ADVANCED
ENGINEERING
MATHEMATICS

ADVANCED ENGINEERING MATHEMATICS

FOURTH EDITION

C. RAY WYLIE

William R. Kenan, Jr., Professor of Mathematics
Chairman, Department of Mathematics
Furman University

McGRAW-HILL BOOK COMPANY

New York   St. Louis   San Francisco   Auckland   Dusseldorf
Johannesburg   Kuala Lumpur   London   Mexico   Montreal
New Delhi   Panama   Paris   São Paulo
Singapore   Sydney   Tokyo   Toronto

ADVANCED
ENGINEERING
MATHEMATICS

34567890 DODO 7987

This book was set in Times New Roman. The editors were A. Anthony
Arthur and Michael LaBarbera; the designer was Jo Jones; the
production supervisor was Sam Ratkewitch. New drawings were done
by Eric G. Hieber Associates Inc.

Library of Congress Cataloging in Publication Data

Wylie, Clarence Raymond, date
    Advanced engineering mathematics.

    1. Mathematics—1961–   I. Title.
QA401.W9 1975   510   74-14523
ISBN 0-07-072180-7

CONTENTS

PREFACE

The first edition of this book was written to provide an introduction to those branches of postcalculus mathematics with which the average analytical engineer or physicist needs to be reasonably familiar in order to carry on his own work effectively and keep abreast of current developments in his field. In the present edition, as in the second and third, although the material has been largely rewritten, the various additions, deletions, and refinements have been made only because they seemed to contribute to the achievement of this goal.

Because ordinary differential equations are probably the most immediately useful part of postcalculus mathematics for the student of applied science, and because the techniques for solving simple ordinary differential equations stem naturally from the techniques of calculus, this book begins with a chapter on ordinary differential equations of the first order and their applications. This is followed by two other chapters on ordinary differential equations, which develop the theory and applications of linear equations and systems of linear equations with constant coefficients. In particular, a new section on Green's function and its interpretation as an influence function has been added to the first of these two chapters. Following these is a chapter on finite differences containing the usual applications to interpolation, numerical differentiation and integration, and the step-by-step solution of differential equations using both Milne's method and the Runge-Kutta method. There is also a section on linear difference equations with constant coefficients closely paralleling the preceding development for differential equations and a section on the method of least squares and the related topic of orthogonal polynomials. It is hoped that the material in this chapter will provide a useful background in classical finite differences, on which a more extensive course in computer-oriented numerical analysis can be based. Chapter 5 is devoted to the application of the preceding ideas to mechanical and electrical systems, and, as in the first three editions, the mathematical identity of these fields is emphasized. The next two chapters deal first with partial differential equations and boundary-value problems and second with Bessel functions and Legendre polynomials, very much as in the third edition though with a number of new examples and exercises.

Chapters 10 and 11 deal with determinants and matrices as far as the Cayley-Hamilton theorem, Sylvester's identity, and infinite series of matrices and their use in solving matric differential equations. Chapter 12 is a new chapter on the calculus of variations, covering such topics as the maxima and minima of functions of several variables, Lagrange's multipliers, the extremal properties of the eigenvalues of matric equations, Euler's equation, Hamilton's principle, and Lagrange's equation. Chapters

13 and 14 deal with vector and tensor analysis. The last four chapters provide an introduction to the theory of functions of a complex variable, with applications to the evaluation of real definite integrals, the complex inversion integral, stability criteria, conformal mapping, and the Schwarz-Christoffel transformation.

This book falls naturally into three major subdivisions. The first nine chapters constitute a reasonably self-contained treatment of ordinary and partial differential equations and their applications. The next five chapters cover the related areas of linear algebra, the calculus of variations, and vector and tensor analysis. The last four chapters cover the elementary theory and applications of functions of a complex variable. With this organization, the book, which contains enough material for a 2-year postcalculus course in applied mathematics, is well adapted for use as a text for any of several shorter courses.

In this edition, as in the first three, every effort has been made to keep the presentation detailed and clear while at the same time maintaining acceptable standards of precision and accuracy. To achieve this, more than the usual number of worked examples and carefully drawn figures have been included, and in every development there has been a conscious attempt to make the transitions from step to step so clear that a student with no more than a good background in calculus, working with paper and pencil, should seldom be held up more than momentarily. Over 750 new exercises of varying degrees of difficulty have been added to the 1,387 problems which appeared in the third edition. Many of these involve extensions of topics presented in the text or related topics which could not be treated because of limitations of space. Hints are included in many of the exercises, and answers to the odd-numbered ones are given at the end of the book. As in the first three editions, words and phrases defined informally in the body of the text are set in boldface. Illustrative examples are set.in type of a different size from that used for the main body of the text.

The indebtedness of the author to his colleagues, students, and former teachers is too great to catalog, and to all who have given help and encouragement in the preparation of this book, I can offer here only a most inadequate acknowledgment of my appreciation. In particular, I am deeply grateful to those users of this book, both teachers and students, who have been kind enough to write me their impressions and criticisms of the first three editions and their suggestions for an improved fourth edition. Finally, I must express my gratitude to my wife Ellen and my student Moffie Hills, who have shared with me the task of proofreading the manuscript in all its stages.

C. Ray Wylie

TO THE STUDENT

This book has been written to help you in your development as an applied scientist, whether engineer, physicist, chemist, or mathematician. It contains material which you will find of great use, not only in the technical courses you have yet to take, but also in your profession after graduation as long as you deal with the analytical aspects of your field.

I have tried to write a book which you will find not only useful but also easy to study from, at least as easy as a book on advanced mathematics can be. There is a good deal of theory in it, for it is the theoretical portion of a subject which is the basis for the nonroutine applications of tomorrow. But nowhere will you find theory for its own sake, interesting and legitimate as this may be to a pure mathematician. Our theoretical discussions are designed to illuminate principles, to indicate generalizations, to establish limits within which a given technique may or may not safely be used, or to point out pitfalls into which one might otherwise stumble. On the other hand, there are many applications illustrating, with the material at hand, the usual steps in the solution of a physical problem: formulation, manipulation, and interpretation. These examples are, without exception, carefully set up and completely worked, with all but the simplest steps included. Study them carefully, with paper and pencil at hand, for they are an integral part of the text. If you do this, you should find the exercises, though challenging, still within your ability to work.

Terms defined informally in the body of the text are always indicated by the use of boldface type. Italic type is used for emphasis. It is suggested that you read each section through for the main ideas before you concentrate on filling in any of the details. You will probably be surprised at how many times a detail which seems to hold you up in one paragraph is explained in the next as the discussion unfolds.

Because this book is long and contains material suitable for various courses, your instructor may begin with any of a number of chapters. However, the overall structure of the book is the following: The first nine chapters are devoted to the general theme of ordinary and partial differential equations and related topics. Here you will find basic analytical techniques for solving the equations in which physical problems must be formulated when continuously changing quantities are involved. Chapters 10 through 14 deal with the somewhat related topics of matrix theory and linear algebra, the calculus of variations, vector analysis, and an introduction to generalized coordinates and tensor analysis. Finally, Chapters 15 to 18 provide an introduction to the theory and applications of functions of a complex variable. (Chapter 4, in particular, is worthy of note because it provides an introduction to numerical analysis, the modern field which deals with techniques for obtaining numerical answers to problems too complicated to be solved by exact analytic methods.)

It has been gratifying to receive letters from students who have used this book, giving me their reactions to it, pointing out errors and misprints in it, and offering suggestions for its improvement. Should you be inclined to do so, I should be happy to hear from you also. And now good luck and every success.

C. Ray Wylie

ADVANCED
ENGINEERING
MATHEMATICS

CHAPTER 1
Ordinary Differential Equations of the First Order

1.1 Introduction

An equation involving one or more derivatives of a function is called a **differential equation**. By a **solution** of a differential equation is meant a relation between the dependent and independent variables which is free of derivatives and which, when substituted into the given equation, reduces it to an identity. The study of the existence, nature, and determination of solutions of differential equations is of fundamental importance not only to the pure mathematician but also to anyone engaged in the mathematical analysis of natural phenomena.

In general, a mathematician considers it a triumph if he is able to prove that a given differential equation possesses a solution and if he can deduce a few of the more important properties of that solution. A physicist or engineer, on the other hand, is usually greatly disappointed if a specific expression for the solution cannot be exhibited. The usual compromise is to find some practical procedure by means of which the required solution can be approximated with satisfactory accuracy.

Not all differential equations are difficult enough to make this necessary, however, and there are several large and very important classes of equations for which solutions can readily be found. For instance, an equation such as

$$\frac{dy}{dx} = f(x)$$

is really a differential equation, and the integral

$$y = \int f(x)\, dx + c$$

is a solution. More generally, the equation

$$\frac{d^n y}{dx^n} = g(x)$$

is a differential equation whose solution can be found by n successive integrations. Except in name, the process of integration is actually an example of a process for solving differential equations.

In this and the following two chapters we shall consider differential equations which are next in difficulty after those which can be solved by direct integration. These equations form only a very small part of the class of all differential equations, and yet with a knowledge of them a scientist is equipped to handle a great variety of applications. To get so much for so little is indeed remarkable.

1.2 Fundamental Definitions

If the derivatives which appear in a differential equation are total derivatives, the equation is called an **ordinary differential equation**; if partial derivatives occur, the equation is called a **partial differential equation**. By the **order** of a differential equation is meant the order of the highest derivative which appears in the equation.

EXAMPLE 1

The equation $x^2 y'' + xy' + (x^2 - 4)y = 0$ is an *ordinary* differential equation of the *second* order connecting the dependent variable y with its first and second derivatives and with the independent variable x.

EXAMPLE 2

The equation

$$\frac{\partial^4 u}{\partial x^4} + \frac{\partial^4 u}{\partial x^2\,\partial y^2} + \frac{\partial^4 u}{\partial y^4} = 0$$

is a *partial* differential equation of the *fourth* order.

At present we shall be concerned exclusively with ordinary differential equations.

An equation which is linear, i.e., of the first degree, in the *dependent* variable and its derivatives is called a **linear differential equation**. From this definition it follows that the most general (ordinary) linear differential equation of order n is of the form

(1) $$p_0(x)y^{(n)} + p_1(x)y^{(n-1)} + \cdots + p_{n-1}(x)y' + p_n(x)y = r(x)$$

A differential equation which is not linear, i.e., cannot be put in the form (1), is said to be **nonlinear**. In general, linear equations are much easier to solve than nonlinear ones, and most elementary applications involve linear equations.

EXAMPLE 3

The equation $y'' + 4xy' + 2y = \cos x$ is a *linear* equation of the second order. The presence of the terms xy' and $\cos x$ does not alter the fact that the equation is linear, because, by definition, linearity is determined solely by the way the *dependent* variable y and its derivatives enter into combination among themselves.

EXAMPLE 4

The equation $y'' + 4yy' + 2y = \cos x$ is a *nonlinear* equation because of the occurrence of the product of y and one of its derivatives.

EXAMPLE 5

The equation $y'' + \sin y = 0$ is *nonlinear* because of the presence of $\sin y$, which is a nonlinear function of y.

As illustrated by the simple differential equation

$$\frac{dy}{dx} = e^{-x^2}$$

and its solution

$$y = \int e^{-x^2}\,dx + c$$

the solution of a differential equation may depend upon integrals which cannot be evaluated in terms of elementary functions. This example also illustrates the fact that a solution of a differential equation usually involves one or more arbitrary constants.

A detailed treatment of the question of the maximum number of *essential* arbitrary constants that a solution of a differential equation may contain or even of what is meant by essential constants is quite difficult.† For our purposes, if an expression contains n arbitrary constants, we shall consider them essential if they cannot, through formal rearrangement of the expression, be replaced by any smaller number of constants. For example,

(2) $$a \cos^2 x + b \sin^2 x + c \cos 2x$$

contains three arbitrary constants. However, since

$$\cos 2x = \cos^2 x - \sin^2 x$$

the expression (2) can be written in the form

$$a \cos^2 x + b \sin^2 x + c(\cos^2 x - \sin^2 x) = (a + c) \cos^2 x + (b - c) \sin^2 x$$
$$= d \cos^2 x + e \sin^2 x$$

where $d = a + c$ and $e = b - c$. The fact that the three arbitrary constants a, b, and c can be replaced by the two constants d and e shows that the former are not all essential. On the other hand, since $\cos^2 x$ and $\sin^2 x$ are linearly independent‡ (whereas $\cos^2 x$, $\sin^2 x$, and $\cos 2x$ are linearly dependent), it follows that there is no further rearrangement of the given expression that will permit d and e to be combined into, and replaced by, a single new arbitrary constant. Hence d and e are essential.

It is frequently the case (especially with linear equations) that a differential equation of order n possesses solutions containing n essential arbitrary constants, or parameters, but none containing more. However, there are equations such as

$$\left| \frac{dy}{dx} \right| + |y| = 0$$

(which clearly has only the single solution $y = 0$) and

$$\left| \frac{dy}{dx} \right| + 1 = 0$$

(which has no solutions at all) which possess *no* solutions containing *any* arbitrary constants. Moreover, there are also simple differential equations which possess solutions containing more essential parameters than the order of the equation. For instance it is easy to verify that the arc of the family $y = c_1 x^2$ $(x \leq 0)$ (Fig. 1.1a) corresponding to any value of c_1 can be paired with the arc of the family $y = c_2 x^2$ $(x \geq 0)$ (Fig. 1.1b) corresponding to any value of c_2, to give a function which satisfies the differential equation

(3) $$xy' = 2y$$

---

† See, for instance, R. P. Agnew, "Differential Equations," 2d ed., pp. 103–105, McGraw-Hill, New York, 1960.
‡ See Definitions 1 and 2, Sec. 10.5.

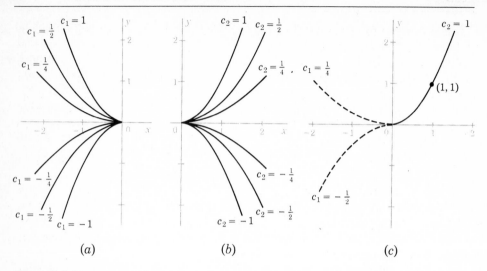

Figure 1.1
Arcs of different parabolas of the family $y = cx^2$ pieced together to give solutions of the differential equation $xy' = 2y$.

for all values of x (Fig. 1.1c). In other words, for all choices of the *two* essential constants c_1 and c_2, the rule

$$y = \begin{cases} c_1 x^2 & x \leq 0 \\ c_2 x^2 & x > 0 \end{cases}$$

defines a function which is continuous and differentiable for all values of x and which satisfies the equation (3) over the entire x axis. A still more striking example of this sort appears in Exercise 44, where a first-order equation with a solution containing infinitely many essential parameters is given.

As the foregoing suggests, it is difficult, if not impossible, to make statements valid for all differential equations. The theory of differential equations is essentially a body of theorems concerning particular classes of equations defined by such considerations as the order and linearity of the equation and the continuity and boundedness of its coefficients. Typical of these theorems is the following result,† which is of fundamental importance in the theory of the equations we shall consider in this chapter, namely, equations of the first order.

THEOREM 1   Let (x_0, y_0) be a point of the xy plane; let R be the rectangular region defined by the inequalities $|x - x_0| \leq a$, $|y - y_0| \leq b$; let $f(x,y)$ and $f_y(x,y) \equiv \partial f(x,y)/\partial y$ be single-valued and continuous at all points of R; let M be a constant such that $|f(x,y)| < M$ at all points of R; and let h be the smaller of the numbers a and b/M. Then, on the interval $|x - x_0| < h$, there is a unique continuous function y which satisfies the equation $y' = f(x,y)$ and takes on the value y_0 when $x = x_0$.

---

† See, for instance, M. Golomb and M. E. Shanks, "Elements of Ordinary Differential Equations," 2d ed., pp. 63–78, McGraw-Hill, New York, 1965.

It is instructive to reconsider Eq. (3) in the light of Theorem 1. For this equation we have $f(x,y) = 2y/x$, and, clearly, neither f nor f_y exists when $x = 0$. Hence, it follows from Theorem 1 that over an interval containing $x = 0$ neither the existence nor the uniqueness of a solution of Eq. (3) can be guaranteed. Actually, as our earlier discussion pointed out, Eq. (3) does have solutions which are valid for all values of x. However, as Fig. 1.1c illustrates, over any interval which contains $x = 0$, the solution curve which passes through a given point (x_0, y_0), for example, $(1,1)$, is not unique. On the other hand, according to Theorem 1, over any interval which contains x_0 but does not contain $x = 0$, the solution curve which passes through a given point (x_0, y_0) is unique.

Almost all applications of differential equations involve equations which possess solutions containing at least one arbitrary constant, and for such equations it is convenient to introduce the following definitions. A solution which contains at least one arbitrary constant is called a **general solution**. A solution obtained from a general solution by assigning particular values to the arbitrary constants which appear in it is called a **particular solution**. Solutions which cannot be obtained from any general solution by assigning specific values to its arbitrary constants are called **singular solutions**. If a general solution has the property that *every* solution of the differential equation can be obtained from it by assigning suitable values to its arbitrary constants, it is said to be a **complete solution**. A general solution can thus be thought of as a description of some family of particular solutions, and a complete solution can be thought of as a description of the set of all solutions of the given equation.

It is important to note that we speak of *a* general solution and *a* complete solution of a differential equation and not of *the* general solution and *the* complete solution. If an equation has a general solution or a complete solution, it has many such solutions, and these may differ significantly in form. Moreover, in particular problems involving differential equations, the choice of which complete solution to use often has an important bearing on the ease with which the problem can be solved.

EXAMPLE 6

Verify that $y = ae^{-x} + be^{2x}$ is a solution of the equation $y'' - y' - 2y = 0$ for all values of the constants a and b.

By differentiating y, substituting into the differential equation as indicated, and then collecting terms on a and b, we obtain

$$(ae^{-x} + 4be^{2x}) - (-ae^{-x} + 2be^{2x}) - 2(ae^{-x} + be^{2x})$$

$$= (e^{-x} + e^{-x} - 2e^{-x})a + (4e^{2x} - 2e^{2x} - 2e^{2x})b$$

$$= 0a + 0b = 0$$

for all values of a and b. Thus, $y = ae^{-x} + be^{2x}$ is a general solution of $y'' - y' - 2y = 0$. In fact, as we shall see in Sec. 2.2, it is a complete solution of this equation.

It is interesting to note that although $y_1 = ae^{-x}$ and $y_2 = be^{2x}$ also satisfy the equation $yy'' - (y')^2 = 0$, the sum $y = y_1 + y_2 = ae^{-x} + be^{2x}$ is *not* a solution of $yy'' - (y')^2 = 0$. In fact, differentiating, substituting, and simplifying, we have

$$(ae^{-x} + be^{2x})(ae^{-x} + 4be^{2x}) - (-ae^{-x} + 2be^{2x})^2 \equiv 9abe^x$$

and this cannot vanish identically unless either a or b is zero, i.e., unless the sum y consists of just one or the other of the two individual solutions. Roughly speaking, the reason for this difference in behavior is that the equation $y'' - y' - 2y = 0$ is linear, whereas the equation $yy'' - (y')^2 = 0$ is nonlinear. More precisely, as we shall see in

Theorem 1, Sec. 2.1, for linear equations in which y or one of its derivatives appears in every term, the sum of two solutions is also a solution, whereas, in general, the sum of two solutions of a nonlinear equation is *not* a solution of that equation.

Occasionally it is necessary to determine a differential equation of order n which has a given function containing n arbitrary constants as a general solution. This can be done (at least theoretically) by differentiating the given expression n times and then eliminating the arbitrary constants by algebraic manipulation of the original equation and these derived equations.

EXAMPLE 7

If a and b are arbitrary constants, find a second-order equation which has

$$(4) \qquad y = ae^x + b \cos x$$

as a general solution.

By differentiating the given expression, we find

$$(5) \qquad y' = ae^x - b \sin x$$

$$(6) \qquad y'' = ae^x - b \cos x$$

Then, by adding and subtracting Eqs. (4) and (6), we obtain

$$a = \frac{y + y''}{2e^x} \qquad b = \frac{y - y''}{2 \cos x}$$

Substitution of these into (5) gives

$$y' = \frac{y + y''}{2e^x} e^x - \frac{y - y''}{2 \cos x} \sin x$$

and finally

$$(7) \qquad (1 + \tan x)y'' - 2y' + (1 - \tan x)y = 0$$

Although (7), except for its obvious multiples, is the only second-order differential equation having (4) as a general solution, it is by no means the only equation of which (4) is a general solution. For instance, if (6) is differentiated twice more, we obtain

$$y^{iv} = ae^x + b \cos x$$

and comparing this with (4), we see that the given function also satisfies the very simple equation

$$(8) \qquad y^{iv} = y$$

Since Eq. (8) is of the fourth order, it presumably possesses general solutions containing four arbitrary constants, and it is easy to verify that

$$y = ae^x + b \cos x + ce^{-x} + d \sin x$$

does in fact satisfy Eq. (8) for all values of a, b, c, and d.

EXERCISES

Describe each of the following equations, giving its order and telling whether it is ordinary or partial and linear or nonlinear:

1 $y'' + 3y' + 2y = x^4$ 2 $y'' + (a + b \cos 2x)y = 0$

3 $y''' + 6y'' + 11y' + 6y = e^x$ 4 $y^{iv} + xy'' + y^2 = 0$

5 $\dfrac{d(xy')}{dx} + xy = 0$ 6 $(x + y)\,dy = (x - y)\,dx$

7 $a^2 \dfrac{\partial^2 u}{\partial x^2} = \dfrac{\partial^2 u}{\partial t^2}$

8 $\dfrac{\partial^2 (x^2 \, \partial^2 u/\partial x^2)}{\partial x^2} = \dfrac{\partial^2 u}{\partial t^2}$

9 $\dfrac{\partial^2 u}{\partial x^2} = u \dfrac{\partial u}{\partial t}$

10 $\dfrac{\partial^2 u}{\partial x^2} + \dfrac{\partial^2 u}{\partial y^2} + \dfrac{\partial^2 u}{\partial z^2} = \phi(x, y, z)$

Show that not all the constants which appear in the following expressions are essential, and in each case rearrange the expression so that all the constants which remain are essential:

11 Ae^{x+k}

12 $a + \ln bx$

13 $a \ln x^b$

14 $\dfrac{ax + b}{cx + d}$

15 $A \sin (x + b) + C \sin (x + d)$

16 $A[\cos (x + a) + \cos (x - a)]$

17 $a \cosh^2 \theta + b \sinh^2 \theta + c \cosh 2\theta$

18 $a \sin 3x + b \sin x + c \sin^3 x$

19 $\dfrac{A}{x + 1} + \dfrac{B}{x + 2} + \dfrac{C}{x^2 + 3x + 2}$

20 $a(x - 6y - 7) + b(3x + 4y + 5) + c(5x + 3y + 4)$

Verify that each of the following equations has the indicated solution for all values of the constants a and b:

21 $y'' + 4y = 0$ $y = a \cos 2x + b \sin 2x$

22 $y'' - 4y = 0$ $y = ae^{2x} + be^{-2x}$

23 $y'' + 3y' + 2y = 12e^{2x}$ $y = ae^{-x} + be^{-2x} + e^{2x}$

24 $y'' - 6y' + 9y = 0$ $y = ae^{3x} + bxe^{3x}$

25 $(\cos 2x)y' + (2 \sin 2x)y = 2$ $y = a \cos 2x + \sin 2x$

26 $2xy \, dy = (y^2 - x) \, dx$ $y^2 = ax - x \ln x$

27 $y'' + (y')^2 + 1 = 0$ $y = \ln \cos (x - a) + b$

28 $\dfrac{\partial^2 u}{\partial x^2} = \dfrac{\partial u}{\partial t}$ $u = ae^{-9t} \cos (3x + b)$

29 $4 \dfrac{\partial^2 u}{\partial x^2} = \dfrac{\partial^2 u}{\partial t^2}$ $u = af(x + 2t) + bg(x - 2t)$

If a and b are arbitrary constants, find a differential equation of minimum order of which each of the following expressions is a general solution:

30 $y = ae^{-2x} + be^x$ **31** $y = ae^{-2t} + bte^{-2t}$

32 $y = ae^{-t} + be^t + ce^{2t}$ **33** $y = 2ax + bx^2$

34 $y = a \cosh 2x + b \sinh 2x$ **35** $y = \sin (ax + b)$

36   Find a differential equation which has as a general solution the expression which defines the family of all parabolas which touch the x axis and have their axes vertical.

37   Find a differential equation which has as a general solution the expression which defines the family of all lines which touch the parabola $2y = x^2$. Verify that the equation of the given parabola defines a function which is a singular solution of the required differential equation.

38   Verify that for all values of m the function $y = mx + f(m)$ is a solution of the differential equation $y = xy' + f(y')$. [This differential equation is called **Clairaut's equation**, after the French mathematician A. C. Clairaut (1713–1765).]

39   Verify that for all values of the arbitrary constants a and b both $y_1 = a$ and $y_2 = bx^2$ satisfy each of the differential equations

$$xy'' = y' \quad \text{and} \quad 2yy'' = (y')^2$$

but that $y = a + bx^2$ will satisfy only the first of these equations. Explain.

40   Verify that for all values of the arbitrary constants a and b both $y_1 = a$ and $y_2 = b\sqrt{x}$ satisfy each of the differential equations

$$2xy'' + y' = 0 \quad \text{and} \quad 8x^3(y'')^2 - yy' = 0$$

but that $y = a + b\sqrt{x}$ will satisfy only the first of these equations. Explain.

41 Verify that for all values of the arbitrary constants a and b both $y_1 = a(x - 1)^2$ and $y_2 = b(x + 1)^2$ satisfy each of the differential equations

$$(x^2 - 1)y'' - 2xy' + 2y = 0 \quad \text{and} \quad 2yy'' - (y')^2 = 0$$

but that $y = a(x - 1)^2 + b(x + 1)^2$ will satisfy only the first of these equations. Explain.

42 Verify that for all values of the arbitrary constants c_1 and c_2 the differential equation $xy' = 2y + x$ is satisfied by the function

$$y = \begin{cases} c_1 x^2 - x & x \le 0 \\ c_2 x^2 - x & x > 0 \end{cases}$$

Explain.

43 Verify that for all values of the arbitrary constants c_1, c_2, and c_3 the differential equation

$$(x^2 - 1)y' = 4xy$$

is satisfied by the function

$$y = \begin{cases} c_1(x^2 - 1)^2 & x < -1 \\ c_2(x^2 - 1)^2 & -1 \le x \le 1 \\ c_3(x^2 - 1)^2 & x > 1 \end{cases}$$

Explain.

44 Verify that for all values of the arbitrary constants $\{c_n\}$ $(n = \ldots, -2, -1, 0, 1, 2, \ldots)$ the differential equation

$$(1 - \cos x)y' = (\sin x)y$$

is satisfied by the function

$$y = c_n(1 - \cos x) \quad 2n\pi \le x < 2(n + 1)\pi$$

Explain.

1.3 Separable First-Order Equations

In many cases a first-order differential equation can be reduced by algebraic manipulations to the form

(1) $$f(x)\, dx = g(y)\, dy$$

Such an equation is said to be **separable** because the variables x and y can be *separated* from each other in such a way that x appears only in the coefficient of dx and y appears only in the coefficient of dy. An equation of this type can be solved at once by integration, and we have the general solution

(2) $$\int f(x)\, dx = \int g(y)\, dy + c$$

where c is an arbitrary constant of integration. It must be borne in mind, however, that the integrals which appear in (2) may be impossible to evaluate in terms of elementary functions, and numerical or graphical integration may be required before this solution can be put to practical use.

Other forms which should be recognized as being separable are

(3) $$f(x)G(y)\, dx = F(x)g(y)\, dy$$

(4) $$\frac{dy}{dx} = M(x)N(y)$$

A general solution of Eq. (3) can be found by first dividing by the product $F(x)G(y)$ to separate the variables and then integrating:

$$\int \frac{f(x)}{F(x)}\, dx = \int \frac{g(y)}{G(y)}\, dy + c$$

Similarly, a general solution of Eq. (4) can be found by first multiplying by dx and dividing by $N(y)$ and then integrating:

$$\int \frac{dy}{N(y)} = \int M(x)\, dx + c$$

Clearly, the process of solving a separable equation will often involve division by one or more expressions. In such cases the results are valid where the divisors are not equal to zero but may or may not be meaningful for values of the variables for which the division is impossible. Such values require special consideration and, as we shall see in the next example, may lead us to singular solutions.

EXAMPLE 1

Solve the differential equation $dx + xy\, dy = y^2\, dx + y\, dy$.

It is not immediately evident that this equation is separable. In any case, however, the best first step in solving an equation of this sort is to collect terms on dx and dy. This gives

$$(1 - y^2)\, dx = y(1 - x)\, dy$$

which is of the form (3). Hence, division by the product $(1 - x)(1 - y^2)$ will separate the variables and reduce the equation to the standard form (1):

$$\frac{dx}{1 - x} = \frac{y\, dy}{1 - y^2}$$

Thm 1 does not apply

Now, multiplying by -2 and integrating, we obtain the following equation defining y as an implicit function of x:

$$2 \ln |1 - x| = \ln |1 - y^2| + c$$

In this case, as in many problems of this sort, it is possible to write the solution in a more convenient form by first combining the logarithmic terms and then taking antilogarithms:

$$\ln \frac{|1 - x|^2}{|1 - y^2|} = c \qquad \frac{|1 - x|^2}{|1 - y^2|} = e^c = k^2$$

where $k^2 = e^c$ is necessarily positive. Finally, clearing of fractions and eliminating the absolute values, we have

$$(1 - x)^2 = \pm k^2(1 - y^2) \qquad k \neq 0$$

The two possibilities here can, of course, be combined into one by writing

$$(1 - x)^2 = \lambda(1 - y^2)$$

where now λ can take on any real value, positive or negative, except 0. The solution of the differential equation thus defines the family of conics

(5)
$$\frac{(x - 1)^2}{\lambda} + y^2 = 1 \qquad \lambda \neq 0$$

typical members of which are shown in Fig. 1.2. If $\lambda > 0$, the solution curves are all ellipses; if $\lambda < 0$, the solution curves are all hyperbolas.

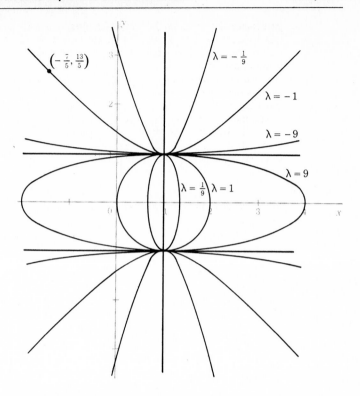

Figure 1.2
Typical members of the solution family $[(x - 1)^2/\lambda] + y^2 = 1$ of the
differential equation $(1 - y^2)\, dx = y(1 - x)\, dy$.

In most practical problems a general solution of a differential equation is required
to satisfy specific conditions which permit its arbitrary constants to be uniquely
determined. For instance, in the present problem we might ask for the particular
solution curve which passes through the point $(-\frac{7}{5}, \frac{13}{5})$. Substituting these values of
x and y, we then have

$$\frac{(-\frac{7}{5} - 1)^2}{\lambda} + \left(\frac{13}{5}\right)^2 = 1$$

from which we find the value $\lambda = -1$ and thence the specific solution

(6)
$$y^2 = 1 + (x - 1)^2$$

Equation (6) defines the unique member of the family of curves (5) which passes
through the point $(-\frac{7}{5}, \frac{13}{5})$. However, over any interval which contains both $x = -\frac{7}{5}$
and $x = 1$, there are many functions which satisfy the given differential equation and
are such that $y = \frac{13}{5}$ when $x = -\frac{7}{5}$. In fact, the upper branch of *any* curve of the
family (5) for $x > 1$ can be associated with the upper branch of the curve (6) for $x \leq 1$
to give a function which satisfies the given equation and fulfills the condition that
$y = \frac{13}{5}$ when $x = -\frac{7}{5}$. This is, of course, consistent with the fact that according to
Theorem 1, Sec. 1.2, the uniqueness of the solution for which $y = \frac{13}{5}$ when $x = -\frac{7}{5}$
can be guaranteed only over an interval around $x = -\frac{7}{5}$ which does not contain $x = 1$,
since y' is undefined at $x = 1$.

It should be noted that in separating variables in the given differential equation it
was necessary to divide by $1 - x$ and by $1 - y^2$; hence, the possibility that $x = 1$ and
the possibility that $y = \pm 1$ were implicitly ruled out. Therefore, had we desired the

particular solution curve which passed through any point with coordinates of the form $(1, y_0)$, $(x_0, 1)$, or $(x_0, -1)$, we could not have found that curve, even if it existed, by starting with the general solution (5) and particularizing the arbitrary constant λ. Instead, it would have been necessary to return to the differential equation and search for the required solution by some method other than separation of variables. In this case it is obvious that the linear equations $x = 1$, $y = 1$, and $y = -1$ all define solutions of the given differential equation and, moreover, satisfy, respectively, the conditions $(1, y_0)$, $(x_0, 1)$, and $(x_0, -1)$. None of these can be *obtained* from our general solution, although $x = 1$ can be *included* in the first form of it by permitting λ to take on the (previously excluded) value zero. In this case, then, only $y = 1$ and $y = -1$ appear as singular solutions of the given equation.

EXERCISES

Find a general solution of each of the following equations:

1   $y' = -2xy$

2   $(\sin x)\, dy = 2y\, (\cos x)\, dx$

3   $y' = 3x^2(1 + y^2)$

4   $x\, dy = 3y\, dx$

5   $2(xy + x)y' = y$

6   $x\, dy = (y^2 - 3y + 2)\, dx$

7   $(y + x^2 y)\, dy = (xy^2 - x)\, dx$

8   $y\, dx - x\, dy = x(dy - y\, dx)$

9   $yy' = 2(xy + x)$

10   $dx + y\, dy = x^2 y\, dy$

11   $ye^{x+y}\, dy = dx$

12   $xe^{x^2+y}\, dx = y\, dy$

13   $y' = \left(\dfrac{y + 1}{x + 1}\right)^2$

14   $y' = \dfrac{2(y^2 + y - 2)}{x^2 + 4x + 3}$

15   $y'' + (y')^2 + 1 = 0$ *Hint:* Observe that $y'' = dy'/dx$.

16   $xy'' = y'$

17   $yy'' = (y')^2$

Find the particular solution of each of the following equations which satisfies the indicated conditions:

18   $2xy' + y = 0$ $x = 4,\ y = 1$

19   $y' + 2y = 0$ $x = 0,\ y = 100$

20   $2x\, dx - dy = x(x\, dy - 2y\, dx)$ $x = -3,\ y = 1$

21   $dy = x(2y\, dx - x\, dy)$ $x = 1,\ y = 4$

22   Is there a solution of the equation $x\, dy = 3(y - 1)\, dx$ satisfying the two conditions $y = 3$ when $x = 1$ and $y = 9$ when $x = 2$? Is there a solution of this equation which satisfies the two conditions $y = 3$ when $x = -1$ and $y = 9$ when $x = 2$? Explain.

23   Find a solution of the equation $(1 - x^2)\, dy + 4xy\, dx = 0$ with the property that $y = 9$ when $x = -2$, $y = 2$ when $x = 0$, and $y = 0$ when $x = 2$.

24   Show that every solution of the equation $y' = ky$ is of the form $y = Ae^{kx}$. *Hint:* Let y be any solution of the given equation, and consider the derivative of the fraction y/e^{kx}.

25   A critical student watching his professor integrate the separable equation $f(x)\, dx = g(y)\, dy$ objected that the procedure was incorrect, since one side was integrated with repect to x while the other side was integrated with respect to y. How would you answer the student's objection?

26   Show that there is no loss of generality if the arbitrary constant added when a separable equation is integrated is written in the form $\ln c$ rather than just c. Do you think this would ever be a convenient thing to do? Is there any loss of generality if the integration constant is written in the form c^2? $\tan c$? $\sin c$? e^c? $\sinh c$? $\cosh c$?

27   Show that the change of dependent variable from y to v defined by the substitution $v = ax + by + c$ will always transform the equation $y' = f(ax + by + c)$ into a separable equation.

Using the substitution described in Exercise 27, find a general solution of each of the following equations:

28   $y' = (x - y)^2$

29   $y' = e^{2x+y-1} - 2$

30   $y' = (x + y - 3)^2 - 2(x + y - 3)$

31   $y' = (x - y + 1)^2 + x - y$

1.4 Homogeneous First-Order Equations

If all terms in the coefficient functions $M(x,y)$ and $N(x,y)$ in the general first-order differential equation

$$(1) \qquad\qquad M(x,y) \, dx = N(x,y) \, dy$$

are of the same total degree in the variables x and y, then either of the substitutions $y = ux$ and $x = vy$ will reduce the equation to one which is separable.

More generally, if $M(x,y)$ and $N(x,y)$ have the property that for all positive values of λ the substitution of λx for x and λy for y converts them, respectively, into the expressions

$$\lambda^n M(x,y) \qquad \text{and} \qquad \lambda^n N(x,y)$$

then Eq. (1) can always be reduced to a separable equation by either of the substitutions $y = ux$ and $x = vy$.

Functions with the property that the substitutions

$$x \to \lambda x \qquad \text{and} \qquad y \to \lambda y \qquad \lambda > 0$$

merely reproduce the original forms multiplied by λ^n are called **homogeneous functions of degree** n. As a direct extension of this terminology, the differential equation (1) is said to be **homogeneous** when $M(x,y)$ and $N(x,y)$ are homogeneous functions *of the same degree.*

EXAMPLE 1

Is the function

$$F(x,y) = x(\ln \sqrt{x^2 + y^2} - \ln y) + ye^{x/y}$$

homogeneous?

To decide this question, we replace x by λx and y by λy, getting

$$\begin{aligned}
F(\lambda x, \lambda y) &= \lambda x(\ln \sqrt{\lambda^2 x^2 + \lambda^2 y^2} - \ln \lambda y) + \lambda y e^{\lambda x/\lambda y} \\
&= \lambda x[(\ln \sqrt{x^2 + y^2} + \ln \lambda) - (\ln y + \ln \lambda)] + \lambda y e^{x/y} \\
&= \lambda[x(\ln \sqrt{x^2 + y^2} - \ln y) + ye^{x/y}] \\
&= \lambda F(x,y)
\end{aligned}$$

The given function is therefore homogeneous of degree 1.

If Eq. (1), assumed now to be homogeneous, is written in the form

$$\frac{dy}{dx} = \frac{M(x,y)}{N(x,y)}$$

it is evident that the fraction on the right is a homogeneous function of degree 0, since the same power of λ will multiply both numerator and denominator when the test substitutions $x \to \lambda x$ and $y \to \lambda y$ are made. But if

$$\frac{M(\lambda x, \lambda y)}{N(\lambda x, \lambda y)} = \frac{M(x,y)}{N(x,y)}$$

it follows, by assigning to the arbitrary symbol λ the value $1/x$ if x is positive and the value $-1/x$ if x is negative, that

$$\frac{M(x,y)}{N(x,y)} = \frac{M(\lambda x, \lambda y)}{N(\lambda x, \lambda y)} = \begin{cases} \dfrac{M(1, y/x)}{N(1, y/x)} & x > 0 \\[2mm] \dfrac{M(-1, -y/x)}{N(-1, -y/x)} & x < 0 \end{cases}$$

In either case it is clear that the result is a function of the fractional argument y/x. Thus, an alternative standard form for a homogeneous first-order differential equation is

$$(2) \qquad\qquad \frac{dy}{dx} = R\left(\frac{y}{x}\right)$$

Although in practice it is not necessary to reduce a homogeneous equation to the form (2) in order to solve it, the theory of the substitution $y = ux$, or $u = y/x$, is most easily developed when the equation is written in this form.

Now if $y = ux$, then $dy/dx = u + x\, du/dx$ (or, equivalently, $dy = u\, dx + x\, du$). Hence under this substitution, Eq. (2) becomes

$$u + x\frac{du}{dx} = R(u)$$

or

$$(3) \qquad\qquad x\, du = [R(u) - u]\, dx$$

If $R(u) \equiv u$, then Eq. (2) is simply

$$\frac{dy}{dx} = \frac{y}{x}$$

and this is separable at the outset. If $R(u) \not\equiv u$, then we can divide (3) by the product $x[R(u) - u]$, getting

$$\frac{du}{R(u) - u} = \frac{dx}{x}$$

The variables have now been separated, and the equation can be integrated at once. Finally, by replacing u by its value y/x, we can obtain the equation defining the original dependent variable y as a function of x.

EXAMPLE 2

Solve the equation $(x^2 + 3y^2)\, dx - 2xy\, dy = 0$

By inspection, this equation is homogeneous, since all terms in the coefficient of each differential are of the second degree. Hence we substitute $y = ux$ and $dy = u\, dx + x\, du$, getting

$$(x^2 + 3u^2 x^2)\, dx - 2x^2 u(u\, dx + x\, du) = 0$$

or, dividing by x^2 and collecting terms,

$$(1 + u^2)\, dx - 2xu\, du = 0$$

Separating variables, we obtain

$$\frac{dx}{x} - \frac{2u\,du}{1 + u^2} = 0$$

and then, by integrating, we find

$$\ln |x| - \ln |1 + u^2| = c$$

This can be written as

$$\ln \left| \frac{x}{1 + u^2} \right| = \ln e^c = \ln k \qquad \text{where } k = e^c > 0$$

Hence $|x/(1 + u^2)| = k$; or, replacing u by y/x and dropping absolute values,

$$\frac{x}{1 + (y/x)^2} = \pm k$$

Finally, clearing of fractions, we have

$$x^3 = K(x^2 + y^2) \qquad K = \pm k$$

From the preceding steps, it appears that K must be different from zero. However, it is easy to verify by direct substitution that the function corresponding to $K = 0$, namely, $x = 0$, is also a solution of the given equation. Hence, in the general solution we have just obtained, K is actually unrestricted.

EXERCISES

Determine which, if any, of the following functions are homogeneous:

1 $\dfrac{x}{x^2 + y^2}$ **2** $\sin \dfrac{x}{x^2 + y^2}$

3 $x \left[\ln \dfrac{2x^2 + y^2}{x} - \ln (x + y) \right] + y^2 \tan \dfrac{x + 2y}{3x - y}$ **4** $\dfrac{x^2 + y^2 + 1}{xy + 2}$

5 Prove that the substitution $x = vy$ will also transform any homogeneous first-order differential equation into one which is separable.

6 Under what conditions, if any, do you think that the substitution $x = vy$ would be more convenient than the substitution $y = ux$?

7 Show that the product of a homogeneous function of degree m and a homogeneous function of degree n is a homogeneous function of degree $m + n$.

8 Show that the quotient of a homogeneous function of degree m by a homogeneous function of degree n is a homogeneous function of degree $m - n$.

9 If $f(x,y,c_1) = 0$ and $f(x,y,c_2) = 0$ are two solution curves of a homogeneous first-order differential equation, and if P_1 and P_2 are, respectively, the points of intersection of these curves and an arbitrary line, $y = mx$, through the origin, prove that the slopes of these two curves at P_1 and P_2 are equal.

Find a general solution of each of the following differential equations:

10 $(x^2 + y^2)\,dx = 2xy\,dy$ **11** $2xy' = y - x$

12 $xy' - y = \sqrt{x^2 - y^2}$ **13** $x^2\,dy = (xy - y^2)\,dx$

Solve each of the following equations and discuss the family of solution curves:

14 $\dfrac{dy}{dx} = \dfrac{2x - y}{x - 2y}$ **15** $\dfrac{dy}{dx} = \dfrac{x - y}{x + 3y}$

16 $\dfrac{dy}{dx} = \dfrac{x + 2y}{2x + y}$ **17** $\dfrac{dy}{dx} = \dfrac{x + y}{x - y}$

Find the particular solution of each of the following equations which satisfies the given conditions:

18 $x^2 y \, dx = (x^3 - y^3) \, dy$ $x = 1, y = 1$
19 $xy' = y + \sqrt{x^2 + y^2}$ $x = 4, y = 3$
20 $(3y^3 - x^3) \, dx = 3xy^2 \, dy$ $x = 1, y = 2$
21 $(x^4 + y^4) \, dx = 2x^3 y \, dy$ $x = 1, y = 0$
22 $y' = \sec \dfrac{y}{x} + \dfrac{y}{x}$ $x = 2, y = \pi$
23 $(x^3 + y^3) \, dx = 2xy^2 \, dy$ $x = 1, y = 0$
24 If $aB \neq bA$, show that by choosing d and D suitably the equation

$$\frac{dy}{dx} = \frac{ax + by + c}{Ax + By + C}$$

can be reduced to a homogeneous equation in the new variables t and z by the substitutions

$$x = t + d \quad \text{and} \quad y = z + D$$

Using the substitutions described in Exercise 24, find a general solution of each of the following equations:

25 $y' = \dfrac{x - y + 5}{x + y - 1}$ 26 $y' = \dfrac{2x + 2y + 1}{3x + y - 2}$
27 Discuss Exercise 24 in the case when $aB = bA$. *Hint:* Recall Exercise 27, Sec. 1.3.
28 . Prove that $b + c = 0$ is a sufficient condition for all solutions of the equation $y' = (ax + by)/(cx + ey)$ to be conics. Prove further that when this is the case, the conics are all ellipses if $c^2 + ae < 0$ and are all hyperbolas if $c^2 + ae > 0$.
29 Extend the result of Exercise 28 by showing that $b + c = 0$ is also a necessary condition for all solutions of the equation $y' = (ax + by)/(cx + ey)$ to be conics.
30 If $M(x,y) \, dx = N(x,y) \, dy$ is a homogeneous equation, prove that if it is expressed in terms of the polar coordinates r and θ by means of the substitutions $x = r \cos \theta$ and $y = r \sin \theta$, it becomes separable.

Solve each of the following equations, using the method described in Exercise 30:

31 $y' = \dfrac{x + y}{x - y}$ 32 $y' = \dfrac{x + 2y}{2x - y}$
33 Give an example of a function which is homogeneous according to our definition but is not homogeneous if the condition $f(\lambda x, \lambda y) = \lambda^n f(x,y)$ is required to hold for all real values of λ.
34 If $f(x,y)$ is a homogeneous function of degree n, show that

$$x \frac{\partial f}{\partial x} + y \frac{\partial f}{\partial y} = nf$$

What is the generalization of this result to functions of more than two variables? (This result is commonly referred to as **Euler's theorem for homogeneous functions**.)

1.5 Exact First-Order Equations

Associated with each suitably differentiable function of two variables $f(x,y)$ is an expression called its **total differential**, namely,

$$df = \frac{\partial f}{\partial x} \, dx + \frac{\partial f}{\partial y} \, dy$$

Conversely, if the differential equation

$$M(x,y)\, dx + N(x,y)\, dy = 0$$

has the property that

$$M(x, y) = \frac{\partial f}{\partial x} \quad \text{and} \quad N(x, y) = \frac{\partial f}{\partial y}$$

then it can be written in the form

$$\frac{\partial f}{\partial x}\, dx + \frac{\partial f}{\partial y}\, dy = df = 0$$

from which it follows that $f(x,y) = k$ is a solution for all values of the constant k. An equation of this sort is said to be *exact* since, as it stands, its left member is an *exact* differential.

When $M(x,y)$ and $N(x,y)$ are sufficiently simple, it is possible to tell by inspection whether or not there exists a function f with the property that

$$\frac{\partial f}{\partial x} = M(x, y) \quad \text{and} \quad \frac{\partial f}{\partial y} = N(x, y)$$

In general, however, this cannot be done, and it is desirable to have a straightforward test to determine when a given first-order equation is exact. Such a criterion is provided by the following theorem.

THEOREM 1   If $\partial M/\partial y$ and $\partial N/\partial x$ are continuous, then the differential equation $M(x,y)\, dx + N(x,y)\, dy = 0$ is exact if and only if

$$\frac{\partial M}{\partial y} = \frac{\partial N}{\partial x}$$

Proof   To prove the theorem, let us assume first that the given equation is exact. Under this assumption there exists a function f such that

$$M = \frac{\partial f}{\partial x} \quad \text{and} \quad N = \frac{\partial f}{\partial y}$$

Hence, $$\frac{\partial M}{\partial y} = \frac{\partial^2 f}{\partial y\, \partial x} \quad \text{and} \quad \frac{\partial N}{\partial x} = \frac{\partial^2 f}{\partial x\, \partial y}$$

Moreover, $\partial^2 f/(\partial y\, \partial x)$ and $\partial^2 f/(\partial x\, \partial y)$ are continuous since we have just found them to be equal, respectively, to $\partial M/\partial y$ and $\partial N/\partial x$, which are continuous by hypothesis. Therefore it follows, from the familiar properties of partial derivatives, that the order of differentiation is immaterial and

$$\frac{\partial^2 f}{\partial y\, \partial x} = \frac{\partial^2 f}{\partial x\, \partial y}$$

Hence $\partial M/\partial y = \partial N/\partial x$, and the "only if" part of the theorem is established.

To complete the proof we must now show that if $\partial M/\partial y = \partial N/\partial x$, then there is a function f such that $\partial f/\partial x = M$ and $\partial f/\partial y = N$. To do this, let us first integrate

$M(x,y)$ with respect to x, holding y fixed. This gives us the expression

(1)
$$f(x,y) = \int_a^x M(x,y)\,dx + c(y) \qquad a \text{ arbitrary}$$

in which, since the integration is done with respect to x while y is held constant, the integration "constant" is actually a function of y to be determined. Clearly, $\partial f/\partial x = M(x,y)$; and our proof will be complete if we can determine $c(y)$ so that $\partial f/\partial y = N(x,y)$.

Now, observing that under the hypothesis that $\partial M/\partial y$ is continuous the operations of integrating with respect to x and differentiating with respect to y can legitimately be interchanged, and recalling our current assumption that $\partial M/\partial y = \partial N/\partial x$, we have, from (1),

$$\frac{\partial f}{\partial y} = \frac{\partial}{\partial y}\int_a^x M(x,y)\,dx + c'(y)$$

$$= \int_a^x \frac{\partial M}{\partial y}\,dx + c'(y)$$

$$= \int_a^x \frac{\partial N}{\partial x}\,dx + c'(y)$$

$$= N(x,y) - N(a,y) + c'(y)$$

Thus, $\partial f/\partial y$ will equal $N(x,y)$, as required, if $c(y)$ is determined so that $c'(y) = N(a,y)$, that is, if

$$c(y) = \int_b^y N(a,y)\,dy \qquad b \text{ arbitrary}$$

We have thus shown that if $\partial M/\partial y = \partial N/\partial x$, then

(2)
$$f(x,y) = \int_a^x M(x,y)\,dx + \int_b^y N(a,y)\,dy$$

is a function such that

$$df = \frac{\partial f}{\partial x}\,dx + \frac{\partial f}{\partial y}\,dy = M(x,y)\,dx + N(x,y)\,dy$$

This establishes the "if" assertion of the theorem, and our proof is complete.

Since the proof of the preceding theorem tells us that when the equation $M(x,y)\,dx + N(x,y)\,dy = 0$ is exact, its left member is, in fact, the total differential of the function f defined by (2), it follows that the solution in this case can be found at once by integration. Thus we have the following corollary.

COROLLARY 1 If the differential equation $M(x,y)\,dx + N(x,y)\,dy = 0$ is exact, then, for all values of the constant k,

$$\int_a^x M(x,y)\,dx + \int_b^y N(a,y)\,dy = k$$

is a solution of the equation.

EXAMPLE 1

Show that the equation $(2x + 3y - 2)\, dx + (3x - 4y + 1)\, dy = 0$ is exact, and find a general solution.

Applying the test provided by Theorem 1, we find

$$\frac{\partial M}{\partial y} = \frac{\partial(2x + 3y - 2)}{\partial y} = 3 \quad \text{and} \quad \frac{\partial N}{\partial x} = \frac{\partial(3x - 4y + 1)}{\partial x} = 3$$

Since the two partial derivatives are equal, the equation is exact. Its solution can therefore be found by means of Corollary 1, Theorem 1:

$$\int_a^x (2x + 3y - 2)\, dx + \int_b^y (3a - 4y + 1)\, dy = k$$

$$(x^2 + 3xy - 2x)\Big|_a^x + (3ay - 2y^2 + y)\Big|_b^y = k$$

$$x^2 + 3xy - 2y^2 - 2x + y = k + a^2 + 3ab - 2b^2 - 2a + b = K$$

Occasionally a differential equation which is not exact can be made exact by multiplying it by some simple expression. For example, if the (exact) equation $2xy^3\, dx + 3x^2y^2\, dy = 0$ is simplified by the natural process of dividing out the common factor xy^2, the resulting equation, namely, $2y\, dx + 3x\, dy = 0$, is *not* exact. Conversely, however, the last equation can be restored to its original exact form by multiplying it through by xy^2. This illustrates the general result† that every first-order equation which possesses a family of solutions can be made exact by multiplying it by a suitable factor, called an **integrating factor**. In general, the determination of an integrating factor for a given equation is difficult. However, as the following examples show, in particular cases an integrating factor can often be found by inspection.

EXAMPLE 2

Show that $1/(x^2 + y^2)$ is an integrating factor for the equation $(x^2 + y^2 - x)\, dx - y\, dy = 0$, and then solve the equation.

The test provided by Theorem 1 shows that in its present form the given equation is not exact. However, if it is multiplied by the indicated factor, it can be rewritten in the form

$$\left(1 - \frac{x}{x^2 + y^2}\right) dx - \frac{y}{x^2 + y^2}\, dy = 0$$

Testing again, we find that the last equation is exact, and we can now use Corollary 1 to find a general solution of it. However, it is simpler to observe that it can also be written

$$dx - \frac{x\, dx + y\, dy}{x^2 + y^2} = 0 \quad \text{or} \quad dx - \tfrac{1}{2}\, d[\ln (x^2 + y^2)] = 0$$

Hence, integrating, we have

$$x - \ln \sqrt{x^2 + y^2} = k$$

EXAMPLE 3

Find an integrating factor for the equation $y\, dx + (x^2y^3 + x)\, dy = 0$, and solve the equation.

Since this equation can be rewritten in the form

$$(y\, dx + x\, dy) + x^2y^3\, dy = 0$$

---

† See, for instance, Golomb and Shanks, op. cit., pp. 52–53.

and since $y\,dx + x\,dy = d(xy)$, it is natural to multiply the equation by $1/x^2y^2$, getting

$$\frac{d(xy)}{(xy)^2} + y\,dy = 0$$

This equation can now be integrated by inspection, and we have

$$-\frac{1}{xy} + \frac{y^2}{2} = k$$

EXAMPLE 4

Find an integrating factor for the equation $x\,dy - y\,dx = (4x^2 + y^2)\,dy$, and solve the equation.

In this equation, the terms on the left seem related equally well to

$$d\left(\frac{y}{x}\right) = \frac{x\,dy - y\,dx}{x^2} \qquad \text{or} \qquad d\left(\frac{x}{y}\right) = \frac{y\,dx - x\,dy}{y^2}$$

If we pursue the first suggestion and multiply the equation by $1/x^2$, we obtain

$$d\left(\frac{y}{x}\right) = \left(4 + \frac{y^2}{x^2}\right)dy$$

This equation is still not exact, but it is separable, and division by $4 + y^2/x^2$ gives us

$$\frac{d(y/x)}{4 + (y/x)^2} = dy$$

Integrating this, we have finally

$$\tfrac{1}{2}\,\mathrm{Tan}^{-1}\frac{y}{2x} = y + k$$

The results of the last three examples suggest the following observations, which are often helpful:

a. If a first-order differential equation contains the combination $x\,dx + y\,dy = \frac{1}{2}\,d(x^2 + y^2)$, try some function of $x^2 + y^2$ as a multiplier to make the equation integrable.

b. If a first-order differential equation contains the combination $x\,dy + y\,dx = d(xy)$, try some function of xy as a multiplier to make the equation integrable.

c. If a first-order differential equation contains the combination $x\,dy - y\,dx$, try $1/x^2$ or $1/y^2$ as a multiplier to make the equation integrable.

EXERCISES

Show that the following equations are exact and integrate each one:

1 $(y^2 - 1)\,dx + (2xy - \sin y)\,dy = 0$
2 $(2xy + x^3)\,dx + (x^2 + y^2)\,dy = 0$
3 $(3x^2 - 6xy)\,dx - (3x^2 + 2y)\,dy = 0$
4 $(x\sqrt{x^2 + y^2} + y)\,dx + (y\sqrt{x^2 + y^2} + x)\,dy = 0$
5 $(2xy^4 + \sin y)\,dx + (4x^2y^3 + x\cos y)\,dy = 0$

Find a general solution of each of the following equations by first multiplying by a suitable factor and then integrating:

6 $y(1 + xy)\,dx + (2y - x)\,dy = 0$　　　　**7** $3(y^4 + 1)\,dx + 4xy^3\,dy = 0$
8 $(xy^2 + y)\,dx + (x - x^2y)\,dy = 0$　　　**9** $(x^2 + y^2 + 2x)\,dy = 2y\,dx$
10 $x\,dy + 3y\,dx = xy\,dy$

Solve each of the following equations by two methods:

11 $2y\,dx + (3y - 2x)\,dy = 0$ **12** $(x^2 - y^2)\,dy = 2xy\,dx$

13 $x\,dy + y\,dx = \dfrac{dx}{y} - \dfrac{dy}{x}$ **14** $2x \ln y\,dx + \dfrac{1 + x^2}{y}\,dy = 0$

15 $\sqrt{x^2 + y^2}\,dx = x\,dy - y\,dx$

16 Solve the equation $(xy^2 - y)\,dx + (x^2 y - x)\,dy = 0$ first by integrating it as an exact equation and then by multiplying it by $1/x^2 y^2$ before integrating it. Reconcile your results.

17 Show that if the equation of Exercise 16 is multiplied by any differentiable function of the product xy, it is still exact.

18 If $\phi(x,y)$ is an integrating factor of the differential equation $M(x,y)\,dx + N(x,y)\,dy = 0$, show that ϕ satisfies the partial differential equation

$$M\frac{\partial \phi}{\partial y} - N\frac{\partial \phi}{\partial x} + \left(\frac{\partial M}{\partial y} - \frac{\partial N}{\partial x}\right)\phi = 0$$

19 Show that $f(x,y) = k$ is a general solution of the differential equation $M(x,y)\,dx + N(x,y)\,dy = 0$ if and only if

$$M\frac{\partial f}{\partial y} - N\frac{\partial f}{\partial x} \equiv 0$$

20 Using the result of the preceding exercise, show that if the equation $M(x,y)\,dx + N(x,y)\,dy = 0$ is both homogeneous and exact, its solution is $xM(x,y) + yN(x,y) = k$. *Hint:* Recall the result of Exercise 34, Sec. 1.4.

21 Show that if $\phi(x,y)$ is an integrating factor leading to the solution $f(x,y) = k$ for the differential equation $M(x,y)\,dx + N(x,y)\,dy = 0$, then $\phi F(f)$ is also an integrating factor, where F is an arbitrary differentiable function.

22 If the equation $M(x,y)\,dx + N(x,y)\,dy = 0$ is homogeneous, show that $1/(xM + yN)$ is an integrating factor. *Hint:* Observe that

$$\frac{M\,dx + N\,dy}{xM + yN} = \frac{dx}{x} + \frac{(x\,dy - y\,dx)N}{x(xM + yN)}$$

$$= \frac{dx}{x} + \frac{(x\,dy - y\,dx)/x^2}{M/N + y/x}$$

23 Prove the "if" part of Theorem 1 by first integrating $N(x,y)$ with respect to y.

24 Show that the arbitrary constants a and b which appear in the formula of Corollary 1, Theorem 1, add no generality to the solution. *Hint:* Consider the partial derivatives with respect to a and b of the left-hand side of the formula.

1.6 Linear First-Order Equations

First-order equations which are linear form an important class of differential equations which can always be routinely solved by the use of an integrating factor. By definition, a linear first-order differential equation cannot contain products, powers, or other nonlinear combinations of y or y'. Hence its most general form is

$$F(x)\frac{dy}{dx} + G(x)y = H(x)$$

If we divide this equation by $F(x)$ and rename the coefficients, it appears in the more usual form

(1) $$\frac{dy}{dx} + P(x)y = Q(x)$$

The presence of two terms on the left side of (1) involving, respectively, dy/dx and y suggests strongly that this expression is in some way related to the derivative of a product, say $\phi(x)y$, having y as one factor. Now the derivative of $\phi(x)y$ is

(2)
$$\frac{d[\phi(x)y]}{dx} = \phi(x)\frac{dy}{dx} + \frac{d\phi(x)}{dx}y$$

and the left member of (1) can be made identically equal to this if we first multiply Eq. (1) by $\phi(x)$, getting

(3)
$$\phi(x)\frac{dy}{dx} + \phi(x)P(x)y = \phi(x)Q(x)$$

and then make the second terms in (2) and (3) equal by choosing $\phi(x)$ so that

$$\frac{d\phi(x)}{dx} = \phi(x)P(x)$$

This is a simple separable equation, any nontrivial solution of which will meet our requirements. Hence, we can write, in particular,

$$\frac{d\phi(x)}{\phi(x)} = P(x)\,dx$$

$$\ln|\phi(x)| = \int P(x)\,dx$$

$$\phi(x) = \exp\left[\int P(x)\,dx\right]^\dagger$$

Thus, after Eq. (1) is multiplied by the factor

$$\phi(x) = \exp\left[\int P(x)\,dx\right]$$

it can be written in the form

$$\frac{d}{dx}\left\{\exp\left[\int P(x)\,dx\right]y\right\} = Q(x)\exp\left[\int P(x)\,dx\right]$$

The left-hand side is now an exact derivative and hence can be integrated at once. Moreover, the right-hand side is a function of x only and can therefore be integrated also with at most practical difficulties requiring numerical integration. Thus we have, on performing these integrations,

$$y\exp\left[\int P(x)\,dx\right] = \int Q(x)\exp\left[\int P(x)\,dx\right]dx + c$$

and finally, after dividing by $\exp\left[\int P(x)\,dx\right]$,

(4) $\quad y = \exp\left[-\int P(x)\,dx\right]\int Q(x)\exp\left[\int P(x)\,dx\right]dx + c\exp\left[-\int P(x)\,dx\right]$

---

† The notation $\exp f(x)$ is frequently used in place of $e^{f(x)}$, especially when $f(x)$ is a complicated expression.

The factor exp $[\int P(x)\, dx]$, which converts the general linear first-order differential equation (1) into one which can be integrated by inspection, is, of course, an integrating factor for (1).

Equation (4) should *not* be remembered as a formula for the solution of Eq. (1). Instead, a linear first-order equation should be solved by actually carrying out the steps we have described:

a. Compute the integrating factor exp $[\int P(x)\, dx]$.
b. Multiply both sides of the given equation by this factor.
c. Integrate both sides of the resulting equation, taking advantage of the fact that the integral of the left member is *always* just y times the integrating factor.
d. Solve the integrated equation for y.

EXAMPLE 1

Find the solution of the equation $(1 + x^2)(dy - dx) = 2xy\, dx$ for which $y = 1$ when $x = 0$.

Dividing the given equation by $(1 + x^2)\, dx$ and transposing, we have

$$(5) \qquad \frac{dy}{dx} - \frac{2x}{1 + x^2}\, y = 1$$

which is a linear first-order equation. In this case $P(x) = -2x/(1 + x^2)$; hence the integrating factor is

$$\exp\left(\int \frac{-2x}{1 + x^2}\, dx\right) = \exp\left[-\ln(1 + x^2)\right] = \exp\left[\ln(1 + x^2)^{-1}\right] = \frac{1}{1 + x^2} \,\dagger$$

Multiplying Eq. (5) by this factor gives the equation

$$\frac{1}{1 + x^2}\frac{dy}{dx} - \frac{2x}{(1 + x^2)^2}\, y = \frac{1}{1 + x^2}$$

Integrating this, remembering that the integral of the left member is just y times the integrating factor, we have

$$\frac{y}{1 + x^2} = \text{Tan}^{-1} x + k \qquad \text{or} \qquad y = (1 + x^2)\, \text{Tan}^{-1} x + k(1 + x^2)$$

To find the specific solution required, we substitute the given conditions $x = 0$, $y = 1$ into the general solution, getting $1 = 0 + k$. The required solution is, therefore,

$$y = (1 + x^2)\, \text{Tan}^{-1} x + (1 + x^2)$$

EXERCISES

Find a general solution of each of the following equations:

1 $(2y + x^2)\, dx = x\, dy$
2 $y' + 2xy + x = \exp(-x^2)$
3 $y' + y \tan x = \sec x$
4 $y' + y \cot x = \sin 2x$
5 $x^2\, dy + (2xy - x + 1)\, dx = 0$
6 $(1 - x^2)y' + xy = 2x$
7 $y' + \dfrac{y}{1 - x} = x^2 - x$
8 $y' = \dfrac{2y}{x + 1} + (x + 1)^3$
9 $xy' + (1 + x)y = e^{-x}$
10 $xy' + 2(1 - x^2)y = 1$

$\dagger$ Note that $\exp(\ln u) \equiv e^{\ln u} = u$, for any expression u.

Find the particular solution of each of the following equations which satisfies the indicated conditions:

11 $y' + y = e^x$ $x = 0, y = 2$

12 $y' + y = e^{-x}$ $x = 0, y = 3$

13 $(x^2 + 1) \, dy = (x^3 - 2xy + x) \, dx$ $x = 1, y = 1$

14 $y' + (1 + 2x)y = \exp(-x^2)$ $x = 0, y = 3$

15 $(1 + x^2) \, dy = (1 + xy) \, dx$ $x = 1, y = 0$

16 Find a solution of the equation $x \, dy + (x^2 - 3y) \, dx = 0$ which simultaneously meets the condition $x = -1, y = 1$ and the condition $x = 1, y = -1$.

17 Find a general solution of the equation $y^2 \, dx + (3xy - 4y^3) \, dy = 0$.   *Hint:* Consider x as the dependent variable.

18 Find a general solution of the equation $y'' + [y'/(x - 1)] = x - 1$.   *Hint:* Consider y' as the dependent variable.

19 Show that the substitution $z = y^{1-n}$ will reduce the **Bernoulli equation**†

$$y' + P(x)y = Q(x)y^n$$

to a linear first-order equation.

Using the substitution described in Exercise 19, find a general solution of each of the following equations:

20 $y' + y = xy^2$ **21** $y' + y = \dfrac{x}{y}$

22 $dy = (xy^2 + 3xy) \, dx$ **23** $3xy' + y + x^2y^4 = 0$

24 Show that if one solution, say $y = u(x)$, of the **Riccati equation**‡

$$y' = P(x)y^2 + Q(x)y + R(x)$$

is known, then the substitution $y = u + (1/z)$ will transform this equation into a linear first-order equation in the new dependent variable z.

Using the substitution described in Exercise 24, find a general solution of each of the following equations:

25 $y' = xy^2 + (1 - 2x)y + x - 1$.   *Hint:* Observe that $y = 1$ is a solution of this equation.

26 $y' = (y - 1)[y + (1/x)]$

27 $y' = (y + x)(y + x - 2)$.   *Hint:* A particular solution of this equation should be evident by inspection.

28 $y' = (y + x + 3)(y + x + \frac{1}{2})$

29 Prove that no extra generality in the final answer results from using

$$\exp\left[\int P(x) \, dx + c\right]$$

instead of just $\exp[\int P(x) \, dx]$ as an integrating factor for the equation

$$y' + P(x)y = Q(x)$$

1.7 Applications of First-Order Differential Equations

The mathematical formulation of physical problems involving continuously changing quantities often leads to differential equations of the first order. The following examples show how such equations arise and how they are handled.

---

† Named for the Swiss mathematician Jakob Bernoulli (1654–1705), a member of a family which in little more than a century produced eight distinguished mathematicians.

‡ Named for the Italian mathematician J. F. Riccati (1676–1754).

EXAMPLE 1

A tank is initially filled with 100 gal of salt solution containing 1 lb of salt per gallon. Fresh brine containing 2 lb of salt per gallon runs into the tank at the rate of 5 gal/min, and the mixture, assumed to be kept uniform by stirring, runs out at the same rate. Find the amount of salt in the tank at any time t, and determine how long it will take for this amount to reach 150 lb.

Let Q lb be the total amount of salt in solution in the tank at any time t, and let dQ be the increase in this amount during the infinitesimal interval of time dt. At any time t, the amount of salt per gallon of solution is therefore $Q/100$ (lb/gal). Now the change dQ in the total amount of salt in the tank is clearly the net gain in the interval dt due to the fresh brine running into and the mixture running out of the tank. The rate at which salt enters the tank is

$$5 \text{ gal/min} \times 2 \text{ lb/gal} = 10 \text{ lb/min}$$

Hence in the interval dt the gain in salt from this source is

$$10 \text{ lb/min} \times dt \text{ min} = 10 \, dt \text{ lb}$$

Likewise, since the concentration of salt in the mixture as it leaves the tank is the same as the concentration $Q/100$ in the tank itself, the amount of salt leaving the tank in the interval dt is

$$5 \text{ gal/min} \times \frac{Q}{100} \text{ lb/gal} \times dt \text{ min} = \frac{Q}{20} \, dt \text{ lb}$$

Therefore,

$$dQ = 10 \, dt - \frac{Q}{20} \, dt = \left(10 - \frac{Q}{20}\right) dt$$

This equation can be written in the form

(1) $$\frac{dQ}{200 - Q} = \frac{dt}{20}$$

and solved as a separable equation, or it can be written

(2) $$\frac{dQ}{dt} + \frac{Q}{20} = 10$$

and treated as a linear equation.

Considering it as a linear equation, we must first compute the integrating factor

$$\exp\left(\int P \, dt\right) = \exp\left(\int \frac{dt}{20}\right) = e^{t/20}$$

Multiplying Eq. (2) by this factor gives

$$e^{t/20}\left(\frac{dQ}{dt} + \frac{Q}{20}\right) = 10e^{t/20}$$

From this, by integration, we obtain

$$Qe^{t/20} = 200e^{t/20} + c \quad \text{or} \quad Q = 200 + ce^{-t/20}$$

Substituting the initial conditions $t = 0$, $Q = 100$, we find

$$100 = 200 + c \quad \text{or} \quad c = -100$$

Hence,

$$Q = 200 - 100e^{-t/20}$$

To find how long it will be before there is 150 lb of salt in the tank, we must find the value of t such that

$$150 = 200 - 100e^{-t/20} \quad \text{or} \quad e^{-t/20} = \tfrac{1}{2}$$

From this we have at once

$$-\frac{t}{20} = \ln \tfrac{1}{2} = -\ln 2 = -0.693 \quad \text{and} \quad t = 13.9 \text{ min}$$

EXAMPLE 2

A hemispherical tank of radius R is initially filled with water. At the bottom of the tank there is a hole of radius r through which the water drains under the influence of gravity. Find the depth of the water at any time t, and determine how long it will take the tank to drain completely.

Let the origin be chosen at the lowest point of the tank, let y be the instantaneous depth of the water, and let x be the instantaneous radius of the free surface of the water (Fig. 1.3). Then in the infinitesimal interval dt the water level will fall by the amount dy, and the resultant decrease in the volume of water in the tank will be

$$dV = \pi x^2 \, dy$$

This, of course, must equal the volume of water that leaves the orifice during the same interval dt. Now from **Torricelli's law**† the velocity with which a liquid issues from an orifice is

$$v = \sqrt{2gh}$$

where g is the acceleration of gravity and h is the instantaneous height, or **head**, of the liquid above the orifice. In the interval dt, then, a stream of water of length $v \times dt = \sqrt{2gy} \, dt$ and of cross-sectional area πr^2‡ will emerge from the outlet. The volume of this amount of water is

$$dV = \text{area} \times \text{length} = \pi r^2 \sqrt{2gy} \, dt$$

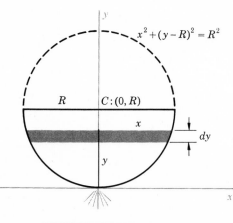

$$x^2 + (y - R)^2 = R^2$$

Figure 1.3
A vertical plane section through the center of a hemispherical tank.

---

† Named for the Italian mathematician and physicist Evangelista Torricelli (1608–1647).
‡ This neglects the fact that the stream contracts as it leaves the orifice. How much the cross section of the stream decreases depends in a very complicated way upon the size and shape of both the tank and the orifice and also upon the head. However, in most practical problems reasonably accurate answers can be obtained by assuming that the cross section of the stream just after it leaves the orifice is 0.6 times the area of the orifice.

Hence, equating the two expressions for dV, we obtain the differential equation

$$(3) \qquad \pi x^2 \, dy = -\pi r^2 \sqrt{2gy} \, dt$$

the minus sign indicating that as t increases, the depth y decreases.

Before this equation can be solved, x must be expressed in terms of y. This is easily done through the use of the equation of the circle which describes the vertical cross section of the tank:

$$x^2 + (y - R)^2 = R^2 \qquad \text{or} \qquad x^2 = 2yR - y^2$$

With this, the differential equation (3) can be written

$$\pi(2yR - y^2) \, dy = -\pi r^2 \sqrt{2gy} \, dt$$

This is a simple separable equation which can be solved without difficulty:

$$(2Ry^{1/2} - y^{3/2}) \, dy = -r^2 \sqrt{2g} \, dt$$

$$\tfrac{4}{3} R y^{3/2} - \tfrac{2}{5} y^{5/2} = -r^2 \sqrt{2g} \, t + c$$

Since $y = R$ when $t = 0$, we find

$$\tfrac{14}{15} R^{5/2} = c$$

and thus

$$\tfrac{4}{3} R y^{3/2} - \tfrac{2}{5} y^{5/2} = -r^2 \sqrt{2g} \, t + \tfrac{14}{15} R^{5/2}$$

To find how long it will take the tank to empty, we must determine the value of t when $y = 0$:

$$0 = -r^2 \sqrt{2g} \, t + \tfrac{14}{15} R^{5/2}$$

$$t = \frac{14}{15} \frac{R^{5/2}}{r^2 \sqrt{2g}}$$

EXAMPLE 3

Under certain conditions it is observed that the rate at which a solid substance dissolves varies directly as the product of the amount of undissolved solid present in the solvent and the difference between the instantaneous concentration and the saturation concentration of the substance. If 20 lb of solute is dumped into a tank containing 120 lb of solvent and at the end of 12 min the concentration is observed to be 1 part in 30, find the amount of solute in solution at any time t if the saturation concentration is 1 part of solute in 3 parts of solvent.

If Q is the amount of the material in solution at time t, then $20 - Q$ is the amount of undissolved material present at that time and $Q/120$ is the corresponding concentration. Hence, according to the given law,

$$\frac{dQ}{dt} = k(20 - Q)\left(\frac{1}{3} - \frac{Q}{120}\right) = \frac{k}{120}(20 - Q)(40 - Q)$$

This is a simple separable equation, and we have at once

$$\frac{dQ}{(20 - Q)(40 - Q)} = \frac{k}{120} \, dt$$

To integrate the left member it is convenient to use the method of partial fractions and write

$$\frac{1}{(20 - Q)(40 - Q)} = \frac{A}{20 - Q} + \frac{B}{40 - Q} = \frac{A(40 - Q) + B(20 - Q)}{(20 - Q)(40 - Q)}$$

This will be an identity if and only if

$$1 = A(40 - Q) + B(20 - Q)$$

Setting $Q = 20$ and $Q = 40$, in turn, we find from this that

$$A = \tfrac{1}{20} \quad \text{and} \quad B = -\tfrac{1}{20}$$

Hence the differential equation can be written

$$\frac{1}{20}\left(\frac{1}{20 - Q} - \frac{1}{40 - Q}\right) dQ = \frac{k}{120}\, dt$$

and, integrating, we have

(4) $$-\ln(20 - Q) + \ln(40 - Q) = \frac{k}{6}\, t + c$$

To determine the integration constant c we observe that when $t = 0$, the amount $Q = Q_0$ of dissolved material is zero. Hence

$$-\ln 20 + \ln 40 = c \quad \text{or} \quad c = \ln 2$$

and Eq. (4) can be written

(5) $$\ln \frac{40 - Q}{2(20 - Q)} = \frac{k}{6}\, t$$

To find the physical constant k we use the fact that when $t = 12$, the concentration $Q/120$ is $\tfrac{1}{30}$, or $Q = 4$. Substituting these values gives

$$\ln \tfrac{36}{32} = 2k \quad \text{or} \quad k = \tfrac{1}{2}\ln \tfrac{9}{8} = 0.05889$$

Passing to exponential form from Eq. (5), in order to solve for Q, we have

$$\frac{40 - Q}{40 - 2Q} = e^{0.0098t}$$

and finally

$$Q = \frac{40 - 40e^{0.0098t}}{1 - 2e^{0.0098t}} = \frac{40(1 - e^{-0.0098t})}{2 - e^{-0.0098t}}$$

EXAMPLE 4

According to **Fourier's law**[†] **of heat conduction**, the amount of heat in Btu per unit time flowing through an area is proportional to the area and to the temperature gradient, in degrees per unit length, in the direction of the perpendicular to the area. On the basis of this law, obtain a formula for the steady-state heat loss per unit time from a unit length of pipe of radius r_0 carrying steam at temperature T_0 if the pipe is covered with insulation of thickness w, the outer surface of which remains at the constant temperature T_1. What is the temperature distribution through the insulation, i.e., what is the temperature in the insulation as a function of the radius?

Since the problem tells us that steady-state conditions have been reached, it follows that the heat loss per unit time from a unit length of the pipe is a constant independent of time, say Q. Furthermore it is reasonable to suppose that heat conduction through the insulation in the direction of the length of the pipe is negligible in comparison with the heat flow in the radial direction; and this we shall assume to be the case. We shall also make the obvious assumption that the heat flow through the insulation has circular symmetry; i.e., we shall assume that the temperature in the insulation depends only on the radial distance r. Let us now consider a typical cross section of the pipe and insulation, as suggested in Fig. 1.4. Clearly, under the assumption that all heat flow through the insulation is radial, it follows that for the unit length of pipe we are considering, all the heat that passes into the insulation through its inner surface will eventually pass into the air through its outer surface. Moreover, on the way, this same

† See footnote p. 215.

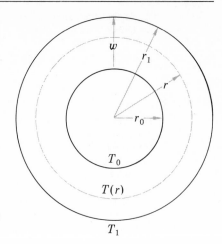

Figure 1.4
A typical cross section of an insulated pipe.

amount of heat Q will also pass through every coaxial cylindrical area between r and $r_1 = r_0 + w$. Now if we let T denote the temperature in the insulation at the radius r, it follows that dT/dr is the temperature gradient, or temperature change per unit length, in the direction perpendicular to the cylindrical area of radius r. Hence, by Fourier's law, we have for the (as yet unknown) amount of heat Q flowing through this general area per unit time

$$Q = \text{thermal conductivity} \times \text{area} \times \text{temperature gradient}$$

$$= k(1 \times 2\pi r)\frac{dT}{dr}$$

We thus have the exceedingly simple separable equation

$$dT = \frac{Q}{2\pi k}\frac{dr}{r}$$

Hence

$$T = \frac{Q}{2\pi k}\ln r + c$$

To determine the integration constant c, we use the fact that $T = T_0$ when $r = r_0$; whence

$$T_0 = \frac{Q}{2\pi k}\ln r_0 + c \qquad \text{or} \qquad c = T_0 - \frac{Q}{2\pi k}\ln r_0$$

and, substituting and collecting terms,

(6) $$T = T_0 + \frac{Q}{2\pi k}(\ln r - \ln r_0)$$

Furthermore, $T = T_1$ when $r = r_0 + w = r_1$. Hence

$$T_1 = T_0 + \frac{Q}{2\pi k}(\ln r_1 - \ln r_0)$$

from which we find easily that

(7) $$Q = \frac{(T_1 - T_0)2\pi k}{\ln r_1 - \ln r_0}$$

Since k is the (presumably) known thermal conductivity of the insulation, this formula gives the heat loss per unit time, as required.

To find the temperature distribution through the insulation, we merely substitute for $Q/2\pi k$ from Eq. (7) into Eq. (6), getting

$$T = T_0 + (T_1 - T_0)\frac{\ln r - \ln r_0}{\ln r_1 - \ln r_0}$$

EXAMPLE 5

The curves of a family C are said to be **orthogonal trajectories** of the curves of a family K, and vice versa, if at every intersection of a curve of C with a curve of K the two curves are perpendicular. Using this definition, find the equation of the family of orthogonal trajectories of the curves of the family defined by the equation $y^2 = 4cx$.

To find the equation of the orthogonal trajectories of the curves of the given family, we must first find the expression for the slope of a general curve of this family at a general point. This, of course, is just the process of finding the differential equation satisfied by a given family of functions, which we illustrated in Example 7, Sec. 1.2. In this case, differentiating the equation $y^2 = 4cx$, we have

(8) $$2yy' = 4c$$

Then, by eliminating c, we find $y^2 = 2yy'x$, and from this we obtain the required slope formula

$$y' = \frac{y}{2x}$$

Since the curves $y^2 = 4cx$ and their orthogonal trajectories are to be perpendicular at every intersection, it follows that at every point the slopes of the members of the two families which pass through that point must be negative reciprocals. Thus the required curves must be the solution curves of the differential equation

$$y' = -\frac{2x}{y}\dagger$$

This is a homogeneous equation, but it is simpler to treat it as a separable equation:

$$y\,dy + 2x\,dx = 0$$

$$\frac{y^2}{2} + x^2 = k^2$$

where the integration constant has been written as a square to emphasize the fact that it cannot be negative, since it is the sum of two squares. Typical curves of the two families are shown in Fig. 1.5.

Although the concept of orthogonal trajectories appears to be essentially a geometric one, it actually is intimately related to many important physical problems. For instance, in what are known in physics as two-dimensional field problems, the equipotential

---

† Note that it would be incorrect to take the slope formula given by Eq. (8), namely,

$$y' = \frac{2c}{y}$$

and use its negative reciprocal

$$y' = -\frac{y}{2c}$$

as the differential equation to be satisfied by the required orthogonal trajectories. The curves obtained by integrating this simple separable equation, namely, $y = ke^{-x/2c}$, depend on the two constants k and c, and, in general, a curve of this family and a curve of the given family will intersect at right angles only if they correspond to the same value of c.

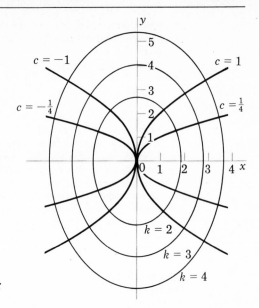

Figure 1.5
The curves $y^2 = 4cx$ and their
orthogonal trajectories.

lines and the lines of flux are orthogonal trajectories. More specifically, if one knows the family of isothermal curves, or curves joining points at the same temperature, in some problem in heat flow, then their orthogonal trajectories are the lines in which the heat flows. Thus in Fig. 1.5, if the ellipses are thought of as isothermal curves, the parabolas are the paths along which the heat flow takes place.

EXERCISES

1 Radium disintegrates at a rate proportional to the amount of radium instantaneously present. If one-half of any given amount of radium will disappear in 1,590 years, what fraction will disintegrate during the first century? During the tenth century? How long will it be before one-fourth of the original amount has disintegrated?

2 Living tissues, both plant and animal, contain carbon, derived ultimately from the carbon dioxide in the air. Most of this carbon is the stable isotope ^{12}C, but a small fixed percentage of it is unstable radioactive ^{14}C. It appears that there is little or no segregation of these two forms of carbon in living organisms, and the ratio $^{14}C/^{12}C$ is essentially constant for all types of tissue. When the tissue dies, the vital processes, of course, end, no more carbon of either form is added to the tissue, and the amount of ^{14}C present at the time of death decreases at a rate proportional to the amount instantaneously present, with a half-life of approximately 5,500 years.
(a) If x_0 is the amount of ^{14}C in a given specimen of tissue at the moment of death (determined as a known percentage of the unchanged amount of ^{12}C calculated by chemical analysis of the specimen), and if x is the amount present in the tissue t years after death, express x as a function of t.
(b) A piece of charcoal found in the Lascaux Cave in France (the cave with the remarkable paintings of prehistoric animals) contained 14.8 percent of the original amount of ^{14}C. Date the occupation of the cave that produced the charcoal.
(c) A charred branch of a tree killed by the eruption that formed Crater Lake, Oregon, contained 44.5 percent of the original amount of ^{14}C. Date the eruption.

3 For a certain population, both the birth rate and the death rate are constant multiples of the number of individuals instantaneously present. Find the population as a function of time.

4 Work Exercise 3 if the proportionality factor in the birth rate, instead of remaining constant, decreases uniformly with time.

5 Work Exercise 3 if the proportionality factors in both the birth rate and the death rate are exponentially decreasing functions of time.

6 It is a fact of common experience that when a rope is wound around a rough cylinder, a small force at one end can resist a much larger force at the other. Quantitatively, it is found that throughout the portion of the rope in contact with the cylinder, the change in tension per unit length is proportional to the tension, the numerical value of the proportionality constant being the coefficient of friction between the rope and the cylinder divided by the radius of the cylinder. Assuming a coefficient of friction of 0.35, how many times must a rope be snubbed around a post 1 ft in diameter for a man holding one end to be able to resist a force 200 times greater than he can exert?

7 According to **Lambert's law of absorption**,† when light passes through a transparent medium, the amount absorbed by any thin layer of the material perpendicular to the direction of the light is proportional to the amount incident on that layer and to the thickness of that layer. In his underwater explorations off Bermuda, Beebe observed that at a depth of 50 ft the intensity of illumination was 10 cd/ft², and that at 250 ft it had fallen to 0.2 cd/ft². Find the law connecting intensity with depth in this case.

8 (a) If p lb/in² is the atmospheric pressure and ρ lb/in³ is the density of air at a height of h in above the surface of the earth, show that $(dp/dh) + \rho = 0$.
(b) If $p = 14.7$ lb/in² at sea level, and if it has fallen to half this value at 18,000 ft, find the formula for the pressure at any height, given that p is proportional to ρ (isothermal conditions). What is the predicted height of the atmosphere under this assumption?
(c) Work part (b), given that p is proportional to $\rho^{1.4}$ (adiabatic conditions).

9 Although water is often assumed to be incompressible, it actually is not. In fact, using pounds and feet as units, the weight of 1 ft³ of water under pressure p is approximately $w(1 + kp)$, where $w = 64$, $k = 2 \times 10^{-8}$, and p is measured from standard atmospheric pressure as an origin. Using this information, find the pressure at any depth y below the surface of the ocean. At a depth of 6 mi, by what factor does the actual pressure exceed the pressure computed on the assumption that water is incompressible?

10 When ethyl acetate in dilute aqueous solution is heated in the presence of a small amount of acid, it decomposes according to the equation

$$\underset{\text{Ethyl acetate}}{CH_3COOC_2H_5} + \underset{\text{Water}}{H_2O} \rightarrow \underset{\text{Acetic acid}}{CH_3COOH} + \underset{\text{Ethyl alcohol}}{C_2H_5OH}$$

Since this reaction takes place in dilute solution, the quantity of water present is so great that the loss of the small amount which combines with the ethyl acetate produces no appreciable change in the total amount. Hence of the reacting substances only the ethyl acetate suffers a measurable change in concentration. A chemical reaction of this sort, in which the concentration of only one reacting substance changes, is called a **first-order reaction**. It is a law of physical chemistry that the rate at which a substance is being used up, i.e., transformed, in a first-order reaction is proportional to the amount of that substance instantaneously present. If the initial concentration of ethyl acetate is C_0, find the expression for its concentration at any time t.

11 In some chemical reactions where two substances combine to form a third, the amount of each of the reacting substances changes appreciably. In such cases it is observed that the rate at which the resulting compound is formed is proportional to the product of the untransformed amounts of the two reacting substances. If two substances combine in the ratio 1:2, by weight, to form a third substance, and

---

† Named for the German mathematician and astronomer Johann Heinrich Lambert (1728–1777).

if it is observed that 10 min after 10 g of the first substance and 20 g of the second are mixed the amount of the product which has been formed is 5 g, find an expression for the amount of the product present at any time t. How long will it be before one-half the final amount of the product has been formed?

12 (a) Work Exercise 11 given that instead of 10 g of the first substance and 20 g of the second, 20 g of each substance are mixed.
(b) Work Exercise 11 given that 10 g of the first substance and 30 g of the second are mixed.

13 Work Example 3 given that the amount of solute dumped into the tank is 40 lb instead of 20 lb.

14 Work Example 3 given that the saturation concentration is 1 part of solute to 12 parts of solvent.

15 Work Example 3 given that the saturation concentration is 1 part of solute to 6 parts of solvent.

16 Work Example 3 with *concentration* defined as the ratio of solute to solution instead of solute to solvent.

17 Work Exercise 16 if the saturation concentration is 1 part of solute to 7 parts of solution.

18 A tank contains 100 gal of brine in which 50 lb of salt is dissolved. Brine containing 2 lb/gal of salt runs into the tank at the rate of 3 gal/min, and the mixture, assumed to be kept uniform by stirring, runs out of the tank at the rate of 2 gal/min. Assuming the tank sufficiently large to avoid overflow, find the amount of salt in the tank as a function of t. When will the concentration of salt in the tank reach $\frac{3}{2}$ lb/gal?

19 Work Exercise 18 with the rates of influx and efflux interchanged. When does the tank contain the maximum amount of salt, and what is this amount?

Find the orthogonal trajectories of the curves of each of the following families:

20 $x^2 - y^2 = c$ **21** $y^2 = cx^3$

22 $y = (x - c)^2$ **23** $x^2 + 2y^2 = cy$

24 $y^2 = x^2 + cx$ **25** $x^2 + y^2 = cx$

26 Families of orthogonal trajectories are not always found through the use of differential equations. In particular, according to Property 2, Sec. 15.6, if any analytic, i.e., differentiable, function of the complex variable $z = x + iy$ is reduced to the standard form $u(x,y) + iv(x,y)$, where $u(x,y)$ and $v(x,y)$ are real functions, then the curves $v(x,y) = k$ are the orthogonal trajectories of the curves $u(x,y) = c$, and vice versa. Verify this fact for the function $1/z$ and compare your results with the results of Exercise 25.

27 Verify the property described in Exercise 26 for the function z^3.

28 According to **Newton's law of cooling**, the rate at which the temperature of a body decreases is proportional to the difference between the instantaneous temperature of the body and the temperature of the surrounding medium. If a body whose temperature is initially 100°C is allowed to cool in air which remains at the constant temperature 20°C, and if it is observed that in 10 min the body has cooled to 60°C, find the temperature of the body as a function of time.

29 Obtain a formula for the amount of heat lost under steady-state conditions from 1 ft² of furnace wall h ft thick if the temperature in the furnace is T_0, the temperature of the air outside the furnace is T_1, and the thermal conductivity of the material of the furnace wall is k. What is the temperature distribution through the wall?

30 The inner and outer surfaces of a hollow sphere are maintained at the respective temperatures T_0 and T_1. If the inner and outer radii of the spherical shell are r_0 and r_1, and if the thermal conductivity of the material of the shell is k, find the amount of heat lost from the sphere per unit time. What is the temperature distribution through the shell?

31 A tank and its contents weigh 100 lb. The average heat capacity of the system is 0.5 Btu/(lb)(°F). The liquid in the tank is heated by an immersion heater which delivers 100 Btu/min. Heat is lost from the system at a rate proportional to the difference between the temperature of the system, assumed constant throughout at any instant, and the temperature of the surrounding air, the proportionality constant being 2 Btu/(min)(°F). If the air temperature remains constant at 70°F, and if the initial temperature of the tank and its contents is 55°F, find the temperature of the tank as a function of time.

32 A perfectly insulated tank of negligible heat capacity contains P lb of brine at T_0°F. Hot brine at T_a°F runs into the tank at the rate of a lb/min, and the brine in the tank, brought instantly to a uniform temperature throughout by vigorous stirring, runs out at the same rate. If the specific heat of the brine is 1 Btu/(lb)(°F), find the temperature of the brine in the tank as a function of time. *Hint:* If h_0 is the amount of heat in 1 lb of brine at the temperature T_0, the total amount of heat in the brine in the tank when its temperature is T is $H = P[(T - T_0) + h_0]$. The change in the heat content of the brine in the tank during the time interval dt is then dH.

33 Work Exercise 32 if the brine, instead of running out at the influx rate of a lb/min, runs out at the rate of b lb/min, where $b \neq a$. Verify that the solution to this exercise approaches the solution to Exercise 32 as $b \to a$. *Hint:* With the notation introduced in the hint to Exercise 32, the heat content of the brine in the tank in the present problem is $H = [P + (a - b)t][(T - T_0) + h_0]$.

34 A perfectly insulated tank of negligible heat capacity contains 1,000 lb of brine at 60°F. It is necessary that in exactly 20 min the tank contains 2,000 lb of brine at 150°F. To accomplish this, hot brine at 160°F is run into the tank and mixed brine, kept at a uniform temperature by stirring, is run off. What should be the rates of influx and efflux to achieve the desired conditions?

35 A metal ball at temperature 100°C is placed in a tank of negligible heat capacity containing water at 40°C. The tank is perfectly insulated so that no heat is lost from the system, and the only transfer of heat is that from the ball to the water. After 15 min it is observed that the temperature of the water is 50°C and the temperature of the ball is 80°C. Assuming that the temperature of the ball at any instant is the same at all points and that the temperature of the water at any instant is the same at all points, find the temperature of the ball as a function of time. When will the temperature of the ball be 75°C? *Hint:* Contrary to our first impression, numerical values for the mass of the water, the mass of the ball, and the specific heats of the water and the material of the ball are not needed. Introducing convenient symbols for these quantities, compute the amount of heat in the system at $t = 0$, $t = 15$, $t = \infty$, and $t = t$. Then, observing that these are all equal, show that the temperature of the water is a linear function of the temperature of the ball. Finally, use Newton's law of cooling (Exercise 28) to set up the appropriate differential equation.

36 Work Example 2 if the tank has the shape of an inverted right circular cone of radius R and height h.

37 Work Example 2 if the tank is a sphere initially filled with water, instead of just a hemisphere. How long will it take for half the water to run out? How long will it take for the tank to drain? (Assume that there is a small hole at the top of the sphere where air can enter the tank.)

38 Work Example 2 if, instead of draining through an orifice of constant area, the tank drains through an orifice whose area is controlled by a float valve in such a way that it is proportional to the instantaneous depth of the water in the tank.

39 A tank having the shape of a right circular cylinder of radius R and height h is filled with water. The tank drains through an orifice whose area is controlled by a float valve in such a way that it is proportional to the instantaneous depth of the water. Express the depth of the water as a function of time. How long will it take the tank to drain?

40 Work Example 2 if the tank has the shape of a right circular cylinder of radius R and height h and if in addition to a hole of radius r in the bottom of the tank there is also a hole of radius r in the side at a height of $h/2$ above the base.

41 A cylindrical tank of length l has semicircular end sections of radius R. The tank is placed with its axis horizontal and is initially filled with water. How long will it take the tank to drain through a hole of radius r in the bottom of the tank?

42 What is the shape of a tank which is a surface of revolution if the tank drains so that the water level falls at a constant rate?

43 What is the shape of a typical perpendicular cross section of a horizontal trough of constant cross section which drains so that the water level falls at a constant rate?

44 Water flows into a vertical cylindrical tank of cross sectional-area A ft² at the rate of Q ft³/min. At the same time the water flows out under the influence of gravity through a hole of area a ft² in the base of the tank. If the water is initially h ft deep, find the instantaneous depth as a function of time. What is the limiting depth of the water as time increases indefinitely?

45 A vertical cylindrical tank of radius R and height h has a narrow crack of width w running vertically from top to bottom. If the tank is initially filled with water and allowed to drain through the crack under the influence of gravity, find the instantaneous depth of the water as a function of time. How long will it take the tank to empty? *Hint:* First imagine the crack to be a series of adjacent orifices, and integrate to find the total efflux from the crack in the infinitesimal time interval dt.

46 A mothball loses mass by evaporation at a rate proportional to its instantaneous surface area. If half the mass is lost in 100 days, how long will it be before the radius has decreased to one-half its initial value? How long will it be before the mothball disappears completely?

47 When a volatile substance is placed in a sealed container, molecules leave its surface at a rate proportional to the area of the surface and return at a rate proportional to the amount which has evaporated. If a volatile material is spread evenly to a depth h over the bottom of a closed box, find the depth of the material at any time. Under what conditions, if any, will all the material eventually evaporate?

48 A sphere of volatile material is suddenly introduced into a closed container. Assuming the same laws of evaporation and condensation described in Exercise 47, set up the differential equation whose solution gives the radius of the sphere as a function of time.

49 A barge is being towed at 16 ft/s when the towline breaks. It continues thereafter in a straight line but slows down at a rate proportional to the square root of its instantaneous velocity. If 2 min after the towline breaks, the velocity of the barge is observed to be 9 ft/s, how far does it move before it comes to rest?

50 A rapidly rotating flywheel, after power is shut off, coasts to rest under the influence of a friction torque which is proportional to the instantaneous angular velocity ω. If the moment of inertia of the flywheel is I, and if its initial angular velocity is ω_0, find its instantaneous angular velocity as a function of time. How long will it take the flywheel to come to rest? *Hint:* Use **Newton's law in torsional form:**

$$\text{Moment of inertia} \times \text{angular acceleration} = \text{torque}$$

to set up the differential equation describing the motion.

51 The friction torque acting to slow down a flywheel is actually not proportional to the first power of the angular velocity at all speeds. As a more realistic example than Exercise 50, suppose that a flywheel of moment of inertia $I = 7.5$ lb-ft/s² coasts to rest from an initial speed of 1,000 rad/min under the influence of a retarding torque T estimated to be

$$T = \begin{cases} \dfrac{\sqrt{\omega}}{10} \text{ ft-lb} & 0 < \omega < 100 \text{ rad/min} \\[2ex] \dfrac{1}{10}\left(7.5 + \dfrac{\omega^2}{4{,}000}\right) \text{ ft-lb} & 100 < \omega < 1{,}000 \text{ rad/min} \end{cases}$$

Find ω as a function of time, and determine how long it will take the flywheel to come to rest.

52 A body weighing w lb falls from rest under the influence of gravity and a retarding force due to air resistance, assumed to be proportional to the velocity of the body. Find the equations expressing the velocity of fall and the distance fallen as functions of time, and verify that these reduce to the ideal laws

$$v = gt \quad \text{and} \quad s = \tfrac{1}{2}gt^2$$

when the coefficient of air resistance approaches zero. *Hint:* Use **Newton's law,**

$$\text{mass} \times \text{acceleration} = \text{force}$$

to set up the differential equation which describes the motion.

53 An object is projected upward from the surface of the earth with velocity 64 ft/s. Air resistance is proportional to the first power of the velocity, the proportionality constant being such that if the body were to fall freely, its velocity would approach the limiting value $v_\infty = 256$ ft/s. How high does the body rise? When and with what velocity does it strike the ground?

54 Work Exercise 52 given that the retarding force due to air resistance is proportional to the square of the velocity of fall.

55 A body of weight w falls from rest under the influence of gravity and a retarding force proportional to the nth power of the velocity. Show that the velocity of the body approaches the limiting value $v_\infty = \sqrt[n]{w/k}$, where k is the proportionality constant in the law of air resistance. Show, also, that the time τ that it takes the body to reach one-half its limiting velocity is given by the equation

$$\tau = \alpha \frac{v_\infty}{g}$$

and compute the value of α for:

(a) $n = \tfrac{1}{4}$ (b) $n = \tfrac{1}{3}$ (c) $n = \tfrac{1}{2}$ (d) $n = 1$
(e) $n = 2$ (f) $n = 3$ (g) $n = 4$

56 A particle of mass m moves along the x-axis under the influence of a force which is directed toward the origin and proportional to the distance of the particle from the origin. If the body starts from rest at the point where $x = x_0$, find the equations which express its velocity and its distance from the origin as functions of time. *Hint:*

$$\frac{dv}{dt} = \frac{dv}{dx}\frac{dx}{dt} = v\frac{dv}{dx}$$

57 Work Exercise 56 (a) if the particle starts from the origin with velocity $v = v_0$; (b) if the body starts from the point $x = x_0$ with velocity $v = v_0$.

58 Work Exercise 56 if the force, instead of being directed toward the origin, is directed away from the origin.

59 Work Exercise 56 (a) if the particle moves under the influence of a force which is directed toward the origin and inversely proportional to the square of the distance of the particle from the origin; (b) if the particle moves under the influence of a force which is directed away from the origin and inversely proportional to the square of the distance of the particle from the origin.

60 A body falls from rest from a height so great that the fact that the force of gravity varies inversely as the square of the distance from the center of the earth cannot be neglected. Find the equations expressing the velocity of fall and the distance fallen as functions of time in the ideal case in which air resistance is neglected. (Note the hint to Exercise 56.)

61 Under the conditions of Exercise 60, determine the minimum initial velocity with which a body must be projected upward if it is to leave the earth and never return.

62 A cylinder of mass m and radius r rolls, without sliding, down an inclined plane of inclination angle α. Neglecting friction, express the distance which the cylinder has rolled down the plane as a function of time. *Hint:* Observe first that since friction is neglected, the principle of the conservation of energy implies that the sum of the kinetic energy and the potential energy of the cylinder remains constant throughout the motion. Then note that the kinetic energy of the cylinder consists of two parts: that due to the translation of the cylinder and that due to its rotation.

63 Work Exercise 62 for a rolling sphere of mass m and radius r.

64 When a capacitor of capacity C is being charged through a resistance R by a battery which supplies a constant voltage E, the instantaneous charge Q on the capacitor satisfies the differential equation

$$R\frac{dQ}{dt} + \frac{Q}{C} = E\,\dagger$$

Find Q as a function of time if the capacitor is initially uncharged, i.e., if $Q_0 = 0$. How long will it be before the charge on the capacitor is one-half its final value?

65 Work Exercise 64 if the battery is replaced by a generator which supplies an alternating voltage equal to $E_0 \sin \omega t$.

66 Determine how an initially charged capacitor will discharge through a resistance; i.e., work Exercise 64 if there is no voltage source in the circuit and the charge on the capacitor is initially some nonzero value Q_0.

67 When a switch is closed in a circuit containing a resistance R, an inductance L, and a battery which supplies a constant voltage E, the current i builds up at a rate defined by the equation

$$L\frac{di}{dt} + Ri = E\,\ddagger$$

Find the current i as a function of time. How long will it take i to reach one-half its final value?

68 Work Exercise 67 if the battery is replaced by a generator which supplies an alternating voltage equal to $E_0 \cos \omega t$.

69 What is the equation of the curve in which a perfectly flexible cable, of uniform weight per unit length w, will hang when it is suspended between two points at the same height? *Hint:* Consider a section of the cable, such as that shown in Fig. 1.6, and note that if H is the horizontal tension in the cable at its lowest point A,

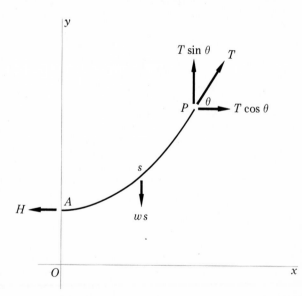

Figure 1.6

---

† For a derivation of this equation see Sec. 5.2.
‡ For a derivation of this equation see Sec. 5.2.

if T is the tension in the cable at a general point P, and if s is the length of the cable between A and P, then

$$T \sin \theta = ws \qquad T \cos \theta = H$$

These imply that

$$\tan \theta = y' = \frac{ws}{H} \quad \text{and hence} \quad \frac{dy'}{dx} = \frac{w}{H} \frac{ds}{dx}$$

After ds/dx is expressed in terms of y' by means of the familiar formula for the differential of arc length, the integrations required to find y' and then y will be simplified if they are carried out in terms of hyperbolic functions.

70 Find the equation of the curve in which a perfectly flexible cable hangs when instead of bearing a uniform load per unit length, as in Exercise 69, it bears a uniform horizontal load. (This is approximately the case when the cable is part of a suspension bridge carrying a horizontal roadbed whose weight is much greater than the weight of the cable.)

71 Find the equation of a curve with the property that light rays emanating from the origin will all be reflected from it into rays which are parallel to the x-axis. *Hint:* Note in Fig. 1.7 that if t is the tangent to the required curve at a general point P, then the angles 1, 2, and 3 all have the same measure, say α. Then from the exterior-angle theorem $\alpha = \beta - \alpha$, which implies that $\tan \alpha = \tan(\beta - \alpha)$. Expanding $\tan(\beta - \alpha)$ and observing that $\tan \alpha = y'$ and $\tan \beta = y/x$ leads to the differential equation $y = 2xy' + y(y')^2$. This can be solved by first solving for y' and then using an integrating factor, or it can be reduced to a Clairaut equation by multiplying it by y and setting $y^2 = u$.

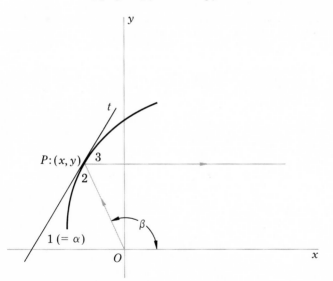

Figure 1.7

72 A weight W is to be supported by a column having the shape of a solid of revolution. If the material of the column weighs ρ lb/ft^3, and if the radius of the upper base of the column is to be r_0, determine how the radius of the column should vary if at all cross sections the load per unit area is to be the same.

73 A vertical cylindrical tank of radius r is filled with liquid to a depth h. When the tank is rotated about its axis, centrifugal force tends to drive the liquid outward from the axis of the tank. Under steady conditions of rotation with constant angular velocity ω, find the equation of the curve in which the free surface of the liquid is intersected by a plane through the axis of the cylinder, assuming the tank to be sufficiently deep to prevent liquid from spilling over the edge. *Hint:* The net force acting on a particle in the surface of the liquid is the resultant of the weight of the particle, acting downward, and the centrifugal force on the particle, acting radially outward; and this resultant force is perpendicular to the free surface of the liquid.

74 A coast guard vessel is pursuing a smuggler in a dense fog. The fog lifts momentarily, and the smuggler is seen d mi away; then the fog descends and the smuggler can no longer be seen. It is known, however, that in his attempt to escape, the smuggler will set off on a straight course of unknown direction with constant speed v_1. If the velocity of the coast guard boat is v_2 ($> v_1$), what course should it follow to be sure of intercepting and capturing the smuggler? *Hint:* Choose the origin at the point where the smuggler was seen, and observe that if the coast guard boat spirals the origin in a path such that its distance from the origin is always equal to $v_1 t$, it will necessarily intercept the smuggler. Of course before the coast guard boat can begin its spiral, it must first reach a point where its distance from the origin is $v_1 t_0$, where t_0 is the elapsed time since the smuggler was seen. In implementing these suggestions it will be convenient to use the polar-coordinate equations $x = r(t) \cos \theta(t)$ and $y = r(t) \sin \theta(t)$ to describe the path of the coast guard vessel.

75 One winter morning it began snowing heavily and continued at a constant rate all day. At noon a snowplow, able to clear c ft^3 of snow per hour with a blade w ft in width, started plowing. At 1:00 it had gone 2 mi and at 2:00 it had gone one additional mile. When did it start snowing? *Hint:* If t_0 is the number of hours before noon when it began to snow, if t is the time measured in hours after noon, if s is the constant rate at which the snow is falling, if $y(t)$ is the depth of the snow, and if $x(t)$ is the distance the snowplow has traveled, then

$$y = s(t_0 + t) \qquad \text{and} \qquad wy\, dx = c\, dt$$

CHAPTER 2
Linear Differential Equations with Constant Coefficients

2.1 The General Linear Second-Order Equation

The general linear differential equation of the second order can be written in the standard form

$$(1) \qquad y'' + P(x)y' + Q(x)y = R(x)$$

where P, Q, and R are known functions. Clearly, no loss of generality results from assuming the coefficient of y'' to be unity, since this can always be accomplished by division. Because of the presence of the term $R(x)$, which is unlike the other terms in that it does not contain the dependent variable y or any of its derivatives, Eq. (1) is said to be **nonhomogeneous**. If $R(x)$ is identically zero, we have the so-called **homogeneous** equation†

$$(2) \qquad y'' + P(x)y' + Q(x)y = 0$$

The existence of solutions of Eqs. (1) and (2) is guaranteed by the following theorem.‡

THEOREM 1 Let (a,b) be an interval in which $P(x)$, $Q(x)$, and $R(x)$ are continuous, let x_0 be an arbitrary point of (a,b), and let y_0 and y'_0 be arbitrary numbers. Then over the interval (a,b) the equation $y'' + P(x)y' + Q(x)y = R(x)$ has one and only one solution $y = y(x)$ with the property that $y(x_0) = y_0$ and $y'(x_0) = y'_0$.

In general, neither Eq. (1) nor Eq. (2) can be solved in terms of known functions. The theory associated with such special cases as have been studied at length is, for the most part, very difficult. At this stage we shall consider in detail only the simple, though highly important, case in which $P(x)$ and $Q(x)$ are constants. However, both as an illustration of how certain properties of the solutions of a differential equation can be established even though the form of those solutions is unknown and also because we shall have need of the results themselves, we shall begin by proving three fundamental theorems pertaining to the solutions of the general equations (1) and (2).

---

† It is regrettable that in describing linear equations of all orders the word homogeneous should be used in a manner totally unlike its use in describing homogeneous equations of the first order (Sec. 1.4). The usage is universal, however, and must be accepted.

‡ See, for instance, E. L. Ince, "Ordinary Differential Equations," pp. 73–75, Dover, New York, 1944.

THEOREM 2 If y_1 and y_2 are any solutions of the homogeneous equation $y'' + P(x)y' + Q(x)y = 0$, then $y_3 = c_1y_1 + c_2y_2$, where c_1 and c_2 are arbitrary constants, is also a solution.

Proof To establish this theorem, it is only necessary to substitute the expression for y_3 into the given differential equation and verify that it is satisfied:

$$
\begin{aligned}
y_3'' + P(x)y_3' + Q(x)y_3 &= (c_1y_1 + c_2y_2)'' + P(x)(c_1y_1 + c_2y_2)' \\
&\quad + Q(x)(c_1y_1 + c_2y_2) \\
&= (c_1y_1'' + c_2y_2'') + P(x)(c_1y_1' + c_2y_2') \\
&\quad + Q(x)(c_1y_1 + c_2y_2) \\
&= [y_1'' + P(x)y_1' + Q(x)y_1]c_1 \\
&\quad + [y_2'' + P(x)y_2' + Q(x)y_2]c_2 \\
&= 0c_1 + 0c_2 = 0
\end{aligned}
$$

where the coefficients of c_1 and c_2 vanish identically because, by hypothesis, both y_1 and y_2 are solutions of the homogeneous equation (2).

Theorem 2 assures us that if we have two solutions of the homogeneous equation (2), then we can obtain infinitely many other solutions simply by forming linear combinations of these two. However, it leaves completely unanswered the important question of whether or not *all* solutions of (2) can be obtained from the pair (y_1, y_2) in this fashion. To decide this point we need the stronger result contained in the next theorem.

THEOREM 3 If y_1 and y_2 are two solutions of the homogeneous equation $y'' + P(x)y' + Q(x)y = 0$ for which

$$
W(y_1, y_2)\dagger \equiv \begin{vmatrix} y_1 & y_2 \\ y_1' & y_2' \end{vmatrix} = y_1y_2' - y_2y_1' \neq 0
$$

then there exist constants c_1 and c_2 such that any solution y_3 of the homogeneous equation can be expressed in the form $y_3 = c_1y_1 + c_2y_2$.

Proof To prove this theorem, it is convenient to show first that any pair of solutions of Eq. (2), say y_i and y_j, satisfies the relation

$$
(3) \qquad W(y_i, y_j) = y_iy_j' - y_jy_i' = k_{ij} \exp\left[-\int P(x)\,dx \right]
$$

where k_{ij} is a suitable constant. To establish this, we begin with the hypothesis that both y_i and y_j are solutions of (2) and hence that

$$
y_i'' + P(x)y_i' + Q(x)y_i = 0
$$
$$
y_j'' + P(x)y_j' + Q(x)y_j = 0
$$

---

† The notation $W(y_1, y_2)$ is customarily used to denote this combination of two functions, in honor of Hoene Wronsky (1778–1853), Polish philosopher and mathematician who was one of the first to study determinants of this type. Such determinants are usually referred to as **wronskians**.

If the first of these equations is multiplied by y_j and subtracted from y_i times the second, we obtain

(4)
$$(y_i y_j'' - y_j y_i'') + P(x)(y_i y_j' - y_j y_i') = 0$$

Now,

$$\frac{dW(y_i, y_j)}{dx} \equiv \frac{d(y_i y_j' - y_j y_i')}{dx} = (y_i y_j'' + y_i' y_j') - (y_j y_i'' + y_j' y_i')$$

$$= y_i y_j'' - y_j y_i''$$

Hence, Eq. (4) can be written

$$\frac{dW(y_i, y_j)}{dx} + P(x)W(y_i, y_j) = 0$$

This is a very simple, separable differential equation whose solution can be written down immediately:

$$W(y_i, y_j) = k_{ij} \exp\left[-\int P(x)\,dx\right]^\dagger$$

where k_{ij} is an integration constant. This establishes the relation (3), which is usually known as **Abel's identity**, after the Norwegian mathematician Niels Abel (1802–1829).

Now consider the two pairs of solutions (y_3, y_1) and (y_3, y_2), where y_3 is any solution whatsoever of the homogeneous equation (2). Applying Abel's identity (3) to each of these pairs in turn, we have

$$y_3 y_1' - y_1 y_3' = k_{31} \exp\left[-\int P(x)\,dx\right]$$

$$y_3 y_2' - y_2 y_3' = k_{32} \exp\left[-\int P(x)\,dx\right]$$

These can be regarded as two simultaneous linear equations in the quantities y_3 and y_3' from which, in general, y_3 can easily be found:

$$y_3 = \frac{y_1 k_{32} \exp\left[-\int P(x)\,dx\right] - y_2 k_{31} \exp\left[-\int P(x)\,dx\right]}{y_1 y_2' - y_2 y_1'}$$

If we now apply Abel's identity to the denominator of the last expression, we obtain

$$y_3 = \frac{y_1 k_{32} \exp\left[-\int P(x)\,dx\right] - y_2 k_{31} \exp\left[-\int P(x)\,dx\right]}{k_{12} \exp\left[-\int P(x)\,dx\right]}$$

$$= \frac{k_{32}}{k_{12}} y_1 - \frac{k_{31}}{k_{12}} y_2$$

---

† Since an exponential function can never vanish, it follows that wherever the exponent $-\int P(x)\,dx$ is defined, the wronskian of y_i and y_j is either never zero or identically zero, according as $k_{ij} \neq 0$ or $k_{ij} = 0$.

Interpreting k_{32}/k_{12} as c_1 and $-k_{31}/k_{12}$ as c_2, we have thus succeeded in exhibiting *any* solution y_3 as a linear combination $c_1 y_1 + c_2 y_2$ of the two particular solutions y_1 and y_2, provided only that the expression

$$y_1 y_2' - y_2 y_1' = W(y_1, y_2)$$

by which we had to divide in order to solve for y_3, does not vanish. Theorem 3 is thus established.

From Theorem 3 it is clear that to find a complete solution of Eq. (2) we must first find two particular solutions that have a nonvanishing wronskian, and then we must form a linear combination of these solutions with arbitrary coefficients. Such a pair of solutions is called a **basis** for the complete solution. We must remember, however, that although there are infinitely many pairs of particular solutions y_1 and y_2 which can be used as a basis for constructing a complete solution of Eq. (2),† neither Theorem 2 nor Theorem 3 tells us how to find them. In fact there is *no* general method for solving Eq. (2)‡ and the only procedure applicable in all cases is one which permits us to determine a second, nonproportional solution when one solution is known.

To develop this procedure, let us suppose that $y_1(x) \not\equiv 0$ is a solution of Eq. (2), and let us attempt to find a function $\phi(x)$ with the property that $\phi(x) y_1(x)$ is also a solution of (2). Substituting $y = \phi(x) y_1(x)$ into Eq. (2), we have

$$(y_1'' \phi + 2y_1' \phi' + y_1 \phi'') + P(x)(y_1' \phi + y_1 \phi') + Q(x)(y_1 \phi)$$
$$= [y_1'' + P(x)y_1' + Q(x)y_1]\phi + [2y_1' + P(x)y_1]\phi' + y_1 \phi'' \overset{?}{=} 0$$

Now, the coefficient of ϕ in the last expression is identically zero, since, by hypothesis, y_1 is a solution of Eq. (2). Hence the last equation will be satisfied provided ϕ is chosen such that

$$y_1 \phi'' + [2y_1' + P(x)y_1]\phi' = 0$$

Since $\phi'' = d\phi'/dx$, this is a simple separable equation in ϕ', and we have

$$\frac{d\phi'}{\phi'} + \left[\frac{2y_1'}{y_1} + P(x) \right] dx = 0$$

or, integrating,

$$\ln |\phi'| + 2 \ln |y_1| + \int P(x) \, dx = \ln |c|$$

Hence, combining the logarithms and taking antilogarithms gives

$$\phi' = \frac{c \exp\left[-\int P(x) \, dx\right]}{y_1^2}$$

---

† See Exercise 7.
‡ The closest thing to a general solution procedure is the use of infinite series, described in Sec. 9.1.

Integrating again, we find

$$\phi = c \int \frac{\exp\left[-\int P(x)\,dx\right]}{y_1{}^2}\,dx + k$$

from which we obtain, for all values of c and k, the solution

(5) $$\phi y_1 = cy_1(x) \int \frac{\exp\left[-\int P(x)\,dx\right]}{y_1{}^2(x)}\,dx + ky_1(x)$$

Since this contains two arbitrary constants, it is actually a complete solution, provided that the two particular solutions from which it is constructed, namely, the given solution y_1 and the one we have just obtained

$$y_1(x) \int \frac{\exp\left[-\int P(x)\,dx\right]}{y_1{}^2(x)}\,dx$$

have a nonvanishing wronskian. It is not difficult to show that this is always the case, although we shall leave the proof as an exercise.†

EXAMPLE 1

Find a complete solution of the equation $x^2y'' + xy' - 4y = 0$, given that $y = x^2$ is one solution.

Substituting the assumed solution $y = x^2\phi$ into the given differential equation, we have

$$x^2(2\phi + 4x\phi' + x^2\phi'') + x(2x\phi + x^2\phi') - 4x^2\phi = 0$$

or, simplifying,

$$x\phi'' + 5\phi' = 0$$

Separating variables, we obtain

$$\frac{d\phi'}{\phi'} + \frac{5}{x}\,dx = 0$$

Then, integrating, we get

$$\ln|\phi'| + 5\ln|x| = \ln|c| \qquad \text{or} \qquad \phi' = \frac{c}{x^5}$$

and finally, integrating again,

$$\phi = -\frac{c}{4x^4} + k$$

Therefore the complete solution is equally well

$$y = x^2\phi = -\frac{c}{4x^2} + kx^2 \qquad \text{or} \qquad y = \frac{a}{x^2} + kx^2$$

where $a = -c/4$.

The solution of the nonhomogeneous equation (1) is based on the following theorem.

---

† See Exercise 19.

THEOREM 4 If Y is any solution of the nonhomogeneous equation

$$y'' + P(x)y' + Q(x)y = R(x)$$

and if $c_1 y_1 + c_2 y_2$ is a complete solution of the homogeneous equation obtained from this by deleting the term $R(x)$, then $y = c_1 y_1 + c_2 y_2 + Y$ is a complete solution of the nonhomogeneous equation.

Proof Let $\bar{y}$ be any solution whatsoever of the nonhomogeneous equation (1). Then

$$\bar{y}'' + P(x)\bar{y}' + Q(x)\bar{y} = R(x)$$

and, similarly, since Y is also a solution of (1),

$$Y'' + P(x)Y' + Q(x)Y = R(x)$$

If we subtract these two equations, we obtain

$$(\bar{y}'' - Y'') + P(x)(\bar{y}' - Y') + Q(x)(\bar{y} - Y) = 0$$

or $\qquad (\bar{y} - Y)'' + P(x)(\bar{y} - Y)' + Q(x)(\bar{y} - Y) = 0$

Thus the quantity $\bar{y} - Y$ satisfies the homogeneous equation (2) and hence, by Theorem 3, must be expressible in the form

$$\bar{y} - Y = c_1 y_1 + c_2 y_2$$

provided that $W(y_1, y_2) \neq 0$, that is, provided that $c_1 y_1 + c_2 y_2$ is a complete solution of (2), as we assumed. Therefore, transposing, we have

$$\bar{y} = c_1 y_1 + c_2 y_2 + Y$$

Since $\bar{y}$ was *any* solution of the nonhomogeneous equation (1), Theorem 4 is thus established.

The term Y, which can be any solution of (1) no matter how special, is called a **particular integral** of the nonhomogeneous equation. The expression $c_1 y_1 + c_2 y_2$, which is a complete solution of the homogeneous equation corresponding to (1), is called the **complementary function** of the nonhomogeneous equation. The steps to be carried out in solving an equation of the form (1) can now be summarized as follows:

a. Delete the term $R(x)$ from the given equation and find two solutions of the resulting homogeneous equation which have a nonvanishing wronskian. Then combine these to form the complementary function $c_1 y_1 + c_2 y_2$ of the given equation.

b. Find one particular solution Y of the nonhomogeneous equation itself.

c. Add the complementary function $c_1 y_1 + c_2 y_2$ found in step **a** to the particular integral found in step **b** to obtain the complete solution $y = c_1 y_1 + c_2 y_2 + Y$ of the given equation.

In the following sections we shall investigate how these theoretical steps can be carried out when $P(x)$ and $Q(x)$ are constants, i.e., when we have the so-called **linear differential equation with constant coefficients**.

EXERCISES

Using the one solution indicated, find a complete solution of each of the following equations:

1 $y'' + y = 0$ $y_1 = \sin x$
2 $y'' - y' - 2y = 0$ $y_1 = e^{-x}$
3 $x^2y'' + 4xy' - 4y = 0$ $y_1 = x$
4 $y'' + 2y' + y = 0$ $y_1 = e^{-x}$
5 $(1 - 2x)y'' + 2y' + (2x - 3)y = 0$ $y_1 = e^x$
6 $(2x - x^2)y'' + 2(x - 1)y' - 2y = 0$ $y_1 = x - 1$
7 If y_1 and y_2 have a nonvanishing wronskian, show that $y_3 = c_1 y_1 + c_2 y_2$ and $y_4 = k_1 y_1 + k_2 y_2$ also have a nonvanishing wronskian provided $c_1 k_2 - c_2 k_1 \neq 0$.

Verify that each of the following equations has the indicated solutions, and in each case construct two different complete solutions using two different bases:

8 $y'' - y = 0$ $y_1 = e^x, y_2 = e^{-x}$
9 $y'' - 3y' + 2y = 0$ $y_1 = e^x, y_2 = e^{2x}$
10 $y'' + y = 0$ $y_1 = \sin\left(x + \dfrac{\pi}{4}\right), \ y_2 = \sin\left(x - \dfrac{\pi}{4}\right)$
11 $x^2y'' + xy' - y = 0$ $y_1 = x, y_2 = \dfrac{1}{x}$

12 Show that if two differentiable functions are proportional, their wronskian is identically zero.
13 Show that the converse of the assertion of Exercise 12 is false. *Hint:* Consider the following pair of functions:
$$y_1 = x^2 \qquad y_2 = \begin{cases} -x^2 & -\infty < x \le 0 \\ x^2 & 0 < x < \infty \end{cases}$$

14 Show that the converse of the assertion of Exercise 12 is true over any interval in which neither function takes on the value zero.
15 If the wronskian of two functions is different from zero at every point of an interval, show that there is no point of the interval at which either function has a repeated zero.
16 Show that there is no point at which two nonproportional solutions of Eq. (2) are simultaneously zero, except possibly at a point where $P(x)$ is undefined.
17 Show that there is no point at which two nonproportional solutions of Eq. (2) simultaneously take on extreme values, except possibly at a point where $P(x)$ is undefined.
18 Given two nonproportional solutions of Eq. (2), show that between any two consecutive zeros of either solution there is exactly one zero of the other solution. *Hint:* Let y_1 and y_2 be the two solutions, let a and b be two consecutive zeros of y_1, apply Rolle's theorem to the quotient y_1/y_2, and note the contradiction unless $y_2 = 0$ at some point between a and b.
19 Show that the two solutions
$$y_1 \qquad \text{and} \qquad y_1 \int \frac{\exp\left[-\int P(x)\,dx\right]}{y_1^2}\,dx$$
of Eq. (2) have a nonvanishing wronskian.
20 Explain how Abel's identity can be used to find a second solution of the equation $y'' + P(x)y' + Q(x)y = 0$ when one solution is known. Illustrate this method by applying it to Exercises 1, 2, and 3.
21 If the quotient of two nonproportional solutions of the differential equation $y'' + P(x)y' + Q(x)y = 0$ exists at all points of an interval, prove that it is either an increasing function at all points of the interval or else is a decreasing function at all points of the interval.

22 Let $y = f(x)$ and $y = g(x)$ be nontrivial solutions of the differential equations $y'' + Q_1(x)y = 0$ and $y'' + Q_2(x)y = 0$, respectively, on an interval at all points of which $Q_1(x) \geq Q_2(x)$. Assuming that $Q_1(x) \not\equiv Q_2(x)$, show that $f(x)$ vanishes at least once between any two consecutive zeros of $g(x)$. *Hint:* Let a and b be two consecutive zeros of $g(x)$, and assume that $f(x)$ is not zero for any value of x between a and b. Show first that $f(x)$ and $g(x)$ can both be supposed positive on (a,b). Next show that $W(f,g)|_{x=a} \geq 0$ and $W(f,g)|_{x=b} \leq 0$. Finally, by combining the differential equations, show that $dW(f,g)/dx \geq 0$ on (a,b), and from this derive a contradiction which will complete the proof. This result and the result of Exercise 18 are examples of what are known as **separation theorems,** and are due to the Swiss mathematician J. C. F. Sturm (1803–1855).

23 If $Q(x) \leq 0$, show that no nontrivial solution of $y'' + Q(x)y = 0$ can have more than one real zero. *Hint:* Apply the result of Exercise 22 to the pair of equations $y'' + Q(x)y = 0$ and $y'' = 0$.

2.2 The Homogeneous Linear Equation with Constant Coefficients

When $P(x)$ and $Q(x)$ are constants, the general linear second-order differential equation can be written in the standard form

(1) $$ay'' + by' + cy = f(x)$$

A second standard form which is also common is based on the so-called **operator notation**. In this, the symbol of differentiation d/dx is replaced by D, so that, by definition,

$$\frac{dy}{dx} = Dy\dagger$$

As an immediate extension, the second derivative, which of course is obtained by a repetition of the process of differentiation, is written

$$\frac{d^2y}{dx^2} = D(Dy) = D^2y$$

Similarly,

$$\frac{d^3y}{dx^3} = D(D^2y) = D^3y$$

$$\frac{d^4y}{dx^4} = D(D^3y) = D^4y$$

. . . .

Evidently, positive integral powers of D (which are the only ones we have defined) obey the usual laws of exponents.

If due care is taken to see that variables are not moved across the sign of differentiation by a careless interchange of the order of factors containing variable coefficients,‡ the operator D can be handled in many respects as though it were a simple

---

† Just as the prime notation, $y', y'', \ldots$, may in specific instances indicate derivatives with respect to x, t, or any other independent variable, so the operator notation, $Dy, D^2y, \ldots$, may also indicate derivatives with respect to an independent variable other than x, depending on the context.

‡ See Exercise 3.

algebraic quantity. For instance, after defining $(aD^2 + bD + c)\phi(x)$ to mean $aD^2\phi(x) + bD\phi(x) + c\phi(x)$, we have for the polynomial operator $3D^2 - 10D - 8$ and its factored equivalents, applied to the particular function $\phi(x) = x^2$,

$$(3D^2 - 10D - 8)x^2 = 3(2) - 10(2x) - 8(x^2) = 6 - 20x - 8x^2$$

$$(3D + 2)(D - 4)x^2 = (3D + 2)(2x - 4x^2)$$
$$= (6 - 24x) + (4x - 8x^2) = 6 - 20x - 8x^2$$

$$(D - 4)(3D + 2)x^2 = (D - 4)(6x + 2x^2)$$
$$= (6 + 4x) - (24x + 8x^2) = 6 - 20x - 8x^2$$

These results illustrate how algebraically equivalent forms of an operator yield identical results when applied to the same function.

Using the operator D, we can evidently write Eq. (1) in the alternative standard form

$$(1a) \qquad\qquad (aD^2 + bD + c)y = f(x)$$

Many writers base the solution of Eq. (1) upon the operational properties of the symbol D. However, we shall postpone all operational methods until the chapter on the Laplace transformation, where operational calculus can be developed easily and efficiently in its proper setting.

Following the theory of the last section, we first attempt to find a complete solution of the homogeneous equation

$$(2) \qquad\qquad ay'' + by' + cy = 0$$

or

$$(2a) \qquad\qquad (aD^2 + bD + c)y = 0$$

obtained from (1) or (1a) by deleting $f(x)$. In searching for particular solutions of (2) to be combined into a complete solution, it is natural to try

$$y = e^{mx}$$

where m is a constant to be determined, because all derivatives of this function are alike except for a numerical coefficient. Substituting into Eq. (2) and then factoring e^{mx} from every term, we have

$$e^{mx}(am^2 + bm + c) = 0$$

as the condition to be satisfied if $y = e^{mx}$ is to be a solution. Since e^{mx} can never be zero, it is thus necessary that

$$(3) \qquad\qquad am^2 + bm + c = 0$$

This purely algebraic equation is known as the **characteristic** or **auxiliary** equation of either Eq. (1) or Eq. (2), since its roots determine or *characterize* the only possible solutions of the assumed form $y = e^{mx}$. In practice it is obtained not by substituting $y = e^{mx}$ into the given differential equation and then simplifying, but rather by substituting m^2 for y'', m for y', and 1 for y in the given equation or, still more simply, by equating to zero the operational coefficient of y and then letting the symbol D play the role of m:

$$aD^2 + bD + c = 0$$

The characteristic equation (3) is a simple quadratic which will in general be satisfied by two values of m:

$$m = \frac{-b \pm \sqrt{b^2 - 4ac}}{2a}$$

Using these values, say m_1 and m_2, two solutions

$$y_1 = e^{m_1 x} \quad \text{and} \quad y_2 = e^{m_2 x}$$

can be constructed. From this pair, according to Theorem 2, Sec. 2.1, an infinite family of solutions

(4)
$$y = c_1 y_1 + c_2 y_2 = c_1 e^{m_1 x} + c_2 e^{m_2 x}$$

can be formed. Moreover, by Theorem 3, Sec. 2.1, if the wronskian of these solutions is different from zero, then (4) is a complete solution of Eq. (2); i.e., it contains all possible solutions of the homogeneous equation. Accordingly we compute

$$W(y_1, y_2) = y_1 y_2' - y_2 y_1' = e^{m_1 x}(m_2 e^{m_2 x}) - e^{m_2 x}(m_1 e^{m_1 x})$$
$$= (m_2 - m_1)e^{(m_1 + m_2)x}$$

Since $e^{(m_1 + m_2)x}$ can never vanish, it is clear that *a complete solution of Eq. (2) is always given by (4) except in the special case when $m_1 = m_2$ and the wronskian vanishes identically.*

EXAMPLE 1

Find a complete solution of the differential equation $y'' + 7y' + 12y = 0$.
The characteristic equation in this case is

$$m^2 + 7m + 12 = 0$$

and its roots are

$$m_1 = -3 \quad \text{and} \quad m_2 = -4$$

Since these values are different, a complete solution is

$$y = c_1 e^{-3x} + c_2 e^{-4x}$$

EXAMPLE 2

Find a complete solution of the equation $y'' + 2y' + 5 = 0$.
The characteristic equation in this case is

$$m^2 + 2m + 5 = 0$$

and its roots are

$$m_1 = -1 + 2i \quad \text{and} \quad m_2 = -1 - 2i$$

Since these are distinct, a complete solution is

$$y = c_1 e^{(-1+2i)x} + c_2 e^{(-1-2i)x}$$

Although the last expression is undeniably a complete solution of the given equation, it is unsatisfactory for many practical purposes because it involves complex exponentials, which are awkward to handle and are not tabulated. It is therefore a matter of considerable importance to construct a more convenient complete solution when m_1 and m_2 are conjugate complex numbers.

To do this, let us suppose that

$$m_1 = p + iq \qquad \text{and} \qquad m_2 = p - iq$$

so that a complete solution as first constructed is

$$y = c_1 e^{(p+iq)x} + c_2 e^{(p-iq)x} = c_1 e^{px} e^{iqx} + c_2 e^{px} e^{-iqx}$$

By factoring out e^{px} this can be written as

$$y = e^{px}(c_1 e^{iqx} + c_2 e^{-iqx})$$

The expression in parentheses can be simplified by using the **Euler formulas** (Sec. 15.7)

$$e^{i\theta} = \cos\theta + i\sin\theta \qquad \text{and} \qquad e^{-i\theta} = \cos\theta - i\sin\theta$$

taking $\theta = qx$. The result of these substitutions is

$$y = e^{px}[c_1(\cos qx + i\sin qx) + c_2(\cos qx - i\sin qx)]$$
$$= e^{px}[(c_1 + c_2)\cos qx + i(c_1 - c_2)\sin qx]$$

If we now define two new arbitrary constants by the equations

$$A = c_1 + c_2 \qquad \text{and} \qquad B = i(c_1 - c_2)$$

the complete solution can finally be put in the purely real form

$$y = e^{px}(A\cos qx + B\sin qx)$$

Of course it is not difficult to verify by direct substitution that both $y_1 = e^{px}\cos qx$ and $y_2 = e^{px}\sin qx$ are particular solutions (with nonvanishing wronskian) of the homogeneous equation (2). For a completely satisfactory derivation this should now be done, since we do not yet know that our formal treatment of complex exponentials, as though they obeyed the same laws as real exponentials, is justified.

EXAMPLE 2 (continued)

Applying the preceding reasoning to Example 2 makes it clear that $p = -1$ and $q = 2$. Hence the complete solution can be written

$$y = e^{-x}(A\cos 2x + B\sin 2x)$$

When the characteristic equation has equal roots, the two independent solutions normally arising from the substitution of $y = e^{mx}$ become identical and, as pointed out above, we do not have an adequate basis for constructing a complete solution. To find a second, nonproportional solution in this case, we use the method developed in the last section.

By hypothesis, the characteristic equation in this case has equal roots, say $m_1 = m_2 = r$. Hence it must be of the form

$$m^2 - 2rm + r^2 = 0$$

which implies that the differential equation itself is

$$y'' - 2ry' + r^2 y = 0$$

In particular, we observe from this that the coefficient $P(x)$ in this case is the number $-2r$. Clearly, $y_1 = e^{m_1 x} = e^{rx}$ is one solution of the differential equation, and, from Eq. (5), Sec. 2.1, a second, independent solution is given by

$$y_1 \int \frac{e^{-\int P(x)\,dx}}{y_1{}^2}\,dx = e^{rx} \int \frac{e^{2rx}}{(e^{rx})^2}\,dx = xe^{rx} = xe^{m_1 x}$$

Thus, *in the exceptional case in which the characteristic equation has equal roots, a complete solution of Eq. (2) is*

$$y = c_1 e^{m_1 x} + c_2 x e^{m_1 x}$$

EXAMPLE 3

Find a complete solution of the equation $(D^2 + 6D + 9)y = 0$.

In this case the characteristic equation $m^2 + 6m + 9 = 0$ is a perfect square, with roots $m_1 = m_2 = -3$. Hence, by our last remark, a complete solution of the given equation is

$$y = c_1 e^{-3x} + c_2 x e^{-3x}$$

The complete process for solving the homogeneous equation (2) in all possible cases is summarized in Table 2.1.

Table 2.1

Differential equation: $ay'' + by' + cy = 0$ or $(aD^2 + bD + c)y = 0$		
Characteristic equation: $am^2 + bm + c = 0$ or $aD^2 + bD + c = 0$		
Nature of the roots of the characteristic equation	Condition on the coefficients of the characteristic equation	Complete solution of the differential equation
Real and unequal $m_1 \neq m_2$	$b^2 - 4ac > 0$	$y = c_1 e^{m_1 x} + c_2 e^{m_2 x}$
Real and equal $m_1 = m_2$	$b^2 - 4ac = 0$	$y = c_1 e^{m_1 x} + c_2 x e^{m_1 x}$
Conjugate complex $m_1 = p + iq$ $m_2 = p - iq$	$b^2 - 4ac < 0$	$y = e^{px}(A \cos qx + B \sin qx)$

In particular applications, the two arbitrary constants in the complete solution must usually be determined to fit given initial conditions on y and y', or their equivalent. The following examples will clarify the procedure.

EXAMPLE 4

Find the solution of the equation $y'' - 4y' + 4y = 0$ for which $y = 3$ and $y' = 4$ when $x = 0$.

The characteristic equation of the differential equation is

$$m^2 - 4m + 4 = 0$$

Its roots are $m_1 = m_2 = 2$; hence a complete solution is

$$y = c_1 e^{2x} + c_2 x e^{2x}$$

By differentiating this, we find

$$y' = 2c_1e^{2x} + c_2(e^{2x} + 2xe^{2x}) = (2c_1 + c_2)e^{2x} + 2c_2xe^{2x}$$

Substituting the given data into the equations for y and y', respectively, we have

$$3 = c_1 \quad \text{and} \quad 4 = 2c_1 + c_2$$

Hence, $c_1 = 3$, $c_2 = -2$, and the required solution is

$$y = 3e^{2x} - 2xe^{2x}$$

EXAMPLE 5

Find the solution of the equation $(4D^2 + 16D + 17)y = 0$ for which $y = 1$ when $t = 0$ and $y = 0$ when $t = \pi$.

In this case, the statement of the problem makes it clear that the independent variable is not x but t. This does not affect the characteristic equation $4m^2 + 16m + 17 = 0$ or its roots $m = 2 \pm \frac{1}{2}i$. However, the solution which we construct from these roots must be expressed in terms of t rather than x:

$$y = e^{-2t}\left(A\cos\frac{t}{2} + B\sin\frac{t}{2}\right)$$

Substituting the given conditions into this equation, we find

$$1 = A \quad \text{and} \quad 0 = e^{-2\pi}B \quad \text{or} \quad B = 0$$

Hence, the required solution is $y = e^{-2t}\cos(t/2)$.

EXERCISES

1 What is the difference between Dy and yD?
2 Verify that

$$(D + 1)(D^2 + 2)\sin 3x = (D^2 + 2)(D + 1)\sin 3x$$
$$= (D^3 + D^2 + 2D + 2)\sin 3x$$

3 Is $(D + 1)(D + x)e^x = (D + x)(D + 1)e^x$? Explain.

Find a complete solution of each of the following equations:

4 $y'' + y' - 2y = 0$ 5 $y'' + 5y' + 4y = 0$
6 $y'' - 5y = 0$ 7 $y'' + 5y' = 0$
8 $(4D^2 + 4D + 1)y = 0$ 9 $(9D^2 - 12D + 4)y = 0$
10 $10y'' + 6y' + y = 0$ 11 $y'' + 10y' + 26y = 0$

Find the particular solution of each of the following equations which satisfies the given conditions:

12 $y'' + 3y' - 4y = 0$ $y = 4, y' = -2$ when $x = 0$
13 $y'' + 4y = 0$ $y = 2, y' = 6$ when $x = 0$
14 $y'' - 4y = 0$ $y = 1, y' = -1$ when $x = 0$
15 $25y'' + 20y' + 4y = 0$ $y = y' = 0$ when $x = 0$
16 $(D^2 + 6D + 9)y = 0$ $y = 0, y' = 3$ when $x = 0$
17 $(D^2 + 2D + 5)y = 0$ $y = 1$ when $x = 0, y = 0$ when $x = \pi$
18 $(D^2 + 2D + 5)y = 0$ $y = 1$ when $x = 0, y' = 0$ when $x = \pi$
19 Show that there is always a unique solution of the equation $y'' + y = 0$ satisfying given conditions of the form $y = y_0$ when $x = x_0$ and $y = y_1$ when $x = x_1$ unless $x_1 = x_0 + n\pi$. What is the situation when $x_1 = x_0 + n\pi$?
20 For what values of λ, if any, are there nontrivial solutions of the equation $y'' + \lambda^2y = 0$ which satisfy the conditions $y(0) = 0$ and $y(\pi) = 0$? What are these solutions?

21 For what values of λ, if any, are there nontrivial† solutions of the equation $y'' + \lambda^2 y = 0$ which satisfy the conditions $y'(0) = 0$ and $y'(\pi) = 0$? What are these solutions?

22 For what values of λ, if any, are there nontrivial solutions of the equation $y'' + \lambda^2 y = 0$ which satisfy the conditions $y(0) = 0$ and $y'(\pi) = 0$? What are these solutions?

23 Show that the only values of λ for which there exist nontrivial solutions of the equation $y'' + \lambda^2 y = 0$ satisfying the conditions $y(0) = 0$ and $y(\pi) = y'(\pi)$ are the roots of the equation $\tan \pi\lambda = \lambda$. Show that this equation has infinitely many roots.

24 Work Exercise 20 for the equation $y'' - \lambda^2 y = 0$.

25 Work Exercise 22 for the equation $y'' - \lambda^2 y = 0$.

26 (a) Show that if the roots of the characteristic equation of a differential equation are $m = p \pm iq$, then the differential equation itself is $y'' - 2py' + (p^2 + q^2)y = 0$.
(b) Verify by direct substitution that $y_1 = e^{px} \cos qx$ and $y_2 = e^{px} \sin qx$ are solutions of the equation

$$y'' - 2py' + (p^2 + q^2)y = 0$$

(c) Verify that the particular solutions indicated in part (b) have a nonvanishing wronskian.

27 Show that if $k \neq 0$, both $y = A \cos (kx + B)$ and $y = G \sin (kx + H)$ are complete solutions of the equation $y'' + k^2 y = 0$.

28 Show that if $k \neq 0$, $y = A \cosh kx + B \sinh kx$ is a complete solution of the equation $y'' - k^2 y = 0$.

29 Show that $y = e^{px}(A \cosh qx + B \sinh qx)$ is a complete solution of the equation $ay'' + by' + cy = 0$ when the roots of the characteristic equation are $m = p \pm q$ and $q \neq 0$.

30 If the roots of its characteristic equation are real, show that no nontrivial solution of the equation $ay'' + by' + cy = 0$ can have more than one real zero.

31 If the characteristic equation of the differential equation $ay'' + by' + cy = 0$ has distinct roots m_1 and m_2, show that

$$y = \frac{e^{m_1 x} - e^{m_2 x}}{m_1 - m_2}$$

is a particular solution of the equation. Determine the limit of this expression as $m_2 \to m_1$ and discuss its relation to the solution of the differential equation when the characteristic equation has equal roots.

32 (a) Show that $(D - a)^2 f(x) = e^{ax} D^2 [e^{-ax} f(x)]$.
(b) Use the formula of part (a) to show that $(D - a)^2 [(c_1 + c_2 x)e^{ax}] = 0$.
(c) Explain how the formula of part (b) can be used to obtain the complete solution of a differential equation whose characteristic equation has equal roots.

33 (a) Show that $D^2(xe^{mx}) = m^2 xe^{mx} + 2me^{mx}$.
(b) Using the result of part (a), show that if $p(D)$ is a quadratic polynomial in D, then $p(D)(xe^{mx}) = p(m)xe^{mx} + p'(m)e^{mx}$.
(c) Explain how the formula of part (b) can be used to obtain a second, non-proportional solution of a differential equation whose characteristic equation has equal roots. *Hint:* Recall that if a polynomial equation $p(x) = 0$ has a double root $x = r$, then $x = r$ is also a root of the equation $p'(x) = 0$.

34 What meaning, if any, do you think can be assigned to D^0? D^{-1}? D^{-2}?

35 Show that the change of dependent variable defined by the substitution $y = -z'/zP$ changes the **Riccati equation** $y' = P(x)y^2 + Q(x)y + R(x)$ into the linear second-order equation $z'' - [Q + (P'/P)]z' + PRz = 0$.

---

† By a **nontrivial** solution of an equation we mean a solution of the equation which is not identically zero. A **trivial solution**, of course, is one which is identically zero.

Using the result of Exercise 35, solve each of the following equations:

36 $xy' = x^2y^2 - y + 1$ **37** $x^2y' = x^4y^2 + (3x^2 - 2x)y + 2$

38 $(\cos x)y' = (\cos^2 x)y^2 + (\sin x - 2\cos x)y + 5$

2.3 The Nonhomogeneous Equation

In the last section we learned how to solve the homogeneous equation $ay'' + by' + cy = 0$, and with this knowledge we can now obtain the complementary function of the nonhomogeneous equation

$$(1) \qquad\qquad ay'' + by' + cy = f(x)$$

However, we must also have a particular integral, i.e., a particular solution, of Eq. (1) before we can construct its complete solution, namely,

$$y = \text{complementary function} + \text{particular integral}$$

Various procedures are available for the determination of particular solutions of Eq. (1), some applicable no matter what $f(x)$ may be, others useful only when $f(x)$ belongs to some suitably specialized class of functions. It should be borne in mind, however, that in applying Theorem 4, Sec. 2.1, the important thing is not *how* we obtain a particular solution of Eq. (1) but merely *that* we have one such solution. Any method, from outright guessing to the most sophisticated theoretical technique, is legitimate, provided that it leads to a solution that can be checked in (1). In this section we shall introduce the so-called **method of undetermined coefficients**, which appears initially to be based on little more than guesswork but which can readily be formalized into a well-defined procedure applicable to a well-defined and very important class of cases.

To illustrate the method, suppose that we wish to find a particular integral of the equation

$$(2) \qquad\qquad y'' + 4y' + 3y = 5e^{2x}$$

Since differentiating an exponential of the form e^{kx} merely reproduces that function with, at most, a change in its numerical coefficient, it is natural to "guess" that it may be possible to determine A so that

$$Y = Ae^{2x}$$

will be a solution of (2). To check this, we substitute $Y = Ae^{2x}$ into the given equation, getting

$$4Ae^{2x} + 8Ae^{2x} + 3Ae^{2x} \overset{?}{=} 5e^{2x} \qquad \text{or} \qquad 15Ae^{2x} \overset{?}{=} 5e^{2x}$$

which will be an identity if and only if $A = \frac{1}{3}$. Thus, the required particular integral is

$$Y = \tfrac{1}{3}e^{2x}$$

Now suppose that the right-hand side of (2), instead of being $5e^{2x}$, were $5\sin 2x$. Guided by our previous success in determining a particular solution, we might perhaps be led to try

$$Y = A\sin 2x$$

as a particular integral. Substituting this to check whether or not it can be a solution, we obtain

$$-4A \sin 2x + 8A \cos 2x + 3A \sin 2x \overset{?}{=} 5 \sin 2x$$

$$-A \sin 2x + 8A \cos 2x \overset{?}{=} 5 \sin 2x$$

and this cannot be an identity unless, simultaneously, $A = -5$ and $A = 0$, which is absurd. The difficulty here, of course, is that differentiating $\sin 2x$ introduces the new function $\cos 2x$, which must also be eliminated identically from the equation resulting from the substitution of $Y = A \sin 2x$. Since the one arbitrary constant A cannot satisfy two independent conditions, it is clear that we must arrange to incorporate *two* arbitrary constants in our tentative choice for Y without, at the same time, introducing new terms which will lead to still more conditions. This is easily done by assuming

$$Y = A \sin 2x + B \cos 2x$$

which contains the necessary second parameter yet cannot introduce any further new functions since it already is a linear combination of *all* the independent terms that can be obtained from $\sin 2x$ by repeated differentiation. The actual determination of A and B is a simple matter, for substitution into the new version of (2) yields

$$(-4A \sin 2x - 4B \cos 2x) + 4(2A \cos 2x - 2B \sin 2x)$$

$$+ 3(A \sin 2x + B \cos 2x) \overset{?}{=} 5 \sin 2x$$

$$(-A - 8B) \sin 2x + (8A - B) \cos 2x \overset{?}{=} 5 \sin 2x$$

and for this to be an identity requires that

$$-A - 8B = 5 \quad \text{and} \quad 8A - B = 0$$

from which we find immediately that $A = -\frac{1}{13}$ and $B = -\frac{8}{13}$. Hence, finally,

$$Y = -\frac{\sin 2x + 8 \cos 2x}{13}$$

With these illustrations in mind we are now in a position to describe more precisely the use of the method of undetermined coefficients for finding particular integrals.

RULE 1 If $f(x)$ is a function for which repeated differentiation yields only a finite number of linearly independent expressions, then, in general, a particular Y for the nonhomogeneous equation $ay'' + by' + cy = f(x)$ can be found by

    a. Assuming Y to be an arbitrary linear combination of all the linearly independent terms which arise from $f(x)$ by repeated differentiation
    b. Substituting Y into the given differential equation
    c. Determining the arbitrary constants in Y so that the equation resulting from the substitution is identically satisfied.

The class of functions $f(x)$ possessing only a finite number of linearly independent derivatives consists of the simple functions

k

x^n n a positive integer

e^{kx}

$\cos kx$

$\sin kx$

and any others obtainable from these by a finite number of additions, subtractions, and multiplications. If $f(x)$ possesses infinitely many independent derivatives, as is the case, for instance, with the simple function $1/x$, it is occasionally convenient to assume for Y an infinite series whose terms are the respective derivatives of $f(x)$ each multiplied by an arbitrary constant. However, the use of the method of undetermined coefficients in such cases involves questions of convergence which never arise when $f(x)$ has only a finite number of independent derivatives.

When $f(x)$ is the sum of several terms, say $f(x) = R_1(x) + R_2(x)$, we can find a particular integral Y of Eq. (1) in either of two slightly different ways. If we wish, we can find Y by applying Rule 1 to $f(x)$ in its entirety. On the other hand, we can also find Y by solving two shorter problems, one involving just $R_1(x)$, the other involving just $R_2(x)$. The basis for this method is provided by the following theorem, whose proof is left as an exercise.

THEOREM 1 If Y_1 is a solution of the equation $ay'' + by' + cy = R_1(x)$, and if Y_2 is a solution of the equation $ay'' + by' + cy = R_2(x)$, then $Y = Y_1 + Y_2$ is a solution of the equation

$$ay'' + by' + cy = R_1(x) + R_2(x)$$

EXAMPLE 1

Find a particular integral for the equation

$$(3) \qquad\qquad y'' + 3y' + 2y = 10e^{3x} + 4x^2$$

If we wish, we can find Y by beginning with the expression $Y = Ae^{3x} + Bx^2 + Cx + D$, which means that we are going to handle the various terms in $f(x)$ at the same time. On the other hand, according to Theorem 1, we can also find Y by first finding a particular integral Y_1 for the equation

$$(3a) \qquad\qquad y'' + 3y' + 2y = 10e^{3x}$$

then finding a particular integral Y_2 for the equation

$$(3b) \qquad\qquad y'' + 3y' + 2y = 4x^2$$

and finally taking Y to be the sum $Y_1 + Y_2$.

Using the second method (which means that the expressions we have to substitute are not quite so lengthy and the subsequent collection of terms is not quite so involved), we assume $Y_1 = Ae^{3x}$, substitute into Eq. (3a), and determine A so that the resulting equation will be an identity:

$$9Ae^{3x} + 3(3Ae^{3x}) + 2(Ae^{3x}) = 10e^{3x}$$
$$20Ae^{3x} = 10e^{3x}$$

which implies that $A = \frac{1}{2}$ and $Y_1 = \frac{1}{2}e^{3x}$. Then we assume $Y_2 = Bx^2 + Cx + D$, substitute into Eq. (3b), and determine B, C, and D so that, again, the resulting equation will be an identity:

$$2B + 3(2Bx + C) + 2(Bx^2 + Cx + D) = 4x^2$$
$$2Bx^2 + (6B + 2C)x + (2B + 3C + 2D) = 4x^2$$
$$2B = 4$$
$$6B + 2C = 0$$
$$2B + 3C + 2D = 0$$

Solving these simultaneously, we find at once that

$$B = 2 \qquad C = -6 \qquad D = 7$$

Hence $Y_2 = 2x^2 - 6x + 7$ and, finally,

$$Y = Y_1 + Y_2 = \frac{e^{3x}}{2} + 2x^2 - 6x + 7$$

There is one important exception to the procedure we have just been outlining, which we must now investigate. Suppose, for example, that we wish to find a particular integral for the equation

$$(4) \qquad\qquad y'' + 5y' + 6y = e^{-3x}$$

Proceeding in the way we have just described, we would start with

$$Y = Ae^{-3x}$$

and substitute, getting

$$9Ae^{-3x} + 5(-3Ae^{-3x}) + 6(Ae^{-3x}) \overset{?}{=} e^{-3x}$$

$$0 \overset{?}{=} e^{-3x}$$

This is obviously an impossibility, and it is important that we be able to recognize and handle such cases. The source of the difficulty is easily identified. For the characteristic equation of Eq. (4) is

$$m^2 + 5m + 6 = 0$$

and since its roots are $m_1 = -3$ and $m_2 = -2$, the complementary function of Eq. (4) is

$$y = c_1 e^{-3x} + c_2 e^{-2x}$$

Thus, the term on the right-hand side of (4) is proportional to a term in the complementary function; i.e., it is a solution of the related homogeneous equation and hence can yield only 0 when it is substituted into the left member.

One way† in which we might attempt to avoid this difficulty would be to find a particular integral of the equation

$$(5) \qquad\qquad y'' + 5y' + 6y = e^{kx}$$

with $k \neq -3$, and then take the limit of this solution as $k \to -3$. Thus, substituting $Y = Ae^{kx}$, as usual, we have

$$k^2 Ae^{kx} + 5kAe^{kx} + 6Ae^{kx} = e^{kx}$$

whence $\qquad A = \dfrac{1}{k^2 + 5k + 6} \qquad$ and $\qquad Y = \dfrac{e^{kx}}{k^2 + 5k + 6}$

Unfortunately, the limit of this as $k \to -3$ is infinite, and so we must look further. To do this, we note that for all values of c_1 and c_2

$$y = c_1 e^{-3x} + c_2 e^{-2x} + \frac{e^{kx}}{k^2 + 5k + 6}$$

---

† Another way is indicated in Exercise 30.

is a solution of Eq. (5). Hence, taking $c_1 = -1/(k^2 + 5k + 6)$ and $c_2 = 0$, we obtain

$$Y = \frac{e^{kx} - e^{-3x}}{k^2 + 5k + 6}$$

as another particular integral of the nonhomogeneous equation (5). Now, as $k \to -3$ [and Eq. (5) approaches the given equation (4)], the last expression becomes an indeterminate of the form 0/0. Evaluating it by L'Hospital's rule, we find that the limit is

$$-xe^{-3x}$$

and by direct substitution it is easily verified that this actually is a solution of Eq. (4).

It is not necessary to go through this limiting process in particular cases where $f(x)$ contains a term duplicating a term in the complementary function, for we have the following extension of Rule 1.

RULE 2 In the differential equation $ay'' + by' + cy = f(x)$, let $f(x)$ be a sum $R_1(x) + \cdots + R_n(x)$, and let Y_i be the group of terms normally included in the trial particular integral Y because of $R_i(x)$. If any term in Y_i duplicates a term in the complementary function of the differential equation, then before it is substituted into the equation, Y_i must be multiplied by the lowest positive integral power of x which will eliminate all such duplications. The results of our discussion are summarized in Table 2.2.

Table 2.2

Differential equation: $ay'' + by' + cy = f(x)$ or $(aD^2 + bD + c)y = f(x)$	
$f(x)$†	Necessary choice for the trial particular integral Y‡
1. α	A
2. αx^n (n a positive integer)	$A_0 x^n + A_1 x^{n-1} + \cdots + A_{n-1}x + A_n$
3. αe^{rx} (r either real or complex)	Ae^{rx}
4. $\alpha \cos kx$§ 5. $\alpha \sin kx$	$A \cos kx + B \sin kx$
6. $\alpha x^n e^{rx} \cos kx$ 7. $\alpha x^n e^{rx} \sin kx$	$(A_0 x^n + \cdots + A_{n-1}x + A_n)e^{rx} \cos kx$ $+ (B_0 x^n + \cdots + B_{n-1}x + B_n)e^{rx} \sin kx$

† When $f(x)$ consists of a sum of several terms, the appropriate choice for Y is the sum of the Y expressions corresponding to these terms individually.

‡ Whenever a term in any of the Y's listed in this column duplicates a term in the complementary function, all terms in that Y expression must be multiplied by the lowest positive integral power of x sufficient to eliminate all such duplications.

§ The hyperbolic functions $\cosh kx$ and $\sinh kx$ can be handled either by expressing them in terms of exponentials or by using formulas entirely analogous to those in lines 4 to 7.

EXAMPLE 2

Find a complete solution of the equation $y'' + 5y' + 6y = 3e^{-2x} + e^{3x}$.

The roots of the characteristic equation

$$m^2 + 5m + 6 = 0$$

are $m_1 = -2$ and $m_2 = -3$. Hence the complementary function is

$$c_1 e^{-2x} + c_2 e^{-3x}$$

For the trial solution Y_1 corresponding to the term $3e^{-2x}$ we would normally use Ae^{-2x}. However, e^{-2x} is a part of the complementary function, and thus, following the second footnote to Table 2.2, we must multiply this by x before including it in Y_1. For the term e^{3x} the normal choice for a trial solution, namely, $Y_2 = Be^{3x}$, is satisfactory as it stands, since e^{3x} is not contained in the complementary function. Hence we assume

$$Y = Y_1 + Y_2 = Axe^{-2x} + Be^{3x}$$

Substituting this into the differential equation, we have

$$(4Axe^{-2x} - 4Ae^{-2x} + 9Be^{3x}) + 5(-2Axe^{-2x} + Ae^{-2x} + 3Be^{3x})$$
$$+ 6(Axe^{-2x} + Be^{3x}) = 3e^{-2x} + e^{3x}$$

or

$$Ae^{-2x} + 30Be^{3x} = 3e^{-2x} + e^{3x}$$

Equating coefficients of like terms, we find $A = 3$ and $B = \frac{1}{30}$. Hence

$$Y = 3xe^{-2x} + \frac{e^{3x}}{30}$$

and a complete solution is

$$y = c_1 e^{-2x} + c_2 e^{3x} + 3xe^{-2x} + \frac{e^{3x}}{30}$$

EXAMPLE 3

Find a complete solution of the equation $y'' - 2y' + y = xe^x - e^x$.

The characteristic equation in this case is

$$m^2 - 2m + 1 = 0$$

and its roots are $m_1 = m_2 = 1$. Hence, the complementary function is

$$c_1 e^x + c_2 xe^x$$

According to line 6 of Table 2.2 (with $n = r = 1$, $k = 0$) we would normally try

$$Y_1 = (A_0 x + A_1)e^x$$

as the particular integral required for the term xe^x. Moreover, since the particular integral for the term $-e^x$, namely, $Y_2 = Ae^x$, is already a part of Y_1, it need not be included separately. However, the terms in Y_1 are both in the complementary function. Hence, following the second footnote in Table 2.2, we must multiply Y_1 by the lowest positive integral power of x which will eliminate all duplication of terms between Y_1 and the complementary function. This means that Y_1 must be multiplied by x^2, since multiplying it by x would still leave the term xe^x common to Y_1 and the complementary function. Thus we continue with the modified trial solution

$$Y = (A_0 x^3 + A_1 x^2)e^x$$

Substituting this into the differential equation, we obtain

$$[A_0 x^3 + (6A_0 + A_1)x^2 + (6A_0 + 4A_1)x + 2A_1]e^x$$
$$- 2[A_0 x^3 + (3A_0 + A_1)x^2 + 2A_1 x]e^x + (A_0 x^3 + A_1 x^2)e^x = xe^x - e^x$$

or

$$6A_0 xe^x + 2A_1 e^x = xe^x - e^x$$

This will be identically true if and only if $A_0 = \frac{1}{6}$ and $A_1 = -\frac{1}{2}$. Hence

$$Y = \left(\frac{x^3}{6} - \frac{x^2}{2}\right) e^x$$

and so a complete solution is

$$y = c_1 e^x + c_2 x e^x - \frac{x^2 e^x}{2} + \frac{x^3 e^x}{6}$$

EXERCISES

Find a complete solution of each of the following equations:

1 $y'' + 4y' + 5y = 2e^x$ 2 $y'' + 4y' + 3y = x - 1$
3. $y'' + y' = x + 2$ 4 $y'' - y = e^x + 2e^{2x}$
5 $y'' + y = \cos x + 3 \sin 2x$ 6 $y'' + 4y' + 13y = \cos 3x - \sin 3x$
7 $y'' + 3y' = \sin x + 2 \cos x$ 8 $y'' + 2y' + 10y = 25x^2 - 3e^{-x}$
9 $(D^2 + 4D + 4)y = xe^{-x}$ 10 $(D^2 + 1)y = e^x \sin x$
11 $y'' + 2y' + y = \cos^2 x$ *Hint:* Recall that $\cos^2 x = (1 + \cos 2x)/2$.
12 $10y'' - 6y' + y = 30 \sin x \cos x$ 13 $y'' - 5y' + 6y = \cosh x$
14 $y'' - 5y' + 4y = \cosh x$
15 **(a)** Show that $Y = -\cosh x$ is a particular integral of the equation

$$y'' + y' - 2y = e^{-x}$$

(b) Determine A so that $Y = A \sinh x$ will be a particular integral of this equation.

Find the solution of each of the following equations which satisfies the given conditions:

16 $y'' + 4y' + 5y = 20e^x$ $y = y' = 0$ when $x = 0$
17 $y'' + 4y' + 4y = 8x - 10$ $y = 2, y' = 0$ when $x = 0$
18 $y'' + 4y' + 3y = 4e^{-x}$ $y = 0, y' = 2$ when $x = 0$
19 $y'' + 2y' + 5y = 10 \cos x$ $y = 5, y' = 6$ when $x = 0$
20 Show that $Y = (\sin \lambda t - \sin kt)/(k^2 - \lambda^2)$ is a particular integral of the equation $y'' + k^2 y = \sin \lambda t$ and investigate the limiting case when $\lambda \to k$.
21 Construct a solution of the equation $y'' - 2ay' + a^2 y = e^{\lambda t}$ which will approach a solution of the equation $y'' - 2ay' + a^2 y = e^{at}$ as $\lambda \to a$.
22 If y_1 and y_2 are two solutions of the nonhomogeneous equation $y'' + P(x)y' + Q(x)y = R(x)$, determine for what values of c_1 and c_2, if any, $y = c_1 y_1 + c_2 y_2$ is a solution of this equation.
23 If y_1 and y_2 are, respectively, solutions of the equations $y'' + P(x)y' + Q(x)y = R_1(x)$ and $y'' + P(x)y' + Q(x)y = R_2(x)$, show that $y = y_1 + y_2$ is always a solution of the equation $y'' + P(x)y' + Q(x)y = R_1(x) + R_2(x)$.
24 Using the method of undetermined coefficients, find a particular integral of the equation $y'' - y = 1/x$. For what values of x, if any, is this solution meaningful?
25 Using the method of undetermined coefficients, find a particular integral of the equation $y'' + y = x^{1/2}$. For what values of x, if any, is this solution meaningful?
26 If y_1 is a solution of the homogeneous equation $y'' + P(x)y' + Q(x)y = 0$, show that the substitution $y = \phi y_1$ will reduce the problem of finding a particular integral of the nonhomogeneous equation $y'' + P(x)y' + Q(x)y = R(x)$ to the solution of a linear first-order equation in which ϕ' is the independent variable.

Using the method of Exercise 26, and the given solution y_1, find a particular integral of each of the following equations:

27 **(a)** $y'' - 3y' + 2y = e^x$ $y_1 = e^x$
    (b) $y'' - 3y' + 2y = e^x$ $y_1 = e^{2x}$
28 $y'' - 2y' + y = xe^x$ $y_1 = e^x$
29 **(a)** $y'' + y = \cos x$ $y_1 = \sin x$
    (b) $y'' + y = \cos x$ $y_1 = \cos x$

30 (a) Explain how the result of part **(b)** of Exercise 33, Sec. 2.2, can be used to obtain a particular integral of the equation $ay'' + by' + cy = e^{rx}$ when $m = r$ is a simple root of the characteristic equation.

(b) Generalize the results of parts **(a)** and **(b)** of Exercise 33, Sec. 2.2, to show that if $p(D)$ is a quadratic polynomial in D, then $p(D)(x^2 e^{mx}) = p(m)x^2 e^{mx} + 2p'(m)xe^{mx} + p''(m)e^{mx}$.

(c) Explain how the result of part **(b)** can be used to obtain a particular integral of the equation $ay'' + by' + cy = xe^{rx}$ when $m = r$ is a double root of the characteristic equation.

2.4 Particular Integrals by the Method of Variation of Parameters

For certain theoretical purposes and occasionally in applications it is desirable to be able to find a particular integral of the equation

$$(1) \qquad\qquad ay'' + by' + cy = f(x)$$

in cases where the method of undetermined coefficients will not work, i.e., when $f(x)$ is not one of the simple functions possessing only a finite number of independent derivatives. A procedure known as **variation of parameters** will do this for all linear equations, including the general equation with variable coefficients,

$$(2) \qquad\qquad y'' + P(x)y' + Q(x)y = R(x)$$

regardless of the form of $R(x)$, provided that a complete solution of the corresponding homogeneous equation is known. This method differs from the method of undetermined coefficients in that integration rather than differentiation is involved,† which means that the price we pay for greater generality is usually the inconvenience of integrals which cannot be evaluated in terms of familiar functions.

The fundamental idea behind the process is this. Instead of using two arbitrary *constants* c_1 and c_2 to combine two independent solutions of the homogeneous equation

$$(3) \qquad\qquad y'' + P(x)y' + Q(x)y = 0$$

as we do in constructing the complementary function, we attempt to find two *functions* of x, say u_1 and u_2, such that

$$Y = u_1 y_1 + u_2 y_2$$

will be a solution of the nonhomogeneous equation (2). Having two unknown functions u_1 and u_2, we require two equations for their determination. One of these will be obtained by substituting Y into the given differential equation (2); the other remains at our disposal. As the analysis proceeds it will become clear what this second condition should be.

From $Y = u_1 y_1 + u_2 y_2$ we have, by differentiation,

$$Y' = (u_1 y_1' + u_1' y_1) + (u_2 y_2' + u_2' y_2)$$

---

† This is the origin of the name *particular integral*.

Another differentiation will clearly introduce second derivatives of the unknown functions u_1 and u_2, with attendant complications, unless we arrange to eliminate the first-derivative terms u_1' and u_2' from Y'. This can be done if we make

(4) $$u_1' y_1 + u_2' y_2 = 0$$

which thus becomes the necessary second condition on u_1 and u_2.

Proceeding now with the simplified expression

$$Y' = u_1 y_1' + u_2 y_2'$$

we find $$Y'' = (u_1 y_1'' + u_1' y_1') + (u_2 y_2'' + u_2' y_2')$$

Substituting Y, Y', and Y'' into Eq. (2), we obtain

$$(u_1 y_1'' + u_1' y_1' + u_2 y_2'' + u_2' y_2') + P(x)(u_1 y_1' + u_2 y_2')$$
$$+ Q(x)(u_1 y_1 + u_2 y_2) = R(x)$$

or

$$u_1[y_1'' + P(x)y_1' + Q(x)y_1] + u_2[y_2'' + P(x)y_2' + Q(x)y_2]$$
$$+ u_1' y_1' + u_2' y_2' = R(x)$$

The expressions in brackets vanish because, by hypothesis, both y_1 and y_2 are solutions of the homogeneous equation (3). Hence, we find for the other condition on u_1 and u_2

(5) $$u_1' y_1' + u_2' y_2' = R(x)$$

Solving Eqs. (4) and (5) for u_1' and u_2', we obtain

(6) $$u_1' = -\frac{y_2}{y_1 y_2' - y_2 y_1'} R(x) \quad \text{and} \quad u_2' = \frac{y_1}{y_1 y_2' - y_2 y_1'} R(x)$$

The functions y_1, y_2, y_1', y_2', and $R(x)$ are all known. Hence, u_1 and u_2 can be found by a single integration. With u_1 and u_2 known, the particular integral

$$Y = u_1 y_1 + u_2 y_2$$

is completely determined.

We should notice, of course, that if $y_1 y_2' - y_2 y_1' = 0$, the solution for u_1' and u_2' cannot be carried out. However, $y_1 y_2' - y_2 y_1'$ is precisely the wronskian of the two solutions y_1 and y_2, and if these are independent, as we suppose them to be, then their wronskian cannot vanish.

EXAMPLE 1

Find a complete solution of the equation $y'' + y = \sec x$

By inspection, we see that the complementary function in this case is

$$A \cos x + B \sin x$$

Hence, taking $y_1 = \cos x$ and $y_2 = \sin x$, we have from Eq. (6)

$$u_1' = -\frac{\sin x}{\cos x (\cos x) - \sin x (-\sin x)} \sec x = -\tan x$$

$$u_2' = \frac{\cos x}{\cos x (\cos x) - \sin x (-\sin x)} \sec x = 1$$

Therefore,

$$u_1 = -\int \tan x \, dx = \ln |\cos x| \quad \text{and} \quad u_2 = \int dx = x$$

and thus,

$$Y = u_1 y_1 + u_2 y_2 = (\ln |\cos x|) \cos x + x \sin x$$

Finally,

$$y = A \cos x + B \sin x + (\ln |\cos x|) \cos x + x \sin x$$

EXERCISES

Find a complete solution of each of the following equations:

1 $y'' - 2y' + y = x^3 e^x$ 2 $y'' + y = \sin x$
3 $y'' - y = \cosh x$ 4 $y'' + 2y' + 2y = e^{-x} \cos x$
5 $y'' + 2y' + y = \dfrac{e^{-x}}{x}$ 6 $y'' + y = \tan x$
7 $y'' + y = \dfrac{1}{1 + \sin x}$ 8 $y'' + 4y' + 4y = \dfrac{e^{-2x}}{x^2}$
9 $y'' + 2y' + y = e^{-x} \ln x$ 10 $4y'' + 8y' + 5y = e^{-x} \sec \dfrac{x}{2}$

11 Find a particular integral of the equation $x^2 y'' + xy' - y = 1/(1 + x)$, given that $y_1 = x$ and $y_2 = 1/x$ are two solutions of the related homogeneous equation.
12 Find a particular integral of the equation $x^2 y'' - xy' + y = 1/x$, given that $y_1 = x$ and $y_2 = x \ln x$ are two solutions of the related homogeneous equation.
13 Find a particular integral of the equation $x^2 y'' - 2xy' + 2y = x \ln x$, given that $y_1 = x$ and $y_2 = x^2$ are two solutions of the related homogeneous equation.
14 Using the method of variation of parameters, show that the expression

$$y = A \cos kx + B \sin kx + \frac{1}{k} \int_0^x \sin k(x - s) f(s) \, ds$$

is a complete solution of the equation $y'' + k^2 y = f(x)$. *Hint:* Introduce the dummy variable s in the integrals which define u_1 and u_2. Then move $y_1(x)$ and $y_2(x)$ into the respective integrals and combine the two integrals.

Find a complete solution of each of the following equations:

15 $y'' + 2ay' + (a^2 - b^2)y = f(x)$ 16 $y'' + 2ay' + a^2 y = f(x)$
17 $y'' + 2ay' + (a^2 + b^2)y = f(x)$
18 Using the method of variation of parameters, find a particular integral of the equation $y'' - y = 1/x$. How does this result compare with the result of Exercise 24, Sec. 2.3?
19 Using the method of variation of parameters, find a particular integral of the equation $y'' + y = x^{1/2}$. How does this result compare with the result of Exercise 25, Sec. 2.3?
20 (a) Use the result of Exercise 15 to show that if $0 < b < a$ and if $|f(x)|$ is bounded for $x \geq x_0$, then the absolute value of every solution of the equation $y'' + 2ay' + (a^2 - b^2)y = f(x)$ is bounded for $x \geq x_0$.
    (b) If $0 < b < a$, if $|f(x)|$ is bounded for $x \geq x_0$, and if $\lim_{x \to \infty} f(x) = L$, show that if y is any solution of the equation $y'' + 2ay' + (a^2 - b^2)y = f(x)$, then

$$\lim_{x \to \infty} y = \frac{L}{a^2 - b^2}$$

21 (a) Using the result of Exercise 17, show that if $0 < a$ and if $|f(x)|$ is bounded for $x \geq x_0$, then the absolute value of every solution of the equation $y'' + 2ay' + (a^2 + b^2)y = f(x)$ is bounded for $x \geq x_0$.
    (b) If $0 < a$, if $|f(x)|$ is bounded for $x \geq x_0$, if $\lim_{x \to \infty} f(x) = L$, and if y is any solution of the equation

$$y'' + 2ay' + (a^2 + b^2)y = f(x)$$

what is $\lim_{x \to \infty} y$?

2.5 Equations of Higher Order

The theory of the linear differential equation of order higher than 2,

(1) $$y^{(n)} + P_1(x)y^{(n-1)} + \cdots + P_{n-1}(x)y' + P_n(x)y = R(x)$$

parallels the second-order case in all significant details. In particular, with the obvious changes required by the fact that $n > 2$, the three fundamental theorems of Sec. 2.1 hold for linear equations of all orders.† For the especially important case of the homogeneous, linear, constant-coefficient equation of order higher than 2,

(2) $$a_0 y^{(n)} + a_1 y^{(n-1)} + \cdots + a_{n-1}y' + a_n y = 0$$

the substitution $y = e^{mx}$ leads, as before, to the characteristic equation

(3) $$a_0 m^n + a_1 m^{n-1} + \cdots + a_{n-1}m + a_n = 0$$

which can be obtained in a specific problem simply by replacing each derivative by the corresponding power of m. The degree of this algebraic equation will be the same as the order of the differential equation (2); hence, counting repeated roots the appropriate number of times, we find that the number of roots $m_1, m_2, \ldots$ will equal the order of the differential equation. From these roots, the solution of the homogeneous equation can be constructed by adding together the terms that were listed in Table 2.1 as corresponding to each of the various root types. The only extension necessary is required when the characteristic equation (3) has roots of multiplicity greater than 2: *If y_1 is the solution normally corresponding to a root m_1, and if this root occurs k (>2) times, then not only are y_1 and xy_1 solutions (as in the second-order case) but x^2y_1, $x^3y_1, \ldots, x^{k-1}y_1$ are also solutions and must be included in the complementary function.*

For the nonhomogeneous, constant-coefficient equation

(4) $$a_0 y^{(n)} + a_1 y^{(n-1)} + \cdots + a_{n-1}y' + a_n y = f(x)$$

it is still true that the complete solution is the sum of the complementary function, obtained by solving the associated homogeneous equation, and a particular integral. In the important case when $f(x)$ is a function possessing only a finite number of independent derivatives the particular integral can be found just as before by using the tentative choices for Y listed in Table 2.2. Variation of parameters can be extended to those problems which the method of undetermined coefficients cannot handle. An example or two will make these ideas clear.

EXAMPLE 1

Find a complete solution of the equation $y''' + 3y'' + 3y' + y = 0$.

In this case the characteristic equation is

$$m^3 + 3m^2 + 3m + 1 \equiv (m + 1)^3 = 0$$

---

† Before Theorem 3, Sec. 2.1, can be extended to equations of higher order, the wronskian of more than two functions must be defined. The appropriate generalization is

$$W(y_1, y_2, \ldots, y_n) = \begin{vmatrix} y_1 & y_2 & \cdots & y_n \\ y_1' & y_2' & \cdots & y_n' \\ \cdots\cdots\cdots\cdots\cdots\cdots\cdots \\ y_1^{(n-1)} & y_2^{(n-1)} & \cdots & y_n^{(n-1)} \end{vmatrix}$$

which clearly reduces to the definition of Sec. 2.1 if $n = 2$. Various properties of the wronskian when $n > 2$ will be found among the exercises.

Hence $m = -1$ is a triple root, and not only are e^{-x} and xe^{-x} solutions of the differential equation but so too is x^2e^{-x}. Since these three solutions have a nonvanishing wronskian (see Exercise 24), it follows that a complete solution of the equation is

$$y = c_1e^{-x} + c_2xe^{-x} + c_3x^2e^{-x}$$

EXAMPLE 2

Find a complete solution of the equation $y''' + 5y'' + 9y' + 5y = 3e^{2x}$.
The characteristic equation in this case is

$$m^3 + 5m^2 + 9m + 5 = 0$$

By inspection $m = -1$ is seen to be a root. Hence c_1e^{-x} must be one term in the complementary function. When the factor corresponding to this root is divided out of the characteristic equation, there remains the quadratic equation

$$m^2 + 4m + 5 = 0$$

Its roots are $m = -2 \pm i$; thus the complementary function must also contain

$$e^{-2x}(c_2 \cos x + c_3 \sin x)$$

The entire complementary function is therefore

$$c_1e^{-x} + e^{-2x}(c_2 \cos x + c_3 \sin x)$$

For a particular integral we try, as usual, $Y = Ae^{2x}$. Substituting this into the differential equation gives

$$(8Ae^{2x}) + 5(4Ae^{2x}) + 9(2Ae^{2x}) + 5(Ae^{2x}) = 3e^{2x} \quad \text{or} \quad 51Ae^{2x} = 3e^{2x}$$

Hence $\qquad\qquad\qquad\qquad A = \frac{1}{17} \qquad Y = \frac{1}{17}e^{2x}$

and therefore $\qquad y = c_1e^{-x} + e^{-2x}(c_2 \cos x + c_3 \sin x) + \frac{1}{17}e^{2x}$

EXAMPLE 3

Find a complete solution of the equation $(D^4 + 8D^2 + 16)y = -\sin x$.
The characteristic equation here is

$$m^4 + 8m^2 + 16 = 0 \quad \text{or} \quad (m^2 + 4)^2 = 0$$

The roots of this equation are $m = \pm 2i, \pm 2i$. Hence, the complementary function contains not only the terms

$$\cos 2x \quad \text{and} \quad \sin 2x$$

but also these terms multiplied by x and therefore is

$$c_1 \cos 2x + c_2 \sin 2x + c_3x \cos 2x + c_4x \sin 2x$$

To find a particular integral we try $Y = A \cos x + B \sin x$, which, on substitution into the differential equation, gives

$$(A \cos x + B \sin x) + 8(-A \cos x - B \sin x) + 16(A \cos x + B \sin x) = -\sin x$$

or $\qquad\qquad\qquad\qquad 9A \cos x + 9B \sin x = -\sin x$

This will be an identity if and only if $A = 0$† and $B = -\frac{1}{9}$. Therefore

$$Y = -\frac{\sin x}{9}$$

---

† Since the given differential equation contains only derivatives of even order, we could have foreseen that Y would need to contain only a sine term in order to match the sine term on the right-hand side and that $Y = B \sin x$ would therefore be a satisfactory trial solution. This simplification should be clearly understood, for it can often be applied to the analysis of vibrating mechanical systems with negligible friction or electric circuits with negligible resistance.

and the complete solution is

$$y = c_1 \cos 2x + c_2 \sin 2x + c_3 x \cos 2x + c_4 x \sin 2x - \frac{\sin x}{9}$$

EXAMPLE 4

For what nonzero values of λ, if any, does the equation $y^{iv} - \lambda^4 y = 0$ have solutions which satisfy the four conditions $y(0) = y''(0) = y(l) = y''(l) = 0$ and are not identically zero? What are these solutions if they exist?

The characteristic equation in this case is $m^4 - \lambda^4 = 0$, and its roots are $m = \pm \lambda$, $\pm i\lambda$. Hence, a complete solution is

$$y = c_1 \cos \lambda x + c_2 \sin \lambda x + c_3 e^{\lambda x} + c_4 e^{-\lambda x}$$

It is more convenient, however, to introduce hyperbolic functions and work with a complete solution of the following form:

$$y = A \cos \lambda x + B \sin \lambda x + C \cosh \lambda x + E \sinh \lambda x$$

Differentiating this twice gives us

$$y'' = \lambda^2(-A \cos \lambda x - B \sin \lambda x + C \cosh \lambda x + E \sinh \lambda x)$$

Hence, substituting the first two of the given conditions, we obtain the relations

$$A + C = 0$$
$$\lambda^2(-A + C) = 0$$

which imply that $A = C = 0$. Using this information and the last two conditions, we have, further,

$$B \sin \lambda l + E \sinh \lambda l = 0$$
$$\lambda^2(-B \sin \lambda l + E \sinh \lambda l) = 0$$

Dividing out λ^2 and then adding these equations, we find that

$$2E \sinh \lambda l = 0$$

Now the hyperbolic sine is zero if and only if its argument is zero. Moreover, by the statement of the problem we are restricted to nonzero values of λ. Hence, $\sinh \lambda l \neq 0$, and therefore $E = 0$, which implies that

$$B \sin \lambda l = 0$$

We have already been forced to the conclusion that $A = C = E = 0$; hence if $B = 0$, the solution would be identically zero, contrary to the requirements of the problem. Thus we must have

$$\sin \lambda l = 0 \qquad \text{or} \qquad \lambda = \frac{n\pi}{l}$$

For these values of λ, and for these only, there are solutions meeting the requirements of the problem. Clearly, these solutions are all of the form $y = B \sin \lambda x$, or, more specifically, $y = B_n \sin (n\pi x/l)$.

EXERCISES

Find a complete solution of each of the following equations:

1   $(D^3 + 6D^2 + 11D + 6)y = 6x - 7$ 2 $(D^4 - 16)y = e^x$
3   $y''' - 2y'' - 3y' + 10y = 40 \cos x$ 4 $y^{iv} + 10y'' + 9y = \cos 2x$

5 $(D^4 + 8D^2 - 9)y = 9x^2 + 5 \sin 2x$ **6** $(D^3 + D^2 + 3D - 5)y = e^x$
7 $(D^4 + 2D^3 - 3D^2 - 4D + 4)y = e^x$ **8** $(D^3 - 7D + 6)y = e^x - 5e^{2x}$
9 $y^{iv} + 4y = \cos x + \sin 2x$. *Hint:* By adding and subtracting the appropriate term on the left-hand side of the characteristic equation, rewrite it as the difference of two squares.

Find the solution of each of the following equations which satisfies the given conditions:

10 $y''' - 3y' + 2y = 0$ $y = 0, y' = 2, y'' = 0$ when $x = 0$
11 $(D^3 + 2D^2 - D - 2)y = 10 \sin x$ $y = 1, y' = -2, y'' = -1$ when $x = 0$
12 $(D^3 - 2D^2 + D - 2)y = 0$ $y = y' = y'' = 1$ when $x = 0$
13 $y^{iv} + 5y'' + 4y = 0$ $y = y'' = 0$ when $x = 0$; $y = 1, y' = 2$
                                          when $x = \pi/2$

14 Using the method of variation of parameters, find a particular integral of the equation $y''' - 3y'' + 3y' - y = e^x/x$.

Using the method of variation of parameters, obtain a formula for a particular integral of each of the following equations:

15 $(D^3 - 6D^2 + 11D - 6)y = f(x)$ **16** $y''' - y'' + y' - y = f(x)$
17 $y''' - y'' - y' + y = f(x)$ **18** $y^{iv} - 5y'' + 4y = f(x)$
19 For what nonzero values of λ, if any, does the equation $y^{iv} - \lambda^4 y = 0$ have solutions which satisfy the conditions $y(0) = y''(0) = y(1) = y'(1) = 0$ and are not identically zero? What are these solutions if they exist?
20 Work Exercise 19 if the given conditions are $y(0) = y''(0) = y''(1) = y'''(1) = 0$.
21 (a) Generalize the results of Exercise 33, Sec. 2.2, and Exercise 30, Sec. 2.3, by showing that if $p(D)$ is a cubic polynomial in D, then

$$p(D)(xe^{mx}) = p(m)xe^{mx} + p'(m)e^{mx}$$

$$p(D)(x^2e^{mx}) = p(m)x^2e^{mx} + 2p'(m)xe^{mx} + p''(m)e^{mx}$$

$$p(D)(x^3e^{mx}) = p(m)x^3e^{mx} + 3p'(m)x^2e^{mx} + 3p''(m)xe^{mx} + p'''(m)e^{mx}$$

(b) Discuss the application of these formulas to the solution of the third-order linear differential equation with constant coefficients.
22 If three functions are linearly dependent, prove that their wronskian is identically zero.
23 Prove that the wronskian of the functions $e^{m_1 x}$, $e^{m_2 x}$, and $e^{m_3 x}$ is different from zero if and only if m_1, m_2, and m_3 are all different.
24 Prove that the wronskian of the functions e^{mx}, xe^{mx}, and x^2e^{mx} is never equal to zero.
25 Prove that if a and m are real and if a is different from zero, the wronskian of the functions e^{mx}, $\cos ax$, and $\sin ax$ is never equal to zero.
26 Prove that if $\lambda \neq 0$, the wronskian of the functions $\cos \lambda x$, $\sin \lambda x$, $\cosh \lambda x$, and $\sinh \lambda x$ is never equal to zero.
27 State and prove the generalization of Theorem 2, Sec. 2.1, for the case $n = 3$.
28 State and prove the generalization of Theorem 4, Sec. 2.1, for the case $n = 3$.
29 If y_1, y_2, and y_3 are three solutions of the equation $y''' + P_1(x)y'' + P_2(x)y' + P_3(x)y = 0$, prove the following generalization of Abel's identity

$$W(y_1, y_2, y_3) = k_{123} \exp\left[-\int P_1(x)\, dx\right]$$

Hint: Write out the three equations expressing the hypothesis that y_1, y_2, and y_3 are solutions of the given equation, multiply them by u_1, u_2, and u_3, respectively, add the resulting equations, and then determine u_1, u_2, and u_3 so that y_1, y_2, y_3 and their first derivatives are eliminated from the result.
30 Using the result of Exercise 29, state and prove the generalization of Theorem 3, Sec. 2.1, for the case $n = 3$.

2.6 Applications

Linear differential equations with constant coefficients find their most important application in the study of electric circuits and vibrating mechanical systems. So useful to engineers are the results of this analysis that we shall devote an entire chapter to its major features. However, there are also other applications of considerable interest, and although we cannot discuss them at length, we shall present a few typical examples in this section.

One important field in which linear differential equations often arise is the study of the bending of beams. When a beam is bent, it is obvious that the fibers near the concave surface of the beam are compressed whereas those near the convex surface are stretched. Somewhere between these regions of compression and tension there must, from considerations of continuity, be a surface of fibers which are neither compressed nor stretched. This is known as the **neutral surface** of the beam, and the curve of any particular fiber in this surface is known as the **elastic curve** or **deflection curve** of the beam. The line in which the neutral surface is cut by any plane cross section of the beam is known as the **neutral axis** of that cross section (Fig. 2.1).

The loads which cause a beam to bend may be of two sorts: they may be concentrated at one or more points along the beam, or they may be continuously distributed with a density $w(x)$ known as the **load per unit length**. In either case we have two important related quantities. One is the **shear** $V(x)$ at any point along the beam, which is defined to be the algebraic sum of all the transverse forces† which act on the beam on the positive side of the point in question (Fig. 2.2). The other is the **moment** $M(x)$, which is defined as the total moment produced at a general point along the beam by all the forces, transverse or not, which act on the beam on one side or the other of the point in question. We shall consider the load per unit length and the shear to be positive if they act in the direction of the negative y axis (the direction in which loads usually act on a beam). The moment we shall take to be positive if it acts to bend the beam so that it is concave toward the positive y axis. With these conventions of sign (which are not universally adopted) it is shown in the study of the strength of materials that the deflection curve of the beam satisfies the second-order differential equation

(1) $$EIy'' = M$$

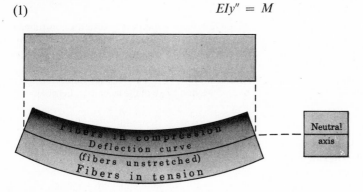

Figure 2.1
A beam before and after bending.

---

† A **transverse force** is one whose direction is perpendicular to the length of the beam.

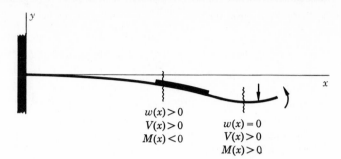

Figure 2.2
The conventions for the signs of the moment, shear, and load per unit length at a general point of a beam.

where E is the modulus of elasticity of the material of the beam, and I, which may be a function of x, is the moment of inertia of the cross-sectional area of the beam about the neutral axis of the cross section. If the beam bears only transverse loads, it can be shown further that we have the two additional relations

(2) $$\frac{dM}{dx} = \frac{d(EIy'')}{dx} = V$$

(3) $$\frac{d^2M}{dx^2} = \frac{dV}{dx} = \frac{d^2(EIy'')}{dx^2} = -w$$

In most elementary applications the moment M is an explicit function of x; hence Eq. (1) can be solved and the deflection $y(x)$ determined simply by performing two integrations. However in problems in which the load has a component in the direction of the length of the beam, M depends on y, and Eq. (1) can be solved only through the use of techniques from the field of differential equations. An interesting example of this sort is provided by the classic problem of the buckling of a slender column.

EXAMPLE 1

A long slender column of length L and uniform cross section whose ends are constrained to remain in the same vertical line but are otherwise free, i.e., able to turn, is compressed by a load F. Determine the possible deflection curves of the column and the loads required to produce each one.

Let coordinates be chosen as shown in Fig. 2.3. Then, clearly, the moment arm of the load F about a general point P on the deflection curve of the beam is y; hence Eq. (1) becomes

(4) $$EIy'' = -Fy$$

the minus sign indicating that when y is positive (as shown), the moment is negative since it has produced a deflection curve which is convex toward the positive y axis.

By hypothesis, the column is of uniform cross section; hence the moment of inertia I is a constant. Therefore (4) is a constant-coefficient differential equation and can be solved by the methods of Sec. 2.2. Accordingly, we set up the characteristic equation

$$EIm^2 + F = 0$$

and solve it, getting

$$m = \pm\sqrt{\frac{F}{EI}}\,i$$

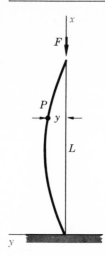

Figure 2.3
A slender column buckling under
a vertical load.

A complete solution of (4) is therefore

(5)
$$y = A \cos \sqrt{\frac{F}{EI}}\, x + B \sin \sqrt{\frac{F}{EI}}\, x$$

To determine the constants A and B, we have the information that $y = 0$ when $x = 0$ and also when $x = L$. Substituting the first of these into Eq. (5), we see at once that $A = 0$. Substituting the second, we obtain

$$0 = B \sin \sqrt{\frac{F}{EI}}\, L$$

Since $\sin \sqrt{(F/EI)}\,L$ in general is not equal to zero, it follows that $B = 0$, which, since we have already found $A = 0$, means that $y \equiv 0$. However, if the load F has just the right value to make $\sqrt{(F/EI)}\,L = n\pi$, then the last equation will be satisfied without B being 0 and equilibrium is then possible in a deflected position defined by

$$y = B \sin \frac{n\pi x}{L}$$

Since n can take on any of the values $1, 2, 3, \ldots$, there are thus infinitely many different critical loads

$$F_n = \left(\frac{n\pi}{L}\right)^2 EI$$

each with its own particular deflection curve. For values of F below the lowest critical load, the column will remain in its undeflected vertical position or if displaced slightly from it, will return to it as an equilibrium configuration. For values of F above the lowest critical load and different from the higher critical loads, the column can theoretically remain in a vertical position, but the equilibrium is unstable, and if the column is deflected slightly, it will not return to a vertical position but will continue to deflect until it collapses. Thus only the lowest critical load is of much practical significance.

In many physical systems vibratory motion is possible but undesirable. In such cases it is important to know the frequency at which vibration *can* take place in order to avoid periodic external influences that might be in resonance with the natural frequency of the system. For simple linear systems in which (as is usually the case) friction is neglected, the underlying differential equation is eventually reducible to the form

$$y'' + \omega^2 y = 0$$

Since the complete solution of this equation is

$$y = A \cos \omega t + B \sin \omega t$$

and since both cos ωt and sin ωt represent periodic behavior of frequency

$$\omega \text{ rad/unit time} \qquad \text{or} \qquad \frac{\omega}{2\pi} \text{ cycles/unit time}$$

it is clear that the frequency can be read just as well from the differential equation itself as from any of its solutions, general or particular. The important part of such a frequency calculation, then, is the formulation of the differential equation and not its solution.

EXAMPLE 2

A weight W_2 is suspended from a pulley of weight W_1 and radius R, as shown in Fig. 2.4. Constraints, which need not be specified, prevent any swinging of the system and permit it to move only in the vertical direction. If a spring of **modulus** k, that is, a spring requiring k units of force to stretch it 1 unit of length, is inserted in the otherwise inextensible cable which supports the pulley, find the frequency with which the system will vibrate in the vertical direction if it is displaced slightly from its equilibrium position. Friction between the cable and the pulley prevents any slippage, but all other frictional effects are to be neglected.

As coordinate to describe the system we choose the vertical displacement y of the center of the pulley, the downward direction being taken as positive. Now when the center of the pulley moves a distance y, the length of the spring must change by $2y$. Moreover, as this happens, the pulley must rotate through an angle

$$\theta = \frac{y}{R} \qquad \text{and} \qquad \frac{d\theta}{dt} = \frac{1}{R}\frac{dy}{dt}$$

It will be convenient to formulate the differential equation governing this problem through the use of the so-called **energy method**. From the fundamental law of the conservation of energy, it follows that *if no energy is lost through friction or other irreversible changes, then in a mechanical system the sum of the instantaneous potential and kinetic energies must remain constant.* In the present problem, the potential energy consists of two parts: (*a*) the potential energy in the weights W_1 and W_2 due to their

Figure 2.4
An unusual spring-suspended weight in equilibrium and after vertical displacement.

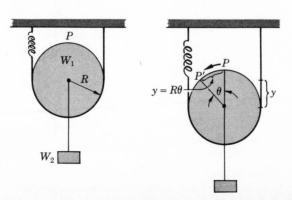

position in the gravitational field and (b) the potential energy stored in the stretched spring. Taking the equilibrium position of the system as the reference level for potential energy we have for (a)

$$(6) \qquad (PE)_a = -(W_1 + W_2)y$$

the minus sign indicating that a positive y corresponds to a lowering of the weights and hence a decrease in the potential energy. The potential energy stored in the spring is simply the amount of work required to stretch the spring from its equilibrium elongation, say δ, to its instantaneous elongation $\delta + 2y$. Since the force in the spring at any time is

$$F = \text{elongation} \times \text{force per unit elongation} = sk$$

we have for the potential energy of type b

$$(7) \qquad (PE)_b = \int_{s_1}^{s_2} F\,ds = \int_{\delta}^{\delta+2y} ks\,ds = k\,\frac{s^2}{2}\bigg|_{\delta}^{\delta+2y} = 2ky^2 + 2k\,\delta y$$

The kinetic energy likewise consists of two parts: (a) the energy of translation of the weights W_1 and W_2, namely,

$$(8) \qquad (KE)_a = \frac{1}{2}\,\frac{W_1 + W_2}{g}\,(\dot{y})^2 \; \dagger$$

and (b) the energy of rotation of the pulley, namely,

$$(KE)_b = \tfrac{1}{2}I(\dot{\theta})^2$$

or, recalling that the polar moment of inertia of a circular disk is

$$I = M\,\frac{R^2}{2} = \frac{W_1}{g}\,\frac{R^2}{2}$$

$$(9) \qquad (KE)_b = \frac{1}{2}\,\frac{W_1}{g}\,\frac{R^2}{2}\,(\dot{\theta})^2 = \frac{1}{2}\,\frac{W_1}{g}\,\frac{R^2}{2}\left(\frac{\dot{y}}{R}\right)^2 = \frac{W_1}{4g}\,(\dot{y})^2$$

The conservation of energy now requires that

$$\text{Kinetic energy} + \text{potential energy} = \text{constant}$$

or, substituting from Eqs. (6) to (9)

$$\frac{W_1}{4g}\,(\dot{y})^2 + \frac{W_1 + W_2}{2g}\,(\dot{y})^2 + (2ky^2 + 2k\,\delta y) - (W_1 + W_2)y = C$$

Differentiating this with respect to time, we have

$$\frac{W_1}{2g}\,\dot{y}\ddot{y} + \frac{W_1 + W_2}{g}\,\dot{y}\ddot{y} + 4ky\dot{y} + 2k\,\delta\dot{y} - (W_1 + W_2)\dot{y} = 0$$

Dividing out $\dot{y}$ (which surely cannot be identically zero when the system is in motion) and collecting terms gives

$$\frac{3W_1 + 2W_2}{2g}\,\ddot{y} + 4ky = (W_1 + W_2) - 2k\delta = 0$$

the terms on the right equaling zero since the elongation δ of the spring in its equilibrium position, when it is supporting one-half the total weight $W_1 + W_2$, is

$$\delta = \frac{W_1 + W_2}{2k}$$

---

† In problems in dynamics, first and second derivatives *with respect to time* are often indicated by placing one and two dots, respectively, over the variable in question.

The differential equation describing the vertical movement of the system is therefore

$$\ddot{y} + \frac{8kg}{3W_1 + 2W_2}\, y = 0$$

From this, as we pointed out above, we can immediately read the natural frequency of the system, namely,

$$\frac{1}{2\pi}\sqrt{\frac{8kg}{3W_1 + 2W_2}} \qquad \text{cycles/unit time}$$

In general, differential equations with variable coefficients are very difficult to solve and rarely can be solved in terms of elementary functions. However, there is one important linear differential equation with variable coefficients which can always be reduced by a suitable substitution to a linear equation with constant coefficients and hence solved without difficulty. This is the so-called **equation of Euler.**†

$$(10) \qquad a_0 x^n y^{(n)} + a_1 x^{n-1} y^{(n-1)} + \cdots + a_{n-1} x y' + a_n y = 0$$

in which the coefficient of each derivative is proportional to the corresponding power of the independent variable. If we change the independent variable from x to z by means of the substitution

$$x = e^z \qquad \text{or} \qquad z = \ln x$$

Eq. (10) becomes an equation in y and z with constant coefficients which can then be solved by the methods of Sec. 2.5. Finally, replacing z by $\ln x$ in the solution of the transformed equation, we obtain the solution of the original differential equation.

EXAMPLE 3

Find a complete solution of the differential equation

$$x^3 \frac{d^3 y}{dx^3} + 4x^2 \frac{d^2 y}{dx^2} - 5x \frac{dy}{dx} - 15y = 0$$

Under the transformation $x = e^z$ or $z = \ln x$ we have by a straightforward application of the chain rule

$$\frac{dy}{dx} = \frac{dy}{dz}\frac{dz}{dx} = \frac{1}{x}\frac{dy}{dz}$$

$$\frac{d^2 y}{dx^2} = \frac{d}{dx}\left(\frac{1}{x}\frac{dy}{dz}\right) = -\frac{1}{x^2}\frac{dy}{dz} + \frac{1}{x}\frac{d^2 y}{dz^2}\frac{dz}{dx} = -\frac{1}{x^2}\frac{dy}{dz} + \frac{1}{x^2}\frac{d^2 y}{dz^2}$$

$$\frac{d^3 y}{dx^3} = \frac{d}{dx}\left[\frac{1}{x^2}\left(-\frac{dy}{dz} + \frac{d^2 y}{dz^2}\right)\right] = -\frac{2}{x^3}\left(-\frac{dy}{dz} + \frac{d^2 y}{dz^2}\right)$$

$$+ \frac{1}{x^2}\left(-\frac{d^2 y}{dz^2} + \frac{d^3 y}{dz^3}\right)\frac{dz}{dx} = \frac{2}{x^3}\frac{dy}{dz} - \frac{3}{x^3}\frac{d^2 y}{dz^2} + \frac{1}{x^3}\frac{d^3 y}{dz^3}$$

Substituting these into the given differential equation, we have

$$x^3 \left[\frac{1}{x^3}\left(2\frac{dy}{dz} - 3\frac{d^2 y}{dz^2} + \frac{d^3 y}{dz^3}\right)\right] + 4x^2 \left[\frac{1}{x^2}\left(-\frac{dy}{dz} + \frac{d^2 y}{dz^2}\right)\right]$$

$$- 5x\left(\frac{1}{x}\frac{dy}{dz}\right) - 15y = 0$$

---

† Also called **Cauchy's equation**, after the French mathematician Augustin Louis Cauchy (1789–1857).

or, simplifying and collecting terms,

$$\frac{d^3y}{dz^3} + \frac{d^2y}{dz^2} - 7\frac{dy}{dz} - 15y = 0$$

The characteristic equation of the last equation is

$$m^3 + m^2 - 7m - 15 \equiv (m - 3)(m^2 + 4m + 5) = 0$$

From its roots, $m_1 = 3$, $m_2, m_3 = -2 \pm i$, we obtain the complete solution

$$y = c_1 e^{3z} + e^{-2z}(c_2 \cos z + c_3 \sin z)$$

Finally, replacing z by $\ln x$, we have

$$y = c_1 \exp(3 \ln x) + \exp(-2 \ln x)[c_2 \cos(\ln x) + c_3 \sin(\ln x)]$$

$$= c_1 x^3 + \frac{1}{x^2}[c_2 \cos(\ln x) + c_3 \sin(\ln x)]$$

EXERCISES

Find a complete solution of each of the following equations:

1   $x^2y'' + xy' - y = 0$ 2 $x^2y'' - 6y = 1 + \ln x$

3   $x^2y'' - xy' + y = x^5$ 4 $x^3y''' - 3x^2y'' + 7xy' - 8y = 0$

5   $2x^2y'' + 5xy' + y = 3x + 2$ 6 $x^3y''' + 2x^2y'' - xy' + y = 0$

7   $x^4y^{\text{iv}} + 6x^3y''' + 15x^2y'' + 9xy' - 9y = 3 \ln x + \dfrac{1}{x^2}$

8   Since e^z is always positive, does the use of the substitution $x = e^z$ mean that an Euler equation can be solved only for positive values of x? If not, how can a solution be obtained which will be valid for negative values of x?

9   If $x = e^z$ and if $d/dz \equiv D$, establish the following operational equivalences:

$$x\frac{d}{dx} = D$$

$$x^2\frac{d^2}{dx^2} = D(D - 1)$$

$$x^3\frac{d^3}{dx^3} = D(D - 1)(D - 2)$$

....

Explain how these formulas can be used to shorten the work of solving an Euler equation.

10   (a) Show that the substitution $Ax + B = e^z$, or $z = \ln(Ax + B)$, will reduce the equation

$$a(Ax + B)^2\frac{d^2y}{dx^2} + b(Ax + B)\frac{dy}{dx} + y = 0$$

to a linear equation with constant coefficients. For $n > 2$, can the equation

$$a_0(Ax + B)^n\frac{d^ny}{dx^n} + a_1(Ax + B)^{n-1}\frac{d^{n-1}y}{dx^{n-1}} + \cdots + a_{n-1}(Ax + B)y' + a_ny = 0$$

be solved in a similar fashion?

(b) Find a complete solution of the equation

$$(x - 2)^2\frac{d^2y}{dx^2} + 2(x - 2)\frac{dy}{dx} - 6y = 0$$

11   (a) A weight W hangs from a spring of modulus k. Neglecting friction, and assuming the weight to be guided so that it can move only in the vertical direction, set up the differential equation which describes its vertical motion and determine its natural frequency.

(b) In part (a), suppose that $W = 16$ lb, $k = 6$ lb/in, and suppose that the weight is pulled down 2 in below its equilibrium position and released from rest at that point. Find the equation describing its subsequent motion. Take the value of g to be 384 in/s².

(c) Work part (b) if the weight begins to move from its equilibrium position with an initial velocity of 3 in/s in the positive direction.

(d) Work part (b) if the weight begins to move with initial velocity 8 in/s in the positive direction from a point 4 in above its equilibrium position.

12 A circular cylinder of radius r and height h, made of material weighing w lb/in, floats in water in such a way that its axis is always vertical. Neglecting all forces except gravity and the buoyant force of the water, as given by the principle of Archimedes, determine the period with which the cylinder will vibrate in the vertical direction if it is depressed slightly from its equilibrium position and released.

13 A cylinder weighing 50 lb floats in water with its axis vertical. When depressed slightly and released, it vibrates with period 2 s. Neglecting all frictional effects, find the diameter of the cylinder.

14 A straight hollow tube rotates with constant angular velocity ω about a vertical axis which is perpendicular to the tube at its midpoint. A pellet of mass m slides without friction in the interior of the tube. Find the equation describing the radial motion of the pellet until it emerges from the tube, assuming that it starts from rest at a radial distance a from the midpoint of the tube.

15 A straight hollow tube rotates with constant angular velocity ω about a horizontal axis which is perpendicular to the tube at its midpoint. Show that if the initial conditions are suitably chosen, a pellet sliding without friction in the tube will never be ejected but will execute simple harmonic motion within the tube.

16 A long slender column of uniform cross section is built-in rigidly at its base. Its upper end, which is free to move out of line, bears a vertical load F. Determine the possible deflection curves of the column and the load required to produce each one. *Hint:* Unlike Example 1, it is convenient here to take the origin at the free end of the column. Then $y(0) = 0$ and $y'(L) = 0$.

17 When the effect of its own weight is neglected, show that the deflection of a uniform cantilever beam at the point $x = x_0$ due to a unit load at the point $x = x_1$ is equal to the deflection at $x = x_1$ due to a unit load at $x = x_0$. *Hint:* Between the free end of the beam and the point where the load is applied, the moment is identically zero, but this is not the case between the fixed end of the beam and the point where the load is applied. Hence two differential equations must be solved to obtain the entire deflection curve.

18 A uniform cantilever beam of length L is subjected to an oblique tensile force F at the free end. Find the tip deflection of the beam as a function of the angle θ between the direction of the force and the initial direction of the beam.

19 A cantilever beam has the shape of a solid of revolution whose radius varies as $\sqrt{x}$, where x is the distance from the free end of the beam. A tensile force F is applied at the free end of the beam at an angle of 45° with the initial direction of the beam. Find the deflection curve of the beam.

20 A uniform shaft of length L rotates about its axis with constant angular velocity ω. The ends of the shaft are held in bearings which are free to swing out of line, as shown in Fig. 2.5, if the shaft deflects from its neutral position. Show that there are infinitely many critical speeds at which the shaft can rotate in a deflected position, and find these speeds and the associated deflection curves. *Hint:* During rotation, centrifugal force applies a load per unit length given by

$$w(x) = -\frac{\rho A \omega^2}{g} y$$

where A is the cross-sectional area of the shaft and ρ is the density of the material of the shaft. Substitute this into Eq. (3), solve the resulting differential equation, and then impose the conditions that at $x = 0$ and at $x = L$ both the deflection of the shaft and the moment are zero. It will be convenient to use hyperbolic rather than exponential functions in taking account of the real roots of the characteristic equation.

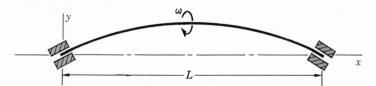

Figure 2.5

21 Work Exercise 20 if the bearings are fixed in position and cannot swing out of line.

22 Under the assumption of very small motions (so that the end of the spring and the weight W may be considered to move in a purely vertical direction) and neglecting friction, determine the natural frequency of the system shown in Fig. 2.6 if the bar

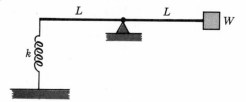

Figure 2.6

is of uniform cross section, absolutely rigid, and of weight w. *Hint:* Use the energy method to obtain the differential equation of the system, recalling that the moment of inertia of a uniform bar of length l and mass m about its midpoint is $\frac{1}{12}ml^2$.

23 Under the assumption of very small motions and neglecting friction, determine the natural frequency of the system shown in Fig. 2.7 if the bar is of uniform cross section, absolutely rigid, and of weight w.

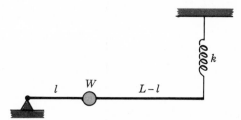

Figure 2.7

24 A weight W hangs by an inextensible cord from the circumference of a pulley of radius R and moment of inertia I. The pulley is prevented from rotating freely by a spring of modulus k, attached as shown in Fig. 2.8. Considering only displace-

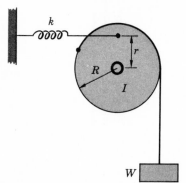

Figure 2.8

ments so small that the departure of the spring from the horizontal can be neglected, and neglecting all friction, determine the natural frequency of the oscillations that occur when the system is slightly disturbed.

25 A perfectly flexible cable of length $2L$ ft, weighing w lb/ft, hangs over a frictionless peg of negligible diameter. At $t = 0$ the cable is released from rest in a position in which the portion hanging on one side is a ft longer than that on the other. Find the equation of motion of the cable as it slips over the peg.

26 A perfectly flexible cable of length L ft, weighing w lb/ft, lies in a straight line on a frictionless table top, a ft of the cable hanging over the edge. At $t = 0$ the cable is released and begins to slide off the edge of the table. Assuming that the height of the table is greater than L, determine the motion of the cable until it leaves the table top.

27 A perfectly flexible cable of length L, weighing w lb/ft, hangs over a pulley as shown in Fig. 2.9. The radius of the pulley is R, and its moment of inertia is I.

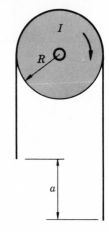

Figure 2.9

Friction between the cable and the pulley prevents any relative slipping, although the pulley is free to turn without appreciable friction. At $t = 0$ the cable is released from rest in a position in which the portion hanging on one side is a ft longer than that hanging on the other. Determine the motion of the cable until the short end first makes contact with the pulley.

28 Neglecting friction and assuming angular displacements θ so small that θ is a satisfactory approximation to $\sin \theta$ and $\theta^2/2$ is a satisfactory approximation to

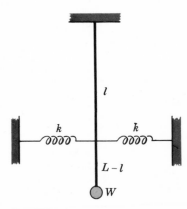

Figure 2.10

$1 - \cos \theta$, find the natural frequency of the system shown in Fig. 2.10 if the bar is of uniform cross section, absolutely rigid, and of weight w.

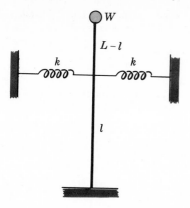

Figure 2.11

29 Neglecting friction and assuming angular displacements θ so small that θ is a satisfactory approximation to $\sin \theta$ and $\theta^2/2$ is a satisfactory approximation to $1 - \cos \theta$, find the natural frequency of the system shown in Fig. 2.11 if the bar is of uniform cross section, absolutely rigid, and of weight w. In what significant respect does this system differ from that in Exercise 28?

30 (a) Two disks each of moment of inertia I are connected by an elastic shaft of modulus k, that is, a shaft which requires k units of torque to twist one end through an angle of 1 rad with respect to the other end. The system is mounted in frictionless bearings, as shown in Fig. 2.12. Neglecting the moment of inertia of the shaft, find the natural frequency with which the disks will oscillate if they are twisted through equal but opposite angles and then released.

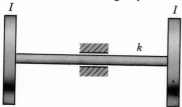

Figure 2.12

(b) What is the natural frequency of the system if the moments of inertia of the disks are respectively I_1 and I_2? *Hint:* Not only does the total energy of the system remain constant, but so does the total angular momentum.

31 A pendulum, consisting of a mass m at the end of an inextensible cord of length l and negligible mass, swings between maximum angular displacements of $\pm \alpha$. If θ is the instantaneous angular displacement of the pendulum from the vertical, use the energy method to show that

$$(\dot\theta)^2 = 2\frac{g}{l}(\cos \theta - \cos \alpha)$$

From this, determine the period of the pendulum if it swings through an angle small enough to make θ a satisfactory approximation to $\sin \theta$.

32 (a) In the expression for $\dot\theta$ from Exercise 31, use the half-angle formula $\cos u = 1 - 2\sin^2(u/2)$ to show that

$$(\dot\theta)^2 = \frac{4g}{l}\left(\sin^2\frac{\alpha}{2} - \sin^2\frac{\theta}{2}\right)$$

Integrate this differential equation, assuming that $\theta = 0$ when $t = 0$.

(b) In the integral obtained in part **(a)** change the variable of integration from θ to ϕ by the substitution

$$\sin\frac{\theta}{2} = \sin\frac{\alpha}{2}\sin\phi$$

and show that

$$t = \sqrt{\frac{l}{g}}\int_0^\phi \frac{d\phi}{\sqrt{1 - k^2\sin^2\phi}} \qquad \text{where } k^2 = \sin^2\frac{\alpha}{2}$$

This integral is known as an **elliptic integral of the first kind** and is commonly denoted $F(\phi,k)$. The function $F(\phi,k)$ is tabulated in most elementary handbooks.

33 Using the results of Exercises 31 and 32, together with tables of $F(\phi,k)$, compare the true period of a pendulum swinging through an angle of $90°$ on each side of the vertical with the period computed under the simplifying assumption that $\sin\theta \doteq \theta$.

2.7 Green's Functions

Theory for its own sake is a luxury we can rarely, if ever, afford in this book. However, when a point of theory is intimately related to practical applications, its inclusion may strengthen not only our theoretical insight but our intuitive understanding as well. In particular, the mathematical objects known as *Green's functions* are of this nature. They are of fundamental importance in the theory of differential equations, and they are also immediate generalizations of important physical quantities. To introduce them, via their physical counterparts, it is convenient to begin with the simple problem of determining the deflection curve of a stretched elastic string.

Consider a perfectly flexible elastic string stretched to a length L under tension T. We assume, as a first possibility, that the string bears a distributed load per unit length $w(x)$. Moreover, we shall suppose that under this load all deflections are perpendicular to the length of the string (Fig. 2.13a). Hence, given any two values of x, the load acting on the portion of the string between these values is the same before and after the string deflects.

On an arbitrary element of the string we then have the forces shown in Fig. 2.13b. Since the deflected string is in equilibrium, the net horizontal force and the net vertical force on the element must both be zero. Hence

(1) $$F_1\cos\alpha_1 = F_2\cos\alpha_2$$

(2) $$F_2\sin\alpha_2 = F_1\sin\alpha_1 - w(x)\,\Delta x$$

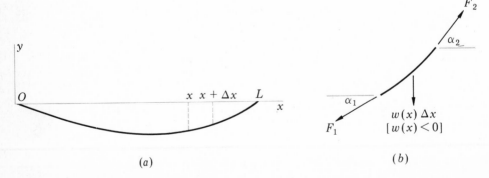

(a)   (b)

Figure 2.13
A stretched string deflected by a distributed load.

The first of these equations tells us that the horizontal component of the force in the string is a constant, and we shall further assume that the deflections are so small that this constant horizontal component does not differ appreciably from the tension T in the string before it is loaded. Then, dividing the respective terms in Eq. (2) by the equal quantities $F_2 \cos \alpha_2$, $F_1 \cos \alpha_1$, and T, we have

$$(3) \qquad\qquad \tan \alpha_2 = \tan \alpha_1 - \frac{w(x)\,\Delta x}{T}$$

Now $\tan \alpha_2$ is the slope of the deflection curve at the point $x + \Delta x$, and $\tan \alpha_1$ is the slope of the deflection curve at the point x. Hence Eq. (3) can be rewritten in the form

$$\frac{y'(x + \Delta x) - y'(x)}{\Delta x} = -\frac{w(x)}{T}$$

In the limit as $\Delta x \to 0$, we thus obtain

$$(4) \qquad\qquad Ty'' = -w(x)$$

as the differential equation satisfied by the deflection curve of the string.

Let us now determine the deflection of the string under the influence of a concentrated rather than a distributed load. A concentrated load is, of course, a mathematical fiction which cannot be realized physically, since any nonzero load concentrated at a single point implies an infinite pressure which would immediately cut through the string. Nonetheless, the use of concentrated loads in analyzing physical systems, such as beams and strings, is both common and fruitful.

At the outset, we note from Eq. (4) that y'' is zero at all points of the string where there is no distributed load. Hence, since $y'' = 0$ implies that y is a linear function, it follows that the deflection curve of the string under the influence of a single concentrated load P consists of two linear sections, as shown in Fig. 2.14. As in our earlier discussion, the equilibrium of the string implies the following conditions

$$F_1 \cos \alpha_1 = F_2 \cos \alpha_2 = T$$

$$F_1 \sin \alpha_1 + F_2 \sin \alpha_2 = -P$$

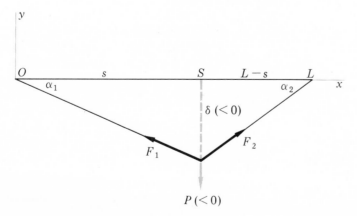

Figure 2.14
A stretched string deflected by a concentrated load.

From these we obtain, as above, the equation

$$\tan \alpha_1 + \tan \alpha_2 = \frac{-P}{T}$$

or $$\frac{-\delta}{s} + \frac{-\delta}{L-s} = \frac{-P}{T} \quad \text{and} \quad \delta = \frac{P(L-s)s}{TL}$$

With the deflection δ known, it is a simple matter to use similar triangles to find the deflection of the string at any point x. The results are

$$(5) \qquad\qquad y(x) = \begin{cases} \dfrac{P(L-s)x}{TL} & 0 \le x \le s \\[2ex] \dfrac{P(L-x)s}{TL} & s \le x \le L \end{cases}$$

From this formula, it follows immediately that *the deflection of a string at a point x due to a concentrated load P applied at a point s is the same as the deflection produced at the point s by an equal load applied at the point x.* When P is a unit load, it is customary to use the notation $G(x,s)$ as a new name for the function defined by (5), to indicate that it depends on both the point s where the load is applied and the point x where the deflection is observed. The preceding observation then asserts that $G(x,s)$ is symmetric in the two variables x and s; that is,

$$G(x,s) = G(s,x)$$

$G(x,s)$ is often referred to as an **influence function**, since it describes the *influence* which a unit load concentrated at the point s has at any point x of the string. The symmetry of $G(x,s)$ is an illustration of the important **Maxwell-Rayleigh reciprocity law**, which holds in many physical systems, both mechanical and electrical.

It is interesting and important to note that by means of the influence function $G(x,s)$, an expression for the deflection of a string under an arbitrary distributed load can be found without solving Eq. (4). To do this, we reason as follows. Let the interval $[0,L]$ be subdivided into n subintervals by the points $s_0 = 0, s_1, s_2, \ldots, s_n = L$, let $\Delta s_i = s_i - s_{i-1}$, and let ξ_i be an arbitrary point in Δs_i. Further, let the portion of the distributed load acting on the subinterval Δs_i, namely, $w(\xi_i)\,\Delta s_i$, be regarded as concentrated at the point $s = \xi_i$. The deflection produced at the point x by this load is the magnitude of the load multiplied by the deflection produced at x by a *unit* load at the point $s = \xi_i$, namely,

$$[w(\xi_i)\,\Delta s_i]G(x,\xi_i)$$

If we add up the infinitesimal deflections produced at the point x by the various infinitesimal concentrated forces which together are equivalent to the actual distributed load, we obtain the sum

$$\sum_{i=1}^{n} w(\xi_i)G(x,\xi_i)\,\Delta s_i$$

In the limit, as each $\Delta s_i \to 0$, this sum becomes an integral and for the deflection at an arbitrary point x we have the formula

$$(6) \qquad\qquad y(x) = \int_0^L w(s)G(x,s)\,ds$$

We have already observed that the influence function

(7)
$$G(x,s) = \begin{cases} \dfrac{(L - s)x}{TL} & 0 \le x \le s \\[2mm] \dfrac{(L - x)s}{TL} & s \le x \le L \end{cases}$$

associated with the differential equation $Ty'' = w(x)$ is a symmetric function of the two arguments x and s. Other properties of $G(x,s)$ are also worth noting. In the first place, it is obvious that $G(x,s)$ satisfies the boundary conditions of the problem; i.e., just as the string satisfies the conditions that its deflection is zero when $x = 0$ and when $x = L$, so it is also true that $G(0,s) = G(L,s) = 0$ for all values of s such that $0 \le s \le L$. It is also easy to verify that $G(x,s)$ is a continuous function of x on the interval $[0,L]$. This is obvious, except possibly at the point $x = s$, where we have for the left- and right-hand limits of $G(x,s)$

$$\lim_{x \to s^-} G(x,s) = \lim_{x \to s^-} \frac{(L - s)x}{TL} = \frac{(L - s)s}{TL}$$

$$\lim_{x \to s^+} G(x,s) = \lim_{x \to s^+} \frac{(L - x)s}{TL} = \frac{(L - s)s}{TL}$$

and these are clearly equal and equal to $G(s,s)$. On the other hand, the derivative of $G(x,s)$ with respect to x is discontinuous at the point $x = s$, where, in fact, it has a (downward) jump of $-1/T$. To verify this, we note first that $G(x,s)$ is obviously differentiable at all points of the interval $[0,L]$ except possibly at $x = s$. There we observe that

$$\lim_{x \to s^-} G_x(x,s) = \lim_{x \to s^-} \frac{L - s}{TL} = \frac{L - s}{TL}$$

$$\lim_{x \to s^+} G_x(x,s) = \lim_{x \to s^+} \left(-\frac{s}{TL}\right) = -\frac{s}{TL}$$

These limiting values are not equal, and their difference is

$$-\frac{s}{TL} - \frac{L - s}{TL} = -\frac{1}{T}$$

as asserted. Finally, we note that since $G(x,s)$ consists of two linear expressions, it satisfies the homogeneous differential equation $Ty'' = 0$ at all points of the interval $[0,L]$ except at $x = s$. In fact, at $x = s$ the second derivative $G_{xx}(x,s)$ does not exist since, as we have just observed, $G_x(x,s)$ is discontinuous at that point.

The properties which we have just noted are not accidental characteristics of the influence function of one particular problem. Instead, as the following definition makes clear, they identify an important class of functions associated with linear differential equations with variable as well as constant coefficients.

DEFINITION 1 Given the differential equation

$$a_0(x)y'' + a_1(x)y' + a_2(x)y = 0$$

and the boundary conditions $y(a) = y(b) = 0$. A function $G(x,s)$ with the property that

  a. $G(x,s)$ satisfies the given differential equation for $a \leq x < s$ and for $s < x \leq b$
  b. $G(a,s) = G(b,s) = 0$ for $a \leq s \leq b$
  c. $G(x,s)$ is a continuous function of x for $a \leq x \leq b$
  d. $G_x(x,s)$ is continuous for $a \leq x < s$ and for $s < x \leq b$ but has a step discontinuity of magnitude $-1/a_0(s)$ at $x = s$

is called the Green's function of the problem defined by the given differential equation and its boundary conditions.†

EXAMPLE 1

     Using Definition 1, construct the Green's function for the differential equation $y'' + k^2 y = 0$ with the boundary conditions $y(0) = y(b) = 0$.

     Since any solution of the equation $y'' + k^2 y = 0$ is of the form $y = A \cos kx + B \sin kx$, it follows from property **a** that the required function $G(x,s)$ must be defined by expressions of the form

$$G(x,s) = \begin{cases} A_1 \cos kx + B_1 \sin kx & 0 \leq x \leq s \\ A_2 \cos kx + B_2 \sin kx & s \leq x \leq b \end{cases}$$

In order for the left-hand boundary condition to be met, as required by property **b**, it is necessary that $A_1 = 0$. Similarly, in order for the boundary condition at $x = b$ to be met, it is necessary that $A_2 \cos kb + B_2 \sin kb = 0$, which will be the case if we take $A_2 = C \sin kb$ and $B_2 = -C \cos kb$, where C is arbitrary. Thus $G(x,s)$ is restricted to the form

$$G(x,s) = \begin{cases} B_1 \sin kx & 0 \leq x \leq s \\ C(\sin kb \cos kx - \cos kb \sin kx) = C \sin k(b - x) & s \leq x \leq b \end{cases}$$

     Further, in order for $G(x,s)$ to be continuous at $x = s$, as required by property **c**, it is necessary that $B_1 \sin ks = C \sin k(b - s)$, or $B_1 = E \sin k(b - s)$ and $C = E \sin ks$, where E is arbitrary. Thus $G(x,s)$ is further restricted to the form

$$G(x,s) = \begin{cases} E \sin k(b - s) \sin kx & 0 \leq x \leq s \\ E \sin ks \sin k(b - x) & s \leq x \leq b \end{cases}$$

Finally, to satisfy property **d**, we must have

$$\lim_{x \to s^+} G_x(x,s) - \lim_{x \to s^-} G_x(x,s) = -1$$

or $\quad \displaystyle\lim_{x \to s^+} [-kE \sin ks \cos k(b - x)] - \lim_{x \to s^-} [kE \sin k(b - s) \cos kx] = -1$

$$-kE[\sin ks \cos k(b - s) + \sin k(b - s) \cos ks] = -1$$

$$-kE \sin kb = -1$$

$$E = \frac{1}{k \sin kb}$$

---

† The property of symmetry, i.e., the property that $G(x,s) = G(s,x)$ is not a part of the definition of a Green's function but in a large class of cases is a consequence of the four properties required by the definition; see Exercise 11.

With E known, the function $G(x,s)$ is completely determined, and we have

$$G(x,s) = \begin{cases} \dfrac{\sin kx \sin k(b-s)}{k \sin kb} & 0 \le x \le s \\[2ex] \dfrac{\sin ks \sin k(b-x)}{k \sin kb} & s \le x \le b \end{cases}$$

provided, of course, that $kb \ne n\pi$.† It is interesting to note that $G(x,s) = G(s,x)$ even though we did not impose this condition in the course of our derivation.

Green's functions are not only closely related to the influence functions which arise in many practical problems but also have much in common with the results of the method of variation of parameters, discussed in Sec. 2.5. To explore this matter, let us consider again the differential equation

$$a_0(x)y'' + a_1(x)y' + a_2(x)y = f(x)$$

or, equivalently,

$$y'' + \frac{a_1(x)}{a_0(x)} y' + \frac{a_2(x)}{a_0(x)} y = \frac{f(x)}{a_0(x)} \qquad a_0(x) \ne 0 \text{ for } a \le x \le b$$

If $y_1(x)$ and $y_2(x)$ are two linearly independent solutions of the related homogeneous equation, then, according to the method of variation of parameters, $y = u_1 y_1 + u_2 y_2$ will be a solution of the given nonhomogeneous equation provided

$$u_1' = -\frac{y_2}{y_1 y_2' - y_1' y_2} \frac{f}{a_0} \qquad u_2' = \frac{y_1}{y_1 y_2' - y_1' y_2} \frac{f}{a_0}$$

Let us now solve for u_1 and u_2 by integrating their derivatives between x and a and between x and b, respectively. Then, recalling that $y_1 y_2' - y_1' y_2$ is the wronskian W of the two solutions y_1 and y_2, we have

$$u_1 = \int_x^a \frac{-y_2(s)}{W(s)} \frac{f(s)}{a_0(s)} \, ds = \int_a^x \frac{y_2(s)}{W(s)} \frac{f(s)}{a_0(s)} \, ds$$

$$u_2 = \int_x^b \frac{y_1(s)}{W(s)} \frac{f(s)}{a_0(s)} \, ds$$

and $\quad y = u_1 y_1 + u_2 y_2 = y_1(x) \int_a^x \frac{y_2(s)}{W(s)} \frac{f(s)}{a_0(s)} \, ds + y_2(x) \int_x^b \frac{y_1(s)}{W(s)} \frac{f(s)}{a_0(s)} \, ds$

We now ask under what conditions, if any, the solution defined by the last equation can satisfy the boundary conditions $y(a) = y(b) = 0$. Putting $x = a$ makes the first

---

† It is worth noting that if $kb = n\pi$, there is a nontrivial function meeting the boundary conditions $y(0) = y(b) = 0$ and satisfying the equation $y'' + k^2 y = 0$ at *all* points of the interval $[0,b]$. In fact, beginning with the general solution

$$y = A \cos kx + B \sin kx$$

it is clear that the conditions $y(0) = y(b) = 0$ will be met if and only if $A = 0$ and $B \sin kb = 0$. Since $kb = n\pi$, the second condition is satisfied for all values of B; that is, B need not be zero. This illustrates the important general result that *for equations containing a parameter* (such as k in the equation $y'' + k^2 y = 0$) *Green's function fails to exist for any value of the parameter for which there is a nontrivial solution of the differential equation which satisfies the boundary conditions of the problem.*

integral vanish; hence $y(a)$ will be zero for arbitrary f if and only if the particular solution $y_2(x)$ is zero when $x = a$. At $x = b$, the second integral is zero, and y will be zero for arbitrary f if and only if $y_1(x) = 0$ when $x = b$. Assuming that y_1 and y_2 are chosen to satisfy these conditions, we can write

$$y = \int_a^x \frac{y_1(x) y_2(s)}{W(s)a_0(s)} f(s) \, ds + \int_x^b \frac{y_1(s) y_2(x)}{W(s)a_0(s)} f(s) \, ds$$

which is of the form

$$(8) \qquad\qquad\qquad y = \int_a^b G(x,s) f(s) \, ds$$

where

$$G(x,s) = \begin{cases} \dfrac{y_1(s) y_2(x)}{W(s)a_0(s)} & x \le s \le b, \text{ that is, } a \le x \le s \\[2ex] \dfrac{y_1(x) y_2(s)}{W(s)a_0(s)} & a \le s \le x, \text{ that is, } s \le x \le b \end{cases}$$

From the way in which $y_1(x)$ and $y_2(x)$ were selected, it is clear that $G(x,s)$ satisfies the boundary conditions of the problem. It is also evident that $G(x,s)$ is a continuous function of x for each value of s, since $y_1(x)$ and $y_2(x)$ are continuous functions of x and at $x = s$ the right- and left-hand limits $G(s^+,s)$ and $G(s^-,s)$ are equal. Furthermore, except at $x = s$, $G(x,s)$ satisfies the homogeneous form of the given differential equation since $y_1(x)$ and $y_2(x)$ are solutions of this equation. Finally, for $G_x(x,s)$ we have

$$G_x(x,s) = \begin{cases} \dfrac{y_1(s) y_2'(x)}{W(s)a_0(s)} & a \le x < s \\[2ex] \dfrac{y_1'(x) y_2(s)}{W(s)a_0(s)} & s < x \le b \end{cases}$$

Hence

$$G_x(s^+,s) - G_x(s^-,s) = \frac{y_1'(s) y_2(s)}{W(s)a_0(s)} - \frac{y_1(s) y_2'(s)}{W(s)a_0(s)}$$

$$= -\frac{y_1(s) y_2'(s) - y_1'(s) y_2(s)}{W(s)a_0(s)} = -\frac{1}{a_0(s)}$$

which shows that $G_x(x,s)$ has a jump of the required amount at $x = s$. Thus $G(x,s)$ has the four properties necessary to make it the Green's function for the given problem. Moreover, from Eq. (8) it is clear that the solution of the general nonhomogeneous boundary-value problem is given by a formula just like the one we derived for the particular case of the loaded string, Eq. (6).

As a final example of a more sophisticated application of a Green's function, let us consider the following problem. A flexible elastic string of weight per unit length $\rho(x)$ is stretched under tension T between two points a distance L apart. Determine the frequencies at which the string can perform free vibrations and the shape of the curve of maximum deflection corresponding to each natural frequency.

This problem is just like the one we discussed earlier except that now the load per unit length instead of being a known static load is the (unknown) dynamic load arising from the inertia forces of the moving string itself. To obtain an expression for this load, consider an element of the string of length Δx. The mass of this element is

$[\rho(x)\,\Delta x]/g$, and when the string is vibrating, the acceleration of this element is $\ddot{y}$. Hence, the inertia force due to the motion of this element is

$$-\frac{\rho(x)\,\Delta x}{g}\,\ddot{y}\dagger$$

and the corresponding load *per unit length* is

$$-\frac{\rho(x)}{g}\,\ddot{y}$$

Now if the string is vibrating at a single natural frequency in the absence of any damping forces, all points of the string must move harmonically with that frequency and with phase differences which are either 0 or 180°. In other words, during the vibration, the curve of the string is defined by an equation of the form

$$y(x,t) = \phi(x)\sin\omega t$$

where ω is the (as yet unknown) frequency of the vibrations and $\phi(x)$ is the (as yet unknown) curve of maximum displacements. Thus the load per unit length of the string is of the form

$$w(x,t) = -\frac{\rho(x)}{g}\left[-\omega^2\phi(x)\sin\omega t\right]$$

and Eq. (6) gives us

$$y(x,t) \equiv \phi(x)\sin\omega t = \int_0^L G(x,s)\omega^2\,\frac{\rho(s)}{g}\,\phi(s)\sin\omega t\,ds$$

The factor $\sin\omega t$ is independent of the variable of integration s and hence can be removed from the integral and canceled from the equation. Thus the amplitude function $\phi(x)$ satisfies the *integral equation*‡

$$\phi(x) = \frac{\omega^2}{g}\int_0^L G(x,s)\rho(s)\phi(s)\,ds$$

In a typical problem, this equation would probably have to be solved approximately in the following way. Make some reasonable guess of the nature of the function $\phi(x)$, say $\phi_1(x)$, substitute this into the integrand, and compute the integral as a function of x. If the result is proportional to the "input" $\phi_1(x)$, we have a solution and the unknown frequency ω can be inferred from the proportionality constant. Of course, it is highly unlikely that we would ever hit upon the solution by guessing, and so the integrated result, or "output," say $\phi_2(x)$, will not be proportional to $\phi_1(x)$ and we shall not have found a solution. However, the process can be repeated, $\phi_2(x)$ can be substituted into the integral to give a new output $\phi_3(x)$, and so on. It can be shown (compare Exercises 43 to 45, Sec. 8.6) that in a large class of cases the sequence

---

† The minus sign here indicates that when the acceleration is negative, the string is deflected as though it carried a positive load, and vice versa.

‡ An equation involving the integral of an unknown function is called an **integral equation**. As this problem illustrates, Green's functions play an important role in connecting the theory of differential equations with the theory of integral equations.

of functions $\{\phi_n(x)\}$ determined by this procedure will converge to a function $\phi(x)$ which is a solution and the ratio $\phi_{n-1}(x)/\phi_n(x)$ will approach a value which is independent of x. Thus by repeating the iteration a sufficient number of times, $\phi(x)$ and ω can be approximated with satisfactory accuracy.

In particular, if $\rho(x)$ is a constant, say ρ, then for a uniform string of length L we have the integral equation

$$\phi(x) = \frac{\omega^2 \rho}{g} \int_0^L G(x,s)\phi(s)\, ds$$

or, substituting the Green's function for a string from Eq. (7),

$$\phi(x) = \frac{\omega^2 \rho}{g} \int_0^x \frac{(L-x)s}{TL}\phi(s)\, ds + \frac{\omega^2 \rho}{g} \int_x^L \frac{(L-s)x}{TL}\phi(s)\, ds$$

$$= \frac{\omega^2 \rho (L-x)}{gTL} \int_0^x s\phi(s)\, ds + \frac{\omega^2 \rho x}{gTL} \int_x^L (L-s)\phi(s)\, ds$$

A reasonable guess for $\phi(x)$ might be $\phi_1(x) = A \sin(n\pi x/L)$, since this is a simple function with the property that for each value of n it is zero for $x = 0$ and for $x = L$, as the deflection of a string should be. Using this, we have for the integrals on the right,

$$\frac{A\omega^2 \rho (L-x)}{gTL} \int_0^x s \sin \frac{n\pi s}{L}\, ds + \frac{A\omega^2 \rho x}{gTL} \int_x^L (L-s) \sin \frac{n\pi s}{L}\, ds$$

$$= \frac{A\omega^2 \rho (L-x)}{gTL} \left[\frac{L^2}{n^2 \pi^2} \sin \frac{n\pi s}{L} - \frac{Ls}{n\pi} \cos \frac{n\pi s}{L} \right]_0^x$$

$$+ \frac{A\omega^2 \rho x}{gTL} \left[-\frac{(L-s)L}{n\pi} \cos \frac{n\pi s}{L} - \frac{L^2}{n^2 \pi^2} \sin \frac{n\pi s}{L} \right]_x^L$$

$$= \frac{A\omega^2 \rho}{gT} \frac{L^2}{n^2 \pi^2} \sin \frac{n\pi x}{L}$$

This will be equal to the input, $A \sin(n\pi x/L)$, if and only if

$$\frac{\omega^2 \rho}{gT} \frac{L^2}{n^2 \pi^2} = 1$$

Thus there are infinitely many natural frequencies, given by the formula

$$\omega_n = \frac{n\pi}{L} \sqrt{\frac{gT}{\rho}}$$

with corresponding deflection curves

$$y_n(x,t) = A_n \sin \frac{n\pi x}{L} \sin \omega_n t$$

It is interesting to note that we shall obtain these results by an entirely different method in Sec. 8.4.

EXERCISES

1 Construct the Green's function for a string of length L using the method of variation of parameters.

2 Determine the deflection curve of a string of length L bearing equal concentrated loads P at $x = L/4$ and at $x = L/2$.

3 Determine the deflection curve of a string of length L bearing concentrated loads P at $x = L/4$, $-2P$ at $x = L/2$, and $3P$ at $x = 3L/4$.

4 Find the deflection curve of a string of length L bearing a load per unit length $w(x) = -x$, first by solving the differential equation (4) and then by using the Green's function for the string.

5 Work Exercise 4 if the load per unit length is

$$w(x) = \begin{cases} 0 & 0 \le x < L/2 \\ -1 & L/2 < x \le L \end{cases}$$

6 Find the deflection curve of a string of length L bearing a load per unit length $w(x) = -x$ and a concentrated load P at $x = L/4$.

7 Construct the Green's function for the equation $y'' + 2y' + 2y = 0$ with the boundary conditions $y(0) = 0$, $y(\pi/2) = 0$. Is this Green's function symmetric? What is the Green's function for the equation $e^{2x}y'' + 2e^{2x}y' + 2e^{2x}y = 0$? Is this Green's function symmetric?

8 The boundary conditions appearing in Definition 1, namely, $y(a) = y(b) = 0$, can be replaced by the more general conditions

$$\alpha_1 y(a) = \alpha_2 y'(a) \qquad \text{and} \qquad \beta_1 y(b) = \beta_2 y'(b)$$

and a Green's function for the corresponding problem can be obtained by requiring that $G(x,s)$ also satisfy the new boundary conditions at $x = a$ and at $x = b$. Find the Green's function for the equation $y'' + k^2 y = 0$ if the boundary conditions are

(a) $y(0) = y'(b) = 0$ (b) $y'(0) = y(b) = 0$
(c) $y'(a) = y'(b) = 0$ (d) $y(a) = y'(a), y(b) = 0$

9 Work Exercise 8 for the equation $y'' - k^2 y = f(x)$.

10 Find the Green's function for the equation $(y'/x^2)' + 2y/x^4 = 0$ with the boundary conditions $y(0) = y'(1) = 0$. *Hint:* The given differential equation is an Euler equation, after the indicated differentiation is carried out and the resulting equation cleared of fractions.

11 Show that the Green's function for an equation of the form $[r(x)y']' + p(x)y = 0$ is symmetric. *Hint:* Recall Abel's identity for the wronskian of a second-order differential equation.

12 The angle of twist θ produced in a uniform shaft of length L by a torque T is given by the formula $\theta = TL/E_s J$, where E_s is the modulus of elasticity in shear of the material of the shaft and J is the polar moment of inertia of the cross-sectional area of the shaft about the center of gravity of the cross section. Using this formula, find the influence function which gives the angle of twist at a point x due to a unit torque applied at a point s of a shaft rigidly clamped at the left end and free at the right end.

13 (a) By considering the limit of the angle of twist produced in a shaft of infinitesimal length by a torque applied to its ends, show that the torque transmitted *through* any cross section of a twisted shaft is given by the formula

$$T = E_s J \frac{d\theta}{dx}$$

(b) Using the result of part (a), show that the angle of twist produced in a uniform shaft by a torque per unit length equal to $t(x)$ satisfies the differential equation

$$E_s J \theta'' = -t(x)$$

14 Using the result of part **(b)** of Exercise 13, find the Green's function of a uniform shaft of length L if each end of the shaft is clamped so that it cannot twist.

15 Using the result of Exercise 12, set up the integral equation satisfied by the maximum-deflection function $\phi(x)$ for the free vibrations of a uniform shaft of length L which is clamped at the left end and free at the right. Show that $\phi(x) = A \sin [(2n + 1)\pi x/2L]$ is a solution of this equation for each value of n, and find the corresponding natural frequencies.

16 Using the result of Exercise 14, set up the integral equation satisfied by the maximum-deflection function $\phi(x)$ for the free torsional vibrations of a uniform shaft of length L which is clamped at both ends. Show that $\phi(x) = A \sin (n\pi x/L)$ is a solution of this equation for each value of n, and find the corresponding natural frequencies.

17 Find the influence function which gives the deflection of a uniform cantilever beam at a point x due to a unit load applied at the point s. Verify that this influence function $G(x,s)$ has the following properties, the primes denoting differentiation with respect to x:
(a) $G(x,s)$ satisfies the differential equation $EIy^{iv} = 0$ for $0 \le x < s$ and $s < x \le L$.
(b) $G(0,s) = G'(0,s) = G''(L,s) = G'''(L,s) = 0$; that is, $G(x,s)$ satisfies the end conditions for a cantilever beam.
(c) $G(x,s)$, $G'(x,s)$, and $G''(x,s)$ are continuous for $0 \le x \le L$.
(d) $G'''(x,s)$ has a jump of $-1/EI$ at $x = s$.
Is $G(x,s)$ a symmetric function of x and s?

18 Using the influence function obtained in Exercise 17, find the deflection curve of a uniform cantilever beam bearing concentrated loads P at $x = L/2$ and at $x = L$.

19 Find the deflection curve of a uniform cantilever beam of length L bearing a load per unit length $w(x) = x$, first by solving the differential equation $EIy^{iv} = -w(x)$ and then by using the influence function obtained in Exercise 17.

20 Taking the following properties, suggested in Exercise 17, as the definition of the Green's function for a fourth-order differential equation whose leading coefficient is a_0, find the Green's function for the equation $EIy^{iv} - y = 0$ with the boundary conditions $y(0) = y''(0) = y(L) = y''(L) = 0$:
(a) $G(x,s)$ satisfies the given differential equation except at $x = s$.
(b) $G(x,s)$ satisfies the boundary conditions accompanying the differential equation.
(c) $G(x,s)$, $G'(x,s)$, and $G''(x,s)$ are continuous at all points of the interval of the problem.
(d) $G'''(x,s)$ has a jump of $-1/a_0$ at the point s.

21 Verify carefully the italicized observation which follows Eq. (5).

CHAPTER 3
Simultaneous Linear Differential Equations

3.1 Introduction

In many problems in applied mathematics there are not one but several dependent variables, each a function of a single independent variable, usually time. The formulation of such a problem in mathematical terms frequently leads to a system of simultaneous differential equations, as many equations as there are dependent variables.

Often these equations are nonlinear and exceedingly difficult to solve, even with the aid of a computer. In many important cases, however, they are not only linear but have constant coefficients and are relatively easy to solve. There are various methods of solving such systems. In one, which bears a strong resemblance to the solution of systems of simultaneous algebraic equations, the system is reduced by successive elimination of the unknowns until a single differential equation remains. This is solved and then, working backward, the solutions for the other variables are found, one by one, until the problem is completed. A second method, which amounts to considering the system as a single matric differential equation, generalizes the ideas of complementary function and particular integral and through their use obtains solutions for all the variables at the same time. Finally, the use of the Laplace transformation provides a straightforward operational procedure for solving systems of linear differential equations with constant coefficients which is probably preferable in most applications to either of the other methods.

In this chapter we shall attempt through examples to present the first two methods, leaving the third to Chap. 7, where we discuss the Laplace transformation and its applications in detail.

3.2 The Reduction of a System to a Single Equation

Consider the following system of equations:

$$\frac{dx}{dt} + 2x + \frac{dy}{dt} + 6y = 2e^t$$

$$2\frac{dx}{dt} + 3x + 3\frac{dy}{dt} + 8y = -1$$

If we first eliminate dx/dt by subtracting the second equation from twice the first, we obtain

(1)
$$x - \frac{dy}{dt} + 4y = 4e^t + 1$$

If we next eliminate x by subtracting 3 times the first equation from twice the second, we obtain

$$(2) \qquad \frac{dx}{dt} + 3\frac{dy}{dt} - 2y = -6e^t - 2$$

Finally, if we differentiate Eq. (1) and subtract the result from Eq. (2), all occurrences of x and its derivatives will be eliminated and we shall have an equation in y alone:

$$\frac{d^2 y}{dt^2} - \frac{dy}{dt} - 2y = -10e^t - 2$$

It is now a simple matter to solve this equation by the methods of Chap. 2, and we find without difficulty

$$(3) \qquad y = c_1 e^{-t} + c_2 e^{2t} + 5e^t + 1$$

Various possibilities are available for finding x. By far the simplest is to use Eq. (1), which gives x directly in terms of y and its first derivative. Thus

$$x = \frac{dy}{dt} - 4y + 4e^t + 1$$

$$= (-c_1 e^{-t} + 2c_2 e^{2t} + 5e^t) - 4(c_1 e^{-t} + c_2 e^{2t} + 5e^t + 1) + 4e^t + 1$$

$$(4) \qquad = -5c_1 e^{-t} - 2c_2 e^{2t} - 11e^t - 3$$

On the other hand, we might (less efficiently) substitute y and dy/dt into either of the given equations and solve the resulting first-order differential equation for x, or we might substitute into Eq. (2) and then find x by a single integration. Pursuing the latter possibility, we have

$$\frac{dx}{dt} = -3\frac{dy}{dt} + 2y - 6e^t - 2$$

$$= -3(-c_1 e^{-t} + 2c_2 e^{2t} + 5e^t) + 2(c_1 e^{-t} + c_2 e^{2t} + 5e^t + 1) - 6e^t - 2$$

$$= 5c_1 e^{-t} - 4c_2 e^{2t} - 11e^t$$

and, integrating,

$$(5) \qquad x = -5c_1 e^{-t} - 2c_2 e^{2t} - 11e^t + c_3$$

This appears to be a more general solution for x than the one given by Eq. (4) since it contains c_3 as a new *arbitrary* integration constant, whereas Eq. (4) contains only the *specific* additive constant -3. Which is the correct expression for x, Eq. (4) or Eq. (5)?

The need to resolve this question becomes even more pressing if we attempt to find x by using one of the original equations. Specifically, substituting y and dy/dt into the first of the given equations, we have

$$\frac{dx}{dt} + 2x = -\frac{dy}{dt} - 6y + 2e^t$$

$$= -(-c_1 e^{-t} + 2c_2 e^{2t} + 5e^t) - 6(c_1 e^{-t} + c_2 e^{2t} + 5e^t + 1) + 2e^t$$

$$= -5c_1 e^{-t} - 8c_2 e^{2t} - 33e^t - 6$$

This is a nonhomogeneous linear differential equation whose characteristic equation is $m + 2 = 0$ and whose complementary function is therefore

$$c_4 e^{-2t}$$

For a particular integral we substitute, as usual,

$$X = Ae^{-t} + Be^{2t} + Ce^t + E$$

getting

$$(-Ae^{-t} + 2Be^{2t} + Ce^t) + 2(Ae^{-t} + Be^{2t} + Ce^t + E)$$
$$= -5c_1 e^{-t} - 8c_2 e^{2t} - 33e^t - 6$$

whence, collecting terms and then equating the coefficients of like terms, we find

$$A = -5c_1 \qquad B = -2c_2 \qquad C = -11 \qquad E = -3$$

and finally

(6)
$$x = -5c_1 e^{-t} - 2c_2 e^{2t} - 11e^t + c_4 e^{-2t} - 3$$

This expression agrees with neither of the previous ones, and although it contains the same additive constant as the solution given by Eq. (4), it also contains a variable term $c_4 e^{-2t}$ contained in neither Eq. (4) nor Eq. (5). Now our question must be: Which is the correct expression for x, Eq. (4), Eq. (5), *or* Eq. (6)?

To settle this matter, let us do precisely what we would do at this stage in a simple problem involving simultaneous algebraic equations: let us check our answers by substituting them into the given equations. To check the solution apparently given by Eq. (3) and Eq. (6) we need substitute only into the second of the original equations, since (6) was derived directly from the first equation and so, perforce, (3) and (6) must satisfy it. Thus, substituting as indicated, we have

$$2(\quad 5c_1 e^{-t} - 4c_2 e^{2t} - 11e^t - 2c_4 e^{-2t})$$
$$+ 3(-5c_1 e^{-t} - 2c_2 e^{2t} - 11e^t + \quad c_4 e^{-2t} - 3)$$
$$+ 3(-\quad c_1 e^{-t} + 2c_2 e^{2t} + \quad 5e^t)$$
$$+ 8(\quad c_1 e^{-t} + \quad c_2 e^{2t} + \quad 5e^t \qquad\qquad + 1) = -1$$

or
$$-c_4 e^{-2t} - 1 = -1$$

This will be an identity if and only if $c_4 = 0$, which means that the expression for x in (6) reduces to the one given by (4). Similarly, if the solution given by (3) and (5) is checked in the second equation of the given system, the result will be an identity if and only if $c_3 = -3$. Thus, a complete solution of the problem is given by the pair of expressions (3) and (4).

The general situation illustrated by the preceding work can be summarized as follows: if after one of the unknowns has been determined, the other is found by integration or the solution of a differential equation, extraneous constants (such as c_3 and c_4 in the preceding discussion) will always be introduced. The values of these constants or their relation to other constants already present must be determined by substituting the solutions for the two dependent variables into the original equations and making sure that they are identically satisfied.

In general, the steps in the reduction of a system of equations to a single equation are not so obvious as they were in the example we have just worked. For this reason it is frequently convenient to rewrite the given equations in the D notation. Then, if we regard the operational coefficients of the variables as ordinary algebraic coefficients, the method of elimination will usually be apparent. Still more systematically, determinants can be used to obtain the single equation satisfied by any one of the unknowns, very much as in the case of linear algebraic equations.

Suppose, for definiteness, that we have the second-order system

$$(a_{11}D^2 + b_{11}D + c_{11})x + (a_{12}D^2 + b_{12}D + c_{12})y = \phi_1(t)$$
$$(a_{21}D^2 + b_{21}D + c_{21})x + (a_{22}D^2 + b_{22}D + c_{22})y = \phi_2(t)$$

or, more compactly,

$$P_{11}(D)x + P_{12}(D)y = \phi_1(t)$$
$$P_{21}(D)x + P_{22}(D)y = \phi_2(t)$$

where the P's denote the polynomial operators which act on x and y. If these were, as indeed they appear to be, two algebraic equations in x and y, we could eliminate x at once by subtracting $P_{21}(D)$ times the first equation from $P_{11}(D)$ times the second equation, getting

(7) $$[P_{11}(D)P_{22}(D) - P_{12}(D)P_{21}(D)]y = P_{11}(D)\phi_2(t) - P_{21}(D)\phi_1(t)$$

Moreover, this procedure is clearly valid even though the system consists of differential equations rather than algebraic equations. For "multiplying" the first equation by

$$P_{21}(D) \equiv a_{21}D^2 + b_{21}D + c_{21}$$

is simply a way of performing in one step the operations of adding a_{21} times the second derivative of the equation and b_{21} times the first derivative of the equation to c_{21} times the equation itself; and these steps are individually well defined and completely correct. Similarly, multiplying the second equation by

$$P_{11}(D) \equiv a_{11}D^2 + b_{11}D + c_{11}$$

merely furnishes in one step the sum of a_{11} times the second derivative of the equation, b_{11} times the first derivative of the equation, and c_{11} times the equation itself. Finally, the subtraction of the two equations obtained by the multiplications we have just described eliminates x and each of its derivatives because these operations produce in each equation exactly the same combination of x and its various derivatives. Similarly, of course, y can be eliminated from the system by subtracting $P_{12}(D)$ times the second equation from $P_{22}(D)$ times the first equation, leaving a differential equation from which x can be found at once.

The preceding observations can easily be formulated in determinant notation. In fact, the (operational) coefficient of y in Eq. (7) is simply the determinant of the (operational) coefficients of the unknowns in the original system, namely,

$$\begin{vmatrix} P_{11}(D) & P_{12}(D) \\ P_{21}(D) & P_{22}(D) \end{vmatrix}$$

Furthermore, the right-hand side of (7) can be identified as the expanded form of the determinant

$$\begin{vmatrix} P_{11}(D) & \phi_1(t) \\ P_{21}(D) & \phi_2(t) \end{vmatrix}$$

Thus, Eq. (7) can be written in the form

(8)
$$\begin{vmatrix} P_{11}(D) & P_{12}(D) \\ P_{21}(D) & P_{22}(D) \end{vmatrix} y = \begin{vmatrix} P_{11}(D) & \phi_1(t) \\ P_{21}(D) & \phi_2(t) \end{vmatrix}$$

which is precisely what Cramer's rule (Theorem 8, Sec. 10.5) would yield if applied to the given system as though it were purely algebraic. In just the same way, the result of eliminating y from the original system, namely,

$$[P_{11}(D)P_{22}(D) - P_{12}(D)P_{21}(D)]x = P_{22}(D)\phi_1(t) - P_{12}(D)\phi_2(t)$$

can be written

(9)
$$\begin{vmatrix} P_{11}(D) & P_{12}(D) \\ P_{21}(D) & P_{22}(D) \end{vmatrix} x = \begin{vmatrix} \phi_1(t) & P_{12}(D) \\ \phi_2(t) & P_{22}(D) \end{vmatrix}$$

provided we keep in mind that in the determinant on the right the operators $P_{12}(D)$ and $P_{22}(D)$ must operate on $\phi_2(t)$ and $\phi_1(t)$, respectively, and hence the diagonal products must be interpreted to mean

$$P_{22}(D)\phi_1(t) \qquad \text{and} \qquad P_{12}(D)\phi_2(t)$$

and not $\phi_1(t)P_{22}(D)$ and $\phi_2(t)P_{12}(D)$

The use of Cramer's rule to obtain the differential equations satisfied by the individual dependent variables is in no way restricted to the case of two equations in two unknowns. Exactly the same procedure can be applied to systems of any number of equations, regardless of the degree of the polynomial operators which appear as the coefficients of the unknowns. Moreover, as Eqs. (8) and (9) illustrate, the polynomial operators appearing in the left members of the equations which result when the original system is "solved" for the various unknowns are identical. Hence the characteristic equations of these differential equations are identical, and therefore, except for the presence of different arbitrary constants, the complementary functions in the solutions for the various unknowns are all the same. The constants in these complementary functions are not all independent, however, and relations will always exist between them serving to reduce their number to the figure required by the following theorem.†

THEOREM 1   If the determinant of the operational coefficients of a system of n linear differential equations with constant coefficients is not identically zero, then the total number of independent arbitrary constants in any complete solution of the system is equal to the degree of the determinant of the operational coefficients, regarded as a polynomial in D. In particular cases in which the determinant of the operational coefficients is identically zero, the system may have no solution or it may have solutions containing any number of independent constants.

The necessary relations between the constants appearing initially in the solutions for the unknowns can always be found by substituting these solutions into the n

---

† For a proof of this result see, for instance, E. L. Ince, "Ordinary Differential Equations," pp. 144–150, Dover, New York, 1944.

equations in the original system and then equating to zero the net coefficients of the terms that occur in each of these equations.†

EXAMPLE 1

Find a complete solution of the system

$$(10) \qquad \begin{aligned} (2D^2 + 3D - 9)x + (D^2 + 7D - 14)y &= 4 \\ (D + 1)x + (D + 2)y &= -8e^{2t} \end{aligned}$$

From the preceding discussion we know that the question satisfied by x is

$$\begin{vmatrix} 2D^2 + 3D - 9 & D^2 + 7D - 14 \\ D + 1 & D + 2 \end{vmatrix} x = \begin{vmatrix} 4 & D^2 + 7D - 14 \\ -8e^{2t} & D + 2 \end{vmatrix}$$

or, expanding the determinants and operating, as required, on the known functions 4 and $-8e^{2t}$,‡

$$(D^3 - D^2 + 4D - 4)x = 8 + 32e^{2t}$$

The roots of the characteristic equation of this differential equation are $\pm 2i$, 1. Hence the complementary function is

$$c_1 \cos 2t + c_2 \sin 2t + c_3 e^t$$

It is easy to see that

$$X = -2 + 4e^{2t}$$

is a particular integral, and therefore

$$(11) \qquad x = c_1 \cos 2t + c_2 \sin 2t + c_3 e^t - 2 + 4e^{2t}$$

The solution for y can now be found by substituting the last expression into either of the original equations and solving the resulting differential equation for y. However, it is usually a little simpler to use Cramer's rule again. Doing this, we find that y must satisfy the equation

$$\begin{vmatrix} 2D^2 + 3D - 9 & D^2 + 7D - 14 \\ D + 1 & D + 2 \end{vmatrix} y = \begin{vmatrix} 2D^2 + 3D - 9 & 4 \\ D + 1 & -8e^{2t} \end{vmatrix}$$

or

$$(D^3 - D^2 + 4D - 4)y = -4 - 40e^{2t}$$

The solution of this equation presents no difficulty, and we find at once that

$$(12) \qquad y = k_1 \cos 2t + k_2 \sin 2t + k_3 e^t + 1 - 5e^{2t}$$

However, Eqs. (11) and (12) do not yet constitute the solution of the given system since, collectively, they contain six independent arbitrary constants, whereas, according to Theorem 1, a complete solution of (10) can contain only three arbitrary constants.

---

† In most problems the necessary relations between the constants can be found by substituting the initial expressions for the variables into all but one of the given differential equations (though not necessarily into each set of $n - 1$ equations). However, there are systems for which natural solution procedures require substitution into *all* of the given equations to obtain the necessary relations between the arbitrary constants. Examples of this sort will be found in Exercises 16 and 28.

‡ In carrying out these expansions it must be borne in mind that the operational elements in the determinant on the right operate on the algebraic elements 4 and $-8e^{2t}$, whereas in the determinant on the left these elements all operate on the variable x and not on each other. This is the reason why in expanding the determinant on the right we have the reduction $D(4) = 0$, whereas in expanding the determinant on the left we have only formal multiplications such as $2D^2 2 = 4D^2$ and $3D2 = 6D$.

To accomplish the necessary reduction in the number of constants we must now substitute from (11) and (12) into either one or the other, i.e., into all but one, of the original equations, say the second since it is the simpler:

$$(D + 1)(c_1 \cos 2t + c_2 \sin 2t + c_3 e^t - 2 + 4e^{2t})$$
$$+ (D + 2)(k_1 \cos 2t + k_2 \sin 2t + k_3 e^t + 1 - 5e^{2t}) = -8e^{2t}$$

or, performing the indicated differentiations and collecting terms,

$$(c_1 + 2c_2 + 2k_1 + 2k_2) \cos 2t + (-2c_1 + c_2 - 2k_1 + 2k_2) \sin 2t$$
$$+ (2c_3 + 3k_3)e^t - 8e^{2t} = -8e^{2t}$$

As it stands, with all six constants completely arbitrary, this equation is not identically satisfied. It will be an identity if and only if

$$c_1 + 2c_2 + 2k_1 + 2k_2 = 0$$
$$-2c_1 + c_2 - 2k_1 + 2k_2 = 0$$
$$2c_3 + 3k_3 = 0$$

From these we find (among many equivalent possibilities)

$$k_1 = \frac{-3c_1 - c_2}{4} \qquad k_2 = \frac{c_1 - 3c_2}{4} \qquad k_3 = -\tfrac{2}{3}c_3$$

Hence a complete solution to our problem is given by the pair of functions

$$x = c_1 \cos 2t + c_2 \sin 2t + c_3 e^t - 2 + 4e^{2t}$$
$$y = -\tfrac{1}{4}(3c_1 + c_2) \cos 2t + \tfrac{1}{4}(c_1 - 3c_2) \sin 2t - \tfrac{2}{3}c_3 e^t + 1 - 5e^{2t}$$

Though tedious, it is perfectly straightforward to verify that these expressions satisfy the first of the original pair of equations without additional restrictions on the constants.

EXAMPLE 2

Solve the system of equations

$$Dx + (D - 1)y + (D + 2)z = 2e^t$$
$$(D - 1)x + Dy + (D - 2)z = ae^t$$
$$(D + 1)x + (D - 2)y + (D + 6)z = e^t$$

From the preceding discussion we expect that the differential equation satisfied by z is

$$\begin{vmatrix} D & D - 1 & D + 2 \\ D - 1 & D & D - 2 \\ D + 1 & D - 2 & D + 6 \end{vmatrix} z = \begin{vmatrix} D & D - 1 & 2e^t \\ D - 1 & D & ae^t \\ D + 1 & D - 2 & e^t \end{vmatrix}$$

However, expanding the determinants and operating, as required, on the known functions $2e^t$, ae^t, and e^t, we obtain

$$0z = (a - 3)e^t$$

Clearly, unless $a = 3$, this equation, and hence the system itself, has no solution. On the other hand, if $a = 3$, this equation is satisfied by any function z. In fact, if $a = 3$, it is easy to verify that the third equation in the given system is equal to twice the first equation minus the second. Hence, when $a = 3$, the last equation is dependent upon the first two and is automatically satisfied by any functions $x(t), y(t), z(t)$ which satisfy them. Thus, considering only the first two equations, we can write

$$Dx + (D - 1)y = 2e^t - (D + 2)z$$
$$(D - 1)x + Dy = 3e^t - (D - 2)z$$

and for every differentiable function z this system can be solved for x and y. Specifically,

$$\begin{vmatrix} D & D-1 \\ D-1 & D \end{vmatrix} x = \begin{vmatrix} 2e^t - (D+2)z & D-1 \\ 3e^t - (D-2)z & D \end{vmatrix}$$

or

(13) $(2D - 1)x = 2e^t - (5D - 2)z$

and

$$\begin{vmatrix} D & D-1 \\ D-1 & D \end{vmatrix} y = \begin{vmatrix} D & 2e^t - (D+2)z \\ D-1 & 3e^t - (D-2)z \end{vmatrix}$$

or

(14) $(2D - 1)y = 3e^t + (3D - 2)z$

From Eqs. (13) and (14), x and y can be found in terms of z. Moreover, since z is subject only to the restriction that it be differentiable, it may contain any number of arbitrary constants, and hence, when $a = 3$, but not otherwise, the solution of the original system may also contain any number of independent arbitrary constants, as asserted by Theorem 1.

EXERCISES

With the understanding that $D \equiv d/dt$, find a complete solution of each of the following systems of equations:

1 $(D + 5)x + (D + 4)y = e^{-t}$
   $(D + 2)x + (D + 1)y = 3$

2 $(D + 5)x + (D + 3)y = e^{-t}$
   $(D + 2)x + (D + 1)y = 3$

3 $(D + 5)x + (D + 3)y = e^{-t}$
   $(2D + 1)x + (D + 1)y = 3$

4 $(D - 1)x + (D - 2)y = 0$
   $(D - 5)x + (2D - 7)y = e^{-t}$

5 $(D - 1)x + (D + 9)y = t$
   $(D - 2)x + (2D + 9)y = 4$

6 $(2D + 5)x - (2D + 3)y = t$
   $(D - 2)x + (D + 2)y = 0$

7 $(2D + 3)x + (D + 4)y = -\cos t$
   $(D + 1)x + (D + 2)y = 2 \sin t$

8 $(2D^2 + 1)x + (D + 2)y = 5$
   $(D^2 - 16)x + (D - 4)y = 4$

9 $(D^2 + 1)x + (D^2 + 3)y = 0$
   $(3D + 1)x + (2D + 6)y = 0$

10 $(9D^2 + 8)x + (3D^2 + 4)y = 0$
    $(2D^2 + 1)x + (D^2 + 2)y = 120 \cos 3t$

11 $(D - 1)x - y = 0$
    $-2x + (D - 1)y - z = 0$
    $-2y + (D - 1)z = 0$

12 $(D + 1)x + y + 2z = 0$
    $x + (D + 2)y + z = e^{-t}$
    $5x + y + (D - 2)z = 5e^{-t}$

13 $(D + 1)x + (D + 5)y + (2D + 5)z = 15e^t$
    $(2D + 1)x + (D + 2)y + (3D + 1)z = 10e^t$
    $(D + 3)x + (3D + 4)y + (4D + 6)z = 21e^t$

14 $(D + 1)x + (D + 3)y + (2D + 3)z = e^t$
    $(2D + 1)x + (D + 2)y + (3D + 1)z = 0$
    $(D + 3)x + (3D + 11)y + (4D + 13)z = 0$

15 $(D + 1)x + (D + 1)y + (2D + 3)z = 0$
    $(2D + 1)x + (D + 2)y + (3D + 5)z = 0$
    $(D + 3)x + (3D + 1)y + (4D + 5)z = 0$

16 $(2D^2 - D - 1)x + (D - 1)y = 0$
    $(D^2 - 1)x + (D - 1)y = 0$

17 $(3D^2 + 3D + 2)x + (D^2 + 2D + 3)y = 0$
    $(2D^2 - D - 2)x + (D^2 + D + 1)y = 8$

Find the solution of each of the following systems which satisfies the indicated conditions:

18 $(2D + 1)x + (D + 2)y = 0$ **(a)** $x_0 = 0,\ y_0 = 7$
      $(D - 1)x + (D + 3)y = 5$ **(b)** $x_0 = 7,\ y'_0 = 8$

19 $(2D^2 + 3D - 9)x + (D^2 + 7D - 14)y = 4$ **(a)** $x_0 = -2,\ y_0 = 0,\ y'_0 = -6$
      $(D + 1)x + \quad\quad (D + 2)y = 0$ **(b)** $x_0 = 3,\ y_0 = -3,\ x'_0 = 7$

20   Find a system of differential equations having

$$x = c_1 e^t + c_2 e^{2t} \qquad y = c_1 e^t + 3c_2 e^{2t}$$

as a complete solution.

21   Find a system of differential equations having

$$x = c_1 e^{-t} + c_2 e^t + c_3 e^{2t} \qquad y = c_1 e^{-t} - c_2 e^t + 2c_3 e^{2t}$$

as a complete solution.

22   If (x_1, y_1) and (x_2, y_2) are two solutions of the system

$$P_{11}(D)x + P_{12}(D)y = 0$$
$$P_{21}(D)x + P_{22}(D)y = 0$$

prove that $(c_1 x_1 + c_2 x_2,\ c_1 y_1 + c_2 y_2)$ is also a solution of the system for all values of the constants c_1 and c_2.

23   In Exercise 17, determine (operational) multiples of the two equations which, when added, will yield an equation expressing y directly in terms of x and its various derivatives. Can this be done for the system of equations in Exercise 19? Do you think that this can be done in general?

24   A system consists of two tanks each containing V gal of brine. The brine in the first tank initially contains s_1 lb of salt per gallon, the brine in the second tank initially contains s_2 lb of salt per gallon. Fresh brine containing s lb of salt per gallon flows into the first tank at the rate of a gal/min and the mixture, kept uniform by stirring, runs into the second tank at the same rate. From the second tank, the mixture, kept uniform by stirring, runs out at the same rate. Find the amounts of salt in each tank as functions of time. Under what conditions, if any, will the amount of salt in the second tank reach a relative maximum or minimum value?

25   Two tanks are connected as shown in Fig. 3.1. The first tank contains 100 gal of

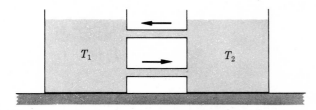

Figure 3.1

pure water; the second contains 100 gal of brine containing 2 lb of salt per gallon. Liquid circulates through the tanks at a constant rate of 5 gal/min. If the brine in each tank is kept uniform by stirring, find the amount of salt in each tank as a function of time.

26   Three tanks are connected as shown in Fig. 3.2. The first tank contains 100 gal of pure water; the second contains 100 gal of brine containing 1 lb of salt per gallon; the third contains 100 gal of brine containing 2 lb of salt per gallon. Liquid circulates through the tanks at a constant rate of 5 gal/min. If the brine in each tank is kept uniform by stirring, find the amount of salt in each tank as a function of time.

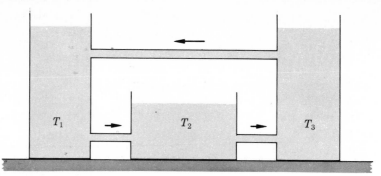

Figure 3.2

27 **(a)** Consider (as a simplified mathematical model of a hot-water radiator) a tank containing p_1 lb of liquid of specific heat c Btu/(lb)(°F) and a second tank, containing p_2 lb of the same liquid, connected to the first by pipes of negligible volume through which the liquid circulates at the rate of q lb/min. The first tank loses heat to air of constant temperature T_1 according to Newton's law of cooling with proportionality constant k_1; the second tank loses heat to air of constant temperature T_2 according to Newton's law with proportionality constant k_2. Initially, the temperature of the liquid in both tanks is T_0. If a heat source supplies heat to the first tank at h Btu/min, and if the temperature of the liquid in each tank is assumed to be kept uniform throughout that tank, set up the system of differential equations which give the temperature of the liquid in each tank as a function of time.

(b) Solve the differential equations derived in part **(a)** if $c = 1$ Btu/(lb)(°F), $p_1 = 900$ lb, $p_2 = 100$ lb, $k_1 = 3$, $k_2 = 11$, $q = 15$ lb/min, $T_0 = 50$°F, $T_1 = 80$°F, $T_2 = 70$°F, and $h = 100$ Btu/min.

(c) Solve the differential equations derived in part **(a)** if $c = 1$ Btu/(lb)(°F), $p_1 = p_2 = 200$ lb, $k_1 = 2$, $k_2 = 50$, $q = 10$ lb/min, $T_0 = 60$°F, $T_1 = 90$°F, $T_2 = 70$°F, and $h = 150$ Btu/min.

28 If the system

$$(2D^2 - D - 1)x + (2D^2 + 4D - 6)y = 0$$

$$(D^2 + 2D - 3)x + (D^2 + 7D - 8)y = 0$$

is solved by determining both x and y from the differential equations obtained by using Cramer's rule, show that x and y must be substituted into *each* of the given equations to obtain the necessary relations between the arbitrary constants.

29 If each of the operational coefficients in the system

$$P_{11}(D)x + P_{12}(D)y = 0$$

$$P_{21}(D)x + P_{22}(D)y = 0$$

contains $D - a$ as a simple factor and if the determinant of the operational coefficients contains $D - a$ only as a double factor, show that neither x nor y contains a term of the form te^{at} even though the characteristic equation of the differential equations for both x and y has a as a double root. *Hint:* Factor $D - a$ from each operational coefficient, set $(D - a)x = u$ and $(D - a)y = v$, and note that neither u nor v can contain a term of the form e^{at}.

3.3 Complementary Functions and Particular Integrals for Systems of Equations

The concepts of *characteristic equation*, *complementary function*, and *particular integral* can be extended from a single differential equation to a system of equations. The generalization is not entirely obvious, however, and is most effectively carried out using the notation and some of the ideas of matrix algebra. On the other hand, when the roots of the characteristic equation of the system are all distinct, and when there is no duplication between the terms on the right-hand sides of the various equations and the terms in the complementary function of the system, the analogy between a single equation and a system of equations is very close and can fruitfully be discussed at this stage. Problems arising from repeated roots and from duplicating terms are suggested in the exercises and explored more fully in Sec. 10.6.

To illustrate these ideas in their simplest form, let us consider the following system of equations:

$$
\begin{aligned}
(D + 1)x + (D + 2)y + (D + 3)z &= -e^{-t} \\
(D + 2)x + (D + 3)y + (2D + 3)z &= e^{-t} \\
(4D + 6)x + (5D + 4)y + (20D - 12)z &= 7e^{-t}
\end{aligned}
$$
(1)

As in the case of a single equation, we shall first make the system homogeneous by neglecting the terms on the right, getting

$$
\begin{aligned}
(D + 1)x + (D + 2)y + (D + 3)z &= 0 \\
(D + 2)x + (D + 3)y + (2D + 3)z &= 0 \\
(4D + 6)x + (5D + 4)y + (20D - 12)z &= 0
\end{aligned}
$$
(2)

Guided by our experience in solving single equations, let us now attempt to find solutions of this system of the form

$$
x = ae^{mt} \qquad y = be^{mt} \qquad z = ce^{mt}
$$
(3)

Substituting these into the equations (2) and dividing out the common factor e^{mt} leads to the set of algebraic equations

$$
\begin{aligned}
(m + 1)a + (m + 2)b + (m + 3)c &= 0 \\
(m + 2)a + (m + 3)b + (2m + 3)c &= 0 \\
(4m + 6)a + (5m + 4)b + (20m - 12)c &= 0
\end{aligned}
$$
(4)

If nontrivial solutions for x, y, and z, that is, solutions that do not vanish identically, are to be obtained, it is necessary that a, b, and c not all be zero. However, the values $a = b = c = 0$ obviously satisfy the system (4) and in general will be the only solution of this set of equations. In fact, from college algebra (or from Corollary 1, Theorem 8, Sec. 10.5) we know that no other solutions are possible unless the determinant of the coefficients in (4) is equal to zero. Thus we must have

$$
\begin{vmatrix}
m + 1 & m + 2 & m + 3 \\
m + 2 & m + 3 & 2m + 3 \\
4m + 6 & 5m + 4 & 20m - 12
\end{vmatrix} = 0
$$

or, expanding and collecting terms,

$$
-(m - 1)(m - 2)(m - 3) = 0
$$

This equation, which defines all the values of m for which nontrivial solutions of (4), and hence of (2), can exist, is the **characteristic equation** of the system. It is, of course, nothing but the determinant of the operational coefficients of the system equated to zero, with D replaced by m; and its expanded form is precisely the characteristic equation of the differential equations obtained for the individual variables by the elimination process described in Sec. 3.2.

From the roots of the last equation, namely, $m_1 = 1$, $m_2 = 2$, $m_3 = 3$, we can construct three particular solutions

(5)
$$x_1 = a_1 e^t \qquad x_2 = a_2 e^{2t} \qquad x_3 = a_3 e^{3t}$$
$$y_1 = b_1 e^t \qquad y_2 = b_2 e^{2t} \qquad y_3 = b_3 e^{3t}$$
$$z_1 = c_1 e^t \qquad z_2 = c_2 e^{2t} \qquad z_3 = c_3 e^{3t}$$

provided that we establish the proper relations between the constants in each of the three sets.

To do this, we note that the constants a_i, b_i, c_i must satisfy the equations of the system (4) for the corresponding value $m = m_i$. Thus for $m_1 = 1$, we must have

$$2a_1 + 3b_1 + 4c_1 = 0$$
$$3a_1 + 4b_1 + 5c_1 = 0$$
$$10a_1 + 9b_1 + 8c_1 = 0$$

We know, of course, that the determinant of the coefficients of this system is equal to zero. Hence these equations are nontrivially solvable, i.e., have a solution other than $a_1 = b_1 = c_1 = 0$, and it is easy to verify that for all values of k_1 they are satisfied by

$$a_1 = -k_1 \qquad b_1 = 2k_1 \qquad c_1 = -k_1$$

Therefore, for each value of k_1,

(6)
$$x_1 = -k_1 e^t$$
$$y_1 = 2k_1 e^t$$
$$z_1 = -k_1 e^t$$

is a particular solution of (2) corresponding to the characteristic root $m_1 = 1$.

Similarly, for $m_2 = 2$, we have from (4)

$$3a_2 + 4b_2 + 5c_2 = 0$$
$$4a_2 + 5b_2 + 7c_2 = 0$$
$$14a_2 + 14b_2 + 28c_2 = 0$$

and it is easy to verify that for all values of k_2 these are satisfied by

$$a_2 = 3k_2 \qquad b_2 = -k_2 \qquad c_2 = -k_2$$

Therefore a second family of particular solutions of (2) is

(7)
$$x_2 = 3k_2 e^{2t}$$
$$y_2 = -k_2 e^{2t}$$
$$z_2 = -k_2 e^{2t}$$

Finally, for $m_3 = 3$, we have from (4)

$$4a_3 + 5b_3 + 6c_3 = 0$$
$$5a_3 + 6b_3 + 9c_3 = 0$$
$$18a_3 + 19b_3 + 48c_3 = 0$$

and, from these,

$$a_3 = 9k_3 \qquad b_3 = -6k_3 \qquad c_3 = -k_3$$

A third family of particular solutions of (2) is therefore

(8)
$$x_3 = 9k_3 e^{3t}$$
$$y_3 = -6k_3 e^{3t}$$
$$z_3 = -k_3 e^{3t}$$

Since the equations of the homogeneous system (2) are all linear, sums of solutions will also be solutions (see Exercise 22, Sec. 3.2). Hence we can combine the three families of particular solutions (6), (7), and (8) into a complete solution of (2):

(9)
$$x = x_1 + x_2 + x_3 = -k_1 e^t + 3k_2 e^{2t} + 9k_3 e^{3t}$$
$$y = y_1 + y_2 + y_3 = 2k_1 e^t - k_2 e^{2t} - 6k_3 e^{3t}$$
$$z = z_1 + z_2 + z_3 = -k_1 e^t - k_2 e^{2t} - k_3 e^{3t}$$

This is the **complementary function** of the original nonhomogeneous system (1). We note that it contains precisely three independent arbitrary constants, as required by Theorem 1, Sec. 3.2. The relations between the nine constants originally present in the three particular solutions (5) could also have been found by substituting those solutions into the equations of the homogeneous system (2) and equating coefficients, as we did in Example 1, Sec. 3.2.

To complete the problem we now need to find a particular solution, or "integral," of the nonhomogeneous system (1). When, as we are assuming in the present section, the roots of the characteristic equation are all distinct and there is no duplication between the nonhomogeneous terms and the terms in the complementary function, we choose for x, y, and z individual trial solutions exactly as described in Table 2.2. Thus in the present case we assume

$$X = Ae^{-t} \qquad Y = Be^{-t} \qquad Z = Ce^{-t}$$

Substituting these into (1) and collecting terms, we find

$$(B + 2C)e^{-t} = -e^{-t}$$
$$(A + 2B + C)e^{-t} = e^{-t}$$
$$(2A - B - 32C)e^{-t} = 7e^{-t}$$

Hence $A = 3$, $B = -1$, $C = 0$, and

$$X = 3e^{-t} \qquad Y = -e^{-t} \qquad Z = 0$$

The complete solution of the original system (1) is therefore

$$x = -k_1 e^t + 3k_2 e^{2t} + 9k_3 e^{3t} + 3e^{-t}$$
$$y = 2k_1 e^t - k_2 e^{2t} - 6k_3 e^{3t} - e^{-t}$$
$$z = -k_1 e^t - k_2 e^{2t} - k_3 e^{3t}$$

EXERCISES

Find a complete solution of each of the following systems:

1 $(D + 2)x + (D + 4)y = 1$
$(D + 1)x + (D + 5)y = 2$

2 $(2D + 1)x + (D + 2)y = 0$
$(D + 3)x + (D + 6)y = -3e^t$

3 $(D + 1)x + (4D - 2)y = t - 1$
$(D + 2)x + (5D - 2)y = 2t - 1$

4 $(D + 5)x + (D + 7)y = 4e^{2t}$
$(2D + 1)x + (3D + 1)y = 0$

5 $(2D + 1)x + (D + 2)y = 6e^t$
$(D + 2)x + (D + 4)y = 4e^{-t}$

6 $(2D + 1)x + (D - 1)y = -3 \cos t$
$(D + 2)x + (D + 3)y = 5 \sin t$

7 $\quad (2D + 1)x + \quad\quad (D + 2)y = 8e^{-t}$
$(D^2 + D + 9)x + (D^2 - 2D + 12)y = 6$

8 $(2D + 1)x + (D^2 + 6D + 1)y = 0$
$(D + 2)x + (D^2 + 2D + 5)y = 6e^{2t}$

9 $(2D^2 + 5)x + (D^2 + 3)y = -8 \sin 3t$
$(D^2 + 7)x + (D^2 + 5)y = 8 \sin 3t$

Hint: Noting that only derivatives of even order appear in the two equations, assume first $x = a \cos mt$, $y = b \cos mt$ and then $x = c \sin mt$, $y = d \sin mt$, where m is a parameter to be determined.

10 $(2D + 1)x + (D + 1)y = 0$
$(D - 2)x + (D - 1)y = 0$

11 $(2D + 11)x + (D + 3)y + \quad (D - 2)z = \quad 14e^t$
$(D - 2)x + (D - 1)y + \quad\quad\quad Dz = \quad -2e^t$
$(D + 1)x + (D - 3)y + (2D - 4)z = \quad 4e^t$

12 Verify that the characteristic equation of the system

$$(2D + 1)x + (D + 1)y = 0$$

$$(D - 4)x + (D - 3)y = 0$$

has the repeated root $m = 1$. Show, further, that although a_1 and b_1 can be related so that $x = a_1e^t$ and $y = b_1e^t$ will constitute a particular solution of the system, it is impossible to determine a_2 and b_2 so that $x = a_2te^t$ and $y = b_2te^t$ will form a particular solution of the system. Show, however, that a complete solution can be found by considering simultaneously the two tentative particular solutions and assuming $x = a_1e^t + a_2te^t$ and $y = b_1e^t + b_2te^t$.

13 Verify that in the system of equations

$$(2D + 1)x + (D + 1)y = e^t$$

$$(D - 7)x + (D - 5)y = 0$$

the nonhomogeneous term e^t duplicates a term in the complementary function of the system. Verify, further, that it is impossible to determine A and B so that $X = Ate^t$ and $Y = Bte^t$ will form a particular integral of the system but that A, B, C, and E can be determined so that $X = Ate^t + Ce^t$ and $Y = Bte^t + Ee^t$ will constitute a particular integral of the system.

14 Two particles, each of weight w, are attached to a perfectly flexible, weightless, elastic string stretched under tension T as shown in Fig. 3.3. The particles vibrate in a direction perpendicular to the length of the string through amplitudes so small that

Figure 3.3

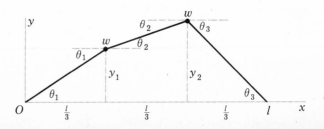

(a) The tension in the string remains constant, and

(b) The angles shown in Fig. 3.3 are so small that their sines can with satisfactory accuracy be approximated by their tangents.

Neglecting all forces but the elastic forces supplied by the string, set up the differential equations describing the behavior of the system, find the natural frequencies of the system and the ratios of the amplitudes of the two particles at each frequency.

15 Work Exercise 14 for the system of three particles shown in Fig. 3.4.

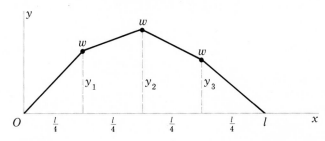

Figure 3.4

CHAPTER 4
Finite Differences

4.1 The Differences of a Function

In the last three chapters we have developed methods for the solution of several large and important classes of differential equations. There are, of course, other families of equations for which exact solutions can be found, but, in general, differential equations more complicated than the simple ones we have been considering must be solved by approximate, numerical methods. Among the most important of these are what are known as *finite-difference methods*. Since finite differences also occur in other branches of mathematical analysis, such as interpolation, numerical differentiation and integration, curve fitting, and the smoothing of data, an applied mathematician should have some familiarity with them.

Suppose that we have a function $y = f(x)$ given in tabular form for a sequence of values of x:

x	$f(x)$
x_0	$f(x_0)$
x_1	$f(x_1)$
x_2	$f(x_2)$
x_3	$f(x_3)$
$\dots$	$\dots$

If $f(x_i)$ and $f(x_j)$ are any two values of $f(x)$, then the **first divided differences** of $f(x)$ are defined by the formula†

$$(1) \qquad f(x_i,x_j) = \frac{f(x_i) - f(x_j)}{x_i - x_j}$$

Similarly, if $f(x_i,x_j)$ and $f(x_j,x_k)$ are two first divided differences of $f(x)$ having one argument, x_j, in common, then the **second divided differences** of $f(x)$ are defined by the formula

$$(2) \qquad f(x_i,x_j,x_k) = \frac{f(x_i,x_j) - f(x_j,x_k)}{x_i - x_k}$$

Proceeding inductively, we define a divided difference of any order as the difference between two divided differences of the next lower order, overlapping in all but one of

---

† In most applications the subscripts of the arguments x_i and x_j will be consecutive integers, but this is not a necessary restriction on the definition.

their arguments, divided by the difference between the extreme, or nonoverlapping, arguments appearing in these differences.† From these definitions it is clear that divided differences have the following properties.

PROPERTY 1   Any divided difference of the sum (or difference) of two functions is equal to the sum (or difference) of the divided differences of the individual functions.

PROPERTY 2   Any divided difference of a constant times a function is equal to the constant times the divided difference of the function.

In many applications it is convenient to have the divided differences of a function prominently displayed. This is usually done by constructing a **difference table** in which each difference is entered in the appropriate column, midway between the elements in the preceding column from which it is constructed:

x	$f(x)$			
x_0	$f(x_0)$			
		$f(x_0,x_1)$		
x_1	$f(x_1)$		$f(x_0,x_1,x_2)$	
		$f(x_1,x_2)$		$f(x_0,x_1,x_2,x_3)$
x_2	$f(x_2)$		$f(x_1,x_2,x_3)$	
		$f(x_2,x_3)$		$\cdots$
x_3	$f(x_3)$		$\cdots$	
		$\cdots$		
$\cdots$	$\cdots$			

or in a specific numerical example,

x	x^3			
0	0			
		1		
1	1		4	
		13		1
3	27		8	0
		37		1
4	64		14	0
		93		1
7	343		20	
		193		
9	729			

Usually the values of x in a table of data will be equally spaced, and the differences of the function will be based on sets of consecutive functional values. When this is the case, the denominators in the divided differences of any given order are all equal, and it is customary to omit them. This leads to a modified set of quantities known

---

† Though obvious only for divided differences of the first order, it is true (see Exercises 22 and 23) that divided differences of all orders are symmetric functions of their arguments. Thus

$$f(x_i,x_j,x_k) = f(x_i,x_k,x_j) = f(x_j,x_i,x_k) = \cdots$$

simply as the **differences** of the function. If the constant difference between successive values of x is h, so that the general value of x in the table is

$$x_k = x_0 + kh \qquad k = \ldots, -2, -1, 0, 1, 2, \ldots$$

and the corresponding functional value is

$$y_k = f(x_k) = f(x_0 + kh) = f_k$$

then the first differences of f are defined by the formula

$$(3) \qquad\qquad \Delta f_k = f_{k+1} - f_k$$

Differences of higher order are defined in the same way, the **second differences** being

$$(4) \qquad\qquad \Delta^2 f_k = \Delta(\Delta f_k) = \Delta f_{k+1} - \Delta f_k$$

and, in general, for positive integral values of n,

$$(5) \qquad\qquad \Delta^n f_k = \Delta(\Delta^{n-1} f_k) = \Delta^{n-1} f_{k+1} - \Delta^{n-1} f_k$$

These differences are also displayed in difference tables just like divided differences.

Evidently the **difference operator** Δ has the characteristic properties of a linear operator, for

$$\begin{aligned}
\Delta(f_k \pm g_k) &= (f_{k+1} \pm g_{k+1}) - (f_k \pm g_k) \\
&= (f_{k+1} - f_k) \pm (g_{k+1} - g_k) \\
&= \Delta f_k \pm \Delta g_k
\end{aligned}$$

and if c is a constant,

$$\Delta(c f_k) = c f_{k+1} - c f_k = c(f_{k+1} - f_k) = c\, \Delta f_k$$

Moreover, Δ obeys the usual law of exponents

$$\Delta^m(\Delta^n f_k) = \Delta^{m+n} f_k$$

provided both m and n are positive integers.

When the values of the independent variable are equally spaced, the divided differences of a function can easily be expressed in terms of the ordinary differences and vice versa. Specifically,

$$f(x_0, x_1) = \frac{f(x_0) - f(x_1)}{x_0 - x_1} = \frac{f_0 - f_1}{-h} = \frac{\Delta f_0}{h}$$

$$f(x_0, x_1, x_2) = \frac{f(x_0, x_1) - f(x_1, x_2)}{x_0 - x_2} = -\frac{1}{2h}\left(\frac{\Delta f_0}{h} - \frac{\Delta f_1}{h}\right) = \frac{\Delta^2 f_0}{2!\, h^2}$$

and, in general,

$$(6) \qquad\qquad f(x_0, x_1, \ldots, x_n) = \frac{\Delta^n f_0}{n!\, h^n}$$

More generally, if the points used in constructing an nth divided difference are the $n + 1$ equally spaced points between $x_0 - kh$ and $x_0 + (n - k)h$, inclusive, it is easy to show that

$$(7) \qquad\qquad f(x_{-k}, x_{-k+1}, \ldots, x_{n-k}) = \frac{\Delta^n f_{-k}}{n!\, h^n}$$

The Δ symbolism for the differences of a function is known as the **advancing-difference notation.** In some applications, however, another notation, known as the **central-difference notation,** is more convenient. In this, the symbol δ is used instead of Δ, and the subscript appearing in the symbol for any difference is the average of the subscripts already assigned by this convention to the elements which are subtracted in forming that difference. Thus,

$$\Delta f_k = f_{k+1} - f_k = \delta f_{k+1/2} \qquad \Delta f_{k+1} = f_{k+2} - f_{k+1} = \delta f_{k+3/2}$$

$$\Delta^2 f_k = \Delta f_{k+1} - \Delta f_k = \delta f_{k+3/2} - \delta f_{k+1/2} = \delta^2 f_{k+1}$$

. . . .

The following difference tables show the relation between the advancing- and the central-difference notations:

x	f			
x_0	f_0			
		Δf_0		
x_1	f_1		$\Delta^2 f_0$	
		Δf_1		$\Delta^3 f_0$
x_2	f_2		$\Delta^2 f_1$		$\Delta^4 f_0$
		Δf_2		$\Delta^3 f_1$
x_3	f_3		$\Delta^2 f_2$	
		Δf_3		
x_4	f_4			

x	f			
x_0	f_0			
		$\delta f_{1/2}$		
x_1	f_1		$\delta^2 f_1$	
		$\delta f_{3/2}$		$\delta^3 f_{3/2}$
x_2	f_2		$\delta^2 f_2$		$\delta^4 f_2$
		$\delta f_{5/2}$		$\delta^3 f_{5/2}$
x_3	f_3		$\delta^2 f_3$	
		$\delta f_{7/2}$		
x_4	f_4			

In the first table, elements with the same subscript lie on lines sloping downward, or *advancing* into the table. In the second, elements with the same subscript lie on lines extending horizontally, or *centrally*, into the table.

Closely associated with Δ and δ is the operator E, defined as the operator which increases the argument of a function by one tabular interval. Thus

$$E(f_k) = f(x_k + h) = f(x_{k+1}) = f_{k+1}$$

Applying E a second time again increases the argument of f by h; that is,

$$E^2 f(x_k) = E[Ef(x_k)] = Ef(x_k + h) = f(x_k + 2h) = f(x_{k+2}) = f_{k+2}$$

and, in general, we define

(8) $$E^r f(x_k) = f(x_k + rh) = f(x_{k+r}) = f_{k+r}$$

for *any* real number r. Clearly, E obeys the laws

$$E(f_k \pm g_k) = Ef_k \pm Eg_k$$

$$E(cf_k) = cEf_k \qquad c \text{ a constant}$$

$$E^r(E^s f_k) = E^{r+s} f_k$$

Two operators with the property that when they are applied to the same function they yield the same result are said to be **operationally equivalent.** Now from the definition of Δf_k, we have

$$\Delta f_k = f_{k+1} - f_k = Ef_k - f_k$$

or, symbolically, $$\Delta f_k = (E - 1)f_k$$

Hence we have the operational equivalences

(9) $$\Delta = E - 1$$

(10) $$E = 1 + \Delta$$

(11) $$E - \Delta = 1$$

Moreover, by definition

$$\Delta f_k = \delta f_{k+1/2} = \delta E^{1/2} f_k$$

Hence we have the further equivalences

(12) $$\Delta = \delta E^{1/2}$$

(13) $$\delta = \Delta E^{-1/2}$$

Also, substituting from (12) into (9) and solving for δ, we have

(14) $$\delta = E^{1/2} - E^{-1/2}$$

By means of (9) we can express the various differences of a function in terms of successive entries in the table of the function. For we can write

$$\Delta^n f_k = (E - 1)^n f_k$$

and then, using the binomial expansion,

$$
\begin{aligned}
\Delta^n f_k &= \left[E^n - \binom{n}{1} E^{n-1} + \binom{n}{2} E^{n-2} + \cdots \right. \\
&\qquad \left. + (-1)^{n-1} \binom{n}{n-1} E + (-1)^n \binom{n}{n} \right] f_k\dagger \\
&= E^n f_k - n E^{n-1} f_k + \frac{n(n-1)}{2} E^{n-2} f_k + \cdots \\
&\qquad + (-1)^{n-1} n E f_k + (-1)^n f_k
\end{aligned}
$$

(15)
$$
\begin{aligned}
&= f_{k+n} - n f_{k+n-1} + \frac{n(n-1)}{2} f_{k+n-2} + \cdots \\
&\qquad + (-1)^{n-1} n f_{k+1} + (-1)^n f_k
\end{aligned}
$$

Specifically, taking $k = 0$ and $n = 1, 2, 3, 4, \ldots$, we have

(15a)
$$
\begin{aligned}
\Delta f_0 &= f_1 - f_0 \\
\Delta^2 f_0 &= f_2 - 2f_1 + f_0 \\
\Delta^3 f_0 &= f_3 - 3f_2 + 3f_1 - f_0 \\
\Delta^4 f_0 &= f_4 - 4f_3 + 6f_2 - 4f_1 + f_0
\end{aligned}
$$

$\cdots\cdots\cdots\cdots\cdots\cdots\cdots\cdots\cdots$

$\dagger$ The quantities $\binom{n}{j}$ are the so-called **binomial coefficients**, defined by the formula

$$\binom{n}{j} = \frac{n!}{j!\,(n-j)!}$$

The fact that the first divided difference of a function is precisely the difference quotient whose limit defines the derivative of the function suggests that in some respects the properties of the differences of a function and the properties of the derivatives of a function may be analogous. This is actually the case, and among other interesting results we have the following theorem.

THEOREM 1   The nth divided differences of a polynomial of degree n are constant.

Proof   To prove this theorem, it is clearly sufficient to establish the asserted property for the special polynomial x^n. To do this, we observe that for x^n the first divided difference is simply

$$f(x_i,x_j) = \frac{x_i^n - x_j^n}{x_i - x_j} = x_i^{n-1} + x_i^{n-2} x_j + \cdots + x_i x_j^{n-2} + x_j^{n-1}$$

which is a homogeneous and symmetric function of x_i and x_j of degree $n - 1$. For the second divided difference we have, of course,

$$f(x_i,x_j,x_k) = \frac{f(x_i,x_j) - f(x_j,x_k)}{x_i - x_k}$$

Moreover, since divided differences of all orders are symmetric functions of the arguments (see the footnote on p. 105), it follows that the numerator of the last fraction vanishes when $x_i = x_k$. Hence it must contain $x_i - x_k$ as a factor and therefore, as we verified explicitly for the first divided difference, the indicated division is exact. Thus the second divided difference of x^n is a homogeneous and symmetric expression of degree $n - 2$ in x_i, x_j, and x_k. Continuing in this way, it is evident that after differencing n times, the degree of the resultant expression will be zero; that is, x^n will have been reduced to a constant, independent of $x_i, x_j, x_k, \ldots$, as asserted.

Since ordinary differences are proportional to the corresponding divided differences, it is clear that Theorem 1 also holds for these differences, whether they are expressed in terms of the advancing- or the central-difference notation. Thus we have the following corollary of Theorem 1.

COROLLARY 1   The nth (ordinary) differences of a polynomial of degree n are constant.

The analogy between differences and derivatives becomes even more striking if we consider the operator Δ and introduce the so-called **factorial polynomials**:

(16) $$(x)^{(n)} = x(x - 1) \cdots (x - n + 1)$$

(17) $$(x)^{-(n)} = \frac{1}{(x + 1)(x + 2) \cdots (x + n)}$$

In general, these play the same role in the calculus of finite differences that the power functions x^n and x^{-n} play in ordinary calculus. In particular, we have the important formulas

(18) $$\Delta(x)^{(n)} = n(x)^{(n-1)}$$

(19) $$\Delta(x)^{-(n)} = -n(x)^{-(n+1)}$$

whose resemblance to the formulas for differentiating x^n and x^{-n} is unmistakable. The proofs of these involve only a little elementary algebra, and we shall leave them as exercises.

In view of formulas (18) and (19), it is a matter of some interest to be able to express an arbitrary polynomial $p(x)$ of degree n in terms of factorial polynomials. One way this can be done is to write, by analogy with Maclaurin's expansion,

$$(20) \qquad p(x) = a_0 + a_1(x)^{(1)} + a_2(x)^{(2)} + \cdots + a_n(x)^{(n)}$$

where, clearly, since $p(x)$ is of degree n, no terms beyond $a_n(x)^{(n)}$ need be included. If we set $x = 0$ in Eq. (20), every term after the first becomes zero, since each contains x as a factor. Hence

$$a_0 = p(0)$$

Now if we use (18) to take the first difference of $p(x)$, as given by Eq. (20), we get

$$\Delta p(x) = a_1 + 2a_2(x)^{(1)} + 3a_3(x)^{(2)} + \cdots + na_n(x)^{(n-1)}$$

and if we set $x = 0$ in this expression, we obtain

$$a_1 = \Delta p(0)$$

Differencing again, we find

$$\Delta^2 p(x) = 2!\, a_2 + 3 \cdot 2a_3(x)^{(1)} + \cdots + n(n-1)a_n(x)^{(n-2)}$$

and, evaluating at $x = 0$,

$$a_2 = \frac{\Delta^2 p(0)}{2!}$$

Continuing this process of differencing and evaluating at $x = 0$, we obtain the general formula

$$(21) \qquad a_j = \frac{\Delta^j p(0)}{j!} \qquad j = 0, 1, 2, \ldots, n$$

which obviously resembles closely the familiar formula for the coefficients in Maclaurin's expansion. This procedure is especially convenient when we are given a difference table of the polynomial rather than the polynomial itself.

When we are given the polynomial itself, it is usually inefficient to construct a difference table and apply the preceding method. Instead, it is better to proceed in the following way. If we divide $p(x)$ by x, we get a remainder r_0 (which is just the constant term in p) and a quotient $q_0(x)$, so that we can write

$$(22) \qquad p(x) = r_0 + xq_0(x)$$

Now, if we divide $q_0(x)$ by $x - 1$, we get a remainder r_1 and a quotient $q_1(x)$ such that

$$q_0(x) = r_1 + (x - 1)q_1(x)$$

Hence, substituting into (22),

$$(23) \quad p(x) = r_0 + x[r_1 + (x - 1)q_1(x)] = r_0 + r_1(x)^{(1)} + x(x - 1)q_1(x)$$

If, further, we divide $q_1(x)$ by $x - 2$, we obtain a remainder r_2 and a quotient $q_2(x)$ such that

$$q_1(x) = r_2 + (x - 2)q_2(x)$$

and, substituting into (23),

$$p(x) = r_0 + r_1(x)^{(1)} + x(x - 1)[r_2 + (x - 2)q_2(x)]$$
$$= r_0 + r_1(x)^{(1)} + r_2(x)^{(2)} + x(x - 1)(x - 2)q_2(x)$$

Each application of this procedure leads to a new quotient whose degree is 1 less than the degree of the preceding quotient. Hence the process must terminate, and yield the required expansion, after $n + 1$ steps:

(24) $$p(x) = r_0 + r_1(x)^{(1)} + r_2(x)^{(2)} + \cdots + r_{n-1}(x)^{(n-1)} + r_n(x)^{(n)}$$

Obviously, the required divisions can easily be carried out by the elementary process of synthetic division. Moreover, it is clear from Eqs. (20), (21), and (24) that

$$r_j = a_j = \frac{\Delta^j p(0)}{j!}$$

or

(25) $$\Delta^j p(0) = j! \, r_j$$

Hence, this method provides a convenient way of constructing the difference table of a polynomial in the important case when $h = 1$, since it furnishes us with the leading entry in each column of the table and from these the table can be extended as far as desired by simple addition, using the identity

$$\Delta^{j-1} f_{k+1} = \Delta^{j-1} f_k + \Delta^j f_k$$

EXAMPLE 1

Express $p(x) = x^4 - 5x^3 + 3x + 4$ in terms of factorial polynomials and construct the difference table of the function for $h = 1$.

Using synthetic division, we have at once

1	1	-5	0	3	4
		1	-4	-4	
2	1	-4	-4	$\underline{-1}$	
		2	-4		
3	1	-2	$\underline{-8}$		
		3			
	1	1			

The remainders r_0, r_1, r_2, r_3, r_4 are the underscored numbers 4, -1, -8, 1, 1. Hence

$$p(x) \equiv x^4 - 5x^3 + 3x + 4 = 4 - (x)^{(1)} - 8(x)^{(2)} + (x)^{(3)} + (x)^{(4)}$$

as can be verified by direct expansion.

Now from (25) we have

$$p(0) = 4 \qquad \Delta p(0) = -1 \qquad \Delta^2 p(0) = -16 \qquad \Delta^3 p(0) = 6 \qquad \Delta^4 p(0) = 24$$

Hence we have the leading entries in the difference table for $p(x)$, and by crisscross addition, as indicated, the table can be extended and the values of $p(x)$ determined as far as desired.

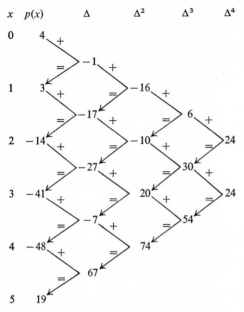

| x | $p(x)$ | Δ | Δ^2 | Δ^3 | Δ^4 |

Once a function has been expressed as a series of factorial polynomials, it is a simple matter to apply Eq. (18) or (19) to obtain its various differences. Conversely, when a function has been expressed as a series of factorial polynomials, it is easy to use these equations "in reverse" and find a new function having the given function as its first difference. By analogy with the terminology of calculus, we shall refer to such a function as an **antidifference**.

EXAMPLE 2

What is the general antidifference of the polynomial

$$p(x) = x^4 - 5x^3 + 3x + 4$$

i.e., what polynomial has $p(x)$ as its first difference?

From the results of Example 1 we know that

$$p(x) = (x)^{(4)} + (x)^{(3)} - 8(x)^{(2)} - (x)^{(1)} + 4$$

Hence, from Eq. (18), it is clear that the required antidifference, which is often denoted by the symbol $\Delta^{-1}p(x)$, is

$$\Delta^{-1}p(x) = \frac{(x)^{(5)}}{5} + \frac{(x)^{(4)}}{4} - \frac{8(x)^{(3)}}{3} - \frac{(x)^{(2)}}{2} + 4(x)^{(1)} + c$$

where c is an arbitrary constant† which can, and in general must, be added, since the difference of any constant is obviously zero. The analogy between antidifferences and indefinite integrals, or antiderivatives, is clear.

---

† Actually, c need not be a constant but can be an arbitrary periodic function of period $h = 1$ since the values of such a function at $x = 0, 1, 2, \ldots$ are all the same and hence have zero differences.

The determination of antidifferences is not just a mathematical curiosity but is intimately related to the important problem of finding the sums of series. To see this, consider any two consecutive columns in a difference table:

$$\Delta^k f_1$$
$$\Delta^{k+1} f_1$$
$$\Delta^k f_2$$
$$\Delta^{k+1} f_2$$
$$\dots$$
$$\dots$$
$$\Delta^k f_n$$
$$\Delta^{k+1} f_n$$
$$\Delta^k f_{n+1}$$

Now, from the definition of a difference, we have

$$\sum_{i=1}^{n} \Delta^{k+1} f_i = (\Delta^k f_2 - \Delta^k f_1) + (\Delta^k f_3 - \Delta^k f_2) + \cdots$$
$$+ (\Delta^k f_n - \Delta^k f_{n-1}) + (\Delta^k f_{n+1} - \Delta^k f_n)$$

or, canceling the common terms in the series on the right,

$$(26) \qquad \sum_{i=1}^{n} \Delta^{k+1} f_i = \Delta^k f_{n+1} - \Delta^k f_1$$

Since the kth difference of a function is obviously an antidifference of the $(k + 1)$st difference, it is clear that Eq. (26) is equivalent to the following theorem.

THEOREM 2 If $F(i)$ is any antidifference of $f(i)$, then the sum from $i = 1$ to $i = n$ of the series whose general term is $f(i)$ is $F(n + 1) - F(1)$.

EXAMPLE 3

What is the sum of the squares of the first n odd integers?

To facilitate finding the necessary antidifference, we first express the general term of the given series, namely, $(2i - 1)^2$, in terms of factorial polynomials:

$$(2i - 1)^2 = 4i(i - 1) + 1 = 4(i)^{(2)} + 1$$

Then, by the last theorem,

$$\sum_{i=1}^{n} (2i - 1)^2 = \sum_{i=1}^{n} [4(i)^{(2)} + 1] = \left[\frac{4(i)^{(3)}}{3} + (i)^{(1)} \right]_{i=1}^{i=n+1}$$
$$= \frac{4(n + 1)^{(3)}}{3} + (n + 1)^{(1)} - \frac{4(1)^{(3)}}{3} - (1)^{(1)}$$
$$= \frac{4(n + 1)n(n - 1)}{3} + (n + 1) - 0 - 1$$
$$= \frac{4n^3 - n}{3}$$

EXERCISES

1 Prove Properties 1 and 2.
2 Prove formulas (6) and (7).   **3** Prove formulas (18) and (19).

4 (a) Show that the divided differences of a function are unchanged if the arguments are all increased by a constant c while the corresponding functional values are left unchanged.

(b) How are the divided differences of a function changed if each argument is multiplied by a constant c while the corresponding functional values are left unchanged?

5 Express the following polynomials in terms of factorial polynomials, and construct a difference table, with $h = 1$, for each function:

(a) $x^3 - x + 1$ (b) $x^4 - 2x^3 - x$ (c) $x^5 - 2x^4 + 4x^3 - x + 6$

6 For each of the following difference tables, find the polynomial of minimum degree which yields the given data:

(a)

x	y				
0	-1				
		-1			
1	-2		6		
		5		6	
2	3		12		0
		17		6	
3	20		18		
		3			
4	55				

(b)

x	y					
0	6					
		-5				
1	1		2			
		-3		-6		
2	-2		-4		24	
		-7		18		0
3	-9		14		24	
		7		42		
4	-2		56			
		63				
5	61					

7 Can the use of factorial polynomials illustrated in Example 1 be extended to the construction of the difference table of a polynomial for a tabular interval other than $h = 1$? *Hint:* Note that the values of $p(x)$ for $x = 0, h, 2h, \ldots$ are the same as the values of $p(hz)$ for $z = 0, 1, 2, \ldots$.

8 Using the results of Exercise 7, construct the difference table of each of the following polynomials for the indicated value of h:

(a) $x^3 + 2x^2 - 1$ $h = 0.5$

(b) $x^2 + 3x - 10$ $h = 0.1$

(c) $x^3 + 3x^2 - x + 11$ $h = 2$

9 If $h = 1$, show that for all values of the constants a and b, each of the following functions satisfies the indicated relation:

(a) $y = a2^x + b3^x$ $(E^2 - 5E + 6)y = 0$

(b) $y = a2^x + bx2^x$ $(E^2 - 4E + 4)y = 0$

(c) $y = a3^x + b(-2)^x$ $(\Delta^2 + \Delta - 6)y = 0$

10 Find the sum of the cubes of the first n integers.

11 Evaluate   (a) $\sum_{i=1}^{n} i(i + 1)(i + 2)$ (b) $\sum_{i=1}^{n} i(i + 1)(i + 3)$

12 If $h = 1$, show that $\Delta a^x = a^x(a - 1)$.

13 If $h = 1$, show that

(a) $\Delta \cos (ax + b) = -2 \sin \dfrac{a}{2} \sin \left(ax + b + \dfrac{a}{2}\right)$

(b) $\Delta \sin (ax + b) = 2 \sin \dfrac{a}{2} \cos \left(ax + b + \dfrac{a}{2}\right)$

14 Using the results of Exercise 13, find the sum of each of the following series:

(a) $\displaystyle\sum_{x=1}^{n} \sin \dfrac{\pi}{n} x$ (b) $\displaystyle\sum_{x=1}^{n} \cos (x\theta + \alpha)$ (c) $\displaystyle\sum_{x=1}^{n} \sin (x\theta + \alpha)$

15 Express each of the following in terms of factorial functions of the form $(x)^{-(k)}$:

(a) $\dfrac{x}{(x + 1)(x + 2)}$ *Hint:* Add and subtract 2 in the numerator and then separate the resulting fraction into two parts.

(b) $\dfrac{1}{(x + 2)(x + 3)}$ (c) $\dfrac{x - 1}{(x + 1)(x + 3)}$

16 What is $\displaystyle\sum_{k=1}^{n} \dfrac{k}{(k + 1)(k + 2)(k + 3)}$?

17 Show that $\displaystyle\sum_{k=1}^{n} y_k = \dfrac{E^n - 1}{E - 1} y_1$, and then, by putting $E + 1 = \Delta$ show that

$$\sum_{k=1}^{n} y_k = \left[n + \dfrac{n(n - 1)}{2!} \Delta + \dfrac{n(n - 1)(n - 2)}{3!} \Delta^2 + \cdots\right] y_1$$

18 Using the result of Exercise 17, evaluate

(a) $\displaystyle\sum_{k=1}^{n} k^2$ (b) $\displaystyle\sum_{k=1}^{n} k^3$ (c) $\displaystyle\sum_{k=1}^{n} k^4$

19 Show that $\Delta(f_k g_k) = f_{k+1} \Delta g_k + g_k \Delta f_k = g_{k+1} \Delta f_k + f_k \Delta g_k$.

20 Explain how the formula of Exercise 19 can be used in a process of "summation by parts" analogous to integration by parts.

21 Illustrate the result of Exercise 20 by finding the sum of each of the following series:

(a) $\displaystyle\sum_{k=1}^{n} k2^k$ *Hint:* In the formula of Exercise 19, let $g_k = k$ and $\Delta f_k = 2^k$.

(b) $\displaystyle\sum_{k=1}^{n} k^2 3^k$ (c) $\displaystyle\sum_{k=1}^{n} k \cos k$ (d) $\displaystyle\sum_{k=1}^{n} k \sin ak$

22 Show that

$$f(x_0, x_1) = \dfrac{f(x_0)}{x_0 - x_1} + \dfrac{f(x_1)}{x_1 - x_0}$$

and

$$f(x_0, x_1, x_2) = \dfrac{f(x_0)}{(x_0 - x_1)(x_0 - x_2)} + \dfrac{f(x_1)}{(x_1 - x_0)(x_1 - x_2)} + \dfrac{f(x_2)}{(x_2 - x_0)(x_2 - x_1)}$$

What is the generalization of these results to differences of higher order?

23 Show that

$$f(x_0, x_1) = \dfrac{\begin{vmatrix} f(x_0) & f(x_1) \\ 1 & 1 \end{vmatrix}}{\begin{vmatrix} x_0 & x_1 \\ 1 & 1 \end{vmatrix}} \quad \text{and} \quad f(x_0, x_1, x_2) = \dfrac{\begin{vmatrix} f(x_0) & f(x_1) & f(x_2) \\ x_0 & x_1 & x_2 \\ 1 & 1 & 1 \end{vmatrix}}{\begin{vmatrix} x_0^2 & x_1^2 & x_2^2 \\ x_0 & x_1 & x_2 \\ 1 & 1 & 1 \end{vmatrix}}$$

24 Show that $(x)^{(a)}(x)^{(b)} \neq (x)^{(a+b)}$, but that $(x + a)^{(a)}(x)^{(b)} = (x + a)^{(a+b)}$.

25 If we define

$$f(x_0, x_1, \ldots, x_{n-1}, x_n, x_n) = \lim_{x \to x_n} f(x_0, x_1, \ldots, x_{n-1}, x_n, x),$$

show that

$$f(x_0, x_1, \ldots, x_{n-1}, x_n, x_n) = \dfrac{df(x_0, x_1, \ldots, x_{n-1}, x)}{dx}\bigg|_{x = x_n}$$

4.2 Interpolation Formulas

One of the most important applications of finite differences is to the problem of interpolation. In courses such as algebra and trigonometry, where tables of the elementary functions must occasionally be used, it is customary to obtain values between adjacent entries by the method of proportional parts or linear interpolation. As is well known, this procedure amounts to replacing the arc of the tabulated function over one tabular interval by its chord and then reading the required functional value from the chord rather than from the arc itself (Fig. 4.1a). In this case the formula for the interpolated value turns out to be

$$(1) \qquad f(x_0 + rh) = f(x_0) + r[f(x_0 + h) - f(x_0)] = f_0 + r\,\Delta f_0$$

Obviously, if h is relatively large, or if the graph of $f(x)$ is changing direction rapidly, the chord may not be a good approximation to the arc and linear interpolation may involve a substantial error. One way to overcome this difficulty would be to approximate the graph of $f(x)$ by some curve which would "fit" the true arc more closely than a straight line could and then read the interpolated value from this approximating curve rather than from the chord (Fig. 4.1b). If, specifically, the graph of $f(x)$ is approximated over two successive tabular intervals by a parabola of the form $y = a + bx + cx^2$ chosen to pass through the three points

$$[x_0, f(x_0)] \qquad [x_0 + h, f(x_0 + h)] \qquad [x_0 + 2h, f(x_0 + 2h)]$$

the formula for the interpolated value is found without difficulty to be

$$f(x_0 + rh) = f(x_0) + r[f(x_0 + h) - f(x_0)]$$
$$+ \frac{r(r-1)}{2!}[f(x_0 + 2h) - 2f(x_0 + h) + f(x_0)]$$
$$(2) \qquad = f_0 + r\,\Delta f_0 + \frac{r(r-1)}{2!}\,\Delta^2 f_0$$

Proceeding in this fashion, using polynomial curves of higher and higher order to approximate the graph of $f(x)$, one could derive a succession of interpolation formulas

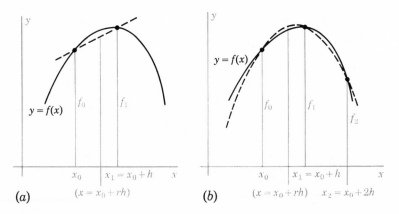

Figure 4.1
Straight-line and parabolic approximations to a given function.

involving higher and higher differences of the tabulated function and providing in general higher and higher accuracy in the interpolated values. In this section we shall obtain several important interpolation formulas, though we shall derive them by methods more general than the geometric approach we have just suggested.

Probably the most fundamental interpolation formula is **Newton's divided-difference formula**

$$(3) \quad f(x) = f(x_0) + (x - x_0)f(x_0,x_1) + (x - x_0)(x - x_1)f(x_0,x_1,x_2) + \cdots$$
$$+ (x - x_0)(x - x_1)\cdots(x - x_{n-1})f(x_0,x_1,\ldots,x_n)$$
$$+ (x - x_0)(x - x_1)\cdots(x - x_n)f(x,x_0,x_1,\ldots,x_n)$$

From this all the other interpolation formulas of interest to us can easily be derived by suitably specializing the points $x_0, x_1, \ldots, x_n$, which need not be equally spaced or taken in consecutive order. For convenience in establishing (3) we shall restrict our discussion to some special, though adequately typical, value of n, say $n = 2$. Then, beginning with the third difference

$$f(x,x_0,x_1,x_2) = \frac{f(x,x_0,x_1) - f(x_0,x_1,x_2)}{x - x_2}$$

and solving for $f(x,x_0,x_1)$, we have

$$(4) \qquad f(x,x_0,x_1) = f(x_0,x_1,x_2) + (x - x_2)f(x,x_0,x_1,x_2)$$

But

$$f(x,x_0,x_1) = \frac{f(x,x_0) - f(x_0,x_1)}{x - x_1}$$

and substituting this into (4) and solving for $f(x,x_0)$, we find

$$(5) \quad f(x,x_0) = f(x_0,x_1) + (x - x_1)f(x_0,x_1,x_2) + (x - x_1)(x - x_2)f(x,x_0,x_1,x_2)$$

Finally, since

$$f(x,x_0) = \frac{f(x) - f(x_0)}{x - x_0}$$

we have, on substituting this into (5) and solving for $f(x)$

$$f(x) = f(x_0) + (x - x_0)f(x_0,x_1) + (x - x_0)(x - x_1)f(x_0,x_1,x_2)$$
$$+ (x - x_0)(x - x_1)(x - x_2)f(x,x_0,x_1,x_2)$$

which is precisely (3) in the special case $n = 2$. The extension of the preceding argument to any value of n is obvious.

The last term in (3) differs from the other terms in that the divided difference appearing in it contains x as one of its arguments and hence is not to be found among the entries in the difference table of $f(x)$. For this reason the last term is usually referred to as the **remainder after $n + 1$ terms** or simply as the **error term**, and the interpolation series is often written in the form

$$(6) \qquad\qquad f(x) = p_n(x) + r_{n+1}(x)$$

where, of course, $p_n(x)$ is the nth-degree polynomial

(7) $f(x_0) + (x - x_0)f(x_0,x_1) + (x - x_0)(x - x_1)f(x_0,x_1,x_2) + \cdots$

$$+ (x - x_0)(x - x_1) \cdots (x - x_{n-1})f(x_0,x_1, \ldots .x_n)$$

and $r_{n+1}(x)$ is the function

(8) $$(x - x_0)(x - x_1) \cdots (x - x_n)f(x,x_0,x_1, \ldots ,x_n)$$

Using (6), (7), and (8), it is possible to obtain an interesting alternative expression for an nth divided difference and ultimately a somewhat more tractable form of the remainder term in (3). To do this, we observe that $r_{n+1}(x)$ vanishes at least $n + 1$ times on the closed interval between the largest and smallest values of the set $(x_0,x_1, \ldots ,x_n)$, since, in fact, it vanishes when $x = x_0, x_1, \ldots , x_n$. Therefore, assuming that the necessary derivatives exist, it follows from Rolle's theorem that $r'_{n+1}(x)$ must vanish at least n times on this interval, $r''_{n+1}(x)$ must vanish at least $n - 1$ times on this interval, and, continuing in this fashion, $r^{(n)}_{n+1}(x)$ must vanish at least once on this interval. That is, there must exist at least one value of x, say $x = \xi$, between the largest and smallest values of the set $(x_0,x_1, \ldots ,x_n)$ such that $r^{(n)}_{n+1}(\xi) = 0$. Hence, differentiating (6) n times and evaluating the result for $x = \xi$, we have

(9) $$f^{(n)}(\xi) - p_n{}^{(n)}(\xi) = r^{(n)}_{n+1}(\xi) = 0$$

Now, from (7), the coefficient of x^n in the nth-degree polynomial $p_n(x)$ is $f(x_0,x_1, \ldots ,x_n)$ Therefore

$$p_n{}^{(n)}(\xi) = n! \, f(x_0,x_1, \ldots ,x_n)$$

and hence, from (9), we have for each value of n,

(10) $$f(x_0,x_1, \ldots ,x_n) = \frac{f^{(n)}(\xi)}{n!}$$

where ξ is somewhere between the largest and smallest values of the set $(x_0,x_1, \ldots ,x_n)$.

Applying (10) to the $(n + 1)$st divided difference appearing in the expression for $r_{n+1}(x)$ in (8), we have, as an alternative form of $r_{n+1}(x)$,

(8a) $$r_{n+1}(x) = (x - x_0)(x - x_1) \cdots (x - x_n)\frac{f^{(n+1)}(\xi)}{(n + 1)!}$$

where now ξ is somewhere between the largest and the smallest values in the set $(x,x_0,x_1, \ldots ,x_n)$. The error term $r_{n+1}(x)$ is of great importance in theoretical studies of the convergence of the interpolation series (3), but the difficulty of estimating the factor

$$f(x,x_0,x_1, \ldots ,x_n) = \frac{f^{(n+1)}(\xi)}{(n + 1)!}$$

often limits its usefulness in numerical work. Of course, if $f(x)$ is a polynomial of degree m, say, its divided differences of order greater than m are all exactly zero, and if we extend the series (3) sufficiently far, the error term will be zero. In our calculations we shall neglect the error term on the assumption that we have extended the interpolation series to the point where the error term is either zero or at least negligibly small.

EXAMPLE 1

Find $f(2)$ from the following data:

x	$f(x)$	$f(x_i,x_j)$	$f(x_i,x_j,x_k)$	$f(x_i,x_j,x_k,x_i)$
-1.0	3.000			
		-5.000		
0.0	-2.000		5.500	
		3.250		-1.000
0.5	-0.375		3.500	
		6.750		-1.000
1.0	3.000		1.000	
		8.750		-1.000
2.5	16.125		-1.500	
		5.750		
3.0	19.000			

The construction of the difference table of the given function presents no problem, and using Newton's formula with $x_0 = 0$, we can write at once

$$f(2) = -2.000 + (2 - 0)(3.250) + (2 - 0)(2 - 0.5)(3.500)$$
$$+ (2 - 0)(2 - 0.5)(2 - 1)(-1.000)$$
$$= 12.000$$

In passing, we note that the ordinary process of linear interpolation yields the value $f(2) = 13.750$.

Closely associated with Newton's divided-difference formula is **Lagrange's interpolation formula**†

(11)
$$f(x) = \frac{(x - x_1)(x - x_2)\cdots(x - x_n)}{(x_0 - x_1)(x_0 - x_2)\cdots(x_0 - x_n)} f(x_0)$$
$$+ \frac{(x - x_0)(x - x_2)\cdots(x - x_n)}{(x_1 - x_0)(x_1 - x_2)\cdots(x_1 - x_n)} f(x_1) + \cdots$$
$$+ \frac{(x - x_0)(x - x_1)\cdots(x - x_{n-1})}{(x_n - x_0)(x_n - x_1)\cdots(x_n - x_{n-1})} f(x_n)$$

Like Newton's divided-difference formula, this formula provides the equation of a polynomial of degree n (or less) which takes on the $n + 1$ prescribed functional values $f(x_0), f(x_1), \ldots, f(x_n)$ when x takes on the values $x_0, x_1, \ldots, x_n$. Equation (11) can easily be derived from Eq. (3), but it is simpler merely to verify its properties. Clearly, it is a polynomial of degree n (or less), since each term on the right is a polynomial of degree n. Moreover, when $x = x_0$, every fraction except the first vanishes because of the factor $x - x_0$, and at the same time the first fraction reduces to 1, leaving just $f(x) = f(x_0)$, as required when $x = x_0$. In the same way, when $x = x_1$, every fraction except the second becomes zero, and we have $f(x) = f(x_1)$. Similarly, we can verify without difficulty that $f(x)$ reduces to $f(x_2), f(x_3), \ldots, f(x_n)$ when $x = x_2, x_3, \ldots, x_n$, as required.

When the points $x_0, x_1, \ldots, x_n$ on which Newton's divided-difference formula is

---

† Named for the French mathematician Joseph Louis Lagrange (1736–1813).

based are regularly spaced with tabular interval h, say, it is generally more convenient to express formula (3) in terms of ordinary differences. To do this we observe that if

$$x = x_0 + rh \qquad \text{and} \qquad x_k = x_0 + kh$$

then
$$x - x_k = h(r - k) \qquad k = 0, 1, 2, \ldots, n$$

and

$$(12) \qquad (x - x_0)(x - x_1) \cdots (x - x_j) = h^{j+1} r(r - 1) \cdots (r - j)$$

Also, from Eq. (6), Sec. 4.1, we have

$$(13) \qquad f(x_0, x_1, \ldots, x_{j+1}) = \frac{\Delta^{j+1} f_0}{(j + 1)! \, h^{j+1}}$$

Hence, substituting from (12) and (13) into (3), we find

$$f(x) \equiv f(x_0 + rh)$$

$$(14) \qquad = f_0 + r \Delta f_0 + \frac{r(r - 1)}{2!} \Delta^2 f_0 + \frac{r(r - 1)(r - 2)}{3!} \Delta^3 f_0 + \cdots$$

which is known as the **forward Gregory-Newton interpolation formula**.† Obviously this is a direct generalization of the formulas of linear and parabolic interpolation [Eqs. (1) and (2)]. Of course, the error term in (3) can be transformed into a corresponding error term for the series (14), but we shall leave this as an exercise.

For tables of limited extent, formula (14) is especially adapted to interpolation near the upper end, i.e., for smaller values of x, and cannot conveniently be used near the lower end. For the latter case it would be desirable to have a formula using differences located above rather than below the point of interpolation. Such a formula can easily be derived by choosing the points $x_0, x_1, \ldots, x_n$ used in the divided-difference formula (3) to be the points

$$x_0, x_0 - h, x_0 - 2h, \ldots, x_0 - nh$$

Then
$$x - x_k = h(r + k) \qquad k = 0, 1, 2, \ldots, n$$

and

$$(15) \qquad (x - x_0)(x - x_1) \cdots (x - x_j) = h^{j+1} r(r + 1) \cdots (r + j)$$

Moreover, in this case the typical difference $f(x_0, x_1, \ldots, x_{j+1})$ becomes $f(x_0, x_{-1}, \ldots, x_{-j-1})$, and from the symmetry of divided differences this is equal to $f(x_{-j-1}, x_{-j}, \ldots, x_{-1}, x_0)$. Hence, using Eq. (7), Sec. 4.1 (with $n = k = j + 1$), we have for our current choice of points

$$(16) \qquad f(x_0, x_1, \ldots, x_{j+1}) = \frac{\Delta^{j+1} f_{-j-1}}{(j + 1)! \, h^{j+1}}$$

---

† Conamed for the Scottish mathematician James Gregory (1638–1675).

Finally, substituting from (15) and (16) into (3), we find

$$f(x) \equiv f(x_0 + rh)$$

$$(17) \qquad = f_0 + r\,\Delta f_{-1} + \frac{r(r+1)}{2!}\,\Delta^2 f_{-2} + \frac{r(r+1)(r+2)}{3!}\,\Delta^3 f_{-3} + \cdots$$

which is known as the **backward Gregory-Newton interpolation formula**.

EXAMPLE 2

Compute $f(1.03)$ from the following data:

x	$f(x)$	Δ	Δ^2	Δ^3
1.00	1.000000			
		0.257625		
1.05	1.257625		0.015750	
		0.273375		0.000750
1.10	1.531000		0.016500	
		0.289875		0.000750
1.15	1.820875		0.017250	
		0.307125		
1.20	2.128000			

The construction of the difference table presents no difficulty, and we need merely identify $x_0 = 1.00$, $h = 0.05$, $r = 0.6$ and then substitute into formula (14):

$$f(1.03) = f[1.00 + (0.6)(0.05)]$$

$$= 1.000000 + (0.6)(0.257625) + \frac{(0.6)(0.6 - 1)}{2!}(0.015750)$$

$$+ \frac{(0.6)(0.6 - 1)(0.6 - 2)}{3!}(0.000750)$$

$$= 1.152727$$

Linear interpolation uses only the first two terms of the last series and hence yields the (presumably) less accurate value $f(1.03) = 1.154575$.

There are various ways of obtaining central-difference interpolation formulas. For instance, we can choose the points used in Newton's divided-difference formula in the following order:

$$x_0 = x_0, \ x_1 = x_0 + h, \ x_2 = x_0 - h, \ x_3 = x_0 + 2h, \ x_4 = x_0 - 2h, \ldots$$

Then substituting into (3) and using Eq. (7), Sec. 4.1, to simplify the various divided differences, we find

$$(18) \qquad f(x) \equiv f(x_0 + rh)$$

$$= f_0 + r\,\Delta f_0 + \frac{r(r-1)}{2!}\,\Delta^2 f_{-1} + \frac{r(r-1)(r+1)}{3!}\,\Delta^3 f_{-1}$$

$$+ \frac{r(r-1)(r+1)(r-2)}{4!}\,\Delta^4 f_{-2} + \cdots$$

or, introducing the central-difference operator δ by means of the operational equivalence $\Delta = \delta E^{1/2}$ [Eq. (12), Sec. 4.1],

$$(18a) \quad f(x_0 + rh) = f_0 + r\,\delta f_{1/2} + \frac{r(r-1)}{2!}\,\delta^2 f_0 + \frac{(r+1)r(r-1)}{3!}\,\delta^3 f_{1/2}$$

$$+ \frac{(r+1)r(r-1)(r-2)}{4!}\,\delta^4 f_0 + \cdots$$

This is known as the **forward Newton-Gauss interpolation formula**.

In exactly the same way, by choosing the points $x_0, x_1, x_2, \ldots$ in the order

$$x_0 = x_0,\ x_1 = x_0 - h,\ x_2 = x_0 + h,\ x_3 = x_0 - 2h,\ x_4 = x_0 + 2h, \ldots$$

and again substituting into (3) we obtain, after introducing the central-difference notation,

$$(19) \quad f(x_0 + rh) = f_0 + r\,\delta f_{-1/2} + \frac{(r+1)r}{2!}\,\delta^2 f_0 + \frac{(r+1)r(r-1)}{3!}\,\delta^3 f_{-1/2}$$

$$+ \frac{(r+2)(r+1)r(r-1)}{4!}\,\delta^4 f_0 + \cdots$$

which is usually referred to as the **backward Newton-Gauss interpolation formula**.

If we take the average of Eqs. (18a) and (19), we obtain a useful result known as **Stirling's interpolation formula**:†

$$(20) \quad f(x_0 + rh) = f_0 + \frac{r}{1!}\,\frac{\delta f_{1/2} + \delta f_{-1/2}}{2} + \frac{r^2}{2!}\,\delta^2 f_0$$

$$+ \frac{r(r^2-1)}{3!}\,\frac{\delta^3 f_{1/2} + \delta^3 f_{-1/2}}{2} + \frac{r^2(r^2-1)}{4!}\,\delta^4 f_0 + \cdots$$

Another useful formula can be obtained by eliminating the differences of odd order from Eq. (18a) by means of the formulas $\delta f_{1/2} = f_1 - f_0$, $\delta^3 f_{1/2} = \delta^2 f_1 - \delta^2 f_0, \ldots$. This gives

$$f(x_0 + rh) = f_0 + r(f_1 - f_0) + \frac{r(r-1)}{2!}\,\delta^2 f_0 + \frac{(r+1)r(r-1)}{3!}\,(\delta^2 f_1 - \delta^2 f_0)$$

$$+ \frac{(r+1)r(r-1)(r-2)}{4!}\,\delta^4 f_0$$

$$+ \frac{(r+2)(r+1)r(r-1)(r-2)}{5!}\,(\delta^4 f_1 - \delta^4 f_0) + \cdots$$

or, collecting terms,

$$f(x_0 + rh) = -(r-1)f_0 - \frac{r(r-1)(r-2)}{3!}\,\delta^2 f_0$$

$$- \frac{(r+1)r(r-1)(r-2)(r-3)}{5!}\,\delta^4 f_0 - \cdots + rf_1$$

$$+ \frac{(r+1)r(r-1)}{3!}\,\delta^2 f_1 + \frac{(r+2)(r+1)r(r-1)(r-2)}{5!}\,\delta^4 f_1 + \cdots$$

---

† Named for the Scottish mathematician James Stirling (1692–1770).

Finally, if we set $1 - r = s$ in the coefficients of the differences of f_0, we obtain the symmetric form

$$(21) \quad f(x_0 + rh) = sf_0 + \frac{s(s^2 - 1)}{3!} \delta^2 f_0 + \frac{s(s^2 - 1)(s^2 - 4)}{5!} \delta^4 f_0 + \cdots$$

$$+ rf_1 + \frac{r(r^2 - 1)}{3!} \delta^2 f_1 + \frac{r(r^2 - 1)(r^2 - 4)}{5!} \delta^4 f_1 + \cdots$$

which is known as the **Laplace-Everett interpolation formula**.

EXERCISES

1 Establish Eq. (2) by finding the equation of the approximating parabola and evaluating it at $x = x_0 + rh$. *Hint:* Take x_0, x_1, x_2 to be 0, h, $2h$, respectively.

2 Compute (a) $f(1.3)$ and (b) $f(1.95)$ from the following data:

x	1.1	1.2	1.5	1.7	1.8	2.0
$f(x)$	1.112	1.219	1.636	2.054	2.323	3.011

3 Compute (a) $\sqrt{50.2}$ and (b) $\sqrt{55.9}$ from the following data:

x	$\sqrt{x}$
50	7.07107
51	7.14143
52	7.21110
53	7.28011
54	7.34847
55	7.41620
56	7.48331

4 Fit a polynomial of minimum degree to the data of Example 1.

5 Fit a polynomial of minimum degree to the following data:

x	-1	1	2	4	5
$f(x)$	13	15	13	33	67

6 (a) Supply the details required to complete the derivation of Eq. (18).
   (b) Supply the details required to complete the derivation of Eq. (19).

7 Give an operational derivation of the forward Gregory-Newton interpolation formula by observing that $f(x) \equiv f(x_0 + rh) = E^r f_0$ and then replacing E by its operational equivalent $1 + \Delta$.

8 Give an operational derivation of the backward Gregory-Newton interpolation formula by observing that $f(x) \equiv f(x_0 + rh) = E^r f_0$ and then replacing E by its operational equivalent

$$\frac{E}{E - \Delta} = \frac{1}{1 - \Delta E^{-1}} = (1 - \Delta E^{-1})^{-1}$$

9 If y_0, y_1, y_2, y_3 are the values of a function at the equally spaced values x_0, x_1, x_2, x_3, show that the best estimate of the value of y corresponding to the value of x midway between x_1 and x_2 is

$$\frac{y_1 + y_2}{2} + \frac{(y_1 + y_2) - (y_0 + y_3)}{16}$$

10 Three readings are taken at equally spaced points $x = 0, h, 2h$ near the maximum (or minimum) of a function $y = f(x)$. Show that the abscissa of the maximum (or minimum) is approximately

$$\left(\frac{1}{2} - \frac{\Delta y_0}{\Delta^2 y_0}\right) h$$

and that the maximum (or minimum) ordinate is approximately

$$y_1 - \frac{(\Delta y_1 + \Delta y_0)^2}{8\Delta^2 y_0}$$

11 Work Exercise 10, given that the three points where readings are taken are not equally spaced.

12 Derive Lagrange's interpolation formula from Eq. (3) for the case $n = 2$. *Hint:* Use the results of Exercise 22, Sec. 4.1.

13 Derive Lagrange's interpolation formula by expanding

$$\frac{f(x)}{(x - x_0)(x - x_1)\cdots(x - x_n)}$$

into partial fractions.

14 Obtain the error terms in the forward and backward Gregory-Newton formulas from the error term in Newton's divided-difference formula.

15 Estimate the errors made in the calculations required in Exercise 3.

4.3 Numerical Differentiation and Integration

Any of the interpolation formulas we obtained in the last section can be used to find the derivative of a tabular function. For instance, if we consider the forward Gregory-Newton formula

$$f(x_0 + rh) = f_0 + r\,\Delta f_0 + \frac{r(r-1)}{2!}\Delta^2 f_0 + \frac{r(r-1)(r-2)}{3!}\Delta^3 f_0$$

$$+ \frac{r(r-1)(r-2)(r-3)}{4!}\Delta^4 f_0 + \cdots$$

and differentiate with respect to r, we find

(1) $$hf'(x_0 + rh) = \Delta f_0 + \frac{2r-1}{2}\Delta^2 f_0 + \frac{3r^2 - 6r + 2}{6}\Delta^3 f_0$$

$$+ \frac{2r^3 - 9r^2 + 11r - 3}{12}\Delta^4 f_0 + \cdots$$

(2) $$h^2 f''(x_0 + rh) = \Delta^2 f_0 + (r-1)\Delta^3 f_0 + \frac{6r^2 - 18r + 11}{12}\Delta^4 f_0 + \cdots$$

(3) $$h^3 f'''(x_0 + rh) = \Delta^3 f_0 + \frac{2r-3}{2}\Delta^4 f_0 + \cdots$$

(4) $$h^4 f^{iv}(x_0 + rh) = \Delta^4 f_0 + \cdots$$

Specifically, if we put $r = 0$, we find for the successive derivatives at the tabular point x_0

(5) $\quad f'(x_0) = \dfrac{1}{h}(\Delta f_0 - \tfrac{1}{2}\Delta^2 f_0 + \tfrac{1}{3}\Delta^3 f_0 - \tfrac{1}{4}\Delta^4 f_0 + \cdots)$

(6) $\quad f''(x_0) = \dfrac{1}{h^2}(\Delta^2 f_0 - \Delta^3 f_0 + \tfrac{11}{12}\Delta^4 f_0 - \cdots)$

(7) $\quad f'''(x_0) = \dfrac{1}{h^3}(\Delta^3 f_0 - \tfrac{3}{2}\Delta^4 f_0 + \cdots)$

(8) $\quad f^{iv}(x_0) = \dfrac{1}{h^4}(\Delta^4 f_0 - \cdots)$

Similarly, from the backward Gregory-Newton formula we obtain

(9) $\quad hf'(x_0 + rh) = \Delta f_{-1} + \dfrac{2r+1}{2}\Delta^2 f_{-2} + \dfrac{3r^2 + 6r + 2}{6}\Delta^3 f_{-3}$

$$+ \dfrac{2r^3 + 9r^2 + 11r + 3}{12}\Delta^4 f_{-4} + \cdots$$

(10) $\quad h^2 f''(x_0 + rh) = \Delta^2 f_{-2} + (r+1)\Delta^3 f_{-3} + \dfrac{6r^2 + 18r + 11}{12}\Delta^4 f_{-4} + \cdots$

(11) $\quad h^3 f'''(x_0 + rh) = \Delta^3 f_{-3} + \dfrac{2r+3}{2}\Delta^4 f_{-4} + \cdots$

(12) $\quad h^4 f^{iv}(x_0 + rh) = \Delta^4 f_{-4} + \cdots$

and, at the point x_0,

(13) $\quad f'(x_0) = \dfrac{1}{h}(\Delta f_{-1} + \tfrac{1}{2}\Delta^2 f_{-2} + \tfrac{1}{3}\Delta^3 f_{-3} + \tfrac{1}{4}\Delta^4 f_{-4} + \cdots)$

(14) $\quad f''(x_0) = \dfrac{1}{h^2}(\Delta^2 f_{-2} + \Delta^3 f_{-3} + \tfrac{11}{12}\Delta^4 f_{-4} + \cdots)$

(15) $\quad f'''(x_0) = \dfrac{1}{h^3}(\Delta^3 f_{-3} + \tfrac{3}{2}\Delta^4 f_{-4} + \cdots)$

(16) $\quad f^{iv}(x_0) = \dfrac{1}{h^4}(\Delta^4 f_{-4} + \cdots)$

For a complete development, an error term analogous to Eq. (8a), Sec. 4.2, should be found for any formula of numerical differentiation. This can be done, but the results are of relatively little use in routine calculations, and we shall not take them into account. However, it should be borne in mind that unless we are dealing with a polynomial, numerical differentiation may involve errors of considerable magnitude, the errors increasing significantly as derivatives of higher order are computed.

EXAMPLE 1

Find the first and second derivatives of $\sqrt{x}$ at $x = 2.5$ from the table

x	$\sqrt{x}$	Δ	Δ^2
2.50	1.58114		
		0.01573	
2.55	1.59687		-0.00015
		0.01558	
2.60	1.61245		-0.00015
		0.01543	
2.65	1.62788		-0.00014
		0.01529	
2.70	1.64317		-0.00015
		0.01514	
2.75	1.65831		

Using Eqs. (5) and (6) with $x_0 = 2.50$ and $h = 0.05$, we find at once

$$f'(2.5) = \frac{1}{0.05} [0.01573 - \tfrac{1}{2}(-0.00015)] = 0.3160$$

$$f''(2.5) = \frac{1}{(0.05)^2} (-0.00015) = -0.0600$$

The correct values to four decimal places are, of course,

$$f'(2.5) = \frac{1}{2\sqrt{x}}\bigg|_{x=2.5} = 0.3162$$

$$f''(2.5) = \frac{-1}{4x\sqrt{x}}\bigg|_{x=2.5} = -0.0632$$

To obtain formulas for numerical integration, it is convenient to begin by considering the related problem of the summation of series, a topic on which we touched briefly at the end of Sec. 4.1. In doing this it will be convenient to use certain additional operational equivalences which we shall now develop. We begin with Maclaurin's expansion,

$$(17) \qquad f(x + h) = f(x) + hf'(x) + \frac{h^2}{2!} f''(x) + \frac{h^3}{3!} f'''(x) + \cdots$$

or, introducing the operators E and $D \equiv d/dx$,

$$(18) \qquad Ef(x) = \left(1 + hD + \frac{h^2 D^2}{2!} + \frac{h^3 D^3}{3!} + \cdots\right) f(x)$$

Now the series on the right is simply the expansion of the exponential e^{hD}. Hence, we can write (18) in the form

$$Ef(x) = e^{hD} f(x)$$

from which we infer the operational equivalences

$$(19) \qquad E = e^{hD}$$

$$(20) \qquad \Delta = E - 1 = e^{hD} - 1$$

Next, we introduce the integration operator I by writing

$$If(x) = \int_x^{x+h} f(x)\, dx$$

Then

$$IDf(x) = If'(x) = \int_x^{x+h} f'(x)\, dx = f(x + h) - f(x) = \Delta f(x)$$

and if $F(x)$ is any antiderivative of $f(x)$,

$$DIf(x) = D \int_x^{x+h} f(x)\, dx = D[F(x + h) - F(x)]$$

$$= f(x + h) - f(x) = \Delta f(x)$$

Hence D and I commute with each other, and we have the further equivalences

(21) $$ID = DI = \Delta$$

We are now in a position to give an operational derivation of the famous **Euler-Maclaurin summation formula**:

(22) $$\sum_{i=0}^{n} f_i = \frac{1}{h} \int_{x_0}^{x_n} f(x)\, dx + \tfrac{1}{2}(f_0 + f_n) + \sum_{i=1}^{\infty} \frac{B_{2i}}{(2i)!} h^{2i-1}(f_n^{(2i-1)} - f_0^{(2i-1)})$$

where the B's are the Bernoulli numbers, $B_2 = \tfrac{1}{6}$, $B_4 = -\tfrac{1}{30}, \ldots$, to be defined below. We begin by writing, with the aid of Eq. (21),

$$h\, \Delta f(x) = hDIf(x)$$

or, replacing Δ by its equivalent from Eq. (20),

$$h(e^{hD} - 1)f(x) = hDIf(x)$$

or further,

(23) $$hf(x) = \frac{hD}{e^{hD} - 1} If(x)$$

It is now necessary to expand the fractional operator $hD/(e^{hD} - 1)$ in a power series in hD. This can be done in various ways, but perhaps the simplest is to replace e^{hD} by its series equivalent and then make use of the method of undetermined coefficients. Thus we have

$$\frac{hD}{(1 + hD + h^2D^2/2! + h^3D^3/3! + \cdots) - 1}$$

$$= a_0 + a_1 hD + \frac{a_2}{2!} h^2D^2 + \frac{a_3}{3!} h^3D^3 + \cdots$$

where the a's are coefficients to be determined to make this expression an identity. Simplifying the fraction on the left and then clearing of fractions gives us, further,

$$1 = \left(1 + \frac{hD}{2!} + \frac{h^2D^2}{3!} + \frac{h^3D^3}{4!} + \cdots\right)\left(a_0 + a_1 hD + \frac{a_2}{2!} h^2D^2 + \frac{a_3}{3!} h^3D^3 + \cdots\right)$$

Now, multiplying the two series and equating the coefficients of like powers of hD on the two sides of this identity, we obtain

$$a_0 = 1 \qquad \frac{a_0}{2!} + a_1 = 0 \qquad \frac{a_0}{3!} + \frac{a_1}{2!} + \frac{a_2}{2!} = 0$$

$$\frac{a_0}{4!} + \frac{a_1}{3!} + \frac{a_2}{2!\,2!} + \frac{a_3}{3!} = 0$$

$$\frac{a_0}{5!} + \frac{a_1}{4!} + \frac{a_2}{3!\,2!} + \frac{a_3}{2!\,3!} + \frac{a_4}{4!} = 0$$

· ·

from which we find without difficulty

$$a_0 = 1,\ a_1 = -\tfrac{1}{2},\ a_2 = \tfrac{1}{6},\ a_3 = 0,\ a_4 = -\tfrac{1}{30},\ a_5 = 0,\ \dots$$

The function $z/(e^z - 1)$ occurs in numerous applications, and the coefficients $\{a_i\}$ in its expansion have many interesting properties. These coefficients are ordinarily referred to as the **Bernoulli numbers** $\{B_i\}$,† and formulas have been developed which give them explicitly for any value of i. Except for B_1, all B's with odd subscripts are zero. The first few nonzero B's are

$$B_0 = 1 \qquad B_1 = -\tfrac{1}{2} \qquad B_2 = \tfrac{1}{6} \qquad B_4 = -\tfrac{1}{30} \qquad B_6 = \tfrac{1}{42} \qquad B_8 = -\tfrac{1}{30}$$

$$B_{10} = \frac{5}{66} \qquad B_{12} = -\frac{691}{2{,}730} \qquad B_{14} = \frac{7}{6} \qquad B_{16} = -\frac{3{,}617}{510}$$

$$B_{18} = \frac{43{,}867}{798} \qquad B_{20} = -\frac{174{,}611}{330}$$

and, as these values suggest, B_{2n} becomes infinite as $n \to \infty$.

Returning now to Eq. (23), we can write

$$f(x) = \frac{1}{h}\frac{hD}{e^{hD} - 1}\, If(x) = \frac{1}{h}\left[\sum_{i=0}^{\infty} \frac{B_i}{i!}(hD)^i\right] If(x)$$

or, detaching the first term from the series and factoring hD from the remaining terms,

$$f(x) = \frac{1}{h}\left[1 + hD \sum_{i=1}^{\infty} \frac{B_i}{i!}(hD)^{i-1}\right] If(x)$$

$$= \frac{1}{h} If(x) + \left[\sum_{i=1}^{\infty} \frac{B_i}{i!}(hD)^{i-1}\right] DIf(x)$$

(24)
$$= \frac{1}{h}\int_{x}^{x+h} f(x)\,dx + \left[\sum_{i=1}^{\infty} \frac{B_i}{i!}(hD)^{i-1}\right] \Delta f(x)$$

---

† The notation for the Bernoulli numbers is not completely standardized. Some writers use the symbol B_n to denote the absolute, rather than the algebraic, value of the coefficient of z^n in the expansion of $z/(e^z - 1)$. Others define B_{2n-1} to be the absolute value of what we call B_{2n}.

Now let us evaluate Eq. (24) for $x = x_0, x_1, \ldots, x_{n-1}$ and add the results, recalling that $\Delta f_0 + \Delta f_1 + \cdots + \Delta f_{n-1} = f_n - f_0$:

$$f_0 = \frac{1}{h} \int_{x_0}^{x_1} f(x)\, dx + \left[\sum_{i=1}^{\infty} \frac{B_i}{i!} (hD)^{i-1} \right] \Delta f_0$$

$$f_1 = \frac{1}{h} \int_{x_1}^{x_2} f(x)\, dx + \left[\sum_{i=1}^{\infty} \frac{B_i}{i!} (hD)^{i-1} \right] \Delta f_1$$

$$\cdots\cdots\cdots\cdots\cdots\cdots\cdots\cdots\cdots\cdots\cdots\cdots\cdots\cdots\cdots$$

$$f_{n-1} = \frac{1}{h} \int_{x_{n-1}}^{x_n} f(x)\, dx + \left[\sum_{i=1}^{\infty} \frac{B_i}{i!} (hD)^{i-1} \right] \Delta f_{n-1}$$

$$\sum_{i=0}^{n-1} f_i = \frac{1}{h} \int_{x_0}^{x_n} f(x)\, dx + \left[\sum_{i=1}^{\infty} \frac{B_i}{i!} (hD)^{i-1} \right] (f_n - f_0)$$

Since $B_1 = -\frac{1}{2}$ and $B_3 = B_5 = B_7 = \cdots = 0$, the last formula can be simplified somewhat by detaching the first term from the sum on the right-hand side and then setting $i = 2j$ in the rest of the series:

$$\sum_{i=0}^{n-1} f_i = \frac{1}{h} \int_{x_0}^{x_n} f(x)\, dx - \tfrac{1}{2}(f_n - f_0) + \left[\sum_{j=1}^{\infty} \frac{B_{2j}}{(2j)!} h^{2j-1} D^{2j-1} \right] (f_n - f_0)$$

$$= \frac{1}{h} \int_{x_0}^{x_n} f(x)\, dx - \tfrac{1}{2}(f_n - f_0) + \sum_{j=1}^{\infty} \frac{B_{2j}}{(2j)!} h^{2j-1} (f_n^{(2j-1)} - f_0^{(2j-1)})$$

Finally, if we add f_n to both sides of this identity, we obtain formula (22), as required.†
If Eq. (22) is solved for the integral, we obtain

$$(25) \qquad \int_{x_0}^{x_n} f(x)\, dx = h \sum_{i=0}^{n} f_i - \frac{h}{2}(f_0 + f_n) - \sum_{j=1}^{\infty} \frac{B_{2j}}{(2j)!} h^{2j} (f_n^{(2j-1)} - f_0^{(2j-1)})$$

which is a fundamental formula of numerical integration. Equation (25) is especially adapted to the integration of functions defined by analytic expressions which can conveniently be differentiated. For functions defined only by a table of values it is usually more convenient to have an integration formula in which the "correction terms" are expressed as differences rather than as derivatives. To obtain such a formula

---

† All operational derivations are suggestive rather than rigorous, and ours is no exception. For a rigorous development of the Euler-Maclaurin formula, including the necessary remainder term, see, for example, Kenneth S. Miller, "An Introduction to the Calculus of Finite Differences and Difference Equations," pp. 105–110, Holt, New York, 1960.

from Eq. (25), we need only replace the derivatives $f_0', f_0''', \dots$ by means of Eqs. (5), (7), $\dots$ and the derivatives $f_n', f_n''', \dots$ by means of Eqs. (13), (15), $\dots$. This gives us

$$\int_{x_0}^{x_n} f(x)\,dx = h\left(\frac{f_0}{2} + f_1 + \cdots + f_{n-1} + \frac{f_n}{2}\right)$$

$$- \frac{h^2}{12}\left[\frac{1}{h}\left(\Delta f_{n-1} + \frac{\Delta^2 f_{n-2}}{2} + \frac{\Delta^3 f_{n-3}}{3} + \frac{\Delta^4 f_{n-4}}{4} + \cdots\right)\right.$$

$$\left. - \frac{1}{h}\left(\Delta f_0 - \frac{\Delta^2 f_0}{2} + \frac{\Delta^3 f_0}{3} - \frac{\Delta^4 f_0}{4} + \cdots\right)\right]$$

$$+ \frac{h^4}{720}\left[\frac{1}{h^3}(\Delta^3 f_{n-3} + \tfrac{3}{2}\Delta^4 f_{n-4} + \cdots)\right.$$

$$\left. - \frac{1}{h^3}(\Delta^3 f_0 - \tfrac{3}{2}\Delta^4 f_0 + \cdots)\right]$$

$$+ \cdots$$

(26)
$$= h(\tfrac{1}{2}f_0 + f_1 + \cdots + f_{n-1} + \tfrac{1}{2}f_n) - \frac{h}{12}(\Delta f_{n-1} - \Delta f_0)$$

$$- \frac{h}{24}(\Delta^2 f_{n-2} + \Delta^2 f_0) - \frac{19h}{720}(\Delta^3 f_{n-3} - \Delta^3 f_0)$$

$$- \frac{3h}{160}(\Delta^4 f_{n-4} + \Delta^4 f_0) - \cdots$$

which is known as **Gregory's formula of numerical integration**. In passing, we note that both (25) and (26) reduce to the well-known trapezoidal rule of integration if the correction terms are neglected.

EXAMPLE 2

Compute $\int_0^1 f(x)\,dx$ for the function defined by the following table:

x	$f(x)$	Δ	Δ^2	Δ^3	Δ^4
0.0	0.4698220				
		0.0144778			
0.2	0.4842998		−0.0004670		
		0.0140108		0.0000290	
0.4	0.4983106		−0.0004380		−0.0000024
		0.0135728		0.0000266	
0.6	0.5118834		−0.0004114		−0.0000023
		0.0131614		0.0000243	
0.8	0.5250448		−0.0003871		
		0.0127743			
1.0	0.5378191				

Using Eq. (26), with $n = 5$ and $h = 0.2$, we have at once

$$\int_0^1 f(x)\,dx = \frac{1}{5}\left(\frac{0.4698220}{2} + 0.4842998 + 0.4983106 + 0.5118834\right.$$

$$\left. + 0.5250448 + \frac{0.5378191}{2}\right)$$

$$- \frac{1}{60}(0.0127743 - 0.0144778) - \frac{1}{120}(-0.0003871 - 0.0004670)$$

$$- \frac{19}{3,600}(0.0000243 - 0.0000290) - \cdots$$

$$= 0.5047073$$

The integral in this problem is actually $\int_0^1 \log (2.95 + 0.5x)\,dx$ and its exact value is easily found to be 0.5047074 correct to seven decimal places. The approximate value of the integral is therefore in error by only 0.0000001.

1 From the data in the following table compute the first three derivatives of $\ln x$ at $x = 200$ and at $x = 205$, and check them against the exact values:

x	$\ln x$
200	5.29831737
201	5.30330491
202	5.30826770
203	5.31320598
204	5.31811999
205	5.32300998

2 Using the data of Exercise 1, compute the first three derivatives of $\ln x$ at $x = 200.5$ and at $x = 204.5$, and check them against the exact values.

3 From the data in the following table compute $f'(x)$ at $x = 0$ and at $x = 10$:

x	$f(x)$
0	1.000
1	1.221
3	1.822
4	2.226
7	4.055
10	7.389

4 Compute $\int_0^1 \sin x^2\,dx$ by the trapezoidal rule with $h = 0.1$, and compare your answer with the results given by Eq. (25):
  (a) When only the first-derivative correction terms are included.
  (b) When the first- and third-derivative correction terms are included.

5 Work Exercise 4 using Eq. (26) and the corresponding difference correction terms.

6 Use Eq. (22) to obtain a formula for the sum of the cubes of the first n integers.

7 Use Eq. (22) to obtain a formula for the sum of the fourth powers of the first n integers.

8 Using the trapezoidal rule, compute $\int_0^x e^{-x^2}\,dx$ for values of x at intervals of $h = 0.1$ from 0 to 1.

9 Using the trapezoidal rule, compute $\int_x^1 (\sin x)/x\,dx$ for values of x at intervals of $h = 0.1$ from 0 to 1.

10 Show that when $n = 2$ and differences of order higher than the second are neglected, Eq. (26) reduces to

$$\int_{x_0}^{x_2} f(x)\, dx = \frac{h}{3}\,(f_0 + 4f_1 + f_2)$$

By applying this to successive pairs of tabular intervals, establish **Simpson's rule**:

$$\int_{x_0}^{x_n} f(x)\, dx = \frac{h}{3}\,(f_0 + 4f_1 + 2f_2 + \cdots + 2f_{n-2} + 4f_{n-1} + f_n) \qquad n \text{ even}$$

11 By making two applications of the formula of Exercise 10, establish the following formula for the numerical evaluation of double integrals:

$$\int_{y_0}^{y_2} \int_{x_0}^{x_2} f(x,y)\, dx\, dy = \frac{hk}{9}\,[(f_{00} + f_{02} + f_{20} + f_{22})$$

$$+\, 4(f_{01} + f_{10} + f_{12} + f_{21}) + 16f_{11}]$$

where h and k are, respectively, the intervals at which x and y are tabulated, and $f_{ij} = f(x_0 + ih, y_0 + jk)$. Give the generalization of this result to integrals of the form

$$\int_{y_0}^{y_{2n}} \int_{x_0}^{x_{2m}} f(x,y)\, dx\, dy$$

12 Using Eq. (22), establish the relation

$$\sum_{i=c}^{\infty} \frac{1}{i^2} \sim \frac{1}{c} + \frac{1}{2c^2} + \frac{1}{6c^3} - \frac{1}{30c^5} + \cdots + \frac{B_{2i}}{c^{2i+1}} + \cdots$$

(Note that the equality sign is not used between the two members of the given relation. In fact, although B_{2n} at first decreases with n, it increases more rapidly than c^{2n+1} for any c as $n \to \infty$; hence the series on the right diverges and certainly cannot be said to equal the sum on the left. Interestingly enough, however, the series on the right is a good approximation to the sum if it is terminated at the point where its terms are a minimum. Divergent series with this property are usually referred to as **asymptotic series**.)

13 By writing

$$\sum_{i=1}^{\infty} \frac{1}{i^2} = \sum_{i=1}^{c-1} \frac{1}{i^2} + \sum_{i=c}^{\infty} \frac{1}{i^2}$$

use the result of Exercise 12 to approximate $\sum_{i=1}^{\infty} (1/i^2)$ if (**a**) $c = 5$ and (**b**) $c = 10$. Check your answers against the exact value of the sum, which is $\pi^2/6$.

14 Using the methods outlined in Exercises 12 and 13, evaluate the sum

$$1 + \frac{1}{3^2} + \frac{1}{5^2} + \frac{1}{7^2} + \cdots$$

and check your answer against the exact value of $\pi^2/8$.

15 What is $\sum_{i=1}^{1,000} 1/i$?

16 Establish the operational equivalence $D = (1/h) \ln (1 + \Delta)$, and then by expanding the right member into an infinite series, establish formulas (5) to (8). *Hint:* Begin with Eq. (20).

17 Establish the operational equivalence $D = (2/h) \sinh^{-1} (\delta/2)$, and use it to develop a central-difference formula for numerical differentiation. *Hint:* Begin with Eq. (14), Sec. 4.1, and Eq. (19).

18 If the forward Gregory-Newton interpolation formula is integrated from $r = 0$ to $r = n$, a formula of numerical integration of the form

$$\int_{x_0}^{x_n} f(x)\, dx = c_0 f_0 + c_1 \Delta f_0 + c_2 \Delta^2 f_0 + \cdots + c_n \Delta^n f_0$$

is obtained. Determine the values of the c's when $n = 2, 3$, and 4.

19 If Lagrange's interpolation formula is integrated from x_0 to x_n, a formula of numerical integration of the form

$$\int_{x_0}^{x_n} f(x)\, dx = c_0 f_0 + c_1 f_1 + \cdots + c_{n-1} f_{n-1} + c_n f_n$$

is obtained. Determine the values of the c's when $n = 2$. If $n = 2$ and the x's are equally spaced, show that this formula becomes Simpson's rule. (For arbitrary n the formula obtained by this method is known as the **Newton-Cotes integration formula**.)

20 Obtain the values of the c's in the formula of Exercise 19 if the x's are equally spaced and (a) $n = 3$, (b) $n = 4$.

21 Show that with the exception of B_1, all the Bernoulli numbers with odd subscripts are zero. *Hint:* show that the function

$$\frac{z}{e^z - 1} + \frac{z}{2}$$

is an even function.

4.4 The Numerical Solution of Differential Equations

One of the most important applications of finite differences is to the numerical solution of differential equations which because of their complexity cannot be solved by exact methods. Many procedures are available for doing this,[†] some of considerable generality, others especially adapted to equations of a particular form. Of the many methods which have been devised, we shall present only the *method of Milne* and the *Runge-Kutta method*. These can be applied to simultaneous differential equations as well as to single equations of any order and are therefore adequate for almost any problem one is likely to encounter.

The fundamental problem is to find the solution of the first-order differential equation

$$(1) \qquad \frac{dy}{dx} = f(x, y)$$

which satisfies the initial condition $y = y_0$ when $x = x_0$. We do not, of course, expect to find an equation for the solution. Instead, our object is merely to plot or tabulate the solution curve point by point, beginning at (x_0, y_0) and continuing thereafter at selected values of x, usually equally spaced, until the solution has been extended over the required range.

---

† See, for instance, H. Levy and E. A. Baggott, "Numerical Studies in Differential Equations," vol. 1, Watts, London, 1934, and W. E. Milne, "Numerical Solutions of Differential Equations," Wiley, New York, 1953.

To develop **Milne's method** we begin with Eq. (1), Sec. 4.3, written in terms of y rather than f, and evaluate it for $r = 1, 2, 3$, and 4, getting

$$y'_1 = \frac{1}{h}(\Delta y_0 + \tfrac{1}{2}\Delta^2 y_0 - \tfrac{1}{6}\Delta^3 y_0 + \tfrac{1}{12}\Delta^4 y_0 + \cdots)$$

$$y'_2 = \frac{1}{h}(\Delta y_0 + \tfrac{3}{2}\Delta^2 y_0 + \tfrac{1}{3}\Delta^3 y_0 - \tfrac{1}{12}\Delta^4 y_0 + \cdots)$$

$$y'_3 = \frac{1}{h}(\Delta y_0 + \tfrac{5}{2}\Delta^2 y_0 + \tfrac{11}{6}\Delta^3 y_0 + \tfrac{1}{4}\Delta^4 y_0 + \cdots)$$

$$y'_4 = \frac{1}{h}(\Delta y_0 + \tfrac{7}{2}\Delta^2 y_0 + \tfrac{13}{3}\Delta^3 y_0 + \tfrac{25}{12}\Delta^4 y_0 + \cdots)$$

or, neglecting differences beyond the fourth and replacing the remaining differences by their equivalent expressions in terms of the successive functional values [Eqs. (15a), Sec. 4.1],

(2)

$$y'_1 = \frac{1}{12h}(-3y_0 - 10y_1 + 18y_2 - 6y_3 + y_4)$$

$$y'_2 = \frac{1}{12h}(y_0 - 8y_1 + 8y_3 - y_4)$$

$$y'_3 = \frac{1}{12h}(-y_0 + 6y_1 - 18y_2 + 10y_3 + 3y_4)$$

$$y'_4 = \frac{1}{12h}(3y_0 - 16y_1 + 36y_2 - 48y_3 + 25y_4)$$

Now if we subtract the second equation in the set (2) from twice the sum of the first and third equations and solve the result for y_4, we obtain

$$y_4 = y_0 + \frac{4h}{3}(2y'_1 - y'_2 + 2y'_3)$$

or, in more general terms,

(3)
$$y_{n+1} = y_{n-3} + \frac{4h}{3}(2y'_{n-2} - y'_{n-1} + 2y'_n)$$

If we know the values of y and y' down to and including their values at x_n, Eq. (3) enables us to "reach out" one step further and compute y_{n+1}. With y_{n+1} known, we can then return to the given differential equation (1) and compute y'_{n+1}. Then using Eq. (3) again, with n increased by 1 throughout, we can find y_{n+2}, and so on, step by step, until the solution has been extended over the desired range. All that remains is to devise a means of finding enough y's and y''s to get the process started.

One possibility is to begin the tabulation of y by expanding it in a Taylor series about the point $x = x_0$:

(4)
$$y = y_0 + y'_0(x - x_0) + \frac{y''_0}{2!}(x - x_0)^2 + \frac{y'''_0}{3!}(x - x_0)^3 + \cdots$$

The value of y_0 is, of course, given. The value of y_0' can be found at once by substituting x_0 and y_0 into the given differential equation (1). To find the second derivative we need only differentiate the given equation, getting

$$(5) \qquad y'' = \frac{\partial f}{\partial x} + \frac{\partial f}{\partial y} y'$$

Since $f(x,y)$ is a given function, its partial derivatives are known and become definite numbers when x_0 and y_0 are substituted into them. Moreover, the value of y' at (x_0,y_0) has already been found, and thus (5) furnishes the value of y_0''. Similarly, differentiating (5) and evaluating the result at (x_0,y_0) will give y_0''', and so on. In this way the first few terms of the expansion of y around the point (x_0,y_0) can be constructed. In especially favorable cases the general term of the series (4) can be found and the region of convergence established. When this happens, (4) is the required solution, and we need look no further. In general, however, successive differentiation of $f(x,y)$ becomes too complicated to continue, or the resulting series converges too slowly to be of practical value, and we must fall back on Milne's or some similar method.

With (4) available as a representation of y in the neighborhood of $x = x_0$, we can set $x = x_0 + h \equiv x_1$ and calculate y_1. Similarly, setting $x = x_0 + 2h$ and $x_0 + 3h$, we can find y_2 and y_3. Then substituting (x_1,y_1), (x_2,y_2), and (x_3,y_3) into the given differential equation, we can compute y_1', y_2', and y_3' without difficulty. With these values we are then in a position to begin the step-by-step solution of the differential equation by means of Eq. (3).

From the preceding discussion it is clear that Eq. (3) is in general adequate for the step-by-step solution of $y' = f(x,y)$. However, as a precaution against errors of various kinds, it is desirable to have a second, independent formula into which y_{n+1} can be substituted as a check. To obtain such an equation we return to (2) and add 4 times the third equation to the sum of the second and fourth and solve the resulting equation for y_4, getting

$$y_4 = y_2 + \frac{h}{3}(y_2' + 4y_3' + y_4')$$

or, in more general terms,

$$(6) \qquad y_{n+1} = y_{n-1} + \frac{h}{3}(y_{n-1}' + 4y_n' + y_{n+1}')$$

This formula cannot be used as a formula of extrapolation, since it involves y_{n+1}', which cannot be found unless y_{n+1} is already known. However, after y_{n+1} has been found by means of (3), y_{n+1}' can be calculated, and enough information is then available to permit the use of (6). If the value of y_{n+1}, as given by (6), agrees with the value found from (3), we are ready to move on to the calculation of y_{n+2}. On the other hand, if the two values of y_{n+1} do not agree, we must use the second value of y_{n+1} to compute a new value of y_{n+1}', substitute these into (6), and continue the process until two successive values of y_{n+1} are in agreement. When this happens, we are ready to continue the tabulation of y by returning to (3) and determining an initial estimate of y_{n+2}.

Formulas like (3), which express a new value exclusively in terms of quantities already found, are known as **open formulas** or **predictor formulas**. Those, like (6), which express a new value in terms of one or more additional new quantities and which, therefore, can be used only for purposes of checking and refining are known as **closed formulas** or **corrector formulas**.

The method of Milne is readily extended to the solution of simultaneous and higher-order equations. For instance, if we have the two equations

$$y' = f(x,y,z) \qquad \text{and} \qquad z' = g(x,y,z)$$

with the initial conditions $y = y_0$, $z = z_0$ when $x = x_0$, and if by independent means we have calculated (y_1,y_2,y_3), (z_1,z_2,z_3), and the related quantities (y'_1,y'_2,y'_3) and (z'_1,z'_2,z'_3), then, using Eq. (3) and an identical version of it with z replacing y, we can compute y_4 and z_4. After that, we can compute y'_4 and z'_4 from the differential equations and again use (3) to obtain y_5 and z_5, and so on, as far as desired. Of course, the closed formula (6) can be used to check and correct both y_{n+1} and z_{n+1} if and when this is deemed necessary.

The application of Milne's method to equations of higher order is now immediate, since such an equation can always be replaced by a system of simultaneous first-order equations. For instance, $y'' = g(x,y,y')$ is equivalent to the system

$$y' = z \qquad z' = g(x,y,z)$$

which is just a special case, with $f(x,y,z) \equiv z$, of the general problem of two simultaneous first-order equations.

The Runge-Kutta method differs from Milne's method in several significant respects. In the first place, it requires no predetermination of a set of starting values and hence is completely self-contained. Second, it does not require the values of x at which the solution is being tabulated to be equally spaced; hence, the interval between successive values of x can be varied throughout the process, as time and accuracy may require.

The Runge-Kutta method can be thought of as a generalization of the following extremely simple (and quite inaccurate) procedure. If one is given the first-order differential equation

$$(1) \qquad \frac{dy}{dx} = f(x,y)$$

and the initial conditions $y = y_0$ when $x = x_0$, the value of y at $x_1 = x_0 + \Delta x$, say $y_1 = y_0 + \Delta y$, can be approximated by using the usual differential estimate of the increment Δy, namely,

$$\Delta y \doteq \frac{dy}{dx}\bigg|_{x_0, y_0} \Delta x = f(x_0, y_0)\, \Delta x$$

With this value for Δy available, an approximate value for $y_1 = y_0 + \Delta y$, namely

$$(7) \qquad y_1 = y_0 + \frac{dy}{dx}\bigg|_{x_0, y_0} \Delta x = y_0 + f(x_0, y_0)\, \Delta x$$

is determined, and the process can be repeated to obtain $y_2, y_3, \ldots$.

On the other hand, having a first approximation to y_1, one can compute dy/dx at the new point (x_1, y_1) and then use the *average* of the slopes at (x_0, y_0) and (x_1, y_1) to compute a (presumably) more accurate estimate of Δy, and hence of y_1, before attempting to calculate y_2. This method yields the value

$$(8) \quad y_1 = y_0 + \frac{1}{2}\left(\frac{dy}{dx}\Big|_{x_0, y_0} + \frac{dy}{dx}\Big|_{x_1, y_1}\right)\Delta x = y_0 + \tfrac{1}{2}[f(x_0, y_0) + f(x_1, y_1)]\,\Delta x$$

Still another possibility is to recompute Δy and y_1 by using not the slope at (x_0, y_0) or the average of the slopes at (x_0, y_0) and (x_1, y_1) but the slope at the midpoint of the segment determined by (x_0, y_0) and (x_1, y_1), namely, $(x_0 + \Delta x/2, y_0 + \Delta y/2)$. This yields the (presumably) refined value

$$(9) \qquad y_1 = y_0 + \frac{dy}{dx}\Big|_{\substack{x_0 + \Delta x/2 \\ y_0 + \Delta y/2}} \Delta x = y_0 + f\left(x_0 + \frac{\Delta x}{2}, y_0 + \frac{\Delta y}{2}\right)\Delta x$$

$$= y_0 + f\left[x_0 + \frac{\Delta x}{2}, y_0 + \tfrac{1}{2}f(x_0, y_0)\,\Delta x\right]\Delta x$$

The procedure based on Eq. (7) is known as **Euler's method**; that based on Eq. (8) is known as the **modified Euler method**; and that based on Eq. (9) is known as **Runge's method**.

In the Runge-Kutta method, three (or more) preliminary estimates of Δy are computed and then the value of Δy which is finally used to determine y_1 is taken to be a linear combination of all of these. Specifically, in **Kutta's third-order method** the three estimates of Δy are

$$k_1 \equiv \Delta_1 y = f(x_0, y_0)\,\Delta x \equiv f(x_0, y_0)h$$

which is just the estimate of Δy used in Euler's method;

$$k_2 \equiv \Delta_2 y = f(x_0 + p\,\Delta x, y_0 + p\,\Delta_1 y)\,\Delta x \equiv f(x_0 + ph, y_0 + pk_1)h$$

which is just like the estimate used in Runge's method except that instead of being evaluated at the midpoint of the segment determined by (x_0, y_0) and (x_1, y_1), the slope is evaluated at a point of the segment yet to be determined; and

$$k_3 \equiv \Delta_3 y = f(x_0 + q\,\Delta x, y_0 + r\,\Delta_2 y + \overline{q - r}\,\Delta_1 y)\,\Delta x$$
$$\equiv f(x_0 + qh, y_0 + rk_2 + \overline{q - r}\,k_1)h$$

Finally, the value of Δy which is actually used in the determination of y_1 is taken to be

$$\Delta y = ak_1 + bk_2 + ck_3$$

where a, b, c are parameters which, like the parameters p, q, r, are to be chosen to ensure the highest possible accuracy in Δy.

The determination of a, b, c, p, q, r is accomplished in the following way.† First, the expansion of $\Delta y = y - y_0$ in terms of powers of $\Delta x \equiv h$ is obtained, using implicit differentiation of the given equation (1) to compute the derivatives which must be evaluated to find the coefficients. Then $k_1 \equiv \Delta_1 y$, $k_2 \equiv \Delta_2 y$, and $k_3 \equiv \Delta_3 y$ are also expanded in series of powers of h, and, these expansions are used to express

---

† The details of this development can be found in the third edition of this book, pp. 112–114.

$\Delta y = ak_1 + bk_2 + ck_3$ as a power series in h. Finally a, b, c, p, q, r are chosen to make the two series for Δy agree as far as the terms involving $h^3 \equiv (\Delta x)^3$, so that the difference between the true value of Δy (given by the first series) and the estimated value of y (given by the second series) will be of the order of $h^4 \equiv (\Delta x)^4$. Equating the constant terms and the coefficients of h, h^2, and h^3 in the two series gives the conditions

(10)
$$a + b + c = 1 \qquad bp + cq = \tfrac{1}{2}$$
$$\frac{bp^2 + cq^2}{2} = \tfrac{1}{6} \qquad cpr = \tfrac{1}{6}$$

The first three of these equations are linear in a, b, and c and can easily be solved to express a, b, and c in terms of p and q. The fourth equation can then be used to express r in terms of p and q also:

(11)
$$a = \frac{6pq - 3(p + q) + 2}{6pq} \qquad b = \frac{2 - 3q}{6p(p - q)}$$
$$c = \frac{2 - 3p}{6q(q - p)} \qquad r = \frac{q(q - p)}{p(2 - 3p)}$$

Since p and q are arbitrary, we thus have a two-parameter family of formulas which can be used for the step-by-step solution of the equation $y' = f(x,y)$ with an error which is of the order of $h^4 \equiv (\Delta x)^4$.

The following particular cases are worthy of note:

Case I: $a = \tfrac{1}{4}$ $b = 0$ $c = \tfrac{3}{4}$ $p = \tfrac{1}{3}$ $q = r = \tfrac{2}{3}$
$$\Delta y = \tfrac{1}{4}(k_1 + 3k_3)$$

where
$$k_1 = f(x_0, y_0)h$$
$$k_2 = f(x_0 + \tfrac{1}{3}h, y_0 + \tfrac{1}{3}k_1)h$$
$$k_3 = f(x_0 + \tfrac{2}{3}h, y_0 + \tfrac{2}{3}k_2)h$$

Case II: $a = \tfrac{1}{4}$ $b = c = \tfrac{3}{8}$ $p = q = r = \tfrac{2}{3}$
$$\Delta y = \tfrac{1}{8}(2k_1 + 3k_2 + 3k_3)$$

where
$$k_1 = f(x_0, y_0)h$$
$$k_2 = f(x_0 + \tfrac{2}{3}h, y_0 + \tfrac{2}{3}k_1)h$$
$$k_3 = f(x_0 + \tfrac{2}{3}h, y_0 + \tfrac{2}{3}k_2)h$$

The values of the parameters in Case II cannot be obtained from Eqs. (11), since $p = q$, but can be checked directly in Eqs. (10).

The foregoing analysis can be extended without difficulty to yield step-by-step solution procedures in which the error is of the order of $h^5 \equiv (\Delta x)^5$. In particular, the following two sets of formulas are quite useful:

Case III: $\Delta y = \tfrac{1}{6}(k_1 + 2k_2 + 2k_3 + k_4)$

where
$$k_1 = f(x_0, y_0)h$$
$$k_2 = f(x_0 + \tfrac{1}{2}h, y_0 + \tfrac{1}{2}k_1)h$$
$$k_3 = f(x_0 + \tfrac{1}{2}h, y_0 + \tfrac{1}{2}k_2)h$$
$$k_4 = f(x_0 + h, y_0 + k_3)h$$

Case IV: $$\Delta y = \tfrac{1}{8}(k_1 + 3k_2 + 3k_3 + k_4)$$

where
$$k_1 = f(x_0, y_0)h$$
$$k_2 = f(x_0 + \tfrac{1}{3}h,\ y_0 + \tfrac{1}{3}k_1)h$$
$$k_3 = f(x_0 + \tfrac{2}{3}h,\ y_0 + k_2 - \tfrac{1}{3}k_1)h$$
$$k_4 = f(x_0 + h,\ y_0 + k_3 - k_2 + k_1)h$$

The solution process based on III is often referred to specifically as *the* **Runge-Kutta method**.

Any of the Runge-Kutta formulas can be used to solve simultaneous and hence higher-order differential equations. For instance, using III, we can tabulate the solution of the system of equations

$$\frac{dy}{dx} = f(x, y, z) \qquad \frac{dz}{dx} = g(x, y, z)$$

at intervals of $\Delta x \equiv h$ by computing

$$k_1 \equiv \Delta_1 y = f(x_0, y_0, z_0)h$$
$$l_1 \equiv \Delta_1 z = g(x_0, y_0, z_0)h$$
$$k_2 \equiv \Delta_2 y = f(x_0 + \tfrac{1}{2}h,\ y_0 + \tfrac{1}{2}k_1,\ z_0 + \tfrac{1}{2}l_1)h$$
$$l_2 \equiv \Delta_2 z = g(x_0 + \tfrac{1}{2}h,\ y_0 + \tfrac{1}{2}k_1,\ z_0 + \tfrac{1}{2}l_1)h$$
$$k_3 \equiv \Delta_3 y = f(x_0 + \tfrac{1}{2}h,\ y_0 + \tfrac{1}{2}k_2,\ z_0 + \tfrac{1}{2}l_2)h$$
$$l_3 \equiv \Delta_3 z = g(x_0 + \tfrac{1}{2}h,\ y_0 + \tfrac{1}{2}k_2,\ z_0 + \tfrac{1}{2}l_2)h$$
$$k_4 \equiv \Delta_4 y = f(x_0 + h,\ y_0 + k_3,\ z_0 + l_3)h$$
$$l_4 \equiv \Delta_4 z = g(x_0 + h,\ y_0 + k_3,\ z_0 + l_3)h$$

and then using the formulas

$$\Delta y = \tfrac{1}{6}(k_1 + 2k_2 + 2k_3 + k_4)$$
$$\Delta z = \tfrac{1}{6}(l_1 + 2l_2 + 2l_3 + l_4)$$

If the various increments are computed in the indicated order, each involves only quantities which have previously been calculated.

EXAMPLE 1

Tabulate the solution of $y' = x^2 + y$ at intervals of $h = 0.1$ if $y = -1$ when $x = 0$. Using the Runge-Kutta formulas III for the first increment, we have

$$k_1 = -0.1000 \qquad k_2 = -0.1048 \qquad k_3 = -0.1050 \qquad k_4 = -0.1095$$

and $$\Delta y = -0.1048$$

Hence $$y_1 = y_0 + \Delta y = -1.1048$$

For the second increment we have, similarly,

$$k_1 = -0.1095 \qquad k_2 = -0.1137 \qquad k_3 = -0.1139 \qquad k_4 = -0.1179$$

and $$\Delta y = -0.1138$$

Hence $$y_2 = y_1 + \Delta y = -1.2186$$

For the third increment, we have

$$k_1 = -0.1179 \quad\quad k_2 = -0.1215 \quad\quad k_3 = -0.1217 \quad\quad k_4 = -0.1250$$

and

$$\Delta y = -0.1216$$

Hence

$$y_3 = y_2 + \Delta y = -1.3402$$

This process can, of course, be continued as far as desired. However, we shall calculate just one more value of y, this time using Milne's method, which can now be applied since we have values for y_0, y_1, y_2, and y_3. To do this, we must first compute y_0', y_1', y_2', y_3' from the differential equation, getting

$$y_0' = -1.0000 \quad\quad y_1' = -1.0948 \quad\quad y_2' = -1.1786 \quad\quad y_3' = -1.2502$$

With these values, we can now use the open formula, Eq. (3), to obtain

$$y_4 = -1.4682$$

Using this value, we find from the differential equation that

$$y_4' = -1.3082$$

With this we can use Eq. (3) again to find y_5, or we can first use the closed formula, Eq. (6), to check y_4 before continuing. In this case Eq. (6) also gives us $y_4 = -1.4682$, which is a good check on the accuracy of our calculations.

The differential equation $y' = x^2 + y$ is so simple that it can be solved exactly without recourse to numerical methods, and by the methods of Chap. 1 or 2 we find at once that the required solution is

$$y = e^x - x^2 - 2x - 2$$

For $x = 1, 2, 3, 4$ this gives us the correct values

$$y_1 = -1.1048 \quad\quad y_2 = -1.2186 \quad\quad y_3 = -1.3401 \quad\quad y_4 = -1.4682$$

The values we computed for y_1, y_2, and y_4 agree with these to four decimal places, and the value we computed for y_3 differs from the correct value by only 1 in the fourth place.

EXERCISES

1   Explain how Milne's method can be used to solve a differential equation of the form $y''' = f(x,y,y',y'')$.

2   Using the Runge-Kutta method III, find y_6 and y_9 in Example 1 without finding y_5, y_7, or y_8. How do these values compare with the correct values?

3   Using Kutta's third-order approximation I, tabulate the solution of the equation $y' = x - y$ at intervals of $h = 0.1$ if $y = 1$ when $x = 1$.

4   Using Kutta's third-order approximation II, tabulate the solution of the equation $y' = x + y$ at intervals of $h = 0.1$ if $y = 1$ when $x = 0$.

5   Using the Runge-Kutta method III, tabulate the function $y = e^{-x^2}$ for $x = 0.0$, 0.1, 0.2, 0.3, 0.4, and 0.5. How do your answers compare with the exact values of this function? *Hint:* Find a differential equation satisfied by y.

6   Using the Runge-Kutta method III, evaluate $\int_0^x e^{-t^2}\, dt$ for $x = 0.0, 0.2, 0.4, 0.6$, 0.8, and 1.0.

7   Using the Runge-Kutta method III, find y_1, y_2, y_3, z_1, z_2, z_3 for the solution of the system

$$\frac{dy}{dx} = x + z \quad\quad\quad \frac{dz}{dx} = x - y$$

given $h = 0.1$, and $y = 0$, $z = 1$ when $x = 0$.

8   Using Milne's method, tabulate the solution of the equation $y' = x + y$ at intervals of $h = 0.1$ if $y = 1$ when $x = 0$.

9 Work Exercise 3 using the open formula

$$y_{n+1} = y_n + h(y'_n + \tfrac{1}{2}\Delta y'_{n-1} + \tfrac{5}{12}\Delta^2 y'_{n-2} + \tfrac{3}{8}\Delta^3 y'_{n-3} + \tfrac{251}{720}\Delta^4 y'_{n-4})$$

and the closed formula

$$y_{n+1} = y_n + h(y'_{n+1} - \tfrac{1}{2}\Delta y'_n - \tfrac{1}{12}\Delta^2 y'_{n-1} - \tfrac{1}{24}\Delta^3 y'_{n-2} - \tfrac{19}{720}\Delta^4 y'_{n-3})$$

(These equations constitute the so-called **Adams-Bashforth method** for the numerical solution of differential equations.)

10 Using Milne's method, tabulate the solution of the system

$$\frac{dy}{dx} = y^2 + xz \qquad \frac{dz}{dx} = x^2 + yz$$

at intervals of $h = 0.1$, given $y_0 = 0$, $z_0 = 1$.

11 Explain how the Adams-Bashforth method described in Exercise 9 can be extended to systems of differential equations and to differential equations of higher order.

12 Eliminate the differences from the formulas of Exercise 9 and express y_{n+1} directly in terms of y_n and the various values of y'.

13 Using the open formula

$$y_{n+1} = 2y_n - y_{n-1} + h^2(y''_n + \tfrac{1}{12}\Delta^2 y''_{n-2})$$

and the closed formula

$$y_{n+1} = 2y_n - y_{n-1} + h^2(y''_n + \tfrac{1}{12}\Delta^2 y''_{n-1})$$

tabulate the solution of the equation $y'' = x + y$ at intervals of $h = 0.1$, given that $y_0 = 1$, $y'_0 = 0$. How do your results compare with the exact solution?

14 Work Exercise 4 using the open formula

$$y_{n+1} = y_n + \frac{h}{12}(23y'_n - 16y'_{n-1} + 5y'_{n-2})$$

and the closed formula

$$y_{n+1} = y_n + \frac{h}{12}(5y'_{n+1} + 8y'_n - y'_{n-1})$$

(These equations constitute the so-called **Adams-Moulton method** for the numerical solution of differential equations.)

15 Set up the Kutta third-order approximation corresponding to the values $p = \tfrac{1}{2}$, $q = 1$ and show that it reduces to Simpson's rule when $f(x, y)$ is independent of y.

16 By expanding each term in Eq. (3) around the point $x = x_{n-3}$, show that the principal part of the error in Milne's open formula is $\tfrac{14}{45}h^5 y^v_{n-3}$. What is the principal part of the error in Milne's closed formula?

17 Find the equation of the polynomial of minimum degree for which y and y' take on prescribed values (y_0, y'_0) and (y_1, y'_1) at $x = 0$ and at $x = h$. What is the value of y_2 given by this polynomial? How might this result be used to carry out the step-by-step integration of a differential equation of the form $y' = f(x, y)$? How might an accompanying closed formula be obtained?

18 Find the equation of the polynomial of minimum degree for which y and y'' take on prescribed values (y_0, y''_0) and (y_1, y''_1) at $x = 0$ and at $x = h$. What is the value of y_2 given by this polynomial? How might this result be used to carry out the step-by-step integration of a differential equation of the form $y'' = f(x, y)$? How might an accompanying closed formula be obtained?

4.5 Difference Equations

The many similarities we have already observed between the calculus of finite differences and the ordinary, or infinitesimal, calculus suggest that there should be a theory of difference equations roughly paralleling the theory of differential equations, and this is indeed the case. However, in the study of difference equations we do not ordinarily consider equations of the form

(1) $$f(\Delta)y = \phi(x)$$

as might be expected by analogy with the differential equation

(2) $$f(D)y = \phi(x)$$

but rather equations of the form

(3) $$f(E)y = \phi(x)$$

This, of course, is simply a matter of notational convenience, since, using the operational equivalence $\Delta = E - 1$, any function of Δ can be transformed at once into a function of E, and vice versa. In this section we shall restrict ourselves to the case of a single **linear, constant-coefficient difference equation**

(4) $$(a_0 E^r + a_1 E^{r-1} + \cdots + a_{r-1} E + a_r)y = \phi(x)$$

where $\phi(x)$ is a linear combination of terms or products of terms from the set

$$k^x \quad \cos kx \quad \sin kx \quad k \text{ a constant}$$

and $\qquad\qquad x^n \qquad n$ a nonnegative integer

Since the substitution $t = hx$ will transform a function of t tabulated at intervals of h into a function of x tabulated at unit intervals, it is clearly no restriction to assume $h = 1$, so that invariably $Ef(x) = f(x + 1)$, and we shall do this throughout the present section. We shall base our solution of Eq. (4) primarily on analogy with linear, constant-coefficient differential equations, and such theoretical results as we may need we shall cite without proof.

In Eq. (4), if both a_0 and a_r are different from zero, as we shall henceforth suppose, the positive integer r is called the **order** of the equation. If $\phi(x)$ is identically zero, Eq. (4) is said to be **homogeneous**; if $\phi(x)$ is not identically zero, Eq. (4) is said to be **nonhomogeneous**. By a solution of (4) we mean a function of x with the property that when it is substituted into (4), it reduces the equation to an identity. From a theoretical point of view both x and y should be regarded as continuous variables related by Eq. (4) on a set of equally spaced values of x. However, in practical problems we are almost always interested in y only for the discrete values $x = \ldots, -3, -2, -1, 0, 1, 2, 3, \ldots$, and in our work we shall attempt no more than the determination of solutions defined on this domain set.

For the second-order linear difference equation, with either variable or constant coefficients, we have three theorems completely analogous to the fundamental theorems of Sec. 2.1.

THEOREM 1 If $y_1(x)$ and $y_2(x)$ are any two solutions of the homogeneous equation $(a_0 E^2 + a_1 E + a_2)y = 0$, then $c_1 y_1(x) + c_2 y_2(x)$, where c_1 and c_2 are arbitrary constants, is also a solution.

THEOREM 2 If $y_1(x)$ and $y_2(x)$ are two solutions of the homogeneous equation $(a_0E^2 + a_1E + a_2)y = 0$ for which

$$C[y_1(x), y_2(x)]\dagger = \begin{vmatrix} y_1(x) & y_2(x) \\ Ey_1(x) & Ey_2(x) \end{vmatrix} \neq 0$$

then any solution $y_3(x)$ of the homogeneous equation can be written in the form $y_3(x) = c_1y_1(x) + c_2y_2(x)$ where c_1 and c_2 are suitable constants.

As a consequence of Theorem 2, the expression $c_1y_1(x) + c_2y_2(x)$ is called a **complete solution** of the homogeneous equation when the particular solutions $y_1(x)$ and $y_2(x)$ satisfy the condition $C[y_1(x), y_2(x)] \neq 0$.

THEOREM 3 If $Y(x)$ is any solution of the nonhomogeneous equation

$$(a_0E^2 + a_1E + a_2)y = \phi(x)$$

and if $c_1y_1(x) + c_2y_2(x)$ is a complete solution of the homogeneous equation obtained from this by deleting the term $\phi(x)$, then any solution $y(x)$ of the nonhomogeneous equation can be written in the form

$$y(x) = c_1y_1(x) + c_2y_2(x) + Y(x)$$

where c_1 and c_2 are suitable constants.

As in the theory of differential equations, any complete solution of the related homogeneous equation is usually called a **complementary function** of the nonhomogeneous equation. The extension of these theorems to difference equations of order greater than 2 is obvious.

To find particular solutions of the homogeneous equation

$$(5) \qquad (a_0E^2 + a_1E + a_2)y = 0$$

when the coefficients a_0, a_1, a_2 are constants, we might try, as with the analogous differential equation, the substitution

$$(6) \qquad y = e^{mx}$$

However, it is more convenient to assume

$$(7) \qquad y = M^x$$

which is clearly equivalent to (6) with $M = e^m$. Substituting this into (5), recalling our agreement that $Ef(x) = f(x + 1)$, we obtain

$$a_0M^{x+2} + a_1M^{x+1} + a_2M^x = 0$$

or, dividing out M^x,

$$(8) \qquad a_0M^2 + a_1M + a_2 = 0$$

---

† The function $C[y_1(x), y_2(x)]$ is customarily referred to as **Casorati's determinant**, after the Italian mathematician Felice Casorati (1835–1890). Its resemblance to the wronskian $W[y_1(x), y_2(x)]$ (see Sec. 2.1) is apparent.

Naturally enough, this is called the **characteristic equation** of the difference equation (5).

If the roots M_1 and M_2 of (8) are distinct, then

$$C(M_1{}^x, M_2{}^x) = M_1{}^x M_2^{x+1} - M_2{}^x M_1^{x+1} = M_1{}^x M_2{}^x(M_2 - M_1) \neq 0\dagger$$

and hence, by Theorem 2, a complete solution of Eq. (5) is

$$(9) \qquad\qquad y = c_1 M_1{}^x + c_2 M_2{}^x$$

If M_1 and M_2 are real, this is a completely acceptable form of the solution. However, if M_1 and M_2 are complex, then (9) is inconvenient for most purposes, and it is desirable that we reduce it to a more useful form. To do this, let the roots be

$$M_1, M_2 = p \pm iq = re^{\pm i\theta}$$

where $\qquad\qquad r = \sqrt{p^2 + q^2} \qquad$ and $\qquad \tan\theta = \dfrac{q}{p}\ddagger$

Then we can write

$$\begin{aligned}
y &= c_1(re^{i\theta})^x + c_2(re^{-i\theta})^x \\
&= r^x(c_1 e^{i\theta x} + c_2 e^{-i\theta x}) \\
&= r^x[c_1(\cos\theta x + i\sin\theta x) + c_2(\cos\theta x - i\sin\theta x)] \\
&= r^x[(c_1 + c_2)\cos\theta x + i(c_1 - c_2)\sin\theta x]
\end{aligned}$$

or, renaming the constants,

$$(10) \qquad\qquad y = r^x(A\cos\theta x + B\sin\theta x)$$

If $M_1 = M_2$, clearly $C(M_1{}^x, M_2{}^x) = 0$, and we must find a second, independent solution before we can construct the complete solution of (5). Again, by analogy with differential equations, we are led to try

$$y = xM_1{}^x$$

and we find by direct substitution that this is indeed a solution when the characteristic equation (8) has equal roots. For we have

$$\begin{aligned}
a_0(x + 2)M_1^{x+2} &+ a_1(x + 1)M_1^{x+1} + a_2 x M_1{}^x \\
&= xM_1{}^x(a_0 M_1{}^2 + a_1 M_1 + a_2) + M_1^{x+1}(2a_0 M_1 + a_1) = 0
\end{aligned}$$

since the coefficient of $xM_1{}^x$ vanishes, because in any case M_1 satisfies the characteristic equation (8); and the coefficient of M_1^{x+1} vanishes because when the characteristic equation has equal roots, their common value is $M_1 = -a_1/2a_0$. Moreover, for the solutions $M_1{}^x$ and $xM_1{}^x$ we have

$$C(M_1{}^x, xM_1{}^x) = M_1{}^x(x + 1)M_1^{x+1} - xM_1{}^x M_1^{x+1} = M_1^{2x+1} \neq 0$$

---

† Since $a_2 \neq 0$, or else the difference equation would be of order less than 2, contrary to hypothesis, it is clear that neither M_1 nor M_2 can be zero.

‡ For a discussion of the exponential form of a complex number, see Sec. 15.7.

Hence, according to Theorem 2, a complete solution when the characteristic equation has equal roots is

(11) $$y = c_1 M_1{}^x + c_2 x M_1{}^x$$

The results of the preceding discussion are summarized in Table 4.1.

EXAMPLE 1

Find a complete solution of the difference equation

$$(E^2 + 2E + 4)y = 0$$

The characteristic equation in this case is $M^2 + 2M + 4 = 0$, and its roots are $M_1, M_2 = -1 \pm i\sqrt{3}$. Since

$$r = \sqrt{(-1)^2 + (\sqrt{3})^2} = 2 \quad \text{and} \quad \theta = \tan^{-1} \frac{\sqrt{3}}{-1} = \frac{2\pi}{3}$$

we have as a complete solution

$$y = 2^x \left(A \cos \frac{2\pi x}{3} + B \sin \frac{2\pi x}{3} \right)$$

To solve the nonhomogeneous equation

(12) $$(a_0 E^2 + a_1 E + a_2)y = \phi(x)$$

we must, according to Theorem 3, add a particular solution of (12) to a complete solution of the related homogeneous equation (5). To find the necessary particular solution Y, we use the method of undetermined coefficients, starting with an arbitrary linear combination of all the independent terms which arise from $\phi(x)$ by repeatedly applying the operator E. As in the case of differential equations, if any term in the initial choice for Y duplicates a term in the complementary function, it and all associated terms must be multiplied by the lowest positive integral power of x which will eliminate all duplications. The procedure is summarized in Table 4.2.

Table 4.1

Difference equation: $(a_0 E^2 + a_1 E + a_2)y = 0 \quad a_0, a_2 \neq 0$
Characteristic equation: $a_0 M^2 + a_1 M + a_2 = 0$

Nature of the roots of the characteristic equation	Condition on the coefficients of the characteristic equation	Complete solution of the difference equation
Real and unequal $M_1 \neq M_2$	$a_1{}^2 - 4a_0 a_2 > 0$	$y = c_1 M_1{}^x + c_2 M_2{}^x$
Real and equal $M_1 = M_2$	$a_1{}^2 - 4a_0 a_2 = 0$	$y = c_1 M_1{}^x + c_2 x M_1{}^x$
Conjugate complex $M_1 = p + iq$ $M_2 = p - iq$	$a_1{}^2 - 4a_0 a_2 < 0$	$y = r^x(A \cos \theta x + B \sin \theta x)$ $r = \sqrt{p^2 + q^2}$ $\tan \theta = q/p$

EXAMPLE 2

Find a complete solution of the difference equation

$$(E^2 - 5E + 6)y = x + 2^x$$

The characteristic equation in this case is $M^2 - 5M + 6 = 0$, and from its roots, $M_1 = 2$ and $M_2 = 3$, we can immediately construct the complementary function $y = c_1 2^x + c_2 3^x$. For a particular solution we would ordinarily try $Y = Ax + B + C2^x$. However, it is clear that $C2^x$ duplicates a term in the complementary function. Hence, according to the second footnote to Table 4.2, we must multiply $C2^x$ by x before incorporating it in our choice for Y. [The expression $Ax + B$ is *not* to be multiplied by x because it arises from a term, x, in $\phi(x)$ which does not duplicate anything in the complementary function.] Thus we substitute $Y = Ax + B + Cx2^x$ into the difference equation, getting

$$[A(x + 2) + B + C(x + 2)2^{x+2}] - 5[A(x + 1) + B + C(x + 1)2^{x+1}]$$
$$+ 6(Ax + B + Cx2^x) = x + 2^x$$

or
$$2Ax + (-3A + 2B) - 2C2^x = x + 2^x$$

which will be an identity if and only if $A = \frac{1}{2}$, $B = \frac{3}{4}$, and $C = -\frac{1}{2}$. A complete solution is therefore

$$y = c_1 2^x + c_2 3^x + \frac{2x + 3}{4} - \frac{x2^x}{2}$$

EXAMPLE 3

Find the sum of the series

$$s = \sum_{x=1}^{n} xk^x \qquad k \neq 1$$

Clearly, s satisfies the first-order difference equation

$$s_{n+1} - s_n \equiv (E - 1)s_n = (n + 1)k^{n+1}$$

Table 4.2

Difference equation: $(a_0 E^2 + a_1 E + a_2)y = \phi(x)$	
$\phi(x)$†	Necessary choice for particular solution Y‡
1. α (constant)	A
2. αx^k (k a positive integer)	$A_0 x^k + A_1 x^{k-1} + \cdots + A_{k-1}x + A_k$
3. αk^x	Ak^x
4. $\alpha \cos kx$	$A \cos kx + B \sin kx$
5. $\alpha \sin kx$	
6. $\alpha x^k l^x \cos mx$	$(A_0 x^k + \cdots + A_{k-1}x + A_k)l^x \cos mx$
7. $\alpha x^k l^x \sin mx$	$\quad + (B_0 x^k + \cdots + B_{k-1}x + B_k)l^x \sin mx$

† When $\phi(x)$ consists of a sum of several terms, the appropriate choice for Y is the sum of the Y expressions corresponding to these terms individually.

‡ Whenever a term in any of the Y's listed in this column duplicates a term in the complementary function, all terms in that Y expression must be multiplied by the lowest positive integral power of x sufficient to eliminate all such duplications.

Moreover, although Tables 4.1 and 4.2 appear to be concerned only with second-order equations, the procedures they outline are correct for difference equations of any order. Hence we set up the characteristic equation $M - 1 = 0$ and from its root construct the complementary function

$$s = c_1(1)^n = c_1$$

To find a particular "integral," we assume

$$S = (An + B)k^{n+1}$$

Then, substituting, we must have

$$[A(n + 1) + B]k^{n+2} - (An + B)k^{n+1} = (n + 1)k^{n+1}$$

or, dividing out k^{n+1} and collecting terms,

$$n(Ak - A) + (Ak + Bk - B) = n + 1$$

This will be an identity in the variable n if and only if

$$A(k - 1) = 1 \quad \text{and} \quad Ak + B(k - 1) = 1$$

or

$$A = \frac{1}{k - 1} \quad \text{and} \quad B = -\frac{1}{(k - 1)^2}$$

Hence

$$S = \left[\frac{n}{k - 1} - \frac{1}{(k - 1)^2}\right] k^{n+1}$$

and a complete solution is

$$s = c_1 + S = c_1 + \frac{n(k - 1) - 1}{(k - 1)^2} k^{n+1}$$

To determine c_1 we use the obvious fact that $s = k$ when $n = 1$. Thus we must have

$$k = c_1 + \frac{k - 2}{(k - 1)^2} k^2 \quad \text{or} \quad c_1 = \frac{k}{(k - 1)^2}$$

Hence, finally,

$$s = \frac{k + [n(k - 1) - 1]k^{n+1}}{(k - 1)^2}$$

EXAMPLE 4

In the system shown in Fig. 4.2a the point P_0 is kept at the constant potential V_0 with respect to the ground. What is the potential at each of the points $P_1, P_2, \ldots, P_{n-1}$?

According to Kirchhoff's first law, the sum of the currents flowing toward any junction in a network must equal the sum of the currents flowing away from that junction. Hence at a general point P_{x+1} (Fig. 4.2b) we have

$$i_x = i_{x+1} + I_{x+1}$$

or, replacing each current by its equivalent according to Ohm's law, $I = E/R$,

$$\frac{V_x - V_{x+1}}{r} = \frac{V_{x+1} - V_{x+2}}{r} + \frac{V_{x+1}}{2r}$$

or, finally,

(13) $$V_{x+2} - \tfrac{5}{2}V_{x+1} + V_x = 0$$

This equation holds for $x = 2, \ldots, n - 2$, that is, at all but the points P_1 and P_{n-1}, where Eq. (13) reduces to the respective conditions

(14) $$-V_2 + \tfrac{5}{2}V_1 = V_0 \quad \text{since } V_0 \text{ is given}$$

(15) $$-\tfrac{5}{2}V_{n-1} + V_{n-2} = 0 \quad \text{since } V_n = 0$$

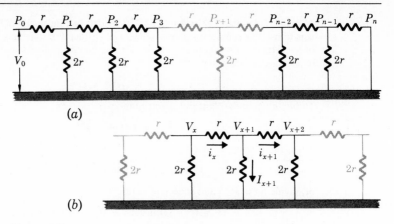

Figure 4.2
A ladder-type network with identical loops. (Although the network shown in
Fig. 4.2a appears to contain exactly seven loops, the number of loops is
actually indefinite. This is implied by the fact that the central portion of the
figure is drawn with lighter lines; this convention will be used throughout the
book to suggest a configuration of indefinite extent.)

Equations (13), (14), and (15) constitute a system of $n - 1$ linear algebraic equations
from which the unknown potentials $V_1, V_2, \ldots, V_{n-1}$ can be found by completely
elementary though very tedious steps for any particular value of n. However, it is much
simpler and much more elegant to regard Eq. (13) as a second-order difference equation
subject to the end conditions (14) and (15), which will serve to determine the values of
the arbitrary constants appearing in any complete solution of (13).

Taking this point of view, we first set up the characteristic equation of Eq. (13):

$$M^2 - \tfrac{5}{2}M + 1 = 0$$

From its roots, $M_1 = \tfrac{1}{2}$ and $M_2 = 2$, we then construct a complete solution of Eq. (13),
namely,

$$V_x = A(\tfrac{1}{2})^x + B2^x$$

Substituting this into Eqs. (14) and (15), we have

$$-\left(\frac{A}{4} + 4B\right) + \frac{5}{2}\left(\frac{A}{2} + 2B\right) = V_0$$

$$-\frac{5}{2}\left(\frac{A}{2^{n-1}} + B2^{n-1}\right) + \left(\frac{A}{2^{n-2}} + B2^{n-2}\right) = 0$$

or $A + B = V_0$ and $\dfrac{A}{2^n} + B2^n = 0$

from which we find at once

$$A = \frac{2^{2n}}{2^{2n} - 1}\, V_0 \qquad B = -\frac{1}{2^{2n} - 1}\, V_0$$

The final solution is therefore

$$V_x = \left(\frac{2^{2n}}{2^x} - 2^x\right)\frac{V_0}{2^{2n} - 1}$$

That this reduces to V_0 when $x = 0$ and reduces to 0 when $x = n$ is easily verified.

One of the many uses of difference equations is in analyzing the accuracy of methods
for the numerical solution of differential equations. Although a detailed discussion of

this topic must be left to more advanced texts on numerical analysis, we can make an introductory investigation of this problem by considering the solution of the simple equation

(16) $$y' = Ay$$

by any of the predictor-corrector methods we presented in the last section, say Milne's.

Milne's method, as we know, makes use of the predictor formula [Eq. (3), Sec. 4.4]

$$y_{n+1} = y_{n-3} + \frac{4h}{3}(2y'_{n-2} - y'_{n-1} + 2y'_n)$$

to obtain a first estimate of y_{n+1} and then iterates the corrector formula

$$y_{n+1} = y_{n-1} + \frac{h}{3}(y'_{n-1} + 4y'_n + y'_{n+1})$$

to determine the value of y_{n+1} which is finally accepted. For the differential equation (16), Milne's corrector-formula becomes

$$y_{n+1} = y_{n-1} + \frac{h}{3}(Ay_{n-1} + 4Ay_n + Ay_{n+1})$$

Hence, *the numerical solution given by Milne's method is just a particular solution of the second-order difference equation*

(17) $$\left(1 - \frac{Ah}{3}\right)y_{n+1} - \frac{4Ah}{3}y_n - \left(1 + \frac{Ah}{3}\right)y_{n-1} = 0$$

When we put $Ah/3 = \lambda$, for convenience, the characteristic equation of Eq. (17) becomes

$$(1 - \lambda)M^2 - 4\lambda M - (1 + \lambda) = 0$$

and its roots are

$$M_1, M_2 = \frac{2\lambda \pm \sqrt{1 + 3\lambda^2}}{1 - \lambda}$$

Hence, the tabular function y_n, which is our approximation to the solution of (16), is contained in the complete solution

$$y = c_1 M_1{}^n + c_2 M_2{}^n$$

and as $h \to 0$, this particular solution should, presumably, approach the exact solution of (16), namely, $y = y_0 e^{A(x - x_0)}$.

To determine under what conditions, if any, this will be the case, it is convenient to expand M_1 and M_2 in terms of powers of λ and then retain only the lowest power of λ in the expansion of each root. Doing this, we have

$$M_1, M_2 = [2\lambda \pm (1 + 3\lambda^2)^{1/2}](1 - \lambda)^{-1}$$
$$= [2\lambda \pm (1 + \tfrac{3}{2}\lambda^2 + \cdots)](1 + \lambda + \lambda^2 + \cdots)$$
$$= (2\lambda \pm 1 + \cdots)(1 + \lambda + \cdots)$$

and $$M_1 = 1 + 3\lambda + \cdots \qquad M_2 = -1 + \lambda + \cdots$$

Hence

$$y_n \doteq c_1(1 + 3\lambda)^n + c_2(-1 + \lambda)^n$$

(18)
$$= c_1(1 + Ah)^{(x_n-x_0)/h} + c_2(-1)^n \left(1 - \frac{Ah}{3}\right)^{(x_n-x_0)/h}$$

since $\lambda = Ah/3$ and $n = (x_n - x_0)/h$. As $h \to 0$, each term in (18) contains an indeterminate of the form 1^∞. Evaluating these by L'Hospital's rule, or recalling from calculus that

$$\lim_{z \to 0} (1 + z)^{b/z} = e^b$$

we see that for small values of h we have the approximation

(19)
$$y_n = c_1 e^{A(x_n-x_0)} + c_2(-1)^n e^{-A(x_n-x_0)/3}$$

If our initial data were perfectly accurate, and if there were no roundoff errors in our calculations, the correct values of c_1 and c_2 would be y_0 and 0, respectively, and (19) would reduce to the required solution of the differential equation $y' = Ay$. However, because of inevitable errors of various kinds, c_2 will in general not be zero. Thus the numerical solution we obtain is actually the sum of an approximation to the exact solution plus a so-called **parasitic solution** which for small values of h is approximately $c_2(-1)^n e^{-A(x_n-x_0)/3}$.

If $A > 0$, the parasitic solution is approximately a decaying exponential function which soon becomes vanishingly small. In this case, the procedure is said to be **numerically stable**. On the other hand, if $A < 0$, the numerical value of the parasitic solution increases exponentially (while continually alternating in sign) and sooner or later, depending on the size of c_2 and A, it will become the principal part of the numerical solution we obtain. In this case, the procedure is said to be **numerically unstable**.

The analysis, leading to essentially the same conclusions, which can be made for the equation $y' = Ay + f(x)$ is similar except that the difference equation arising from Milne's corrector formula is nonhomogeneous rather than homogeneous.

For the general first-order differential equation $y' = f(x,y)$ Milne's corrector formula is not expressible as a simple difference equation. However, if $f(x,y)$ is replaced by its Taylor's expansion around the point (x_0,y_0), namely,

$$y' = f(x,y) = f(x_0,y_0) + \left[\frac{\partial f}{\partial x}\bigg|_{x_0,y_0} (x - x_0) + \frac{\partial f}{\partial y}\bigg|_{x_0,y_0} (y - y_0)\right]$$

$$+ \frac{1}{2!}\left[\frac{\partial^2 f}{\partial x^2}\bigg|_{x_0,y_0} (x - x_0)^2 + 2\frac{\partial^2 f}{\partial x \, \partial y}\bigg|_{x_0,y_0} (x - x_0)(y - y_0)\right.$$

$$\left. + \frac{\partial^2 f}{\partial y^2}\bigg|_{x_0,y_0} (y - y_0)^2\right] + \cdots$$

and if powers of $x - x_0$ and $y - y_0$ higher than the first are neglected, then in the neighborhood of (x_0,y_0), the differential equation $y' = f(x,y)$ becomes, approximately,

$$y' \equiv (y - y_0)' = \frac{\partial f}{\partial y}\bigg|_{x_0,y_0} (y - y_0) + \left[f(x_0,y_0) + \frac{\partial f}{\partial x}\bigg|_{x_0,y_0} (x - x_0)\right]$$

which is of the form we have just discussed. By the preceding discussion, the solution given by Milne's corrector formula will be stable if the coefficient of $y - y_0$, namely $\partial f / \partial y|_{x_0, y_0}$, is positive and unstable if it is negative.

Clearly, if any other of the predictor-corrector methods we described in the last section (see Exercises 9, 14, and 17, Sec. 4.4) is applied to Eq. (16), the corrector formula becomes a homogeneous, constant-coefficient difference equation and the preceding analysis can be repeated. The chief difference is that the related difference equation may be of order greater than 2. This means that its characteristic equation will, in general, have more than two roots and its complementary function will be of the form

$$y_n = c_1 M_1{}^n + c_2 M_2{}^n + \cdots + c_k M_k{}^n$$

As with Milne's method, one of these terms will approach the exact solution of Eq. (16) as $h \to 0$; the others will be parasitic solutions. If the absolute values of the roots from which the parasitic solutions arise are all less than 1, then as n, or x, increases, each will decay exponentially and the process will yield an approximation to the exact solution. On the other hand, if the absolute value of even one of the extraneous roots is greater than 1, then the corresponding parasitic solution will increase exponentially and the procedure will be numerically unstable.

Since numerical instability is commonly observed in predictor-corrector methods of solving differential equations, many workers prefer the Runge-Kutta method, which involves no numerical instability, at least for sufficiently small values of the tabular interval h.

EXERCISES

1 Find a complete solution of each of the following equations:
   (a) $(E^2 + 7E + 12)y = 0$ (b) $(E^2 + 6E + 9)y = 0$
   (c) $(E^2 + 2E + 2)y = 0$ (d) $(\Delta^2 - 3\Delta + 2)y = 0$
2 Find a complete solution of each of the following equations:
   (a) $(E^2 - E - 6)y = x^2$ (b) $(4E^2 - 4E + 1)y = x + 2 + 2^x$
   (c) $(E^2 + 4)y = \cos x$ (d) $(\Delta^2 + 6\Delta + 18)y = 2^{-x}$
   (e) $(E^2 - 3E + 2)y = 2^x + 2^{-x}$ (f) $(E^2 - 4E + 4)y = 2^x$
3 Find a complete solution of each of the following equations:
   (a) $(E^2 - E - 6)y = x + 3^x$ (b) $(E^2 + 1)y = \sin x$
4 Find a complete solution of each of the following equations:
   (a) $(E^3 - 6E^2 + 11E - 6)y = 0$ (b) $(E^4 - 16)y = x + 3^x$
   (c) $(E^4 + 10E^2 + 9)y = 0$ (d) $(E^4 + 8E^2 - 9)y = 5$
5 Show that the difference equation $(E^2 - 2\lambda E + 1)y = 0$ has the indicated solution in each of the following special cases:

   $\lambda < -1$: $y = A(-1)^x \cosh \mu x + B(-1)^x \sinh \mu x$ $\cosh \mu = -\lambda$

   $\lambda = -1$: $y = A(-1)^x + Bx(-1)^x$

   $-1 < \lambda < 1$: $y = A \cos \mu x + B \sin \mu x$ $\cos \mu = \lambda$

   $\lambda = 1$: $y = A + Bx$

   $1 < \lambda$: $y = A \cosh \mu x + B \sinh \mu x$ $\cosh \mu = \lambda$

6 Work Example 4 with both P_0 and P_n maintained at the constant potential V_0.
7 Work Example 4 given that the common value of the resistances in the vertical branches is kr.
8 A system consists of n spring-connected masses, as shown in Fig. 4.3. What is the displacement of each mass from its original position when the system is again in equilibrium after a force F_0 is applied to the right-hand end?

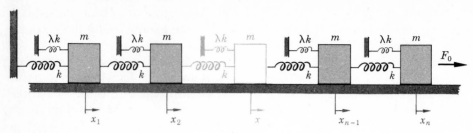

Figure 4.3
(See explanation of convention used in Fig. 4.2.)

9 Show that the nth-order determinant

$$D_n = \begin{vmatrix} a & 1 & 0 & \cdots & 0 & 0 \\ 1 & a & 1 & \cdots & 0 & 0 \\ 0 & 1 & a & \cdots & 0 & 0 \\ \cdots & \cdots & \cdots & \cdots & \cdots & \cdots \\ 0 & 0 & 0 & \cdots & a & 1 \\ 0 & 0 & 0 & \cdots & 1 & a \end{vmatrix}$$

satisfies the difference equation $(E^2 - aE + 1)D = 0$. Hence show that when $a > 2$,

$$D_n = \frac{\sinh (n + 1)\mu}{\sinh \mu} \qquad \text{where } \cosh \mu = \frac{a}{2}$$

What is D_n if $a = 2$? $-2 < a < 2$? $a = -2$? $a < -2$?

10 If $y_1(x)$ and $y_2(x)$ are any two solutions of the general linear second-order difference equation

$$[a_0(x)E^2 + a_1(x)E + a_2(x)]y = 0$$

show that Casorati's determinant $C[y_1(x), y_2(x)]$ satisfies the relation

$$[a_0(x)E - a_2(x)]C = 0$$

Hint: Write down the conditions that both $y_1(x)$ and $y_2(x)$ satisfy the given equation; then eliminate the terms in $Ey_1(x)$ and $Ey_2(x)$.

11 Prove Theorem 1.

12 Prove Theorem 2 in the special case where the coefficients are constants. *Hint:* Recall the proof of Theorem 2, Sec. 2.1, and use the result of Exercise 10.

13 Prove Theorem 3.

14 Show that the integral

$$I_n = \int_0^\pi \frac{\cos nt - \cos n\lambda}{\cos t - \cos \lambda} \, dt$$

satisfies the equation $[E^2 - (2 \cos \lambda)E + 1]I = 0$. Solve this equation, and find an explicit expression for I_n.

15 Discuss the solution of each of the following equations:
(a) $Ey = 0$ (b) $Ey = \phi(x)$
(c) $(a_0E^2 + a_1E)y = 0$ (d) $(a_0E^2 + a_1E)y = \phi(x)$

16 (a) Show that Euler's method for the numerical solution of differential equations [Eq. (7), Sec. 4.4] is numerically stable for the differential equation $y' = Ay$ for all values of A.
(b) Verify that as $h \to 0$, the general solution of the difference equation involved in part (a) converges to the general solution of $y' = Ay$ for all values of A.

17 Work Exercise 16 for the modified Euler method [Eq. (8), Sec. 4.4].

18 Discuss the stability of Milne's corrector formula as applied to the equations
(a) $y' = Ay + 1$ (b) $y' = Ay + x$ (c) $y' = Ay + 2^x$

19 (a) Discuss the stability of the closed formula given in Exercise 14, Sec. 4.4, in relation to the equation $y' = Ay$.
(b) Verify that as $h \to 0$, one of the solutions of the difference equation involved in part (a) converges to the general solution of $y' = Ay$.

20 Show that the corrector formula

$$y_{n+1} = y_n + \frac{h}{24}(9y'_{n+1} + 19y'_n - y'_{n-1} + y'_{n-2})$$

is stable in relation to the equation $y' = Ay$ if $\gamma \equiv Ah/24$ is a sufficiently small negative number. *Hint:* Note that from the characteristic equation of the related difference equation, γ can easily be plotted as a function of M.

21 Discuss the stability of Milne's predictor formula [Eq. (3), Sec. 4.4] in relation to the equation $y' = Ay$. *Hint:* Note the hint given in Exercise 20.

22 (a) Show that the closed formula given in Exercise 13, Sec. 4.4, is numerically stable for the equation $y'' = Ay$ for all values of A.
(b) Verify that as $h \to 0$, the general solution of the difference equation involved in part (a) converges to the general solution of $y'' = Ay$.

4.6 The Method of Least Squares

The problem of fitting a curve to a set of points admits of two somewhat different interpretations. In the first place, we may ask for the equation of a curve of prescribed type which passes exactly through each point of the given set. For polynomial curves this is most easily accomplished by interpolation formulas like those developed in Sec. 4.2. On the other hand, we may weaken these requirements and ask for some simpler curve whose equation contains too few parameters to permit it to pass exactly through each given point but which comes "as close as possible" to each point. For instance, given a set of points as in Fig. 4.4a, a straight line passing as close as possible to each point may well be more useful than some complicated curve passing exactly through each point. This will certainly be the case with experimental data which theoretically should fall along a straight line but which fail to do so because of errors of observation. For most purposes the necessary measure of "as close as possible" is taken to be the least-squares criterion,† and the process of applying this criterion is known as the **method of least squares**, which we now develop.

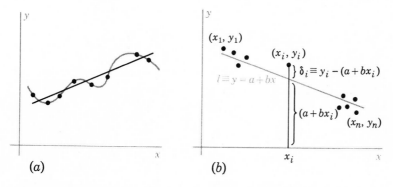

(a)   **(b)**

Figure 4.4
The approximate fitting of a straight line to a set of points.

---

† A brief discussion of the reasons for this will be found in A. M. Mood, "Introduction to the Theory of Statistics," p. 311, McGraw-Hill, New York, 1950.

Let us begin by supposing that we wish to fit a straight line l whose equation is

(1) $$y = ax + b$$

to the n points $(x_1, y_1), (x_2, y_2), \ldots, (x_n, y_n)$. Since two points completely determine a straight line, it will in general be impossible for the required line to pass through more than two of the given points, and it may not pass through any. Hence, the coordinates of the general point (x_i, y_i) will not satisfy Eq. (1). That is, when we substitute x_i into Eq. (1), we get not y_i but the ordinate of l which, as we see from Fig. 4.4b, differs from y_i by δ_i. In other words,

(2) $$y_i - (a + bx_i) = \delta_i \neq 0$$

If we compute the discrepancy δ_i for each point of the set and form the sum of the squares of these quantities (in order to prevent large positive and large negative δ's canceling each other and thereby giving an unwarranted impression of accuracy), we obtain

(3) $$E = \sum_{i=1}^{n} \delta_i{}^2 = (y_1 - a - bx_1)^2 + (y_2 - a - bx_2)^2 + \cdots + (y_n - a - bx_n)^2$$

The quantity E is obviously a measure of how well the line fits the set of points as a whole. For E will be zero if and only if each of the points lies on l, and the larger E is, the farther the points are, on the average, from l. The least-square or "squares" criterion is now simply this: *that the parameters a and b should be chosen so as to make the sum of the squares of the deviations E as small as possible.*

To do this, we apply the usual conditions for minimizing a function of several variables and equate to zero the two first derivatives, $\partial E/\partial a$ and $\partial E/\partial b$. This gives us the two equations

$$\frac{\partial E}{\partial a} = 2(y_1 - a - bx_1)(-1) + 2(y_2 - a - bx_2)(-1) + \cdots$$
$$+ 2(y_n - a - bx_n)(-1) = 0$$
$$\frac{\partial E}{\partial b} = 2(y_1 - a - bx_1)(-x_1) + 2(y_2 - a - bx_2)(-x_2) + \cdots$$
$$+ 2(y_n - a - bx_n)(-x_n) = 0$$

or, dividing by 2 and collecting terms on the unknown coefficients a and b,

(4) $$na + b \sum_{i=1}^{n} x_i = \sum_{i=1}^{n} y_i$$

(5) $$a \sum_{i=1}^{n} x_i + b \sum_{i=1}^{n} x_i{}^2 = \sum_{i=1}^{n} x_i y_i$$

Equations (4) and (5) are two simultaneous linear equations whose solution for a and b presents no difficulty.

If we reconsider Eq. (2) from a purely algebraic point of view, it appears that as i varies from 1 to n, it defines a system of n equations in two unknowns, a and b, which should, ideally, be satisfied but which actually are not. Moreover, minimizing E, as given by (3), is nothing more than minimizing the sum of the squares of the amounts by which these n equations fail to be satisfied. These observations suggest a somewhat more general interpretation of the method of least squares, namely, that it is simply a process for finding the best possible values for a set of m unknowns, say $z_1, z_2, \ldots,$ z_m, connected by n linear equations

$$\alpha_{11}z_1 + \alpha_{12}z_2 + \cdots + \alpha_{1m}z_m = \beta_1$$
$$\alpha_{21}z_1 + \alpha_{22}z_2 + \cdots + \alpha_{2m}z_m = \beta_2$$
$$\cdots\cdots\cdots\cdots\cdots\cdots\cdots\cdots\cdots$$
$$\alpha_{n1}z_1 + \alpha_{n2}z_2 + \cdots + \alpha_{nm}z_m = \beta_n$$

where $n > m$.

Since the number of equations in the last set exceeds the number of unknowns, the system presumably does not admit of an exact solution; i.e., there is no set of values for $z_1, z_2, \ldots, z_m$ for which each equation is exactly satisfied. Hence, letting δ_i be the amount by which the ith equation fails to be satisfied, we consider the discrepancies

$$\delta_i = \alpha_{i1}z_1 + \alpha_{i2}z_2 + \cdots + \alpha_{im}z_m - \beta_i \quad (\neq 0) \quad i = 1, 2, \ldots, n$$

and attempt to find values of $z_1, z_2, \ldots, z_m$ which will make

$$E = \sum_{i=1}^{n} \delta_i^2 = \sum_{i=1}^{n} (\alpha_{i1}z_1 + \alpha_{i2}z_2 + \cdots + \alpha_{im}z_m - \beta_i)^2$$

as small as possible.

To minimize E we must equate to zero each of its first partial derivatives

$$\frac{\partial E}{\partial z_1}, \frac{\partial E}{\partial z_2}, \ldots, \frac{\partial E}{\partial z_m}$$

For $\partial E/\partial z_1$ this gives the equation

$$\frac{\partial E}{\partial z_1} = \sum_{i=1}^{n} 2(\alpha_{i1}z_1 + \alpha_{i2}z_2 + \cdots + \alpha_{im}z_m - \beta_i)\alpha_{i1} = 0$$

or
$$z_1 \sum_{i=1}^{n} \alpha_{i1}\alpha_{i1} + z_2 \sum_{i=1}^{n} \alpha_{i2}\alpha_{i1} + \cdots + z_m \sum_{i=1}^{n} \alpha_{im}\alpha_{i1} = \sum_{i=1}^{n} \alpha_{i1}\beta_i$$

and similarly, for the other partial derivatives,

$$z_1 \sum_{i=1}^{n} \alpha_{i1}\alpha_{i2} + z_2 \sum_{i=1}^{n} \alpha_{i2}\alpha_{i2} + \cdots + z_m \sum_{i=1}^{n} \alpha_{im}\alpha_{i2} = \sum_{i=1}^{n} \alpha_{i2}\beta_i$$
$$\cdots\cdots\cdots\cdots\cdots\cdots\cdots\cdots\cdots\cdots\cdots\cdots\cdots\cdots$$
$$z_1 \sum_{i=1}^{n} \alpha_{i1}\alpha_{im} + z_2 \sum_{i=1}^{n} \alpha_{i2}\alpha_{im} + \cdots + z_m \sum_{i=1}^{n} \alpha_{im}\alpha_{im} = \sum_{i=1}^{n} \alpha_{im}\beta_i$$

We have thus obtained a system of m linear equations in m unknowns $z_1, z_2, \ldots, z_m$, whose solution is now a routine matter. As a practical detail, it is worthy of note that these minimizing conditions, or **normal equations** as they are usually called, can be written down at once according to the following rule.

RULE 1 If each of n linear equations in the m unknowns

$$z_1, z_2, \ldots, z_m \quad n > m$$

is multiplied by the coefficient of z_i in that equation, the sum of the resulting equations is the ith normal equation in the least-squares solution of the system.

EXAMPLE 1

By the method of least squares, fit a parabolic equation $y = a + bx + cx^2$ to the data.

x	-3	-2	0	3	4
y	18	10	2	2	5

Substituting these pairs of values into the equation $y = a + bx + cx^2$, we find that a, b, and c (which play the roles of z_1, z_2, and z_3 in the preceding general discussion) should satisfy the conditions

$$
\begin{aligned}
a - 3b + 9c &= 18 \\
a - 2b + 4c &= 10 \\
a \quad\quad\quad &= 2 \\
a + 3b + 9c &= 2 \\
a + 4b + 16c &= 5
\end{aligned}
$$

In general, three unknowns cannot be made to satisfy more than three conditions; hence, the most we can do is to determine values of a, b, and c which will satisfy these conditions as nearly as possible.

To set up the first of the three normal equations required by the method of least squares, we must multiply each of the equations of condition by the coefficient of a in that equation and add, getting in this case simply the sum of the five equations:

$$5a + 2b + 38c = 37$$

To set up the second normal equation, we multiply each equation by the coefficient of b in that equation and add, getting

$$
\begin{array}{r}
-3a + 9b - 27c = -54 \\
-2a + 4b - 8c = -20 \\
0 + 0 + 0 = 0 \\
3a + 9b + 27c = 6 \\
4a + 16b + 64c = 20 \\
\hline
2a + 38b + 56c = -48
\end{array}
$$

In the same way, multiplying each equation by the coefficient of c in that equation and adding, we get the third normal equation:

$$
\begin{array}{r}
9a - 27b + 81c = 162 \\
4a - 8b + 16c = 40 \\
0 + 0 + 0 = 0 \\
9a + 27b + 81c = 18 \\
16a + 64b + 256c = 80 \\
\hline
38a + 56b + 434c = 300
\end{array}
$$

The solution of the three normal equations is a simple matter, and we find

$$a = 1.82 \qquad b = -2.65 \qquad c = 0.87$$

The required solution is therefore

$$y = 1.82 - 2.65x + 0.87x^2$$

When, as is often the case, the abscissas of the points to which we wish to fit a polynomial curve are equally spaced, the labor involved in the least-squares procedure we have just described can be significantly reduced by using what are known as **orthogonal polynomials**.

DEFINITION 1 If $n + 1$ polynomials $P_{nm}(x)$ of respective degrees $m = 0, 1, 2, \ldots, n$ have the property that

$$(6) \qquad\qquad \sum_{x=0}^{n} P_{nj}(x)P_{nk}(x) = 0 \qquad j \neq k$$

they are said to be orthogonal polynomials.

By methods which need not concern us here,† it has been shown that for each value of n there exists a set of $n + 1$ orthogonal polynomials, and the general formula for them has been obtained:

$$(7) \qquad P_{nm}(x) = \sum_{i=0}^{m} (-1)^i \binom{m}{i}\binom{m+i}{i}\frac{(x)^{(i)}}{(n)^{(i)}} \qquad m = 0, 1, 2, \ldots, n$$

In particular

$$P_{n0}(x) = 1$$

$$P_{n1}(x) = 1 - 2\frac{x}{n}$$

$$P_{n2}(x) = 1 - 6\frac{x}{n} + 6\frac{x(x-1)}{n(n-1)}$$

$$P_{n3}(x) = 1 - 12\frac{x}{n} + 30\frac{x(x-1)}{n(n-1)} - 20\frac{x(x-1)(x-2)}{n(n-1)(n-2)}$$

Clearly, for each $m \leq n$, any polynomial of degree m can be expressed as a linear combination of the polynomials

$$P_{n0}(x), P_{n1}(x), \ldots, P_{nm}(x)$$

since the expression

$$(8) \qquad\qquad P(x) = a_0 P_{n0}(x) + a_1 P_{n1}(x) + \cdots + a_m P_{nm}(x)$$

is obviously a polynomial of degree m containing the maximum number, $m + 1$, of independent, arbitrary constants which can appear in the general polynomial of this degree. Moreover, the coefficients $a_0, a_1, \ldots, a_m$ in (8) can easily be found; for if we multiply both sides of this identity by $P_{ni}(x)$, say, and then sum from $x = 0$ to $x = n$, we get

$$\sum_{x=0}^{n} P(x)P_{ni}(x) = a_0 \sum_{x=0}^{n} P_{n0}(x)P_{ni}(x) + \cdots + a_i \sum_{x=0}^{n} P_{ni}^2(x) + \cdots$$

$$+ a_m \sum_{x=0}^{n} P_{nm}(x)P_{ni}(x)$$

---

† See, for instance, W. E. Milne, "Numerical Analysis," pp. 265–275 and 375–381, Princeton University Press, Princeton, N.J., 1949.

But, from the so-called **orthogonality property** of the polynomials $\{P_{nm}(x)\}$, which is expressed by Eq. (6), it follows that every term on the right-hand side of the last expression is zero except the sum

$$a_i \sum_{x=0}^{n} P_{ni}^2(x)$$

Hence, solving for a_i, we obtain the formula

$$(9) \qquad a_i = \frac{\displaystyle\sum_{x=0}^{n} P(x)P_{ni}(x)}{\displaystyle\sum_{x=0}^{n} P_{ni}^2(x)} \qquad i = 0, 1, \ldots, m$$

The property described by Eq. (6) is very important, and we shall encounter it again in Sec. 11.2 when we attempt, in a manner analogous to the expansion (8), to express an arbitrary vector as a linear combination of certain given, independent vectors. Also, in Chaps. 6, 8, and 9 we shall study expansion problems resembling (8), in which the coefficients will be determined through the use of orthogonality properties involving integrals rather than sums, as in (6).

Clearly, an expansion of the form (8) can be created for any function $f(x)$, polynomial or not, merely by using the coefficient formula (9) with $f(x)$ replacing $P(x)$. Such expansions are of great importance for, although it is obvious that they cannot represent $f(x)$ exactly unless $f(x)$ is a polynomial of degree n or less, they provide the best polynomial approximations to $f(x)$ in the least-squares sense. To prove this, suppose that we have a function $f(x)$ defined for the $n + 1$ equally spaced values $x = 0, 1, \ldots, n$, which we wish to approximate with a polynomial of degree m ($< n$). If we assume the polynomial to be written in the form (8), the discrepancy at the general point x is

$$f(x) - a_0 P_{n0}(x) - \cdots - a_i P_{ni}(x) - \cdots - a_m P_{nm}(x)$$

and the principle of least squares requires that we minimize the sum

$$(10) \qquad E = \sum_{x=0}^{n} [f(x) - a_0 P_{n0}(x) - \cdots - a_i P_{ni}(x) - \cdots - a_m P_{nm}(x)]^2$$

If we equate to zero the derivative of E with respect to a_i, say, we obtain the general minimizing condition

$$\frac{\partial E}{\partial a_i} = \sum_{x=0}^{n} 2[f(x) - a_0 P_{n0}(x) - \cdots - a_i P_{ni}(x) - \cdots - a_m P_{nm}(x)]P_{ni}(x) = 0$$

or, breaking up the sum,

$$(11) \quad \sum_{x=0}^{n} f(x)P_{ni}(x) - a_0 \sum_{x=0}^{n} P_{n0}(x)P_{ni}(x) - \cdots$$

$$- a_i \sum_{x=0}^{n} P_{ni}^2(x) - \cdots - a_m \sum_{x=0}^{n} P_{nm}(x)P_{ni}(x) = 0$$

But from the orthogonality of the P's, the sums involving two different P's are all zero, and Eq. (11) reduces to

$$\sum_{x=0}^{n} f(x)P_{ni}(x) - a_i \sum_{x=0}^{n} P_{ni}^{2}(x) = 0$$

or

(12)
$$a_i = \frac{\displaystyle\sum_{x=0}^{n} f(x)P_{ni}(x)}{\displaystyle\sum_{x=0}^{n} P_{ni}^{2}(x)} \qquad i = 0, 1, \ldots, m$$

which is exactly the same as (9) with $P(x)$ replaced by $f(x)$.

The advantage of using orthogonal polynomials instead of the general least-squares procedure is now clear. In the first place, through their use the coefficients in the least-squares polynomial approximation to a function $f(x)$ defined for the $n + 1$ equally spaced values $x = 0, 1, \ldots, n$ can be found one at a time by formula (12) without the necessity of solving any simultaneous equations. In the second place, since formula (12) for a_i does not involve m, the degree of the polynomial we are fitting to the data, it follows that if we wish to increase m, that is, add another term to the approximating polynomial, all previously calculated coefficients remain unchanged and only the coefficient of the new term need be computed.

The sum appearing in the denominator of (12) need not be calculated directly because a general formula for it is available, namely,

(13)
$$\sum_{x=0}^{n} P_{ni}^{2}(x) = \frac{(n + i + 1)^{(i+1)}\dagger}{(2i + 1)(n)^{(i)}}$$

In particular,

$$\sum_{x=0}^{n} P_{n0}^{2}(x) = n + 1$$

$$\sum_{x=0}^{n} P_{n1}^{2}(x) = \frac{(n + 1)(n + 2)}{3n}$$

$$\sum_{x=0}^{n} P_{n2}^{2}(x) = \frac{(n + 1)(n + 2)(n + 3)}{5n(n - 1)}$$

$$\sum_{x=0}^{n} P_{n3}^{2}(x) = \frac{(n + 1)(n + 2)(n + 3)(n + 4)}{7n(n - 1)(n - 2)}$$

To determine the accuracy with which the polynomial approximation fits the data it is not necessary to compute E from (10), since it can be shown that in general

(14)
$$E = \sum_{x=0}^{n} f^{2}(x) - \sum_{i=0}^{m} \left[a_i^{2} \sum_{x=0}^{n} P_{ni}^{2}(x) \right]$$

---

† See, for instance, Milne, loc. cit.

EXAMPLE 2

Using orthogonal polynomials, fit equations of the form $y = a_0 + a_1 t$ and $y = a_0 + a_1 t + a_2 t^2$ to the data

t	0.00	0.25	0.50	0.75	1.00
y	0.00	0.06	0.20	0.60	0.90

As a first step we must introduce an auxiliary variable $x = 4t$, which will take on the values 0, 1, 2, 3, 4 when t takes on the given values 0.00, 0.25, 0.50, 0.75, 1.00. Then, because there are five given points to which the required curves are to be fitted, we observe that $n + 1 = 5$ or $n = 4$. Next, lacking tables of the orthogonal polynomials, we must compute the values of $P_{40}(x)$, $P_{41}(x)$, and $P_{42}(x)$ for the five values $x = 0, 1, 2, 3, 4$. This is a simple matter, of course, and the values shown in Table 4.3 can be calculated at once. It is then necessary to compute the sums of the products of the respective values of the y's and each of the P's. These products are shown in the last three columns of the table. The coefficients a_0, a_1, and a_2 are then given by Eq. (12):

$$a_0 = \frac{1.760}{5.000} = 0.3520 \qquad a_1 = \frac{-1.170}{2.500} = -0.4680 \qquad a_2 = \frac{0.370}{3.500} = 0.1057$$

In terms of x, the line of best fit is therefore

$$y = a_0 P_{40}(x) + a_1 P_{41}(x) = 0.3520 - 0.4680 \left(1 - \frac{x}{2}\right)$$

$$= -0.116 + 0.234x$$

and the parabola of best fit is

$$y = a_0 P_{40}(x) + a_1 P_{41}(x) + a_2 P_{42}(x)$$

$$= 0.3520 - 0.4680 \left(1 - \frac{x}{2}\right) + 0.1057 \left(1 - \frac{3x}{2} + \frac{x^2 - x}{2}\right)$$

$$= -0.0103 + 0.0226x + 0.0529x^2$$

Then, by setting $x = 4t$, we obtain the curves of best fit for the data as originally given:

$$y = -0.116 + 0.936t \qquad \text{and} \qquad y = -0.0103 + 0.0904t + 0.8464t^2$$

Table 4.3

t	x	y	P_{40}	P_{41}	P_{42}	yP_{40}	yP_{41}	yP_{42}
0.00	0	0.00	1.000	1.000	1.000	0.000	0.000	0.000
0.25	1	0.06	1.000	0.500	-0.500	0.060	0.030	-0.030
0.50	2	0.20	1.000	0.000	-1.000	0.200	0.000	-0.200
0.75	3	0.60	1.000	-0.500	-0.500	0.600	-0.300	-0.300
1.00	4	0.90	1.000	-1.000	1.000	0.900	-0.900	0.900

$$\sum_{x=0}^{4} P_{40}{}^2 = 5.000 \qquad \sum_{x=0}^{4} yP_{40} = 1.760$$

$$\sum_{x=0}^{4} P_{41}{}^2 = 2.500 \qquad \sum_{x=0}^{4} yP_{41} = -1.170$$

$$\sum_{x=0}^{4} P_{42}{}^2 = 3.500 \qquad \sum_{x=0}^{4} yP_{42} = 0.370$$

Using Eq. (14), we find that the sum of the squares of the departures of the points from the line and from the parabola of best fit are, respectively, $E_1 = 0.0465$ and $E_2 = 0.0074$. From the relative size of E_1 and E_2 we conclude that the parabola fits the data significantly better than the straight line does.

In various applications the position of a moving object is observed at a series of equally spaced times, and its velocity and acceleration are required at these times. Clearly, these can be estimated by using the formulas for numerical differentiation which we obtained in Sec. 4.3. However, since the interpolation polynomials from which these formulas of numerical differentiation were derived fit the raw data *exactly*, they and any formulas based on them are seriously influenced by even small errors in the data. On the other hand, a polynomial curve fitted to the data or a portion of the data by the method of least squares will be less influenced by random errors and will represent more nearly the underlying, presumably smooth trend of the data. Hence, derivatives computed from such approximating functions will in general be more accurate than those computed by differentiating interpolation polynomials. These ideas are illustrated geometrically in Fig. 4.5, where it is clear that the slope of the interpolation polynomial fluctuates markedly from point to point, whereas the slope of the least-squares approximation changes in a smooth fashion which is almost certainly a more reliable description of the trend the data would exhibit if free from random errors.

In applying these ideas to the analysis of observed positional data, it is customary to fit a polynomial curve of relatively low degree, say a parabola, to successive sets of five, seven, or nine observations and then take the ordinate and the first and second derivatives of this approximating polynomial at the central point of the set as the corrected, or "smoothed," position, velocity, and acceleration of the body at that instant.

To illustrate this technique, let us use the method of orthogonal polynomials to fit a parabola to the five points

t	x	y	
$-2h$	0	y_{-2}	
$-h$	1	y_{-1}	
0	2	y_0	$x = 2 + \dfrac{t}{h}$
h	3	y_1	
$2h$	4	y_2	

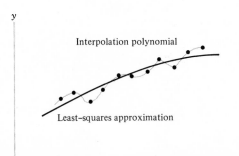

y

Interpolation polynomial

Least–squares approximation

x

Figure 4.5
The relative smoothness of an interpolation polynomial and a least-squares approximation to a set of points.

To do this, we use the coefficient formula (12) and the values of $P_{40}(x)$, $P_{41}(x)$, $P_{42}(x)$, together with the sums of the squares of these values, as tabulated in Example 2. The results are found immediately to be

$$a_0 = \tfrac{1}{5}(y_{-2} + y_{-1} + y_0 + y_1 + y_2)$$
$$a_1 = \tfrac{2}{5}(y_{-2} + \tfrac{1}{2}y_{-1} - \tfrac{1}{2}y_1 - y_2)$$
$$a_2 = \tfrac{2}{7}(y_{-2} - \tfrac{1}{2}y_{-1} - y_0 - \tfrac{1}{2}y_1 + y_2)$$

and thus the formula for the approximating polynomial is

$$(15) \quad a_0 P_{40}(x) + a_1 P_{41}(x) + a_2 P_{42}(x) = \tfrac{1}{5}(y_{-2} + y_{-1} + y_0 + y_1 + y_2)$$
$$+ \tfrac{2}{5}(y_{-2} + \tfrac{1}{2}y_{-1} - \tfrac{1}{2}y_1 - y_2)\left(1 - \frac{x}{2}\right)$$
$$+ \tfrac{2}{7}(y_{-2} - \tfrac{1}{2}y_{-1} - y_0 - \tfrac{1}{2}y_1 + y_2)\left(1 - \tfrac{3}{2}x + \frac{x^2 - x}{2}\right)$$

Putting $x = 2$, we find for the smoothed midordinate

$$(16) \qquad Y_0 = \frac{-3y_{-2} + 12y_{-1} + 17y_0 + 12y_1 - 3y_2}{35}$$

To approximate dy/dt, we recall that $x = 2 + t/h$; hence

$$\frac{dy}{dt} = \frac{dy}{dx}\frac{dx}{dt} = \frac{1}{h}\frac{dy}{dx}$$

Therefore, differentiating (15) with respect to x, dividing by h, and then setting $x = 2$, we find for the smoothed value of the first derivative at the central point of the set

$$(17) \qquad Y_0' = \frac{-2y_{-2} - y_{-1} + y_1 + 2y_2}{10h}$$

Similarly, a second differentiation would yield

$$(18) \qquad Y_0'' = \frac{2y_{-2} - y_{-1} - 2y_0 - y_1 + 2y_2}{7h^2}$$

as the smoothed value of the second derivative at the central point of the set. However, it is better to find Y_0'' by applying formula (17) to the table of the smoothed first-derivative values, getting

$$(19) \qquad Y_0'' = \frac{-2Y_{-2}' - Y_{-1}' + Y_1' + 2Y_2'}{10h}$$

or, replacing each derivative by its expression in terms of the appropriate y values from (17),

$$(20) \quad Y_0'' = \frac{4y_{-4} + 4y_{-3} + y_{-2} - 4y_{-1} - 10y_0 - 4y_1 + y_2 + 4y_3 + 4y_4}{100h^2}$$

Since Eqs. (16) and (17) require a knowledge of two ordinates on each side of those being smoothed, it is evident that these equations can be used only for points after the second and before the $(n - 1)$st in a table of data. Similarly, formula (20) for the second derivative can be used only between the fifth and the $(n - 4)$th points, inclusive.

To smooth to the ends of a table, we must derive auxiliary formulas from Eq. (15) by evaluating it and its derivatives at $x = 0, 1, 3$, and 4 as well as at $x = 2$. These results will be found among the exercises at the end of this section. In general, central formulas, i.e., formulas in which the element being smoothed is as near as possible to the central member of the set of data appearing in the smoothing formula, should be used wherever possible.

The method of least squares is not limited in its application to problems in which the equations to be satisfied are linear. Sometimes, by a suitable transformation, the problem can be converted into one in which the parameters do enter linearly. For instance, to fit an equation of the important type $y = ae^{bx}$, we can take the natural logarithm of each side, getting

$$\ln y = \ln a + bx$$

Then, considering x and $\ln y$ as new variables, say X and Y, and $\ln a$ and b as new parameters, say A and B, we can regard the problem as requiring the determination of A and B such that the *linear* equation

$$Y = A + BX$$

gives the best possible fit to the known pairs of values $X \, (=x)$ and $Y \, (= \ln y)$. Once A has been found it is, of course, a simple matter to find the actual parameter a, since $A = \ln a$.

Similarly, fitting a function $y = kx^n$ can be reduced to a linear problem by first taking logarithms (preferably to the base 10), getting

$$\log y = \log k + n \log x$$

This equation is linear in the parameters $K = \log k$ and $N = n$. Hence the determination of the parameters can be carried out as outlined above.

On the other hand, it is not possible to make a rigorous linearization of general systems of nonlinear equations of condition. But if a reasonable approximation to a solution of such a system is available, an approximate linearization of the problem can be achieved in the following way.

Let the equations to be satisfied (as nearly as possible) be

$$(21) \qquad f_1(x,y) = 0, f_2(x,y) = 0, \ldots, f_n(x,y) = 0$$

and suppose that (x_0, y_0) is known, by inspection or otherwise, to be an approximate solution of this system. Then we can expand each function in a generalized Taylor series about the point (x_0, y_0), getting, for $i = 1, 2, \ldots, n$,

$$f_i(x,y) = f_i(x_0, y_0) + \frac{\partial f_i}{\partial x}\bigg|_{x_0, y_0} (x - x_0) + \frac{\partial f_i}{\partial y}\bigg|_{x_0, y_0} (y - y_0)$$

$$+ \frac{1}{2}\left[\frac{\partial^2 f_i}{\partial x^2}\bigg|_{x_0, y_0} (x - x_0)^2 + 2 \frac{\partial^2 f_i}{\partial x \, \partial y}\bigg|_{x_0, y_0} (x - x_0)(y - y_0) \right.$$

$$\left. + \frac{\partial^2 f_i}{\partial y^2}\bigg|_{x_0, y_0} (y - y_0)^2 \right] + \cdots$$

Now, if (x_0, y_0) is a reasonable approximation to the required solution, the quantities $x - x_0$ and $y - y_0$ will be small, and hence their squares, products, and higher powers will be negligible in comparison with the quantities themselves. Omitting these quantities thus reduces the set (21) to the system

$$(22) \quad f_i(x,y) = f_i(x_0,y_0) + \frac{\partial f_i}{\partial x}\bigg|_{x_0,y_0} (x - x_0) + \frac{\partial f_i}{\partial y}\bigg|_{x_0,y_0} (y - y_0) = 0$$

$$i = 1, 2, \ldots, n$$

which is linear in the unknown corrections $x - x_0$ and $y - y_0$. The method of least squares can now be applied to the system (22) in a straightforward way, following which the preliminary estimate (x_0,y_0) can be appropriately corrected. Of course, if desired, the functions $f_i(x,y)$ can be expanded about the corrected solution (x_1,y_1) and the process repeated. The extension to systems with more than two unknowns

$$f_1(x,y,z,\ldots) = 0, f_2(x,y,z,\ldots) = 0, \ldots, f_n(x,y,z,\ldots) = 0$$

is immediate.

EXAMPLE 3

Fit an equation of the form $y = kx^n$ to the data

x	1	2	3	4
y	2.500	8.000	19.000	50.000

and compute the value of E.

Let us first work the problem by using the logarithmic equivalent

$$\log y = \log k + n \log x$$

of the function we are trying to fit to the data. The equations of condition are then

$$0.3979 = \log k$$
$$0.9031 = \log k + 0.3010n$$
$$1.2788 = \log k + 0.4771n$$
$$1.6990 = \log k + 0.6021n$$

and from these, by the usual process, we obtain the normal equations

$$4.0000 \log k + 1.3802n = 4.2788$$
$$1.3802 \log k + 0.6807n = 1.9049$$

Solving these simultaneously, we find $\log k = 0.3472$ and $n = 2.096$. Hence, $k = 2.224$, and the required function is

$$y = 2.224x^{2.096}$$

To find E we must evaluate the function $y = 2.224x^{2.096}$ for $x = 1, 2, 3, 4$; subtract these results from the corresponding values of y as originally given; square these differences; and add them. The work is shown in Table 4.4.

Table 4.4

x	$y \, (= 2.224x^{2.096})$	y (given)	δ	δ^2
1	2.224	2.500	0.276	0.076
2	9.510	8.000	−1.510	2.280
3	22.243	19.000	−3.243	10.517
4	40.655	50.000	9.345	87.329
				$E = 100.202$

Although we have no real basis for such a conviction, this value of E should strike us as discouragingly large, especially in view of the fact that we have tried to choose the parameters k and n to make it as small as possible. To explore the matter further, let us reconsider the problem in a more elementary way and determine k and n so that the curve will pass exactly through the points $(3,19)$ and $(4,50)$ without regard to the remaining pair of points. This requires that

$$19 = k3^n \quad \text{and} \quad 50 = k4^n$$

Dividing the second equation by the first gives us $(\frac{4}{3})^n = \frac{50}{19}$. Hence, taking logarithms,

$$n = \frac{\log 50 - \log 19}{\log 4 - \log 3} = 3.36$$

With n known, it is easy to find k from the equation $19 = k3^n$:

$$\log k = \log 19 - 3.36 \log 3 = 9.67563 - 10 \quad \text{and} \quad k = 0.474$$

Now, for the function $y = 0.474x^{3.36}$, the calculation of E leads to the following results:

x	$y \, (= 0.474x^{3.36})$	y (given)	δ	δ^2
1	0.474	2.500	2.026	4.105
2	4.865	8.000	3.135	9.828
3	19.000	19.000	0.000	0.000
4	50.000	50.000	0.000	0.000
				$E = \overline{13.933}$ (!)

This is a remarkable improvement in the closeness of fit, which surely requires explanation.

The question will become clearer if we consider the sums of the squares of the errors associated with the respective functions $y = 2.224x^{2.096}$ and $y = 0.474x^{3.36}$ when they are written in logarithmic form. These are:

x	$\log y \, (= \log 2.224 + 2.096 \log x)$	$\log y$ (given)	δ	δ^2
1	0.3471	0.3979	0.0508	0.00258
2	0.9782	0.9031	-0.0751	0.00564
3	1.3472	1.2788	-0.0684	0.00468
4	1.6091	1.6990	0.0899	0.00808
				$E = \overline{0.02098}$

and

x	$\log y \, (= \log 0.474 + 3.36 \log x)$	$\log y$ (given)	δ	δ^2
1	-0.3244	0.3979	0.7233	0.52172
2	0.6871	0.9031	0.2160	0.04666
3	1.2788	1.2788	0.0000	0.00000
4	1.6990	1.6990	0.0000	0.00000
				$E = \overline{0.56838}$

The function $y = 2.224x^{2.096}$ which we fitted logarithmically by the method of least squares fits the logarithms of the data much better than does the second function we derived. Moreover, it does this by keeping the discrepancies δ_i about equally small. However, a given difference δ in the logarithms of two numbers represents only a small difference in the numbers if the logarithms are near zero but represents a large difference if the logarithms themselves are large. For instance, with a difference of 0.10000 in the logarithms of two numbers we might have either

$$
\begin{aligned}
0.10000 &= \log \text{ of } 1.259 \\
0.00000 &= \log \text{ of } 1.000
\end{aligned}
$$

$$\text{Difference of numbers} = \quad 0.259$$

or

$$
\begin{aligned}
1.60000 &= \log \text{ of } 39.811 \\
1.50000 &= \log \text{ of } 31.623
\end{aligned}
$$

$$\text{Difference of numbers} = \quad 8.188$$

Hence, the average approximation to the original data of our problem is significantly improved by keeping the errors in the larger logarithms as small as possible, even at the expense of considerably larger errors in the smaller logarithms. And, clearly, there is no reason to believe that the function which best fits the logarithms of the data will necessarily give the best approximation to the data themselves.

As a final approach to the problem, let us now try the general method of handling nonlinear equations of condition. Assuming again an equation of the form $y = kx^n$, and substituting the four given sets of values, we find that k and n should satisfy the conditions

$$
\begin{aligned}
2.5 &= k \\
8.0 &= k2^n \\
19.0 &= k3^n \\
50.0 &= k4^n
\end{aligned}
$$

As an initial estimate of the values of k and n, let us use the values $k = 0.474$ and $n = 3.36$, which we obtained by passing the curve exactly through the points $(3,19)$ and $(4,50)$. Then expanding each of the equations of condition in a Taylor series around the initial approximation $(0.474, 3.36)$, we find

$$f_1(k,n) \equiv k - 2.500 = -2.026 + (k - 0.474) = 0$$

$$f_2(k,n) \equiv k2^n - 8.000 \doteq (4.865 - 8.000) + 2^n|_{0.474,3.36}(k - 0.474)$$

$$+ k2^n \ln 2|_{0.474,3.36}(n - 3.36)$$

$$= -3.135 + 10.267(k - 0.474) + 3.372(n - 3.36) = 0$$

$$f_3(k,n) \equiv k3^n - 19.000 \doteq (19.000 - 19.000) + 3^n|_{0.474,3.36}(k - 0.474)$$

$$+ k3^n \ln 3|_{0.474,3.36}(n - 3.36)$$

$$= 40.098(k - 0.474) + 20.874(n - 3.36) = 0$$

$$f_4(k,n) \equiv k4^n - 50.000 \doteq (50.000 - 50.000) + 4^n|_{0.474,3.36}(k - 0.474)$$

$$+ k4^n \ln 4|_{0.474,3.36}(n - 3.36)$$

$$= 105.411(k - 0.474) + 69.314(n - 3.36) = 0$$

When we let $u = k - 0.474$ and $v = n - 3.36$, the approximate equations of condition are therefore

$$
\begin{aligned}
u &= 2.026 \\
10.267u + 3.372v &= 3.135 \\
40.098u + 20.874v &= 0.000 \\
105.411u + 69.314v &= 0.000
\end{aligned}
$$

The construction of the normal equations, by multiplying each equation of condition first by the coefficient of u and then by the coefficient of v in that equation and adding, is a routine matter, and we find without difficulty

$$12{,}825.740u + 8{,}178.084v = 34.213$$
$$8{,}178.084u + 5{,}251.525v = 10.571$$

Hence $u = 0.197$ and $v = -0.305$

and the corrected estimates of k and n are

$$k = 0.474 + 0.197 = 0.671$$
$$n = 3.36 - 0.305 = 3.055$$

For the function $y = 0.671x^{3.055}$ a straightforward calculation yields $E = 22.628$, which is still not as small as the value we found for the curve that passed exactly through the points (3,19) and (4,50). However, a second application, based upon expanding the equations of condition around $k = 0.671$ and $n = 3.055$, yields the improved values

$$k = 0.733 \quad \text{and} \quad n = 3.039$$

and $E = 10.052$, which is the smallest value of E we have yet found. Another repetition of the process would no doubt improve this slightly.

EXERCISES

1 Show that when a polynomial curve is fitted to a set of points by the method of least squares, the sum of the vertical distances from the points to the curve is zero.

2 Compute E in Example 1 and compare it with the sum of the squares of the vertical distances from the given points to each of the following curves:
(a) $y = 1.80 - 2.60x + 0.90x^2$ (b) $y = 1.85 - 2.70x + 0.85x^2$

3 Fit a straight line to the data

x	1	3	6	7	9
y	1	5	6	10	12

(a) By minimizing the sum of the squares of the vertical distances from the points to the line.
(b) By minimizing the sum of the squares of the horizontal distances from the points to the line.

4 Fit an equation of the form $y = a + bx + cx^2$ to the data

x	-1	0	2	3	5
y	-4	4	8	9	7

5 Find the most plausible values of x and y from the system of equations

$$x + y = 2$$
$$2x - 3y = 9$$
$$20x + 16y = 4$$

(a) Without dividing out the factor 4 from the last equation.
(b) After dividing out the factor 4 from the last equation.
Explain.

6 Fit equations of each of the forms
(a) $ax + by - 1 = 0$ (b) $ax + y - c = 0$ (c) $x + by - c = 0$
to the data

x	0	1	2	3
y	1.1	1.9	3.0	3.9

by minimizing the sum of the squares of the amounts by which each of the equations, in turn, fails to be satisfied. Compare the results and explain the differences.

7 **(a)** Verify that

$$\sum_{x=0}^{n} P_{n0}(x)P_{n1}(x) = 0 \qquad \sum_{x=0}^{n} P_{n0}(x)P_{n2}(x) = 0 \qquad \sum_{x=0}^{n} P_{n1}(x)P_{n2}(x) = 0$$

(b) Verify that

$$\sum_{x=0}^{n} P_{n0}^{2}(x) = n + 1 \qquad \sum_{x=0}^{n} P_{n1}^{2}(x) = \frac{(n+1)(n+2)}{3n}$$

$$\sum_{x=0}^{n} P_{n2}^{2}(x) = \frac{(n+1)(n+2)(n+3)}{5n(n-1)}$$

8 Using orthogonal polynomials, fit functions of each of the forms $y = a + bx$ and $y = a + bx + cx^2$ to the data

x	0.50	1.00	1.50	2.00	2.50	3.00
y	1.01	1.08	1.16	1.25	1.29	1.30

and compute the value of E for each approximation.

9 Work Exercise 8 using the data

x	0.00	0.20	0.40	0.60	0.80	1.00
y	1.00	1.25	1.50	2.00	2.00	3.20

10 Fit an equation of the form $y = kx^n$ to the data

x	1	2	3	4
y	0.10	0.80	4.00	13.00

(a) By first taking logarithms and then working with the linearized equation $\log y = \log k + n \log x$.
(b) By first obtaining approximate values of k and n and then linearizing by expanding the equations of condition in Taylor's series around these values and retaining only the linear terms.

11 Fit an equation of the form $y = Ae^{ax}$ to the data

x	1	2	3	4
y	1.65	2.70	4.50	7.35

(a) By first taking natural logarithms and then working with the linearized equation $\ln y = \ln A + ax$.
(b) By first obtaining approximate values of A and a and then linearizing by expanding the equations of condition in Taylor's series around these values and retaining only the linear terms.

12 Derive the following modifications of formulas (16) and (17):

$$Y_0 = \tfrac{1}{35}(31y_0 + 9y_1 - 3y_2 - 5y_3 + 3y_4)$$

$$Y_0' = \frac{1}{70h}(-54y_0 + 13y_1 + 40y_2 + 27y_3 - 26y_4)$$

and

$$Y_0 = \tfrac{1}{35}(9y_{-1} + 13y_0 + 12y_1 + 6y_2 - 5y_3)$$

$$Y_0' = \frac{1}{70h}(-34y_{-1} + 3y_0 + 20y_1 + 17y_2 - 6y_3)$$

13 Derive formula (14).

14 A circular arc is to be fitted to a set of points $(x_1, y_1), \ldots, (x_n, y_n)$. Discuss the relative merits of doing this by minimizing the sum of the squares of the vertical distances from the points to the circular arc and by taking the equation of the circle in the form $x^2 + y^2 + ax + by + c = 0$ and minimizing the sum of the squares of the amounts by which the coordinates of the points fail to satisfy this equation.

15 An equation of the form $y = Ae^{ax}$ is to be fitted to a set of points $(1, y_1), (2, y_2), \ldots,$ (n, y_n). By observing that y must satisfy a certain linear constant-coefficient, first-order difference equation, obtain the following equations of condition:

$$y_2 - e^a y_1 = 0$$
$$y_3 - e^a y_2 = 0$$
$$\cdots \cdots \cdots \cdots$$
$$y_n - e^a y_{n-1} = 0$$

Show how A can be found after the best least-squares approximation to a has been found from these equations. Discuss the advantages of this method relative to the method of linearizing by taking logarithms and the general method of using Taylor's series to linearize the equations of condition in a problem in which the parameters enter nonlinearly.

16 Show how the method of Exercise 15 can be adapted to the problem of fitting a curve of the form $y = Ae^{ax}$ to a set of points $(x_1, y_1), (x_2, y_2), \ldots, (x_n, y_n)$ when the x's are equally spaced at an arbitrary interval h.

17 Apply the method of Exercise 15 to the data of Exercise 11. Compute E and compare it with the values of E for parts a and b of Exercise 11.

18 Explain how the method of Exercise 15 can be extended to fitting functions of the form

$$y = Ae^{ax} + Be^{bx} + \cdots + Ke^{kx}$$

19 Explain how a continuous function can be approximated over an interval (p,q) by minimizing the integral of the square of the difference between the given function and the chosen approximation. Illustrate by:
(a) Approximating the function $y = \cos x$ over the interval $(0, \pi/2)$ with a function of the form $y = a - bx^2$.
(b) Approximating the function $y = \sin x$ over the interval $(0, \pi/2)$ with a function of the form $y = ax - bx^3$. *Hint:* Note that the derivatives of the integral with respect to a and b can be found, according to Leibnitz' rule, p. 313, by differentiating the integrand before the integration is done.

20 Compute the first derivative of $\ln x$ at $x = 2.10, 2.20,$ and 2.30 from the table

x	$\ln x$
2.00	0.69315
2.05	0.71784
2.10	0.74194
2.15	0.76547
2.20	0.78846
2.25	0.81093
2.30	0.83291
2.35	0.85442
2.40	0.87547
2.45	0.89609
2.50	0.91629

(a) by using formula (17); (b) by using formula (5), Sec. 4.3. In each case compare the results with the correct answer

$$\frac{d \ln x}{dx} = \frac{1}{x}$$

21 The following table was obtained from the data of Exercise 20 by giving to each entry a random error with numerical value between 100 and 200 in the fifth decimal place.

x	$\ln x$ ("corrupted")
2.00	0.69168
2.05	0.71599
2.10	0.74016
2.15	0.76691
2.20	0.78957
2.25	0.80943
2.30	0.83095
2.35	0.85308
2.40	0.87684
2.45	0.89723
2.50	0.91813

Using this table, compute the first derivative of $\ln x$ at $x = 2.10, 2.20,$ and 2.30 (a) by using formula (17); (b) by using formula (5), Sec. 4.3. In each case compare the results with the correct answer

$$\frac{d \ln x}{dx} = \frac{1}{x}$$

22 (a) Approximate the solution of the equation $y'' + y = 0$ for which $y_0 = 1$ and $y_0' = 0$ by assuming $y = 1 - ax^2$ and minimizing the integral from 0 to 1 of the square of the amount by which this function fails to satisfy the differential equation.
(b) Approximate the solution of the equation $y'' + y = 0$ for which $y_0 = 0$ and $y_0' = 1$ by assuming $y = x - ax^3$ and minimizing the integral from 0 to 1 of the square of the amount by which this function fails to satisfy the differential equation.

23 Approximate the solution of the equation $y'' + x^2 y = \sin x$ for which $y(0) = y(\pi) = 0$ by assuming $y = A \sin x$ and choosing A to minimize the integral from 0 to π of the square of the amount by which $A \sin x$ fails to satisfy the differential equation.

24 Show that Eqs. (4) and (5) can always be solved for a and b by showing that the determinant of the coefficients of the system is always different from zero. [Consider the discriminant of the equation

$$(\lambda x_1 + 1)^2 + (\lambda x_2 + 1)^2 + \cdots + (\lambda x_n + 1)^2 = 0$$

thought of as an equation in λ with no real roots.]

25 Show that if a line with equation $x \cos \theta + y \sin \theta - p = 0$ is fitted to a set of points $(x_1, y_1), \ldots, (x_n, y_n)$ by minimizing the sum of the squares of the perpendicular distances from the points to the line, the value of θ is given by the formula

$$\tan 2\theta = \frac{2 r_{xy} \sigma_x \sigma_y}{\sigma_x{}^2 - \sigma_y{}^2}$$

where σ_x and σ_y are, respectively, the so-called **standard deviations** of the x values and the y values,

$$\sigma_x = \frac{1}{n} \sqrt{n \sum_{i=1}^{n} x_i{}^2 - \left(\sum_{i=1}^{n} x_i\right)^2} \quad \text{and} \quad \sigma_y = \frac{1}{n} \sqrt{n \sum_{i=1}^{n} y_i{}^2 - \left(\sum_{i=1}^{n} y_i\right)^2}$$

and r_{xy} is the **coefficient of correlation** between the x values and the y values,

$$r_{xy} = \frac{n \sum_{i=1}^{n} x_i y_i - \sum_{i=1}^{n} x_i \sum_{i=1}^{n} y_i}{n^2 \sigma_x \sigma_y}$$

What is the value of p in the equation of the line of best fit?

CHAPTER 5
Mechanical and Electric Circuits

5.1 Introduction

An examination of the application of differential equations to mechanical and electrical systems is valuable for at least two reasons. First, it will furnish us with useful information about the behavior of certain physical systems of great practical interest. Second, and perhaps more important, it will provide a striking example of the role which mathematics plays in unifying widely differing phenomena. For instance, we shall see that merely by a renaming of the variables, the analysis of the motion of a weight vibrating on a spring becomes the analysis of a simple electric circuit. Moreover, this correspondence is not just qualitative or descriptive. It is quantitative, in the sense that if one is given any of a wide variety of vibrating mechanical systems, an electric circuit can be constructed whose currents or voltages, as we choose, will give the *exact* values of the displacements in the mechanical system when suitable scale factors are introduced. Since electric circuits are easy to assemble, and since currents and voltages are easy to measure, this affords a practical method of studying the vibration of complicated mechanical configurations, such as engine crankshafts, which are expensive to make and modify and whose motions are difficult to record accurately.†

5.2 Systems with One Degree of Freedom

A system which can be described by one coordinate, i.e., by one physical datum such as a displacement, an angle, a current, or a voltage, is called a **system of one degree of freedom**. A system requiring more than one coordinate for its complete description is called a **system of several degrees of freedom**. A single differential equation suffices for the mathematical description of a system of one degree of freedom. A set of simultaneous differential equations, as many equations as there are degrees of freedom, is necessary for the analysis of systems of more than one degree of freedom. We shall begin our investigations by considering, as prototypes, each of the systems shown in Fig. 5.1. In each case we assume that all the elements of the system are concentrated, or **lumped**. In other words, such things as the distributed mass of the spring in Fig. 5.1*a*, the distributed moment of inertia of the shaft in Fig. 5.1*b*, and the resistance of the leads in Fig. 5.1*c* and *d* we assume to be either negligible or taken

---

† Of course, mechanical models of electric circuits can also be constructed, but there is little practical reason for doing this.

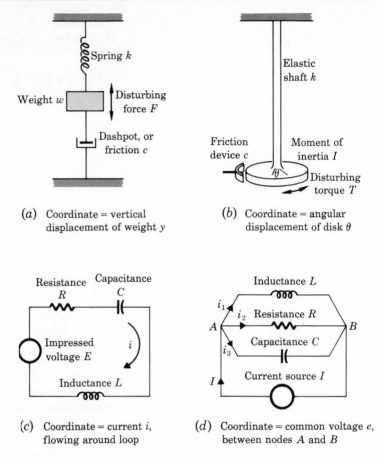

(a) Coordinate = vertical
     displacement of weight y

(b) Coordinate = angular
     displacement of disk θ

(c) Coordinate = current i,
     flowing around loop

(d) Coordinate = common voltage e,
     between nodes A and B

Figure 5.1
Four simple systems of one degree of freedom: (a) translational mechanical;
(b) torsional mechanical; (c) series electrical; (d) parallel electrical.

into account through suitable corrections added to the corresponding major elements.†

In Fig. 5.1a, we assume that the weight is guided, so that only vertical motion, without swinging, is possible. As indicated, the effect of friction is not neglected. Instead, we suppose that a retarding force proportional to the velocity acts at all times. Friction of this sort is known as **viscous friction** or **viscous damping**. Its existence is well established for moderate velocities, although for large velocities the resistance may be more nearly proportional to the square or even the cube of the velocity.

The analysis of this system is based upon Newton's law

(1)   Mass × acceleration = force

---

† In many problems these assumptions are not sufficiently accurate, and the continuous distribution of the components of the system must be considered. As we shall see in Chap. 8, this leads to *partial* rather than *ordinary* differential equations.

Measuring the displacement y from the equilibrium position of the weight, with the positive direction upward, we have

(2) $$\text{Acceleration} = \frac{d^2 y}{dt^2}$$

The most obvious force acting on the mass is the attraction of gravity

(3) $$\text{Gravitational force} = -w$$

the minus sign indicating that this force acts downward. To compute the elastic force, we note first that the spring is assumed to be of modulus k; that is, k units of force are required to stretch the spring 1 unit of length. This means that when the weight w is hung on the spring, it stretches the spring a distance equal to w/k. Hence, when the weight moves from this equilibrium level during the course of its motion, the instantaneous elongation of the spring is $w/k - y$. If this quantity is positive, the spring is stretched and therefore applies to the weight a force in the upward, or positive, direction. If this quantity is negative, the spring is compressed and therefore applies to the weight a force which acts in the downward, or negative, direction. The force the spring exerts on the weight at any time is therefore

$$\text{Force per unit elongation} \times \text{instantaneous elongation}$$

or

(4) $$\text{Elastic force} = k\left(\frac{w}{k} - y\right) = w - ky$$

To determine the frictional force, we observe that the velocity of the weight is dy/dt; hence, from the assumption of viscous damping,

(5) $$\text{Frictional force} = -c\frac{dy}{dt}$$

the minus sign indicating that the resistance always acts in opposition to the velocity. Finally, through some external agency, a disturbing force, usually periodic, may act on the system, upsetting its condition of equilibrium. We shall consider specifically the important case in which

(6) $$\text{Impressed force} = F_0 \cos \omega t \qquad F_0 \text{ a constant}$$

Substituting from Eqs. (2) to (6) into Newton's law, Eq. (1), we thus have

$$\frac{w}{g}\frac{d^2 y}{dt^2} = -w + (w - ky) - c\frac{dy}{dt} + F_0 \cos \omega t$$

or

(7) $$\frac{w}{g}\frac{d^2 y}{dt^2} + c\frac{dy}{dt} + ky = F_0 \cos \omega t$$

We note from this equation that the gravitational force on the weight is canceled by that part of the elastic force due to the initial elongation of the spring. Therefore in analyzing problems like this in the future we shall from the outset neglect both the gravitational force and the initial, or static, elongation of the spring which it produces.

Equation (7) is a typical nonhomogeneous, linear differential equation of the second order with constant coefficients, whose complete solution we can easily find by the methods of Chap. 2. Presumably it will be accompanied by given initial conditions

$$y(0) = y_0 \qquad \text{and} \qquad \frac{dy}{dt}\bigg|_{t=0} = y_0'$$

and by using these, the constants in any complete solution can be determined. However, before continuing with the solution of Eq. (7) we shall derive the equations governing the other systems shown in Fig. 5.1.

The analysis of the system in Fig. 5.1*b* is based on Newton's law in torsional form:

(8) $\qquad\qquad$ Moment of inertia $\times$ angular acceleration $=$ torque

In this case the various torques are

(9) $\qquad\qquad$ Elastic torque due to twisting of shaft $= -k\theta$

(10) $\qquad\qquad$ Viscous damping torque $= -c\dfrac{d\theta}{dt}$

(11) $\qquad\qquad$ Impressed torque $= T_0 \cos \omega t \qquad T_0$ a constant

Since the angular acceleration is $d^2\theta/dt^2$, we have, on substituting into Newton's law, Eq. (8),

$$I\frac{d^2\theta}{dt^2} = -k\theta - c\frac{d\theta}{dt} + T_0 \cos \omega t$$

or

(12) $$I\frac{d^2\theta}{dt^2} + c\frac{d\theta}{dt} + k\theta = T_0 \cos \omega t$$

This, too, is a completely familiar differential equation, and when it is accompanied by the initial conditions

$$\theta(0) = \theta_0 \qquad \text{and} \qquad \frac{d\theta}{dt}\bigg|_{t=0} = \theta_0'$$

it can easily be solved for the function describing the behavior of any particular system.

The analysis of the series, or one-loop, electric circuit shown in Fig. 5.1*c* is based on **Kirchhoff's second law**: *the algebraic sum of the potential differences around any closed loop in an electric network is zero,* or *the voltage impressed on a closed loop is equal to the sum of the voltage drops in the rest of the loop.* Using well-known electrical laws, we have

(13) $\qquad\qquad$ Voltage drop across resistance $= iR$

(14) $\qquad\qquad$ Voltage drop across capacitor $= \dfrac{1}{C}\displaystyle\int^t i\, dt$

(15) $\qquad\qquad$ Voltage drop across inductance $= L\dfrac{di}{dt}$

Thus, assuming the important case in which

(16) $\qquad\qquad$ Impressed voltage $= E_0 \cos \omega t \qquad E_0$ a constant

we have, on substituting from Eqs. (13) to (16) into Kirchhoff's second law,

$$(17) \qquad L\frac{di}{dt} + iR + \frac{1}{C}\int^t i\, dt = E_0 \cos \omega t$$

Strictly speaking, this is not a differential equation but an **integrodifferential equation**, since an integral as well as a derivative of the dependent variable i appears in it. The operational methods we shall develop in Chap. 7 will handle it directly, but before we can apply the techniques we have available at this stage, we must convert it into a pure differential equation. There are several ways of doing so. The first is to regard not i but $\int^t i\, dt$ as the dependent variable of the problem. This is not merely a mathematical strategem, for the quantity

$$Q = \int^t i\, dt$$

i.e., the integrated flow of current into the capacitor, is precisely the quantity of electricity, or electric charge, instantaneously on the capacitor. In terms of Q, then, we have the equation

$$(18a) \qquad L\frac{d^2Q}{dt^2} + R\frac{dQ}{dt} + \frac{1}{C}Q = E_0 \cos \omega t$$

subject, of course, to the given initial conditions

$$Q(0) \equiv \int^{t=0} i\, dt = Q_0 \qquad \text{and} \qquad \frac{dQ}{dt}\bigg|_{t=0} = i(0) = i_0$$

On the other hand, we can also convert Eq. (17) into a differential equation simply by differentiating it with respect to time, getting

$$(18b) \qquad L\frac{d^2i}{dt^2} + R\frac{di}{dt} + \frac{1}{C}i = -\omega E_0 \sin \omega t$$

The initial conditions required for an equation of this form are

$$i(0) = i_0 \qquad \text{and} \qquad \frac{di}{dt}\bigg|_{t=0} = i_0'$$

The first of these was given for the original equation. The second can be found from the original equation, since

$$\frac{di}{dt} = \frac{1}{L}\left(E_0 \cos \omega t - iR - \frac{1}{C}\int^t i\, dt\right)$$

and the right-hand side is completely known at $t = 0$.

To establish the differential equation describing the behavior of the parallel, or one-node-pair, electric circuit shown in Fig. 5.1d, we must use **Kirchhoff's first law:** *the algebraic sum of the currents flowing toward any point in an electrical network is zero.* Solving for i in Eqs. (13), (14), and (15) in terms of the unknown potential difference e between the points A and B we obtain, respectively,

$$(19) \qquad \text{Current through resistance} = \frac{e}{R}$$

(20) $$\text{Current (apparently) through capacitor} = C\frac{de}{dt}$$

(21) $$\text{Current through inductance} = \frac{1}{L}\int^t e\, dt$$

Thus, assuming the important case of a current source such that

(22) $$\text{Impressed current} = I_0 \cos \omega t \qquad I_0 \text{ a constant}$$

we have, on substituting from Eqs. (19) to (22) into Kirchhoff's first law,

(23) $$C\frac{de}{dt} + \frac{1}{R}e + \frac{1}{L}\int^t e\, dt = I_0 \cos \omega t$$

Again, our derivation has led us to an integrodifferential equation. To convert it into a pure differential equation we can consider $\int^t e\, dt = U$, say, as a new dependent variable, getting

(24a) $$C\frac{d^2 U}{dt^2} + \frac{1}{R}\frac{dU}{dt} + \frac{1}{L}U = I_0 \cos \omega t$$

subject to initial conditions of the form

$$U(0) \equiv \int^{t=0} e\, dt = U_0 \qquad \text{and} \qquad \frac{dU}{dt}\bigg|_{t=0} = e_0$$

On the other hand, we can simply differentiate Eq. (23) with respect to time, getting

(24b) $$C\frac{d^2 e}{dt^2} + \frac{1}{R}\frac{de}{dt} + \frac{1}{L}e = -\omega I_0 \sin \omega t$$

subject to initial conditions of the form

$$e(0) = e_0 \qquad \text{and} \qquad \frac{de}{dt}\bigg|_{t=0} = e_0'$$

When we collect the differential equations we have derived

(7) $$\frac{w}{g}\frac{d^2 y}{dt^2} + c\frac{dy}{dt} + ky = F_0 \cos \omega t \qquad \text{translational mechanical}$$

(12) $$I\frac{d^2\theta}{dt^2} + c\frac{d\theta}{dt} + k\theta = T_0 \cos \omega t \qquad \text{torsional mechanical}$$

(18a) $$L\frac{d^2 Q}{dt^2} + R\frac{dQ}{dt} + \frac{Q}{C} = E_0 \cos \omega t$$

series electrical

(18b) $$L\frac{d^2 i}{dt^2} + R\frac{di}{dt} + \frac{i}{C} = -\omega E_0 \sin \omega t$$

(24a) $$C\frac{d^2 U}{dt^2} + \frac{1}{R}\frac{dU}{dt} + \frac{U}{L} = I_0 \cos \omega t$$

parallel electrical

(24b) $$C\frac{d^2 e}{dt^2} + \frac{1}{R}\frac{de}{dt} + \frac{e}{L} = -\omega I_0 \sin \omega t$$

their essential mathematical identity becomes apparent. Moreover, we can see the possibility of various physical analogies. For instance, if we compare the translational mechanical and the series electrical systems, we find that

$$\text{Mass } \frac{w}{g} \leftrightarrow \text{inductance } L$$

$$\text{Friction } c \leftrightarrow \text{resistance } R$$

$$\text{Spring modulus } k \leftrightarrow \text{elastance } \frac{1}{C}$$

$$\text{Impressed force } F \leftrightarrow \begin{cases} \text{impressed voltage } E & \text{using (18}a) \\ \dfrac{dE}{dt} & \text{using (18}b) \end{cases}$$

$$\text{Displacement } y \leftrightarrow \begin{cases} \text{charge } Q & \text{using (18}a) \\ \text{current } i & \text{using (18}b) \end{cases}$$

and, if we compare the translational mechanical and the parallel electrical systems, we have the correspondences

$$\text{Mass } \frac{w}{g} \leftrightarrow \text{capacitance } C$$

$$\text{Friction } c \leftrightarrow \text{conductance } \frac{1}{R}$$

$$\text{Spring modulus } k \leftrightarrow \text{susceptance } \frac{1}{L}$$

$$\text{Impressed force } F \leftrightarrow \begin{cases} \text{impressed current } I & \text{using (24}a) \\ \dfrac{dI}{dt} & \text{using (24}b) \end{cases}$$

$$\text{Displacement } y \leftrightarrow \begin{cases} \displaystyle\int^{t} e\, dt & \text{using (24}a) \\ \text{voltage } e & \text{using (24}b) \end{cases}$$

Except briefly in the exercises, we shall not pursue these analogies further. Instead we shall investigate one or two of the systems in detail.

EXERCISES

1 A weight w hangs from a spring of modulus k_1, which in turn hangs from a spring of modulus k_2. What is the modulus of the suspension formed by the two springs together?

2 (a) A weight w hangs from two springs of equal length but different moduli, k_1 and k_2, each of which is connected directly to the weight. What is the modulus of the suspension formed by the two springs together?

   (b) A weight w hangs from a spring of modulus k_1 and at the same time is supported from below by a spring of modulus k_2. What is the modulus of the suspension formed by the two springs together?

3 One reasonable way to take into account the (relatively small) distributed moment of inertia of the elastic shaft in the system shown in Fig. 5.1*b* is to compute the kinetic energy of the shaft as it vibrates and then add to the moment of inertia of the disk a correction sufficient to increase its kinetic energy by an amount equal to the kinetic energy of the shaft. Following this procedure, show that if the polar moment of inertia of the entire shaft is I_s, the correction to be added to I is $\frac{1}{3}I_s$. What correction should be added to the mass in the system shown in Fig. 5.1*a* to take into account the distributed mass of the spring?

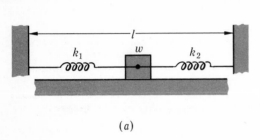

(*a*)

Figure 5.2

(*b*)

4 In Fig. 5.2*a* the unstretched lengths of the springs are l_1 and l_2, respectively. What is the equilibrium position of the weight? Set up the differential equation describing the motion of the weight and show that it is independent of the initial elongations of the springs.

5 In Fig. 5.2*b* the unstretched lengths of the springs are l_1 and l_2, respectively. What is the equilibrium position of the weight? Set up the differential equation describing the motion of the weight and show that it is independent of the initial elongations of the springs. (Because of this result and the result of Exercise 4, it is customary in setting up mass-spring problems to assume that all springs are initially unstretched.)

6 Show that the differential equation of the series electric circuit can be written in the form

$$LC \frac{d^2 e_c}{dt^2} + RC \frac{de_c}{dt} + e_c = E_0 \cos \omega t$$

where e_c is the potential difference across the capacitor, and also in the form

$$LC \frac{d^2 e_r}{dt^2} + RC \frac{de_r}{dt} + e_r = -\omega RC E_0 \sin \omega t$$

where e_r is the potential difference across the resistance. What is the form of this equation if the dependent variable is the potential difference e_i across the inductance?

7 What is the form of the differential equation of the parallel electric circuit if the dependent variable is the current through the inductance? The current through the resistance? The current (apparently) through the capacitor?

8 Let v_1 be an arbitrary frequency and let σ be an arbitrary length. Express the differential equation of the translational mechanical system in terms of the dimensionless variables

$$X = \frac{y}{\sigma} \qquad T = v_1 t$$

and show that this equation can be simplified to a form in which each coefficient is a dimensionless quantity.

9 Let v_2 be an arbitrary frequency and let α be an arbitrary angle. Express the differential equation of the torsional mechanical system in terms of the dimensionless variables

$$X = \frac{\theta}{\alpha} \qquad T = v_2 t$$

and show that this equation can be simplified to a form in which each coefficient is a dimensionless quantity.

10 Let v_3 be an arbitrary frequency and let γ be an arbitrary charge. Express the differential equation of the series electric circuit in terms of the dimensionless variables

$$X = \frac{Q}{\gamma} \qquad T = v_3 t$$

and show that this equation can be simplified to a form in which each coefficient is a dimensionless quantity.

11 Define the appropriate dimensionless variables and reduce the differential equation of the parallel electric circuit to dimensionless form.

12 Explain how the results of Exercises 8 to 11 can be used to construct electric circuits that will be exact scale models of given translational or torsional systems.

5.3 The Translational Mechanical System

In the last section we saw that the displacement y of the weight in the translational mechanical system (Fig. 5.1a) satisfies the differential equation

(1)
$$\frac{w}{g}\frac{d^2y}{dt^2} + c\frac{dy}{dt} + ky = F_0 \cos \omega t$$

In accordance with the general theory of Chap. 2, y must therefore consist of two parts. One is the *complementary function*, obtained by solving Eq. (1) when the term representing the impressed force is deleted; the other is the *particular integral*. The complementary function describes the motion of the weight in the absence of any external disturbance. This intrinsic, or natural, behavior of the system is called the **free motion**. The particular integral† describes the response of the system to a specific influence external to the system, the driving force $F_0 \cos \omega t$ in the case we are here considering. The behavior which it represents is called the **forced motion**.

The nature of the free motion of the system will depend upon the roots of the characteristic equation

$$\frac{w}{g} m^2 + cm + k = 0$$

namely,
$$m_1, m_2 = -\frac{cg}{2w} \pm \frac{g}{2w}\sqrt{c^2 - \frac{4kw}{g}}$$

Since g, w, and k are all positive and c is nonnegative, and since the radical, when real, is certainly less than c, it follows that the roots m_1 and m_2, if real, are always negative

---

† From our work in Chap. 2, we know that a nonhomogeneous linear differential equation has infinitely many particular integrals. When we speak here of *the* particular integral, we mean the unique solution containing no terms which are also in the complementary function; that is, the solution obtained by the method of undetermined coefficients.

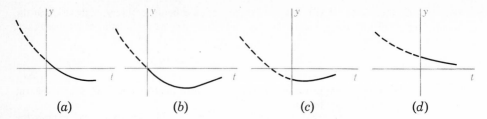

Figure 5.3
Displacement-time plots for free, overdamped, and critically damped motion.

and, if complex, have real parts which are negative or zero. We must now consider three possibilities:

$$c^2 - \frac{4kw}{g} \begin{cases} > 0 \\ = 0 \\ < 0 \end{cases}$$

If $c^2 - 4kw/g > 0$, or $c^2 > 4kw/g$, there is a relatively large amount of friction, and, naturally enough, the system is said to be **overdamped**. In this case the roots of the characteristic equation are real and unequal, and the free motion, i.e., the motion described by the complementary function, is given by

$$y = Ae^{m_1 t} + Be^{m_2 t}$$

where, as we pointed out above, both m_1 and m_2 are negative. Thus y approaches zero as time increases indefinitely. This, of course, is perfectly consistent with the familiar observation that if a system upon which no external force is acting is displaced from its equilibrium position, it will eventually return to that position as friction causes the motion to subside.

If we set $y = 0$, we obtain the equation

$$Ae^{m_1 t} + Be^{m_2 t} = 0 \qquad \text{or} \qquad e^{(m_1 - m_2)t} = -\frac{B}{A} \qquad A \neq 0$$

If A and B, which will of course be determined by the initial conditions of the problem, are of opposite sign, there is one and only one value of t which satisfies the last equation. On the other hand, since a real exponential function must always be positive, it follows that when A and B have the same sign or when one or the other of them is zero, there is no time when $y = 0$. A plot of the displacement y during the free motion of an overdamped system must therefore resemble one of the curves shown in Fig. 5.3 or the reflection of one of these curves in the t axis. Figure 5.3a, b, and c illustrates the possibilities when A and B are of opposite sign and y vanishes once and only once. Assuming that the weight starts its motion when $t = 0$, the zero of y may occur at $t = 0$ or in the physically irrelevant interval $-\infty < t < 0$. Figure 5.3d illustrates both the case when A and B are of like sign and the case when either A or B is zero and y can never vanish.

If $c^2 - 4kw/g = 0$, we have the borderline case in which the roots of the characteristic equation are real and equal:

$$m_1 = m_2 = -\frac{cg}{2w}$$

When this occurs, the motion is said to be **critically damped,** and the exact value of the damping which produces it, namely,

$$(2) \qquad\qquad\qquad c_c = 2\sqrt{\frac{kw}{g}}$$

is known as the **critical damping**. In this case the free motion is given by

$$y = Ae^{m_1 t} + Bte^{m_1 t}$$

If we set $y = 0$, we obtain

$$Ae^{m_1 t} + Bte^{m_1 t} = 0 \qquad \text{or} \qquad t = -\frac{A}{B} \qquad B \neq 0$$

If $B = 0$, there is no value of t for which $y = 0$, but in all other cases there is one and only one value of t for which $y = 0$. This may be in the physically irrelevant interval $-\infty < t < 0$, however, and so it is possible that y will not vanish in the actual motion even when $B \neq 0$. Clearly, there is no essential difference in the character of the motion in the overdamped and critically damped cases, and the possible plots of the displacement y in the critically damped cases are also represented by the curves of Fig. 5.3 and their reflections in the t axis.

If $c^2 - 4kw/g < 0$, the motion is said to be **underdamped**. The roots of the characteristic equation in this case are the conjugate complex numbers

$$m_1, m_2 = -\frac{cg}{2w} \pm i\,\frac{g}{2w}\sqrt{\frac{4kw}{g} - c^2} = -p \pm iq$$

where

$$(3) \qquad\qquad p = \frac{cg}{2w} \qquad \text{and} \qquad q = \frac{g}{2w}\sqrt{\frac{4kw}{g} - c^2}$$

The free motion is therefore described by

$$(4a) \qquad\qquad y = e^{-pt}(A \cos qt + B \sin qt)$$

or equally well by

$$(4b) \qquad\qquad y = Ge^{-pt} \cos (qt - H)$$

or by

$$(4c) \qquad\qquad y = Ke^{-pt} \sin (qt - L)$$

where A, B, G, H, K, and L are arbitrary constants.

The motion described by either (4a), (4b), or (4c) is known as a **damped oscillation,** and its general appearance is shown in Fig. 5.4. It is not periodic, since the factor e^{-pt} which multiplies the trigonometric terms is continuously decreasing. However, there are regularly spaced passages through the equilibrium position at intervals of π/q. In fact, from the description of the motion provided by Eq. (4b) it is clear that $y = 0$ whenever

$$\cos (qt - H) = 0$$

i.e., when $qt - H = \pi/2 + n\pi$ or

$$t = \frac{1}{q}\left(H + \frac{\pi}{2}\right) + \frac{n\pi}{q} \qquad n = 0, 1, 2, \dots$$

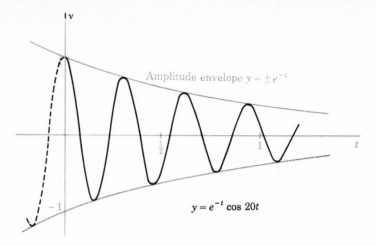

Figure 5.4
A typical displacement-time plot for an underdamped system.

Hence, we can speak of the **pseudo period** $2\pi/q$ and of the **pseudo frequency** or **frequency with damping** ω_d, defined by

$$(5) \qquad \frac{\omega_d}{2\pi} = \frac{q}{2\pi} = \frac{1}{2\pi}\frac{g}{2w}\sqrt{\frac{4kw}{g} - c^2} = \frac{1}{2\pi}\sqrt{\frac{kg}{w} - \frac{c^2 g^2}{4w^2}} \qquad \text{cycles/unit time}$$

If $c = 0$, that is, if there is no damping in the system, the motion is strictly periodic and its frequency, which we shall call the **undamped natural frequency** ω_n, is, from (5), given by

$$(6) \qquad\qquad \frac{\omega_n}{2\pi} = \frac{1}{2\pi}\sqrt{\frac{kg}{w}} \qquad \text{cycles/unit time}$$

Clearly, the frequency when damping is present is always less than the undamped natural frequency. The ratio of the two frequencies is

$$\frac{\omega_d}{\omega_n} = \frac{\sqrt{kg/w - c^2 g^2/4w^2}}{\sqrt{kg/w}}$$

$$= \sqrt{1 - \frac{c^2 g}{4kw}} = \sqrt{1 - \frac{c^2}{c_c^2}}$$

since, from Eq. (2), $c_c^2 = 4kw/g$. Figure 5.5 shows a plot of ω_d/ω_n vs. c/c_c. Evidently, if the actual damping is only a small fraction of the critical damping, as it often is, $\omega_d/\omega_n \doteq 1$ and the effect of friction on the frequency of the motion is very small. This explains why friction is usually neglected in natural-frequency calculations.

Still using Eq. (4b), it is clear that the extreme values of y occur when

$$\frac{dy}{dt} = G[-pe^{-pt}\cos(qt - H) - qe^{-pt}\sin(qt - H)] = 0$$

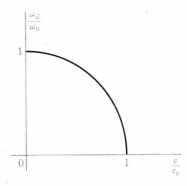

Figure 5.5
The effect of friction on frequency in an underdamped system.

i.e., when $\tan{(qt - H)} = -p/q$ or, finally, when

$$t = \frac{H}{q} - \frac{1}{q} \, \mathrm{Tan}^{-1} \frac{p}{q} + \frac{n\pi}{q} = T + \frac{n\pi}{q}$$

where T denotes the constant $H/q - (1/q) \, \mathrm{Tan}^{-1} (p/q)$.

The ratio of successive extreme displacements on the same side of the equilibrium position is a quantity of considerable importance. Its value is

$$\frac{y_n}{y_{n+2}} = \frac{y\left(T + \dfrac{n\pi}{q}\right)}{y\left[T + \dfrac{(n+2)\pi}{q}\right]}$$

$$= \frac{G \exp\left[-p\left(T + \dfrac{n\pi}{q}\right)\right] \cos\left[q\left(T + \dfrac{n}{q}\pi\right) - H\right]}{G \exp\left\{-p\left[T + \dfrac{(n+2)\pi}{q}\right]\right\} \cos\left[q\left(T + \dfrac{n+2}{q}\pi\right) - H\right]}$$

$$= e^{2\pi p/q} \, \frac{\cos{(qT + n\pi - H)}}{\cos{(qT + n\pi - H + 2\pi)}}$$

(7) $= e^{2\pi p/q}$

Since this result depends only on the parameters of the system and not on n, we have thus established the following remarkable result.

THEOREM 1   The ratio of successive maximum (or minimum) displacements remains constant throughout the entire free motion of an underdamped system.

If we take the natural logarithm of the expression in (7), we have

(8) $$\ln \frac{y_n}{y_{n+2}} = \frac{2\pi p}{q}$$

This quantity, known as the **logarithmic decrement** δ, is a convenient measure, in **nepers per cycle**, of the rate at which the motion dies away.† Substituting for p and q from (3) into (8), we find

$$\delta = \frac{2\pi p}{q} = 2\pi \frac{cg/2w}{(g/2w)\sqrt{4kw/g - c^2}} = 2\pi \frac{c}{\sqrt{c_c^2 - c^2}}$$

Solved for c/c_c, this becomes

$$(9) \qquad \frac{c}{c_c} = \frac{\delta}{\sqrt{\delta^2 + 4\pi^2}}$$

Since y_n and y_{n+2} are quantities which are relatively easy to measure, δ can easily be computed. Then from Eq. (9) the fraction of critical damping present in a given system can be found at once.

Now that we have investigated the free motion of the translational mechanical system in the overdamped, critically damped, and underdamped cases, it remains for us to consider the forced motion. To do this we must, of course, find a particular integral of Eq. (1):

$$(1) \qquad \frac{w}{g}\frac{d^2y}{dt^2} + c\frac{dy}{dt} + ky = F_0 \cos \omega t$$

Assuming, as usual,

$$Y = A \cos \omega t + B \sin \omega t$$

and substituting into (1), collecting terms, and equating the coefficients of $\cos \omega t$ and $\sin \omega t$ on each side of the equation, we obtain the two conditions

$$\left(k - \omega^2 \frac{w}{g}\right) A + \omega c B = F_0$$

$$-\omega c A + \left(k - \omega^2 \frac{w}{g}\right) B = 0$$

from which we find immediately

$$A = \frac{k - \omega^2(w/g)}{[k - \omega^2(w/g)]^2 + (\omega c)^2} F_0$$

$$B = \frac{\omega c}{[k - \omega^2(w/g)]^2 + (\omega c)^2} F_0$$

Hence

$$Y = F_0 \frac{[k - \omega^2(w/g)] \cos \omega t + \omega c \sin \omega t}{[k - \omega^2(w/g)]^2 + (\omega c)^2}$$

$$= \frac{F_0}{\sqrt{[k - \omega^2(w/g)]^2 + (\omega c)^2}}$$

$$\times \left\{\frac{k - \omega^2(w/g)}{\sqrt{[k - \omega^2(w/g)]^2 + (\omega c)^2}} \cos \omega t + \frac{\omega c}{\sqrt{[k - \omega^2(w/g)]^2 + (\omega c)^2}} \sin \omega t\right\}$$

---

† Equivalently, though less conventionally, the rate of attenuation could be expressed in **decibels per cycle** by means of the definition

$$\text{Decibels} = 20 \ln \frac{y_n}{y_{n+2}}$$

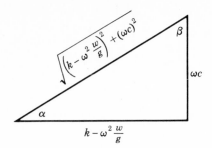

Figure 5.6
The triangle defining the phase angles appearing in Eqs. (10a) and (10b).

Referring to the triangle shown in Fig. 5.6, it is evident that Y can be written in either of the equivalent forms

$$Y = \frac{F_0}{\sqrt{[k - \omega^2(w/g)]^2 + (\omega c)^2}}(\cos \omega t \cos \alpha + \sin \omega t \sin \alpha)$$

$$(10a) \qquad = \frac{F_0}{\sqrt{[k - \omega^2(w/g)]^2 + (\omega c)^2}} \cos(\omega t - \alpha)$$

or

$$Y = \frac{F_0}{\sqrt{[k - \omega^2(w/g)]^2 + (\omega c)^2}}(\cos \omega t \sin \beta + \sin \omega t \cos \beta)$$

$$(10b) \qquad = \frac{F_0}{\sqrt{[k - \omega^2(w/g)]^2 + (\omega c)^2}} \sin(\omega t + \beta)$$

The first of these equations is the more convenient because it involves the same function (the cosine) as the excitation term in the differential equation. Hence, the phase relation between the response of the system and the disturbing force can easily be inferred. Accordingly, we shall continue with the first expression for Y.

If we divide the numerator and denominator by k and rearrange slightly, we obtain

$$Y = \frac{\dfrac{F_0}{k}}{\sqrt{\left(1 - \omega^2 \dfrac{w}{kg}\right)^2 + \left(\dfrac{\omega c}{k}\right)^2}} \cos(\omega t - \alpha)$$

$$= \frac{\dfrac{F_0}{k}}{\sqrt{\left(1 - \dfrac{\omega^2}{kg/w}\right)^2 + \left(\dfrac{\omega}{\sqrt{kg/w}} \dfrac{2c}{\sqrt{4kw/g}}\right)^2}} \cos(\omega t - \alpha)$$

$$= \frac{\delta_{st}}{\sqrt{\left(1 - \dfrac{\omega^2}{\omega_n^2}\right)^2 + \left(2 \dfrac{\omega}{\omega_n} \dfrac{c}{c_c}\right)^2}} \cos(\omega t - \alpha)$$

where $\delta_{st} = F_0/k$ is the **static deflection** which a *constant* force of magnitude F_0 would produce in a spring of modulus k and, as before, $\omega_n{}^2 = kg/w$ and $c_c{}^2 = 4kw/g$.

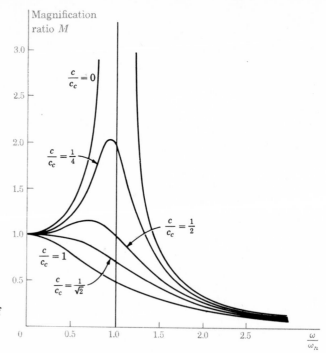

Figure 5.7
Curves of the magnification ratio M as a function of the impressed frequency ratio ω/ω_n for various amounts of damping.

The quantity

$$(11) \qquad M = \frac{1}{\sqrt{\left(1 - \dfrac{\omega^2}{\omega_n{}^2}\right)^2 + \left(2\dfrac{\omega}{\omega_n}\dfrac{c}{c_c}\right)^2}}$$

is called the **magnification ratio**. It is the factor by which the static deflection produced in a spring of modulus k by a steady force F_0 must be multiplied in order to give the amplitude of the vibrations which result when the same force acts dynamically with frequency ω. Curves of the magnification ratio M plotted against the **frequency ratio** ω/ω_n for various values of the **damping ratio** c/c_c are shown in Fig. 5.7. An inspection of Fig. 5.7 reveals the following interesting facts:

a. $M = 1$, regardless of the amount of damping, if $\omega/\omega_n = 0$.

b. If $0 < c/c_c < 1/\sqrt{2}$, M rises to a maximum as ω/ω_n increases from 0, the peak value of M occurring in all cases before the impressed frequency reaches the undamped natural frequency ω_n.

c. The smaller the amount of friction, the larger the maximum of M, until for conditions of undamped resonance, namely, $c/c_c = 0$ and $\omega/\omega_n = 1$, infinite magnification, i.e., a response of infinite amplitude, occurs.

d. If $c/c_c \geq 1/\sqrt{2}$, the magnification ratio decreases steadily as ω/ω_n increases from 0.

e. For all values of c/c_c, M approaches zero as the impressed frequency is raised indefinitely above the undamped natural frequency of the system.

The angle $\qquad \alpha = \tan^{-1}\dfrac{\omega c}{k - \omega^2(w/g)} \qquad 0 \leq \alpha \leq \pi$

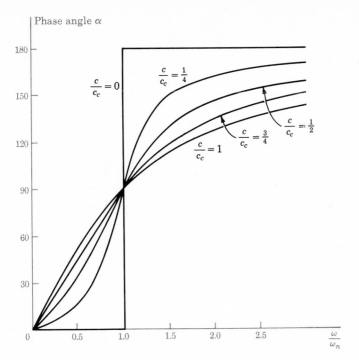

Figure 5.8
Curves of the phase angle α as a function of the impressed frequency ratio ω/ω_n for various amounts of damping.

which appears in Eq. (10a) and is shown in Fig. 5.6, is known as the **phase angle** or **angle of lag** of the response. Like the magnification ratio, it can easily be expressed in terms of the dimensionless parameters ω/ω_n and c/c_c. To do this we need only divide the numerator and denominator of the right-hand side of the last expression by k and rearrange slightly:

$$\alpha = \tan^{-1}\frac{\omega c/k}{1 - \omega^2(w/kg)}$$

$$= \tan^{-1}\frac{\dfrac{\omega}{\sqrt{kg/w}}\dfrac{2c}{\sqrt{4kw/g}}}{1 - \dfrac{\omega^2}{kg/w}}$$

(12)
$$= \tan^{-1}\frac{2(\omega/\omega_n)(c/c_c)}{1 - (\omega/\omega_n)^2}$$

It is important to note that α is *not* to be read from the principal-value branch of the arctangent relation, for it is evident from Fig. 5.6 that $\sin \alpha$ is always positive, whereas $\cos \alpha$ can be either positive or negative. Hence, α must be an angle between 0 and π and not an angle in the principal-value range $(-\pi/2, \pi/2)$. Plots of α vs. the frequency ratio ω/ω_n for various values of the damping ratio c/c_c are shown in Fig. 5.8.

The physical significance of α is shown in Fig. 5.9. The displacement Y reaches its

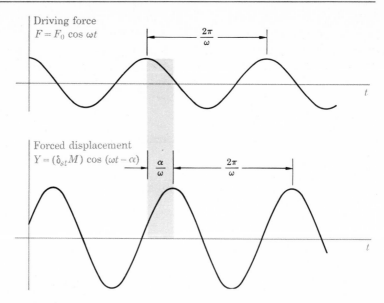

Figure 5.9
The significance of the phase angle as a measure of the time by which the response lags the excitation in a mechanical system.

maxima α/ω units of time *after* or *later than* the driving force reaches its corresponding peak values. When the frequency of the disturbing force is well below the undamped natural frequency of the system, α is small and the forced vibrations lag only slightly behind the driving force. When the impressed frequency is equal to the natural frequency, $\alpha = \pi/2$ and the response of the system lags the excitation by one-fourth cycle. As ω increases indefinitely, the lag of the response approaches half a cycle, or, in other words, the response becomes 180° out of phase with respect to the driving force.

The results of our detailed study of the vibrating weight can now be summarized. The complete motion of the system consists of two parts. The first is described by the complementary function of the underlying differential equation and may be either oscillatory or nonoscillatory, according as the amount of friction in the system is less than or more than the critical damping figure for the system. In any case, however, this part of the solution contains factors which decay exponentially, and it soon becomes vanishingly small. For this reason it is known as the **transient**. The general expression for the transient contains two arbitrary constants, which, after the complete solution has been constructed, must be determined to fit the initial conditions of displacement and velocity. The second part of the solution is described by the particular integral. In the highly important case in which the system is acted upon by a pure harmonic disturbing force (we considered only $F = F_0 \cos \omega t$, but without exception all our conclusions are equally valid for $F = F_0 \sin \omega t$), this term represents a harmonic displacement of the same frequency as the excitation but lagging behind the latter. The amplitude of this displacement is a determinate multiple of the steady deflection which would be produced in the system by a constant force of the same magnitude as the actual, alternating force. This factor of magnification, like the amount of lag, depends only on two dimensionless parameters, ω/ω_n, which is the ratio of the impressed frequency to the undamped natural frequency of the system, and c/c_c, which is the ratio of the actual amount of damping to the critical damping of the system. The motion described by the particular integral does not decay as time goes on but continues its periodic behavior indefinitely. For this reason, although it is obviously not independent of time, it is known as the **steady state**.

EXAMPLE 1

A 50-lb weight is suspended from a spring of modulus 20 lb/in. When the system is vibrating freely, it is observed that in consecutive cycles the maximum displacement decreases by 40 percent. If a force equal to 10 cos ωt acts upon the system, find the amplitude and phase lag of the resultant steady-state motion if (a) $\omega = 6$, (b) $\omega = 12$, and (c) $\omega = 18$ rad/s.

The first step here is to determine the amount of damping present in the system. From the given data it is clear that

$$y_{n+2} = 0.60 y_n$$

and thus that

$$\delta = \ln \frac{y_n}{y_{n+2}} = \ln \frac{1}{0.60} = 0.511$$

Hence, by Eq. (9),

$$\frac{c}{c_c} = \frac{\delta}{\sqrt{\delta^2 + 4\pi^2}} = \frac{0.511}{\sqrt{(0.511)^2 + 4\pi^2}} = 0.081$$

Next we must compute the undamped natural frequency of the system. Using Eq. (6), we have

$$\omega_n = \sqrt{\frac{kg}{w}} = \sqrt{\frac{20 \times 384}{50}} = 12.4 \text{ rad/s}$$

Knowing c/c_c and ω_n, we can now use Eq. (11) to compute the magnification ratio and Eq. (12) to compute the phase shift for $\omega = 6$, 12, and 18. Direct substitution gives the values

ω	M	α
6	1.30	0.10
12	5.94	1.19
18	0.88	2.93

Finally, it is clear that a 10-lb force, acting statically, will stretch a spring of modulus 20 lb/in a distance

$$\delta_{st} = \tfrac{10}{20} = 0.5 \text{ in}$$

Hence, multiplying this static deflection by the appropriate values of the magnification ratio, we find for the amplitude A of the steady-state motion the values

ω	6	12	18
A	0.65	2.97	0.44

Using these values and the corresponding values of α, we can now write the equations describing the steady-state motion in the three given cases:

ω	Y
6	0.65 cos (6t − 0.10)
12	2.97 cos (12t − 1.19)
18	0.44 cos (18t − 2.93)

The amplitude corresponding to the impressed frequency $\omega = 12$ is much larger than either of the others because this frequency very nearly coincides with the natural frequency of the system, $\omega_n = 12.4$.

EXAMPLE 2

A system containing a negligible amount of damping is disturbed from its equilibrium position by the sudden application at $t = 0$ of a force equal to $F_0 \sin \omega t$. Discuss the subsequent motion of the system if ω is close to the natural frequency ω_n.

The differential equation to be solved here is

$$\frac{w}{g} \frac{d^2y}{dt^2} + ky = F_0 \sin \omega t$$

The complementary function is, clearly,

$$A \cos \sqrt{\frac{kg}{w}} t + B \sin \sqrt{\frac{kg}{w}} t = A \cos \omega_n t + B \sin \omega_n t$$

and it is easy to verify that a particular integral is

$$Y = \frac{F_0}{k - \omega^2(w/g)} \sin \omega t = \frac{\omega_n^2 \delta_{st}}{\omega_n^2 - \omega^2} \sin \omega t$$

Hence a complete solution can be written

$$y = A \cos \omega_n t + B \sin \omega_n t + \frac{\omega_n^2 \delta_{st}}{\omega_n^2 - \omega^2} \sin \omega t$$

Since $y = 0$ when $t = 0$, we must have $A = 0$, leaving

(13)
$$y = B \sin \omega_n t + \frac{\omega_n^2 \delta_{st}}{\omega_n^2 - \omega^2} \sin \omega t$$

and
$$v = \frac{dy}{dt} = \omega_n B \cos \omega_n t + \frac{\omega_n^2 \delta_{st}}{\omega_n^2 - \omega^2} \omega \cos \omega t$$

Substituting $v = 0$ and $t = 0$ in the last equation, we obtain

$$0 = \omega_n B + \frac{\omega \omega_n^2 \delta_{st}}{\omega_n^2 - \omega^2} \qquad \text{or} \qquad B = \frac{\omega \omega_n \delta_{st}}{\omega^2 - \omega_n^2}$$

Hence, substituting into (13), we find for the required solution

$$y = \frac{\omega_n \delta_{st}}{\omega^2 - \omega_n^2} (\omega \sin \omega_n t - \omega_n \sin \omega t)$$

If the impressed frequency ω is very close to the natural frequency ω_n, we may, for descriptive purposes, substitute ω_n for ω in the first term in the expression in parentheses (although of course we cannot do this in the denominator of the coefficient fraction). This gives us

$$y \doteq \frac{\omega_n^2 \delta_{st}}{\omega^2 - \omega_n^2} (\sin \omega_n t - \sin \omega t)$$

If we now convert the difference of the sine terms into a product, we get

$$y \doteq -\omega_n^2 \delta_{st} \frac{2 \cos [(\omega + \omega_n)/2]t \, \sin [(\omega - \omega_n)/2]t}{(\omega + \omega_n)(\omega - \omega_n)}$$

If we denote the small quantity $\omega - \omega_n$ by 2ϵ and note that $\omega + \omega_n$ is approximately equal to both 2ω and $2\omega_n$, the last expression can be written in the form

$$y \doteq -\omega_n \delta_{st} \frac{\sin \epsilon t}{2\epsilon} \cos \omega t$$

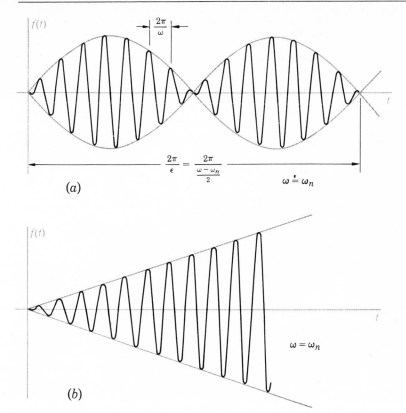

Figure 5.10
The phenomenon of beats.

Since ϵ is a small quantity, the period $2\pi/\epsilon$ of the factor $\sin \epsilon t$ is large. Hence, the form of the last expression shows that y can be regarded as essentially a periodic function, $\cos \omega t$, of frequency ω, with slowly varying amplitude

$$\omega_n \delta_{st} \frac{\sin \epsilon t}{2\epsilon}$$

Figure 5.10 shows the general nature of this behavior when ω is nearly but not quite equal to ω_n and in the limiting case when $\omega = \omega_n$ and conditions of **pure resonance** exist.

This is one of the simplest illustrations of the phenomenon of **beats**, which occurs whenever an impressed frequency is close to a natural frequency of a system or whenever two slightly different frequencies are impressed upon a system regardless of what its natural frequencies may be. A waveform of variable amplitude, like that shown in Fig. 5.10a, is said to be **amplitude-modulated**, and the lighter curves to which the actual wave periodically rises and falls are called its **envelope**.

EXERCISES

1   If friction is neglected, show that the natural frequency of a system consisting of a mass on an elastic suspension is approximately equal to $3.13/\sqrt{\delta_{st}}$ Hz,† where δ_{st} is the deflection, in inches, produced in the suspension when the mass hangs in static equilibrium.

---

† The unit *cycles per second* is now given the name hertz (Hz) in honor of the German physicist Heinrich Hertz (1857–1894).

2   A motor of unknown weight is set on a felt mounting pad of unknown spring constant. What is the natural frequency of the system if the motor is observed to compress the pad $\frac{1}{16}$ in?

3   Show that the logarithmic decrement δ can also be computed by the formula

$$\delta = \frac{1}{k} \ln \frac{y_n}{y_{n+2k}} \qquad k = 1, 2, 3, \ldots$$

4   Prove that the logarithmic decrement δ is equal to the natural logarithm of the ratio of *any* nonzero displacement to the displacement one full cycle later.

5   For a given value of c/c_c, determine the minimum number of cycles required to produce a reduction of at least 50 percent in the maxima of a damped oscillation.

6   Investigate the motion of a weight hanging on a spring when the disturbing force is equal to $F_0 \sin \omega t$ instead of $F_0 \cos \omega t$. In particular, show that Eqs. (11) and (12) for the magnification ratio and phase shift, respectively, are still the same.

7   A weight of 54 lb hangs from a spring of modulus 36 lb/in. During the free motion of the system it is observed that the maximum displacement of the weight decreases to one-tenth of its value in five complete cycles of the motion. Find the equation describing the steady-state motion produced by a force equal to $6 \sin 15t$.

8   A weight of 96 lb hangs from a spring of modulus 25 lb/in. The damping in the system is 60 percent of critical. Determine the motion of the weight if it is pulled downward 1 in from its equilibrium position and released with an upward velocity of 2 in/s.

9   Work Exercise 8 if a constant force of 50 lb is suddenly applied to the system when it is at rest in its equilibrium position.

10 A weight of 128 lb hangs from a spring of modulus 75 lb/in. The damping in the system is 28 percent of critical. Determine the motion of the weight if it is pulled downward 2 in from its equilibrium position and suddenly released.

11 Work Exercise 10 if a force equal to $F_0 e^{-10t}$ is suddenly applied to the system when it is at rest in its equilibrium position.

12 If c/c_c is small, show that the logarithmic decrement is approximately

$$\delta = \frac{y_n - y_{n+2}}{y_n} = \frac{\Delta y_n}{y_n}$$

13 Show that the energy dissipated during the nth cycle of a damped oscillation is equal to $(k/2)(y_n{}^2 - y_{n+2}^2)$. Hence, using the result of Exercise 12, show that when c/c_c is small, the energy loss during the nth cycle is approximately $k\, y_n{}^2 \delta$.

14 If the roots of the characteristic equation in the overdamped case are $m = -r \pm s$, show that in general the complementary function can be written

$$y = Ae^{-rt} \cosh (st + B) \quad \text{or} \quad y = Ce^{-rt} \sinh (st + D)$$

according as y has no real zero or one real zero. Are there any exceptions?

15 Show that a complete description of the motion of an underdamped system acted upon by a force $F_0 \cos \omega t$ can be written in the form

$$y = \exp\left[-\left(\frac{c}{c_c}\right) \omega_n t \right] \left[A \cos \sqrt{1 - \left(\frac{c}{c_c}\right)^2} \, \omega_n t + B \sin \sqrt{1 - \left(\frac{c}{c_c}\right)^2} \, \omega_n t \right]$$
$$+ \delta_{st} M \cos (\omega t - \alpha)$$

16 (a) Obtain a result comparable to that of Exercise 15 if the system is critically damped.

    (b) Obtain a result comparable to that of Exercise 15 if the system is overdamped.

17 In many applications involving forces arising from rotating parts that have become unbalanced, the amplitude of the sinusoidal disturbing force acting on a system is not constant but varies as the square of the frequency. If a weight suspended from a spring is acted upon by a force of this character, determine the steady-state motion. In particular, determine the form of the magnification ratio and the formula for the angle of lag.

18 Show that maxima of the plots of the magnification ratio vs. the frequency ratio under the conditions of Exercise 17 always occur at values of the impressed frequency ω which are greater than the undamped natural frequency of the system ω_n.

19 A uniform bar of length l and weight w rests on two parallel rollers which rotate about fixed axes as shown in Fig. 5.11. Friction between the bar and each roller is

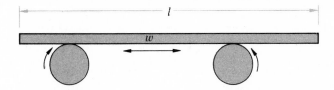

Figure 5.11

assumed to be "dry," or coulomb, i.e., proportional to the normal force between the bar and the roller, the proportionality constant being the so-called **coefficient of friction** μ. When the bar, which always remains in a line perpendicular to the axes of the rollers, is displaced slightly from a symmetrical position, it executes small oscillations in the horizontal direction. Determine the period of these oscillations, and show how the value of μ can thus be found experimentally.

20 A particle of weight w moves along the x axis under the influence of a force equal to $-kx$. Friction in the system is assumed to be dry rather than viscous; i.e., it is proportional to the normal force between the particle and the surface on which it moves and does not depend on the velocity. Show that the motion of the particle is governed by the differential equations

$$\frac{w}{g}\frac{d^2x}{dt^2} + kx = \begin{cases} \mu w & \text{particle moving to left} \\ -\mu w & \text{particle moving to right} \end{cases}$$

If the particle starts from rest at the point $x = x_0$, find x as a function of t. What is the decrease in amplitude per cycle? When will the particle come to rest?

21 Show that the maxima of the curves of the magnification ratio vs. the frequency ratio occur when

$$\frac{\omega}{\omega_n} = \sqrt{1 - 2\left(\frac{c}{c_c}\right)^2}$$

22 If y_0 and v_0 are, respectively, the initial displacement and initial velocity with which an overdamped system begins its motion, show that

$$\frac{w}{g}v_0^2 + cv_0 y_0 + ky_0^2 > 0$$

is the condition that the complementary function have a real zero.

23 In addition to the condition of Exercise 22, what further requirement is necessary to ensure that the zero of the complementary function will be nonnegative, i.e., will occur during the actual motion?

24 An overdamped system begins to move from its equilibrium position with velocity v_0. Show that its maximum displacement occurs when

$$t = \frac{1}{\omega_n\sqrt{(c/c_c)^2 - 1}}\tanh^{-1}\sqrt{1 - \left(\frac{c_c}{c}\right)^2}$$

Hint: Use the result of Exercise 14.

25 In Exercise 24, show that the maximum displacement is

$$y_{\text{max}} = \frac{v_0}{\omega_n}\left(\tan\frac{\gamma}{2}\right)^{\sec\gamma} \qquad \text{where } \gamma = \text{Sin}^{-1}\frac{c_c}{c}$$

26   Investigate the answers to Exercises 24 and 25 in the limit when c/c_c approaches 1. Check your conclusions by working directly with the equation for the transient in the critically damped case.

27   Show that the maximum displacements during the free motion of an underdamped system do not occur midway between the zeros of the displacement but precede the midpoints by the constant amount

$$\frac{\text{Sin}^{-1} (c/c_c)}{\omega_n\sqrt{1 - (c/c_c)^2}}$$

28   Work Example 2 if the impressed force is $F_0 \cos \omega t$.

29   A 48-lb weight hangs from a spring of modulus 50 lb/in. In 10 cycles of the motion it is observed that the maximum displacement decreases by 50 percent. Determine the steady-state motion of the system if it is acted upon simultaneously by forces equal to $F_0 \cos 15t$ and $F_0 \cos 16t$. Do you think that these two forces will produce the phenomenon of beats? Why?

30   A system is acted upon by two forces

$$F_1 \sin \omega_1 t \qquad \text{and} \qquad F_2 \sin \omega_2 t$$

Friction, though present in the system, is so small that it can be neglected in determining the forced motion. Discuss the steady-state behavior of the system if ω_1 and ω_2 are nearly equal but neither is close to the natural frequency of the system. In particular, show that the response consists of a term of frequency $(\omega_1 + \omega_2)/2$ whose amplitude is modulated by a factor of frequency $(\omega_1 - \omega_2)/2$, and determine the limits between which the amplitude varies. *Hint:* Note first that the assumption of at least a little friction in the system implies that only the particular integrals contribute to the steady-state motion. (Why?) Then after the particular integrals have been determined, note that the expression $K_1 \sin \omega_1 t + K_2 \sin \omega_2 t$ can be written

$$\frac{K_1 + K_2}{2} (\sin \omega_1 t + \sin \omega_2 t) + \frac{K_1 - K_2}{2} (\sin \omega_1 t - \sin \omega_2 t)$$

31   Show that Eq. (1) can be written in the form

$$\frac{d^2y}{dt^2} + 2 \frac{c}{c_c} \omega_n \frac{dy}{dt} + \omega_n^2 y = \delta_{st}\omega_n^2 \cos \omega t$$

32   In the critically damped case, show that the common value of the roots of the characteristic equation of Eq. (1) is $-\omega_n$.

5.4 The Series Electric Circuit

All the results we obtained in the last section can, after a suitable change in terminology, be applied to any of the other systems we have considered. However, the concepts central in one field are not always of equal importance in related fields, and it seems desirable to illustrate the minor differences in the application of our general theory to various classes of systems by considering one of the electrical circuits in some detail.

For the simple series circuit with an alternating impressed voltage, we derived (among several equivalent forms) the equation

(1) $$L \frac{d^2Q}{dt^2} + R \frac{dQ}{dt} + \frac{1}{C} Q = E_0 \cos \omega t$$

and on comparing this with the differential equation of the vibrating weight,

$$\frac{w}{g}\frac{d^2y}{dt^2} + c\frac{dy}{dt} + ky = F_0 \cos \omega t$$

we noted the correspondences

$$\text{Mass } \frac{w}{g} \leftrightarrow \text{inductance } L$$

$$\text{Friction } c \leftrightarrow \text{resistance } R$$

$$\text{Spring modulus } k \leftrightarrow \text{elastance } \frac{1}{C}$$

$$\text{Impressed force } F \leftrightarrow \text{impressed voltage } E$$

$$\text{Displacement } y \leftrightarrow \text{charge } Q$$

$$\text{Velocity } v \leftrightarrow \text{current } i$$

Extending this correspondence to the derived results by making the appropriate substitutions, we infer from the undamped natural frequency of the mechanical system

$$\omega_n = \sqrt{\frac{kg}{w}}$$

that the electric circuit has a natural frequency

$$\Omega_n = \sqrt{\frac{1}{LC}}$$

when no resistance is present. Furthermore, the concept of critical damping

$$c_c = 2\sqrt{\frac{kw}{g}}$$

leads to the concept of critical resistance

$$R_c = 2\sqrt{\frac{L}{C}}$$

which determines whether the free behavior of the electrical system will be oscillatory or nonoscillatory.

The notion of magnification ratio can also be extended to the electrical case, but it is not customary to do so because the extension would relate to Q (the analog of the displacement y), whereas in most electrical problems it is not Q but i which is the variable of interest. To see how a related concept arises in the electrical case, let us convert the particular integral Y given by Eq. (10b), Sec. 5.3, into its electrical equivalent. By direct substitution (using Fig. 5.6 to obtain the phase angle β) the result is found to be

$$Q = \frac{E_0 \sin(\omega t + \beta)}{\sqrt{(1/C - \omega^2 L)^2 + (\omega R)^2}} \qquad \beta = \mathrm{Tan}^{-1}\frac{1/C - \omega^2 L}{\omega R}$$

To obtain the current i, we differentiate this, getting

$$\frac{dQ}{dt} = i = \frac{E_0\omega \cos(\omega t + \beta)}{\sqrt{(1/C - \omega^2 L)^2 + (\omega R)^2}}$$

or, dividing numerator and denominator by ω in the expressions for both i and β and then introducing a new phase angle $\delta = -\beta$,

(2)
$$i = \frac{E_0 \cos(\omega t - \delta)}{\sqrt{R^2 + (\omega L - 1/\omega C)^2}}$$

where

(3)
$$\delta = -\beta = \text{Tan}^{-1}\frac{\omega L - 1/\omega C}{R}$$

From Eq. (2) we infer that the steady-state current produced by an alternating voltage is of the same frequency as the voltage but differs from it in phase by

$$\frac{\delta}{\omega} \text{ units of time} \quad \text{or} \quad \frac{\delta/\omega}{2\pi/\omega} = \frac{\delta}{2\pi} \text{ cycles}$$

Moreover, from Eq. (3) it is clear that the numerator of tan δ (which is proportional to sin δ) can be either positive or negative, whereas the denominator of tan δ (which is proportional to cos δ) is always positive. Hence δ must be an angle between $-\pi/2$ and $\pi/2$, and so the principal-value designation in Eq. (3) is appropriate. If δ is positive, the steady-state current *lags* the voltage; if δ is negative, the steady-state current *leads* the voltage.

Furthermore, from Eq. (2) we see that the amplitude of the steady-state current is obtained by dividing the amplitude of the impressed voltage E_0 by the expression

(4)
$$\sqrt{R^2 + \left(\omega L - \frac{1}{\omega C}\right)^2}$$

By analogy with Ohm's law, $I = E/R$, the quantity (4) thus appears as a generalized resistance, although it is actually called the **impedance** of the circuit. While not the analog of the magnification ratio, the impedance is clearly a similar concept. Since impedance is defined as

$$\frac{\text{Voltage}}{\text{Current}}$$

the mechanical quantity corresponding to this is the ratio

$$\frac{\text{Force}}{\text{Velocity}}$$

This is called the **mechanical impedance** by some writers and in certain mechanical problems has proved a useful notion.†

There is another approach to the problem of determining the steady-state current produced by a harmonic voltage that is well worth investigating. Suppose that given

---

† See, for instance, T. von Kármán and M. A. Biot, "Mathematical Methods in Engineering," pp. 370–378, McGraw-Hill, New York, 1940.

either $E = E_0 \cos \omega t$ *or* $E_0 \sin \omega t$, we write the basic differential equation (1) in the form

(5) $$L \frac{d^2Q}{dt^2} + R \frac{dQ}{dt} + \frac{1}{C} Q = E_0 e^{j\omega t} = E_0 (\cos \omega t + j \sin \omega t)\dagger$$

This includes both possibilities for the voltage, and if the real and the imaginary terms retain their identity throughout the analysis, then the real part of the particular integral corresponding to $E_0 e^{j\omega t}$ will be the particular integral for $E_0 \cos \omega t$ and the imaginary part will be the particular integral for $E_0 \sin \omega t$.

To see that this is actually the case, we must first find a particular integral of Eq. (5). As usual, we do this by assuming

$$Q = Ae^{j\omega t}$$

and substituting into the differential equation. This gives

$$L(-\omega^2 Ae^{j\omega t}) + R(j\omega Ae^{j\omega t}) + \frac{1}{C}(Ae^{j\omega t}) = E_0 e^{j\omega t}$$

which will be an identity if and only if

$$A = \frac{E_0}{-\omega^2 L + j\omega R + 1/C}$$

Hence $$Q = \frac{E_0}{j\omega R - \omega^2 L + 1/C} e^{j\omega t}$$

From this, by differentiation, we find that

$$\frac{dQ}{dt} = i = \frac{j\omega E_0}{j\omega R - \omega^2 L + 1/C} e^{j\omega t} = \frac{E_0}{R + j(\omega L - 1/\omega C)} e^{j\omega t}$$

To find the real and imaginary parts of this expression, it is convenient to use the fact (Sec. 15.7) that any complex number $a + jb$ can be written in the form $a + jb = re^{j\delta}$, where the magnitude r and the angle δ of the complex number are related to the components a and b as shown in Fig. 5.12. Applied to the denominator of the second expression for i, this gives

$$R + j\left(\omega L - \frac{1}{\omega C}\right) = \sqrt{R^2 + \left(\omega L - \frac{1}{\omega C}\right)^2} \, e^{j\delta}$$

where $$\delta = \text{Tan}^{-1} \frac{\omega L - 1/\omega C}{R}$$

Hence we can rewrite i in the form

$$i = \frac{E_0}{\sqrt{R^2 + (\omega L - 1/\omega C)^2} \, e^{j\delta}} e^{j\omega t}$$

$$= \frac{E_0}{\sqrt{R^2 + (\omega L - 1/\omega C)^2}} e^{j(\omega t - \delta)}$$

$$= E_0 \frac{\cos(\omega t - \delta) + j \sin(\omega t - \delta)}{\sqrt{R^2 + (\omega L - 1/\omega C)^2}}$$

---

† To avoid confusing $i = \sqrt{-1}$ with $i =$ current, we shall throughout the rest of this chapter follow the practice, standard in electrical engineering, of writing $\sqrt{-1} = j$.

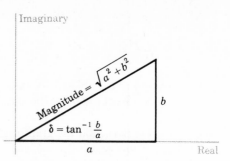

Figure 5.12
The relations between the magnitude, angle, and components of a general complex number $a + jb$.

Comparing this with Eqs. (2) and (3) makes it clear that the real part here is exactly the particular integral corresponding to $E_0 \cos \omega t$, as we derived it directly. Similarly, had we taken the trouble to work it out explicitly, we would have found for the particular integral corresponding to $E_0 \sin \omega t$ precisely the imaginary part of the last expression. Since it is much easier to find the particular integral corresponding to an exponential term than it is to find the particular integral corresponding to a cosine or sine term, the advantage of using $E_0 e^{j\omega t}$ in place of $E_0 \cos \omega t$ or $E_0 \sin \omega t$ is obvious.

The expression

$$R + j\left(\omega L - \frac{1}{\omega C}\right) \quad \text{or} \quad j\omega L + R + \frac{1}{j\omega C}$$

is called the **complex impedance** Z. Its magnitude is the quantity (4) which we referred to simply as the impedance. Its angle δ is the **phase shift**. The real part of Z is clearly a resistance. The imaginary part of Z is called the **reactance**. The reciprocal of Z is called the **admittance**. The real part of the admittance is called the **conductance**, and the imaginary part is called the **susceptance**.

The most striking property of the complex impedance is that when any electric elements are connected in series or in parallel, the corresponding impedances combine just as simple resistances do. Thus the steady-state current through a series of Z's (Fig. 5.13) can be found by dividing the impressed voltage by the single impedance

$$Z = Z_1 + Z_2 + \cdots + Z_n$$

Similarly, the current through a set of elements connected in parallel (Fig. 5.14) can be found by dividing the impressed voltage by the single impedance Z defined by the relation

$$\frac{1}{Z} = \frac{1}{Z_1} + \frac{1}{Z_2} + \cdots + \frac{1}{Z_n}$$

This makes it unnecessary to use differential equations in determining the *steady-state* behavior of an electric network (or of a mechanical system, if the concept of mechan-

Figure 5.13
Impedances connected in series.

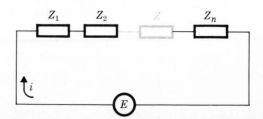

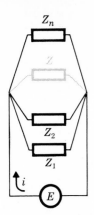

Figure 5.14
Impedances connected in parallel.

ical impedance is used). For the *transient* behavior, however, this is not true until the impedance concept is generalized through the use of the Laplace transformation (Chap. 7).

EXAMPLE 1

A series circuit in which both the charge and the current are initially zero contains the elements $L = 1$ H, $R = 1,000$ Ω, $C = 6.25 \times 10^{-6}$ F. If a constant voltage $E = 24$ V is suddenly switched into the circuit, find the peak value of the resultant current.

The differential equation we must solve is

$$\frac{d^2Q}{dt^2} + 1,000 \frac{dQ}{dt} + \frac{Q}{6.25 \times 10^{-6}} = 24$$

subject to the conditions that $Q = i = 0$ when $t = 0$. The characteristic equation in this case is

$$m^2 + 1,000m + 160,000 = 0$$

and its roots are $m_1 = -200$, $m_2 = -800$. Hence the complementary function is

$$c_1 e^{-200t} + c_2 e^{-800t}$$

To find a particular integral, we assume $Q = A$ and substitute into the differential equation, getting immediately

$$A = 150 \times 10^{-6}$$

The complete solution is therefore

$$Q = c_1 e^{-200t} + c_2 e^{-800t} + 150 \times 10^{-6}$$

Differentiating gives

$$\frac{dQ}{dt} = i = -200c_1 e^{-200t} - 800c_2 e^{-800t}$$

Substituting the initial conditions for Q and i gives the pair of equations

$$c_1 + c_2 + 150 \times 10^{-6} = 0 \quad \text{and} \quad c_1 + 4c_2 = 0$$

from which we find at once

$$c_1 = -200 \times 10^{-6} \qquad c_2 = 50 \times 10^{-6}$$

and
$$i = 0.04(e^{-200t} - e^{-800t})$$

To find the time when i is a maximum, we must equate to zero the time derivative of i:

$$0.04(-200e^{-200t} + 800e^{-800t}) = 0$$

Dividing out $0.04 \times 800e^{-200t}$ and transposing, we have $e^{-600t} = \frac{1}{4}$. From this, taking logarithms, we find

$$-600t = -\ln 4 = -1.386 \quad \text{and} \quad t_{max} = 0.0023 \text{ s}$$

The maximum value of i can now be found by substituting the value of t_{max} into the general expression for i. The result is

$$i_{max} = 0.019 \text{ A}$$

EXERCISES

1 In Example 1, find the potential difference across each element as a function of time.
2 Find the steady-state current produced in the circuit of Example 1 by a voltage $E = 120 \cos 120\pi t$ V.
3 An open series circuit contains the elements $L = 0.01$ H, $R = 250\ \Omega$, $C = 10^{-6}$ F. At $t = 0$, with the capacitor charged to the value $Q_0 = 10^{-5}$ C, the circuit is closed. Find the resultant current as a function of time.
4 Work Exercise 3, given that the circuit elements are $L = 6.4 \times 10^{-3}$ H, $R = 1.6 \times 10^2\ \Omega$, $C = 10^{-6}$ F.
5 Work Exercise 3, given that the circuit elements are $L = 0.01$ H, $R = 120\ \Omega$, $C = 10^{-6}$ F.
6 A series circuit in which $Q_0 = i_0 = 0$ contains the elements $L = 0.15$ H, $R = 800\ \Omega$, $C = 4 \times 10^{-6}$ F. If a constant voltage $E = 25$ V is suddenly switched into the circuit, find the resultant current as a function of time.
7 Work Exercise 6, given that the circuit elements are $L = 0.16$ H, $R = 800\ \Omega$, $C = 10^{-6}$ F.
8 Work Exercise 7, given that $E = e^{-500t}$ V.
9 A series circuit in which $Q_0 = i_0 = 0$ contains the elements $L = 1$ H, $R = 1,000\ \Omega$, $C = 4 \times 10^{-6}$ F. A voltage $E = 110 \sin 50\pi t$ V is suddenly switched into the circuit. Find the resultant current as a function of time.
10 Work Exercise 9 if $L = 1$ H, $R = 800\ \Omega$, $C = 4 \times 10^{-6}$ F, $E = 120 \cos 600t$ V.
11 A series circuit in which $Q_0 = i_0 = 0$ contains the elements $L = 0.02$ H, $R = 250\ \Omega$, $C = 2 \times 10^{-6}$ F. A constant voltage $E = 28$ V is suddenly switched into the circuit. Find the time it takes for the potential difference across the capacitor to build up to one-half its final value.
12 A capacitor $C = 4 \times 10^{-6}$ F, a resistance $R = 250\ \Omega$, and an inductance $L = 1$ H are connected in parallel. A current source delivering a constant current $I = 0.01$ A is suddenly connected across the common terminals of the elements. Find the resultant voltage as a function of time.
13 Find the steady-state voltage in Exercise 12 if $I = 0.01 \sin 150t$ A.
14 For the series electric circuit, what is the analog of the static deflection δ_{st}?
15 Show that Eq. (1) can also be written in the form

$$\frac{d^2Q}{dt^2} + 2\frac{R}{R_c}\Omega_n\frac{dQ}{dt} + \Omega_n^2 Q = Q_{st}\Omega_n^2 \cos \omega t$$

where Q_{st} is the quantity identified in Exercise 14 as the analog of the static deflection δ_{st}.
16 Find the particular integral of Eq. (1) corresponding to the driving voltage $E_0 \sin \omega t$, and verify that it produces a current equal to the imaginary part of the particular integral corresponding to the complex voltage $E_0 e^{j\omega t}$.
17 (a) Prove that if a set of elements with impedance Z_1 is connected in series with a set of elements with impedance Z_2, the impedance of the resultant combination is $Z_1 + Z_2$.

(b) Prove that if a set of elements with impedance Z_1 is connected in parallel with a set of elements with impedance Z_2, the impedance of the resultant combination is given by

$$\frac{1}{Z} = \frac{1}{Z_1} + \frac{1}{Z_2}$$

18 A constant voltage is suddenly switched into a nonoscillatory series circuit in which $Q_0 = i_0 = 0$. Show that the potential difference across the capacitor can never overshoot its final value.

19 For what value(s) of ω is the impedance $\sqrt{R^2 + (\omega L - 1/\omega C)^2}$ a minimum? Compare this with the corresponding property of the magnification ratio. Explain.

20 If the frequency of the voltage $E_0 \cos \omega t$ impressed on a series circuit is the same as the natural frequency of the circuit in the absence of any resistance, show that the amplitudes of the steady-state potential differences across the inductance and the capacitance are each equal to $E_0 R_c/2R$.

21 Instead of using the ratio R/R_c as a dimensionless parameter in circuit analysis, it is customary to use the so-called **quality factor** Q (not to be confused with the charge Q) defined to be $R_c/2R$. Express the impedance and the phase angle for a simple series circuit in terms of the resistance R, the frequency ratio Ω/Ω_n, and the quality factor Q.

22 Find the steady-state voltage between the terminals A and B of the network shown in Fig. 5.1d by solving Eq. (24a), Sec. 5.2, and verify that your answer is equal to the steady-state voltage calculated directly from the impedances ωL, R, and $1/\omega C$ of the three elements in the parallel combination.

23 A 50-lb weight hangs from a spring of modulus 25 lb/in. Friction in the system is equal to one-tenth of critical, and the weight is driven by a force equal to $10 \cos 3t$ lb. Determine the appropriate scale factors and the components of an equivalent series circuit under the restriction that the orders of magnitude of L, R, C, E_0, and ω_2 are, respectively, 0.1 H, 100 Ω, 10^{-6} F, 0.1 V, and 600 Hz.

24 Work Exercise 23 if an equivalent parallel circuit is required.

5.5 Systems with Several Degrees of Freedom

The laws of Newton and Kirchhoff, together with the theory of simultaneous linear differential equations developed in Chap. 3 and the theory of difference equations developed in Sec. 4.5, form the basis for the analysis of large classes of systems with more than one degree of freedom. The details of such applications can best be made clear through examples.

EXAMPLE 1

Assuming friction to be negligible, find the natural frequencies of the mass-spring system shown in Fig. 5.15. What are the relative amplitudes with which the two masses vibrate at each of the natural frequencies? If a force $F = 40 \sin 3t$ is suddenly applied to M_1 when the system is in equilibrium, find the displacements of M_1 and M_2 as functions of time.

As usual, we suppose the masses to be guided, by constraints which need not be specified, so that they can move only in the vertical direction. The instantaneous displacements of the masses from their equilibrium positions we shall use as coordinates to describe the system, displacements above the equilibrium positions being considered positive. Since friction is assumed to be negligible, the only forces acting on the masses besides the attraction of gravity are those transmitted to them by the attached springs. Moreover, as suggested by the derivation of Eq. (7), Sec. 5.2, and confirmed for systems with two degrees of freedom by Exercise 13 at the end of this section, the force of gravity can be neglected provided we also neglect the initial elongation of the springs and assume that each is unstretched when the system is in equilibrium.

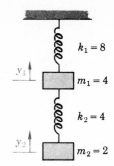

Figure 5.15
A simple mass-spring system.

When the displacements of the masses M_1 and M_2 are y_1 and y_2, respectively, the upper spring is changed in length by the amount y_1 and the lower spring is changed in length by the amount $y_1 - y_2$. Because of these changes in length, the springs exert forces equal to

$$8y_1 \quad \text{and} \quad 4(y_1 - y_2)$$

respectively. Hence, applying Newton's law to each mass and taking due account of the direction of the forces applied to each mass by the attached springs, we have

$$4\frac{d^2y_1}{dt^2} = -8y_1 - 4(y_1 - y_2) \quad 2\frac{d^2y_2}{dt^2} = 4(y_1 - y_2)$$

or

(1)
$$(4D^2 + 12)y_1 - 4y_2 = 0$$
$$-4y_1 + (2D^2 + 4)y_2 = 0$$

From these equations we find that the equation satisfied by y_1 is

$$\begin{vmatrix} 4D^2 + 12 & -4 \\ -4 & 2D^2 + 4 \end{vmatrix} y_1 = 0 \quad \text{or} \quad 8(D^4 + 5D^2 + 4)y_1 = 0$$

The characteristic equation of this differential equation is

$$m^4 + 5m^2 + 4 = 0$$

and its roots are $m = \pm j, \pm 2j$. Hence

(2)
$$y_1 = c_1 \cos t + c_2 \sin t + c_3 \cos 2t + c_4 \sin 2t$$

Since the system (1) is homogeneous, it is evident that y_2 satisfies the same differential equation that y_1 satisfies. Therefore,

(3)
$$y_2 = d_1 \cos t + d_2 \sin t + d_3 \cos 2t + d_4 \sin 2t$$

At the present stage, we have no information about the values of the constants $c_1, c_2, \ldots, d_4$. However, even without this information we can infer the natural frequencies of the system. In fact, no matter what the values of $c_1, c_2, \ldots, d_4$ may be, it is clear that Eqs. (2) and (3) represent sums of periodic displacements with frequencies $\omega_1 = 1$ and $\omega_2 = 2$. Moreover, since Eqs. (2) and (3) (when their constants are suitably related) constitute a complete solution of the system (1) and hence a complete description of all possible motions of the given physical system, it follows that free vibrations at frequencies other than $\omega_1 = 1$ and $\omega_2 = 2$ are impossible.

To find the relative amplitudes with which the masses vibrate at the respective frequencies, we must determine the values of $c_1, c_2, \ldots, d_4$. As usual, we do this by substituting the general expressions for y_1 and y_2 into one of the differential equations, say the second, and then relating the c's and d's so that the resulting equation is identically satisfied. Doing this, we first obtain the equation

$$- 4(\quad c_1 \cos t + c_2 \sin t + \quad c_3 \cos 2t + \quad c_4 \sin 2t)$$
$$+ 2(-d_1 \cos t - d_2 \sin t - 4d_3 \cos 2t - 4d_4 \sin 2t)$$
$$+ 4(\quad d_1 \cos t + d_2 \sin t + \quad d_3 \cos 2t + \quad d_4 \sin 2t) = 0$$

from which we infer that

$$-4c_1 + 2d_1 = 0 \qquad -4c_3 - 4d_3 = 0$$
$$-4c_2 + 2d_2 = 0 \qquad -4c_4 - 4d_4 = 0$$

Thus $\qquad d_1 = 2c_1 \qquad d_2 = 2c_2 \qquad d_3 = -c_3 \qquad d_4 = -c_4$

and, specifically,

(4) $\qquad y_1 = c_1 \cos t + c_2 \sin t + c_3 \cos 2t + c_4 \sin 2t$

(5) $\qquad y_2 = 2(c_1 \cos t + c_2 \sin t) - (c_3 \cos 2t + c_4 \sin 2t)$

From these expressions, it is clear that when the system is vibrating at the single frequency $\omega_1 = 1$, that is, when $c_3 = c_4 = 0$, the masses move up and down together but M_2 vibrates with twice the amplitude of M_1. Similarly, when the system is vibrating at the single frequency $\omega_2 = 2$, that is, when $c_1 = c_2 = 0$, the masses vibrate with equal amplitudes but in opposite directions.

To determine the motion of the system when the force $F = 40 \sin 3t$ acts on M_1 we must first find a particular integral for the system

(1a)
$$(4D^2 + 12)y_1 - \qquad\qquad 4y_2 = 40 \sin 3t$$
$$-4y_1 + (2D^2 + 4)y_2 = 0$$

obtained from (1) by adding the term $40 \sin 3t$ to the right-hand side of the first equation. Since only derivatives of even order occur in (1a) (because of our assumption that friction was negligible) we need only assume

$$Y_1 = A_1 \sin 3t$$
$$Y_2 = A_2 \sin 3t$$

Then substituting into (1a), dividing out the common factor $\sin 3t$, and simplifying, we obtain the conditions

$$-6A_1 - \quad A_2 = 10$$
$$-2A_1 - 7A_2 = 0$$

from which we find $A_1 = -\frac{7}{4}$ and $A_2 = \frac{2}{4}$. The required particular integral is then

$$Y_1 = -\tfrac{7}{4} \sin 3t \qquad Y_2 = \tfrac{2}{4} \sin 3t$$

Adding these to the expressions (4) and (5), respectively, gives us the complete solution of the nonhomogeneous system:

$$y_1 = c_1 \cos t + c_2 \sin t + c_3 \cos 2t + c_4 \sin 2t - \tfrac{7}{4} \sin 3t$$
$$y_2 = 2c_1 \cos t + 2c_2 \sin t - c_3 \cos 2t - c_4 \sin 2t + \tfrac{2}{4} \sin 3t$$

Imposing now the first of our initial conditions, namely, $y_1 = y_2 = 0$ when $t = 0$, we find

$$c_1 + c_3 = 0$$
$$2c_1 - c_3 = 0$$

from which it follows that $c_1 = c_3 = 0$. Differentiating y_1 and y_2 to obtain expressions for the velocities $\dot{y}_1$ and $\dot{y}_2$ and then imposing the condition that $\dot{y}_1 = \dot{y}_2 = 0$ when $t = 0$, we obtain, similarly, the equations

$$c_2 + 2c_4 - \tfrac{21}{4} = 0$$

$$2c_2 - 2c_4 + \tfrac{6}{4} = 0$$

whence $c_2 = \tfrac{5}{4}$ and $c_4 = 2$. The required solution for y_1 and y_2 when M_1 is acted upon by the force $F = 40 \sin 3t$ and the system begins to move from rest in its equilibrium position is thus

$$y_1 = \tfrac{5}{4} \sin t + 2 \sin 2t - \tfrac{7}{4} \sin 3t$$

$$y_2 = \tfrac{5}{2} \sin t - 2 \sin 2t + \tfrac{1}{2} \sin 3t$$

From these expressions it appears that the terms from the complementary function, i.e., the terms describing vibrations at the natural frequencies $\omega_1 = 1$ and $\omega_2 = 2$, persist indefinitely, and this is correct under our assumption that there is no friction in the system. However, any amount of friction, no matter how small, would actually cause these terms to die away, and only the terms of frequency $\omega = 3$ would appear in the steady state.

EXAMPLE 2

In the circuit shown in Fig. 5.16, find the current in each loop as a function of time, given that all charges and currents are zero when the switch is closed at $t = 0$.

We take as variables the currents i_1 and i_2 flowing in the respective loops, noting that the current in the common branch is therefore $i_1 - i_2$. Applying Kirchhoff's second law to each loop, we obtain the equations

$$0.5 \frac{di_1}{dt} + 200(i_1 - i_2) = 50$$

$$300i_2 + 200(i_2 - i_1) + \frac{1}{50 \times 10^{-6}} \int_0^t i_2 \, dt = 0$$

or, letting $Q_2 = \int^t i_2 \, dt$,

$$\frac{di_1}{dt} + 400i_1 - 400 \frac{dQ_2}{dt} = 100$$

(6)

$$-2i_1 + 5 \frac{dQ_2}{dt} + 200Q_2 = 0$$

The characteristic equation of this system is

$$\begin{vmatrix} m + 400 & -400m \\ -2 & 5m + 200 \end{vmatrix} = 5\,(m^2 + 280m + 16{,}000) = 0$$

From its roots, $m_1 = -80$ and $m_2 = -200$, we can construct the expressions

$$i_1 = a_1 e^{-80t} + b_1 e^{-200t}$$

$$Q_2 = a_2 e^{-80t} + b_2 e^{-200t}$$

which, after the constants a_1, a_2, b_1, b_2 are properly related, will constitute the complementary function of the system.

Figure 5.16
A simple two-loop electric circuit.

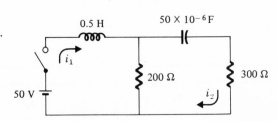

Substituting these expressions into the second of the two differential equations (6), we obtain

$$-2(a_1e^{-80t} + b_1e^{-200t}) + 5(-80a_2e^{-80t} - 200b_2e^{-200t})$$
$$+ 200(a_2e^{-80t} + b_2e^{-200t}) = 0$$

This will be an identity if and only if $a_1 = -100a_2$ and $b_1 = -400b_2$. Therefore the complementary function is

$$i_1 = -100a_2e^{-80t} - 400b_2e^{-200t}$$
$$Q_2 = a_2e^{-80t} + b_2e^{-200t}$$

To find a particular integral, we assume $i_1 = A_1$ and $Q_2 = A_2$. Substituting these into the nonhomogeneous system of differential equations (6), we obtain

$$400A_1 = 100$$
$$-2A_1 + 200A_2 = 0$$

Hence $A_1 = \frac{1}{4}$ and $A_2 = \frac{1}{400}$

and therefore a complete solution of the system is

$$i_1 = -100a_2e^{-80t} - 400b_2e^{-200t} + \frac{1}{4}$$
$$Q_2 = a_2e^{-80t} + b_2e^{-200t} + \frac{1}{400}$$

Since $i_1 = 0$ and $Q_2 = 0$ when $t = 0$, we must have

$$0 = -100a_2 - 400b_2 + \frac{1}{4}$$
$$0 = a_2 + b_2 + \frac{1}{400}$$

From these we find without difficulty that $a_2 = -\frac{1}{240}$ and $b_2 = \frac{1}{600}$. The required currents are therefore

$$i_1 = \frac{5}{12}e^{-80t} - \frac{2}{3}e^{-200t} + \frac{1}{4}$$
$$i_2 = \frac{dQ_2}{dt} = \frac{1}{3}e^{-80t} - \frac{1}{3}e^{-200t}$$

Evidently, $i_2 = 0$ when $t = 0$, as required.

EXAMPLE 3

Find the natural frequencies of the network shown in Fig. 5.17.

By applying Kirchhoff's second law to each loop in turn, we obtain the equations

$$L\frac{di_1}{dt} + \frac{1}{C}\int^t (i_1 - i_2)\, dt = 0$$

$$\frac{1}{C}\int^t (i_2 - i_1)\, dt + L\frac{di_2}{dt} + \frac{1}{C}\int^t (i_2 - i_3)\, dt = 0$$

$$\cdots\cdots\cdots\cdots\cdots\cdots\cdots\cdots\cdots\cdots\cdots\cdots\cdots\cdots$$

$$\frac{1}{C}\int^t (i_{k+1} - i_k)\, dt + L\frac{di_{k+1}}{dt} + \frac{1}{C}\int^t (i_{k+1} - i_{k+2})\, dt = 0$$

$$\cdots\cdots\cdots\cdots\cdots\cdots\cdots\cdots\cdots\cdots\cdots\cdots\cdots\cdots$$

$$\frac{1}{C}\int^t (i_{n-1} - i_{n-2})\, dt + L\frac{di_{n-1}}{dt} + \frac{1}{C}\int^t (i_{n-1} - i_n)\, dt = 0$$

$$\frac{1}{C}\int^t (i_n - i_{n-1})\, dt + L\frac{di_n}{dt} + \frac{1}{C}\int^t i_n\, dt = 0$$

or, introducing new variables via the substitutions

$$\int^t i_k\, dt = Q_k \qquad i_k = \frac{dQ_k}{dt} = DQ_k \qquad \frac{di_k}{dt} = \frac{d^2Q_k}{dt^2} = D^2Q_k$$

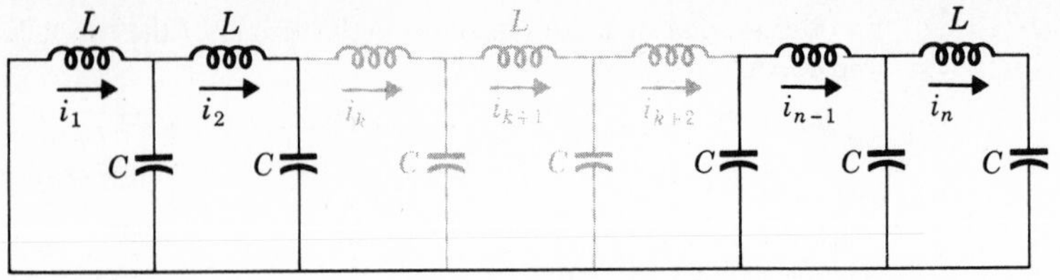

Figure 5.17
An oscillatory ladder-type network. (See explanation of convention used in Fig. 4.2.)

and rearranging slightly,

$$(LCD^2 + 1)Q_1 - Q_2 = 0$$

$$-Q_1 + (LCD^2 + 2)Q_2 - Q_3 = 0$$

$$\cdots\cdots\cdots\cdots\cdots\cdots\cdots\cdots\cdots\cdots$$

(7) $$-Q_k + (LCD^2 + 2)Q_{k+1} - Q_{k+2} = 0$$

$$\cdots\cdots\cdots\cdots\cdots\cdots\cdots\cdots\cdots\cdots$$

$$-Q_{n-2} + (LCD^2 + 2)Q_{n-1} - Q_n = 0$$

$$-Q_{n-1} + (LCD^2 + 2)Q_n = 0$$

Only derivatives of even orders appear in these equations because, by hypothesis, there are no resistances anywhere in the network. This means that the response of the system to any set of nonzero initial conditions of charge and current will be purely oscillatory. Hence, *as far as the determination of the natural frequencies of the system is concerned*, it is sufficient to assume solutions of the form

$$Q_k = A_k \cos \omega t$$

where ω is the unknown frequency of the response and the A's are arbitrary constants. Substituting into the equations of the set (7), dividing each equation by $-\cos \omega t$, and setting

$$LC\omega^2 = \alpha^2$$

we obtain the algebraic equations

$$-(1 - \alpha^2)A_1 + A_2 = 0$$

$$A_1 - (2 - \alpha^2)A_2 + A_3 = 0$$

$$\cdots\cdots\cdots\cdots\cdots\cdots\cdots\cdots\cdots$$

(8) $$A_k - (2 - \alpha^2)A_{k+1} + A_{k+2} = 0$$

$$\cdots\cdots\cdots\cdots\cdots\cdots\cdots\cdots\cdots$$

$$A_{n-2} - (2 - \alpha^2)A_{n-1} + A_n = 0$$

$$A_{n-1} - (2 - \alpha^2)A_n = 0$$

In order for these equations to have a nontrivial solution for the A's, that is, in order for the network to exhibit a response that is not identically zero, it is necessary that the determinant of the coefficients in these equations be zero. However, in this case the determinant of the coefficients is of the nth order, and to expand it and then solve the resulting nth-degree equation in the unknown quantity $\alpha^2 = LC\omega^2$ would be prohibitively time-consuming. Hence it is much better to proceed in the following way. With the exception of the first and last equations, each equation of the system (8) is of the form

$$A_k - (2 - \alpha^2)A_{k+1} + A_{k+2} = 0$$

In other words, for $k = 1, 2, \ldots, n - 2$, the A's satisfy the linear, constant-coefficient, second-order difference equation†

$$(9) \qquad \left[E^2 - 2 \left(1 - \frac{\alpha^2}{2} \right) E + 1 \right] A_k = 0$$

The first and last equations, which clearly do not fit into the pattern of Eq. (9), are, of course, the two boundary conditions necessary for the determination of the two arbitrary constants which appear in the complete solution of this difference equation.

Following the theory of Sec. 4.5, the first step in the solution of Eq. (9) is to solve its characteristic equation

$$(10) \qquad m^2 - 2 \left(1 - \frac{\alpha^2}{2} \right) m + 1 = 0$$

getting

$$m_1, m_2 = 1 - \frac{\alpha^2}{2} \pm \sqrt{ \left(1 - \frac{\alpha^2}{2} \right)^2 - 1 }$$

The continuation now involves an investigation of the special cases which arise according as $1 - \alpha^2/2$ is equal to or greater than 1, between -1 and 1, or equal to or less than -1.

First, we can immediately reject the possibility that $1 - \alpha^2/2 \geq 1$, since this implies that $\alpha^2 \leq 0$, which is impossible because $\alpha^2 \equiv LC\omega^2$ is an intrinsically positive quantity.

We next consider the possibility that $1 - \alpha^2/2 = -1$, that is, that $\alpha^2 = 4$. In this case $m_1 = m_2 = -1$ and so, according to Table 4.1, the complete solution of Eq. (9) is

$$A_k = (c_1 + c_2 k)(-1)^k$$

Imposing the boundary conditions on the coefficient function A_k by substituting A_1 and A_2 into the first of the equations (8) and substituting A_{n-1} and A_n into the last of the equations (8), we have

$$-(-3)[(c_1 + c_2)(-1)] + [(c_1 + 2c_2)(-1)^2] \equiv -2c_1 - c_2 = 0$$
$$\{[c_1 + (n - 1)c_2](-1)^{n-1}\} - (-2)[(c_1 + nc_2)(-1)^n] \equiv (-1)^n[c_1 + (n + 1)c_2] = 0$$

But these two equations obviously have only the trivial solution $c_1 = c_2 = 0$. This implies that $A_k \equiv 0$; hence the possibility that $1 - \alpha^2/2 = -1$ must also be rejected.

If $1 - \alpha^2/2 < -1$, it is convenient to introduce a new parameter μ by writing

$$(11) \qquad 1 - \frac{\alpha^2}{2} = -\cosh \mu \qquad \text{or} \qquad \alpha^2 = 2 + 2 \cosh \mu \qquad \mu \neq 0$$

In terms of μ, the roots of the characteristic equation (10) become in this case

$$- \cosh \mu \pm \sqrt{\cosh^2 \mu - 1} = -\cosh \mu \pm \sinh \mu = -e^{\pm \mu}$$

Hence, a complete solution of (9) can be written

$$A_k = c_1(-e^\mu)^k + c_2(-e^{-\mu})^k = (-1)^k(c_1 e^{\mu k} + c_2 e^{-\mu k})$$
$$= (-1)^k[c_1(\cosh \mu k + \sinh \mu k) + c_2(\cosh \mu k - \sinh \mu k)]$$
$$= (-1)^k(d_1 \cosh \mu k + d_2 \sinh \mu k)$$

---

† This is true, of course, only because the loops of the network, with the exception of the first and last, are all identical. In general, the possibility of using difference equations should always be considered in studying systems, both mechanical and electrical, which consist essentially of a number of identical components identically connected.

where $d_1 = c_1 + c_2$ and $d_2 = c_1 - c_2$. Again imposing the boundary conditions by substituting into the first and last of the equations (8), we have

$$-(1 + 2 \cosh \mu)(d_1 \cosh \mu + d_2 \sinh \mu) + (d_1 \cosh 2\mu + d_2 \sinh 2\mu) = 0$$

$$(-1)^{n-1}[d_1 \cosh (n - 1)\mu + d_2 \sinh (n - 1)\mu]$$
$$+ (-1)^n(2 \cosh \mu)(d_1 \cosh n\mu + d_2 \sinh n\mu) = 0$$

From these, by collecting terms and then simplifying through the use of the identities

$$2 \cosh^2 \mu = \cosh 2\mu + 1$$
$$2 \sinh \mu \cosh \mu = \sinh 2\mu$$
$$2 \cosh n\mu \cosh \mu = \cosh (n + 1)\mu + \cosh (n - 1)\mu$$
$$2 \sinh n\mu \cosh \mu = \sinh (n + 1)\mu + \sinh (n - 1)\mu$$

we obtain
$$(1 + \cosh \mu)d_1 + (\sinh \mu)d_2 = 0$$
$$[\cosh (n + 1)\mu]d_1 + [\sinh (n + 1)\mu]d_2 = 0$$

These equations will have a nontrivial solution for d_1 and d_2 if and only if

$$\begin{vmatrix} 1 + \cosh \mu & \sinh \mu \\ \cosh (n + 1)\mu & \sinh (n + 1)\mu \end{vmatrix} = \sinh (n + 1)\mu + \sinh n\mu$$
$$= 2 \sinh \frac{2n + 1}{2} \mu \cosh \frac{\mu}{2} = 0$$

This can vanish only if $\mu = 0$, which is impossible since, from (11), $\mu = 0$ implies $1 - \alpha^2/2 = -1$, and this possibility has already been considered and rejected. Hence the assumption that $1 - \alpha^2/2 < -1$ also leads only to a trivial solution and must therefore be rejected.

It remains now to consider the possibility that $-1 < 1 - \alpha^2/2 < 1$. To investigate this case, let us put

(12) $$1 - \frac{\alpha^2}{2} = \cos \mu \quad \text{or} \quad \alpha^2 = 2 - 2 \cos \mu$$

Then the roots of the characteristic equation (10) are

$$m_1, m_2 = \cos \mu \pm \sqrt{\cos^2 \mu - 1} = \cos \mu \pm j \sin \mu = e^{\pm j\mu}$$

and a complete solution of the difference equation (9) is now

$$A_k = c_1(e^{j\mu})^k + c_2(e^{-j\mu})^k = c_1 e^{j\mu k} + c_2 e^{-j\mu k}$$
$$= c_1(\cos \mu k + j \sin \mu k) + c_2(\cos \mu k - j \sin \mu k)$$
$$= d_1 \cos \mu k + d_2 \sin \mu k$$

where $d_1 = c_1 + c_2$ and $d_2 = j(c_1 - c_2)$. Again imposing the boundary conditions on A_k, we have

$$-(2 \cos \mu - 1)(d_1 \cos \mu + d_2 \sin \mu) + (d_1 \cos 2\mu + d_2 \sin 2\mu) = 0$$

$$[d_1 \cos (n - 1)\mu + d_2 \sin (n - 1)\mu] - (2 \cos \mu)(d_1 \cos n\mu + d_2 \sin n\mu) = 0$$

From these, by collecting terms and then simplifying through the use of the identities

$$2 \cos^2 \mu = 1 + \cos 2\mu$$
$$2 \sin \mu \cos \mu = \sin 2\mu$$
$$2 \cos n\mu \cos \mu = \cos (n + 1)\mu + \cos (n - 1)\mu$$
$$2 \sin n\mu \cos \mu = \sin (n + 1)\mu + \sin (n - 1)\mu$$

we obtain
$$(\cos \mu - 1)d_1 + (\sin \mu)d_2 = 0$$
$$[\cos (n + 1)\mu]d_1 + [\sin (n + 1)\mu]d_2 = 0$$

These two equations will have a nontrivial solution for d_1 and d_2 if and only if

$$\begin{vmatrix} \cos \mu - 1 & \sin \mu \\ \cos (n + 1)\mu & \sin (n + 1)\mu \end{vmatrix} = \sin n\mu - \sin (n + 1)\mu = -2 \sin \frac{\mu}{2} \cos \frac{2n + 1}{2} \mu = 0$$

Now $\sin (\mu/2)$ can be zero only if μ is a multiple of 2π, which is impossible since $\mu = 2N\pi$ implies, from (12), that $1 - \alpha^2/2 = 1$, which has already been considered and rejected. Hence we must have

$$\cos \frac{2n + 1}{2} \mu = 0$$

Therefore

$$\frac{2n + 1}{2} \mu = \frac{\pi}{2} + N\pi = \frac{2N + 1}{2} \pi$$

and

$$\mu = \frac{2N + 1}{2n + 1} \pi \qquad N = 0, 1, 2, \ldots, n - 1$$

The values $N = 0, 1, 2, \ldots, n - 1$ lead to distinct values of μ which in turn define the n natural frequencies of the network, since, from (12),

$$\sqrt{LC\omega^2} \equiv \alpha = \sqrt{2(1 - \cos \mu)} = 2 \sin \frac{\mu}{2}$$

Hence the required frequencies are given by the formula

$$\omega_N = \frac{2}{\sqrt{LC}} \sin \left(\frac{2N + 1}{2n + 1} \frac{\pi}{2} \right) \qquad N = 0, 1, 2, \ldots, n - 1$$

EXERCISES

1 Work Example 1 if the force $40 \sin 3t$ is suddenly applied to M_2 when the system is in equilibrium.

2 In Example 1 find the particular integral associated with a force equal to $F_0 \sin \omega t$ acting on M_1, and discuss the corresponding steady-state motion as a function of ω.

3 Work Exercise 2 if the force $F_0 \sin \omega t$ acts on M_2.

4 If $M_1 = 1$, $M_2 = 2$, $k_1 = 1$, $k_2 = k_3 = 2$ for the system shown in Fig. 5.18 and if friction is negligible, find the natural frequencies of the system. What are the

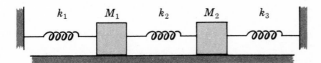

k_1 M_1 k_2 M_2 k_3

Figure 5.18

relative amplitudes of the two masses when the system is vibrating at each of these frequencies?

5 Work Exercise 4 if $M_1 = 1$, $M_2 = 3$, $k_1 = 1$, $k_2 = k_3 = 3$.

6 Work Exercise 4 if $M_1 = M_2 = 1$, $k_1 = 1$, $k_2 = 3$, $k_3 = 9$.

7 Find the displacements of M_1 and M_2 as functions of t in Exercise 4 if the system starts from rest with $x_1 = 1$ and $x_2 = 0$.

8 Prove that for no values of the parameters M_1, M_2, k_1, k_2, k_3 can the two natural frequencies of the system shown in Fig. 5.18 be equal.

9 In the system shown in Fig. 5.19 the parameters M_1, k_1, and ω are assumed to be known. Determine k_2 and M_2 so that in the steady-state forced motion of the system the mass M_1 will remain at rest. What is the resultant amplitude of the mass M_2? If this amplitude is deemed to be too large, can viscous friction be introduced into the **dynamic damper**, i.e., the (M_2, k_2) subsystem, to reduce it?

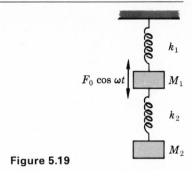

Figure 5.19

10   Assuming that there is no friction, find the natural frequencies of the system shown
in Fig. 5.20. What are the relative amplitudes of the two masses when the system is
vibrating at each of these frequencies? How does the quantity $x_1 + x_2$ vary with
time? How does the quantity $x_1 - x_2$ vary with time? Under what conditions, if
any, would you expect this system to exhibit the phenomenon of beats?

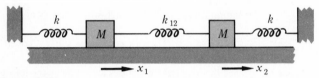

Figure 5.20

11   Find the equations of motion of the system shown in Fig. 5.20 if $M = \frac{1}{9}$, $k = 9$,
$k_{12} = \frac{20}{9}$ and if the system starts from rest in the position where $x_1 = 2$ and $x_2 = 0$.
Analyze this motion for the existence of beats.

12   Find the steady-state motion produced in the system discussed in Example 1 by a
force $F_0 \cos \omega t$ acting on M_1 if a frictional force equal to $-4\, dy_1/dt$ acts on M_1
and a frictional force equal to $-2\, dy_2/dt$ acts on M_2. What is the lag of the response
of M_1 and M_2 with respect to the driving force? *Hint:* Replace $F_0 \cos \omega t$ by $F_0 e^{j\omega t}$.

13   Set up the differential equations governing the motion of the system shown in Fig. 5.21,

Figure 5.21

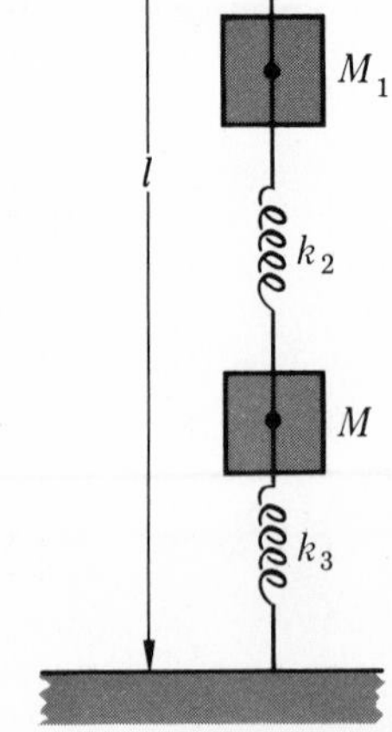

taking into account the force of gravity and the forces due to the initial elongations of the springs, and verify that these forces cancel each other identically.

14 A mass M_1 hanging from a spring of modulus k_1 constitutes a system with natural frequency $\omega = \sqrt{k_1/M_1}$. If a mass M_2 hangs from M_1 by a spring of modulus k_2, prove that the two natural frequencies of the resulting system are such that one is always greater than ω and one is always less than ω. Is this true if M_2 is also connected to the ground by a spring of modulus k_3?

15 A uniform bar 4 ft long and weighing 16 lb/ft is supported as shown in Fig. 5.22 on springs of modulus 24 and 15 lb/in, respectively. If the springs are guided so that only vertical displacement of the center of the bar is possible, and if friction is negligible, find the natural frequencies of the system. *Hint:* As coordinates, use the

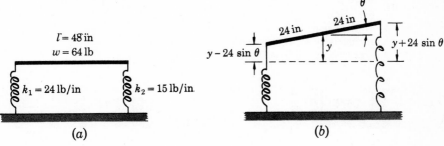

(a)   (b)

Figure 5.22

displacement y of the center of the bar and the angle of rotation θ of the bar about its center. Assume displacements so small that $\cos\theta$ can be replaced by 1 and $\sin\theta$ can be replaced by θ.

16 In the network shown in Fig. 5.23 the current and the charge on the capacitor in the

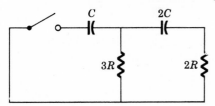

Figure 5.23

closed loop are both zero, but the capacitor in the open loop bears a charge Q_0. Find the current in each loop as a function of time after the switch is closed.

17 Work Exercise 16 for the network shown in Fig. 5.24.

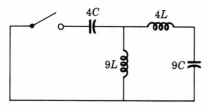

Figure 5.24

18 Find the current in each loop of the network shown in Fig. 5.25 if the switch is closed at an instant when all charges and currents are zero.

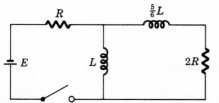

Figure 5.25

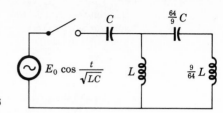

Figure 5.26

19 Work Exercise 18 for the network shown in Fig. 5.26.
20 In Example 3, find the normal modes, i.e., the sets of A's, for each of the natural frequencies.
21 Work Example 3 with the capacitor in series with the inductance in the last loop removed.
22 Work Example 3 with the inductances and capacitances in each loop interchanged.
23 Assuming that friction is negligible, find the natural frequencies of the system of n equal masses shown in Fig. 5.27.

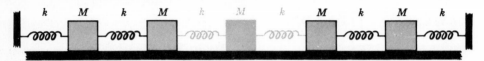

Figure 5.27

24 Assuming that friction is negligible, find the natural frequencies of the system of n identical disks connected by identical lengths of elastic shafting shown in Fig. 5.28.

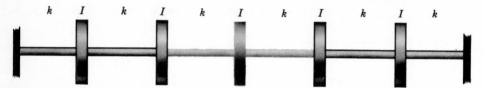

Figure 5.28

25 Work Exercise 23 if the spring connecting the right-hand mass to the wall is removed.
26 Work Exercise 24 if the two lengths of shafting which connect the system to the walls are removed.
27 If a voltage $E_0 \cos \omega t$ is inserted in series with the inductance in the first loop of the network in Example 3, find expressions for the steady-state charges on the various capacitors if

(a) $0 < \omega < \dfrac{2}{\sqrt{LC}}$ (b) $\omega > \dfrac{2}{\sqrt{LC}}$

28 If the capacitances and inductances in the network in Example 3 are interchanged, and if a voltage $E_0 \cos \omega t$ is then inserted in series with the capacitance in the first loop, find expressions for the steady-state charges on the various capacitors if

(a) $0 < \omega < \dfrac{1}{2\sqrt{LC}}$ (b) $\omega > \dfrac{1}{2\sqrt{LC}}$

29 Show that the electrical and mechanical systems shown in Fig. 5.29 are governed by differential equations which are identical in form. Set up these differential equations and reduce each to a dimensionless form.

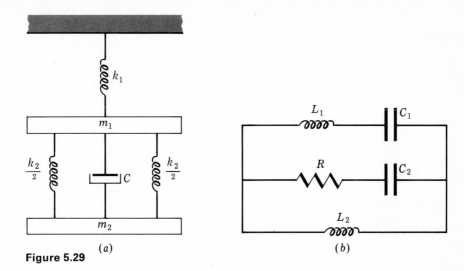

(a)　　　　　　　　　　　　　　(b)

Figure 5.29

30　Construct an electric network which will be governed by differential equations of exactly the same form as those which govern the mechanical system shown in Fig. 5.30. Set up these differential equations and reduce each to a dimensionless form.

Figure 5.30

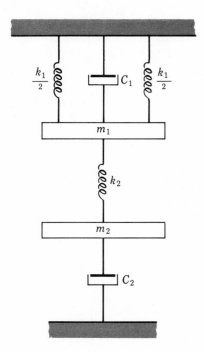

CHAPTER 6
Fourier Series and Integrals

6.1 Introduction

In Chap. 2 we learned that nonhomogeneous, linear, constant-coefficient differential equations containing terms of the form

$$A \cos \omega t \quad \text{and} \quad B \sin \omega t$$

can easily be solved for all values of ω. Then in Chap. 5 we discovered that such differential equations are fundamental in the study of physical systems subjected to periodic disturbances. In many cases, however, the forces, torques, voltages, or currents which act on a system, although periodic, are by no means so simple as pure sine and cosine waves. For instance, the voltage impressed on an electric circuit might consist of a series of pulses, as shown in Fig. 6.1a, or the disturbing influence acting on a mechanical system might be a force of constant magnitude whose direction is periodically and instantaneously reversed, as in Fig. 6.1b.

This raises the question whether a general periodic function† can be expressed as a series of sine and cosine terms. Specifically, since for all integral values of n

$$\cos \frac{n\pi(t + 2p)}{p} = \cos \frac{n\pi t}{p} \quad \text{and} \quad \sin \frac{n\pi(t + 2p)}{p} = \sin \frac{n\pi t}{p}$$

Figure 6.1
Typical periodic forcing functions.

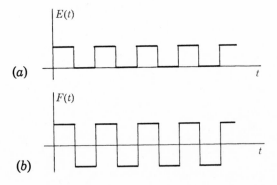

(a)

(b)

---

† A function $f(t)$ is said to be **periodic** if there exists a constant $2p$ with the property that $f(t + 2p) = f(t)$ for all t. If $2p$ is the smallest number for which this identity holds, $2p$ is called the **period** of the function.

it is natural to ask whether an arbitrary function $f(t)$ of period $2p$ can be represented by a series of the form

$$\frac{a_0\dagger}{2} + a_1 \cos \frac{\pi t}{p} + a_2 \cos \frac{2\pi t}{p} + \cdots + a_n \cos \frac{n\pi t}{p} + \cdots$$

$$+ b_1 \sin \frac{\pi t}{p} + b_2 \sin \frac{2\pi t}{p} + \cdots + b_n \sin \frac{n\pi t}{p} + \cdots$$

If this is the case, then the methods of Chap. 5, applied to the individual terms of such a series, will enable us, in fact, to analyze the behavior of systems acted upon by general periodic disturbances. The possibility of such expansions and their determination when they exist are the subject matter of **Fourier analysis,**‡ to which we devote this chapter.

6.2 The Euler Coefficients

To obtain formulas for the coefficients a_n and b_n in the expansion

(1) $\quad f(t) = \dfrac{a_0}{2} + a_1 \cos \dfrac{\pi t}{p} + a_2 \cos \dfrac{2\pi t}{p} + \cdots + a_n \cos \dfrac{n\pi t}{p} + \cdots$

$$+ b_1 \sin \frac{\pi t}{p} + b_2 \sin \frac{2\pi t}{p} + \cdots + b_n \sin \frac{n\pi t}{p} + \cdots$$

assuming, of course, that it exists, we shall need the following definite integrals, which are valid for all values of d, provided m and n are integers satisfying the given restrictions

(2) $$\int_d^{d+2p} \cos \frac{n\pi t}{p} \, dt = 0 \qquad n \neq 0$$

(3) $$\int_d^{d+2p} \sin \frac{n\pi t}{p} \, dt = 0$$

(4) $$\int_d^{d+2p} \cos \frac{m\pi t}{p} \cos \frac{n\pi t}{p} \, dt = 0 \qquad m \neq n$$

(5) $$\int_d^{d+2p} \cos^2 \frac{n\pi t}{p} \, dt = p \qquad n \neq 0$$

(6) $$\int_d^{d+2p} \cos \frac{m\pi t}{p} \sin \frac{n\pi t}{p} \, dt = 0$$

(7) $$\int_d^{d+2p} \sin \frac{m\pi t}{p} \sin \frac{n\pi t}{p} \, dt = 0 \qquad m \neq n$$

(8) $$\int_d^{d+2p} \sin^2 \frac{n\pi t}{p} \, dt = p \qquad n \neq 0$$

---

† The introduction of the factor $\frac{1}{2}$ in this term is a conventional device to render the final formulas for the coefficients more symmetric.

‡ Named for J. B. J. Fourier (1768–1830), French mathematician and confidant of Napoleon, who first undertook the systematic study of such expansions in a memorable monograph, "Théorie analytique de la chaleur," published in 1822. The use of such series in particular problems, however, dates from the time of Daniel Bernoulli (1700–1782), who used them to solve certain problems connected with vibrating strings.

With these integrals available, the determination of a_n and b_n proceeds as follows.[†]

To find a_0, we assume that the series (1) can legitimately be integrated term by term from $t = d$ to $t = d + 2p$.[‡] Then

$$\int_d^{d+2p} f(t)\, dt = \frac{a_0}{2} \int_d^{d+2p} dt + a_1 \int_d^{d+2p} \cos \frac{\pi t}{p}\, dt + \cdots$$

$$+ a_n \int_d^{d+2p} \cos \frac{n\pi t}{p}\, dt + \cdots + b_1 \int_d^{d+2p} \sin \frac{\pi t}{p}\, dt + \cdots$$

$$+ b_n \int_d^{d+2p} \sin \frac{n\pi t}{p}\, dt + \cdots$$

The integral on the left can always be evaluated, since $f(t)$ is a known function which is assumed to be integrable. At worst, some method of approximate integration, like those we discussed in Sec. 4.3, will be required. The first term on the right is simply

$$\frac{1}{2}a_0 t \Big|_d^{d+2p} = pa_0$$

By Eq. (2) all integrals with a cosine in the integrand vanish, and by Eq. (3) all integrals containing a sine vanish. Hence the integrated result reduces to

$$\int_d^{d+2p} f(t)\, dt = pa_0$$

or

(9)
$$a_0 = \frac{1}{p} \int_d^{d+2p} f(t)\, dt$$

To find a_n $(n = 1, 2, 3, \ldots)$, we multiply each side of (1) by $\cos (n\pi t/p)$ and then integrate from d to $d + 2p$, assuming again that term-by-term integration is justified. This gives

$$\int_d^{d+2p} f(t) \cos \frac{n\pi t}{p}\, dt = \frac{1}{2}a_0 \int_d^{d+2p} \cos \frac{n\pi t}{p}\, dt + a_1 \int_d^{d+2p} \cos \frac{\pi t}{p} \cos \frac{n\pi t}{p}\, dt + \cdots$$

$$+ a_n \int_d^{d+2p} \cos^2 \frac{n\pi t}{p}\, dt + \cdots + b_1 \int_d^{d+2p} \sin \frac{\pi t}{p} \cos \frac{n\pi t}{p}\, dt + \cdots$$

$$+ b_n \int_d^{d+2p} \sin \frac{n\pi t}{p} \cos \frac{n\pi t}{p}\, dt + \cdots$$

Again, the integral on the left is completely determined. By Eqs. (2) and (4), all integrals on the right containing only cosine terms vanish except the one involving

---

[†] The procedure here is analogous to the procedure we used in Sec. 4.6 to express an arbitrary polynomial as a linear combination of orthogonal polynomials. The most obvious difference is that in Sec. 4.6 the orthogonality condition involved the *summation* of products of two different functions over a discrete set of values, while here the orthogonality condition involves the *integration* of products of two continuous functions.

[‡] A sufficient condition for a series of integrable functions to be integrable term by term is that it be uniformly convergent (Theorem 6, Sec. 15.1).

$\cos^2 (n\pi t/p)$, which, by Eq. (5), is equal to p. Finally, by Eq. (6), every integral which contains a sine is zero. Hence

$$\int_d^{d+2p} f(t) \cos \frac{n\pi t}{p} \, dt = pa_n$$

or

(10)
$$a_n = \frac{1}{p} \int_d^{d+2p} f(t) \cos \frac{n\pi t}{p} \, dt$$

To determine b_n, we continue essentially the same procedure. We multiply (1) by $\sin (n\pi t/p)$ and then integrate from d to $d + 2p$, getting

$$\int_d^{d+2p} f(t) \sin \frac{n\pi t}{p} \, dt = \tfrac{1}{2}a_0 \int_d^{d+2p} \sin \frac{n\pi t}{p} \, dt + a_1 \int_d^{d+2p} \cos \frac{\pi t}{p} \sin \frac{n\pi t}{p} \, dt + \cdots$$

$$+ a_n \int_d^{d+2p} \cos \frac{n\pi t}{p} \sin \frac{n\pi t}{p} \, dt + \cdots + b_1 \int_d^{d+2p} \sin \frac{\pi t}{p} \sin \frac{n\pi t}{p} \, dt + \cdots$$

$$+ b_n \int_d^{d+2p} \sin^2 \frac{n\pi t}{p} \, dt + \cdots$$

As before, every integral on the right vanishes but one, leaving

$$\int_d^{d+2p} f(t) \sin \frac{n\pi t}{p} \, dt = pb_n$$

or

(11)
$$b_n = \frac{1}{p} \int_d^{d+2p} f(t) \sin \frac{n\pi t}{p} \, dt$$

Formulas (9), (10), and (11) are known as the **Euler** or **Euler-Fourier formulas**, and the series (1), when its coefficients have these values, is known as the **Fourier series** of $f(t)$. In most applications, the interval over which the coefficients are computed is either $(-p,p)$ or $(0,2p)$; so the value of d in the Euler formulas is usually either $-p$ or 0. Actually, the formula for a_0 need not be listed, since it can be obtained from the general expression for a_n by putting $n = 0$.† It was to achieve this that we wrote the constant term as $\tfrac{1}{2}a_0$ in the original expansion.

We must be careful at this stage not to delude ourselves with the belief that we have proved that every periodic function $f(t)$ has a Fourier expansion which converges to it. What our analysis has shown is merely that *if* a function $f(t)$ has an expansion of the form (1) for which term-by-term integration is valid, *then* the coefficients in that series must be given by the Euler formulas. Questions concerning the convergence of Fourier series and (if they converge) the conditions under which they will represent the functions which generated them are many and difficult. These problems are primarily of theoretical interest, however, for almost any conceivable practical application is covered by the famous **theorem of Dirichlet**.‡

---

† It is not necessarily the case, however, that the value of a_0 in a particular problem can be obtained by putting $n = 0$ in the *integrated* formula for a_n. For instance, in Example 2 the integrated formula for a_n is indeterminate when $n = 0$, and evaluation of the indeterminacy yields -3 instead of the correct value 3, which is obtained by putting $n = 0$ *before* integrating.
‡ Named for the German mathematician Peter Gustave Lejeune Dirichlet (1805–1859). For a proof of this theorem see, for instance, Philip Franklin, A Simple Discussion of the Representation of Functions by Fourier Series, pp. 357–361 in "Selected Papers on Calculus," Mathematical Association of America, 1969.

THEOREM 1   If $f(t)$ is a bounded periodic function which in any one period has at most a finite number of local maxima and minima and a finite number of points of discontinuity, then the Fourier series of $f(t)$ converges to $f(t)$ at all points where $f(t)$ is continuous and converges to the average of the right- and left-hand limits of $f(t)$ at each point where $f(t)$ is discontinuous.

   The conditions of Theorem 1, usually referred to as the **Dirichlet conditions**, make it clear that a function need not be continuous in order to possess a valid Fourier expansion. This means that a function may have a graph consisting of a number of disjointed arcs of different curves, each defined by a different formula, and still be representable by a Fourier series. In using the Euler formulas to find the coefficients in the expansion of such a function it will therefore be necessary to break up the range of integration $(d, d + 2p)$ to correspond to the various segments of the function. Thus in Fig. 6.2, the function $f(t)$ is defined by three different expressions, $f_1(t), f_2(t),$ $f_3(t)$ on successive portions of the period interval $d \leq t \leq d + 2p$. Hence it is necessary to write the Euler formulas as

$$a_n = \frac{1}{p} \int_d^{d+2p} f(t) \cos \frac{n\pi t}{p} \, dt$$

$$= \frac{1}{p} \int_d^r f_1(t) \cos \frac{n\pi t}{p} \, dt + \frac{1}{p} \int_r^s f_2(t) \cos \frac{n\pi t}{p} \, dt + \frac{1}{p} \int_s^{d+2p} f_3(t) \cos \frac{n\pi t}{p} \, dt$$

$$b_n = \frac{1}{p} \int_d^{d+2p} f(t) \sin \frac{n\pi t}{p} \, dt$$

$$= \frac{1}{p} \int_d^r f_1(t) \sin \frac{n\pi t}{p} \, dt + \frac{1}{p} \int_r^s f_2(t) \sin \frac{n\pi t}{p} \, dt + \frac{1}{p} \int_s^{d+2p} f_3(t) \sin \frac{n\pi t}{p} \, dt$$

Incidentally, according to Theorem 1, the Fourier series of the function shown in Fig. 6.2 will converge to the average values, indicated by dots, at the discontinuities at d, r, and $d + 2p$, regardless of the definition (or lack of definition) of the function at these points.

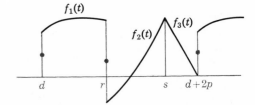

Figure 6.2
A periodic function defined by different
formulas over different portions of a period.

EXAMPLE 1

   What is the Fourier expansion of the periodic function whose definition in one period is

$$f(t) = \begin{cases} 0 & -\pi < t < 0 \\ \sin t & 0 < t < \pi \end{cases}$$

In this case the half period of the given function is $p = \pi$. Hence, taking $d = -\pi$ in the Euler formulas, we have

$$a_n = \frac{1}{\pi} \int_{-\pi}^{\pi} f(t) \cos nt \, dt = \frac{1}{\pi} \int_{-\pi}^{0} 0 \cos nt \, dt + \frac{1}{\pi} \int_{0}^{\pi} \sin t \cos nt \, dt$$

$$= \frac{1}{\pi} \left[-\frac{1}{2} \left\{ \frac{\cos (1-n)t}{1-n} + \frac{\cos (1+n)t}{1+n} \right\} \right]_0^{\pi}$$

$$= -\frac{1}{2\pi} \left[\frac{\cos (\pi - n\pi)}{1-n} + \frac{\cos (\pi + n\pi)}{1+n} - \left(\frac{1}{1-n} + \frac{1}{1+n} \right) \right]$$

$$= -\frac{1}{2\pi} \left(\frac{-\cos n\pi}{1-n} + \frac{-\cos n\pi}{1+n} - \frac{2}{1-n^2} \right)$$

$$= \frac{1 + \cos n\pi}{\pi(1 - n^2)} \qquad n \neq 1$$

$$a_1 = \frac{1}{\pi} \int_{0}^{\pi} \sin t \cos t \, dt = \left. \frac{\sin^2 t}{2\pi} \right|_0^{\pi} = 0$$

$$b_n = \frac{1}{\pi} \int_{-\pi}^{\pi} f(t) \sin nt \, dt = \frac{1}{\pi} \int_{-\pi}^{0} 0 \sin nt \, dt + \frac{1}{\pi} \int_{0}^{\pi} \sin t \sin nt \, dt$$

$$= \frac{1}{\pi} \left[\frac{1}{2} \left\{ \frac{\sin (1-n)t}{1-n} - \frac{\sin (1+n)t}{1+n} \right\} \right]_0^{\pi} = 0 \qquad n \neq 1$$

$$b_1 = \frac{1}{\pi} \int_{0}^{\pi} \sin^2 t \, dt = \frac{1}{\pi} \left[\frac{t}{2} - \frac{\sin 2t}{4} \right]_0^{\pi} = \frac{1}{2}$$

Hence, evaluating the coefficients for $n = 0, 1, 2, \ldots$, we have

$$f(t) = \frac{1}{\pi} + \frac{\sin t}{2} - \frac{2}{\pi} \left(\frac{\cos 2t}{3} + \frac{\cos 4t}{15} + \frac{\cos 6t}{35} + \frac{\cos 8t}{63} + \cdots \right)$$

Plots showing the accuracy with which the first n terms of this series represent the given function are shown in Fig. 6.3 for $n = 1, 2, 3$. For $n = 4, 5, 6, \ldots$ the graphs of the partial sums are almost indistinguishable from the graph of $f(t)$.

Interesting numerical series can often be obtained from Fourier series by evaluating them at particular points. For instance, if we set $t = \pi/2$ in the above expansion, we find

$$f\left(\frac{\pi}{2} \right) = 1 = \frac{1}{\pi} + \frac{1}{2} - \frac{2}{\pi} \left(-\frac{1}{3} + \frac{1}{15} - \frac{1}{35} + \frac{1}{63} - \cdots \right)$$

or

$$\frac{1}{1 \cdot 3} - \frac{1}{3 \cdot 5} + \frac{1}{5 \cdot 7} - \frac{1}{7 \cdot 9} + \cdots = \frac{\pi - 2}{4}$$

EXAMPLE 2

Find the expansion of the periodic function whose definition in one period is

$$f(t) = \begin{cases} -t & -3 < t < 0 \\ t & 0 < t < 3 \end{cases}$$

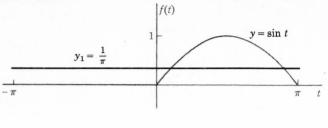

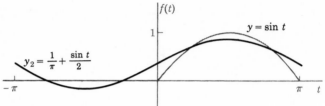

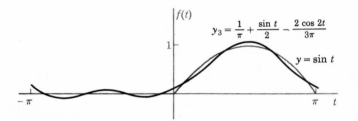

Figure 6.3
The approximation of a function by the first few terms of its
Fourier expansion.

In this case the period of the function is 6. Hence $p = 3$, and, from (10) and (11),
taking $d = -3$, we have

$$a_n = \frac{1}{3} \int_{-3}^{0} -t \cos \frac{n\pi t}{3} \, dt + \frac{1}{3} \int_{0}^{3} t \cos \frac{n\pi t}{3} \, dt$$

$$= -\frac{1}{3} \left[\frac{9}{n^2\pi^2} \cos \frac{n\pi t}{3} + \frac{3t}{n\pi} \sin \frac{n\pi t}{3} \right]_{-3}^{0} + \frac{1}{3} \left[\frac{9}{n^2\pi^2} \cos \frac{n\pi t}{3} + \frac{3t}{n\pi} \sin \frac{n\pi t}{3} \right]_{0}^{3}$$

$$= -\frac{3}{n^2\pi^2} (1 - \cos n\pi) + \frac{3}{n^2\pi^2} (\cos n\pi - 1)$$

$$= \frac{6}{n^2\pi^2} (\cos n\pi - 1) \qquad n \neq 0$$

$$a_0 = \frac{1}{3} \int_{-3}^{0} -t \, dt + \frac{1}{3} \int_{0}^{3} t \, dt = -\frac{t^2}{6} \Big|_{-3}^{0} + \frac{t^2}{6} \Big|_{0}^{3} = \frac{3}{2} + \frac{3}{2} = 3$$

$$b_n = \frac{1}{3} \int_{-3}^{0} -t \sin \frac{n\pi t}{3} \, dt + \frac{1}{3} \int_{0}^{3} t \sin \frac{n\pi t}{3} \, dt$$

$$= -\frac{1}{3} \left[\frac{9}{n^2\pi^2} \sin \frac{n\pi t}{3} - \frac{3t}{n\pi} \cos \frac{n\pi t}{3} \right]_{-3}^{0} + \frac{1}{3} \left[\frac{9}{n^2\pi^2} \sin \frac{n\pi t}{3} - \frac{3t}{n\pi} \cos \frac{n\pi t}{3} \right]_{0}^{3}$$

$$= \frac{3}{n\pi} \cos (-n\pi) - \frac{3}{n\pi} \cos n\pi = 0$$

Substituting these coefficients into the series (1), we obtain

$$f(t) = \frac{3}{2} - \frac{12}{\pi^2} \left(\frac{1}{1} \cos \frac{\pi t}{3} + \frac{1}{9} \cos \frac{3\pi t}{3} + \frac{1}{25} \cos \frac{5\pi t}{3} + \cdots \right)$$

Determine the Fourier expansions of the periodic functions whose definitions in one period are:

1 $f(t) = \begin{cases} 1 & 0 < t < 1 \\ 0 & 1 < t < 2 \end{cases}$

2 $f(t) = \begin{cases} 1 & 0 < t < 1 \\ -1 & 1 < t < 2 \end{cases}$

3 $f(t) = \begin{cases} 0 & -2 < t < -1 \\ 1 & -1 < t < 0 \\ -1 & 0 < t < 1 \\ 0 & 1 < t < 2 \end{cases}$

4 $f(t) = \begin{cases} 1 & -2 < t < -1 \\ 0 & -1 < t < 0 \\ -1 & 0 < t < 1 \\ 0 & 1 < t < 2 \end{cases}$

5 $f(t) = t \qquad 0 < t < 1$

6 $f(t) = t \qquad -1 < t < 1$

7 $f(t) = \begin{cases} 2 & 0 < t < 2\pi/3 \\ 1 & 2\pi/3 < t < 4\pi/3 \\ 0 & 4\pi/3 < t < 2\pi \end{cases}$

8 $f(t) = \begin{cases} 0 & -2 < t < -1 \\ 1 + t & -1 < t < 0 \\ 1 - t & 0 < t < 1 \\ 0 & 1 < t < 2 \end{cases}$

9 $f(t) = e^{-t} \qquad 0 < t < 2$

10 $f(t) = \sin t \qquad 0 < t < \pi$

11 $f(t) = \begin{cases} 0 & -\pi < t < 0 \\ t & 0 < t < \pi \end{cases}$

12 $f(t) = \begin{cases} \cos t & -\pi < t < 0 \\ \sin t & 0 < t < \pi \end{cases}$

13 $f(t) = t - t^3 \qquad -1 < t < 1$

14 $f(t) = \begin{cases} 0 & -\pi < t < 0 \\ t^2 & 0 < t < \pi \end{cases}$

15 Establish the following numerical results:

$$1 + \frac{1}{2^2} + \frac{1}{3^2} + \frac{1}{4^2} + \frac{1}{5^2} + \cdots = \frac{\pi^2}{6}$$

$$1 - \frac{1}{2^2} + \frac{1}{3^2} - \frac{1}{4^2} + \frac{1}{5^2} - \cdots = \frac{\pi^2}{12}$$

$$1 + \frac{1}{3^2} + \frac{1}{5^2} + \frac{1}{7^2} + \frac{1}{9^2} + \cdots = \frac{\pi^2}{8}$$

Hint: Use the results of Exercise 14.

6.3 Half-Range Expansions

When $f(t)$ possesses certain symmetry properties, the coefficients in its Fourier expansion become especially simple. This was illustrated in Example 2 of the last section, where the given function was symmetric in the y axis and its expansion contained only cosine terms, i.e., only terms which themselves were symmetric in the y axis. In this section we shall investigate in detail just what effect the symmetry of $f(t)$ has on the coefficients in the Fourier series for $f(t)$.

Suppose first of all that $f(t)$ is an **even function**, i.e., suppose that $f(-t) = f(t)$ for all t or, geometrically, that the graph of $f(t)$ is symmetric in the vertical axis. Taking $d = -p$ in the formula for a_n [Eq. (10), Sec. 6.2] we can write

$$a_n = \frac{1}{p} \int_{-p}^{p} f(t) \cos \frac{n\pi t}{p}\, dt = \frac{1}{p} \int_{-p}^{0} f(t) \cos \frac{n\pi t}{p}\, dt + \frac{1}{p} \int_{0}^{p} f(t) \cos \frac{n\pi t}{p}\, dt$$

Now, in the integral from $-p$ to 0 let us make the substitution

$$t = -s \qquad dt = -ds$$

Then since $t = -p$ implies $s = p$, and since $t = 0$ implies $s = 0$, this integral becomes

(1) $$\frac{1}{p} \int_{p}^{0} f(-s) \cos \frac{-n\pi s}{p} (-ds)$$

But $f(-s) = f(s)$, from the hypothesis that $f(t)$ is an even function. Moreover, the cosine is also an even function; i.e.,

$$\cos \frac{-n\pi s}{p} = \cos \frac{n\pi s}{p}$$

Finally, the negative sign associated with ds in (1) can be eliminated by changing the limits back to the normal order, 0 to p. The integral (1) then becomes

$$\frac{1}{p} \int_0^p f(s) \cos \frac{n\pi s}{p} \, ds$$

and thus a_n can be written

$$a_n = \frac{1}{p} \int_0^p f(s) \cos \frac{n\pi s}{p} \, ds + \frac{1}{p} \int_0^p f(t) \cos \frac{n\pi t}{p} \, dt$$

$$= \frac{2}{p} \int_0^p f(t) \cos \frac{n\pi t}{p} \, dt$$

since the two integrals are identical, except for the dummy variable of integration, which is immaterial.

Similarly, we can write

$$b_n = \frac{1}{p} \int_{-p}^0 f(t) \sin \frac{n\pi t}{p} \, dt + \frac{1}{p} \int_0^p f(t) \sin \frac{n\pi t}{p} \, dt$$

Again, putting $t = -s$ and $dt = -ds$ in the first integral, we find

$$b_n = \frac{1}{p} \int_p^0 f(-s) \sin \frac{-n\pi s}{p} \, (-ds) + \frac{1}{p} \int_0^p f(t) \sin \frac{n\pi t}{p} \, dt$$

But, by hypothesis, $f(-s) = f(s)$; and $\sin(-n\pi s/p) = -\sin(n\pi s/p)$ from the familiar properties of the sine function. Hence, reversing the limits on the first integral, as before, we have

$$b_n = -\frac{1}{p} \int_0^p f(s) \sin \frac{n\pi s}{p} \, ds + \frac{1}{p} \int_0^p f(t) \sin \frac{n\pi t}{p} \, dt = 0$$

since, except for the irrelevant dummy variable of integration, the two integrals are identical in all but sign. Thus we have established the following useful result.

THEOREM 1 If $f(t)$ is an even periodic function, the coefficients in the Fourier series of $f(t)$ are given by the formulas

$$a_n = \frac{2}{p} \int_0^p f(t) \cos \frac{n\pi t}{p} \, dt \qquad b_n \equiv 0$$

If $f(t)$ is an **odd function**, i.e., if $f(t)$ has the property that $f(-t) = -f(t)$ for all values of t, then by an almost identical argument we can establish the following companion result.

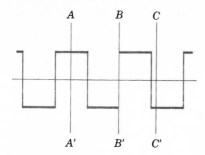

Figure 6.4
How oddness and evenness depend
on choice of axes.

THEOREM 2 If $f(t)$ is an odd periodic function, the coefficients in the Fourier series of $f(t)$ are given by the formulas

$$a_n \equiv 0 \qquad b_n = \frac{2}{p} \int_0^p f(t) \sin \frac{n\pi t}{p}\, dt$$

It should be emphasized here that oddness and evenness are not intrinsic properties of a graph but depend upon its relation to the axes of the coordinate system. For instance, in Fig. 6.4, if the line AA' is chosen as the vertical axis, the graph represents an even function whose Fourier series, in accordance with Theorem 1, will contain only cosine terms. On the other hand, if BB' is chosen as the vertical axis, the graph represents an odd function and, by Theorem 2, only sine terms will appear in its expansion. Finally, if a general line, such as CC', is chosen as the vertical axis, the graph represents a function which is neither odd nor even, and both sines and cosines will appear in its Fourier series.

The observations we have just made about the Fourier coefficients of odd and even functions serve to reduce by half the labor of expanding such functions. However, their chief value is that they allow us to meet the requirements of certain problems†️ in which expansions containing *only* cosine terms or expansions containing *only* sine terms must be constructed.

Let us suppose that the conditions of a problem require us to consider the values of a function *only* in the interval from 0 to p. In other words, conditions of periodicity are irrelevant to the problem, and what the function may be outside the interval $(0,p)$ is completely immaterial. This being the case, we can define the function *in any way we please* over the interval $(-p,0)$, then define it over the rest of the t axis by the simple requirement that it be of period $2p$, and finally use the Euler formulas to determine the coefficients in the Fourier series of the periodic extension we have thus created. Between $-p$ and 0 this series will, of course, converge to whatever extension we created over this interval, but *irrespective of this extension* the series will represent the given function between 0 and p, as required.

In particular, if we extend the function from 0 to $-p$ by reflecting it in the vertical axis, so that $f(-t) = f(t)$, the original function together with its extension is even; hence the Fourier expansion of its periodic continuation will contain only cosine terms

---

† Examples of such problems will be found in Sec. 8.4.

[including, of course, the constant term $a_0/2 = (a_0/2) \cos (0\pi t/p)$], whose coefficients as we showed above, will be given by

$$a_n = \frac{2}{p} \int_0^p f(t) \cos \frac{n\pi t}{p} \, dt$$

On the other hand, if we extend the same function from 0 to $-p$ by reflecting it in the origin, so that $f(-t) = -f(t)$, the extended function is odd and hence the Fourier series of its periodic continuation will contain only sine terms, whose coefficients will be given by

$$b_n = \frac{2}{p} \int_0^p f(t) \sin \frac{n\pi t}{p} \, dt$$

Thus, simply by imagining the appropriate extension of a function originally defined only for $0 < t < p$ we can obtain expansions representing the function on this interval and containing only cosine terms or only sine terms, as we please. Such series are known as **half-range expansions.**

EXAMPLE 1

Find the half-range expansions of the function

$$f(t) = t - t^2 \qquad 0 < t < 1$$

The half-range cosine expansion is obtained by first extending $t - t^2$ from the given interval (0,1) to the interval $(-1,0)$ by reflection in the y axis and then taking the function thus defined from -1 to 1 as one period of a periodic function of period $2p = 2$. However, once we understand the reasoning underlying the procedure we need give no thought to the extension but can write immediately, on the basis of Theorem 1,

$$b_n \equiv 0$$

$$a_n = \frac{2}{1} \int_0^1 (t - t^2) \cos \frac{n\pi t}{1} \, dt$$

$$= 2 \left[\left(\frac{\cos n\pi t}{n^2 \pi^2} + \frac{t}{n\pi} \sin n\pi t \right) - \left(\frac{2t}{n^2 \pi^2} \cos n\pi t - \frac{2}{n^3 \pi^3} \sin n\pi t + \frac{t^2}{n\pi} \sin n\pi t \right) \right]_0^1$$

$$= 2 \left(\frac{\cos n\pi - 1}{n^2 \pi^2} - \frac{2 \cos n\pi}{n^2 \pi^2} \right)$$

$$= -\frac{2(1 + \cos n\pi)}{n^2 \pi^2} \qquad n \neq 0$$

$$a_0 = \frac{2}{1} \int_0^1 (t - t^2) \, dt = 2 \left[\frac{t^2}{2} - \frac{t^3}{3} \right]_0^1 = \frac{1}{3}$$

Hence it is possible to represent $f(t) = t - t^2$ for $0 < t < 1$ by the series

$$(2) \qquad f(t) = \frac{1}{6} - \frac{4}{\pi^2} \left(\frac{\cos 2\pi t}{4} + \frac{\cos 4\pi t}{16} + \frac{\cos 6\pi t}{36} + \frac{\cos 8\pi t}{64} + \cdots \right)$$

Similarly, the half-range sine expansion is obtained by first extending the given function, $t - t^2$, to the interval $(-1,0)$ by reflection in the origin and then extending periodically the function thus defined over $(-1,1)$. However, all we actually need to do to obtain the expansion is to note that, according to Theorem 2,

$$a_n \equiv 0$$

and

$$b_n = \frac{2}{1} \int_0^1 (t - t^2) \sin \frac{n\pi t}{1} \, dt$$

$$= 2 \left[\left(\frac{\sin n\pi t}{n^2 \pi^2} - \frac{t}{n\pi} \cos n\pi t \right) - \left(\frac{2t}{n^2 \pi^2} \sin n\pi t + \frac{2}{n^3 \pi^3} \cos n\pi t - \frac{t^2}{n\pi} \cos n\pi t \right) \right]_0^1$$

$$= 2 \left[\left(-\frac{\cos n\pi}{n\pi} \right) - \left(\frac{2(\cos n\pi - 1)}{n^3 \pi^3} - \frac{\cos n\pi}{n\pi} \right) \right]$$

$$= \frac{4(1 - \cos n\pi)}{n^3 \pi^3}$$

Hence it is also possible to represent $f(t)$ for $0 < t < 1$ by the series

$$(3) \qquad f(t) = \frac{8}{\pi^3} \left(\frac{\sin \pi t}{1} + \frac{\sin 3\pi t}{27} + \frac{\sin 5\pi t}{125} + \frac{\sin 7\pi t}{343} + \cdots \right)$$

Series (2) and (3) are by no means the only Fourier series that will represent $t - t^2$ on the interval $(0,1)$. They are merely the most convenient or most useful ones. In fact, with every possible extension of $t - t^2$ from 0 to -1 there is associated a series yielding $t - t^2$ for $0 < t < 1$. For instance, a third such series might be obtained by letting the extension be simply the one defined by $t - t^2$ itself for $-1 < t < 0$. In this case

$$a_n = \frac{1}{1} \int_{-1}^1 (t - t^2) \cos \frac{n\pi t}{1} \, dt$$

$$= \left[\left(\frac{\cos n\pi t}{n^2 \pi^2} + \frac{t}{n\pi} \sin n\pi t \right) - \left(\frac{2t}{n^2 \pi^2} \cos n\pi t - \frac{2}{n^3 \pi^3} \sin n\pi t + \frac{t^2}{n\pi} \sin n\pi t \right) \right]_{-1}^1$$

$$= -\frac{4 \cos n\pi}{n^2 \pi^2} \qquad n \neq 0$$

$$a_0 = \frac{1}{1} \int_{-1}^1 (t - t^2) \, dt = \left[\frac{t^2}{2} - \frac{t^3}{3} \right]_{-1}^1 = -\frac{2}{3}$$

$$b_n = \frac{1}{1} \int_{-1}^1 (t - t^2) \sin \frac{n\pi t}{1} \, dt$$

$$= \left[\left(\frac{1}{n^2 \pi^2} \sin n\pi t - \frac{t}{n\pi} \cos n\pi t \right) \right.$$

$$\left. - \left(\frac{2t}{n^2 \pi^2} \sin n\pi t + \frac{2}{n^3 \pi^3} \cos n\pi t - \frac{t^2}{n\pi} \cos n\pi t \right) \right]_{-1}^1$$

$$= -\frac{2 \cos n\pi}{n\pi}$$

Hence, for $0 < t < 1$ it is also possible to write

$$(4) \qquad f(t) = -\frac{1}{3} + \frac{4}{\pi^2} \left(\frac{\cos \pi t}{1} - \frac{\cos 2\pi t}{4} + \frac{\cos 3\pi t}{9} - \frac{\cos 4\pi t}{16} + \cdots \right)$$

$$+ \frac{2}{\pi} \left(\frac{\sin \pi t}{1} - \frac{\sin 2\pi t}{2} + \frac{\sin 3\pi t}{3} - \frac{\sin 4\pi t}{4} + \cdots \right)$$

Figure 6.5a, b, and c shows the extended periodic functions represented, respectively, by the series (2), (3), and (4).

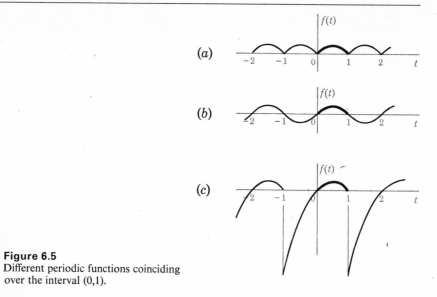

Figure 6.5
Different periodic functions coinciding
over the interval (0,1).

Figure 6.5 and the associated expansions illustrate another interesting and important fact. In Fig. 6.5c the graph as a whole is not continuous but has jumps at $t = \pm 1, \pm 3, \ldots$. In the corresponding series (4), the coefficients (of the sine terms) decrease only at a rate proportional to $1/n$. On the other hand, the graph in Fig. 6.5a is everywhere continuous but has corners, or points, where the tangent changes direction discontinuously. In the corresponding series (2), the coefficients become small much more rapidly than in (4); in fact they decrease at a rate proportional to $1/n^2$. Finally, the graph in Fig. 6.5b not only is continuous but also has a continuous tangent; i.e., there are no points where the tangent changes direction abruptly. This smoother behavior of the function is reflected in the coefficients in the corresponding series (3), which in this case approach zero at a rate proportional to $1/n^3$. These observations are summed up and generalized in the following theorem, which we cite without proof.†

THEOREM 3   As n becomes infinite, the coefficients a_n and b_n in the Fourier expansion of a periodic function satisfying the Dirichlet conditions always approach zero at least as rapidly as c/n, where c is a constant independent of n. If the function has one or more points of discontinuity, then either a_n or b_n, and in general both, can decrease no faster than this. In general, if a function $f(t)$ and its first $k - 1$ derivatives satisfy the Dirichlet conditions and are everywhere continuous, then as n becomes infinite the coefficients a_n and b_n in the Fourier series of $f(t)$ tend to zero at least as rapidly as c/n^{k+1}. If, in addition, the kth derivative of $f(t)$ is not everywhere continuous, then either a_n or b_n, and in general both, can tend to zero no faster than c/n^{k+1}.

More concisely, though less accurately, Theorem 3 asserts that the smoother the function, the faster its Fourier expansion converges. Conversely, by observing the

---

† See, for instance, H. S. Carslaw, "Fourier Series," pp. 269–271, Dover, New York, 1930.

rate at which the terms in the Fourier series of an otherwise unknown function approach zero, we can obtain useful information about the degree of smoothness of the function.

Closely associated with the last result are the following observations, which we also state without proof.

THEOREM 4†   The integral of any periodic function which satisfies the Dirichlet conditions can be found by term-by-term integration of the Fourier series of the function.

THEOREM 5‡   If $f(t)$ is a periodic function which satisfies the Dirichlet conditions and is everywhere continuous, and if $f'(t)$ also satisfies the Dirichlet conditions, then wherever it exists, $f'(t)$ can be found by term-by-term differentiation of the Fourier series of $f(t)$.

EXERCISES

1   By considering the identity

$$f(t) = \frac{f(t) + f(-t)}{2} + \frac{f(t) - f(-t)}{2}$$

show that any function defined over an interval which is symmetric with respect to the origin can be written as the sum of an even function and an odd function.

Obtain the half-range sine and cosine expansions of each of the following functions:

2 $f(t) = \begin{cases} 1 & 0 < t < 1 \\ 0 & 1 < t < 3 \end{cases}$ 3 $f(t) = t \quad 0 < t < p$

4 $f(t) = t^2 \quad 0 < t < p$ 5 $f(t) = \cos t \quad 0 < t < 2\pi$

6 $f(t) = \sin t \quad 0 < t < 2\pi$ 7 $f(t) = e^{-at} \quad 0 < t < 1$

8 $f(t) = \sin at \quad 0 < t < \pi$, a not an integer

9 $f(t) = \cos at \quad 0 < t < \pi$, a not an integer

10 By setting $t = \pi$ in the half-range cosine expansion of $\cos at$ obtained in Exercise 9, show that

$$\cot a\pi = \frac{1}{a\pi} - \frac{2a}{\pi} \sum_{n=1}^{\infty} \frac{1}{n^2 - a^2} = \frac{1}{\pi} \sum_{n=-\infty}^{\infty} \frac{1}{n + a}$$

11 Using the results of Exercise 8, show that

$$\csc a\pi = \frac{1}{a\pi} - \frac{2a}{\pi} \sum_{n=1}^{\infty} \frac{(-1)^n}{n^2 - a^2} = \frac{1}{\pi} \sum_{n=-\infty}^{\infty} \frac{(-1)^n}{n + a}$$

Hint: At the appropriate point, recall the series obtained in Exercise 10.

12 By integrating the half-range cosine series for t^2 obtained in Exercise 4, obtain the half-range sine series for t^3. *Hint:* Recall from Exercise 3 the half-range sine series for t.

† See, for instance, E. C. Titchmarsh, "Theory of Functions," pp. 419–421, Oxford, New York, 1939.

‡ See, for instance, E. T. Whittaker and G. N. Watson, "Modern Analysis," pp. 168–169, Macmillan, New York, 1943.

13 By integrating the half-range sine series for t^2 obtained in Exercise 4 and evaluating the result for $t = p/2$, show first that

$$1 + \frac{1}{3^4} + \frac{1}{5^4} + \frac{1}{7^4} + \cdots = \frac{\pi^4}{96}$$

and then find the half-range cosine series for t^3. *Hint:* Recall from Exercise 3 the half-range sine series for t.

14 By integrating the result of Exercise 13, find the half-range sine expansion of t^4.

15 Show that

$$1 - \frac{1}{3^3} + \frac{1}{5^3} - \frac{1}{7^3} + \cdots = \frac{\pi^3}{32}$$

16 Show that

$$\frac{1}{1^2 3^2} - \frac{3}{5^2 7^2} + \frac{5}{9^2 11^2} - \frac{7}{13^2 15^2} + \cdots = \frac{\sqrt{2}\,\pi^2}{128}$$

Hint: Evaluate the half-range cosine expansion of t^2 for $t = p/4$.

17 Obtain a series, different from the half-range sine expansion, which will represent $t - t^2$ for $0 < t < 1$ and which has coefficients decreasing as $1/n^3$.

18 Is it possible to obtain a series representing $t - t^2$ for $0 < t < 1$ whose coefficients will decrease as $1/n^4$?

19 Find a function whose half-range cosine series will have coefficients which decrease as $1/n^4$. Determine the expansion.

20 Without calculating the coefficients, determine how fast the coefficients in the half-range expansions of each of the following functions will decrease:

(a) $f(t) = t(1 - t)^2 \qquad 0 < t < 1$

(b) $f(t) = \begin{cases} 10t^4 - 15t^3 + 6t & 0 < t < 1 \\ 1 & 1 < t < 2 \end{cases}$

(c) $f(t) = \begin{cases} 10t^4 - 15t^3 + 6t & 0 < t < 1 \\ 2t - t^2 & 1 < t < 2 \end{cases}$

(d) $f(t) = 1/(2 + \cos t) \qquad 0 < t < 2\pi$

21 If $f(t)$, originally defined only for $0 < t < p$, is extended from p to $2p$ by reflection in the line $t = p$, show that the half-range sine expansion of the extended function contains no terms of the form

$$\sin \frac{n\pi t}{2p} \qquad n \text{ even}$$

and show that the coefficients of the terms of the form

$$\sin \frac{n\pi t}{2p} \qquad n \text{ odd}$$

are given by the formula

$$b_n = \frac{2}{p} \int_0^p f(t) \sin \frac{n\pi t}{2p}\, dt$$

22 Determine how $f(t)$, originally defined only for $0 < t < p$, must be extended from p to $2p$ if the half-range cosine expansion of the extended function is to contain no terms of the form

$$\cos \frac{n\pi t}{2p} \qquad n \text{ even}$$

Derive a formula for the nonzero coefficients.

23 If

$$
f(t) = \begin{cases} 1 & 0 < t < a \\ \dfrac{t-1}{a-1} & a < t < 1 \\ 0 & 1 < t < 2 \end{cases}
$$

and if a is only slightly less than 1, discuss the behavior of the coefficients in the half-range cosine expansion of $f(t)$ for small and medium values of n as well as for $n \to \infty$.

24 If

$$
f(t) = \begin{cases} 2 & 0 < t < 1 - \dfrac{1}{2a} \\ 1 - \sin a\pi t & 1 - \dfrac{1}{2a} < t < 1 + \dfrac{1}{2a} \\ 0 & 1 + \dfrac{1}{2a} < t < 2 \end{cases}
$$

and if a is a large even integer, discuss the behavior of the coefficients in the half-range cosine expansion of $f(t)$ for small and medium values of n as well as for $n \to \infty$.

25 Prove Theorem 3 for the special case $k = 1$. Under what conditions, if any, can either a_n or b_n decrease faster than c/n^2? *Hint:* Assuming that $f'(t)$ has a single point of discontinuity in each period, apply integration by parts to the formulas for a_n and b_n, taking u in the formula $\int u\, dv = uv - \int v\, du$ to be $f(t)$.

6.4 Alternative Forms of Fourier Series

The original form of the Fourier series of a function, as derived in Sec. 6.2, can be converted into several other trigonometric forms and into one in which imaginary exponentials appear instead of real trigonometric functions. For instance, in the series

$$
f(t) = \frac{a_0}{2} + \sum_{n=1}^{\infty} \left(a_n \cos \frac{n\pi t}{p} + b_n \sin \frac{n\pi t}{p} \right)
$$

we can apply to each pair of terms of the same frequency the usual procedure for reducing the sum of a sine and a cosine of the same angle to a single term:

$$
f(t) = \frac{a_0}{2} + \sum_{n=1}^{\infty} \sqrt{a_n^2 + b_n^2} \left(\frac{a_n}{\sqrt{a_n^2 + b_n^2}} \cos \frac{n\pi t}{p} + \frac{b_n}{\sqrt{a_n^2 + b_n^2}} \sin \frac{n\pi t}{p} \right)
$$

If we now define the angles γ_n and δ_n from the triangle shown in Fig. 6.6 and set

$$
A_0 = \frac{a_0}{2} \quad \text{and} \quad A_n = \sqrt{a_n^2 + b_n^2}
$$

the last series can be written

$$
f(t) = A_0 + \sum_{n=1}^{\infty} A_n \left(\cos \frac{n\pi t}{p} \cos \gamma_n + \sin \frac{n\pi t}{p} \sin \gamma_n \right)
$$

$$
= A_0 + \sum_{n=1}^{\infty} A_n \cos \left(\frac{n\pi t}{p} - \gamma_n \right)
$$

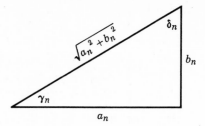

Figure 6.6
The triangle defining the phase angles
γ_n and δ_n for the resultant of the terms
of frequency $n\pi/p$ in a Fourier series.

or, equally well,

$$f(t) = A_0 + \sum_{n=1}^{\infty} A_n \left(\cos \frac{n\pi t}{p} \sin \delta_n + \sin \frac{n\pi t}{p} \cos \delta_n \right)$$

$$= A_0 + \sum_{n=1}^{\infty} A_n \sin \left(\frac{n\pi t}{p} + \delta_n \right)$$

In either of these forms, the quantity $A_n = \sqrt{a_n{}^2 + b_n{}^2}$ is the resultant amplitude of the components of frequency $n\pi/p$, that is, the amplitude of the **nth harmonic** in the expansion. The phase angles

$$\gamma_n = \tan^{-1} \frac{b_n}{a_n} \quad \text{and} \quad \delta_n = \tan^{-1} \frac{a_n}{b_n} = \frac{\pi}{2} - \gamma_n$$

measure the lag or lead of the nth harmonic with reference to a standard cosine or standard sine wave of the same frequency.

The complex exponential form of a Fourier series is obtained by substituting the exponential equivalents of the cosine and sine terms into the original form of the series:

$$f(t) = \frac{a_0}{2} + \sum_{n=1}^{\infty} \left(a_n \frac{e^{ni\pi t/p} + e^{-ni\pi t/p}}{2} + b_n \frac{e^{ni\pi t/p} - e^{-ni\pi t/p}}{2i} \right)$$

Collecting terms on the various exponentials and noting that $1/i = -i$, we obtain

$$f(t) = \frac{a_0}{2} + \sum_{n=1}^{\infty} \left(\frac{a_n - ib_n}{2} e^{ni\pi t/p} + \frac{a_n + ib_n}{2} e^{-ni\pi t/p} \right)$$

If we now define

$$c_0 = \frac{a_0}{2} \qquad c_n = \frac{a_n - ib_n}{2} \qquad c_{-n} = \frac{a_n + ib_n}{2}$$

the last series can be written in the more symmetric form

(1) $$f(t) = \sum_{n=-\infty}^{\infty} c_n e^{ni\pi t/p}$$

When it is used at all, this exponential form is used as a basic form in its own right; i.e., it is not obtained by transformation from the trigonometric form but is constructed directly from the given function. To do this requires that expressions be available for the direct evaluation of the coefficients c_n. These can easily be found from the definitions of c_0, c_n, and c_{-n}. For

$$c_0 = \tfrac{1}{2}a_0 = \frac{1}{2p} \int_d^{d+2p} f(t) \, dt$$

$$c_n = \frac{a_n - ib_n}{2} = \frac{1}{2}\left[\frac{1}{p} \int_d^{d+2p} f(t) \cos \frac{n\pi t}{p} \, dt - i\frac{1}{p} \int_d^{d+2p} f(t) \sin \frac{n\pi t}{p} \, dt\right]$$

$$= \frac{1}{2p} \int_d^{d+2p} f(t) \left(\cos \frac{n\pi t}{p} - i \sin \frac{n\pi t}{p}\right) dt$$

$$= \frac{1}{2p} \int_d^{d+2p} f(t) e^{-ni\pi t/p} \, dt$$

$$c_{-n} = \frac{a_n + ib_n}{2} = \frac{1}{2}\left[\frac{1}{p} \int_d^{d+2p} f(t) \cos \frac{n\pi t}{p} \, dt + i\frac{1}{p} \int_d^{d+2p} f(t) \sin \frac{n\pi t}{p} \, dt\right]$$

$$= \frac{1}{2p} \int_d^{d+2p} f(t) \left(\cos \frac{n\pi t}{p} + i \sin \frac{n\pi t}{p} \, dt\right)$$

$$= \frac{1}{2p} \int_d^{d+2p} f(t) e^{ni\pi t/p} \, dt$$

Clearly, whether the index n is positive, negative, or zero, c_n is correctly given by the single formula

(2) $$c_n = \frac{1}{2p} \int_d^{d+2p} f(t) e^{-ni\pi t/p} \, dt$$

As usual, d will almost always be either $-p$ or 0.

In the complex representation defined by (1) or (2), a certain symmetry between the formula for a function and the formula for its Fourier coefficients is evident. In fact, taking $d = -p$, the expressions

$$f(t) = \sum_{n=-\infty}^{\infty} c_n e^{ni\pi t/p} \quad \text{and} \quad c_n = \frac{1}{2p} \int_{-p}^{p} f(t) e^{-ni\pi t/p} \, dt$$

are of essentially the same structure, as the following correlation of symbols and operations reveals:

$$t \sim n$$

$$f(t) \sim c_n \equiv c(n)$$

$$e^{ni\pi t/p} \sim e^{-ni\pi t/p}$$

$$\sum_{n=-\infty}^{\infty} (\cdot) \sim \frac{1}{2p} \int_{-p}^{p} (\cdot) \, dt$$

This duality is worthy of note, and as our development proceeds to the Fourier integral and thence to the Laplace transform, it will become still more striking and fundamental.

EXAMPLE 1

Find the complex form of the Fourier series of the periodic function whose definition in one period is $f(t) = e^{-t}, -1 < t < 1$.

Since $p = 1$, we have from (2), taking $d = -1$,

$$c_n = \frac{1}{2} \int_{-1}^{1} e^{-t} e^{-n i \pi t}\, dt = \frac{1}{2} \left[\frac{e^{-(1 + n i \pi)t}}{-(1 + n i \pi)} \right]_{-1}^{1}$$

$$= \frac{e^{-(1 + n i \pi)} - e^{(1 + n i \pi)}}{-2(1 + n i \pi)}$$

$$= \frac{e e^{n i \pi} - e^{-1} e^{-n i \pi}}{2(1 + n i \pi)}$$

Now $e^{i\pi} = \cos \pi + i \sin \pi = -1$, and thus $e^{n i \pi} = e^{-n i \pi} = (-1)^n$. Hence

$$c_n = \frac{(-1)^n}{(1 + n i \pi)} \frac{e - e^{-1}}{2} = \frac{(-1)^n (1 - n i \pi) \sinh 1}{1 + n^2 \pi^2}$$

The expansion of $f(t)$ is therefore

$$f(t) = \sum_{n = -\infty}^{\infty} (-1)^n \frac{(1 - n i \pi) \sinh 1}{1 + n^2 \pi^2} e^{n i \pi t}$$

This, of course, can be converted into the real trigonometric form without difficulty, for we have, by definition,

$$c_n = \frac{a_n - i b_n}{2} \quad \text{and} \quad c_{-n} = \frac{a_n + i b_n}{2}$$

and thus, by adding and subtracting these expressions,

$$a_n = c_n + c_{-n} \quad \text{and} \quad b_n = i(c_n - c_{-n})$$

Therefore in this problem

$$a_n = \frac{(-1)^n (1 - n i \pi) \sinh 1}{1 + n^2 \pi^2} + \frac{(-1)^n (1 + n i \pi) \sinh 1}{1 + n^2 \pi^2} = \frac{(-1)^n 2 \sinh 1}{1 + n^2 \pi^2}$$

$$b_n = i \left[\frac{(-1)^n (1 - n i \pi) \sinh 1}{1 + n^2 \pi^2} - \frac{(-1)^n (1 + n i \pi) \sinh 1}{1 + n^2 \pi^2} \right] = \frac{(-1)^n 2 n \pi \sinh 1}{1 + n^2 \pi^2}$$

Hence we can also write

$$f(t) = \sinh 1 - 2 \sinh 1 \left(\frac{\cos \pi t}{1 + \pi^2} - \frac{\cos 2\pi t}{1 + 4\pi^2} + \frac{\cos 3\pi t}{1 + 9\pi^2} - \cdots \right)$$

$$- 2\pi \sinh 1 \left(\frac{\sin \pi t}{1 + \pi^2} - \frac{2 \sin 2\pi t}{1 + 4\pi^2} + \frac{3 \sin 3\pi t}{1 + 9\pi^2} - \cdots \right)$$

EXERCISES

What is the amplitude of the resultant term of frequency $n\pi/p$ in the Fourier series of the periodic functions whose definitions in one period are the following?

1 $f(t) = t + t^2$ $-1 < t < 1$

2 $f(t) = \begin{cases} -t & -1 < t < 0 \\ t^2 - t & 0 < t < 1 \end{cases}$

3 $f(t) = e^{-t}$ $-1 < t < 1$

4 Show that the amplitude of the nth harmonic in the Fourier series of a function $f(t)$ is equal to $2\sqrt{c_n c_{-n}}$.

Find the complex form of the Fourier series of the periodic functions whose definitions in one period are

5 $f(t) = \begin{cases} 1 & 0 < t < 1 \\ 0 & 1 < t < 2 \end{cases}$ **6** $f(t) = \begin{cases} 1 & 0 < t < 1 \\ -1 & 1 < t < 2 \end{cases}$

7 $f(t) = t \quad 0 < t < 1$ **8** $f(t) = t \quad -1 < t < 1$
9 $f(t) = \cos t \quad -\pi/2 < t < \pi/2$. *Hint:* Use the fact that $\cos \theta = \frac{1}{2}(e^{i\theta} + e^{-i\theta})$.
10 $f(t) = \sin t \quad 0 < t < \pi$ **11** $f(t) = \cosh t \quad -1 < t < 1$
12 $f(t) = \sinh t \quad -1 < t < 1$

6.5 Applications

Although we shall see other uses of Fourier series in later chapters, their most important application at the present stage of our work is in the analysis of the behavior of physical systems subjected to periodic disturbances.

EXAMPLE 1

If the **root-mean-square**, or **rms**, **value** of a function $f(t)$ over an interval (a,b) is defined to be

(1)
$$\sqrt{\frac{\int_a^b f^2(t)\, dt}{b - a}}$$

express the rms value of a periodic function over one period in terms of the coefficients in its Fourier expansion.

If $f(t)$ is of period $2p$, we can write, as usual,

$$f(t) = \frac{a_0}{2} + a_1 \cos \frac{\pi t}{p} + a_2 \cos \frac{2\pi t}{p} + \cdots + a_n \cos \frac{n\pi t}{p} + \cdots$$
$$+ b_1 \sin \frac{\pi t}{p} + b_2 \sin \frac{2\pi t}{p} + \cdots + b_n \sin \frac{n\pi t}{p} + \cdots$$

Hence, $f^2(t)$ will consist exclusively of squared terms of the form

$$\tfrac{1}{4} a_0^2 \qquad a_n^2 \cos^2 \frac{n\pi t}{p} \qquad b_n^2 \sin^2 \frac{n\pi t}{p}$$

and cross-product terms of the form

$$a_0 a_n \cos \frac{n\pi t}{p} \qquad a_0 b_n \sin \frac{n\pi t}{p} \qquad 2a_m a_n \cos \frac{m\pi t}{p} \cos \frac{n\pi t}{p}$$
$$2a_m b_n \cos \frac{m\pi t}{p} \sin \frac{n\pi t}{p} \qquad 2b_m b_n \sin \frac{m\pi t}{p} \sin \frac{n\pi t}{p}$$

As in the original derivation of the Euler formulas in Sec. 6.2, the integral of every cross-product term, taken over one period of $f(t)$, is zero. Moreover, for the squared terms we have

$$\frac{a_0^2}{4} \int_{-p}^{p} dt = \frac{a_0^2 p}{2} \qquad a_n^2 \int_{-p}^{p} \cos^2 \frac{n\pi t}{p}\, dt = a_n^2 p \qquad b_n^2 \int_{-p}^{p} \sin^2 \frac{n\pi t}{p}\, dt = b_n^2 p$$

Hence, dividing the sum of the nonzero terms by the length of the period, $2p$, and then taking the square root of the result, we obtain for the required rms value

(2)
$$f(t)\Big|_{\text{rms}} = \sqrt{\frac{a_0^2}{4} + \frac{1}{2} \sum_{n=1}^{\infty} (a_n^2 + b_n^2)}$$

Since the coefficients in the complex exponential form of the Fourier series of $f(t)$ are related to the coefficients in the real trigonometric form by the equations

$$c_0 = \frac{a_0}{2} \qquad c_n = \frac{a_n - ib_n}{2} \qquad c_{-n} = \frac{a_n + ib_n}{2} = \bar{c}_n\dagger$$

it follows that Eq. (2) can also be written

$$f(t)\Big|_{\text{rms}} = \sqrt{c_0{}^2 + 2 \sum_{n=1}^{\infty} c_n \bar{c}_n}$$

In particular, if $I = f(t)$ is a periodic electric current flowing through a resistance R, the average power dissipated per cycle is, since power = voltage $\times$ current,

$$\frac{1}{2p} \int_{-p}^{p} EI\, dt = \frac{1}{2p} \int_{-p}^{p} I^2 R\, dt = R \left(\frac{1}{2p} \int_{-p}^{p} I^2\, dt \right) = RI^2 \Big|_{\text{rms}}$$

$$= \left[\frac{a_0{}^2}{4} + \frac{1}{2} \sum_{n=1}^{\infty} (a_n{}^2 + b_n{}^2) \right] R = \left(c_0{}^2 + 2 \sum_{n=1}^{\infty} c_n \bar{c}_n \right) R$$

EXAMPLE 2

Determine the steady-state forced vibrations of the system shown in Fig. 6.7a if the applied force $F(t)$ is that shown in Fig. 6.7b.

Since the concepts we developed in our study of forced vibrations in Chap. 5 are applicable only to periodic functions which are simple sines and cosines, our first step must be to express the driving force $F(t)$ in terms of such functions; i.e., our first step must be to determine the Fourier expansion of $F(t)$. Since $F(t)$ is clearly an odd function of t, no cosine terms can be present, and thus we need only compute b_n:

$$b_n = \frac{2}{\frac{1}{2}} \int_0^{1/2} 20 \sin \frac{n\pi t}{\frac{1}{2}}\, dt = 80 \left[-\frac{\cos 2n\pi t}{2n\pi} \right]_0^{1/2}$$

$$= 40 \frac{1 - \cos n\pi}{n\pi}$$

$$= \begin{cases} 0 & n \text{ even} \\ \dfrac{80}{n\pi} & n \text{ odd} \end{cases}$$

Hence $F(t) = \dfrac{80}{\pi} \left(\sin 2\pi t + \dfrac{\sin 6\pi t}{3} + \dfrac{\sin 10\pi t}{5} + \dfrac{\sin 14\pi t}{7} + \cdots \right)$

and this is the expression that would appear on the right-hand side of the differential equation describing the motion of the system if we were to set up the equation. Now we are asked only to find the *steady-state* forced motion of the system; hence we need to determine only the particular integral corresponding to $F(t)$. Moreover, since the relevant differential equation (even though we have not set it up) is obviously linear, the required particular integral can be found very simply by using the ideas of Sec. 5.3. In fact, all we need do is to apply the proper magnification and phase shift to each component of the driving force $F(t)$ and add the results. Preparatory to this, we must determine the static deflections that would be produced in the system by steady forces having the magnitudes of the various terms of $F(t)$. These are equal to

$$(\delta_{st})_n = \frac{80/n\pi \text{ lb}}{100 \text{ lb/in}} = \frac{4}{5n\pi} \text{ in} \qquad n \text{ odd}$$

$\dagger$ As usual, a bar over a complex number denotes the **conjugate** of that number, that is, the number obtained by changing the sign of its imaginary component.

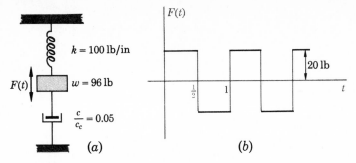

Figure 6.7
A spring-mass system acted upon by an alternating square-wave force.

Then we must calculate the undamped natural frequency of the system:

$$\omega_N\dagger = \sqrt{\frac{kg}{w}} = \sqrt{\frac{100 \times 384}{96}} = 20 \text{ rad/s}$$

The rest of the work can best be presented in tabular form (see Table 6.1).

Table 6.1

Term	δ_{st}	$\dfrac{\omega}{\omega_N}$	$M = \dfrac{1}{\sqrt{\left(1 - \dfrac{\omega^2}{\omega_N{}^2}\right)^2 + \left(2\dfrac{c}{c_c}\dfrac{\omega}{\omega_N}\right)^2}}$	$\alpha = \tan^{-1}\dfrac{2\dfrac{c}{c_c}\dfrac{\omega}{\omega_N}}{1 - \dfrac{\omega^2}{\omega_N{}^2}}$	Steady-state term = $\delta_{st}M\sin(\omega t - \alpha)$
1	$\dfrac{4}{5\pi}$	$\dfrac{2\pi}{20}$	1.11	0.035 rad	$0.28\sin(2\pi t - 0.035)$
2	$\dfrac{4}{15\pi}$	$\dfrac{6\pi}{20}$	6.83	0.701 rad	$0.58\sin(6\pi t - 0.701)$
3	$\dfrac{4}{25\pi}$	$\dfrac{10\pi}{20}$	0.68	3.035 rad	$0.03\sin(10\pi t - 3.035)$
4	$\dfrac{4}{35\pi}$	$\dfrac{14\pi}{20}$	0.26	3.084 rad	$0.01\sin(14\pi t - 3.084)$
...	...	...	...	...	...

Figure 6.8 shows the steady-state motion of the system plotted as a function of time.

This example illustrates an exceedingly important but sometimes misunderstood characteristic of forced vibrations. If the driving force is not a pure sine or cosine function, its Fourier expansion will contain terms whose frequencies are above the fundamental, or apparent, frequency of the excitation. If the frequency of one of these harmonics happens to be close to the natural, or resonant, frequency of the system, and if the amount of friction in the system is small, the corresponding magnification ratio will be large and its value may offset many times the smaller amplitude of that harmonic and make the resultant term the dominant part of the entire response. If and when this happens, the response will appear to be of a higher frequency than the

---

† We must remember that here the subscript N in ω_N stands for *natural* and is in no way connected with the parameter n which identifies the general term in the Fourier expansion of $F(t)$.

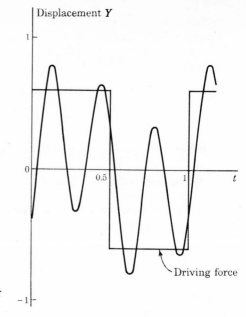

Figure 6.8
A response of apparent frequency greater than that of the excitation producing it.

force which produces it. Figure 6.8 shows this clearly, for, although the force alternates only once per second, the weight is seen to move up and down 3 times per second.

It is interesting to note that although the driving force in this example is discontinuous, both the displacement and the velocity it produces are continuous. This is suggested by the plot of the displacement shown in Fig. 6.8 and confirmed by an application of Theorem 3, Sec. 6.3. In fact, since the frequency of the general term in the Fourier expansion of the driving force $F(t)$ is $(2n - 1)2\pi \approx 4n\pi$, it follows, by neglecting all but the highest power of n in the denominator of the magnification ratio M, that for n sufficiently large, M is arbitrarily close to

$$\frac{1}{\omega^2/\omega_N{}^2} \approx \frac{1}{(4n\pi)^2/(20)^2} = \frac{25}{n^2\pi^2}$$

Therefore, since the static deflection corresponding to the general term in the expansion of $F(t)$ is

$$(\delta_{st})_n = \frac{4}{5(2n - 1)\pi} \approx \frac{2}{5n\pi}$$

it follows that as n becomes infinite, the coefficient of the general term in the expansion of the steady-state displacement, namely, $(\delta_{st})_n M$, tends to zero as

$$\frac{2}{5n\pi} \frac{25}{n^2\pi^2} = \frac{10}{n^3\pi^3}$$

Thus, according to Theorem 3, Sec. 6.3, the steady-state displacement $Y(t)$ and the steady-state velocity $\dot{Y}(t)$ are continuous, but the acceleration $\ddot{Y}(t)$ is discontinuous.

EXAMPLE 3

Find the steady-state current produced in the circuit shown in Fig. 6.9a by the voltage shown in Fig. 6.9b.

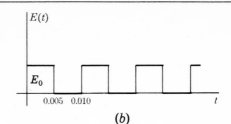

Figure 6.9
A series circuit driven by a square-wave voltage.

As in Example 2, our first step here is to determine the Fourier expansion of the driving force, i.e., the voltage. However, since we plan to use the complex impedance to find the steady-state current, we must use the complex exponential rather than the real trigonometric form of the Fourier series. Hence we compute

$$c_n = \frac{1}{0.01} \int_0^{0.005} E_0 e^{-ni\pi t/0.005} \, dt = 100 E_0 \left. \frac{e^{-ni\pi t/0.005}}{-ni\pi/0.005} \right|_0^{0.005}$$

$$= E_0 \frac{1 - e^{-ni\pi}}{2ni\pi}$$

$$= \begin{cases} 0 & n \text{ even, } n \neq 0 \\ \dfrac{E_0}{ni\pi} = -\dfrac{iE_0}{n\pi} & n \text{ odd} \end{cases}$$

$$c_0 = \frac{1}{0.01} \int_0^{0.005} E_0 \, dt = \frac{E_0}{2}$$

Therefore

$$E(t) = E_0 \left(\cdots + \frac{ie^{-600i\pi t}}{3\pi} + \frac{ie^{-200i\pi t}}{\pi} + \frac{1}{2} - \frac{ie^{200i\pi t}}{\pi} - \frac{ie^{600i\pi t}}{3\pi} - \cdots \right)$$

In Sec. 5.4 we showed that the steady-state current produced by a voltage of the form $Ae^{i\omega t}$ can be found simply by dividing the voltage by the complex impedance

$$Z(\omega) = R + i\left(\omega L - \frac{1}{\omega C}\right)$$

Using the data of the present problem, we have

$$Z(\omega) = 250 + i\left(0.02\omega - \frac{10^6}{2\omega}\right)$$

or, since $\omega = 200 n\pi \qquad n \text{ odd}$

we have $Z(\omega) \equiv Z_n = 250 + i\left(4n\pi - \frac{2{,}500}{n\pi}\right) \qquad n \text{ odd}$

Hence, dividing each term in the expansion of the voltage $E(t)$ by the value of Z for the corresponding frequency, i.e., the corresponding value of n, we find

$$I(t) = \sum_{n=-\infty}^{\infty} D_n e^{200ni\pi t} \qquad n \text{ odd}†$$

where

$$D_n = \frac{c_n}{Z_n} = -\frac{iE_0}{n\pi} \frac{1}{250 + i(4n\pi - 2{,}500/n\pi)} = \frac{-iE_0}{250n\pi + i(4n^2\pi^2 - 2{,}500)}$$

If we want the real trigonometric form of this expansion, namely,

$$I(t) = \frac{a_0}{2} + a_1 \cos 200\pi t + a_3 \cos 600\pi t + \cdots + b_1 \sin 200\pi t + b_3 \sin 600\pi t + \cdots$$

we have at once

$$a_n = D_n + D_{-n} = -iE_0 \left[\frac{1}{250n\pi + i(4n^2\pi^2 - 2{,}500)} + \frac{1}{-250n\pi + i(4n^2\pi^2 - 2{,}500)} \right]$$

$$= -\frac{2E_0(4n^2\pi^2 - 2{,}500)}{(250n\pi)^2 + (4n^2\pi^2 - 2{,}500)^2} \qquad n \text{ odd}$$

$$b_n = i(D_n - D_{-n}) = E_0 \left[\frac{1}{250n\pi + i(4n^2\pi^2 - 2{,}500)} - \frac{1}{-250n\pi + i(4n^2\pi^2 - 2{,}500)} \right]$$

$$= \frac{500n\pi E_0}{(250n\pi)^2 + (4n^2\pi^2 - 2{,}500)^2} \qquad n \text{ odd}$$

EXERCISES

1   In Example 3, show that the current $I(t)$ is continuous even though the voltage $E(t)$ is discontinuous. Is the charge $Q(t)$ on the condenser continuous?

2   Show that if a periodic function $f(t)$ is expressed in either of the forms

$$A_0 + A_1 \cos(\omega t - \gamma_1) + A_2 \cos(2\omega t - \gamma_2) + A_3 \cos(3\omega t - \gamma_3) + \cdots$$
$$A_0 + A_1 \sin(\omega t + \delta_1) + A_2 \sin(2\omega t + \delta_2) + A_3 \sin(3\omega t + \delta_3) + \cdots$$

its rms value over one period is equal to

$$\sqrt{A_0^2 + \frac{1}{2} \sum_{n=1}^{\infty} A_n^2}$$

3   If $f(t) = \sum_{n=-\infty}^{\infty} c_n e^{ni\pi t/p}$ and $g(t) = \sum_{n=-\infty}^{\infty} d_n e^{ni\pi t/p}$ are two functions of period $2p$, show that the average value of the product $f(t)g(t)$ over one period is $\sum_{n=-\infty}^{\infty} c_n d_{-n}$.

4   In Example 2, determine the steady-state motion if the amount of friction is doubled and the spring is changed to one of modulus 144 lb/in.

5   In Example 3, determine the steady-state current if

$$E(t) = \begin{cases} E_0 & 0 < t < 0.005 \\ -E_0 & 0.005 < t < 0.01 \end{cases}$$

6   Determine the steady-state motion of the system shown in Fig. 6.10.

---

† Because of the presence of the capacitor, the impedance Z_0 for the dc component, or component of zero frequency, is infinite. Hence the term $E_0/2$ in the expansion of $E(t)$ makes no contribution to the steady-state current.

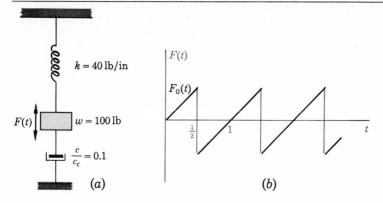

Figure 6.10

7   Work Exercise 6 if the amount of friction is reduced to $c/c_c = 0.02$.

8   Determine the steady-state motion of the system shown in Fig. 6.11.

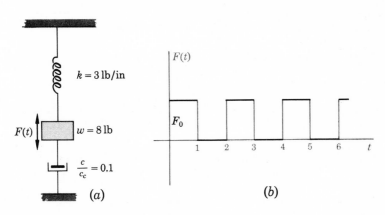

Figure 6.11

9   Work Exercise 8 if $k = 60$ lb/in, $w = 160$ lb, and $c/c_c = 0.04$.

10   Determine the steady-state current in the circuit shown in Fig. 6.12.

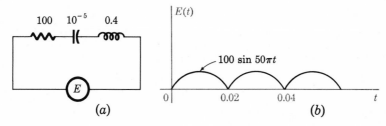

Figure 6.12

11   Determine the steady-state current in the circuit shown in Fig. 6.13.

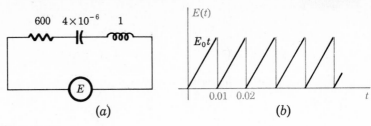

$$(a) \qquad\qquad\qquad\qquad (b)$$

Figure 6.13

12   Determine the steady-state motion produced in the system shown in Fig. 5.15 if:
(a) The force $F(t)$ shown in Fig. 6.7b acts on M_1.
(b) The force $F(t)$ shown in Fig. 6.7b acts on M_2.

13   In Example 2, discuss the problem of determining the complete motion, transient as well as steady-state.

14   If $F(t)$ is the periodic function whose definition in one period is

$$F(t) = \begin{cases} 1 & 0 < t < \pi \\ 0 & \pi < t < 2\pi \end{cases}$$

find the solution of each of the following equations which satisfies the indicated conditions:
(a) $y'' - y = F(t)$ $y_0 = y_0' = 0$
(b) $y'' + y = F(t)$ $y_0 = y_0' = 0$
(c) $y'' - y = F(t)$ $y_0 = 1, y_0' = 0$
(d) $y'' - 3y' + 2y = F(t)$ $y_0 = y_0' = 0$

15   If $F(t)$ is the periodic function whose definition in one period is $F(t) = |t|$, $-\pi < t < \pi$, find the solution of each of the following equations which satisfies the indicated conditions:
(a) $y'' - y = F(t)$ $y_0 = y_0' = 0$
(b) $y'' + 4y = F(t)$ $y_0 = y_0' = 0$
(c) $y'' + 9y = F(t)$ $y_0 = y_0' = 0$

6.6 The Fourier Integral as the Limit of a Fourier Series

The properties of Fourier series developed thus far are adequate to accomplish the expansion of any periodic function satisfying the Dirichlet conditions and, in conjunction with the theory of Chap. 5, enable us to find the response of numerous mechanical and electrical systems to general periodic disturbances. On the other hand, in many problems the impressed force or voltage is nonperiodic rather than periodic, a single unrepeated pulse, for instance. Functions of this sort cannot be handled directly through the use of Fourier series, since such series necessarily define only periodic functions. However, by investigating the limit (if any) which is approached by a Fourier series as the period of the given function becomes infinite, a suitable representation for nonperiodic functions can perhaps be obtained. An example is probably the best way to introduce the theory of this procedure.

Consider, then, the function $f_p(t)$ shown in Fig. 6.14, namely,

$$f_p(t) = \begin{cases} 0 & -p < t < -1 \\ 1 & -1 < t < 1 \\ 0 & 1 < t < p \end{cases}$$

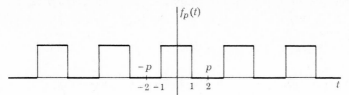

Figure 6.14
A periodic function of period $2p = 4$.

in the limit as $p \to \infty$, as suggested by Fig. 6.15. This function is clearly even, and thus its Fourier expansion contains only cosine terms, i.e.,

$$(1) \qquad f_p(t) = \frac{a_0}{2} + a_1 \cos \frac{\pi t}{p} + a_2 \cos \frac{2\pi t}{p} + \cdots + a_n \cos \frac{n\pi t}{p} + \cdots$$

where

$$(2a) \qquad a_0 = \frac{2}{p} \int_0^1 1 \, dt = \frac{2}{p}$$

and

$$(2b) \qquad a_n = \frac{2}{p} \int_0^1 1 \cos \frac{n\pi t}{p} \, dt = \frac{2}{p} \left[\frac{\sin (n\pi t/p)}{n\pi/p} \right]_0^1 = \frac{2}{p} \frac{\sin (n\pi/p)}{n\pi/p}$$

It will help us now to understand what happens as $p \to \infty$ if we plot the spectrum

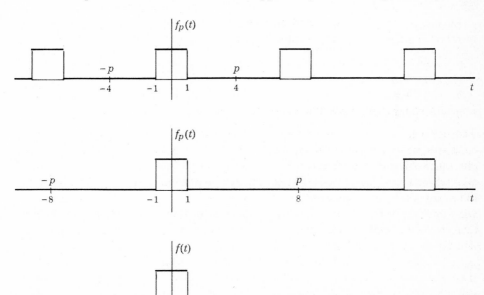

Figure 6.15
The nonperiodic limit of a sequence of periodic functions whose periods become infinite.

of $f_p(t)$, that is, plot the general coefficient, or amplitude, a_n as a function of the frequency†

$$\omega_n = \frac{n\pi}{p} \text{ rad/unit time}$$

for different values of p. Introducing the symbol ω_n in Eq. (2b), we then have

(3)
$$a_n = \frac{2}{p} \frac{\sin \omega_n}{\omega_n}$$

where successive values of n correspond to values of ω_n which differ by the constant amount

$$\Delta\omega = \omega_{n+1} - \omega_n = \frac{(n+1)\pi}{p} - \frac{n\pi}{p} = \frac{\pi}{p}$$

If we now consider the curve

(4)
$$y = \frac{2}{p}\frac{\sin x}{x} = \frac{2}{\pi}\frac{\sin x}{x}\frac{\pi}{p} = \frac{2}{\pi}\frac{\sin x}{x}\Delta\omega$$

suggested by the form of Eq. (3), it is clear that the values of a_n for any particular value of p are simply the ordinates of this curve at the equally spaced points $x = 0$, $\omega_1 (= \pi/p)$, $\omega_2 (= 2\pi/p)$, ... Figure 6.16 shows plots of a_n vs. ω_n, and the corresponding **amplitude envelope**

$$y = \frac{2}{p}\frac{\sin x}{x}$$

for $p = 2, 4$, and 8.

As suggested by Fig. 6.16 and confirmed by Eq. (4), these coefficient plots, or spectra, for different values of p differ in only two respects:

a. In the vertical scale, which is inversely proportional to p (or directly proportional to $\Delta\omega$)
b. In the horizontal interval between successive ordinates, which is also inversely proportional to p (or equal to $\Delta\omega$)

The fact that as $p \to \infty$ (or $\Delta\omega \to 0$), the frequencies $\omega_n = n\pi/p$ of the terms in (1) become more and more closely spaced while the coefficients approach zero suggests that this particular series, thought of as a function of p, is actually a sum of infinitesimals whose limit is an integral. Indeed, this is true of Fourier series in general, as the following outline of steps reveals.

For convenience, we begin with the complex exponential form of a Fourier series [Eqs. (1) and (2), Sec. 6.4]

$$f_p(t) = \sum_{n=-\infty}^{\infty} c_n e^{ni\pi t/p}$$

$$c_n = \frac{1}{2p}\int_{-p}^{p} f_p(t)e^{-ni\pi/p}\, dt \equiv \frac{1}{2p}\int_{-p}^{p} f_p(\tau)e^{-ni\pi\tau/p}\, d\tau$$

---

† In this case, the subscript n does *not* stand for *natural*, as N did in Example 2 in the last section, but is used to identify the frequency of the nth harmonic in the expansion of $f_p(t)$.

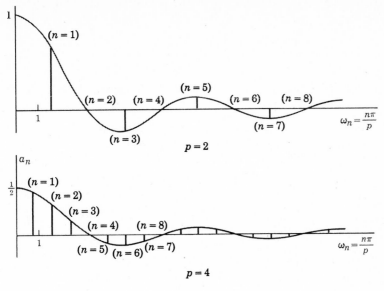

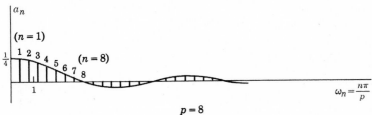

Figure 6.16
The behavior of the Fourier coefficients of a function as the period of the function becomes infinite.

Substitution of the second expression for c_n into $f_p(t)$ then gives

$$f_p(t) = \sum_{n=-\infty}^{\infty} \left[\frac{1}{2p} \int_{-p}^{p} f_p(\tau) e^{-ni\pi\tau/p} \, d\tau \right] e^{ni\pi t/p}$$

$$= \sum_{n=-\infty}^{\infty} \left[\frac{1}{2\pi} \int_{-p}^{p} f_p(\tau) e^{-ni\pi\tau/p} \, d\tau \right] e^{ni\pi t/p} \frac{\pi}{p}$$

Now, as above, let us denote the frequency of the general term by

$$\omega_n = \frac{n\pi}{p}$$

and the difference in frequency between successive terms by

$$\Delta\omega = \frac{\pi}{p}$$

Then $f_p(t)$ can be written

(5) $$f_p(t) = \sum_{n=-\infty}^{\infty} \left[\frac{1}{2\pi} e^{i\omega_n t} \int_{-p}^{p} f_p(\tau) e^{-i\omega_n \tau} \, d\tau \right] \Delta\omega$$

If we now define

(6)
$$F(\omega) = \frac{e^{i\omega t}}{2\pi} \int_{-p}^{p} f_p(\tau)e^{-i\omega \tau} \, d\tau$$

Eq. (5) becomes simply

(7)
$$f_p(t) = \sum_{n=-\infty}^{\infty} F(\omega_n) \, \Delta\omega$$

where ω_n is a point (the left-hand end point, in fact) in the nth subinterval $\Delta\omega$. Under very general conditions, the limit of a sum of the form (7) is the integral

$$\int_{-\infty}^{\infty} F(\omega) \, d\omega$$

Hence, since $p \to \infty$ implies $\Delta\omega \to 0$, it follows that there is good reason to believe that as $p \to \infty$, the nonperiodic limit of $f_p(t)$, say $f(t)$, can be written as the integral

(8)
$$f(t) = \int_{-\infty}^{\infty} \left[\frac{1}{2\pi} e^{i\omega t} \int_{-\infty}^{\infty} f(\tau)e^{-i\omega \tau} \, d\tau \right] d\omega$$

Though our derivation of it has been far from complete,† Eq. (8) is actually a valid representation of the nonperiodic limit function $f(t)$, provided that

a. In every finite interval $f(t)$ satisfies the Dirichlet conditions.
b. The improper integral $\int_{-\infty}^{\infty} |f(t)| \, dt$ exists.

Under these conditions, the so-called **Fourier integral** (8) gives the value of $f(t)$ at all points where $f(t)$ is continuous and gives the average of the right- and left-hand limits of $f(t)$ at all points where $f(t)$ is discontinuous.

The Fourier integral can be written in various forms. For instance, we can write Eq. (8) as

(9)
$$f(t) = \int_{-\infty}^{\infty} g(\omega)e^{i\omega t} \, d\omega$$

where

(10)
$$g(\omega) = \frac{1}{2\pi} \int_{-\infty}^{\infty} f(\tau)e^{-i\omega \tau} \, d\tau$$

These two expressions, in which the symmetry between $f(t)$ and its **coefficient function** $g(\omega)$ is unmistakable, constitute what is known as a **Fourier transform pair**.‡ The

---

† The situation is actually not so simple as we have made it appear, for from (6) it is clear that the structure of the function $F(\omega)$ is not fixed but depends on p. Hence, as p increases, the function we are evaluating changes and the elementary theory of the definite integral is not strictly applicable. Moreover, the fact that the summation extends over an infinite range makes additional investigation of the limiting process necessary. The modifications required for a rigorous justification of our conclusions can be found in more advanced texts, such as R. V. Churchill, "Fourier Series and Boundary Value Problems," 2d ed., pp. 113–117, McGraw-Hill, New York, 1963.
‡ It is sometimes more convenient to associate the factor $1/2\pi$ with the integral for $f(t)$ instead of with the integral for $g(\omega)$. It is also possible to achieve a still more symmetric form by associating the factor $1/\sqrt{2\pi}$ with each of the integrals.

coefficient function $g(\omega)$ is, of course, completely equivalent to $f(t)$, since when it is known, $f(t)$ is determined through Eq. (9). In effect, we thus have two different representations of the function of our discussion: $f(t)$ in the time domain and $g(\omega)$ in the frequency domain. In passing, we note that elaborate tables of Fourier transform pairs have been prepared for engineering use.†

If we choose, we can, of course, move the factor $e^{i\omega t}$ into the integrand of the inner integral in (8), since it does not involve the variable τ of that integration. This gives

$$(11) \qquad f(t) = \frac{1}{2\pi} \int_{-\infty}^{\infty} \int_{-\infty}^{\infty} f(\tau) e^{-i\omega(\tau - t)} \, d\tau \, d\omega$$

In this, we can replace the exponential by its trigonometric equivalent, getting

$$f(t) = \frac{1}{2\pi} \int_{-\infty}^{\infty} \int_{-\infty}^{\infty} f(\tau)[\cos \omega(\tau - t) - i \sin \omega(\tau - t)] \, d\tau \, d\omega$$

If we break this up into two integrals, we get

$$f(t) = \frac{1}{2\pi} \int_{-\infty}^{\infty} \int_{-\infty}^{\infty} f(\tau) \cos \omega(\tau - t) \, d\tau \, d\omega$$

$$- \frac{i}{2\pi} \int_{-\infty}^{\infty} \int_{-\infty}^{\infty} f(\tau) \sin \omega(\tau - t) \, d\tau \, d\omega$$

The fact that $\sin \omega(\tau - t)$ is an odd function of ω makes it seem plausible that the second integral is always zero, and indeed this must be the case since by hypothesis $f(t)$ is purely real. Thus we obtain the real trigonometric representation

$$(12a) \qquad f(t) = \frac{1}{2\pi} \int_{-\infty}^{\infty} \int_{-\infty}^{\infty} f(\tau) \cos \omega(\tau - t) \, d\tau \, d\omega$$

Since the integrand of (12a) is an even function of ω, we need perform the ω integration only between 0 and ∞, provided we multiply the result by 2. This gives us the modified form

$$(12b) \qquad f(t) = \frac{1}{\pi} \int_{0}^{\infty} \int_{-\infty}^{\infty} f(\tau) \cos \omega(\tau - t) \, d\tau \, d\omega$$

If $f(t)$ is either an odd function or an even function, further simplifications are possible. To see this, we first expand the factor $\cos \omega(\tau - t)$ in the integrand of (12a) getting

$$f(t) = \frac{1}{2\pi} \int_{-\infty}^{\infty} \int_{-\infty}^{\infty} f(\tau) \cos \omega\tau \cos \omega t \, d\tau \, d\omega$$

$$+ \frac{1}{2\pi} \int_{-\infty}^{\infty} \int_{-\infty}^{\infty} f(\tau) \sin \omega\tau \sin \omega t \, d\tau \, d\omega$$

---

† G. A. Campbell and R. M. Foster, "Fourier Integrals for Practical Applications," Van Nostrand, Princeton, N.J., 1948.

and then write the inner integrals as the sums of integrals over $(-\infty,0)$ and $(0,\infty)$. Then

$$f(t) = \frac{1}{2\pi} \int_{-\infty}^{\infty} \left[\int_{-\infty}^{0} f(\tau) \cos \omega\tau \cos \omega t \, d\tau + \int_{0}^{\infty} f(\tau) \cos \omega\tau \cos \omega t \, d\tau \right] d\omega$$

$$+ \frac{1}{2\pi} \int_{-\infty}^{\infty} \left[\int_{-\infty}^{0} f(\tau) \sin \omega\tau \sin \omega t \, d\tau + \int_{0}^{\infty} f(\tau) \sin \omega\tau \sin \omega t \, d\tau \right] d\omega$$

Next we make the substitutions $\tau = -z$ and $d\tau = -dz$ in the integrals from $-\infty$ to 0:

(13) $\quad f(t) = \dfrac{1}{2\pi} \int_{-\infty}^{\infty} \left[\int_{\infty}^{0} f(-z) \cos (-\omega z) \cos \omega t \, (-dz) \right.$

$$\left. + \int_{0}^{\infty} f(\tau) \cos \omega\tau \cos \omega t \, d\tau \right] d\omega$$

$$+ \frac{1}{2\pi} \int_{-\infty}^{\infty} \left[\int_{\infty}^{0} f(-z) \sin (-\omega z) \sin \omega t \, (-dz) \right.$$

$$\left. + \int_{0}^{\infty} f(\tau) \sin \omega\tau \sin \omega t \, d\tau \right] d\omega$$

Now if $f(t)$ is an even function, so that $f(-z) = f(z)$, the inner integrals in the first term in (13) become identical when the minus sign associated with dz is used to reverse the limits in the first of these integrals. Similarly, the inner integrals in the second term turn out to be negatives of each other. Hence we have simply

(14) $\qquad f(t) = \dfrac{1}{\pi} \int_{-\infty}^{\infty} \int_{0}^{\infty} f(\tau) \cos \omega\tau \cos \omega t \, d\tau \, d\omega \qquad f(t)$ even

This is called the **Fourier cosine integral** and is analogous to the half-range cosine expansion of a periodic function which is even.

If $f(t)$ is an odd function, so that $f(-z) = -f(z)$, then the inner integrals in the first term in (13) are negatives of each other and the inner integrals in the second term are equal, and we have

(15) $\qquad f(t) = \dfrac{1}{\pi} \int_{-\infty}^{\infty} \int_{0}^{\infty} f(\tau) \sin \omega\tau \sin \omega t \, d\tau \, d\omega \qquad f(t)$ odd

This is the **Fourier sine integral**, the analog of the half-range sine expansion of an odd periodic function.

For some purposes it is convenient to have the Fourier cosine- and sine-integral representations displayed as transform pairs. Thus we can write (14) in the form

$$f(t) = \int_{-\infty}^{\infty} g(\omega) \cos \omega t \, d\omega$$

(14a) $\qquad\qquad\qquad\qquad\qquad\qquad\qquad\qquad f(t)$ even

$$g(\omega) = \frac{1}{\pi} \int_{0}^{\infty} f(\tau) \cos \omega\tau \, d\tau$$

and (15) in the form

$$f(t) = \int_{-\infty}^{\infty} g(\omega) \sin \omega t \, d\omega$$

(15a) $\qquad\qquad\qquad\qquad\qquad\qquad\qquad\qquad f(t)$ odd

$$g(\omega) = \frac{1}{\pi} \int_{0}^{\infty} f(\tau) \sin \omega\tau \, d\tau$$

Of course, Eqs. (14), (14a), (15), and (15a) can all be modified by performing the ω integrations only from 0 to ∞ and multiplying the results by 2.

To illustrate the Fourier integral representation of a nonperiodic function, let us return to the isolated pulse which we used to introduce the ideas developed in this section (Fig. 6.15). Since this function is clearly even, we can use (14), getting

$$f(t) = \frac{1}{\pi} \int_{-\infty}^{\infty} \int_0^1 1 \cos \omega\tau \cos \omega t \, d\tau \, d\omega = \frac{1}{\pi} \int_{-\infty}^{\infty} \cos \omega t \left[\frac{\sin \omega\tau}{\omega} \right]_0^1 d\omega$$

$$= \frac{1}{\pi} \int_{-\infty}^{\infty} \frac{\cos \omega t \sin \omega}{\omega} \, d\omega$$

$$(16) \qquad = \frac{2}{\pi} \int_0^{\infty} \frac{\cos \omega t \sin \omega}{\omega} \, d\omega$$

where the last step follows because the integrand is an even function of ω. Thus, although it is impossible to find an elementary antiderivative for the integrand of the last integral, we know that as a definite integral it must be equal to 1 if t is between -1 and 1, must equal $\frac{1}{2}$ if $t = \pm 1$, and must be zero if t is numerically greater than 1.

In the Fourier series representation of a periodic function it was a matter of some interest to determine how well the first few terms of the expansion represented the function (Fig. 6.3). The corresponding problem in the nonperiodic case is to investigate how well the Fourier integral represents a function when only the components in the lower portion of the (continuous) frequency range are taken into account. Suppose, therefore, that we consider only the frequencies below ω_0. In this case, from (16) we have, as an approximation to $f(t)$, the finite integral

$$\frac{2}{\pi} \int_0^{\omega_0} \frac{\cos \omega t \sin \omega}{\omega} \, d\omega$$

Now
$$\cos a \sin b = \frac{\sin (a + b) - \sin (a - b)}{2}$$

and thus we can write the last integral as

$$\frac{1}{\pi} \int_0^{\omega_0} \frac{\sin \omega(t + 1)}{\omega} \, d\omega - \frac{1}{\pi} \int_0^{\omega_0} \frac{\sin \omega(t - 1)}{\omega} \, d\omega$$

In the first of these integrals let $\omega(t + 1) = u$, and in the second let $\omega(t - 1) = u$. Then, for our approximation to $f(t)$ we have

$$\frac{1}{\pi} \int_0^{\omega_0(t+1)} \frac{\sin u}{u} \, du - \frac{1}{\pi} \int_0^{\omega_0(t-1)} \frac{\sin u}{u} \, du$$

Although integrals of this form cannot be expressed in terms of elementary functions, they occur so often in applied mathematics that they have been named and tabulated. Specifically,

$$\text{Si}(x) \equiv \int_0^x \frac{\sin u}{u} \, du$$

is known as the **sine-integral function** of x and is tabulated in numerous handbooks.†

---

† See, for instance, E. Jahnke, F. Emde, and F. Lösch, "Tables of Higher Functions," 6th ed., McGraw-Hill, New York, 1960.

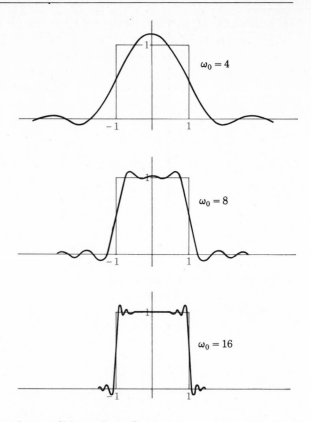

Figure 6.17
The approximation of a function
by its Fourier integral taken
only over frequencies less than ω_0.

With this notation the approximation to $f(t)$ can be written

$$\frac{1}{\pi} \operatorname{Si}\left[\omega_0(t + 1)\right] - \frac{1}{\pi} \operatorname{Si}\left[\omega_0(t - 1)\right]$$

Figure 6.17 shows this approximation for $\omega_0 = 4$, 8, and 16 rad per unit time. Physically speaking, these curves represent the output of an ideal low-pass filter, cutting off all frequencies above ω_0 when the input signal is an isolated rectangular pulse.

An examination of the curves shown in Fig. 6.17 in the neighborhood of the discontinuities at $t = \pm 1$ reveals an interesting and important characteristic of both Fourier integrals and Fourier series. As ω_0 increases and the approximation to $f(t)$ becomes better and better, it appears that the amount of the overshoot on the right and left of each discontinuity does *not* approach zero, as we might expect, but is wiped out in the limit by being pinched into a vanishingly small interval on each side of each discontinuity. This is indeed the case. In fact, it can be shown† that no matter how large the value of ω_0 in the approximation of a Fourier integral and no matter how large the value of n in the approximation of a Fourier series by its partial sums, at each finite jump of the function the approximating curve will overshoot the right- and left-hand limits of the function by an amount which approaches

$$\frac{1}{2} - \frac{1}{\pi} \operatorname{Si}(\pi) \doteq 0.09$$

---

† See, for example, Louis Brand, "Advanced Calculus," pp. 536–538, Wiley, New York, 1955.

times the actual amount of the jump. The convergence guaranteed by the Dirichlet theorem is achieved because the noninfinitesimal overshoot occurs over an interval whose length approaches zero as ω_0 and n become infinite. This curious behavior is known as the **Gibbs' phenomenon**, after the American mathematical physicist Josiah Willard Gibbs (1839–1903), who first explained it.

The Fourier integral representation of a nonperiodic function can be used in essentially the same way as the Fourier series representation of a periodic function in applications like those we considered in Sec. 6.5. For instance, if an electric circuit is acted upon by a nonperiodic voltage $f(t)$ whose Fourier integral representation is [Eqs. (9) and (10)]

$$f(t) = \int_{-\infty}^{\infty} g(\omega)e^{i\omega t}\, d\omega \qquad \text{where} \qquad g(\omega) = \frac{1}{2\pi} \int_{-\infty}^{\infty} f(\tau)e^{-i\omega \tau}\, d\tau$$

we can still, for purposes of analysis, think of $f(t)$ as being the sum of an infinite number of complex voltages $e^{i\omega t}$. In fact, the only practical distinction between the periodic and nonperiodic cases is that in the latter the spectrum of $f(t)$ contains terms of *all* frequencies and the amplitude, or intensity, of the component associated with any particular frequency ω is infinitesimal, namely,

$$g(\omega)\, d\omega$$

In Sec. 5.4 we saw that the current produced in a circuit of impedance $Z(\omega)$ by a complex voltage $E_0 e^{i\omega t}$ is simply

$$\frac{E_0 e^{i\omega t}}{Z(\omega)}$$

Hence, to find the current produced by the nonperiodic voltage $f(t)$, we need only divide the infinitesimal voltage

$$[g(\omega)\, d\omega]e^{i\omega t}$$

associated with the general frequency ω by the value of the impedance $Z(\omega)$ at that frequency and then "add," i.e., integrate, all the infinitesimal currents thus obtained. The result is simply

$$I(t) = \int_{-\infty}^{\infty} \frac{g(\omega)}{Z(\omega)} e^{i\omega t}\, d\omega \qquad \text{where} \qquad g(\omega) = \frac{1}{2\pi} \int_{-\infty}^{\infty} f(\tau)e^{-i\omega \tau}\, d\tau$$

and $Z(\omega)$ is the impedance of the circuit.

EXERCISES

1   Make an amplitude-frequency plot for $p = 2$, 4, and 8 for the periodic function whose definition in one period is

(a) $f(t) = \begin{cases} 0 & -p < t < -1 \\ -1 & -1 < t < 0 \\ 1 & 0 < t < 1 \\ 0 & 1 < t < p \end{cases}$ (b) $f(t) = \begin{cases} 0 & -p < t < -1 \\ 1 + t & -1 < t < 0 \\ 1 - t & 0 < t < 1 \\ 0 & 1 < t < p \end{cases}$

(c) $f(t) = \begin{cases} 0 & -p < t < -1 \\ 1 - t^2 & -1 < t < 1 \\ 0 & 1 < t < p \end{cases}$

(d) $f(t) = \begin{cases} 0 & -p < t < -1 \\ \sin \pi t & -1 < t < 1 \\ 0 & 1 < t < p \end{cases}$

2 Make an amplitude-frequency plot for $p = 2, 4,$ and 8 for the periodic function whose definition in one period is

(a) $f(t) = \begin{cases} 0 & -p < t < -1 \\ e^t & -1 < t < 0 \\ e^{-t} & 0 < t < 1 \\ 0 & 1 < t < p \end{cases}$ (b) $f(t) = \begin{cases} 0 & -p < t < -1 \\ -e^t & -1 < t < 0 \\ e^{-t} & 0 < t < 1 \\ 0 & 1 < t < p \end{cases}$

3 Make an amplitude-frequency plot for $p = 2, 4,$ and 8 for the periodic function whose definition in one period is

(a) $f(t) = \begin{cases} e^t & -p < t < 0 \\ e^{-t} & 0 < t < p \end{cases}$ (b) $f(t) = \begin{cases} -e^t & -p < t < 0 \\ e^{-t} & 0 < t < p \end{cases}$

In what significant way do the results of this exercise differ from the results of Exercise 2?

4 If the shape and duration of the pulse in one period of a periodic function remain fixed as $p \to \infty$, show that the amplitude envelopes corresponding to different values of p differ only in their vertical scale.

5 Find the Fourier integral representation of each of the following functions:

(a) $f(t) = \begin{cases} e^{at} & t < 0 \\ e^{-at} & t > 0 \end{cases}$ (b) $f(t) = \begin{cases} 0 & t < 0 \\ e^{-at} & t > 0 \end{cases}$

(c) $f(t) = \begin{cases} \sin t & t^2 < \pi^2 \\ 0 & t^2 > \pi^2 \end{cases}$ (d) $f(t) = \begin{cases} \cos t & t^2 < \pi^2/4 \\ 0 & t^2 > \pi^2/4 \end{cases}$

(e) $f(t) = \begin{cases} 0 & -\infty < t < 0 \\ 1 & 0 < t < 1 \\ 0 & 1 < t < \infty \end{cases}$ (f) $f(t) = \begin{cases} 1 - t^2 & t^2 < 1 \\ 0 & t^2 > 1 \end{cases}$

6 Express each of the following definite integrals in terms of the sine-integral function:

(a) $\int_a^b \dfrac{\sin x^n}{x}\, dx$ *Hint:* Let $x^n = u$.

(b) $\int_a^b \sin e^{\lambda x}\, dx$

(c) $\int_a^b \dfrac{\sin x}{x^3}\, dx$ *Hint:* Use integration by parts.

(d) $\int_1^2 \dfrac{\sin x \sin 2x \sin 3x}{x}\, dx$ *Hint:* Convert the products in the numerator into sums.

7 Find the Fourier integral representation of the function

$$f(t) = \begin{cases} 0 & -\infty < t < -1 \\ -1 & -1 < t < 0 \\ 1 & 0 < t < 1 \\ 0 & 1 < t < \infty \end{cases}$$

and express the integral which approximates this function for frequencies between 0 and ω_0 in terms of the sine-integral function.

8 Work Exercise 7 for the function

$$f(t) = \begin{cases} 0 & -\infty < t < -1 \\ 1 + t & -1 < t < 0 \\ 1 - t & 0 < t < 1 \\ 0 & 1 < t < \infty \end{cases}$$

9   Work Exercise 7 for the function

$$f(t) = \begin{cases} t & t^2 < 1 \\ 0 & t^2 > 1 \end{cases}$$

10   If the **cosine-integral function** Ci (x) is defined by the formula

$$\text{Ci}\,(x) = -\int_x^\infty \frac{\cos u}{u}\,du$$

evaluate each of the following definite integrals

(a) $\displaystyle\int_1^2 \frac{\cos x^n}{x}\,dx$ **(b)** $\displaystyle\int_a^b \cos e^{\lambda x}\,dx$

(c) $\displaystyle\int_a^b \frac{\cos \lambda x}{1 - x^2}\,dx$ $a, b > 1$ *Hint:* First use partial fractions.

(d) $\displaystyle\int_1^2 \frac{\cos x \sin 2x \sin 3x}{x}\,dx$

11   Determine the coefficient function $g(\omega)$ for the finite wave train

$$f(t) = \begin{cases} \sin \omega_0 t & t^2 < (N\pi/\omega_0)^2 \\ 0 & t^2 > (N\pi/\omega_0)^2 \end{cases}$$

Discuss $g(\omega)$ as a function of ω for different values of the number N of waves in the train, and show that the longer the train, the more the coefficient function concentrates around the frequency ω_0, that is, the more accurately ω_0 can be determined by scanning the coefficient function through a narrow bandpass filter.

12   Show that

$$\frac{2}{\pi}\int_0^\infty \frac{(2 - \omega^2)\cos \omega t + 3\omega \sin \omega t}{(2 - \omega^2)^2 + 9\omega^2} \frac{\sin \omega}{\omega}\,d\omega$$

is a particular integral of the equation $y'' + 3y' + 2y = f(t)$ where

$$f(t) = \begin{cases} 1 & t^2 < 1 \\ 0 & t^2 > 1 \end{cases}$$

13   Find a particular integral of the equation $y'' + ay' + by = f(t)$ where $f(t)$ is the function described in Exercise 12.

14   Work Exercise 13 if $f(t)$ is the function described in Exercise 8.

15   Work Exercise 13 if $f(t)$ is the function described in Exercise 9.

16   **(a)** Using the Fourier integral representation (12a) and the concepts of magnification ratio and phase angle, obtain a formula for the steady-state response of a mechanical system of one degree of freedom to a nonperiodic driving force.

   (b) Show that the steady-state response of an electric circuit of impedance $Z(\omega)$ to a nonperiodic voltage

$$E(t) = \int_{-\infty}^\infty g(\omega)e^{i\omega t}\,d\omega$$

can be written

$$I(t) = 2\int_0^\infty \left\{ \mathscr{R}\left[\frac{g(\omega)}{Z(\omega)}\right] \cos \omega t - \mathscr{I}\left[\frac{g(\omega)}{Z(\omega)}\right] \sin \omega t \right\} d\omega$$

where $\mathscr{R}[g(\omega)/Z(\omega)]$ and $\mathscr{I}[g(\omega)/Z(\omega)]$ denote, respectively, the real and the imaginary parts of the complex expression $g(\omega)/Z(\omega)$.

17 If we call $g(\omega)$ in Eq. (10) the **Fourier transform** of $f(t)$, show that if the various transforms exist, then:

(a) The Fourier transform of $e^{\pm i\omega_0 t} f(t)$ is $g(\omega \mp \omega_0)$.

(b) The Fourier transform of $f'(t)$ is $i\omega g(\omega)$.

(c) The Fourier transform of $\displaystyle\int_{-\infty}^{t} f(t)\, dt$ is $g(\omega)/i\omega$.

(d) The Fourier transform of $f_1(t) f_2(t)$ is

$$\int_{-\infty}^{\infty} g_1(u) g_2(\omega - u)\, du = \int_{-\infty}^{\infty} g_1(\omega - v) g_2(v)\, dv$$

where $g_1(\omega)$ and $g_2(\omega)$ are, respectively, the Fourier transforms of $f_1(t)$ and $f_2(t)$.

18 Let $f(t)$ be a function which is identically zero outside the interval $(-1,1)$, so that the Fourier transform of $f(t)$ is

$$T(f) = \frac{1}{2\pi} \int_{-1}^{1} f(t) e^{-i\omega t}\, dt$$

By repeated differentiation of $T(1)$ with respect to ω, show that

$$T(t^n) = \frac{i^n}{\pi} \frac{d^n S(\omega)}{d\omega^n}$$

where $S(\omega) = (\sin \omega)/\omega$. Explain how this result can be used to obtain the Fourier transform of a single pulse defined between -1 and 1 by a convergent power series.†

19 Using the definitions of Exercise 18, show that

$$T(e^{ni\pi t}) = \frac{1}{\pi} S(\omega - n\pi)$$

Explain how this result can be used to obtain the Fourier transform of a single pulse defined between -1 and 1 by a Fourier series in either complex exponential or real trigonometric form.

20 If $f(t)$ is a pulse defined between -1 and 1 by either of the equivalent series

$$\sum_{n=0}^{\infty} (a_n \cos n\pi t + b_n \sin n\pi t) \qquad \sum_{n=-\infty}^{\infty} c_n e^{ni\pi t}$$

use the results of Exercise 19 to show that

$$a_n = \pi[\phi(n\pi) + \phi(-n\pi)]$$
$$b_n = i\pi[\phi(n\pi) - \phi(-n\pi)]$$
$$c_n = \pi \phi(n\pi)$$

where $\phi(\omega)$ is the Fourier transform of the pulse.

21 Justify the observation that in Eqs. (14), (14a), (15), and (15a) the ω integration need be performed only from 0 to ∞. *Hint:* Consider the oddness and evenness of $g(\omega)$.

6.7 From the Fourier Integral to the Laplace Transform

In many applications of the Fourier integral the function to be represented is identically zero before some instant, usually $t = 0$. When this is the case, the general

---

† In particular problems, "Tables of the Function $(\sin u)/u$ and Its First Eleven Derivatives," Harvard University Press, Cambridge, Mass., will be of considerable help.

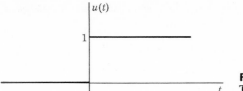

Figure 6.18
The unit step function $u(t)$.

Fourier transform pair, given by Eqs. (9) and (10), Sec. 6.6, becomes the **unilateral Fourier transform pair**

$$(1) \qquad f(t) = \frac{1}{2\pi} \int_{-\infty}^{\infty} g(\omega)e^{i\omega t}\, d\omega \qquad g(\omega) = \int_{0}^{\infty} f(\tau)e^{-i\omega\tau}\, d\tau\dagger$$

Useful as this is in many applications, it is still inadequate to represent such a simple function as the important **unit step function** (Fig. 6.18):

$$u(t) = \begin{cases} 0 & t < 0 \\ 1 & t > 0 \end{cases}$$

In fact, for this function we have from (1)

$$g(\omega) = \int_{0}^{\infty} u(\tau)e^{-i\omega\tau}\, d\tau = \int_{0}^{\infty} 1e^{-i\omega\tau}\, d\tau = \frac{e^{-i\omega\tau}}{-i\omega}\Big|_{0}^{\infty} = \frac{\cos \omega\tau - i \sin \omega\tau}{-i\omega}\Big|_{0}^{\infty}$$

and this is completely meaningless, since both the cosine and sine oscillate without limit as their arguments become infinite.

As an artifice to handle this case, and others like it, the function e^{-at} is sometimes inserted in place of the unit step function. Now as we shall soon see, e^{-at} has a unilateral Fourier transform when a is positive. Moreover, when a approaches zero, e^{-at}, considered for $t > 0$, approaches the unit step function (Fig. 6.19). Hence, it is natural to hope that the order of the operations of letting a approach zero and taking the Fourier transform can be interchanged. *If* this is the case, then we can postpone letting $a \to 0$ until *after* the transform has been taken, and all will be well.

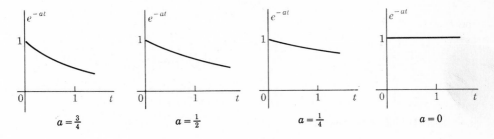

Figure 6.19
The approach of e^{-at} ($a, t > 0$) to the unit step function as a approaches zero.

---

† Note that, for later convenience, we have chosen to incorporate the factor $1/2\pi$ in the integral for $f(t)$ rather than in the integral for $g(\omega)$, as we did in the last section.

In the present problem the development proceeds as follows. Instead of transforming $u(t)$ we transform e^{-at}, getting

$$g(\omega) = \int_0^\infty e^{-a\tau}e^{-i\omega\tau}\,d\tau = \frac{e^{-(a+i\omega)\tau}}{-(a+i\omega)}\bigg|_0^\infty = \frac{e^{-a\tau}e^{-i\omega\tau}}{-(a+i\omega)}\bigg|_0^\infty = \frac{1}{a+i\omega}$$

since the factor $e^{-a\tau}$ $(a > 0)$ now ensures that the antiderivative vanishes at the upper limit. Thus, from (1),

$$f(t) = \frac{1}{2\pi}\int_{-\infty}^\infty g(\omega)e^{i\omega t}\,d\omega$$

$$= \frac{1}{2\pi}\int_{-\infty}^\infty \frac{e^{i\omega t}}{a+i\omega}\,d\omega$$

$$= \frac{1}{2\pi}\int_{-\infty}^\infty \frac{\cos\omega t + i\sin\omega t}{a+i\omega}\frac{a-i\omega}{a-i\omega}\,d\omega$$

$$= \frac{1}{2\pi}\int_{-\infty}^\infty \frac{(a\cos\omega t + \omega\sin\omega t) + i(a\sin\omega t - \omega\cos\omega t)}{a^2+\omega^2}\,d\omega$$

Now, the imaginary part of the integrand, namely,

$$\frac{a\sin\omega t - \omega\cos\omega t}{a^2+\omega^2}$$

is an odd function of ω and hence will vanish when integrated between the limits $-\infty$ and ∞. On the other hand, the real part of the integrand is an even function of ω, and thus, integrating over just the positive half of the range, we can write

$$f(t) = \frac{1}{\pi}\int_0^\infty \frac{a\cos\omega t + \omega\sin\omega t}{a^2+\omega^2}\,d\omega = \frac{1}{\pi}\int_0^\infty \frac{a\cos\omega t}{a^2+\omega^2}\,d\omega + \frac{1}{\pi}\int_0^\infty \frac{\omega\sin\omega t}{a^2+\omega^2}\,d\omega$$

In the first integral in the right member, let $\omega = az$. Then

$$f(t) = \frac{1}{\pi}\int_0^\infty \frac{\cos azt}{1+z^2}\,dz + \frac{1}{\pi}\int_0^\infty \frac{\omega\sin\omega t}{a^2+\omega^2}\,d\omega$$

We are now in a position to let a approach zero. As this happens,

$$f(t) \equiv e^{-at} \to u(t)$$

and thus we obtain

$$u(t) = \frac{1}{\pi}\int_0^\infty \frac{dz}{1+z^2} + \frac{1}{\pi}\int_0^\infty \frac{\sin\omega t}{\omega}\,d\omega = \frac{1}{2} + \frac{1}{\pi}\int_0^\infty \frac{\sin\omega t}{\omega}\,d\omega$$

Incidentally, since $u(t) = 1$ for $t > 0$, it follows that if we put $t = 1$ in this result, we have

$$1 = \frac{1}{2} + \frac{1}{\pi}\int_0^\infty \frac{\sin\omega}{\omega}\,d\omega \qquad \text{and hence} \qquad \int_0^\infty \frac{\sin\omega}{\omega}\,d\omega \equiv \text{Si}(\infty) = \frac{\pi}{2}$$

Thus we have established the value of another definite integral for which there is no elementary antiderivative.

The factor e^{-at} $(a > 0)$ is called a **convergence factor** since, as we have just seen, when it is inserted in the integrands of certain divergent infinite integrals, it decreases with increasing t at a rate sufficient to make them converge. The use we have just made of this convergence factor is both artificial and clumsy, however, and it would be desirable to make this procedure more systematic. To do this, let us define an auxiliary function

$$F(t) = \begin{cases} 0 & t < 0 \\ e^{-at}f(t) & t > 0 \end{cases}$$

where $f(t)$ is the function of actual interest. Then, applying the unilateral Fourier transformation (1) to $F(t)$, which surely satisfies the necessary conditions if $f(t)$ does, we have for $t > 0$,

$$(2) \qquad F(t) \equiv e^{-at}f(t) = \frac{1}{2\pi} \int_{-\infty}^{\infty} g(\omega)e^{i\omega t}\, d\omega$$

where

$$(3) \quad g(\omega) = \int_{0}^{\infty} F(\tau)e^{-i\omega\tau}\, d\tau = \int_{0}^{\infty} \left[e^{-a\tau}f(\tau)\right]e^{-i\omega\tau}\, d\tau = \int_{0}^{\infty} f(\tau)e^{-(a+i\omega)\tau}\, d\tau$$

If we now multiply both sides of Eq. (2) by e^{at}, we obtain

$$f(t) = \frac{e^{at}}{2\pi} \int_{-\infty}^{\infty} g(\omega)e^{i\omega t}\, d\omega = \frac{1}{2\pi} \int_{-\infty}^{\infty} g(\omega)e^{(a+i\omega)t}\, d\omega$$

Moreover, from the last form of the expression for $g(\omega)$ in Eq. (3), it is clear that ω enters the analysis only through the binomial $a + i\omega$. To emphasize this fact, we shall write $g(a + i\omega)$ instead of $g(\omega)$. Then the equations of the transform pair become

$$f(t) = \frac{1}{2\pi} \int_{-\infty}^{\infty} g(a + i\omega)e^{(a+i\omega)t}\, d\omega$$

$$g(a + i\omega) = \int_{0}^{\infty} f(\tau)e^{-(a+i\omega)\tau}\, d\tau$$

Finally, let us put $a + i\omega = s$, noting that

$$d\omega = \frac{d(a + i\omega)}{i} = \frac{ds}{i}$$

and that $s = a - i\infty$ when $\omega = -\infty$ and $s = a + i\infty$ when $\omega = \infty$. Thus we obtain the pair of equations

$$f(t) = \frac{1}{2\pi i} \int_{a-i\infty}^{a+i\infty} g(s)e^{st}\, ds \qquad g(s) = \int_{0}^{\infty} f(\tau)e^{-s\tau}\, d\tau = \int_{0}^{\infty} f(t)e^{-st}\, dt$$

These equations for $f(t)$ and $g(s)$ constitute a **Laplace transform pair**.† The function $g(s)$ is known as the **Laplace transform** of $f(t)$. The integral for $f(t)$ is known as the **complex inversion integral**.

We have thus naturally and inevitably encountered the Laplace transformation through our attempt to provide the unilateral Fourier transformation with a built-in convergence factor. This transformation is the foundation of the modern form of the

---

† Named for the French mathematician Pierre Simon de Laplace (1749–1827), who used such transforms in his researches in the theory of probability.

operational calculus, which was originated in quite another form by the English electrical engineer Oliver Heaviside (1850–1925) around 1890. In the next chapter we shall develop an extensive list of formulas for the use of the Laplace transform itself, although the meaning and use of the inversion integral we must leave to the chapters on complex-variable theory.

EXERCISES

1 Show that $f(t) = t$ does not have a unilateral Fourier transform but that $F(t) = e^{-at}t$ $(a > 0)$ does.

2 Does t^2 have a unilateral Fourier transform? Does $e^{-at}t^2$?

3 Does $\cos kt$ have a unilateral Fourier transform? Does $e^{-at} \cos kt$?

4 Does $\sin kt$ have a unilateral Fourier transform? Does $e^{-at} \sin kt$?

CHAPTER 7
The Laplace Transformation

7.1 Theoretical Preliminaries

In the last chapter we traced the evolution of the Laplace transformation from the unilateral Fourier integral. Our development made it clear that for the Laplace transform of $f(t)$ to exist and for $f(t)$ to be recoverable from its transform, it is sufficient that

    a. In every interval of the form $0 \leq t_1 \leq t \leq t_2$, $f(t)$ be bounded and have at most a finite number of maxima and minima and a finite number of finite discontinuities.

    b. There exist a real constant a such that the improper integral $\int_0^\infty |e^{-at}f(t)|\, dt = \int_0^\infty e^{-at}|f(t)|\, dt$ is convergent.

Functions satisfying condition **a** we shall henceforth describe as **piecewise regular**.

    Condition **b** is frequently replaced by the stronger, i.e., more restrictive, condition that

    c. There is a constant α with the property that $e^{-\alpha t}|f(t)|$ remains bounded as t becomes infinite, i.e., there are constants α, M, and T such that

$$e^{-\alpha t}|f(t)| < M \qquad \text{for all } t > T$$

Functions which satisfy condition **c** are usually described as being of **exponential order**.

    Since $e^{-\alpha t}$ is a monotonically decreasing function of α, it is clear that if

$$e^{-\alpha t}|f(t)| < M \qquad \text{for all } t > T$$

then for all $\alpha_1 > \alpha$ it is also true that $e^{-\alpha_1 t}|f(t)| < M$ for $t > T$. Thus the α required by condition **c** is not unique. The greatest lower bound α_0 of the set of all α's which can be used in condition **c** is often called the **abscissa of convergence** of $f(t)$.

    The abscissa of convergence α_0 of a function $f(t)$ may or may not itself be one of the α's which will serve in condition **c**. For instance, if $f(t) = t$, then for every positive α, and no others, $e^{-\alpha t}|f(t)| = |t|e^{-\alpha t}$ remains bounded, and in fact approaches zero, as t becomes infinite. Obviously the greatest lower bound of the set of all positive numbers is the number zero. Hence, in this case, $\alpha_0 = 0$, although for α_0 itself, $|t|e^{-\alpha_0 t} = |t|$ increases beyond all bounds as $t \to \infty$. Thus for the function $f(t) = t$, α_0 is *not* one of the α's that can be used in condition **c**.

    On the other hand, if $f(t) = e^{2t}$, then for every α greater than or *equal* to 2, $e^{-\alpha t}|f(t)| = e^{-\alpha t}e^{2t} = e^{-(\alpha-2)t}$ is bounded as $t \to \infty$. Since the greatest lower bound of all numbers equal to or greater than 2 is 2, it is clear that in this case the abscissa of convergence α_0 is 2 and moreover is a value of α which will serve in condition **c**.

Since $e^{-\alpha t}|f(t)| < M$ implies only that $|f(t)| < Me^{\alpha t}$, it is obvious that if a function is of exponential order, its absolute value need not remain bounded as $t \to \infty$ but it must not increase more rapidly than some constant multiple of a simple exponential function of t. As the particular function $f(t) = \sin e^{t^2}$ shows, *the derivative of a function of exponential order is not necessarily of exponential order*. On the other hand, it is not difficult to show that *if $f(t)$ is piecewise regular and of exponential order, then $\int_0^t f(t)\, dt$ is also of exponential order*.

With a function $f(t)$ satisfying either conditions **a** and **b** or conditions **a** and **c**, the Laplace transformation associates a function of s, which we shall denote by $\mathscr{L}\{f(t)\}$† or simply by $\mathscr{L}\{f\}$. This is defined by the formula

$$(1) \qquad \mathscr{L}\{f(t)\} = \int_0^\infty f(t)e^{-st}\, dt\ddagger$$

The function $f(t)$ whose Laplace transform is a given function of s, say $\phi(s)$, we shall call the **inverse** of $\phi(s)$ and denote by $\mathscr{L}^{-1}\{\phi(s)\}$. From the concluding discussion of the last chapter we have good reason to believe that the function having $\phi(s)$ for its transform is given by the complex inversion integral

$$(2) \qquad f(t) = \frac{1}{2\pi i}\int_{a-i\infty}^{a+i\infty} \phi(s)e^{st}\, ds$$

where s is the complex variable $a + i\omega$, but we shall make no use of this fact in the present chapter. Indeed, in this chapter we shall regard s as a real-valued parameter.

It is obvious that the derivation of the fundamental properties of the Laplace transform will involve manipulation of the definitive integral (1). This integral is clearly improper, since its upper limit is infinite, and it may also be improper because of discontinuities of $f(t)$ at one or more points in the range of integration. However, inasmuch as $f(t)$ is assumed to be piecewise regular, these discontinuities can be at worst finite jumps which can easily be handled by breaking up the range of integration into subranges whose end points are the points of discontinuity. We shall therefore usually not pay explicit attention to the possible jumps of $f(t)$. Questions associated with the infinite upper limit in (1) are more serious, however, and cannot be passed over so lightly.

At the outset, we recall that by an integral of the form

$$(3) \qquad \int_a^\infty h(s,t)\, dt$$

we mean

$$\lim_{b \to \infty}\int_a^b h(s,t)\, dt$$

---

† Many writers consistently use only small letters to denote functions of t and use the corresponding capital letters to denote the transforms of these functions. Thus what we shall write as $\mathscr{L}\{f(t)\}$ is often written as $F(s)$.

‡ Clearly, the variable of integration t is a **dummy variable** and can be replaced at pleasure by any other symbol. From time to time we shall find it convenient to do this in our work.

and that for this limit to exist for a particular value of s, say $s = s_1$, it must be possible to show that for any $\varepsilon > 0$ there exists a number B such that

$$\left| \int_a^\infty h(s_1,t) \, dt - \int_a^b h(s_1,t) \, dt \right| = \left| \int_b^\infty h(s_1,t) \, dt \right| < \varepsilon$$

for all values of $b > B$. The number B will, of course, depend on ε and in general will also depend on s_1, the particular value of s under consideration. It may happen, however, that when ε is given, one and the same value of B will serve equally well, or *uniformly*, for all members of some set of s values. When this is the case, i.e., when B is a function of ε but not of s, the integral (3) is said to **converge uniformly** or to be **uniformly convergent** over that particular set of s values.

The importance of uniform convergence is apparent from the following theorems, which we shall have to use in this chapter but whose proofs we leave to more advanced texts.†

THEOREM 1 If $g(s,t)$ is a continuous function of s and t for $\alpha \leq s \leq \beta$ and $t \geq a$, if $f(t)$ is at least piecewise regular for $t \geq a$, and if the integral $G(s) = \int_a^\infty f(t)g(s,t) \, dt$ converges uniformly over the interval $\alpha \leq s \leq \beta$, then $G(s)$ is a continuous function of s for $\alpha \leq s \leq \beta$.

Since the definitive property of a continuous function is that

$$\lim_{s \to s_0} G(s) = G(s_0)$$

this theorem states, in effect, that under the appropriate conditions the limit of $G(s)$ can be found by taking the limit inside the t-integral sign or, equivalently, that the order of integrating with respect to t and taking the limit with respect to s can be interchanged.

THEOREM 2 If $g(s,t)$ is a continuous function of s and t for $\alpha \leq s \leq \beta$ and $t \geq a$, if $f(t)$ is at least piecewise regular for $t \geq a$, and if the integral $G(s) = \int_a^\infty f(t)g(s,t) \, dt$ converges uniformly over the interval $\alpha \leq s \leq \beta$, then

$$\int_\alpha^\beta G(s) \, ds \equiv \int_\alpha^\beta \left[\int_a^\infty f(t)g(s,t) \, dt \right] ds = \int_a^\infty \left[\int_\alpha^\beta f(t)g(s,t) \, ds \right] dt$$

In words, Theorem 2 asserts that under the appropriate conditions the integral of $G(s)$ can be found by integrating inside the t-integral sign or, equivalently, that the order of integrating with respect to t and with respect to s can be interchanged.

THEOREM 3 If $g(s,t)$ and $g_s(s,t) \equiv \partial g(s,t)/\partial s$ are continuous functions of s and t for $\alpha \leq s \leq \beta$ and $t \geq a$, if $f(t)$ is at least piecewise regular for $t \geq a$, if the integral

$$G(s) = \int_a^\infty f(t)g(s,t) \, dt$$

---

† See, for instance, H. S. Carslaw, "Fourier Series," pp. 198–201, Dover, New York, 1930.

converges, and if $\int_a^\infty f(t)g_s(s,t)\,dt$ converges uniformly over the interval $\alpha \le s \le \beta$, then

$$G'(s) \equiv \frac{d}{ds}\int_a^\infty f(t)g(s,t)\,dt = \int_a^\infty f(t)\,\frac{\partial g(s,t)}{\partial s}\,dt = \int_a^\infty f(t)g_s(s,t)\,dt$$

In words, Theorem 3 asserts that under the appropriate conditions, the derivative of $G(s)$ can be found by differentiating inside the t-integral sign or, equivalently, that the order of integrating with respect to t and differentiating with respect to s can be interchanged.

Obviously, if we take $g(s,t)$ to be the continuous function e^{-st} and take $a = 0$, the integral $G(s)$ referred to in the last three theorems is precisely the Laplace transform of the function $f(t)$. However, before we can apply these theorems to our work we must determine under what conditions the Laplace transform integral converges uniformly. We begin by proving the following weaker result.

THEOREM 4   If $f(t)$ is piecewise regular and of exponential order, then $\mathscr{L}\{f(t)\} = \int_0^\infty f(t)e^{-st}\,dt$ converges absolutely for any value of s greater than the abscissa of convergence α_0 of $f(t)$.

Proof   To establish this theorem, we must show that

$$(4) \qquad \lim_{b\to\infty}\int_0^b |f(t)e^{-st}|\,dt = \lim_{b\to\infty}\int_0^b |f(t)|e^{-st}\,dt$$

exists, and to do this it is necessary that we have an upper bound for $|f(t)|$ over the entire range of integration $t \ge 0$. Now, by hypothesis, $f(t)$ is of exponential order and therefore has an abscissa of convergence α_0. Hence, there exist numbers M_1 and T such that for all $t > T$ and any α greater than, but bounded from α_0, that is, any α such that $\alpha > \alpha_1 > \alpha_0$, we have

$$|f(t)| < M_1 e^{\alpha t}$$

Moreover, since $f(t)$ is piecewise regular, it is bounded over the finite interval $0 \le t \le T$; that is, there exists a positive number M_2 such that

$$|f(t)| < M_2 = (M_2 e^{-\alpha t})(e^{\alpha t}) \qquad \text{for } 0 \le t \le T$$

Now if $\alpha \ge 0$, then for $0 \le t \le T$ we have $M_2 e^{-\alpha t} \le M_2$. On the other hand, if $\alpha < 0$, then for $0 \le t \le T$ we have $M_2 e^{-\alpha t} \le M_2 e^{-\alpha T}$. Thus if we let M be the largest of the three numbers M_1, M_2, $M_2 e^{-\alpha T}$, it is clear that

$$|f(t)| < M e^{\alpha t} \qquad \text{for } all\ t \ge 0$$

Hence, returning to the integral in (4) and replacing $|f(t)|$ by its upper bound, we have

$$I \equiv \int_0^b |f(t)|e^{-st}\,dt \le \int_0^b Me^{\alpha t}e^{-st}\,dt = \frac{Me^{-(s-\alpha)t}}{-(s-\alpha)}\bigg|_0^b = \frac{M}{s-\alpha}\left(1 - e^{-(s-\alpha)b}\right)$$

Now if $s > \alpha$, the exponential in the last expression decreases monotonically and hence the expression itself increases monotonically and approaches $M/(s - \alpha)$ as b becomes infinite. Therefore

$$I \leq \frac{M}{s - \alpha} \qquad s > \alpha > \alpha_0$$

Since the integrand of I is everywhere nonnegative, it is clear that I is a monotonically increasing function of b. Hence, being bounded above, as we have just shown, it must approach a limit as b becomes infinite. Since $s > \alpha > \alpha_0$ is clearly equivalent to the condition $s > \alpha_0$, the theorem is established.

Since the absolute value of an integral is always equal to or less than the integral of the absolute value, it follows from the preceding discussion that

$$\left| \int_0^b f(t)e^{-st}\, dt \right| \leq \int_0^b |f(t)|e^{-st}\, dt \equiv I \leq \frac{M}{s - \alpha}$$

Hence, letting $b \to \infty$, we have the following important result.

THEOREM 5 If $f(t)$ is piecewise regular and of exponential order with abscissa of convergence α_0, then for all values of s and α such that $s > \alpha > \alpha_0$

$$|\mathscr{L}\{f(t)\}| \leq \frac{M}{s - \alpha} \qquad \text{where } M \text{ is independent of } s$$

Finally, from Theorem 5 we draw the following interesting conclusions.

COROLLARY 1 If $f(t)$ is piecewise regular and of exponential order, then $\mathscr{L}\{f(t)\}$ approaches zero as s becomes infinite.

COROLLARY 2 If $f(t)$ is piecewise regular and of exponential order, then $s\mathscr{L}\{f(t)\}$ is bounded as s becomes infinite.

Corollaries 1 and 2 make it clear that not all functions of s are Laplace transforms—or at least not Laplace transforms of functions of the "respectable" type defined by conditions **a** and **c**. For instance, $\phi(s) = s/(s - 1)$ does not approach zero as s becomes infinite; hence it is not the Laplace transform of any "respectable" function. Also, although $\phi(s) = 1/\sqrt{s}$ does approach zero as s becomes infinite, it is not the transform of any "respectable" function, since $s\phi(s) = \sqrt{s}$ is not bounded as s becomes infinite.

We are now in a position to establish the uniform convergence of the integral defining $\mathscr{L}\{f(t)\}$.

THEOREM 6 If $f(t)$ is piecewise regular and of exponential order with abscissa of convergence α_0, then for any number $s_0 > \alpha_0$,

$$\mathscr{L}\{f(t)\} = \int_0^\infty f(t)e^{-st}\, dt$$

converges uniformly for all values of s such that $s \geq s_0$.

Proof   To prove this theorem, we must show that given any $\varepsilon > 0$, there exists a number B, depending on ε but not on s, such that

$$\left| \int_b^\infty f(t)e^{-st}\, dt \right| < \varepsilon \qquad \text{for all } b > B \text{ and all } s \geq s_0 > \alpha_0$$

Now

$$\left| \int_b^\infty f(t)e^{-st}\, dt \right| \leq \int_b^\infty |f(t)|e^{-st}\, dt$$

and we know that for $s > \alpha_0$ the integral on the right approaches zero as b becomes infinite, since this is implied by the fact that

$$\int_0^\infty |f(t)|e^{-st}\, dt$$

is convergent for $s > \alpha_0$ (Theorem 4). In other words, given any $\varepsilon > 0$ and any $s_0 > \alpha_0$, there exists a number B such that

$$\int_b^\infty f(t)e^{-s_0 t}\, dt < \varepsilon \qquad \text{for all } b > B$$

Now if $s \geq s_0$, it is obvious that $e^{-st} \leq e^{-s_0 t}$. Hence

$$\left| \int_b^\infty f(t)e^{-st}\, dt \right| \leq \int_b^\infty |f(t)|e^{-st}\, dt \leq \int_b^\infty |f(t)|e^{-s_0 t}\, dt$$

and so for any $s \geq s_0$ the integral on the left is less than ε for all values of b greater than the particular B which suffices for the integral on the right. This value of B is clearly independent of s (since it arises from the *specific* value $s = s_0$), and so the proof of the theorem is complete.

In succeeding sections we shall find that many relatively complicated operations upon $f(t)$, such as differentiation and integration, for instance, can be replaced by simple algebraic operations such as multiplication or division by s upon the transform of $f(t)$. This is analogous to the way in which such operations as multiplication and division of numbers are replaced by the simpler processes of addition and subtraction when we work not with the numbers themselves but with their logarithms. Our primary purpose in this chapter is to develop rules of transformation and tables of transforms which can be used, like tables of logarithms, to facilitate the manipulation of functions and by means of which we can recover the proper function from its transform at the end of a problem.

EXERCISES

1   Which of the following functions are of exponential order?

(a) t^n (b) $\tan t$ (c) e^{t^2} (d) $\cosh t$ (e) $1/t$ (f) $t^2 e^{3t}$

2   Show by an example that it is possible for the abscissa of convergence of a function to be negative.

3   What is the abscissa of convergence of each of the following functions?

(a) $\cos kt$ (b) $\sin kt$ (c) t^2 (d) $\cosh kt$ (e) $\sinh kt$ (f) $\ln (1 + t)$

4   For which of the functions in Exercise 3 is α_0 a value of α which will serve in condition c?

5 Show that each of the following integrals converges uniformly over the indicated set of s values:

(a) $\int_0^\infty \dfrac{\sin st}{1 + t^2}\, dt$ all real values of s

(b) $\int_1^\infty \dfrac{1}{s^4 + t^4}\, dt$ all real values of s

(c) $\int_0^\infty e^{-st^2}\, dt$ all values of $s > s_0 > 0$

(d) $\int_0^\infty \dfrac{\sin t}{1 + st}\, dt$ all values of $s > s_0 > 0$

Hint: Consider the alternating series which results when the integration is performed over the successive subintervals $(0,\pi)$, $(\pi,2\pi)$, ...

6 Prove that if a piecewise regular function satisfies condition **b**, it does not necessarily satisfy condition **c**. *Hint:* Consider the function

$$f(t) = \begin{cases} e^{n^2} & t = n \\ 0 & t \neq n \end{cases} \quad n = 0, 1, 2, \ldots$$

7 Prove that if a piecewise regular function satisfies condition **c**, it also satisfies condition **b**. *Hint:* The proof of this is very much like the proof of Theorem 4.

8 Prove that if $f(t)$ is piecewise regular and of exponential order, then $\int_0^t f(t)\, dt$ is also piecewise regular and of exponential order. Show further that if α_0 and α_1 are, respectively, the abscissas of convergence of $f(t)$ and $\int_0^t f(t)\, dt$ and if $\alpha_0 \geq 0$, then $\alpha_1 \leq \alpha_0$. Is it necessarily true that $\alpha_1 \leq \alpha_0$ if $\alpha_0 < 0$?

9 In the proof of Theorem 4, why must α be bounded from α_0, that is, why cannot it simply be assumed that $\alpha > \alpha_0$? *Hint:* Consider the function $f(t) = t$ as a counterexample.

10 Prove that a function $f(t)$ is of exponential order if and only if s can be chosen so that $\lim_{t \to \infty} e^{-st} f(t) = 0$. If $f(t)$ is of exponential order, show that its abscissa of convergence α_0 is the greatest lower bound of all values of s such that

$$\lim_{t \to \infty} e^{-st} f(t) = 0.$$

7.2 The General Method

The utility of the Laplace transformation is based primarily upon the following three theorems.

THEOREM 1 $\mathscr{L}\{c_1 f_1(t) \pm c_2 f_2(t)\} = c_1 \mathscr{L}\{f_1(t)\} \pm c_2 \mathscr{L}\{f_2(t)\}$

Proof   To prove this, we have by definition

$$\mathscr{L}\{c_1 f_1(t) \pm c_2 f_2(t)\} = \int_0^\infty [c_1 f_1(t) \pm c_2 f_2(t)] e^{-st}\, dt$$

$$= c_1 \int_0^\infty f_1(t) e^{-st}\, dt \pm c_2 \int_0^\infty f_2(t) e^{-st}\, dt$$

$$= c_1 \mathscr{L}\{f_1(t)\} \pm c_2 \mathscr{L}\{f_2(t)\}$$

as asserted. The extension of Theorem 1 to linear combinations of more than two functions is obvious.

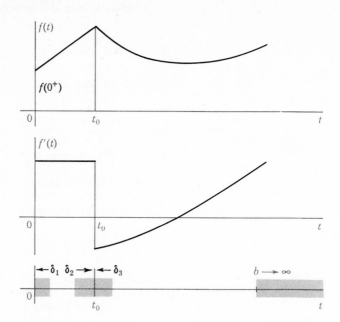

Figure 7.1
A continuous function whose
derivative has a point of
discontinuity.

THEOREM 2 If $f(t)$ is a continuous, piecewise regular function of exponential order whose derivative is also piecewise regular and of exponential order, and if $f(t)$ approaches the limit $f(0^+)$ as t approaches zero from the right, then the Laplace transform of $f(t)$ is given by the formula

$$\mathcal{L}\{f'(t)\} = s\mathcal{L}\{f(t)\} - f(0^+)$$

provided s is greater than the abscissa of convergence of $f(t)$.

Proof To prove this, let us suppose for definiteness that there is a single point, say $t = t_0$, where though $f(t)$ is continuous, its derivative has a finite jump, as suggested by Fig. 7.1. Then, by definition,

$$\mathcal{L}\{f'(t)\} = \int_0^\infty f'(t)e^{-st}\,dt$$

$$= \lim_{\substack{\delta_1,\delta_2,\delta_3 \to 0 \\ b \to \infty}} \left[\int_{\delta_1}^{t_0-\delta_2} f'(t)e^{-st}\,dt + \int_{t_0+\delta_3}^{b} f'(t)e^{-st}\,dt \right]$$

If we use integration by parts on these integrals, choosing

$$u = e^{-st} \qquad\qquad dv = f'(t)\,dt$$
$$du = -se^{-st}\,dt \qquad v = f(t)$$

we have

$$\mathcal{L}\{f'(t)\} = \lim_{\substack{\delta_1,\delta_2,\delta_3 \to 0 \\ b \to \infty}} \left[e^{-st}f(t)\Big|_{\delta_1}^{t_0-\delta_2} + s\int_{\delta_1}^{t_0-\delta_2} f(t)e^{-st}\,dt \right.$$

$$\left. + e^{-st}f(t)\Big|_{t_0+\delta_3}^{b} + s\int_{t_0+\delta_3}^{b} f(t)e^{-st}\,dt \right]$$

In the limit the two integrals which remain combine to give precisely

$$s \int_0^\infty f(t)e^{-st}\, dt = s\mathscr{L}\{f(t)\}$$

Similarly, the first evaluated portion yields

$$e^{-st_0}f(t_0-) - f(0^+)$$

and the second yields simply

$$0 - e^{-st_0}f(t_0+)$$

because, since $f(t)$ is of exponential order, s can be chosen sufficiently large, i.e., greater than the abscissa of convergence of $f(t)$, for the contribution from the upper limit to be zero. Now $f(t)$ was assumed to be continuous. Hence at t_0 (as at all other points) its right- and left-hand limits must be equal. Therefore, the terms

$$e^{-st_0}f(t_0-) \qquad \text{and} \qquad -e^{-st_0}f(t_0+)$$

cancel, leaving finally

$$\mathscr{L}\{f'(t)\} = s\mathscr{L}\{f(t)\} - f(0^+)$$

as asserted. The extension of the preceding proof to functions whose derivatives have more than one finite jump is obvious. The extension of the theorem to the relatively unimportant case in which $f(t)$ itself is permitted to have finite jumps is indicated in Exercise 3.

COROLLARY 1 If both $f(t)$ and $f'(t)$ are continuous, piecewise regular functions of exponential order, and if $f''(t)$ is piecewise regular and of exponential order, then

$$\mathscr{L}\{f''(t)\} = s^2\mathscr{L}\{f(t)\} - sf(0^+) - f'(0^+)$$

where $f(0^+)$ and $f'(0^+)$ are, respectively, the values that $f(t)$ and $f'(t)$ approach as t approaches zero from the right.

Proof This result follows immediately by applying Theorem 2 twice to $f''(t)$:

$$\begin{aligned}
\mathscr{L}\{f''(t)\} = \mathscr{L}\{[f'(t)]'\} &= s\mathscr{L}\{f'(t)\} - f'(0^+) \\
&= s[s\mathscr{L}\{f(t)\} - f(0^+)] - f'(0^+) \\
&= s^2\mathscr{L}\{f(t)\} - sf(0^+) - f'(0^+)
\end{aligned}$$

as asserted. The extension of this result to derivatives of higher order is obvious (Exercise 1).

THEOREM 3 If $f(t)$ is piecewise regular and of exponential order, then the Laplace transform of $\int_a^t f(t)\, dt$ is given by the formula

$$\mathscr{L}\left\{\int_a^t f(t)\, dt\right\} = \frac{1}{s}\mathscr{L}\{f(t)\} + \frac{1}{s}\int_a^0 f(t)\, dt$$

Proof To prove this theorem, we have by definition

$$\mathscr{L}\left\{\int_a^t f(t)\, dt\right\} = \int_0^\infty \left[\int_a^t f(x)\, dx\right]e^{-st}\, dt$$

where the dummy variable x has been introduced in the inner integral for convenience. If we integrate the last integral by parts, with

$$u = \int_a^t f(x)\, dx \qquad dv = e^{-st}\, dt$$

$$du = f(t)\, dt \qquad v = \frac{e^{-st}}{-s}$$

we have

$$\mathscr{L}\left\{\int_a^t f(t)\, dt\right\} = \left[\frac{e^{-st}}{-s}\int_a^t f(x)\, dx\right]_0^\infty + \frac{1}{s}\int_0^\infty f(t)e^{-st}\, dt$$

Since $f(t)$ is of exponential order, so too is its integral (Exercise 8, Sec. 7.1). Hence s can be chosen sufficiently large for the integrated portion to vanish at the upper limit, leaving

$$\mathscr{L}\left\{\int_a^t f(t)\, dt\right\} = \frac{1}{s}\int_a^0 f(x)\, dx + \frac{1}{s}\mathscr{L}\{f(t)\}$$

as asserted. The extension of this result to repeated integrals of $f(t)$ is obvious (Exercise 2).

Although we need many more formulas before the Laplace transformation can be applied effectively to specific problems, Theorems 1 to 3 allow us to outline all the essential steps in the usual application of this method to the solution of differential equations. Suppose that we are given the equation

$$ay'' + by' + cy = f(t) \quad a,\, b,\, c \text{ constants}$$

If we take the Laplace transform of both sides, we have by Theorem 1

$$a\mathscr{L}\{y''\} + b\mathscr{L}\{y'\} + c\mathscr{L}\{y\} = \mathscr{L}\{f(t)\}$$

Now applying Theorem 2 and its corollary, we have

$$a(s^2\mathscr{L}\{y\} - sy_0 - y_0') + b(s\mathscr{L}\{y\} - y_0) + c\mathscr{L}\{y\} = \mathscr{L}\{f(t)\}$$

where y_0 and y_0' are the given initial values of y and y'. Collecting terms on $\mathscr{L}\{y\}$ and then solving for $\mathscr{L}\{y\}$, we obtain finally

$$\mathscr{L}\{y\} = \frac{\mathscr{L}\{f(t)\} + (as + b)y_0 + ay_0'}{as^2 + bs + c}$$

Now $f(t)$ is a given function of t; hence its Laplace transform (if it exists) is a perfectly definite function of s (although as yet we have no specific formulas for finding it). Moreover, y_0 and y_0' are definite numbers known from the data of the problem. Hence the transform of y is a completely known function of s. Thus if we had available a sufficiently extensive table of transforms, we could find in it the function $y(t)$ having the right-hand side of the last equation for its transform, *and this function would be the formal solution to our problem, initial conditions and all*. The formal solution could then be substituted into the differential equation to verify that it was indeed the genuine solution.

This brief discussion illustrates the two great advantages of the Laplace transformation in solving linear constant-coefficient differential equations: it reduces the problem to one in algebra, and it takes care of initial conditions without the necessity of first constructing a complete solution and then specializing the arbitrary constants it contains. Clearly, our immediate task is to implement this process by establishing an adequate table of transforms.

EXERCISES

1 Show that under the appropriate conditions

$$\mathcal{L}\{f'''\} = s^3\mathcal{L}\{f\} - s^2 f_0 - s f_0' - f_0''$$

What is $\mathcal{L}\{f^{(n)}\}$?

2 Show that

$$\mathcal{L}\left\{\int_a^t \int_a^t f(t)\, dt\, dt\right\} = \frac{1}{s^2}\mathcal{L}\{f\} + \frac{1}{s^2}\int_a^0 f(t)\, dt + \frac{1}{s}\int_a^0 \int_a^t f(t)\, dt\, dt$$

3 If $f(t)$ satisfies all the conditions of Theorem 2 except that it has an upward jump of magnitude J_0 at $t = t_0$, show that

$$\mathcal{L}\{f'(t)\} = s\mathcal{L}\{f\} - f_0 - J_0 e^{-s t_0}$$

4 Is the proof of Theorem 3 valid if $f(t)$ has a finite jump at $t = t_0$?

5 Devise a proof of Theorem 3 based upon the use of Theorem 2.

6 Show that

$$\mathcal{L}\{f(at)\} = \frac{1}{a}\mathcal{L}\{f(t)\}\Big|_{s\to s/a}$$

7 (a) Given $\mathcal{L}\{\cos t\} = s/(s^2 + 1)$, use the result of Exercise 6 to determine $\mathcal{L}\{\cos bt\}$.

(b) Given $\mathcal{L}\{\sin t\} = 1/(s^2 + 1)$, use the result of Exercise 6 to determine $\mathcal{L}\{\sin bt\}$.

8 (a) Given $\mathcal{L}\{\sin t\} = 1/(s^2 + 1)$, use Theorem 2 to obtain $\mathcal{L}\{\cos t\}$.

(b) Given $\mathcal{L}\{\cos t\} = s/(s^2 + 1)$, use Theorem 2 to obtain $\mathcal{L}\{\sin t\}$.

(c) Use Theorem 3 to obtain $\mathcal{L}\{\sin t\}$ from $\mathcal{L}\{\cos t\}$.

(d) Use Theorem 3 to obtain $\mathcal{L}\{\cos t\}$ from $\mathcal{L}\{\sin t\}$.

9 Explain how the Laplace transform can be used to solve a system of simultaneous linear differential equations with constant coefficients. In particular, given that $y = y_0$ and $z = z_0$ when $t = 0$, obtain formulas for the Laplace transforms of y and z if

$$a_1 \frac{dy}{dt} + b_1 y + c_1 \frac{dz}{dt} + d_1 z = f_1(t)$$

$$a_2 \frac{dy}{dt} + b_2 y + c_2 \frac{dz}{dt} + d_2 z = f_2(t)$$

10 The function of n defined by the equation

$$S\{f(t)\} = \int_0^\pi f(t) \sin nt\, dt \qquad n = 1, 2, 3, \ldots$$

is called the **sine transform** of $f(t)$. Show that

$$S\{f''\} = -n^2 S\{f\} + n[f(0) - (-1)^n f(\pi)]$$

11 The function of n defined by the equation

$$C\{f(t)\} = \int_0^\pi f(t) \cos nt\, dt \qquad n = 0, 1, 2, \ldots$$

is called the **cosine transform** of $f(t)$. Obtain a formula expressing $C\{f''\}$ in terms of $C\{f\}$.

12 Obtain formulas expressing $S\{f'\}$ and $C\{f'\}$ in terms of $S\{f\}$ and $C\{f\}$.

13 (a) Noting that the values of $S\{f\}$ are just $\pi/2$ times the coefficients in the half-range sine expansion of $f(t)$, defined for $0 < t < \pi$, use the result of Exercise 10 to obtain the half-range sine expansion of $f(t) = t^2$, $0 < t < \pi$, from the half-range sine expansion of $f(t) = 1$ Is this method easier than the direct calculation of the expansion of t^2 by the formulas of Sec. 6.3?

(b) Employing ideas similar to those in part (a), use the result of Exercise 11 to obtain the half-range cosine expansion of $f(t) = t^3$, $0 < t < \pi$, from the half-range cosine expansion of $f(t) = t$.

14 Let $T\{f(t)\}$ be a general integral transform

$$T\{f(t)\} = \int_a^b f(t)K(s,t)\, dt$$

where $K(s,t)$ is the so-called **kernel** of the transformation. Obtain conditions on $K(s,t)$ so that $T\{f'\}$ and $T\{f''\}$ contain no terms involving the evaluation of f or any of its derivatives. Find at least one kernel satisfying these conditions.

15 If both $f(t)$ and $f'(t)$ are piecewise regular and of exponential order, and if $f(t)$ is continuous, which implies that $f(0^+) = 0$, show that as s becomes infinite, $\mathscr{L}\{f(t)\}$ tends to zero at least as rapidly as c/s^2. Can this result be generalized?

7.3 The Transforms of Special Functions

Among all the functions whose transforms we might now think of tabulating, the most important are the simple ones

$$e^{-at} \qquad \cos bt \qquad \sin bt \qquad t^n$$

and the unit step function,

$$u(t) = \begin{cases} 0 & t < 0 \\ 1 & t > 0 \end{cases}$$

shown in Fig. 7.2. Once we know the transforms of these functions, nearly all the formulas we shall need can be obtained through the use of a few additional theorems which we shall establish in the next section. The specific results are the following.

FORMULA 1 $\mathscr{L}\{e^{-at}\} = \dfrac{1}{s + a}$

FORMULA 2 $\mathscr{L}\{\cos bt\} = \dfrac{s}{s^2 + b^2}$

FORMULA 3 $\mathscr{L}\{\sin bt\} = \dfrac{b}{s^2 + b^2}$

FORMULA 4 $\mathscr{L}\{t^n\} = \begin{cases} \dfrac{\Gamma(n + 1)}{s^{n+1}} & n > -1 \\[2mm] \dfrac{n!}{s^{n+1}} & n \text{ a positive integer} \end{cases}$

FORMULA 5 $\mathscr{L}\{u(t)\} = \dfrac{1}{s}$

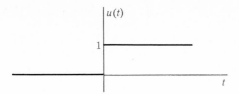

Figure 7.2
The unit step function $u(t)$.

To prove Formula 1, we have simply

$$\mathscr{L}\{e^{-at}\} = \int_0^\infty e^{-at}e^{-st}\, dt = \frac{e^{-(s+a)t}}{-(s+a)}\Big|_0^\infty = \frac{1}{s+a} \qquad \text{if } s + a > 0$$

To prove Formula 2, we have

$$\mathscr{L}\{\cos bt\} = \int_0^\infty \cos bt\, e^{-st}\, dt = \frac{e^{-st}}{s^2+b^2}(-s \cos bt + b \sin bt)\Big|_0^\infty$$

$$= \frac{s}{s^2+b^2} \qquad \text{if } s > 0$$

To prove Formula 3, we have

$$\mathscr{L}\{\sin bt\} = \int_0^\infty \sin bt\, e^{-st}\, dt = \frac{e^{-st}}{s^2+b^2}(-s \sin bt - b \cos bt)\Big|_0^\infty$$

$$= \frac{b}{s^2+b^2} \qquad \text{if } s > 0$$

Before we can prove Formula 4 it will be necessary for us to investigate briefly the so-called **gamma function** or **generalized factorial function**, defined by the equation

$$(1) \qquad \Gamma(x) = \int_0^\infty e^{-t}t^{x-1}\, dt$$

This improper integral can be shown to be convergent for all $x > 0$.

To determine the simple properties of the gamma function and its relation to the familiar factorial function

$$n! = n(n-1)\cdots 3\cdot 2\cdot 1$$

defined in elementary algebra for positive integral values of n, let us apply integration by parts to the definitive integral (1), taking

$$u = e^{-t} \qquad\qquad dv = t^{x-1}\, dt$$

$$du = -e^{-t}\, dt \qquad v = \frac{t^x}{x}$$

Then

$$\Gamma(x) = \frac{t^x e^{-t}}{x}\Big|_0^\infty + \frac{1}{x}\int_0^\infty e^{-t}t^x\, dt$$

Under the restriction $x > 0$, the integrated portion vanishes at both limits. By comparison with (1), it is clear that the integral which remains is simply $\Gamma(x + 1)$. Thus we have established the important recurrence relation

(2)
$$\Gamma(x) = \frac{\Gamma(x + 1)}{x} \qquad x > 0$$

or

(2a)
$$x\Gamma(x) = \Gamma(x + 1)$$

Moreover, we have specifically

$$\Gamma(1) = \int_0^\infty e^{-t}\, dt = -e^{-t}\Big|_0^\infty = 1$$

Therefore, using (2a),

$$\Gamma(2) = 1 \cdot \Gamma(1) = 1$$
$$\Gamma(3) = 2 \cdot \Gamma(2) = 2 \cdot 1 = 2!$$
$$\Gamma(4) = 3 \cdot \Gamma(3) = 3 \cdot 2! = 3!$$

and in general

(3)
$$\Gamma(n + 1) = n! \qquad n = 1, 2, 3, \ldots$$

The connection between the gamma function and ordinary factorials is now clear. However, the gamma function constitutes an essential extension of the idea of a factorial, since its argument x is not restricted to positive integral values but can vary continuously.

From (2) and the fact that $\Gamma(1) = 1$, it is evident that $\Gamma(x)$ becomes infinite as x approaches zero. It is thus clear that $\Gamma(x)$ cannot be defined for $x = 0, -1, -2, \ldots$ in a way consistent with Eq. (2); hence we shall leave it undefined for these values of x. For all other values of x, however, $\Gamma(x)$ is well defined, the use of the recurrence formula (2a) effectively removing the restriction that x be positive, which the integral definition (1) requires. By methods which need not concern us here, tables of $\Gamma(x)$ have been constructed and can be found, usually as tables of $\log \Gamma(x)$, in most elementary handbooks. Because of the recurrence formula which the gamma function satisfies, these tables ordinarily cover only a unit interval on x, usually the interval $1 \le x \le 2$. A plot of $\Gamma(x)$ is shown in Fig. 7.3.

EXAMPLE 1

What is the value of $I = \int_0^\infty \sqrt{z}\, e^{-z^3}\, dz$?

This integral is typical of many which can be reduced to the standard form of the gamma function by a suitable substitution. In this case it is clear on comparing the given integral with (1) that we should let

$$z = t^{1/3} \qquad dz = \tfrac{1}{3}t^{-2/3}\, dt$$

getting
$$I = \int_0^\infty \sqrt{t^{1/3}}\, e^{-t}(\tfrac{1}{3}t^{-2/3}\, dt) = \frac{1}{3}\int_0^\infty e^{-t}t^{1/2-1}\, dt = \tfrac{1}{3}\Gamma(\tfrac{1}{2})$$

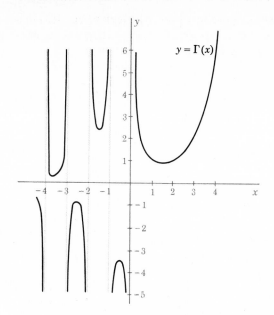

Figure 7.3
The function $y = \Gamma(x)$.

Since $\Gamma(\frac{1}{2})$ cannot be found in the usual table, which lists $\Gamma(x)$ only for $1 \le x \le 2$, it is necessary to use the recurrence relation (2) to bring the argument of the gamma function into this interval:

$$I = \tfrac{1}{3}\Gamma(\tfrac{1}{2}) = \frac{1}{3}\frac{\Gamma(\tfrac{3}{2})}{\tfrac{1}{2}} = \tfrac{2}{3}(0.86623) = 0.59082\dagger$$

Returning now to Formula 4, we have

$$\mathscr{L}\{t^n\} = \int_0^\infty t^n e^{-st}\, dt$$

In an attempt to reduce this integral to the standard form of the gamma function, let us make the substitution

$$t = \frac{z}{s} \qquad dt = \frac{dz}{s}$$

Then $\qquad \mathscr{L}\{t^n\} = \int_0^\infty \left(\frac{z}{s}\right)^n e^{-z}\frac{dz}{s} = \frac{1}{s^{n+1}}\int_0^\infty e^{-z}z^n\, dz = \frac{\Gamma(n+1)}{s^{n+1}}$

Since $\Gamma(n+1) = n!$ when n is a positive integer, this establishes the second part of Formula 4 also.

It is interesting to note that if n is negative,

$$s\mathscr{L}\{t^n\} = \frac{\Gamma(n+1)}{s^n}$$

---

† Actually, the value of $\Gamma(\frac{1}{2})$ is known exactly and in fact is equal to $\sqrt{\pi}$ (Exercise 14). Hence, in this example $I = \sqrt{\pi}/3$.

is not bounded as $s \to \infty$. Hence, according to Corollary 2, Theorem 5, Sec. 7.1, this function of s is not the Laplace transform of a piecewise regular function of exponential order. This, of course, is obvious, since when n is negative, t^n, though of exponential order (with abscissa of convergence $\alpha_0 = 0$), is not bounded in the neighborhood of the origin and so is not piecewise regular. It can be shown, however, that the improper integral defining $\mathscr{L}\{t^n\}$ exists for $n > -1$ although it does not exist for $n \le -1$. Formula 4 must therefore be qualified by the restriction $n > -1$.

Formula 5 can be obtained immediately by taking $n = 0$ in Formula 4.

EXAMPLE 2

What is the Laplace transform of $\sinh bt$?
Since $\sinh bt = (e^{bt} - e^{-bt})/2$, we have

$$\mathscr{L}\{\sinh bt\} = \mathscr{L}\left\{\frac{e^{bt} - e^{-bt}}{2}\right\} = \frac{1}{2}\left(\frac{1}{s-b} - \frac{1}{s+b}\right) = \frac{b}{s^2 - b^2}$$

The analogy with Formula 3 for the transform of $\sin bt$ is apparent.

EXAMPLE 3

If
$$\mathscr{L}\{y\} = \frac{s+1}{s^2 + s - 6}$$

what is $y(t)$?
None of our formulas yields a transform resembling this one. However, using the method of partial fractions, we can write

$$\frac{s+1}{s^2 + s - 6} = \frac{s+1}{(s-2)(s+3)} = \frac{A}{s-2} + \frac{B}{s+3} = \frac{A(s+3) + B(s-2)}{(s-2)(s+3)}$$

For this to be an identity, we must have

$$s + 1 = A(s+3) + B(s-2)$$

Setting $s = 2$ and $s = -3$ in turn, we find from this that $A = \frac{3}{5}$, $B = \frac{2}{5}$. Hence

$$\mathscr{L}\{y(t)\} = \frac{1}{5}\left(\frac{3}{s-2} + \frac{2}{s+3}\right)$$

Formula 1 can now be applied to the individual terms, and we find

$$y(t) = \tfrac{1}{5}(3e^{2t} + 2e^{-3t})$$

EXAMPLE 4

Solve for $y(t)$ from the simultaneous equations

$$y' + 2y + 6\int_0^t z\, dt = -2u(t)$$

$$y' + z' + z = 0$$

if $y_0 = -5$ and $z_0 = 6$.
We begin by taking the Laplace transform of each equation, term by term, using Theorems 2 and 3, Sec. 7.2:

$$(s\mathscr{L}\{y\} + 5) + 2\mathscr{L}\{y\} + \frac{6}{s}\mathscr{L}\{z\} = -\frac{2}{s}$$

$$(s\mathscr{L}\{y\} + 5) + (s\mathscr{L}\{z\} - 6) + \mathscr{L}\{z\} = 0$$

Obvious simplifications then lead to the following pair of linear algebraic equations in the transforms of the unknown functions $y(t)$ and $z(t)$:

$$(s^2 + 2s)\mathscr{L}\{y\} + 6\mathscr{L}\{z\} = -2 - 5s$$
$$s\mathscr{L}\{y\} + (s + 1)\mathscr{L}\{z\} = 1$$

Since it is $y(t)$ that we are asked to find, we solve these simultaneous equations for $\mathscr{L}\{y\}$, getting

$$\mathscr{L}\{y\} = \frac{\begin{vmatrix} -2 - 5s & 6 \\ 1 & s + 1 \end{vmatrix}}{\begin{vmatrix} s^2 + 2s & 6 \\ s & s + 1 \end{vmatrix}} = \frac{-5s^2 - 7s - 8}{s^3 + 3s^2 - 4s}$$

Applying the method of partial fractions to this expression, we have

$$\mathscr{L}\{y\} = \frac{-5s^2 - 7s - 8}{s^3 + 3s^2 - 4s} = \frac{2}{s} - \frac{4}{s - 1} - \frac{3}{s + 4}$$

Finally, taking the inverse of each of these terms, we find

$$y(t) = 2u(t) - 4e^t - 3e^{-4t}$$

EXERCISES

1 Plot each of the following functions:

(a) $u(t - 2)$ (b) $u(t^2)$ (c) $u(t^2 - 1)$ (d) $u(t^3 - 6t^2 + 11t - 6)$

2 Plot each of the following functions:

(a) $u(t - 2) - u(t - 1)$ (b) $u(t) + u(t - 1) + u(t - 2) + u(t - 3) + \cdots$
(c) $u(\sin t)$ (d) $2u(\sin \pi t) - 1$

3 What is $\mathscr{L}\{\cosh kt\}$?
4 What is $\mathscr{L}\{\cos (at + b)\}$? *Hint:* First express cos $(at + b)$ as the difference of two terms.
5 What is $\mathscr{L}\{\cos^2 bt\}$? *Hint:* First express cos^2 bt as a function of $2bt$.
6 What is $\mathscr{L}\{(t + 1)^2\}$?
7 Find the inverse of each of the following functions:

(a) $\dfrac{1}{s + 3}$ (b) $\dfrac{1}{s^4}$ (c) $\dfrac{1}{s^2 + 9}$ (d) $\dfrac{2s + 3}{s^2 + 9}$ (e) $\dfrac{s + 3}{(s + 1)(s - 3)}$

8 Find the solution of each of the following differential equations which satisfies the given conditions:

(a) $y'' + 4y' - 5y = 0$ $y_0 = 1, y_0' = 0$
(b) $y'' - 4y = 0$ $y_0 = -1, y_0' = 1$
(c) $4y'' + y = 0$ $y_0 = 2, y_0' = 1$
(d) $y'' + 4y = u(t)$ $y_0 = y_0' = 0$
(e) $y'' + 3y' + 2y = e^t$ $y_0 = 1, y_0' = 0$
(f) $y''' + 6y'' + 11y' + 6y = 0$ $y_0 = 2, y_0' = 1, y_0'' = -1$
(g) $y''' - y'' + 4y' - 4y = t$ $y_0 = y_0' = 0, y_0'' = 1$

9 Solve for $z(t)$ in Example 4.
10 Find the solution of the following system of equations:

$$y' + y + 2z' + 3z = e^{-t}$$
$$3y' - y + 4z' + z = 0$$
     $y_0 = -1, z_0 = 0$

11 Find the solution of the following system of equations:

$$(D + 1)y + (2D + 3)z = 0$$
$$(D - 4)y + (3D - 8)z = \sin t$$
     $y_0 = 2, z_0 = -1$

12   Evaluate each of the following integrals:

     (a) $\displaystyle\int_0^\infty \frac{e^{-x}}{\sqrt{x}}\, dx$ **(b)** $\displaystyle\int_0^\infty \exp\left(-\sqrt{x}\right) dx$ **(c)** $\displaystyle\int_0^\infty (x + 1)^2 e^{-x^3}\, dx$

13   Evaluate each of the following integrals:

     (a) $\displaystyle\int_0^\infty \frac{x^c}{c^x}\, dx$ *Hint:* Recall that $c^x = \exp(x \ln c)$.

     (b) $\displaystyle\int_0^1 \frac{dx}{\sqrt{\ln(1/x)}}$ *Hint:* Let $\ln \dfrac{1}{x} = z$.

     (c) $\displaystyle\int_0^1 x^m \left(\ln \frac{1}{x}\right)^n dx$

14   Show that $\Gamma(\frac{1}{2}) = \sqrt{\pi}$.   *Hint:* First show that

$$\Gamma(\tfrac{1}{2}) = 2\int_0^\infty e^{-x^2}\, dx = 2\int_0^\infty e^{-y^2}\, dy$$

Then multiply these integrals and evaluate the resulting double integral by changing to polar coordinates.

15   A particle of mass m moves along the x axis under the influence of a force which varies inversely as the distance from the origin. If the particle begins to move from rest at the point $x = a$, find the time it takes it to reach the origin. *Hint:* After the equation of motion is set up, recall the hint given in Exercise 56, Sec. 1.7.

16   Show that

$$\int_0^{\pi/2} \cos^{2m-1}\theta \sin^{2n-1}\theta\, d\theta = \frac{\Gamma(m)\Gamma(n)}{2\Gamma(m+n)} \qquad m, n > 0$$

Hint: First show that $\Gamma(m) = \int_0^\infty 2x^{2m-1}e^{-x^2}\, dx$ and $\Gamma(n) = \int_0^\infty 2y^{2n-1}e^{-y^2}\, dy$. Then multiply these integrals and convert the resulting repeated integral into polar coordinates. Next, note that the repeated integral in polar coordinates can be written as the product of two integrals, one of which is the integral asked for in the exercise. Finally, establish the required result by showing that the other integral is equal to $\Gamma(m+n)$.

17   By setting $2m - 1 = k$ and $n = \frac{1}{2}$ in the result of Exercise 16, show that

$$\int_0^{\pi/2} \cos^k\theta\, d\theta = \frac{\sqrt{\pi}}{2}\frac{\Gamma[(k+1)/2]}{\Gamma[(k/2)+1]} \qquad k > -1$$

What is $\int_0^{\pi/2} \sin^k\theta\, d\theta$?

18   Put $x = \cos^2\theta$ in the integral in Exercise 16 and show that

$$\int_0^1 x^{m-1}(1-x)^{n-1}\, dx = \frac{\Gamma(m)\Gamma(n)}{\Gamma(m+n)} \qquad m, n > 0$$

This integral is usually referred to as the **beta function** of m and n, $B(m,n)$.

19   Evaluate each of the following integrals:

     (a) $\displaystyle\int_0^{\pi/2} \sqrt{\cos\theta}\, d\theta$ **(b)** $\displaystyle\int_0^{\pi/2} \sqrt{\tan\theta}\, d\theta$

     (c) $\displaystyle\int_0^1 \frac{dz}{\sqrt{1-z^4}}$ *Hint:* Let $z^4 = x$.

     (d) $\displaystyle\int_0^1 (1-z^k)^{1/k}\, dz$ **(e)** $\displaystyle\int_0^a \sqrt{a^n - z^n}\, dz$

20 (a) Show that $B(m,n) = B(n,m)$.

(b) Show that

$$B(m,n) = \int_0^\infty \frac{z^{m-1}}{(1 + z)^{m+n}}\, dz$$

Hint: Transform the integral which defines $B(m,n)$ by the substitution $x = z/(1 + z)$.

(c) Show that

$$B(m,m) = \frac{\sqrt{\pi}\, \Gamma(m)}{2^{2m-1}\Gamma(m + \frac{1}{2})}$$

Hint: Use the double-angle sine formula in the integral in Exercise 16.

7.4 Further General Theorems

We are now in a position to derive a number of theorems that will be of considerable use in the application of the Laplace transformation to practical problems. We begin with a result which allows us to infer the behavior of a function $f(t)$ for small positive values of t from the behavior of $\mathscr{L}\{f(t)\}$ for large positive values of s.

THEOREM 1   If both $f(t)$ and $f'(t)$ are piecewise regular and of exponential order, then

$$\lim_{s \to \infty} s\mathscr{L}\{f(t)\} = \lim_{t \to 0^+} f(t) = f(0^+)$$

Proof   For convenience we shall prove this under the additional assumption that $f(t)$ is continuous, leaving as an exercise the proof under the less restrictive conditions of the theorem as stated. We may thus begin with the result of Theorem 2, Sec. 7.2, namely,

$$\mathscr{L}\{f'(t)\} = s\mathscr{L}\{f(t)\} - f(0^+)$$

Hence, taking the limit of each side, we have

(1)
$$\lim_{s \to \infty} \mathscr{L}\{f'(t)\} = \lim_{s \to \infty} s\mathscr{L}\{f(t)\} - f(0^+)$$

However, under the conditions of the theorem, it follows from Corollary 1, Theorem 5, Sec. 7.1, that

$$\lim_{s \to \infty} \mathscr{L}\{f'(t)\} = 0$$

Therefore, from (1),

$$\lim_{s \to \infty} s\mathscr{L}\{f(t)\} = f(0^+)$$

as asserted.

An analogous result which allows us to infer the behavior of a function $f(t)$ for large positive values of t from the behavior of $\mathscr{L}\{f(t)\}$ for small values of s is contained in the following theorem.

THEOREM 2   If both $f(t)$ and $f'(t)$ are piecewise regular and of exponential order, and if the abscissa of convergence of $f'(t)$ is negative, then

$$\lim_{s \to 0} s\mathscr{L}\{f(t)\} = \lim_{t \to \infty} f(t)$$

provided these limits exist.

Proof Here, as in the proof of Theorem 1, we shall base our argument on the additional assumption that $f(t)$ is continuous. Then again we may take limits in the result of Theorem 2, Sec. 7.2, getting

$$\lim_{s \to 0} \mathscr{L}\{f'(t)\} = \lim_{s \to 0} s\mathscr{L}\{f(t)\} - f(0^+)$$

or

(2)
$$\lim_{s \to 0} s\mathscr{L}\{f(t)\} = \lim_{s \to 0} \mathscr{L}\{f'(t)\} + f(0^+)$$

But
$$\lim_{s \to 0} \mathscr{L}\{f'(t)\} = \lim_{s \to 0} \int_0^\infty f'(t)e^{-st}\, dt$$

and under the conditions of the present theorem we can invoke Theorems 6 and 1, Sec. 7.1, and take the limit on the right inside the integral sign. Thus

$$\lim_{s \to 0} \mathscr{L}\{f'(t)\} = \int_0^\infty f'(t)\left(\lim_{s \to 0} e^{-st}\right) dt = \int_0^\infty f'(t)\, dt = f(t)\Big|_0^\infty$$
$$= \lim_{t \to \infty} f(t) - f(0^+)$$

Substituting this into (2), we have finally

$$\lim_{s \to 0} s\mathscr{L}\{f(t)\} = \left[\lim_{t \to \infty} f(t) - f(0^+)\right] + f(0^+)$$
$$= \lim_{t \to \infty} f(t)$$

as asserted.†

When the Laplace transform of an unknown function $f(t)$ contains the factor s,‡ it is often convenient to find $f(t)$ by means of the following theorem.

THEOREM 3 If $f(t)$ is piecewise regular and of exponential order, if $\mathscr{L}\{f(t)\} = s\phi(s)$, and if the inverse of the factor $\phi(s)$ is continuous for $t > 0$, then

$$f(t) = \frac{d}{dt}\, \mathscr{L}^{-1}\{\phi(s)\}$$

Proof To prove this theorem, let $F(t) \equiv \mathscr{L}^{-1}\{\phi(s)\}$ be the function which has the factor $\phi(s)$ for its transform. If $F(t)$ is continuous, as assumed, then, by Theorem 2, Sec. 7.2,

$$\mathscr{L}\{F'(t)\} = s\mathscr{L}\{F(t)\} - F(0^+) = s\phi(s) - F(0^+)$$

---

† In realistic applications of this theorem, $\mathscr{L}\{f(t)\}$ will be known, but $f(t)$ and its abscissa of convergence will be unknown. Hence it is desirable that conditions for the use of the theorem be expressed in terms of $\mathscr{L}\{f(t)\}$ rather than $f(t)$. This can be done, since it is possible to show that Theorem 2 holds if and only if $\mathscr{L}\{f(t)\}$ is bounded for all nonnegative real values of s and all complex values of s with nonnegative real parts. Thus, for example, even though $\lim_{s \to 0} [s/(s^2 + 1)]$ exists, Theorem 2 cannot be applied to $\mathscr{L}\{f(t)\} = 1/(s^2 + 1)$, since this is unbounded for the values $s = \pm i \equiv 0 \pm i$. In this case, of course, $f(t) = \sin t$, and clearly $\lim_{t \to \infty} (\sin t)$ does not exist.

‡ This can always be arranged, of course, by multiplying and dividing the transform by s; that is, $\Phi(s) \equiv s\Phi(s)/s = s\phi(s)$.

But, by Theorem 1,

$$F(0^+) = \lim_{s \to \infty} s\mathscr{L}\{F(t)\} = \lim_{s \to \infty} s\phi(s) = 0$$

the last step following from Corollary 1, Theorem 5, Sec. 7.1, since $s\phi(s)$ is the transform of the function $f(t)$, which, though unknown, is assumed to be "respectable." Hence

$$\mathscr{L}\{f(t)\} = \mathscr{L}\{F'(t)\}$$

since each is equal to $s\phi(s)$. Therefore†

$$f(t) = \frac{dF(t)}{dt} = \frac{d}{dt} \mathscr{L}^{-1}\{\phi(s)\}$$

as asserted.

EXAMPLE 1

What is $\mathscr{L}^{-1}\{s/(s^2 + 4)\}$?

By Formula 2, Sec. 7.3, we see immediately that the required inverse is $f(t) = \cos 2t$. However, it is interesting that we can also obtain this result by suppressing the factor s, finding the inverse $F(t)$ of the remaining portion of the transform, namely

$$\frac{1}{s^2 + 4}$$

and then differentiating this inverse according to Theorem 3:

$$f(t) = \frac{d}{dt} \mathscr{L}^{-1}\left\{\frac{1}{s^2 + 4}\right\} = \frac{d}{dt}\left(\frac{\sin 2t}{2}\right) = \cos 2t$$

as before. The usual applications of this theorem are, of course, not of this trivial character.

When the Laplace transform of an unknown function $f(t)$ contains the factor $1/s$,‡ it is often convenient to find $f(t)$ by means of the following theorem.

THEOREM 4 If $f(t)$ is piecewise regular and of exponential order, and if $\mathscr{L}\{f(t)\} = \phi(s)/s$, then

$$f(t) = \int_0^t \mathscr{L}^{-1}\{\phi(s)\} \, dt$$

Proof To prove this theorem, let $F(t) \equiv \mathscr{L}^{-1}\{\phi(s)\}$ be the function which has the factor $\phi(s)$ for its transform. Then by Theorem 3, Sec. 7.2,

$$\mathscr{L}\left\{\int_0^t F(t) \, dt\right\} = \frac{1}{s} \mathscr{L}\{F(t)\} + \frac{1}{s}\int_0^0 F(t) \, dt = \frac{1}{s} \mathscr{L}\{F(t)\} = \frac{\phi(s)}{s}$$

Thus both $f(t)$ and $\int_0^t F(t) \, dt \equiv \int_0^t \mathscr{L}^{-1}\{\phi(s)\} \, dt$ have $\phi(s)/s$ for their Laplace transform and so must be equal, as asserted.

---

† This, of course, assumes the "obvious" theorem that if two functions have the same transform, they are identical. This is strictly true if the functions are continuous. If discontinuities are permitted, the most we can say is that two functions with the same transform cannot differ over any interval of positive length, although they may differ at various isolated points. A detailed discussion of this result (Lerch's theorem) would take us too far afield.

‡ This can always be arranged, of course, by multiplying and dividing the transform by s; that is, $\Phi(s) \equiv s\Phi(s)/s = \phi(s)/s$.

EXAMPLE 2

What is

$$\mathcal{L}^{-1}\left(\frac{1}{s(s^2 + 4)}\right)$$

Here, using the last theorem, we first suppress the factor $1/s$, getting

$$\phi(s) = \frac{1}{s^2 + 4}$$

By Formula 3, Sec. 7.3, the inverse of this is $F(t) = \frac{1}{2}\sin 2t$. Finally, we obtain $f(t)$ by integrating $F(t)$ from 0 to t:

$$f(t) = \int_0^t \frac{\sin 2t}{2}\, dt = -\frac{\cos 2t}{4}\Big|_0^t = \frac{1 - \cos 2t}{4}$$

One of the most useful properties of the Laplace transformation is contained in the so-called **first shifting theorem**.

THEOREM 5 $\mathcal{L}\{e^{-at}f(t)\} = \mathcal{L}\{f(t)\}_{s\to s+a}$

Proof   By definition,

$$\mathcal{L}\{e^{-at}f(t)\} = \int_0^\infty e^{-at}f(t)e^{-st}\, dt = \int_0^\infty f(t)e^{-(s+a)t}\, dt$$

and the last integral is in structure exactly the Laplace transform of $f(t)$ itself, except that $s + a$ takes the place of s.

In words, Theorem 5 says that *the transform of e^{-at} times a function of t is equal to the transform of the function itself, with s replaced by $s + a$.* Conversely, as a tool for finding inverses this theorem states that if we reverse the substitution $s \to s + a$, that is, if we replace $s + a$ by s or s by $s - a$ in the transform of a function, then the inverse of the modified transform $\phi(s - a)$ must be multiplied by e^{-at} to obtain the inverse of the original transform $\phi(s)$. This procedure is summarized in the following result.

COROLLARY 1 $\mathcal{L}^{-1}\{\phi(s)\} = e^{-at}\mathcal{L}^{-1}\{\phi(s - a)\}$

By means of Theorem 5 we can easily establish the following important formulas.

FORMULA 1 $\mathcal{L}\{e^{-at}\cos bt\} = \dfrac{s + a}{(s + a)^2 + b^2}$

FORMULA 2 $\mathcal{L}\{e^{-at}\sin bt\} = \dfrac{b}{(s + a)^2 + b^2}$

FORMULA 3 $\mathcal{L}\{e^{-at}t^n\} = \begin{cases} \dfrac{\Gamma(n + 1)}{(s + a)^{n+1}} & n > -1 \\[3mm] \dfrac{n!}{(s + a)^{n+1}} & n \text{ a positive integer} \end{cases}$

EXAMPLE 3

If
$$\mathcal{L}\{y\} = \frac{2s + 5}{s^2 + 4s + 13}$$

what is y?

By obvious manipulations we obtain

$$\mathcal{L}\{y\} = \frac{2(s + 2) + 1}{(s + 2)^2 + 3^2} = 2\frac{s + 2}{(s + 2)^2 + 3^2} + \frac{1}{3}\frac{3}{(s + 2)^2 + 3^2}$$

Hence, by Formulas 1 and 2,

$$y = 2e^{-2t}\cos 3t + \tfrac{1}{3}e^{-2t}\sin 3t$$

EXAMPLE 4

What is the solution of the differential equation

$$y'' + 2y' + y = te^{-t}$$

for which $y_0 = 1$ and $y_0' = -2$?

Transforming both sides of the given equation, we have

$$(s^2\mathcal{L}\{y\} - s + 2) + 2(s\mathcal{L}\{y\} - 1) + \mathcal{L}\{y\} = \frac{1}{(s + 1)^2}$$

$$(s^2 + 2s + 1)\mathcal{L}\{y\} = \frac{1}{(s + 1)^2} + s$$

$$\mathcal{L}\{y\} = \frac{1}{(s + 1)^4} + \frac{s}{(s + 1)^2}$$

By Formula 3, the inverse of the first fraction in $\mathcal{L}\{y\}$ is

$$\frac{t^3 e^{-t}}{3!}$$

To find the inverse of the second fraction we can write it in the form

$$\frac{s + 1 - 1}{(s + 1)^2} = \frac{1}{s + 1} - \frac{1}{(s + 1)^2}$$

and take the inverse of each term, or we can suppress the factor s, take the inverse of what remains, and differentiate this result, according to Theorem 3. By either method we obtain immediately $e^{-t} - te^{-t}$. Hence

$$y = \frac{t^3 e^{-t}}{3!} + e^{-t} - te^{-t}$$

In this example the characteristic equation of the differential equation has repeated roots, and moreover the term on the right is a part of the complementary function; yet neither of these features requires any special treatment in the operational solution of the problem. This is another of the many advantages of the Laplace transform method of solving linear differential equations with constant coefficients.

In some problems a system which becomes active at $t = 0$ because of some initial disturbance, is subsequently acted upon by another disturbance beginning at a later time, say $t = a$. The analytical representation of such functions and the nature of their Laplace transforms are therefore matters of some importance. To illustrate, suppose that we wish an expression describing the function whose graph is shown in

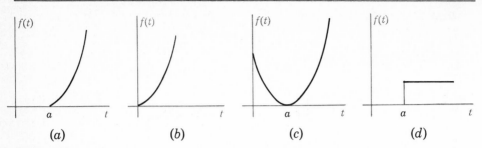

Figure 7.4
Plot describing the graph of a function which has been translated and cut off.

Fig. 7.4*a*, the curve being congruent to the right half of the parabola $y = t^2$ shown in Fig. 7.4*b*. It is not enough to recall the translation formula from analytic geometry and write $f(t) = (t - a)^2$, because this equation, even with the usual qualification that $f(t) \equiv 0$ for $t < 0$, defines the curve shown in Fig. 7.4*c* and not the required graph. However, if we take the unit step function and translate it a units to the right by writing $u(t - a)$, we obtain the function shown in Fig. 7.4*d*. Since this vanishes for $t < a$ and is equal to 1 for $t > a$, the product $(t - a)^2 u(t - a)$ will be identically zero for $t < a$ and will be identically equal to $(t - a)^2$ for $t > a$ and hence will define precisely the arc we want. More generally, the expression

$$f(t - a)u(t - a)$$

represents the function obtained by translating $f(t)$ a units to the right and cutting it off, i.e., making it vanish identically to the left of a.

EXAMPLE 5

What is the equation of the function whose graph is shown in Fig. 7.5*a*?

Clearly we can regard this function as the sum of the two translated step functions shown in Fig. 7.5*b*. Hence its equation is

$$u(t - a) - u(t - b)$$

Although the function shown in Fig. 7.5*a* is not ordinarily given a name, it could appropriately be referred to as a **filter function**; for when any other function is multiplied by this filter function, it is annihilated completely, i.e., reduced identically to zero, outside the "passband" $a < t < b$, and reproduced without any change whatsoever for values of t within the passband.

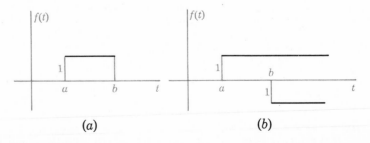

Figure 7.5
The construction of a rectangular pulse, or filter function, from two step functions.

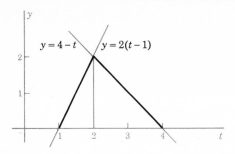

Figure 7.6
A graph consisting of straight-line segments.

EXAMPLE 6

What is the equation of the function whose graph is shown in Fig. 7.6?

To obtain the segment of this function between 1 and 2 we must multiply the expression $2(t - 1)$ by a factor which will be zero to the left of 1, unity between 1 and 2, and zero to the right of 2. By Example 5, such a function is $u(t - 1) - u(t - 2)$. Hence

$$2(t - 1)[u(t - 1) - u(t - 2)]$$

defines the segment of the given function between 1 and 2 and vanishes elsewhere. Similarly

$$(-t + 4)[u(t - 2) - u(t - 4)]$$

defines the segment of the given function between 2 and 4 and vanishes elsewhere. The complete representation of the function is therefore

$$2(t - 1)[u(t - 1) - u(t - 2)] + (-t + 4)[u(t - 2) - u(t - 4)]$$

$$= 2(t - 1)u(t - 1) - 3(t - 2)u(t - 2) + (t - 4)u(t - 4)$$

The transforms of functions that have been translated and cut off are given by the so-called **second shifting theorem.**

THEOREM 6 $\mathscr{L}\{f(t - a)u(t - a)\} = e^{-as}\mathscr{L}\{f(t)\}$ $a \geq 0$

Proof   To prove this, we have by definition

$$\mathscr{L}\{f(t - a)u(t - a)\} = \int_0^\infty f(t - a)u(t - a)e^{-st}\, dt = \int_a^\infty f(t - a)e^{-st}\, dt$$

the last step following since the integration effectively starts not at $t = 0$ but at $t = a$ because $f(t - a)u(t - a)$ vanishes identically to the left of $t = a$. Now let $t - a = T$ and $dt = dT$. Then the last integral becomes

$$\int_0^\infty f(T)e^{-s(T+a)}\, dT = e^{-as}\int_0^\infty f(T)e^{-sT}\, dT = e^{-as}\mathscr{L}\{f(t)\}$$

as asserted.

Before Theorem 6 can be applied, the function being transformed must be expressed in terms of the binomial argument which appears in the unit step function. This will not often be the case; so it will frequently be necessary to alter the form of the function, as originally given, before it can be conveniently transformed. In many cases this can be done by inspection. On the other hand, we can always proceed in the following systematic way. Suppose we wish to transform

$$f(t)u(t - a)$$

As it stands, this cannot be handled by Theorem 6; so we rewrite it in the form

$$f[(t - a) + a]u(t - a) \equiv F(t - a)u(t - a)$$

where, by definition, $F(t - a) = f[(t - a) + a] = f(t)$, or

$$F(t) = f(t + a)$$

Theorem 6 can now be applied, and we have

$$\mathscr{L}\{f(t)u(t - a)\} = \mathscr{L}\{F(t - a)u(t - a)\} = e^{-as}\mathscr{L}\{F(t)\} = e^{-as}\mathscr{L}\{f(t + a)\}$$

Thus we have established the following useful result.

COROLLARY 1 $\mathscr{L}\{f(t)u(t - a)\} = e^{-as}\mathscr{L}\{f(t + a)\}$ $a > 0$

As a tool for finding inverses, it is convenient to restate Theorem 6 in the following form.

COROLLARY 2   If $\mathscr{L}^{-1}\{\phi(s)\} = f(t)$, then $\mathscr{L}^{-1}\{e^{-as}\phi(s)\} = f(t - a)u(t - a)$.

In words, Corollary 2 states that *suppressing the factor e^{-as} in a transform requires that the inverse of what remains be translated a units to the right and cut off to the left of the point $t = a$.*

EXAMPLE 7

What is the transform of the function whose graph is shown in Fig. 7.7?
The equation of this function is obviously

$$g(t) = -(t^2 - 3t + 2)[u(t - 1) - u(t - 2)]$$
$$= -f(t)u(t - 1) + f(t)u(t - 2)$$

where $f(t) = t^2 - 3t + 2$. However, the form of $f(t)$ is such that Theorem 6 cannot be applied directly to either term in the expression for $g(t)$. Hence we use Corollary 1, observing that

$$f(t + 1) = (t + 1)^2 - 3(t + 1) + 2 = t^2 - t$$

and $$f(t + 2) = (t + 2)^2 - 3(t + 2) + 2 = t^2 + t$$

The required transform is therefore

$$-e^{-s}\mathscr{L}\{t^2 - t\} + e^{-2s}\mathscr{L}\{t^2 + t\} = -e^{-s}\left(\frac{2}{s^3} - \frac{1}{s^2}\right) + e^{-2s}\left(\frac{2}{s^3} + \frac{1}{s^2}\right)$$

Figure 7.7
A parabolic pulse.

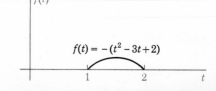

$$f(t) = -(t^2 - 3t + 2)$$

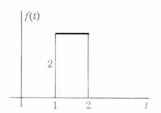

Figure 7.8
A rectangular pulse.

EXAMPLE 8

Find the solution of the equation $y' + 3y + 2\int_0^t y\, dt = f(t)$ for which $y_0 = 1$ if $f(t)$ is the function whose graph is shown in Fig. 7.8.

In this case $f(t) = 2u(t - 1) - 2u(t - 2)$, and thus the differential equation can be written

$$y' + 3y + 2\int_0^t y\, dt = 2u(t - 1) - 2u(t - 2)$$

Taking transforms, we have

$$(s\mathscr{L}\{y\} - 1) + 3\mathscr{L}\{y\} + \frac{2}{s}\mathscr{L}\{y\} = \frac{2e^{-s}}{s} - \frac{2e^{-2s}}{s}$$

or

$$(s^2 + 3s + 2)\mathscr{L}\{y\} = 2e^{-s} - 2e^{-2s} + s$$

and

$$\mathscr{L}\{y\} = \frac{s}{(s + 1)(s + 2)} + \frac{2e^{-s}}{(s + 1)(s + 2)} - \frac{2e^{-2s}}{(s + 1)(s + 2)}$$

The first term in $\mathscr{L}\{y\}$ can be written

$$\frac{2}{s + 2} - \frac{1}{s + 1}$$

Hence its inverse is $2e^{-2t} - e^{-t}$. If the exponential factors are suppressed in the second and third terms in $\mathscr{L}\{y\}$, the algebraic portion which remains can be written

$$2\left(\frac{1}{s + 1} - \frac{1}{s + 2}\right)$$

and the inverse of this is $2e^{-t} - 2e^{-2t}$. However, because the factors e^{-s} and e^{-2s} were neglected, it is necessary to take the last expression, translate it 1 unit to the right and cut it off to the left of $t = 1$, and also translate it 2 units to the right and cut it off to the left of $t = 2$ in order to obtain the inverses of the original terms. This gives for y

$$y = (2e^{-2t} - e^{-t}) + 2(e^{-(t-1)} - e^{-2(t-1)})u(t - 1) - 2(e^{-(t-2)} - e^{-2(t-2)})u(t - 2)$$

Plots of these three terms and of their sum, that is, y itself, are shown in Fig. 7.9.

We have already made repeated use of Theorems 2 and 3, Sec. 7.2, on the transforms of derivatives and integrals. On the other hand, it is sometimes convenient or necessary to consider the derivatives and integrals of transforms. The basis for this is contained in the next two theorems.

THEOREM 7 If $f(t)$ is piecewise regular and of exponential order, and if $\mathscr{L}\{f(t)\} = \phi(s)$, then $\mathscr{L}\{tf(t)\} = -\phi'(s)$.

Proof By definition we have

$$\mathscr{L}\{f(t)\} = \int_0^\infty f(t)e^{-st}\, dt$$

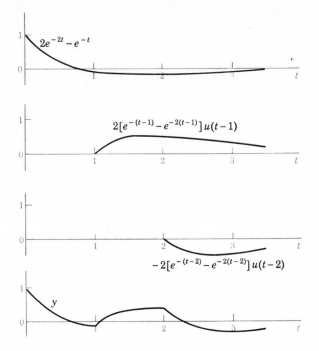

Figure 7.9
The solution of Example 8.

and differentiating this with respect to s, we obtain

$$\frac{d}{ds} \int_0^\infty f(t) e^{-st}\, dt = \phi'(s)$$

Now under our usual assumption that $f(t)$ is piecewise regular and of exponential order, the product $tf(t)$ also satisfies these conditions. Hence by Theorem 6, Sec. 7.1, the integral which results when $\mathscr{L}\{f(t)\}$ is differentiated partially with respect to s, namely,

$$\int_0^\infty -tf(t) e^{-st}\, dt$$

converges uniformly. Therefore according to Theorem 3, Sec. 7.1, the integral for $\mathscr{L}\{f(t)\}$ can legitimately be differentiated with respect to s inside the integral sign. Thus, performing the differentiation, we have

$$\phi'(s) = \int_0^\infty f(t)(-t e^{-st})\, dt$$

or

$$\int_0^\infty [tf(t)] e^{-st}\, dt \equiv \mathscr{L}\{tf(t)\} = -\phi'(s)$$

as asserted.

By taking inverses in the assertion of Theorem 7 and then solving for $f(t)$ we obtain the following useful result.

COROLLARY 1   If $\mathscr{L}\{f(t)\} = \phi(s)$, then

$$f(t) \equiv \mathscr{L}^{-1}\{\phi(s)\} = -\frac{1}{t}\, \mathscr{L}^{-1}\{\phi'(s)\}$$

Corollary 1 is often helpful when the inverse of a transform cannot conveniently be found but the inverse of the derivative of the transform is known. The extension of Theorem 7 and its corollary to repeated differentiation of transforms is obvious.

EXAMPLE 9

What is $\mathscr{L}\{t^2 \sin 2t\}$?

By a repeated application of Theorem 7, we have

$$\mathscr{L}\{t^2 \sin 2t\} = (-1)^2 \frac{d^2 \mathscr{L}\{\sin 2t\}}{ds^2} = \frac{d^2}{ds^2}\left(\frac{2}{s^2 + 4}\right) = \frac{12s^2 - 16}{(s^2 + 4)^3}$$

EXAMPLE 10

What is y if $\mathscr{L}\{y\} = \ln[(s + 1)/(s - 1)]$?

Using Corollary 1 of Theorem 7, we have immediately

$$y = -\frac{1}{t}\mathscr{L}^{-1}\left\{\frac{d}{ds}\left(\ln\frac{s + 1}{s - 1}\right)\right\} = -\frac{1}{t}\mathscr{L}^{-1}\left\{\frac{1}{s + 1} - \frac{1}{s - 1}\right\}$$

$$= \frac{e^{-t} - e^t}{-t} = \frac{2\sinh t}{t}$$

THEOREM 8　If $f(t)$ is piecewise regular and of exponential order, if $\mathscr{L}\{f(t)\} = \phi(s)$, and if $f(t)/t$ has a limit as t approaches zero from the right, then

$$\mathscr{L}\left\{\frac{f(t)}{t}\right\} = \int_s^\infty \phi(s)\, ds$$

Proof　By definition

$$\phi(s) = \mathscr{L}\{f(t)\} = \int_0^\infty f(t)e^{-st}\, dt$$

Hence, integrating from s to ∞, we obtain

$$\int_s^\infty \phi(s)\, ds = \int_s^\infty\left[\int_0^\infty f(t)e^{-st}\, dt\right] ds$$

Now under the assumption that $\lim_{t\to 0^+}[f(t)/t]$ exists and that $f(t)$ itself is piecewise regular and of exponential order, it follows from Theorems 6 and 2, Sec. 7.1, that the integration with respect to s can be performed inside the t integral, i.e., that the order of integration in the repeated integral can be reversed. Hence, performing the integration, we have

$$\int_s^\infty \phi(s)\, ds = \int_0^\infty \int_s^\infty f(t)e^{-st}\, ds\, dt = \int_0^\infty f(t)\left[\frac{e^{-st}}{-t}\right]_s^\infty dt$$

$$= \int_0^\infty \frac{f(t)}{t} e^{-st}\, dt$$

$$= \mathscr{L}\left\{\frac{f(t)}{t}\right\}$$

as asserted.

By taking inverses in the assertion of Theorem 8 and then solving for $f(t)$, we obtain the following useful result.

COROLLARY 1   If $\mathscr{L}\{f(t)\} = \phi(s)$, then

$$f(t) \equiv \mathscr{L}^{-1}\{\phi(s)\} = t\mathscr{L}^{-1}\left\{\int_s^\infty \phi(s)\,ds\right\}$$

Corollary 1 is often useful in finding inverses when the integral of a transform is simpler to work with than the transform itself. The extension of Theorem 8 and its corollary to repeated integration of transforms is immediate.

EXAMPLE 11

What is $\mathscr{L}\{(\sin kt)/t\}$?
By Theorem 8, we have

$$\mathscr{L}\left\{\frac{\sin kt}{t}\right\} = \int_s^\infty \mathscr{L}\{\sin kt\}\,ds = \int_s^\infty \frac{k}{s^2 + k^2}\,ds = \left.\mathrm{Tan}^{-1}\frac{s}{k}\right|_s^\infty$$

$$= \frac{\pi}{2} - \mathrm{Tan}^{-1}\frac{s}{k} = \mathrm{Cot}^{-1}\frac{s}{k}$$

EXAMPLE 12

What is y if $\mathscr{L}\{y\} = s/(s^2 - 1)^2$?
Using Corollary 1, Theorem 8, we have immediately

$$y = t\mathscr{L}^{-1}\left\{\int_s^\infty \frac{s}{(s^2 - 1)^2}\,ds\right\} = t\mathscr{L}^{-1}\left\{\left.\frac{-1}{2(s^2 - 1)}\right|_s^\infty\right\}$$

$$= t\mathscr{L}^{-1}\left\{\frac{1}{2(s^2 - 1)}\right\}$$

$$= t\mathscr{L}^{-1}\left\{\frac{1}{4}\left(\frac{1}{s - 1} - \frac{1}{s + 1}\right)\right\}$$

$$= \frac{t}{4}(e^t - e^{-t}) = \frac{t \sinh t}{2}$$

EXERCISES

Find the Laplace transform of each of the following functions:

1 $u(t - a)$

2 $\cos(t - 1)u(t - 1)$

3 $u(1 - e^{-t})$

4 $u(t^3 - 1)$

5 $t^2 u(t - 2)$

6 $(t^2 - 1)u(t - 1)$

7 $\cos t\, u(t - 1)$

8 $e^{2t}u(t - 2)$

9 $f(t) = \begin{cases} \sin t & 0 < t < \pi \\ 0 & \pi < t \end{cases}$

10 $f(t) = \begin{cases} t & 0 < t < 2 \\ 2 & 2 < t \end{cases}$

11 The function graphed in Fig. 7.10.

Figure 7.10

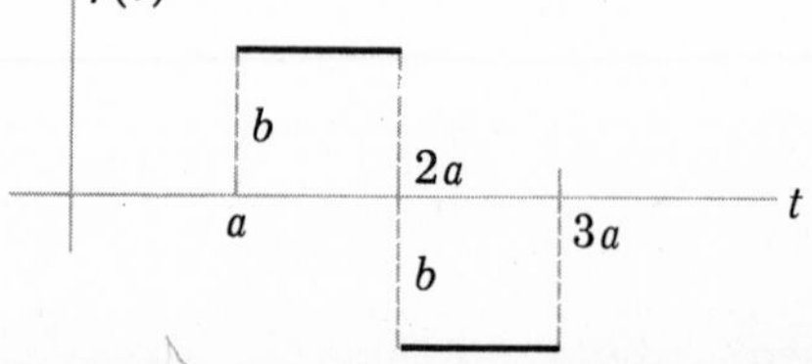

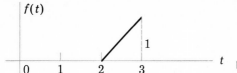

Figure 7.11

12 The function graphed in Fig. 7.11.

13 $te^{-3t} \sin 2t$

14 $t \int_0^t e^{-3t} \sin 2t \, dt$

15 $e^{-3t} \int_0^t t \sin 2t \, dt$

16 $\int_0^t te^{-3t} \sin 2t \, dt$

17 $\dfrac{1 - \cos 3t}{t}$

18 $\dfrac{e^{2t} - 1}{t}$

19 $\dfrac{e^{-3t} \sin 2t}{t}$

20 $\int_0^t \dfrac{e^t - \cos 2t}{t} \, dt$

21 $e^{-3t} \int_0^t \dfrac{\sin 2t}{t} \, dt$

22 $\int_0^t \dfrac{e^{-3t} \sin 2t}{t} \, dt$

Find the inverse of each of the following transforms:

23 $\dfrac{1}{(s + 2)^4}$

24 $\dfrac{s}{(s + 2)^4}$

25 $\dfrac{s + 1}{9s^2 + 6s + 5}$

26 $\dfrac{1}{s(s + 2)^2}$

27 $\dfrac{1}{s^2(s + 1)}$

28 $\dfrac{1}{(s + 1)(s^2 + 2s + 5)}$

29 $\dfrac{1}{s^4 + 5s^2 + 4}$

30 $\dfrac{e^{-2s}}{s^2 + 4}$

31 $\dfrac{e^{-3s}}{s^2 - 9}$

32 $\dfrac{e^{-s}}{(s + 1)^3}$

33 $\dfrac{e^{-s} + e^{-2s}}{s^2 - 3s + 2}$

34 $\ln \dfrac{s + a}{s + b}$

35 $\ln \dfrac{s^2 - 1}{s^2}$

36 $\ln \dfrac{s^2 + 1}{s(s + 1)}$

37 $s \ln [(s - 1)/(s + 1)] + 2$. What role, if any, does the additive constant 2 play in this problem?

38 $\dfrac{2}{(s^2 + 4)^2}$ *Hint:* Multiply and divide the transform by s.

39 $\dfrac{s + 2}{(s^2 + 4s + 5)^2}$

40 $\dfrac{2s + 3}{(s^2 + 3s + 2)^2}$

41 $\dfrac{1}{s} \mathrm{Tan}^{-1} \dfrac{1}{s}$

42 $\mathrm{Tanh}^{-1} \dfrac{1}{s}$

43 $\dfrac{1}{(s^2 + 2s + 2)^2}$

44 $\dfrac{s}{(s^2 + a^2)^3}$

Find the solution of each of the following differential equations which satisfies the indicated conditions:

45 $y'' + 4y' + 3y = e^{-t}$ $y_0 = y_0' = 1$

46 $y'' + 4y = \cos 2t$ $y_0 = -2, y_0' = 1$

47 $y'' + 3y' + 2y = u(t - 1)$ $y_0 = 0, y_0' = 1$

48 $y'' + 4y' + 4y = (t - 2)e^{-(t-2)}u(t - 2)$ $y_0 = 1, y'_0 = -1$

49 $y^{iv} + 2y'' + y = 0$ $y_0 = y'_0 = y''_0 = 0, y'''_0 = 1$

50   Find the values of $f(0^+)$ and of $\lim_{t \to \infty} f(t)$, if it exists, if $\mathscr{L}\{f(t)\}$ is:

(a) $\dfrac{s^2 + 1}{s^3 + 6s^2 + 11s + 6}$

(b) $\dfrac{s + 3}{2s^3 - 3s^2 - 2s}$

(c) $\dfrac{s^2 + s + 1}{s^3 - s^2 + 2}$

(d) $\dfrac{s + 2}{s^3 + 3s^2 + 4s + 2}$

51   Show that under appropriate conditions

$$\lim_{s \to \infty} s[s\mathscr{L}\{f(t)\} - f(0^+)] = f'(0^+)$$

and that $$\lim_{s \to \infty} s[s^2\mathscr{L}\{f(t)\} - sf(0^+) - f'(0^+)] = f''(0^+)$$

What conditions beyond those of Theorem 1 are necessary for the validity of these results? Can the value of $f^{(n)}(0^+)$ be obtained by an extension of these formulas?

52   Show that under appropriate conditions

$$\lim_{s \to 0} s[s\mathscr{L}\{f(t)\} - f(0^+)] = \lim_{t \to \infty} f'(t)$$

What conditions beyond those of Theorem 2 are necessary for the validity of this result? Can this result be generalized to the determination of $\lim_{t \to \infty} f^{(n)}(t)$ from $\mathscr{L}\{f(t)\}$?

53   What is $\displaystyle\int_0^\infty \frac{e^{-t} - e^{-2t}}{t} \, dt$? *Hint:* Use Theorem 1, Sec. 7.1, to justify the observation that

$$\int_0^\infty \frac{e^{-t} - e^{-2t}}{t} \, dt = \int_0^\infty \frac{e^{-t} - e^{-2t}}{t} \left(\lim_{s \to 0} e^{-st} \right) dt = \lim_{s \to 0} \mathscr{L} \left\{ \frac{e^{-t} - e^{-2t}}{t} \right\}$$

54   What is $\displaystyle\int_0^\infty e^{-t} \frac{1 - \cos t}{t} \, dt$?

55   What is $\displaystyle\int_0^\infty \frac{e^{-at} \cos bt - e^{-pt} \cos qt}{t} \, dt$?

56   (a) Prove Theorem 1 without assuming that $f(t)$ is continuous. *Hint:* Use the result of Exercise 3, Sec. 7.2.

(b) Prove Theorem 2 without assuming that $f(t)$ is continuous.

(c) Where in the proof of Theorem 2 is use made of the hypothesis that the abscissa of convergence of $f(t)$ is negative?

57   Use the Laplace transformation to solve the variable-coefficient linear differential equation

$$ty'' + 2(t - 1)y' + (t - 2)y = 0$$

Hint: Use Theorem 7 in transforming the equation. Then note that the result is a linear first-order differential equation in $\mathscr{L}\{y\}$ which can be solved by the method of Sec. 1.6.

58   Using the hint of Exercise 57, solve each of the following differential equations:

(a) $ty'' + 2(2t - 1)y' + 4(t - 1)y = 0$

(b) $ty'' - 2(2t + 1)y' + (3t + 4)y = 0$

(c) $ty'' - 2y' + ty = 0$

(d) $ty'' + 2(2t - 1)y' - 4y = 0$

59   Using the hint of Exercise 57, show that the equation

$$(at + b)y'' + (ct + d)y' + (et + f)y = 0$$

can be solved in terms of elementary functions if

$$ad - bc = 2a\lambda^2 \qquad \text{and} \qquad af - be = ac\lambda$$

where λ is a nonpositive integer.

60 Discuss the following problem as a possible application of the Laplace transformation. The behavior of a certain system is governed by a linear second-order differential equation with constant, though unknown, coefficients. The response of the system to a specific test disturbance, a unit step function, say, can be recorded. Is it possible, using numerical or graphical integration, to calculate the Laplace transform of such a response for several values of s and thus obtain a set of linear algebraic equations from which the coefficients in the differential equation can be obtained?

7.5 The Heaviside Expansion Theorems

The frequent use we have had to make of partial fractions indicates clearly the importance of this technique in operational calculus. It is therefore highly desirable to have the procedure systematized as much as possible. The following theorems, usually associated with the name of Heaviside, are very useful in this connection.

THEOREM 1 If $f(t) = \mathcal{L}^{-1}\{p(s)/q(s)\}$, where $p(s)$ and $q(s)$ are polynomials and the degree of $q(s)$ is greater than the degree of $p(s)$, then the term in $f(t)$ corresponding to an unrepeated linear factor $s - a$ of $q(s)$ is equally well

$$\frac{p(a)}{q'(a)} e^{at} \quad \text{or} \quad \frac{p(a)}{Q(a)} e^{at}$$

where $Q(s)$ is the product of all the factors of $q(s)$ except $s - a$.

Proof In the familiar partial fraction decomposition of $p(s)/q(s)$, an unrepeated linear factor $s - a$ of $q(s)$ gives rise to a single fraction of the form $A/(s - a)$. Hence, if we denote by $h(s)$ the sum of all the fractions corresponding to the other factors of $q(s)$, we can write

$$\frac{p(s)}{q(s)} = \frac{A}{s - a} + h(s)$$

where, since $s - a$ is an *unrepeated* factor of $q(s)$, the term $h(s)$ does not contain $s - a$ as a factor of its denominator and hence remains finite as s approaches a. Multiplying this identity by $s - a$ then gives

$$\frac{(s - a)p(s)}{q(s)} = \frac{p(s)}{q(s)/(s - a)} = A + (s - a)h(s)$$

If we now let s approach a, the second term on the right vanishes and we have

$$A = \lim_{s \to a} \frac{p(s)}{q(s)/(s - a)}$$

The limit of the numerator here is evidently $p(a)$. The denominator appears as an indeterminate of the form $0/0$. However, if we evaluate it as usual according to L'Hospital's rule, by differentiating numerator and denominator with respect to s and then letting s approach a, we obtain just $q'(a)$. Hence

$$A = \frac{p(a)}{q'(a)}$$

On the other hand, we could have eliminated the indeterminacy before passing to the limit simply by canceling $s - a$ into $q(s)$, which by hypothesis contains this factor. Doing this, we obtain the equivalent form of A:

$$A = \frac{p(a)}{Q(a)}$$

Finally, taking inverses, we see that the fraction $A/(s - a)$ gives rise to the term

$$A e^{at} = \frac{p(a)}{q'(a)} e^{at} = \frac{p(a)}{Q(a)} e^{at}$$

in the inverse $f(t)$, as asserted.

If $q(s)$ contains only unrepeated linear factors, then by applying Theorem 1 to each factor in turn, we obtain the following useful result.

COROLLARY 1   If $f(t) = \mathcal{L}^{-1}\{p(s)/q(s)\}$, and if $q(s)$ is completely factorable into unrepeated real linear factors

$$(s - a_1), (s - a_2), \ldots, (s - a_n)$$

then

$$f(t) = \sum_{i=1}^{n} \frac{p(a_i)}{q'(a_i)} e^{a_i t} = \sum_{i=1}^{n} \frac{p(a_i)}{Q_i(a_i)} e^{a_i t}$$

where $Q_i(s)$ is the product of all the factors of $q(s)$ except the factor $s - a_i$.

THEOREM 2   If $f(t) = \mathcal{L}^{-1}\{p(s)/q(s)\}$, where $p(s)$ and $q(s)$ are polynomials and the degree of $q(s)$ is greater than the degree of $p(s)$, then the terms in $f(t)$ corresponding to a repeated linear factor $(s - a)^r$ in $q(s)$ are

$$\left[\frac{\phi^{(r-1)}(a)}{(r - 1)!} + \frac{\phi^{(r-2)}(a)}{(r - 2)!} \frac{t}{1!} + \cdots + \frac{\phi'(a)}{1!} \frac{t^{r-2}}{(r - 2)!} + \phi(a) \frac{t^{r-1}}{(r - 1)!} \right] e^{at}$$

where $\phi(s)$ is the quotient of $p(s)$ and all the factors of $q(s)$ except $(s - a)^r$.

Proof   From the familiar theory of partial fractions we recall that a repeated linear factor $(s - a)^r$ of $q(s)$ gives rise to the component fractions

$$\frac{A_1}{s - a} + \frac{A_2}{(s - a)^2} + \cdots + \frac{A_{r-1}}{(s - a)^{r-1}} + \frac{A_r}{(s - a)^r}$$

If, as before, we let $h(s)$ denote the sum of the fractions corresponding to all the other factors of $q(s)$, we have

$$\frac{p(s)}{q(s)} \equiv \frac{\phi(s)}{(s - a)^r} = \frac{A_1}{s - a} + \frac{A_2}{(s - a)^2} + \cdots + \frac{A_{r-1}}{(s - a)^{r-1}} + \frac{A_r}{(s - a)^r} + h(s)$$

Multiplying this identity by $(s - a)^r$ gives

$$\phi(s) = A_1(s - a)^{r-1} + A_2(s - a)^{r-2} + \cdots + A_{r-1}(s - a) + A_r + (s - a)^r h(s)$$

If we put $s = a$ in this identity, we obtain

$$\phi(a) = A_r$$

If we now differentiate $\phi(s)$, we have

$$\phi'(s) = A_1(r - 1)(s - a)^{r-2} + A_2(r - 2)(s - a)^{r-3} + \cdots$$
$$+ A_{r-1} + [r(s - a)^{r-1}h(s) + (s - a)^r h'(s)]$$

Again setting $s = a$, we find this time

$$\phi'(a) = A_{r-1}$$

Continuing this process of differentiation and evaluation, and noting that the first $r - 1$ derivatives of the product $(s - a)^r h(s)$ will all vanish when $s = a$, we obtain successively

$$\phi''(a) = 2! \, A_{r-2}$$
$$\phi'''(a) = 3! \, A_{r-3}$$
$$\cdots\cdots\cdots\cdots\cdots$$
$$\phi^{(r-1)}(a) = (r - 1)! \, A_1$$

or
$$A_{r-k} = \frac{\phi^{(k)}(a)}{k!} \qquad k = 0, 1, 2, \ldots, r - 1$$

The terms in the expansion of $p(s)/q(s)$ which correspond to the factor $(s - a)^r$ are therefore

$$\frac{\phi^{(r-1)}(a)}{(r - 1)!} \frac{1}{s - a} + \frac{\phi^{(r-2)}(a)}{(r - 2)!} \frac{1}{(s - a)^2} + \cdots + \frac{\phi'(a)}{1!} \frac{1}{(s - a)^{r-1}} + \phi(a) \frac{1}{(s - a)^r}$$

When we recall that

$$\mathscr{L}^{-1}\left\{\frac{1}{(s - a)^n}\right\} = \frac{t^{n-1}e^{at}}{(n - 1)!}$$

it is evident that the terms in $f(t)$ which arise from these fractions are

$$\frac{\phi^{(r-1)}(a)}{(r - 1)!} e^{at} + \frac{\phi^{(r-2)}(a)}{(r - 2)!} \frac{te^{at}}{1!} + \cdots + \frac{\phi'(a)}{1!} \frac{t^{r-2}e^{at}}{(r - 2)!} + \phi(a) \frac{t^{r-1}e^{at}}{(r - 1)!}$$

Finally, if we factor out e^{at} from this expression, we have precisely the assertion of the theorem.

THEOREM 3   If $f(t) = \mathscr{L}^{-1}\{p(s)/q(s)\}$, where $p(s)$ and $q(s)$ are polynomials and the degree of $q(s)$ is greater than the degree of $p(s)$, then the terms in $f(t)$ which correspond to an unrepeated, irreducible quadratic factor $(s + a)^2 + b^2$ of $q(s)$ are

$$\frac{e^{-at}}{b} (\phi_i \cos bt + \phi_r \sin bt)$$

where ϕ_r and ϕ_i are, respectively, the real and the imaginary parts of $\phi(-a + ib)$ and where $\phi(s)$ is the quotient of $p(s)$ and all the factors of $q(s)$ except the factor $(s + a)^2 + b^2$.

Proof From the familiar theory of partial fractions, we recall that an unrepeated irreducible quadratic factor $(s + a)^2 + b^2$ of $q(s)$ gives rise to a single fraction of the form

$$\frac{As + B}{(s + a)^2 + b^2}$$

in the partial-fraction expansion of $p(s)/q(s)$. If again we let $h(s)$ denote the sum of the fractions corresponding to all the other factors of $q(s)$, we can therefore write

$$\frac{p(s)}{q(s)} \equiv \frac{\phi(s)}{(s + a)^2 + b^2} = \frac{As + B}{(s + a)^2 + b^2} + h(s)$$

Multiplying this identity by $(s + a)^2 + b^2$, we obtain

$$\phi(s) = As + B + [(s + a)^2 + b^2]h(s)$$

Now put $s = -a + ib$. This value, of course, makes $(s + a)^2 + b^2$ vanish; hence the last product drops out, leaving

$$\phi(-a + ib) = (-a + ib)A + B$$

or, reducing $\phi(-a + ib)$ to its standard complex form $\phi_r + i\phi_i$,

$$\phi_r + i\phi_i = (-aA + B) + ibA$$

Equating real and imaginary terms, respectively, in this equality, we find

$$\phi_r = -aA + B \qquad \text{and} \qquad \phi_i = bA$$

or, solving for A and B,

$$A = \frac{\phi_i}{b} \qquad B = \frac{b\phi_r + a\phi_i}{b}$$

Thus the partial fraction which corresponds to the quadratic factor $(s + a)^2 + b^2$ is

$$\frac{As + B}{(s + a)^2 + b^2} = \frac{1}{b}\frac{\phi_i s + (b\phi_r + a\phi_i)}{(s + a)^2 + b^2} = \frac{1}{b}\left[\frac{(s + a)\phi_i}{(s + a)^2 + b^2} + \frac{b\phi_r}{(s + a)^2 + b^2}\right]$$

By Formulas 1 and 2, Sec. 7.4, the inverse of the last expression is

$$\frac{1}{b}(\phi_i e^{-at} \cos bt + \phi_r e^{-at} \sin bt)$$

Finally, factoring out e^{-at}, we have the assertion of the theorem.

There is a fourth theorem dealing with repeated irreducible quadratic factors but because of its complexity and limited usefulness, we shall not develop it here. Fortunately, many of the simpler transforms involving repeated quadratic factors can be handled by other means, e.g., the convolution theorem of Sec. 7.7.

EXAMPLE 1

If $\mathscr{L}\{f(t)\} = (s^2 + 2)/s(s + 1)(s + 2)$, what is $f(t)$?

The roots of the denominator in this case are $s = 0, -1, -2$. Hence we must compute the values of

$$p(s) = s^2 + 2 \qquad \text{and} \qquad q'(s) = 3s^2 + 6s + 2$$

for these values of s. The results are

$$p(0) = 2 \qquad p(-1) = 3 \qquad p(-2) = 6$$
$$q'(0) = 2 \qquad q'(-1) = -1 \qquad q'(-2) = 2$$

From the corollary of Theorem 1 we now have at once

$$f(t) = \frac{2}{2} e^{0t} + \frac{3}{-1} e^{-t} + \frac{6}{2} e^{-2t} = 1 - 3e^{-t} + 3e^{-2t}$$

Equally well, of course, we could have obtained the coefficients in the inverse by suppressing, in turn, each of the factors of the denominator and evaluating the rest of the fraction at the root corresponding to the suppressed factor. In particular, we note that

$$Q_1(0) = (s + 1)(s + 2)\Big|_{s=0} = 2 = q'(0)$$

$$Q_2(-1) = s(s + 2)\Big|_{s=-1} = -1 = q'(-1)$$

$$Q_3(-2) = s(s + 1)\Big|_{s=-2} = 2 = q'(-2)$$

EXAMPLE 2

If $\mathcal{L}\{y\} = s/(s + 2)^2(s^2 + 2s + 10)$, what is y?

Considering first the repeated linear factor, we identify

$$\phi(s) = \frac{s}{s^2 + 2s + 10} \qquad \text{and} \qquad \phi'(s) = \frac{-s^2 + 10}{(s^2 + 2s + 10)^2}$$

Evaluating these for the root $s = -2$, we obtain

$$\phi(-2) = -\tfrac{1}{5} \qquad \text{and} \qquad \phi'(-2) = \tfrac{3}{50}$$

Hence, by Theorem 2, the terms in y corresponding to $(s + 2)^2$ are

$$\left(\frac{3}{50} - \frac{t}{5}\right) e^{-2t} = \frac{(3 - 10t)e^{-2t}}{50}$$

For the quadratic factor $s^2 + 2s + 10 \equiv (s + 1)^2 + 3^2$, we have

$$\phi(s) = \frac{s}{(s + 2)^2}$$

Hence $$\phi(-a + ib) = \phi(-1 + 3i) = \frac{-1 + 3i}{[(-1 + 3i) + 2]^2}$$

$$= \frac{-1 + 3i}{(1 + 3i)^2} = \frac{-1 + 3i}{-8 + 6i} = \frac{13 - 9i}{50}$$

and thus $\phi_r = \tfrac{13}{50}$, $\phi_i = -\tfrac{9}{50}$. The term in y corresponding to the factor $s^2 + 2s + 10$ is therefore

$$\frac{1}{3} \frac{e^{-t}(-9 \cos 3t + 13 \sin 3t)}{50}$$

Adding the two partial inverses, we have finally

$$y = \frac{(3 - 10t)e^{-2t}}{50} + \frac{e^{-t}(-9 \cos 3t + 13 \sin 3t)}{150}$$

Find the functions which have the following transforms:

1 $\dfrac{s^2 - s + 3}{s^3 + 6s^2 + 11s + 6}$
    2 $\dfrac{s + 2}{(s + 1)(s^2 + 4)}$

3 $\dfrac{s}{(s + 2)^2(s^2 + 1)}$
    4 $\dfrac{s}{(s + 1)(s + 2)^3}$

5 $\dfrac{s}{s^4 - 2s^2 + 1}$
    6 $\dfrac{s + 1}{(s^2 + 1)(s^2 + 4s + 13)}$

7 $\dfrac{s + 2}{s^4 + 4s^3 + 4s^2 - 4s - 5}$
    8 $\dfrac{s^2 + 2}{(s^2 + 4s + 5)(s^2 + 6s + 10)}$

9 $\dfrac{1}{(s^2 - s - 6)^4}$
    10 $\dfrac{s + 2}{s^4 - 16s^2 + 100}$

Solve the following differential equations:

11 $y''' - 2y'' - y' + 2y = u(t - 2)$ $y_0 = y_0' = 0,\ y_0'' = 1$
12 $y''' + 3y'' + 3y' + y = \cosh t$ $y_0 = y_0' = y_0'' = 0$
13 $y^{\mathrm{iv}} + 2y''' + 2y'' + 2y' + y = e^{-t}$ $y_0 = y_0' = y_0'' = y_0''' = 0$

Solve each of the following systems of equations:

14 $x'' + 2x' + \displaystyle\int_0^t y\,dt = 0$

$4x'' - x' + y = e^{-t}$ $x_0 = 0,\ x_0' = 1$

15 $(D^2 + D + 1)x\ + (D - 1)y = u(t)$
$(D^2 + 2D + 3)x + (3D^2 + 4D - 3)y = 0$ $x_0 = x_0' = y_0 = y_0' = 0$

16 $y' - 3z\ = 5$
$y + z' - w = 3 - 2t$ $y_0 = 1,\ z_0 = 0,\ w_0 = -1$
$z + w' = -1$

17 In the proof of Theorem 3, verify that if the identity

$$\phi(s) = As + B + [(s + a)^2 + b^2]h(s)$$

is evaluated for $s = -a - ib$, instead of for $s = -a + ib$, the same inverse is obtained.

18 Find the inverse of $s/[(s + 1)(s^2 + 2s + 5)]$ by factoring the irreducible quadratic factor $s^2 + 2s + 5$† into the unrepeated linear factors $s + 1 + 2i$ and $s + 1 - 2i$ and then applying Theorem 1 to these factors as well as to the real factor $s + 1$. Does your answer agree with the result obtained by using Theorem 3 to handle the quadratic factor? Do you think that this alternative procedure could be used to handle irreducible quadratic factors in general?

19 Using the procedure suggested in Exercise 18, find the inverse of $s/[(s^2 + 4)^2]$. Does your answer agree with the result obtained by using Corollary 1, Theorem 8, Sec. 7.4?

20 If $q(s)$ is a polynomial of degree n containing only unrepeated real linear factors, show that the sum of the numerators of the fractions in the partial fraction decomposition of $p(s)/q(s)$ is equal to the coefficient of s^{n-1} in $p(s)$. Is there a comparable result if $q(s)$ contains only unrepeated real linear factors and unrepeated quadratic factors?

---

† There is no contradiction in factoring an expression previously described as irreducible because *irreducible* means, technically, "having no *real* factors."

7.6 The Transforms of Periodic Functions

The application of the Laplace transformation to the important case of general periodic functions is based upon the following theorem.

THEOREM 1 If $f(t)$ is a piecewise regular function of exponential order which is periodic with period k, then

$$\mathscr{L}\{f(t)\} = \frac{\int_0^k f(t)e^{-st} \, dt}{1 - e^{-ks}}$$

Proof By definition

$$\mathscr{L}\{f(t)\} = \int_0^\infty f(t)e^{-st} \, dt$$

$$= \int_0^k f(t)e^{-st} \, dt + \int_k^{2k} f(t)e^{-st} \, dt + \int_{2k}^{3k} f(t)e^{-st} \, dt + \cdots$$

Now, in the second integral, let $t = T + k$; in the third integral let $t = T + 2k$; and, in general, let $t = T + nk$ in the $(n + 1)$st integral. In each case $dt = dT$, and the new limits become 0 and k. Hence

$$\mathscr{L}\{f(t)\} = \int_0^k f(t)e^{-st} \, dt + \int_0^k f(T + k)e^{-s(T+k)} \, dT$$

$$+ \int_0^k f(T + 2k)e^{-s(T+2k)} \, dT + \cdots$$

$$= \int_0^k f(T)e^{-sT} \, dT + e^{-ks} \int_0^k f(T + k)e^{-sT} \, dT$$

$$+ e^{-2ks} \int_0^k f(T + 2k)e^{-sT} \, dT + \cdots$$

However, $f(T + k) = f(T + 2k) = \cdots = f(T + nk) = \cdots = f(T)$ for all values of T, since, by hypothesis, $f(t)$ is of period k. Thus we have

$$\mathscr{L}\{f(t)\} = \int_0^k f(T)e^{-sT} \, dT + e^{-ks} \int_0^k f(T)e^{-sT} \, dT + e^{-2ks} \int_0^k f(T)e^{-sT} \, dT + \cdots$$

$$= (1 + e^{-ks} + e^{-2ks} + \cdots) \int_0^k f(T)e^{-sT} \, dT$$

Finally, if the infinite geometric progression which multiplies the integral is explicitly summed, using the familiar formula $S = 1/(1 - r)$, where the common ratio r is e^{-ks}, we obtain the result of the theorem.

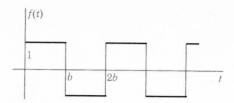

Figure 7.12
An alternating rectangular wave.

EXAMPLE 1

Find the transform of the rectangular wave shown in Fig. 7.12.
The period of the given function is $2b$. Hence, by Theorem 1,

$$\mathscr{L}\{f(t)\} = \frac{1}{1 - e^{-2bs}} \int_0^{2b} f(t)e^{-st}\, dt$$

$$= \frac{1}{1 - e^{-2bs}} \left(\int_0^b 1e^{-st}\, dt + \int_b^{2b} -1e^{-st}\, dt \right)$$

$$= \frac{1}{1 - e^{-2bs}} \left(\frac{e^{-st}}{-s}\Big|_0^b - \frac{e^{-st}}{-s}\Big|_b^{2b} \right)$$

$$= \frac{1}{1 - e^{-2bs}} \frac{1 - 2e^{-bs} + e^{-2bs}}{s} = \frac{(1 - e^{-bs})^2}{s(1 - e^{-bs})(1 + e^{-bs})}$$

$$= \frac{1 - e^{-bs}}{s(1 + e^{-bs})} = \frac{e^{bs/2} - e^{-bs/2}}{s(e^{bs/2} + e^{-bs/2})} = \frac{1}{s} \tanh \frac{bs}{2}$$

EXAMPLE 2

Find the transform of the sawtooth wave shown in Fig. 7.13.
In this case, the period of the function is k. Hence

$$\mathscr{L}\{f(t)\} = \frac{1}{1 - e^{-ks}} \int_0^k t e^{-st}\, dt = \frac{1}{1 - e^{-ks}} \left[\frac{e^{-st}}{s^2}(-st - 1) \right]_0^k$$

$$= \frac{1 - (1 + ks)e^{-ks}}{s^2(1 - e^{-ks})} = \frac{(1 + ks) - (1 + ks)e^{-ks} - ks}{s^2(1 - e^{-ks})}$$

$$= \frac{(1 + ks)(1 - e^{-ks}) - ks}{s^2(1 - e^{-ks})}$$

$$= \frac{1 + ks}{s^2} - \frac{k}{s(1 - e^{-ks})}$$

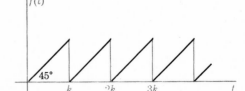

Figure 7.13
A sawtooth wave.

EXAMPLE 3

What is the Laplace transform of the **staircase function**

$$f(t) = n + 1 \qquad nk < t < (n + 1)k \qquad n = 0, 1, 2, \dots$$

shown in Fig. 7.14a?
The required transform can easily be found by direct calculation. However, it is

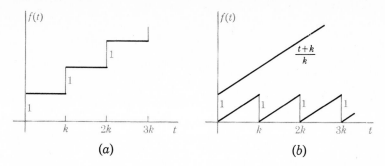

Figure 7.14
The staircase function and its synthesis.

even simpler to obtain it by considering $f(t)$ to be the difference between the two functions shown in Fig. 7.14b. The transform of the linear function $(t + k)/k$ can be found at once by Formula 4, Sec. 7.3. Except for the obvious coefficient $1/k$, the transform of the sawtooth function was obtained in the last example. Hence

$$\mathscr{L}\{f(t)\} = \frac{1}{k}\left(\frac{1}{s^2} + \frac{k}{s}\right) - \frac{1}{k}\left[\frac{1 + ks}{s^2} - \frac{k}{s(1 - e^{-ks})}\right] = \frac{1}{s(1 - e^{-ks})}$$

EXAMPLE 4

What is $f(t)$ if $\mathscr{L}\{f(t)\} = 1/[(s + a)(1 - e^{-ks})]$?

Although $\mathscr{L}\{f(t)\}$ somewhat resembles the transform of the staircase function obtained in the last example, the correspondence is not sufficiently close to provide us with the required inverse. Moreover, we cannot successfully employ the result of the last example after first using the corollary of Theorem 5, Sec. 7.4, for if we replace s by $s - a$ (intending to multiply the inverse of the resulting fraction by e^{-at}), the given transform becomes

$$\frac{1}{s(1 - e^{-k(s-a)})} = \frac{1}{s(1 - e^{ak}e^{-ks})}$$

and now, because of the factor e^{ak}, which is not equal to 1 except in the trivial cases $a = 0$ or $k = 0$, we still do not have the transform of the staircase function. It appears, therefore, that we must make a direct attack upon the problem. To do this, let us reverse the derivation of Theorem 1 and replace $1/(1 - e^{-ks})$ by the infinite geometric series of which it is the sum:

$$f(t) = \frac{1}{(s + a)(1 - e^{-ks})} = \frac{1}{s + a}(1 + e^{-ks} + e^{-2ks} + e^{-3ks} + \cdots)$$

$$= \frac{1}{s + a} + \frac{e^{-ks}}{s + a} + \frac{e^{-2ks}}{s + a} + \frac{e^{-3ks}}{s + a} + \cdots$$

Now let us assume that we can take the inverse of this infinite series term by term. If we neglect the exponential factor in, say, the $(n + 1)$st term, the inverse of what remains is obvious, namely,

$$e^{-at}$$

But having neglected the exponential e^{-nks}, we must, according to Corollary 2 of Theorem 6, Sec. 7.4, translate the function e^{-at} to the right a distance of nk and then cut it off to the left of $t = nk$. When this is done for each term, we have

$$f(t) = e^{-at} + e^{-a(t-k)}u(t - k) + e^{-a(t-2k)}u(t - 2k) + e^{-a(t-3k)}u(t - 3k) + \cdots$$

When we take into account the cutoff properties of the various translated step functions, it is clear that $f(t)$ is equal to

e^{-at}	over interval $(0,k)$
$e^{-at} + e^{ak}e^{-at}$	over interval $(k,2k)$
$e^{-at} + e^{ak}e^{-at} + e^{2ak}e^{-at}$	over interval $(2k,3k)$
. . . .	. . . .
$e^{-at} + e^{ak}e^{-at} + e^{2ak}e^{-at} + \cdots + e^{nak}e^{-at}$	over interval $[nk,(n + 1)k]$

In order to obtain a more convenient expression for $f(t)$ over the general interval $nk < t < (n + 1)k$, we can sum the finite geometric progression defining $f(t)$ in this range. Since this progression contains $n + 1$ terms and has the common ratio $r = e^{ak}$, it follows that over this interval we have

$$f(t) = e^{-at}(1 + e^{ak} + e^{2ak} + \cdots + e^{nak}) = e^{-at}\frac{(e^{ak})^{n+1} - 1}{e^{ak} - 1}$$

$$= \frac{e^{-a[t-(n+1)k]}}{e^{ak} - 1} - \frac{e^{-at}}{e^{ak} - 1} \qquad nk < t < (n + 1)k$$

Now, to achieve a more symmetric form, let us define $\tau = t - (n + 1)k$. Clearly, $t = nk$ corresponds to $\tau = -k$, and $t = (n + 1)k$ corresponds to $\tau = 0$, so that, for each value of n, the parameter τ ranges from $-k$ to 0 as t ranges from nk to $(n + 1)k$. If we make this substitution in the first fraction only, $f(t)$ assumes the form

$$f(t) = \frac{e^{-a\tau}}{e^{ak} - 1} - \frac{e^{-at}}{e^{ak} - 1} \qquad \begin{matrix} -k < \tau < 0 \\ nk < t < (n + 1)k \end{matrix}$$

The second term describes a continuous function, which decreases steadily if $a > 0$. The first term is completely independent of n, that is, yields the same set of values over each interval, because no matter what n may be, as t ranges from nk to $(n + 1)k$, τ always ranges from $-k$ to 0. Moreover, the first term is discontinuous, since at the left end of any interval, where $\tau = -k$, its value is

$$\frac{e^{-a(-k)}}{e^{ak} - 1}$$

while at the right end, where $\tau = 0$, its value is

$$\frac{1}{e^{ak} - 1}$$

The periodic function which it represents therefore has an upward jump of

$$\frac{e^{ak}}{e^{ak} - 1} - \frac{1}{e^{ak} - 1} = 1$$

at each of the points $t = k, 2k, 3k, \ldots$.

In Fig. 7.15 the discontinuous periodic function represented by the first term in $f(t)$, the continuous transient term represented by the second fraction, and $f(t)$ itself are shown for $a = \frac{1}{2}$ and $k = 2$.

EXAMPLE 5

What is the solution of the equation $y' + 3y + 2\int_0^t dt = f(t)$ if $y_0 = 1$ and if $f(t)$ is the function shown in Fig. 7.16?

Taking the transform of each side of the given equation, using the result of Example 2 to transform $f(t)$, we have

$$s(\mathscr{L}\{y\} - 1) + 3\mathscr{L}\{y\} + \frac{2}{s}\mathscr{L}\{y\} = \frac{1 + s}{s^2} - \frac{1}{s(1 - e^{-s})}$$

or $\qquad \mathscr{L}\{y\} = \dfrac{s^2 + s + 1}{s(s + 1)(s + 2)} - \dfrac{1}{(s + 1)(s + 2)(1 - e^{-s})}$

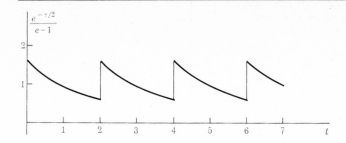

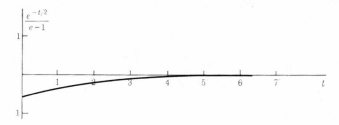

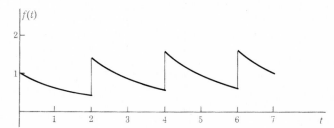

Figure 7.15

Plot showing the inverse of $\phi(s) = \dfrac{1}{(s + \frac{1}{2})(1 - e^{-2s})}$.

The inverse of the first fraction can be found immediately by the corollary of the first Heaviside theorem:

$$\tfrac{1}{2} - e^{-t} + \tfrac{3}{2}e^{-2t}$$

To find the inverse of the second fraction we must write

$$\frac{1}{(s + 1)(s + 2)(1 - e^{-s})} = \left(\frac{1}{s + 1} - \frac{1}{s + 2}\right)\frac{1}{1 - e^{-s}}$$

$$= \frac{1}{(s + 1)(1 - e^{-s})} - \frac{1}{(s + 2)(1 - e^{-s})}$$

Figure 7.16
A sawtooth wave.

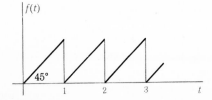

and then use the results of Example 4. In this case $k = 1$, and thus the inverse over the interval $n < t < n + 1$ is

$$\left(\frac{e^{-t}}{e-1} - \frac{e^{-t}}{e-1}\right) - \left(\frac{e^{-2t}}{e^2-1} - \frac{e^{-2t}}{e^2-1}\right)$$

or $\qquad \left(\frac{e^{-t}}{e-1} - \frac{e^{-2t}}{e^2-1}\right) - \left(\frac{e^{-t}}{e-1} - \frac{e^{-2t}}{e^2-1}\right) \qquad -1 < \tau < 0$

The second term is obviously a continuous function of t and is simply an additional contribution to the transient of the system. The periodic function defined by the first term is also continuous in this case, because the unit jumps exhibited by each of the fractions at $t = 1, 2, 3, \ldots$ are of opposite sign and hence cancel each other. The entire solution for y is therefore

$$y = \frac{1 - 2e^{-t} + 3e^{-2t}}{2} + \left(\frac{e^{-t}}{e-1} - \frac{e^{-2t}}{e^2-1}\right) - \left(\frac{e^{-t}}{e-1} - \frac{e^{-2t}}{e^2-1}\right)$$

$$= \underbrace{\left[-\frac{e-2}{e-1}e^{-t} + \frac{3e^2-5}{2(e^2-1)}e^{-2t}\right]}_{\text{Transient}} + \underbrace{\left(\frac{1}{2} - \frac{e^{-t}}{e-1} + \frac{e^{-2t}}{e^2-1}\right)}_{\text{Steady-state}} \qquad -1 < \tau < 0$$

Figure 7.17 shows a plot of the component terms and of Y itself.

The analysis of equations like the one we considered in Example 5 is so important that a table of additional results similar to the one we obtained in Example 4 would be highly desirable. Using for the most part only the procedure illustrated in Example 4, we can easily develop such a table, as we shall now show.

Figure 7.17
The solution of Example 5.

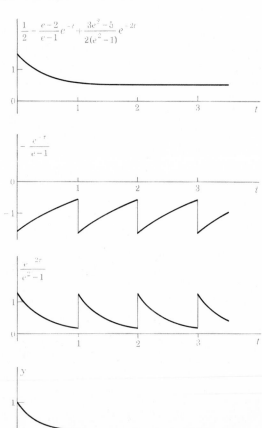

To eliminate unnecessary writing, it will be convenient to introduce the functions defined in Table 7.1, where k is an arbitrary positive number, n is a nonnegative integer, and x is a variable which is to be replaced by t or τ, as indicated in Table 7.2. The functions $\phi_1(x,k)$ and $\phi_2(x,k)$ are, respectively, the staircase function and the **Morse dot function**. The functions $\phi_3(x,k)$ and $\phi_4(x,k)$ are the integrals from 0 to x of $\phi_1(x,k)$ and $\phi_2(x,k)$, respectively. The function $\phi_5(x,a,k)$ is precisely the one we encountered in the solution of Examples 4 and 5. The others, though somewhat more complicated, arise in the same way and can be plotted just as easily when the parameters a, b, and k are known.

Table 7.1a

Definition of functional symbol	Definition of function over general interval $nk < x < (n + 1)k$
$\phi_1(x,k)$	$n + 1$
$\phi_2(x,k)$	$\dfrac{(-1)^n + 1}{2}$
$\phi_3(x,k)$	$(n + 1)x - \dfrac{n(n + 1)k}{2}$
$\phi_4(x,k)$	$\dfrac{(-1)^n + 1}{2} x + \dfrac{k}{4} [1 - (-1)^n(2n + 1)]$

Table 7.1b

Definition of functional symbol	Definition of function
$\phi_5(x,a,k)$	$\dfrac{e^{-ax}}{e^{ak} - 1}$
$\phi_6(x,a,k)$	$\dfrac{e^{-ax}}{e^{ak} + 1}$
$\phi_7(x,a,b,k)$	$\dfrac{e^{-ax} \cos b(x + k) - e^{-a(x+k)} \cos bx}{2(\cosh ak - \cos bk)}$
$\phi_8(x,a,b,k)$	$\dfrac{e^{-ax} \cos b(x + k) + e^{-a(x+k)} \cos bx}{2(\cosh ax + \cos bx)}$
$\phi_9(x,a,b,k)$	$\dfrac{e^{-ax} \sin b(x + k) - e^{-a(x+k)} \sin bx}{2(\cosh ak - \cos bk)}$
$\phi_{10}(x,a,b,k)$	$\dfrac{e^{-ax} \sin b(x + k) + e^{-a(x+k)} \sin bx}{2(\cosh ak + \cos bk)}$
$\phi_{11}(x,a,k)$	$\dfrac{(x + k)e^{-ax} - xe^{-a(x+k)}}{2(\cosh ak - 1)}$
$\phi_{12}(x,a,k)$	$\dfrac{(x + k)e^{-ax} + xe^{-a(x+k)}}{2(\cosh ak + 1)}$

Table 7.2 lists the inverses of all elementary periodic-type transforms which are likely to be encountered. Of course, as Example 5 illustrated, it is usually necessary to employ the method of partial fractions before the results of Table 7.2 can be applied.

Formulas 1 to 4 are obtained by obvious applications of Theorem 1 and of Theorem 3, Sec. 7.2. Formula 5 was derived in detail in Example 4, and the derivations of Formulas 6 to 10 follow almost exactly the same pattern. All that is necessary is to

Table 7.2

Laplace transform	Inverse over general interval $nk < t < (n + 1)k$ $-k < \tau < 0$
1. $\dfrac{1}{s(1 - e^{-ks})}$	$\phi_1(t,k)$
2. $\dfrac{1}{s(1 + e^{-ks})}$	$\phi_2(t,k)$
3. $\dfrac{1}{s^2(1 - e^{-ks})}$	$\phi_3(t,k)$
4. $\dfrac{1}{s^2(1 + e^{-ks})}$	$\phi_4(t,k)$
5. $\dfrac{1}{(s + a)(1 - e^{-ks})} \quad a \neq 0$	$\phi_5(\tau,a,k) - \phi_5(t,a,k)$
6. $\dfrac{1}{(s + a)(1 + e^{-ks})} \quad a \neq 0$	$(-1)^n \phi_6(\tau,a,k) + \phi_6(t,a,k)$
7. $\dfrac{s + a}{[(s + a)^2 + b^2](1 - e^{-ks})}$	$\phi_7(\tau,a,b,k) - \phi_7(t,a,b,k)$†
8. $\dfrac{s + a}{[(s + a)^2 + b^2](1 + e^{-ks})}$	$(-1)^n \phi_8(\tau,a,b,k) + \phi_8(t,a,b,k)$‡
9. $\dfrac{b}{[(s + a)^2 + b^2](1 - e^{-ks})}$	$\phi_9(\tau,a,b,k) - \phi_9(t,a,b,k)$†
10. $\dfrac{b}{[(s + a)^2 + b^2](1 + e^{-ks})}$	$(-1)^n \phi_{10}(\tau,a,b,k) + \phi_{10}(t,a,b,k)$‡
11. $\dfrac{1}{(s + a)^2(1 - e^{-ks})} \quad a \neq 0$	$\phi_{11}(\tau,a,k) - \phi_{11}(t,a,k)$
12. $\dfrac{1}{(s + a)^2(1 + e^{-ks})} \quad a \neq 0$	$(-1)^n \phi_{12}(\tau,a,k) + \phi_{12}(t,a,k)$

† The possibility that, simultaneously, a is zero and bk is an even multiple of π is to be ruled out.

‡ The possibility that, simultaneously, a is zero and bk is an odd multiple of π is to be ruled out.

express as complex exponentials the sines and cosines which appear in the inverses of the individual terms after the transform has been converted to an infinite series. The expression for $f(t)$ over any interval $nk < t < (n + 1)k$ is then, as in Example 4, just a finite geometric progression which can be summed and converted to a purely real form without difficulty.

The derivation of Formulas 11 and 12 are somewhat different because of the repeated factors in the denominators of the transforms. Over the general interval $nk < t < (n + 1)k$, these lead to expressions for $f(t)$ which are series of the form

$$\sum_{j=0}^{n} (t - jk)e^{-a(t - jk)} = te^{-at} \sum_{j=0}^{n} (e^{ak})^j - ke^{-at} \sum_{j=0}^{n} j(e^{ak})^j$$

in the case of Formula 11, and

$$\sum_{j=0}^{n} (-1)^j(t - jk)e^{-a(t - jk)} = te^{-at} \sum_{j=0}^{n} (-e^{ak})^j - ke^{-at} \sum_{j=0}^{n} j(-e^{ak})^j$$

in the case of Formula 12. In each instance, the second series is not a geometric progression and must be summed by other means. Fortunately, the results of Example 3, Sec. 4.5, are applicable, and through their use the inverses given in Table 7.2 can easily be established.

The transient, or t-evaluated, terms in the inverses in Table 7.2 are all continuous for all $t \geq 0$. This is true of the periodic, or τ-evaluated, terms if and only if the degree of the polynomial part of the denominator of the transform exceeds the degree of the numerator by more than 1. If this is not the case, there is an upward jump of 1 at each of the points $t = k, 2k, 3k, \ldots, nk, \ldots$ if the denominator of the transform contains $1 - e^{-ks}$ and a jump of $(-1)^n$ if the denominator of the transform contains $1 + e^{-ks}$.

EXAMPLE 6

A simple series circuit contains the elements $R = 400 \,\Omega$, $L = 0.2$ H, $C = 10^{-6}$ F. At $t = 0$, while the circuit is completely passive, an exponential sawtooth voltage wave, equal to $E_0e^{-5,000t}$ V throughout one period and repeating itself every 0.002 s, is switched into the circuit. Find the total current and the steady-state current which result.

The differential equation to be solved is

$$0.2\frac{di}{dt} + 400i + 10^6 \int_0^t i\, dt = E(t)$$

Taking the Laplace transform of both sides, we obtain

$$\mathscr{L}\{i\}\left(0.2s + 400 + \frac{10^6}{s}\right) = \frac{E_0 \int_0^{0.002} e^{-5,000t}e^{-st}\, dt}{1 - e^{-0.002s}}$$

$$= \frac{E_0}{1 - e^{-0.002s}}\left[\frac{e^{-t(s + 5,000)}}{-(s + 5,000)}\right]_0^{0.002}$$

or

$$\mathscr{L}\{i\}\frac{s^2 + 2,000s + 5 \times 10^6}{5s} = E_0 \frac{1 - e^{-0.002s - 10}}{(s + 5,000)(1 - e^{-0.002s})}$$

$$= E_0 \frac{(1 - e^{-10}) + e^{-10}(1 - e^{-0.002s})}{(s + 5,000)(1 - e^{-0.002s})}$$

$$= E_0 \frac{e^{-10}}{s + 5,000} + E_0 \frac{1 - e^{-10}}{(s + 5,000)(1 - e^{-0.002s})}$$

Hence,

$$\mathscr{L}\{i\} = \frac{5E_0 e^{-10}s}{(s + 5,000)[(s + 1,000)^2 + (2,000)^2]}$$

$$+ \frac{5E_0(1 - e^{-10})s}{(s + 5,000)[(s + 1,000)^2 + (2,000)^2](1 - e^{-0.002s})}$$

Now by simple partial-fraction manipulations we find

$$\frac{s}{(s + 5,000)[(s + 1,000)^2 + (2,000)^2]}$$

$$= \frac{1}{4,000}\left[-\frac{1}{s + 5,000} + \frac{s + 1,000}{(s + 1,000)^2 + (2,000)^2} \right]$$

From this point the entire solution can be written down at once:

$$i = \frac{5E_0 e^{-10}}{4,000}(-e^{-5,000t} + e^{-1,000t}\cos 2,000t)$$

$$- \frac{5E_0(1 - e^{-10})}{4,000}[\phi_5(\tau, 5,000, 0.002) - \phi_5(t, 5,000, 0.002)]$$

$$+ \frac{5E_0(1 - e^{-10})}{4,000}[\phi_7(\tau, 1,000, 2,000, 0.002) - \phi_7(t, 1,000, 2,000, 0.002)$$

The steady-state current is described by the terms in τ:

$$i_{ss} = -\frac{5E_0(1 - e^{-10})}{4,000}[\phi_5(\tau, 5,000, 0.002) - \phi_7(\tau, 1,000, 2,000, 0.002)]$$

or written out at length:

$$i_{ss} = -\frac{E_0(1 - e^{-10})}{800}$$

$$\times \left[\frac{e^{-5,000\tau}}{e^{10} - 1} - \frac{e^{-1,000\tau}\cos 2,000(\tau + 0.002) - e^{-1,000(\tau + 0.002)}\cos 2,000\tau}{2(\cosh 2 - \cos 4)} \right]$$

This function, plotted for $-0.002 < \tau < 0$, defines one complete cycle of the steady-state current. Of course, the unit jumps in ϕ_5 and ϕ_7 at the ends of each period just cancel, leaving the steady-state current continuous, as it must be.

The operational solution of a problem like this, leading as it does to a relatively simple finite expression for the response, is in general to be preferred to the use of Fourier series, which leaves the answer in the form of an infinite series, as in Example 3, Sec. 6.5.

The many points of similarity between differential equations and difference equations, noted in Sec. 4.5, suggest that the Laplace transformation may be useful in solving difference equations as well as differential equations. This is indeed the case, and some of the ideas of the present section are helpful in this connection. To investigate this matter, let us first recall that a constant-coefficient linear difference equation (of the second order) is an equation of the form

$$(aE^2 + bE + c)y = f(t)$$

or, more explicitly,

$$ay(t + 2) + by(t + 1) + cy(t) = f(t)$$

which is to hold between the values of y and f at *integral* values of the independent variable t. This implies that both y and f are functions whose values change only at the integral values $t = 0, 1, 2, \ldots$; that is, y and f are what are often called **jump**

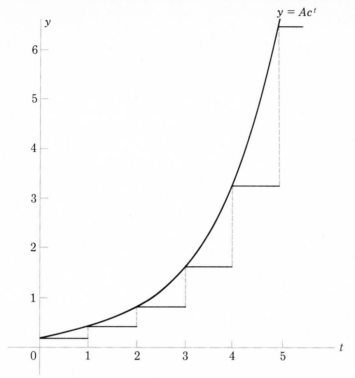

Figure 7.18
The jump function associated with the function Ac^t.

functions. For instance, although a plot of Ac^t, thought of as a function of the continuous variable t, is a familiar exponential curve, its plot, when its value changes only at the integral values $t = n$, is the jump function shown in Fig. 7.18. We shall denote by the symbol $f_{(j)}$ the jump function associated with a given continuous function f. Thus for any t, the value of $f_{(j)}(t)$ is $f(n)$, where n is the largest integer equal to or less than t.

To transform the jump function associated with a function $f(t)$, it is convenient to recall from Examples 5 and 6 of Sec. 7.4 that the filter function $u(t - a) - u(t - b)$ takes on the value 1 if t is between a and b and takes on the value 0 if t is less than a or greater than b. Then, using the filter function

$$u(t - n) - u(t - \overline{n + 1})$$

we can write the required jump function as

$$f_{(j)}(t) = \sum_{n=0}^{\infty} f(n)[u(t - n) - u(t - \overline{n + 1})]$$

and the Laplace transform of $f_{(j)}(t)$ can now be found without difficulty.

EXAMPLE 7

If $f(t) = c^t$, what is $\mathcal{L}\{f_{(j)}(t)\}$?

In this case the jump function $f_{(j)}(t)$ is described by the expression

$$\sum_{n=0}^{\infty} c^n[u(t - n) - u(t - \overline{n + 1})]$$

Hence the required transform is

$$\int_0^\infty f_{(j)}(t)e^{-st}\, dt = \int_0^\infty \sum_{n=0}^\infty c^n[u(t - n) - u(t - \overline{n + 1})]e^{-st}\, dt$$

$$= \sum_{n=0}^\infty c^n[\mathscr{L}\{u(t - n)\} - \mathscr{L}\{u(t - \overline{n + 1})\}]$$

$$= \sum_{n=0}^\infty c^n \left(\frac{e^{-ns}}{s} - \frac{e^{-(n+1)s}}{s}\right) = \sum_{n=0}^\infty c^n \frac{1 - e^{-s}}{s} e^{-ns}$$

$$= \frac{1 - e^{-s}}{s} \sum_{n=0}^\infty (ce^{-s})^n$$

Finally, noting that the last series is just a geometric progression with common ratio $r = ce^{-s}$, we have

$$\mathscr{L}\{f_{(j)}(t)\} = \frac{1 - e^{-s}}{s} \frac{1}{1 - ce^{-s}} = \frac{e^s - 1}{s(e^s - c)}$$

Before a difference equation can be solved by Laplace transform methods, formulas for the transforms of $y(t + 1)$ and $y(t + 2)$ must be available. These are given in the next two theorems.

THEOREM 2   If $y(t)$ is a jump function, then

$$\mathscr{L}\{y(t + 1)\} = e^s \mathscr{L}\{y(t)\} - y_0 \frac{e^s - 1}{s}$$

Proof   In the integral defining $\mathscr{L}\{y(t + 1)\}$, namely,

$$\mathscr{L}\{y(t + 1)\} = \int_0^\infty y(t + 1)e^{-st}\, dt$$

let us put $t + 1 = T$. Then

$$\mathscr{L}\{y(t + 1)\} = \int_1^\infty y(T)e^{-s(T - 1)}\, dT$$

$$= e^s \int_1^\infty y(T)e^{-sT}\, dT$$

$$= e^s \left[\int_0^\infty y(T)e^{-sT}\, dT - \int_0^1 y(T)e^{-sT}\, dT\right]$$

However, since $y(t)$ is a jump function, its value between 0 and 1 is constant and equal, in fact, to its value at $t = 0$, namely, y_0. Hence

$$\mathscr{L}\{y(t + 1)\} = e^s \left[\int_0^\infty y(T)e^{-sT}\, dT - y_0 \int_0^1 e^{-sT}\, dT\right]$$

$$= e^s \left[\mathscr{L}\{y(t)\} - y_0 \frac{1 - e^{-s}}{s}\right]$$

$$= e^s \mathscr{L}\{y(t)\} - y_0 \frac{e^s - 1}{s}$$

as asserted.

THEOREM 3 If $y(t)$ is a jump function, then

$$\mathscr{L}\{y(t + 2)\} = e^{2s}\mathscr{L}\{y(t)\} - (y_0 e^s + y_1)\frac{e^s - 1}{s}$$

EXAMPLE 8

What is the solution of the difference equation

$$y(t + 2) - 5y(t + 1) + 6y(t) = 5^t$$

for which $y_0 = 1$ and $y_1 = 3$?

Taking the Laplace transform of this equation, with the help of Theorems 2 and 3 and Example 7, we have

$$\left[e^{2s}\mathscr{L}\{y\} - (e^s + 3)\frac{e^s - 1}{s}\right] - 5\left(e^s\mathscr{L}\{y\} - \frac{e^s - 1}{s}\right) + 6\mathscr{L}\{y\} = \frac{e^s - 1}{s(e^s - 5)}$$

or

$$(e^{2s} - 5e^s + 6)\mathscr{L}\{y\} = \frac{e^s - 1}{s}\left(e^s - 2 + \frac{1}{e^s - 5}\right)$$

Hence

$$\mathscr{L}\{y\} = \frac{e^s - 1}{s}\left[\frac{1}{e^s - 3} + \frac{1}{(e^s - 2)(e^s - 3)(e^s - 5)}\right]$$

By a straightforward application of familiar partial-fraction techniques (treating e^s as the variable) we find that

$$\frac{1}{(e^s - 2)(e^s - 3)(e^s - 5)} = \frac{1}{6}\left(\frac{2}{e^s - 2} - \frac{3}{e^s - 3} + \frac{1}{e^s - 5}\right)$$

Hence, substituting and combining terms in $\mathscr{L}\{y\}$, we find

$$\mathscr{L}\{y\} = \frac{1}{6}\frac{e^s - 1}{s}\left(\frac{2}{e^s - 2} + \frac{3}{e^s - 3} + \frac{1}{e^s - 5}\right)$$

Therefore, by the result of Example 7,

$$y = \tfrac{1}{6}(2 \cdot 2^n + 3 \cdot 3^n + 5^n) = \tfrac{1}{6}(2^{n+1} + 3^{n+1} + 5^n)$$

EXERCISES

1 Using Theorem 1, verify that:

(a) $\mathscr{L}\{\sin bt\} = \dfrac{b}{s^2 + b^2}$ (b) $\mathscr{L}\{\cos bt\} = \dfrac{s}{s^2 + b^2}$

2 Obtain the Laplace transform of the staircase function (Fig. 7.14a) by direct evaluation of the Laplace transform integral.

Find the Laplace transforms of the periodic functions whose definitions in one period are

3 $f(t) = \sin t \quad 0 < t < \pi$

4 $f(t) = \begin{cases} \sin t & 0 < t < \pi \\ 0 & \pi < t < 2\pi \end{cases}$

5 $f(t) = \begin{cases} t & 0 < t < a \\ 0 & a < t < 2a \end{cases}$

6 $f(t) = \begin{cases} 1 & 0 < t < a \\ 0 & a < t < 2a \\ -1 & 2a < t < 3a \\ 0 & 3a < t < 4a \end{cases}$

Find the inverse of each of the following transforms:

7 $\dfrac{s}{(s + 1)(s + 2)(1 - e^{-2s})}$ **8** $\dfrac{e^{-s}}{s(s^2 + 2s + 5)(1 + e^{-s})}$

9 $\dfrac{1}{(s + 1)(s + 2)^2(1 + e^{-s})}$ **10** $\dfrac{3s + 5}{(s + 1)(s^2 + 4s + 5)(1 - e^{-3s})}$

Solve the following differential equations and explain your answers, $f(t)$ being in each case a periodic function defined over one period as indicated:

11 $y' + 4y + 3\displaystyle\int_0^t y\, dt = f(t)$ $f(t) = \begin{cases} 1 & 0 < t < 2 \\ -1 & 2 < t < 4 \end{cases}$ $y_0 = 1$

12 $y'' + 4y' + 4y = f(t)$ $f(t) = \begin{cases} 1 & 0 < t < 1 \\ 0 & 1 < t < 2 \end{cases}$ $y_0 = y_0' = 0$

13 $y'' + y = f(t)$ $f(t) = \begin{cases} 1 & 0 < t < \pi \\ 0 & \pi < t < 2\pi \end{cases}$ $y_0 = y_0' = 0$

14 In the proof of Theorem 1, the geometric progression which had to be summed is meaningful only if the absolute value of the common ratio $r = e^{-as}$ is less than 1. Does this cause any problems?

15 According to the footnotes to Table 7.2, certain values of a, b, and k cannot be allowed to occur simultaneously in Formulas 7 to 10 of Table 7.2 because the formulas become meaningless for these values. Why is this?

16 Plot the jump function $f_{(J)}(t)$ associated with each of the following functions:
   (a) $f(t) = t^2$ (b) $f(t) = t^3 - 3t^2 + 2t$
   (c) $f(t) = \sin t$ (d) $f(t) = \sin \pi t$

17 If $y(t)$ is a jump function, show that the Laplace transform of $\Delta y = y(t + 1) - y(t)$ is $(e^s - 1)\mathcal{L}\{y(t)\} - y_0[(e^s - 1)/s]$.

18 If $f(t) = t^2$, show that the Laplace transform of the associated jump function is

$$\frac{e^s + 1}{s(e^s - 1)^2}$$

Hint: Note that $\Delta f \equiv (t + 1)^2 - t^2$, and apply the result of Exercise 17.

19 (a) If $f(t) = tc^{t-1}$, show that the Laplace transform of the associated jump function is

$$\frac{e^s - 1}{s(e^s - c)^2}$$

Hint: Compute the first difference of $f(t)$ and use the result of Exercise 17.
   (b) If $f(t) = t(t - 1)c^{t-2}$, what is the Laplace transform of the associated jump function?

20 Establish the following results:
   (a) If $f(t) = \sin bt$, then

$$\mathcal{L}\{f_{(J)}(t)\} = \frac{\sin b}{e^{2s} - 2e^s \cos b + 1} \frac{e^s - 1}{s}$$

   (b) If $f(t) = \cos bt$, then

$$\mathcal{L}\{f_{(J)}(t)\} = \frac{e^s - \cos b}{e^{2s} - 2e^s \cos b + 1} \frac{e^s - 1}{s}$$

Hint: Compute the first differences of $\sin bt$ and $\cos bt$, apply the result of Exercise 17 to each of these equations, and then obtain $\mathcal{L}\{(\sin bt)_{(J)}\}$ and $\mathcal{L}\{(\cos bt)_{(J)}\}$ by solving these equations simultaneously.

21 (a) If $f(t) = e^{at} \sin bt$, what is the Laplace transform of the associated jump function?
   (b) If $f(t) = e^{at} \cos bt$, what is the Laplace transform of the associated jump function?

22 (a) If $f(t) = \sinh bt$, what is $\mathcal{L}\{f_{(J)}(t)\}$?
   (b) If $f(t) = \cosh bt$, what is $\mathcal{L}\{f_{(J)}(t)\}$?

23 (a) Show that $\mathscr{L}^{-1}\{1/s(e^s - 1)^2\}$ is the jump function corresponding to the function $(t^2 - t)/2$.

(b) Show that $\mathscr{L}^{-1}\{1/s(e^s - c)\}$, $c \neq 1$, is the jump function corresponding to the function $(c^t - 1)/(c - 1)$.

(c) Show that $\mathscr{L}^{-1}\{1/s(e^s - c)^2\}$, $c \neq 1$, is the jump function corresponding to the function

$$\frac{tc^{t-1}}{c - 1} - \frac{c^t}{(c - 1)^2} + \frac{1}{(c - 1)^2}$$

24 Solve each of the following difference equations:

(a) $y(t + 2) - 4y(t + 1) + 3y(t) = 3^t$ $\qquad y_0 = y_1 = 0$

(b) $y(t + 2) - 4y(t + 1) + 4y(t) = u(t)$ $\qquad y_0 = 0, y_1 = 1$

(c) $y(t + 2) - 4y(t + 1) + 3y(t) = t$ $\qquad y_0 = y_1 = 0$

(d) $y(t + 2) - y(t + 1) + y(t) = 0$ $\qquad y_0 = 1, y_1 = 2$

Hint: Use the results of Exercise 20.

(e) $y(t + 2) + 6y(t + 1) + 10y(t) = 0$ $\qquad y_0 = 1, y_1 = -6$

Hint: Use the results of Exercise 21.

(f) $y(t + 2) - 2y(t + 1) + 10y(t) = 2^t$ $\qquad y_0 = 0, y_1 = 1$

25 Derive the following formulas of Table 7.2:

(a) Formula 6 $\qquad$ (b) Formula 7 $\qquad$ (c) Formula 8

(d) Formula 9 $\qquad$ (e) Formula 10 $\qquad$ (f) Formula 11

(g) Formula 12

7.7 Convolution and the Duhamel Formulas

We conclude this chapter by establishing a result concerning the product of transforms which is of considerable theoretical as well as practical interest.

THEOREM 1 $\quad \mathscr{L}\{f(t)\}\mathscr{L}\{g(t)\} = \mathscr{L}\left\{\int_0^t f(t - \lambda)g(\lambda)\,d\lambda\right\}$

$$= \mathscr{L}\left\{\int_0^t f(\lambda)g(t - \lambda)\,d\lambda\right\}$$

Proof Working with the term on the right in the first equality, we have by definition

(1) $\qquad \mathscr{L}\left\{\int_0^t f(t - \lambda)g(\lambda)\,d\lambda\right\} = \int_0^\infty \left[\int_0^t f(t - \lambda)g(\lambda)\,d\lambda\right] e^{-st}\,dt$

Now $\qquad\qquad u(t - \lambda) = \begin{cases} 1 & \lambda < t \\ 0 & \lambda > t \end{cases}$

and thus $\qquad f(t - \lambda)g(\lambda)u(t - \lambda) = \begin{cases} f(t - \lambda)g(\lambda) & \lambda < t \\ 0 & \lambda > t \end{cases}$

Since this product vanishes for all values of λ greater than the upper limit t, the inner integration in (1) can be extended to infinity if the factor $u(t - \lambda)$ is inserted in the integrand. Hence,

(2) $\qquad \mathscr{L}\left\{\int_0^t f(t - \lambda)g(\lambda)\,d\lambda\right\} = \int_0^\infty \left[\int_0^\infty f(t - \lambda)g(\lambda)u(t - \lambda)\,d\lambda\right] e^{-st}\,dt$

Now our usual assumptions about the functions we transform are sufficient to permit the order of integration in (2) to be interchanged:

$$(3) \quad \mathscr{L}\left\{\int_0^t f(t - \lambda)g(\lambda)\, d\lambda\right\} = \int_0^\infty \left[\int_0^\infty f(t - \lambda)g(\lambda)u(t - \lambda)e^{-st}\, dt\right] d\lambda$$

$$= \int_0^\infty g(\lambda)\left[\int_0^\infty f(t - \lambda)u(t - \lambda)e^{-st}\, dt\right] d\lambda$$

Because of the presence of $u(t - \lambda)$, the integrand of the inner integral is identically zero for all $t < \lambda$. Hence, the inner integration effectively starts not at $t = 0$ but at $t = \lambda$. Therefore

$$(4) \quad \mathscr{L}\left\{\int_0^t f(t - \lambda)g(\lambda)\, d\lambda\right\} = \int_0^\infty g(\lambda)\left[\int_\lambda^\infty f(t - \lambda)e^{-st}\, dt\right] d\lambda$$

Now, in the inner integral on the right of (4), let $t - \lambda = \tau$ and $dt = d\tau$. Then

$$\mathscr{L}\left\{\int_0^t f(t - \lambda)g(\lambda)\, d\lambda\right\} = \int_0^\infty g(\lambda)\left[\int_0^\infty f(\tau)e^{-s(\tau+\lambda)}\, d\tau\right] d\lambda$$

$$= \int_0^\infty g(\lambda)e^{-s\lambda}\left[\int_0^\infty f(\tau)e^{-s\tau}\, d\tau\right] d\lambda$$

$$= \left[\int_0^\infty f(\tau)e^{-s\tau}\, d\tau\right]\left[\int_0^\infty g(\lambda)e^{-s\lambda}\, d\lambda\right]$$

$$= \mathscr{L}\{f(t)\}\mathscr{L}\{g(t)\}$$

as asserted.

The **convolution**, or **Faltung**,† integral

$$\int_0^t f(t - \lambda)g(\lambda)\, d\lambda$$

is frequently denoted simply by $f(t) * g(t)$. In this notation Theorem 1 becomes

$$\mathscr{L}\{f(t)\}\mathscr{L}\{g(t)\} = \mathscr{L}\{f(t) * g(t)\}$$

EXAMPLE 1

If $\mathscr{L}\{f(t)\} = 1/(s^2 + 4s + 13)^2$, what is $f(t)$?
Clearly, we can write $\mathscr{L}\{f(t)\}$ in the form

$$\frac{1}{[(s + 2)^2 + 3^2]^2}$$

and then use the corollary of the first shifting theorem (Theorem 5, Sec. 7.4) to obtain

$$(5) \quad f(t) = \mathscr{L}^{-1}\left\{\frac{1}{[(s + 2)^2 + 3^2]^2}\right\} = e^{-2t}\mathscr{L}^{-1}\left\{\frac{1}{(s^2 + 3^2)^2}\right\}$$

Now $\dfrac{1}{(s^2 + 3^2)^2} = \mathscr{L}\left\{\dfrac{\sin 3t}{3}\right\} \mathscr{L}\left\{\dfrac{\sin 3t}{3}\right\}$

---

† German for *folding*.

Hence, by the convolution theorem,

$$\mathcal{L}^{-1}\left\{\frac{1}{(s^2 + 3^2)^2}\right\} = \frac{1}{9}\int_0^t \sin 3(t - \lambda) \sin 3\lambda \, d\lambda$$

$$= \frac{1}{9}\int_0^t \frac{\cos(6\lambda - 3t) - \cos 3t}{2} \, d\lambda$$

$$= \frac{1}{18}\left[\frac{\sin(6\lambda - 3t)}{6} - \lambda \cos 3t\right]_0^t$$

$$= \frac{1}{18}\left(\frac{\sin 3t}{3} - t \cos 3t\right)$$

Therefore, from (5), $f(t) = \dfrac{e^{-2t}(\sin 3t - 3t \cos 3t)}{54}$

This example illustrates how in certain cases the convolution theorem can be used in place of a fourth Heaviside theorem to handle repeated quadratic factors in the denominator of a transform.

EXAMPLE 2

Find a particular integral of the differential equation

$$y'' + 2ay' + (a^2 + b^2)y = f(t)$$

Taking the Laplace transform of the given equation and assuming $y_0 = y_0' = 0$ since we desire only a *particular* solution, we find

$$\mathcal{L}\{y\} = \frac{1}{(s + a)^2 + b^2} \mathcal{L}\{f(t)\}$$

Now

$$\frac{1}{(s + a)^2 + b^2} = \mathcal{L}\left\{\frac{e^{-at}\sin bt}{b}\right\}$$

Hence

$$\mathcal{L}\{y\} = \mathcal{L}\{f(t)\}\mathcal{L}\left\{\frac{e^{-at}\sin bt}{b}\right\}$$

and thus, by the convolution theorem,

$$y = \frac{1}{b}\int_0^t f(t - \lambda)e^{-a\lambda}\sin b\lambda \, d\lambda$$

or, equally well,

$$y = \frac{1}{b}\int_0^t f(\lambda)e^{-a(t-\lambda)}\sin b(t - \lambda) \, d\lambda = \frac{e^{-at}}{b}\int_0^t f(\lambda)e^{a\lambda}\sin b(t - \lambda) \, d\lambda$$

It is interesting to compare this procedure with the method of variation of parameters (Sec. 2.4) for the determination of particular integrals of linear differential equations. The two give identical results in the case of constant-coefficient differential equations.

An especially important application of the convolution theorem makes it possible to determine the response of a system to a general excitation if its response to a unit step function is known. To develop this idea we shall need the concepts of **transfer function** and **indicial admittance**.

Any physical system capable of responding to an excitation can be thought of as a device by means of which an input function is transformed into an output function. If we assume that all initial conditions are zero at the moment when a single excitation, or **input,** $f(t)$ begins to act, then by setting up the differential equations describing the system, taking Laplace transforms, and solving for the transform of the **output** $y(t)$, we obtain a relation of the form

(6)
$$\mathscr{L}\{y(t)\} = \frac{\mathscr{L}\{f(t)\}}{Z(s)}$$

where $Z(s)$ is a function of s whose coefficients depend solely on the parameters of the system. Moreover, in the usual applications to linear systems, $Z(s)$ will be just the quotient of two polynomials in s.

In electrical problems where the input is an applied voltage $E_0 e^{j\omega t}$ and the output is the resultant current, the function $Z(s)$, except for the fact that the frequency variable $j\omega$ is replaced by the Laplace transform parameter s, is just the impedance of the network. However, the importance of $Z(s)$ is not restricted to electric circuits, and for systems of all sorts the function

$$\frac{1}{Z(s)} = \frac{\mathscr{L}\{y(t)\}}{\mathscr{L}\{f(t)\}} = \frac{\mathscr{L}\{\text{output}\}}{\mathscr{L}\{\text{input}\}}$$

is an exceedingly important quantity, usually called the **transfer function**. In particular, after s has been replaced by $j\omega$, the transfer function can be used to determine the effect of any system on the phase and amplitude of a sinusoidal input of arbitrary frequency, just as in the electrical case.

If a unit step function is applied to a system with transfer function $1/Z(s)$, then from (6) we have

$$\mathscr{L}\{y(t)\} = \frac{\mathscr{L}\{u(t)\}}{Z(s)} = \frac{1}{sZ(s)}$$

The response in this particular case is called the **indicial admittance** $A(t)$; that is,

(7)
$$\mathscr{L}\{A(t)\} = \frac{1}{sZ(s)}$$

Using (7), we can now rewrite (6) in the form

$$\mathscr{L}\{y(t)\} = \frac{\mathscr{L}\{f(t)\}}{Z(s)} = \frac{s\mathscr{L}\{f(t)\}}{sZ(s)} = s\mathscr{L}\{A(t)\}\mathscr{L}\{f(t)\}$$

Hence, by the convolution theorem,

$$\mathscr{L}\{y(t)\} = s\mathscr{L}\left\{\int_0^t A(t-\lambda)f(\lambda)\,d\lambda\right\} = s\mathscr{L}\left\{\int_0^t A(\lambda)f(t-\lambda)\,d\lambda\right\}$$

Because of the factor s, it follows from Theorem 3, Sec. 7.4, that

$$y(t) = \frac{d}{dt}\left[\int_0^t A(t-\lambda)f(\lambda)\,d\lambda\right] = \frac{d}{dt}\left[\int_0^t A(\lambda)f(t-\lambda)\,d\lambda\right]$$

Therefore, performing the indicated differentiations,† we have equivalently,

(8)
$$y(t) = \int_0^t A'(t - \lambda)f(\lambda)\, d\lambda + A(0)f(t)$$

and

(9)
$$y(t) = \int_0^t A(\lambda)f'(t - \lambda)\, d\lambda + A(t)f(0)$$

Since $A(t)$ is by definition the response of a system which is initially passive, it follows that $A(0) = 0$. Hence Eq. (8) becomes simply

(10)
$$y(t) = \int_0^t A'(t - \lambda)f(\lambda)\, d\lambda$$

Finally, by making the change of variable $\tau = t - \lambda$ in the integrals in (9) and (10) we obtain the related expressions

(11)
$$y(t) = \int_0^t A'(\tau)f(t - \tau)\, d\tau$$

(12)
$$y(t) = \int_0^t A(t - \tau)f'(\tau)\, d\tau + A(t)f(0)$$

Formulas (9) to (12) all serve to express the response of a system to a general driving function $f(t)$ in terms of the experimentally accessible response of the system to a unit step function. They are often referred to, collectively, as **Duhamel's formulas**, after the French mathematician J. M. C. Duhamel (1797–1872).

It is possible to interpret these integrals in physical terms as follows. Let the driving function $f(t)$ be given, and imagine it approximated by a series of step functions, as shown in Fig. 7.19. The first step function is of noninfinitesimal magnitude $f(0)$. All later step functions in the approximation are of infinitesimal magnitude, and their contributions in the limit will have to be taken into account by integration. Specifically, since

$$\frac{\Delta f}{\Delta \lambda} \doteq \frac{df}{dt}\bigg|_{t = \lambda} = f'(\lambda)$$

we have for the height Δf_i of the general infinitesimal step function the approximate expression

$$\Delta f_i \doteq f'(\lambda_i)\, \Delta \lambda_i$$

Now if $A(t)$ is the indicial admittance of the system, the first step function $f(0)u(t)$ produces a response equal to

$$f(0)A(t)$$

from the very definition of the indicial admittance as the response of the system to a unit step input beginning to act at $t = 0$. For the second step function $\Delta f_1\, u(t - \lambda_1)$,

---

† According to **Leibnitz' rule**, if $F(t) = \int_{a(t)}^{b(t)} \phi(x,t)\, dx$, where a and b are differentiable functions of t and where $\phi(x,t)$ and $\partial\phi(x,t)/\partial t$ are continuous in x and t, then

$$\frac{dF}{dt} = \int_{a(t)}^{b(t)} \frac{\partial\phi(x,t)}{\partial t}\, dx + \phi[b(t),t]\frac{db(t)}{dt} - \phi[a(t),t]\frac{da(t)}{dt}$$

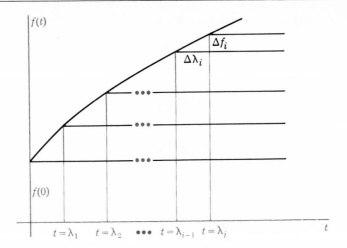

Figure 7.19
The synthesis of a general function by means of step functions.

there is a lag of $t = \lambda_1$ units of time before it begins to act. Hence the infinitesimal response which it produces is

$$\Delta f_1[A(t - \lambda_1)] \qquad \text{or} \qquad f'(\lambda_1)\, \Delta \lambda_1[A(t - \lambda_1)]$$

Similarly, the third step function produces the response

$$f'(\lambda_2)\, \Delta \lambda_2[A(t - \lambda_2)]$$

and in general the $(i + 1)$st step function produces the response

$$f'(\lambda_i)\, \Delta \lambda_i[A(t - \lambda_i)]$$

If these contributions to the total response are added, we obtain for the response at a general time t

$$y(t) = f(0)A(t) + f'(\lambda_1)\, \Delta \lambda_1[A(t - \lambda_1)] + f'(\lambda_2)\, \Delta \lambda_2[A(t - \lambda_2)] + \cdots$$
$$+ f'(\lambda_i)\, \Delta \lambda_i[A(t - \lambda_i)] + \cdots$$
$$= f(0)A(t) + \sum f'(\lambda_i)A(t - \lambda_i)\, \Delta \lambda_i$$

the summation extending over all step-function inputs which have begun to act up to the instant t. In the limit when each $\Delta \lambda$ approaches zero and the height of each step function after the first, $f(0)u(t)$, approaches zero, the sum in the last expression becomes an integral, and, except for the dummy variable, we have Eq. (12).

To give a physical interpretation of Eq. (10), we must first determine the significance of the derivative of the indicial admittance $A'(t)$. To do this, we shall need the concept of a **unit impulse**.

Suppose that we have the function shown in Fig. 7.20. This consists of a suddenly applied excitation of constant magnitude acting for a certain period of time and then suddenly ceasing, the product of duration and magnitude being unity. If a is very small, the period of application is correspondingly small but the magnitude of the excitation is very great. It is sometimes convenient to pursue this idea to the limit and imagine

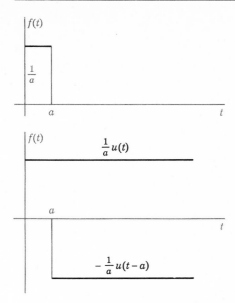

Figure 7.20
Plot suggesting the nature of a unit impulse.

an input function of arbitrarily large magnitude acting for an infinitesimal time, the product of duration and intensity remaining unity as $a \to 0$. The resulting "function" is usually referred to as the **unit impulse** $I(t)$ or the **δ function** $\delta(t)$.†

In somewhat different terms, the δ function $\delta(t - t_0)$ is often described by the following purported definition:

$$(13a) \qquad \delta(t - t_0) = \begin{cases} 0 & t \neq t_0 \\ \infty & t = t_0 \end{cases}$$

$$(13b) \qquad \int_{-\infty}^{\infty} \delta(t - t_0)\, dt = 1$$

Taken literally this is nonsense, for the area under a curve which coincides with the t axis at every point but one must surely be zero and not unity, as (13b) suggests. However, if (13) is considered to be merely suggestive of the limiting process by which we first described the unit impulse, then whatever its shortcomings as a definition, it is at least as meaningful as certain other useful and reasonably respectable concepts in applied mathematics.

Consider, for instance, the familiar concept of a concentrated load on a beam (Fig. 7.21a). Clearly, such a load is physically unrealizable and must be viewed as an idealization of the following nature. Imagine that over the interval $(x_0, x_0 + a)$ the beam bears a distributed load whose magnitude per unit length is P/a (Fig. 7.21b). Then no matter how small a may be, the total load on the beam, being equal to the product of the intensity P/a and the interval length a, is just P. As $a \to 0$, the ideal concept of a concentrated load thus emerges as a limiting form of a realizable

---

† More specifically, $\delta(t)$ is often called the **Dirac δ function**, after the British theoretical physicist P. A. M. Dirac (1902–   ).

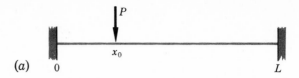

Load per unit length $= \dfrac{P}{a}$

Figure 7.21
Plot suggesting the interpretation of
a concentrated load on a beam as a
unit impulse.

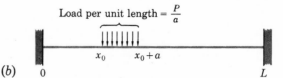

distributed load. If one were now asked to describe the load per unit length $w(x)$ in the limiting case, one would probably give the following "definition":

$$(14a) \qquad w(x) = \begin{cases} 0 & x \neq x_0 \\ \infty & x = x_0 \end{cases}$$

$$(14b) \qquad \int_0^L w(x)\,dx = P$$

which corresponds in all essential respects to the description of the δ function provided by (13).

One important formal property of the δ function is its ability to isolate or reproduce a particular value of a function $f(t)$ according to the following formula:

$$(15) \qquad \int_{-\infty}^{\infty} f(t)\delta(t - t_0)\,dt = f(t_0)$$

To justify this we revert to the preliminary approximation to the δ function and use it in place of $\delta(t - t_0)$ in (15). This gives us the approximating integral

$$\int_{t_0}^{t_0+a} f(t)\,\frac{1}{a}\,dt$$

By the **law of the mean for integrals,**† this integral is equal to

$$(16) \qquad a\,\frac{f(\xi)}{a} = f(\xi) \qquad t_0 < \xi < t_0 + a$$

Now as $a \to 0$, perforce $\xi \to t_0$, and so, from (16), the integral approaches $f(t_0)$, as asserted.

The unit impulse is only the first of an infinite sequence of so-called **singularity functions.** As a direct generalization of the unit impulse we have the **unit doublet** (Fig. 7.22), defined (loosely) as

$$\lim_{a \to 0} \frac{u(t) - 2u(t - a) + u(t - 2a)}{a^2}$$

---

† This asserts that if $\phi(t)$ is continuous over the closed range of integration $b \leq t \leq c$, then there exists at least one value of t, say $t = \xi$, between b and c such that

$$\int_b^c \phi(t)\,dt = (c - b)\phi(\xi)$$

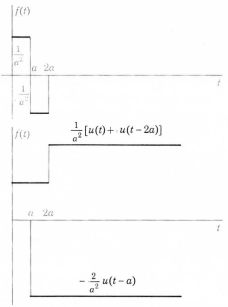

Figure 7.22
Plot suggesting the nature of a
unit doublet.

the unit triplet, defined similarly as

$$\lim_{a \to 0} \frac{u(t) - 3u(t - a) + 3u(t - 2a) - u(t - 3a)}{a^3}$$

and so on, indefinitely. Some of the properties of these "functions" will be found among the exercises at the end of this section.

Whatever its theoretical limitations, the δ function is a useful device which, at worst, can be used formally provided the answers to which it leads are subsequently checked experimentally or by independent analysis. It is interesting and important that in many applications the use of the δ function can be rigorously justified by arguments based on what is known as the **Stieltjes integral**,† a generalization of the familiar Riemann integral. More generally, all the singularity functions are examples of mathematical objects known as **generalized functions**, or **distributions**, which are studied in the recently developed **theory of distributions**.‡

To determine the Laplace transform of a unit impulse, we return to the prelimiting approximation

$$\frac{u(t) - u(t - a)}{a}$$

shown in Fig. 7.20. Transforming this expression, we have for all $a > 0$,

$$\frac{1}{a}\left(\frac{1}{s} - \frac{e^{-as}}{s}\right) = \frac{1 - e^{-as}}{as}$$

---

† Named for the Dutch mathematician T. J. Stieltjes (1856–1894).

‡ An introductory account of the theory of distributions can be found in Athanasios Papoulis, "The Fourier Integral and Its Applications," pp. 269–282, McGraw-Hill, New York, 1962.

As $a \to 0$, this transform assumes the indeterminate form 0/0, but evaluating it in the usual way by L'Hospital's rule, we obtain immediately the limiting value 1. In the same way we can show that the transforms of the unit doublet and the unit triplet are, respectively, s and s^2, and the transforms of the other singularity functions follow exactly the same pattern. Since these transforms do not approach zero as s becomes infinite, we know from Corollary 1 of Theorem 5, Sec. 7.1, that they are not the transforms of piecewise regular functions of exponential order. This, of course, is obvious, for although the singularity functions are all of exponential order, they are limiting forms involving unbounded behavior in the neighborhood of the origin and hence are not piecewise regular.

We are now in a position to resume our attempt to give a physical interpretation to formula (10). For convenience let us denote by $h(t)$ the response of the system under discussion when the input is a unit impulse. We have already seen [Eq. (6)] that

$$\mathcal{L}\{y(t)\} = \frac{\mathcal{L}\{f(t)\}}{Z(s)}$$

Hence, if $f(t)$ is a unit impulse, so that $\mathcal{L}\{f(t)\} = 1$ and $y(t) = h(t)$, we have

$$\mathcal{L}\{h(t)\} = \frac{1}{Z(s)} = s\,\frac{1}{sZ(s)} = s\mathcal{L}\{A(t)\}$$

Thus, from Theorem 3, Sec. 7.4, it follows that

$$h(t) = \frac{dA(t)}{dt} = A'(t)$$

or in words, *the response of a system to a unit impulse is the derivative of the response of the system to a unit step function.*

Now let a general input $f(t)$ be approximated by a series of infinitesimal impulses, as suggested in Fig. 7.23. For the first impulse, whose magnitude, by definition, is the product

$$f(0)\,\Delta\lambda_0 \equiv f(\lambda_0)\,\Delta\lambda_0$$

the infinitesimal response is $[f(\lambda_0)\,\Delta\lambda_0]A'(t)$, since $A'(t) \equiv h(t)$ is the response to a unit impulse acting at $t = 0$. The second impulse $f(\lambda_1)\,\Delta\lambda_1$ does not occur until

Figure 7.23
The synthesis of a general function
by means of impulses.

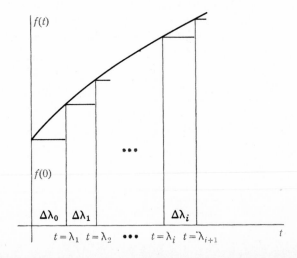

$t = \lambda_1$; hence the response which it produces is $[f(\lambda_1) \, \Delta\lambda_1] A'(t - \lambda_1)$. And in general, the response produced by the $(i + 1)$st impulse is

$$[f(\lambda_i) \, \Delta\lambda_i] A'(t - \lambda_i)$$

If these contributions to the total response are added, we obtain for the response at a general time t

$$y(t) = \sum f(\lambda_i) A'(t - \lambda_i) \, \Delta\lambda_i$$

the summation extending over all impulses which have acted on the system up to the time t. In the limit when each $\Delta\lambda \to 0$, the last sum becomes an integral, and we have formula (10).

EXERCISES

Find the inverse of each of the following transforms:

1 $\dfrac{1}{(s^2 + 4)^2}$

2 $\dfrac{s}{(s^2 + 9)^3}$

3 $\dfrac{s}{s + 2}$

4 $\dfrac{s^4 + 2s + 3}{s^2 + 4}$

5 $\dfrac{s^2 + 4s + 4}{(s^2 + 4s + 13)^2}$

6 $\dfrac{s^4 + 2s^2 - s}{(s + 1)(s^2 + 1)^2}$

7 Using the convolution theorem, find a particular integral of the equation

$$y'' + 2ay' + a^2 y = f(t)$$

8 Using the convolution theorem, find a particular integral of the equation

$$y'' + (a + b)y' + aby = f(t)$$

9 (a) Show that

$$\mathcal{L}^{-1} \left\{ \frac{1}{\sqrt{s}\,(s - 1)} \right\} = \frac{2e^t}{\sqrt{\pi}} \int_0^{\sqrt{t}} e^{-\tau^2} \, d\tau$$

 Hint: Use the convolution theorem, and then in the resulting integral let $\sqrt{\lambda} = \tau$.
 (b) What is $\mathcal{L}^{-1}\{1/(s\sqrt{s} + 1)\}$?

10 (a) What is $\mathcal{L}^{-1}\{1/[\sqrt{s}\,(s^2 + 1)]\}$?
 (b) What is $\mathcal{L}^{-1}\{s/[\sqrt{s}\,(s^2 + 1)]\}$?

11 Verify that the Laplace transform of the unit doublet is s and that the transform of the unit triplet is s^2.

12 (a) What is $\mathcal{L}\{I(t - t_0)\}$?
 (b) If $D(t)$ denotes the unit doublet function, what is $\mathcal{L}\{D(t - t_0)\}$?

13 (a) Find the transform of the triangular pulse shown in Fig. 7.24a. What is the limit of the transform of this pulse as $a \to 0$?

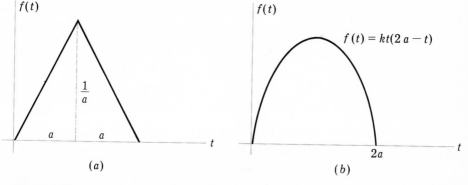

(a)   (b)

Figure 7.24

(b) Determine k so that the parabolic pulse shown in Fig. 7.24b will have unit area. What is the transform of this unit pulse? What is the limit of the transform of this pulse as $a \to 0$?

14 If $D(t)$ denotes the unit doublet function, show that $\int_{-\infty}^{\infty} f(t)D(t - t_0)\, dt = -f'(t_0)$. If $T(t)$ denotes the unit triplet function, what is $\int_{-\infty}^{\infty} f(t)T(t - t_0)\, dt$?

15 Find $A(t)$ and $h(t)$ for the equation $y'' + 3y' + 2y = 0$. Verify that $h(t) = A'(t)$, and then verify formulas (10) and (12) when this equation is driven by the function $f(t) = e^t$.

16 **(a)** Find $A(t)$ and $h(t)$ for the system shown in Fig. 7.25 if the input is applied to

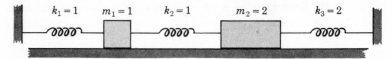

Figure 7.25

M_1 and if the output is the response, i.e., displacement, of M_2. Verify that $h(t) = A'(t)$. What is the response of M_2 to an arbitrary force applied to M_1 when the system is in its equilibrium position?

(b) Work part **(a)** if the input is applied to M_2 and the output is the response of M_1.

17 Find $A(t)$ and $h(t)$ for the system shown in Fig. 7.26 if the input voltage is applied

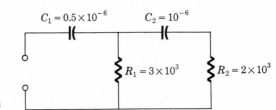

Figure 7.26

across the indicated terminals and if the output is the current through R_2. Verify that $h(t) = A'(t)$. What is the current through R_2 due to an arbitrary voltage $E(t)$ applied across the terminals when all charges and currents in the system are zero?

18 Show that the solution of the equation $ay'' + by' + cy = 0$ when $y_0 = 0$ and $y_0' = 1$ is exactly the same as the solution of the equation $ay'' + by' + cy = a\delta(t)$ when $y_0 = y_0' = 0$. Does this fact have a physical interpretation? With what combination of singularity functions must an initially passive second-order equation be driven in order to have the same solution as the undriven equation with initial conditions $y(0) = y_0$ and $y'(0) = y_0'$?

19 Show that $f(t) * [g(t) * h(t)] = \int_0^t \int_0^\lambda f(t - \lambda)g(\lambda - \mu)h(\mu)\, d\mu\, d\lambda$.

20 **(a)** Show that $f(t) * [g(t) \pm h(t)] = [f(t) * g(t)] \pm [f(t) * h(t)]$.

(b) Show that $f(t) * [g(t) * h(t)] = [f(t) * g(t)] * h(t)$.

21 Show that $\mathscr{L}\{f(t)\}\mathscr{L}\{g(t)\}\mathscr{L}\{h(t)\} = \mathscr{L}\{f(t) * g(t) * h(t)\}$.

22 Show that $1 * 1 = t$ and that $1 * 1 * 1 = t^2/2$. What is the generalization of these results to n factors?

23 Evaluate:

(a) $\delta(t - a) * f(t)$ **(b)** $u(t - a) * f(t)$

(c) $t^m * t^n$ m, n nonnegative integers.

24 If $f(0) = g(0) = 0$, show that $f'(t) * g(t) = f(t) * g'(t)$ and that

$$[f(t) * g(t)]' = \frac{f'(t) * g(t) + f(t) * g'(t)}{2}$$

25 If $f(t)$ and $g(t)$ are given functions, is it possible to solve the equation

$$f(t) * x(t) = g(t) \qquad \text{for } x(t)$$

CHAPTER 8
Partial Differential Equations

8.1 Introduction

In our previous work we have seen how the analysis of mechanical and electrical systems containing lumped parameters often leads to ordinary differential equations. However, assumptions to the effect that all masses exist as mass points, that all springs are weightless, or that the elements of an electric circuit are concentrated in ideal resistances, inductances, and capacitances rather than continuously distributed are frequently not sufficiently accurate. In such cases a more realistic approach usually leads to one or more partial differential equations which must be solved to obtain a description of the behavior of the system. In this chapter we shall discuss such equations as they commonly arise in engineering and in physics. We shall begin our study by examining in detail the derivation from physical principles of certain important partial differential equations. Then, knowing the forms of most frequent occurrence, we shall investigate methods of solution and their application to specific problems.

8.2 The Derivation of Equations

One of the first problems to be attacked through the use of partial differential equations was that of the vibration of a stretched, flexible string. Today, after nearly 250 years, it is still an excellent initial example.

Let us consider, then, an elastic string, stretched under a tension T between two points on the x axis (Fig. 8.1a). The weight of the string per unit length after it is stretched we suppose to be a known function $w(x)$. Besides the elastic and inertia forces inherent in the system, the string may also be acted upon by a distributed load whose magnitude per unit length we assume to be a known function of x, y, t, and the transverse velocity $\dot{y}$, say $f(x, y, \dot{y}, t)$. In formulating the problem we assume that

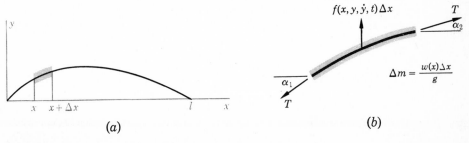

Figure 8.1
A typical element of a vibrating string.

a. The motion takes place entirely in one plane, and in this plane each particle moves at right angles to the equilibrium position of the string.

b. The deflection of the string during the motion is so small that the resulting change in length of the string has no effect on the tension T.

c. The string is perfectly flexible, i.e., can transmit force only in the direction of its length.

d. The slope of the deflection curve of the string is at all points and at all times so small that with satisfactory accuracy $\sin \alpha$ can be replaced by $\tan \alpha$, where α is the inclination angle of the tangent to the deflection curve.

Gravitational and frictional forces, if any, we suppose to be taken into account in the expression for the load per unit length $f(x,y,\dot{y},t)$.

With these assumptions in mind, let us consider a general infinitesimal segment of the string as a free body (Fig. 8.1b). By assumption **a**, the mass of such an element is $\Delta m = w(x)\,\Delta x/g$. By assumption **b**, the forces which act at the ends of the element are the same, namely, T. By assumption **c**, these forces are directed along the respective tangents to the deflection curve; and, by assumption **d**, their transverse components are

$$T \sin \alpha_2 = T \sin \alpha|_{x+\Delta x} \doteq T \tan \alpha|_{x+\Delta x}$$

and
$$T \sin \alpha_1 = T \sin \alpha|_x \doteq T \tan \alpha|_x$$

The acceleration produced in Δm by these forces and by the portion of the distributed load $f(x,y,\dot{y},t)\,\Delta x$ which acts over the interval Δx is approximately $\partial^2 y/\partial t^2$, where y is the ordinate of an arbitrary point of the element. The time derivative is here written as a partial derivative because obviously y depends not only upon t but upon x as well. Applying Newton's law to the element, we can thus write

(1)
$$\frac{w(x)\,\Delta x}{g}\frac{\partial^2 y}{\partial t^2} = T \tan \alpha|_{x+\Delta x} - T \tan \alpha|_x + f(x,y,\dot{y},t)\,\Delta x$$

or, dividing by Δx,

$$\frac{w(x)}{g}\frac{\partial^2 y}{\partial t^2} = T\left(\frac{\tan \alpha|_{x+\Delta x} - \tan \alpha|_x}{\Delta x}\right) + f(x,y,\dot{y},t)$$

The fraction on the right consists of the difference between the values of $\tan \alpha$ at $x + \Delta x$ and at x, divided by the difference Δx. In other words, it is precisely the difference quotient for the function $\tan \alpha$. Hence its limit as $\Delta x \to 0$ is the derivative of $\tan \alpha$ with respect to x, that is, $(\partial \tan \alpha)/\partial x$. But since $\tan \alpha = \partial y/\partial x$, this can be written simply as $\partial^2 y/\partial x^2$. Our final result, then, is that the deflection $y(x,t)$ of a stretched string satisfies the partial differential equation†

(2)
$$\frac{\partial^2 y}{\partial t^2} = \frac{Tg}{w(x)}\frac{\partial^2 y}{\partial x^2} + \frac{g}{w(x)} f(x, y, \dot{y}, t)$$

---

† The question of what constitutes a satisfactory derivation of the partial differential equation describing a given physical system is not a simple one. To attempt to give a careful limiting argument is, in effect, "to strain at a gnat and swallow a camel," since, being ultimately atomic, no physical system is continuous. Perhaps our purported derivations should be regarded merely as plausibility arguments suggesting that certain partial differential equations be accepted as the axioms of a theoretical or "rational" study of applied mathematics, whose practical importance, in contrast to its purely mathematical interest, is to be judged by how well its conclusions describe past observations and predict new ones.

In most applications the weight of the string per unit length $w(x)$ is a constant, and external forces are negligible; i.e., with satisfactory accuracy $f(x,y,\dot{y},t)$ can be assumed to be identically zero. When this is the case, Eq. (2) reduces to the **one-dimensional wave equation**

(3)
$$\frac{\partial^2 y}{\partial t^2} = a^2 \frac{\partial^2 y}{\partial x^2} \qquad a^2 = \frac{Tg}{w}$$

The dimensions of a^2 are

$$\frac{\text{Force} \times \text{acceleration}}{\text{Weight/unit length}} = \frac{(ML/T^2)(L/T^2)}{(ML/T^2)(1/L)} = \frac{L^2}{T^2}$$

that is, a has the dimensions of velocity. The significance of this will become apparent in Sec. 8.3 when we discuss the d'Alembert solution of the wave equation.

Closely related to the vibrating string is the vibrating membrane. To obtain the partial differential equation describing its behavior, we suppose that it is stretched across some closed curve C in the xy plane and that when it vibrates, each particle moves in a direction perpendicular to the xy plane. We assume, further, that the weight per unit area of the membrane after it is stretched is a known function $w(x,y)$ and that the tension per unit length is the same at all points and in all directions in the membrane. Finally, we suppose that the membrane is acted upon by a known distributed force whose magnitude per unit area is $f(x,y,z,\dot{z},t)$. Then by computing the transverse, or z, components of the tensile forces acting across the boundaries of a typical two-dimensional element of the membrane (Fig. 8.2) and applying Newton's law to the mass of such an element, we find without difficulty that the deflection of the membrane $z(x,y,t)$ satisfies the equation

(4)
$$\frac{\partial^2 z}{\partial t^2} = \frac{Tg}{w(x,y)} \left(\frac{\partial^2 z}{\partial x^2} + \frac{\partial^2 z}{\partial y^2} \right) + \frac{g}{w(x,y)} f(x,y,z,\dot{z},t)$$

If the weight per unit area of the membrane is the same at all points, and if there are no external forces, i.e., if $w(x,y)$ is a constant and if $f(x,y,z,\dot{z},t) \equiv 0$, then Eq. (4) reduces to the **two-dimensional wave equation**

(5)
$$\frac{\partial^2 z}{\partial t^2} = a^2 \left(\frac{\partial^2 z}{\partial x^2} + \frac{\partial^2 z}{\partial y^2} \right) \qquad a^2 = \frac{Tg}{w}$$

Here, as in the case of the vibrating string, the parameter a has the dimensions of velocity.

Figure 8.2
A typical element of a vibrating membrane.

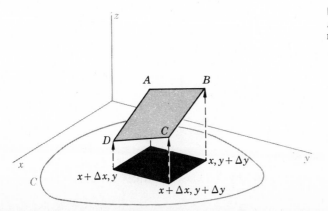

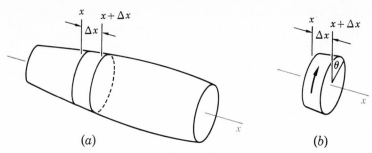

Figure 8.3
A typical element of a vibrating shaft.

As a third problem leading to a partial differential equation, let us consider a shaft vibrating torsionally (Fig. 8.3a). The material of the shaft we assume to have a modulus of elasticity in shear E_s and to be of uniform weight per unit volume ρ. The cross-sectional area of the shaft at a distance x from one end of the shaft we suppose to be a known function of x, say $A(x)$. The polar moment of inertia $J(x)$ of a general cross section about its center of gravity we also suppose known. In addition to the obvious elastic and inertia torques, the shaft may also be acted upon by a distributed external torque whose magnitude per unit length is a known function, say $f(x,\theta,\dot\theta,t)$, where θ is the angle through which a general cross section has rotated from its equilibrium position and $\dot\theta$ is the angular velocity with which that cross section rotates while the shaft is vibrating. We assume further that

a. All cross sections of the shaft remain plane during rotation.
b. Each cross section rotates about its center of gravity.
c. The shape of a general cross section does not differ greatly from a circle.

Frictional torques, if any, we suppose to be taken into account in the expression for the distributed torque per unit length $f(x,\theta,\dot\theta,t)$.

We begin by considering as a free body an infinitesimal segment of the shaft bounded by two cross sections a distance Δx apart (Fig. 8.3b). The mass of such a disk is approximately

$$\Delta m = \frac{\rho A(x)\,\Delta x}{g}$$

and, by definition, its radius of gyration is

$$k = \sqrt{\frac{J(x)}{A(x)}}$$

Hence its polar moment of inertia is approximately

$$\Delta I = k^2\,\Delta m = \frac{J(x)}{A(x)}\frac{\rho A(x)\,\Delta x}{g} = \frac{J(x)\rho\,\Delta x}{g}$$

The rotation of such an element is produced by the portion of the distributed torque $f(x,\theta,\dot\theta,t)\,\Delta x$ which acts on it and by the elastic torques T transmitted to it through its end sections by the adjacent portions of the shaft. Therefore, applying Newton's law in torsional form, we have

$$\frac{J(x)\rho\,\Delta x}{g}\frac{\partial^2\theta}{\partial t^2} = T|_{x+\Delta x} - T|_x + f(x,\theta,\dot\theta,t)\,\Delta x$$

or, dividing by Δx and then letting $\Delta x \to 0$,

(6) $$\frac{J(x)\rho}{g}\frac{\partial^2\theta}{\partial t^2} = \frac{\partial T}{\partial x} + f(x,\theta,\dot\theta,t)$$

Now from the strength of materials, we recall that the torque transmitted through any cross section of a twisted shaft is proportional to the twist per unit length, i.e., the slope of the (θ,x) curve, at that cross section:

$$T = \tau\frac{\partial\theta}{\partial x}$$

The proportionality constant τ is known as the **torsional rigidity**. For shafts which are solids of revolution it can be shown that

$$\tau = E_sJ(x)$$

and this result can be used with satisfactory accuracy whenever the cross sections of a shaft are approximately circular. Hence in such cases Eq. (6) becomes

(7) $$\frac{J(x)\rho}{g}\frac{\partial^2\theta}{\partial t^2} = \frac{\partial[E_sJ(x)\,\partial\theta/\partial x]}{\partial x} + f(x,\theta,\dot\theta,t)$$

However, for configurations whose cross sections differ appreciably from circles, e.g., propeller blades or aircraft wings, it is necessary to determine the torsional rigidity τ by experimental means and continue the solution of Eq. (6) by numerical rather than analytical methods.

In most elementary applications the shafts are of uniform circular cross section, and there are no external distributed torques. In such cases $J(x)$ is a constant, $f(x,\theta,\dot\theta,t)$ is identically zero, and Eq. (7) reduces to

(8) $$\frac{\partial^2\theta}{\partial t^2} = a^2\frac{\partial^2\theta}{\partial x^2} \qquad a^2 = \frac{E_sg}{\rho}$$

which is, again, just the one-dimensional wave equation.

Another vibration problem of considerable practical interest concerns the transverse vibrations of a beam. To obtain the partial differential equation describing these vibrations, let us first choose a coordinate system such that the beam in its undeflected position coincides with a portion of the x axis and the deflections occur in the direction of the y axis. A general cross section of the beam we assume to be of known area $A(x)$ and known moment of inertia $I(x)$ about its neutral axis. The material of the beam we suppose to be of weight per unit volume ρ and modulus of elasticity E. In addition to the intrinsic elastic and inertia forces, the beam may also be acted upon by a distributed load of known intensity $f(x,y,\dot y,t)$. Gravitational forces and frictional forces, if any, we suppose included in this distributed load. Finally, we assume that all particles of the beam move in a purely transverse direction, i.e., that the slight rotation of the cross sections as the beam vibrates is negligible.

Now from the discussion in Sec. 2.6 we recall the following formulas of beam flexure:

$$M(x) = EI(x)\frac{d^2y}{dx^2} \qquad \frac{dM(x)}{dx} = V(x) \qquad \frac{dV(x)}{dx} = -w(x)$$

where $M(x)$ = bending moment at general cross section
$\quad\quad V(x)$ = shear, or net transverse force, to right of general cross section
$\quad\quad w(x)$ = load per unit length at general cross section

Hence, combining these relations into a single equation, we have

$$(9) \qquad w(x) = -\frac{\partial V(x)}{\partial x} = -\frac{\partial^2 M(x)}{\partial x^2} = -\frac{\partial^2[EI(x)\,\partial^2 y/\partial x^2]}{\partial x^2}$$

where the derivatives are now written as partial derivatives, since in a vibration problem y depends upon t as well as upon x.

During vibration the load per unit length on the beam consists of two parts: the external load $f(x,y,\dot{y},t)$ and the inertia load due to the motion of the beam itself. Now the mass of an infinitesimal segment of the beam of length Δx is approximately $[\rho A(x)\,\Delta x]/g$, and the transverse acceleration of such a mass element is $\partial^2 y/\partial t^2$. Hence the inertia load per unit length is

$$\frac{\dfrac{\rho A(x)\,\Delta x}{g}\dfrac{\partial^2 y}{\partial t^2}}{\Delta x} = \frac{\rho A(x)}{g}\frac{\partial^2 y}{\partial t^2}†$$

and therefore the total load per unit length is

$$w(x) = \frac{\rho}{g}A(x)\frac{\partial^2 y}{\partial t^2} + f(x,y,\dot{y},t)$$

Substituting this into Eq. (9), we have finally

$$(10) \qquad \frac{\partial^2[EI(x)\,\partial^2 y/\partial x^2]}{\partial x^2} = -\frac{\rho}{g}A(x)\frac{\partial^2 y}{\partial t^2} - f(x,y,\dot{y},t)$$

In many important applications the beam under consideration is of constant cross section, and there is no external load; that is, A and I are constants, and $f(x,y,\dot{y},t) \equiv 0$. Under these conditions Eq. (10) reduces to the simpler form

$$(11) \qquad a^2\frac{\partial^4 y}{\partial x^4} = -\frac{\partial^2 y}{\partial t^2} \qquad a^2 = \frac{EIg}{A\rho}$$

In this case the parameter a does *not* have the dimensions of velocity.

An entirely different class of problems leading to partial differential equations is encountered in the study of the flow of heat in thermally conducting regions. To ob-

---

† The sign of the inertia load per unit length can be checked by observing that when the beam is instantaneously concave toward the positive y axis, its elements are either losing velocity in the negative y direction or gaining velocity in the positive y direction and hence have positive acceleration. Therefore the inertia load per unit length is positive, as it should be under the convention we established in Sec. 2.6 (Fig. 2.2). Similarly, when the beam is instantaneously convex toward the positive y axis, the acceleration of its particles is negative and so too is the inertia load per unit length.

tain the equation governing this phenomenon we must make use of the following experimental facts:

a. Heat flows in the direction of decreasing temperature.

b. The rate at which heat flows through an area is proportional to the temperature gradient, in degrees per unit distance, in the direction perpendicular to the area.

c. The quantity of heat gained or lost by a body when its temperature changes is proportional to the mass of the body and to the temperature change.

The proportionality constant in **b** is called the **thermal conductivity** of the material k. The proportionality constant in **c** is called the **specific heat** of the material c.

Let us now consider the thermal conditions in an infinitesimal element of a conducting solid (Fig. 8.4). If the weight of the conducting material per unit volume is ρ, the mass of such an element is

$$\Delta m = \frac{\rho \, \Delta x \, \Delta y \, \Delta z}{g}$$

Furthermore, if u is the temperature at any point at any time, and if Δu is the temperature change which occurs in the element in the infinitesimal time interval Δt, then the quantity of heat stored in the element in this time is, by **c**,

$$\Delta H = c \, \Delta m \, \Delta u = \frac{c\rho \, \Delta x \, \Delta y \, \Delta z \, \Delta u}{g}$$

and the rate at which heat is being stored is approximately

$$(12) \qquad \frac{\Delta H}{\Delta t} = \frac{c\rho}{g} \, \Delta x \, \Delta y \, \Delta z \, \frac{\Delta u}{\Delta t}$$

The heat which produces the temperature change Δu comes from two sources. First, heat may be generated throughout the body, by electrical or chemical means for instance, at a known rate per unit volume, say $f(x,y,z,t)$. The rate at which heat is being received by the element from this source is then

$$(13) \qquad f(x,y,z,t) \, \Delta x \, \Delta y \, \Delta z$$

Second, the element may also gain heat by virtue of heat transfer through its various faces.

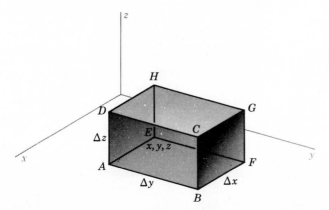

Figure 8.4
A typical volume element in a region of three-dimensional heat flow.

In particular, the rate at which heat flows into the element through the rear face *EFGH* is, by **b**, approximately

$$-k\,\Delta y\,\Delta z\,\frac{\partial u}{\partial x}\bigg|_{\substack{x\\ y+(1/2)\Delta y\\ z+(1/2)\Delta z}}$$

where, as an average figure, we have used the temperature gradient $\partial u/\partial x$ at the center $(x,\ y+\frac{1}{2}\Delta y,\ z+\frac{1}{2}\Delta z)$ of the face *EFGH*. The minus sign is necessary because the element *gains* heat through the rear face if the normal temperature gradient, i.e., the rate of change of temperature in the x direction, is *negative*. Similarly, the element gains heat through the front face *ABCD* at the approximate rate

$$k\,\Delta y\,\Delta z\,\frac{\partial u}{\partial x}\bigg|_{\substack{x+\Delta x\\ y+(1/2)\Delta y\\ z+(1/2)\Delta z}}$$

The sum of these two expressions is the net rate at which the element gains heat because of heat flow in the x direction.

In the same way we find that the rates at which the element gains heat because of flow in the y and z directions are, respectively,

$$-k\,\Delta x\,\Delta z\,\frac{\partial u}{\partial y}\bigg|_{\substack{x+(1/2)\Delta x\\ y\\ z+(1/2)\Delta z}} \qquad +\,k\,\Delta x\,\Delta z\,\frac{\partial u}{\partial y}\bigg|_{\substack{x+(1/2)\Delta x\\ y+\Delta y\\ z+(1/2)\Delta z}}$$

and

$$-k\,\Delta x\,\Delta y\,\frac{\partial u}{\partial z}\bigg|_{\substack{x+(1/2)\Delta x\\ y+(1/2)\Delta y\\ z}} \qquad +\,k\,\Delta x\,\Delta y\,\frac{\partial u}{\partial z}\bigg|_{\substack{x+(1/2)\Delta x\\ y+(1/2)\Delta y\\ z+\Delta z}}$$

Now the rate at which heat is being stored in the element, Eq. (12), must equal the rate at which heat is being produced in the element, Eq. (13), plus the rate at which heat is flowing into the element from the rest of the region. Hence we have the approximate relation

$$\frac{c\rho}{g}\,\Delta x\,\Delta y\,\Delta z\,\frac{\Delta u}{\Delta t} = f(x,y,z,t)\,\Delta x\,\Delta y\,\Delta z + k\,\Delta y\,\Delta z\left(\frac{\partial u}{\partial x}\bigg|_{\substack{x+\Delta x\\ y+(1/2)\Delta y\\ z+(1/2)\Delta z}} - \frac{\partial u}{\partial x}\bigg|_{\substack{x\\ y+(1/2)\Delta y\\ z+(1/2)\Delta z}}\right)$$

$$+\,k\,\Delta x\,\Delta z\left(\frac{\partial u}{\partial y}\bigg|_{\substack{x+(1/2)\Delta x\\ y+\Delta y\\ z+(1/2)\Delta z}} - \frac{\partial u}{\partial y}\bigg|_{\substack{x+(1/2)\Delta x\\ y\\ z+(1/2)\Delta z}}\right)$$

$$+\,k\,\Delta x\,\Delta y\left(\frac{\partial u}{\partial z}\bigg|_{\substack{x+(1/2)\Delta x\\ y+(1/2)\Delta y\\ z+\Delta z}} - \frac{\partial u}{\partial z}\bigg|_{\substack{x+(1/2)\Delta x\\ y+(1/2)\Delta y\\ z}}\right)$$

Finally, dividing by $k\,\Delta x\,\Delta y\,\Delta z$ and letting Δx, Δy, Δz, and Δt approach zero, we obtain the equation of heat conduction

$$(14) \qquad a^2\,\frac{\partial u}{\partial t} = \frac{\partial^2 u}{\partial x^2} + \frac{\partial^2 u}{\partial y^2} + \frac{\partial^2 u}{\partial z^2} + \frac{1}{k}f(x,y,z,t) \qquad a^2 = \frac{c\rho}{kg}$$

The parameter a in this equation does not have the dimensions of velocity.

In many important cases, heat is neither generated nor lost in the body, and we are interested only in the limiting steady-state temperature distribution which exists when all change of temperature with time has ceased. Under these conditions both $f(x,y,z,t)$ and $\partial u/\partial t$ are identically zero, and Eq. (14) becomes simply

$$(15) \qquad \frac{\partial^2 u}{\partial x^2} + \frac{\partial^2 u}{\partial y^2} + \frac{\partial^2 u}{\partial z^2} = 0$$

This exceedingly important equation, which arises in many applications besides steady-state heat flow, is known as **Laplace's equation**, and is often written in the abbreviated form

$$(16) \qquad \nabla^2 u = 0$$

As a final example of the derivation of partial differential equations from physical principles, we consider the flow of electricity in a long cable or transmission line. We assume the cable to be imperfectly insulated so that there is both capacitance and current leakage to ground (Fig. 8.5). Specifically, let

$$x = \text{distance from sending end of cable}$$
$$e(x,t) = \text{potential at any point on cable at any time}$$
$$i(x,t) = \text{current at any point on cable at any time}$$
$$R = \text{resistance of cable } \textit{per unit length}$$
$$L = \text{inductance of cable } \textit{per unit length}$$
$$G = \text{conductance to ground } \textit{per unit length of cable}$$
$$C = \text{capacitance to ground } \textit{per unit length of cable}$$

Since the potential at Q is equal to the potential at P minus the drop in potential along the element PQ, we see from the equivalent circuit shown in Fig. 8.5b that

$$e(x + \Delta x) = e(x) - (R \, \Delta x)i - (L \, \Delta x)\frac{\partial i}{\partial t}\dagger$$

or

$$e(x + \Delta x) - e(x) \equiv \Delta e = -(R \, \Delta x)i - (L \, \Delta x)\frac{\partial i}{\partial t}$$

Finally, dividing by Δx and then letting Δx approach zero gives

$$(17) \qquad \frac{\partial e}{\partial x} = -Ri - L\frac{\partial i}{\partial t}$$

Likewise, the current at Q is equal to the current at P minus the current loss through leakage to ground and the apparent current loss due to the varying charge stored on the element. Hence, referring again to Fig. 8.5b, we have

$$i(x + \Delta x) = i(x) - (G \, \Delta x)e - (C \, \Delta x)\frac{\partial e}{\partial t}$$

or

$$i(x + \Delta x) - i(x) \equiv \Delta i = -(G \, \Delta x)e - (C \, \Delta x)\frac{\partial e}{\partial t}$$

---

† See Eqs. (13), (15), (19), and (20), Sec. 5.2, for an explicit statement of the circuit laws which are used in this derivation.

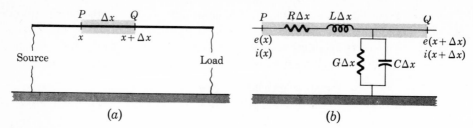

Figure 8.5
A typical element of a transmission line.

Finally, dividing by Δx and letting Δx approach zero gives

(18)
$$\frac{\partial i}{\partial x} = -Ge - C\frac{\partial e}{\partial t}$$

If we differentiate Eq. (17) with respect to x and Eq. (18) with respect to t, we obtain

$$\frac{\partial^2 e}{\partial x^2} = -R\frac{\partial i}{\partial x} - L\frac{\partial^2 i}{\partial x\,\partial t}$$

$$\frac{\partial^2 i}{\partial t\,\partial x} = -G\frac{\partial e}{\partial t} - C\frac{\partial^2 e}{\partial t^2}$$

If we eliminate the term $\partial^2 i/(\partial t\,\partial x)[\equiv \partial^2 i/(\partial x\,\partial t)]$ between these two equations and then substitute for $\partial i/\partial x$ from (18), we find that e satisfies the equation

(19)
$$\frac{\partial^2 e}{\partial x^2} = LC\frac{\partial^2 e}{\partial t^2} + (RC + GL)\frac{\partial e}{\partial t} + RGe$$

By differentiating Eq. (17) with respect to t and Eq. (18) with respect to x and then eliminating the derivatives of e, we obtain a similar equation for i:

(20)
$$\frac{\partial^2 i}{\partial x^2} = LC\frac{\partial^2 i}{\partial t^2} + (RC + GL)\frac{\partial i}{\partial t} + RGi$$

Equations (19) and (20) are known as the **telephone equations**.

Two special cases of the telephone equations are worthy of note:

a. If leakage and inductance are negligible, i.e., if $G = L = 0$, as they are, for example, for coaxial cables, Eqs. (19) and (20) reduce, respectively, to

(21a)
$$\frac{\partial^2 e}{\partial x^2} = RC\frac{\partial e}{\partial t}$$

(21b)
$$\frac{\partial^2 i}{\partial x^2} = RC\frac{\partial i}{\partial t}$$

These are known as the **telegraph equations**. Mathematically, they are identical with the one-dimensional heat equation, i.e., the equation to which (14) reduces when there are no heat sources in the conducting region and the temperature depends only on one space coordinate.

b. At high frequencies the factor introduced by the time differentiation is large. Hence the terms involving e and $\partial e/\partial t$ or i and $\partial i/\partial t$ are insignificant in comparison with the terms containing the corresponding second derivatives $\partial^2 e/\partial t^2$ and $\partial^2 i/\partial t^2$. In this case Eqs. (19) and (20) reduce, respectively, to

$$(22a) \qquad \frac{\partial^2 e}{\partial x^2} = LC \frac{\partial^2 e}{\partial t^2}$$

$$(22b) \qquad \frac{\partial^2 i}{\partial x^2} = LC \frac{\partial^2 i}{\partial t^2}$$

Each of these is an example of the one-dimensional wave equation [Eq. (3)], $1/\sqrt{LC}$ having, in fact, the dimensions of velocity. These equations are obtained at any frequency, of course, if $R = G = 0$.

It is interesting to note that nowhere in the derivation of any of the preceding equations was any use made of boundary conditions. In other words, the same partial differential equation is satisfied by the deflections of a membrane whether the membrane is round or square; and the same equation is satisfied by the deflections of a vibrating beam whether the beam is built-in at one end and free at the other, built-in at both ends, or simply supported at both ends. Likewise, the flow of heat in a body is described by the same equation whether the surface is maintained at a constant temperature, insulated against heat loss, or allowed to cool freely by radiation to the surrounding medium. In general, as we shall soon see, the role of boundary conditions, e.g., permanent conditions of constraint or of temperature, is to determine the *form* of those solutions of a partial differential equation which are relevant to a particular problem. Subsequent to this, initial conditions of displacement, velocity, or temperature, say, determine specific values for the arbitrary constants appearing in these solutions.

EXERCISES

1 Supply the details of the derivation of Eq. (4) for the transverse vibrations of a membrane.

2 What is the form of the heat equation if the thermal conductivity k and the specific heat c vary from point to point in the body?

3 Consider the telephone equations in the so-called distortionless case when $RC = LG$, and put $a^2 = RG$ and $v^2 = 1/LC$. Prove that if $e(x,t)$ [or, equally well, $i(x,t)$] is written in the form $e(x,t) = \varepsilon^{-avt}y(x,t)$, then the function y satisfies the wave equation

$$v^2 \frac{\partial^2 y}{\partial x^2} = \frac{\partial^2 y}{\partial t^2}$$

Note: To avoid confusion with the voltage, ε is used here in place of e to denote the base of natural logarithms.

4 Derive the partial differential equation satisfied by the concentration u of a liquid diffusing through a porous solid. *Hint:* The rate at which liquid diffuses through an area is proportional to the area and to the concentration gradient in the direction perpendicular to the area.

5 Consider a region of space filled with a moving fluid. Let the density of the fluid at the point (x,y,z) at time t be $\rho(x,y,z,t)$, and let the particle instantaneously at the point (x,y,z) have velocity components v_x, v_y, and v_z, respectively, in the directions of the coordinate axes. By considering the flow through the boundaries of an infinitesimal region of dimensions Δx, Δy, Δz, show that the velocity components satisfy the so-called **equation of continuity**

$$\frac{\partial(\rho v_x)}{\partial x} + \frac{\partial(\rho v_y)}{\partial y} + \frac{\partial(\rho v_z)}{\partial z} + \frac{\partial \rho}{\partial t} = 0$$

6　If $u(x,t)$ is the displacement of a general cross section of a bar which is vibrating longitudinally, i.e., in the direction of its length, show that

$$A(x)\frac{\partial^2 u}{\partial t^2} = \frac{Eg}{\rho}\frac{\partial[A(x)\,\partial u/\partial x]}{\partial x}$$

where $A(x)$ = cross-sectional area of bar
　　　E = modulus of elasticity of material of bar
　　　ρ = weight per unit volume of material

Hint: Use the definition of the modulus of elasticity,

$$E = \frac{\text{stress}}{\text{strain}} = \frac{\text{force/unit area}}{\text{stretch/unit length}}$$

to obtain the expression

$$F = EA\frac{\partial u}{\partial x}$$

for the force transmitted through a general cross section of a stretched bar.

7　Modify the derivation of Eq. (10) to take into account the rotational inertia of the cross sections of the beam. *Hint:* Consider the portion of the beam between x and $x + \Delta x$ and compute its moment of inertia about the neutral axis of its end section. Then note that as the beam vibrates, this slice of the beam remains perpendicular to the deflection curve of the beam and hence rotates through an angle equal to the inclination angle of the deflection curve at the corresponding point. Next, assuming deflections so small that the inclination angle α can be replaced by its tangent $\partial y/\partial x$, compute the angular acceleration $\ddot{\alpha}$ of the slice and the inertial moment which this angular acceleration implies. Then note that between x and $x + \Delta x$ the total moment decreases by an amount equal to the inertial moment of the slice. Finally, include this change in the expression for $\partial M/\partial x$ and complete the derivation as in the text.

8　Verify that each of the following equations has the indicated solution:

(a) $\dfrac{\partial^2 u}{\partial x\,\partial y} = 0$　　$u = f(x) + g(y)$　　(b) $u\dfrac{\partial^2 u}{\partial x\,\partial y} = \dfrac{\partial u}{\partial x}\dfrac{\partial u}{\partial y}$　　$u = f(x)g(y)$

(c) $a\dfrac{\partial u}{\partial x} + b\dfrac{\partial u}{\partial y} = 0$　　$u = f(ay - bx)$

(d) $\dfrac{\partial^2 y}{\partial x^2} = \dfrac{\partial^2 y}{\partial t^2}$　　$y = f(x - t) + g(x + t)$

(e) $\dfrac{\partial^3 u}{\partial x^3} - 6\dfrac{\partial^3 u}{\partial y\,\partial x^2} + 11\dfrac{\partial^3 u}{\partial y^2\,\partial x} - 6\dfrac{\partial^3 u}{\partial y^3} = 0$

$$u = f(x + y) + g(2x + y) + h(3x + y)$$

9　Explain how the method of undetermined coefficients can be used to obtain a particular solution of the equation

$$a\frac{\partial^2 u}{\partial x^2} + b\frac{\partial^2 u}{\partial x\,\partial y} + c\frac{\partial^2 u}{\partial y^2} = \phi(x,y)$$

if

(a) $\phi(x,y) = e^{mx + ny}$　　　　　　　(b) $\phi(x,y) = \sin(mx + ny)$
(c) $\phi(x,y) = \cos(mx + ny)$　　　　　(d) $\phi(x,y) = px^2 + qxy + ry^2$
(e) $\phi(x,y)$ is a homogeneous polynomial of degree k in x and y.

10　Find a particular solution of each of the following equations:

(a) $\dfrac{\partial^2 u}{\partial x^2} + \dfrac{\partial^2 u}{\partial y^2} = \cos(x + 2y)$　　　　(b) $\dfrac{\partial^2 u}{\partial x^2} - \dfrac{\partial u}{\partial y} = 2e^{2x + 3y}$

(c) $\dfrac{\partial^2 u}{\partial x^2} + 3\dfrac{\partial^2 u}{\partial x\,\partial y} + 2\dfrac{\partial^2 u}{\partial y^2} = 2x - y$

11 **(a)** Show that Laplace's equation in three dimensions

$$\frac{\partial^2 u}{\partial x^2} + \frac{\partial^2 u}{\partial y^2} + \frac{\partial^2 u}{\partial z^2} = 0$$

is satisfied by the function

$$u = \frac{1}{\sqrt{(x - a)^2 + (y - b)^2 + (z - c)^2}}$$

for all values of the constants a, b, c.

(b) Determine whether or not Laplace's equation in two dimensions

$$\frac{\partial^2 u}{\partial x^2} + \frac{\partial^2 u}{\partial y^2} = 0$$

is satisfied by the function

$$u = \frac{1}{\sqrt{(x - a)^2 + (y - b)^2}}$$

12 Show that Laplace's equation in two dimensions is satisfied by the function $u = \ln\left[(x - a)^2 + (y - b)^2\right]$ for all values of the constants a and b.

13 Show that under the substitution $u(x,y) = w(x,y)e^{-(bx + ay)}$ the equation

$$\frac{\partial^2 u}{\partial x\,\partial y} + a\frac{\partial u}{\partial x} + b\frac{\partial u}{\partial y} + cu = 0$$

becomes

$$\frac{\partial^2 w}{\partial x\,\partial y} + (c - ab)w = 0$$

14 By assuming $u(x,t) = \phi(x)\sin t$, find a solution of the equation

$$\frac{\partial^2 u}{\partial x^2} - \frac{\partial^2 u}{\partial t^2} = x \sin t$$

satisfying the conditions $u(0,t) = 0$ and $u(l,t) = 0$.

15 Show that if $u_1(x,y)$ and $u_2(x,y)$ are solutions of the equation

$$p_1(x,y)\frac{\partial^2 u}{\partial x^2} + p_2(x,y)\frac{\partial^2 u}{\partial x\,\partial y} + p_3(x,y)\frac{\partial^2 u}{\partial y^2}$$

$$+ q_1(x,y)\frac{\partial u}{\partial x} + q_2(x,y)\frac{\partial u}{\partial y} + r_1(x,y)u = 0$$

then for all values of the constants c_1 and c_2, the expression $c_1 u_1(x,y) + c_2 u_2(x,y)$ is also a solution.

16 Show that when Laplace's equation in cartesian coordinates

$$\frac{\partial^2 u}{\partial x^2} + \frac{\partial^2 u}{\partial y^2} + \frac{\partial^2 u}{\partial z^2} = 0$$

is transformed into cylindrical coordinates by means of the substitutions $x = r\cos\theta$, $y = r\sin\theta$, $z = z$, it becomes

$$\frac{\partial^2 u}{\partial r^2} + \frac{1}{r}\frac{\partial u}{\partial r} + \frac{1}{r^2}\frac{\partial^2 u}{\partial\theta^2} + \frac{\partial^2 u}{\partial z^2} = 0$$

17 Show that when Laplace's equation in cartesian coordinates is transformed into spherical coordinates by means of the substitutions $x = r\sin\theta\cos\phi$, $y = r\sin\theta\sin\phi$, $z = r\cos\theta$, it becomes

$$r^2\sin\theta\,\frac{\partial^2 u}{\partial r^2} + 2r\sin\theta\,\frac{\partial u}{\partial r} + \sin\theta\,\frac{\partial^2 u}{\partial\theta^2} + \cos\theta\,\frac{\partial u}{\partial\theta} + \frac{1}{\sin\theta}\frac{\partial^2 u}{\partial\phi^2} = 0$$

18 By considering the resultant horizontal force on the element of the string shown in Fig. 8.1b, show that the assumption that all particles move at right angles to the equilibrium position of the string and the assumption that the tension is the same at all points of the string are, strictly speaking, incompatible.

19 Show that the assumed compatibility of the two assumptions referred to in Exercise 18 is equivalent to the assumption that the product $(\partial y/\partial x)(\partial^2 y/\partial x^2)$ is negligibly small. *Hint:* Recall that $\cos \alpha = 1/\sqrt{1 + \tan^2 \alpha}$.

20 Derive Eq. (3) by approximating the string by a weightless elastic cord bearing n equally spaced mass particles each equal to $(1/n)$th of the mass of the string and then letting n become infinite. *Hint:* Recall the relation between the second difference and the second derivative of a function.

8.3 The d'Alembert Solution of the Wave Equation

Each of the partial differential equations in the last section can be solved by a method of considerable generality known as **separation of variables**. For the one-dimensional wave equation, however, there is also an elegant, special method known as **d'Alembert's solution,**† which, because of the importance of this equation, we shall examine in some detail before developing more general techniques.

The whole matter is very simple. In fact, if f is a function possessing a second derivative, then, by the chain rule,

$$\frac{\partial f(x - at)}{\partial t} = -af'(x - at) \qquad \frac{\partial f(x - at)}{\partial x} = f'(x - at)$$

$$\frac{\partial^2 f(x - at)}{\partial t^2} = a^2 f''(x - at) \qquad \frac{\partial^2 f(x - at)}{\partial x^2} = f''(x - at)$$

and from these results it is evident that $y = f(x - at)$ satisfies the equation

(1)
$$\frac{\partial^2 y}{\partial t^2} = a^2 \frac{\partial^2 y}{\partial x^2}$$

It is an equally simple matter to prove that if g is an arbitrary twice-differentiable function, then $g(x + at)$ is likewise a solution of (1). Hence, since (1) is a linear equation, it follows that the sum

(2)
$$y = f(x - at) + g(x + at)$$

is also a solution. In fact, it can be shown (see Exercise 13) that if f and g are arbitrary twice-differentiable functions, then (2) is a *complete* solution of (1), that is, *any* solution of (1) can be expressed in the form (2).

This form of the solution of the wave equation is especially useful for revealing the significance of the parameter a and its dimensions of velocity. Suppose, specifically,

---

† Named for the French mathematician Jean Le Rond d'Alembert (1717–1783). The d'Alembert solution is actually not a special method but a special application of a general procedure known as the **method of characteristics**. Unfortunately, this cannot be applied with comparable simplicity to problems involving the heat equation and Laplace's equation, and so, despite its theoretical interest, we shall not discuss it here. An introduction to the theory can be found in Arnold Sommerfeld, "Partial Differential Equations in Physics," pp. 36–43, Academic, New York, 1949.

that we consider the vibrations of a uniform string† stretching from $-\infty$ to ∞. If its transverse displacement is given by (2), we have in fact two waves traveling in opposite directions along the string, each with velocity a. For consider the function $f(x - at)$. At $t = 0$, it defines the curve $y = f(x)$, and at any later time $t = t_1$, it defines the curve $y = f(x - at_1)$. But these curves are identical except that the latter is translated to the right a distance equal to at_1. Thus the entire configuration moves along the string without distortion a distance of at_1 in t_1 units of time. The velocity with which the wave is propagated is therefore

$$v = \frac{at_1}{t_1} = a$$

Similarly, the function $g(x + at)$ defines a configuration which moves to the left along the string with constant velocity a. The total displacement of the string is, of course, the algebraic sum of these two traveling waves.

To carry the solution through in detail, let us suppose that the initial displacement of the string at any point x is given by $\phi(x)$ and that the initial velocity of the string at any point is $\theta(x)$. Then, as conditions to determine the form of f and g, we have, from (2) and its first derivative with respect to t,

(3) $y(x,0) = \phi(x) = [f(x - at) + g(x + at)]_{t=0} = f(x) + g(x)$

(4) $\left.\dfrac{\partial y}{\partial t}\right|_{x,0} = \theta(x) = [-af'(x - at) + ag'(x + at)]_{t=0} = -af'(x) + ag'(x)$

Dividing Eq. (4) by a and then integrating with respect to x, we find

$$-f(x) + g(x) = \frac{1}{a} \int_{x_0}^{x} \theta(x) \, dx$$

Combining this with Eq. (3) and introducing the dummy variable s in the integrals, we obtain

$$f(x) = \frac{1}{2}\left[\phi(x) - \frac{1}{a}\int_{x_0}^{x} \theta(s)\, ds\right] \qquad g(x) = \frac{1}{2}\left[\phi(x) + \frac{1}{a}\int_{x_0}^{x}\theta(s)\, ds\right]$$

With the forms of f and g known, we can now write

$$y = f(x - at) + g(x + at)$$
$$= \left[\frac{\phi(x - at)}{2} - \frac{1}{2a}\int_{x_0}^{x-at}\theta(s)\, ds\right] + \left[\frac{\phi(x + at)}{2} + \frac{1}{2a}\int_{x_0}^{x+at}\theta(s)\, ds\right]$$

or, combining the integrals,

(5) $$y(x,t) = \frac{\phi(x - at) + \phi(x + at)}{2} + \frac{1}{2a}\int_{x-at}^{x+at}\theta(s)\, ds$$

---

† The use of the string as an illustration is purely a matter of convenience, and *any* quantity satisfying the wave equation possesses the mathematical properties developed for the string.

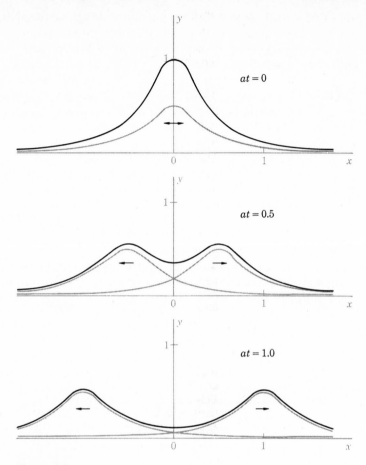

Figure 8.6
The propagation of a disturbance along a two-way infinite string.

EXAMPLE 1

A string stretching to infinity in both directions is given the initial displacement

$$\phi(x) = \frac{1}{1 + 8x^2} \quad †$$

and released from rest. Determine its subsequent motion.

Since $\theta(x) \equiv 0$, we have from (5) simply

$$y(x,t) = \frac{\phi(x - at) + \phi(x + at)}{2} = \frac{1}{2}\left[\frac{1}{1 + 8(x - at)^2} + \frac{1}{1 + 8(x + at)^2}\right]$$

The deflection of the string when $at = 0.0, 0.5,$ and 1.0 is shown in Fig. 8.6.

---

† The initial deflection curve $y = \phi(x)$ clearly violates assumption **d**, Sec. 8.2, since at $x = -\frac{1}{4}$ (for instance), $\phi'(x) \equiv \tan \alpha = \frac{16}{9} = 1.78$ while $\sin \alpha = 0.87$. This difficulty can easily be overcome, however, by assuming instead of $\phi(x)$ a new deflection curve

$$\phi^*(x) = \frac{\phi(x)}{k}$$

where k is a sufficiently large constant, say $k = 10,000$. Using $\phi(x)$ instead of $\phi^*(x)$ in this and similar problems is just a convenient way of eliminating the constant factor $1/k$ at each step of our work.

The motion of a semi-infinite string whose finite end is fixed is completely equivalent to the motion of one half of a two-way infinite string having a fixed point, or **node**, located at some finite point, say the origin. To capitalize on this fact we need only imagine the actual string, stretching from 0 to ∞, to be extended in the opposite direction to $-\infty$. The initial conditions of velocity and displacement for the new portion of the string we define to be equal in magnitude but opposite in sign to those given for the actual string.† The solution for the resulting two-way infinite string can be written down at once, using Eq. (5). In the nature of the extended initial conditions, the displacement at the origin due to the wave traveling to the right from the left half of the string will always be equal but opposite in sign to the displacement at the origin due to the wave traveling to the left from the right half of the string. Hence the string will always remain at rest at the origin, and the solution for the right half of the extended string will be precisely the solution of the original problem.

EXAMPLE 2

A semi-infinite string is given the displacement shown in Fig. 8.7a and released from rest. Determine its subsequent motion.

We first imagine the string extended to $-\infty$ and released from rest in the extended initial configuration shown in Fig. 8.7b. Since $\theta(x) \equiv 0$, we have, from (5),

$$y(x,t) = \frac{\phi(x - at) + \phi(x + at)}{2}$$

where $\phi(x)$ is the displacement function shown in Fig. 8.7b.‡ We thus have two displacement waves, each of shape defined by $\frac{1}{2}\phi(x)$, one traveling to the right and one traveling to the left along the string. Plots of these waves are shown in Fig. 8.8. An inspection of these configurations reveals the important fact that a displacement wave is reflected from a fixed or "closed" end without distortion but with reversal of sign.

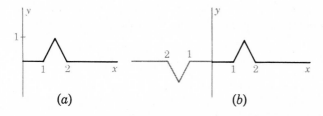

(a)   (b)

Figure 8.7
A semi-infinite string and its conceptual extension.

---

† This method of extending the initial conditions is sufficient but not necessary (see Exercise 9).
‡ If, as suggested by Fig. 8.7a, the graph of $\phi(x)$ has one or more corner points, then, strictly speaking, $\phi(x)$ does not describe an admissible initial displacement function. In fact, in the derivation of Eq. (5) both $f(x)$ and $g(x)$ were assumed to be twice differentiable, and therefore $\phi(x)$ must also be twice differentiable, which is not the case if there are points where the derivative of $\phi(x)$ is undefined. The apparent solutions obtained from Eq. (5) by overlooking this fact are therefore at best only formal solutions and are to be viewed with suspicion unless and until it is verified directly that they satisfy the given partial differential equation and its accompanying boundary and initial conditions. Questions concerning the existence and uniqueness of solutions of partial differential equations are quite difficult, and in our work we shall be concerned mainly with techniques for obtaining formal solutions. For an extended discussion of the problem of establishing the validity of solutions derived by purely formal means see, for instance, R. V. Churchill, "Fourier Series and Boundary Value Problems," 2d ed., pp. 126–163, McGraw-Hill, New York, 1963.

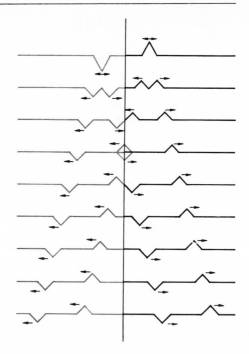

Figure 8.8
The propagation of a disturbance along a
semi-infinite string.

The motion of a finite string can be obtained as the motion of a segment of an
infinite string with suitably defined initial displacement and velocity. If the string is
given between 0 and l, say, we first imagine that it is extended from 0 to $-l$ with initial
conditions which are equal but opposite in sign to those for the actual string. Then we
extend the string to infinity in each direction subject to initial conditions which
duplicate with period $2l$ the initial configuration between $-l$ and l.† Finally, after we
obtain the solution for the infinite string we have thus created, its behavior for
$0 \le x \le l$ will be a complete description of the motion of the actual finite string in
which we are really interested.

EXAMPLE 3

A string of length l is given the initial displacement shown in Fig. 8.9 and released from
rest. Determine its subsequent motion.
The necessary extension of the string and one-half cycle of its motion are shown in

Figure 8.9
A finite string with initial displacement.

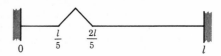

$$0 \qquad \frac{l}{5} \qquad \frac{2l}{5} \qquad\qquad\qquad l$$

---

† This, of course, is essentially the procedure we used in Sec. 6.3 to obtain the half-range sine
expansion of a function originally defined only over a finite portion of the real axis. The
relation of Fourier series to the problem of the vibrating string, and to the solution of the
wave equation in general, will become clear in the next section when we develop the method
of separation of variables.

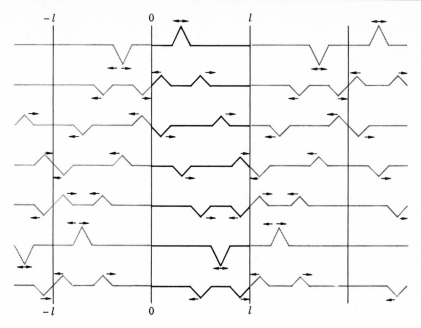

Figure 8.10
One-half cycle of the motion of a finite string.

Fig. 8.10. An inspection of Fig. 8.10 shows that the period of the motion, i.e., the least time for its return to its initial state, is just the time for either of the traveling waves to traverse a distance $2l$. In other words, since the velocity of the waves is a, the period is $2l/a$. The frequency of the vibrations is therefore $a/2l$. We shall encounter this formula again when we solve the wave equation by the method of separation of variables in the next section.

EXERCISES

1 A uniform string stretching from $-\infty$ to ∞ is given the initial displacement

$$y(x,0) = \begin{cases} 1 - |x| & x^2 < 1 \\ 0 & x^2 \geq 1 \end{cases}$$

and released from rest. Find the displacement of the string as a function of x and t and plot the displacement curves for $at = \frac{1}{4}, \frac{1}{2}, \frac{3}{4}, 1$. What is the transverse velocity of the string at $x = 0$? *Hint:* It will be convenient to express the initial displacement condition in terms of the unit step function.

2 In Exercise 1, plot the displacement of the string at $x = 3$ as a function of at.

3 A uniform string stretching from $-\infty$ to ∞ is given the initial displacement

$$y(x,0) = \begin{cases} \cos x & x^2 < \pi^2/4 \\ 0 & x^2 \geq \pi^2/4 \end{cases}$$

and released from rest. Find the displacement of the string as a function of x and t and plot the displacement curves for $at = \pi/4, \pi/2, 3\pi/4$. What is the transverse velocity of the string at $x = 0$?

4 In Exercise 3, plot the displacement of the string at $x = 2\pi$ as a function of at.

5 A uniform string stretching from $-\infty$ to ∞ while at rest in its equilibrium position is struck in such a way that the portion of the string between $x = -1$ and $x = 1$ is given a velocity of 1. Find the displacement of the string as a function of x and t, and plot the displacement curves for $at = 1$ and 2.

6 In Exercise 5, plot the displacement of the string at $x = 3$ as a function of at.

7 A uniform string stretching from 0 to ∞ is initially displaced into the curve $y = xe^{-x}$ and released from rest in that position. Find its displacement as a function of x and t.

8 A uniform string stretching from 0 to ∞ while at rest in its equilibrium position is struck in such a way that the portion of the string between $x = 1$ and $x = 4$ is given the velocity 2. Plot the velocity and displacement of the string at $x = 1$, $x = 4$, and $x = 10$ as functions of at.

9 A uniform string stretching from 0 to ∞ begins its motion with initial displacement $\phi(x)$ and initial velocity $\theta(x)$. Show that its motion can be found as the motion of the right half of a two-way infinite string provided merely that the intial displacement $\phi(-x)$ and the initial velocity $\theta(-x)$ for the negative extension of the string satisfy the condition

$$\phi(x) + \phi(-x) = -\frac{1}{a}\int_{-x}^{x} \theta(s)\, ds$$

10 If a semi-infinite string begins its motion with initial displacement $\phi(x) = (\sin x)/a$ and initial velocity $\theta(x) = 1$, and if the negative extension of the string is imagined to have the initial displacement $\phi(-x) = 0$, find the necessary initial velocity for the extended portion of the string.

11 The initial displacement of a two-way infinite string is

$$y(x,0) = \frac{1}{1 + x^2} \qquad x^2 < \infty$$

With what velocity must the string start to move if its subsequent motion is to consist solely of a wave traveling to the right?

12 The initial velocity of a two-way infinite string is

$$\dot{y}(x,0) = \begin{cases} \sin x & x^2 < \pi^2 \\ 0 & x^2 \geq \pi^2 \end{cases}$$

From what initial displacement must the string start to move if its subsequent motion is to consist solely of a wave traveling to the right?

13 Show that under the substitutions $u = x - at$ and $v = x + at$, the equation

$$\frac{\partial^2 y}{\partial t^2} = a^2 \frac{\partial^2 y}{\partial x^2}$$

becomes $\partial^2 y/(\partial u\, \partial v) = 0$. Hence show that $y = f(x - at) + g(x + at)$ is the most general solution of the one-dimensional wave equation.

14 In the d'Alembert solution of the wave equation, suppose that $f(x - at)$ and $g(x + at)$ are plotted as surfaces above the xt plane.
(a) Show that for all values of h and k and for all $t \geq 0$ the height of $f(x - at)$ is constant at all points above the line $x - at = h$ and the height of $g(x + at)$ is constant at all points above the line $x + at = k$.
(b) Show how the displacement and velocity at the position and time associated with the point (x,t) can be found by drawing the lines of the families $x - at = h$ and $x + at = k$ which pass through this point and observing where they intersect the x axis.
(c) Suppose that the disturbance which sets a two-way infinite string in motion is confined to the interval (x_1,x_2), and consider the four lines

$$x - at = x_1,\ x - at = x_2,\ x + at = x_1,\text{ and } x + at = x_2$$

Discuss the behavior of the string at a general "point" (x,t) in each of the six regions into which these lines divide the upper half of the xt plane.

15 Suppose that the disturbance which sets a one-way infinite string in motion is confined to the interval $0 < x_1 \leq x \leq x_2$ and consider the four lines $x - at = x_1$, $x - at = x_2$, $x + at = x_1$, $x + at = x_2$ and the two lines of the family $x - at = h$ which pass through the intersections of the lines $x + at = x_1$ and $x + at = x_2$ with the t axis. Discuss the behavior of the string at a general "point" (x,t) in each of the nine regions into which these six lines divide the first quadrant of the xt plane.

16 Discuss the possibility of finding solutions of the form $u = f(x + \lambda y)$ for the equation

$$A \frac{\partial^2 u}{\partial x^2} + 2B \frac{\partial^2 u}{\partial x \, \partial y} + C \frac{\partial^2 u}{\partial y^2} = 0$$

and show that according as $B^2 - AC$ is greater than, equal to, or less than zero, there will be two, one, or no (real) values of λ for which such solutions exist. (The given equation is said to be **hyperbolic**, **parabolic**, or **elliptic** in the respective cases, and the nature of its solutions and their properties is significantly different in each case.)

17 The equation

$$A(x,y) \frac{\partial^2 u}{\partial x^2} + 2B(x,y) \frac{\partial^2 u}{\partial x \, \partial y} + C(x,y) \frac{\partial^2 u}{\partial y^2} = f\left(u, \frac{\partial u}{\partial x}, \frac{\partial u}{\partial y}, x, y\right)$$

is said to be **hyperbolic**, **parabolic**, or **elliptic at a point** (x,y) according as $B^2(x,y) - A(x,y)C(x,y)$ is, respectively, greater than, equal to, or less than zero at that point. For what values of x and y is each of the following equations hyperbolic? Parabolic? Elliptic?

(a) $(y + 1) \dfrac{\partial^2 u}{\partial x^2} + 2x \dfrac{\partial^2 u}{\partial x \, \partial y} + \dfrac{\partial^2 u}{\partial y^2} = x + y$

(b) $(x + 1) \dfrac{\partial^2 u}{\partial x^2} + 2x \dfrac{\partial^2 u}{\partial x \, \partial y} + y \dfrac{\partial^2 u}{\partial y^2} = u$

(c) $(1 - y) \dfrac{\partial^2 u}{\partial x^2} + 2(1 - x) \dfrac{\partial^2 u}{\partial x \, \partial y} + (1 + y) \dfrac{\partial^2 u}{\partial y^2} - x \dfrac{\partial u}{\partial y} + y \dfrac{\partial u}{\partial x} = 0$

18 Let

$$A(x,y) \frac{\partial^2 u}{\partial x^2} + 2B(x,y) \frac{\partial^2 u}{\partial x \, \partial y} + C(x,y) \frac{\partial^2 u}{\partial y^2} = f\left(u, \frac{\partial u}{\partial x}, \frac{\partial u}{\partial y}, x, y\right)$$

and let $\phi(x,y) = c_1$ and $\psi(x,y) = c_2$ be the families of functions obtained from the values of y' which satisfy the equation

$$A(x,y)(y')^2 - 2B(x,y)y' + C(x,y) = 0$$

These functions are called the **characteristics** of the given partial differential equation, and if $B^2(x,y) - A(x,y)C(x,y) \geq 0$, they define families of (real) curves which are called **characteristic curves**.

(a) What are the characteristic curves of the one-dimensional wave equation?

(b) If the given partial differential equation is hyperbolic, show that the change of independent variables defined by the substitutions $r = \phi(x,y)$, $s = \psi(x,y)$ will reduce it to the standard form

$$\frac{\partial^2 u}{\partial r \, \partial s} = F\left(u, \frac{\partial u}{\partial r}, \frac{\partial u}{\partial s}, r, s\right)$$

(c) If the given partial differential equation is parabolic, show that the change of independent variables defined by the substitutions $r = x$, $s = \phi(x,y)$ will reduce it to the standard form

$$\frac{\partial^2 u}{\partial r^2} = F\left(u, \frac{\partial u}{\partial r}, \frac{\partial u}{\partial s}, r, s\right)$$

(d) If the given partial differential equation is elliptic, show that the change of independent variables defined by the substitutions $r + is = \phi(x,y), r - is = \psi(x,y)$ will reduce it to the standard form

$$\frac{\partial^2 u}{\partial r^2} + \frac{\partial^2 u}{\partial s^2} = F\left(u, \frac{\partial u}{\partial r}, \frac{\partial u}{\partial s}, r, s\right)$$

19 Using the substitutions described in the last exercise, solve each of the following equations:

(a) $\dfrac{\partial^2 u}{\partial x^2} + 3\dfrac{\partial^2 u}{\partial x\,\partial y} + 2\dfrac{\partial^2 u}{\partial y^2} = 0$ (b) $\dfrac{\partial^2 u}{\partial x^2} + 4\dfrac{\partial^2 u}{\partial x\,\partial y} + 4\dfrac{\partial^2 u}{\partial y^2} = 0$

(c) $\dfrac{\partial^2 u}{\partial x^2} + 4\dfrac{\partial^2 u}{\partial x\,\partial y} + 5\dfrac{\partial^2 u}{\partial y^2} = 0$ (d) $x\dfrac{\partial^2 u}{\partial x\,\partial y} + y\dfrac{\partial^2 u}{\partial y^2} = 0$

(e) $x\dfrac{\partial^2 u}{\partial x\,\partial y} - y\dfrac{\partial^2 u}{\partial y^2} = \dfrac{\partial u}{\partial y}$ (f) $\dfrac{\partial^2 u}{\partial x^2} - 4x^2\dfrac{\partial^2 u}{\partial y^2} = \dfrac{1}{x}\dfrac{\partial u}{\partial x}$

20 (a) Discuss the possibility of extending the d'Alembert solution to the two-dimensional wave equation

$$a^2\left(\frac{\partial^2 u}{\partial x^2} + \frac{\partial^2 u}{\partial y^2}\right) = \frac{\partial^2 u}{\partial t^2}$$

(b) Discuss the possibility of finding solutions of the form $e^{\lambda x + \mu y}$ for the equation

$$A\frac{\partial^2 u}{\partial x^2} + B\frac{\partial^2 u}{\partial x\,\partial y} + C\frac{\partial^2 u}{\partial y^2} + D\frac{\partial u}{\partial x} + E\frac{\partial u}{\partial y} + F = 0$$

where A, B, C, D, E, F are constants.

8.4 Separation of Variables

We are now ready to consider the solution of partial differential equations by the method of separation of variables. Although this method is not universally applicable, it suffices for most of the partial differential equations encountered in elementary applications in engineering and physics and leads directly to the heart of the branch of mathematics which deals with *boundary-value problems*.

The idea behind the method is the familiar mathematical stratagem of reducing a new problem to dependence upon an old one. In this case we attempt to convert the given partial differential equation into several ordinary differential equations, hoping that what we know about the latter will prove adequate for a successful continuation.

To illustrate the details of the procedure, let us again consider the wave equation, this time taking the undamped torsionally vibrating shaft of finite length as a specific representation:†

$$\frac{\partial^2 \theta}{\partial t^2} = a^2\frac{\partial^2 \theta}{\partial x^2}$$

We assume, as a working hypothesis, that solutions for the angle of twist θ exist as products of a function of x alone and a function of t alone:

$$\theta(x,t) = X(x)T(t)$$

---

† We do this not primarily for the sake of variety but because different end conditions, requiring somewhat different mathematical treatment, are possible for a torsionally vibrating shaft, whereas a stretched string must necessarily have each end fixed.

If this is the case, partial differentiation of θ amounts to total differentiation of one or the other of the factors of θ and we have

$$\frac{\partial^2 \theta}{\partial x^2} = X''T \quad \text{and} \quad \frac{\partial^2 \theta}{\partial t^2} = XT''$$

Substituting these into the wave equation, we obtain

$$XT'' = a^2 X''T$$

Dividing by XT then gives

(1)
$$\frac{T''}{T} = a^2 \frac{X''}{X}$$

as a necessary condition that $\theta(x,t) = X(x)T(t)$ should be a solution.

Now the left member of (1) is clearly independent of x. Hence (in spite of its appearance) the right-hand side of (1) must also be independent of x, since it is identically equal to the expression on the left. Similarly, each member of (1) must be independent of t. Therefore, being independent of both x and t, each side of (1) must be a constant, say μ, and we can write

$$\frac{T''}{T} = a^2 \frac{X''}{X} = \mu$$

Thus the determination of solutions of the original partial differential equation has been reduced to the determination of solutions of the two ordinary differential equations

$$T'' = \mu T \quad \text{and} \quad X'' = \frac{\mu}{a^2} X$$

Assuming that we need consider only real values of μ, there are three cases to investigate:

$$\mu > 0 \quad \mu = 0 \quad \mu < 0$$

If $\mu > 0$, we can write $\mu = \lambda^2$. In this case the two differential equations and their solutions are

$$T'' = \lambda^2 T \qquad\qquad X'' = \frac{\lambda^2}{a^2} X$$

$$T = Ae^{\lambda t} + Be^{-\lambda t} \qquad X = Ce^{\lambda x/a} + De^{-\lambda x/a}$$

But a solution of the form

$$\theta(x,t) = X(x)T(t) = (Ce^{\lambda x/a} + De^{-\lambda x/a})(Ae^{\lambda t} + Be^{-\lambda t})$$

cannot describe the undamped vibrations of a system because it is not periodic, i.e., does not repeat itself periodically as time increases. Hence, although product solutions of the differential equation exist for $\mu > 0$, they have no significance in relation to the problem we are considering.

If $\mu = 0$, the equations and their solutions are

$$T'' = 0 \qquad\qquad X'' = 0$$

$$T = At + B \qquad X = Cx + D$$

But, again, a solution of the form

$$\theta(x,t) = X(x)T(t) = (Cx + D)(At + B)$$

cannot describe a periodic motion. Hence the alternative $\mu = 0$ must also be rejected.

Finally, if $\mu < 0$, we can write $\mu = -\lambda^2$. Then the component differential equations and their solutions are

$$T'' = -\lambda^2 T \qquad\qquad X'' = -\frac{\lambda^2}{a^2} X$$

$$T = A \cos \lambda t + B \sin \lambda t \qquad X = C \cos \frac{\lambda}{a} x + D \sin \frac{\lambda}{a} x$$

In this case the solution

$$(2) \qquad \theta(x,t) = X(x)T(t) = \left(C \cos \frac{\lambda}{a} x + D \sin \frac{\lambda}{a} x\right)(A \cos \lambda t + B \sin \lambda t)$$

is clearly periodic, repeating itself every time t increases by $2\pi/\lambda$. In other words, $\theta(x,t)$ represents a vibratory motion with period $2\pi/\lambda$ or frequency $\lambda/2\pi$.

It remains now to find the value or values of λ and the constants A, B, C, and D. Since the admissible values of λ are determined by the boundary conditions of the problem, the continuation now varies in some respects, depending upon how the shaft is constrained at its ends. We shall discuss in turn the following simple cases (Fig. 8.11):

a. Both ends of the shaft are built-in, i.e., are constrained so that no twisting can take place.

b. Both ends of the shaft are free to twist.

c. One end of the shaft is built-in; the other is free to twist.

If both ends of the shaft are held fixed, we have the following conditions to impose upon the general expression for $\theta(x,t)$, assuming the x axis chosen along the shaft so that the left end of the shaft is at $x = 0$ and the right end is at $x = l$:

$$\theta(0,t) = \theta(l,t) = 0 \qquad \text{identically in } t$$

Substituting $x = 0$ into expression (2), we find

$$\theta(0,t) \equiv 0 = C(A \cos \lambda t + B \sin \lambda t)$$

This condition will obviously be fulfilled for all values of t if both A and B are zero. In this case, however, $\theta(x,t)$ is zero at all times, and the shaft remains motionless, a

Figure 8.11
End conditions for a shaft vibrating torsionally:
(a) fixed-fixed; (b) free-free; (c) fixed-free.

(a)

(b)

(c)

possible but trivial solution in which we have no interest. Hence we are driven to the other alternative, $C = 0$, which reduces (2) to the form

$$\theta(x,t) = D \sin \frac{\lambda}{a} x \, (A \cos \lambda t + B \sin \lambda t)$$

The second boundary condition, namely, that the right end of the shaft remains motionless at all times, requires that

$$\theta(l,t) \equiv 0 = D \sin \frac{\lambda l}{a} (A \cos \lambda t + B \sin \lambda t)$$

As before, we reject the possibility that $A = B = 0$, since it leads only to a trivial solution. Moreover, we cannot permit $D = 0$, since that too, with C already zero, leads to the trivial case. The only possibility which remains is that

$$\sin \frac{\lambda l}{a} = 0 \qquad \text{or} \qquad \frac{\lambda l}{a} = n\pi$$

From the continuous infinity of values of the parameter λ for which periodic product solutions of the wave equation exist, we have thus been forced to reject all but the values

$$(3) \qquad\qquad\qquad \lambda_n = \frac{n\pi a}{l} \qquad n = 1, 2, 3, \ldots$$

These and only these values of λ (still infinite in number, however) yield solutions which, in addition to being periodic, also satisfy the end, or boundary, conditions of the problem at hand. With these solutions, one for each admissible value of λ, we must now attempt to construct a solution which will satisfy the remaining conditions of the problem, namely, that the shaft starts its motion at $t = 0$ with a known angle of twist $\theta(x,0) = f(x)$ and a known angular velocity $\partial\theta/\partial t|_{x,0} = g(x)$ at every section.

Now the wave equation is linear, and thus if we have several solutions, their sum is also a solution. Hence writing the solution associated with the nth value of λ in the form

$$\theta_n(x,t) = \sin \frac{\lambda_n}{a} x \, (A_n \cos \lambda_n t + B_n \sin \lambda_n t)$$

$$= \sin \frac{n\pi x}{l} \left(A_n \cos \frac{n\pi a t}{l} + B_n \sin \frac{n\pi a t}{l} \right)^{\dagger}$$

it is natural enough (though perhaps optimistic, in view of the questions of convergence that are raised) to ask if an *infinite* series of *all* the θ_n's, say

$$(4) \qquad \theta(x,t) = \sum_{n=1}^{\infty} \theta_n(x,t) = \sum_{n=1}^{\infty} \sin \frac{n\pi x}{l} \left(A_n \cos \frac{n\pi a t}{l} + B_n \sin \frac{n\pi a t}{l} \right)$$

can be made to yield a solution fitting the initial conditions of angular displacement and velocity.

---

† The constants A and B now bear subscripts to indicate that they are not necessarily the same in the solutions associated with the different values of λ. The constant D can, of course, be absorbed into the constants A and B and need not be explicitly included.

This can be done, and in fact in this case the determination of the coefficients A_n and B_n requires nothing more than a simple application of Fourier series, as developed in Chap. 6. For if we set $t = 0$ in $\theta(x,t)$, we obtain from Eq. (4) and the given initial displacement condition the requirement that

$$\theta(x,0) \equiv f(x) = \sum_{n=1}^{\infty} A_n \sin \frac{n\pi x}{l}$$

The problem of determining the A_n's so that this will be true is nothing but the problem of expanding a given function $f(x)$ in a half-range sine series over the interval $(0,l)$. Using Theorem 2, Sec. 6.3, we have explicitly

$$A_n = \frac{2}{l} \int_0^l f(x) \sin \frac{n\pi x}{l} \, dx$$

To determine the B_n's, we note, further, that

$$\frac{\partial \theta}{\partial t} = \sum_{n=1}^{\infty} \sin \frac{n\pi x}{l} \left(-A_n \sin \frac{n\pi a t}{l} + B_n \cos \frac{n\pi a t}{l} \right) \frac{n\pi a}{l}$$

Hence, putting $t = 0$, we have, from the initial velocity condition,

$$\left. \frac{\partial \theta}{\partial t} \right|_{x,0} \equiv g(x) = \sum_{n=1}^{\infty} \left(\frac{n\pi a}{l} B_n \right) \sin \frac{n\pi x}{l}$$

This, again, merely requires that the B_n's be determined so that the quantities

$$\frac{n\pi a}{l} B_n$$

will be the coefficients in the half-range sine expansion of the known function $g(x)$. Thus

$$\frac{n\pi a}{l} B_n = \frac{2}{l} \int_0^l g(x) \sin \frac{n\pi x}{l} \, dx \quad \text{or} \quad B_n = \frac{2}{n\pi a} \int_0^l g(x) \sin \frac{n\pi x}{l} \, dx$$

Aside from convergence questions, our problem is now completely solved. We know that a uniform shaft with both ends restrained against twisting can vibrate torsionally at any of an infinite number of natural frequencies,

$$\frac{\lambda_n}{2\pi} = \frac{na}{2l} \quad \text{cycles/unit time} \quad n = 1, 2, 3, \ldots$$

If and when the shaft vibrates at a single one of these frequencies, we know that the angular displacements along the shaft vary periodically between extreme values proportional to

$$\sin \frac{n\pi x}{l}$$

Finally, assuming any initial conditions of velocity and displacement which satisfy the Dirichlet conditions, we know how to construct, at least formally,† the instantaneous deflection curve as an infinite series of the deflection curves associated with the respective natural frequencies λ_n.

---

† See the footnote to Example 2, Sec. 8.3.

The treatment of the shaft with both ends free follows closely the preceding analysis, once we obtain the proper analytic formulation of the end conditions. To obtain this formulation, we observe that at a free end, although we do not know the amount of twist, we do know that there is no torque acting through the end section since there is no shaft material beyond the end section. Recalling from the discussion of Sec. 8.2 the expression for the torque transmitted through a general cross section of a twisted shaft, we thus find that the free ends are characterized by the requirement that

$$E_s J \left.\frac{\partial \theta}{\partial x}\right|_{end} = 0$$

Since E_s is a nonzero constant of the material of the shaft, and since J cannot vanish for a shaft of uniform section such as we are considering, it follows that at a free end $\partial \theta / \partial x = 0$.

Returning to the original product solution (2), we find that

$$\frac{\partial \theta}{\partial x} = \left(-C \frac{\lambda}{a} \sin \frac{\lambda}{a} x + D \frac{\lambda}{a} \cos \frac{\lambda}{a} x\right)(A \cos \lambda t + B \sin \lambda t)$$

Substituting $x = 0$ and equating the result to zero, we obtain the condition

$$\frac{\lambda}{a} D(A \cos \lambda t + B \sin \lambda t) = 0 \qquad \text{for all } t$$

and from this we conclude that $D = 0$. Similarly, imposing the right-hand end condition by substituting $x = l$ and again equating the result to zero, we find

$$-C \frac{\lambda}{a} \sin \frac{\lambda l}{a} (A \cos \lambda t + B \sin \lambda t) = 0 \qquad \text{for all } t$$

Since we cannot permit $C = 0$, we must have

$$\sin \frac{\lambda l}{a} = 0 \qquad \text{or} \qquad \frac{\lambda l}{a} = n\pi$$

Thus, as in the last example, to have the end conditions of the problem fulfilled, λ must be restricted to one of the discrete set of values

$$\lambda_n = \frac{n\pi a}{l} \qquad n = 1, 2, 3, \ldots$$

Again we construct the product solution for each admissible value of λ, getting

$$\theta_n(x,t) = \left(\cos \frac{\lambda_n}{a} x\right)(A_n \cos \lambda_n t + B_n \sin \lambda_n t)$$

$$= \cos \frac{n\pi x}{l} \left(A_n \cos \frac{n\pi a t}{l} + B_n \sin \frac{n\pi a t}{l}\right)$$

and attempt to form an infinite series of these solutions,

$$\theta(x,t) = \sum_{n=1}^{\infty} \theta_n(x,t) = \sum_{n=1}^{\infty} \cos \frac{n\pi x}{l} \left(A_n \cos \frac{n\pi a t}{l} + B_n \sin \frac{n\pi a t}{l}\right)$$

which will satisfy the initial displacement condition $\theta(x,0) = f(x)$ and the initial velocity condition $\partial \theta / \partial t |_{x,0} = g(x)$.

To satisfy the initial displacement condition we must have

$$(5) \qquad \theta(x,0) \equiv f(x) = \sum_{n=1}^{\infty} A_n \cos \frac{n\pi x}{l}$$

which requires that the A_n's be the coefficients in the half-range cosine expansion† of the known function $f(x)$ over the interval $(0,l)$, that is, that

$$A_n = \frac{2}{l} \int_0^l f(x) \cos \frac{n\pi x}{l} \, dx$$

To satisfy the initial velocity condition, we must have

$$(6) \qquad \frac{\partial \theta}{\partial t}\bigg|_{x,0} \equiv g(x) = \sum_{n=1}^{\infty} \left(\frac{n\pi a}{l} B_n \right) \cos \frac{n\pi x}{l}$$

which requires that the quantities

$$\frac{n\pi a}{l} B_n$$

be the coefficients in the half-range cosine series for $g(x)$ over the interval $(0,l)$, that is, that

$$\frac{n\pi a}{l} B_n = \frac{2}{l} \int_0^l g(x) \cos \frac{n\pi x}{l} \, dx \qquad \text{or} \qquad B_n = \frac{2}{n\pi a} \int_0^l g(x) \cos \frac{n\pi x}{l} \, dx$$

We note, in passing, that since the admissible λ's are the same for the free-free shaft and the fixed-fixed shaft, the natural frequencies of the two systems are the same. The amplitudes through which they vibrate are not the same, however. In fact, for the fixed-fixed shaft we found that the distribution of amplitudes along the shaft was given by the function $\sin (n\pi x/l)$, whereas for the free-free shaft the amplitudes are given by $\cos (n\pi x/l)$.

　　The case of the shaft with one end fixed and the other free can be disposed of quickly. Taking the fixed end at $x = 0$ and the free end at $x = l$, we have the two conditions

$$\theta(0,t) = 0 \qquad \text{and} \qquad \frac{\partial \theta}{\partial x}\bigg|_{l,t} = 0 \qquad \text{for all } t$$

Imposing the first of these upon the general product solution (2) gives

$$C(A \cos \lambda t + B \sin \lambda t) = 0$$

whence it follows that $C = 0$. Imposing the second then gives

$$\frac{\lambda}{a} D \cos \frac{\lambda l}{a} (A \cos \lambda t + B \sin \lambda t) = 0$$

from which we conclude that

$$\cos \frac{\lambda l}{a} = 0 \qquad \frac{\lambda l}{a} = \frac{(2n-1)\pi}{2} \qquad \text{and} \qquad \lambda_n = \frac{(2n-1)\pi a}{2l}$$

---

† In general, the half-range cosine expansion of a function begins with a constant term. This series does not, because we rejected earlier the possibility $\mu = 0$, which would have led to such a term. Had there been an acceptable product solution corresponding to $\mu = 0$, we would, of course, have had to add it to the solutions arising from the assumption $\mu = -\lambda^2$ when we constructed the infinite series for $\theta(x,t)$ (see Exercise 1).

The general solution of the problem, formed by adding together the product solutions corresponding to each λ_n, is therefore

$$\theta(x,t) = \sum_{n=1}^{\infty} \sin \frac{\lambda_n}{a} x \, (A_n \cos \lambda_n t + B_n \sin \lambda_n t)$$

$$= \sum_{n=1}^{\infty} \sin \frac{(2n-1)\pi x}{2l} \left[A_n \cos \frac{(2n-1)\pi a t}{2l} + B_n \sin \frac{(2n-1)\pi a t}{2l} \right]$$

To fit the initial displacement condition $\theta(x,0) = f(x)$, we must have

$$f(x) = \sum_{n=1}^{\infty} A_n \sin \frac{(2n-1)\pi x}{2l}$$

This is not quite the usual half-range sine-expansion problem, since the arguments of the various terms are not integral multiples of the fundamental argument $\pi x/l$. It is, however, the special half-range sine expansion over $(0,l)$ discussed in Exercise 21, Sec. 6.3, where the formula for the coefficients was shown to be

$$A_n = \frac{2}{l} \int_0^l f(x) \sin \frac{(2n-1)\pi x}{2l} \, dx$$

Similarly, to fit the initial velocity condition $\partial\theta/\partial t|_{x,0} = g(x)$, we must have

$$g(x) = \sum_{n=1}^{\infty} \left[\frac{(2n-1)\pi a}{2l} B_n \right] \sin \frac{(2n-1)\pi x}{2l}$$

which requires that

$$B_n = \frac{4}{(2n-1)\pi a} \int_0^l g(x) \sin \frac{(2n-1)\pi x}{2l} \, dx$$

EXERCISES

1 Discuss the restrictions implicitly imposed on $f(x)$ and $g(x)$ by the absence of constant terms in the series in Eqs. (5) and (6). What is the physical significance of these restrictions?

2 Verify that the solutions of the wave equation obtained in this section can all be written in the form

$$\theta(x,t) = f(x - at) + g(x + at)$$

as required by the d'Alembert theory.

3 Which of the following equations can be solved by the method of separation of variables?

(a) $a\dfrac{\partial^2 u}{\partial x \, \partial y} + bu = 0$

(b) $x^2 \dfrac{\partial^2 u}{\partial x^2} + y \dfrac{\partial^2 u}{\partial y^2} = 0$

(c) $a\dfrac{\partial^2 u}{\partial x^2} + b\dfrac{\partial^2 u}{\partial x \, \partial y} + c\dfrac{\partial u}{\partial y} = 0$

(d) $a\dfrac{\partial^2 u}{\partial x^2} + b\dfrac{\partial^2 u}{\partial y^2} + c\dfrac{\partial^2 u}{\partial z^2} = 0$

(e) $a\dfrac{\partial^2 u}{\partial x^2} + b\dfrac{\partial^2 u}{\partial x \, \partial y} + c\dfrac{\partial^2 u}{\partial y^2} = 0$

(f) $a\dfrac{\partial^2 u}{\partial x^2} + b\dfrac{\partial^2 u}{\partial y^2} + c\dfrac{\partial u}{\partial x} + d\dfrac{\partial u}{\partial y} = 0$

4 A uniform shaft fixed at one end and free at the other is twisted so that each cross section rotates through an angle proportional to the distance from the fixed end. If the shaft is released from rest in this position, find its subsequent angular displacement as a function of x and t.

5 A uniform shaft fixed at each end is twisted so that each cross section rotates through an angle proportional to $x(l - x)$, where l is the length of the shaft and x is the distance from the left end. If the shaft is released from rest in this position, find its subsequent angular displacement as a function of x and t.

6 A uniform shaft free at each end is twisted so that each cross section rotates through an angle proportional to $(2x - l)/2$, where l is the length of the shaft and x is the distance from the left end. If the shaft is released from rest in this position, find its subsequent angular displacement as a function of x and t.

7 Show that the natural frequencies of a uniform string are given by

$$f_n = \frac{n}{2l} \sqrt{\frac{Tg}{w}} \qquad \text{cycles/unit time}$$

where l = length of the string
        T = tension under which string is stretched
        w = weight of string per unit length

How does doubling the tension affect the pitch of the fundamental tone of the string? Why is it that the strings of most string instruments are either of different lengths or have their lengths changed by the performer as he plays?

8 A uniform string stretched between the points $(0,0)$ and $(l,0)$ is given the initial displacement

$$y(x,0) = f(x) = \begin{cases} x & 0 < x < l/2 \\ l - x & l/2 < x < l \end{cases}$$

and released from rest. Find its subsequent displacement as a function of x and t.

9 In Exercise 8, show that the displacement of the midpoint of the string varies between its successive maximum and minimum values as a linear function of at. *Hint:* Recall the result of Example 2, Sec. 6.2.

10 While in its equilibrium position, a uniform string stretched between the points $(0,0)$ and $(l,0)$ is given the initial velocity

$$\dot{y}(x,0) = g(x) = \begin{cases} ax/l & 0 < x < l/2 \\ a(l - x)/l & l/2 < x < l \end{cases}$$

Find its subsequent displacement as a function of x and t.

11 In Exercise 10, show that the displacement of the midpoint of the string is a periodic function of t of period $2l/a$ whose definition in one period is $f(t) = (lat - a^2 t^2)/4l$.

12 While in its equilibrium position, a uniform string stretched between the points $(0,0)$ and $(l,0)$ is given the initial velocity

$$\dot{y}(x,0) = g(x) = \begin{cases} 0 & 0 < x < (l - k)/2 \\ a/k & (l - k)/2 < x < (l + k)/2 \\ 0 & (l + k)/2 < x < l \end{cases}$$

Find its subsequent displacement as a function of x and t. Does your answer appear to have a meaningful limit as $k \to 0$? If so, to what problem do you think it is the answer?

13 A uniform string stretched between the points $(0,0)$ and $(l,0)$ is given the following initial displacement and initial velocity:

$$y(x,0) = f(x) = \sin \pi x/l \qquad 0 < x < l$$

$$\dot{y}(x,0) = g(x) = \begin{cases} 0 & 0 < x < l/4 \\ a & l/4 < x < 3l/4 \\ 0 & 3l/4 < x < l \end{cases}$$

Find its subsequent displacement as a function of x and t.

14   The curved surface of a rod of length l is perfectly insulated against the flow of heat. The rod, which is so thin that heat flow in it can be assumed to be one-dimensional, is initially at the uniform temperature $u = 100°C$. Find the temperature at any point in the rod at any subsequent time if at $t = 0$ the temperature at each end of the rod is suddenly reduced to $0°C$ and maintained at that temperature thereafter. *Hint:* For heat flow in one dimension, the heat equation reduces to

$$\frac{\partial^2 u}{\partial x^2} = a^2 \frac{\partial u}{\partial t}$$

15   In Exercise 14, find the temperature at the midpoint of the rod and show that for arbitrarily small positive values of t it is different from zero. What does this appear to say about the rate at which thermal disturbances are propagated? Do you agree with this conclusion? *Hint:* Consider $u(l/2, t)$ as a power series in the quantity

$$z = \exp\left(-\pi^2 t/a^2 l^2\right)$$

and recall that a power series cannot converge to a constant over any interval unless it converges to that same constant at *every* point of its interval of convergence.

16   (a) Work Exercise 14 if the initial temperature distribution in the rod is $u(x,0) = 0°C$ and if the left and right ends are maintained, after $t = 0$, at the respective temperatures $u(0,t) = 0°C$ and $u(l,t) = 100°C$.
       (b) Work part (a) if the left and right ends of the rod are maintained at the respective temperatures $u(0,t) = 50°C$ and $u(l,t) = 100°C$.

17   (a) Work Exercise 14 if the initial temperature distribution in the rod is $u(x,0) = 100°C$, the left end of the rod is perfectly insulated, and the right end of the rod is maintained, after $t = 0$, at the constant temperature $u(l,t) = 0°C$. *Hint:* The temperature gradient through an insulated surface must be zero.
       (b) Work Exercise 14 if the initial temperature distribution in the rod is $u(x,0) = 0°C$, the left end of the rod is perfectly insulated, and the right end is maintained, after $t = 0$, at the constant temperature $u(l,t) = 100°C$.

18   Work Exercise 14 with both ends of the rod insulated and the initial temperature distribution in the rod given by

$$u(x,0) = f(x) = u_0 \frac{x}{l} \qquad 0 < x < l$$

19   (a) In Exercise 14, determine the quantity of heat remaining in the rod when $t = \alpha a^2 l^2$ and verify that it is equal to the quantity of heat in the rod at $t = 0$ minus the heat lost through the two ends between $t = 0$ and $t = \alpha a^2 l^2$. *Hint:* Recall the laws of heat flow given in Sec. 8.2 in connection with the derivation of the heat equation. Then, at the appropriate point, use the results of Exercise 15, Sec. 6.2.
       (b) Set up and check the heat balance at $t = \alpha a^2 l^2$ for the rod in part (a) of Exercise 16.

20   Show that the torsional vibrations of any uniform fixed-free shaft of length l are always the same as those of the left half of a suitably chosen fixed-fixed shaft of length $2l$. Is the converse true? That is, does the motion of the left half of a fixed-fixed shaft of length $2l$ always represent a possible motion of a fixed-free shaft of length l?

8.5 Orthogonal Functions and the General Expansion Problem

The three examples we considered in the last section embody all the significant features of the general boundary-value problem, but they give an exaggerated picture of the role of Fourier series in the final expansion required to fit the initial conditions. In general, a knowledge of Fourier series, as such, will not suffice to obtain the necessary expansion. Hence before we attempt to summarize the major characteristics of boundary-value problems, as illustrated in our examples, we shall consider an additional example or two in which Fourier series play no part.

EXAMPLE 1

A slender curved rod of length l has its curved surface perfectly insulated against the flow of heat. Its left end is maintained at the constant temperature $u = 0$, and its right end radiates freely into air of constant temperature $u = 0$. If the initial temperature distribution in the rod is given by

$$u(x,0) = f(x)$$

find the temperature at any point of the rod at any subsequent time.

Since the rod is very thin, and since its lateral surface is perfectly insulated, we shall assume that all points of any given cross section are at the same temperature and that the flow of heat in the rod is therefore entirely in the x direction. Thus we have to solve the heat equation [Eq. (14), Sec. 8.2] specialized to one-dimensional flow without heat sources:

(1)
$$\frac{\partial^2 u}{\partial x^2} = a^2 \frac{\partial u}{\partial t}$$

At the left end of the rod we have the obvious fixed-temperature condition $u(0,t) = 0$. At the right end we have a radiation condition which must be formulated analytically before we can proceed with our solution.

Now, according to **Stefan's law**, the amount of heat radiated from a given area dA in a given time interval dt is

$$dQ = \sigma(U^4 - U_0{}^4) \, dA \, dt$$

where U = absolute temperature of radiating surface

U_0 = absolute temperature of surrounding medium

σ = proportionality constant

This quantity of heat must have come to the surface by conduction from the interior of the body; hence as a second estimate for dQ we have the expression

$$dQ = -k \frac{\partial U}{\partial n} \, dA' \, dt$$

where k = thermal conductivity

$\partial U/\partial n$ = temperature gradient in direction perpendicular to dA

dA' = element of area congruent to dA and parallel to dA an infinitesimal distance below dA in the body

Therefore, equating the two expressions for dQ, we have

$$-k \frac{\partial U}{\partial n} \, dA' \, dt = \sigma(U^4 - U_0{}^4) \, dA \, dt$$

or, canceling the common factors and expanding $U^4 - U_0{}^4$ in powers of $(U - U_0)/U_0$,

$$-k \frac{\partial U}{\partial n} = \sigma(U^4 - U_0{}^4)$$

$$= \sigma\{[U_0 + (U - U_0)]^4 - U_0{}^4\}$$

$$= \sigma U_0{}^4 \left[\left(1 + \frac{U - U_0}{U_0}\right)^4 - 1\right]$$

$$= \sigma U_0{}^4 \left[4 \frac{U - U_0}{U_0} + 6 \left(\frac{U - U_0}{U_0}\right)^2 + \cdots\right]$$

Finally, if $U - U_0$ is small in comparison with U_0, as we shall suppose, we can neglect everything on the right except the first term, getting

$$-\frac{\partial U}{\partial n} = h(U - U_0) \qquad h = \frac{4\sigma U_0{}^3}{k}$$

In our problem, the normal to the surface from which radiation takes place, i.e., the right end of the rod, is the x axis. Hence, if we set $u = U - U_0$, which means that we measure temperatures from U_0 as a reference value, our second boundary condition becomes

(2)
$$-\frac{\partial u}{\partial n}\bigg|_{l,t} \equiv -\frac{\partial u}{\partial x}\bigg|_{l,t} = hu(l,t)$$

As before, we begin by assuming a product solution $u = XT$ and substituting it into the heat equation (1):

$$X''T = a^2 XT'$$

Dividing by XT, we have

$$\frac{X''}{X} = a^2 \frac{T'}{T}$$

from which, since x and t are independent variables, we conclude that

$$\frac{X''}{X} \qquad \text{and} \qquad a^2 \frac{T'}{T}$$

must equal the same constant, say μ.

If $\mu > 0$, say $\mu = \lambda^2$, we have, from the fraction involving T,

$$T' = \frac{\lambda^2}{a^2} T \qquad \text{and} \qquad T = C \exp \frac{\lambda^2 t}{a^2}$$

But this is absurd, since it indicates that T, and hence the temperature $u = XT$, increases beyond all bounds as t increases. Hence we must reject the possibility that $\mu > 0$.

If $\mu = 0$, we have simply

$$X'' = 0 \qquad\qquad T' = 0$$
$$X = Ax + B \qquad T = C$$

whence, letting $C = 1$, as we can without loss of generality since A and B are arbitrary, we obtain

$$u = XT = Ax + B$$

For this to be relevant to our problem it must reduce to zero when $x = 0$; hence $B = 0$. Moreover, it must satisfy Eq. (2) when $x = l$; hence $-A = hAl$, which implies that $A = 0$, since both h and l are positive constants. Thus $\mu = 0$ leads only to a trivial solution and must also be rejected.

Finally, if $\mu < 0$, say $\mu = -\lambda^2$, the component differential equations and their solutions are

$$X'' = -\lambda^2 X \qquad\qquad T' = -\frac{\lambda^2}{a^2} T$$

$$X = A \cos \lambda x + B \sin \lambda x \qquad T = C \exp\left(-\frac{\lambda^2 t}{a^2}\right)$$

and, again letting $C = 1$,

$$u = XT = (A \cos \lambda x + B \sin \lambda x) \exp [-\lambda^2 t/a^2]$$

To fit the left end condition we must have $u(0,t) \equiv 0 = A \exp[-\lambda^2 t/a^2]$. Hence $A = 0$, and u reduces to

$$u = B \exp [-\lambda^2 t/a^2] \sin \lambda x$$

To fit the right end condition (2) we must have

$$-B \exp\left[-\lambda^2 t/a^2\right] \lambda \cos \lambda l = hB \exp\left[-\lambda^2 t/a^2\right] \sin \lambda l$$

or, dividing out the exponential and collecting terms,

$$B(h \sin \lambda l + \lambda \cos \lambda l) = 0$$

If $B = 0$, the solution is trivial, since A is already known to be zero. Hence we must have

$$h \sin \lambda l + \lambda \cos \lambda l = 0$$

or

$$\tan \lambda l = -\frac{\lambda}{h} = -\frac{\lambda l}{hl}$$

or finally

$$\tan z = -\alpha z$$

where

$$z = \lambda l \quad \text{and} \quad \alpha = \frac{1}{hl}$$

This equation is not like the simple equations

$$\sin \lambda l = 0 \quad \text{and} \quad \cos \lambda l = 0$$

which determined the admissible values of λ in the examples of the last section, and its roots cannot be found by inspection. To determine them, it is convenient to consider the graphs of the two functions

$$y_1 = \tan z \quad \text{and} \quad y_2 = -\alpha z$$

The abscissas of the points of intersection of these curves (Fig. 8.12), being values of z for which $y_1 = y_2$, are then the solutions of the equation $\tan z = -\alpha z$. Obviously, there are an infinite number of roots z_n. However, unlike the roots of $\sin \lambda l = 0$ and $\cos \lambda l = 0$, they are not evenly spaced, although, as Fig. 8.12 indicates, the interval between successive values of z_n *approaches* π as n becomes infinite.

From each root z_n we obtain at once the corresponding value of λ

$$\lambda_n = \frac{z_n}{l}$$

and the associated product solution

$$u_n(x,t) = T_n(t)X_n(x) = B_n \exp\left(-\frac{\lambda_n^2 t}{a^2}\right) \sin \lambda_n x$$

Then we form a series of these particular solutions

(3) $$u(x,t) = \sum_{n=1}^{\infty} u_n(x,t) = \sum_{n=1}^{\infty} B_n \exp\left(-\frac{\lambda_n^2 t}{a^2}\right) \sin \lambda_n x$$

and attempt to determine the constants B_n so that the function defined by the series will satisfy the initial condition

$$u(x,0) = f(x)$$

Finally, putting $t = 0$ in (3), we find that this requires

(4) $$u(x,0) \equiv f(x) = \sum_{n=1}^{\infty} B_n \sin \lambda_n x$$

Thus, as in the examples in the last section, to satisfy the initial conditions we must be able to expand an arbitrary function in an infinite series of known functions determined by a differential equation and a set of boundary conditions. However, although the functions in terms of which the expansion is to be carried out are sines, the values of λ appearing in their arguments are spaced at incommensurable intervals and so the required series is *not* a Fourier series. Clearly, something is involved which includes Fourier series as a special case but is itself more general and more fundamental.

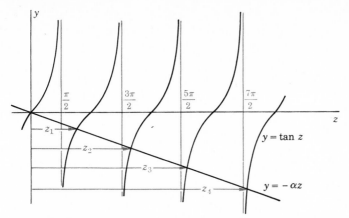

Figure 8.12
The graphical solution of the equation $\tan z = -\alpha z$.

If we review thoughtfully our earlier discussion of Fourier series (Sec. 6.2), it should be apparent that the decisive property of the set of functions $\{\cos (n\pi x/l),\ \sin (n\pi x/l)\}$ which made it possible to determine one by one the coefficients in the assumed expansion

$$f(x) = \tfrac{1}{2}a_0 + a_1 \cos \frac{\pi x}{l} + a_2 \cos \frac{2\pi x}{l} + \cdots + b_1 \sin \frac{\pi x}{l} + b_2 \sin \frac{2\pi x}{l} + \cdots$$

was that the integral of the product of any two members of the set taken over the appropriate interval is zero. For it was this which enabled us to multiply the series for $f(x)$ by $\cos (n\pi x/l)$ or $\sin (n\pi x/l)$ and eliminate all but one of the unknown coefficients simply by integrating from d to $d + 2l$.

Now sines and cosines are by no means the only functions from which sets can be constructed having the property that the integral between suitable limits of the product of two distinct members of the set is zero. In fact, the trigonometric functions which appear in Fourier expansions are merely one of the simplest examples of infinitely many such systems of functions, whose existence we shall soon establish.

DEFINITION 1   If a sequence of real functions

$$\{\phi_n(x)\} \qquad n = 1, 2, 3, \ldots$$

which are defined over some interval (a,b), finite or infinite, has the property that

$$\int_a^b \phi_m(x)\phi_n(x)\,dx \begin{cases} = 0 & m \neq n \\ \neq 0 & m = n \end{cases}$$

then the functions are said to form an orthogonal set on that interval.

DEFINITION 2   If the functions of an orthogonal set $\{\phi_n(x)\}$ have the property that

$$\int_a^b \phi_n^2(x)\,dx = 1 \qquad \text{for all values of } n$$

then the functions are said to be orthonormal on the interval (a,b).

Any set of orthogonal functions can easily be converted into an orthonormal set. In fact, if the functions of the set $\{\phi_n(x)\}$ are orthogonal, and if k_n is the (necessarily positive) value of $\int_a^b \phi_n^2(x)\, dx$, then the functions

$$\frac{\phi_1(x)}{\sqrt{k_1}}, \ \frac{\phi_2(x)}{\sqrt{k_2}}, \ \frac{\phi_3(x)}{\sqrt{k_3}}, \dots$$

are clearly orthonormal. It is therefore no specialization to assume that an orthogonal set of functions is also orthonormal.

DEFINITION 3 If a sequence of real functions $\{\phi_n(x)\}$ has the property that over some interval (a,b), finite or infinite,

$$\int_a^b p(x)\phi_m(x)\phi_n(x)\, dx \begin{cases} = 0 & m \neq n \\ \neq 0 & m = n \end{cases}$$

then the functions are said to be orthogonal with respect to the weight function $p(x)$ on that interval.

Any set of functions orthogonal with respect to a weight function $p(x)$ can be converted into a set of functions orthogonal in the first sense (Definition 1) simply by multiplying each member of the set by $\sqrt{p(x)}$ if, as we shall suppose, $p(x) \geq 0$ on the interval of orthogonality.

With respect to any set of functions $\{\phi_n(x)\}$ orthogonal over an interval (a,b), an arbitrary function $f(x)$ has a formal expansion analogous to a Fourier expansion, for we can write

$$(5) \qquad f(x) = a_1\phi_1(x) + a_2\phi_2(x) + a_3\phi_3(x) + \cdots + a_n\phi_n(x) + \cdots$$

Then, multiplying by $\phi_n(x)$ and integrating formally between the appropriate limits, a and b, we have

$$\int_a^b f(x)\phi_n(x)\, dx = a_1 \int_a^b \phi_1(x)\phi_n(x)\, dx + a_2 \int_a^b \phi_2(x)\phi_n(x)\, dx + \cdots$$
$$+ a_n \int_a^b \phi_n^2(x)\, dx + \cdots$$

From the property of orthogonality, all integrals on the right are zero except the one which contains a square in its integrand. Hence we can solve at once for a_n as the quotient of two known integrals:

$$a_n = \frac{\int_a^b f(x)\phi_n(x)\, dx}{\int_a^b \phi_n^2(x)\, dx}$$

However, although the orthogonality of the ϕ's makes it possible to determine the coefficients in the expansion (5), this property is not sufficient to guarantee that this series converges to $f(x)$ or even converges at all.

To pursue this matter a little further,† it is convenient to introduce the notion of a **null function**.

---

† An extended discussion of these ideas requires a knowledge of the Lebesgue integral, which we do not assume in this book.

DEFINITION 4 A real function $f(x)$ is said to be a null function on the interval (a,b) if $\int_a^b f^2(x)\, dx = 0$.

If $f(x)$ is identically zero, it is obviously a null function. However, a null function need not be identically zero. In fact, since the area under a curve is not altered when the ordinate of the curve at one or more isolated points is changed, it is clear that we can have $\int_a^b f^2(x)\, dx = 0$ even though $f(x)$ has nonzero values at a finite or countably infinite number of points between a and b. On the other hand, if there is any subinterval of (a,b), no matter how short, at all points of which $f(x)$ is different from zero, then $\int_a^b f^2(x)\, dx \neq 0$, and $f(x)$ is not a null function. From this it is not difficult to show that *a null function is zero at every point where it is continuous.* Prove

Clearly, if an arbitrary function is multiplied by a null function, the product is also a null function and hence the integral of such a product is zero. In particular, this means that any null function is orthogonal to every member of an orthogonal set $\{\phi_n(x)\}$. It is also conceivable that a nonnull function $g(x)$ might be orthogonal to every ϕ, that is, that we might have

$$\int_a^b g(x)\phi_n(x)\, dx = 0 \qquad \text{for all values of } n$$

In such a case, every coefficient in the expansion of $g(x)$ in terms of the ϕ's would be zero, and the series (5) would converge to zero at all points of (a,b) even though $g(x)$ was not a null function. That this is actually possible is easily shown by example. For instance, although the functions $\{\sin nx\}$ are readily shown to be orthogonal over the interval $(-\pi,\pi)$, not every function can be represented on this interval by a series of the form

$$a_1 \sin x + a_2 \sin 2x + \cdots + a_n \sin nx + \cdots$$

In particular, if $g(x) = x^2$, we have for the coefficients in its formal expansion

$$a_n = \frac{\int_{-\pi}^{\pi} x^2 \sin nx\, dx}{\int_{-\pi}^{\pi} \sin^2 nx\, dx}$$

$$= \frac{1}{\pi}\left[\frac{2x}{n^2} \sin nx - \left(\frac{x^2}{n} - \frac{2}{n^3} \right) \cos nx \right]_{-\pi}^{\pi} = 0$$

for all values of n. More generally, since every member of the set $\{\sin nx\}$ is odd, it is clear that no series of these functions can represent *any* even function on the interval $(-\pi,\pi)$.

Evidently, important as it is, orthogonality is not the whole story, and the functions in our orthogonal systems must possess some further property before the expansion (5) can be used with confidence. What is required is that the set of functions $\{\phi_n(x)\}$, in addition to being orthogonal, should also possess the property of **completeness** described in the following definition.

DEFINITION 5 A set of orthogonal functions $\{\phi_n(x)\}$ is said to be complete if the relation $\int_a^b f(x)\phi_n(x)\, dx = 0$ can hold for all values of n only if $f(x)$ is a null function.

If $\{\phi_n(x)\}$ is a complete orthogonal set, then clearly not all the coefficients in the expansion of a nonnull function $g(x)$ can be zero and thus no nontrivial function can have a trivial expansion. In fact, we have the following theorem.

THEOREM 1   If the formal expansion

$$a_1\phi_1(x) + a_2\phi_2(x) + \cdots + a_n\phi_n(x) + \cdots$$

of a function $g(x)$ in terms of the members of a complete orthonormal set $\{\phi_n(x)\}$ converges and can be integrated term by term, then the sum of the series differs from $g(x)$ by at most a null function; i.e., the sum of the series cannot differ from $g(x)$ over any interval of finite length.

Proof   By hypothesis, the series $\sum\limits_{n=1}^{\infty} a_n\phi_n(x)$ converges to some function; hence it is meaningful to consider the difference

$$h(x) = g(x) - \sum_{n=1}^{\infty} a_n\phi_n(x)$$

If we can prove that $h(x)$ is a null function, the assertion of the theorem will be established. To do this, consider

$$\int_a^b \phi_m(x)h(x)\,dx = \int_a^b \phi_m(x)\left[g(x) - \sum_{n=1}^{\infty} a_n\phi_n(x)\right]dx$$

$$= \int_a^b \phi_m(x)g(x)\,dx - \int_a^b \phi_m(x)\left[\sum_{n=1}^{\infty} a_n\phi_n(x)\right]dx$$

$$= \int_a^b \phi_m(x)g(x)\,dx - \sum_{n=1}^{\infty} a_n \int_a^b \phi_m(x)\phi_n(x)\,dx$$

The first integral in the last expression is simply the coefficient a_m. Furthermore, in the series, every integral is zero except the one for which $n = m$, which is equal to 1 since the ϕ's form an orthonormal set. Hence, we have finally,

$$\int_a^b \phi_m(x)h(x)\,dx = a_m - a_m = 0 \qquad m = 1, 2, 3, \ldots$$

Thus $h(x)$ is orthogonal to every one of the ϕ's. Therefore, since the ϕ's form a complete set, $h(x)$ must be a null function, and the theorem is established.

Closely associated with the concept of completeness is the concept of **closure,†** described in the following definitions.

DEFINITION 6   If $\lim\limits_{n\to\infty} \int_a^b [f(x) - S_n(x)]^2\,dx = 0$, the sequence of functions $S_n(x)$ is said to converge in the mean to $f(x)$.

DEFINITION 7   If $S_n(x) = a_1\phi_1(x) + a_2\phi_2(x) + \cdots + a_n\phi_n(x)$ is the nth partial sum of the expansion of $f(x)$ in terms of the members of an orthonormal set $\{\phi_n(x)\}$, and if $S_n(x)$ converges in the mean to $f(x)$ for every $f(x)$, then the set $\{\phi_n(x)\}$ is said to be closed.

---

† What we have called *completeness* some authors call *closure* and vice versa.

One important property of closed orthonormal sets is contained in the **theorem of Parseval**.

THEOREM 2 If $a_1\phi_1(x) + a_2\phi_2(x) + \cdots$ is the expansion of a function $f(x)$ in terms of the members of a closed orthonormal set $\{\phi_n(x)\}$, then

$$\sum_{n=1}^{\infty} a_n^2 = \int_a^b f^2(x)\, dx$$

Proof From the definition of closure, we have

$$\lim_{m\to\infty} \int_a^b \left[f(x) - \sum_{n=1}^{m} a_n\phi_n(x) \right]^2 dx = 0$$

or $$\lim_{m\to\infty} \int_a^b \left[f^2(x) - 2f(x)\sum_{n=1}^{m} a_n\phi_n(x) + \left\{ \sum_{n=1}^{m} a_n\phi_n(x) \right\}^2 \right] dx = 0$$

If we now perform the indicated integration, remembering that

$$\int_a^b f(x)\phi_n(x)\, dx = a_n$$

and observing that in the integral of the sum in the last term

$$\int_a^b \phi_m(x)\phi_n(x)\, dx = \begin{cases} 0 & m \neq n \\ 1 & m = n \end{cases}$$

we obtain $$\lim_{m\to\infty} \left[\int_a^b f^2(x)\, dx - 2\sum_{n=1}^{m} a_n^2 + \sum_{n=1}^{m} a_n^2 \right] = 0$$

or $$\sum_{n=1}^{\infty} a_n^2 = \int_a^b f^2(x)\, dx$$

as asserted.

As an immediate consequence of the last theorem we have the following result.

THEOREM 3 A closed orthonormal system $\{\phi_n(x)\}$ is also complete.

Proof To prove this, let us suppose, contrary to the theorem, that the closed orthonormal system $\{\phi_n(x)\}$ is not complete. This implies that there is at least one nonnull function $f(x)$ which is orthogonal to each of the ϕ's and which therefore has the property that every coefficient in its expansion in terms of the ϕ's is zero. However, since the set $\{\phi_n(x)\}$ is closed, we have from Parseval's theorem,

$$\int_a^b f^2(x)\, dx = \sum_{n=1}^{\infty} a_n^2$$

Hence, since each a_n is zero, as we have just observed, it follows that $f(x)$ is a null function, contrary to our assumption. This contradiction forces us to abandon the supposition that the closed set $\{\phi_n(x)\}$ is incomplete, and the theorem is established.

The converse of Theorem 3 is also true, but the proof of this fact is difficult and we shall not attempt it.

A great deal of important advanced mathematics deals with the properties of special orthogonal systems and with the validity of the formal expansion we have just created. In the next chapter we shall examine in some detail two such systems, the Bessel functions and the Legendre polynomials. Questions concerning the convergence of the generalized Fourier series (5), however, we shall not discuss, and in our work we shall assume not only that all the expansions we obtain converge but that they actually represent the functions which generated them.

Orthogonal functions arise naturally and inevitably in many types of problems in pure and applied mathematics.† Their existence in problems like those we have been considering is guaranteed by the following important theorem.‡

THEOREM 4   Given the differential equation

$$\frac{d[r(x)y']}{dx} + [q(x) + \lambda p(x)]y = 0$$

where $r(x)$ and $p(x)$ are continuous on the closed interval $a \le x \le b$ and $q(x)$ is continuous at least over the open interval $a < x < b$. If $\lambda_1, \lambda_2, \lambda_3, \ldots$ are the values of the parameter λ for which there exist nontrivial solutions of this equation possessing continuous first derivatives and satisfying the boundary conditions

$$a_1 y(a) - a_2 y'(a) = 0$$
$$b_1 y(b) - b_2 y'(b) = 0$$

where a_1, a_2, b_1, b_2 are any constants such that a_1 and a_2 are not both zero and b_1 and b_2 are not both zero, and if $y_1, y_2, y_3, \ldots$ are the solutions corresponding to these values of λ, then the functions $\{y_n(x)\}$ form a system orthogonal with respect to the weight function $p(x)$ over the interval (a,b).

Proof   To prove this, let y_m and y_n be the solutions associated with two distinct values of λ, say λ_m and λ_n. This means that

$$\frac{d(ry_m')}{dx} + (q + \lambda_m p)y_m = 0$$

$$\frac{d(ry_n')}{dx} + (q + \lambda_n p)y_n = 0$$

---

† It is interesting and instructive in this connection to reread the discussion of orthogonal polynomials in Sec. 4.6 and to refer to the discussion of the orthogonality of vectors in Sec. 11.2.

‡ This theorem and the boundary-value problem with which it deals are usually associated with the names of the Swiss mathematician J. C. F. Sturm (1803–1855) and the French mathematician Joseph Liouville (1809–1882).

Now multiply the first of these equations by y_n and the second by y_m and then subtract the second equation from the first. The result, after transposing, is

$$(\lambda_m - \lambda_n) p y_m y_n = y_m \frac{d(ry_n')}{dx} - y_n \frac{d(ry_m')}{dx}$$

or, integrating between a and b,

$$(\lambda_m - \lambda_n) \int_a^b p y_m y_n \, dx = \int_a^b \left[y_m \frac{d(ry_n')}{dx} \right] dx - \int_a^b \left[y_n \frac{d(ry_m')}{dx} \right] dx$$

If we can prove that the integral on the left vanishes whenever m and n are different, we shall have established the orthogonality property of the functions of the set $\{y_n(x)\}$. This we shall prove by showing that the right-hand side of the last equation is zero. To do this we begin by integrating the terms on the right by parts:

$$\int_a^b \left[y_m \frac{d(ry_n')}{dx} \right] dx \xrightarrow[\substack{u = y_m, \\ du = y_m' \, dx}]{\substack{dv = d(ry_n') \\ v = ry_n'}} ry_m y_n' \Big|_a^b - \int_a^b ry_n' y_m' \, dx$$

$$\int_a^b \left[y_n \frac{d(ry_m')}{dx} \right] dx \xrightarrow[\substack{u = y_n, \\ du = y_n' \, dx}]{\substack{dv = d(ry_m') \\ v = ry_m'}} ry_n y_m' \Big|_a^b - \int_a^b ry_m' y_n' \, dx$$

When we subtract these expressions, the integrals which remain on the right cancel and we have

(6) $$\int_a^b \left[y_m \frac{d(ry_n')}{dx} \right] dx - \int_a^b \left[y_n \frac{d(ry_m')}{dx} \right] dx = r(y_m y_n' - y_m' y_n) \Big|_a^b$$

Now y_m and y_n are not merely solutions of the given differential equation. For every m and n, they also satisfy the boundary conditions

$$a_1 y(a) = a_2 y'(a) \qquad \text{and} \qquad b_1 y(b) = b_2 y'(b)$$

Substituting for $y'(a)$ and $y'(b)$ from these expressions into the evaluated antiderivative in (6), we obtain

$$\int_a^b \left[y_m \frac{d(ry_n')}{dx} \right] dx - \int_a^b \left[y_n \frac{d(ry_m')}{dx} \right] dx$$

$$= r(b) \left[y_m(b) \frac{b_1}{b_2} y_n(b) - \frac{b_1}{b_2} y_m(b) y_n(b) \right]$$

$$- r(a) \left[y_m(a) \frac{a_1}{a_2} y_n(a) - \frac{a_1}{a_2} y_m(a) y_n(a) \right] \equiv 0$$

If a_2 or b_2, or both, should be zero, this result can still be established by substituting for $y(a)$ or $y(b)$, or both, instead of for their derivatives, since a_1 and a_2 cannot both vanish nor can b_1 and b_2. Moreover, if $r(a) = 0$, then the first boundary condition becomes irrelevant; i.e., the integrated terms vanish at $x = a$ without the need of any condition on the solutions y_m and y_n. Likewise, if $r(b) = 0$, the second boundary condition is irrelevant. We have thus shown that under the conditions of the theorem

$$(\lambda_m - \lambda_n) \int_a^b p y_m y_n \, dx = 0$$

Since λ_m and λ_n were any two *distinct* values of λ, the difference $\lambda_m - \lambda_n$ cannot vanish. Hence

$$\int_a^b p y_m y_n \, dx = 0$$

and the theorem is established.

In each of the torsional vibration problems we considered in Sec. 8.4, the functions in terms of which we had to expand the initial conditions satisfied a differential equation and a set of boundary conditions included under Theorem 4. This (and not the coincidental fact that Fourier series were involved) explains why the final expansion could be carried out in each case.

EXAMPLE 1 (continued)

When we left Example 1 in order to develop the theory necessary to complete its solution, we were faced with the necessity of expanding the initial temperature $u(x,0) = f(x)$ in a series of the form (4),

$$f(x) = \sum_{n=1}^{\infty} B_n \sin \lambda_n x$$

where the functions in the set $\{\sin \lambda_n x\}$ were the solutions of the differential equation

$$X'' + \lambda^2 X = 0$$

which satisfied the conditions

$$X(0) = 0$$

$$hX(l) + X'(l) = 0$$

We now note that this equation and the accompanying boundary conditions are in all respects a special case covered by Theorem 4. In fact, with λ^2 written in place of λ, we have

$$r(x) = 1 \qquad q(x) = 0 \qquad p(x) = 1$$
$$a = 0 \qquad b = l$$
$$a_1 = 1 \qquad a_2 = 0 \qquad b_1 = h \qquad b_2 = -1$$

Hence, by Theorem 4, the functions $\{\sin \lambda_n x\}$ form a set orthogonal with respect to the weight function $p(x) = 1$ on the interval $(0,l)$.

To determine B_n we now multiply Eq. (4) by $\sin \lambda_n x$ and integrate term by term from 0 to l. Because of the orthogonality of the functions $\{\sin \lambda_n x\}$, every integral on the right vanishes except the one whose integrand contains $\sin^2 \lambda_n x$. Therefore

$$B_n = \frac{\int_0^l f(x) \sin \lambda_n x \, dx}{\int_0^l \sin^2 \lambda_n x \, dx}$$

or, evaluating the integral in the denominator and recalling that $z_n \equiv \lambda_n l$ satisfies the equation $\sin z_n = -\alpha z_n \cos z_n$, we have

$$B_n = \frac{2}{l(1 + \alpha \cos^2 z_n)} \int_0^l f(x) \sin \lambda_n x \, dx$$

With B_n determined, the formal solution of our heat-flow problem is now complete.

Problems involving second-order differential equations are not the only ones in which orthogonal functions arise. In particular, we have the following important theorem covering fourth-order systems, of which the vibrating beam is a special case.

THEOREM 5   Given the differential equation

$$\frac{d^2[r(x)y'']}{dx^2} + [q(x) + \lambda p(x)]y = 0$$

where $r(x)$ and $p(x)$ are continuous on the closed interval (a,b) and $q(x)$ is continuous at least on the open interval (a,b). If $\lambda_1, \lambda_2, \lambda_3, \ldots$ are the values of the parameter λ for which there exist nontrivial solutions of this equation possessing continuous third derivatives and satisfying the boundary conditions

$$a_1 y(a) - \alpha_1(ry'')'|_{x=a} = 0 \qquad a_2 y'(a) - \alpha_2(ry'')|_{x=a} = 0$$
$$b_1 y(b) - \beta_1(ry'')'|_{x=b} = 0 \qquad b_2 y'(b) - \beta_2(ry'')|_{x=b} = 0$$

where neither a_i and α_i nor b_i and β_i are both zero, and if $y_1, y_2, y_3, \ldots$ are the non-trivial solutions corresponding to these values of λ, then the functions $\{y_n(x)\}$ form a system orthogonal with respect to the weight function $p(x)$ over the interval (a,b).

EXAMPLE 2

A uniform cantilever of length l begins to vibrate with initial displacement $y(x,0) = f(x)$ and initial velocity $\partial y/\partial t|_{x,0} = g(x)$. Find its displacement at any point at any subsequent time.

     For definiteness let us assume that the built-in end of the beam is at the origin. Then, since the beam is of uniform cross section and bears no external load, we have to solve Eq. (11), Sec. 8.2,

$$a^2 \frac{\partial^4 y}{\partial x^4} = -\frac{\partial^2 y}{\partial t^2}$$

subject to the boundary conditions

$$y(0,t) = 0 \qquad \text{i.e., initial displacement at built-in end} = 0$$

$$\left.\frac{\partial y}{\partial x}\right|_{0,t} = 0 \qquad \text{i.e., slope at built-in end} = 0$$

$$\left.\frac{\partial^2 y}{\partial x^2}\right|_{l,t} = 0 \qquad \text{i.e., moment } EI\frac{\partial^2 y}{\partial x^2} \text{ at free end} = 0$$

$$\left.\frac{\partial^3 y}{\partial x^3}\right|_{l,t} = 0 \qquad \text{i.e., shear } \frac{\partial[EI(\partial^2 y/\partial x^2)]}{\partial x} \text{ at free end} = 0$$

     As usual, we begin by assuming a product solution $y(x,t) = X(x)T(t)$, substituting it into the given partial differential equation, getting $a^2 X^{iv}T = -XT''$, and then separating variables,

$$a^2 \frac{X^{iv}}{X} = -\frac{T''}{T}$$

Since x and t are independent variables, these two fractions must have a common constant value, say μ. If $\mu \le 0$, the solution for T cannot be periodic, as we know it must be to represent undamped vibrations. Hence we restrict μ to be positive† and write $\mu = \lambda^2$. This leads to the component differential equations

$$T'' = -\lambda^2 T \qquad \text{and} \qquad X^{iv} = \frac{\lambda^2}{a^2} X$$

---

† In vibration problems where it is clear that only periodic solutions are possible, engineers often take their initial assumption to be $y(x,t) = X(x)(A \cos \lambda t + B \sin \lambda t)$ as, in effect, we did in Example 3, Sec. 5.5, in studying the undamped vibrations of an electric network with only a finite number of degrees of freedom.

and the respective solutions

(7) $$T = A \cos \lambda t + B \sin \lambda t$$

(8) $$X = C \cos \left(\sqrt{\frac{\lambda}{a}} x \right) + D \sin \left(\sqrt{\frac{\lambda}{a}} x \right) + E \cosh \left(\sqrt{\frac{\lambda}{a}} x \right) + F \sinh \left(\sqrt{\frac{\lambda}{a}} x \right)$$

Imposing the first boundary condition, namely,

$$y(0,t) \equiv X(0)T(t) = 0$$

we find

$$(C + E)T(t) = 0$$

Since we cannot permit $T(t)$ to be identically zero, for then the entire solution would be trivial, we conclude that

$$C + E = 0$$

Imposing the second boundary condition, namely,

$$\left. \frac{\partial y}{\partial x} \right|_{0,t} \equiv X'(0)T(t) = 0$$

we find $(D + F) \sqrt{\lambda/a}\, T(t) = 0$. Hence $D + F = 0$.

From the third and fourth boundary conditions, namely,

$$\left. \frac{\partial^2 y}{\partial x^2} \right|_{l,t} \equiv X''(l)T(t) = 0 \quad \text{and} \quad \left. \frac{\partial^3 y}{\partial x^3} \right|_{l,t} \equiv X'''(l)T(t) = 0$$

we obtain, respectively,

$$\left[-C \cos \left(\sqrt{\frac{\lambda}{a}} l \right) - D \sin \left(\sqrt{\frac{\lambda}{a}} l \right) \right.$$
$$\left. + E \cosh \left(\sqrt{\frac{\lambda}{a}} l \right) + F \sinh \left(\sqrt{\frac{\lambda}{a}} l \right) \right] \frac{\lambda}{a} T(t) = 0$$

and

$$\left[C \sin \left(\sqrt{\frac{\lambda}{a}} l \right) - D \cos \left(\sqrt{\frac{\lambda}{a}} l \right) \right.$$
$$\left. + E \sinh \left(\sqrt{\frac{\lambda}{a}} l \right) + F \cosh \left(\sqrt{\frac{\lambda}{a}} l \right) \right] \left(\frac{\lambda}{a} \right)^{3/2} T(t) = 0$$

Hence, setting

(9) $$z = \sqrt{\frac{\lambda}{a}} l$$

we must have

$$-C \cos z - D \sin z + E \cosh z + F \sinh z = 0$$
$$C \sin z - D \cos z + E \sinh z + F \cosh z = 0$$

If we eliminate C and D from these equations by using the conditions $C + E = 0$ and $D + F = 0$, we obtain the system

(10)
$$E(\cosh z + \cos z) + F(\sinh z + \sin z) = 0$$
$$E(\sinh z - \sin z) + F(\cosh z + \cos z) = 0$$

These equations will have a solution other than the obvious trivial solution $E = F = 0$ if and only if the determinant of the coefficients is equal to zero. Thus we must have

$$\begin{vmatrix} \cosh z + \cos z & \sinh z + \sin z \\ \sinh z - \sin z & \cosh z + \cos z \end{vmatrix} = 0$$

or, expanding and simplifying,

$$\cosh z \cos z = -1$$

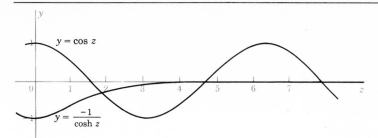

Figure 8.13
The graphical solution of the equation $\cos z = -1/\cosh z$.

The existence of infinitely many roots of this equation, i.e., $\cos z = -1/\cosh z$, can be inferred from Fig. 8.13, where the graphs of

$$y = \cos z \quad \text{and} \quad y = -\frac{1}{\cosh z}$$

are plotted.

From these roots, $z_1, z_2, z_3, \ldots$, we can find the relevant values of λ at once from Eq. (9):

$$\lambda_1 = \frac{az_1{}^2}{l^2}, \ \lambda_2 = \frac{az_2{}^2}{l^2}, \ \lambda_3 = \frac{az_3{}^2}{l^2}, \ldots$$

When z has any one of the values $z_1, z_2, z_3, \ldots$ (and E and F have the corresponding values E_n and F_n), Eqs. (10) become dependent and we can write either

$$\frac{E_n}{F_n} = -\frac{\sinh z_n + \sin z_n}{\cosh z_n + \cos z_n} \quad \text{or} \quad \frac{E_n}{F_n} = -\frac{\cosh z_n + \cos z_n}{\sinh z_n - \sin z_n}$$

as we choose. Using the former, we have

$$E_n = -C_n = -(\sinh z_n + \sin z_n)K_n$$

$$F_n = -D_n = (\cosh z_n + \cos z_n)K_n$$

where K_n is an arbitrary constant. Therefore, substituting into Eq. (8), we have

$$X_n(x) = C_n \cos\left(\sqrt{\frac{\lambda_n}{a}}\,x\right) + D_n \sin\left(\sqrt{\frac{\lambda_n}{a}}\,x\right) + E_n \cosh\left(\sqrt{\frac{\lambda_n}{a}}\,x\right) + F_n \sinh\left(\sqrt{\frac{\lambda_n}{a}}\,x\right)$$

$$= K_n(\sinh z_n + \sin z_n)\left[\cos\left(z_n\frac{x}{l}\right) - \cosh\left(z_n\frac{x}{l}\right)\right]$$

$$- K_n(\cosh z_n + \cos z_n)\left[\sin\left(z_n\frac{x}{l}\right) - \sinh\left(z_n\frac{x}{l}\right)\right]$$

Hence, absorbing K_n in the (still arbitrary) coefficients A_n and B_n in the expression for T_n given by Eq. (7) and redefining X_n to be the completely determined function

$$(11) \quad X_n = (\sinh z_n + \sin z_n)\left[\cos\left(z_n\frac{x}{l}\right) - \cosh\left(z_n\frac{x}{l}\right)\right]$$

$$- (\cosh z_n + \cos z_n)\left[\sin\left(z_n\frac{x}{l}\right) - \sinh\left(z_n\frac{x}{l}\right)\right]$$

we have

$$y(x,t) = \sum_{n=1}^{\infty} X_n(x)T_n(t) = \sum_{n=1}^{\infty} X_n(x)(A_n \cos \lambda_n t + B_n \sin \lambda_n t)$$

as the formal solution of the partial differential equation which meets the four given boundary conditions.

To satisfy the initial displacement condition, we must have

(12)
$$y(x,0) \equiv f(x) = \sum_{n=1}^{\infty} A_n X_n(x)$$

and to satisfy the initial velocity condition, we must have

(13)
$$\left.\frac{\partial y}{\partial t}\right|_{x,0} \equiv g(x) = \sum_{n=1}^{\infty} (\lambda_n B_n) X_n(x)$$

Thus, again, to satisfy the initial conditions we must be able to expand an arbitrary function in an infinite series of known functions, in this case the functions of the set $\{X_n(x)\}$ defined by Eq. (11). These bear little or no resemblance to the terms of a Fourier series, but the required expansions can easily be carried out using the orthogonality of the X's which is guaranteed by Theorem 5 (and of course their completeness, which, as usual, we must assume). In fact, with λ^2/a^2 written in place of λ, our problem is just the special case of Theorem 5 for which

$$r(x) = 1 \qquad q(x) = 0 \qquad p(x) = 1$$
$$a = 0 \qquad b = l$$
$$a_1 = 1 \qquad \alpha_1 = 0 \qquad b_1 = 0 \qquad \beta_1 = -1$$
$$a_2 = 1 \qquad \alpha_2 = 0 \qquad b_2 = 0 \qquad \beta_2 = -1$$

Hence, the functions of the set $\{X_n(x)\}$ are orthogonal with respect to the weight function $p(x) = 1$ over the interval $(0,l)$.

With the orthogonality of the X_n's now established, we can determine A_n and B_n immediately by multiplying Eqs. (12) and (13) by $X_n(x)$ and integrating from 0 to l. The results are

$$A_n = \frac{\int_0^l f(x)X_n(x)\,dx}{\int_0^l X_n^2(x)\,dx} \qquad \text{and} \qquad B_n = \frac{\int_0^l g(x)X_n(x)\,dx}{\lambda_n \int_0^l X_n^2(x)\,dx}$$

We are now in a position to summarize the main features of a simple boundary-value problem: by assuming that solutions for the dependent variable exist in the form of products of functions of the respective independent variables, the original partial differential equation is broken down into several ordinary differential equations, each involving a parameter λ which ranges over a continuous infinity of values.

When the boundary conditions of the problem are imposed upon the product solutions obtained by solving the component ordinary differential equations, it is necessary, in order to avoid solutions which are identically zero, that the parameter λ satisfy a certain equation. This equation is known as the **characteristic equation** of the problem, and its roots, in general infinite in number, are known as the **characteristic values, eigenvalues**, or **Eigenwerte**† of the problem. Only for them can nontrivial solutions be found satisfying both the partial differential equation and the given boundary conditions. In a vibration problem, the characteristic values determine the natural frequencies of the system, and the characteristic equation is therefore usually called the **frequency equation**. The solutions which correspond to the respective characteristic values are known as the **characteristic functions** or **eigenfunctions** of the problem. In a vibration problem, they are usually called the **normal modes**, since they define the relative amplitudes of the extreme positions between which the system oscillates when it is vibrating at a single natural frequency, i.e., in a "normal" manner.

To satisfy the initial conditions of the problem it is necessary to be able to express an arbitrary function as an infinite series of the characteristic functions of the problem.

---

† German for *characteristic values*.

This can be done in most cases of interest because under very general conditions the characteristic functions of a boundary-value problem form an orthogonal set over the particular interval related to the problem.

EXERCISES

1 If $r(a) = r(b)$, show that the conclusion of Theorem 4 follows equally well if the boundary conditions are of the form $y(a) = y(b)$ and $y'(a) = y'(b)$.

2 In the proof of Theorem 4, verify that the integrated result also vanishes in each of the remaining cases, $a_2 = 0$, $b_2 \neq 0$; $a_2 \neq 0$, $b_2 = 0$; $a_2 = b_2 = 0$.

3 Verify by direct integration that the characteristic functions of Example 1 are orthogonal over the interval (a,b).

4 In Example 1, if $\alpha = 1$, compute the values of z_1, z_2, and z_3, and determine the first three coefficients in the expansion of the initial condition $u(x,0) \equiv f(x) = x$.

5 In Example 1, if the left end of the rod is perfectly insulated, determine the characteristic equation, show that it has infinitely many roots, and prove that $z_{n+1} - z_n$ approaches π as n becomes infinite.

6 Find the temperature $u(x,t)$ in a slender rod of length l whose curved surface and left end are perfectly insulated and whose right end radiates freely into air of constant temperature $0°C$ if the rod is initially at the temperature $100°C$ throughout.

7 In Example 1, if both ends of the rod radiate freely into air of constant temperature, $0°C$, determine the characteristic equation, show that it has infinitely many roots, and prove that $z_{n+1} - z_n$ approaches π as n becomes infinite.

8 A slender rod of length l has its curved surface and left end perfectly insulated. Its right end radiates freely into air of constant temperature $70°C$. Initially the temperature throughout the rod is $100°C$. Find the temperature of the rod at any point at any subsequent time. *Hint:* Let $U = u - 70$ be the dependent variable.

9 Work Example 1 if the left end of the rod is maintained at the constant temperature $100°C$.

10 A uniform shaft of length l has its left end fixed and its right end elastically restrained, i.e., attached to a torsional spring which applies a restoring torque proportional to the angle through which the end section has rotated. Find the frequency equation for the shaft and verify that the nth natural frequency of the shaft is between the nth natural frequencies of a fixed-free and a fixed-fixed shaft of the same dimensions. Using Theorem 4, verify also that the normal modes of the shaft are orthogonal.

11 Find the frequency equation for a uniform shaft of length l whose left end is free and whose right end is elastically restrained. Verify that the nth natural frequency of the shaft is between the nth natural frequencies of a free-free shaft and a fixed-free shaft of the same dimensions. Using Theorem 4, verify also that the normal modes of the shaft are orthogonal.

12 Find the frequency equation for a uniform shaft of length l whose ends are elastically restrained by torsional springs of the same modulus, show that the nth natural frequency of the shaft is between the nth natural frequencies of a free-free shaft and a fixed-fixed shaft of the same dimensions, and verify that the normal modes of the shaft are orthogonal.

13 In Exercises 10 and 11 verify the orthogonality of the normal modes by direct integration.

14 What is the lowest natural frequency of a steel shaft 100 in long and 2 in in diameter which is restrained as in
    (a) Exercise 10   **(b)** Exercise 11   **(c)** Exercise 12
    if the modulus of the torsional springs is 1,000 in-lb/rad. Take $E_s = 12,000,000$ lb/in^2 and $\rho = 0.25$ lb/in^3.
    (d) What value of k will make the lowest natural frequency of the shaft in Exercise 10 halfway between the lowest natural frequencies of a fixed-free shaft and a fixed-fixed shaft of the same dimensions?

15 Prove Theorem 5.

16 Prove that the general linear second-order differential equation

$$p_0(x)y'' + p_1(x)y' + p_2(x)y = \lambda y$$

can be reduced to an equation of the Sturm-Liouville form (see Theorem 4) by multiplying it by the factor

$$\frac{1}{p_0(x)} \exp \left[\int_{x_0}^x \frac{p_1(x)}{p_0(x)} \, dx \right]$$

17 Show that the solutions of the equation

$$xy'' + y' + \frac{\mu}{x} y = 0$$

which satisfy each of the following sets of conditions form an orthogonal set on the indicated interval. In each case, solve the equation, impose the given boundary conditions, and verify the orthogonality property by direct integration:

(a) $y(1) = 0,\ y(e) = 0$ (b) $y'(1) = 0,\ y'(e) = 0$

(c) $y(1) = 0,\ y'(\sqrt{e}) = 0$ (d) $y(1) = 0,\ y(e) = y'(e)$

Hint: Note the result of Exercise 16, and then recall the discussion of Euler's equation in Example 3, Sec.2.6.

18 Work Exercise 17 for each of the following boundary-value problems:

(a) $x^2y'' + 2xy' + \mu y = 0$ $y(1) = 0,\ y(e) = 0$

(b) $x^2y'' + 3xy' + \mu y = 0$ $y(1) = 0,\ y'(\sqrt{e}) = 0$

19 Find the frequency equation and normal modes for the transverse vibration of a uniform beam whose ends are

(a) Hinged-hinged. *Hint:* A **hinged end** is one where a beam, though constrained so it cannot deflect, is still free to turn; i.e., an end where both the displacement and the moment are zero at all times. A hinged end is often referred to as a **simply supported end**.

(b) Fixed-fixed   (c) Free-free

(d) Fixed-hinged   (e) Free-hinged

20 Find the frequency equation for the transverse vibration of a uniform cantilever bearing a concentrated mass at the free end. *Hint:* At the free end, one boundary condition is that the shear, instead of being zero, is equal to the inertia force of the attached mass.

21 Find the frequency equation for the torsional vibrations of a shaft of length $2l$ which is clamped at $x = 0$ and free at $x = 2l$ if the radius of the portion of the shaft between $x = 0$ and $x = l$ is r_1 and the radius of the portion of the shaft between $x = l$ and $x = 2l$ is r_2. *Hint:* Solve the problem separately for the interval $(0,l)$ and the interval $(l,2l)$ and then, in addition to the two end conditions, impose the condition that at $x = l$ both the angle of twist and the transmitted torque are continuous.

22 Show that the system $\{\cos nx\},\ n = 0, 1, 2, 3, \ldots$, is orthogonal but not complete over the interval $(-\pi,\pi)$.

23 Find the frequency equation for a uniform torsional cantilever if a disk of polar moment of inertia I_p is attached to the free end of the shaft. *Hint:* At the free end of the shaft the boundary condition is that the torque, instead of being zero, is equal to the inertia torque of the disk.

24 Show that the normal modes in Exercise 23 are not orthogonal.

25 Find the frequency equation for the transverse vibrations of a uniform hinged-hinged beam bearing a concentrated mass at its midpoint.

26 Show that for an orthonormal system $\{\phi_n(x)\}$, whether closed or not, we have **Bessel's inequality**

$$\sum_{n=1}^{\infty} a_n{}^2 \le \int_a^b f^2(x) \, dx$$

where the a's are the coefficients in the generalized Fourier expansion of $f(x)$ in terms of the ϕ's and (a,b) is the interval of orthogonality. Using this result, show that

$$\lim_{n \to \infty} \int_a^b \phi_n(x)f(x)\, dx = 0$$

27 If $\{\phi_n(x)\}$ is an orthonormal set over the interval (a,b), show that the values of the c's which make

$$\int_a^b [f(x) - c_1\phi_1(x) - c_2\phi_2(x) - \cdots - c_n\phi_n(x)]^2\, dx$$

a minimum are the corresponding coefficients in the generalized Fourier expansion of $f(x)$ in terms of the ϕ's.

28 What is the minimum value of the integral in Exercise 27?

29 Specialize the antiderivative relation

$$\int p y_m y_n\, dx = \frac{1}{\lambda_m - \lambda_n}\, r(y_m y_n' - y_m' y_n)$$

obtained in the proof of Theorem 4, to the work of Example 1, evaluate the indeterminacy arising when $\lambda_m \to \lambda_n$, and use this result to check the value of $\int_0^l \sin^2 \lambda_n x\, dx$ obtained by direct integration.

30 Show that the substitution $t = \lambda x$ eliminates the parameter λ from the equation

$$\frac{d(x^u y')}{dx} + (cx^{u-2} + \lambda^2 x^u)y = 0$$

Hence show that the solutions of this equation are functions of the product λx. Then, applying L'Hospital's rule to the antiderivative relation obtained in the proof of Theorem 4, evaluate

$$\lim_{m \to n} \int p y_m y_n\, dx \equiv \lim_{\lambda_m \to \lambda_n} \int p(x)y(\lambda_m x)y(\lambda_n x)\, dx$$

Hint: Note that

$$\frac{dy(\lambda x)}{dx} = \lambda\, \frac{dy(\lambda x)}{d(\lambda x)} \quad \text{and} \quad \frac{dy(\lambda x)}{d\lambda} = x\, \frac{dy(\lambda x)}{d(\lambda x)}$$

8.6 Further Applications

Many problems in partial differential equations involve features not found in the simple examples we have used to explain the standard elementary theory. We cannot here investigate in detail the variations and extensions of this theory, but as illustrations we shall present several additional examples exhibiting techniques of practical interest. In the first example, we shall see how the analysis of the forced vibrations of a continuous system leads to a nonhomogeneous rather than a homogeneous partial differential equation. In the second, we shall see how Fourier integrals, rather than Fourier series, enter into problems where the boundary conditions fail to provide a characteristic equation and λ remains a continuous parameter. In the third, we shall see that though a partial differential equation may be separable, it may be impossible to make its product solutions fit the boundary conditions and so other methods must be used to solve it. In the fourth, we shall see how a partial differential equation involving three rather than two independent variables leads to a *double* series of characteristic functions and *two* separate expansion problems. The important matter of the application of Laplace transform methods to the solution of partial differential equations we shall consider in the next section.

EXAMPLE 1

A uniform string of length l is acted upon by a distributed periodic force

$$f(x,t) = \frac{w}{g}\, \phi(x) \sin \omega t$$

If the string is initially at rest in its equilibrium position, determine its subsequent motion, given that frictional effects are negligible.

From Eq. (2), Sec. 8.2, it is clear that the deflection of the string satisfies the partial differential equation

(1)
$$\frac{\partial^2 y}{\partial t^2} = a^2 \frac{\partial^2 y}{\partial x^2} + \phi(x) \sin \omega t$$

As in the case of a system with a single degree of freedom, the motion of the string consists of two parts, one described by the solution of the homogeneous equation

(2)
$$\frac{\partial^2 y}{\partial t^2} = a^2 \frac{\partial^2 y}{\partial x^2}$$

and the other described by a particular solution corresponding to the nonhomogeneous term $\phi(x) \sin \omega t$.

To find the solution of the homogeneous equation (2), i.e., to determine the free motion of the string, we assume a product solution

$$y_H(x,t) = X(x)T(t)$$

and proceed *exactly* as we did in solving the wave equation for the torsional vibrations of a fixed-fixed shaft of uniform cross section in Sec. 8.4. The result is

(3)
$$y_H(x,t) = \sum_{n=1}^{\infty} \sin \frac{n\pi x}{l} \left(A_n \cos \frac{n\pi a t}{l} + B_n \sin \frac{n\pi a t}{l} \right)$$

To find a particular solution of the nonhomogeneous equation (1), we observe that from physical considerations, the motion produced by the applied force must be periodic with the same period as the force. Moreover, since the system is assumed to be frictionless, it is clear that the motion of the string must be in phase with the force. Hence it is reasonable to assume a solution of the form

$$Y(x,t) = \Phi(x) \sin \omega t$$

We can now proceed in either of two ways. In the first place, we can substitute $Y(x,t)$ into the nonhomogeneous equation (1), divide out the common factor $\sin \omega t$, solve the resulting nonhomogeneous ordinary differential equation, namely,

$$-\omega^2 \Phi = a^2 \Phi'' + \phi(x)$$

and impose upon it the boundary conditions that

$$Y(0,t) = Y(l,t) = 0$$

i.e., the conditions that

$$\Phi(0) = \Phi(l) = 0$$

When $\Phi(x)$ has been determined so that these conditions are fulfilled, we can then construct the complete solution

(4)
$$y(x,t) = y_H(x,t) + Y(x,t)$$

$$= \sum_{n=1}^{\infty} \sin \frac{n\pi x}{l} \left(A_n \cos \frac{n\pi a t}{l} + B_n \sin \frac{n\pi a t}{l} \right) + \Phi(x) \sin \omega t$$

The initial conditions can now be imposed, giving

(5)
$$y(x,0) \equiv 0 = \sum_{n=1}^{\infty} A_n \sin \frac{n\pi x}{l}$$

and

(6)
$$\dot{y}(x,0) \equiv 0 = \sum_{n=1}^{\infty} \frac{n\pi a}{l} B_n \sin \frac{n\pi x}{l} + \omega\Phi(x)$$

From (5) we conclude that the A's are the coefficients in the half-range sine expansion of 0; hence $A_n = 0$ for all values of n. From (6) we conclude, similarly, that the quantities $\{(n\pi a/l)B_n\}$ are the coefficients in the half-range sine expansion of $-\omega\Phi(x)$; hence

$$B_n = -\frac{2\omega}{n\pi a} \int_0^l \Phi(x) \sin \frac{n\pi x}{l} \, dx$$

provided that ω is not equal to one of the natural frequencies $\{\omega_n\} = \{n\pi a/l\}$ of the system, i.e., provided that the system is not being driven at resonance.

On the other hand, before substituting $Y(x,t) = \Phi(x) \sin \omega t$ into the nonhomogeneous equation (1), we can expand $\phi(x)$ into a series of the characteristic functions of the homogeneous problem, i.e., we can express $\phi(x)$ as a half-range sine series with known coefficients, say

$$\phi(x) = \sum_{n=1}^{\infty} C_n \sin \frac{n\pi x}{l} \qquad \text{where} \qquad C_n = \frac{2}{l} \int_0^l \phi(x) \sin \frac{n\pi x}{l} \, dx$$

Then, assuming for the corresponding particular integral $\Phi(x)$ a half-range sine series with undetermined coefficients, say

$$\Phi(x) = \sum_{n=1}^{\infty} D_n \sin \frac{n\pi x}{l}$$

we have, on substituting

$$Y(x,t) = \left(\sum_{n=1}^{\infty} D_n \sin \frac{n\pi x}{l}\right) \sin \omega t$$

into Eq. (1) and then dividing out $\sin \omega t$,

$$-\omega^2 \sum_{n=1}^{\infty} D_n \sin \frac{n\pi x}{l} = -a^2 \sum_{n=1}^{\infty} D_n \left(\frac{n\pi}{l}\right)^2 \sin \frac{n\pi x}{l} + \sum_{n=1}^{\infty} C_n \sin \frac{n\pi x}{l}$$

Making this relation an identity by collecting terms and then equating to zero the coefficient of $\sin (n\pi x/l)$ for each value of n, we find

$$D_n = \frac{C_n}{(n\pi a/l)^2 - \omega^2} = \frac{C_n}{\omega_n^2 - \omega^2}$$

Hence

$$\Phi(x) = \sum_{n=1}^{\infty} D_n \sin \frac{n\pi x}{l} = \sum_{n=1}^{\infty} \frac{C_n}{\omega_n^2 - \omega^2} \sin \frac{n\pi x}{l}$$

and
$$Y(x,t) = \Phi(x) \sin \omega t = \left(\sum_{n=1}^{\infty} \frac{C_n}{\omega_n^2 - \omega^2} \sin \frac{n\pi x}{l}\right) \sin \omega t$$

Thus the complete formal solution of the nonhomogeneous equation becomes

$$y(x,t) = \sum_{n=1}^{\infty} \sin \frac{n\pi x}{l} \left(A_n \cos \frac{n\pi a t}{l} + B_n \sin \frac{n\pi a t}{l}\right)$$

$$+ \left(\sum_{n=1}^{\infty} \frac{C_n}{\omega_n^2 - \omega^2} \sin \frac{n\pi x}{l}\right) \sin \omega t$$

$$= \sum_{n=1}^{\infty} \sin \frac{n\pi x}{l} \left(A_n \cos \frac{n\pi a t}{l} + B_n \sin \frac{n\pi a t}{l} + \frac{C_n}{\omega_n^2 - \omega^2} \sin \omega t\right)$$

To satisfy the initial conditions, we must have

$$y(x,0) \equiv 0 = \sum_{n=1}^{\infty} A_n \sin \frac{n\pi x}{l}$$

whence $A_n \equiv 0$; and

$$\dot{y}(x,0) \equiv 0 = \sum_{n=1}^{\infty} \sin \frac{n\pi x}{l} \left(\frac{n\pi a}{l} B_n + \frac{\omega C_n}{\omega_n^2 - \omega^2} \right)$$

whence $\qquad \omega_n B_n + \dfrac{\omega C_n}{\omega_n^2 - \omega^2} = 0 \qquad$ or $\qquad B_n = - \dfrac{\omega C_n}{\omega_n(\omega_n^2 - \omega^2)}$

From the expression for B_n it is clear that the frequency ω of the impressed force must not coincide with any natural frequency ω_n of the string unless the corresponding coefficient C_n is equal to zero, i.e., unless the term

$$\sin \frac{n\pi x}{l} \equiv \sin \omega_n \frac{x}{a}$$

is missing from the half-range expansion of $\phi(x)$. If $\omega = \omega_n$ and $C_n \neq 0$ for some particular value of n, then the string is effectively being driven at a condition of resonance and displacements of arbitrarily large amplitudes will be built up (see Exercise 3).

EXAMPLE 2

A slender rod whose curved surface is perfectly insulated stretches from $x = 0$ to $x = \infty$. Find the temperature in the rod as a function of x and t if the left end of the rod is maintained at the constant temperature 0°C and if initially the temperature along the rod is given by $u(x,0) = f(x)$.

Exactly as in Example 1, Sec. 8.5, we find that the function

$$u = B \exp \left[-\lambda^2 t/a^2 \right] \sin \lambda x$$

satisfies the heat equation $\qquad \dfrac{\partial^2 u}{\partial x^2} = a^2 \dfrac{\partial u}{\partial t}$

and the boundary condition at the left end of the rod,

$$u(0,t) \equiv 0$$

Lacking a second boundary condition, however, we have no further restriction on λ. Therefore, instead of having an infinite set of *discrete* characteristic values λ_n, with corresponding solutions

$$u_n(x,t) = B_n \exp \left[-\lambda_n^2 t/a^2 \right] \sin \lambda_n x$$

we have a *continuous* family of solutions

$$u_\lambda(x,t) = B(\lambda) \exp \left[-\lambda^2 t/a^2 \right] \sin \lambda x$$

where the arbitrary constant B is now associated not with n but with the continuous parameter λ, which can assume *any* real value.

We cannot speak of an infinite series of particular solutions in this case. Instead of *adding* the product solutions for each value of n we therefore try *integrating* them over all values of λ, getting

(7) $\qquad u(x,t) = \displaystyle\int_{-\infty}^{\infty} B(\lambda) \exp \left(- \frac{\lambda^2 t}{a^2} \right) \sin \lambda x \, d\lambda$

By direct substitution it is easily verified that this integral is a solution of the heat equation.

It is now necessary to impose the initial condition $u(x,0) = f(x)$ on the solution $u(x,t)$. When we set $t = 0$ in Eq. (7), it is clear that this requires that

$$f(x) = \int_{-\infty}^{\infty} B(\lambda) \sin \lambda x \, d\lambda$$

This we recognize as just an instance of the Fourier integral considered in Sec. 6.6. There, in discussing what we called the Fourier sine integral, we saw [Eq. (15a), Sec. 6.6] that if

$$f(x) = \int_{-\infty}^{\infty} B(\lambda) \sin \lambda x \, d\lambda$$

then the coefficient function $B(\lambda)$ is given by the formula

$$B(\lambda) = \frac{1}{\pi} \int_{0}^{\infty} f(x) \sin \lambda x \, dx$$

Introducing the dummy variable s for x in the integral defining $B(\lambda)$ and then substituting for $B(\lambda)$ in Eq. (7), we have

$$u(x,t) = \int_{-\infty}^{\infty} \exp\left(-\frac{\lambda^2 t}{a^2}\right)\left[\frac{1}{\pi}\int_{0}^{\infty} f(s) \sin \lambda s \, ds\right] \sin \lambda x \, d\lambda$$

$$= \frac{1}{\pi} \int_{-\infty}^{\infty} \int_{0}^{\infty} \exp\left(-\frac{\lambda^2 t}{a^2}\right) f(s) \sin \lambda s \sin \lambda x \, ds \, d\lambda$$

which is the required solution.

EXAMPLE 3

Find the steady-state potential at any point of an infinitely long transmission line if a signal voltage $E_0 \cos \omega t$ is applied at the sending end $x = 0$.

Here we have to solve the telephone equation

(8) $$\frac{\partial^2 e}{\partial x^2} = LC \frac{\partial^2 e}{\partial t^2} + (RC + GL)\frac{\partial e}{\partial t} + RGe$$

subject to the boundary conditions

(9) $$e(0,t) = E_0 \cos \omega t \qquad e(x,t) \text{ bounded as } x \to \infty$$

If we assume a product solution $e(x,t) = X(x)T(t)$ and separate variables, we obtain

$$\frac{X''}{X} = \frac{LCT'' + (RC + GL)T' + RGT}{T} = \mu$$

Thus the factor T satisfies the differential equation

$$LCT'' + (RC + GL)T' + (RG - \mu)T = 0$$

and hence T must be of one or the other of the forms (see footnote on p. 385)

$$\varepsilon^{pt}(A \cos qt + B \sin qt)$$
$$\varepsilon^{pt}(At + B)$$
$$A\varepsilon^{p_1 t} + B\varepsilon^{p_2 t}$$

Now the steady-state behavior of a system which is acted upon by a periodic disturbance must be periodic. Under no circumstances, however, can the last two expressions represent periodic behavior. Moreover, the first of the three expressions can represent periodic behavior only if $p = 0$, which is impossible, since $p \equiv -(RC + GL)/2LC \neq 0$. Hence no product solution of Eq. (8) is capable of describing what we know the steady-state response of the line must be.

If we reconsider the problem, in an attempt to find an alternative method of solution, it seems reasonable to expect that under the given conditions the voltage along the line will vary harmonically with time while exhibiting attenuation and phase shift depending on the distance from the sending end. Hence, as an alternative approach, we are led to try an expression of the form

$$(10) \qquad e(x,t) = E_0 \varepsilon^{-ax} \cos{(\omega t + bx)}$$

If $a > 0$, this obviously satisfies each of the boundary conditions (9), and perhaps the constants a and b can be determined so that it will satisfy the differential equation (8) also.

If we substitute the tentative solution (10) into the telephone equation (8), divide out $E_0 e^{-ax}$, and collect terms, we obtain without difficulty

$$(a^2 - b^2 + LC\omega^2 - RG) \cos{(\omega t + bx)}$$
$$+ [2ab + \omega(RC + GL)] \sin{(\omega t + bx)} = 0$$

This will be an identity if and only if

$$(11) \qquad a^2 - b^2 = RG - LC\omega^2$$

$$(12) \qquad 2ab = -(RC + GL)\omega$$

Now by adding the square of Eq. (12) to the square of Eq. (11), we obtain

$$(a^2 + b^2)^2 = (RG - LC\omega^2)^2 + (RC + GL)^2 \omega^2$$

or

$$(13) \qquad a^2 + b^2 = \sqrt{(RG - LC\omega^2)^2 + (RC + GL)^2 \omega^2}$$

Finally, by solving (11) and (13) simultaneously, we find

$$a^2 = \tfrac{1}{2}[\sqrt{(RG - LC\omega^2)^2 + (RC + GL)^2 \omega^2} + (RG - LC\omega^2)]$$
$$b^2 = \tfrac{1}{2}[\sqrt{(RG - LC\omega^2)^2 + (RC + GL)^2 \omega^2} - (RG - LC\omega^2)]$$

From the form of these equations it is clear that a^2 and b^2 are both positive. Hence a and b are real, and, with their values now determined, Eq. (10) becomes the required solution. In a similar manner, of course, the steady-state response to a signal voltage of the form $E_0 \sin{\omega t}$ can be found.

By means of these results it is now possible to find the steady-state voltage corresponding to *any* periodic signal. For if $e(0,t) = f(t)$ is a periodic function with period $2p$, it can be expanded in a Fourier series

$$f(t) = \frac{1}{2}a_0 + a_1 \cos{\frac{\pi t}{p}} + a_2 \cos{\frac{2\pi t}{p}} + \cdots + b_1 \sin{\frac{\pi t}{p}} + b_2 \sin{\frac{2\pi t}{p}} + \cdots$$

and the steady-state solution for each of these terms can be found. Then, since the telephone equation is linear, the sum of the steady-state responses to each of these terms will be the steady-state response of the line to the entire signal $f(t)$. Moreover, if the input signal is not periodic, the steady-state response can still be found by a similar analysis after first expressing the input signal as a Fourier integral instead of a Fourier series.

EXAMPLE 4

A very thin sheet of metal coincides with the square in the xy plane whose vertices are the points $(0,0)$, $(1,0)$, $(1,1)$, and $(0,1)$. The upper and lower faces of the sheet are perfectly insulated, so that heat flow in the sheet is purely two-dimensional. Initially, the temperature distribution in the sheet is $u(x,y,0) = f(x,y)$, where $f(x,y)$ is a known function. If there are no sources of heat in the sheet, find the temperature at any point at any subsequent time, given that the edges parallel to the x axis are perfectly insulated and that the edges parallel to the y axis are maintained at the constant temperature $0°C$.

Here we have to solve the two-dimensional form of the heat equation [Eq. (14), Sec. 8.2],

$$(14) \qquad \frac{\partial^2 u}{\partial x^2} + \frac{\partial^2 u}{\partial y^2} = a^2 \frac{\partial u}{\partial t}$$

subject to the boundary conditions

$$(15) \qquad u(0,y,t) = 0 \qquad u(1,y,t) = 0$$

$$(16) \qquad \left.\frac{\partial u}{\partial y}\right|_{x,0,t} = 0 \qquad \left.\frac{\partial u}{\partial y}\right|_{x,1,t} = 0$$

and the initial condition

$$(17) \qquad u(x,y,0) = f(x,y)$$

Because we now have three independent variables, we begin with a product solution of the form

$$(18) \qquad u(x,y,t) = X(x)Y(y)T(t)$$

Then, substituting this into Eq. (14) and attempting to separate variables, we get

$$(19) \qquad \frac{X''}{X} = a^2 \frac{T'}{T} - \frac{Y''}{Y}$$

Although y and t enter together on the right-hand side of (19), they are both independent of x, and so each side of this equation must be a constant, say μ. Thus the factor X satisfies the equation

$$X'' = \mu X$$

If $\mu > 0$, say $\mu = \lambda^2$, we have

$$X = A \cosh \lambda x + B \sinh \lambda x$$

Now from the first of the boundary conditions (15), namely,

$$u(0,y,t) \equiv X(0)Y(y)T(t) = AY(y)T(t) = 0$$

it follows that $A = 0$. Likewise, from the second of the conditions (15), namely,

$$u(1,y,t) \equiv X(1)Y(y)T(t) = (B \sinh \lambda)Y(y)T(t) = 0$$

it follows that $B = 0$. Hence, when $\mu > 0$, the factor $X(x)$ vanishes identically and only a trivial solution is possible.

If $\mu = 0$, we have

$$X = Ax + B$$

and again the boundary conditions (15) can be satisfied only if $A = B = 0$.

Finally, if $\mu < 0$, say $\mu = -\lambda^2$, we have

$$X = A \cos \lambda x + B \sin \lambda x$$

From the first of the boundary conditions (15) we conclude that $A = 0$. The second condition requires that $(B \sin \lambda)Y(y)T(t) = 0$, and since we cannot permit B to be zero, we must have

$$\sin \lambda = 0 \qquad \text{and} \qquad \lambda = m\pi \qquad m = 1, 2, 3, \ldots$$

Therefore

$$(20) \qquad X_m(x) = \sin m\pi x \qquad m = 1, 2, 3, \ldots$$

Continuing with the other equation arising from (19), we now have

$$a^2 \frac{T'}{T} - \frac{Y''}{Y} = -m^2\pi^2$$

or

$$(21) \qquad \frac{Y''}{Y} = a^2 \frac{T'}{T} + m^2\pi^2$$

Since y and t are also independent, each member of the last equation must be a constant, say η. Thus the factor Y satisfies the equation

$$Y'' = \eta Y$$

If $\eta > 0$, say $\eta = v^2$, we have

$$Y = C \cosh vy + D \sinh vy$$

and
$$Y' = vC \sinh vy + vD \cosh vy$$

From the first of the boundary conditions (16), namely,

$$\left.\frac{\partial u}{\partial y}\right|_{x,0,t} \equiv X(x)Y'(0)T(t) = X(x)(vD)T(t) = 0$$

it follows that $D = 0$. Likewise, from the second of the conditions (16), namely,

$$\left.\frac{\partial u}{\partial y}\right|_{x,1,t} \equiv X(x)Y'(1)T(t) = X(x)(vC \sinh v)T(t) = 0$$

it follows that $C = 0$. Hence, when $\eta > 0$, the factor $Y(y)$ vanishes identically and only a trivial solution is possible.

If $\eta = 0$, we have
$$Y = Cy + D$$

and this time the boundary conditions (16) require that $C = 0$ but do not restrict D. Hence $Y = D$ is a possible solution for the factor Y.

Finally, if $\eta < 0$, say $\eta = -v^2$, we have

$$Y = C \cos vy + D \sin vy$$

and
$$Y' = -vC \sin vy + vD \cos vy$$

From the first of the conditions (16) we conclude again that $D = 0$. The second of the conditions (16) requires that

$$X(x)(-vC \sin v)T(t) = 0$$

and since we cannot permit $C = 0$, we must have

$$\sin v = 0 \quad\quad \text{and} \quad\quad v = n\pi \quad\quad n = 1, 2, 3, \ldots$$

Therefore
$$Y_n(y) = \cos n\pi y \quad\quad n = 1, 2, 3, \ldots$$

or including the solution $Y = \text{constant}$ obtained when $\eta = 0$,

$$(22) \quad\quad\quad\quad Y_n(y) = \cos n\pi y \quad\quad n = 0, 1, 2, 3, \ldots$$

From (21) it is now clear that the factor T satisfies the equation

$$T' = -\frac{(m^2 + n^2)\pi^2}{a^2} T$$

and hence that

$$(23) \quad\quad\quad\quad T = E_{mn} \exp\left[-(m^2 + n^2)\pi^2 t/a^2\right]$$

Therefore, combining (20) and (22) with (23), we can write the product solution (18) explicitly as

$$(24) \quad\quad u_{mn}(x,y,t) = E_{mn} \sin m\pi x \cos n\pi y \exp\left[-(m^2 + n^2)\pi^2 t/a^2\right]$$

None of the product solutions (24), by itself, can reduce to the required initial temperature distribution (17). Hence we must form a series of them and attempt to make this series satisfy the initial temperature condition. But now, since we have *two* independent parameters m and n in the product solutions, the general solution for u will be a *double* series:

$$u(x,y,t) = \sum_{m,n} u_{mn}(x,y,t) = \sum_{n=0}^{\infty} \sum_{m=1}^{\infty} E_{mn} \sin m\pi x \cos n\pi y \exp\left[-(m^2 + n^2)\pi^2 t/a^2\right]$$

When $t = 0$, this must reduce to $f(x,y)$; that is,

$$(25) \qquad f(x,y) = \sum_{n=0}^{\infty} \left(\sum_{m=1}^{\infty} E_{mn} \sin m\pi x \right) \cos n\pi y$$

Now the inner summation in (25) is a function only of n and x, say $G_n(x)$, and hence (25) can be written

$$f(x,y) = \sum_{n=0}^{\infty} G_n(x) \cos n\pi y$$

But for any particular value of x this is just the Fourier half-range cosine expansion of $f(x,y)$, thought of now as a function of y for $0 \le y \le 1$. Hence, by familar theory we can write

$$(26) \qquad G_n(x) = 2 \int_0^1 f(x,y) \cos n\pi y \, dy$$

But, by definition,

$$G_n(x) = \sum_{m=1}^{\infty} E_{mn} \sin m\pi x$$

and this is just the half-range sine expansion of the now known function $G_n(x)$ for $0 \le x \le 1$. Hence

$$(27) \qquad E_{mn} = 2 \int_0^1 G_n(x) \sin m\pi x \, dx$$

If we wish, we can now substitute for $G_n(x)$ from (26) into (27), getting

$$E_{mn} = 2 \int_0^1 \left[2 \int_0^1 f(x,y) \cos n\pi y \, dy \right] \sin m\pi x \, dx$$

$$= 4 \int_0^1 \int_0^1 f(x,y) \cos n\pi y \sin m\pi x \, dy \, dx$$

With E_{mn} determined for all values of m and n, the formal solution of the problem is now complete.

EXERCISES

1   Work Example 1 if

$$\phi(x) = \begin{cases} 0 & 0 < x < l/4 \\ 1 & l/4 < x < 3l/4 \\ 0 & 3l/4 < x < l \end{cases}$$

2   Work Example 1 if $\phi(x) = \sin (\pi x/l)$.

3   Work Example 1 if

$$\phi(x) = \begin{cases} x & 0 < x < l/2 \\ l - x & l/2 < x < l \end{cases}$$

4   Work Example 1 if $\phi(x) = x(l - x)$.

5   In Example 1, where does the first of the two methods break down if ω is one of the natural frequencies of the string?

6   Can the procedure illustrated in Example 1 be modified to obtain a description of the motion of the string when the impressed force is of the form $f(x,t) = \phi(x)\theta(t)$ where:

    (a) $\theta(t)$ is a general periodic function? How?

    (b) $\theta(t)$ is not periodic? How?

7   Work Example 1 if the impressed force is $(w/g)\phi(x)\theta(t)$, where $\phi(x) = \sin (\pi x/l)$ and $\theta(t)$ is the periodic function whose definition in one period is

$$\phi(t) = \begin{cases} 1 & 0 < t < p \\ -1 & p < t < 2p \end{cases}$$

8 Work Example 1 if the impressed force is $(w/g) \sin (\pi x/l) e^{-t}$.

9 Can the procedure illustrated in Example 1 be modified to obtain a description of the motion of the string when the frequency of the impressed force is one of the natural frequencies of the string? How?

10 Work Example 1 if the impressed force is

(a) $\dfrac{w}{g} \sin \dfrac{\pi x}{l} \sin \dfrac{2\pi a t}{l}$ (b) $\dfrac{w}{g} \sin \dfrac{\pi x}{l} \sin \dfrac{\pi a t}{l}$

11 Work Example 1 if the impressed force is $(w/g) \phi(x)\theta(t)$, where

$$\phi(x) = \begin{cases} x & 0 < x < l/2 \\ l - x & l/2 < x < l \end{cases}$$

and

(a) $\theta(t) = \sin \dfrac{2\pi a t}{l}$ (b) $\theta(t) = \sin \dfrac{3\pi a t}{l}$

12 A uniform string of length l is acted upon by a distributed frictional force equal at each point to

$$-\frac{cw}{g} \frac{\partial y}{\partial t}$$

where c is an arbitrary proportionality constant. Discuss the subsequent motion of the string, given that it starts with initial displacement $y(x,0) = g(x)$ and initial velocity $\dot{y}(x,0) = h(x)$. In particular, show that certain frequencies in the spectrum of the string may be overdamped while others are underdamped, and determine which ones are of each type.

13 A string subject to viscous damping as described in the last exercise is acted upon by a force per unit length equal to $(wa^2/gl) \phi(x) \sin \omega t$. Discuss the possibility of finding the steady-state motion of the string by assuming a solution of the form $Y_{ss}(x,t) = A(x) \sin \omega t + B(x) \cos \omega t$, substituting it into the appropriate differential equation, and determining $A(x)$ and $B(x)$ so that the equation is identically satisfied. Are the concepts of magnification ratio and phase shift relevant here?

14 A string subject to viscous damping as described in Exercise 12 is acted upon by a force per unit length equal to $(wa^2/gl) \sin (n\pi x/l) \sin \omega t$. Discuss the possibility of finding the steady-state motion of the string by assuming a solution of the form $\sin (n\pi x/l) (A \sin \omega t + B \cos \omega t)$, where A and B are constants. Are the concepts of magnification ratio and phase shift relevant here? Can the procedure described here be adapted to the determination of the steady-state response of the string to an impressed force per unit length equal to $(wa^2/gl) \phi(x) \sin \omega t$?

15 A uniform shaft of length l with both ends free, vibrating torsionally, is acted upon by a periodic impressed torque per unit length equal to $(J\rho a^2/gl^2) \phi(x) \sin \omega t$. If the initial angular displacement of the shaft is $\theta(x,0) = g(x)$, and if its initial angular velocity is $\dot{\theta}(x,0) = h(x)$, discuss the problem of determining its subsequent motion.

16 Work Exercise 15 for a uniform shaft of length l fixed at $x = 0$ and free at $x = l$.

17 Assuming that the function defined by formula (7) can legitimately be differentiated inside the integral sign, verify that it satisfies the one-dimensional heat equation.

18 Work Example 2 if the left end of the rod is perfectly insulated.

19 A slender rod of infinite length has its curved surface perfectly insulated. Find the steady-state temperature distribution in the rod if the temperature at the finite end of the rod varies according to the law $u(0,t) = \sin \omega t$. Explain how this result can be used to determine the steady-state temperature distribution produced by an arbitrary periodic temperature condition at the finite end of the rod.

20 A slender rod of length l has its curved surface perfectly insulated. Its right end is maintained at the constant temperature $u(l,t) = 0°C$. At the left end the temperature varies according to the law $u(0,t) = \sin \omega t$. Determine the steady-state temperature distribution in the rod. Explain how this result can be used to determine the steady-state temperature distribution produced in the rod by an arbitrary periodic temperature condition at the left end. *Hint:* Verify that λ can be chosen so that

$$u_1(x,t) = \sin \omega t \cos \frac{\lambda x}{l} \cosh \lambda \left(2 - \frac{x}{l}\right) - \cos \omega t \sin \frac{\lambda x}{l} \sinh \lambda \left(2 - \frac{x}{l}\right)$$

and

$$u_2(x,t) = \sin \omega t \cos \lambda \left(2 - \frac{x}{l}\right) \cosh \frac{\lambda x}{l} - \cos \omega t \sin \lambda \left(2 - \frac{x}{l}\right) \sinh \frac{\lambda x}{l}$$

are solutions of the one-dimensional heat equation. Then determine A_1 and A_2 so that $u(x,t) = A_1 u_1(x,t) + A_2 u_2(x,t)$ satisfies the boundary conditions of the problem.

21 A slender rod of length l has its curved surface perfectly insulated. Heat is generated within the rod at a rate per unit volume equal to $\phi(x)$. Find the temperature in the rod as a function of x and t if both ends of the rod are maintained at the constant temperature $u(0,t) = u(l,t) = 0°C$ and if the initial temperature distribution in the rod is $u(x,0) = g(x)$.

22 Work Exercise 21 if the left end of the rod is perfectly insulated while the right end is maintained at the constant temperature $u(l,t) = 0°C$.

23 Work Example 3 by replacing the signal voltage $E_0 \cos \omega t$ by $E_0 \varepsilon^{j\omega t}$ and assuming a solution of the form

$$e(x,t) = E_0 \varepsilon^{j\omega t + (a+jb)x}$$

24 If the transmission line in Example 3 is initially dead, i.e., if at $t = 0$ the potential and current along the line are identically zero, determine the complete response of the line, transient as well as steady-state, to the signal voltage $E_0 \cos \omega t$. *Hint:* show that if $-p \pm iq$ are the roots of the equation

$$LCm^2 + (RC + GL)m + (RG + \lambda^2) = 0$$

then

$$e_\lambda = \varepsilon^{-pt}[A(\lambda) \cos qt + B(\lambda) \sin qt] \sin \lambda x$$

is a solution of the telephone equation which is bounded as $x \to \infty$ and is zero for all values of t when $x = 0$. Then show that the steady-state solution found in Example 3 plus the integral of e_λ over all values of λ is a solution which satisfies both boundary conditions (9). Finally, determine $A(\lambda)$ and $B(\lambda)$ so that both e and $\partial e/\partial t$ are zero when $t = 0$.

25 Work Example 4, given that the edges from (0,0) to (0,1) and (1,0) are maintained at the constant temperature $0°C$ and that the other two edges are insulated.

26 Determine E_{mn} in Example 4 if $f(x,y) = x + y$.

27 A thin sheet of metal coincides with the square in the xy plane whose vertices are the points (0,0), (1,0), (1,1), and (0,1). Along the edge from (0,0) to (1,0) the temperature distribution $u(x,0) = f(x)$ is maintained. The other three edges are maintained at the temperature $0°C$. Find the steady-state temperature as a function of x and y.

28 Work Exercise 27 if the boundary conditions are

(a) $u(x,0) = f(x)$ $\quad \left.\dfrac{\partial u}{\partial y}\right|_{x,1} = \left.\dfrac{\partial u}{\partial x}\right|_{0,y} = \left.\dfrac{\partial u}{\partial x}\right|_{1,y} = 0$

(b) $u(x,0) = u(x,1) = 0$ $\quad \left.\dfrac{\partial u}{\partial x}\right|_{0,y} = 0 \quad u(1,y) = f(y)$

29 If an arbitrary temperature distribution exists along *each* of the edges of a square sheet of metal, how can the steady-state temperature distribution in the sheet be found?

30 A solid coincides with the cube whose vertices are the points (0,0,0), (1,0,0), (1,1,0), (0,1,0), (0,0,1), (1,0,1), (1,1,1), and (0,1,1). The temperature distribution $u(x,y,0) = f(x,y)$ is maintained over the face of the solid which lies in the xy plane. The other five faces are maintained at the constant temperature $0°C$. Find the steady-state temperature in the solid as a function of x, y, and z.

31 Work Exercise 30 if:

(a) The upper horizontal face of the solid is perfectly insulated.

(b) The four vertical faces of the solid are perfectly insulated.

(c) The front and back faces of the solid are perfectly insulated.

32 A thin sheet of metal bounded by the x axis and the lines $x = 0$ and $x = 1$ and stretching to infinity in the y direction has its upper and lower faces perfectly insulated and its vertical edges maintained at the constant temperature 0°C. Over its base, the temperature distribution $u(x,0) = 100$°C is maintained. Find the steady-state temperature at any point in the sheet.

33 Work Exercise 32 if the boundary conditions are

(a) $\dfrac{\partial u}{\partial x}\bigg|_{0,y} = \dfrac{\partial u}{\partial x}\bigg|_{1,y} = 0 \qquad u(x,0) = 100$

(b) $u(0,y) = 0 \qquad\qquad u(1,y) = 100 \qquad\qquad u(x,0) = 100x$

(c) $u(0,y) = 0 \qquad\qquad \dfrac{\partial u}{\partial x}\bigg|_{1,y} = 0 \qquad\qquad u(x,0) = 100$

(d) $u(0,y) = 0 \qquad\qquad u(x,0) = 0 \qquad\qquad u(1,y) = 100$

34 Work Exercise 32, given that the left edge and lower edge of the sheet are maintained at the constant temperature 0°C and the known distribution $u(1,y) = f(y)$ is maintained along the right edge.

35 In Example 4, find the temperature as a function of x, y, and t if the upper edge is maintained at the temperature distribution $u(x,1) = f(x)$, the other edges are maintained at the temperature 0°C, and at $t = 0$ the entire sheet is at the temperature $u(x,y,0) = 0$°C.

36 In Exercise 32, find the temperature as a function of x, y, and t if at $t = 0$ the entire sheet is at the temperature $u(x,y,0) = 0$°C.

37 Find the steady-state motion of a uniform beam of length l which is simply supported at each end and which is acted upon by a distributed load whose magnitude per unit length is $x(l - x) \sin \omega t$.

38 A uniform cantilever beam is built-in at $x = 0$ and free at $x = l$. Find the steady-state motion produced in the beam by a distributed load whose magnitude per unit length is $x \sin \omega t$.

39 Find the steady-state motion of a uniform cantilever beam of length l if the free end of the beam is forced to move so that its displacement is equal to $A \sin \omega t$.

40 Find the steady-state motion of a uniform cantilever beam of length l if a force equal to $F \sin \omega t$ acts at the free end of the beam.

41 Find the frequency equation of the uniform cantilever beam shown in Fig. 8.14a.

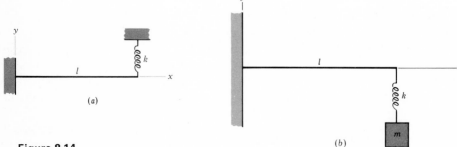

Figure 8.14

42 Assuming that the attached mass is guided so that it can only move in a direction perpendicular to the length of the beam, find the frequency equation of the uniform cantilever shown in Fig. 8.14b.

43 By substituting $\theta(x,t) = X(x) \sin \omega t$ into Eq. (7), Sec. 8.2, dividing out $\sin \omega t$, and integrating the resulting equation between the appropriate limits, show that the characteristic functions for the torsional vibrations of a nonuniform shaft fixed at $x = 0$ and free at $x = l$ satisfy the integral equation

$$X(x) = \frac{\omega^2 \rho}{E_s g} \int_0^x \frac{1}{J(r)} \int_r^l J(s)X(s)\, ds\, dr$$

44 Find the integral equation satisfied by the characteristic functions for the torsional vibrations of a nonuniform shaft which is:
(a) Fixed at $x = 0$ and at $x = l$
(b) Free at $x = 0$ and at $x = l$

45 The integral equations obtained in Exercises 43 and 44 are often used in the following way to approximate the natural frequencies of a nonuniform shaft vibrating torsionally when an exact solution is impossible. Assume a reasonable expression $X_1(x)$ as a first approximation to the characteristic function $X(x)$, substitute this into the right-hand side of the appropriate integral equation, and perform the indicated integrations, getting a second function, say $X_2(x)$. Repeat the process by substituting $X_2(x)$ and integrating to obtain $X_3(x)$, and so on. If the ratio $X_n(x)/X_{n+1}(x)$ becomes (approximately) the same for all values of x, then this ratio is (approximately) equal to $\omega^2 \rho / E_s g$ and $X_n(x)$ is (approximately) the characteristic function, or normal mode, corresponding to this frequency. This procedure is often referred to as the **method of Stodola**. Prove that this iterative process does indeed converge to the lowest natural frequency of the shaft and the corresponding characteristic function. *Hint:* Express $X_1(x)$ as an infinite series of the characteristic functions $\{\phi_i(x)\}$ of the shaft and observe that $\phi_i(x)$ satisfies the integral equation when ω has the corresponding value ω_i.

46 How can the method of Stodola be modified to obtain other natural frequencies besides the lowest? *Hint:* Observe that after ϕ_1 has been obtained, the orthogonality property of the ϕ's can be used to determine a function $X_1(x)$ in whose expansion the term ϕ_1 will not appear.

47 Obtain the integral equation for the problem of the torsional vibrations of a nonuniform shaft built-in at $x = 0$ and bearing a disk of polar moment of inertia I at the free end. *Hint:* At the free end of the shaft, the torque transmitted through the end section is equal to the inertia torque of the disk, $-I\, \partial^2\theta/\partial t^2|_{x=l}$.

48 Obtain the integral equation for the problem of the transverse vibration of a nonuniform string of length l.

49 Obtain the integral equation for the problem of the transverse vibration of a nonuniform cantilever beam of length l, and prove that the method of Stodola converges to the lowest natural frequency of the cantilever.

50 Determine the natural frequencies of a uniform square drumhead.

8.7 Laplace Transform Methods

In Chap. 7 we observed how the Laplace transformation converted an ordinary linear constant-coefficient differential equation into a linear algebraic equation from which the transform of the dependent variable could readily be found. In much the same way, the Laplace transformation can often be used to advantage in solving linear constant-coefficient partial differential equations in two independent variables. In such cases it leads not to an *algebraic* equation but to an *ordinary differential* equation in the transform of the dependent variable. The general procedure is as follows.

The given partial differential equation, with its accompanying boundary conditions and initial conditions, is transformed with respect to one of its independent variables, usually t. Partial derivatives with respect to this variable are, of course, transformed by the familiar formulas of Theorem 2 of Sec. 7.2 and its corollary. For partial derivatives with respect to the other independent variable we assume† that the operations

---

† This is justified by Theorems 3 and 6, Sec. 7.1.

of differentiating and taking the Laplace transform can be interchanged. Then, if the independent variables are x and t, say, we have

$$\mathscr{L}\left\{\frac{\partial f(x,t)}{\partial x}\right\} = \int_0^\infty \frac{\partial f(x,t)}{\partial x} e^{-st}\, dt = \frac{\partial}{\partial x}\int_0^\infty f(x,t)e^{-st}\, dt$$

$$= \frac{d}{dx}\,\mathscr{L}\{f(x,t)\}$$

the derivative in the term being a total derivative because $\mathscr{L}\{f(x,t)\}$ is not a function of t. Similar formulas hold, of course, for x derivatives of higher orders. Thus, the result of the transformation is an ordinary differential equation in $\mathscr{L}\{f(x,t)\}$ in which x is the independent variable and s enters as a parameter. Because s occurs in the coefficients of the differential equation, the arbitrary constants appearing in its complete solution will in general be functions of s which must be determined by imposing the transformed boundary conditions on the complete solution of the transformed differential equation. After this has been done, the inverse transformation is carried out and the solution to the original problem is obtained. The details of this process can best be made clear through examples.

EXAMPLE 1

A semi-infinite string is initially at rest in a position coinciding with the positive half of the x axis. At $t = 0$, the left end of the string begins to move along the y axis in a manner described by $y(0,t) = f(t)$, where $f(t)$ is a known function. Find the displacement $y(x,t)$ of the string at any point at any subsequent time.

The partial differential equation to be solved is, of course, the one-dimensional wave equation

$$(1) \qquad \frac{\partial^2 y}{\partial t^2} = a^2\,\frac{\partial^2 y}{\partial x^2}$$

subject to the boundary conditions

$$(2) \qquad y(0,t) = f(t)$$

$$(3) \qquad y(x,t) \text{ bounded as } x \to \infty$$

and the initial conditions

$$(4) \qquad y(x,0) = 0$$

$$(5) \qquad \left.\frac{\partial y}{\partial t}\right|_{x,0} = 0$$

If we take the Laplace transform of Eq. (1) *with respect to* t, we obtain

$$s^2\mathscr{L}\{y(x,t)\} - sy(x,0) - \left.\frac{\partial y}{\partial t}\right|_{x,0} = a^2\mathscr{L}\left\{\frac{\partial^2 y(x,t)}{\partial x^2}\right\} = a^2\,\frac{d^2}{dx^2}\,\mathscr{L}\{y(x,t)\}$$

or, using the initial conditions (4) and (5),

$$(6) \qquad \frac{d^2\mathscr{L}\{y(x,t)\}}{dx^2} - \frac{s^2}{a^2}\,\mathscr{L}\{y(x,t)\} = 0$$

Solving this ordinary differential equation for $\mathscr{L}\{y(x,t)\}$, we find without difficulty that

$$(7) \qquad \mathscr{L}\{y(x,t)\} = A(s)e^{-(s/a)x} + B(s)e^{(s/a)x}$$

To determine the coefficient functions $A(s)$ and $B(s)$, we observe first that if $y(x,t)$ remains finite as $x \to \infty$ [condition (3)], so must $\mathscr{L}\{y(x,t)\}$. Hence, $B(s)$ must be zero. Furthermore, putting $x = 0$ in (7), after $B(s)$ is set equal to zero, we have $\mathscr{L}\{y(0,t)\} = A(s)$, and from the boundary condition (2) we have $\mathscr{L}\{y(0,t)\} = \mathscr{L}\{f(t)\}$. Therefore (7) becomes

$$\mathscr{L}\{y(x,t)\} = \mathscr{L}\{f(t)\}e^{-(s/a)x} = \mathscr{L}\{f(t)\}e^{-(x/a)s}$$

The inverse of this can be found at once by suppressing the exponential factor and using Corollary 2 of Theorem 6, Sec. 7.4. The solution to our problem is therefore

$$y(x,t) = f\left(t - \frac{x}{a}\right) u\left(t - \frac{x}{a}\right)$$

which represents a wave traveling to the right along the string with velocity a. Evidently, the effect of this wave is to give the string at a general point the same displacement that the left end of the string had x/a units of time earlier.

EXAMPLE 2

A semi-infinite string is initially at rest in a position coinciding with the positive half of the x axis. A concentrated transverse force of magnitude F_0 moves along the string with constant velocity v, beginning at $t = 0$ at the point $x = 0$. Find the displacement $y(x,t)$ of the string at any point at any subsequent time.

In this problem, since there is an external force applied to the string, we must use the nonhomogeneous wave equation [Eq. (2), Sec. 8.2]

$$\frac{\partial^2 y}{\partial t^2} = a^2 \frac{\partial^2 y}{\partial x^2} + \frac{g}{w} F(x,t)$$

To obtain $F(x,t)$, we observe that a single concentrated load F_0 acting at the point $x = vt$ corresponds to a load per unit length which is infinite at $x = vt$ and zero everywhere else. Hence, since F_0 is assumed to act on the string in the negative y direction, we can write

$$F(x,t) = -F_0 \delta\left(t - \frac{x}{v}\right)$$

where $\delta(t - x/v)$ is the unit impulse, or δ function, discussed in Sec. 7.7. Our problem, therefore, is to solve the equation

(8) $$\frac{\partial^2 y}{\partial t^2} = a^2 \frac{\partial^2 y}{\partial x^2} - \frac{g}{w} F_0 \delta\left(t - \frac{x}{v}\right)$$

subject to the boundary conditions

(9) $$y(0,t) = 0$$

(10) $$y(x,t) \text{ bounded as } x \to \infty$$

and the initial conditions

(11) $$y(x,0) = 0$$

(12) $$\left.\frac{\partial y}{\partial t}\right|_{x,0} = 0$$

If we take the Laplace transform of Eq. (8) with respect to t and use the initial conditions (11) and (12), we obtain, just as in Example 1,

$$s^2 \mathscr{L}\{y(x,t)\} = a^2 \frac{d^2}{dx^2} \mathscr{L}\{y(x,t)\} - \frac{g}{w} F_0 e^{-(x/v)s}$$

or

(13) $$\frac{d^2}{dx^2} \mathscr{L}\{y(x,t)\} - \frac{s^2}{a^2} \mathscr{L}\{y(x,t)\} = \frac{gF_0}{a^2 w} e^{-(s/v)x}$$

The complete solution of this equation by the methods of Chap. 2 presents no difficulty, and we find for the complete solution

$$(14)\quad \mathscr{L}\{y(x,t)\} = A(s)e^{-(s/a)x} + B(s)e^{(s/a)x} + \begin{cases} \dfrac{gv^2 F_0}{w(a^2 - v^2)s^2}\, e^{-(s/v)x} & v \neq a \\[2ex] -\dfrac{gF_0}{2was}\, xe^{-(s/v)x} & v = a \end{cases}$$

In each case we must have $B(s) = 0$ in order that $\mathscr{L}\{y(x,t)\}$ should remain finite as $x \to \infty$. To determine $A(s)$ we have, from the boundary condition (9), the information that when $x = 0$,

$$\mathscr{L}\{y(x,t)\} \equiv \mathscr{L}\{y(0,t)\} = 0$$

Hence, substituting into Eq. (14), we obtain

$$A(s) = \begin{cases} -\dfrac{gv^2 F_0}{w(a^2 - v^2)s^2} & v \neq a \\[2ex] 0 & v = a \end{cases}$$

and therefore

$$\mathscr{L}\{y(x,t)\} = \begin{cases} \dfrac{gv^2 F_0}{w(a^2 - v^2)s^2}\, (e^{-(x/v)s} - e^{-(x/a)s}) & v \neq a \\[2ex] -\dfrac{gF_0}{2was}\, xe^{-(x/a)s} & v = a \end{cases}$$

Taking inverses, we have finally

$$(15)\quad y(x,t) = \frac{gv^2 F_0}{w(a^2 - v^2)}\left[\left(t - \frac{x}{v}\right)u\left(t - \frac{x}{v}\right) - \left(t - \frac{x}{a}\right)u\left(t - \frac{x}{a}\right)\right] \quad v \neq a$$

and

$$(16)\qquad\qquad y(x,t) = -\frac{gF_0}{2wa}\, x\, u\left(t - \frac{x}{a}\right) \quad v = a$$

Plots of (15) in the subsonic case $v = \frac{3}{4}a$, and the supersonic case $v = \frac{5}{4}a$, and of the transonic case $v = a$ described by (16) are shown in Fig. 8.15 for a typical time t. The discontinuity in $y(x,t)$ in the transonic case when the disturbance travels with exactly the propagation velocity a is interesting.

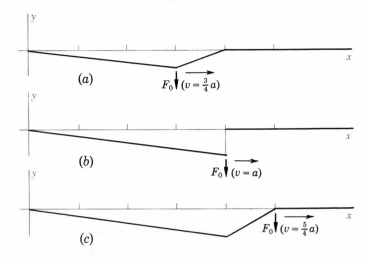

Figure 8.15
The displacement of a semi-infinite string produced by a concentrated force moving along the string with a velocity of (a) $\frac{3}{4}$, (b) 1, and (c) $\frac{5}{4}$ times the propagation velocity for the string.

EXAMPLE 3

A semi-infinite cable of negligible leakage and inductance is initially dead. At $t = 0$ an arbitrary signal voltage $E(t)$ is suddenly applied at the sending end. Find the potential $e(x,t)$ at any point on the cable at any subsequent time.

In this problem we have to solve the telegraph equation [Eq. (21a), Sec. 8.2],

$$(17) \qquad \frac{\partial^2 e}{\partial x^2} = a^2 \frac{\partial e}{\partial t} † \qquad a^2 = RC$$

subject to the boundary conditions

$$(18) \qquad e(0,t) = E(t)$$

$$(19) \qquad e(x,t) \text{ bounded as } x \to \infty$$

and the initial condition

$$(20) \qquad e(x,0) = 0$$

Taking the Laplace transform of (17) with respect to t and using the initial condition (20), we obtain

$$\frac{d^2}{dx^2} \mathscr{L}\{e(x,t)\} = a^2 s \mathscr{L}\{e(x,t)\}$$

as the ordinary differential equation satisfied by the transform of the potential. Solving this for $\mathscr{L}\{e(x,t)\}$, we find without difficulty that

$$(21) \qquad \mathscr{L}\{e(x,t)\} = A(s) \exp(-a\sqrt{s}\,x) + B(s) \exp(a\sqrt{s}\,x)$$

Since $e(x,t)$ and hence $\mathscr{L}\{e(x,t)\}$ are to remain finite as $x \to \infty$, it is necessary that $B(s) = 0$. To determine $A(s)$ we observe that when $x = 0$,

$$\mathscr{L}\{e(x,t)\} = \mathscr{L}\{E(t)\}$$

Hence, substituting into Eq. (21), we find

$$A(s) = \mathscr{L}\{E(t)\}$$

and

$$(22) \qquad \mathscr{L}\{e(x,t)\} = \mathscr{L}\{E(t)\} \exp(-ax\sqrt{s})$$

To determine $e(x,t)$ it will be necessary to use the convolution theorem, but before this can be done we must know the inverse of $\exp(-ax\sqrt{s})$. Up to this point in our work we have not encountered any function of t having this function of s for its transform. However, it can be shown (see Exercises 1 and 2) that

$$\mathscr{L}\left\{\frac{b\varepsilon^{-b^2/4t}}{2\sqrt{\pi}\,t^{3/2}}\right\}‡ = \exp(-b\sqrt{s})$$

Hence, taking $b = ax$ and setting up the convolution integral, we obtain from (22)

$$(23) \qquad e(x,t) = \frac{ax}{2\sqrt{\pi}} \int_0^t E(t-\lambda) \frac{\varepsilon^{-a^2 x^2/4\lambda}}{\lambda^{3/2}} \, d\lambda$$

---

† This is identical with the one-dimensional heat equation, and so all our conclusions apply equally well to the flow of heat in a slender, insulated, semi-infinite rod whose left end is maintained at an arbitrary time-dependent temperature $u_0(t)$.

‡ Here, as usual, to avoid confusion with the voltage, the symbol ε is used in place of e to denote the base of natural logarithms.

In particular, if $E(t)$ is a unit step voltage, we have, since $u(t - \lambda) = 1$ for $\lambda < t$ and $u(t - \lambda) = 0$ for $\lambda > t$,

$$e(x,t) = \frac{ax}{2\sqrt{\pi}} \int_0^t \frac{\varepsilon^{-a^2x^2/4\lambda}}{\lambda^{3/2}} \, d\lambda$$

If we now change the variable of integration from λ to z by the substitution $a^2x^2/4\lambda = z^2$, then $\lambda = a^2x^2/4z^2$, $d\lambda = -a^2x^2/2z^3 \, dz$, and the last integral becomes

$$e(x,t) = \frac{ax}{2\sqrt{\pi}} \int_\infty^{ax/2\sqrt{t}} \varepsilon^{-z^2} \frac{8z^3}{a^3x^3} \left(-\frac{a^2x^2}{2z^3} \, dz \right)$$

$$= \frac{2}{\sqrt{\pi}} \int_{ax/2\sqrt{t}}^\infty \varepsilon^{-z^2} \, dz$$

(24)
$$= \frac{2}{\sqrt{\pi}} \int_0^\infty \varepsilon^{-z^2} \, dz - \frac{2}{\sqrt{\pi}} \int_0^{ax/2\sqrt{t}} \varepsilon^{-z^2} \, dz$$

Under the substitution $z^2 = v$, the first integral becomes

$$\frac{1}{\sqrt{\pi}} \int_0^\infty \varepsilon^{-v} v^{(1/2)-1} \, dv = \frac{1}{\sqrt{\pi}} \Gamma(\tfrac{1}{2}) = 1$$

since $\Gamma(\tfrac{1}{2}) = \sqrt{\pi}$. Hence Eq. (24) can be written

$$e(x,t) = 1 - \frac{2}{\sqrt{\pi}} \int_0^{ax/2\sqrt{t}} \varepsilon^{-z^2} \, dz$$

(25)
$$= 1 - \text{erf} \, \frac{ax}{2\sqrt{t}}$$

where

(26)
$$\text{erf} \, \theta = \frac{2}{\sqrt{\pi}} \int_0^\theta \varepsilon^{-z^2} \, dz$$

This is the so-called **error function**, a tabulated function which can be found in most handbooks of mathematical tables.†

EXERCISES

1 If

$$f(\lambda) = \int_0^\infty \frac{e^{-z} e^{-\lambda/z}}{\sqrt{z}} \, dz$$

show by means of the substitution $u = \lambda/z$ that

$$f(\lambda) = \sqrt{\lambda} \int_0^\infty \frac{e^{-u} e^{-\lambda/u}}{u^{3/2}} \, du$$

---

† Actually, most handbooks list not the error function as here defined and used in physics and engineering but the so-called **probability integral** of mathematical statistics:

$$\Phi(\theta) = \frac{1}{\sqrt{2\pi}} \int_0^\theta \varepsilon^{-w^2/2} \, dw$$

If the substitution $z = w/\sqrt{2}$ is made in the error function (26), it becomes

$$\frac{2}{\sqrt{2\pi}} \int_0^{\sqrt{2}\,\theta} \varepsilon^{-w^2/2} dw$$

and we obtain the relation

$$\text{erf} \, \theta = 2\Phi(\sqrt{2}\,\theta)$$

Hence, by differentiating the first expression for $f(\lambda)$, show that

$$f'(\lambda) = -\frac{f(\lambda)}{\sqrt{\lambda}}$$

Solve this differential equation, using the fact that

$$f(0) = \Gamma(\tfrac{1}{2}) = \sqrt{\pi}$$

and show that

$$f(\lambda) = \sqrt{\pi}\,\exp\,(-2\sqrt{\lambda})$$

Finally, use this result to show that

$$\mathscr{L}\left\{\frac{e^{-b^2/4t}}{\sqrt{\pi t}}\right\} = \frac{\exp\,(-b\sqrt{s})}{\sqrt{s}}$$

2   Use the results of the last exercise, together with Theorem 8, Sec. 7.4, to show that

$$\mathscr{L}\left\{\frac{be^{-b^2/4t}}{2\sqrt{\pi}\,t^{3/2}}\right\} = \exp\,(-b\sqrt{s})$$

3   (a) In Example 3, what is the response of the line if $E(t)$ is a unit impulse voltage? *Hint:* Recall from Sec. 7.7 the relation between the response of a system to a unit step function and to a unit impulse.
(b) Using Eq. (25) and the appropriate Duhamel formula, obtain a formula different from Eq. (23) for the response of the line in Example 3 to a general voltage.

4   Using Laplace transform methods, determine the motion of a uniform string of length l whose initial displacement and initial velocity are, respectively, $y(x,0) = \sin\,(m\pi x/l)$ and $\dot{y}(x,0) = \sin\,(n\pi x/l)$, where m and n are integers. Can these results be used to obtain the motion of the string produced by arbitrary initial conditions? How?

5   Using Laplace transform methods, determine the response of a uniform string of length l to a distributed force $f(x,t) = \sin\,(n\pi x/l) \sin \omega t$ if the string is initially at rest in its equilibrium position. Explain how these results can be used to determine the response of the string to a distributed force $f(x,t) = g(x) \sin \omega t$, where $g(x)$ is defined arbitrarily on the interval $0 < x < l$, and to a distributed force $\sin\,(n\pi x/l)h(t)$, where $h(t)$ is an arbitrary periodic function whose frequency is distinct from each natural frequency of the string.

6   Work Example 2 given that the transverse force which moves along the string is a rectangular pulse of height F_0 initially acting on the portion of the string between $x = 0$ and $x = 1$.

7   A semi-infinite string whose weight per unit length is w has its left end fixed at the origin. The infinite end is fastened to a ring which slides without friction along a vertical rod. Initially, the string is at rest in a position coinciding with the positive x axis. At $t = 0$ the support which maintained the string in its horizontal position is removed and the string begins to fall freely under the influence of gravity. Determine its subsequent position as a function of x and t.

8   A shaft of uniform cross section is built-in at $x = 0$ and free at $x = l$. At $t = 0$, while the shaft is at rest in its equilibrium position, a constant torque T_0 is suddenly applied to the free end. Find the Laplace transform of the resultant angular displacement. What is the angular displacement of the free end as a function of time? *Hint:* The boundary condition at $x = l$ is $E_sJ\,(\partial\theta/\partial x) = T_0$.

9   Work Exercise 8, given that the torque applied at the free end is a unit impulse instead of a step function.

10 A semi-infinite string initially at rest in a position coinciding with the positive x axis is acted upon by a concentrated force $F_0 \sin \omega t$ applied at the point $x = b$. Find the Laplace transform of the resultant displacement of the string. What is the displacement of the string at the point $x = b$ as a function of time?

CHAPTER 9
Bessel Functions and Legendre Polynomials

9.1 Theoretical Preliminaries

In solving partial differential equations by the method of separation of variables, we are often led to ordinary differential equations with variable coefficients which cannot be solved in terms of familiar functions. The usual procedure in such cases is to obtain solutions in the form of infinite series, which can be taken as the definitions of new functions to be studied in detail and eventually tabulated if they prove of sufficient importance. In this section we shall discuss the general problem of obtaining series solutions of the form

$$(1) \qquad y = (x - a)^r[a_0 + a_1(x - a) + a_2(x - a)^2 + \cdots]$$

for the general linear second-order differential equation

$$(2) \qquad y'' + P(x)y' + Q(x)y = 0$$

We shall not require the exponent r to be a positive integer, and in general it will not be. Hence the solutions we obtain will usually not be Taylor expansions.

The analysis involves a consideration of several cases, depending upon the behavior of the coefficient functions $P(x)$ and $Q(x)$ at the point $x = a$ around which we propose to expand the solution y. In most of our work, the variables x and y and the coefficient functions $P(x)$ and $Q(x)$ will all be real. However, this is not a necessary restriction, and in the basic definitions and theorems we shall introduce in this section x, y, $P(x)$, and $Q(x)$ may be either real or complex.

In the first place, both $P(x)$ and $Q(x)$ may be analytic at $x = a$; that is, they may possess Taylor expansions around the point $x = a$. When this happens, $x = a$ is said to be an **ordinary point** of the differential equation. A point which is not an ordinary point of the differential equation is called a **singular point**. At a singular point, although $P(x)$ and $Q(x)$ do not both possess Taylor expansions, it may be that the products

$$(x - a)P(x) \qquad \text{and} \qquad (x - a)^2 Q(x)$$

do have Taylor expansions. A singular point at which this is the case is said to be **regular**; otherwise it is called **irregular**. In our work we shall be concerned exclusively with the expansion of solutions of Eq. (2) around ordinary points and regular singular points.

EXAMPLE 1

For the differential equation

$$y'' + \frac{2}{x}y' + \frac{3}{x(x - 1)^3}y = 0$$

$x = 0$ and $x = 1$ are singular points, since at $x = 0$ both $P(x)$ and $Q(x)$ become infinite, while at $x = 1$, although $P(x)$ is analytic, $Q(x)$ becomes infinite. All other points are ordinary points. The point $x = 0$ is a regular singular point, since each of the products

$$xP(x) = 2$$

and $\qquad x^2 Q(x) = \dfrac{3x}{(x-1)^3} = -3x(1-x)^{-3} = -3x(1 + 3x + 6x^2 + \cdots)$

is analytic at $x = 0$, that is, can be expanded in a series of positive integral powers of x. The point $x = 1$ is an irregular singular point, however, because although the product

$$(x - 1)P(x) = \frac{2(x-1)}{x} = 2(x - 1)[1 + (x - 1)]^{-1}$$

$$= 2(x - 1)[1 - (x - 1) + (x - 1)^2 - \cdots]$$

is analytic at $x = 1$, the product

$$(x - 1)^2 Q(x) = \frac{3}{x(x-1)}$$

becomes infinite there and hence is not analytic.

The importance of the classification of values of x into ordinary and singular points is apparent from the following theorems, which are proved in more advanced treatments of the theory of differential equations.†

THEOREM 1   At an ordinary point $x = a$ of the differential equation

$$y'' + P(x)y' + Q(x)y = 0$$

every solution is analytic, i.e., can be represented by a series of the form

$$y = a_0 + a_1(x - a) + a_2(x - a)^2 + \cdots$$

Moreover, the radius of convergence of each series solution is equal to the distance from a to the nearest singular point of the equation.

THEOREM 2   At a regular singular point $x = a$ of the differential equation

$$y'' + P(x)y' + Q(x)y = 0$$

there is at least one solution which possesses an expansion of the form

$$y = (x - a)^r[a_0 + a_1(x - a) + a_2(x - a)^2 + \cdots]$$

and this series will converge for $0 < |x - a| < R$, where R is the distance from a to the nearest of the other singular points of the equation.

THEOREM 3   At an irregular singular point $x = a$ of the differential equation

$$y'' + P(x)y' + Q(x)y = 0$$

there are in general no solutions with expansions consisting solely of powers of $x - a$.

---

† See, for instance, E. T. Whittaker and G. N. Watson, "Modern Analysis," pp. 194–203, Macmillan, New York, 1943.

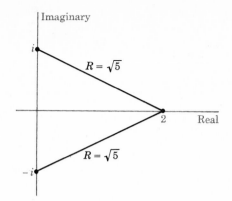

Figure 9.1
The radius of convergence as the distance
to a complex singular point.

In using Theorems 1 and 2 to infer the radius of convergence of power-series solutions of Eq. (2), it must be borne in mind that the singular point nearest to, but distinct from, the point of expansion may be complex, even though the point around which we are expanding is real. For instance, for the differential equation

$$y'' + \frac{1}{1 + x^2} y' + y = 0$$

the coefficient functions

$$P(x) = \frac{1}{1 + x^2} \quad \text{and} \quad Q(x) = 1$$

are analytic for all real values of x. However, $P(x)$ fails to be analytic at $x = \pm i$; hence, these two points are singular points of the differential equation. Therefore, a series solution around the ordinary point $x = 2$, say, would have radius of convergence $R = \sqrt{5}$, since in the complex plane the distance from the point of expansion $x = 2$ to the nearest singular point $x = i$ (or $x = -i$) is $\sqrt{5}$ (Fig. 9.1).

To obtain series solutions of Eq. (2) around an ordinary point or a regular singular point we use the so-called **method of Frobenius.**† First, for convenience, we translate axes, if necessary, so that the point of expansion becomes the point $x = 0$. Now, if $x = 0$ is either an ordinary or a regular singular point, both $xP(x)$ and $x^2Q(x)$ are analytic, and hence we can write

$$xP(x) = b_0 + b_1x + b_2x^2 + \cdots$$
$$x^2Q(x) = c_0 + c_1x + c_2x^2 + \cdots$$

Therefore, multiplying Eq. (2) by x^2 and then substituting for $xP(x)$ and $x^2Q(x)$, we have

(3) $x^2y'' + x(b_0 + b_1x + b_2x^2 + \cdots)y' + (c_0 + c_1x + c_2x^2 + \cdots)y = 0$

Next we assume a series of the desired form

(4) $$y = x^r(a_0 + a_1x + a_2x^2 + \cdots)$$

---

† Named for the German mathematician F. G. Frobenius (1849–1917).

where, without loss of generality, we can suppose that $a_0 \neq 0$. If we substitute this series into Eq. (3), we have

$$x^2[a_0 r(r-1)x^{r-2} + a_1(r+1)rx^{r-1} + a_2(r+2)(r+1)x^r + \cdots]$$
$$+ x(b_0 + b_1 x + b_2 x^2 + \cdots)[a_0 r x^{r-1} + a_1(r+1)x^r + a_2(r+2)x^{r+1} + \cdots]$$
$$+ (c_0 + c_1 x + c_2 x^2 + \cdots)(a_0 x^r + a_1 x^{r+1} + a_2 x^{r+2} + \cdots) = 0$$

or, collecting terms on the various powers of x,

$$(5) \qquad a_0[r(r-1) + b_0 r + c_0]x^r$$
$$+ \{a_1[(r+1)r + b_0(r+1) + c_0] + a_0(b_1 r + c_1)\}x^{r+1}$$
$$+ \{a_2[(r+2)(r+1) + b_0(r+2) + c_0]$$
$$+ a_1[b_1(r+1) + c_1] + a_0(b_2 r + c_2)\}x^{r+2} + \cdots = 0$$

Equation (5) will be an identity if and only if the coefficient of each power of x is zero, and thus we obtain the set of equations

$$a_0[r(r-1) + b_0 r + c_0] = 0$$
$$(6) \qquad a_1[(r+1)r + b_0(r+1) + c_0] + a_0(b_1 r + c_1) = 0$$
$$a_2[(r+2)(r+1) + b_0(r+2) + c_0] + a_1[b_1(r+1) + c_1] + a_0(b_2 r + c_2) = 0$$
$$\cdots\cdots\cdots\cdots\cdots\cdots\cdots\cdots\cdots\cdots\cdots\cdots\cdots\cdots\cdots\cdots\cdots\cdots$$

Since $a_0 \neq 0$, it follows from the first of these equations that

$$(7) \qquad r^2 + (b_0 - 1)r + c_0 = 0$$

This quadratic equation in r is known as the **indicial equation** of the differential equation relative to the point of expansion, and its roots r_1 and r_2 are known as the **exponents** of the differential equation at that point. For each of these values there is, in general, a series solution of the form (4). Moreover, the coefficients in these expansions can be determined, one by one, from the successive equations in the set (6), which express each of the a's in turn in terms of the a's which precede it in the series (4).

If the indicial equation has a double root, it is obvious that two series solutions cannot be obtained by the present method. It is also true (though not obvious) that if the roots of the indicial equation differ by an integer, this method fails, in general, to provide a second series solution.† In either of these cases, however, a second solution can be found by the method of Sec. 2.1, that is, by assuming $y = \phi(x)y_1(x)$, where $y_1(x)$ is the first series solution, and then determining $\phi(x)$ so that the product $\phi(x)y_1(x)$ will satisfy the given differential equation. Another method, usually preferable to the one described in Sec. 2.1, is illustrated in the following example.

EXAMPLE 2

Find series solutions around the origin for the equation $xy'' + y' + y = 0$.

Since the given equation has $P(x) = Q(x) = 1/x$, it follows that the origin is a regular singular point. Hence, by Theorem 2, there exists at least one solution with an expansion of the form

$$y = x^r(a_0 + a_1 x + a_2 x^2 + \cdots) = \sum_{n=0}^{\infty} a_n x^{n+r}$$

---

† See Exercise 19.

Substituting this series and its first two derivatives gives us the equation

$$x \sum_{n=0}^{\infty} a_n(n + r)(n + r - 1)x^{n+r-2} + \sum_{n=0}^{\infty} a_n(n + r)x^{n+r-1} + \sum_{n=0}^{\infty} a_n x^{n+r} = 0$$

From this, by collecting terms on the various powers of x, we obtain

$$[a_0 r(r - 1) + a_0 r]x^{r-1} + [a_1(r + 1)r + a_1(r + 1) + a_0]x^r + \cdots$$
$$+ [a_n(n + r)(n + r - 1) + a_n(n + r) + a_{n-1}]x^{n+r-1} + \cdots = 0$$

This will be an identity, and the assumed series will be a solution, if and only if the coefficient of each power of x vanishes. Hence we have the following system of equations to determine the a's:

$$a_0 r^2 = 0$$
$$a_1(r + 1)^2 + a_0 = 0$$
$$\cdots\cdots\cdots\cdots\cdots\cdots$$
$$a_n(n + r)^2 + a_{n-1} = 0$$
$$\cdots\cdots\cdots\cdots\cdots\cdots$$

Since $a_0 \neq 0$, it follows from the first of these equations that $r = 0$. However, for reasons that will soon become clear, it is convenient to continue a little further with r as a parameter before we replace it with the value zero. Doing this, we find in succession

$$a_0 = a_0 \qquad \text{that is, } a_0 \text{ is arbitrary}$$

$$a_1 = -\frac{a_0}{(r + 1)^2}$$

$$a_2 = -\frac{a_1}{(r + 2)^2} = \frac{a_0}{(r + 1)^2(r + 2)^2}$$

and, by an obvious induction,

$$a_n = -\frac{a_{n-1}}{(r + n)^2} = (-1)^n \frac{a_0}{(r + 1)^2(r + 2)^2 \cdots (r + n)^2}$$

Thus, taking $a_0 = 1$ for convenience, we have obtained the series

$$\bar{y} = x^r \left[1 - \frac{x}{(r + 1)^2} + \frac{x^2}{(r + 1)^2(r + 2)^2} - \frac{x^3}{(r + 1)^2(r + 2)^2(r + 3)^2} + \cdots \right]$$

which has the property that when it is substituted into the given differential equation, the coefficient of x^{r-1} is r^2 and the coefficients of all other powers of x are zero. In other words, $\bar{y}$ satisfies the equation

(8) $$x\bar{y}'' + \bar{y}' + \bar{y} = r^2 x^{r-1}$$

Clearly, if $r = 0$, this becomes the original equation and $\bar{y}$ reduces to the solution

$$y_1 = 1 - \frac{x}{(1!)^2} + \frac{x^2}{(2!)^2} - \frac{x^3}{(3!)^2} + \cdots$$

Since the indicial equation, namely, $r^2 = 0$, has only the one root $r = 0$, it is evident that a second power-series solution cannot be found. In an attempt to find a second solution of some other form, let us now regard $\bar{y}$ as a function of r† and differentiate Eq. (8) partially with respect to r, remembering, of course, that x is independent of r. The result is

$$x \frac{\partial(\bar{y}'')}{\partial r} + \frac{\partial(\bar{y}')}{\partial r} + \frac{\partial(\bar{y})}{\partial r} = 2rx^{r-1} + r^2 x^{r-1} \ln x$$

---

† It was for this purpose that we decided earlier to retain r as a parameter rather than replace it with the value zero.

When the order of differentiating with respect to x and r is interchanged, this becomes

$$x\left(\frac{\partial \bar{y}}{\partial r}\right)'' + \left(\frac{\partial \bar{y}}{\partial r}\right)' + \frac{\partial \bar{y}}{\partial r} = r(2 + r \ln x)\, x^{r-1}$$

If we set $r = 0$ in this equation, the right-hand side becomes zero, which shows that $\partial \bar{y}/\partial r\,|_{r=0}$ satisfies the original differential equation and is, presumably, a second, independent solution.

Carrying out the indicated differentiation, we have

$$\frac{\partial \bar{y}}{\partial r} = x^r \ln x \left[1 - \frac{x}{(r+1)^2} + \frac{x^2}{(r+1)^2(r+2)^2} - \frac{x^3}{(r+1)^2(r+2)^2(r+3)^2} + \cdots \right]$$

$$+ x^r \left\{ -x \frac{-2}{(r+1)^3} + x^2 \left[\frac{-2}{(r+1)^3} \frac{1}{(r+2)^2} + \frac{1}{(r+1)^2} \frac{-2}{(r+2)^3} \right] \right.$$

$$- x^3 \left[\frac{1}{(r+2)^2(r+3)^2} \frac{-2}{(r+1)^3} + \frac{1}{(r+1)^2(r+3)^2} \frac{-2}{(r+2)^3} \right.$$

$$\left. + \frac{1}{(r+1)^2(r+2)^2} \frac{-2}{(r+3)^3} \right] + \cdots \left. \right\}$$

Finally, letting $r = 0$, we have as a second solution of the original equation

$$y_2 = \frac{\partial \bar{y}}{\partial r}\bigg|_{r=0} = \ln x \left[1 - \frac{x}{(1!)^2} + \frac{x^2}{(2!)^2} - \frac{x^3}{(3!)^2} + \cdots \right]$$

$$+ 2 \left[\frac{x}{(1!)^2} - \frac{x^2}{(2!)^2}\left(1 + \frac{1}{2}\right) + \frac{x^3}{(3!)^2}\left(1 + \frac{1}{2} + \frac{1}{3}\right) - \cdots \right]$$

Since y_2 contains the term $\ln x$ while y_1 does not, it is clear that y_1 and y_2 cannot be proportional. In other words, they are independent particular solutions, and a complete solution of the original equation is $y = c_1 y_1 + c_2 y_2$.

EXERCISES

1 Find the singular points of each of the following equations, and determine whether they are regular or irregular:
(a) $y'' + xy' + y = 0$ (b) $e^x y'' + 2y' - xy = 0$
(c) $x^2 y'' - \lambda^2 y = 0$ (d) $x^2 y'' + y' + y = 0$
(e) $(1 - x^2)y'' + y' + y = 0$ (f) $x^2(1 - x)y'' + (1 - x)y' + y = 0$

2 Find the indicial equation relative to each of the singular points of each of the equations in Exercise 1.

Find two independent power-series solutions around the origin for each of the following equations:

3 $y'' + y = 0$ 4 $2y'' + y' - y = 0$
5 $y'' + xy = 0$ 6 $9x^2 y'' + (x + 2)y = 0$
7 $2x^2 y'' + 3xy' + (x^2 - 1)y = 0$
8 $2x^2 y'' + (2x^2 + 3x)y' + (x - 1)y = 0$

Show that each of the following equations has a single independent power-series solution around the origin, and find this solution. If possible, find a second, independent solution by the method illustrated in Example 2.

9 $x^2 y'' + (x - 2)y = 0$ 10 $4x^2 y'' + (4x + 1)y = 0$

11 What do you think would be appropriate definitions for the terms *ordinary point*, *regular singular point*, *irregular singular point*, and *indicial equation* relative to the general linear third-order differential equation

$$y''' + P(x)y'' + Q(x)y' + R(x)y = 0$$

12 What is the indicial equation relative to the origin for the third-order differential equation

$$y''' + P(x)y'' + Q(x)y' + R(x)y = 0$$

given that $xP(x)$, $x^2Q(x)$, and $x^3R(x)$ are all analytic at the origin?

For each of the following equations, find all the power-series solutions around the origin:

13 $x^2y'' + y = 0$ **14** $xy'' + y = 0$

15 The **point at infinity** is said to be an ordinary point or a singular point of the differential equation

$$y'' + P(x)y' + Q(x)y = 0$$

according as the equation obtained from this by the substitution $x = 1/u$ has an ordinary point or a singular point at $u = 0$. Show that under this transformation the original equation becomes

$$u^4 \frac{d^2y}{du^2} + \left[2u^3 - u^2 P\left(\frac{1}{u}\right) \right] \frac{dy}{du} + Q\left(\frac{1}{u}\right) y = 0$$

and use this result to determine the nature of the point at infinity for the equation $(x^2 + 1)y'' + y' + y = 0$.

16 Verify that under the change of dependent variable defined by the substitution

$$y = z \exp \left(-\tfrac{1}{2} \int P(x)\, dx \right)$$

the differential equation $y'' + P(x)y' + Q(x)y = 0$ becomes

$$\frac{d^2z}{dx^2} + R(x)z = 0$$

where $$R(x) = Q(x) - \frac{1}{2}\frac{dP(x)}{dx} - \tfrac{1}{4}P^2(x)$$

17 Using the result of Exercise 16, determine conditions on a_1, b_1, a_2, b_2, and c_2 which will ensure that the equation

$$y'' + (a_1 + b_1x)y' + (a_2 + b_2x + c_2x^2)y = 0$$

can be solved in terms of elementary functions.

18 Show that if the origin is an irregular singular point of the differential equation $y'' + P(x)y' + Q(x)y = 0$, then the indicial equation relative to the origin is of the first degree at most.

19 Verify that if the roots of the indicial equation differ by unity, then, in general, the two roots lead to the same series solution of the equation $y'' + P(x)y' + Q(x)y = 0$. Under what conditions will this not be the case? Is this true if the roots differ by an integer greater than 1?

9.2 The Series Solution of Bessel's Equation

One of the most important of all variable-coefficient differential equations is

(1) $$x^2 \frac{d^2y}{dx^2} + x\frac{dy}{dx} + (\lambda^2x^2 - \nu^2)y = 0$$

which is known as **Bessel's equation†** of order ν with a parameter λ. This arises in a

---

† Named for the German mathematician and astronomer Friedrich Wilhelm Bessel (1784–1846), although special cases of this equation had been studied earlier by Jakob Bernoulli (1703), Daniel Bernoulli (1732), and Leonhard Euler (1764).

great variety of problems, including almost all applications involving partial differential equations, such as the wave equation or the heat equation, in regions possessing circular symmetry.

As a preliminary step in the solution of Eq. (1), let us change the independent variable from x to t by means of the substitution

$$(2) \qquad t = \lambda x$$

Since $dy/dx = \lambda\, dy/dt$ and $d^2y/dx^2 = \lambda^2\, d^2y/dt^2$, Eq. (1) becomes

$$(3) \qquad t^2 \frac{d^2y}{dt^2} + t\frac{dy}{dt} + (t^2 - v^2)y = 0$$

which is known simply as **Bessel's equation of order** v.

For Eq. (3) it is clear that

$$P(t) = \frac{1}{t} \quad \text{and} \quad Q(t) = \frac{t^2 - v^2}{t^2}$$

Hence, the origin is a regular singular point of the equation, and all other values of t are ordinary points. At the origin, where we propose to obtain series solutions of (3), the indicial equation [Eq. (7), Sec. 9.1] is $r^2 - v^2 = 0$, and therefore, by the theory of the preceding section, we are led to try a series solution of the form

$$(4) \qquad y = \sum_{k=0}^{\infty} a_k t^{v+k}$$

Substituting this into Eq. (3), we obtain

$$t^2 \sum_{k=0}^{\infty} a_k(v+k)(v+k-1)t^{v+k-2} + t \sum_{k=0}^{\infty} a_k(v+k)t^{v+k-1}$$

$$+ (t^2 - v^2) \sum_{k=0}^{\infty} a_k t^{v+k} = 0$$

or, bringing the coefficients into the respective sums and then combining the sums,

$$\sum_{k=0}^{\infty} a_k[(v+k)(v+k-1) + (v+k) - v^2]t^{v+k} + \sum_{k=0}^{\infty} a_k t^{v+k+2} = 0$$

This will be an identity, and y will be a solution, if and only if the coefficient of every power of t is zero. Since v is a root of the indicial equation, it follows that the coefficient of the lowest power of t, namely, t^v, is zero. For the next power, t^{v+1}, the coefficient is $a_1[(v+1)v + (v+1) - v^2] = a_1(2v+1)$, and for $k \geq 2$, the coefficient of t^{v+k} is

$$a_k[(v+k)(v+k-1) + (v+k) - v^2] + a_{k-2} \qquad \text{or} \qquad a_k k(2v+k) + a_{k-2}$$

Thus to determine the a's we have the conditions

$$(5) \qquad a_1(2v+1) = 0$$

and

$$a_k k(2v+k) + a_{k-2} = 0$$

or

$$(6) \qquad a_k = -\frac{a_{k-2}}{k(2v+k)}$$

From (5) it is clear that $a_1 = 0$ for all values of v except possibly $v = -\frac{1}{2}$; and even in this case we can assume $a_1 = 0$, since we are interested only in conditions *sufficient* for the existence of solutions of the form (4). Moreover, from (6) it is apparent that any coefficient a_k is a multiple of the second preceding coefficient a_{k-2}. Hence, beginning with a_1, it follows that every coefficient with an odd subscript must be zero.

On the other hand, starting with a_0, which is still perfectly arbitrary, and taking $k = 2, 4, 6, \ldots$ successively in the recurrence formula (6), we have

$$a_0 = a_0$$

$$a_2 = -\frac{a_0}{2(2v + 2)} = -\frac{a_0}{2^2 \cdot 1! \, (v + 1)}$$

$$a_4 = -\frac{a_2}{4(2v + 4)} = -\frac{a_2}{2^2 \cdot 2(v + 2)} = \frac{a_0}{2^4 \cdot 2! \, (v + 2)(v + 1)}$$

$$a_6 = -\frac{a_4}{6(2v + 6)} = -\frac{a_4}{2^2 \cdot 3(v + 3)} = -\frac{a_0}{2^6 \cdot 3! \, (v + 3)(v + 2)(v + 1)}$$

and, in general,

$$a_{2m} = \frac{(-1)^m a_0}{2^{2m} m! \, (v + m)(v + m - 1) \cdots (v + 3)(v + 2)(v + 1)}$$

Now a_{2m} is the coefficient of t^{v+2m} in the series (4) for y. Hence it would be convenient if a_{2m} contained the factor 2^{v+2m} in its denominator instead of just 2^{2m}. To achieve this, we write

$$a_{2m} = \frac{(-1)^m}{2^{v+2m} m! \, (v + m) \cdots (v + 2)(v + 1)} (2^v a_0)$$

Furthermore, the factors $(v + m)(v + m - 1) \cdots (v + 3)(v + 2)(v + 1)$ in the denominator of a_{2m} suggest a factorial. In fact, if v were an integer, a factorial could be created by multiplying the numerator and the denominator of a_{2m} by $v!$. However, since v is not necessarily an integer, we must use not $v!$ but its generalization, $\Gamma(v + 1)$ (Sec. 7.3) for this purpose. Then, except for the values $v = -1, -2, -3, \ldots$, for which $\Gamma(v + 1)$ is not defined, we can write

$$a_{2m} = \frac{(-1)^m}{2^{v+2m} m! \, (v + m) \cdots (v + 2)(v + 1)\Gamma(v + 1)} [2^v \Gamma(v + 1)a_0]$$

Since the gamma function satisfies the recurrence relation

$$(v + j)\Gamma(v + j) = \Gamma(v + j + 1)$$

the factors $(v + 1), (v + 2), \ldots, (v + m)$ can be successively telescoped into the gamma function, and the expression for a_{2m} can be written

$$a_{2m} = \frac{(-1)^m}{2^{v+2m} m! \, \Gamma(v + m + 1)} [2^v \Gamma(v + 1)a_0]$$

In this formula a_0 is still arbitrary, and since we are looking only for particular solutions, it is convenient to choose

$$a_0 = \frac{1}{2^v \Gamma(v + 1)}$$

so that, finally,

$$a_{2m} = \frac{(-1)^m}{2^{v+2m} m! \, \Gamma(v + m + 1)}$$

The series for y is therefore, from (4),

(7)
$$y = t^v \left[\frac{1}{2^v \Gamma(v + 1)} - \frac{t^2}{2^{v+2}\Gamma(v + 2)} + \frac{t^4}{2^{v+4}2! \, \Gamma(v + 3)} - \cdots \right]$$

$$= \sum_{m=0}^{\infty} \frac{(-1)^m t^{v + 2m}}{2^{v+2m}m! \, \Gamma(v + m + 1)}$$

The function defined by this infinite series is known as the **Bessel function of the first kind of order** v and is denoted $J_v(t)$. Since Bessel's equation of order v has no finite singular points except the origin, it follows from Theorem 2, Sec. 9.1, that the series for $J_v(t)$ converges for all values of t if $v \geq 0$. The graphs of $J_0(t)$ and $J_1(t)$ are shown in Fig. 9.2. Their resemblance to the graphs of $\cos t$ and $\sin t$ is interesting. In particular, they illustrate the important fact that for every value of v the equation $J_v(t) = 0$ has infinitely many real roots.†

Let us now consider the series arising from the other root of the indicial equation, namely, $r = -v$. We could, of course, begin again with a series analogous to (4) and determine its coefficients one by one, just as we did for $J_v(t)$, but there is no need to go to this trouble. In fact, since v enters into Bessel's equation only in the form of a square, it follows that the series obtained from (7) by replacing v by $-v$ will satisfy Bessel's equation, provided only that the gamma functions appearing in the denominators of the various terms are all defined. This is necessarily the case unless v is an integer; hence when v is not an integer, the function

(8)
$$J_{-v}(t) = \sum_{m=0}^{\infty} \frac{(-1)^m t^{-v + 2m}}{2^{-v+2m}m! \, \Gamma(-v + m + 1)}$$

is a second particular solution of Bessel's equation of order v. Moreover, since $J_{-v}(t)$ contains negative powers of t while $J_v(t)$ does not, it is obvious that in the neighborhood of the origin $J_{-v}(t)$ is unbounded while $J_v(t)$ remains finite. Hence $J_v(t)$ and $J_{-v}(t)$ cannot be proportional and therefore are two independent solutions of Bessel's equation. According to Theorem 2, Sec. 2.1, a complete solution of Bessel's equation when v is not an integer is then

(9)
$$y(t) = c_1 J_v(t) + c_2 J_{-v}(t)$$

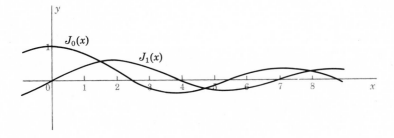

Figure 9.2
The Bessel functions of the first kind $J_0(x)$ and $J_1(x)$.

---

† For a proof of this fact, see Exercise 23, Sec. 9.4.

Instead of $J_{-v}(t)$, some writers take the linear combination

$$(10) \qquad\qquad Y_v(t) = \frac{\cos v\pi \, J_v(t) - J_{-v}(t)}{\sin v\pi}$$

as a second, independent solution of Bessel's equation. Using $Y_v(t)$, which is known as the **Bessel function of the second kind of order** v, we can write a complete solution of Bessel's equation

$$(11) \qquad\qquad y(t) = c_1 J_v(t) + c_2 Y_v(t) \qquad v \text{ not an integer}$$

In some applications it is convenient to use still another form of the general solution of Bessel's equation based upon the two particular solutions

$$(12) \qquad\qquad \begin{aligned} H_v^{(1)}(t) &= J_v(t) + iY_v(t) \\ H_v^{(2)}(t) &= J_v(t) - iY_v(t) \end{aligned}$$

These are known as **Hankel functions**† or **Bessel functions of the third kind of order** v, and in terms of them a complete solution of Eq. (3) can be written

$$(13) \qquad\qquad y(t) = c_1 H_v^{(1)}(t) + c_2 H_v^{(2)}(t) \qquad v \text{ not an integer}$$

It is interesting to note that (9), (11), and (13) are correct expressions for the general solution of Eq. (3) even when v is an odd multiple of $\frac{1}{2}$ and the roots of the indicial equation $r^2 - v^2 = 0$ differ by an integer. In the last section we pointed out that when this happens, a second, independent series solution of the form (4) will usually not exist. It *may* exist, however, and this is one of the instances when it actually does.‡

If v is an integer, say $v = n$, the situation is somewhat different. Again the roots of the indicial equation differ by an integer, namely, $2n$, and it is to be expected that a second solution of the form (4) will not exist. In fact, when we consider $J_{-v}(t)$ as the limit of $J_v(t)$ as v approaches $-n$ and remember that when its argument approaches any nonpositive integer, the gamma function becomes infinite, it follows that as v approaches $-n$, the first n terms in the series (7) approach zero and the series effectively begins with the term for which $m = n$:

$$J_{-n}(t) = \sum_{m=n}^{\infty} \frac{(-1)^m t^{-n+2m}}{2^{-n+2m} m! \, \Gamma(-n + m + 1)}$$

In this, let the variable of summation be changed from m to j by the substitution $m = j + n$. Then

$$\begin{aligned} J_{-n}(t) &= \sum_{j=0}^{\infty} \frac{(-1)^{j+n} t^{-n+2(j+n)}}{2^{-n+2(j+n)}(j + n)! \, \Gamma[-n + (j + n) + 1]} \\ &= \sum_{j=0}^{\infty} \frac{(-1)^n (-1)^j t^{n+2j}}{2^{n+2j}(j + n)! \, \Gamma(j + 1)} \\ &= (-1)^n \sum_{j=0}^{\infty} \frac{(-1)^j t^{n+2j}}{2^{n+2j} \Gamma(n + j + 1) j!} = (-1)^n J_n(t) \end{aligned}$$

---

† Named for the German mathematician Hermann Hankel (1839–1873).
‡ See Exercise 13.

Thus, when v is an integer, the function $J_{-v}(t)$ is proportional to $J_v(t)$. These two solutions are therefore not independent, and the linear combination $c_1 J_v(t) + c_2 J_{-v}(t)$ is no longer a complete solution of Bessel's equation. Moreover, without additional definitions, neither (11) nor (13) provides a complete solution, since $Y_v(t)$, as defined by (10), assumes the indeterminate form $0/0$ when v is an integer.

A complete solution when v is an integer can be found in either of several ways. One is to use the method developed in Sec. 2.1 for finding a second solution of a linear second-order differential equation when one solution is known. The result, as given by Eq. (5) of Sec. 2.1, with $y_1(t) = J_n(t)$ and $P(t) = 1/t$, is

$$y(t) = c_1 J_n(t) + c_2 J_n(t) \int \frac{dt}{t J_n^2(t)}$$

Another procedure is to obtain a second, independent solution by evaluating the limit of $Y(t)$ as $v \to n$. The details are somewhat involved, and we shall not present them here. The limit function, which exists and is independent of $J_n(t)$ for all values of n, is commonly denoted by $Y_n(t)$; that is,

$$(14) \qquad Y_n(t) = \lim_{v \to n} Y_v(t) = \lim_{v \to n} \frac{\cos v\pi\, J_v(t) - J_{-v}(t)}{\sin v\pi}$$

More explicitly, this limiting process leads to the following expression for $Y_n(t)$:

$$(15) \quad Y_n(t) = \frac{2}{\pi} \left(\ln \frac{t}{2} + \gamma\dagger \right) J_n(t) - \frac{1}{\pi} \sum_{k=0}^{n-1} \frac{2^{n-2k}(n-k-1)!\ddagger}{t^{n-2k}k!}$$

$$- \frac{1}{\pi} \sum_{k=0}^{\infty} \frac{(-1)^k t^{n+2k}}{2^{n+2k}k!\,(n+k)!} \times$$

$$\left[\left(1 + \frac{1}{2} + \cdots + \frac{1}{k} \right)^{\S} + \left(1 + \frac{1}{2} + \cdots + \frac{1}{n+k} \right) \right]$$

The corresponding specializations of the Hankel functions (12) are defined in the obvious way in terms of $Y_n(t)$:

$$(16) \qquad H_n^{(1)}(t) = J_n(t) + iY_n(t) \qquad H_n^{(2)}(t) = J_n(t) - iY_n(t)$$

With Formulas (14) and (16), we can now eliminate from (11) and (13) the restriction that v is not an integer and use these results for all values of v, integral as well as nonintegral. Plots of $Y_0(t)$ and $Y_1(t)$ are shown in Fig. 9.3. Among other things, they illustrate the important facts that for all values of v, $Y_v(t)$ is unbounded in the neighborhood of the origin and, like $J_v(t)$ and $J_{-v}(t)$, has infinitely many real zeros.

Reversing the transformation (2) which we used to eliminate the parameter λ from the general form of Bessel's equation (1), we can now summarize the results of the preceding discussion in the following theorem.

---

† The symbol γ denotes the following limit:

$$\lim_{m \to \infty} \left(1 + \frac{1}{2} + \frac{1}{3} + \cdots + \frac{1}{m} - \ln m \right) = 0.5772\cdots$$

This important number is known as **Euler's constant**.
‡ When $n = 0$, this sum is to be taken equal to zero.
§ When $k = 0$, this sum is to be taken equal to zero.

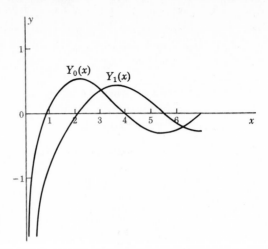

Figure 9.3
The Bessel functions of the second
kind $Y_0(x)$ and $Y_1(x)$.

THEOREM 1   For all values of v, a complete solution of Bessel's equation of order v with a parameter λ,

$$x^2 y'' + xy' + (\lambda^2 x^2 - v^2)y = 0$$

can be written in either of the forms

$$y(x) = c_1 J_v(\lambda x) + c_2 Y_v(\lambda x)$$

or $$y(x) = c_1 H_v^{(1)}(\lambda x) + c_2 H_v^{(2)}(\lambda x)$$

If v is not an integer, a complete solution can also be written

$$y(x) = c_1 J_v(\lambda x) + c_2 J_{-v}(\lambda x)$$

$J_v(\lambda x)$, $J_{-v}(\lambda x)$, and $Y_v(\lambda x)$ all have infinitely many real zeros. If $v \geq 0$, $J_v(\lambda x)$ is finite for all values of x but $J_{-v}(\lambda x)$ and $Y_v(\lambda x)$ are unbounded in the neighborhood of the origin. $H_v^{(1)}(\lambda x)$ and $H_v^{(2)}(\lambda x)$ are complex-valued functions when x is real.

EXERCISES

1   If y_1 and y_2 are any two solutions of Bessel's equation of order v, show that $y_1 y_2' - y_1' y_2 = c/x$, where c is a suitable constant. *Hint:* Recall Abel's identity from Sec. 2.1.

2   If y_1 and y_2 are two independent solutions of Bessel's equation of order v, show that there is no value of x for which y_1 and y_2 are simultaneously zero. *Hint:* Use the result of Exercise 1.

3   By determining the coefficient of $1/x$ on the left-hand side, show that if v is not an integer

$$J_v(x) J_{-v}'(x) - J_v'(x) J_{-v}(x) = -\frac{2}{\pi x} \sin v\pi$$

Is this result correct if v is an integer? *Hint:* Use the result of Exercise 1 and the fact that $\Gamma(x)\Gamma(1 - x) = \pi/(\sin \pi x)$ if x is not an integer.

4   Show that $J_n(x) Y_n'(x) - J_n'(x) Y_n(x) = 2/\pi x$.

5   Show that except possibly at the origin, no nontrivial solution of Bessel's equation can have a double root.

6 If v is not an integer, show that

$$Y = \frac{\pi}{2 \sin v\pi} \left[J_v(x) \int_a^x f(s)J_{-v}(s) \, ds - J_{-v}(x) \int_a^x f(s)J_v(s) \, ds \right]$$

is a particular integral of the nonhomogeneous Bessel equation

$$x^2 y'' + xy' + (x^2 - v^2)y = xf(x)$$

What is the corresponding result if v is an integer? *Hint:* Use the method of variation of parameters and the result of Exercise 3.

7 Show that under the change of dependent variable defined by the substitution $y = u/\sqrt{t}$, Bessel's equation of order v becomes

$$\frac{d^2 u}{dt^2} + \left(1 + \frac{1 - 4v^2}{4t^2}\right) u = 0$$

Hence show that for large values of t, solutions of Bessel's equation are described approximately by expressions of the form

$$c_1 \frac{\sin t}{\sqrt{t}} + c_2 \frac{\cos t}{\sqrt{t}}$$

[More precisely, it can be shown that

$$J_v(t) \sim \sqrt{\frac{2}{\pi t}} \cos \left(t - \frac{\pi}{4} - \frac{v\pi}{2}\right)$$

$$Y_v(t) \sim \sqrt{\frac{2}{\pi t}} \sin \left(t - \frac{\pi}{4} - \frac{v\pi}{2}\right)$$

where the symbol $\sim$ means that the limit of the ratio of the two quantities connected by it approaches 1 as t becomes infinite.]

8 Show that if $v_1 > v_2$, then between any two consecutive zeros of $J_{v_1}(t)$ there is at least one zero of $J_{v_2}(t)$. *Hint:* Recall the separation theorem established in Exercise 22, Sec. 2.1, and apply it to the transformed equation obtained in the last exercise.

9 Show that in every interval of length π on the t axis there is at least one zero of $J_0(t)$ and at least one zero of $Y_0(t)$. *Hint:* Apply the separation theorem of Exercise 22, Sec. 2.1, to the equations $u'' + (1/4t^2)u = 0$ and $u'' + u = 0$.

10 Show that

$$\int \frac{dx}{xJ_v^2(x)} = \frac{\pi}{2} \frac{Y_v(x)}{J_v(x)}$$

What is

$$\int \frac{dx}{xY_v^2(x)}$$

11 What is

$$\frac{d}{dx} \left[\ln \frac{Y_v(x)}{J_v(x)}\right]$$

12 Show that the smallest positive root of the equation $J_v(x) = 0$ is equal to or greater than $\frac{1}{2}\sqrt{4v^2 - 1}$. *Hint:* Recall the result of Exercise 23, Sec. 2.1.

13 Using the result of Exercise 19, Sec. 9.1, verify that when $v = \frac{1}{2}, \frac{3}{2}, \ldots$, Bessel's equation should have two independent power-series solutions but that when $v = 1, 2, \ldots$, it should not.

9.3 Modified Bessel Functions

Certain equations closely resembling Bessel's equation occur so often that their solutions are also named and studied as functions in their own right. The most important of these is

$$(1) \qquad x^2 y'' + xy' - (x^2 + v^2)y = 0$$

which is known as the **modified Bessel equation of order** v. Since this can be written in the form

$$x^2 y'' + xy' + (i^2 x^2 - v^2)y = 0$$

it is evident that this is nothing but Bessel's equation of order v with the imaginary parameter $\lambda = i$. However, in actual applications, to write the complete solution of (1) in the form

$$y = c_1 J_v(ix) + c_2 Y_v(ix)$$

and retain the imaginaries would be about as awkward as to take the solution of

$$y'' - y = 0$$

to be

$$y = c_1 \cos ix + c_2 \sin ix$$

and use this complex expression instead of resorting to real exponentials or hyperbolic functions. Accordingly, we seek modifications of $J_v(ix)$ and $Y_v(ix)$ which will be real functions of real variables.

Now,
$$J_v(ix) = \sum_{k=0}^{\infty} \frac{(-1)^k (ix)^{v+2k}}{2^{v+2k} k! \, \Gamma(v+k+1)}$$

$$= i^v \sum_{k=0}^{\infty} \frac{x^{v+2k}}{2^{v+2k} k! \, \Gamma(v+k+1)}$$

Moreover, $J_v(ix)$ multiplied by any constant will also be a solution of the equation we are considering. Hence, in particular, we can multiply it by i^{-v}, getting

$$i^{-v} J_v(ix) = \sum_{k=0}^{\infty} \frac{x^{v+2k}}{2^{v+2k} k! \, \Gamma(v+k+1)}$$

This is a completely real function, identical with $J_v(x)$ except that its terms, instead of alternating in sign, are all positive. This new function, which is related to $J_v(x)$ in the same way that $\cosh x$ and $\sinh x$ are related to $\cos x$ and $\sin x$, is known as the **modified Bessel function of the first kind of order** v and is customarily denoted by $I_v(x)$.† If v is not an integer, the function $I_{-v}(x)$ obtained from $I_v(x)$ by replacing v by $-v$ throughout is a second, independent solution of Eq. (1), whose complete solution can therefore be written

$$y = c_1 I_v(x) + c_2 I_{-v}(x)$$

On the other hand, instead of using $I_{-v}(x)$, many writers take the second solution of the modified Bessel equation to be the linear combination

$$K_v(x) = \frac{\pi}{2} \frac{I_{-v}(x) - I_v(x)}{\sin v\pi}$$

---

† A few authors and some handbooks continue to use the original designation $i^{-v} J_v(ix)$ to denote this function.

which is known as the **modified Bessel function of the second kind of order** v. If v is not an integer, this is a well-defined solution which is clearly independent of $I_v(x)$. If v is an integer n, this assumes the indeterminate form $0/0$, but a tedious evaluation by L'Hospital's rule leads to a limiting expression

$$K_n(x) = \lim_{v \to n} K_v(x) = \lim_{v \to n} \frac{\pi}{2} \frac{I_{-v}(x) - I_v(x)}{\sin v\pi}$$

$$= (-1)^{n+1} \left(\ln \frac{x}{2} + \gamma\dagger \right) I_n(t) + \frac{1}{2} \sum_{k=0}^{n-1} \frac{(-1)^k (n - k - 1)! \, 2^{n-2k}}{k! \, x^{n-2k}} \ddagger$$

$$+ \frac{(-1)^n}{2} \sum_{k=0}^{\infty} \frac{n + 2k}{k! \, (n + k)! \, 2^{n+k}}$$

$$\times \left\{ \left[1 + \frac{1}{2} + \cdots + \frac{1}{k} \right]^{\S} + \left[1 + \frac{1}{2} + \cdots + \frac{1}{n + k} \right] \right\}$$

which is a solution independent of $I_n(x)$. This is a useful result because, as we might expect, $I_v(x)$ and $I_{-v}(x)$ are not independent when v is an integer. In fact, when $v = n$, we have the identity

$$(-1)^n J_{-n}(ix) = J_n(ix)$$

and then, by obvious steps,

$$(i^2)^n J_{-n}(ix) = J_n(ix)$$

$$i^n J_{-n}(ix) = i^{-n} J_n(ix)$$

$$I_{-n}(x) = I_n(x)$$

Plots of $I_0(x)$ and $I_1(x)$ are shown in Fig. 9.4; plots of $K_0(x)$ and $K_1(x)$ in Fig. 9.5. As these graphs illustrate, the modified Bessel functions have no real zeros except possibly at $x = 0$. They also illustrate that for $v \geq 0$, $I_v(x)$ is finite at the origin but $K_v(x)$, like $I_{-v}(x)$, becomes infinite as x approaches zero.

Like the ordinary Bessel equation, the modified Bessel equation frequently occurs in a form containing a parameter λ:

(2) $$x^2 y'' + xy' - (\lambda^2 x^2 + v^2) y = 0$$

A complete solution of this is, of course,

$$y = c_1 I_v(\lambda x) + c_2 K_v(\lambda x) \qquad v \text{ unrestricted}$$

If v is not an integer, we have the alternative form

$$y = c_1 I_v(\lambda x) + c_2 I_{-v}(\lambda x)$$

A second equation closely related to Bessel's equation is

(3) $$x^2 y'' + xy' + (-ix^2 - v^2) y = 0$$

---

† The symbol γ denotes Euler's constant.
‡ When $n = 0$, this sum is to be taken equal to zero.
§ When $k = 0$, this sum is to be taken equal to zero.

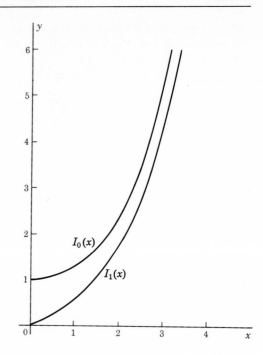

Figure 9.4
The modified Bessel functions of the
first kind $I_0(x)$ and $I_1(x)$.

This can be regarded either as Bessel's equation of order v with parameter $\lambda = \sqrt{-i}$
or as the modified Bessel equation of order v with parameter $\lambda = \sqrt{i}$. From the former
point of view, a complete solution can be written

$$y = c_1 J_v(\sqrt{-i}\, x) + c_2 Y_v(\sqrt{-i}\, x)$$

From the second point of view the solution can be written

$$y = c_1 I_v(\sqrt{i}\, x) + c_2 K_v(\sqrt{i}\, x)$$

Now a complete solution can be constructed from *any* pair of independent particular

Figure 9.5
The modified Bessel functions of the
second kind $K_0(x)$ and $K_1(x)$.

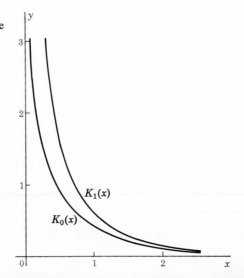

solutions; and it is customary in studying Eq. (3) to select $J_\nu(\sqrt{-i}\,x)$ and $K_\nu(\sqrt{i}\,x)$ for this purpose. Making this choice, and recalling that $-i = i^3$, we have for the complete solution of Eq. (3)

$$y = c_1 J_\nu(i^{3/2}x) + c_2 K_\nu(i^{1/2}x)$$

Now
$$J_\nu(i^{3/2}x) = \sum_{k=0}^{\infty} \frac{(-1)^k (i^{3/2}x)^{\nu+2k}}{2^{\nu+2k} k!\,\Gamma(\nu+k+1)}$$

$$= i^{3\nu/2} \sum_{k=0}^{\infty} \frac{(-1)^k i^{3k} x^{\nu+2k}}{2^{\nu+2k} k!\,\Gamma(\nu+k+1)}$$

Moreover, i^{3k} can take on only one of the four values

$$
\begin{array}{ll}
1 & k = 0,\,4,\,8,\ldots \\
-i & k = 1,\,5,\,9,\ldots \\
-1 & k = 2,\,6,\,10,\ldots \\
i & k = 3,\,7,\,11,\ldots
\end{array}
$$

Hence the first, third, fifth, ... terms in the series for $J_\nu(i^{3/2}x)$ are real, and the second, fourth, sixth, ... are imaginary. Separating the series into its real and imaginary parts, we obtain

$$J_\nu(i^{3/2}x) = i^{3\nu/2}\left[\sum_{j=0}^{\infty} \frac{(-1)^j x^{\nu+4j}}{2^{\nu+4j}(2j)!\,\Gamma(\nu+2j+1)}\right.$$

$$\left. + i\sum_{j=0}^{\infty} \frac{(-1)^j x^{\nu+2+4j}}{2^{\nu+2+4j}(2j+1)!\,\Gamma(\nu+2j+2)}\right]$$

$$= i^{3\nu/2}\left(\sum_r + i\sum_i\right)$$

Furthermore, by Demoivre's theorem (page 739),

$$i^{3\nu/2} = \left(\cos\frac{\pi}{2} + i\sin\frac{\pi}{2}\right)^{3\nu/2} = \cos\frac{3\pi\nu}{4} + i\sin\frac{3\pi\nu}{4}$$

and therefore

$$J_\nu(i^{3/2}x) = \left(\cos\frac{3\pi\nu}{4} + i\sin\frac{3\pi\nu}{4}\right)\left(\sum_r + i\sum_i\right)$$

$$= \left(\cos\frac{3\pi\nu}{4}\sum_r - \sin\frac{3\pi\nu}{4}\sum_i\right)$$

$$+ i\left(\cos\frac{3\pi\nu}{4}\sum_i + \sin\frac{3\pi\nu}{4}\sum_r\right)$$

$J_\nu(i^{3/2}x)$ thus consists of one purely real series plus i times a second purely real series. The series forming the real part of this expression defines the function **ber**$_\nu$ x. The series forming the imaginary part defines the function **bei**$_\nu$ x. The letters *be* suggest the relation between these new functions and the Bessel functions themselves. The

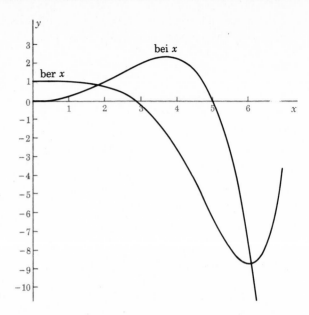

Figure 9.6
The functions ber x and bei x.

terminal letters r and i, of course, suggest the adjectives *real* and *imaginary*. For the important case $v = 0$, we have explicitly

$$\text{ber}_0\, x \equiv \text{ber}\, x = \sum_{j=0}^{\infty} \frac{(-1)^j x^{4j}}{2^{4j}[(2j)!]^2}$$

$$\text{bei}_0\, x \equiv \text{bei}\, x = \sum_{j=0}^{\infty} \frac{(-1)^j x^{4j+2}}{2^{4j+2}[(2j+1)!]^2}$$

Plots of ber x and bei x are shown in Fig. 9.6. The graphs oscillate with increasing amplitudes.

In a similar way the function $K_v(i^{1/2}x)$ can be expressed as a real series plus i times another real series. These series are taken as the definitions of the new functions **ker**$_v\, x$ and **kei**$_v\, x$,† respectively. A complete solution of Eq. (3) can thus be written

$$y = c_1(\text{ber}_v\, x + i\,\text{bei}_v\, x) + c_2(\text{ker}_v\, x + i\,\text{kei}_v\, x)$$

The function ber$_v\, x + i$ bei$_v\, x$ is finite at the origin but becomes infinite as x becomes infinite; ker$_v\, x + i$ kei$_v\, x$ is infinite at the origin but approaches zero as x becomes infinite.

For real values of x the expression ber$_v\, x + i$ bei$_v\, x$ is a complex number and, like any complex number, has a characteristic length, or modulus, $M_v(x)$ and a charac-teristic angle, or amplitude, $\vartheta_v(x)$. Clearly (Fig. 9.7),

$$M_v(x) = \sqrt{\text{ber}_v^2\, x + \text{bei}_v^2\, x} \qquad \text{and} \qquad \vartheta_v(x) = \tan^{-1} \frac{\text{bei}_v\, x}{\text{ber}_v\, x}$$

---

† The letters *ke* are derived from the name of the British mathematical physicist Lord Kelvin (1824–1907).

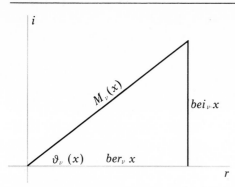

Figure 9.7
The polar representation of
$\text{ber}_v x + i\, \text{bei}_v x$.

In many applications, it is convenient to replace the expression $\text{ber}_v x + i\, \text{bei}_v x$ by its exponential equivalent

$$M_v(x) \exp\left[i\vartheta_v(x)\right]$$

and for this reason the functions $M_v(x)$ and $\vartheta_v(x)$, like the other Bessel functions, have been tabulated for certain important values of v.

Similarly, of course, the expression $\text{ker}_v x + i\, \text{kei}_v x$ can be replaced by an equivalent exponential form

$$N_v(x)e^{i\varphi_v(x)}$$

where $N_v(x) = \sqrt{\text{ker}_v^2 x + \text{kei}_v^2 x}$ and $\varphi_v(x) = \tan^{-1}\dfrac{\text{kei}_v x}{\text{ker}_v x}$

EXERCISES

1 If y_1 and y_2 are any two solutions of the modified Bessel equation of order v, show that $y_1 y_2' - y_1' y_2 = c/x$, where c is a suitable constant.

2 By determining the coefficient of $1/x$ on the left-hand side, show that if v is not an integer,

$$I_v(x)I'_{-v}(x) - I'_v(x)I_{-v}(x) = -\frac{2}{\pi x}\sin v\pi$$

Is this result correct if v is an integer?

3 Show that $I_v(x)K'_v(x) - I'_v(x)K_v(x) = -1/x$.

4 Show that

$$\int \frac{dx}{xI_v^2(x)} = -\frac{K_v(x)}{I_v(x)}$$

5 What is

$$\int \frac{dx}{xK_v^2(x)}$$

6 Show that under the change of dependent variable defined by the substitution $y = u/\sqrt{x}$, the modified Bessel equation of order v becomes

$$\frac{d^2 u}{dx^2} - \left(1 + \frac{4v^2 - 1}{4x^2}\right)u = 0$$

Hence show that for large values of x, solutions of the modified Bessel equation are described approximately by expressions of the form

$$c_1 \frac{e^x}{\sqrt{x}} + c_2 \frac{e^{-x}}{\sqrt{x}}$$

[More precisely, it can be shown that

$$I_\nu(x) \sim \frac{e^x}{\sqrt{2\pi x}} \qquad \text{and} \qquad K_\nu(x) \sim \sqrt{\frac{\pi}{2x}}\, e^{-x}$$

as x becomes infinite.]

7   Show that $J_0(i^{1/2}x) = \text{ber } x - i \text{ bei } x$.

8   Verify that

$$\text{ber}_\nu\, x = \sum_{k=0}^{\infty} \frac{[\cos(3\nu/4 + k/2)\pi]x^{\nu+2k}}{2^{\nu+2k}k!\,\Gamma(\nu+k+1)}$$

9   Verify that

$$\text{bei}_\nu\, x = \sum_{k=0}^{\infty} \frac{[\sin(3\nu/4 + k/2)\pi]x^{\nu+2k}}{2^{\nu+2k}k!\,\Gamma(\nu+k+1)}$$

10   Show that $(x \text{ ber}'\, x)' = -x \text{ bei } x$ and that $(x \text{ bei}'\, x)' = x \text{ ber } x$.

9.4 Equations Solvable in Terms of Bessel Functions

There are many differential equations whose solutions can be expressed in terms of Bessel functions, in particular, the large and important family described in the following theorem.

THEOREM 1   If $(1 - a)^2 \geq 4c$ and if neither d, p, nor q is zero, then, except in the obvious special cases when it reduces to Euler's equation,† the differential equation

$$x^2 y'' + x(a + 2bx^p)y' + [c + dx^{2q} + b(a + p - 1)x^p + b^2 x^{2p}]y = 0$$

has as a complete solution

$$y = x^\alpha e^{-\beta x^p}[c_1 J_\nu(\lambda x^q) + c_2 Y_\nu(\lambda x^q)]$$

where $\alpha = \dfrac{1-a}{2}$ $\beta = \dfrac{b}{p}$ $\lambda = \dfrac{\sqrt{|d|}}{q}$ $\nu = \dfrac{\sqrt{(1-a)^2 - 4c}}{2q}$

If $d < 0$, J_ν and Y_ν are to be replaced by I_ν and K_ν, respectively. If ν is not an integer, Y_ν and K_ν can be replaced by $J_{-\nu}$ and $I_{-\nu}$ if desired.

The proof of this theorem, while straightforward, is lengthy and involved, and we shall not present it here. It consists in transforming the given equation by means of the substitutions

$$y = x^{(1-a)/2}e^{-(b/p)x^p}Y \qquad \text{and} \qquad x = \left(\frac{qX}{\sqrt{|d|}}\right)^{1/q}$$

and verifying that when the parameters are properly identified, the result is precisely Bessel's equation in terms of the new variables X and Y.

One special case of Theorem 1 is useful enough to be stated as a corollary.

COROLLARY 1   If $(1 - r)^2 \geq 4b$, if $a \neq 0$, and if either $r - 2 < s$ or $b = 0$, then, except in the obvious special case when it reduces to Euler's equation,† the differential equation

$$(x^r y')' + (ax^s + bx^{r-2})y = 0$$

---

† Equation (10), Sec. 2.6.

has as a complete solution

$$y = x^{\alpha}[c_1 J_{\nu}(\lambda x^{\gamma}) + c_2 Y_{\nu}(\lambda x^{\gamma})]$$

where

$$\alpha = \frac{1-r}{2} \qquad \gamma = \frac{2-r+s}{2} \qquad \lambda = \frac{2\sqrt{|a|}}{2-r+s} \qquad \nu = \frac{\sqrt{(1-r)^2 - 4b}}{2-r+s}$$

If $a < 0$, J_{ν} and Y_{ν} are to be replaced by I_{ν} and K_{ν}, respectively. If ν is not an integer, Y_{ν} and K_{ν} can be replaced by $J_{-\nu}$ and $I_{-\nu}$ if desired.

EXAMPLE 1

Find a complete solution of the equation

$$x^2 y'' + x(4x^4 - 3)y' + (4x^8 - 5x^2 + 3)y = 0$$

Clearly, this is a special case of the equation of Theorem 1 with $a = -3$, $b = 2$, $p = 4$, $c = 3$, $d = -5$, and $q = 1$. Hence,

$$\alpha = 2 \qquad \beta = \tfrac{1}{2} \qquad \lambda = \sqrt{|-5|} = \sqrt{5} \qquad \text{and} \qquad \nu = 1$$

A complete solution is therefore

$$y = x^2 e^{-x^4/2}[c_1 I_1(\sqrt{5}\, x) + c_2 K_1(\sqrt{5}\, x)]$$

EXAMPLE 2

What is a complete solution of the equation $y'' + y = 0$?
Obviously, one possibility is

$$y = c_1 \cos x + c_2 \sin x$$

However, $y'' + y = 0$ is also a special case of the equation of Corollary 1, with $r = 0$, $s = 0$, $a = 1$, and $b = 0$. Hence

$$\alpha = \tfrac{1}{2} \qquad \gamma = 1 \qquad \lambda = 1 \qquad \nu = \tfrac{1}{2}$$

and so we can also write

$$y = \sqrt{x}\,[d_1 J_{1/2}(x) + d_2 J_{-1/2}(x)]$$

It follows, therefore, from Theorem 2, Sec. 2.1, that for proper choice of the constants c_1 and c_2, each of the particular solutions

$$\sqrt{x}\, J_{1/2}(x) \qquad \text{and} \qquad \sqrt{x}\, J_{-1/2}(x)$$

must be expressible in the form $c_1 \cos x + c_2 \sin x$.
Now, since $\Gamma(\tfrac{3}{2}) = \tfrac{1}{2}\Gamma(\tfrac{1}{2}) = \tfrac{1}{2}\sqrt{\pi}$, the series for $J_{1/2}(x)$ begins with the term

$$\frac{x^{1/2}}{2^{1/2}\,\Gamma(\tfrac{3}{2})} = \sqrt{\frac{2x}{\pi}}$$

Hence the series for $\sqrt{x}\, J_{1/2}(x)$ begins with the term $\sqrt{2/\pi}\, x$. Therefore, if we write

$$\sqrt{x}\, J_{1/2}(x) = \sqrt{\frac{2}{\pi}}\, x - \cdots = c_1 \cos x + c_2 \sin x$$

and put $x = 0$ in this identity, we find $c_1 = 0$. Subsequently, by equating the coefficients of x, we find

$$c_2 = \sqrt{\frac{2}{\pi}}$$

We have thus established the interesting and important result that

$$\sqrt{x}\, J_{1/2}(x) = \sqrt{\frac{2}{\pi}} \sin x \qquad \text{or} \qquad J_{1/2}(x) = \sqrt{\frac{2}{\pi x}} \sin x$$

In a similar fashion it can be shown that

$$J_{-1/2}(x) = \sqrt{\frac{2}{\pi x}} \cos x$$

EXERCISES

1 Verify that Corollary 1 is indeed a special case covered by Theorem 1.
2 Solve the equation of Example 2, Sec. 9.1, in terms of Bessel functions.
3 Solve Exercise 6, Sec. 9.1, in terms of Bessel functions.
4 Solve Exercise 7, Sec. 9.1, in terms of Bessel functions.
5 Can Exercise 8, Sec. 9.1, be solved in terms of Bessel functions by means of Theorem 1?
6 Solve Exercise 9, Sec. 9.1, in terms of Bessel functions.
7 Solve Exercise 10, Sec. 9.1, in terms of Bessel functions.

Find a complete solution of each of the following equations:

8   $(x^2 y')' + [x^2 - n(n - 1)]y = 0$
9   $y'' + x^m y = 0$ 10 $xy'' + 2y' + 4xy = 0$
11 $xy'' - y' + 4x^5 y = 0$ 12 $x^2 y'' + 3xy' + (1 + x)y = 0$
13 $x^2 y'' + 2x^2 y' + (x^4 + x^2 - 2)y = 0$
14 $x^2 y'' + (2x^2 + x)y' + (x^2 + 3x - 1)y = 0$
15 $[(a + bx)y']' - y = 0$. *Hint:* Introduce a new independent variable via the substitution $t = a + bx$.
16 $[(1 + x)^2 y']' + (a + bx)y = 0$ $a - b < \frac{1}{4}$
17 $y'' + ae^{mx}y = 0$. *Hint:* Introduce a new independent variable via the substitution $t = e^{mx/2}$.
18 $y'' + ay' + (b + ce^{mx})y = 0$, $am > 0$ and $a^2 \geq 4b$. *Hint:* Introduce a new independent variable via the substitution $t = e^{ax}$.
19 Show that $I_{1/2}(x) = \sqrt{2/\pi x} \sinh x$ and $I_{-1/2}(x) = \sqrt{2/\pi x} \cosh x$.
20 Show that any solution of

$$(x^{m-1}y')' + kx^{m-2}y = 0 \qquad \text{or} \qquad (x^{m-1}y')' - kx^{m-2}y = 0$$

will also satisfy the equation $(x^m y'')'' - k^2 x^{m-2}y = 0$.

21 Find a complete solution of $(x^2 y'')'' - 9y = 0$.
22 Find a complete solution of $x^2 y^{iv} + 8xy''' + 12y'' - y = 0$.
23 Show that if $b = \frac{1}{4} - v^2$, the function $y = \sqrt{x}\, J_v(x)$ is a solution of the equation $y'' + (1 + b/x^2)y = 0$.. Hence, applying the result of Exercise 22, Sec. 2.1, to this equation and the equation $y'' + k^2 y = 0$ for suitable k, show that $J_v(x)$ has infinitely many zeros.

9.5 Identities for the Bessel Functions

The Bessel functions are related by an amazing array of identities. Fundamental among these are the consequences of the following pair of theorems.

THEOREM 1 $\dfrac{d[x^v J_v(x)]}{dx} = x^v J_{v-1}(x)$

Proof   To prove this theorem, we first multiply the series for $J_v(x)$ by x^v and then differentiate it term by term:

$$J_v(x) = \sum_{k=0}^{\infty} \frac{(-1)^k x^{v+2k}}{2^{v+2k} k! \, \Gamma(v+k+1)}$$

$$x^v J_v(x) = \sum_{k=0}^{\infty} \frac{(-1)^k x^{2v+2k}}{2^{v+2k} k! \, \Gamma(v+k+1)}$$

$$\frac{d[x^v J_v(x)]}{dx} = \sum_{k=0}^{\infty} \frac{(-1)^k 2(v+k) x^{2v+2k-1}}{2^{v+2k} k! \, (v+k) \Gamma(v+k)}$$

$$= \sum_{k=0}^{\infty} \frac{(-1)^k x^v x^{v-1+2k}}{2^{v-1+2k} k! \, \Gamma(v+k)}$$

$$= x^v \sum_{k=0}^{\infty} \frac{(-1)^k x^{v-1+2k}}{2^{v-1+2k} k! \, \Gamma(v-1+k+1)}$$

$$= x^v J_{v-1}(x)$$

as asserted.

THEOREM 2 $\dfrac{d[x^{-v} J_v(x)]}{dx} = -x^{-v} J_{v+1}(x)$

Proof   This theorem can be proved in essentially the same manner as Theorem 1, but it is easier and perhaps more instructive to proceed as follows. By applying Corollary 1 of Theorem 1, Sec. 9.4, to the equation

$$\frac{d(x^{-1-2v} y')}{dx} + x^{-1-2v} y = 0$$

with $r = s = -1 - 2v$ and $a = 1$, $b = 0$, it is clear that $x^{1+v} J_{1+v}(x)$ is a particular solution. Hence

$$\frac{d\{x^{-1-2v}[x^{1+v} J_{1+v}(x)]'\}}{dx} + x^{-1-2v}[x^{1+v} J_{1+v}(x)] = 0$$

Now, using Theorem 1 to compute the derivative of the quantity $x^{1+v} J_{1+v}(x)$, we have further

$$\frac{d\{x^{-1-2v}[x^{1+v} J_v(x)]\}}{dx} = -x^{-v} J_{v+1}(x)$$

or finally,

$$\frac{d[x^{-v} J_v(x)]}{dx} = -x^{-v} J_{v+1}(x)$$

as asserted.

    By using their definitions in terms of $J_v(x)$ and $J_{-v}(x)$, one can readily show that *the Bessel functions of the second kind $Y_v(x)$ and the Hankel functions $H_v^{(1)}(x)$ and $H_v^{(2)}(x)$ also satisfy the identities of Theorems 1 and 2.* Furthermore, by arguments similar to those we have just used, the following theorems can be established.

THEOREM 3 $\dfrac{d[x^\nu I_\nu(x)]}{dx} = x^\nu I_{\nu-1}(x)$

THEOREM 4 $\dfrac{d[x^{-\nu} I_\nu(x)]}{dx} = x^{-\nu} I_{\nu+1}(x)$

THEOREM 5 $\dfrac{d[x^\nu K_\nu(x)]}{dx} = -x^\nu K_{\nu-1}(x)$

THEOREM 6 $\dfrac{d[x^{-\nu} K_\nu(x)]}{dx} = -x^{-\nu} K_{\nu+1}(x)$

Performing the indicated differentiations in the identities of Theorems 1 and 2, we obtain, respectively,

$$x^\nu J_\nu'(x) + \nu x^{\nu-1} J_\nu(x) = x^\nu J_{\nu-1}(x)$$
$$x^{-\nu} J_\nu'(x) - \nu x^{-\nu-1} J_\nu(x) = -x^{-\nu} J_{\nu+1}(x)$$

Dividing the first of these by x^ν and multiplying the second by x^ν and solving for $J_\nu'(x)$ in each case gives

(1) $$J_\nu'(x) = J_{\nu-1}(x) - \frac{\nu}{x} J_\nu(x)$$

(2) $$J_\nu'(x) = \frac{\nu}{x} J_\nu(x) - J_{\nu+1}(x)$$

Adding these and dividing by 2, we obtain a third formula for $J_\nu'(x)$:

(3) $$J_\nu'(x) = \frac{J_{\nu-1}(x) - J_{\nu+1}(x)}{2}$$

Subtracting (2) from (1) gives the important recurrence formula

$$J_{\nu-1}(x) + J_{\nu+1}(x) = \frac{2\nu}{x} J_\nu(x)$$

Written as

(4) $$J_{\nu+1}(x) = \frac{2\nu}{x} J_\nu(x) - J_{\nu-1}(x)$$

this formula serves to express Bessel functions of higher orders in terms of functions of lower orders, frequently a useful manipulation. Written as

(5) $$J_{\nu-1}(x) = \frac{2\nu}{x} J_\nu(x) - J_{\nu+1}(x)$$

it serves similarly to express Bessel functions of large negative orders (for instance) in terms of Bessel functions whose orders are numerically smaller.

EXAMPLE 1

Express $J_4(ax)$ in terms of $J_0(ax)$ and $J_1(ax)$.

Taking $v = 3$ in (4), we first have

$$J_4(ax) = \frac{6}{ax} J_3(ax) - J_2(ax)$$

Applying (4) again to $J_3(ax)$ and then to $J_2(ax)$, we have further

$$J_4(ax) = \frac{6}{ax} \left[\frac{4}{ax} J_2(ax) - J_1(ax) \right] - J_2(ax)$$

$$= \left(\frac{24}{a^2 x^2} - 1 \right) J_2(ax) - \frac{6}{ax} J_1(ax)$$

$$= \left(\frac{24}{a^2 x^2} - 1 \right) \left[\frac{2}{ax} J_1(ax) - J_0(ax) \right] - \frac{6}{ax} J_1(ax)$$

$$= \left(\frac{48}{a^3 x^3} - \frac{8}{ax} \right) J_1(ax) - \left(\frac{24}{a^2 x^2} - 1 \right) J_0(ax)$$

EXAMPLE 2

Show that $d[xJ_v(x)J_{v+1}(x)]/dx = x[J_v{}^2(x) - J_{v+1}^2(x)]$.

Performing the indicated differentiation, we have

$$\frac{d[xJ_v(x)J_{v+1}(x)]}{dx} = J_v(x)J_{v+1}(x) + xJ_v'(x)J_{v+1}(x) + xJ_v(x)J_{v+1}'(x)$$

Then, substituting for $xJ_v'(x)$ from (2) and for $xJ_{v+1}'(x)$ from (1), we have

$$\frac{d[xJ_v(x)J_{v+1}(x)]}{dx} = J_v(x)J_{v+1}(x) + J_{v+1}(x)[vJ_v(x) - xJ_{v+1}(x)]$$

$$+ J_v(x)[xJ_v(x) - (v+1)J_{v+1}(x)]$$

$$= x[J_v{}^2(x) - J_{v+1}^2(x)]$$

The basic differentiation identities of Theorems 1 and 2, when written as integration formulas

(6)
$$\int x^v J_{v-1}(x)\, dx = x^v J_v(x) + c$$

(7)
$$\int x^{-v} J_{v+1}(x)\, dx = -x^{-v} J_v(x) + c$$

suffice for the integration of numerous simple expressions involving Bessel functions. For example, taking $v = 1$ in (6), we have

$$\int x J_0(x)\, dx = x J_1(x) + c$$

Similarly, taking $v = 0$ in (7), we find

$$\int J_1(x)\, dx = -J_0(x) + c$$

Usually, however, integration by parts must be used in addition to formulas (6) and (7).

EXAMPLE 3

What is $\int J_3(x)\, dx$?

If we multiply and divide the integrand by x^2, we have

$$\int x^2 [x^{-2} J_3(x)]\, dx$$

and so, integrating by parts with

$$u = x^2 \qquad dv = x^{-2} J_3(x)\, dx$$
$$du = 2x\, dx \qquad v = -x^{-2} J_2(x) \qquad \text{by (7), with } \nu = 2$$

we have

$$\int J_3(x)\, dx = -J_2(x) + 2 \int x^{-1} J_2(x)\, dx$$
$$= -J_2(x) - 2x^{-1} J_1(x) + c \qquad \text{by (7), with } \nu = 1$$

EXAMPLE 4

What is

$$\int \frac{J_2(3x)}{x^2}\, dx$$

Here it is convenient to multiply the numerator and denominator by $9x^2$, getting

$$\tfrac{1}{9} \int (3x)^2 J_2(3x)\, \frac{dx}{x^4}$$

Now, integrating by parts with

$$u = (3x)^2 J_2(3x) \qquad\qquad dv = \frac{dx}{x^4}$$

$$du = (3x)^2 J_1(3x)\, 3dx \qquad v = -\frac{1}{3x^3}$$

we have

$$\int \frac{J_2(3x)}{x^2}\, dx = \frac{1}{9}\left[-\frac{3J_2(3x)}{x} + 3 \int 3x J_1(3x)\, \frac{dx}{x^2} \right]$$

Again using integration by parts, with

$$u = 3x J_1(3x) \qquad\qquad dv = \frac{dx}{x^2}$$

$$du = 3x J_0(3x)\, 3dx \qquad v = -\frac{1}{x}$$

we have further

$$\int \frac{J_2(3x)}{x^2}\, dx = \frac{1}{9}\left\{ -\frac{3J_2(3x)}{x} + 3\left[-3J_1(3x) + 9 \int J_0(3x)\, dx \right] \right\}$$

$$= -\frac{J_2(3x)}{3x} - J_1(3x) + 3 \int J_0(3x)\, dx$$

The residual integral $\int J_0(3x)\, dx$ cannot be evaluated in finite form in terms of any of the Bessel functions we have encountered.

In general, an integral of the form

$$\int x^m J_n(x)\, dx$$

where m and n are integers such that $m + n \geq 0$, can be completely integrated if $m + n$ is odd but will ultimately depend upon the residual integral $\int J_0(x)\,dx$ if $m + n$ is even. For this reason $\int_0^x J_0(x)\,dx$ has been tabulated.†

Another class of identities of considerable interest can be obtained from the expansion of the function

$$(8) \qquad \exp\left[\frac{x}{2}\left(t - \frac{1}{t}\right)\right] = e^{xt/2}e^{-x/2t}$$

in terms of powers of t. To derive this expansion we first replace the exponentials on the right of (8) by their infinite series, getting

$$\left(\sum_{i=0}^{\infty} \frac{1}{i!}\frac{x^i t^i}{2^i}\right)\left(\sum_{j=0}^{\infty} \frac{(-1)^j}{j!}\frac{x^j t^{-j}}{2^j}\right)$$

When these series are multiplied together, we obtain a term containing t^n $(n \geq 0)$ when and only when the general term in the second series, i.e., the term containing t^{-j}, is multiplied by the term in the first series which contains t^{n+j}, that is, the term for which $i = n + j$. Therefore, taking into account all possible values of j, we find that the total coefficient of t^n in the product of the two series is

$$\sum_{j=0}^{\infty}\left[\frac{1}{(n+j)!}\frac{x^{n+j}}{2^{n+j}}\right]\left[\frac{(-1)^j}{j!}\frac{x^j}{2^j}\right] = \sum_{j=0}^{\infty}\frac{(-1)^j x^{n+2j}}{2^{n+2j}j!\,\Gamma(n+j+1)} = J_n(x)$$

Similarly, a term containing t^{-n} arises when and only when the general term in the first series, i.e., the term containing t^i, is multiplied by the term in the second series which contains t^{-n-i}, that is, the term for which $j = n + i$. Therefore, taking into account all possible values of i, we find that the total coefficient of t^{-n} in the product of the two series is

$$\sum_{i=0}^{\infty}\left[\frac{1}{i!}\frac{x^i}{2^i}\right]\left[\frac{(-1)^{n+i}}{(n+i)!}\frac{x^{n+i}}{2^{n+i}}\right] = (-1)^n\sum_{i=0}^{\infty}\frac{(-1)^i x^{n+2i}}{2^{n+2i}i!\,\Gamma(n+i+1)} = (-1)^n J_n(x)$$

Hence

$$(9) \qquad \exp\left[\frac{x}{2}\left(t - \frac{1}{t}\right)\right] = J_0(x) + \sum_{n=1}^{\infty} J_n(x)[t^n + (-1)^n t^{-n}]$$

Now let $t = e^{i\phi}$, so that

$$\frac{1}{2}\left(t - \frac{1}{t}\right) = \frac{e^{i\phi} - e^{-i\phi}}{2} = i\sin\phi$$

and

$$\exp\left[\frac{x}{2}\left(t - \frac{1}{t}\right)\right] = \exp\,(ix\sin\phi) = \cos\,(x\sin\phi) + i\sin\,(x\sin\phi)$$

In the same way, when n is even, say $n = 2k$, we have

$$t^n + (-1)^n t^{-n} = t^{2k} + (-1)^{2k}t^{-2k} = e^{i2k\phi} + e^{-i2k\phi} = 2\cos 2k\phi$$

---

† "Handbook of Mathematical Functions," Superintendent of Documents, GPO, Washington, D.C., 1965.

and when n is odd, say $n = 2k - 1$, we have

$$t^n + (-1)^n t^{-n} = t^{2k-1} + (-1)^{2k-1} t^{-2k+1} = e^{i(2k-1)\phi} - e^{-i(2k-1)\phi}$$
$$= 2i \sin (2k - 1)\phi$$

Thus Eq. (9) can be written

$$\exp (ix \sin \phi) \equiv \cos (x \sin \phi) + i \sin (x \sin \phi)$$

$$= J_0(x) + 2 \sum_{k=1}^{\infty} J_{2k}(x) \cos 2k\phi + 2i \sum_{k=1}^{\infty} J_{2k-1}(x) \sin (2k - 1)\phi$$

Equating real and imaginary parts in the last expression, we obtain the identities

(10) $$\cos (x \sin \phi) = J_0(x) + 2 \sum_{k=1}^{\infty} J_{2k}(x) \cos 2k\phi$$

(11) $$\sin (x \sin \phi) = 2 \sum_{k=1}^{\infty} J_{2k-1}(x) \sin (2k - 1)\phi$$

The series on the right in (10) and (11) are, of course, just the Fourier expansions of the functions on the left.

Now multiply both sides of (10) by $\cos n\phi$ and both sides of (11) by $\sin n\phi$, and integrate each identity with respect to ϕ from 0 to π. Since

$$\int_0^\pi \cos m\phi \cos n\phi \, d\phi = \int_0^\pi \sin m\phi \sin n\phi \, d\phi = 0 \quad m \neq n$$

$$\int_0^\pi \cos^2 n\phi \, d\phi = \int_0^\pi \sin^2 n\phi \, d\phi = \frac{\pi}{2}$$

this yields

$$\int_0^\pi \cos n\phi \cos (x \sin \phi) \, d\phi = \begin{cases} \pi J_n(x) & n \text{ even} \\ 0 & n \text{ odd} \end{cases}$$

$$\int_0^\pi \sin n\phi \sin (x \sin \phi) \, d\phi = \begin{cases} 0 & n \text{ even} \\ \pi J_n(x) & n \text{ odd} \end{cases}$$

If we add these two expressions and divide by π, we have, for all integral values of n

$$J_n(x) = \frac{1}{\pi} \int_0^\pi [\cos n\phi \cos (x \sin \phi) + \sin n\phi \sin (x \sin \phi)] \, d\phi$$

since for every value of n, one or the other of the integrals vanishes while the remaining one contributes $J_n(x)$. Finally, using the formula for the cosine of the difference of two quantities, we have

(12) $$J_n(x) = \frac{1}{\pi} \int_0^\pi \cos (n\phi - x \sin \phi) \, d\phi \quad n \text{ an integer}$$

EXERCISES

1 Express $J_5(x)$ in terms of $J_0(x)$ and $J_1(x)$.
2 Express $J_{3/2}(x)$ and $J_{-3/2}(x)$ in terms of $\sin x$ and $\cos x$.
3 What is $d[x^2 J_3(2x)]/dx$?   4 What is $d[xJ_0(x^2)]/dx$?
5 Show that

$$\frac{d[x^2 J_{v-1}(x)J_{v+1}(x)]}{dx} = 2x^2 J_v(x) \frac{d[J_v(x)]}{dx}$$

6 Prove Theorem 2 by using the series expansion for $J_\nu(x)$.

7 Show that

(a) $4J_\nu''(x) = J_{\nu-2}(x) - 2J_\nu(x) + J_{\nu+2}(x)$

(b) $8J_\nu'''(x) = J_{\nu-3}(x) - 3J_{\nu-1}(x) + 3J_{\nu+1}(x) - J_{\nu+3}(x)$

8 Show that

$$J_\nu''(x) = \left[\frac{\nu(\nu+1)}{x^2} - 1\right] J_\nu(x) - \frac{1}{x} J_{\nu-1}(x)$$

9 Show that $\cos x = J_0(x) - 2J_2(x) + 2J_4(x) - 2J_6(x) + \cdots$.

10 Show that $\sin x = 2J_1(x) - 2J_3(x) + 2J_5(x) - 2J_7(x) + \cdots$.

11 Show that

$$J_0(x) = \frac{1}{\pi} \int_0^\pi \cos(x \cos \phi) \, d\phi$$

12 By expressing the exponential as an infinite series and then integrating term by term, show that

$$I_0(x) = \frac{1}{\pi} \int_0^\pi \exp(x \cos \phi) \, d\phi$$

Hint: It will be helpful to recall Wallis' formulas

$$\int_0^{\pi/2} \cos^{2k} \phi \, d\phi = \int_0^{\pi/2} \sin^{2k} \phi \, d\phi = \frac{\pi}{2} \frac{(2k)!}{2^{2k}(k!)^2}$$

$$\int_0^{\pi/2} \cos^{2k+1} \phi \, d\phi = \int_0^{\pi/2} \sin^{2k+1} \phi \, d\phi = \frac{2^{2k}(k!)^2}{(2k+1)!}$$

13 By expanding the integrand into an infinite series and integrating term by term, show that

$$\int_0^{\pi/2} J_0(x \cos \phi) \cos \phi \, d\phi = \frac{\sin x}{x}$$

14 Show that

$$\int_0^{\pi/2} J_1(x \cos \phi) \, d\phi = \frac{1 - \cos x}{x}$$

15 What is $\int_0^{\pi/2} J_2(x \cos \phi) \cos \phi \, d\phi$?

16 What is $\int_0^{\pi/2} J_0(x \cos \phi) \cos^3 \phi \, d\phi$?

17 Show that $\int J_0(x) \, dx = 2[J_1(x) + J_3(x) + J_5(x) + \cdots]$. *Hint:* Use Formula (3).

18 Show that

$$\int J_0(x) \, dx = J_1(x) + \int \frac{J_1(x)}{x} \, dx$$

$$= J_1(x) + \frac{J_2(x)}{x} + 1 \cdot 3 \int \frac{J_2(x)}{x^2} \, dx$$

$$= J_1(x) + \frac{J_2(x)}{x} + \frac{1 \cdot 3}{x^2} J_3(x) + 1 \cdot 3 \cdot 5 \int \frac{J_3(x)}{x^3} \, dx$$

$$\cdots\cdots\cdots\cdots\cdots\cdots\cdots\cdots\cdots\cdots\cdots$$

$$= J_1(x) + \frac{J_2(x)}{x} + \frac{1 \cdot 3}{x^2} J_3(x) + \cdots + \frac{(2n-2)!}{2^{n-1}(n-1)!x^{n-1}} J_n(x)$$

$$+ \frac{(2n)!}{2^n n!} \int \frac{J_n(x)}{x^n} \, dx$$

Hint: Use repeated integration by parts, each time taking $dv = x^{k+1} J_k(x) \, dx$.

19 Show that

$$J_\nu(x)J_{-(\nu+1)}(x) - J_{\nu+1}(x)J_{-\nu}(x) = -\frac{2 \sin \nu\pi}{\pi x}$$

Hint: Recall the result of Exercise 3, Sec. 9.2.

20 What is $J_\nu(x)Y_{\nu+1}(x) - J_{\nu+1}(x)Y_\nu(x)$?
21 What is $I_\nu(x)I_{-(\nu+1)}(x) - I_{\nu+1}(x)I_{-\nu}(x)$?
22 What is $I_\nu(x)K_{\nu+1}(x) + I_{\nu+1}(x)K_\nu(x)$?
23 Verify that

$$J_2(x) = x^2 \left(\frac{1}{x}\frac{d}{dx}\right)\left(\frac{1}{x}\frac{d}{dx}\right) J_0(x)$$

24 Verify that

$$J_n(x) = (-1)^n x^n \left(\frac{1}{x}\frac{d}{dx}\right)^n J_0(x)$$

25 Show that $\int J_0(x)\cos x\,dx = xJ_0(x)\cos x + xJ_1(x)\sin x + c$.
26 Show that $\int J_0(x)\sin x\,dx = xJ_0(x)\sin x - xJ_1(x)\cos x + c$.
27 What is $\int J_1(x)\cos x\,dx$?   **28** What is $\int J_1(x)\sin x\,dx$?
29 Show that

$$\int xJ_n{}^2(x)\,dx = \frac{x^2}{2}\,[J_n{}^2(x) - J_{n-1}(x)J_{n+1}(x)] + \cdot c$$

Hint: After integrating by parts, the result of Exercise 5 may be helpful.
30 Show that

$$\int [J_{\nu-1}^2(x) - J_{\nu+1}^2(x)]\,x\,dx = 2\nu J_\nu{}^2(x) + c$$

31 What is:
  (a) $\int xJ_0(x)\cos x\,dx$ (b) $\int xJ_1(x)\sin x\,dx$
32 What is:
  (a) $\int xJ_1(x)\cos x\,dx$ (b) $\int xJ_0(x)\sin x\,dx$
33 Show that:
  (a) $\int xJ_0(x)\,dx = xJ_1(x) + c$
  (b) $\int x^2 J_0(x)\,dx = x^2 J_1(x) + xJ_0(x) - \int J_0(x)\,dx + c$
  (c) $\int x^3 J_0(x)\,dx = (x^3 - 4x)J_1(x) + 2x^2 J_0(x) + c$
  (d) $\int x^4 J_0(x)\,dx = (x^4 - 9x^2)J_1(x) + (3x^3 - 9x)J_0(x) + 9\int J_0(x)\,dx + c$
34 Show that:
  (a) $\int \dfrac{J_1(x)}{x}\,dx = -J_1(x) + \int J_0(x)\,dx + c$
  (b) $\int J_1(x)\,dx = -J_0(x) + c$
  (c) $\int xJ_1(x)\,dx = -xJ_0(x) + \int J_0(x)\,dx + c$
  (d) $\int x^2 J_1(x)\,dx = 2xJ_1(x) - x^2 J_0(x) + c$
  (e) $\int x^3 J_1(x)\,dx = 3x^2 J_1(x) - (x^3 - 3x)J_0(x) - 3\int J_0(x)\,dx + c$
  (f) $\int x^4 J_1(x)\,dx = (4x^3 - 16x)J_1(x) - (x^4 - 8x^2)J_0(x) + c$
35 By replacing $J_0(\lambda x)$ by its infinite series and then integrating termwise, show that
  $\int_0^\infty e^{-ax}J_0(\lambda x)\,dx = 1/\sqrt{\lambda^2 + a^2}$.
36 Using the result of Exercise 35, determine the value of $\int_0^\infty e^{-ax}J_1(\lambda x)\,dx$.
37 What is:

  (a) $\displaystyle\int_0^\infty xe^{-ax}J_1(\lambda x)\,dx$ (b) $\displaystyle\int_0^\infty xe^{-ax}J_0(\lambda x)\,dx$

38 What is:

  (a) $\displaystyle\int_0^\infty e^{-ax}I_0(\lambda x)\,dx$ (b) $\displaystyle\int_0^\infty e^{-ax}I_1(\lambda x)\,dx$

39 What is:
  (a) $\int J_0(\sqrt{x})\,dx$ (b) $\int J_2(\sqrt{x})\,dx$
40 What is $\int xJ_2(1 - x)\,dx$?
41 What is $\int x\ln x\,J_0(\lambda x)\,dx$?
42 What is:
  (a) $\int xJ_0(\sqrt{x})\,dx$ (b) $\int xJ_2(\sqrt{x})\,dx$

43 Show that:

(a) $I_\nu'(x) = I_{\nu-1}(x) - \dfrac{\nu}{x} I_\nu(x)$ (b) $I_\nu'(x) = \dfrac{\nu}{x} I_\nu(x) + I_{\nu+1}(x)$

(c) $I_\nu'(x) = \frac{1}{2}[I_{\nu-1}(x) + I_{\nu+1}(x)]$ (d) $I_{\nu-1}(x) - I_{\nu+1}(x) = \dfrac{2\nu}{x} I_\nu(x)$

44 What is:

(a) $\int x I_0(x)\, dx$ (b) $\int x^2 I_0(x)\, dx$

(c) $\int x I_1(x)\, dx$ (d) $\int x^2 I_1(x)\, dx$

45 Show that

$$x^2 J_2(x) = \frac{1}{2} \int_0^x t(x^2 - t^2) J_0(t)\, dt$$

Hint: Observe that $x^2 J_2(x) = \int_0^x t^2 J_1(t)\, dt = \int_0^x t \int_0^t s J_0(s)\, ds\, dt$ and then change the order of integration in the double integral.

46 Show that

$$x^3 J_3(x) = \frac{1}{2^2 2!} \int_0^x t(x^2 - t^2)^2 J_0(t)\, dt$$

47 Show that

$$x^{n+1} J_{n+1}(x) = \frac{1}{2^n n!} \int_0^x t(x^2 - t^2)^n J_0(t)\, dt$$

48 By substituting $x \sin\theta$ for t in the formula of Exercise 47, show that

$$J_{n+1}(x) = \frac{x^{n+1}}{2^n n!} \int_0^{\pi/2} \sin\theta \cos^{2n+1}\theta\, J_0(x \sin\theta)\, d\theta$$

49 Verify that $y_1 = J_0(\lambda x)$, $y_2 = Y_0(\lambda x)$, $y_3 = x J_1(\lambda x)$, and $y_4 = x Y_1(\lambda x)$ are four particular solutions of the equation

$$\left(\frac{d^2}{dx^2} + \frac{1}{x}\frac{d}{dx} + \lambda^2\right)\left(\frac{d^2}{dx^2} + \frac{1}{x}\frac{d}{dx} + \lambda^2\right) y = 0$$

50 Find a complete solution of the equation

$$\left(\frac{d^2}{dx^2} + \frac{1}{x}\frac{d}{dx} - \lambda^2\right)\left(\frac{d^2}{dx^2} + \frac{1}{x}\frac{d}{dx} - \lambda^2\right) y = 0$$

9.6 The Orthogonality of the Bessel Functions

If we write Bessel's equation of order ν in the form

$$x \frac{d^2 y}{dx^2} + \frac{dy}{dx} + \left(\lambda^2 x - \frac{\nu^2}{x}\right) y \equiv \frac{d(xy')}{dx} + \left(-\frac{\nu^2}{x} + \lambda^2 x\right) y = 0$$

it is clear that it is a special case of the equation covered by Theorem 4, Sec. 8.5, provided we make the identifications

$$p(x) = x \qquad q(x) = -\frac{\nu^2}{x} \qquad r(x) = x$$

and write λ^2 in place of λ. Therefore, if the solutions of Bessel's equation satisfy boundary conditions of the form

(1)
$$A_i y_\nu(\lambda x_i) - B_i \left.\frac{dy_\nu(\lambda x)}{dx}\right|_{x=x_i} = 0 \qquad i = 1, 2$$

they must be orthogonal with respect to the weight function $p(x) = x$ over the interval (x_1,x_2).†

It is not enough, however, to know that the characteristic functions of a problem are orthogonal. In order to carry out the expansions required at the final stage of a typical boundary-value problem, it is also necessary to know the value of the integral of the product of the weight function $p(x)$ and the square of the general characteristic function, taken over the interval of the problem.

We begin this calculation by considering the indefinite integral $\int ty_v^2(t)\,dt$, where $y_v(t)$ is *any* solution of Bessel's equation; i.e.,

(2) $$t^2 y_v'' + t y_v' + (t^2 - v^2)y_v = 0$$

If Eq. (2) is multiplied by y_v' and then integrated, we obtain

(3) $$\int t^2 y_v' y_v''\,dt + \int t(y_v')^2\,dt + \int t^2 y_v y_v'\,dt - v^2 \int y_v y_v'\,dt = 0$$

Now, evaluating the first and third integrals by parts, we have

$$\int t^2 y_v' y_v''\,dt \underset{\substack{u=t^2 \\ du=2t\,dt}}{\xrightarrow{\hspace{1.2cm}}} \underset{\substack{dv=y_v'y_v''dt \\ v=\frac{1}{2}(y_v')^2}}{\xrightarrow{\hspace{1.2cm}}} \tfrac{1}{2}t^2(y_v')^2 - \int t(y_v')^2\,dt$$

$$\int t^2 y_v y_v'\,dt \underset{\substack{u=t^2 \\ du=2t\,dt}}{\xrightarrow{\hspace{1.2cm}}} \underset{\substack{dv=y_vy_v'dt \\ v=\frac{1}{2}y_v^2}}{\xrightarrow{\hspace{1.2cm}}} \tfrac{1}{2}t^2 y_v^2 - \int ty_v^2\,dt$$

Then, substituting these results into Eq. (3), we find

$$\left[\tfrac{1}{2}t^2(y_v')^2 - \int t(y_v')^2\,dt\right] + \int t(y_v')^2\,dt + \left(\tfrac{1}{2}t^2 y_v^2 - \int ty_v^2\,dt\right) - \tfrac{1}{2}v^2 y_v^2 = 0$$

or, collecting terms and solving for $\int ty_v^2\,dt$,

$$\int ty_v^2(t)\,dt = \tfrac{1}{2}(t^2 - v^2)y_v^2(t) + \tfrac{1}{2}t^2\left[\frac{dy_v(t)}{dt}\right]^2$$

If we now put $t = \lambda_m x$, where λ_m is any one of the characteristic values for which solutions satisfying the boundary conditions (1) exist, and then divide by λ_m^2, we obtain the integral in which we are actually interested:

(4) $$\int xy_v^2(\lambda_m x)\,dx = \frac{1}{2\lambda_m^2}\left\{(\lambda_m^2 x^2 - v^2)y_v^2(\lambda_m x) + x^2\left[\frac{dy_v(\lambda_m x)}{dx}\right]^2\right\}$$

The evaluation of (4) between specific limits x_1 and x_2 requires the consideration of several special cases, according as B_i in the boundary conditions (1) is or is not equal to zero. If $B_i = 0$, then (1) becomes simply

(5) $$y_v(\lambda_m x_i) = 0$$

and the antiderivative on the right of (4) reduces to

(6) $$\frac{1}{2\lambda_m^2}x_i^2\left[\frac{dy_v(\lambda_m x)}{dx}\right]^2\Bigg|_{x=x_i}$$

---

† Since $r(x) \equiv x$ vanishes when $x = 0$, it follows from the proof of Theorem 4, Sec. 8.5, that if $x_1 = 0$, no boundary condition of the form (1) will be needed (and none will be available) at $x = x_1$. The relevant condition in this case is that the solutions be bounded at $x = x_1 = 0$.

This can be further simplified by recalling from the last section that all solutions of Bessel's equation, J_v, J_{-v}, Y_v, $H_v^{(1)}$, and $H_v^{(2)}$, as well as arbitrary linear combinations of these functions, satisfy the identity [Eq. (2), Sec. 9.5]

$$t \frac{dy_v(t)}{dt} = vy_v(t) - ty_{v+1}(t)$$

or

$$x \frac{dy_v(\lambda_m x)}{dx} = vy_v(\lambda_m x) - \lambda_m x y_{v+1}(\lambda_m x)$$

Evaluating this at $x = x_i$, and using (5), we find

$$x_i \frac{dy_v(\lambda_m x)}{dx}\bigg|_{x=x_i} = -\lambda_m x_i y_{v+1}(\lambda_m x_i)$$

and so (6) becomes simply

$$\tfrac{1}{2} x_i^2 y_{v+1}^2(\lambda_m x_i)$$

On the other hand, if $B_i \neq 0$, we can solve for the derivative term in Eq. (1) and substitute the result into the right-hand member of Eq. (4), getting

$$\frac{y_v^2(\lambda_m x_i)}{2\lambda_m^2} \left[(\lambda_m x_i)^2 - v^2 + \left(\frac{x_i A_i}{B_i}\right)^2 \right]$$

At each end of the interval (x_1, x_2) we thus have two possibilities for the evaluated integral (4). These lead to four cases, which are summarized in the following important theorem.

THEOREM 1 The solutions of Bessel's equation of order v which satisfy the boundary conditions

$$A_i y_v(\lambda x_i) - B_i \frac{dy_v(\lambda x)}{dx}\bigg|_{x=x_i} = 0 \qquad i = 1, 2$$

form an orthogonal system with respect to the weight function x over the interval (x_1, x_2). The integral of the product of the weight function and the square of any solution of the system $\{y_v(\lambda_m x)\}$, that is,

$$\int_{x_1}^{x_2} x y_v^2(\lambda_m x)\, dx$$

is equal to

$$\frac{x_2^2}{2} y_{v+1}^2(\lambda_m x_2) - \frac{x_1^2}{2} y_{v+1}^2(\lambda_m x_1) \qquad B_1 = B_2 = 0$$

$$\frac{x_2^2}{2} y_{v+1}^2(\lambda_m x_2) - \frac{y_v^2(\lambda_m x_1)}{2\lambda_m^2} \left[(\lambda_m x_1)^2 - v^2 + \left(\frac{x_1 A_1}{B_1}\right)^2 \right] \qquad \begin{matrix} B_1 \neq 0 \\ B_2 = 0 \end{matrix}$$

$$\frac{y_v^2(\lambda_m x_2)}{2\lambda_m^2} \left[(\lambda_m x_2)^2 - v^2 + \left(\frac{x_2 A_2}{B_2}\right)^2 \right] - \frac{x_1^2}{2} y_{v+1}^2(\lambda_m x_1) \qquad \begin{matrix} B_1 = 0 \\ B_2 \neq 0 \end{matrix}$$

$$\frac{y_v^2(\lambda_m x_2)}{2\lambda_m^2} \left[(\lambda_m x_2)^2 - v^2 + \left(\frac{x_2 A_2}{B_2}\right)^2 \right]$$

$$- \frac{y_v^2(\lambda_m x_1)}{2\lambda_m^2} \left[(\lambda_m x_1)^2 - v^2 + \left(\frac{x_1 A_1}{B_1}\right)^2 \right] \qquad \begin{matrix} B_1 \neq 0 \\ B_2 \neq 0 \end{matrix}$$

If $x_1 = 0$, the only boundary condition at $x = x_1$ is that of boundedness, and the contribution to the integral from the lower limit is zero.

EXAMPLE 1

Expand $f(x) = 4x - x^3$ over the interval $(0,2)$ in terms of the Bessel functions of the first kind of order 1 which satisfy the boundary condition

$$J_1(\lambda x)\Big|_{x=2} = 0$$

In this case the characteristic values are the values of λ determined by the roots of the equation $J_1(2\lambda) = 0$. Now the roots of the equation $J_1(z) = 0$ are†

$$z_0 = 0, \; z_1 = 3.832, \; z_2 = 7.016, \; z_3 = 10.174, \; z_4 = 13.324, \ldots$$

Hence,

$$\lambda_0 = 0, \; \lambda_1 = 1.916, \; \lambda_2 = 3.508, \; \lambda_3 = 5.087, \; \lambda_4 = 6.662, \ldots$$

Therefore, since $J_1(\lambda_0 x) = J_1(0) = 0$, the characteristic functions in terms of which the expansion is to be carried out are

$$J_1(\lambda_1 x), \; J_1(\lambda_2 x), \; J_1(\lambda_3 x), \; J_1(\lambda_4 x), \ldots$$

As in the simpler case of Fourier expansions, we begin by writing

$$f(x) = 4x - x^3 = C_1 J_1(\lambda_1 x) + C_2 J_1(\lambda_2 x) + \cdots + C_m J_1(\lambda_m x) + \cdots$$

Multiplying both sides of this expression by $x J_1(\lambda_m x)$, integrating from 0 to 2, and using the results of Theorem 1 (noting that $x_1 = 0$ and $B_2 = 0$), we have

$$\int_0^2 (4x - x^3) x J_1(\lambda_m x) \, dx = C_m \int_0^2 x J_1{}^2(\lambda_m x) \, dx = 2 C_m J_2{}^2(2\lambda_m)$$

Hence

$$C_m = \frac{\int_0^2 (4x^2 - x^4) J_1(\lambda_m x) \, dx}{2 J_2{}^2(2\lambda_m)}$$

For the first term in the integral for C_m, namely,

$$4 \int_0^2 x^2 J_1(\lambda_m x) \, dx = \frac{4}{\lambda_m{}^3} \int_0^2 (\lambda_m x)^2 J_1(\lambda_m x) \, d(\lambda_m x) = \frac{4}{\lambda_m{}^3} \int_0^{2\lambda_m} t^2 J_1(t) \, dt$$

we have immediately, from Eq. (6), Sec. 9.5,

(7)
$$\frac{4}{\lambda_m{}^3} t^2 J_2(t) \Big|_0^{2\lambda_m} = \frac{16}{\lambda_m} J_2(2\lambda_m)$$

To evaluate the integral of the second term, namely,

$$\int_0^2 x^4 J_1(\lambda_m x) \, dx = \frac{1}{\lambda_m{}^5} \int_0^2 (\lambda_m x)^4 J_1(\lambda_m x) \, d(\lambda_m x) = \frac{1}{\lambda_m{}^5} \int_0^{2\lambda_m} t^4 J_1(t) \, dt$$

we use integration by parts, with

$$u = t^2 \qquad dv = t^2 J_1(t) \, dt$$
$$du = 2t \, dt \qquad v = t^2 J_2(t)$$

This gives

$$\int_0^2 x^4 J_1(\lambda_m x) \, dx = \frac{1}{\lambda_m{}^5} \left[t^4 J_2(t) \Big|_0^{2\lambda_m} - 2 \int_0^{2\lambda_m} t^3 J_2(t) \, dt \right]$$

$$= \frac{1}{\lambda_m{}^5} \left[t^4 J_2(t) - 2 t^3 J_3(t) \right]_0^{2\lambda_m}$$

(8)
$$= \frac{16}{\lambda_m{}^2} \left[\lambda_m J_2(2\lambda_m) - J_3(2\lambda_m) \right]$$

---

† See, for instance, Eugene Jahnke, Fritz Emde, and Friedrich Lösch, "Tables of Higher Functions," 6th ed., p. 193, McGraw-Hill, New York, 1960.

Thus, from (7) and (8),

$$C_m = \frac{1}{2J_2{}^2(2\lambda_m)} \left\{ \frac{16}{\lambda_m} J_2(2\lambda_m) - \frac{16}{\lambda_m{}^2} [\lambda_m J_2(2\lambda_m) - J_3(2\lambda_m)] \right\} = \frac{8J_3(2\lambda_m)}{\lambda_m{}^2 J_2{}^2(2\lambda_m)}$$

However, by Eq. (4), Sec. 9.5,

$$J_3(2\lambda_m) = \frac{4}{2\lambda_m} J_2(2\lambda_m) - J_1(2\lambda_m) = \frac{2J_2(2\lambda_m)}{\lambda_m}$$

since the λ's were determined by the condition that $J_1(2\lambda_m) = 0$. Therefore, C_m can be further simplified to

$$C_m = \frac{16}{\lambda_m{}^3 J_2(2\lambda_m)}$$

The same reduction can be repeated for $J_2(2\lambda_m)$, since

$$J_2(2\lambda_m) = \frac{2}{2\lambda_m} J_1(2\lambda_m) - J_0(2\lambda_m) = -J_0(2\lambda_m)$$

Hence, finally, $$C_m = -\frac{16}{\lambda_m{}^3 J_0(2\lambda_m)}$$

The required expansion is therefore

$$4x - x^3 = -16 \sum_{m=1}^{\infty} \frac{1}{\lambda_m{}^3 J_0(2\lambda_m)} J_1(\lambda_m x)$$

Plots showing the degree to which the first term and the first two terms of this series approximate the graph of $4x - x^3$ are shown in Fig. 9.8.

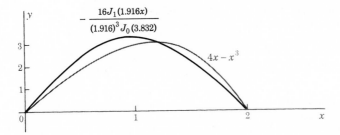

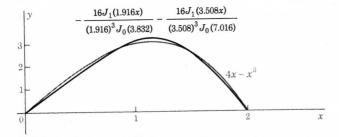

Figure 9.8
The approximation of a function by the first two terms of a Bessel function expansion.

1 Expand $f(x) = 4x - x^3$ over the interval $(0,2)$ in terms of the Bessel functions of the first kind of order 1 which satisfy the boundary condition

$$\frac{dJ_1(\lambda x)}{dx}\bigg|_{x=2} = 0$$

2 Expand $f(x) = 4x - x^3$ over the interval $(0,2)$ in terms of the Bessel functions of the first kind of order 1 which satisfy the boundary condition

$$J_1(\lambda x)\bigg|_{x=2} - 2\frac{dJ_1(\lambda x)}{dx}\bigg|_{x=2} = 0$$

3 Expand $f(x) = 1$ over the interval $(0,3)$ in terms of the Bessel functions of the first kind of order 0 which satisfy the boundary condition $J_0(\lambda x) = 0$ at $x = 3$.

4 Expand $f(x) = 1$ over the interval $(0,3)$ in terms of the Bessel functions of the first kind of order 0 which satisfy the boundary condition

$$J_0(\lambda x)\bigg|_{x=3} - \frac{dJ_0(\lambda x)}{dx}\bigg|_{x=3} = 0$$

5 Expand $f(x) = x^2$ over the interval $(0,2)$ in terms of the Bessel functions of the first kind of order 0 which satisfy the boundary condition

$$\frac{dJ_0(\lambda x)}{dx}\bigg|_{x=2} = 0$$

6 Expand $f(x) = 1$ over the interval $(0,1)$ in terms of the Bessel functions of the first kind of order 1 which satisfy the boundary condition $J_1(\lambda x)|_{x=1} = 0$.

7 Expand $f(x) = x^2$ over the interval $(0,5)$ in terms of the Bessel functions of the first kind of order 2 which satisfy the boundary condition $J_2(\lambda x)|_{x=5} = 0$.

8 Expand $f(x) = \sqrt{x}$ over the interval $(0,\pi)$ in terms of the Bessel functions of the first kind of order $\frac{1}{2}$ which satisfy the boundary condition $J_{1/2}(\lambda x)|_{x=\pi} = 0$. Is this a Fourier series?

9 If the boundary conditions in Theorem 1 are of the form $y_\nu(\lambda x) = 0$ at $x = 1$ and $x = 2$, what is the characteristic equation satisfied by the characteristic values $\{\lambda_m\}$?

10 If $\nu = 0$, calculate the first two roots of the characteristic equation in Exercise 9, correct to two decimal places.

11 If the boundary conditions in Theorem 1 are of the form $dy_\nu(\lambda x)/dx = 0$ at $x = 1$ and at $x = 2$, what is the equation satisfied by the characteristic values?

12 If $\nu = 0$, calculate the first two roots of the characteristic equation in Exercise 11, correct to two decimal places.

13 If the boundary conditions in Theorem 1 are of the form $y_\nu(\lambda x) = 0$ at $x = 1$ and $dy_\nu(\lambda x)/dx = 0$ at $x = 2$, what is the characteristic equation?

14 If $\nu = 0$, calculate the first two roots of the characteristic equation in Exercise 13, correct to two decimal places.

15 Expand $f(x) = 1$ in terms of the Bessel functions of order 0 which satisfy the boundary condition $y_0(\lambda x) = 0$ at $x = 1$ and at $x = 3$.

16 Expand $f(x) = 1$ in terms of the Bessel functions of order 0 which satisfy the boundary conditions $y_0(\lambda x) = 0$ at $x = 1$ and $dy_0(\lambda x)/dx = 0$ at $x = 2$.

17 Expand $f(x) = x$ over the interval $(0,1)$ in terms of the solutions of the equation $(x^{-1}y')' + (4\lambda^2 x - 3x^{-3})y = 0$ which satisfy the boundary condition $y(\lambda x) = 0$ at $x = 1$.

18 Expand $f(x) = \sqrt{x}$ over the interval $(0,1)$ in terms of the solutions of the equation $y'' + (4\lambda^2 x^2 - 1/4x^2)y = 0$ which satisfy the boundary condition $y(\lambda x) = 0$ at $x = 1$.

19 If $\lambda_1, \lambda_2, \ldots$ are the zeros of the function $J_0(x)$, show that

$$\sum_{i=1}^{\infty} \frac{1}{\lambda_i^2} = \frac{1}{4}$$

Hint: Assume that $J_0(x)$ can be written in the factored form

$$J_0(x) = \left(1 - \frac{x}{\lambda_1}\right)\left(1 + \frac{x}{\lambda_1}\right) \cdots \left(1 - \frac{x}{\lambda_i}\right)\left(1 + \frac{x}{\lambda_i}\right) \cdots = \prod_{i=1}^{\infty} \left(1 - \frac{x^2}{\lambda_i^2}\right)$$

20 In Exercise 19, find the value of the sum $\sum_{i=1}^{\infty} \frac{1}{\lambda_i^4}$. *Hint:* Consider the infinite product expansion of both $J_0(x)$ and $J_0^2(x)$.

9.7 Applications of Bessel Functions

Bessel functions occur in a great many practical problems. In principle they are always to be expected when partial differential equations are used in the study. of configurations with circular symmetry. On the other hand, they also arise in numerous applications where neither circular symmetry nor partial differential equations are involved. In this section we shall conclude our treatment of Bessel functions by discussing a variety of problems where their use is required.

EXAMPLE 1

A uniform, perfectly flexible cable of length l and weight per unit length w hangs by one end from a frictionless hook. At $t = 0$, while the cable is at rest in a vertical position, a uniform horizontal velocity v is imparted to the portion of the cable between $x = 0$ and $x = \alpha l$. With coordinates chosen as indicated in Fig. 9.9, find the expression describing the subsequent motion of the cable.

This is essentially the problem of the vibrating string discussed in Sec. 8.2 except for one important difference. Here, instead of being constant, the tension T at a general point of the cable is equal to the weight wx of the portion of the cable below that point. Hence, in this case Eq. (1), Sec. 8.2, becomes in the limit

$$\frac{w}{g}\frac{\partial^2 y}{\partial t^2} = \frac{\partial(wx\,\partial y/\partial x)}{\partial x}$$

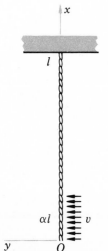

Figure 9.9
A hanging cable given an initial velocity over part of its length.

As usual, we assume a product solution $y = X(x)T(t)$ and attempt to separate variables. Then, substituting, we have

$$T''X = gT(xX')' \qquad \text{or} \qquad \frac{(xX')'}{X} = \frac{T''}{gT}$$

The common value of these two fractions must be a negative constant, say $-\lambda^2$, for otherwise T will not be a periodic function of the time t, as we know it must. Hence

$$T = A \cos (\lambda\sqrt{g}\, t) + B \sin (\lambda\sqrt{g}\, t)$$

and

(1) $$(xX')' + \lambda^2 X = 0$$

Using Corollary 1 of Theorem 1, Sec. 9.4, we find at once that the solution for X is

$$X = CJ_0(2\lambda\sqrt{x}) + DY_0(2\lambda\sqrt{x})$$

Since the displacement of the free end of the cable will obviously be finite, whereas $Y_0(2\lambda\sqrt{x})$ becomes infinite as x approaches zero, it is clear that D must be zero. Moreover, for all values of t, the displacement y is zero when $x = l$. Hence $X(l) = 0$; that is,

(2) $$J_0(2\lambda\sqrt{l}) = 0$$

Mathematically speaking, this is the characteristic equation of the problem, but, physically speaking, it is also the frequency equation. From its infinite set of roots,

$$2\lambda\sqrt{l} = 2.4048,\ 5.5201,\ 8.6537, \ldots$$

the natural frequencies of the cable, namely, $\omega_n = \lambda_n\sqrt{g}$, can be immediately determined:

$$\omega_1 = 1.2024 \sqrt{\frac{g}{l}}, \ \omega_2 = 2.7600 \sqrt{\frac{g}{l}}, \ \omega_3 = 4.3268 \sqrt{\frac{g}{l}}, \ldots$$

We have now been led to an infinite sequence of product solutions,

$$y_m(x,t) = X_m(x)T_m(t) = J_0(2\lambda_m\sqrt{x})[A_m \cos (\lambda_m\sqrt{g}\, t) + B_m \sin (\lambda_m\sqrt{g}\, t)]$$

None of these, by itself, can satisfy the initial conditions

$$y(x,0) \equiv 0 \qquad \text{and} \qquad \frac{\partial y}{\partial t}\bigg|_{x,0} \equiv f(x) = \begin{cases} v & 0 < x < \alpha l \\ 0 & \alpha l < x < l \end{cases}$$

Hence, as usual, we form an infinite series of the individual product solutions,

(3) $$y(x,t) = \sum_{m=1}^{\infty} J_0(2\lambda_m\sqrt{x})[A_m \cos (\lambda_m\sqrt{g}\, t) + B_m \sin (\lambda_m\sqrt{g}\, t)]$$

and attempt to make it fit the initial conditions.

Now Eq. (1) with its accompanying boundary condition $X(l) = 0$ meets all the conditions of Theorem 4, Sec. 8.5. Hence the X's are orthogonal with respect to the weight function $p(x) = 1$ over the interval $(0,l)$, and thus the A's and B's can be determined by the familiar generalized Fourier procedure. To find A_m we put $t = 0$ and $y = 0$ in (3), getting

$$0 = \sum_{m=1}^{\infty} A_m J_0(2\lambda_m\sqrt{x})$$

from which it is obvious that

$$A_m = 0 \qquad m = 1, 2, 3, \ldots$$

To find B_m we differentiate (3) with respect to t and then put $t = 0$ and $\partial y/\partial t|_{t=0} = f(x)$, getting

$$(4) \qquad\qquad f(x) = \sum_{m=1}^{\infty} \sqrt{g}\, \lambda_m B_m J_0(2\lambda_m \sqrt{x})$$

Next we multiply (4) by $J_0(2\lambda_m\sqrt{x})$ and integrate from 0 to l. From the orthogonality of the J_0's, every term on the right but one becomes zero, and we have

$$\int_0^l f(x) J_0(2\lambda_m\sqrt{x})\,dx \equiv \int_0^{\alpha l} v J_0(2\lambda_m\sqrt{x})\,dx = \sqrt{g}\,\lambda_m B_m \int_0^l J_0{}^2(2\lambda_m\sqrt{x})\,dx$$

or

$$B_m = \frac{v \int_0^{\alpha l} J_0(2\lambda_m\sqrt{x})\,dx}{\sqrt{g}\,\lambda_m \int_0^l J_0{}^2(2\lambda_m\sqrt{x})\,dx}$$

To evaluate these integrals we make the obvious substitutions $x = u^2$ and $dx = 2u\,du$, getting

$$B_m = \frac{v \int_0^{\sqrt{\alpha l}} u J_0(2\lambda_m u)\,du}{\sqrt{g}\,\lambda_m \int_0^{\sqrt{l}} u J_0{}^2(2\lambda_m u)\,du}$$

The integral in the numerator is precisely

$$\left.\frac{u J_1(2\lambda_m u)}{2\lambda_m}\right|_0^{\sqrt{\alpha l}} = \frac{\sqrt{\alpha l}\, J_1(2\lambda_m\sqrt{\alpha l})}{2\lambda_m}$$

For the integral in the denominator, the antiderivative is given by Eq. (4), Sec. 9.6. Evaluating this between 0 and $\sqrt{l}$ and using the boundary condition (2), we obtain

$$\frac{l J_1{}^2(2\lambda_m\sqrt{l})}{2}$$

Hence, finally,

$$B_m = \frac{v}{\lambda_m{}^2}\sqrt{\frac{\alpha}{gl}}\,\frac{J_1(2\lambda_m\sqrt{\alpha l})}{J_1{}^2(2\lambda_m\sqrt{l})}$$

With A_m and B_m determined for all values of m, the solution is now complete.

It is interesting to note that since the λ's are incommensurable, there are no two times when the terms $\sin(\lambda_m\sqrt{g}\,t)$ are respectively the same. Hence the cable never returns to a position coinciding exactly with an earlier one unless it is vibrating in one of its normal modes, i.e., unless all but one of the B's are zero, which is not the case for the given initial velocity distribution $f(x)$. This is in sharp contrast to the behavior of a string stretched under uniform tension, which repeats *any* configuration exactly, after intervals of $2l/a$, where a is the propagation velocity for the string.

EXAMPLE 2

A metal fin of triangular cross section is attached to a plane surface to help carry off heat from the latter. Assuming dimensions and coordinates as shown in Fig. 9.10, find the steady-state temperature distribution along the fin if the wall temperature is u_w and if the fin cools freely into air of constant temperature u_0.

We shall base our analysis upon a unit length of the fin and shall assume that the fin is so thin that temperature variations parallel to the base can be neglected. Now consider the heat balance in the element of the fin between x and $x + \Delta x$. This element gains heat by internal flow through its right face and loses heat by internal flow through its left face and also by cooling through its upper and lower surfaces. Through the right face the gain of heat per unit time is

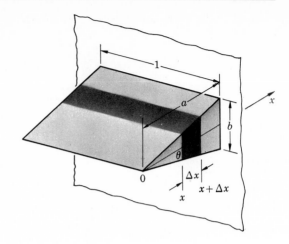

Figure 9.10
A portion of a triangular cooling fin
attached to a flat wall.

$$\text{Area} \times \text{thermal conductivity} \times \text{temperature gradient}$$

or

$$\left[\left(1\frac{bx}{a}\right)k\frac{du}{dx}\right]_{x+\Delta x} = \left(\frac{bkx}{a}\frac{du}{dx}\right)_{x+\Delta x}$$

Through the left face the element loses heat at the rate

$$\left(\frac{bkx}{a}\frac{du}{dx}\right)_x$$

Through the surfaces exposed to the air the element loses heat at the rate

$$\text{Area} \times \text{surface conductivity} \times (\text{surface temperature} - \text{air temperature})$$

or

$$2\left(1\frac{\Delta x}{\cos\theta}\right)h(u - u_0) = \frac{2h(u - u_0)\,\Delta x}{\cos\theta}$$

Under steady-state conditions the rate of gain of heat must equal the rate of loss, and thus we have

$$\left(\frac{bkx}{a}\frac{du}{dx}\right)_{x+\Delta x} = \left(\frac{bkx}{a}\frac{du}{dx}\right)_x + \frac{2h(u - u_0)\,\Delta x}{\cos\theta}$$

Writing this as

$$\frac{(x\,du/dx)_{x+\Delta x} - (x\,du/dx)_x}{\Delta x} - \frac{2ah}{bk\cos\theta}(u - u_0) = 0$$

and letting $\Delta x \to 0$, we obtain the differential equation

$$\frac{d(xu')}{dx} - \frac{2ah}{bk\cos\theta}(u - u_0) = 0$$

If we set $U = u - u_0$ and $\alpha^2 = 2ah/(bk\cos\theta) = [h/(k\sin\theta)]$, this becomes

$$\frac{d(xU')}{dx} - \alpha^2 U = 0$$

This can be solved immediately by means of the corollary of Theorem 1, Sec. 9.4, and we have

$$U = u - u_0 = c_1 I_0(2\alpha\sqrt{x}) + c_2 K_0(2\alpha\sqrt{x})$$

Since $K_0(2\alpha\sqrt{x})$ is infinite when $x = 0$, c_2 must be zero, leaving

$$u - u_0 = c_1 I_0(2\alpha\sqrt{x})$$

Furthermore, $u = u_w$ when $x = a$; hence

$$u_w - u_0 = c_1 I_0(2\alpha\sqrt{a}) \qquad \text{or} \qquad c_1 = \frac{u_w - u_0}{I_0(2\alpha\sqrt{a})}$$

Therefore

$$u = u_0 + (u_w - u_0)\frac{I_0(2\alpha\sqrt{x})}{I_0(2\alpha\sqrt{a})}$$

EXAMPLE 3

A solid consists of one-half of a right circular cylinder of radius b and height h (Fig. 9.11). The lower base, the curved surface, and the vertical plane face are maintained at the constant temperature $u = 0$. Over the upper base the temperature is a known function of position, that is, $u(r,\theta,h) = f(r,\theta)$. Assuming steady-state conditions, find the temperature at any point in the solid.

Because of the nature of the boundaries of the solid it will be highly inconvenient to use the heat equation in the cartesian form in which we derived it in Sec. 8.2. Instead, we use it as expressed in cylindrical coordinates by means of the change of variables

$$x = r \cos\theta \qquad y = r \sin\theta \qquad z = z$$

namely,

$$\frac{\partial^2 u}{\partial r^2} + \frac{1}{r}\frac{\partial u}{\partial r} + \frac{1}{r^2}\frac{\partial^2 u}{\partial\theta^2} + \frac{\partial^2 u}{\partial z^2} = a^2\frac{\partial u}{\partial t}$$

or, more specifically, for steady-state conditions, under which $\partial u/\partial t = 0$,

(5)
$$\frac{\partial^2 u}{\partial r^2} + \frac{1}{r}\frac{\partial u}{\partial r} + \frac{1}{r^2}\frac{\partial^2 u}{\partial\theta^2} + \frac{\partial^2 u}{\partial z^2} = 0$$

Our first step is to assume a product solution $u(r,\theta,z) = R(r)\Theta(\theta)Z(z)$ and substitute it into (5) in an attempt to separate the variables. This gives

$$R''\Theta Z + \frac{1}{r}R'\Theta Z + \frac{1}{r^2}R\Theta''Z + R\Theta Z'' = 0$$

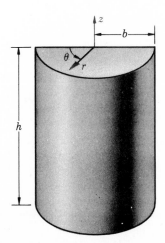

Figure 9.11
A half cylinder in which heat flow occurs because of surface temperature conditions.

or, multiplying by r^2 and dividing by the product $R\Theta Z$,

$$\frac{r^2 R''}{R} + r\frac{R'}{R} + r^2\frac{Z''}{Z} = -\frac{\Theta''}{\Theta} = \mu_1$$

where the common value μ_1 is necessarily a constant, since the variables appearing on the respective sides of the equation are independent of each other.

If $\mu_1 < 0$, say $\mu_1 = -v^2$, then $\Theta''/\Theta = v^2$ and

(6) $\Theta = A \cosh v\theta + B \sinh v\theta$

Now, by hypothesis,

$$u(r,0,z) = R(r)\Theta(0)Z(z) = 0 \quad \text{and} \quad u(r,\pi,z) = R(r)\Theta(\pi)Z(z) = 0$$

and these can hold for all values of r and z, as required by the given boundary conditions, only if $\Theta(0) = \Theta(\pi) = 0$. From (6) we see that the condition $\Theta(0) = 0$ will be satisfied only if $A = 0$. To satisfy the condition $\Theta(\pi) = 0$, it is necessary that

$$B \sinh v\pi = 0$$

which, since $v \neq 0$, is possible only if $B = 0$. Thus the possibility $\mu_1 < 0$ leads only to a trivial solution and hence must be rejected.

If $\mu_1 = 0$, then $\Theta'' = 0$ and

$$\Theta = A + B\theta$$

Again imposing the conditions $\Theta(0) = \Theta(\pi) = 0$, we find, as before, that $A = B = 0$. Hence the possibility $\mu_1 = 0$ must also be rejected, since it, too, leads only to a trivial solution.

Finally, if $\mu_1 > 0$, say $\mu_1 = v^2$, we have

$$\frac{\Theta''}{\Theta} = -v^2 \quad \text{and} \quad \Theta = A \cos v\theta + B \sin v\theta$$

For this to vanish when $\theta = 0$, we must have $A = 0$. For it to vanish when $\theta = \pi$, it is necessary that

$$B \sin v\pi = 0$$

Since we cannot permit B to be zero, because that would again lead to a trivial solution, we must have

$$\sin v\pi = 0$$

Hence $v = 1, 2, 3, \ldots, n, \ldots$

and so for Θ we have the family of solutions

$$\Theta_n(\theta) = \sin n\theta$$

With $\mu_1 = v^2$ now known to be n^2, the differential equation for R and Z becomes

$$r^2\frac{R''}{R} + r\frac{R'}{R} + r^2\frac{Z''}{Z} = n^2$$

or, rearranging,

$$\frac{Z''}{Z} = \frac{n^2}{r^2} - \frac{R''}{R} - \frac{1}{r}\frac{R'}{R} = \mu_2$$

where, again, since r and z are independent variables, it follows that the common value μ_2 must be a constant.

If $\mu_2 < 0$, say $\mu_2 = -\lambda^2$, we have†

$$\frac{R''}{R} + \frac{1}{r}\frac{R'}{R} - \lambda^2 - \frac{n^2}{r^2} = 0 \qquad \text{or} \qquad r^2R'' + rR' - (\lambda^2r^2 + n^2)R = 0$$

which is precisely the modified Bessel equation of order n. Hence

$$R = CI_n(\lambda r) + DK_n(\lambda r)$$

Now $K_n(\lambda r)$ is infinite when $r = 0$; hence, to keep the temperature finite on the axis of the cylinder, it is necessary that $D = 0$. Also, from the given boundary conditions,

$$u(b,\theta,z) = R(b)\,\Theta(\theta)Z(z) = 0$$

and this is possible for all values of θ and z only if

$$R(b) = CI_n(\lambda b) = 0$$

But the modified Bessel function I_n is never zero except possibly at the origin. Therefore the last condition can hold only if $C = 0$. But with C and D both zero, the solution is trivial, and so the possibility that $\mu_2 < 0$ must be rejected.

If $\mu_2 = 0$, then

$$\frac{R''}{R} + \frac{1}{r}\frac{R'}{R} - \frac{n^2}{r^2} = 0 \qquad \text{or} \qquad r^2R'' + rR' - n^2R = 0$$

This is not a Bessel-type equation but is instead an example of the Euler equation (Example 3, Sec. 2.6). By the usual change of independent variable

$$r = e^v \qquad \text{or} \qquad v = \ln r$$

it becomes

$$\frac{d^2R}{dv^2} - n^2R = 0$$

so that

$$R = Ce^{nv} + De^{-nv} = Cr^n + Dr^{-n}$$

To keep the temperature finite on the axis of the cylinder, where $r = 0$, it is necessary that $D = 0$. To keep the temperature zero when $r = b$, it is necessary that

$$R(b) \equiv Cb^n = 0$$

which will be the case only if $C = 0$. This means that again the solution is trivial, and $\mu_2 = 0$ must also be rejected.

Finally, if $\mu_2 > 0$, say $\mu_2 = \lambda^2$, we have

$$\frac{R''}{R} + \frac{1}{r}\frac{R'}{R} + \lambda^2 - \frac{n^2}{r^2} = 0 \qquad \text{or} \qquad r^2R'' + rR' + (\lambda^2r^2 - n^2)R = 0$$

and

$$R = CJ_n(\lambda r) + DY_n(\lambda r)$$

Since $Y_n(\lambda r)$ is infinite when $r = 0$, we must have $D = 0$. To keep the temperature zero on the curved surface of the cylinder, we must have

$$R(b) \equiv CJ_n(\lambda b) = 0$$

---

† Since the group of terms involving Z is much simpler than the group of terms involving R, one might think that it would be better to work next with the equation $Z''/Z = \mu_2$. However, it is not the simplicity of the terms themselves but the nature of the boundary conditions they must satisfy that determines the continuation. Clearly, the condition $R(b) = 0$ is one that can be imposed on R once it is found; whereas the condition $u = f(r,\theta)$ when $z = h$ cannot be imposed on the factor Z by itself but must finally be satisfied by a series expansion involving all the product solutions.

Since $C = 0$ leads to a trivial solution, it is thus necessary that

$$J_n(\lambda b) = 0$$

that is, λ is restricted to the set of values

$$\left\{ \frac{\rho_{nm}}{b} \right\}$$

where ρ_{nm} is the mth one of the roots of the equation $J_n(x) = 0$. Thus, for every value of n, there are infinitely many particular solutions for R, namely,

$$R_{nm}(r) = J_n(\lambda_{nm} r)$$

Now that we know that $\mu_2 = \lambda_{nm}^2$, it is an easy matter to solve for Z, and we have

$$\frac{Z''}{Z} = \lambda_{nm}^2 \quad \text{and} \quad Z = E \cosh \lambda_{nm} z + F \sinh \lambda_{nm} z$$

Since $u(r,\theta,0) \equiv R(r)\Theta(\theta)Z(0) = 0$, it follows that $Z(0) = 0$, from which we conclude that $E = 0$. The solution for Z associated with R_{nm} is therefore

$$Z_{nm}(z) = \sinh \lambda_{nm} z$$

For *each* n we therefore have infinitely many product solutions, consisting of the same factor $\Theta(\theta) = \sin n\theta$ multiplied by the product of any pair of corresponding R's and Z's:

$$u_{nm} = A_{nm} J_n(\lambda_{nm} r) \sinh \lambda_{nm} z \sin n\theta$$

In other words, we have a double array of product solutions,

$$u_{11}, u_{12}, u_{13}, \ldots, u_{1m}, \ldots$$
$$u_{21}, u_{22}, u_{23}, \ldots, u_{2m}, \ldots$$
$$\cdots\cdots\cdots\cdots\cdots\cdots\cdots$$
$$u_{n1}, u_{n2}, u_{n3}, \ldots, u_{nm}, \ldots$$
$$\cdots\cdots\cdots\cdots\cdots\cdots\cdots$$

Since none of the product solutions by itself is capable of representing the given temperature distribution $f(r,\theta)$ on the upper base, we must construct an infinite series of the u_{nm}'s and try to make it fit the temperature condition when $z = h$. To build up a series for u we first add up all the product solutions associated with a particular value of n, getting

$$u_n = \sum_{m=1}^{\infty} u_{nm} = \sin n\theta \sum_{m=1}^{\infty} A_{nm} J_n(\lambda_{nm} r) \sinh \lambda_{nm} z$$

This, of course, amounts to forming the sums of the elements in each of the rows in the above array. Next we add up all these series for every value of n:

$$(7) \qquad u(r,\theta,z) = \sum_{n=1}^{\infty} u_n = \sum_{n=1}^{\infty} \left[\sin n\theta \sum_{m=1}^{\infty} A_{nm} J_n(\lambda_{nm} r) \sinh \lambda_{nm} z \right]$$

The final step now is to determine the A's so that this double series will reduce to $f(r,\theta)$ when $z = h$:

$$(8) \qquad u(r,\theta,h) = f(r,\theta) = \sum_{n=1}^{\infty} \left[\sin n\theta \sum_{m=1}^{\infty} A_{nm} J_n(\lambda_{nm} r) \sinh \lambda_{nm} h \right]$$

To carry out this expansion, let us imagine that r is held constant and that θ is allowed to vary over the range of the problem $(0,\pi)$. Under these conditions the inner sum in (8) is effectively a constant depending on n, say G_n, or more explicitly $G_n(r)$. That is,

$$f(r,\theta) = \sum_{n=1}^{\infty} G_n \sin n\theta$$

But the determination of the G's is a familiar problem—nothing more in fact than the Fourier sine-expansion problem—and we can write immediately

$$(9) \qquad\qquad G_n \equiv G_n(r) = \frac{2}{\pi} \int_0^\pi f(r,\theta) \sin n\theta \, d\theta$$

Thus $G_n(r)$ is a *known* function of r. But, by definition, $G_n(r)$ was the inner sum in (8); i.e.,

$$G_n(r) = \sum_{m=1}^\infty (A_{nm} \sinh \lambda_{nm}h)J_n(\lambda_{nm}r)$$

Hence, it is clear that the A's must be such that the products $A_{nm} \sinh \lambda_{nm}h$ are the coefficients in a Bessel function expansion of the now known function $G_n(r)$. Therefore, from the theory of the last section, recalling that the λ's were determined by the condition $J_n(\lambda b) = 0$, we can write

$$A_{nm} \sinh \lambda_{nm}h = \frac{\int_0^b rG_n(r)J_n(\lambda_{nm}r) \, dr}{(b^2/2)J_{n+1}^2(\lambda_{nm}b)}$$

Hence,
$$A_{nm} = \frac{\int_0^b rG_n(r)J_n(\lambda_{nm}r) \, dr}{(b^2/2) \sinh \lambda_{nm}h \, J_{n+1}^2(\lambda_{nm}b)}$$

where $G_n(r)$ is given by (9). With the coefficients in the series (7) now determined, the problem is solved.

EXAMPLE 4

A thin circular plate of radius b has its upper and lower faces perfectly insulated against the flow of heat. Around the circumference of the plate the time-dependent temperature distribution $u(b,\theta,t) = f(\theta) \cos \omega t$ is maintained. Find the steady-state temperature distribution throughout the plate.

In this problem, the steady-state temperature distribution we are asked to find is not one which is *independent* of time but is, instead, the limiting *periodic* distribution which the temperature approaches as t increases indefinitely and all transients die away. Thus, even though we are seeking a description of the steady state of the system, we must retain the term $\partial u/\partial t$ in the heat equation. On the other hand, since the plate is very thin and has perfectly insulated faces, we may legitimately assume that the heat flow is two-dimensional, i.e., is independent of z. Hence the equation we must solve is

$$\frac{\partial^2 u}{\partial r^2} + \frac{1}{r}\frac{\partial u}{\partial r} + \frac{1}{r^2}\frac{\partial^2 u}{\partial \theta^2} = a^2 \frac{\partial u}{\partial t}$$

If, as usual, we assume a product solution

$$u(r,\theta,t) = R(r)\,\Theta(\theta)T(t)$$

substitute, and divide by $R\Theta T$, we obtain

$$\frac{R''}{R} + \frac{1}{r}\frac{R'}{R} + \frac{1}{r^2}\frac{\Theta''}{\Theta} = a^2 \frac{T'}{T} = \mu_1$$

Here, however, we are faced with a situation we have not previously encountered. For whether μ_1 is positive, negative, or zero, the solution of the equation

$$a^2 \frac{T'}{T} = \mu_1$$

cannot describe a nonconstant periodic function of t, as we know T must be. The only possible continuation is to assume that T is a complex periodic function of period ω, namely

$$T(t) = e^{i\omega t} = \cos \omega t + i \sin \omega t$$

This means that
$$a^2 \frac{T'}{T} = a^2 \frac{i\omega e^{i\omega t}}{e^{i\omega t}} = a^2 \omega i$$

which implies that $\mu_1 = ia^2\omega$. Thus, for the first time, we have a problem in which a complex separation constant is required.

Using this value of μ_1 and separating again, we have

$$r^2 \frac{R''}{R} + r \frac{R'}{R} - ia^2\omega r^2 = -\frac{\Theta''}{\Theta} = \mu_2$$

Clearly, the boundary condition is a periodic function of θ of period 2π. Hence $\mu_2 = n^2$, and therefore from the equation $-\Theta''/\Theta = \mu_2 = n^2$ we have

$$\Theta_n(\theta) = A_n \cos n\theta + B_n \sin n\theta, \qquad n = 0, 1, 2, \ldots.$$

The factor R is now to be determined from the equation

$$r^2 \frac{R''}{R} + r \frac{R'}{R} - ia^2\omega r^2 = n^2$$

or $$r^2 R'' + rR' - (ia^2\omega r^2 + n^2)R = 0$$

A complete solution of this equation is

$$R(r) = C(\text{ber}_n \, a\sqrt{\omega}\, r + i \, \text{bei}_n \, a\sqrt{\omega}\, r) + D(\text{ker}_n \, a\sqrt{\omega}\, r + i \, \text{kei}_n \, a\sqrt{\omega}\, r)$$

Since $\text{ker}_n \, a\sqrt{\omega}\, r + i \, \text{kei}_n \, a\sqrt{\omega}\, r$ is infinite when $r = 0$, whereas the temperature is obviously finite at the center of the plate, it is necessary that $D = 0$, leaving

$$R(r) = C(\text{ber}_n \, a\sqrt{\omega}\, r + i \, \text{bei}_n \, a\sqrt{\omega}\, r) = CM_n(a\sqrt{\omega}\, r) \exp[i\vartheta_n(a\sqrt{\omega}\, r)]$$

We thus have a family of product solutions,

$$u_n(r,\theta,t) = M_n(a\sqrt{\omega}\, r) \exp[i\vartheta_n(a\sqrt{\omega}\, r)](A_n \cos n\theta + B_n \sin n\theta)e^{i\omega t}$$

Since none of these, by itself, can satisfy the boundary condition, we form an infinite series of them

(10) $$u(r,\theta,t) = \sum_{n=0}^{\infty} M_n(a\sqrt{\omega}\, r) \exp[i\vartheta_n(a\sqrt{\omega}\, r)](A_n \cos n\theta + B_n \sin n\theta)e^{i\omega t}$$

and attempt to determine A_n and B_n so that this series will reduce to the appropriate condition on the boundary where $r = b$. However, before we can do this we must modify the boundary condition, as given. Since we have been forced to take T to be the complex exponential $e^{i\omega t}$, we must also change the boundary condition from $f(\theta) \cos \omega t$ to the complex form

$$f(\theta)e^{i\omega t} = f(\theta)(\cos \omega t + i \sin \omega t)$$

Then, when we have solved the problem for this modified boundary condition, the answer to the actual problem will be just the real part of the resulting complex solution.†
   Putting $r = b$ and $u(b,\theta,t) = f(\theta)e^{i\omega t}$ in (10), we have, after dividing out the factor $e^{i\omega t}$,

$$f(\theta) = \sum_{n=0}^{\infty} M_n(a\sqrt{\omega}\, b) \exp[i\vartheta_n(a\sqrt{\omega}\, b)](A_n \cos n\theta + B_n \sin n\theta)$$

From this, by familiar Fourier theory, we have

$$M_n(a\sqrt{\omega}\, b) \exp[i\vartheta_n(a\sqrt{\omega}\, b)]A_n = \frac{1}{\pi} \int_0^{2\pi} f(\theta) \cos n\theta \, d\theta = \frac{1}{\pi} \int_0^{2\pi} f(s) \cos ns \, ds$$

or $$A_n = \frac{\exp[-i\vartheta_n(a\sqrt{\omega}\, b)]}{M_n(a\sqrt{\omega}\, b)} \frac{1}{\pi} \int_0^{2\pi} f(s) \cos ns \, ds$$

---

† This stratagem is similar to the one we introduced in Sec. 5.4 when we replaced the voltages $E_0 \cos \omega t$ and $E_0 \sin \omega t$ by the complex voltage $E_0 e^{i\omega t}$, solved for the resulting current, and then identified the real and imaginary components of this complex current as the responses to $E_0 \cos \omega t$ and $E_0 \sin \omega t$, respectively.

and

$$M_n(a\sqrt{\omega}\,b)\exp\left[-i\vartheta_n(a\sqrt{\omega}\,b)\right]B_n = \frac{1}{\pi}\int_0^{2\pi} f(\theta)\sin n\theta\,d\theta = \frac{1}{\pi}\int_0^{2\pi} f(s)\sin ns\,ds$$

or
$$B_n = \frac{\exp\left[-i\vartheta_n(a\sqrt{\omega}\,b)\right]}{M_n(a\sqrt{\omega}\,b)}\frac{1}{\pi}\int_0^{2\pi} f(s)\sin ns\,ds$$

Substituting these into the series in (10) and combining the exponentials, we obtain

$$\frac{1}{\pi}\sum_{n=0}^{\infty}\frac{M_n(a\sqrt{\omega}\,r)}{M_n(a\sqrt{\omega}\,b)}\exp\left\{i[\omega t + \vartheta_n(a\sqrt{\omega}\,r) - \vartheta_n(a\sqrt{\omega}\,b)]\right\}\left[\cos n\theta\int_0^{2\pi} f(s)\cos ns\,ds\right.$$

$$\left. + \sin n\theta\int_0^{2\pi} f(s)\sin ns\,ds\right]$$

or, consolidating the integrals,

$$\frac{1}{\pi}\sum_{n=0}^{\infty}\frac{M_n(a\sqrt{\omega}\,r)}{M_n(a\sqrt{\omega}\,b)}\exp\left\{i[\omega t + \vartheta_n(a\sqrt{\omega}\,r) - \vartheta_n(a\sqrt{\omega}\,b)]\right\}\int_0^{2\pi} f(s)\cos n(\theta - s)\,ds$$

Finally, by retaining only the real part of this series, we find that the solution to our problem is

$$u(r,\theta,t)$$

$$= \frac{1}{\pi}\sum_{n=0}^{\infty}\frac{M_n(a\sqrt{\omega}\,r)}{M_n(a\sqrt{\omega}\,b)}\cos\left[\omega t + \vartheta_n(a\sqrt{\omega}\,r) - \vartheta_n(a\sqrt{\omega}\,b)\right]\int_0^{2\pi} f(s)\cos n(\theta - s)\,ds$$

From this, it is evident that, although at every point the temperature varies periodically with frequency ω, there is a phase difference between the temperature at an arbitrary radius and the temperature on the boundary.

EXAMPLE 5

What is $\mathscr{L}\{t^\nu J_\nu(\lambda t)\}$ if $\nu \geq 0$?

It is possible to determine the required transform by expressing $t^\nu J_\nu(\lambda t)$ as an infinite series and then taking the transform term by term. However, it is more instructive to proceed as follows.

From Corollary 1, Theorem 1, Sec. 9.4, it is clear that $y = t^\nu J_\nu(\lambda t)$ is a solution of the differential equation

$$(t^{1-2\nu}y')' + \lambda^2 t^{1-2\nu}y = 0$$

that is, $ty'' + (1 - 2\nu)y' + \lambda^2 ty = 0$

If we take the Laplace transform of this equation, recalling Theorem 7, Sec. 7.4, we obtain

$$-\frac{d}{ds}(s^2\mathscr{L}\{y\} - sy_0 - y_0') + (1 - 2\nu)(s\mathscr{L}\{y\} - y_0) - \lambda^2\frac{d}{ds}\mathscr{L}\{y\}$$

$$= -s^2\frac{d\mathscr{L}\{y\}}{ds} - 2s\mathscr{L}\{y\} + y_0 + (1 - 2\nu)(s\mathscr{L}\{y\} - y_0) - \lambda^2\frac{d\mathscr{L}\{y\}}{ds}$$

$$= -(s^2 + \lambda^2)\frac{d\mathscr{L}\{y\}}{ds} - (1 + 2\nu)s\mathscr{L}\{y\} + 2\nu y_0 = 0$$

Now if $\nu \geq 0$, the term $2\nu y_0$ vanishes identically, because either $\nu = 0$ or else

$$y_0 \equiv t^\nu J_\nu(\lambda t)\Big|_{t=0} = 0.$$

Hence, the last equation reduces to the separable differential equation

$$\frac{d\mathscr{L}\{y\}}{\mathscr{L}\{y\}} + (1 + 2\nu)\frac{s\,ds}{s^2 + \lambda^2} = 0$$

Integrating this, we have

$$\ln \mathscr{L}\{y\} + \frac{1 + 2\nu}{2}\ln(s^2 + \lambda^2) = \ln c$$

and therefore

$$\mathscr{L}\{y\} \equiv \mathscr{L}\{t^\nu J_\nu(\lambda t)\} = \frac{c}{(s^2 + \lambda^2)^{(1+2\nu)/2}}$$

To determine c we consider the leading term on each side of the last equality:

$$\mathscr{L}\left\{t^\nu\left[\frac{\lambda^\nu t^\nu}{2^\nu \Gamma(\nu + 1)} - \cdots\right]\right\} = c(s^2 + \lambda^2)^{-(2\nu+1)/2}$$

$$\frac{\lambda^\nu}{2^\nu \Gamma(\nu + 1)}\mathscr{L}\{t^{2\nu} - \cdots\} = c\left(\frac{1}{s^{2\nu+1}} - \cdots\right)$$

$$\frac{\lambda^\nu}{2^\nu \Gamma(\nu + 1)}\left[\frac{\Gamma(2\nu + 1)}{s^{2\nu+1}} - \cdots\right] = c\left(\frac{1}{s^{2\nu+1}} - \cdots\right)$$

Hence, since this must be an identity, it follows that

$$c = \frac{\lambda^\nu \Gamma(2\nu + 1)}{2^\nu \Gamma(\nu + 1)}$$

and so

(11) $$\mathscr{L}\{t^\nu J_\nu(\lambda t)\} = \frac{\lambda^\nu \Gamma(2\nu + 1)}{2^\nu \Gamma(\nu + 1)(s^2 + \lambda^2)^{(2\nu+1)/2}} \qquad \nu \geq 0$$

Numerous other transforms can be obtained from (11). For instance, since, from (11),

$$\mathscr{L}\{J_0(\lambda t)\} = \frac{1}{\sqrt{s^2 + \lambda^2}} \qquad \text{and} \qquad \frac{dJ_0(\lambda t)}{dt} = -\lambda J_1(\lambda t)$$

it follows that

$$\mathscr{L}\{J_1(\lambda t)\} = -\frac{1}{\lambda}\mathscr{L}\left\{\frac{dJ_0(\lambda t)}{dt}\right\} = -\frac{1}{\lambda}[s\mathscr{L}\{J_0(\lambda t)\} - J_0(0)]$$

$$= -\frac{1}{\lambda}\left(\frac{s}{\sqrt{s^2 + \lambda^2}} - 1\right) = \frac{1}{\lambda}\frac{\sqrt{s^2 + \lambda^2} - s}{\sqrt{s^2 + \lambda^2}}$$

$$= \frac{\lambda}{\sqrt{s^2 + \lambda^2}\,(s + \sqrt{s^2 + \lambda^2})}$$

Other results will be found among the exercises.

EXERCISES

1 What is $\mathscr{L}\{tJ_0(\lambda t)\}$?   **2** What is $\mathscr{L}\{t^2 J_0(\lambda t)\}$?

3 What is $\mathscr{L}\{J_2(\lambda t)\}$? *Hint:* Recall from Eq. (3), Sec. 9.5, that

$$J_2(\lambda t) = J_0(\lambda t) - 2\frac{dJ_1(\lambda t)}{d(\lambda t)}$$

4 What is $\mathscr{L}\{J_n(\lambda t)\}$?

5 Show that $\int_0^\infty J_0(\lambda t)\,dt = 1/\lambda$. *Hint:* Consider the integral defining the Laplace transform of $J_0(\lambda t)$.

6 What is:

(a) $\displaystyle\int_0^\infty J_1(\lambda t)\,dt$ (b) $\displaystyle\int_0^\infty tJ_0(\lambda t)\,dt$ (c) $\displaystyle\int_0^\infty tJ_1(\lambda t)\,dt$

7 What is $\displaystyle\int_0^\infty e^{-at}J_0(\lambda t)\,dt$?   **8** What is $\displaystyle\int_0^\infty J_n(\lambda t)\,dt$?

9 What is $\mathscr{L}^{-1}\{1/\sqrt{s^2 + 4s + 13}\}$?

10 What is $\mathscr{L}^{-1}\{1/(s^2 + 2s + 10)^{3/2}\}$?

11 What is $\mathscr{L}^{-1}\{1/[(s + a)\sqrt{s^2 + b^2}]\}$?

12 What is $\mathscr{L}\{J_n(\lambda t)/t\}$? *Hint:* Recall Theorem 8, Sec. 7.4.

13 Show that $\mathscr{L}\{I_0(\lambda t)\} = 1/\sqrt{s^2 - \lambda^2}$.

14 What is:
   (a) $\mathscr{L}\{tI_0(\lambda t)\}$ **(b)** $\mathscr{L}\{tI_1(\lambda t)\}$

15 What is $\mathscr{L}\{I_1(\lambda t)\}$?

16 What is $\mathscr{L}^{-1}\{1/\sqrt{s(s - 1)}\}$?

17 Show that $\int_0^t J_0(\lambda)J_0(t - \lambda)\,d\lambda = \sin t$. *Hint:* Recall the convolution theorem.

18 What is $\mathscr{L}^{-1}\{1/\sqrt{s^4 - a^4}\}$?

19 What is $\mathscr{L}^{-1}\{1/\sqrt{s^4 + 5s^2 + 4}\}$?

20 Find a particular integral of the equation $y'' + y = J_0(t)$.

21 Show that $\int_0^t \sin(t - \lambda)J_0(\lambda)\,d\lambda = tJ_1(t)$.

22 What is $\int_0^t \sin(t - \lambda)J_1(\lambda)\,d\lambda$?

23 Show that

$$I_0(t) = \frac{e^{-t}}{\pi} \int_0^t \frac{e^{2\lambda}}{\sqrt{\lambda(t - \lambda)}}\,d\lambda$$

Hint: Combine Formula 4, Sec. 7.3, for the case $n = -\frac{1}{2}$ with Theorem 5, Sec. 7.4, and then apply the convolution theorem to the result of Exercise 13.

24 Show that $\int_0^t \cos(t - \lambda)J_0(\lambda)\,d\lambda = tJ_0(t)$.

25 Show that $\int_0^t J_0(t - \lambda)J_1(\lambda)\,d\lambda = J_0(t) - \cos t$.

26 What is $\int_0^t J_1(t - \lambda)J_1(\lambda)\,d\lambda$?

27 Show that $\mathscr{L}\{J_0(2\sqrt{t})\} = (1/s)e^{-1/s}$. *Hint:* Observe that $J_0(2\sqrt{t})$ is a solution of the equation $(ty')' + y = 0$.

28 What is $\mathscr{L}\{I_0(2\sqrt{t})\}$?

29 What is:
   (a) $\mathscr{L}^{-1}\{e^{-1/s} - 1\}$ **(b)** $\mathscr{L}^{-1}\{e^{1/s} - 1\}$

30 What is:
   (a) $\mathscr{L}\{tJ_0(2\sqrt{t})\}$ **(b)** $\mathscr{L}\{tI_0(2\sqrt{t})\}$

31 Show that the function $\phi(x) = \int_0^\infty \exp(-x\cosh\theta)\,d\theta$ $(x > 0)$ satisfies the equation $x\phi'' + \phi' - x\phi = 0$. Hence show that $\phi(x)$ is of the form $CK_0(x)$. (It can be shown that the constant C has the value 1.)

32 Prove that

$$J_{m+n}(t) = n \int_0^t J_m(t - \lambda)J_n(\lambda)\,\frac{d\lambda}{\lambda} = m \int_0^t J_m(t - \lambda)J_n(\lambda)\,\frac{d\lambda}{t - \lambda}$$

33 Prove that

$$J_{m+n}(t) = (m + k) \int_0^t J_{m+k}(\lambda)J_{n-k}(t - \lambda)\,\frac{d\lambda}{\lambda}$$

$$= (n - k) \int_0^t J_{m+k}(\lambda)J_{n-k}(t - \lambda)\,\frac{d\lambda}{t - \lambda}$$

34 In Example 1, if the weight per unit length of the cable varies according to the law $w(x) = (k + 1)x^k$, what is the frequency equation?

35 In Example 2, verify that all the heat that enters the fin is lost from its surface. What fraction of the heat entering the fin is lost from the section between $x = 0$ and $x = \lambda a$?

36 Work Example 2 if the fin, instead of running out to a point at $x = 0$, is cut off at $x = c$ and if heat loss through the end of the fin is neglected. *Hint:* Let u_c be the initially unknown temperature at the end where $x = c$, use u_c, with u_w, to determine the integration constants c_1 and c_2, then determine u_c by equating the heat entering the fin to the heat lost from its surface.

37 Work Example 2 if the fin is of rectangular cross section and if
  (a) Heat loss through the end of the fin is neglected.
  (b) Heat loss through the end of the fin is not neglected.

38 Show that the radial temperature distribution in a thin fin of rectangular cross section and outer radius R which completely encircles a heated cylinder of radius r satisfies the differential equation

$$\frac{d(x\,du/dx)}{dx} - \frac{2hx(u - u_0)}{kw} = 0$$

where x is measured radially outward from the center of the cylinder and the other parameters have the same significance aş in Example 2.

39 Neglecting heat loss through the outer end of the fin, solve the differential equation of Exercise 38 and find the temperature distribution in the fin if the cylinder temperature is u_c.

40 Work Exercise 38 if the fin is of triangular cross section.

41 Find the first two natural frequencies of a steel shaft 20 in long vibrating torsionally if the shaft is built-in at one end and free at the other and if the radius of the shaft at a distance x from the free end is $r(x) = (x/20)^{1/4}$. Steel weighs 0.285 lb/in³, and its modulus of elasticity in shear is $E_s = 12 \times 10^6$ lb/in².

42 Work Exercise 41 if the radius of the shaft varies according to the law:

  (a) $r(x) = \left(\dfrac{x}{20}\right)^{1/2}$ **(b)** $r(x) = \left(\dfrac{x}{20}\right)^{3/4}$ **(c)** $r(x) = \dfrac{x}{20}$

43 In Exercise 41, find the instantaneous deflection of the shaft if it starts to vibrate from rest with initial angular displacement $\theta(x,0) = (20 - x)/20$.

44 The lower end of a long thin rod of uniform cross section is clamped so that the rod is vertical. Determine the values of the parameters of the rod for which buckling will occur if the upper end of the rod is displaced slightly from its neutral position. *Hint:* Choosing axes as in Example 1, Sec. 2.6, the problem can be solved by using the relation $(EIy'')' = V$, where V is the transverse component of the weight of the portion of the rod above a general point x and then noting that $y'' \doteq d\theta/dx$ for inclination angles θ small enough that $\tan\theta \doteq \sin\theta \doteq \theta$.

45 A body whose mass varies according to the law $m(t) = m_0(1 + \alpha t)$ moves along the x axis under the influence of a force of attraction which varies directly as the distance from the origin. Determine the equation of motion of the body if it starts from rest at the point $x = x_0$. *Hint:* Recall the general form of Newton's second law, $d(mv)/dt = F$.

46 Work Exercise 45 if the mass of the body varies according to the law

$$m(t) = m_0(1 + \alpha t)^{-1}$$

47 Work Exercise 45 if the force is directed away from the origin.

48 Work Exercise 46 if the force is directed away from the origin.

49 An elastic string whose weight per unit length is $w_0(1 + \alpha x)$, where x is the distance from one end of the string, is stretched under tension T between two points a distance l apart. Find the equation defining the natural frequencies of the string.

50 When alternating current flows through a long cylindrical conductor, the current density is not uniform over the cross section but increases from the center toward the surface. If the actual current, whether of the form $\cos\omega t$ or $\sin\omega t$, is replaced by the complex current $e^{i\omega t}$, as in Example 4, it can be shown† that the resulting current density σ satisfies the differential equation

$$\frac{d^2\sigma}{dr^2} + \frac{1}{r}\frac{d\sigma}{dr} - ik^2\sigma = 0$$

---

† See, for example, N. W. McLachlan, "Bessel Functions for Engineers," pp. 134–137, Oxford University Press, London, 1934.

where $k^2 = 4\pi\omega/\rho$ and ρ is the resistivity of the conductor. Solve this equation, and find the current density as a function of the radius r in a cylindrical conductor of outer radius a if the current density at the surface is σ_0.

51 Find the frequency equation for the transverse vibrations of a cantilever whose width is constant but whose depth varies directly as the distance from the free end. *Hint:* To solve the differential equation defining the normal modes of the beam, recall Exercise 20, Sec. 9.4.

52 Find the frequency equation for the transverse vibrations of a cantilever which is a solid of revolution whose radius varies directly as the distance from the free end.

53 A bar has the shape of a truncated right circular cone of length l, the radii of the bases being r and R. Find the frequency equation for the torsional vibrations of the bar assuming both ends of the bar free.

54 Determine the limiting form of the frequency equation in Exercise 53 when $r \to R$. Check by comparing your result with the frequency equation derived directly for a uniform bar. *Hint:* Express the Bessel functions in terms of sines and cosines.

55 A cantilever beam of length l and breadth b has its upper surface horizontal. The depth of the beam varies directly as the cube root of the distance from the free end. An oblique tensile force F, whose direction makes an angle θ with the horizontal, acts at the free end of the beam. Find the equation of the deflection curve of the beam.

56 Work Exercise 55 if the force is an oblique compressive force.

57 Determine the natural frequencies of a uniform circular drumhead when it is vibrating in a shape which is independent of the polar angle θ.

58 A uniform membrane spans the annular region between two concentric circles of radii r_1 and r_2. Find the natural frequencies of the membrane for modes of vibration which are independent of θ.

59 Find the equation describing the forced vibrations of a uniform circular membrane driven by a force per unit area equal to $F_0 \cos \omega t$.

60 Determine the natural frequencies of a uniform circular drumhead for modes of vibration which are not necessarily independent of θ. What are the associated nodal lines?

61 The response of a certain network of tuned circuits to a unit impulse applied in the first circuit is described by the set of differential equations

$$LC \frac{d^2 e_1}{dt^2} + RC \frac{de_1}{dt} + e_1 = I(t)\dagger$$

$$LC \frac{d^2 e_2}{dt^2} + RC \frac{de_2}{dt} + e_2 = e_1$$

$$\cdots \cdots \cdots \cdots \cdots \cdots \cdots \cdots \cdots \cdots \cdots \cdots \cdots$$

$$LC \frac{d^2 e_n}{dt^2} + RC \frac{de_n}{dt} + e_n = e_{n-1}$$

Find e_n as a function of t. *Hint:* Use Laplace transform methods.

62 Work Example 3 if the curved surface of the solid is perfectly insulated.

63 Work Example 3 if the upper base is maintained at the temperature $u(r,\theta,h) = 0°C$ while the temperature distribution $u(b,\theta,z) = g(\theta,z)$ is maintained on the curved surface.

64 Could Example 3 be worked if an arbitrary temperature distribution $f(r,\theta)$ is maintained on the upper base *and* an arbitrary temperature distribution $g(\theta,z)$ is maintained on the curved surface? How?

65 A right circular cylinder of radius b and height h has its upper and lower bases maintained at the constant temperature $u = 0°C$. The curved surface is maintained at the temperature distribution $u(b,z) = f(z)$. Determine the steady-state temperature distribution throughout the cylinder.

---

† See Samuel Sabaroff, Impulse Excitation of a Cascade of Series Tuned Circuits, *Proc. IRE*, December 1944, pp. 758–760.

66   Work Exercise 65 if the upper and lower bases of the cylinder are perfectly insulated.

67   A right circular cylinder of radius b and height h has its lower base maintained at the constant temperature $u = 100°C$ and its upper base maintained at the constant temperature $u = 0°C$. If its curved surface is perfectly insulated, find the steady-state temperature distribution in the cylinder.

68   Work Exercise 65 if the lower base is maintained at the constant temperature $u = 100°C$ and the curved surface is maintained at the constant temperature $u = 50°C$.

69   Work Exercise 67 if the curved surface cools into air of constant temperature $u_a = 0°C$.

70   If the temperature throughout the cylinder in Exercise 65 at $t = 0$ is given by the known function $\phi(r,z)$, find the temperature distribution at any subsequent time.

71   A two-dimensional region having the shape of a quarter of a circle is initially at the uniform temperature $u = 100°C$. At $t = 0$ the temperature around the entire boundary is suddenly reduced to $0°C$ and maintained thereafter at that value. Find the temperature at any point of the region at any subsequent time.

72   Find the steady-state temperature distribution in a two-dimensional region having the shape of a semicircle of radius b if the diametral boundary is maintained at the constant temperature $u = u_0$ and the curved boundary is maintained at the temperature distribution $u(b,\theta) = f(\theta)$.

73   A two-dimensional region having the shape of a semicircle of radius b is initially at the uniform temperature $u = 100°C$. At $t = 0$ the temperature along the bounding diameter is reduced to $0°C$ and maintained thereafter at that temperature. The curved boundary is maintained at the temperature $u = 100°C$. Find the temperature at any point of the region at any subsequent time.

74   Find the steady-state temperature distribution in the region between two concentric circles of radii r_1 and r_2 if the portion of the inner boundary between $\theta = 0$ and $\theta = \pi$ is maintained at the temperature $u = 100°C$ while the remainder of the inner boundary and the entire outer boundary is maintained at the temperature $u = 0°C$.

75   The region between two concentric circles of radii r_1 and r_2 is initially at the uniform temperature $u = 0°C$. At $t = 0$ the temperature around the entire inner boundary is suddenly raised to $100°C$ and maintained thereafter at that value. Find the temperature at any point of the region at any subsequent time if the outer boundary is maintained at the temperature $0°C$.

9.8   Legendre Polynomials

In Example 3, Sec. 9.7, in solving the steady-state heat equation, i.e., Laplace's equation, in cylindrical coordinates, we found that one of the ordinary differential equations arising from the separation of variables was Bessel's equation. In very much the same way, it turns out that when we apply the method of separation of variables to Laplace's equation in spherical coordinates, one of the ordinary differential equations which results is **Legendre's equation**.

If the expression

$$\nabla^2 F \equiv \frac{\partial^2 F}{\partial x^2} + \frac{\partial^2 F}{\partial y^2} + \frac{\partial^2 F}{\partial z^2}$$

is transformed from cartesian coordinates to spherical coordinates by means of the relations (Fig. 9.12)

$$x = r \sin \theta \cos \phi \qquad y = r \sin \theta \sin \phi \qquad z = r \cos \theta$$

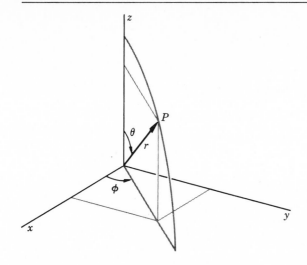

Figure 9.12
The relation between rectangular
and spherical coordinates.

we obtain, after a lengthy but straightforward reduction

$$\nabla^2 F = \frac{1}{r^2 \sin \theta}\left(r^2 \sin \theta \frac{\partial^2 F}{\partial r^2} + 2r \sin \theta \frac{\partial F}{\partial r} + \sin \theta \frac{\partial^2 F}{\partial \theta^2} + \cos \theta \frac{\partial F}{\partial \theta} + \frac{1}{\sin \theta}\frac{\partial^2 F}{\partial \phi^2}\right)$$

Hence, when Laplace's equation $\nabla^2 F = 0$ is expressed in spherical coordinates, it becomes

(1) $\qquad r^2 \sin \theta \dfrac{\partial^2 F}{\partial r^2} + 2r \sin \theta \dfrac{\partial F}{\partial r} + \sin \theta \dfrac{\partial^2 F}{\partial \theta^2} + \cos \theta \dfrac{\partial F}{\partial \theta} + \dfrac{1}{\sin \theta}\dfrac{\partial^2 F}{\partial \phi^2} = 0$

Any solution $F(r,\theta,\phi)$ of this equation is known as a **spherical harmonic**.

In an attempt to solve Eq. (1), we assume a product solution $F(r,\theta,\phi) = R(r)G(\theta,\phi)$. Then, substituting this into (1), we have

$$r^2 \sin \theta \, R''G + 2r \sin \theta \, R'G + \sin \theta \, R \frac{\partial^2 G}{\partial \theta^2} + \cos \theta \, R \frac{\partial G}{\partial \theta} + \frac{R}{\sin \theta}\frac{\partial^2 G}{\partial \phi^2} = 0$$

Dividing through by $RG \sin \theta$ and rearranging gives

$$\frac{r^2 R'' + 2rR'}{R} = -\left(\frac{1}{G}\frac{\partial^2 G}{\partial \theta^2} + \frac{\cos \theta}{G \sin \theta}\frac{\partial G}{\partial \theta} + \frac{1}{G \sin^2 \theta}\frac{\partial^2 G}{\partial \phi^2}\right)$$

This relation can hold only if the common value of these two expressions is a constant. For later convenience we write this constant as $n(n + 1)$; hence we are led to the two equations

(2) $\qquad\qquad\qquad r^2 R'' + 2rR' - n(n + 1)R = 0$

(3) $\qquad\qquad \dfrac{\partial^2 G}{\partial \theta^2} + \dfrac{\cos \theta}{\sin \theta}\dfrac{\partial G}{\partial \theta} + \dfrac{1}{\sin^2 \theta}\dfrac{\partial^2 G}{\partial \phi^2} + n(n + 1)G = 0$

The first of these equations is an instance of Euler's equation (Example 3, Sec. 2.6), and it is easy to verify that its complete solution is

$$R = c_1 r^n + c_2 \frac{1}{r^{n+1}}$$

Solutions $G(\theta,\phi)$ of the second equation, which we shall have to find by a further separation of variables, are known as **surface harmonics**.

If, in (3), we substitute $G(\theta,\phi) = \Theta(\theta)\Phi(\phi)$, we find

$$\Theta''\Phi + \frac{\cos\theta}{\sin\theta}\Theta'\Phi + \frac{1}{\sin^2\theta}\Theta\Phi'' + n(n+1)\Theta\Phi = 0$$

Dividing by $\Theta\Phi/(\sin^2\theta)$ and rearranging slightly gives

$$\sin^2\theta\,\frac{\Theta''}{\Theta} + \sin\theta\cos\theta\,\frac{\Theta'}{\Theta} + n(n+1)\sin^2\theta = -\frac{\Phi''}{\Phi}$$

Again, the common value of the two members of this equation must be a constant, say m^2, and thus we have the pair of equations

(4) $$\Phi'' + m^2\Phi = 0$$

(5) $$\sin^2\theta\,\Theta'' + \sin\theta\cos\theta\,\Theta' + [n(n+1)\sin^2\theta - m^2]\Theta = 0$$

The first of these equations is familiar, and its complete solution can be written down at once:

$$\Phi = c_3\cos m\phi + c_4\sin m\phi$$

The second equation is known as the **associated Legendre equation**,[†] although it is usually studied in the form obtained by setting $x = \cos\theta$.

If $x = \cos\theta$, then

$$\frac{d\Theta}{d\theta} = \frac{d\Theta}{dx}\frac{dx}{d\theta} = -\sin\theta\,\frac{d\Theta}{dx}$$

$$\frac{d^2\Theta}{d\theta^2} = \frac{d}{d\theta}\left(-\sin\theta\,\frac{d\Theta}{dx}\right) = -\cos\theta\,\frac{d\Theta}{dx} - \sin\theta\,\frac{d^2\Theta}{dx^2}\frac{dx}{d\theta}$$

$$= -\cos\theta\,\frac{d\Theta}{dx} + \sin^2\theta\,\frac{d^2\Theta}{dx^2}$$

Hence, substituting these expressions into (5), we obtain

$$\sin^2\theta\left(-\cos\theta\,\frac{d\Theta}{dx} + \sin^2\theta\,\frac{d^2\Theta}{dx^2}\right) + \sin\theta\cos\theta\left(-\sin\theta\,\frac{d\Theta}{dx}\right)$$

$$+ [n(n+1)\sin^2\theta - m^2]\Theta = 0$$

Dividing out $\sin^2\theta$, substituting $x = \cos\theta$ in the coefficients, and simplifying gives

(6) $$(1-x^2)\frac{d^2\Theta}{dx^2} - 2x\frac{d\Theta}{dx} + \left[n(n+1) - \frac{m^2}{1-x^2}\right]\Theta = 0$$

This is the algebraic form of the associated Legendre equation. If $m = 0$, that is, if the solution of the original equation is independent of the longitude angle ϕ, then Eq. (6) reduces to

(7) $$(1-x^2)\frac{d^2\Theta}{dx^2} - 2x\frac{d\Theta}{dx} + n(n+1)\Theta = 0$$

which is known simply as **Legendre's equation**.

---

† Named for the French mathematician Adrien Marie Legendre (1752–1833).

To solve Eq. (7) we use the method of Frobenius and assume a series solution of the form

$$\Theta(x) = \sum_{k=0}^{\infty} a_k x^{k+c}$$

whence
$$\Theta'(x) = \sum_{k=0}^{\infty} a_k (k + c) x^{k+c-1}$$

and
$$\Theta''(x) = \sum_{k=0}^{\infty} a_k (k + c)(k + c - 1) x^{k+c-2}$$

Then, substituting, we have

$$\sum_{k=0}^{\infty} a_k (k + c)(k + c - 1) x^{k+c-2} - \sum_{k=0}^{\infty} a_k (k + c)(k + c - 1) x^{k+c}$$

$$- 2 \sum_{k=0}^{\infty} a_k (k + c) x^{k+c} + n(n + 1) \sum_{k=0}^{\infty} a_k x^{k+c} = 0$$

In this identity, the coefficients of the two lowest powers of x, namely, x^{c-2} and x^{c-1}, are, respectively,

$$a_0 c(c - 1) \qquad \text{and} \qquad a_1(c + 1)c$$

If $c = 0$, each of these is zero without restriction on a_0 and a_1; hence the choice $c = 0$ will lead to two independent solutions of Eq. (7). In this case the coefficient of the general power, x^k, is

$$(k + 2)(k + 1)a_{k+2} + [n(n + 1) - k^2 - k]a_k \qquad \text{or}$$

$$(k + 2)(k + 1)a_{k+2} + (n - k)(n + k + 1)a_k$$

Hence the recurrence relation connecting the successive coefficients is

$$a_{k+2} = - \frac{(n - k)(n + k + 1)}{(k + 1)(k + 2)} a_k$$

and, specifically,

$$a_0 = a_0 \qquad\qquad\qquad a_1 = a_1$$

$$a_2 = - \frac{n(n + 1)}{2!} a_0 \qquad\qquad a_3 = - \frac{(n - 1)(n + 2)}{3!} a_1$$

$$a_4 = \frac{n(n - 2)(n + 1)(n + 3)}{4!} a_0 \qquad a_5 = \frac{(n - 1)(n - 3)(n + 2)(n + 4)}{5!} a_1$$

. . . .

Hence a complete solution of Eq. (7) can be written

$$(8) \quad \Theta(x) = a_0 \left[1 - \frac{n(n + 1)}{2!} x^2 + \frac{n(n - 2)(n + 1)(n + 3)}{4!} x^4 - \cdots \right]$$

$$+ a_1 \left[x - \frac{(n - 1)(n + 2)}{3!} x^3 + \frac{(n - 1)(n - 3)(n + 2)(n + 4)}{5!} x^5 - \cdots \right]$$

These infinite series define what are known as **Legendre functions of the second kind**. Since $x = \pm 1$ are the only singular points of Eq. (7), it follows from Theorem 1, Sec. 9.1, that the radius of convergence of each of these series is 1. It can be shown, however, that neither series converges at either of the end points $x = \pm 1$; that is, the interval of convergence for each series is $-1 < x < 1$.

In many applications the parameter n is a positive integer. If it is odd, then, clearly, the second series in (8) contains only a finite number of terms; if it is even, then the first series contains only a finite number of terms. In either of these cases the series which reduces to a finite sum is known as a **Legendre polynomial** or **zonal harmonic of order** n. The usual standard form for the Legendre polynomials is obtained by assigning to a_0 and a_1 values which will make the coefficient of the highest power of x in each series equal to

$$\frac{(2n)!}{2^n(n!)^2}$$

These values are

$$a_0 = (-1)^{n/2} \frac{1 \cdot 3 \cdot 5 \cdots (n-1)}{2 \cdot 4 \cdot 6 \cdots n} = (-1)^{n/2} \frac{n!}{2^n[(n/2)!]^2}$$

$$a_1 = (-1)^{(n-1)/2} \frac{1 \cdot 3 \cdot 5 \cdots n}{2 \cdot 4 \cdot 6 \cdots (n-1)} = (-1)^{(n-1)/2} \frac{(n+1)!}{2^n[(n-1)/2]! \, [(n+1)/2]!}$$

and the resulting general formula is

$$(9) \quad P_n(x) = \sum_{k=0}^{N} \frac{(-1)^k (2n-2k)!}{2^n k! \, (n-k)! \, (n-2k)!} x^{n-2k} \qquad \begin{matrix} N = n/2 & n \text{ even} \\ N = (n-1)/2 & n \text{ odd} \end{matrix}$$

Specifically, this formula gives

$$P_0(x) = 1 \qquad\qquad P_1(x) = x$$
$$P_2(x) = \tfrac{1}{2}(3x^2 - 1) \qquad\qquad P_3(x) = \tfrac{1}{2}(5x^3 - 3x)$$
$$P_4(x) = \tfrac{1}{8}(35x^4 - 30x^2 + 3) \qquad P_5(x) = \tfrac{1}{8}(63x^5 - 70x^3 + 15x)$$

As these particular results illustrate, $P_n(1) = 1$ and $P_n(-1) = (-1)^n$ for all values of n. Since the infinite series in (8) diverge when $x = \pm 1$, it is clear that *to within an arbitrary constant multiplier, $P_n(x)$ is the only solution of Legendre's equation which is finite on the closed interval* $-1 \le x \le 1$.

One of the fundamental identities involving Legendre polynomials is **Rodrigues' formula.**†

THEOREM 1 $P_n(x) = \dfrac{1}{2^n n!} \dfrac{d^n(x^2 - 1)^n}{dx^n}$

Proof   This result can be proved by direct differentiation and induction, but it is perhaps more interesting to proceed in the following way. If we let

$$v = (x^2 - 1)^n$$

then

$$\frac{dv}{dx} = 2nx(x^2 - 1)^{n-1}$$

---

† Named for the French economist and mathematician Olinde Rodrigues (1794–1851).

Multiplying the last equation by $x^2 - 1$ gives

$$(x^2 - 1) \frac{dv}{dx} = 2nx(x^2 - 1)^n$$

and, finally,

$$(1 - x^2) \frac{dv}{dx} + 2nxv = 0$$

If we differentiate this repeatedly with respect to x, we obtain

$$(1 - x^2)v'' + 2(n - 1)xv' + (2n)v = 0$$
$$(1 - x^2)v''' + 2(n - 2)xv'' + 2(2n - 1)v' = 0$$
$$(1 - x^2)v^{(4)} + 2(n - 3)xv''' + 3(2n - 2)v'' = 0$$

and after $k + 1$ differentiations

$$(1 - x^2)v^{(k+2)} + 2(n - k - 1)xv^{(k+1)} + (k + 1)(2n - k)v^{(k)} = 0$$

If we now take $k = n$ and put $v^{(k)} = u$, the last equation becomes

$$(1 - x^2)u'' - 2xu' + n(n + 1)u = 0$$

which is precisely Legendre's equation. But

$$u = v^{(n)} = \frac{d^n(x^2 - 1)^n}{dx^n}$$

is obviously a polynomial of degree n. Moreover, from (8) it is clear that to within a constant factor there is only one polynomial solution of Legendre's equation, namely, $P_n(x)$. Hence u must be some multiple of $P_n(x)$; that is,

$$\frac{d^n(x^2 - 1)^n}{dx^n} = cP_n(x)$$

Finally, we can determine c by equating the coefficients of x^n in the two members of the last identity. Clearly, the coefficient of x^n on the left-hand side arises solely from the n-fold differentiation of the term of highest degree, namely x^{2n}, and is

$$(2n)(2n - 1) \cdots [2n - (n - 1)] \equiv \frac{(2n)!}{n!}$$

On the other hand, we have already noted that the coefficient of the highest power of x in $P_n(x)$ is

$$\frac{(2n)!}{2^n(n!)^2}$$

Hence we must have

$$\frac{(2n)!}{n!} = c \frac{(2n)!}{2^n(n!)^2} \qquad \text{or} \qquad c = 2^n n!$$

Therefore

$$2^n n! \, P_n(x) = \frac{d^n(x^2 - 1)^n}{dx^n}$$

and

$$P_n(x) = \frac{1}{2^n n!} \frac{d^n(x^2 - 1)^n}{dx^n}$$

as asserted.

Another important identity involving Legendre polynomials is contained in the following theorem.

THEOREM 2 $$\frac{1}{\sqrt{1 - 2xz + z^2}} = P_0(x) + P_1(x)z + P_2(x)z^2 + \cdots$$

$$+ P_n(x)z^n + \cdots$$

Proof To prove this, we expand the radical on the left-hand side by the binomial theorem, getting

$$[1 - z(2x - z)]^{-1/2} = 1 + \tfrac{1}{2}z(2x - z) + \frac{1 \cdot 3}{2^2 2!} z^2(2x - z)^2 + \cdots$$

$$+ \frac{1 \cdot 3 \cdots (2n - 3)}{2^{n-1}(n - 1)!} z^{n-1}(2x - z)^{n-1}$$

$$+ \frac{1 \cdot 3 \cdots (2n - 1)}{2^n n!} z^n(2x - z)^n + \cdots$$

Now z^n can occur only in the terms out to and including the one containing $z^n(2x - z)^n$, and from these, by expanding the various powers of $2x - z$, we find that its total coefficient is

$$\frac{1 \cdot 3 \cdots (2n - 1)}{2^n n!}(2x)^n - \frac{1 \cdot 3 \cdots (2n - 3)}{2^{n-1}(n - 1)!} \frac{n - 1}{1!}(2x)^{n-2}$$

$$+ \frac{1 \cdot 3 \cdots (2n - 5)}{2^{n-2}(n - 2)!} \frac{(n - 2)(n - 3)}{2!}(2x)^{n-4} - \cdots$$

Multiplying and dividing by the factors needed to complete the factorials in the numerators gives

$$\frac{(2n)!}{2^n n! \, n!} x^n - \frac{(2n - 2)!}{2^{n-1} 1! \, (n - 1)! \, (n - 2)!} x^{n-2} + \frac{(2n - 4)!}{2^{n-2} 2! \, (n - 2)! \, (n - 4)!} x^{n-4} - \cdots$$

which is precisely the expanded form of $P_n(x)$, as given by (9). Thus

$$(1 - 2xz + z^2)^{-1/2} = \sum_{n=0}^{\infty} P_n(x)z^n$$

as asserted.

The expression $(1 - 2xz + z^2)^{-1/2}$, in relation to the Legendre polynomials, is an example of a **generating function**: if $\{f_n(x)\}$ is a sequence of functions, and if $\Phi(x,t)$ is a function with the property that when it is expanded in terms of powers of t the coefficient of t^n is $f_n(x)$, then $\Phi(x,t)$ is said to be a generating function for the sequence $\{f_n(x)\}$. We discovered another example of a generating function in Sec. 9.5 when we proved that

$$\exp\left[\frac{x}{2}\left(t - \frac{1}{t}\right)\right] = \sum_{n=-\infty}^{\infty} J_n(x)t^n$$

In many applications the algebraic form of the Legendre polynomials is the more useful. There are problems, however, in which it is essential that they be expressed in terms of θ, the colatitude angle of the spherical coordinate system with which our discussion began. This can easily be done by reversing the transformation $x = \cos \theta$ which led from the trigonometric to the algebraic form of Legendre's equation. However, replacing x by $\cos \theta$ in $P_n(x)$ leads to expressions which are quite inconvenient because of the powers of $\cos \theta$ they contain. Fortunately, using the generating function provided by Theorem 2, we can easily derive more useful forms in which cosines of multiples of θ take the place of powers of $\cos \theta$.

To do this, let us substitute

$$x = \cos \theta = \frac{e^{i\theta} + e^{-i\theta}}{2}$$

into the generating function, getting

$$[1 - z(e^{i\theta} + e^{-i\theta}) + z^2]^{-1/2} = [(1 - ze^{i\theta})(1 - ze^{-i\theta})]^{-1/2} = \sum_{n=1}^{\infty} P_n z^n$$

Now if we use the binomial theorem to expand each of the factors in the middle term of this continued identity, we obtain

$$(1 - ze^{i\theta})^{-1/2} = 1 + \tfrac{1}{2}ze^{i\theta} + \frac{1 \cdot 3}{2 \cdot 4} z^2 e^{2i\theta} + \cdots$$

$$+ \frac{1 \cdot 3 \cdots (2n - 3)}{2 \cdot 4 \cdots (2n - 2)} z^{n-1} e^{(n-1)i\theta} + \frac{1 \cdot 3 \cdots (2n - 1)}{2 \cdot 4 \cdots (2n)} z^n e^{ni\theta} + \cdots$$

and

$$(1 - ze^{-i\theta})^{-1/2} = 1 + \tfrac{1}{2}ze^{-i\theta} + \frac{1 \cdot 3}{2 \cdot 4} z^2 e^{-2i\theta} + \cdots$$

$$+ \frac{1 \cdot 3 \cdots (2n - 3)}{2 \cdot 4 \cdots (2n - 2)} z^{n-1} e^{-(n-1)i\theta} + \frac{1 \cdot 3 \cdots (2n - 1)}{2 \cdot 4 \cdots (2n)} z^n e^{-ni\theta} + \cdots$$

The coefficient of z^n in the product of these two series is easy to determine, and we find for it the expression

$$\frac{1 \cdot 3 \cdots (2n - 1)}{2 \cdot 4 \cdots (2n)} (e^{ni\theta} + e^{-ni\theta}) + \frac{1}{2} \frac{1 \cdot 3 \cdots (2n - 3)}{2 \cdot 4 \cdots (2n - 2)} (e^{(n-2)i\theta} + e^{-(n-2)i\theta})$$

$$+ \frac{1 \cdot 3}{2 \cdot 4} \frac{1 \cdot 3 \cdots (2n - 5)}{2 \cdot 4 \cdots (2n - 4)} (e^{(n-4)i\theta} + e^{-(n-4)i\theta}) + \cdots$$

Hence, replacing the various combinations of exponentials by their cosine equivalents and recalling that the coefficient of z^n in the expansion of the generating function is just P_n, we have finally

$$(10) \quad P_n(\cos \theta) = \frac{1 \cdot 3 \cdots (2n - 1)}{2 \cdot 4 \cdots 2n} 2 \cos n\theta + \frac{1}{2} \frac{1 \cdot 3 \cdots (2n - 3)}{2 \cdot 4 \cdots (2n - 2)} 2 \cos (n - 2)\theta$$

$$+ \frac{1 \cdot 3}{2 \cdot 4} \frac{1 \cdot 3 \cdots (2n - 5)}{2 \cdot 4 \cdots (2n - 4)} 2 \cos (2n - 4)\theta + \cdots$$

If n is odd, the final term in $P_n(\cos \theta)$ contains the factor $\cos \theta$ and is correctly given by the last nonzero term in the series (10). However, if n is even, the final term in $P_n(\cos \theta)$ is a constant which is equal to just half the last nonzero term in the series

(10). This is the case because although the general term in the coefficient of z^n contains both $e^{(n-2k)i\theta}$ and $e^{-(n-2k)i\theta}$, when n is even and $k = n/2$ these terms are identical and arise only once in the product of the series for $(1 - ze^{i\theta})^{-1/2}$ and $(1 - ze^{-i\theta})^{-1/2}$ and not twice. Thus, specifically,

$$P_0(\cos\theta) = 1$$

$$P_1(\cos\theta) = \cos\theta$$

$$P_2(\cos\theta) = \frac{3\cos 2\theta + 1}{4}$$

$$P_3(\cos\theta) = \frac{5\cos 3\theta + 3\cos\theta}{8}$$

$$P_4(\cos\theta) = \frac{35\cos 4\theta + 20\cos 2\theta + 9}{64}$$

$$P_5(\cos\theta) = \frac{63\cos 5\theta + 35\cos 3\theta + 30\cos\theta}{128}$$

. . . .

Since Legendre's equation can be written in the form

$$\frac{d[(1 - x^2)y']}{dx} + n(n + 1)y = 0$$

it is clear that it is a special case, with

$$p(x) = 1 \qquad q(x) = 0 \qquad r(x) = 1 - x^2 \qquad \lambda = n(n + 1)$$

of the equation covered by Theorem 4, Sec. 8.5. Hence, if solutions of Legendre's equation satisfy suitable boundary conditions, they must be orthogonal. In particular, for the important interval $(-1,1)$ no boundary conditions are necessary, since $r(x) \equiv 1 - x^2$ vanishes at each end point; i.e.,

$$(11) \qquad \int_{-1}^{1} P_m(x)P_n(x)\,dx = 0 \qquad m \neq n$$

Before the property of orthogonality can be used to expand an arbitrary function in terms of Legendre polynomials, we must, of course, know the value of the integral of the square of the general Legendre polynomial. This can be obtained in various ways, but perhaps the simplest is to use the generating function provided by Theorem 2. If we square the identity

$$\frac{1}{(1 - 2xz + z^2)^{1/2}} = P_0(x) + P_1(x)z + \cdots + P_n(x)z^n + \cdots$$

and integrate with respect to x from -1 to 1, we obtain

$$\int_{-1}^{1} \frac{dx}{1 - 2xz + z^2} = \int_{-1}^{1} [P_0(x) + P_1(x)z + \cdots + P_n(x)z^n + \cdots]^2\,dx$$

The integral on the left is easily evaluated. On the right, all integrals involving the product of two different P's are zero because of the orthogonality property (11).

Hence

$$(12) \quad -\frac{1}{2z} \ln (1 - 2xz + z^2)^2 \Big|_{-1}^{1} = \int_{-1}^{1} P_0^2(x) \, dx + z^2 \int_{-1}^{1} P_1^2(x) \, dx + \cdots$$

$$+ z^{2n} \int_{-1}^{1} P_n^2(x) \, dx + \cdots$$

Evaluation of the left member leads at once to

$$-\frac{1}{2z} \left[\ln (1 - z)^2 - \ln (1 + z)^2 \right] = \frac{1}{z} \left[\ln (1 + z) - \ln (1 - z) \right]$$

Moreover, if we replace the logarithms by their respective power series, we obtain

$$\frac{1}{z} \left(z - \frac{z^2}{2} + \frac{z^3}{3} - \cdots - \frac{z^{2n}}{2n} + \frac{z^{2n+1}}{2n+1} - \cdots \right)$$

$$-\frac{1}{z} \left(-z - \frac{z^2}{2} - \frac{z^3}{3} - \cdots - \frac{z^{2n}}{2n} - \frac{z^{2n+1}}{2n+1} - \cdots \right)$$

$$= 2 \left(1 + \frac{z^2}{3} + \frac{z^4}{5} + \cdots + \frac{z^{2n}}{2n+1} + \cdots \right)$$

Hence, comparing coefficients in this series and in the right-hand member of (12), we obtain the desired result:

$$(13) \qquad \int_{-1}^{1} P_n^2(x) \, dx = \frac{2}{2n+1}$$

By means of the substitution $x = \cos \theta$, Eqs. (12) and (13) can be transformed at once into corresponding results for the Legendre polynomials in trigonometric form. Hence we can state the following important theorem.

THEOREM 3   The Legendre polynomials in algebraic form satisfy the orthogonality relations

$$\int_{-1}^{1} P_m(x)P_n(x) \, dx = \begin{cases} 0 & m \neq n \\ 2/(2n+1) & m = n \end{cases}$$

In trigonometric form, the Legendre polynomials satisfy the orthogonality relations

$$\int_{0}^{\pi} P_m(\cos \theta)P_n(\cos \theta) \sin \theta \, d\theta = \begin{cases} 0 & m \neq n \\ 2/(2n+1) & m = n \end{cases}$$

EXAMPLE 1

The known temperature distribution $u = f(\theta)$ is maintained over the entire surface of a sphere of radius b. Find the steady-state temperature at any point in the sphere.

Here we have to solve the steady-state heat equation, i.e., Laplace's equation, in spherical coordinates. However, from the obvious circular symmetry of the problem it is clear that u is a function of r and θ only. Hence $\partial^2 u/\partial \phi^2 = 0$, and Eq. (1) reduces to

$$(14) \qquad r^2 \sin \theta \frac{\partial^2 u}{\partial r^2} + 2r \sin \theta \frac{\partial u}{\partial r} + \sin \theta \frac{\partial^2 u}{\partial \theta^2} + \cos \theta \frac{\partial u}{\partial \theta} = 0$$

Assuming a product solution $u = R(r)\Theta(\theta)$ and substituting into (14), we obtain

$$r^2 \sin \theta \, R''\Theta + 2r \sin \theta \, R'\Theta + \sin \theta \, R\Theta'' + \cos \theta \, R\Theta' = 0$$

From this, by dividing by $\sin \theta \, R\Theta$ and transposing, we have

$$\frac{r^2 R''}{R} + \frac{2rR'}{R} = -\frac{\Theta''}{\Theta} - \frac{\cos \theta}{\sin \theta}\frac{\Theta'}{\Theta} = \lambda$$

Since for any λ, the quadratic equation $n^2 + n - \lambda = 0$ is always satisfied by at least one (possibly complex) value of n, it is no specialization to take $\lambda = n(n + 1)$, so that we have the two ordinary differential equations

$$r^2 R'' + 2rR' - n(n + 1)R = 0$$
$$\sin \theta \, \Theta'' + \cos \theta \, \Theta' + n(n + 1) \sin \theta \, \Theta = 0$$

The first of these is just an instance of Euler's equation, and its complete solution is easily found to be

$$R = Ar^n + \frac{B}{r^{n+1}}$$

However, since we require solutions which are finite when $r = 0$, it is clear that we must specialize this by taking $B = 0$. The second equation is Legendre's equation. Since we require solutions of it which are finite over the closed interval $0 \le \theta \le \pi$, and since the only such solutions are the Legendre polynomials $P_n(\cos \theta)$, it is clear that n must be an integer and

$$\Theta = P_n(\cos \theta)$$

Hence we have an infinite sequence of product solutions

$$A_0 P_0 (\cos \theta), \; A_1 r P_1(\cos \theta), \; A_2 r^2 P_2(\cos \theta), \ldots, A_n r^n P_n(\cos \theta), \ldots$$

None of these, by itself can satisfy the given temperature condition

$$u(b,\theta) = f(\theta)$$

on the surface of the sphere. Hence, as usual, we form an infinite series of the individual product solutions and attempt to make it fit the boundary condition. Thus we write

(15) $$u(r,\theta) = \sum_{n=0}^{\infty} A_n r^n P_n(\cos \theta)$$

Then, substituting $r = b$ and $u(b,\theta) = f(\theta)$, we get

$$f(\theta) = \sum_{n=0}^{\infty} A_n b^n P_n(\cos \theta)$$

To find A_n we multiply the last equation by $\sin \theta \, P_n(\cos \theta)$ and integrate from 0 to π. By virtue of the orthogonality properties of the P's, all integrals on the right except one become zero, and we have

$$\int_0^\pi f(\theta) \sin \theta \, P_n(\cos \theta) \, d\theta = A_n b^n \frac{2}{2n + 1}$$

or $$A_n = \frac{2n + 1}{2b^n} \int_0^\pi f(\theta) \sin \theta \, P_n(\cos \theta) \, d\theta$$

With the coefficients in the series (15) determined, the formal solution of the problem is now complete.

EXERCISES

1 Show that

$$\int_{-1}^{1} f(x)P_n(x)\, dx = \frac{(-1)^n}{2^n n!} \int_{-1}^{1} f^{(n)}(x)(x^2 - 1)^n\, dx$$

Hence show that

$$\int_{-1}^{1} x^m P_n(x)\, dx = 0 \qquad \text{if} \qquad m < n$$

Hint: Use Rodrigues' formula and repeated integration by parts.

2 Using the result of Exercise 1, show that

$$\int_{-1}^{1} x^n P_n(x)\, dx = \frac{2^{n+1}(n!)^2}{(2n + 1)!}$$

3 Express x^2, x^3, and x^4 in terms of Legendre polynomials.

4 It can be shown that one solution of the associated Legendre equation is the function $P_n^{(m)}(x)$, defined as follows

$$P_n^{(m)}(x) = \frac{1}{2^n n!}(x^2 - 1)^{m/2}\frac{d^{m+n}(x^2 - 1)^n}{dx^{m+n}} = (x^2 - 1)^{m/2}\frac{d^m P_n(x)}{dx^m}$$

Write out $P_1^{(1)}(x)$, $P_2^{(1)}(x)$, $P_3^{(1)}(x)$, $P_1^{(2)}(x)$, $P_2^{(2)}(x)$, $P_3^{(2)}(x)$.

5 Consider the functions x^i, $i = 0, 1, 2, \ldots$, and let $f_n(x) = \sum_{i=0}^{n} a_{ni}x^i$, where the a's are coefficients to be determined so that $a_{nn} > 0$ and

$$\int_{-1}^{1} f_m(x)f_n(x)\, dx = \begin{cases} 0 & m \neq n \\ \dfrac{2}{2n + 1} & m = n \end{cases}$$

Calculate the a's for $n = 0, 1, 2, 3$, and show that, at least for these values of n, $f_n(x) \equiv P_n(x)$.

6 It is desired to approximate a function $f(x)$ over the interval $(-1,1)$ by a polynomial $P(x)$ of degree n which will make the integral

$$\int_{-1}^{1} [f(x) - P(x)]^2\, dx$$

a minimum. Show that $P(x)$ is the nth partial sum of the expansion of $f(x)$ over the interval $(-1,1)$ in terms of Legendre polynomials. The Legendre polynomials thus play the same role in the least-squares polynomial approximation of continuous functions that the orthogonal polynomials discussed in Sec. 4.6 play in the least-squares approximation of tabular functions. *Hint:* First express $P(x)$ as an arbitrary linear combination of Legendre polynomials.

7 Show that both the Legendre polynomials with even subscripts and the Legendre polynomials with odd subscripts form orthogonal sets over the interval $(0,1)$.

8 By differentiating the generating function for the Legendre polynomials, show that all derivatives of even order of $P_n(x)$ vanish at $x = 0$ if n is odd and that all derivatives of odd order vanish at $x = 0$ if n is even. What are the values of the nonzero derivatives at $x = 0$?

9 Letting $g(x,z)$ denote the generating function for the Legendre polynomials, show that

$$(1 - 2xz + z^2)\frac{\partial g}{\partial z} = (x - z)g(x,z)$$

Then, substituting $\sum_{n=0}^{\infty} P_n(x)z^n$ for $g(x,z)$ in this identity, equate the coefficients of z^n on both sides and establish the recurrence relation

$$(n + 1)P_{n+1}(x) - (2n - 1)xP_n(x) + nP_{n-1}(x) = 0$$

10 Proceeding as in Exercise 9, show that

$$z \frac{\partial g}{\partial z} = (x - z) \frac{\partial g}{\partial x}$$

and then establish the recurrence relation

$$nP_n(x) - xP_n'(x) + P_{n-1}'(x) = 0$$

11 Using Rodrigues' formula, prove that

$$P_{n+1}'(x) - P_{n-1}'(x) = (2n + 1)P_n(x)$$

Hence show that

$$\int_x^1 P_n(x) \, dx = \frac{1}{2n + 1} [P_{n-1}(x) - P_{n+1}(x)]$$

12 Show that

(a) $\displaystyle\int_{-1}^1 P_n(x)P_{n+1}'(x) \, dx = 2$ (b) $\displaystyle\int_{-1}^1 xP_n(x)P_{n-1}(x) \, dx = \frac{2n}{4n^2 - 1}$

Hint: Use the results of Exercises 9 and 11.

13 Introduce a new independent variable in the associated Legendre equation in trigonometric form by means of the substitution $\theta = x/n$, and show that as $n \to \infty$ this equation approaches Bessel's equation.

14 Show that the change of dependent variable defined by the substitution $y = u/\sqrt{\sin \theta}$ transforms Legendre's equation into the equation

$$\frac{d^2u}{d\theta^2} + \left[(n + \tfrac{1}{2})^2 + \frac{1}{4 \sin^2 \theta} \right] u = 0$$

Hence show that for large values of n, $P_n(\cos \theta)$ is described approximately by the expression

$$P_n(\cos \theta) = A_n \cos \left(\frac{2n + 1}{2} \theta + \alpha_n \right)$$

15 By considering $\int_0^\pi P_n^2(\cos \theta) \sin \theta \, d\theta$, show that A_n in the formula of Exercise 14 approaches $\sqrt{2/n\pi}$ as n increases indefinitely.

16 By first putting $x = \cos \theta$ in the generating function for the Legendre polynomials, then setting $\theta = \pi/2$, and finally noting that

$$\binom{-\tfrac{1}{2}}{m} = (-1)^m \frac{(2m)!}{2^{2m}(m!)^2}$$

approaches $(-1)^m(1/\sqrt{m\pi})$ as m increases indefinitely, show that the value of the phase angle α_n in the formula of Exercise 14 is $-\pi/4$.

17 The upper half of the surface of a sphere of radius b is maintained at the temperature $u = 100°C$ and the lower half is maintained at the temperature $u = 0°C$. Find the steady-state temperature distribution in the sphere.

18 If the temperature throughout the sphere in Exercise 17 is initially 0°C, find the temperature at any point in the sphere at any subsequent time.

19 Find the steady-state temperature at any point in a spherical shell of inner radius b_1 and outer radius b_2 if the temperature distributions $u(b_1,\theta) = f_1(\theta)$ and $u(b_2,\theta) = f_2(\theta)$ are maintained over the inner and outer surfaces, respectively.

20 Work Exercise 19 if $f_1(\theta) = 100°C$ and $f_2(\theta) = 0°C$, and check the solution against that obtained in Exercise 30, Sec. 1.7.

21 The temperature distribution $u(b,\theta) = f(\theta)$ is maintained over the curved surface of a hemisphere of radius b. The plane boundary is kept at the temperature $u = 0°C$. Find the steady-state temperature at any point in the hemisphere. *Hint:* The results of Exercise 7 may be of assistance.

22 In Eq. (8), let $u(x)$ denote the series which is a polynomial, and let $v(x)$ denote the series which is not a polynomial. Then a second solution of Legendre's equation which is independent of $P_n(x)$ can be defined as follows:

$$Q_n(x) = \begin{cases} u_n(1)v_n(x) & n \text{ even} \\ -u_n(1)v_n(x) & n \text{ odd} \end{cases}$$

Using the results of Sec. 2.1, show that $Q_n(x)$ is related to $P_n(x)$ by an equation of the form

$$Q_n(x) = A_n P_n(x) \int \frac{dx}{(1 - x^2)[P_n(x)]^2} + B_n$$

23 In Exercise 22, determine (A_0, B_0) and (A_1, B_1), and show that

$$Q_0(x) = \tfrac{1}{2} \ln \frac{1 + x}{1 - x} \quad \text{and} \quad Q_1(x) = \frac{x}{2} \ln \frac{1 + x}{1 - x} - 1 = xQ_0(x) - 1$$

24 Find polynomial solutions of the equation

$$y'' - 2xy' + 2ny = 0 \qquad n \text{ a positive integer}$$

and show that they are orthogonal with respect to the weight function e^{-x^2} over the interval $(-\infty, \infty)$. [This equation is known as **Hermite's equation**, after the French mathematician Charles Hermite (1822–1901), and its polynomial solutions are known as **Hermite polynomials**.]

25 Find polynomial solutions of the equation

$$xy'' + (1 - x)y' + ny = 0 \qquad n \text{ a positive integer}$$

and show that they are orthogonal with respect to the weight function e^{-x} over the interval $(0, \infty)$. [This equation is known as **Laguerre's equation** after the French mathematician Edmond Laguerre (1834–1886), and its polynomial solutions are known as **Laguerre polynomials**.]

CHAPTER 10
Determinants and Matrices

10.1 Determinants

In a restricted sense, at least, the concept of a determinant is already familiar from elementary algebra, where, in solving systems of two and three simultaneous linear equations, we found it convenient to introduce what we called *determinants of the second and third order*. In the work of this book we shall have occasion to generalize these ideas to the solution of systems of more than three linear equations and to other applications not immediately associated with solving equations. For this reason we shall devote this and the following chapter to a review and an extension of our earlier study of determinants and to a discussion of some of the fundamental properties of the related mathematical objects known as *matrices*.

By a **determinant of order** n we mean a certain function of n^2 quantities, which we shall describe more precisely as soon as we have introduced the necessary notation and preliminary definitions. The customary symbol for a determinant consists of a square array of the n^2 quantities enclosed between vertical bars:

$$(1) \qquad |A|\dagger = |a_{ij}| = \begin{vmatrix} a_{11} & a_{12} & \cdots & a_{1n} \\ a_{21} & a_{22} & \cdots & a_{2n} \\ \cdots\cdots\cdots\cdots\cdots\cdots\cdots \\ a_{n1} & a_{n2} & \cdots & a_{nn} \end{vmatrix}$$

For brevity, we shall often use the word *determinant* to refer to this symbol as well as to the expansion‡ for which it stands. Although logically undesirable, this dual usage is quite common and should cause no confusion.

The quantities a_{ij} which appear in (1) are called the **elements** of the determinant. The horizontal lines of elements are called **rows**; the vertical lines of elements are called **columns**. In the convenient **double-subscript notation** illustrated in (1), the first subscript associated with an element identifies the row and the second subscript identifies the column in which the element lies. There is, of course, no reason to suppose that the element in the ith row and jth column is the same as the element in the jth row and ith column, and so in general $a_{ij} \neq a_{ji}$. The sloping line of elements extending from a_{11} to a_{nn} is called the **principal diagonal** of the determinant.

---

† The use of vertical bars in the notation for a determinant and in the notation for the absolute value of a quantity, while perhaps unfortunate, is universal. Which meaning is intended in any particular case should always be clear from the context.

‡ See Definition 1, this section.

The determinant $|M|$ formed by the m^2 elements common to any m rows and any m columns of an nth-order determinant $|A|$ is said to be an **mth-order minor** of $|A|$. The determinant of order $n - m$ formed by the array of elements which remains when the m rows and m columns containing an mth-order minor $|M|$ are deleted from $|A|$ is called the **complementary minor** of $|M|$. If the numbers of the rows and columns of $|A|$ which contain an mth-order minor $|M|$ are, respectively,

$$i_1, i_2, \ldots, i_m \qquad \text{and} \qquad j_1, j_2, \ldots, j_m$$

then $(-1)^{i_1 + i_2 + \cdots + i_m + j_1 + j_2 + \cdots + j_m}$ times the complementary minor of $|M|$ is called the **algebraic complement** of $|M|$. The first-order minors of $|A|$ are, of course, just the elements of $|A|$. Their complementary minors are customarily referred to simply as **minors**, and their algebraic complements are almost universally referred to as **cofactors**. We shall denote the minor of the element a_{ij} by the symbol M_{ij} and its cofactor by the symbol A_{ij}; thus

$$A_{ij} = (-1)^{i+j} M_{ij}$$

Similarly, we shall use the symbols $M_{ij,kl}$ and $A_{ij,kl}$ to denote, respectively, the complementary minor and the algebraic complement of the second-order minor contained in the ith and jth rows and the kth and lth columns of a determinant $|A|$; thus

$$A_{ij,kl} = (-1)^{i+j+k+l} M_{ij,kl}$$

The generalization of this notation is obvious.

EXAMPLE 1

In the fifth-order determinant

$$|A| = \begin{vmatrix} a_{11} & a_{12} & a_{13} & a_{14} & a_{15} \\ a_{21} & a_{22} & a_{23} & a_{24} & a_{25} \\ a_{31} & a_{32} & a_{33} & a_{34} & a_{35} \\ a_{41} & a_{42} & a_{43} & a_{44} & a_{45} \\ a_{51} & a_{52} & a_{53} & a_{54} & a_{55} \end{vmatrix}$$

the minor of the element a_{43} is the fourth-order determinant formed by the elements which remain when the fourth row and third column are deleted from $|A|$, namely,

$$M_{43} = \begin{vmatrix} a_{11} & a_{12} & a_{14} & a_{15} \\ a_{21} & a_{22} & a_{24} & a_{25} \\ a_{31} & a_{32} & a_{34} & a_{35} \\ a_{51} & a_{52} & a_{54} & a_{55} \end{vmatrix}$$

The cofactor A_{43} of the element a_{43} is equal to this minor times $(-1)^{4+3}$; that is,

$$A_{43} = -M_{43}$$

Similarly, the complementary minor of the second-order minor

$$\begin{vmatrix} a_{31} & a_{34} \\ a_{51} & a_{54} \end{vmatrix}$$

contained in the third and fifth rows and the first and fourth columns of $|A|$ is the third-order determinant formed by the elements which remain when these rows and columns are deleted from $|A|$:

$$M_{35,14} = \begin{vmatrix} a_{12} & a_{13} & a_{15} \\ a_{22} & a_{23} & a_{25} \\ a_{42} & a_{43} & a_{45} \end{vmatrix}$$

The algebraic complement $A_{35,14}$ of the given second-order minor is equal to the complementary minor $M_{35,14}$ times $(-1)^{3+5+1+4}$; that is,

$$A_{35,14} = -M_{35,14}$$

For a second-order determinant we have the definition

(2)
$$\begin{vmatrix} a_{11} & a_{12} \\ a_{21} & a_{22} \end{vmatrix} = a_{11}a_{22} - a_{12}a_{21}$$

i.e., a second-order determinant is equal to the difference between the product of the elements on the principal diagonal and the product of the elements on the other diagonal. For a third-order determinant we have the definition

(3)
$$\begin{vmatrix} a_{11} & a_{12} & a_{13} \\ a_{21} & a_{22} & a_{23} \\ a_{31} & a_{32} & a_{33} \end{vmatrix} = a_{11}a_{22}a_{33} + a_{12}a_{23}a_{31} + a_{13}a_{21}a_{32} \\ - a_{13}a_{22}a_{31} - a_{11}a_{23}a_{32} - a_{12}a_{21}a_{33}$$

This expansion can also be obtained by diagonal multiplication, by repeating on the right the first two columns of the determinant and then adding the signed products of the elements on the various diagonals in the resulting array:

The diagonal method of writing out a determinant is correct *only* for determinants of the second and third orders, however, and will in general give incorrect results if applied to determinants of higher order.

We are now in a position to give the general definition of a determinant. This can be done in direct fashion, but since the result is unsuited to the practical evaluation of determinants, we choose to give an inductive definition.

DEFINITION 1   The determinant

$$|A| = \begin{vmatrix} a_{11} & a_{12} & \cdots & a_{1n} \\ a_{21} & a_{22} & \cdots & a_{2n} \\ \cdots\cdots\cdots\cdots\cdots\cdots \\ a_{n1} & a_{n2} & \cdots & a_{nn} \end{vmatrix}$$

is equal to the sum of the products of the elements of any row or column and their respective cofactors; i.e.,

$$|A| = a_{i1}A_{i1} + a_{i2}A_{i2} + \cdots + a_{in}A_{in} = \sum_{j=1}^{n} a_{ij}A_{ij} \qquad \text{row definition}$$

$$= a_{1j}A_{1j} + a_{2j}A_{2j} + \cdots + a_{nj}A_{nj} = \sum_{i=1}^{n} a_{ij}A_{ij} \qquad \text{column definition}$$

Clearly, this definition makes a determinant of order n depend upon n determinants of order $n - 1$, each of which in turn depends upon $n - 1$ determinants of order

$n - 2$, and so on, until finally the expansion involves only second- or third-order determinants which can be written out by the diagonal method. However, before Definition 1 can be accepted and used, it must be shown that the same expansion is obtained no matter which row or column is selected. That this is the case is guaranteed by the following theorem.

THEOREM 1   If the elements of any row or of any column of a determinant are multiplied by their respective cofactors and then added, the sum is the same for all rows and for all columns.

Proof   We shall first prove that the same expansion is obtained no matter which row is chosen. To do this we proceed inductively. Clearly, the theorem is true when $n = 2$; for, expanding the determinant

$$\begin{vmatrix} a_{11} & a_{12} \\ a_{21} & a_{22} \end{vmatrix}$$

in terms of the elements of the first row and their cofactors, we get

$$a_{11}(a_{22}) + a_{12}(-a_{21})$$

and expanding in terms of the elements of the second row and their cofactors, we get

$$a_{21}(-a_{12}) + a_{22}(a_{11})$$

and these two expressions are identical. Let us assume, then, that the assertion of the theorem is true for determinants of order $n - 1$, and let us attempt to prove that it is true for determinants of order n. Specifically, let us expand the nth-order determinant

$$|A| = |a_{ij}|$$

in terms of each of two arbitrary rows, say the ith and the jth, and compare the expansions. In doing this it is, of course, no specialization to assume that $i < j$.

(4a)

$$\begin{vmatrix} a_{11} & \cdots & a_{1k} & \cdots & a_{1l} & \cdots & a_{1n} \\ & & & & & & \\ a_{i1} & \cdots & a_{ik} & \cdots & a_{il} & \cdots & a_{in} \\ & & & & & & \\ a_{j1} & \cdots & a_{jk} & \cdots & \widehat{a_{jl}} & \cdots & a_{jn} \\ & & & & & & \\ a_{n1} & \cdots & a_{nk} & \cdots & a_{nl} & \cdots & a_{nn} \end{vmatrix}$$

$$= \begin{vmatrix} a_{11} & \cdots & a_{1k} & \cdots & a_{1l} & \cdots & a_{1n} \\ a_{i1} & \cdots & \widehat{a_{ik}} & \cdots & a_{il} & \cdots & a_{in} \\ a_{j1} & \cdots & a_{jk} & \cdots & a_{jl} & \cdots & a_{jn} \\ a_{n1} & \cdots & a_{nk} & \cdots & a_{nl} & \cdots & a_{nn} \end{vmatrix} \qquad k < l$$

$$(4b) \quad \begin{vmatrix} a_{11} & \cdots & a_{1l} & \cdots & a_{1k} & \cdots & a_{1n} \\ \hdots & & \hdots & & \hdots & & \hdots \\ a_{i1} & \cdots & a_{il} & \cdots & a_{ik} & \cdots & a_{in} \\ \hdots & & \hdots & & \hdots & & \hdots \\ a_{j1} & \cdots & \widehat{(a_{jl})} & \cdots & a_{jk} & \cdots & a_{jn} \\ \hdots & & \hdots & & \hdots & & \hdots \\ a_{n1} & \cdots & a_{nl} & \cdots & a_{nk} & \cdots & a_{nn} \end{vmatrix}$$

$$= \begin{vmatrix} a_{11} & \cdots & a_{1l} & \cdots & a_{1k} & \cdots & a_{1n} \\ \hdots & & \hdots & & \hdots & & \hdots \\ a_{i1} & \cdots & a_{il} & \cdots & \widehat{(a_{ik})} & \cdots & a_{in} \\ \hdots & & \hdots & & \hdots & & \hdots \\ a_{j1} & \cdots & a_{jl} & \cdots & a_{jk} & \cdots & a_{jn} \\ \hdots & & \hdots & & \hdots & & \hdots \\ a_{n1} & \cdots & a_{nl} & \cdots & a_{nk} & \cdots & a_{nn} \end{vmatrix} \qquad k > l$$

Now, a typical term in the expansion of $|A|$ according to the elements of the ith row is

$$(5) \qquad\qquad a_{ik}A_{ik} = (-1)^{i+k}a_{ik}M_{ik}$$

and this contains the only occurrences of a_{ik} in the entire expansion. Moreover, the $(j - 1)$st row of M_{ik} contains $n - 1$ elements from the jth row of $|A|$, and M_{ik} can legitimately be expanded in terms of these elements, since the hypothesis of our induction is that the theorem in question is true for determinants of order $n - 1$. As a typical term in the expansion of M_{ik} according to the elements from the jth row of $|A|$, we therefore have

$$(6) \qquad\qquad a_{jl}(\text{cofactor of } a_{jl} \text{ in } M_{ik}) \qquad l \ne k$$

and this contains the only occurrences of a_{jl} in the expansion of M_{ik}. Hence, substituting expression (6) into Eq. (5), we find that the expression

$$(7) \qquad\qquad (-1)^{i+k}a_{ik}[a_{jl}(\text{cofactor of } a_{jl} \text{ in } M_{ik})] \qquad l \ne k$$

contains the only occurrences of the product $a_{ik}a_{jl}$ in the expansion of $|A|$ in terms of the elements of the ith row. In exactly the same way, if we first expand $|A|$ in terms of the elements of the jth row and then expand the minor M_{jl} in terms of the $n - 1$ elements from the ith row of $|A|$ which it contains, we conclude that the only occurrences of the product $a_{ik}a_{jl}$ in the expansion of $|A|$ in terms of the elements of the jth row are those in the expression

$$(8) \qquad\qquad (-1)^{j+l}a_{jl}[a_{ik}(\text{cofactor of } a_{ik} \text{ in } M_{jl})] \qquad k \ne l$$

If we can show that (7) and (8) are identical, we shall have completed our proof that under the induction hypothesis all row expansions of $|A|$ are the same, since $a_{ik}a_{jl}$ $(k \ne l)$ is the typical product of an element from the ith row and an element from the jth row, and each term in the expansion of $|A|$ according to the ith row or the jth row must contain one and only one such product.

Now except for the proper power of (-1), both the cofactor of a_{jl} in M_{ik} and the cofactor of a_{ik} in M_{jl} are equal to the determinant of order $n - 2$, say $M_{ij,kl}$, formed by the elements which remain when the ith and jth rows and the kth and lth columns are deleted from $|A|$. In checking the signs of these cofactors, there are two possibilities to consider, according as $k < l$ [see array (4a)] or $k > l$ [see array (4b)]; i.e., according as the kth column precedes or follows the lth column in the determinant $|A|$. Taking due account of the relative positions of the deleted rows and columns, the proper signs are easily determined by inspection, however, and we have, in the respective cases

$$\text{Cofactor of } a_{jl} \text{ in } M_{ik} = (-1)^{(j-1)+(l-1)}M_{ij,kl}$$
$$\text{Cofactor of } a_{ik} \text{ in } M_{jl} = (-1)^{i+k}M_{ij,kl}$$
$$k < l$$

$$\text{Cofactor of } a_{jl} \text{ in } M_{ik} = (-1)^{(j-1)+l}M_{ij,kl}$$
$$\text{Cofactor of } a_{ik} \text{ in } M_{jl} = (-1)^{i+(k-1)}M_{ij,kl}$$
$$k > l$$

Finally, substituting these expressions into (7) and (8), we find that the coefficient of the product $a_{ik}a_{jl}$, as determined by either method of expansion, is

(9a) $$\qquad (-1)^{i+j+k+l}M_{ij,kl} \qquad\qquad\qquad k < l$$

(9b) $$\qquad (-1)^{i+j+k+l-1}M_{ij,kl} = -(-1)^{i+j+k+l}M_{ij,kl} \qquad k > l$$

In exactly the same way, if we expand $|A|$ in terms of two arbitrary columns, say the kth and the lth, we find that the coefficient of the general product $a_{ik}a_{jl}$ is still given by (9a) and (9b). This proves that under the induction hypothesis not only are all column expansions of $|A|$ equal but their common value is the common value of all row expansions of $|A|$. Thus we have completed our proof that *if* the theorem is true for determinants of order $n - 1$, *then* it is true for determinants of order n. Since we have already proved it true for row expansions of second-order determinants and could similarly prove it true for column expansions, our induction is complete; Theorem 1 is established; and Definition 1 is unambiguous.

Since the same expression is obtained whether we expand a determinant in terms of the elements of an arbitrary row or an arbitrary column, we have the following obvious consequence of Theorem 1.

THEOREM 2 If $|A|$ is any determinant, and if $|B|$ is the determinant whose rows are the columns of $|A|$, then $|A| = |B|$.

The proof of Theorem 1 also provides us with the proof of the following important result.

THEOREM 3 Let any two rows (or columns)† be selected from a determinant $|A|$. Then $|A|$ is equal to the sum of the products of all the second-order minors contained in the chosen pair of rows (or columns) each multiplied by its algebraic complement.

---

† Theorems (like this one) whose statements contain words or phrases in parentheses are really two theorems in one. One version is obtained by consistently omitting the word, or words, in parentheses (the word *columns*, in this case); the other is obtained by consistently retaining the words in parentheses and omitting the words to which they are alternatives (the word *rows*, in this case).

Proof   Let the chosen rows be the pth and qth, and, for definiteness, suppose that $p < q$. Now a typical second-order minor from these rows is

(10)
$$\begin{vmatrix} a_{pr} & a_{ps} \\ a_{qr} & a_{qs} \end{vmatrix} = a_{pr}a_{qs} - a_{ps}a_{qr} \qquad r < s$$

and to prove the assertion of the theorem it is sufficient to show that the coefficient of this binomial in the expansion of $|A|$ is

$$A_{pq,rs} = (-1)^{p+q+r+s} M_{pq,rs}$$

To do this, we observe that when the factors in the first product are arranged in the order of magnitude of their first subscripts, that is, $a_{qr}a_{qs}$, the column index r of the first factor is less than the column index s of the second factor. Hence Eq. (9a) can be applied, with

$$i = p \qquad j = q \qquad k = r \qquad \text{and} \qquad l = s$$

giving for the coefficient of $a_{pr}a_{qs}$ the quantity

$$(-1)^{p+q+r+s} M_{pq,rs}$$

Similarly, when the factors in the second product are arranged in the order of magnitude of their first subscripts, that is, $a_{ps}a_{qr}$, the column index s of the first factor is greater than the column index r of the second factor. Hence Eq. (9b) can be applied with

$$i = p \qquad j = q \qquad k = s \qquad \text{and} \qquad l = r$$

giving for the coefficient of $a_{ps}a_{qr}$ the quantity

$$- (-1)^{p+q+s+r} M_{pq,rs}$$

Hence, the expansion of $|A|$ contains the terms

$$a_{pr}a_{qs}[(-1)^{p+q+r+s} M_{pq,rs}] + a_{ps}a_{qr}[-(-1)^{p+q+r+s} M_{pq,rs}]$$

and these are the only occurrences of the products $a_{pr}a_{qs}$ and $a_{ps}a_{qr}$. Finally, from these, by factoring, we obtain

$$(a_{pr}a_{qs} - a_{ps}a_{qr})(-1)^{p+q+r+s} M_{pq,rs} = (a_{pr}a_{qs} - a_{ps}a_{qr})A_{pq,rs}$$

which completes the proof of the theorem in the case where two rows are used for the expansion. An essentially identical argument establishes the assertion of the theorem when two columns are used.

By a somewhat more involved argument the following generalization of Theorem 3 can be established.

THEOREM 4   Let any m rows (or columns) be selected from a determinant $|A|$. Then $|A|$ is equal to the sum of the products of all the mth-order minors contained in the chosen rows (or columns) each multiplied by its algebraic complement.

Both the general result contained in Theorem 4 and the special case $m = 2$ contained in Theorem 3 are usually referred to as **Laplace's expansion** of $|A|$.

EXAMPLE 2

Expand the determinant

$$|A| = \begin{vmatrix} 1 & 2 & 3 & 4 \\ 4 & 3 & 2 & 1 \\ 0 & -1 & 2 & 3 \\ 1 & 6 & 4 & -2 \end{vmatrix}$$

For purposes of illustration we shall obtain the value of this determinant using Definition 1 and also using Theorem 3. According to Definition 1, using the third row because of the presence of the zero element, and taking due account of the alternating signs of the successive cofactors, we have

$$|A| = (0)\begin{vmatrix} 2 & 3 & 4 \\ 3 & 2 & 1 \\ 6 & 4 & -2 \end{vmatrix} - (-1)\begin{vmatrix} 1 & 3 & 4 \\ 4 & 2 & 1 \\ 1 & 4 & -2 \end{vmatrix}$$

$$+ (2)\begin{vmatrix} 1 & 2 & 4 \\ 4 & 3 & 1 \\ 1 & 6 & -2 \end{vmatrix} - (3)\begin{vmatrix} 1 & 2 & 3 \\ 4 & 3 & 2 \\ 1 & 6 & 4 \end{vmatrix}$$

or, expanding the third-order determinants by the diagonal method,

$$|A| = 0 + 75 + 180 - 105 = 150$$

Equivalently, applying Theorem 3 in terms of the first two rows, we have

$$|A| = \begin{vmatrix} 1 & 2 \\ 4 & 3 \end{vmatrix} \cdot \begin{vmatrix} 2 & 3 \\ 4 & -2 \end{vmatrix} + \begin{vmatrix} 1 & 3 \\ 4 & 2 \end{vmatrix}\left(-\begin{vmatrix} -1 & 3 \\ 6 & -2 \end{vmatrix}\right) + \begin{vmatrix} 1 & 4 \\ 4 & 1 \end{vmatrix} \cdot \begin{vmatrix} -1 & 2 \\ 6 & 4 \end{vmatrix}$$

$$+ \begin{vmatrix} 2 & 3 \\ 3 & 2 \end{vmatrix} \cdot \begin{vmatrix} 0 & 3 \\ 1 & -2 \end{vmatrix} + \begin{vmatrix} 2 & 4 \\ 3 & 1 \end{vmatrix}\left(-\begin{vmatrix} 0 & 2 \\ 1 & 4 \end{vmatrix}\right) + \begin{vmatrix} 3 & 4 \\ 2 & 1 \end{vmatrix} \cdot \begin{vmatrix} 0 & -1 \\ 1 & 6 \end{vmatrix}$$

$$= (-5)(-16) + (-10)(16) + (-15)(-16)$$

$$+ (-5)(-3) + (-10)(2) + (-5)(1)$$

$$= 150$$

as before.

With Theorems 1 and 3 a number of other theorems can easily be proved. In particular, we have the following useful results.

THEOREM 5   If all the elements in any row (or in any column) of a determinant are zero, the value of the determinant is zero.

Proof   If we expand the given determinant, according to Definition 1, in terms of the row (or column) of zero elements, each term in the expansion contains a zero factor. Hence, the entire expansion is zero, as asserted.

THEOREM 6   If each element in one row (or in one column) of a determinant is multiplied by c, the value of the determinant is multiplied by c.

Proof   If we expand the given determinant in terms of the row (or column) whose elements have been multiplied by c, each term in the expansion contains c as a factor. If c is then factored from the expansion, the result is just c times the expansion of the original determinant, as asserted.

THEOREM 7   If $|A|$ is any determinant, and if $|B|$ is the determinant obtained from $|A|$ by interchanging any two rows (or any two columns) of $|A|$, then $|B| = -|A|$.

Proof   Let $|A|$ be any determinant, and let $|B|$ be the determinant obtained from $|A|$ by interchanging any two rows (or any two columns) of $|A|$. Now, clearly, if the rows (or columns) of any second-order determinant are interchanged, the resulting determinant is the negative of the original one. Hence, if $|B|$ is expanded in terms of the two rows (or columns) which were interchanged, it follows that each second-order minor occurring as a factor of a term in this expansion is the negative of the corresponding second-order minor from the corresponding pair of rows (or columns) in $|A|$. Therefore, each term in the expansion of $|B|$ is the negative of the corresponding term in the expansion of $|A|$ based on the same two rows (or columns). Thus $|B| = -|A|$, as asserted.

THEOREM 8   If corresponding elements of two rows (or of two columns) of a determinant are proportional, the value of the determinant is zero.

Proof   Clearly, any determinant of the second order whose rows (or columns) are proportional is zero. Hence if we expand the given determinant, according to Theorem 3, in terms of the two rows (or two columns) which are proportional, it follows that each term contains as one factor a second-order minor which is equal to zero. Therefore the entire expansion is zero, as asserted.

THEOREM 9   The sum of the products formed by multiplying the elements of one row (or column) of a determinant by the cofactors of the corresponding elements of another row (or column) is zero.

Proof   Let $|A| = |a_{ik}|$ be the given determinant, and let the elements of some row of $|A|$, say the ith, be multiplied by the cofactors of the corresponding elements in some other row, say the jth, giving the sum

$$\sum_{k=1}^{n} a_{ik}A_{jk}$$

Clearly, according to Definition 1, this can be thought of as the expansion of a determinant whose jth row consists of the elements

$$a_{i1} \quad a_{i2} \quad \cdots \quad a_{in}$$

and whose other rows are identical with the corresponding rows in $|A|$. In this new determinant the ith and jth rows are therefore the same, and hence, by Theorem 8, the determinant is equal to zero. A similar argument leads to the same conclusion if the elements in some column of $|A|$ are multiplied by the cofactors of the corresponding elements in some other column of $|A|$. Thus the theorem is established.

Combining Definition 1 and Theorem 9, we have the following useful result.

COROLLARY 1   If A_{ik} is the cofactor of the element a_{ik} in the determinant $|A| = |a_{ik}|$, then

$$\sum_{k=1}^{n} a_{ik}A_{jk} = \begin{cases} 0 & i \neq j \\ |A| & i = j \end{cases} \quad \text{and} \quad \sum_{i=1}^{n} a_{ik}A_{il} = \begin{cases} 0 & k \neq l \\ |A| & k = l \end{cases}$$

EXAMPLE 3

If we multiply the elements in the first row of the determinant

$$\begin{vmatrix} a_{11} & a_{12} & a_{13} \\ a_{21} & a_{22} & a_{23} \\ a_{31} & a_{32} & a_{33} \end{vmatrix}$$

by the cofactors of the corresponding elements in the third row, say, we obtain the sum

$$a_{11}\begin{vmatrix} a_{12} & a_{13} \\ a_{22} & a_{23} \end{vmatrix} - a_{12}\begin{vmatrix} a_{11} & a_{13} \\ a_{21} & a_{23} \end{vmatrix} + a_{13}\begin{vmatrix} a_{11} & a_{12} \\ a_{21} & a_{22} \end{vmatrix}$$

and this is clearly the expansion of the determinant

$$\begin{vmatrix} a_{11} & a_{12} & a_{13} \\ a_{21} & a_{22} & a_{23} \\ a_{11} & a_{12} & a_{13} \end{vmatrix}$$

according to the third row. Since this determinant has two identical rows, it is therefore equal to zero.

Results such as those of Corollary 1 are often stated more compactly in terms of what is known as the **Kronecker delta**,† usually written δ_{ij}, or sometimes $\delta_j{}^i$, and defined to be 0 or 1 according as $i \neq j$ or $i = j$. When the Kronecker delta is used, the assertions of Corollary 1 can be written in the simpler form

(11) $$\sum_{k=1}^{n} a_{ik}A_{jk} = |A|\delta_{ij}$$

(12) $$\sum_{i=1}^{n} a_{ik}A_{il} = |A|\delta_{kl}$$

THEOREM 10   If the elements in one column of a determinant are expressed as binomials, the determinant can be written as the sum of two determinants, according to the formula

$$\begin{vmatrix} a_{11} & \cdots & a_{1j} + \alpha_{1j} & \cdots & a_{1n} \\ a_{21} & \cdots & a_{2j} + \alpha_{2j} & \cdots & a_{2n} \\ \cdots & & \cdots & & \cdots \\ a_{n1} & \cdots & a_{nj} + \alpha_{nj} & \cdots & a_{nn} \end{vmatrix} = \begin{vmatrix} a_{11} & \cdots & a_{1j} & \cdots & a_{1n} \\ a_{21} & \cdots & a_{2j} & \cdots & a_{2n} \\ \cdots & & \cdots & & \cdots \\ a_{n1} & \cdots & a_{nj} & \cdots & a_{nn} \end{vmatrix}$$

$$+ \begin{vmatrix} a_{11} & \cdots & \alpha_{1j} & \cdots & a_{1n} \\ a_{21} & \cdots & \alpha_{2j} & \cdots & a_{2n} \\ \cdots & & \cdots & & \cdots \\ a_{n1} & \cdots & \alpha_{nj} & \cdots & a_{nn} \end{vmatrix}$$

A similar result holds for a determinant containing a row of elements which are binomials.

---

† Named for the German mathematician Leopold Kronecker (1823–1891).

Proof If we expand the given determinant, according to Definition 1, in terms of the column which contains the binomial elements, we obtain

$$\sum_{i=1}^{n} (a_{ij} + \alpha_{ij})A_{ij} = \sum_{i=1}^{n} a_{ij}A_{ij} + \sum_{i=1}^{n} \alpha_{ij}A_{ij}$$

Since the sums on the right are, respectively, the expansions for the determinants appearing on the right side of the formula in the theorem, the theorem is established.

THEOREM 11 The value of a determinant is unchanged if the elements of any row (or column) are modified by adding to them the same multiple of the corresponding elements in any other row (or column).

Proof If we apply Theorem 10 to the determinant resulting from the given row (or column) modification, we obtain two determinants, one of which is the original determinant and the other of which contains two proportional rows (or columns). By Theorem 8, the second determinant is equal to zero, and the theorem is established.

Theorem 11 is very useful in the practical expansion of determinants for, by its repeated application, one can reduce to zero a number of the elements in a chosen row (or column) of the given determinant. Then, when the determinant is expanded in terms of this row (or column), most of the products involved will be zero and the computation will be appreciably shortened.

EXAMPLE 4

Find the value of the determinant

$$\begin{vmatrix} 3 & 1 & -1 & 2 & 1 \\ 0 & 3 & 1 & 4 & 2 \\ 1 & 4 & 2 & 3 & 1 \\ 5 & -1 & -3 & 2 & 5 \\ -1 & 1 & 2 & 3 & 2 \end{vmatrix}$$

Here, in an attempt to introduce as many zeros as possible into some row, let us add the third column to the second and to the fifth, and let us add twice the third column to the fourth and 3 times the third column to the first. This gives the new but equal determinant

$$\begin{vmatrix} 0 & 0 & -1 & 0 & 0 \\ 3 & 4 & 1 & 6 & 3 \\ 7 & 6 & 2 & 7 & 3 \\ -4 & -4 & -3 & -4 & 2 \\ 5 & 3 & 2 & 7 & 4 \end{vmatrix}$$

Expanding this in terms of the first row, according to Definition 1, we have

$$(-1)(-1)^{1+3}\begin{vmatrix} 3 & 4 & 6 & 3 \\ 7 & 6 & 7 & 3 \\ -4 & -4 & -4 & 2 \\ 5 & 3 & 7 & 4 \end{vmatrix}$$

Now, adding twice the last column to each of the first three, we obtain the equal determinant

$$-\begin{vmatrix} 9 & 10 & 12 & 3 \\ 13 & 12 & 13 & 3 \\ 0 & 0 & 0 & 2 \\ 13 & 11 & 15 & 4 \end{vmatrix}$$

or, expanding in terms of the third row,

$$-(2)(-1)^{3+4}\begin{vmatrix} 9 & 10 & 12 \\ 13 & 12 & 13 \\ 13 & 11 & 15 \end{vmatrix}$$

We can now simplify this by further row or column manipulations, or, since it is of the third order, we can expand it by the diagonal method. The result is -166.

THEOREM 12 The product of two determinants of the same order is a determinant of the same order in which the element in the ith row and jth column is the sum of the products of corresponding elements in the ith row of the first determinant and the jth column of the second determinant.

Proof For simplicity we shall prove this theorem only for determinants of the second order, although for these, direct verification is easier and more natural than the method we shall actually use. The virtue of our proof is that it can be extended immediately to the general case of determinants of any order. We begin by observing that if

$$|A| = \begin{vmatrix} a_{11} & a_{12} \\ a_{21} & a_{22} \end{vmatrix} \quad \text{and} \quad |B| = \begin{vmatrix} b_{11} & b_{12} \\ b_{21} & b_{22} \end{vmatrix}$$

then, by Theorem 3,

$$|A| \cdot |B| = \begin{vmatrix} a_{11} & a_{12} \\ a_{21} & a_{22} \end{vmatrix} \cdot \begin{vmatrix} b_{11} & b_{12} \\ b_{21} & b_{22} \end{vmatrix} = \begin{vmatrix} a_{11} & a_{12} & 0 & 0 \\ a_{21} & a_{22} & 0 & 0 \\ c_{11} & c_{12} & b_{11} & b_{12} \\ c_{21} & c_{22} & b_{21} & b_{22} \end{vmatrix}$$

where c_{11}, c_{12}, c_{21}, and c_{22} are completely arbitrary. In particular, it is convenient to take $c_{11} = c_{22} = -1$ and $c_{12} = c_{21} = 0$, so that we have

$$|A| \cdot |B| = \begin{vmatrix} a_{11} & a_{12} & 0 & 0 \\ a_{21} & a_{22} & 0 & 0 \\ -1 & 0 & b_{11} & b_{12} \\ 0 & -1 & b_{21} & b_{22} \end{vmatrix}$$

Now if we multiply the elements in the first column by b_{11} and the elements in the second column by b_{21} and then add them to the corresponding elements of the third column, we obtain, by Theorem 11, the equal determinant

$$|A| \cdot |B| = \begin{vmatrix} a_{11} & a_{12} & a_{11}b_{11} + a_{12}b_{21} & 0 \\ a_{21} & a_{22} & a_{21}b_{11} + a_{22}b_{21} & 0 \\ -1 & 0 & 0 & b_{12} \\ 0 & -1 & 0 & b_{22} \end{vmatrix}$$

In the same way, if we multiply the elements in the first column by b_{12} and the elements in the second column by b_{22} and then add them to the corresponding elements in the fourth column, we obtain from the last determinant the equal determinant

$$|A| \cdot |B| = \begin{vmatrix} a_{11} & a_{12} & a_{11}b_{11} + a_{12}b_{21} & a_{11}b_{12} + a_{12}b_{22} \\ a_{21} & a_{22} & a_{21}b_{11} + a_{22}b_{21} & a_{21}b_{12} + a_{22}b_{22} \\ -1 & 0 & 0 & 0 \\ 0 & -1 & 0 & 0 \end{vmatrix}$$

If we now expand this determinant by Theorem 3, applied to the last two rows, we obtain

$$|A| \cdot |B| = \begin{vmatrix} a_{11}b_{11} + a_{12}b_{21} & a_{11}b_{12} + a_{12}b_{22} \\ a_{21}b_{11} + a_{22}b_{21} & a_{21}b_{12} + a_{22}b_{22} \end{vmatrix}$$

which is the result asserted by the theorem.

EXERCISES

1 Find the value of each of the following determinants:

(a) $\begin{vmatrix} 1 & 2 & 3 & 4 \\ 2 & 1 & 4 & 3 \\ 3 & 4 & 2 & 1 \\ 4 & 3 & 1 & 2 \end{vmatrix}$ (b) $\begin{vmatrix} 1 & 2 & 3 & 4 \\ 4 & 3 & 2 & 1 \\ 2 & 1 & 4 & 3 \\ 3 & 4 & 1 & 2 \end{vmatrix}$ (c) $\begin{vmatrix} 0 & 1 & 2 & 3 \\ -1 & 0 & 1 & 2 \\ -2 & -1 & 0 & 3 \\ -3 & -2 & -3 & 0 \end{vmatrix}$

2 (a) Find the value(s) of a, if any, for which the diagonal method of expansion yields the correct value for the determinant

$$\begin{vmatrix} 1 & 2 & 3 & 4 \\ -1 & 2 & 0 & 3 \\ 2 & 0 & a & 1 \\ 1 & 4 & -9 & a \end{vmatrix}$$

(b) Show that there is no value of a for which the diagonal method of expansion yields the correct value for the determinant

$$\begin{vmatrix} 1 & 2 & 1 & 1 \\ -1 & 1 & 3 & 2 \\ -1 & 3 & 9 & 1 \\ 2 & 1 & 1 & a \end{vmatrix}$$

3 Show that the number of terms in the expansion of a general determinant of order n is $n!$.

4 If $|A|$ is a second-order determinant each of whose elements is a binomial, how many determinants are there in the sum to which $|A|$ is reduced by repeated applications of Theorem 10? How many are there if $|A|$ is a third-order determinant? If $|A|$ is an nth-order determinant?

5 Calculate the product of each pair of the following determinants, using Theorem 12, and check your results by multiplying the values of the individual determinants:

(a) $\begin{vmatrix} 1 & 1 & -1 \\ 2 & 1 & 3 \\ 1 & 0 & 1 \end{vmatrix}$ (b) $\begin{vmatrix} -2 & 1 & 1 \\ 3 & 1 & 0 \\ -1 & 2 & 4 \end{vmatrix}$ (c) $\begin{vmatrix} 1 & 3 & 4 \\ 2 & -1 & 0 \\ 0 & 1 & 3 \end{vmatrix}$

6 Prove the following generalization of Theorem 12: The product of two determinants of the same order is another determinant of that order in which the element in the ith row and jth column is the sum of the products of corresponding elements in the ith row *or* column of the first determinant and the jth row *or* column of the second determinant, a consistent choice of row or column being maintained for all values of i and j. *Hint:* Use Theorem 2.

7 If $|A| = |a_{ij}|$ is a determinant of order n with the property that $a_{ij} = -a_{ji}$ for all values of i and j such that $1 \le i, j \le n$, prove that $|A| = (-1)^n |A|$. What further conclusion can be drawn if n is odd? *Hint:* Use Theorems 2 and 6.

8 (a) Show that $|A| = \begin{vmatrix} b^2 + ac & bc & c^2 \\ ab & 2ac & bc \\ a^2 & ab & b^2 + ac \end{vmatrix} = 4a^2b^2c^2$

Hint: Verify first that $|A| = \begin{vmatrix} b & c & 0 \\ a & 0 & c \\ 0 & a & b \end{vmatrix}^2$

(b) Find the value of the determinant $\begin{vmatrix} b^2 + c^2 & ab & ca \\ ab & c^2 + a^2 & bc \\ ca & bc & a^2 + b^2 \end{vmatrix}$

9　Prove that

$$\begin{vmatrix} 1 + a_1 & a_2 & a_3 & \cdots & a_n \\ a_1 & 1 + a_2 & a_3 & \cdots & a_n \\ a_1 & a_2 & 1 + a_3 & \cdots & a_n \\ \multicolumn{5}{c}{\cdots\cdots\cdots\cdots\cdots\cdots\cdots} \\ a_1 & a_2 & a_3 & \cdots & 1 + a_n \end{vmatrix} = 1 + a_1 + a_2 + a_3 + \cdots + a_n$$

10　Show that the nth-order determinant

$$\begin{vmatrix} a & b & \cdots & b & b \\ b & a & \cdots & b & b \\ \multicolumn{5}{c}{\cdots\cdots\cdots\cdots\cdots} \\ b & b & \cdots & a & b \\ b & b & \cdots & b & a \end{vmatrix}$$

is equal to $(a - b)^{n-1}[a + (n - 1)b]$.

11　Prove that

$$\begin{vmatrix} 0 & a_1 - a_2 & a_1 - a_3 & \cdots & a_1 - a_n \\ a_1 - a_2 & 0 & a_2 - a_3 & \cdots & a_2 - a_{..} \\ a_1 - a_3 & a_2 - a_3 & 0 & \cdots & a_3 - a_n \\ \multicolumn{5}{c}{\cdots\cdots\cdots\cdots\cdots\cdots\cdots\cdots} \\ a_1 - a_n & a_2 - a_n & a_3 - a_n & \cdots & 0 \end{vmatrix}$$

$$= (-1)^n 2^{n-2}(a_n - a_1) \prod_{i=1}^{n-1} (a_i - a_{i+1}) \qquad n \geq 2$$

12　If D_n is the nth-order determinant

$$\begin{vmatrix} 1 + x^2 & x & 0 & \cdots & 0 & 0 \\ x & 1 + x^2 & x & \cdots & 0 & 0 \\ 0 & x & 1 + x^2 & \cdots & 0 & 0 \\ \multicolumn{6}{c}{\cdots\cdots\cdots\cdots\cdots\cdots\cdots\cdots\cdots} \\ 0 & 0 & 0 & \cdots & 1 + x^2 & x \\ 0 & 0 & 0 & \cdots & x & 1 + x^2 \end{vmatrix}$$

show that $D_n = (1 + x^2)D_{n-1} - x^2 D_{n-2}$. Using this relation, determine the value of D_{10} if $x = 1$; if $x = -1$. Is the value of D_n independent of x?

13　If D_n is the nth-order determinant in which each element on the principal diagonal is a, each element immediately above the principal diagonal is b, each element immediately below the principal diagonal is c, and all other elements are zero, obtain a recurrence relation expressing D_n in terms of D_{n-1} and D_{n-2}. Use this relation to infer the value of D_n if $a = 3$, $b = 2$, $c = 1$.

14　Show that the area of the triangle whose vertices are the points (x_1, y_1), (x_2, y_2), (x_3, y_3) is

$$A = \pm \frac{1}{2} \begin{vmatrix} x_1 & y_1 & 1 \\ x_2 & y_2 & 1 \\ x_3 & y_3 & 1 \end{vmatrix}$$

where the plus or the minus sign is to be chosen according as the vertices of the triangle are numbered consecutively in the counterclockwise or the clockwise direction.

15　If $P_1:(x_1, y_1)$, $P_2:(x_2, y_2)$, and $P_3:(x_3, y_3)$ are three points no two of which lie on the same vertical line, show that the equation of the parabola of the family $y = a + bx + cx^2$ which passes through P_1, P_2, and P_3 can be written in the form

$$\begin{vmatrix} y & 1 & x & x^2 \\ y_1 & 1 & x_1 & x_1^2 \\ y_2 & 1 & x_2 & x_2^2 \\ y_3 & 1 & x_3 & x_3^2 \end{vmatrix} = 0$$

Is this result correct if P_1, P_2, and P_3 are collinear?

16 Show that the equation of the circle which passes through the three points $P_1:(x_1,y_1)$, $P_2:(x_2,y_2)$, and $P_3:(x_3,y_3)$ can be written in the form

$$\begin{vmatrix} x^2 + y^2 & x & y & 1 \\ x_1^2 + y_1^2 & x_1 & y_1 & 1 \\ x_2^2 + y_2^2 & x_2 & y_2 & 1 \\ x_3^2 + y_3^2 & x_3 & y_3 & 1 \end{vmatrix} = 0$$

Is this result correct if the three points are collinear?

17 Show that

$$\begin{vmatrix} a & -b & -a & b \\ b & a & -b & -a \\ c & -d & c & -d \\ d & c & d & c \end{vmatrix} = 4(a^2 + b^2)(c^2 + d^2)$$

18 Evaluate the determinant

$$\begin{vmatrix} 0 & 1 & 2 & 3 & \cdots & n-1 \\ 1 & 0 & 1 & 2 & \cdots & n-2 \\ 2 & 1 & 0 & 1 & \cdots & n-3 \\ 3 & 2 & 1 & 0 & \cdots & n-4 \\ \cdots & \cdots & \cdots & \cdots & \cdots & \cdots \\ n-1 & n-2 & n-3 & n-4 & \cdots & 0 \end{vmatrix}$$

19 (a) If

$$l_1: a_{11}x + a_{12}y + a_{13} = 0$$
$$l_2: a_{21}x + a_{22}y + a_{23} = 0$$
$$l_3: a_{31}x + a_{32}y + a_{33} = 0$$

are three lines no two of which are parallel, show that l_1, l_2, and l_3 are concurrent if and only if

$$|A| = \begin{vmatrix} a_{11} & a_{12} & a_{13} \\ a_{21} & a_{22} & a_{23} \\ a_{31} & a_{32} & a_{33} \end{vmatrix} = 0$$

(b) Show that the area of the triangle determined by the lines l_1, l_2, and l_3 is equal to the absolute value of the expression

$$\frac{1}{2A_{13}A_{23}A_{33}} \begin{vmatrix} A_{11} & A_{12} & A_{13} \\ A_{21} & A_{22} & A_{23} \\ A_{31} & A_{32} & A_{33} \end{vmatrix}$$

where A_{ij} is the cofactor of a_{ij} in $|A|$.

20 Show that

$$\begin{vmatrix} 1 & 1 & 1 & 1 \\ a_1 & a_2 & a_3 & a_4 \\ a_1^2 & a_2^2 & a_3^2 & a_4^2 \\ a_1^3 & a_2^3 & a_3^3 & a_4^3 \end{vmatrix}$$

$$= (a_1 - a_2)(a_1 - a_3)(a_1 - a_4)(a_2 - a_3)(a_2 - a_4)(a_3 - a_4)$$

What is the generalization of this result to determinants of this type of order n? [Determinants of this form are usually referred to as **Vandermonde determinants**, after the French mathematician A. T. Vandermonde (1735–1796).]

21 If $p_1, p_2, \ldots, p_n$ are polynomials, show that the nth-order determinant

$$\begin{vmatrix} p_1(x_1) & p_1(x_2) & \cdots & p_1(x_n) \\ p_2(x_1) & p_2(x_2) & \cdots & p_2(x_n) \\ \cdots & \cdots & \cdots & \cdots \\ p_n(x_1) & p_n(x_2) & \cdots & p_n(x_n) \end{vmatrix}$$

is evenly divisible by

$$\prod_{1 \leq i < j \leq n} (x_i - x_j)$$

22   If $|A|$ is the nth-order determinant $(n > 1)$

$$\begin{vmatrix} a_{11} & a_{12} & \cdots & a_{1n} \\ a_{21} & a_{22} & \cdots & a_{2n} \\ \cdots\cdots\cdots\cdots\cdots \\ a_{n1} & a_{n2} & \cdots & a_{nn} \end{vmatrix}, \text{ show that } \begin{vmatrix} a_{11} & a_{12} & \cdots & a_{1n} & x_1 \\ a_{21} & a_{22} & \cdots & a_{2n} & x_2 \\ \cdots\cdots\cdots\cdots\cdots\cdots\cdots \\ a_{n1} & a_{n2} & \cdots & a_{nn} & x_n \\ y_1 & y_2 & \cdots & y_n & 0 \end{vmatrix}$$

is equal to $-\sum_{i,j=1}^{n} A_{ij}x_i y_j$, where A_{ij} is the cofactor of a_{ij} in $|A|$.

23   If $|A| = |a_{ij}|$, show that $\partial|A|/\partial a_{ij} = A_{ij}$.

24   If each element of a determinant $|A|$ of order n is a differentiable function of t, show that the derivative of $|A|$ with respect to t is equal to the sum of n determinants, the ith one of which is identical with $|A|$ except for the ith row, which consists of the derivatives of the elements of the ith row of $|A|$. *Hint:* Proceed inductively, using Definition 1.

25   If $f_1, f_2, \ldots, f_n$ are suitably differentiable functions of t, show that

$$\frac{d}{dt} \begin{vmatrix} f_1 & \cdots & f_n \\ f_1' & \cdots & f_n' \\ \cdots\cdots\cdots\cdots\cdots \\ f_1^{(n-2)} & \cdots & f_n^{(n-2)} \\ f_1^{(n-1)} & \cdots & f_n^{(n-1)} \end{vmatrix} = \begin{vmatrix} f_1 & \cdots & f_n \\ f_1' & \cdots & f_n' \\ \cdots\cdots\cdots\cdots\cdots \\ f_1^{(n-2)} & \cdots & f_n^{(n-2)} \\ f_1^{(n)} & \cdots & f_n^{(n)} \end{vmatrix}$$

Hint: Use the result of Exercise 24.

10.2 Elementary Properties of Matrices

Closely associated with determinants, yet significantly different and much more fundamental, are the mathematical objects known as **matrices**.

DEFINITION 1   An $m \times n$, or (m,n), matrix is a rectangular array of quantities arranged in m rows and n columns.

When there is no possibility of confusion, matrices are often represented by single capital letters. More explicitly, they are represented by displaying some or all of the constituent quantities between brackets;† thus we write, equivalently,

$$A = [a_{ij}] = \begin{bmatrix} a_{11} & a_{12} & \cdots & a_{1n} \\ a_{21} & a_{22} & \cdots & a_{2n} \\ \cdots\cdots\cdots\cdots\cdots \\ a_{m1} & a_{m2} & \cdots & a_{mn} \end{bmatrix}$$

Two matrices $A = [a_{ij}]$ and $B = [b_{ij}]$ are **equal** if and only if they are identical; i.e., if and only if they contain the same number of rows and the same number of columns and $a_{ij} = b_{ij}$ for all values of i and j.

A matrix consisting of a single row is called a **row matrix**. A matrix consisting of a single column is called a **column matrix**. Both column matrices and row matrices are often referred to as **vectors**. The $n \times m$ matrix obtained from a given $m \times n$ matrix A by writing its rows as columns and its columns as rows is called the **transpose** of A. The transpose of a matrix A is denoted by the symbol A^T. A matrix with the same

---

† Many writers use parentheses or double bars instead of brackets.

number of rows and columns is called a **square matrix**. The determinant whose array of elements is identical with the array of elements of a square matrix is called the **determinant of the matrix**. A square matrix in which every element below the principal diagonal is zero is said to be **upper triangular**. A square matrix in which every element above the principal diagonal is zero is said to be **lower triangular**. A square matrix in which every element not on the principal diagonal is zero is called a **diagonal matrix**. Diagonal matrices are sometimes denoted†

$$
\begin{bmatrix} a_{11} & & 0 \\ & \cdot & \\ 0 & & a_{nn} \end{bmatrix}
$$

A diagonal matrix in which each diagonal element is 1 is called a **unit matrix**. A unit matrix is usually denoted by the symbol I or, more specifically, by the symbol I_n if it is a unit matrix of order n. A matrix, whether square or not, in which every element is zero is called a **null matrix** or a **zero matrix**, and is denoted by the symbol **0**.

If A is an $m \times n$ matrix and if k and l are integers such that $0 < k \le m$ and $0 < l \le n$, then the array of elements common to any k rows and any l columns of A is called a $k \times l$ **submatrix** of A. If A is a square matrix, any square submatrix of A whose principal diagonal is a part of the principal diagonal of A is called a **principal submatrix** of A. A principal submatrix of a square matrix A is thus a submatrix whose rows have the same indices as its columns. The determinants of the square submatrices of any matrix A are called the **minors** of A. The determinant of any principal submatrix of a square matrix A is called a **principal minor** of A.

Most matrices which occur in elementary applications have the property that all their elements are real. However, there are important applications, especially in mathematical physics and quantum mechanics, which involve matrices whose elements are not real. For this reason we shall introduce certain definitions and later state certain fundamental theorems in a form appropriate to matrices whose elements may be general complex quantities.

Recalling that the conjugate of a complex number $z = x + iy$ is the complex number $\bar{z} = x - iy$, we say that the **conjugate** of a matrix A is the matrix $\bar{A}$ whose elements are, respectively, the conjugates of the elements of A. Clearly, a matrix is **real**, i.e., contains only real elements, if and only if A and $\bar{A}$ are the same matrix. Similarly, a matrix A is **imaginary**, i.e., contains only elements which are pure imaginaries, if and only if $A = -\bar{A}$. The transpose of the conjugate of a matrix A is called the **associate** of A.

A matrix equal to its transpose, i.e., a square matrix such that $a_{ij} = a_{ji}$ for $1 \le i,j \le n$ is said to be **symmetric**, since elements *symmetrically* located with respect to the principal diagonal are equal. A matrix equal to the negative of its transpose, i.e., a square matrix such that $a_{ij} = -a_{ji}$ and in which, therefore, $a_{ii} = 0$, is said to be **skew-symmetric**. A matrix equal to its associate, i.e., a square matrix A such that $A = \bar{A}^T$, is said to be **hermitian**.‡ A matrix A such that $A = -\bar{A}^T$ is said to be **skew-hermitian**. Clearly, *a real symmetric matrix is just a hermitian matrix which is real*, and

---

† The boldface zero is a device for indicating that all elements not on the principal diagonal are 0. Diagonal matrices are also written without the boldface zero, i.e., simply with blank space in all off-diagonal positions.

‡ Named for the French mathematician Charles Hermite (1822–1901).

a real skew-symmetric matrix is just a skew-hermitian matrix which is real. Thus, in particular, *any result true for hermitian matrices is automatically true for real symmetric matrices.* For this reason, although real symmetric matrices are of fundamental importance in most of the applications of matrices in this book, we shall as far as possible state our theorems in terms of hermitian matrices.

The concept of a matrix is essentially simpler than the concept of a determinant. For whereas a matrix is just a collection of elements arranged in a particular way, a determinant is a rather complicated function of the elements in a given set. In particular, with every square matrix there is associated a determinant, namely, the determinant whose elements are respectively equal to the elements of the matrix. Thus determinants of order n and $n \times n$ matrices bear to each other the familiar relation of dependent and independent variable, respectively; and it is appropriate to speak of a determinant as a function of a square matrix.

Examples of matrices can be found in many fields. For instance, if m students were given a battery of n different tests, the resulting scores would very probably be displayed in a table containing m rows, one for each student, and n columns, one for each test. The resulting array would, of course, be an $m \times n$ matrix in which the general element a_{ij} was the score which the ith student made on the jth test. Matrices of this sort are of fundamental importance in the branch of mathematical psychology known as *factor analysis*. Similarly, if we had an electric network containing n branches, we might, either experimentally or analytically, determine the current which would flow in the ith branch as a result of inserting a unit voltage in the jth branch. A tabular array of these quantities would also constitute a matrix, the so-called **admittance matrix**, which is of fundamental importance in the theory of electric circuits. Still another example of a matrix is provided by an array of **transition probabilities**: suppose that a system S can exist in any one of n states, say $S_1, S_2, \ldots, S_n$ and that the probability of the system passing from the state S_i to the state S_j by some well-defined random process is p_{ij}. Clearly, the p_{ij}'s can be displayed as an $n \times n$ matrix. Such matrices are of great importance in the theory of probability and its physical applications.

Since, as we pointed out above, both row and column matrices are called vectors, we have, in effect, accepted the following definition.

DEFINITION 2   An n-dimensional vector is an ordered set of n quantities.

This use of the word *vector* requires explanation since at first glance it appears to be somewhat at variance with the familiar usage of elementary physics. In physics, a vector quantity is a quantity, such as a velocity or a force, which possesses both magnitude and direction and hence can be represented by a directed line segment. However, it is clear that such a quantity is uniquely determined by its components in the directions of the three coordinate axes, and conversely. Hence, it can be uniquely associated with an ordered set of three quantities, i.e., with either a row matrix or a column matrix containing three elements. A vector quantity in the physical sense is thus an example of a vector in the matric sense. However, the matric sense includes vectors other than physical vectors. In particular, any set of values of $x_1, x_2, \ldots, x_n$ which satisfies a system of n linear equations in n variables can be thought of as a vector, and in our work we shall often speak of the **solution vectors** of such systems.

By appropriately defining the addition and multiplication of matrices, an algebra of matrices can be developed. As we shall soon see, this is quite different from ordinary algebra, and for this reason, it is convenient to have a term to denote collectively those quantities, such as the variables and constants of our work up to this point, which obey the familiar laws of elementary algebra. These we shall henceforth refer to as **scalars**. For matrices, in contrast to scalars, then, we have the following definitions and rules of operation.

The **sum** or **difference** of two matrices A and B having the same number of rows and the same number of columns is the matrix $A \pm B$ whose elements are the sums or differences of the respective elements of A and B. Obviously, if addition is commutative for the elements of A and B, it is also commutative for A and B themselves, and we have $A + B = B + A$. Similarly, if addition is associative for the elements of the matrices A, B, and C, it is also associative for A, B, and C, and we have $A + (B + C) = (A + B) + C$. Addition and subtraction are not defined for matrices which do not have the same number of rows and the same number of columns.

The **product of a matrix** A **and a scalar** k is the matrix $kA = Ak$ whose elements are the elements of A each† multiplied by k.

The **scalar product of two vectors** having the same number of elements, or **components**, is the sum of the products of corresponding components of the two vectors. The scalar product of two vectors X and Y is also referred to as the **inner product** or **dot product** of X and Y and is often denoted by the symbol $X \cdot Y$. Obviously $X \cdot Y = Y \cdot X$.

The coordinates of a point in three dimensions, say $P{:}(x_1,x_2,x_3)$, form a vector $X = \begin{bmatrix} x_1 & x_2 & x_3 \end{bmatrix}$ in the matric sense, which is completely equivalent to the directed line segment from the origin O to the point P, thought of as a vector in the physical or geometric sense. Now from analytic geometry we know that the square of the length of the segment $\overline{OP}$ is given by the formula

$$(OP)^2 = x_1{}^2 + x_2{}^2 + x_3{}^2$$

But this is simply the scalar product $X \cdot X$ of the vector X with itself. Hence, by analogy, the **length** or **absolute value** of any real vector

$$X = \begin{bmatrix} x_1 & x_2 & \cdots & x_n \end{bmatrix}$$

is defined to be the square root of the scalar product

$$X \cdot X = \sum_{i=1}^{n} x_i{}^2$$

A vector X with the property that $X \cdot X = \sum_{i=1}^{n} x_i{}^2 = 1$ is called a **unit vector**.

From analytic geometry we also know that with every line l there is associated a set of ordered triples

$$(kl_1,kl_2,kl_3) \qquad \begin{matrix} k \neq 0 \\ l_1,\, l_2,\, l_3 \text{ not all zero} \end{matrix}$$

---

† It is important to understand clearly the difference between the product of a matrix A and a scalar k and the product of a determinant $|A|$ and a scalar k. In forming the product kA, as we have just said, *every* element in the matrix A must be multiplied by k. On the other hand, according to Theorem 6, Sec. 10.1, to form the product $k|A|$, we multiply only the elements in some *one* row or some *one* column of the determinant $|A|$ by k.

known as the *direction numbers* of the line. Moreover, if (l_1,l_2,l_3) and (m_1,m_2,m_3) are, respectively, direction numbers of the lines l and m, then l and m are parallel if and only if the sets (l_1,l_2,l_3) and (m_1,m_2,m_3) are proportional. Since the sets (l_1,l_2,l_3) and (m_1,m_2,m_3) can obviously be thought of as vectors

$$L = [l_1 \quad l_2 \quad l_3] \quad \text{and} \quad M = [m_1 \quad m_2 \quad m_3]$$

it is natural to extend these ideas to vectors in general by saying, as a matter of definition, that **two nonzero vectors**

$$X = [x_1 \quad x_2 \quad \cdots \quad x_n] \quad \text{and} \quad Y = [y_1 \quad y_2 \quad \cdots \quad y_n]$$

have the same direction if and only if their components are proportional.

It is also a well-known fact of analytic geometry that if (l_1,l_2,l_3) and (m_1,m_2,m_3) are, respectively, direction numbers of the lines l and m, then l and m are perpendicular if and only if

$$l_1m_1 + l_2m_2 + l_3m_3 = 0$$

But this is simply the condition that the scalar product $L \cdot M$ of the two vectors L and M be equal to zero. Hence, by analogy with this result, we agree that **two nonzero vectors**

$$X = [x_1 \quad x_2 \quad \cdots \quad x_n] \quad \text{and} \quad Y = [y_1 \quad y_2 \quad \cdots \quad y_n]$$

will be called perpendicular or orthogonal if and only if they satisfy the condition

$$X \cdot Y = \sum_{i=1}^{n} x_i y_i = 0$$

This extended concept of orthogonality is of fundamental importance in many applications of matrices.

Two matrices A and B are said to be **conformable in the order** AB if and only if the number of columns in A is equal to the number of rows in B. In other words, if A is an $m \times n$ matrix and if B is a $p \times q$ matrix, A and B are conformable in the order AB if and only if $n = p$. With the idea of conformable matrices introduced, we are now in a position to define the important notion of the **product** of two matrices.

DEFINITION 3 If A is a $p \times q$ matrix and if B is a $q \times r$ matrix, so that A and B are conformable in the order AB, then the product $C = AB$ is the $p \times r$ matrix in which the element c_{ij} in the ith row and jth column is the scalar product of the ith row vector of A and the jth column vector of B; that is, $C = AB$ is the matrix for which $c_{ij} = \sum_{k=1}^{q} a_{ik}b_{kj}$.

Multiplication is not defined for matrices that are not conformable.

EXAMPLE 1

$$\begin{bmatrix} 2 & 3 \\ 1 & -1 \\ 0 & 4 \end{bmatrix} \begin{bmatrix} 5 & -2 & 4 & 7 \\ -6 & 1 & -3 & 0 \end{bmatrix}$$

$$= \begin{bmatrix} 2(5) + 3(-6) & 2(-2) + 3(1) & 2(4) + 3(-3) & 2(7) + 3(0) \\ 1(5) + (-1)(-6) & 1(-2) + (-1)(1) & 1(4) + (-1)(-3) & 1(7) + (-1)(0) \\ 0(5) + 4(-6) & 0(-2) + 4(1) & 0(4) + 4(-3) & 0(7) + 4(0) \end{bmatrix}$$

$$= \begin{bmatrix} -8 & -1 & -1 & 14 \\ 11 & -3 & 7 & 7 \\ -24 & 4 & -12 & 0 \end{bmatrix}$$

From Theorem 12, Sec. 10.1, and Definition 3, it is clear that the way in which we have *defined* the product of two matrices is precisely the way in which we *proved* that the product of two determinants can be formed. Hence we have the following important result.

THEOREM 1   If A and B are square matrices of the same order, the determinant of the product AB is equal to the product of the determinant of A and the determinant of B.

For the multiplication of matrices we have the following important theorems.

THEOREM 2   For suitably conformable matrices, multiplication is associative; that is, $A(BC) = (AB)C$.

Proof   Let A be an $m \times n$ matrix, B an $n \times p$ matrix, and C a $p \times q$ matrix. For convenience, let $BC = D$, $AB = E$, and let

$$A(BC) = AD = F \qquad \text{and} \qquad (AB)C = EC = G$$

Now the element f_{ij} in the ith row and jth column of the matrix $F = AD$ is, by Definition 3,

$$f_{ij} = \sum_{k=1}^{n} a_{ik} d_{kj}$$

Moreover, since $D = BC$, we also have, by Definition 3,

$$d_{kj} = \sum_{l=1}^{p} b_{kl} c_{lj}$$

Hence, substituting,

$$(1) \qquad f_{ij} = \sum_{k=1}^{n} a_{ik} \left(\sum_{l=1}^{p} b_{kl} c_{lj} \right) = \sum_{k=1}^{n} \sum_{l=1}^{p} a_{ik} b_{kl} c_{lj}$$

where the last step follows from the fact that a_{ik} is independent of the index l of the inner summation and hence can be moved across the inner summation sign.

Similarly, the element g_{ij} in the ith row and jth column of the matrix $G = EC$ is

$$g_{ij} = \sum_{l=1}^{p} e_{il} c_{lj}$$

where since $E = AB$, it follows that

$$e_{il} = \sum_{k=1}^{n} a_{ik} b_{kl}$$

Hence, substituting,

$$g_{ij} = \sum_{l=1}^{p} \left(\sum_{k=1}^{n} a_{ik} b_{kl} \right) c_{lj} = \sum_{l=1}^{p} \sum_{k=1}^{n} a_{ik} b_{kl} c_{lj}$$

Finally, interchanging the order of summation in the last double sum, we have

$$(2) \qquad g_{ij} = \sum_{k=1}^{n} \sum_{l=1}^{p} a_{ik} b_{kl} c_{lj}$$

From (1) and (2) it is clear that $f_{ij} = g_{ij}$ for all values of i and j. Hence $F = G$; that is,

$$A(BC) = (AB)C$$

as asserted.

It is interesting and helpful to note that the type symbol of the product of a series of conformable matrices can be obtained by "contracting" the type symbols of the factors by canceling the common interior indices:

$$(m_1,\not{m}_2)(\not{m}_2,\not{m}_3) \cdots (\not{m}_{k-2},\not{m}_{k-1})(\not{m}_{k-1},m_k) \to (m_1,m_k)$$

EXAMPLE 2

$$\begin{bmatrix} 3 & 4 \\ 2 & 1 \end{bmatrix}\left(\begin{bmatrix} 1 & 2 \\ 2 & 5 \end{bmatrix}\begin{bmatrix} 2 & -1 \\ 0 & 3 \end{bmatrix}\right) = \begin{bmatrix} 3 & 4 \\ 2 & 1 \end{bmatrix}\begin{bmatrix} 2 & 5 \\ 4 & 13 \end{bmatrix} = \begin{bmatrix} 22 & 67 \\ 8 & 23 \end{bmatrix}$$

$$\left(\begin{bmatrix} 3 & 4 \\ 2 & 1 \end{bmatrix}\begin{bmatrix} 1 & 2 \\ 2 & 5 \end{bmatrix}\right)\begin{bmatrix} 2 & -1 \\ 0 & 3 \end{bmatrix} = \begin{bmatrix} 11 & 26 \\ 4 & 9 \end{bmatrix}\begin{bmatrix} 2 & -1 \\ 0 & 3 \end{bmatrix} = \begin{bmatrix} 22 & 67 \\ 8 & 23 \end{bmatrix}$$

From the definition of conformable matrices, it is evident that a matrix is conformable to itself if and only if it is square. Hence, a matrix A can be multiplied by itself if and only if it is square. In such a case the product AA is referred to as the **square** of A and is denoted by the symbol A^2. Higher powers of A are defined in similar fashion, Theorem 2 guaranteeing that the definition

$$A^r = \underbrace{AA \cdots A}_{r \text{ factors}}$$

is unambiguous. With A^0 defined as the unit, or identity matrix I, it is obvious that for any nonnegative integers r and s the familiar laws of exponents

$$A^r A^s = A^{r+s} \quad \text{and} \quad (A^r)^s = A^{rs}$$

hold for matric multiplication.

THEOREM 3 For suitably conformable matrices, multiplication is distributive over addition and subtraction; i.e.,

$$A(B \pm C) = AB \pm AC$$

Theorems 2 and 3 are "obvious"; i.e., they assert properties which we know to be true for products in elementary algebra and which, by analogy, we would expect to be true in matric algebra. That these results must be proved and cannot be taken for granted is clear, however, from the next two theorems, which tell us that two other equally simple properties of ordinary algebraic multiplication do not hold for matric multiplication.

THEOREM 4 The product of two nonzero matrices may be a zero matrix; i.e., the fact that $AB = 0$ does not imply either that $A = 0$ or that $B = 0$.

Proof   Clearly, to prove this theorem it is sufficient to exhibit two nonzero matrices whose product is a zero matrix, and we have, among infinitely many possibilities,

$$\begin{bmatrix} 6 & 4 & 2 \\ 9 & 6 & 3 \\ -3 & -2 & -1 \end{bmatrix} \begin{bmatrix} 0 & 1 & -2 \\ -1 & 0 & 3 \\ 2 & -3 & 0 \end{bmatrix} = \begin{bmatrix} 0 & 0 & 0 \\ 0 & 0 & 0 \\ 0 & 0 & 0 \end{bmatrix}$$

THEOREM 5   Even for matrices which are conformable in either order, multiplication is not commutative; i.e., in general $AB \neq BA$.

Proof   To prove this theorem it is sufficient to exhibit two matrices A and B such that $AB \neq BA$, and we have, specifically,

$$\begin{bmatrix} 1 & 2 \\ 3 & 4 \end{bmatrix} \begin{bmatrix} 1 & 1 \\ 4 & 1 \end{bmatrix} = \begin{bmatrix} 9 & 3 \\ 19 & 7 \end{bmatrix} \quad \text{and} \quad \begin{bmatrix} 1 & 1 \\ 4 & 1 \end{bmatrix} \begin{bmatrix} 1 & 2 \\ 3 & 4 \end{bmatrix} = \begin{bmatrix} 4 & 6 \\ 7 & 12 \end{bmatrix}$$

In two special cases, however, the multiplication of matrices is commutative. Though these are simple and obvious, they are of sufficient importance to be stated explicitly.

THEOREM 6   Both unit matrices and zero matrices commute with all suitably conformable matrices; more specifically,

$$AI = IA = A \quad \text{and} \quad A0 = 0A = 0$$

Since matric multiplication is not, in general, commutative, it is desirable to be able to describe concisely the order in which two conformable matrices are to be multiplied. This we shall do by adopting the following terminology. In the product AB we shall say that A **premultiplies** B or B is **premultiplied** by A and B **postmultiplies** A or A is **postmultiplied** by B.

THEOREM 7   The transpose of the product of two conformable matrices is equal to the product of the transposed matrices taken in the other order; that is, $(AB)^T = B^T A^T$.

Proof   Let A be a $p \times q$ matrix and let B be a $q \times r$ matrix. Then from the definition of the transpose of a matrix it follows that the element in the ith row and jth column of $(AB)^T$ is the element in the jth row and ith column of AB, namely,

(3)
$$\sum_{k=1}^{q} a_{jk} b_{ki}$$

On the other hand, the ith row of B^T is, by definition, the ith column of B; that is, the ith row of B^T consists of the elements

$$b_{1i}, b_{2i}, \ldots, b_{qi}$$

Similarly, the jth column of A^T is, by definition, the jth row of A; that is, the jth column of A^T consists of the elements

$$a_{j1}, a_{j2}, \ldots, a_{jq}$$

Hence the element in the ith row and jth column of $B^T A^T$ is

$$b_{1i}a_{j1} + b_{2i}a_{j2} + \cdots + b_{qi}a_{jq}$$

or

$$\sum_{k=1}^{q} b_{ki}a_{jk} = \sum_{k=1}^{q} a_{jk}b_{ki}$$

Since this is the same as the expression (3) for the corresponding element in the matrix $(AB)^T$, the theorem is established.

COROLLARY 1 The transpose of the product of any number of conformable matrices is the product of the transposed matrices taken in the other order; i.e.,

$$(A_1 A_2 \cdots A_n)^T = A_n^T \cdots A_2^T A_1^T$$

The definition of a matrix in no way rules out the possibility that the elements of a matrix are themselves matrices. In fact, it is often convenient to subdivide, or **partition**, a matrix into submatrices and then regard the original matrix as a new matrix having these submatrices as elements. In particular, it is frequently helpful to regard an $m \times n$ matrix $A = [a_{ij}]$ as a row matrix $\begin{bmatrix} C_1 & C_2 & \cdots & C_n \end{bmatrix}$ whose elements are the respective column vectors of A, or as a column matrix

$$\begin{bmatrix} R_1 \\ R_2 \\ \vdots \\ R_m \end{bmatrix}$$

whose elements are the respective row vectors of A.

EXAMPLE 3

For instance, among numerous other possibilities, we can write

$$A = \left[\begin{array}{ccc:c} a_{11} & a_{12} & a_{13} & a_{14} \\ a_{21} & a_{22} & a_{23} & a_{24} \\ \hdashline a_{31} & a_{32} & a_{33} & a_{34} \end{array} \right] = \begin{bmatrix} A_{11} & A_{12} \\ A_{21} & A_{22} \end{bmatrix}$$

where

$$A_{11} = \begin{bmatrix} a_{11} & a_{12} & a_{13} \\ a_{21} & a_{22} & a_{23} \end{bmatrix} \qquad A_{12} = \begin{bmatrix} a_{14} \\ a_{24} \end{bmatrix} \qquad A_{21} = [a_{31} \quad a_{32} \quad a_{33}] \qquad A_{22} = [a_{34}]$$

or equally well

$$A = [A_{11} \quad A_{12} \quad A_{13} \quad A_{14}]$$

where now

$$A_{11} = \begin{bmatrix} a_{11} \\ a_{21} \\ a_{31} \end{bmatrix} \qquad A_{12} = \begin{bmatrix} a_{12} \\ a_{22} \\ a_{32} \end{bmatrix} \qquad A_{13} = \begin{bmatrix} a_{13} \\ a_{23} \\ a_{33} \end{bmatrix} \qquad A_{14} = \begin{bmatrix} a_{14} \\ a_{24} \\ a_{34} \end{bmatrix}$$

In constructing the product of two matrices it is sometimes convenient to partition them before performing the multiplication. This can be done in many ways, but it is of course necessary that the given matrices be conformable and that the various submatrices which must be multiplied together also be conformable. This requirement imposes no restriction on the horizontal partitioning of the first matrix or on the vertical partitioning of the second matrix. It does require, however, that the columns of the first matrix be partitioned into groups such that the number of columns in each group is equal to the number of rows in the corresponding groups into which the rows of the second matrix are partitioned. Matrices for which this is the case are said to be **conformably partitioned**.

EXAMPLE 4

By direct multiplication we have

$$\begin{bmatrix} 1 & 1 & 1 \\ 2 & -1 & 0 \\ -1 & 0 & 2 \end{bmatrix}\begin{bmatrix} 1 & 2 & 3 & -1 \\ 3 & -1 & 1 & 0 \\ 0 & 0 & -2 & 1 \end{bmatrix} = \begin{bmatrix} 4 & 1 & 2 & 0 \\ -1 & 5 & 5 & -2 \\ -1 & -2 & -7 & 3 \end{bmatrix}$$

On the other hand we can write, among various other possibilities,

$$\begin{bmatrix} 1 & 1 & \vdots & 1 \\ 2 & -1 & \vdots & 0 \\ \hdashline -1 & 0 & \vdots & 2 \end{bmatrix} = \begin{bmatrix} A_{11} & A_{12} \\ A_{21} & A_{22} \end{bmatrix} \quad \text{and} \quad \begin{bmatrix} 1 & 2 & 3 & -1 \\ 3 & -1 & 1 & 0 \\ \hdashline 0 & 0 & -2 & 1 \end{bmatrix} = \begin{bmatrix} B_{11} \\ B_{21} \end{bmatrix}$$

and, from this point of view, the product of the two matrices is

$$\begin{bmatrix} A_{11} & A_{12} \\ A_{21} & A_{22} \end{bmatrix}\begin{bmatrix} B_{11} \\ B_{21} \end{bmatrix} = \begin{bmatrix} A_{11}B_{11} + A_{12}B_{21} \\ A_{21}B_{11} + A_{22}B_{21} \end{bmatrix}$$

or, performing the indicated multiplications and additions of the submatrices,

$$\begin{bmatrix} \begin{bmatrix} 4 & 1 & 4 & -1 \\ -1 & 5 & 5 & -2 \end{bmatrix} + \begin{bmatrix} 0 & 0 & -2 & 1 \\ 0 & 0 & 0 & 0 \end{bmatrix} \\ [-1 \quad -2 \quad -3 \quad 1] + [0 \quad 0 \quad -4 \quad 2] \end{bmatrix} = \begin{bmatrix} 4 & 1 & 2 & 0 \\ -1 & 5 & 5 & -2 \\ -1 & -2 & -7 & 3 \end{bmatrix}$$

as before.

Historically, the definition of the product of two matrices was introduced by the English mathematician Arthur Cayley (1821–1895) as a result of his investigations on linear transformations. By a **linear transformation** we mean a relation of the form

$$T_a: \quad \begin{aligned} y_1 &= a_{11}x_1 + a_{12}x_2 + \cdots + a_{1n}x_n \\ y_2 &= a_{21}x_1 + a_{22}x_2 + \cdots + a_{2n}x_n \\ &\cdots\cdots\cdots\cdots\cdots\cdots\cdots\cdots\cdots\cdots\cdots \\ y_n &= a_{n1}x_1 + a_{n2}x_2 + \cdots + a_{nn}x_n \end{aligned}$$

connecting the variables $(x_1,x_2,\ldots,x_n)$ and the variables $(y_1,y_2,\ldots,y_n)$. If $n = 2$, we can think of T_a as a transformation of the cartesian plane which sends a point with coordinates (x_1,x_2) into a point with coordinates (y_1,y_2). Similarly, if $n = 3$, we can think of T_a as a transformation in three dimensions which sends a point with coordinates (x_1,x_2,x_3) into a point with coordinates (y_1,y_2,y_3). If $n > 3$, we can regard T_a as a transformation in a hyperspace of the appropriate number of dimensions, or we may think of it simply as a transformation of an n-component vector X into an n-component vector Y. From the definition of matric multiplication and the equality of two matrices, it is clear that if we introduce the matrices

$$Y = \begin{bmatrix} y_1 \\ y_2 \\ \vdots \\ y_n \end{bmatrix} \qquad A = \begin{bmatrix} a_{11} & a_{12} & \cdots & a_{1n} \\ a_{21} & a_{22} & \cdots & a_{2n} \\ \cdots\cdots\cdots\cdots\cdots \\ a_{n1} & a_{n2} & \cdots & a_{nn} \end{bmatrix} \qquad X = \begin{bmatrix} x_1 \\ x_2 \\ \vdots \\ x_n \end{bmatrix}$$

then the transformation T_a can be written in the form

(4) $$T_a: \quad Y = AX$$

The matrix A in Eq. (4) is usually referred to as the **matrix of the transformation T_a**. It is thus apparent that matrices are intimately related to linear transformations and systems of linear equations.

Suppose, now, that in addition to the transformation T_a which transforms a vector X into a vector Y, we have a second transformation

$$(5) \qquad T_b: \quad Z = BY \qquad Z = \begin{bmatrix} z_1 \\ z_2 \\ \vdots \\ z_n \end{bmatrix} \qquad B = \begin{bmatrix} b_{11} & b_{12} & \cdots & b_{1n} \\ b_{21} & b_{22} & \cdots & b_{2n} \\ \hdotsfor{4} \\ b_{n1} & b_{n2} & \cdots & b_{nn} \end{bmatrix}$$

which transforms a vector Y into a vector Z. If T_a is applied to a vector X and then T_b is applied to the resulting vector Y, the net result is to transform the vector X into the vector Z, and it is a matter of some interest to find the equations of the equivalent transformation connecting Z directly with X. This can easily be done, of course, simply by eliminating the variables $(y_1, y_2, \ldots, y_n)$ between the equations of T_a and T_b. To do this, we observe that the equations of T_a and T_b can be written, respectively,

$$T_a: \quad y_k = \sum_{j=1}^{n} a_{kj} x_j \qquad k = 1, 2, \ldots, n$$

$$T_b: \quad z_i = \sum_{k=1}^{n} b_{ik} y_k \qquad i = 1, 2, \ldots, n$$

Hence, eliminating the y's by substituting for y_k in the equations of T_b, we have

$$z_i = \sum_{k=1}^{n} b_{ik} \left(\sum_{j=1}^{n} a_{kj} x_j \right) = \sum_{j=1}^{n} \left(\sum_{k=1}^{n} b_{ik} a_{kj} \right) x_j \qquad i = 1, 2, \ldots, n$$

Thus the coefficient of x_j in the equation for z_i is

$$\sum_{k=1}^{n} b_{ik} a_{kj}$$

which is precisely the element c_{ij} in the product BA. In other words, the matric form of the single transformation T_{ba} equivalent to following T_a by T_b can be found simply by eliminating Y between Eq. (4) and Eq. (5):

$$Z = BY = B(AX) = (BA)X$$

Thus we have established the following important result.

THEOREM 8 The result of following a linear transformation $T_a: \ Y = AX$ with the linear transformation $T_b: \ Z = BY$ is the single linear transformation† $T_{ba}: \ Z = BAX$, whose matrix is the product BA of the matrices of T_b and T_a.

As a further illustration of the importance of matric multiplication, we return to the idea of transition probabilities which we mentioned earlier in this section. Suppose that a system S can exist in any of n states $S_1, S_2, \ldots, S_n$ and that by some random process the system may pass directly from the ith state to the jth state with probability $p_{ij}^{(1)}$ $(i, j = 1, 2, \ldots, n)$. Naturally, the system may also pass from the ith state to the jth state by first passing to some intermediate state, say the kth, and then passing from the kth state to the jth; and the calculation of these two-step transition

---

† The transformation which results when one transformation is followed by another is usually referred to as the **composition** of the two transformations.

probabilities is a matter of some importance. Now the probability that the system passes from the ith state to the jth state via the kth state is the product of the probability $p_{ik}^{(1)}$ that it passes in one step from S_i to S_k and the probability $p_{kj}^{(1)}$ that it subsequently passes in one step from S_k to S_j. Furthermore, since in any two-step transition from S_i to S_j, the system must pass through *some* intermediate state (including, of course, S_i and S_j themselves) the probability that the system passes in exactly two steps from S_i to S_j can be found by adding the products $p_{ik}^{(1)}p_{kj}^{(1)}$ for all possible intermediate states, i.e., for all possible values of k. Thus

$$p_{ij}^{(2)} = \sum_{k=1}^{n} p_{ik}^{(1)} p_{kj}^{(1)}$$

But the sum on the right in the last formula is precisely the element in the ith row and jth column of the square of the matrix of one-step transition probabilities, $P = [p_{ij}^{(1)}]$. In other words, *the matrix of two-step transition probabilities for any system S is the square of the matrix of one-step transition probabilities for S.* A similar argument shows that the matrix of three-step transition probabilities for S is the cube of the matrix of one-step transition probabilities, and so on.

EXAMPLE 5

Let S be the system consisting of two players A and B who begin with \$2 apiece and match coins until one or the other of them has no more money. If the states of the system are defined by the number of dollars in A's possession, specifically if the system is in the state S_{i+1} whenever A has i dollars ($i = 0, 1, 2, 3, 4$), find the matrix of one-step transition probabilities. Then, by raising this matrix to the second, third, and fourth powers, find the matrices containing the two-, three-, and four-step transition probabilities for S. What is the probability that A will be ruined in at most four turns? What is the probability that A will be ruined in exactly four turns?

Clearly, unless A or B is bankrupt, A must either win a dollar or lose a dollar on each turn, and the probability of each of these events is $\frac{1}{2}$. Hence, if A has i dollars ($i = 1, 2, 3$), that is, if the system is in the state $S_{i+1}(i = 1, 2, 3)$, the probability of a one-step transition to S_i is $\frac{1}{2}$, the probability of a one-step transition to S_{i+2} is $\frac{1}{2}$, and the probability of a one-step transition to any other state is 0. On the other hand, if the system is in the state S_1, that is, if A is bankrupt, the system remains in that state; so the probability of a one-step transition from S_1 to S_1 is 1, and the probability of any other transition from S_1 is 0. Similarly, if the system is in the state S_5, that is, if B is bankrupt, the system remains in that state; hence the probability of a one-step transition from S_5 to S_5 is 1, and the probability of any other transition is 0. Thus the matrix of one-step transition probabilities is

$$P = \begin{bmatrix} 1 & 0 & 0 & 0 & 0 \\ \frac{1}{2} & 0 & \frac{1}{2} & 0 & 0 \\ 0 & \frac{1}{2} & 0 & \frac{1}{2} & 0 \\ 0 & 0 & \frac{1}{2} & 0 & \frac{1}{2} \\ 0 & 0 & 0 & 0 & 1 \end{bmatrix}$$

By multiplying this matrix by itself we find at once that the matrix of two-step transition probabilities is

$$P^2 = \begin{bmatrix} 1 & 0 & 0 & 0 & 0 \\ \frac{1}{2} & \frac{1}{4} & 0 & \frac{1}{4} & 0 \\ \frac{1}{4} & 0 & \frac{1}{2} & 0 & \frac{1}{4} \\ 0 & \frac{1}{4} & 0 & \frac{1}{4} & \frac{1}{2} \\ 0 & 0 & 0 & 0 & 1 \end{bmatrix}$$

Similarly, by computing P^3 and P^4, we find the matrices of three-step and four-step transition probabilities, respectively, to be

$$P^3 = \begin{bmatrix} 1 & 0 & 0 & 0 & 0 \\ \frac{5}{8} & 0 & \frac{1}{4} & 0 & \frac{1}{8} \\ \frac{1}{4} & \frac{1}{4} & 0 & \frac{1}{4} & \frac{1}{4} \\ \frac{1}{8} & 0 & \frac{1}{4} & 0 & \frac{5}{8} \\ 0 & 0 & 0 & 0 & 1 \end{bmatrix} \quad \text{and} \quad P^4 = \begin{bmatrix} 1 & 0 & 0 & 0 & 0 \\ \frac{5}{8} & \frac{1}{8} & 0 & \frac{1}{8} & \frac{1}{8} \\ \frac{3}{8} & 0 & \frac{1}{4} & 0 & \frac{3}{8} \\ \frac{1}{8} & \frac{1}{8} & 0 & \frac{1}{8} & \frac{5}{8} \\ 0 & 0 & 0 & 0 & 1 \end{bmatrix}$$

The probability that A is ruined in at most four turns is simply the probability of a four-step transition from S_3 to S_1, namely, $p_{31}^{(4)} = \frac{3}{8}$, since among such transitions are included those in which the system reaches S_1 in less than four steps and then remains there. The probability that A is ruined in four turns and not before is the probability that S reaches S_1 in four steps but does not reach it in three steps or less, namely, $p_{31}^{(4)} - p_{31}^{(3)} = \frac{3}{8} - \frac{1}{4} = \frac{1}{8}$.

EXERCISES

1 If $A = \begin{bmatrix} 1 & 2 \\ 3 & 4 \end{bmatrix}$, $B = \begin{bmatrix} 1 & -1 \\ 2 & 3 \end{bmatrix}$, $C = \begin{bmatrix} 1 & 4 \\ -1 & 3 \end{bmatrix}$, verify that:

(a) $A(BC) = (AB)C$ (b) $A(B + C) = AB + AC$

(c) $B(A - C) = BA - BC$ (d) $(AB)^T = B^T A^T$

2 Multiply the matrices

$$\begin{bmatrix} 1 & 2 & | & -1 \\ \text{---} & \text{---} & | & \text{---} \\ 3 & 0 & | & 2 \end{bmatrix} \quad \text{and} \quad \begin{bmatrix} 3 & | & 1 \\ 1 & | & 3 \\ \text{---} & | & \text{---} \\ 2 & | & 0 \end{bmatrix}$$

using the indicated partitioning. Check by multiplying without regard to the partitioning.

3 Evaluate the matric polynomial $X^3 - 4X^2 - X + 4I$ for each of the following matrices:

(a) $\begin{bmatrix} 1 & -1 \\ 2 & 0 \end{bmatrix}$ (b) $\begin{bmatrix} 1 & 1 & 2 \\ 1 & 2 & 1 \\ 2 & 1 & 1 \end{bmatrix}$ (c) $\begin{bmatrix} 0 & 1 & 1 \\ -1 & 0 & 1 \\ -1 & -1 & 0 \end{bmatrix}$

4 Verify that $(X - 3I)(X - 2I) = (X - 2I)(X - 3I) = X^2 - 5X + 6I$ for

$$X = \begin{bmatrix} 1 & 2 \\ 2 & -1 \end{bmatrix} \quad \text{and for} \quad X = \begin{bmatrix} 1 & 2 & 0 \\ 0 & 3 & 0 \\ 0 & 0 & 4 \end{bmatrix}$$

5 If $A = \begin{bmatrix} 1 & 2 \\ -1 & 3 \end{bmatrix}$, verify that $A^2 - 4A + 5I = 0$ and use this fact to compute A^3, A^4, and A^5 by first expressing them in terms of the matrices A and I. Check these results by calculating A^3, A^4, and A^5 directly from A.

6 If $A = \begin{bmatrix} 1 & 2 \\ 2 & 4 \end{bmatrix}$, find a nonzero 2×2 matrix X such that AX is a zero matrix. Is $XA = 0$? If $A = \begin{bmatrix} 1 & 2 \\ 2 & 3 \end{bmatrix}$, is it possible to find a nonzero 2×2 matrix X such that AX is a zero matrix?

7 If $A = \begin{bmatrix} 1 & 2 \\ 3 & 4 \end{bmatrix}$, $B = \begin{bmatrix} 2 & 1 \\ 1 & -1 \end{bmatrix}$, $C = \begin{bmatrix} 1 & 2 \\ 2 & 4 \end{bmatrix}$, and $X = \begin{bmatrix} x_1 & x_2 \\ x_3 & x_4 \end{bmatrix}$, solve each of the following equations:

(a) $AX = B$ (b) $BX = A$ (c) $AX = C$ (d) $BX = C$

Is it possible to solve the equation $CX = A$?

8 Show that $\begin{bmatrix} \cos\theta & \sin\theta \\ -\sin\theta & \cos\theta \end{bmatrix}^n = \begin{bmatrix} \cos n\theta & \sin n\theta \\ -\sin n\theta & \cos n\theta \end{bmatrix}$.

9 Show that $\begin{bmatrix} \cosh\theta & \sinh\theta \\ \sinh\theta & \cosh\theta \end{bmatrix}^n = \begin{bmatrix} \cosh n\theta & \sinh n\theta \\ \sinh n\theta & \cosh n\theta \end{bmatrix}$.

10 Under what conditions, if any, is $(AB)^2 = A^2 B^2$?

11 If A and B are symmetric matrices of the same order, prove that the product AB is symmetric if and only if $AB = BA$.

12 If A and B are square matrices which are not symmetric, is it possible for their product to be symmetric?

13 If A and B are square matrices which commute, i.e., square matrices such that $AB = BA$, prove that A^2 and B^2 also commute. Is this true for general positive integral powers of A and B?

14 Prove that $(A + B)^T = A^T + B^T$.

15 Prove Corollary 1, Theorem 7.

16 Prove Theorem 3.

17 Show that if a matrix A commutes with a diagonal matrix D whose diagonal elements are all distinct, then A is a diagonal matrix. Is A necessarily diagonal if the diagonal elements of D are not all distinct?

18 If K is a diagonal matrix whose diagonal elements are all equal to k, prove that the product of K and any conformable matrix A is equal to the product of A and the scalar k. (Because of this property, the matrix K is often referred to as a **scalar matrix**.)

19 By definition, the transpose of the matrix $A = [a_{ij}]$ is the matrix $A^T = [a_{ji}]$. Is this formula correct if the a_{ij}'s are submatrices of A?

20 If A and B are conformable matrices, show that the ith row vector in the product AB is $A_i B$, where A_i is the ith row vector of A. What is the jth column vector in the product AB?

21 By the **derivative of a matrix** A we mean the matrix whose elements are the derivatives of the elements of A. Assuming that the elements of each matrix are differentiable functions of x, use this definition to show that

$$\frac{d(AB)}{dx} = \frac{dA}{dx} B + A \frac{dB}{dx}$$

What is $d(ABC)/dx$? Is $d(A^2)/dx = 2A \, dA/dx$?

22 If $A = \begin{bmatrix} e^{\lambda_1 t} & & & 0 \\ & e^{\lambda_2 t} & & \\ & & \ddots & \\ 0 & & & e^{\lambda_n t} \end{bmatrix}$, show that $\dfrac{dA}{dt} = DA$, where D is the diagonal

matrix $\begin{bmatrix} \lambda_1 & & & 0 \\ & \lambda_2 & & \\ & & \ddots & \\ 0 & & & \lambda_n \end{bmatrix}$. What is $\dfrac{d^2 A}{dt^2}$?

23 If $A = \begin{bmatrix} \sin\lambda_1 t & & & 0 \\ & \sin\lambda_2 t & & \\ & & \ddots & \\ 0 & & & \sin\lambda_n t \end{bmatrix}$, what is $\dfrac{dA}{dt}$? What is $\dfrac{d^2 A}{dt^2}$?

24 If $A = \begin{bmatrix} \cos\lambda_1 t & & & 0 \\ & \cos\lambda_2 t & & \\ & & \ddots & \\ 0 & & & \cos\lambda_n t \end{bmatrix}$, what is $\dfrac{dA}{dt}$? What is $\dfrac{d^2 A}{dt^2}$?

25 Show that in any matrix of transition probabilities, the sum of the elements in any row is 1.

26   Consider the system S consisting of four boxes B_1, B_2, B_3, and B_4 and a single ball, and let the system be in the state S_i ($i = 1, 2, 3, 4$) if the ball is in the box B_i. Transitions from one state to another take place in the following manner. A die is thrown, and if a 1, 2, or 3 turns up, the ball is taken from whichever box it is in and placed in the box bearing the number showing on the die. If 4, 5, or 6 turns up, the ball is taken from wherever it is and placed in box B_4. Find the matrix of one-, two-, and three-step transition probabilities for the system.

27   Consider the system S consisting of three boxes B_1, B_2, and B_3 and a single ball, and let the system be in the state S_i ($i = 1, 2, 3$) if the ball is in the box B_i. Transitions from one state to another take place in the following manner. Three coins are tossed. If no heads turn up, the ball is not moved; if one or more heads turn up, the ball is taken from its box and placed in the box corresponding to the number of heads showing. Find the matrix of one-, two-, three-, and four-step transition probabilities for the system.

28   Show that in computing the product of a $p \times q$ and a $q \times r$ matrix, pqr multiplications and $pr(q - 1)$ additions must be performed. If a $p \times q$ matrix and a $q \times r$ matrix are conformably partitioned into four submatrices by one partition of their rows and one partition of their columns, prove that the same number of multiplications and the same number of additions are required in multiplying the two matrices whether this is done in the original or in the partitioned form.

10.3 Adjoints and Inverses

It is a familiar fact of elementary algebra that any quantity Q which is not equal to zero has a reciprocal

$$Q^{-1} = \frac{1}{Q}$$

with the property that

$$QQ^{-1} = Q^{-1}Q = 1$$

The familiar process of division, which we sometimes inaccurately regard as being essentially independent of multiplication, is nothing but multiplication involving the reciprocal, or multiplicative inverse, of the divisor as one factor. In matric algebra, although we do not define division as such, we can in an important class of cases define the reciprocal, or inverse, of a matrix. With inverses defined, multiplication then serves to accomplish all that we might properly expect to do by division. As usual, our development begins with a number of definitions.

We have already defined the determinant of a square matrix as the determinant whose array of elements is identical with the array of the matrix itself. Clearly, only square matrices have determinants. A square matrix whose determinant is different from zero is said to be **nonsingular**. A square matrix whose determinant is equal to zero is said to be **singular**. Using these notions, we can now give formal definitions of the important concepts of the **adjoint** and the **inverse** of a matrix.

DEFINITION 1   If $A = [a_{ij}]$ is a square matrix, and if A_{ij} is the cofactor of a_{ij} in the determinant of A, then the matrix

$$[A_{ji}] = [A_{ij}]^T = \text{transpose of } [A_{ij}]$$

is called the adjoint of the matrix A.

The adjoint of a square matrix A is sometimes indicated by the notation adj A.

DEFINITION 2   The reciprocal, or inverse, A^{-1} of a nonsingular matrix $A = [a_{ij}]$ is the adjoint of A divided by the determinant of A; that is,

$$A^{-1} = \frac{[A_{ij}]^T}{|A|} = \frac{[A_{ji}]}{|A|}$$

Clearly, although every square matrix has an adjoint, only nonsingular matrices have inverses. The fundamental importance of the inverse of a matrix is apparent from the following theorem.

THEOREM 1   The product of a nonsingular matrix A and its inverse, in either order, is a unit matrix; that is, $A^{-1}A = AA^{-1} = I$.

Proof   Let $A = [a_{ij}]$ be a nonsingular matrix, and consider first the product

$$A^{-1}A = \frac{1}{|A|} \begin{bmatrix} A_{11} & A_{21} & \cdots & A_{n1} \\ A_{12} & A_{22} & \cdots & A_{n2} \\ \cdots & \cdots & \cdots & \cdots \\ A_{1n} & A_{2n} & \cdots & A_{nn} \end{bmatrix} \begin{bmatrix} a_{11} & a_{12} & \cdots & a_{1n} \\ a_{21} & a_{22} & \cdots & a_{2n} \\ \cdots & \cdots & \cdots & \cdots \\ a_{n1} & a_{n2} & \cdots & a_{nn} \end{bmatrix}$$

Clearly, from the definition of matric multiplication, the element in the ith row and jth column of the product of the two matrices on the right is the scalar product

$$\sum_{k=1}^{n} A_{ki}a_{kj}$$

Moreover, from Corollary 1, Theorem 9, Sec. 10.1, this sum is equal to $|A|$ if $i = j$ and is equal to zero if $i \neq j$. Hence

$$A^{-1}A = \frac{1}{|A|} \begin{bmatrix} |A| & 0 & \cdots & 0 \\ 0 & |A| & \cdots & 0 \\ \cdots & \cdots & \cdots & \cdots \\ 0 & 0 & \cdots & |A| \end{bmatrix} = \begin{bmatrix} 1 & 0 & \cdots & 0 \\ 0 & 1 & \cdots & 0 \\ \cdots & \cdots & \cdots & \cdots \\ 0 & 0 & \cdots & 1 \end{bmatrix} = I$$

as asserted. The proof that $AA^{-1} = I$ follows in exactly the same fashion.

COROLLARY 1   For any square matrix, singular or nonsingular,

$$(\text{adj } A)A = A(\text{adj } A) = |A|I$$

COROLLARY 2   If A is a nonsingular matrix, then A^{-1} is also nonsingular and $|A^{-1}| = 1/|A|$.

EXAMPLE 1

If

$$A = \begin{bmatrix} 1 & 2 & 4 \\ -1 & 0 & 3 \\ 3 & 1 & -2 \end{bmatrix}$$

then the determinant of A is

$$\begin{vmatrix} 1 & 2 & 4 \\ -1 & 0 & 3 \\ 3 & 1 & -2 \end{vmatrix} = 7$$

The adjoint of A is the transpose of the cofactor matrix

$$
\begin{bmatrix}
\begin{vmatrix} 0 & 3 \\ 1 & -2 \end{vmatrix} & -\begin{vmatrix} -1 & 3 \\ 3 & -2 \end{vmatrix} & \begin{vmatrix} -1 & 0 \\ 3 & 1 \end{vmatrix} \\[6pt]
-\begin{vmatrix} 2 & 4 \\ 1 & -2 \end{vmatrix} & \begin{vmatrix} 1 & 4 \\ 3 & -2 \end{vmatrix} & -\begin{vmatrix} 1 & 2 \\ 3 & 1 \end{vmatrix} \\[6pt]
\begin{vmatrix} 2 & 4 \\ 0 & 3 \end{vmatrix} & -\begin{vmatrix} 1 & 4 \\ -1 & 3 \end{vmatrix} & \begin{vmatrix} 1 & 2 \\ -1 & 0 \end{vmatrix}
\end{bmatrix}
\quad \text{that is} \quad
\begin{bmatrix}
-3 & 8 & 6 \\
7 & -14 & -7 \\
-1 & 5 & 2
\end{bmatrix}
$$

The inverse of A is therefore

$$
A^{-1} = \frac{1}{7}\begin{bmatrix} -3 & 8 & 6 \\ 7 & -14 & -7 \\ -1 & 5 & 2 \end{bmatrix}
$$

and

$$
A^{-1}A = \frac{1}{7}\begin{bmatrix} -3 & 8 & 6 \\ 7 & -14 & -7 \\ -1 & 5 & 2 \end{bmatrix}\begin{bmatrix} 1 & 2 & 4 \\ -1 & 0 & 3 \\ 3 & 1 & -2 \end{bmatrix} = \frac{1}{7}\begin{bmatrix} 7 & 0 & 0 \\ 0 & 7 & 0 \\ 0 & 0 & 7 \end{bmatrix} = \begin{bmatrix} 1 & 0 & 0 \\ 0 & 1 & 0 \\ 0 & 0 & 1 \end{bmatrix}
$$

From Theorem 1 it is clear that if A is a nonsingular matrix, each of the equations $AX = I$ and $XA = I$ has $X = A^{-1}$ as *one* solution. Actually, we have the following stronger result.

THEOREM 2　If A is a nonsingular matrix, then $X = A^{-1}$ is the unique solution of each of the equations $AX = I$ and $XA = I$.

Proof　Consider first the equation $AX = I$. If both X_1 and X_2 satisfy this equation, then $AX_1 = AX_2$, since each is equal to I. Moreover, since A is nonsingular, it follows that A^{-1} exists. Hence, premultiplying by A^{-1}, we have

$$
A^{-1}AX_1 = A^{-1}AX_2
$$
$$
IX_1 = IX_2
$$
$$
X_1 = X_2
$$

Thus the equation $AX = I$ has in fact just one solution, and from Theorem 1 it follows that this solution is $X = A^{-1}$. A similar argument shows that $X = A^{-1}$ is also the unique solution of the equation $XA = I$, as asserted.

COROLLARY 1　If A is a nonsingular $n \times n$ matrix, if B is an $n \times m$ matrix, and if C is an $m \times n$ matrix, the equation $AX = B$ has the unique solution $X = A^{-1}B$ and the equation $XA = C$ has the unique solution $X = CA^{-1}$.

Various other important theorems follow easily now that the uniqueness of the solution of $AX = I$ for any nonsingular matrix A has been established. In particular, we have the following results.

THEOREM 3　If A is a nonsingular matrix, then $(A^{-1})^{-1} = A$.

Proof　By Theorem 2, $(A^{-1})^{-1}$ is the unique solution of the equation $A^{-1}X = I$. However, it is obvious by inspection that $X = A$ satisfies this equation. Hence $(A^{-1})^{-1} = A$, as asserted.

THEOREM 4   If A and B are nonsingular $n \times n$ matrices, then $(AB)^{-1} = B^{-1}A^{-1}$.

Proof   By Theorem 2, $(AB)^{-1}$ is the unique solution of the equation $(AB)X = I$. However, it is clear that $X = B^{-1}A^{-1}$ satisfies this equation, since

$$(AB)(B^{-1}A^{-1}) = A(BB^{-1})A^{-1} = AIA^{-1} = AA^{-1} = I$$

Hence $(AB)^{-1} = B^{-1}A^{-1}$, as asserted.

COROLLARY 1   If $A, B, \ldots, K$ are nonsingular $n \times n$ matrices, then

$$(AB \cdots K)^{-1} = K^{-1} \cdots B^{-1}A^{-1}$$

With the inverse of a nonsingular matrix defined, it is now possible to define negative integral powers of any nonsingular matrix.

DEFINITION 3   If A is a nonsingular matrix, and if r is a positive integer, then $A^{-r} = (A^{-1})^r$.

Negative powers of singular matrices are not defined. For nonsingular matrices it is now possible to extend the familiar laws of exponents to negative as well as nonnegative integral powers.

THEOREM 5   If A is a nonsingular matrix, then for all integral values of r and s, $A^r A^s = A^{r+s}$ and $(A^r)^s = A^{rs}$.

If in the corollary of Theorem 2 we take B to be the column matrix

$$\begin{bmatrix} b_1 \\ b_2 \\ \vdots \\ b_n \end{bmatrix}$$

the matric equation $AX = B$ is equivalent to the system of nonhomogeneous linear equations

$$a_{11}x_1 + a_{12}x_2 + \cdots + a_{1n}x_n = b_1$$
$$a_{21}x_1 + a_{22}x_2 + \cdots + a_{2n}x_n = b_2$$
$$\cdots\cdots\cdots\cdots\cdots\cdots\cdots\cdots\cdots$$
$$a_{n1}x_1 + a_{n2}x_2 + \cdots + a_{nn}x_n = b_n$$

Hence, it is clear from Corollary 1 of Theorem 2 that the solution of this system exists and is uniquely given by

$$X = A^{-1}B$$

if A is nonsingular.

Similarly, if we take B to be the matrix

$$\begin{bmatrix} y_1 \\ y_2 \\ \vdots \\ y_n \end{bmatrix}$$

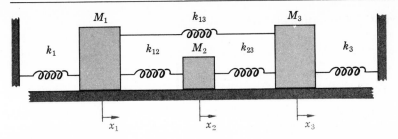

Figure 10.1
A typical system of spring-connected masses.

the matric equation $AX = B$ becomes the linear transformation

$$Y = AX$$

The corollary of Theorem 2 now assures us that if A is nonsingular, the inverse of this transformation, i.e., the transformation that carries us back from the vector Y to the vector X, is unique and has the equation

$$X = A^{-1}Y$$

As a physical illustration of the relation of a matrix and its inverse, let us consider the mass-spring system shown in Fig. 10.1 and determine the forces which act on each of the masses as a result of arbitrary displacements x_1, x_2, and x_3 of the respective masses. The modulus of each spring is the indicated value of k; that is, the force required to stretch each spring a unit distance is the corresponding value of k. If the masses are displaced by the respective amounts x_1, x_2, and x_3, the increases in the lengths of the various springs are

$$\begin{aligned} k_1&: \quad x_1 \\ k_{12}&: \quad x_2 - x_1 \\ k_{13}&: \quad x_3 - x_1 \\ k_{23}&: \quad x_3 - x_2 \\ k_3&: \qquad - x_3 \end{aligned}$$

and the forces represented by these changes in length are

$$\begin{aligned} k_1&: \quad k_1 x_1 \\ k_{12}&: \quad k_{12}(x_2 - x_1) \\ k_{13}&: \quad k_{13}(x_3 - x_1) \\ k_{23}&: \quad k_{23}(x_3 - x_2) \\ k_3&: \quad k_3(-x_3) \end{aligned}$$

a positive force indicating that the spring is stretched and a negative force indicating that the spring is compressed. Hence, taking due account of the direction of the force applied to each mass by each spring attached to it, we find that the forces $f_1, f_2,$ and f_3 which act on the respective masses are

$$\begin{aligned} f_1 &= -k_1 x_1 + k_{12}(x_2 - x_1) + k_{13}(x_3 - x_1) \\ f_2 &= -k_{12}(x_2 - x_1) + k_{23}(x_3 - x_2) \\ f_3 &= -k_{13}(x_3 - x_1) - k_{23}(x_3 - x_2) - k_3 x_3 \end{aligned}$$

(1)

or, collecting terms and rewriting in matric notation,

(2)
$$F = KX$$

where
$$F = \begin{bmatrix} f_1 \\ f_2 \\ f_3 \end{bmatrix} \qquad X = \begin{bmatrix} x_1 \\ x_2 \\ x_3 \end{bmatrix}$$

and
$$K = \begin{bmatrix} -(k_1 + k_{12} + k_{13}) & k_{12} & k_{13} \\ k_{12} & -(k_{12} + k_{23}) & k_{23} \\ k_{13} & k_{23} & -(k_{13} + k_{23} + k_3) \end{bmatrix}$$

Evaluating the first of Eqs. (1) for $x_1 = 1$ and $x_2 = x_3 = 0$, it is clear that $-(k_1 + k_{12} + k_{13})$ is the force applied to the first mass as a result of a unit displacement of that mass. A similar evaluation shows that, in general, the element in the ith row and jth column of K is the force applied to the ith mass as a result of a unit displacement of the jth mass. Because of this property the matrix K is usually referred to as the **stiffness matrix** of the system.

It can easily be verified that for all positive values of the k's the matrix K is nonsingular. Hence, for the physical system shown in Fig. 10.1, K^{-1} exists, and we can solve Eq. (2) for X, getting

$$X = K^{-1}F$$

Now, evaluating the right-hand side of this equation for a force vector F whose jth component is 1 and whose other components are all zero, it follows that the element of K^{-1} in the ith row and jth column is the displacement produced in the ith mass as a result of a unit force applied to the jth mass. Because of this property, the matrix K^{-1} is usually referred to as the **elasticity matrix**† of the system. Our discussion has

Table 10.1

A
$\bar{A}$ = conjugate of A
A^T = transpose of A
$\bar{A}^T$ = associate of A
A^{-1} = inverse or reciprocal of A (A nonsingular)

Condition on A	Type
$A = \bar{A}$	Real
$A = -\bar{A}$	Imaginary
$A = A^T$	Symmetric
$A = -A^T$	Skew-symmetric
$A = \bar{A}^T$	Hermitian
$A = -\bar{A}^T$	Skew-hermitian
$A = (A^T)^{-1}$; that is, $A^{-1} = A^T$ or $AA^T = I$	Orthogonal
$A = (\bar{A}^T)^{-1}$; that is, $A^{-1} = \bar{A}^T$ or $A\bar{A}^T = I$	Unitary

---

† The symmetry of the stiffness and elasticity matrices, which asserts, for instance, that the force acting on the ith mass as a result of a unit displacement of the jth mass is equal to the force acting on the jth mass as a result of a unit displacement of the ith mass, is an illustration of the famous reciprocity theorem of Maxwell, Rayleigh, and Betti. This theorem is the counterpart in mechanics of what is known simply as the *reciprocity theorem* in electric-circuit analysis. It is interesting in this connection to recall the symmetry properties of the Green's function and its interpretation as an influence function, discussed in Sec. 2.7.

thus illustrated the important fact that for any elastic system, the elasticity matrix is the inverse of the stiffness matrix and vice versa.

In the last section we defined a number of special matrices, and now, with the concept of the inverse of a matrix available, our list can be extended to include several additional important types. In Table 10.1 we bring together the types we have already defined, as well as the new ones we are here introducing.

Although we cannot go into details, it is worth noting that orthogonal matrices derive their name from the fact that the matrix of a transformation which is a rotation of mutually perpendicular, or orthogonal, axes in two or three (or more) dimensions is always orthogonal. For instance, it is well known that in the cartesian plane the equations of a general rotation of axes are

$$x_1' = x_1 \cos \alpha + x_2 \sin \alpha$$
$$x_2' = -x_1 \sin \alpha + x_2 \cos \alpha$$
$$\text{or} \quad X' = AX$$

where
$$X' = \begin{bmatrix} x_1' \\ x_2' \end{bmatrix} \quad A = \begin{bmatrix} \cos \alpha & \sin \alpha \\ -\sin \alpha & \cos \alpha \end{bmatrix} \quad X = \begin{bmatrix} x_1 \\ x_2 \end{bmatrix}$$

and it is easy to show that A is orthogonal by verifying that $AA^T = I$.

EXERCISES

1 Find the adjoint of each of the following matrices; when it exists, find the inverse:

(a) $\begin{bmatrix} 1 & 2 \\ 3 & 4 \end{bmatrix}$ (b) $\begin{bmatrix} 2 & -1 & 3 \\ 4 & 0 & -1 \\ 3 & 3 & 2 \end{bmatrix}$ (c) $\begin{bmatrix} 1 & 1 & 1 \\ 1 & 2 & 3 \\ 3 & 2 & 1 \end{bmatrix}$ (d) $\begin{bmatrix} 2 & 3 & 1 \\ 1 & -1 & 2 \\ 1 & 9 & -4 \end{bmatrix}$

2 For each of the following pairs of matrices, verify that $(AB)^{-1} = B^{-1}A^{-1}$:

(a) $A = \begin{bmatrix} 1 & 1 \\ 1 & 2 \end{bmatrix}$ $B = \begin{bmatrix} 1 & 2 \\ 2 & 5 \end{bmatrix}$

(b) $A = \begin{bmatrix} 2 & 0 \\ -1 & 1 \end{bmatrix}$ $B = \begin{bmatrix} 0 & -1 \\ 2 & 3 \end{bmatrix}$

(c) $A = \begin{bmatrix} 1 & 0 & -1 \\ 0 & 2 & 0 \\ 1 & 1 & 3 \end{bmatrix}$ $B = \begin{bmatrix} 1 & 1 & 1 \\ 1 & 2 & 1 \\ 1 & 3 & 2 \end{bmatrix}$

3 Verify the relations $|A^{-1}| = 1/|A|$ and $(A^{-1})^{-1} = A$ for each of the following matrices:

(a) $\begin{bmatrix} 1 & 1 \\ 2 & 3 \end{bmatrix}$ (b) $\begin{bmatrix} 2 & 3 \\ 4 & 5 \end{bmatrix}$ (c) $\begin{bmatrix} 1 & 1 & 0 \\ 2 & 2 & 1 \\ 1 & 2 & 3 \end{bmatrix}$

4 For what values of λ, if any, do the following matrices have inverses?

(a) $\begin{bmatrix} \lambda - 1 & 2 \\ 3 & \lambda \end{bmatrix}$ (b) $\begin{bmatrix} \lambda - 1 & \lambda - 2 \\ \lambda - 3 & \lambda - 4 \end{bmatrix}$ (c) $\begin{bmatrix} 3 & 1 & 0 \\ -4 & 2 & 5 \\ \lambda^2 & \lambda & 1 \end{bmatrix}$

(d) $\begin{bmatrix} \lambda - 1 & \lambda & \lambda + 1 \\ 2 & -1 & 3 \\ \lambda + 3 & \lambda - 2 & \lambda + 7 \end{bmatrix}$ (e) $\begin{bmatrix} 2 - \lambda & -1 & 2\lambda \\ 1 & 2 & 3 \\ 2 & 2 & 1 \end{bmatrix}$

5 Solve the system

$$x_1 - x_2 + 2x_3 = 1$$
$$2x_1 \quad\;\; - \;\; x_3 = 2$$
$$x_1 + x_2 + \;\; x_3 = 3$$

by multiplying both sides of the equivalent matric equation $AX = B$ by the inverse of the matrix of the coefficients, A^{-1}.

6 (a) Under what conditions, if any, does $AB = AC$ imply $B = C$?

(b) If A is a nonsingular matrix, show that $AB = 0$ implies $B = 0$.

7 If D is a nonsingular diagonal matrix, prove that D^{-1} is also a diagonal matrix and that each element on the principal diagonal of D^{-1} is the reciprocal of the corresponding element in D. Does a similar result hold for nonsingular triangular matrices?

8 If A is a singular matrix, prove that the product of A and its adjoint is a null matrix.

9 If A is a nonsingular matrix which commutes with a matrix B, prove that A^{-1} also commutes with B. If B is also nonsingular, do A^{-1} and B^{-1} commute?

10 Prove Corollary 1, Theorem 4.

11 (a) If A is a nonsingular matrix, show that the determinant of the adjoint of A is equal to the $(n-1)$st power of the determinant of A.

(b) If A is a nonsingular matrix, show that the adjoint of the adjoint of A is equal to A multiplied by the $(n-2)$nd power of the determinant of A.

12 Establish the following result: if $A = [a_{ij}]$ is a nonsingular matrix, and if α_{ji} is the cofactor of the element in the jth row and ith column in the determinant of A^{-1}, then

$$|A|\,\alpha_{ji} = a_{ij}$$

13 Prove that the determinant of any orthogonal matrix is either 1 or -1. Is the converse true?

14 Prove that a real matrix is orthogonal if and only if its column vectors are unit vectors which are mutually orthogonal.

15 Prove that if the column vectors of a real matrix A are mutually orthogonal unit vectors, so are the row vectors of A.

16 Show that for all positive values of the k's, the matrix K in Eq. (2) is nonsingular.

17 Find the stiffness matrix for the system shown in Fig. 10.2.

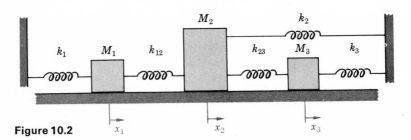

Figure 10.2

18 In mechanics it is shown that if a cantilever beam bears a concentrated load P at a distance s from the free end, then the deflection y at a distance x from the fixed end is given by the formula

$$y = \begin{cases} \dfrac{Px^2(x - 3s)}{6EI} & x \le s \\[2mm] \dfrac{Ps^2(s - 3x)}{6EI} & x \ge s \end{cases}$$

where E and I are physical constants of the beam. Using this formula, obtain the stiffness and elasticity matrices relating the forces and deflections at the positions

$s = L/3, 2L/3, L$ and $x = L/3, 2L/3, L$. In what respect, if any, does this problem differ significantly from the example discussed in the text?

19 Assuming that all springs are initially unstretched, compute the potential energy of the system shown in Fig. 10.1 when the three masses have arbitrary displacements x_1, x_2, and x_3, and show that the force acting on the ith mass is the negative of the partial derivative of the potential energy with respect to x_i.

20 Let K be the stiffness matrix for a frictionless mass-spring system S, let M be the diagonal matrix whose diagonal elements are the respective masses of S, and suppose that X is a vector such that KX is proportional to MX. Show that the square root of the negative of this proportionality constant is one of the natural frequencies of the system. *Hint:* Use Newton's law to set up the equations of motion of the respective masses, and then substitute $x_i = a_i \cos \omega t$.

10.4 Rank and the Equivalence of Matrices

One of the most important characteristics of a matrix is its **rank**.

DEFINITION 1 The rank of a matrix A is the largest value of r for which there exists an $r \times r$ submatrix of A with nonvanishing determinant.

The rank of a matrix A, as we have just defined it, is sometimes referred to more specifically as the **determinant rank** of A. Clearly, as an immediate consequence of Theorem 1, Sec. 10.2, we have the following simple but useful result.

THEOREM 1 If A and B are two $n \times n$ matrices of rank n, then both AB and BA are of rank n.

EXAMPLE 1

The matrix $\begin{bmatrix} 1 & 2 & -1 & 3 \\ 3 & 4 & 0 & -1 \\ -1 & 0 & -2 & 7 \end{bmatrix}$ is of rank 2, since each of the third-order submatrices

$$\begin{bmatrix} 2 & -1 & 3 \\ 4 & 0 & -1 \\ 0 & -2 & 7 \end{bmatrix} \quad \begin{bmatrix} 1 & -1 & 3 \\ 3 & 0 & -1 \\ -1 & -2 & 7 \end{bmatrix} \quad \begin{bmatrix} 1 & 2 & 3 \\ 3 & 4 & -1 \\ -1 & 0 & 7 \end{bmatrix} \quad \begin{bmatrix} 1 & 2 & -1 \\ 3 & 4 & 0 \\ -1 & 0 & -2 \end{bmatrix}$$

is singular, while not all second-order submatrices are singular. Specifically, the determinant of the 2×2 submatrix in the upper left-hand corner is different from zero.

In working with matrices it is frequently necessary to consider the effect of performing upon them certain simple manipulations known as **elementary transformations**.

DEFINITION 2 An elementary transformation of a matrix is any one of the following operations:

a. The multiplication of each element of a row or a column by the same nonzero constant

b. The interchange of two rows or of two columns

c. The addition of any multiple of the elements of one row, or one column, to the corresponding elements of another row, or column, respectively

The most important property of elementary transformations is contained in the following theorem.

THEOREM 2   The rank of a matrix is not altered by any sequence of elementary transformations.

Proof   Let A be an arbitrary matrix, and let r be its rank. Then every minor of order greater than r in A is zero, and at least one minor of order r is different from zero. To prove the theorem it is clearly sufficient to prove that no single elementary transformation can change the rank of A.

Consider first an elementary transformation of type **a**. By Theorem 6, Sec. 10.1, such a transformation cannot affect the vanishing or nonvanishing of any minor of A; hence it cannot alter the rank of A.

A transformation of type **b**, on the other hand, may affect the vanishing or non-vanishing of the minor in some particular position in A. However, after a transformation of type **b** every submatrix in A exists *somewhere* in the resulting matrix, with at most two rows or two columns interchanged. Hence, by Theorem 7, Sec. 10.1, if all minors of A of order greater than r are zero, the same thing will be true after the transformation is performed; and if at least one rth-order minor of A is different from zero, the same thing will be true after the transformation. Thus no transformation of type **b** can alter the rank of A.

Finally, no transformation of type **c** can yield a matrix whose rank is different from the rank of A. For consider the transformation consisting of modifying the elements of the jth row of A by adding to them some multiple, say λ, of the corresponding elements in the ith row. (The case of column modification can be handled by an identical argument.) Clearly, some of the $(r + 1)$st-order minors of A are unaffected by this transformation. Specifically, any $(r + 1)$st-order minor involving neither the ith nor the jth rows, both the ith and the jth rows, or just the ith row will surely be unaffected, i.e., will be left equal to zero. On the other hand, the value of an $(r + 1)$st-order minor involving the jth row but not the ith may conceivably be affected, since one of its rows (the jth) is modified by means of a row of elements (the ith) from out-side that minor. However, by the addition theorem for determinants (Theorem 10, Sec. 10.1) the modified determinant can be written in the form

$$|S_1| + \lambda|S_2|$$

where S_1 and S_2 are square submatrices of A of order $r + 1$ and hence singular, by hypothesis. Thus no vanishing $(r + 1)$st-order minor of A can be transformed into one which is different from zero by a transformation of type **c**; that is, no transformation of type **c** can increase the rank of A. On the other hand, no transformation of type **c** can yield a matrix whose rank is less than the rank of A, either. For if this were the case, then the inverse transformation, i.e., the transformation consisting of adding $-\lambda$ times the elements of the ith row to the corresponding elements of the jth row in the new matrix, would be a transformation which restored the matrix to its original form and hence increased its rank to the original value r; and we have just proved this to be impossible. Thus the rank of A, being neither increased nor decreased by a transformation of type **c**, is invariant under this transformation also, and our proof is complete.

DEFINITION 3 Two matrices one of which (and hence either of which) can be obtained from the other by a series of elementary transformations are said to be equivalent.

It is interesting and important to note that an elementary transformation involving the rows of a matrix A can be accomplished by premultiplying A by a unit matrix on whose rows the same elementary transformation has been performed, and any elementary transformation involving the columns of A can be accomplished by postmultiplying A by a unit matrix on whose columns the same elementary transformation has been performed. More specifically, we have the following theorems, whose proofs follow immediately from the definition of matric multiplication.

THEOREM 3 If A is an arbitrary $m \times n$ matrix, and if M (N)† is the matrix obtained from the identity matrix I_m (I_n) by multiplying the elements in the ith row (column) by λ, then the product MA (AN) is identical with A except for the ith row (column), which consists of the elements of the ith row (column) of A each multiplied by λ.

THEOREM 4 If A is an arbitrary $m \times n$ matrix, and if M (N) is the matrix obtained from the identity matrix I_m (I_n) by interchanging its ith and jth rows (columns), then the product MA (AN) is identical with A except for the ith and jth rows (columns), which are interchanged.

THEOREM 5 If A is an $m \times n$ matrix, and if M (N) is the matrix obtained from the identity matrix I_m (I_n) by adding to the elements of the jth row (column) λ times the corresponding elements in the ith row (column), then the product MA (AN) is identical with A except for the jth row (column), which consists of the elements of the jth row (column) of A plus λ times the corresponding elements in the ith row (column) of A.

From the preceding theorems it is clear that a sequence of elementary transformations $T_1, T_2, \ldots, T_k$ on the rows (columns) of an $m \times n$ matrix A can be accomplished by premultiplying (postmultiplying) A by a sequence of matrices $M_1, M_2, \ldots, M_k$ $(N_1, N_2, \ldots, N_l)$ each obtained from the identity I_m (I_n) by performing on its rows (columns) the same elementary transformation. The product $M_k \cdots M_2 M_1$ $(N_1 N_2 \cdots N_l)$ of the matrices by which A is premultiplied (postmultiplied) can, of course, be expressed as a single matrix P (Q), necessarily of rank m (n) since it is the product of matrices each obtained from I_m (I_n) by elementary transformations and each therefore of rank m (n). We have thus established the following important theorem.

THEOREM 6 If A and B are equivalent matrices, then $B = PAQ$, where P and Q are nonsingular matrices.

---

† See the footnote on p. 459.

In view of the way in which the nonsingular matrices P and Q of the last theorem were obtained, it is interesting to inquire whether, conversely, any nonsingular matrix can be obtained from the corresponding identity matrix by elementary row transformations or elementary column transformations. The answer is yes, as the following pair of theorems makes clear.

THEOREM 7   Any nonsingular $n \times n$ matrix can be reduced to the identity matrix I_n either by elementary row transformations or by elementary column transformations.

Proof   Let $P = [p_{ij}]$ be an arbitrary nonsingular $n \times n$ matrix. Because P is non-singular, at least one element in the first column must be different from zero, and it is no specialization to assume that $p_{11} \neq 0$; for if this is not the case, the interchange of two rows will bring a nonzero element into the leading position. Since $p_{11} \neq 0$, the leading element in the first column can be reduced to 1 by multiplying each of the elements in the first row by $1/p_{11}$. Then by subtracting the appropriate multiple of the first row from each of the other rows, we obtain the matrix

$$\begin{bmatrix} 1 & r_{12} & r_{13} & \cdots & r_{1n} \\ 0 & r_{22} & r_{23} & \cdots & r_{2n} \\ \cdots\cdots\cdots\cdots\cdots \\ 0 & r_{n2} & r_{n3} & \cdots & r_{nn} \end{bmatrix}$$

Since the original matrix, and therefore the last one, is nonsingular, it follows that the submatrix

$$\begin{bmatrix} r_{22} & r_{23} & \cdots & r_{2n} \\ r_{32} & r_{33} & \cdots & r_{3n} \\ \cdots\cdots\cdots\cdots\cdots \\ r_{n2} & r_{n3} & \cdots & r_{nn} \end{bmatrix}$$

is nonsingular. Hence the same reduction can be applied to it, and thus, continuing the process sufficiently, we obtain an upper triangular matrix

$$\begin{bmatrix} 1 & s_{12} & s_{13} & \cdots & s_{1n} \\ 0 & 1 & s_{23} & \cdots & s_{2n} \\ 0 & 0 & 1 & \cdots & s_{3n} \\ \cdots\cdots\cdots\cdots\cdots \\ 0 & 0 & 0 & \cdots & 1 \end{bmatrix}$$

Finally, working upward by row operations similar to those we have just employed, the elements above the diagonal can all be reduced to zero. This proves the assertion of the theorem for reductions involving only elementary row transformations. A similar argument shows that P can also be reduced to the identity by means of elementary column transformations. Thus the theorem is established.

THEOREM 8   Any nonsingular $n \times n$ matrix can be obtained from the identity matrix I_n by a sequence of elementary row transformations or a sequence of elementary column transformations.

Proof   If P is any nonsingular $n \times n$ matrix, we know from the last theorem that a sequence of elementary row transformations can be found which will reduce P to I_n.

Then, by Theorems 3 to 5, we know that there exist corresponding matrices M_1, $M_2, \ldots, M_k$ such that

(1) $$I_n = M_k \cdots M_2 M_1 P$$

Premultiplying this equation by $M_k^{-1}, \ldots, M_2^{-1}, M_1^{-1}$ (which exist, of course, since each of the M's is nonsingular), we obtain

(2) $$M_1^{-1} M_2^{-1} \cdots M_k^{-1} I_n = P$$

Thus if we have a nonsingular matrix P which we wish to obtain from the corresponding identity matrix by a sequence of elementary row transformations, we need only apply to I_n, in reverse order, the inverses of the successive row operations by which P is converted into I_n. A similar argument shows that an arbitrary nonsingular matrix P can also be obtained from the corresponding identity matrix by a sequence of elementary column transformations.

Incidentally, Eq. (1) can be modified to provide an alternative method of determining the inverse of a matrix P. In fact, postmultiplying Eq. (1) by P^{-1}, we have immediately

$$P^{-1} = M_k \cdots M_2 M_1 I_n$$

where, of course, the M's are the matrices corresponding to the elementary row operations by which P itself is reduced to the identity I_n.

EXAMPLE 2

Find a sequence of elementary row transformations which will reduce the matrix

$$P = \begin{bmatrix} 1 & 2 & 0 \\ 2 & 3 & -1 \\ -1 & -1 & 2 \end{bmatrix}$$

to I_3; determine the matrices $M_1, M_2, \ldots$ corresponding to these transformations; and use these results to compute the inverse of P.

By inspection it is clear that P can be reduced to I_3 by the following sequence of row operations:

$T_1:$ row 2 − 2 · row 1 $M_1 = \begin{bmatrix} 1 & 0 & 0 \\ -2 & 1 & 0 \\ 0 & 0 & 1 \end{bmatrix}$ $M_1 P = \begin{bmatrix} 1 & 2 & 0 \\ 0 & -1 & -1 \\ -1 & -1 & 2 \end{bmatrix}$

$T_2:$ row 3 + row 1 $M_2 = \begin{bmatrix} 1 & 0 & 0 \\ 0 & 1 & 0 \\ 1 & 0 & 1 \end{bmatrix}$ $M_2 M_1 P = \begin{bmatrix} 1 & 2 & 0 \\ 0 & -1 & -1 \\ 0 & 1 & 2 \end{bmatrix}$

$T_3:$ − row 2 $M_3 = \begin{bmatrix} 1 & 0 & 0 \\ 0 & -1 & 0 \\ 0 & 0 & 1 \end{bmatrix}$ $M_3 M_2 M_1 P = \begin{bmatrix} 1 & 2 & 0 \\ 0 & 1 & 1 \\ 0 & 1 & 2 \end{bmatrix}$

$T_4:$ row 3 − row 2 $M_4 = \begin{bmatrix} 1 & 0 & 0 \\ 0 & 1 & 0 \\ 0 & -1 & 1 \end{bmatrix}$ $M_4 M_3 M_2 M_1 P = \begin{bmatrix} 1 & 2 & 0 \\ 0 & 1 & 1 \\ 0 & 0 & 1 \end{bmatrix}$

$T_5:$ row 2 − row 3 $M_5 = \begin{bmatrix} 1 & 0 & 0 \\ 0 & 1 & -1 \\ 0 & 0 & 1 \end{bmatrix}$ $M_5 M_4 M_3 M_2 M_1 P = \begin{bmatrix} 1 & 2 & 0 \\ 0 & 1 & 0 \\ 0 & 0 & 1 \end{bmatrix}$

$T_6:$ row 1 − 2 · row 2 $M_6 = \begin{bmatrix} 1 & -2 & 0 \\ 0 & 1 & 0 \\ 0 & 0 & 1 \end{bmatrix}$ $M_6 M_5 M_4 M_3 M_2 M_1 P = \begin{bmatrix} 1 & 0 & 0 \\ 0 & 1 & 0 \\ 0 & 0 & 1 \end{bmatrix}$

The inverse, P^{-1}, of the given matrix can now be found either by using Eq. (2), namely,

$$P^{-1} = M_6 M_5 M_4 M_3 M_2 M_1 I_3$$

or simply by performing on I_3 the same sequence of row transformations used to reduce P to I_3:

$$P^{-1} = T_6 T_5 T_4 T_3 T_2 T_1 I_3$$

The result, by either method, is

$$P^{-1} = \begin{bmatrix} -5 & 4 & 2 \\ 3 & -2 & -1 \\ -1 & 1 & 1 \end{bmatrix}$$

With Theorem 8, we can now prove the converse of Theorem 6.

THEOREM 9 If $B = PAQ$, where P and Q are nonsingular matrices, then A and B are equivalent.

Proof By Theorem 8, P can be obtained from the corresponding identity matrix by a sequence of elementary row transformations, and Q can be obtained from the corresponding identity matrix by a sequence of elementary column transformations. Thus, as in the proof of Theorem 8, there is a set of matrices $M_1, M_2, \ldots, M_k$, each representing some elementary row operation, such that

$$P = M_k \cdots M_2 M_1$$

and a set of matrices $N_1, N_2, \ldots, N_l$, each representing some elementary column operation, such that

$$Q = N_1 N_2 \cdots N_l$$

Hence, $$B = PAQ = (M_k \cdots M_2 M_1) A (N_1 N_2 \cdots N_l)$$

which proves that B is obtained from A by elementary row and column transformations and hence is equivalent to A, as asserted.

By a proof almost identical with the proof of Theorem 7, we can prove the following theorem.

THEOREM 10 Any $m \times n$ matrix of rank r can be reduced by elementary transformations, which in general will involve both rows and columns, to an $m \times n$ matrix in which $a_{ii} = 1$, for $i = 1, 2, \ldots, r$, and all other elements are zero.

From Theorem 2 it is clear that equivalent matrices have the same rank. In view of Theorem 10, it is clear that given two $m \times n$ matrices of the same rank, each can be reduced by elementary transformations to the same standard form, and hence each can be reduced to the other, via the standard form, by elementary transformations. Thus we have established the following important theorem.

THEOREM 11 Two $m \times n$ matrices are equivalent if and only if they have the same rank.

The equivalence relation

$$B = PAQ \qquad P, Q \text{ nonsingular}$$

is a very general one, and many applications involve special cases in which P and Q satisfy additional conditions. These can all be thought of as transformations of a matrix A into a matrix B, and the usual terminology reflects this point of view. Table 10.2 summarizes the various cases of particular interest.

Table 10.2

If P, Q are arbitrary nonsingular matrices	$B = PAQ$	is an equivalence transformation and B is equivalent to A
If $P = Q^{-1}$	$B = Q^{-1}AQ$	is a similarity transformation and B is similar to A
If $P = Q^T$	$B = Q^TAQ$	is a congruence transformation and B is congruent to A
If $P = Q^T = Q^{-1}$	$B = Q^TAQ$ $= Q^{-1}AQ$	is an orthogonal transformation and B is orthogonally similar to A
If $P = \bar{Q}^T = Q^{-1}$	$B = \bar{Q}^TAQ$ $= Q^{-1}AQ$	is a unitary transformation and B is unitarily similar to A

EXERCISES

1   If a matrix is of rank r, is it possible that for some value of ρ, less than r, all minors of order ρ are equal to zero? Why?

2   Determine the rank of each of the following matrices:

(a) $\begin{bmatrix} 3 & -1 & 2 & 4 \\ 6 & 2 & -4 & -8 \end{bmatrix}$ (b) $\begin{bmatrix} 2 & -1 & 3 \\ 1 & -2 & 3 \\ 5 & 0 & 3 \end{bmatrix}$

(c) $\begin{bmatrix} 1 & 2 & 3 & 4 \\ 2 & 1 & 4 & 3 \\ 3 & 0 & 5 & -10 \end{bmatrix}$ (d) $\begin{bmatrix} 1 & 2 & 3 & 4 & 5 \\ 2 & 3 & 4 & 5 & 6 \\ 3 & 4 & 5 & 6 & 7 \\ 4 & 5 & 6 & 7 & 8 \end{bmatrix}$

3   Prove Theorem 3.   4   Prove Theorem 4.

5   Prove Theorem 5.

6   If A is an $m \times 1$ matrix and B is a $1 \times n$ matrix, show that the rank of AB is 1.

7   Prove Theorem 10.

8   Prove that the relation of equivalence has the following properties:
(a) Every matrix is equivalent to itself.
(b) If A is equivalent to B, then B is equivalent to A.
(c) If A is equivalent to B and B is equivalent to C, then A is equivalent to C.
Do the other relations listed in Table 10.2 have these properties?

9   Work Example 2 using only elementary column transformations.

10   Work Example 2 if P is the matrix $\begin{bmatrix} 2 & 1 & 2 \\ 0 & 1 & 1 \\ 1 & 1 & 1 \end{bmatrix}$.

11   Determine the rank of each of the following matrices as a function of λ:

(a) $\begin{bmatrix} 8(1-\lambda) & -2 & 0 \\ -2 & 3-2\lambda & -1 \\ 0 & -1 & 2(1-\lambda) \end{bmatrix}$ (b) $\begin{bmatrix} 1-\lambda & 1 & 1 \\ 1 & 3-\lambda & 3 \\ 2 & 1 & 4-\lambda \end{bmatrix}$

(c) $\begin{bmatrix} 5-\lambda & 4 & -2 \\ 4 & 5-\lambda & -2 \\ -2 & -2 & 3-2\lambda \end{bmatrix}$ (d) $\begin{bmatrix} 3-\lambda & -1 & 1 \\ 6 & -2-\lambda & 3 \\ 2 & -1 & 2-\lambda \end{bmatrix}$

12   Find a sequence of elementary transformations which will reduce the matrix

$$\begin{bmatrix} 0 & 1 & 1 & 3 & 2 \\ 1 & 2 & -1 & 0 & 1 \\ 0 & 1 & 2 & 2 & 1 \\ 1 & 0 & -4 & -5 & -2 \end{bmatrix}$$

to the standard form described in Theorem 10. What is the rank of this matrix?

13   Show that $A = \begin{bmatrix} 0 & 1 & 0 \\ 1 & 2 & 1 \end{bmatrix}$ and $B = \begin{bmatrix} 1 & 0 & 0 \\ 0 & 1 & 1 \end{bmatrix}$ are equivalent, and find non-singular matrices P and Q such that $B = PAQ$.

14   Work Exercise 13 for the matrices

$$A = \begin{bmatrix} 0 & 1 & 0 \\ 1 & 2 & 1 \\ 1 & 1 & 2 \end{bmatrix} \quad \text{and} \quad B = \begin{bmatrix} 1 & 1 & 0 \\ 0 & 1 & 1 \\ 2 & 1 & 1 \end{bmatrix}$$

15   Work Exercise 13 for the matrices

$$A = \begin{bmatrix} 2 & 1 & 0 \\ 1 & 2 & 1 \\ 4 & -1 & -2 \end{bmatrix} \quad \text{and} \quad B = \begin{bmatrix} 2 & 1 & 0 \\ 1 & 1 & 2 \\ 0 & -1 & -4 \end{bmatrix}$$

10.5 Systems of Linear Equations

Determinants and matrices find their most immediate application in the study of **linear dependence** and **independence** and in the closely related problem of the solution of simultaneous linear equations.

DEFINITION 1   The quantities $Q_1, Q_2, \ldots, Q_n$ are said to be linearly dependent if there exists a set of constants $c_1, c_2, \ldots, c_n$, at least one of which is different from zero, such that the equation

$$c_1 Q_1 + c_2 Q_2 + \cdots + c_n Q_n = 0$$

holds identically.

DEFINITION 2   The quantities $Q_1, Q_2, \ldots, Q_n$ are said to be linearly independent if they are not linearly dependent, i.e., if the only linear equation of the form

$$c_1 Q_1 + c_2 Q_2 + \cdots + c_n Q_n = 0$$

which they satisfy identically has $c_1 = c_2 = \cdots = c_n = 0$.

THEOREM 1   If the quantities $Q_1, Q_2, \ldots, Q_n$ are linearly dependent, then at least one (though not necessarily each one) of the quantities can be expressed as a linear combination of the remaining ones.

Proof   Since $Q_1, Q_2, \ldots, Q_n$ are linearly dependent, they necessarily satisfy a linear equation of the form $c_1 Q_1 + c_2 Q_2 + \cdots + c_n Q_n = 0$, in which at least one of the c's, say c_i, is different from zero. This being the case, we can divide by c_i, getting

$$Q_i = -\frac{c_1}{c_i} Q_1 - \frac{c_2}{c_i} Q_2 - \cdots - \frac{c_n}{c_i} Q_n$$

which expresses Q_i as a linear combination of the remaining Q's, as asserted. Since some, though not all, of the c's may be zero, it follows that we may not be able to solve for each of the Q's in this fashion.

EXAMPLE 1

Show that the quantities 1, x, and x^2 are linearly independent.

If 1, x, and x^2 are not linearly independent, they must satisfy identically some linear equation of the form

$$c_1(1) + c_2(x) + c_3(x^2) = 0$$

in which at least one of the c's is different from zero. If this holds identically, it must, in particular, hold for the values $x = -1, 0, 1$. Hence, substituting each of these values, in turn, we obtain the three equations

$$c_1 - c_2 + c_3 = 0$$
$$c_1 \qquad\qquad = 0$$
$$c_1 + c_2 + c_3 = 0$$

By inspection, it is apparent that the only solution of this system is $c_1 = c_2 = c_3 = 0$. Since this contradicts the assumption of linear dependence, the given quantities must be linearly independent, as asserted.

EXAMPLE 2

Show that the vectors

$$V_1 = \begin{bmatrix} 1 \\ 2 \\ 3 \end{bmatrix} \quad V_2 = \begin{bmatrix} 2 \\ -1 \\ 3 \end{bmatrix} \quad V_3 = \begin{bmatrix} 0 \\ 1 \\ -1 \end{bmatrix} \quad V_4 = \begin{bmatrix} 4 \\ -1 \\ 5 \end{bmatrix}$$

are linearly dependent.

These vectors will be linearly dependent if and only if constants c_1, c_2, c_3, and c_4 exist such that

a. At least one of them is different from zero.

$$\textbf{b.} \ c_1 \begin{bmatrix} 1 \\ 2 \\ 3 \end{bmatrix} + c_2 \begin{bmatrix} 2 \\ -1 \\ 3 \end{bmatrix} + c_3 \begin{bmatrix} 0 \\ 1 \\ -1 \end{bmatrix} + c_4 \begin{bmatrix} 4 \\ -1 \\ 5 \end{bmatrix} = \begin{bmatrix} 0 \\ 0 \\ 0 \end{bmatrix}.$$

Condition **b** is, of course, equivalent to the three scalar equations

$$c_1 + 2c_2 \qquad\quad + 4c_4 = 0$$
$$2c_1 - c_2 + c_3 - c_4 = 0$$
$$3c_1 + 3c_2 - c_3 + 5c_4 = 0$$

By combining these equations appropriately it is not difficult to verify that they are satisfied by the values

$$c_1 = 0 \qquad c_2 = 2\lambda \qquad c_3 = \lambda \qquad c_4 = -\lambda \qquad \lambda \text{ arbitrary}$$

and by no others. Hence the four vectors are linearly dependent and, in fact, are connected by the relation

$$0V_1 + 2V_2 + V_3 - V_4 = 0$$

and (except for constant multiples of this) no others. It is therefore obvious that V_2, V_3, and V_4 can each be expressed in terms of the remaining vectors of the set but that V_1 cannot be so expressed.†

On the other hand, the vectors V_1, V_2, and V_3 are linearly independent, since an equation of the form

$$c_1 V_1 + c_2 V_2 + c_3 V_3 = 0$$

implies that

$$c_1 + 2c_2 \quad\quad\;\; = 0$$
$$2c_1 - \;\; c_2 + c_3 = 0$$
$$3c_1 + 3c_2 - c_3 = 0$$

and by direct solution we find that the only values which satisfy this system of equations are $c_1 = c_2 = c_3 = 0$. Similarly, we can verify that V_1, V_2, and V_4 are independent and that V_1, V_3, and V_4 are independent. However, V_2, V_3, and V_4 are dependent, since, as we observed above, they satisfy the relation $2V_2 + V_3 - V_4 = 0$.

As a simple application of the notion of linear independence, we have the following useful result.

THEOREM 2   If $V_1, V_2, \ldots, V_m$ are m vectors each having $n \geq m$ components, and if for $i = 1, 2, \ldots, m$ the first nonzero component of V_i is the ith, then V_1, $V_2, \ldots, V_m$ are linearly independent.

Proof   By hypothesis, the given vectors are of the form

$$V_1 = \begin{bmatrix} v_{11} \\ v_{21} \\ v_{31} \\ \vdots \\ v_{m1} \\ \vdots \\ v_{n1} \end{bmatrix}, V_2 = \begin{bmatrix} 0 \\ v_{22} \\ v_{32} \\ \vdots \\ v_{m2} \\ \vdots \\ v_{n2} \end{bmatrix}, V_3 = \begin{bmatrix} 0 \\ 0 \\ v_{33} \\ \vdots \\ v_{m3} \\ \vdots \\ v_{n3} \end{bmatrix}, \ldots, V_m = \begin{bmatrix} 0 \\ 0 \\ 0 \\ \vdots \\ v_{mm} \\ \vdots \\ v_{nm} \end{bmatrix}$$

where $v_{11}, v_{22}, v_{33}, \ldots, v_{mm}$ are all different from zero. Now, since each vector has n components, the condition $c_1 V_1 + c_2 V_2 + \cdots + c_m V_m = 0$ implies n scalar equations, the first m of which are

$$c_1 v_{11} \quad\quad\quad\quad\quad\quad\quad\quad\quad\quad = 0$$
$$c_1 v_{21} + c_2 v_{22} \quad\quad\quad\quad\quad\quad\;\; = 0$$
$$c_1 v_{31} + c_2 v_{32} + c_3 v_{33} \quad\quad\quad\; = 0$$
$$\cdots\cdots\cdots\cdots\cdots\cdots\cdots\cdots\cdots\cdots\cdots\cdots\cdots$$
$$c_1 v_{m1} + c_2 v_{m2} + c_3 v_{m3} + \cdots + c_m v_{mm} = 0$$

Hence, since $v_{11}, v_{22}, v_{33}, \ldots, v_{mm}$ are all different from zero, it follows that $c_1 = c_2 = c_3 = \cdots = c_m = 0$. Therefore the vectors $V_1, V_2, V_3, \ldots, V_m$ are linearly independent, as asserted.

---

† Of course it is possible for a set of dependent quantities $Q_1, Q_2, \ldots, Q_n$ to satisfy more than one independent linear equation. In problems where this occurs, it may well be that some of the equations can be solved for Q_i, say, while the others cannot. Naturally, if even one equation can be solved for Q_i, then Q_i can be expressed in terms of the other members of the set.

An argument identical to the proof of Theorem 2 establishes the following companion result, covering the case in which the zero components in the various vectors are the last ones rather than the first ones.

THEOREM 3 If $V_1, V_2, \ldots, V_m$ are m vectors each having $n \geq m$ components, and if for $i = 1, 2, \ldots, m$ the last nonzero component of V_i is the $(n - m + i)$th, then $V_1, V_2, \ldots, V_m$ are linearly independent.

From Examples 1 and 2 (as well as Theorems 2 and 3), it is clear that questions concerning linear dependence and independence are closely related to the solution of systems of simultaneous linear equations, and to these we now turn our attention. In the most general case we have a system of the form

$$
\begin{aligned}
a_{11}x_1 + a_{12}x_2 + \cdots + a_{1n}x_n &= b_1 \\
a_{21}x_1 + a_{22}x_2 + \cdots + a_{2n}x_n &= b_2 \\
\cdots\cdots\cdots\cdots\cdots\cdots\cdots\cdots\cdots \\
a_{m1}x_1 + a_{m2}x_2 + \cdots + a_{mn}x_n &= b_m
\end{aligned}
$$

(1)

where m, the number of equations, is not necessarily equal to n, the number of unknowns. If at least one of the m quantities b_i is different from zero, the system is said to be **nonhomogeneous**. If $b_i = 0$ for all values of i, the system is said to be **homogeneous**. If we define the matrices

$$
A = \begin{bmatrix} a_{11} & a_{12} & \cdots & a_{1n} \\ a_{21} & a_{22} & \cdots & a_{2n} \\ \cdots\cdots\cdots\cdots\cdots\cdots \\ a_{m1} & a_{m2} & \cdots & a_{mn} \end{bmatrix} \qquad X = \begin{bmatrix} x_1 \\ x_2 \\ \vdots \\ x_n \end{bmatrix} \qquad B = \begin{bmatrix} b_1 \\ b_2 \\ \vdots \\ b_m \end{bmatrix}
$$

the system can be written in the compact matrix form

(2) $AX = B$

In this form, the matrix A is known as the **coefficient matrix** of the system, and the matrix

$$[AB]$$

obtained by adjoining the column matrix B to the coefficient matrix A as an $(n + 1)$st column is known as the **augmented matrix** of the system.

Before proceeding to the question of the existence and determination of solutions of (2), we shall first prove several important theorems about such solutions on the assumption that they exist.†

THEOREM 4 If X_1 and X_2 are two solutions of the homogeneous matric equation $AX = \mathbf{0}$, then for all values of the constants c_1 and c_2, the vector $c_1X_1 + c_2X_2$ is also a solution of $AX = \mathbf{0}$.

---

† It is interesting to note the striking resemblance between the next three theorems and Theorems 2 to 4 of Sec. 2.1 and Theorems 1 to 3 of Sec. 4.5.

Proof   By direct substitution we have

$$A(c_1X_1 + c_2X_2) = A(c_1X_1) + A(c_2X_2) = 0$$
$$= c_1(AX_1) + c_2(AX_2)$$
$$= c_1 \cdot 0 + c_2 \cdot 0$$
$$= 0$$

where the coefficients of c_1 and c_2 vanish because, by hypothesis, both X_1 and X_2 are solutions of $AX = 0$. Hence $c_1X_1 + c_2X_2$ also satisfies $AX = 0$, as asserted.

THEOREM 5   If k is the maximum number of linearly independent solution vectors of the system $AX = 0$, and if $X_1, X_2, \ldots, X_k$ are k particular linearly independent solution vectors, then any solution vector of $AX = 0$ can be expressed in the form

$$c_1X_1 + c_2X_2 + \cdots + c_kX_k$$

where the c's are scalar constants.

Proof   Let k be the maximum number of linearly independent solution vectors of the equation $AX = 0$; let $X_1, X_2, \ldots, X_k$ be a particular set of k linearly independent solution vectors; and let X_{k+1} be *any* solution vector. If X_{k+1} is one of the vectors in the set $\{X_1, X_2, \ldots, X_k\}$, the assertion of the theorem is obviously true. If X_{k+1} is not a member of the set $\{X_1, X_2, \ldots, X_k\}$, then $X_1, X_2, \ldots, X_k, X_{k+1}$ cannot be linearly independent since, by hypothesis, k is the maximum number of linearly independent solution vectors of $AX = 0$. Hence, the X's must satisfy a linear equation of the form

(3) $$c_1X_1 + c_2X_2 + \cdots + c_kX_k + c_{k+1}X_{k+1} = 0$$

in which at least one c is different from zero. In fact, $c_{k+1} \neq 0$, for otherwise Eq. (3) would reduce to

$$c_1X_1 + c_2X_2 + \cdots + c_kX_k = 0$$

with at least one of the c's different from zero, contrary to the hypothesis that $X_1, X_2, \ldots, X_k$ are linearly independent. But if $c_{k+1} \neq 0$, it is clearly possible to solve Eq. (3) for X_{k+1} and express it in the form asserted by the theorem.

Because of the property guaranteed by the last theorem, a general linear combination of the maximum number of linearly independent solution vectors of $AX = 0$ is usually referred to as a **complete solution** of $AX = 0$.

THEOREM 6   If X_p is a particular solution vector of the nonhomogeneous system $AX = B$, and if $c_1X_1 + c_2X_2 + \cdots + c_kX_k$ is a complete solution of the related homogeneous system $AX = 0$, then any solution of the nonhomogeneous system can be written in the form
$$c_1X_1 + c_2X_2 + \cdots + c_kX_k + X_p$$

Proof   Let X_p be a particular solution vector of the nonhomogeneous equation $AX = B$, and let X_a be any solution vector of this equation. Then $AX_p = B$ and $AX_a = B$; therefore, subtracting these two equations, we have

$$AX_a - AX_p = 0 \quad \text{or} \quad A(X_a - X_p) = 0$$

The last equation shows that the vector $X_a - X_p$ is a solution of the homogeneous equation $AX = 0$. Hence, by Theorem 4, it can be expressed in the form

$$X_a - X_p = c_1 X_1 + c_2 X_2 + \cdots + c_k X_k$$

Transposing gives

$$X_a = c_1 X_1 + c_2 X_2 + \cdots + c_k X_k + X_p$$

Since X_a was *any* solution vector of the equation $AX = B$, the theorem is established.

We now turn our attention to the question of when solutions of the equation $AX = B$ will actually exist. The central result is contained in the following theorem.

THEOREM 7   A system of m simultaneous linear equations in n unknowns, $AX = B$, has a solution if and only if the coefficient matrix A and the augmented matrix $[AB]$ have the same rank, r. When solutions exist, the maximum number of independent arbitrary constants in any general solution, i.e., the maximum number of linearly independent solution vectors of the related homogeneous system $AX = 0$, is $n - r$.

Proof   We shall prove this theorem by applying to the given system

$$a_{11}x_1 + a_{12}x_2 + \cdots + a_{1n}x_n = b_1$$
$$a_{21}x_1 + a_{22}x_2 + \cdots + a_{2n}x_n = b_2$$
$$\cdots\cdots\cdots\cdots\cdots\cdots\cdots\cdots\cdots\cdots\cdots\cdots$$
$$a_{m1}x_1 + a_{m2}x_2 + \cdots + a_{mn}x_n = b_m$$

a procedure, known as the **Gauss reduction**, which closely resembles the method by which we proved Theorem 7, Sec. 10.4. We begin by assuming that $a_{11} \neq 0$, which is no specialization, since at least one of the coefficients in the first equation must be different from zero and by renaming the unknowns, if necessary, it can be brought into the leading position. Next we divide the first equation by a_{11} and then multiply it in turn by $a_{21}, a_{31}, \ldots, a_{m1}$ and subtract it from the second, third, $\ldots$, mth equations, respectively. This gives the equivalent system†

$$x_1 + \alpha_{12}x_2 + \alpha_{13}x_3 + \cdots + \alpha_{1n}x_n = \beta_1$$
$$a'_{22}x_2 + a'_{23}x_3 + \cdots + a'_{2n}x_n = b'_2$$
$$a'_{32}x_2 + a'_{33}x_3 + \cdots + a'_{3n}x_n = b'_3$$
$$\cdots\cdots\cdots\cdots\cdots\cdots\cdots\cdots\cdots\cdots\cdots\cdots$$
$$a'_{m2}x_2 + a'_{m3}x_3 + \cdots + a'_{mn}x_n = b'_m$$

where, explicitly,

$$\alpha_{1j} = \frac{a_{1j}}{a_{11}} \quad \beta_1 = \frac{b_1}{a_{11}} \quad a'_{ij} = a_{ij} - a_{i1}\frac{a_{1j}}{a_{11}} \quad b'_i = b_i - a_{i1}\frac{b_1}{a_{11}}$$

---

† Two equations, or systems of equations, are said to be **equivalent** if every solution of one is a solution of the other, and conversely.

Now we apply the same process to the last $m - 1$ equations, noting that if $a'_{22} = 0$, a renaming of the last $n - 1$ unknowns with possibly a rearrangement of the last $m - 1$ equations will introduce a nonzero coefficient in place of a'_{22} unless all coefficients in the remaining equations are zero, which, of course, may be the case at some stage in the process. The result of this second reduction is the system

$$x_1 + \alpha_{12}x_2 + \alpha_{13}x_3 + \cdots + \alpha_{1n}x_n = \beta_1$$
$$x_2 + \alpha_{23}x_3 + \cdots + \alpha_{2n}x_n = \beta_2$$
$$a''_{33}x_3 + \cdots + a''_{3n}x_n = b''_3$$
$$a''_{43}x_3 + \cdots + a''_{4n}x_n = b''_4$$
$$\cdots\cdots\cdots\cdots\cdots\cdots\cdots\cdots\cdots$$
$$a''_{m3}x_3 + \cdots + a''_{mn}x_n = b''_m$$

We now continue in exactly the same fashion until the process terminates. If $m < n$, that is, if the number of equations is less than the number of unknowns, this may happen because after m steps there are no more equations to which to apply it:

$$x_1 + \alpha_{12}x_2 + \alpha_{13}x_3 + \cdots + \alpha_{1m}x_m + \alpha_{1,m+1}x_{m+1} + \cdots + \alpha_{1n}x_n = \beta_1$$
$$x_2 + \alpha_{23}x_3 + \cdots + \alpha_{2m}x_m + \alpha_{2,m+1}x_{m+1} + \cdots + \alpha_{2n}x_n = \beta_2$$
$$x_3 + \cdots + \alpha_{3m}x_m + \alpha_{3,m+1}x_{m+1} + \cdots + \alpha_{3n}x_n = \beta_3$$
$$\cdots\cdots\cdots\cdots\cdots\cdots\cdots\cdots\cdots\cdots\cdots\cdots\cdots\cdots\cdots$$
$$x_m + \alpha_{m,m+1}x_{m+1} + \cdots + \alpha_{mn}x_n = \beta_m$$

On the other hand, regardless of the relative size of m and n, the process may terminate because before we have been able to transform the ith equation into one beginning with the term x_i for all values of i, say after only k $(<m)$ such reductions, all coefficients in the left members of each of the remaining $m - k$ equations are zero:

$$x_1 + \alpha_{12}x_2 + \alpha_{13}x_3 + \cdots + \alpha_{1k}x_k + \cdots + \cdots + \cdots + \alpha_{1n}x_n = \beta_1$$
$$x_2 + \alpha_{23}x_3 + \cdots + \alpha_{2k}x_k + \cdots + \cdots + \cdots + \alpha_{2n}x_n = \beta_2$$
$$x_3 + \cdots + \alpha_{3k}x_k + \cdots + \cdots + \cdots + \alpha_{3n}x_n = \beta_3$$
$$\cdots + \cdots\cdots + \cdots + \cdots + \cdots + \cdots\cdots\cdots\cdots$$
$$x_k + \cdots + \cdots + \cdots + \alpha_{kn}x_n = \beta_k$$
$$0 + \cdots + \cdots + \quad 0 = \beta_{k+1}$$
$$\cdots + \cdots + \cdots\cdots\cdots\cdots$$
$$0 + \cdots + \quad 0 = \beta_m$$

In the first case, if we transpose all terms containing $x_{m+1}, x_{m+2}, \ldots, x_n$, we have a system of equations from which $x_m, x_{m-1}, x_{m-2}, \ldots, x_2, x_1$ can successively be found in terms of $x_{m+1}, x_{m+2}, \ldots, x_n$, which can be given arbitrary values. In the second case, it may be that

$$\beta_{k+1} = \beta_{k+2} = \cdots = \beta_m = 0$$

so that we have essentially the case we have just discussed except that now it is $x_1, x_2, \ldots, x_k$ which are expressed in terms of the remaining unknowns x_{k+1}, $x_{k+2}, \ldots, x_n$, which can be given arbitrary values. If, however, one or more of the β's after β_k is different from zero, we have a contradiction, and the original system has no solution; in other words, it is **inconsistent**.

Now consider the coefficient matrix and the augmented matrix of the reduced system. Since the augmented matrix contains the coefficient matrix as a submatrix, it is clear that its rank must be at least as great as the rank of the coefficient matrix. Moreover, in the solvable cases it is evident from the definition of rank that the rank of the augmented matrix cannot exceed the rank of the coefficient matrix and hence must be equal to it. Likewise, it is obvious that in the inconsistent case the rank of the augmented matrix is actually greater than the rank of the coefficient matrix. Finally, we observe that the ranks of the coefficient matrix and augmented matrix for the reduced system are equal to the ranks of the respective matrices of the original system, since each step in the Gauss reduction, namely, rearranging the columns of unknowns, rearranging the equations, multiplying and dividing the equations by nonzero constants, and subtracting multiples of one equation from other equations, is an elementary transformation, which, by Theorem 2, Sec. 10.4, cannot change the rank of either matrix. Thus we have established the first assertion of the theorem.

Let us now return to the reduced system in the solvable case. If r is the common value of the rank of the coefficient matrix and the augmented matrix, the reduced system can be written

$$x_1 + \alpha_{12}x_2 + \alpha_{13}x_3 + \cdots + \alpha_{1r}x_r = -\alpha_{1,r+1}x_{r+1} - \cdots - \alpha_{1n}x_n + \beta_1$$
$$x_2 + \alpha_{23}x_3 + \cdots + \alpha_{2r}x_r = -\alpha_{2,r+1}x_{r+1} - \cdots - \alpha_{2n}x_n + \beta_2$$
$$\cdots\cdots\cdots\cdots\cdots\cdots\cdots\cdots\cdots\cdots\cdots\cdots\cdots\cdots\cdots\cdots$$
$$x_r = -\alpha_{r,r+1}x_{r+1} - \cdots - \alpha_{rn}x_n + \beta_r$$

In this form it is clear that the last $n - r$ unknowns, that is, $x_{r+1}, x_{r+2}, \ldots, x_n$, can be given arbitrary values, say

$$x_{r+1} = \lambda_1, x_{r+2} = \lambda_2, \ldots, x_n = \lambda_{n-r}$$

Substituting these values into the last equation in the above system, we obtain x_r immediately. Then, substituting for $x_r, x_{r+1}, \ldots, x_n$ in the next to the last equation, we obtain x_{r-1}, and so on, step by step, until each of the x's is determined. Finally, when the expressions for $x_1, x_2, \ldots, x_r$ are simplified by collecting terms on $\lambda_1, \lambda_2, \ldots, \lambda_{n-r}$, we obtain expressions of the form

$$x_1 = \lambda_1 c_{11} + \lambda_2 c_{12} + \cdots + \lambda_{n-r} c_{1,n-r} + \gamma_1$$
$$\cdots\cdots\cdots\cdots\cdots\cdots\cdots\cdots\cdots\cdots\cdots\cdots\cdots\cdots\cdots$$
$$x_r = \lambda_1 c_{r1} + \lambda_2 c_{r2} + \cdots + \lambda_{n-r} c_{r,n-r} + \gamma_r$$
$$x_{r+1} = \lambda_1$$
$$x_{r+2} = \lambda_2$$
$$\cdots\cdots\cdots\cdots\cdots\cdots\cdots\cdots\cdots\cdots\cdots\cdots\cdots\cdots\cdots$$
$$x_n = \lambda_{n-r}$$

or

$$X = \begin{bmatrix} x_1 \\ \vdots \\ x_r \\ x_{r+1} \\ x_{r+2} \\ \vdots \\ x_m \end{bmatrix} = \lambda_1 \begin{bmatrix} c_{11} \\ \vdots \\ c_{r1} \\ 1 \\ 0 \\ \vdots \\ 0 \end{bmatrix} + \lambda_2 \begin{bmatrix} c_{12} \\ \vdots \\ c_{r2} \\ 0 \\ 1 \\ \vdots \\ 0 \end{bmatrix} + \cdots + \lambda_{n-r} \begin{bmatrix} c_{1,n-r} \\ \vdots \\ c_{r,n-r} \\ 0 \\ 0 \\ \vdots \\ 1 \end{bmatrix} + \begin{bmatrix} \gamma_1 \\ \vdots \\ \gamma_r \\ 0 \\ 0 \\ \vdots \\ 0 \end{bmatrix}$$

The $n - r$ vectors which are multiplied respectively by $\lambda_1, \lambda_2, \ldots, \lambda_{n-r}$ depend only on the a_{ij}'s in the original system and are, in fact, solution vectors of the related homogeneous system $AX = 0$. The vector

$$\begin{bmatrix} \gamma_1 \\ \vdots \\ \gamma_r \\ 0 \\ 0 \\ \vdots \\ 0 \end{bmatrix}$$

depends not only on the a_{ij}'s but also on the b_i's and is clearly a particular solution of the nonhomogeneous system $AX = B$. By Theorem 3, the $n - r$ vectors which are multiplied by the λ's are linearly independent. Hence, the related homogeneous system $AX = 0$ has $n - r$ linearly independent solutions, and the complete solution of the nonhomogeneous system, $AX = B$, contains $n - r$ independent constants, as asserted.

EXAMPLE 3

Find a complete solution of the system

$$\begin{aligned}
x_1 + 2x_2 + x_3 - x_4 + 2x_5 &= 2 \\
x_1 + 4x_2 + 5x_3 - 3x_4 + 8x_5 &= -2 \\
-2x_1 - x_2 + 4x_3 - x_4 + 5x_5 &= -10 \\
3x_1 + 7x_2 + 5x_3 - 4x_4 + 9x_5 &= 4
\end{aligned}$$

Applying the Gauss reduction, we obtain successively

$$\begin{aligned}
x_1 + 2x_2 + x_3 - x_4 + 2x_5 &= 2 \\
2x_2 + 4x_3 - 2x_4 + 6x_5 &= -4 \\
3x_2 + 6x_3 - 3x_4 + 9x_5 &= -6 \\
x_2 + 2x_3 - x_4 + 3x_5 &= -2
\end{aligned}$$

and

$$\begin{aligned}
x_1 + 2x_2 + x_3 - x_4 + 2x_5 &= 2 \\
x_2 + 2x_3 - x_4 + 3x_5 &= -2 \\
0 &= 0 \\
0 &= 0
\end{aligned}$$

Hence, recognizing that x_3, x_4, and x_5 can be given arbitrary values, we have the solutions

$$\begin{aligned}
x_1 &= -2x_2 - x_3 + x_4 - 2x_5 + 2 \\
x_2 &= -2x_3 + x_4 - 3x_5 - 2 \\
x_3 &= x_3 \\
x_4 &= x_4 \\
x_5 &= x_5
\end{aligned}$$

or, taking $x_3 = \lambda_1$, $x_4 = \lambda_2$, $x_5 = \lambda_3$,

$$
\begin{aligned}
x_1 &= \quad 3\lambda_1 - \lambda_2 + 4\lambda_3 + 6 \\
x_2 &= -2\lambda_1 + \lambda_2 - 3\lambda_3 - 2 \\
x_3 &= \quad \lambda_1 \\
x_4 &= \qquad\qquad \lambda_2 \\
x_5 &= \qquad\qquad\qquad\quad \lambda_3
\end{aligned}
$$

A complete solution of the original system is therefore

$$
X = \lambda_1 \begin{bmatrix} 3 \\ -2 \\ 1 \\ 0 \\ 0 \end{bmatrix} + \lambda_2 \begin{bmatrix} -1 \\ 1 \\ 0 \\ 1 \\ 0 \end{bmatrix} + \lambda_3 \begin{bmatrix} 4 \\ -3 \\ 0 \\ 0 \\ 1 \end{bmatrix} + \begin{bmatrix} 6 \\ -2 \\ 0 \\ 0 \\ 0 \end{bmatrix}
$$

where λ_1, λ_2, λ_3 are arbitrary scalars.

The existence of a solution for the nonhomogeneous system implies that the coefficient matrix and the augmented matrix

$$
A = \begin{bmatrix} 1 & 2 & 1 & -1 & 2 \\ 1 & 4 & 5 & -3 & 8 \\ -2 & -1 & 4 & -1 & 5 \\ 3 & 7 & 5 & -4 & 9 \end{bmatrix}
$$

and

$$
[AB] = \begin{bmatrix} 1 & 2 & 1 & -1 & 2 & 2 \\ 1 & 4 & 5 & -3 & 8 & -2 \\ -2 & -1 & 4 & -1 & 5 & -10 \\ 3 & 7 & 5 & -4 & 9 & 4 \end{bmatrix}
$$

have the same rank. The fact that the complete solution contains three independent arbitrary constants implies that the common value of the ranks of the two matrices is 2, since, according to the last theorem

Number of arbitrary constants in complete solution

$$= \text{number of unknowns} - \text{common value of rank}$$

It is, of course, not difficult to verify that A and $[AB]$ are both of rank 2. The vector

$$
X_p = \begin{bmatrix} 6 \\ -2 \\ 0 \\ 0 \\ 0 \end{bmatrix}
$$

is a particular solution of the given nonhomogeneous equation $AX = B$. The vectors

$$
X_1 = \begin{bmatrix} 3 \\ -2 \\ 1 \\ 0 \\ 0 \end{bmatrix} \qquad X_2 = \begin{bmatrix} -1 \\ 1 \\ 0 \\ 1 \\ 0 \end{bmatrix} \qquad \text{and} \qquad X_3 = \begin{bmatrix} 4 \\ -3 \\ 0 \\ 0 \\ 1 \end{bmatrix}
$$

are three linearly independent solutions of the related homogeneous equation $AX = 0$. That $\lambda_1 X_1 + \lambda_2 X_2 + \lambda_3 X_3 + X_p$ is a complete solution of the given system follows, of course, from Theorem 6.

As the last example illustrated, the Gauss reduction provides a practical method for solving systems of simultaneous linear equations in the general case. However, in several important special cases there are other methods which are sometimes more convenient. Specifically, we have the following pair of theorems.

THEOREM 8 (Cramer's Rule)   If the coefficient matrix A of a system $AX = B$ of n linear equations in n unknowns is nonsingular, the system has the unique solution

$$x_1 = \frac{|D_1|}{|A|},\ x_2 = \frac{|D_2|}{|A|}, \ldots, x_n = \frac{|D_n|}{|A|}$$

where D_i is the matrix obtained from A by replacing the ith column of A by the column vector B.

Proof   Since the matrix of coefficients A is nonsingular by hypothesis, it follows that its inverse A^{-1} exists. Hence, premultiplying the given equation $AX = B$ by A^{-1}, we obtain

$$X = A^{-1}B$$

and direct substitution confirms that this actually is a solution. Now, in the column vector $A^{-1}B$, the element in the ith row is simply the scalar product of the ith row vector of A^{-1} and B itself, i.e.,

$$\frac{A_{1i}b_1 + A_{2i}b_2 + \cdots + A_{ni}b_n}{|A|}$$

Moreover, the numerator of this fraction is just the expansion, in terms of the ith column, of the determinant of the matrix D_i obtained from A by replacing the ith column of A by the column vector B. Hence

(4)
$$x_i = \frac{|D_i|}{|A|}$$

as asserted. Since A is nonsingular, the rank of the coefficient matrix and the rank of the augmented matrix are both equal to n. Hence, according to Theorem 7, there can be no arbitrary constant in any complete solution, and the solution we have found is the only one.

If in the $n \times n$ system $AX = B$, the vector B is zero, i.e., if $b_1 = b_2 = \cdots = b_n = 0$, then, clearly, each determinant $|D_i|$ contains a column consisting entirely of zeros and hence is zero. If $|A| \neq 0$, it therefore follows from (4) that $x_i = 0$ for all values of i or, in other words, that only a **trivial solution** is possible. On the other hand, if $B = \mathbf{0}$, the coefficient matrix and the augmented matrix have the same rank, and if $|A| = 0$, the common values of these ranks is at most $r = n - 1$; that is, $r \leq n - 1$. Hence $n - r$ is at least as much as 1, and, by Theorem 7, the equation $AX = \mathbf{0}$ has at least one nontrivial solution vector. Thus we have established the following important corollary of Theorem 8.

COROLLARY 1   A homogeneous system of n linear equations in n unknowns, $AX = \mathbf{0}$, has a nontrivial solution, i.e., a solution other than $x_1 = x_2 = \cdots = x_n = 0$, if and only if the determinant of the coefficients $|A|$ is equal to zero.

More specifically, when the rank of the coefficient matrix of a homogeneous system of n linear equations in n unknowns is $n - 1$, we have the following useful result.

THEOREM 9   If the coefficient matrix of a homogeneous system of n linear equations in n unknowns is of rank $n - 1$, and if the submatrix obtained from A by omitting the kth row is also of rank $n - 1$, then a complete solution of the given system is

$$x_i = cA_{ki} \qquad i = 1, 2, \ldots, n$$

where c is an arbitrary constant.

Proof Since the rank of the $(n - 1) \times n$ matrix remaining when the kth row is deleted from A is $n - 1$, it follows that at least one of the cofactors A_{ki} of the elements in the kth row is different from zero. Hence, not all of the values $x_i = cA_{ki}$ are zero. To verify that these values do indeed satisfy $AX = \mathbf{0}$, we need only substitute them into the general equation of the system, namely,

$$\sum_{i=1}^{n} a_{ji}x_i = 0 \qquad j = 1, 2, \ldots, n$$

and verify that it is satisfied. Doing this, and using Corollary 1 of Theorem 11, Sec. 10.1, to simplify the result, we have

$$\sum_{i=1}^{n} a_{ji}(cA_{ki}) = c \sum_{i=1}^{n} a_{ji}A_{ki} = \begin{cases} 0 & j \neq k \\ |A| & j = k \end{cases}$$

Thus, since $|A| = 0$, by hypothesis, it follows that each equation is satisfied by the given values. Finally, since the rank of both the coefficient matrix and the augmented matrix is $r = n - 1$, it follows by Theorem 7 that the system has just one independent solution vector. Hence the solution given by the formula of the theorem is a complete solution, as asserted.

COROLLARY 1 If the coefficient matrix A of a homogeneous system of $n - 1$ linear equations in n unknowns $AX = \mathbf{0}$ is of rank $n - 1$, then a complete solution of the system is $x_1 = c|M_1|$, $x_2 = -c|M_2|$, $\ldots$, $x_n = (-1)^{n+1}c|M_n|$ where M_j is the $(n - 1) \times (n - 1)$ submatrix obtained from A by deleting the jth column of A.

EXAMPLE 4

Find a complete solution of the system

$$\begin{aligned} x_1 - 2x_2 + x_3 + 3x_4 &= 0 \\ 2x_1 + 2x_2 - x_3 + x_4 &= 0 \\ -x_1 - x_2 + 3x_3 + 2x_4 &= 0 \\ x_1 - 8x_2 - x_3 + 3x_4 &= 0 \end{aligned}$$

It is easy to verify that the determinant of the coefficients of this system, $|A|$, is zero but that the determinant of the 3×3 submatrix in the upper left-hand corner of A is different from zero. Thus the coefficient matrix A and the submatrix remaining when the last row is deleted from A are both of rank 3. Hence, according to the last theorem, the values of the x's which satisfy the given system of equations are proportional to the cofactors of the last row of $|A|$. Thus we have

$$x_1 = -c \begin{vmatrix} -2 & 1 & 3 \\ 2 & -1 & 1 \\ -1 & 3 & 2 \end{vmatrix} = -20c \qquad x_2 = c \begin{vmatrix} 1 & 1 & 3 \\ 2 & -1 & 1 \\ -1 & 3 & 2 \end{vmatrix} = 5c$$

$$x_3 = -c \begin{vmatrix} 1 & -2 & 3 \\ 2 & 2 & 1 \\ -1 & -1 & 2 \end{vmatrix} = -15c \qquad x_4 = c \begin{vmatrix} 1 & -2 & 1 \\ 2 & 2 & -1 \\ -1 & -1 & 3 \end{vmatrix} = 15c$$

or, setting $-5c = k$,

$$x_1 = 4k \qquad x_2 = -k \qquad x_3 = 3k \qquad x_4 = -3k$$

In view of Theorems 8 and 9, it is clearly desirable to have some convenient criterion for determining when a determinant is different from zero. One useful one is provided by the following theorem, whose proof is an interesting application of Corollary 1 of Theorem 8.

THEOREM 10   If in each row of a determinant the absolute value of the element on the principal diagonal is greater than the sum of the absolute values of the remaining elements in that row, the value of the determinant is different from zero.

Proof   Let $|A|$ be an nth-order determinant in which, for each value of i,

$$(5) \qquad |a_{ii}| > \sum_{\substack{j=1 \\ j \neq i}}^{n} |a_{ij}|\dagger$$

To prove that $|A|$ is not equal to zero, let us assume the contrary. Now, if $|A| = 0$, then, by Corollary 1 of Theorem 8, the equations

$$(6) \qquad \begin{aligned} a_{11}x_1 + a_{12}x_2 + \cdots + a_{1n}x_n &= 0 \\ \cdots\cdots\cdots\cdots\cdots\cdots\cdots\cdots\cdots \\ a_{n1}x_1 + a_{n2}x_2 + \cdots + a_{nn}x_n &= 0 \end{aligned}$$

have a nontrivial solution. Let x_k be the component of maximum absolute value of such a solution. Then from the kth equation of the system (6) we have, by transposing,

$$a_{kk}x_k = -\sum_{\substack{j=1 \\ j \neq k}}^{n} a_{kj}x_j$$

Hence, taking absolute values, we have

$$|a_{kk}| \cdot |x_k| \leq \sum_{\substack{j=1 \\ j \neq k}}^{n} |a_{kj}| \cdot |x_j| \leq \sum_{\substack{j=1 \\ j \neq k}}^{n} |a_{kj}| \cdot |x_k|$$

Dividing through by $|x_k|$ gives

$$|a_{kk}| \leq \sum_{\substack{j=1 \\ j \neq k}}^{n} |a_{kj}|$$

But this contradicts some one of the inequalities (5), which hold, by hypothesis. Therefore we must abandon the assumption that $|A| = 0$, and the theorem is established.

If a set of vectors has the property that it contains at least one subset of r vectors which are linearly independent, but if the vectors in every subset containing more than r vectors are linearly dependent, the set is said to be of **dimension** r. Using Theorem 7, it is not difficult to prove the following theorems.

---

† This property of the absolute values of the elements of a determinant is known as **diagonal dominance** and is quite important in the solution of systems of linear equations by iterative methods.

THEOREM 11   A set of row vectors $X_1, X_2, \ldots, X_n$ is of dimension r if and only if the (row-partitioned) matrix

$$\begin{bmatrix} X_1 \\ X_2 \\ \vdots \\ X_n \end{bmatrix}$$

is of rank r.

THEOREM 12   A set of column vectors $X_1, X_2, \ldots, X_n$ is of dimension r if and only if the (column-partitioned) matrix

$$[X_1 \quad X_2 \quad \cdots \quad X_n]$$

is of rank r.

THEOREM 13   If a set of vectors $\{X\}$ is of dimension r, and if $\{X_1, X_2, \ldots, X_r\}$ is a particular subset of r linearly independent vectors, then any vector in the set $\{X\}$ can be expressed as a linear combination of the vectors $X_1, X_2, \ldots, X_r$.

COROLLARY 1   If $X_1, X_2, \ldots, X_n$ are n linearly independent vectors each having n components, then any vector B with n components can be expressed as a linear combination of $X_1, X_2, \ldots, X_n$.

In Sec. 10.4 we observed that what we called simply the *rank* of a matrix is sometimes referred to more specifically as the *determinant rank*. This permits one to distinguish between the rank, as we defined it, and the **row rank** and **column rank**, which are, by definition, the maximum number of linearly independent row and column vectors, respectively, in the matrix. However, the necessity for distinguishing between the three definitions of rank is eliminated by the following theorem, whose proof follows immediately from Theorems 11 and 12.

THEOREM 14   For any matrix A, the determinant rank, the row rank, and the column rank are all equal.

Another interesting consequence of Theorems 11 and 12 is contained in the following theorem.

THEOREM 15   If A and B are conformable matrices of rank r and ρ, respectively, the rank of the product AB is equal to or less than the smaller of the numbers r and ρ.

Proof   Let A be an $m \times p$ matrix, let B be a $p \times n$ matrix, and let the rows of A be $A_1, A_2, \ldots, A_m$. Then the ith row vector in the product AB is (see Exercise 20, Sec. 10.2)

(7) $$A_i B$$

Now, by hypothesis, the rank of A is r. Hence, by Theorem 11, A contains exactly r linearly independent row vectors, which, without loss of generality we may suppose to be the first r, namely, $A_1, A_2, \ldots, A_r$. Therefore, by Theorem 13, it follows that for $i = r + 1, r + 2, \ldots, m$ the row vector A_i must be a linear combination of the first r row vectors:

$$A_i = \lambda_{1i}A_1 + \lambda_{2i}A_2 + \cdots + \lambda_{ri}A_r \qquad i = r + 1, r + 2, \ldots, m$$

Therefore, substituting into (7), we find that for $i = r + 1, r + 2, \ldots, m$, the ith row vector of the product is

$$(\lambda_{1i}A_1 + \lambda_{2i}A_2 + \cdots + \lambda_{ri}A_r)B = \lambda_{1i}A_1B + \lambda_{2i}A_2B + \cdots + \lambda_{ri}A_rB$$

But this shows that each row of AB after the rth is a linear combination of the first r rows, which, in turn, proves that AB contains at most r linearly independent row vectors and hence is of rank at most r. A similar argument, using a column partition of B, shows that the rank of AB is at most equal to ρ. Therefore, the rank of AB is at most equal to the smaller of the numbers r and ρ, as asserted. If A and B are also conformable in the order BA, it is clear that the rank of BA is also equal to or less than the smaller of the pair (r,ρ).

The estimate for the rank of the product AB provided by the last theorem can be supplemented with the following result.†

THEOREM 16 If A is an $m \times p$ matrix of rank r, and if B is a $p \times n$ matrix of rank ρ, the rank of the product AB is equal to or greater than $r + \rho - p$.

As we shall see in later sections, a set of vectors is usually much more convenient to work with if in addition to being linearly independent the vectors are also **orthonormal**, i.e., are of unit length and mutually orthogonal, or perpendicular. A general set of r linearly independent vectors will not ordinarily possess the property of orthonormality, but by an important procedure known as the **Schmidt orthogonalization process**, it is always possible to determine linear combinations of r linearly independent vectors which will be orthonormal as well as independent.

To develop this process, let $V_1, V_2, \ldots, V_r$ be r linearly independent vectors, and let us choose any one of them, say V_1, and reduce it to a unit vector by dividing it by its length $\sqrt{V_1{}^T V_1}$. This gives us the first vector of our orthonormal set:

$$U_1 = \frac{V_1}{\sqrt{V_1{}^T V_1}}$$

We now choose any member of the original set except V_1, say V_2, and write

(8) $$W_2 = V_2 - c_1 U_1$$

where c_1 is a constant to be determined so that W_2 is orthogonal to U_1. This, of course, requires that

$$U_1{}^T W_2 = U_1{}^T(V_2 - c_1 U_1) = 0$$

---

† Both this result and Theorem 15 are due to the English mathematician J. J. Sylvester (1814–1897) and are known together as **Sylvester's law of nullity**. A proof of Theorem 16 can be found in L. Mirsky, "Linear Algebra," p. 162, Oxford, New York, 1955.

From this, since $U_1{}^T U_1 = 1$, we have

$$c_1 = U_1{}^T V_2$$

and, from (8),

$$W_2 = V_2 - (U_1{}^T V_2)U_1$$

We now convert W_2 into a unit vector by dividing it by its length:

$$U_2 = \frac{W_2}{\sqrt{W_2{}^T W_2}} = \frac{V_2 - (U_1{}^T V_2)U_1}{\sqrt{[V_2 - (U_1{}^T V_2)U_1]^T[V_2 - (U_1{}^T V_2)U_1]}}$$

Next, we choose any member of the original set except V_1 and V_2, say V_3, and write

(9) $$W_3 = V_3 - d_1 U_1 - d_2 U_2$$

where d_1 and d_2 are constants to be determined so that W_3 will be orthogonal to both U_1 and U_2. Taking into account the orthonormality of U_1 and U_2 which we have already achieved, this gives us the two conditions

$$U_1{}^T W_3 = U_1{}^T(V_3 - d_1 U_1 - d_2 U_2) = U_1{}^T V_3 - d_1 = 0$$
$$U_2{}^T W_3 = U_2{}^T(V_3 - d_1 U_1 - d_2 U_2) = U_2{}^T V_3 - d_2 = 0$$

Hence, $$d_1 = U_1{}^T V_3 \quad \text{and} \quad d_2 = U_2{}^T V_3$$

and therefore, from (9)

$$W_3 = V_3 - (U_1{}^T V_3)U_1 - (U_2{}^T V_3)U_3$$

W_3 can now be normalized, giving us our third unit vector U_3; and thus the process is continued until the required set of orthonormal vectors is obtained from the original set. It is clear that the process can fail if and only if at some stage

$$W_k = V_k - \sum_{i=1}^{k-1} (U_i{}^T V_k)U_i = 0$$

However, if $W_k = 0$, this implies that

$$V_k = \sum_{i=1}^{k-1} (U_i{}^T V_k)U_i$$

which, if the U's are replaced by their expressions in terms of $V_1, V_2, \ldots, V_{k-1}$, implies that V_k is either zero or a linear combination of the preceding V's. Each of these contradicts the hypothesis that the V's are linearly independent, and hence it follows that W_k cannot be zero for any value of k. An almost identical argument (or the result of Exercise 18) shows that the U's derived by the Schmidt process are linearly independent.

EXERCISES

1 Verify that $\sin x$ and $\cos x$ are linearly independent.
2 (a) Are $\cos^2 x$, $\sin^2 x$, and $\cos 2x$ linearly independent? Why?
  (b) Are $\cos^2 x$, $\sin^2 x$, and $\sin 2x$ linearly independent? Why?
3 Verify that the matrices

$$X_1 = \begin{bmatrix} 1 \\ 1 \\ 0 \end{bmatrix} \quad X_2 = \begin{bmatrix} 1 \\ -1 \\ 1 \end{bmatrix} \quad X_3 = \begin{bmatrix} 2 \\ 1 \\ 3 \end{bmatrix} \quad X_4 = \begin{bmatrix} -1 \\ 4 \\ 5 \end{bmatrix}$$

are linearly dependent and express each matrix as a linear combination of the other three.

4 What conditions must a, b, c, and d satisfy for the matrices $\begin{bmatrix} 1 & 2 \\ -1 & 0 \end{bmatrix}$, $\begin{bmatrix} 2 & 3 \\ -2 & 1 \end{bmatrix}$, and $\begin{bmatrix} a & b \\ c & d \end{bmatrix}$ to be linearly dependent?

5 Determine whether the expressions in the following sets are linearly dependent or linearly independent:

(a) $x_1{}^2 - x_2 x_3$, $x_2{}^2 - x_1 x_3$, $x_3{}^2 - x_1 x_2$

(b) $x_1 + x_2$, $x_2 - x_3$, $x_1 - x_2 - x_3$

6 Show that five or more 2×2 matrices are always linearly dependent. How many 3×3 matrices must we have before we can be sure that they are linearly dependent?

7 Show that if 0 is included in a set of quantities, the members of the set are always linearly dependent.

8 Show that if the quantities $Q_1, Q_2, \ldots, Q_n$ are linearly independent, the members of every proper subset of the Q's are also linearly independent. Is the converse true?

9 If A is a square matrix, and if the equation $AX = 0$ has k linearly independent solution vectors, show that the same is true of the equation $A^T X = 0$. Is this result true if A is not a square matrix?

10 Show that the expressions $a_{11}x + a_{12}y + a_{13}$, $a_{21}x + a_{22}y + a_{23}$, and $a_{31}x + a_{32}y + a_{33}$ are linearly dependent if and only if the three lines

$$a_{11}x + a_{12}y + a_{13} = 0$$

$$a_{21}x + a_{22}y + a_{23} = 0$$

and $\qquad a_{31}x + a_{32}y + a_{33} = 0$

are concurrent. *Hint:* Recall the result of part (a) of Exercise 19, Sec. 10.1. ∘

11 Using the Gauss reduction, find a complete solution of each of the following systems:

(a)
$$\begin{aligned} x_1 + 2x_2 + 4x_3 - x_4 + 2x_5 &= 3 \\ 3x_1 + 4x_2 + 5x_3 - x_4 - 2x_5 &= 7 \\ x_1 + 3x_2 + 4x_3 + 5x_4 - x_5 &= 4 \end{aligned}$$

(b)
$$\begin{aligned} x_1 + x_2 + x_3 - x_4 - x_5 &= 2 \\ x_1 + 2x_2 + 4x_3 - x_4 + 5x_5 &= 3 \\ 3x_1 + 4x_2 + 5x_3 - x_4 - 2x_5 &= 7 \\ x_1 + 3x_2 + 4x_3 + 5x_4 - x_5 &= 4 \\ 2x_1 + 5x_2 + 8x_3 + 4x_4 + x_5 &= 7 \\ x_1 - x_2 - 2x_3 - 7x_4 - x_5 &= 0 \end{aligned}$$

12 Using the Gauss reduction, determine which of the following systems have solutions, and find the solutions when they exist.

(a)
$$\begin{aligned} x + 2y + 4z - w &= 3 \\ 3x + 4y + 5z - w &= 7 \\ x + 3y + 4z + 5w &= 4 \end{aligned}$$

(b)
$$\begin{aligned} x + 2y + 4z + w &= 3 \\ 2x - y + z + 3w &= 7 \\ -4x + 7y + 5z - 7w &= 4 \end{aligned}$$

(c)
$$\begin{aligned} x + 2y + 4z + w &= 0 \\ 2x - y + z + 3w &= 0 \\ -4x + 7y + 5z - 7w &= 0 \end{aligned}$$

(d)
$$\begin{aligned} x + 4y - z + 2w &= 0 \\ 3x + 5y - 2z - w &= 0 \\ x + 5y + 2z + w &= 0 \end{aligned}$$

13 Using Cramer's rule, solve each of the following systems:

(a)
$$\begin{aligned} x - y + 2z &= -5 \\ -x + 3z &= 0 \\ 2x + y &= 1 \end{aligned}$$

(b)
$$\begin{aligned} x - y + 2z + w &= 0 \\ -x + 3z + 2w &= 2 \\ 2x + y - w &= 1 \\ 2x + 2y + z + 3w &= 14 \end{aligned}$$

14 Using Theorem 9, solve each of the following systems:

(a)
$$\begin{aligned} x_1 - 2x_2 + 3x_3 &= 0 \\ 2x_1 + 3x_2 - x_3 &= 0 \\ 4x_1 - x_2 + 5x_3 &= 0 \end{aligned}$$

(b)
$$\begin{aligned} x_1 - 2x_2 + x_3 + 3x_4 &= 0 \\ 2x_1 + x_2 - 3x_3 + x_4 &= 0 \\ 3x_1 + 3x_2 - 2x_3 + x_4 &= 0 \end{aligned}$$

15   Determine the values of λ, if any, for which the following systems have nontrivial solutions, and find such solutions when they exist:

(a) $\begin{aligned} (1-\lambda)x_1 + 2x_2 &= 0 \\ 3x_1 + (2-\lambda)x_2 &= 0 \end{aligned}$ (b) $\begin{aligned} (2-\lambda)x_1 + (1-\lambda)x_2 &= 0 \\ (6-\lambda)x_1 - (5-2\lambda)x_2 &= 0 \end{aligned}$

(c) $\begin{aligned} \lambda x_1 - 2x_2 + x_3 &= 0 \\ \lambda x_1 + (1-\lambda)x_2 + x_3 &= 0 \\ 2x_1 - x_2 + 2\lambda x_3 &= 0 \end{aligned}$

(d) $\begin{aligned} (5-\lambda)x_1 + 4x_2 - 2x_3 &= 0 \\ 4x_1 + (5-\lambda)x_2 - 2x_3 &= 0 \\ -2x_1 - 2x_2 + (3-2\lambda)x_3 &= 0 \end{aligned}$

16   Show by an example that even when A and B are conformable in either order, the rank of AB is in general not equal to the rank of BA.

17   If the rows of a matrix are linearly dependent, are the columns necessarily linearly dependent?

18   Show that if the vectors of a set are mutually orthogonal and nonnull, they are linearly independent.

19   If the quantities $Q_1, Q_2, \ldots, Q_n$ are such that $Q_{r+1}, Q_{r+2}, \ldots, Q_n$ can each be expressed as a linear combination of $Q_1, Q_2, \ldots, Q_r$ show that at most r of the Q's can be linearly independent.

20   Prove Theorem 11.   21   Prove Theorem 12.

22   (a) Prove Theorem 13.   (b) Prove the corollary of Theorem 13.

23   Using the Schmidt process, construct a set of orthonormal vectors from the vectors in each of the following sets:

(a) $V_1 = [3 \quad 4]$ $V_2 = [-1 \quad 1]$

(b) $V_1 = [1 \quad 1 \quad 0]$ $V_2 = [1 \quad 0 \quad 1]$ $V_3 = [0 \quad 1 \quad 1]$

24   Work Exercise 23 for the following sets of vectors:

(a) $V_1 = [1 \quad 2 \quad 2]$ $V_2 = [1 \quad 4 \quad 0]$ $V_3 = [2 \quad 0 \quad 1]$

(b) $V_1 = [1 \quad 1 \quad 1 \quad 1]$ $V_2 = [0 \quad 1 \quad 2 \quad 2]$ $V_3 = [0 \quad 0 \quad 1 \quad 1]$

25   Prove that if an $n \times (n+1)$ matrix A contains a column of elements which are not all zero and if every nth-order determinant in A which contains this column vanishes, then the rank of A is less than n. *Hint:* Expand each of the vanishing determinants in terms of the elements in their common column, consider the determinant of the resulting system of equations, and use the result of Exercise 11, Sec. 10.3.

26   Prove that n vectors $V_1, V_2, \ldots, V_n$ are linearly dependent if and only if the so-called **grammian determinant**, or **grammian**,

$$\begin{vmatrix} V_1^T V_1 & V_1^T V_2 & \cdots & V_1^T V_n \\ V_2^T V_1 & V_2^T V_2 & \cdots & V_2^T V_n \\ \hdotsfor{4} \\ V_n^T V_1 & V_n^T V_2 & \cdots & V_n^T V_n \end{vmatrix}$$

is equal to zero.

27   If A is a square matrix, p a positive integer, and X a vector such that $A^p X \neq 0$ but $A^{p+1} X = 0$, show that the vectors $X, AX, A^2 X, \ldots, A^p X$ are linearly independent.

28   Let A and B be matrices conformable in the order AB. Prove that the rank of AB is equal to the rank of B if and only if $BX = 0$ for every vector X such that $ABX = 0$.

29   If A is a square matrix, and if p is a positive integer such that A^p and A^{p+1} have the same rank r, show that $A^{p+2}, A^{p+3}, A^{p+4}, \ldots$ are also of rank r.

30   Prove that if A is an $n \times n$ matrix, then $A^n, A^{n+1}, A^{n+2}, \ldots$ all have the same rank.

31 (a) Determine the regions in the xy-plane in which Theorem 10 guarantees that the determinant

$$\begin{vmatrix} x & 1 & 1 \\ 1 & y & 1 \\ 1 & 1 & 3 \end{vmatrix}$$

is different from zero.

(b) Expand the determinant in part (a), plot the locus of the pairs of values for which it is equal to zero, and verify that this locus has no point in common with the regions identified in part (a).

32 Prove the following theorem. If a polynomial of degree n vanishes for $n + 1$ distinct values of its variable, it vanishes identically. *Hint*: Consider the system of equations obtained when the $n + 1$ values are substituted into the general polynomial of degree n; then use Corollary 1 of Theorem 8 and the results of Exercise 20, Sec. 10.1.

10.6 Matric Differential Equations

In Sec. 3.3 we saw that the ideas of *complementary function* and *particular integral* which we developed for single linear differential equations could easily be extended to systems of linear differential equations. This analogy is especially striking when we regard a system of linear differential equations with constant coefficients as a single matric equation, much as we regarded a system of linear algebraic equations as a single matric equation in the last section. Moreover, the procedure for handling systems of equations when the characteristic equation has repeated or complex roots or when a term on the right-hand side duplicates a term in the complementary function is best described in the language of matrices. Hence we shall conclude this chapter with a brief discussion of matric differential equations.

Let the system we are given be

$$
\begin{aligned}
p_{11}(D)x_1 + p_{12}(D)x_2 + \cdots + p_{1n}(D)x_n &= f_1(t) \\
p_{21}(D)x_1 + p_{22}(D)x_2 + \cdots + p_{2n}(D)x_n &= f_2(t) \\
&\cdots\cdots\cdots\cdots\cdots\cdots\cdots\cdots\cdots\cdots\cdots \\
p_{n1}(D)x_1 + p_{n2}(D)x_2 + \cdots + p_{nn}(D)x_n &= f_n(t)
\end{aligned}
$$

(1)

where the p_{ij}'s are polynomials in the operator D with constant coefficients. If we define the matrices

$$
P(D) = \begin{bmatrix} p_{11}(D) & p_{12}(D) & \cdots & p_{1n}(D) \\ p_{21}(D) & p_{22}(D) & \cdots & p_{2n}(D) \\ \cdots\cdots\cdots\cdots\cdots\cdots\cdots\cdots \\ p_{n1}(D) & p_{n2}(D) & \cdots & p_{nn}(D) \end{bmatrix} \qquad X = \begin{bmatrix} x_1 \\ x_2 \\ \vdots \\ x_n \end{bmatrix} \qquad F(t) = \begin{bmatrix} f_1(t) \\ f_2(t) \\ \vdots \\ f_n(t) \end{bmatrix}
$$

the system (1) can be written in the compact form

(2) $$P(D)X = F(t)$$

The associated homogeneous equation is, of course,

(3) $$P(D)X = \mathbf{0}$$

The first step in finding the complementary function of Eq. (2) is to assume that solutions of Eq. (3) exist in the form

$$X = Ae^{mt}$$

where the scalar m and the column matrix of constants A have yet to be determined.†
Since

$$D^r(e^{mt}) = m^r e^{mt}$$

it follows that if we substitute the vector $X = Ae^{mt}$ into the homogeneous equation
(3), we obtain just

$$P(m)Ae^{mt} = 0$$

or, dividing out the scalar factor e^{mt},

(4) $$P(m)A = 0‡$$

Now by Corollary 1, Theorem 8, Sec. 10.5, Eq. (4) will have a nontrivial solution if
and only if

(5) $$|P(m)| = 0$$

and for each root m_j of this equation there will be a solution vector A_j of (4) determined to within an arbitrary scalar factor k_j. If the characteristic equation (5) is of
degree N, and if its roots $\{m_j\}$ are all distinct, a complete solution of Eq .(3) [and the
complementary function of Eq. (2)] is then

$$X = k_1 A_1 e^{m_1 t} + k_2 A_2 e^{m_2 t} + \cdots + k_N A_N e^{m_N t}§$$

As in the case of a single scalar differential equation, if the set of roots $\{m_j\}$ includes
one or more pairs of conjugate complex roots, it is desirable to reduce the corresponding complex exponential solution to a purely real form. To see how this can be accomplished, let $p \pm iq$ be a pair of conjugate complex roots of Eq. (5), and let A be a
particular solution vector of (4) corresponding to the root $m = p + iq$; that is, let

$$P(m)A \equiv P(p + iq)A = 0$$

Then, since all the coefficients in (4) are real, it follows by taking conjugates throughout the last equation that

$$P(\bar{m})\bar{A} \equiv P(p - iq)\bar{A} = 0$$

Thus $\bar{A}$ is a solution vector corresponding to the conjugate root $\bar{m} = p - iq$, and
therefore we have the two particular solutions of Eq. (3),

$$Ae^{(p+iq)t} \qquad \text{and} \qquad \bar{A}e^{(p-iq)t}$$

---

† The expressions $x = ae^{mt}$, $y = be^{mt}$, $z = ce^{mt}$ in Eq. (3), Sec. 3.3, are, of course, just the
scalar form of this assumption for the special case $n = 3$.

‡ This is just the matric equivalent of the algebraic system in Eq. (4), Sec. 3.3, which we
obtained in our scalar treatment of the specific system of differential equations considered in
that section.

§ This we should recognize as the matric equivalent of the scalar system (9) of Sec. 3.3, with
$N = 3$ and

$$A_1 = \begin{bmatrix} -1 \\ 2 \\ -1 \end{bmatrix} \qquad A_2 = \begin{bmatrix} 3 \\ -1 \\ -1 \end{bmatrix} \qquad A_3 = \begin{bmatrix} 9 \\ -6 \\ -1 \end{bmatrix}$$

By combining these as follows and applying the Euler formulas, just as we did in handling the complex roots in Sec. 2.2, we obtain the two independent real solutions

$$\frac{Ae^{(p+iq)t} + \bar{A}e^{(p-iq)t}}{2} = e^{pt}\left(\frac{A + \bar{A}}{2}\cos qt - \frac{A - \bar{A}}{2i}\sin qt\right)$$

(6a)
$$= e^{pt}[\mathcal{R}(A)\cos qt - \mathcal{I}(A)\sin qt]$$

$$\frac{Ae^{(p+iq)t} - \bar{A}e^{(p-iq)t}}{2i} = e^{pt}\left(\frac{A - \bar{A}}{2i}\cos qt + \frac{A + \bar{A}}{2}\sin qt\right)$$

(6b)
$$= e^{pt}[\mathcal{I}(A)\cos qt + \mathcal{R}(A)\sin qt]$$

where $\mathcal{R}(A)$ and $\mathcal{I}(A)$ denote the column matrices whose components are, respectively, the real parts of the components of A and the imaginary parts of the components of A. In many cases this method of determining the necessary relations between the coefficients of solutions of (3) of the form

$$x_j = e^{pt}(a_j \cos qt + b_j \sin qt)$$

is simpler than the alternative process of substituting these expressions into the original differential equations, collecting terms, and equating the resulting coefficients to zero.

If $|P(m)| = 0$ has a double root, say $m = r$, we proceed very much as in the case of a single differential equation. If A is a solution of the equation $P(r)A = \mathbf{0}$, then, of course,

$$Ae^{rt}$$

is one solution of (3). However, as a second independent solution we must try not Bte^{rt}, as strict analogy with the scalar case would suggest, but

(7)
$$B_1 te^{rt} + B_2 e^{rt}$$

The term $B_2 e^{rt}$ must be retained in the matric case because in general the matrix B_2 will not be a scalar multiple of A, and hence, in constructing the complete solution, the term $B_2 e^{rt}$ cannot be absorbed in the term Ae^{rt}, as would necessarily be the case with the term $b_2 e^{rt}$ were it included in a trial solution of the form $b_1 te^{rt} + b_2 e^{rt}$ for a single scalar differential equation. It can be shown, however (see Exercises 9 and 10), that to within an arbitrary scalar factor, the matrix B_1 is the same as A. Hence, after (7) has been substituted into the homogeneous system (3), it is only necessary to solve for the ratios of the components of B_2.

Similar observations hold for roots of (5) of higher multiplicity. Thus, for a k-fold root r, the appropriate trial solutions are not Ae^{rt}, Bte^{rt}, $Ct^2 e^{rt}, \ldots, Kt^{k-1}e^{rt}$, but

$$Ae^{rt}$$

$$B_1 te^{rt} + B_2 e^{rt}$$

$$C_1 t^2 e^{rt} + C_2 te^{rt} + C_3 e^{rt}$$

$$\cdots\cdots\cdots\cdots\cdots\cdots\cdots\cdots\cdots\cdots\cdots\cdots$$

$$K_1 t^{k-1}e^{rt} + K_2 t^{k-2}e^{rt} + \cdots + K_{k-1}te^{rt} + K_k e^{rt}$$

In this case, to within arbitrary scalar factors, the matrices $A, B_1, C_1, \ldots, K_1$ are identical (see Exercises 11 to 13).

EXAMPLE 1

Find a complete solution of the system

$$(D^2 + D + 8)x_1 + (D^2 + 6D + 3)x_2 = 0$$

$$(D + 1)x_1 + \quad\quad (D^2 + 1)x_2 = 0$$

In this case the characteristic equation (5) is

$$\begin{vmatrix} m^2 + m + 8 & m^2 + 6m + 3 \\ m + 1 & m^2 + 1 \end{vmatrix} = m^4 + 2m^2 - 8m + 5 = 0$$

with roots $1, 1, -1 \pm 2i$. For the root $-1 + 2i$, Eq. (4) becomes

$$\begin{bmatrix} (-1 + 2i)^2 + (-1 + 2i) + 8 & (-1 + 2i)^2 + 6(-1 + 2i) + 3 \\ (-1 + 2i) + 1 & (-1 + 2i)^2 + 1 \end{bmatrix} \begin{bmatrix} a_1 \\ a_2 \end{bmatrix} = \begin{bmatrix} 0 \\ 0 \end{bmatrix}$$

or

$$\begin{bmatrix} 4 - 2i & -6 + 8i \\ 2i & -2 - 4i \end{bmatrix} \begin{bmatrix} a_1 \\ a_2 \end{bmatrix} = \begin{bmatrix} 0 \\ 0 \end{bmatrix}$$

This is equivalent to the two scalar equations

$$(4 - 2i)a_1 + (-6 + 8i)a_2 = 0$$
$$2ia_1 - (2 + 4i)a_2 = 0$$

Since $m = -1 + 2i$ is a root of the characteristic equation (5), these two equations are dependent and the ratio of a_1 to a_2 can be found equally well from either of them. Using the second, since it is a little simpler, we therefore have

$$\frac{a_1}{a_2} = \frac{1 + 2i}{i} \quad \text{or} \quad A \equiv \begin{bmatrix} a_1 \\ a_2 \end{bmatrix} = \begin{bmatrix} 1 + 2i \\ i \end{bmatrix}$$

Hence,

$$\mathscr{R}(A) = \begin{bmatrix} 1 \\ 0 \end{bmatrix} \quad \text{and} \quad \mathscr{I}(A) = \begin{bmatrix} 2 \\ 1 \end{bmatrix}$$

and thus from (6) we have the two particular solutions

$$X_1 = e^{-t}\left(\begin{bmatrix} 1 \\ 0 \end{bmatrix} \cos 2t - \begin{bmatrix} 2 \\ 1 \end{bmatrix} \sin 2t\right)$$

$$X_2 = e^{-t}\left(\begin{bmatrix} 2 \\ 1 \end{bmatrix} \cos 2t + \begin{bmatrix} 1 \\ 0 \end{bmatrix} \sin 2t\right)$$

For the repeated root $m = 1$, we have one solution of the form Be^t, where, from (4),

$$P(1)B \equiv \begin{bmatrix} 10 & 10 \\ 2 & 2 \end{bmatrix} \begin{bmatrix} b_1 \\ b_2 \end{bmatrix} = \begin{bmatrix} 0 \\ 0 \end{bmatrix}$$

so that we can take

$$B = \begin{bmatrix} b_1 \\ b_2 \end{bmatrix} = \begin{bmatrix} 1 \\ -1 \end{bmatrix}$$

As a second solution we have, from (7),

$$C_1 te^t + C_2 e^t$$

or, since $C_1 = B$ (as we observed above, without proof),

$$\begin{bmatrix} 1 \\ -1 \end{bmatrix} te^t + \begin{bmatrix} c_{12} \\ c_{22} \end{bmatrix} e^t$$

Substituting this into the original system, we obtain two equations, each of which reduces to

$$2c_{12} + 2c_{22} = 1$$

Hence we can take

$$c_{12} = 0 \quad \text{and} \quad c_{22} = \tfrac{1}{2}$$

The solutions associated with the double root $m = 1$ are therefore

$$X_3 = Be^t = \begin{bmatrix} 1 \\ -1 \end{bmatrix} e^t \quad \text{and} \quad X_4 = C_1 te^t + C_2 e^t = \begin{bmatrix} 1 \\ -1 \end{bmatrix} te^t + \begin{bmatrix} 0 \\ \frac{1}{2} \end{bmatrix} e^t \, \dagger$$

The complete solution of the original system is now

$$X = \begin{bmatrix} x_1 \\ x_2 \end{bmatrix} = k_1 X_1 + k_2 X_2 + k_3 X_3 + k_4 X_4$$

or, in scalar form,

$$x_1 = e^{-t}[(k_1 + 2k_2) \cos 2t - (2k_1 - k_2) \sin 2t] + k_3 e^t + k_4 te^t$$

$$x_2 = e^{-t}(k_2 \cos 2t - k_1 \sin 2t) - (k_3 - \tfrac{1}{2}k_4)e^t - k_4 te^t$$

To find a particular integral for the nonhomogeneous system (2) in the usual case in which $F(t)$ is a vector having only a finite number of linearly independent derivatives, we proceed very much as in the case of a single scalar differential equation. At the outset, it is convenient to identify the independent functions $\phi_1(t)$, $\phi_2(t), \ldots,$ $\phi_j(t)$ which appear in the components of $F(t)$ and then express $F(t)$ in the form

$$F(t) = K_1 \phi_1(t) + K_2 \phi_2(t) + \cdots + K_j \phi_j(t)$$

where the K's are appropriate constant column matrices. Then for such terms as do not duplicate vectors already in the complementary function, trial particular integrals can be constructed as described in Table 2.2, provided that the arbitrary scalar constants appearing in the entries in the table are replaced by arbitrary constant vectors. The trial solutions are then substituted into the nonhomogeneous system, and the arbitrary components of the coefficient vectors are determined to make the resulting equations identically true. For terms in the expanded form of $F(t)$ which duplicate vectors in the complementary function, the results of Table 2.2 are still valid with one additional provision: not only must the usual choice for a trial particular integral be multiplied by the lowest positive integral power of the independent variable which will eliminate the duplication, but the products of the normal choice and all lower nonnegative integral powers of the independent variable must also be included in the actual choice. An example should clarify the details of the procedure.

EXAMPLE 2

Find a particular integral for the system of equations in Example 1 if the terms $2e^t$ and $2e^t + e^{-2t}$ appear on the right-hand sides of the respective equations.

At the outset, it is convenient to group together like terms on the right-hand side of the equivalent matric equation by writing

$$P(D)X = F_1 + F_2$$

where

$$F_1 = \begin{bmatrix} 0 \\ e^{-2t} \end{bmatrix} = \begin{bmatrix} 0 \\ 1 \end{bmatrix} e^{-2t} \quad \text{and} \quad F_2 = \begin{bmatrix} 2e^t \\ 2e^t \end{bmatrix} = \begin{bmatrix} 2 \\ 2 \end{bmatrix} e^t$$

Since F_1 does not duplicate any vector in the complementary function, i.e., the solution found in Example 1, we assume as a trial particular integral simply

$$X_p = Ae^{-2t}$$

---

† The most general solution of the equation $2c_{12} + 2c_{22} = 1$, namely, $c_{12} = \lambda$ and $c_{22} = (1 - 2\lambda)/2$, leads to the same expression for X_4 plus a term proportional to X_3, which can be combined with X_3 when the complete solution of the given system is constructed.

where $A = \begin{bmatrix} a \\ \alpha \end{bmatrix}$ is a constant matrix to be determined. Then, substituting, we have

$$P(D)X_p \equiv P(D)Ae^{-2t} = P(-2)Ae^{-2t} = F_1 \equiv \begin{bmatrix} 0 \\ 1 \end{bmatrix} e^{-2t}$$

or, dividing out e^{-2t},

$$P(-2)A \equiv \begin{bmatrix} 10 & -5 \\ -1 & 5 \end{bmatrix} \begin{bmatrix} a \\ \alpha \end{bmatrix} = \begin{bmatrix} 0 \\ 1 \end{bmatrix}$$

This is equivalent to the scalar system

$$10a - 5\alpha = 0$$
$$-a + 5\alpha = 1$$

from which it follows that $a = \frac{1}{9}$, $\alpha = \frac{2}{9}$, $A = \frac{1}{9}\begin{bmatrix} 1 \\ 2 \end{bmatrix}$ Thus $\frac{1}{9}\begin{bmatrix} 1 \\ 2 \end{bmatrix} e^{-2t}$ is one term in the particular integral we are seeking.

To find the terms in the particular integral arising from F_2, we note that since both e^t and te^t occur as terms in the complementary function, the normal choice for a trial particular integral, namely, B_1e^t, must be modified by multiplying it by t^2 *and including the terms* B_2te^t *and* B_3e^t, where

$$B_1 = \begin{bmatrix} b_1 \\ \beta_1 \end{bmatrix} \qquad B_2 = \begin{bmatrix} b_2 \\ \beta_2 \end{bmatrix} \qquad \text{and} \qquad B_3 = \begin{bmatrix} b_3 \\ \beta_3 \end{bmatrix}$$

are constant matrices to be determined. Then, substituting

$$X_p = B_1t^2e^t + B_2te^t + B_3e^t$$

into the equation $P(D)X = F_2$ and using the results of Exercises 9 and 11, we obtain

$$P(D)(B_1t^2e^t + B_2te^t + B_3e^t) = P(1)B_1t^2e^t + 2P'(1)B_1te^t + P''(1)B_1e^t$$
$$+ P(1)B_2te^t + P'(1)B_2e^t + P(1)B_3e^t$$
$$= F_1 = \begin{bmatrix} 2 \\ 2 \end{bmatrix} e^t$$

Hence, equating the coefficients of like terms on the two sides of this equation, we find that

$$P(1)B_1 = 0$$

(8) $$P(1)B_2 + 2P'(1)B_1 = 0$$

$$P(1)B_3 + P'(1)B_2 + P''(1)B_1 = \begin{bmatrix} 2 \\ 2 \end{bmatrix}$$

The first of these equations is simply

$$\begin{bmatrix} 10 & 10 \\ 2 & 2 \end{bmatrix} \begin{bmatrix} b_1 \\ \beta_1 \end{bmatrix} = \begin{bmatrix} 0 \\ 0 \end{bmatrix}$$

which implies that

$$B_1 \equiv \begin{bmatrix} b_1 \\ \beta_1 \end{bmatrix} = \begin{bmatrix} \lambda \\ -\lambda \end{bmatrix} \qquad \lambda \text{ arbitrary}$$

The second of the equations in (8) now becomes

$$\begin{bmatrix} 10 & 10 \\ 2 & 2 \end{bmatrix} \begin{bmatrix} b_2 \\ \beta_2 \end{bmatrix} + 2\begin{bmatrix} 3 & 8 \\ 1 & 2 \end{bmatrix} \begin{bmatrix} \lambda \\ -\lambda \end{bmatrix} = \begin{bmatrix} 0 \\ 0 \end{bmatrix}$$

or $$\begin{bmatrix} 5 & 5 \\ 1 & 1 \end{bmatrix} \begin{bmatrix} b_2 \\ \beta_2 \end{bmatrix} = \begin{bmatrix} 5\lambda \\ \lambda \end{bmatrix}$$

from which, without loss of generality (see Exercise 15) we conclude that

$$B_2 \equiv \begin{bmatrix} b_2 \\ \beta_2 \end{bmatrix} = \begin{bmatrix} \lambda \\ 0 \end{bmatrix}$$

The third equation in the set (8) now becomes

$$\begin{bmatrix} 10 & 10 \\ 2 & 2 \end{bmatrix}\begin{bmatrix} b_3 \\ \beta_3 \end{bmatrix} + \begin{bmatrix} 3 & 8 \\ 1 & 2 \end{bmatrix}\begin{bmatrix} \lambda \\ 0 \end{bmatrix} + \begin{bmatrix} 2 & 2 \\ 0 & 2 \end{bmatrix}\begin{bmatrix} \lambda \\ -\lambda \end{bmatrix} = \begin{bmatrix} 2 \\ 2 \end{bmatrix}$$

or

$$\begin{bmatrix} 10 & 10 \\ 2 & 2 \end{bmatrix}\begin{bmatrix} b_3 \\ \beta_3 \end{bmatrix} = \begin{bmatrix} 2 \\ 2 \end{bmatrix} - \begin{bmatrix} 3\lambda \\ \lambda \end{bmatrix} - \begin{bmatrix} 0 \\ -2\lambda \end{bmatrix} = \begin{bmatrix} 2 - 3\lambda \\ 2 + \lambda \end{bmatrix}$$

which is equivalent to the scalar system

$$10b_3 + 10\beta_3 = 2 - 3\lambda$$
$$2b_3 + 2\beta_3 = 2 + \lambda$$

Since the left member of the first equation is 5 times the left member of the second equation, it follows that the ratio of the right members must also be 5. Hence we must have

$$2 - 3\lambda = 5(2 + \lambda)$$

which implies that $\lambda = -1$. For this value of λ we may without loss of generality (see Exercise 15) take $b_3 = \frac{1}{2}$ and $\beta_3 = 0$. Thus

$$B_3 = \begin{bmatrix} \frac{1}{2} \\ 0 \end{bmatrix} \qquad B_2 = \begin{bmatrix} -1 \\ 0 \end{bmatrix} \qquad \text{and} \qquad B_1 = \begin{bmatrix} -1 \\ 1 \end{bmatrix}$$

Finally, putting our results together, we have the entire particular integral

$$\frac{1}{9}\begin{bmatrix} 1 \\ 2 \end{bmatrix} e^{-2t} + \begin{bmatrix} -1 \\ 1 \end{bmatrix} t^2 e^t + \begin{bmatrix} -1 \\ 0 \end{bmatrix} te^t + \begin{bmatrix} \frac{1}{2} \\ 0 \end{bmatrix} e^t$$

EXERCISES

Find a complete solution of each of the following systems:

1   $(3D + 1)x + (D + 7)y = e^{-t}$
    $(2D + 1)x + (D + 5)y = e^{-t}$

2   $(D + 5)x + (2D + 1)y = e^{-t} + e^t$
    $(D + 7)x + (3D + 1)y = 0$

3   $(D + 5)x + (D + 7)y = 2e^t$
    $(2D + 1)x + (3D + 1)y = e^t$

4   $(D + 2)x + (D + 3)y = -4$
    $(2D - 6)x + (3D - 4)y = 2$

5   $(D + 1)x + (D + 2)y = -e^t$
    $(3D + 1)x + (4D + 7)y = -7e^t$

6   $(D + 1)x + (D + 2)y = -t + 1$
    $(5D + 1)x + (6D + 3)y = -2t + 1$

7   $(2D + 1)x + (D + 2)y = e^{-t}$
    $(3D - 7)x + (3D + 1)y = 0$

8   $(2D + 1)x + (D + 2)y = \sin t$
    $(3D + 1)x + (3D + 5)y = \cos t$

9   Show that $D^r(te^{mt}) = m^r te^{mt} + rm^{r-1}e^{mt}$. Hence show that

$$p(D)te^{mt} = p(m)te^{mt} + p'(m)e^{mt}$$

and

$$P(D)te^{mt} = P(m)te^{mt} + P'(m)e^{mt}$$

where $p(D)$ is a polynomial in the operator D and $P(D)$ is a matrix whose elements are polynomials in D.

10   Using the results of Exercise 9, show that if $m = m_1$ is a double root of the characteristic equation $|P(m)| = 0$, then $X_1 = Ae^{m_1 t}$ and $X_2 = B_1 te^{m_1 t} + B_2 e^{m_1 t}$ are two independent solutions of the system $P(D)X = 0$ provided the matric coefficients A, B_1, and B_2 satisfy the equations

$$P(m_1)A = 0, \; P(m_1)B_1 = 0 \quad \text{and} \quad P(m_1)B_2 + P'(m_1)B_1 = 0$$

11   Show that

$$D^r(t^2 e^{mt}) = m^r t^2 e^{mt} + 2rm^{r-1}te^{mt} + r(r-1)m^{r-2}e^{mt}$$

Hence show that

$$p(D)t^2 e^{mt} = p(m)t^2 e^{mt} + 2p'(m)te^{mt} + p''(m)e^{mt}$$

and

$$P(D)t^2 e^{mt} = P(m)t^2 e^{mt} + 2P'(m)te^{mt} + P''(m)e^{mt}$$

where $p(D)$ is a polynomial in the operator D and $P(D)$ is a matrix whose elements are polynomials in D.

12 Using the results of Exercise 11, show that if $m = m_1$ is a triple root of the characteristic equation $|P(m)| = 0$, then $X_1 = Ae^{m_1 t}$, $X_2 = B_1 t e^{m_1 t} + B_2 e^{m_1 t}$, and $X_3 = C_1 t^2 e^{m_1 t} + C_2 t e^{m_1 t} + C_3 e^{m_1 t}$ are three independent solutions of the system $P(D)X = 0$ provided the matric coefficients A, B_1, B_2, C_1, C_2, C_3 satisfy the equations

$$P(m_1)A = 0 \qquad P(m_1)B_1 = 0 \qquad P(m_1)B_1 + P'(m_1)B_2 = 0$$

$$P(m_1)C_1 = 0 \qquad P(m_1)C_2 + 2P'(m_1)C_1 = 0$$

$$P(m_1)C_3 + P'(m_1)C_2 + P''(m_1)C_1 = 0$$

13 Generalize the results of Exercises 9 and 10 to the function $t^3 e^{mt}$.

14 Generalize the results of Exercises 10 and 12 to the solution of the equation $P(D)X = 0$ arising from a fourfold root of the characteristic equation.

15 (a) In Example 2, verify that no generality is lost in taking $b_2 = \lambda$, $\beta_2 = 0$ by showing that if the general solution $b_2 = g + \lambda$, $\beta_2 = -g$ is used, the term arising from the arbitrary constant g can be absorbed in the term X_4 in the complementary function.

(b) In Example 2, verify that no generality is lost in taking $b_3 = \frac{1}{2}$ and $\beta_3 = 0$.

16 If A, B, and C are constant column matrices and $P(D)$ is a matrix whose elements are polynomials in the operator D, show that

(a) $P(D)A = P(0)A$ (b) $P(D)Bt = P(0)Bt + P'(0)B$

(c) $P(D)Ct^2 = P(0)Ct^2 + 2P'(0)Ct + P''(0)C$

17 In Exercise 10, verify that if $m = m_1$ is a double root of the characteristic equation of a system of two simultaneous differential equations, then the equations

$$P(m_1)B_1 = 0 \qquad \text{and} \qquad P(m_1)B_2 + P'(m_1)B_1 = 0$$

are always solvable for nontrivial vectors B_1 and B_2.

CHAPTER 11
Further Properties of Matrices

11.1 Quadratic Forms

In this section we continue our study of matrices by introducing the important mathematical objects known as *quadratic forms, hermitian forms*, and *bilinear forms*.

By a **quadratic form** we mean a homogeneous second-degree expression in n variables of the form

$$Q(x) = a_{11}x_1^2 + 2a_{12}x_1x_2 + \cdots + 2a_{1n}x_1x_n$$
$$+ \; a_{22}x_2^2 + \cdots + 2a_{2n}x_2x_n$$
$$+ \cdots \cdots \cdots \cdots \cdots \cdots \cdots$$
$$+ \; a_{nn}x_n^2$$

In order that matric notation can be applied to quadratic forms, it is customary to separate each of the cross products into two equal terms and introduce additional coefficients for the new terms by the definition $a_{ji} = a_{ij}$. When this has been done, $Q(x)$ can be expressed in the more symmetric form

$$Q(x) = \quad a_{11}x_1^2 + a_{12}x_1x_2 + \cdots + a_{1n}x_1x_n$$
$$+ a_{21}x_2x_1 + a_{22}x_2^2 + \cdots + a_{2n}x_2x_n$$
$$+ \cdots \cdots \cdots \cdots \cdots \cdots \cdots \cdots \cdots \cdots \cdots \qquad a_{ji} = a_{ij}$$
$$+ a_{n1}x_nx_1 + a_{n2}x_nx_2 + \cdots + a_{nn}x_n^2$$

If we now define the matrices

$$X = \begin{bmatrix} x_1 \\ x_2 \\ \vdots \\ x_n \end{bmatrix} \quad \text{and} \quad A = \begin{bmatrix} a_{11} & a_{12} & \cdots & a_{1n} \\ a_{21} & a_{22} & \cdots & a_{2n} \\ \cdots \cdots \cdots \cdots \cdots \\ a_{n1} & a_{n2} & \cdots & a_{nn} \end{bmatrix} \qquad a_{ji} = a_{ij}$$

it is clear from the definition of matric multiplication that $Q(x)$ can be written in the compact form

$$Q(x) = X^T A X$$

where, since $a_{ji} = a_{ij}$, the **matrix of the quadratic form** A is necessarily symmetric.

If a quadratic form with real coefficients has the property that it is equal to or greater than zero for all real values of its variables, it is said to be **positive**. A positive form which is zero *only* for the values $x_1 = x_2 = \cdots = x_n = 0$ is said to be **positive-definite**. A positive form which is zero for real values other than $x_1 = x_2 = \cdots = x_n = 0$ is said to be **positive-semidefinite**. A quadratic form with real coefficients

Table 11.1

Type of quadratic form	Example
Positive-definite	$x_1^2 + x_2^2$
Negative-definite	$-(x_1^2 + x_2^2)$
Positive-semidefinite	$(x_1 - x_2)^2$
Negative-semidefinite	$-(x_1 - x_2)^2$
Indefinite	$x_1^2 - x_2^2$

which is equal to or less than zero for all real values of its variables is said to be **negative**. A negative form which is zero *only* for the values $x_1 = x_2 = \cdots = x_n = 0$ is said to be **negative-definite**. A negative form which is zero for real values other than $x_1 = x_2 = \cdots = x_n = 0$ is said to be **negative-semidefinite**. Clearly, a negative-definite or negative-semidefinite form can be converted into a positive form of corresponding type by multiplying it by -1. A quadratic form which can take on both positive and negative values for real values of its variables is said to be **indefinite**. Examples of quadratic forms of each type are shown in Table 11.1.

The matrix A of the quadratic form $Q(x) = X^T A X$ is said to be **positive-definite**, **negative-definite**, **positive-semidefinite**, **negative-semidefinite**, or **indefinite** according to the nature of $Q(x)$. Correspondingly, $Q(x)$ is said to be **singular** or **nonsingular** according as its matrix A is singular or nonsingular, i.e., according as $|A|$ is equal to or different from zero.

A quadratic form which is definite is necessarily nonsingular. In fact, if we suppose that $Q(x)$ is both definite and singular, we are led at once to a contradiction, as follows. Let us first write $Q(x)$ in the partially factored form

$$Q(x) = \quad (a_{11}x_1 + a_{12}x_2 + \cdots + a_{1n}x_n)x_1$$
$$+ (a_{21}x_1 + a_{22}x_2 + \cdots + a_{2n}x_n)x_2$$
$$+ \cdots\cdots\cdots\cdots\cdots\cdots\cdots\cdots\cdots$$
$$+ (a_{n1}x_1 + a_{n2}x_2 + \cdots + a_{nn}x_n)x_n$$

Then, assuming that $|A| = 0$, it follows from Corollary 1 of Theorem 8, Sec. 10.5, that there is a nontrivial solution of the system of equations obtained by equating to zero the expressions in parentheses in $Q(x)$. Finally, we observe that for the values of the x's in this (nontrivial) solution, $Q(x)$ itself is equal to zero, contrary to the hypothesis that it is definite.

The converse of the preceding observation is not true. In other words, a nonsingular quadratic form is not necessarily definite. For example, the form $x_1^2 - 4x_1x_2 + 3x_2^2 + 2x_3^2$ is nonsingular since the determinant of its matrix, namely,

$$\begin{vmatrix} 1 & -2 & 0 \\ -2 & 3 & 0 \\ 0 & 0 & 2 \end{vmatrix} = -2$$

is different from zero, yet it is not definite since it is zero for the nontrivial set of values $x_1 = 3$, $x_2 = 1$, $x_3 = 0$.

The complete criterion for the definiteness of a quadratic form is contained in the following theorem, for whose proof we must refer to texts on higher algebra.†

---

† See, for instance, W. L. Ferrar, "Algebra," pp. 138–141, Oxford, New York, 1941.

THEOREM 1 A necessary and sufficient condition that the real quadratic form $X^T A X$ be positive-definite (negative-definite) is that the quantities

$$a_{11}, \quad \begin{vmatrix} a_{11} & a_{12} \\ a_{21} & a_{22} \end{vmatrix}, \quad \begin{vmatrix} a_{11} & a_{12} & a_{13} \\ a_{21} & a_{22} & a_{23} \\ a_{31} & a_{32} & a_{33} \end{vmatrix}, \dots, \quad \begin{vmatrix} a_{11} & \cdots & a_{1n} \\ \cdots & \cdots & \cdots \\ a_{n1} & \cdots & a_{nn} \end{vmatrix}$$

should all be positive (alternate in sign with a_{11} negative).

Clearly, sets of necessary and sufficient conditions for the definiteness of $X^T A X$ equivalent to those given in Theorem 1 can be obtained by first permuting the variables in $X^T A X$ and then applying Theorem 1. Recalling the definition of *principal minors* given in Sec. 10.2, this yields the following more general results.

THEOREM 2 A necessary and sufficient condition that the real quadratic form $X^T A X$ be positive-definite is that every principal minor of A be positive.

THEOREM 3 A necessary and sufficient condition that the real quadratic form $X^T A X$ be negative-definite is that every principal minor of A of odd order be negative and every principal minor of even order be positive.

EXAMPLE 1

The quadratic form

$$[x_1 \quad x_2 \quad x_3] \begin{bmatrix} 1 & 2 & -2 \\ 2 & 5 & -4 \\ -2 & -4 & 5 \end{bmatrix} \begin{bmatrix} x_1 \\ x_2 \\ x_3 \end{bmatrix} = \begin{matrix} x_1{}^2 + 2x_1 x_2 - 2x_1 x_3 \\ +2x_2 x_1 + 5x_2{}^2 - 4x_2 x_3 \\ -2x_3 x_1 - 4x_3 x_2 + 5x_3{}^2 \end{matrix}$$

is positive-definite, since the three quantities

$$1 \qquad \begin{vmatrix} 1 & 2 \\ 2 & 5 \end{vmatrix} = 1 \quad \text{and} \quad \begin{vmatrix} 1 & 2 & -2 \\ 2 & 5 & -4 \\ -2 & -4 & 5 \end{vmatrix} = 1$$

are all positive. Moreover, all the other principal minors, namely, the diagonal elements

$$a_{22} = 5 \qquad \text{and} \qquad a_{33} = 5$$

and the second-order determinants

$$\begin{vmatrix} a_{11} & a_{13} \\ a_{31} & a_{33} \end{vmatrix} = \begin{vmatrix} 1 & -2 \\ -2 & 5 \end{vmatrix} = 1 \quad \text{and} \quad \begin{vmatrix} a_{22} & a_{23} \\ a_{32} & a_{33} \end{vmatrix} = \begin{vmatrix} 5 & -4 \\ -4 & 5 \end{vmatrix} = 9$$

are also positive, in accordance with Theorem 2. In this case the quadratic form can be written equivalently as

$$(x_1 + 2x_2 - 2x_3)^2 + x_2{}^2 + x_3{}^2$$

which, being a sum of squares, can vanish only if

$$x_1 + 2x_2 - 2x_3 = 0 \qquad x_2 = 0 \qquad \text{and} \qquad x_3 = 0$$

and these, in turn, can hold simultaneously only if $x_1 = x_2 = x_3 = 0$.

On the other hand, the quadratic form

$$[x_1 \quad x_2 \quad x_3] \begin{bmatrix} 1 & 2 & -2 \\ 2 & 3 & -4 \\ -2 & -4 & 5 \end{bmatrix} \begin{bmatrix} x_1 \\ x_2 \\ x_3 \end{bmatrix} = \begin{matrix} x_1{}^2 + 2x_1 x_2 - 2x_1 x_3 \\ +2x_2 x_1 + 3x_2{}^2 - 4x_2 x_3 \\ -2x_3 x_1 - 4x_3 x_2 + 5x_3{}^2 \end{matrix}$$

is not definite, since the three quantities

$$1 \qquad \begin{vmatrix} 1 & 2 \\ 2 & 3 \end{vmatrix} = -1 \quad \text{and} \quad \begin{vmatrix} 1 & 2 & -2 \\ 2 & 3 & -4 \\ -2 & -4 & 5 \end{vmatrix} = -1$$

do not fulfill either of the conditions of Theorem 1. In fact, this quadratic form can be written as

$$(x_1 + 2x_2 - 2x_3)^2 - x_2{}^2 + x_3{}^2$$

and since this expression takes on the value 1 when $x_1 = 2$, $x_2 = 0$, $x_3 = 1$ and takes on the value -1 when $x_1 = -2$, $x_2 = 1$, $x_3 = 0$, it is actually indefinite.

In our definition of a quadratic form, neither the matrix of coefficients A nor the matrix of unknowns X was restricted to be real. However, in most elementary applications both A and X will be real, and only for real-valued quadratic forms are such properties as definiteness and indefiniteness defined. Actually, when complex quantities are involved, quadratic forms, as we have defined them, are almost always replaced by related expressions known as **hermitian forms**.

DEFINITION 1 If A is a hermitian matrix, the expression $\overline{X}^T AX$ is known as a hermitian form.

Recalling the definition of a hermitian matrix (Sec. 10.2), it is easy to verify that any hermitian form is equal to its transposed conjugate. Moreover, since such a form is a scalar, i.e., a 1×1 matrix, it is also equal to its transpose. Thus, since its transpose is equal to the conjugate of its transpose, it must be real and we have the following result.

THEOREM 4 The value of a hermitian form is real for all values of its variables.

Because of Theorem 4, positive- and negative-definite, positive- and negative-semidefinite, and indefinite hermitian forms can be defined precisely as the corresponding types of quadratic forms were defined. Moreover, it can be shown that the criteria for definiteness contained in Theorems 1 to 3 hold without change for hermitian forms.

Closely associated with quadratic forms are what are known as **bilinear forms**.

DEFINITION 2 If A is a symmetric matrix, the expression $Y^T AX$ is known as a bilinear form.

Clearly, if $Y = X$, the bilinear form $Y^T AX$ becomes the quadratic form $X^T AX$. If the components of Y are thought of as the coordinates of a "point" in a hyperspace of the appropriate number of dimensions, the bilinear form $Y^T AX$ is sometimes called the **polar** of the point Y with respect to the quadratic form $X^T AX$.

It is interesting to note that the scalar product of two vectors Y and X, namely, $Y^T X$, can be thought of as the bilinear form $Y^T IX$. The condition that Y and X be orthogonal is, then, just the condition that the bilinear form $Y^T IX$ be equal to zero. This suggests that the simple notion of orthogonality introduced in Sec. 10.2, by analogy with the familiar results of solid analytic geometry, be extended to include the concept of **generalized orthogonality**.

DEFINITION 3 Two vectors X and Y are said to be orthogonal with respect to a symmetric matrix A if the bilinear form $Y^T A X$ is equal to zero.

In the spirit of Definition 3, the notion of the length of a vector can also be generalized. In fact, the definition of the length of a vector X introduced in Sec. 10.2, namely, $\sqrt{X^T X}$, can be rewritten $\sqrt{X^T I X}$, and this suggests

$$\sqrt{X^T A X}$$

as the generalized length of the vector X with respect to the symmetric matrix A. This is meaningful, however, only if the quantity under the radical is positive. Hence it is necessary to require further that A be the matrix of a positive-definite quadratic form. A vector whose generalized length with respect to a given symmetric, positive-definite matrix is 1 is said to be **normalized** with respect to that matrix. A vector X can always be normalized with respect to a given symmetric, positive-definite matrix A by dividing it by the positive quantity $\sqrt{X^T A X}$.

Clearly, the Schmidt orthogonalization process, which we discussed at the end of Sec. 10.5, can be carried out equally well using the concepts of generalized orthogonality and generalized length. The notion of orthogonality with respect to a symmetric, nonunit matrix will be of considerable importance in the work of this chapter.

Just as it was convenient in analytic geometry to be able to remove the cross-product term from the equation of a conic, so in many applications involving quadratic forms it is desirable to be able to remove the cross-product terms by a suitable transformation and express the quadratic form as a sum of squares. There are numerous ways of doing this, among which the following, due to Lagrange, is particularly effective. The general idea is first to group together all terms containing one of the variables as a factor, say the terms containing x_1, and then, by suitable manipulations, to convert this part of the expression into a perfect square. Then among the terms not included in this square, those which contain one of the remaining variables, say x_2, are converted into a perfect square; and so on, until the process terminates.

To begin the process, let us assume that $a_{11} \neq 0$ and group together all terms containing x_1 as a factor:

$$(a_{11}x_1{}^2 + 2a_{12}x_1 x_2 + \cdots + 2a_{1n}x_1 x_n) + \sum_{i,j=2}^{n} a_{ij}x_i x_j$$

$$= a_{11}\left(x_1{}^2 + \frac{2a_{12}}{a_{11}}x_1 x_2 + \cdots + \frac{2a_{1n}}{a_{11}}x_1 x_n\right) + \phi_1(x_2, x_3, \ldots, x_n)$$

Now, adding and subtracting the appropriate squares and cross-product terms, none of which involves x_1, we have

$$a_{11}\left[\left(x_1 + \frac{a_{12}}{a_{11}}x_2 + \cdots + \frac{a_{1n}}{a_{11}}x_n\right)^2 - \left(\frac{a_{12}}{a_{11}}\right)^2 x_2{}^2 - \cdots - \left(\frac{a_{1n}}{a_{11}}\right)^2 x_n{}^2\right.$$

$$\left. - 2\frac{a_{12}}{a_{11}}\frac{a_{13}}{a_{11}}x_2 x_3 - \cdots - 2\frac{a_{1,n-1}}{a_{11}}\frac{a_{1n}}{a_{11}}x_{n-1}x_n\right] + \phi_1(x_2, x_3, \ldots, x_n)$$

$$= a_{11}\left(x_1 + \frac{a_{12}}{a_{11}}x_2 + \cdots + \frac{a_{1n}}{a_{11}}x_n\right)^2 + \phi_2(x_2, x_3, \ldots, x_n)$$

The obviously nonsingular transformation

$$T_1: \quad \begin{aligned} y_1 &= x_1 + \frac{a_{12}}{a_{11}} x_2 + \cdots + \frac{a_{1n}}{a_{11}} x_n \\ y_2 &= \qquad\qquad x_2 \\ &\cdots\cdots\cdots\cdots\cdots\cdots\cdots\cdots\cdots \\ y_n &= \qquad\qquad\qquad\qquad\qquad x_n \end{aligned}$$

now reduces $Q(x)$ to the form

$$a_{11} y_1{}^2 + \phi_2(y_2, y_3, \ldots, y_n)$$

where $\phi_2(y_2, y_3, \ldots, y_n)$ is, of course, a quadratic form in the $n - 1$ variables $y_2, y_3, \ldots, y_n$ with coefficients b_{ij}, say.

The same process can, in general, be applied to ϕ_2, and a second nonsingular transformation, of the form

$$T_2: \quad \begin{aligned} z_1 &= y_1 \\ z_2 &= \qquad y_2 + \frac{b_{23}}{b_{22}} y_3 + \cdots + \frac{b_{2n}}{b_{22}} y_n \\ z_3 &= \qquad\qquad\qquad y_3 \\ &\cdots\cdots\cdots\cdots\cdots\cdots\cdots\cdots\cdots \\ z_n &= \qquad\qquad\qquad\qquad\qquad\qquad y_n \end{aligned}$$

extends the reduction to

$$a_{11} z_1{}^2 + b_{22} z_2{}^2 + \phi_3(z_3, z_4, \ldots, z_n)$$

The continuation is now obvious, and the required transformation is, finally, the product, or composition, of the successive transformations $T_1, T_2, \ldots, T_n$.

If at any stage all square terms are missing from the form $\phi_k(u_k, u_{k+1}, \ldots, u_n)$, the process must be modified. If this occurs, either no more terms remain and the reduction is complete, or else there is at least one cross-product term with nonzero coefficient, say $u_i u_j$. If this is the case, the nonsingular transformation

$$\begin{aligned} u_1 &= v_1 \\ &\cdots\cdots\cdots \\ u_{i-1} &= v_{i-1} \\ u_i &= v_i + v_j \\ u_{i+1} &= v_{i+1} \\ &\cdots\cdots\cdots \\ u_{j-1} &= v_{j-1} \\ u_j &= v_i - v_j \\ u_{j+1} &= v_{j+1} \\ &\cdots\cdots\cdots \\ u_n &= v_n \end{aligned}$$

will clearly introduce the square terms $v_i{}^2$ and $v_j{}^2$, after which the process can be continued in its original form.

It is important to note that the linear transformation employed at each stage of the reduction process is rank-preserving. Hence, since the rank of a diagonal matrix is equal to the number of its nonzero diagonal elements, it follows that when X^TAX is transformed into a sum of squares by the Lagrange reduction, the number of square terms present in the final result is equal to the rank of the matrix of the original form. It is also clear that when a positive-definite quadratic form is reduced to a sum of squares by the Lagrange reduction, the final result must consist of the square of each variable with a *positive* coefficient.

EXAMPLE 2

Find a transformation which will reduce to a sum of squares the quadratic form X^TAX, where

$$A = \begin{bmatrix} 1 & -1 & 0 & 2 \\ -1 & 4 & 6 & 4 \\ 0 & 6 & 11 & 8 \\ 2 & 4 & 8 & 8 \end{bmatrix}$$

Following the Lagrange procedure, we first group together the terms containing x_1 as a factor and then complete the square on these terms:

$$(x_1{}^2 - 2x_1x_2 + 4x_1x_4) + (4x_2{}^2 + 12x_2x_3 + 8x_2x_4 + 11x_3{}^2 + 16x_3x_4 + 8x_4{}^2)$$

$$= [(x_1 - x_2 + 2x_4)^2 - x_2{}^2 + 4x_2x_4 - 4x_4{}^2]$$
$$+ (4x_2{}^2 + 12x_2x_3 + 8x_2x_4 + 11x_3{}^2 + 16x_3x_4 + 8x_4{}^2)$$

$$= (x_1 - x_2 + 2x_4)^2 + (3x_2{}^2 + 12x_2x_3 + 12x_2x_4 + 11x_3{}^2 + 16x_3x_4 + 4x_4{}^2)$$

We can now apply the transformation

$$y_1 = x_1 - x_2 + 2x_4 \qquad y_2 = x_2 \qquad y_3 = x_3 \qquad y_4 = x_4$$

getting

$$y_1{}^2 + (3y_2{}^2 + 12y_2y_3 + 12y_2y_4 + 11y_3{}^2 + 16y_3y_4 + 4y_4{}^2)$$

and then continue the process in terms of the y's. However, it is probably a little simpler to continue the reduction in terms of the x's, since doing this makes it unnecessary at the end to determine the composition of the successive transformations we obtain. Hence, incorporating the terms involving x_2, then x_3, and finally x_4 into expressions which are perfect squares, we have

$$(x_1 - x_2 + 2x_4)^2 + 3(x_2{}^2 + 4x_2x_3 + 4x_2x_4) + (11x_3{}^2 + 16x_3x_4 + 4x_4{}^2)$$

$$= (x_1 - x_2 + 2x_4)^2 + [3(x_2 + 2x_3 + 2x_4)^2 - 12x_3{}^2 - 24x_3x_4 - 12x_4{}^2)]$$
$$+ (11x_3{}^2 + 16x_3x_4 + 4x_4{}^2)$$

$$= (x_1 - x_2 + 2x_4)^2 + 3(x_2 + 2x_3 + 2x_4)^2 - (x_3{}^2 + 8x_3x_4 + 8x_4{}^2)$$

$$= (x_1 - x_2 + 2x_4)^2 + 3(x_2 + 2x_3 + 2x_4)^2 - (x_3{}^2 + 8x_3x_4 + 16x_4{}^2) + 8x_4{}^2$$

$$= (x_1 - x_2 + 2x_4)^2 + 3(x_2 + 2x_3 + 2x_4)^2 - (x_3 + 4x_4)^2 + 8x_4{}^2$$

Thus, identifying the squares in the last expression as w_1, w_2, w_3, w_4, respectively, i.e., by applying the transformation

$$T: \quad W = MX \quad \text{where } M = \begin{bmatrix} 1 & -1 & 0 & 2 \\ 0 & 1 & 2 & 2 \\ 0 & 0 & 1 & 4 \\ 0 & 0 & 0 & 1 \end{bmatrix}$$

we change the original quadratic form into the sum of squares

$$w_1{}^2 + 3w_2{}^2 - w_3{}^2 + 8w_4{}^2$$

To verify that this transformation actually reduces $X^T A X$ to a sum of squares, it is necessary that T be solved for X, so that we can substitute for X in the expression $X^T A X$. To do this, we multiply both sides of the equation $W = MX$ by the inverse of M, which surely exists since M is nonsingular. This gives us

$$T^{-1}: \quad X = M^{-1}W \quad \text{where } M^{-1} = \begin{bmatrix} 1 & 1 & -2 & 4 \\ 0 & 1 & -2 & 6 \\ 0 & 0 & 1 & -4 \\ 0 & 0 & 0 & 1 \end{bmatrix}$$

Under T^{-1}, the original quadratic form $X^T A X$ becomes

$$(M^{-1}W)^T A (M^{-1}W) = W^T (M^{-1})^T A M^{-1} W = W^T [(M^{-1})^T A M^{-1}] W$$

and it is easy to verify that $(M^{-1})^T A M^{-1}$ is indeed the diagonal matrix

$$\begin{bmatrix} 1 & 0 & 0 & 0 \\ 0 & 3 & 0 & 0 \\ 0 & 0 & -1 & 0 \\ 0 & 0 & 0 & 8 \end{bmatrix}$$

EXERCISES

1 Classify each of the following quadratic forms:
   (a) $3x_1^2 + 3x_2^2 + 6x_3^2 - 2x_1x_2 - 4x_1x_3$
   (b) $-x_1^2 - 3x_2^2 - 5x_3^2 + 2x_1x_2 + 2x_1x_3 + 2x_2x_3$
   (c) $x_1^2 + 4x_2^2 + 4x_3^2 + 4x_1x_2 + 4x_1x_3 + 6x_2x_3$
   (d) $2x_1^2 + 2x_2^2 + x_3^2 + 2x_1x_3 + 2x_2x_3$
   (e) $x_1^2 + 2x_2^2 + 2x_3^2 + 4x_4^2 - 2x_1x_2 + 2x_2x_3 + 6x_3x_4$

2 Find a transformation which will reduce each of the following quadratic forms to a sum of squares:
   (a) $x_1^2 + 5x_2^2 + 2x_3^2 + 4x_1x_2 + 2x_1x_3 + 6x_2x_3$
   (b) $x_1^2 + x_2^2 + 3x_3^2 + 4x_4^2 + 2x_1x_2 + 4x_1x_3 + 2x_2x_3 + 2x_2x_4 + 6x_3x_4$
   (c) $x_1^2 + 5x_2^2 + 5x_3^2 + 2x_4^2 - 2x_1x_2 + 4x_1x_3 + 2x_1x_4 - 6x_2x_4 + 2x_3x_4$
   (d) $x_1^2 + 4x_2^2 + 2x_3^2 + 4x_4^2 + 4x_1x_2 + 2x_1x_3 + 2x_1x_4 + 2x_2x_3$
   $$\qquad\qquad\qquad\qquad\qquad\qquad\qquad\qquad\qquad\qquad - 2x_2x_4 - 2x_3x_4$$
   (e) $x_1x_2 + x_3x_4$ (f) $x_1x_2 + x_3x_4 + x_4x_5 + x_5x_6$

3 Obtain two more linear transformations each of which will reduce the quadratic form of Example 2 to a sum of squares. Do the sums of squares obtained in the respective cases have any common characteristic?

4 Show that the type of a quadratic form is not altered by the process of reducing it to a sum of squares.

5 Applying the result of Exercise 4 to the results of Exercise 2, determine the type of each of the quadratic forms in Exercise 2.

6 If A is a symmetric matrix, show that $Y^T A X = X^T A Y$.

7 If $f(X) = X^T A X$, where A is a symmetric matrix, show that

$$f(\lambda X + \mu Y) = \lambda^2 X^T A X + 2\lambda\mu X^T A Y + \mu^2 Y^T A Y$$

8 If $X^T A X$ is a positive-definite quadratic form, show that

$$(X^T A Y)^2 \le (X^T A X)(Y^T A Y)$$

the equality sign holding if and only if either X or Y is a null vector or $X = Y$. *Hint:* Use the result of Exercise 7.

9 Prove that a nonsingular quadratic form cannot be semidefinite. *Hint:* Use the result of Exercise 4.

10 If V is a vector for which an indefinite quadratic form $Q = X^T A X$ is equal to zero, and if ε is an arbitrary (small) positive number, show that there are vectors each of whose components differs from the corresponding component of V by less than ε for which Q is positive and also such vectors for which Q is negative.

11   Show that the bilinear form $Y^T A X$, that is, the polar of Y with respect to the quadratic form $Q = X^T A X$, can be obtained by applying the so-called **polar operator**

$$\frac{1}{2}\left(y_1\frac{\partial}{\partial x_1} + y_2\frac{\partial}{\partial x_2} + \cdots + y_n\frac{\partial}{\partial x_n}\right)$$

to Q.

12   (a) Show that regardless of the character of $X^T A X$ and $X^T B X$, there are always singular quadratic forms in the family $\lambda X^T A X + \mu X^T B X$. In general, how many singular forms are there in such a family?

(b) If $A = \begin{bmatrix} 1 & 0 & 0 \\ 0 & 0 & 1 \\ 0 & 1 & 0 \end{bmatrix}$ and $B = \begin{bmatrix} 0 & 1 & 1 \\ 1 & 0 & -1 \\ 1 & -1 & 0 \end{bmatrix}$, determine the singular

quadratic forms in the family

$$\lambda X^T A X + \mu X^T B X$$

13   Find the potential energy stored in the system shown in Fig. 10.1 as a result of the displacements x_1, x_2, and x_3, and show that it is a positive-definite quadratic form in the x's. What is the relation of the matrix of this form to the matrix K discussed in Sec. 10.3 in connection with this system? *Hint:* Recall that the work required to stretch a spring of modulus k a distance s is equal to $\frac{1}{2}ks^2$.

14   Work Exercise 13 for the system shown in Fig. 10.2.

15   From the vectors

$$V_1 = \begin{bmatrix} 1 \\ 0 \\ 0 \end{bmatrix} \quad V_2 = \begin{bmatrix} 1 \\ 1 \\ 0 \end{bmatrix} \quad V_3 = \begin{bmatrix} 1 \\ 1 \\ 1 \end{bmatrix}$$

construct a set of vectors orthonormal with respect to the matrix $\begin{bmatrix} 1 & 0 & 0 \\ 0 & 2 & 0 \\ 0 & 0 & 3 \end{bmatrix}$.

16   Work Exercise 15 for the vectors

$$V_1 = \begin{bmatrix} 1 \\ 0 \\ 0 \end{bmatrix} \quad V_2 = \begin{bmatrix} 0 \\ 1 \\ 0 \end{bmatrix} \quad V_3 = \begin{bmatrix} 0 \\ 0 \\ 1 \end{bmatrix}$$

and the matrix $\begin{bmatrix} 1 & 1 & 0 \\ 1 & 2 & 0 \\ 0 & 0 & 2 \end{bmatrix}$.

11.2 The Characteristic Equation of a Matrix

In studying linear transformations of the form $Y = AX$, where

$$Y = \begin{bmatrix} y_1 \\ y_2 \\ \vdots \\ y_n \end{bmatrix} \quad A = \begin{bmatrix} a_{11} & a_{12} & \cdots & a_{1n} \\ a_{21} & a_{22} & \cdots & a_{2n} \\ \cdots\cdots\cdots\cdots\cdots\cdots \\ a_{n1} & a_{n2} & \cdots & a_{nn} \end{bmatrix} \quad X = \begin{bmatrix} x_1 \\ x_2 \\ \vdots \\ x_n \end{bmatrix}$$

it is an interesting and important problem to determine which vectors, if any, are left unchanged in direction. Since two nontrivial vectors have the same direction if and only if one is a nonzero scalar multiple of the other, this is equivalent to the question of determining those vectors X whose images Y are of the form $Y = \lambda X$, that is, those vectors X such that $AX = \lambda X$ or, equivalently, $AX = \lambda IX$ or, finally,

(1) $(A - \lambda I)X = \mathbf{0}$

Clearly, the matric equation $(A - \lambda I)X = \mathbf{0}$ is equivalent to the scalar system

$$
\begin{aligned}
(a_{11} - \lambda)x_1 + & & a_{12}x_2 + \cdots + & & a_{1n}x_n &= 0 \\
a_{21}x_1 + & (a_{22} - \lambda)x_2 + \cdots + & & & a_{2n}x_n &= 0 \\
& & \cdots\cdots\cdots\cdots\cdots\cdots\cdots\cdots\cdots\cdots & & & \\
a_{n1}x_1 + & & a_{n2}x_2 + \cdots + (a_{nn} - \lambda)x_n &= 0 &
\end{aligned}
$$

(1a)

and, according to Corollary 1 of Theorem 8, Sec. 10.5, a homogeneous system of this sort will have one or more nontrivial solutions if and only if the determinant of the coefficients is equal to zero. This condition, namely,

(2)
$$
|A - \lambda I| = \begin{vmatrix} a_{11} - \lambda & a_{12} & \cdots & a_{1n} \\ a_{21} & a_{22} - \lambda & \cdots & a_{2n} \\ \cdots\cdots\cdots\cdots\cdots\cdots\cdots\cdots\cdots & & & \\ a_{n1} & a_{n2} & \cdots & a_{nn} - \lambda \end{vmatrix} = 0
$$

is obviously a polynomial equation of degree n in the parameter λ with leading coefficient $(-1)^n$, say

(3) $\quad |A - \lambda I| = (-1)^n [\lambda^n - \beta_1 \lambda^{n-1} + \beta_2 \lambda^{n-2} + \cdots$
$$
+ (-1)^{n-1}\beta_{n-1}\lambda + (-1)^n\beta_n] = 0
$$

Both this equation and the equivalent equation obtained by dropping the factor $(-1)^n$ are known as the **characteristic equation** of the matrix A, and the expression in brackets is known as the **characteristic polynomial** of A. For values of λ which satisfy Eq. (3), and for these values only, the matric equation $(A - \lambda I)X = \mathbf{0}$ has nontrivial solution vectors. The n roots of Eq. (3), which of course need not be distinct, are called the **characteristic roots** or **characteristic values** of the matrix A, and the corresponding (nontrivial) solutions of Eq. (1) or Eq. (1a) are called the **characteristic vectors**† of A. Although we have introduced the idea of characteristic values and characteristic vectors of a matrix in connection with the specific problem of determining the vectors left invariant in direction by a linear transformation, these concepts are of fundamental importance in much of matrix theory and in many of its physical applications.

Since most of the applications we have in mind involve matrices which are either real and symmetric or hermitian, we shall for the most part limit the rest of our discussion to the characteristic values and characteristic vectors of such matrices. We begin, however, with several theorems which deal with the characteristic values and characteristic vectors of arbitrary square matrices.

Since the characteristic equation (3) of a square matrix A is a polynomial equation, its roots, say $\lambda_1, \lambda_2, \ldots, \lambda_n$, are connected with its coefficients $-\beta_1, \beta_2, -\beta_3, \ldots,$ $(-1)^n\beta_n$ by the well-known root-coefficient relations

(4)
$$
\begin{aligned}
\beta_1 &= \lambda_1 + \lambda_2 + \cdots + \lambda_n \\
\beta_2 &= \lambda_1\lambda_2 + \lambda_1\lambda_3 + \cdots + \lambda_{n-1}\lambda_n \\
\beta_3 &= \lambda_1\lambda_2\lambda_3 + \cdots + \lambda_{n-2}\lambda_{n-1}\lambda_n \\
& \cdots\cdots\cdots\cdots\cdots\cdots\cdots\cdots\cdots \\
\beta_n &= \lambda_1\lambda_2\lambda_3 \cdots \lambda_n
\end{aligned}
$$

---

† Some writers graft the German word *eigen* meaning *own*, *peculiar*, or *proper*, onto the words *values* and *vectors* and use the hybrid terms *eigenvalues* and *eigenvectors*.

Furthermore, if we set $\lambda = 0$ in Eq. (3), we obtain

$$(5) \qquad\qquad\qquad |A| = (-1)^{2n}\beta_n = \beta_n$$

Hence, from the last of Eqs. (4), we have

$$|A| = \lambda_1\lambda_2\lambda_3 \cdots \lambda_n$$

From this it follows that $|A|$ is zero if and only if at least one of the λ's is zero. Thus we have established the following theorem.

THEOREM 1   A matrix is singular if and only if at least one of its characteristic values is zero.

Equation (5) is only the first of a series of relations connecting the coefficients in Eq. (3) with the principal minors of A. For instance, if $|A - \lambda I|$ is written as the sum of 2^n determinants by repeated use of the addition theorem (Theorem 10, Sec. 10.1), it is clear that the terms containing the first power of λ are obtained by multiplying the term $-\lambda$ in each diagonal element of $|A - \lambda I|$ by the λ-free part of the cofactor of that element. Thus, the coefficient of λ in Eq. (3), namely,

$$(-1)^{2n-1}\beta_{n-1} = -\beta_{n-1}$$

is equal to

$$-(A_{11} + A_{22} + \cdots + A_{nn})$$

Hence it follows that

$$\beta_{n-1} = A_{11} + A_{22} + \cdots + A_{nn}$$

Similarly, the terms containing λ^2 in the expansion of $|A - \lambda I|$ are found by multiplying the terms containing $-\lambda$ in every pair of diagonal elements by the λ-free part of the algebraic complement of the second-order minor containing those diagonal elements. Thus the coefficient of λ^2 in Eq. (3), namely,

$$(-1)^{2n-2}\beta_{n-2} = \beta_{n-2}$$

is equal to

$$A_{12,12} + A_{13,13} + \cdots + A_{(n-1,n),(n-1,n)}$$

The continuation is obvious, and we therefore have the following theorem.

THEOREM 2   If $\lambda^n - \beta_1\lambda^{n-1} + \cdots + (-1)^{n-1}\beta_{n-1} + (-1)^n\beta_n = 0$ is the characteristic equation of a square matrix A, then β_i is equal to the sum of all the principal minors of order i in A.

For $i = 1$, we have, as a special case of Theorem 2, the relation $\beta_1 = \lambda_1 + \lambda_2 + \cdots + \lambda_n = a_{11} + a_{22} + \cdots + a_{nn}$. The quantity $a_{11} + a_{22} + \cdots + a_{nn}$ is called the **trace** of A.

The characteristic polynomial of a matrix A and hence the coefficients $\{\beta_i\}$ and the characteristic roots $\{\lambda_i\}$ have the interesting property of being invariant under any similarity transformation. More precisely, we have the following theorem.

THEOREM 3   If A and B are similar matrices, then A and B have the same characteristic polynomial.

Proof Let $|A - \lambda I|$ be the characteristic polynomial of the matrix A, and let B be a matrix similar to A; that is, let B be any matrix such that $B = S^{-1}AS$ for some nonsingular matrix S. Then the characteristic polynomial of B is

$$|B - \lambda I| = |S^{-1}AS - \lambda I| = |S^{-1}AS - \lambda S^{-1}IS|$$
$$= |S^{-1}(A - \lambda I)S| = |S^{-1}| \cdot |A - \lambda I| \cdot |S|$$

since the determinant of a product of square matrices is equal to the product of the determinants of the individual matrices. Moreover, by Corollary 2 of Theorem 1, Sec. 10.3, $|S^{-1}| \cdot |S| = 1$. Hence $|B - \lambda I| = |A - \lambda I|$, as asserted.

The next three theorems also deal with the characteristic values and characteristic vectors of arbitrary square matrices.

THEOREM 4 A characteristic vector of a square matrix cannot correspond to two distinct characteristic values.

Proof Let λ_1 and λ_2 be distinct characteristic values of a square matrix A, and let X_1 be a characteristic vector of A corresponding, if possible, to both λ_1 and λ_2. Then, simultaneously,

$$(A - \lambda_1 I)X_1 = 0 \quad \text{and} \quad (A - \lambda_2 I)X_1 = 0$$

Hence, subtracting,

$$(6) \qquad (\lambda_2 - \lambda_1)IX_1 = (\lambda_2 - \lambda_1)X_1 = 0$$

However, by hypothesis, $\lambda_1 \neq \lambda_2$. Moreover, a characteristic vector is, by definition, a *nontrivial* solution vector of $(A - \lambda I)X = 0$. Thus $X_1 \neq 0$, and therefore Eq. (6) cannot hold. Hence, the assumption that a characteristic vector can correspond to each of two distinct characteristic values must be abandoned, and the theorem is established.

THEOREM 5 If $X_1, X_2, \ldots, X_m$ ($m \leq n$) are characteristic vectors corresponding respectively to the distinct characteristic values $\lambda_1, \lambda_2, \ldots, \lambda_m$ of an $n \times n$ matrix A, then $X_1, X_2, \ldots, X_m$ are linearly independent.

Proof Let $X_1, X_2, \ldots, X_m$ be characteristic vectors corresponding respectively to the distinct characteristic values $\lambda_1, \lambda_2, \ldots, \lambda_m$ of a square matrix A, and let us suppose, contrary to the theorem, that $X_1, X_2, \ldots, X_m$ are dependent. More specifically, let us suppose that the maximum number of linearly independent vectors in the set is k, where $1 \leq k < m$, and, for convenience, let them be the first k X's. Then any relation of the form

$$\alpha_1 X_1 + \alpha_2 X_2 + \cdots + \alpha_k X_k = 0$$

implies that $\alpha_1 = \alpha_2 = \cdots = \alpha_k = 0$, but there does exist a nontrivial set of γ's, with $\gamma_{k+1} \neq 0$, such that

$$(7) \qquad \gamma_1 X_1 + \gamma_2 X_2 + \cdots + \gamma_k X_k + \gamma_{k+1} X_{k+1} = 0$$

Now multiply Eq. (7) on the left by the matrix A, getting

$$\gamma_1 AX_1 + \gamma_2 AX_2 + \cdots + \gamma_k AX_k + \gamma_{k+1} AX_{k+1} = 0$$

However, $AX_i = \lambda_i X_i$ for each i. Hence the last equation becomes

(8) $$\gamma_1 \lambda_1 X_1 + \gamma_2 \lambda_2 X_2 + \cdots + \gamma_k \lambda_k X_k + \gamma_{k+1} \lambda_{k+1} X_{k+1} = \mathbf{0}$$

If we now multiply Eq. (7) by λ_{k+1} and subtract the result from Eq. (8), we obtain

(9) $$(\lambda_1 - \lambda_{k+1})\gamma_1 X_1 + (\lambda_2 - \lambda_{k+1})\gamma_2 X_2 + \cdots + (\lambda_k - \lambda_{k+1})\gamma_k X_k = \mathbf{0}$$

Since $X_1, X_2, \ldots, X_k$ are linearly independent, by hypothesis, it follows that each coefficient in (9) is equal to zero. Hence, since $\lambda_i - \lambda_{k+1} \neq 0$ $(i = 1, 2, \ldots, k)$, it must be that

$$\gamma_i = 0 \qquad i = 1, 2, \ldots, k$$

But if this is the case, it follows from (7) that

$$\gamma_{k+1} X_{k+1} = \mathbf{0}$$

which is impossible, since neither the scalar γ_{k+1} nor the vector X_{k+1} is zero. This contradiction overthrows the possibility that the characteristic vectors $X_1, X_2, \ldots, X_m$ are linearly dependent, and the theorem is established.

In particular, if an $n \times n$ matrix has n distinct characteristic values, the last theorem tells us that it has n linearly independent characteristic vectors. Hence, using Corollary 1 of Theorem 13, Sec. 10.5, we have the following result.

COROLLARY 1   If the characteristic values of an $n \times n$ matrix A are all distinct, then A has n linearly independent characteristic vectors and any vector with n components can be expressed as a linear combination of the characteristic vectors of A.

Since the characteristic equation of an $n \times n$ matrix is always of degree n, it is obvious that if repeated roots are counted the appropriate number of times, such a matrix always has exactly n characteristic roots. With the same convention one might perhaps be able to say that an $n \times n$ matrix always has exactly n characteristic vectors. However, attempting to assign a multiplicity to a characteristic vector associated with a repeated characteristic root is completely artificial and without significance. The decisive consideration is the number of linearly independent characteristic vectors of a given matrix; hence it is of fundamental importance to know when more than one independent characteristic vector is associated with a repeated characteristic root. The next theorem gives us a partial answer to this question.

THEOREM 6   If λ_1 is a characteristic root of multiplicity r of an $n \times n$ matrix A, then the rank of $A - \lambda_1 I$ is equal to or greater than $n - r$.

Proof   If A is an $n \times n$ matrix, and if $\lambda = \lambda_1$ is a repeated root of multiplicity r of the characteristic equation $|A - \lambda I| = 0$, then when we write $\lambda = \lambda_1 + w$, so that $\lambda = \lambda_1$ corresponds to $w = w_1 = 0$, it is clear that $w_1 = 0$ is a repeated root of multiplicity r of the equation

$$|A - (\lambda_1 + w)I| = |(A - \lambda_1 I) - wI| = 0$$

Hence, the expanded form of the last equation, say

$$(-1)^n[w^n - \sigma_1 w^{n-1} + \sigma_2 w^{n-2} - \cdots + (-1)^{n-1}\sigma_{n-1}w + (-1)^n\sigma_n] = 0$$

must contain w^r as a factor and must therefore reduce to

$$(-1)^n[w^n - \sigma_1 w^{n-1} + \cdots + (-1)^{n-r}\sigma_{n-r}w^r] = 0$$

where $\sigma_{n-r} \neq 0$. Now, by Theorem 2, the coefficient σ_{n-r} is equal to the sum of all principal minors of order $n - r$ of the matrix $A - \lambda_1 I$. Hence, since $\sigma_{n-r} \neq 0$, at least one of these minors must be different from zero. In other words, the rank of $A - \lambda_1 I$ must be at least as great as $n - r$, as asserted.

If, for a particular root λ_1 of multiplicity r of an $n \times n$ matrix A, the equality sign holds in the assertion of Theorem 6, then, according to Theorem 7, Sec. 10.5, there are exactly $n - (n - r) = r$ linearly independent characteristic vectors associated with λ_1. Such a characteristic root is said to be **regular**. However, this is the exception rather than the rule, and in general there will be a single independent characteristic vector associated with a repeated root of any multiplicity. For instance, for the matrix

$$A = \begin{bmatrix} -3 & -7 & -5 \\ 2 & 4 & 3 \\ 1 & 2 & 2 \end{bmatrix}$$

we have

$$|A - \lambda I| = \begin{vmatrix} -3 - \lambda & -7 & -5 \\ 2 & 4 - \lambda & 3 \\ 1 & 2 & 2 - \lambda \end{vmatrix}$$

$$= -\lambda^3 + 3\lambda^2 - 3\lambda + 1 = -(\lambda - 1)^3 = 0$$

Thus A has a single characteristic root, $\lambda = 1$. Moreover, for $\lambda = 1$ the rank of

$$(A - \lambda I)_{\lambda=1} = A - I = \begin{bmatrix} -4 & -7 & -5 \\ 2 & 3 & 3 \\ 1 & 2 & 1 \end{bmatrix}$$

is clearly 2. Hence, according to Theorem 9, Sec. 10.5, the system of equations $(A - I)X = 0$ has a single independent solution, namely,

$$X = \begin{bmatrix} -3 \\ 1 \\ 1 \end{bmatrix}$$

and thus A has just one independent characteristic vector.

Later in this section we shall see that for hermitian matrices, and hence for real symmetric matrices, the assertion of the last theorem can be sharpened to a strict equality; in other words, we shall prove that if $\lambda = \lambda_1$ is a characteristic value of multiplicity r of a hermitian matrix or a real symmetric matrix, then the rank of $A - \lambda_1 I$ is *exactly* $n - r$. Preparatory to this, however, it will be convenient to prove first some other theorems about hermitian matrices.

THEOREM 7   The characteristic values of a hermitian matrix are all real.

Proof   Let A be a hermitian matrix; let λ_1 be any one of its characteristic values; and let X_1 be a characteristic vector corresponding to λ_1. Then

$$(A - \lambda_1 I)X_1 = 0$$

or

(10) $$AX_1 = \lambda_1 X_1$$

and from this, by premultiplying by $\overline{X}_1{}^T$, we obtain .

(11) $$\overline{X}_1{}^T A X_1 = \lambda_1 \overline{X}_1{}^T X_1$$

Now, from the familiar properties of conjugate complex numbers, $\overline{X}_1{}^T X_1$ is real and in fact positive. Furthermore, from Theorem 4, Sec. 11.1, we know that $\overline{X}_1{}^T A X_1$ is also real. Hence, it follows immediately from Eq. (11) that λ_1 is real, as asserted.

Since, as we observed in Sec. 10.2, a real symmetric matrix is just a special case of a hermitian matrix, we have the following corollary of Theorem 7.

COROLLARY 1   The characteristic values of a real symmetric matrix are all real.

Furthermore, since iA is hermitian if A is skew-hermitian, and since $|A - \lambda I| = 0$ implies that $|iA - i\lambda I| = 0$, it follows that if λ_1 is a characteristic value of the skew-hermitian matrix A, then $i\lambda_1$ is a characteristic value of the hermitian matrix iA. Hence, by Theorem 7, $i\lambda_1$ is real, and therefore λ_1 is a pure imaginary. Thus we have established the following result.

COROLLARY 2   The characteristic values of a skew-hermitian matrix are all pure imaginary.

Knowing now that the characteristic roots of a hermitian matrix A are all real, we can return to the characteristic equation of A and prove the following result.

THEOREM 8   If $\lambda^n - \beta_1 \lambda^{n-1} + \beta_2 \lambda^{n-2} - \cdots + (-1)^{n-1}\beta_{n-1}\lambda + (-1)^n\beta_n = 0$ is the characteristic equation of a hermitian matrix A, then the characteristic roots of A are all positive if and only if each β is positive.

Proof   If A is a hermitian matrix, it follows from Theorem 7 that the roots of the characteristic equation

$$\lambda^n - \beta_1 \lambda^{n-1} + \beta_2 \lambda^{n-2} - \cdots + (-1)^{n-1}\beta_{n-1}\lambda + (-1)^n\beta_n = 0$$

are all real. Furthermore, if each β is positive, it follows from Descartes' rule of signs that no root of the characteristic equation can be negative or zero. Hence, all the characteristic roots must be positive. Conversely, if the characteristic roots of A are all positive, then from the root-coefficient relations

$$\beta_1 = \lambda_1 + \lambda_2 + \cdots + \lambda_n$$
$$\beta_2 = \lambda_1\lambda_2 + \lambda_1\lambda_3 + \cdots + \lambda_{n-1}\lambda_n$$
$$\cdots\cdots\cdots\cdots\cdots\cdots\cdots\cdots\cdots$$
$$\beta_n = \lambda_1\lambda_2 \cdots \lambda_n$$

it follows at once that each β is positive, as asserted.

COROLLARY 1   If

$$\lambda^n - \beta_1\lambda^{n-1} + \beta_2\lambda^{n-2} - \cdots + (-1)^{n-1}\beta_{n-1}\lambda + (-1)^n\beta_n = 0$$

is the characteristic equation of a real symmetric matrix A, then the characteristic roots of A are all positive if and only if each β is positive.

One of the most important properties of the characteristic vectors of a hermitian matrix is that of orthogonality. More precisely, we have the following theorem.

THEOREM 9   If X_i and X_j are characteristic vectors corresponding, respectively, to the distinct characteristic values λ_i and λ_j of a hermitian matrix A, then $\overline{X}_i^T X_j = 0$.

Proof   By hypothesis, we have

(12) $$AX_i = \lambda_i X_i$$

(13) $$AX_j = \lambda_j X_j$$

If in the first of these we take the conjugate and then the transpose of each member, we obtain

$$\overline{X}_i^T \overline{A}^T = \overline{\lambda}_i \overline{X}_i^T$$

or, since $\overline{A}^T = A$, by hypothesis, and $\overline{\lambda}_i = \lambda_i$, by Theorem 7,

(14) $$\overline{X}_i^T A = \lambda_i \overline{X}_i^T$$

Now, if we premultiply Eq. (13) by $\overline{X}_i^T$ and postmultiply Eq. (14) by X_j, we obtain, respectively,

$$\overline{X}_i^T A X_j = \lambda_j \overline{X}_i^T X_j \quad \text{and} \quad \overline{X}_i^T A X_j = \lambda_i \overline{X}_i^T X_j$$

Finally, subtracting these equations, we have

$$(\lambda_i - \lambda_j)\overline{X}_i^T X_j = 0$$

or, since $\lambda_i \neq \lambda_j$, by hypothesis,

$$\overline{X}_i^T X_j = 0$$

as asserted.

COROLLARY 1   If X_i and X_j are characteristic vectors corresponding, respectively, to the distinct characteristic values λ_i and λ_j of a hermitian matrix A, then $\overline{X}_i^T A X_j = 0$.

COROLLARY 2   If X_i and X_j are characteristic vectors corresponding, respectively, to the distinct characteristic values λ_i and λ_j of a real symmetric matrix A, then $X_i^T X_j = X_i^T A X_j = 0$.

We are now in a position to return to the question we raised earlier in this section about the rank of the matrix $A - \lambda_i I$ when A is hermitian and λ_i is a characteristic root of A of multiplicity r. As the next theorem shows, every characteristic root of a hermitian matrix is regular; i.e., if A is hermitian, then for every characteristic root λ_i of multiplicity r, the rank of $A - \lambda_i I$ drops to the minimum permitted by Theorem 6, namely $n - r$, and there are r linearly independent characteristic vectors of A corresponding to λ_i.

THEOREM 10   If A is a hermitian matrix, then to every r-fold characteristic root of A there correspond exactly r linearly independent characteristic vectors.

Proof   Let λ_1 be a characteristic root of multiplicity r of a hermitian matrix A, let X_1 be any characteristic vector corresponding to λ_1, and let $V_2, V_3, \ldots, V_n$ be $n - 1$ vectors such that $X_1, V_2, V_3, \ldots, V_n$ are linearly independent. By the Schmidt orthogonalization process it is always possible to convert these vectors into a set of orthonormal vectors $U_1, U_2, U_3, \ldots, U_n$ in which, in particular, U_1 is the normalized form of X_1. The vectors in the set $\{U_i\}$ thus satisfy the relations

$$(15) \qquad \qquad \overline{U}_i^T U_j = \begin{cases} 1 & i = j \\ 0 & i \neq j \end{cases}$$

Now let U be the $n \times n$ matrix whose columns are, respectively, $U_1, U_2, U_3, \ldots, U_n$. Then from (15) it is clear that

$$\overline{U}^T U = I$$

Hence $\overline{U}^T$ is the inverse of U; and therefore from Theorem 3 it follows that the matrix A and the matrix $\overline{U}^T A U \equiv U^{-1} A U$ have the same characteristic polynomial. In other words, the equations

$$|\overline{U}^T A U - \lambda I| = 0 \qquad \text{and} \qquad |A - \lambda I| = 0$$

have the same roots. Now, remembering that $A U_1 = \lambda_1 U_1$, since by hypothesis U_1 is a characteristic vector of A corresponding to $\lambda = \lambda_1$, we have

$$\overline{U}^T A U = \begin{bmatrix} \overline{U}_1^T \\ \overline{U}_2^T \\ \vdots \\ \overline{U}_n^T \end{bmatrix} A [U_1 \quad U_2 \quad \cdots \quad U_n] = \begin{bmatrix} \overline{U}_1^T \\ \overline{U}_2^T \\ \vdots \\ \overline{U}_n^T \end{bmatrix} [A U_1 \quad A U_2 \quad \cdots \quad A U_n]$$

$$= \begin{bmatrix} \overline{U}_1^T \\ \overline{U}_2^T \\ \vdots \\ \overline{U}_n^T \end{bmatrix} [\lambda_1 U_1 \quad A U_2 \quad \cdots \quad A U_n] = \begin{bmatrix} \lambda_1 & \overline{U}_1^T A U_2 & \cdots & \overline{U}_1^T A U_n \\ 0 & \overline{U}_2^T A U_2 & \cdots & \overline{U}_2^T A U_n \\ \multicolumn{4}{c}{\dots\dots\dots\dots\dots\dots\dots} \\ 0 & \overline{U}_n^T A U_2 & \cdots & \overline{U}_n^T A U_n \end{bmatrix}$$

The zeros in the first column in the last matrix appear because of the relation $\overline{U}_i^T U_1 = 0$, $i \neq 1$, guaranteed by (15). The remaining entries in the last matrix are, in general, not equal to zero since the U_i's are not orthogonal with respect to the matrix A. However, because A, and therefore $\overline{U}^T A U$, is hermitian (see Exercise 7), it follows that since the last $n - 1$ elements in the first column are zero, so too are the last $n - 1$ elements in the first row. Thus $\overline{U}^T A U$ is of the form

$$\begin{bmatrix} \lambda_1 & 0 & 0 & \cdots & 0 \\ 0 & \alpha_{22} & \alpha_{23} & \cdots & \alpha_{2n} \\ \multicolumn{5}{c}{\dots\dots\dots\dots\dots\dots\dots\dots} \\ 0 & \alpha_{n2} & \alpha_{n3} & \cdots & \alpha_{nn} \end{bmatrix}$$

and $\qquad [\overline{U}^T A U - \lambda I] = \begin{bmatrix} \lambda_1 - \lambda & 0 & 0 & \cdots & 0 \\ 0 & \alpha_{22} - \lambda & \alpha_{23} & \cdots & \alpha_{2n} \\ 0 & \alpha_{32} & \alpha_{33} - \lambda & \cdots & \alpha_{3n} \\ \multicolumn{5}{c}{\dots\dots\dots\dots\dots\dots\dots\dots\dots} \\ 0 & \alpha_{n2} & \alpha_{n3} & \cdots & \alpha_{nn} - \lambda \end{bmatrix}$

Therefore, if $\lambda = \lambda_1$ is a repeated root of

$$|A - \lambda I| = |\overline{U}^T A U - \lambda I| = 0$$

then $\lambda_1 - \lambda$ must be a factor of the minor of the element in the first row and first column of $|\overline{U}^T A U - \lambda I|$. But if this minor vanishes when $\lambda = \lambda_1$, then the rank of $\overline{U}^T A U - \lambda_1 I$ is at most $n - 2$, since all other minors of order $n - 1$ obviously contain either a row of zeros or a column of zeros. Furthermore, since

$$\overline{U}^T A U - \lambda_1 I \equiv \overline{U}^T (A - \lambda_1 I) U \qquad \text{and} \qquad A - \lambda_1 I$$

are equivalent matrices and therefore have the same rank, it follows that the rank of $A - \lambda_1 I$ is also at most $n - 2$. Hence, by Theorem 7, Sec. 10.5, $(A - \lambda_1 I)X = \mathbf{0}$ has at least two linearly independent solution vectors, and A has at least two linearly independent characteristic vectors corresponding to λ_1.

If the multiplicity of λ_1 is more than 2, the preceding argument can be repeated, using this time any unitary matrix U whose first *two* columns are any two orthonormal characteristic vectors of A corresponding to the characteristic value λ_1. In this case we obtain at once the relation

$$\overline{U}^T A U - \lambda I = \begin{bmatrix} \lambda_1 - \lambda & 0 & 0 & 0 & \cdots & 0 \\ 0 & \lambda_1 - \lambda & 0 & 0 & \cdots & 0 \\ 0 & 0 & \gamma_{33} - \lambda & \gamma_{34} & \cdots & \gamma_{3n} \\ 0 & 0 & \gamma_{43} & \gamma_{44} - \lambda & \cdots & \gamma_{4n} \\ \hdotsfor{6} \\ 0 & 0 & \gamma_{n3} & \gamma_{n4} & \cdots & \gamma_{nn} - \lambda \end{bmatrix}$$

Then, since $\lambda - \lambda_1$ is a characteristic root of multiplicity greater than 2, it follows that $\lambda_1 - \lambda$ must be a factor of the complementary minor of the second-order minor in the first two rows and first two columns of $|\overline{U}^T A U - \lambda I|$. Hence, since all other $(n - 2)$nd-order minors obviously vanish, it is evident that when $\lambda = \lambda_1$, the rank of $\overline{U}^T A U - \lambda I$, and therefore the rank of $A - \lambda I$, is not more than $n - 3$ and A has at least three linearly independent characteristic vectors corresponding to the characteristic value $\lambda = \lambda_1$.

Clearly, this procedure can be continued until we reach the conclusion that if λ_1 is an r-fold characteristic root of A, then the rank of $A - \lambda_1 I$ is at most $n - r$, and hence A has at least r independent characteristic vectors corresponding to λ_1. But by Theorem 6, A can have at most r linearly independent characteristic vectors corresponding to the r-fold root λ_1. Therefore A must have *exactly* r linearly independent characteristic vectors corresponding to λ_1, as asserted.

Since, as we have repeatedly observed, a real symmetric matrix is a special case of a hermitian matrix, it is clear that we also have the following result.

COROLLARY 1   If A is a real symmetric matrix, then to every r-fold characteristic root of A there correspond exactly r linearly independent characteristic vectors.

We are now in a position to prove the following fundamental theorem.

THEOREM 11   Every $n \times n$ hermitian matrix has n linearly independent characteristic vectors.

Proof Let A be an $n \times n$ hermitian matrix. It may, of course, possess one or more repeated characteristic roots, but if it does, we know from the last theorem that to each root of multiplicity r there correspond exactly r linearly independent characteristic vectors. Hence A cannot have more than n linearly independent characteristic vectors. Specifically, let the characteristic roots of A be

$$\lambda_1, \lambda_2, \ldots, \lambda_k \qquad 1 \le k \le n$$

let the multiplicity of λ_i be r_i, where $\sum\limits_{i=1}^{k} r_i = n$; and let

$$X_{i1}, X_{i2}, \ldots, X_{ir_i}$$

be r_i independent characteristic vectors corresponding to λ_i. Suppose, now, contrary to the assertion of the theorem, that these n characteristic vectors of A are not linearly independent. Then there exists a relation of the form

$$(16) \quad (c_{11}X_{11} + \cdots + c_{1r_1}X_{1r_1}) + (c_{21}X_{21} + \cdots + c_{2r_2}X_{2r_2})$$
$$+ (c_{k1}X_{k1} + \cdots + c_{kr_k}X_{kr_k}) = 0$$

in which at least one of the c's is different from zero.

Now consider a typical group of terms, say the ith, in the last expression. By Theorem 4, Sec. 10.5, unless the c's in such a group are all zero, the combination defines a characteristic vector corresponding to the characteristic value $\lambda = \lambda_i$. Thus, Eq. (16) is simply an expression of the form

$$c_1 X_1 + c_2 X_2 + \cdots + c_i X_i + \cdots + c_k X_k$$

in which each c is either 0 or 1 and at least one c is different from zero. But since the X's now correspond to distinct characteristic values, it follows from Theorem 5 that they are linearly independent and hence that each c must be zero. This contradiction establishes the theorem.

COROLLARY 1 Every real symmetric $n \times n$ matrix has n linearly independent characteristic vectors.

If an $n \times n$ matrix has n linearly independent characteristic vectors, then, by means of the Schmidt orthogonalization process applied to the vectors in each of the sets corresponding to a repeated characteristic root, a set of normalized mutually orthogonal characteristic vectors can always be constructed. Hence we have the following important result.

COROLLARY 2 Every $n \times n$ hermitian or real symmetric matrix has a set of n orthonormal characteristic vectors.

An $n \times n$ matrix whose columns are orthonormal characteristic vectors of an $n \times n$ matrix A is said to be a **modal matrix** of A.

In many applications in physics, chemistry, and engineering, it is necessary to consider matric equations of the form $(A - \lambda B)X = 0$ in which A and B are either hermitian or real and symmetric. Such an equation will, of course, have nontrivial solutions if and only if the determinant of the coefficients is equal to zero. Paralleling our earlier terminology for the special case in which $B = I$, the equation $|A - \lambda B| = 0$ is called the **characteristic equation** of the system, its roots are called the **characteristic roots** or **characteristic values** of the system, and the corresponding nontrivial solutions are called the **characteristic vectors** of the system. As one would expect, the theory of the equation $(A - \lambda B)X = \mathbf{0}$ resembles closely the theory of the equation $(A - \lambda I)X = \mathbf{0}$ which we have developed in this section. In particular, we have the following results.

THEOREM 12 The equation $(A - \lambda B)X = \mathbf{0}$ has zero as a characteristic root if and only if A is singular.

Proof This follows immediately from a consideration of the characteristic equation $|A - \lambda B| = 0$ when the left-hand side is expressed as a polynomial in λ.

THEOREM 13 If A and B are hermitian matrices, and if B is definite, then the characteristic values of $(A - \lambda B)X = \mathbf{0}$ are all real.

Proof Let A and B be hermitian matrices, let B be definite, and let X_1 be a characteristic vector of the equation $(A - \lambda B)X = \mathbf{0}$ corresponding to an arbitrary characteristic value λ_1. Then

$$(17) \qquad\qquad AX_1 = \lambda_1 BX_1$$

Hence, premultiplying Eq. (17) by $\bar{X}_1{}^T$, we have

$$(18) \qquad\qquad \bar{X}_1{}^T AX = \lambda_1 \bar{X}_1{}^T BX_1$$

Now from Theorem 4, Sec. 11.1, we know that both $\bar{X}_1{}^T AX_1$ and $\bar{X}_1{}^T BX_1$ are real numbers. Moreover, since B is definite, $\bar{X}_1{}^T BX_1 \neq 0$. Hence it follows from Eq. (18) that λ_1 is a real number, as asserted.

COROLLARY 1 If A and B are real symmetric matrices, and if B is definite, then the characteristic values of $(A - \lambda B)X = \mathbf{0}$ are all real.

By inspection of Eq. (18), the following results are obtained immediately.

COROLLARY 2 If A and B are hermitian (or real symmetric) matrices which are both positive-definite or both negative-definite, then the characteristic values of $(A - \lambda B)X = \mathbf{0}$ are all positive.

COROLLARY 3 If A and B are hermitian (or real symmetric) matrices, and if A is positive-definite and B is negative-definite, or vice versa, then the characteristic values of $(A - \lambda B)X = \mathbf{0}$ are all negative.

THEOREM 14 If A and B are hermitian matrices, and if $X_1, X_2, \ldots, X_k$ are characteristic vectors of the equation $(A - \lambda B)X = \mathbf{0}$ corresponding, respectively, to the distinct characteristic values $\lambda_1, \lambda_2, \ldots, \lambda_k$, then the X's satisfy the generalized orthogonality conditions

$$\overline{X}_i^T B X_j = 0 \quad \text{and} \quad \overline{X}_i^T A X_j = 0 \quad i \neq j$$

Proof   Let A and B be hermitian matrices, let λ_i and λ_j be distinct characteristic values of the equation $(A - \lambda B)X = \mathbf{0}$, and let X_i and X_j be characteristic vectors corresponding respectively to λ_i and λ_j. Then

$$AX_i = \lambda_i B X_i \quad \text{and} \quad AX_j = \lambda_j B X_j$$

If we premultiply the first of these equations by $\overline{X}_j^T$ and the second by $\overline{X}_i^T$, we obtain respectively

(19) $$\overline{X}_j^T A X_i = \lambda_i \overline{X}_j^T B X_i$$

and

(20) $$\overline{X}_i^T A X_j = \lambda_j \overline{X}_i^T B X_j$$

Now, if we take the transpose and then the conjugate of each side of Eq. (19), remembering that both A and B are hermitian, we obtain

(21) $$\overline{X}_i^T A X_j = \lambda_i \overline{X}_i^T B X_j$$

Finally, subtracting Eq. (20) from Eq. (21), we have

$$(\lambda_i - \lambda_j)\overline{X}_i^T B X_j = 0$$

Therefore, since $\lambda_i \neq \lambda_j$, by hypothesis, it follows that

$$\overline{X}_i^T B X_j = 0$$

as asserted. Furthermore, if $\overline{X}_i^T B X_j = 0$, then it follows from Eq. (21) that $\overline{X}_i^T A X_j = 0$, and our proof is complete.

COROLLARY 1   If A and B are real symmetric matrices, and if $X_1, X_2, \ldots, X_k$ are characteristic vectors of the equation $(A - \lambda B)X = \mathbf{0}$, corresponding, respectively, to the distinct characteristic values $\lambda_1, \lambda_2, \ldots, \lambda_k$, then the X's satisfy the generalized orthogonality conditions

$$X_i^T B X_j = 0 \quad \text{and} \quad X_i^T A X_j = 0 \quad i \neq j$$

THEOREM 15   If A and B are hermitian matrices, if B is definite, and if $X_1, X_2, \ldots, X_k$ are characteristic vectors of $(A - \lambda B)X = \mathbf{0}$ corresponding, respectively, to the distinct characteristic values $\lambda_1, \lambda_2, \ldots, \lambda_k$, then $X_1, X_2, \ldots, X_k$ are linearly independent.

Proof   Let A and B be hermitian matrices, let B be definite, and let us suppose, contrary to the theorem, that the characteristic vectors $X_1, X_2, \ldots, X_k$ corresponding respectively to the distinct characteristic values $\lambda_1, \lambda_2, \ldots, \lambda_k$ of $(A - \lambda B)X = \mathbf{0}$ are linearly dependent. Then there exists a relation of the form

$$c_1 X_1 + c_2 X_2 + \cdots + c_k X_k = \mathbf{0}$$

in which at least one of the c's, say c_i, is different from zero. Now if we multiply the last equation through on the left by $\overline{X}_i{}^T B$, we get

$$c_1 \overline{X}_i{}^T B X_1 + c_2 \overline{X}_i{}^T B X_2 + \cdots + c_i \overline{X}_i{}^T B X_i + \cdots + c_k \overline{X}_i{}^T B X_k = 0$$

However, from the orthogonality guaranteed by Theorem 14, it follows that every term in the last equation except $c_i \overline{X}_i{}^T B X_i$ is equal to zero. Moreover, by hypothesis, B is either positive-definite or negative-definite. Hence $\overline{X}_i{}^T B X_i \neq 0$, and, therefore, $c_i = 0$, contrary to the assumption of linear dependence. This contradiction shows that the X's must be linearly independent, and the theorem is established.

COROLLARY 1 If A and B are real symmetric matrices, if B is definite, and if $X_1, X_2, \ldots, X_k$ are characteristic vectors of $(A - \lambda B)X = 0$ corresponding respectively to the characteristic values $\lambda_1, \lambda_2, \ldots, \lambda_k$, then $X_1, X_2, \ldots, X_k$ are linearly independent.

Theorem 15 must not be misinterpreted as asserting that if A and B are hermitian (or real symmetric) $n \times n$ matrices and if B is definite, then $(A - \lambda B)X = 0$ has n linearly independent characteristic vectors. It guarantees that characteristic vectors corresponding to *distinct* characteristic values of $(A - \lambda B)X = 0$ are linearly independent, but it says nothing about how many distinct characteristic values there are or about how many independent characteristic vectors correspond to a repeated characteristic value. If, because of repeated roots, $(A - \lambda B)X = 0$ has fewer than n distinct characteristic values, then, for all we know at present, $(A - \lambda B)X = 0$ may have fewer than n linearly independent characteristic vectors. However, by a proof very much like the proof of Theorem 10, the following result can be established.

THEOREM 16 If A and B are hermitian (or real symmetric) matrices, and if B is positive-definite, then to a repeated characteristic value of $(A - \lambda B)X = 0$ of multiplicity r there correspond exactly r linearly independent characteristic vectors.

With this theorem, it is not difficult to establish the following counterpart of Theorem 11.

THEOREM 17 If A and B are hermitian (or real symmetric) $n \times n$ matrices and if B is positive-definite, then the equation $(A - \lambda B)X = 0$ has exactly n linearly independent characteristic vectors.

By a straightforward application of the Schmidt orthogonalization process applied to the n linearly independent characteristic vectors of $(A - \lambda B)X = 0$ guaranteed by Theorem 17, we can establish the following useful result.

COROLLARY 1 If A and B are hermitian (or real symmetric) matrices, and if B is positive-definite, then $(A - \lambda B)X = 0$ possesses n characteristic vectors which are orthonormal with respect to B.

With Theorem 17 and its corollary available, it is now an easy matter to express an arbitrary vector C with n components as a linear combination of the characteristic vectors of the equation $(A - \lambda B)X = 0$, provided that A and B are hermitian or real symmetric and B is positive-definite. For we can write

$$(22) \qquad C = c_1 X_1 + c_2 X_2 + \cdots + c_n X_n$$

where the X's are characteristic vectors of $(A - \lambda B)X = 0$ mutually orthogonal with respect to B. Then, if we premultiply Eq. (22) by $\bar{X}_i^T B$, we obtain

$$\bar{X}_i^T BC = c_1 \bar{X}_i^T BX_1 + \cdots + c_i \bar{X}_i^T BX_i + \cdots + c_n \bar{X}_i^T BX_n$$

From the orthogonality of the X's, it follows that every term on the right except $c_i \bar{X}_i^T BX_i$ is equal to zero. Moreover, since B is positive-definite, it follows that $\bar{X}_i^T BX_i \neq 0$. Hence we can solve for c_i, getting

$$c_i = \frac{\bar{X}_i^T BC}{\bar{X}_i^T BX_i} \qquad i = 1, 2, \ldots, n$$

If the X's have been normalized with respect to B, that is, if $\bar{X}_i^T BX_i = 1$ ($i = 1, 2, \ldots, n$), the last formula reduces to the simpler expression

$$c_i = \bar{X}_i^T BC \qquad i = 1, 2, \ldots, n$$

The fact that we were able to solve for the coefficients in the expansion (22) without solving any simultaneous equations should make clear the great convenience of working with a set of vectors which are orthogonal.†

EXERCISES

1 Find the characteristic values and the corresponding characteristic vectors of each of the following matrices:

(a) $\begin{bmatrix} 4 & 6 & 6 \\ 1 & 3 & 2 \\ -1 & -5 & -2 \end{bmatrix}$ (b) $\begin{bmatrix} 7 & -2 & -4 \\ 3 & 0 & -2 \\ 6 & -2 & -3 \end{bmatrix}$ (c) $\begin{bmatrix} 2 & 4 & -6 \\ 4 & 2 & -6 \\ -6 & -6 & -15 \end{bmatrix}$

(d) $\begin{bmatrix} 11 & -4 & -7 \\ 7 & -2 & -5 \\ 10 & -4 & -6 \end{bmatrix}$ (e) $\begin{bmatrix} 4 & 6 & 6 \\ 1 & 3 & 2 \\ -1 & -4 & -3 \end{bmatrix}$ (f) $\begin{bmatrix} -4 & 5 & 5 \\ -5 & 6 & 5 \\ -5 & 5 & 6 \end{bmatrix}$

For which of these, if any, are the characteristic vectors orthogonal?

2 Find the characteristic values and the corresponding characteristic vectors for the equation $(A - \lambda B)X = 0$ if

(a) $A = \begin{bmatrix} 2 & -1 & 0 \\ -1 & 2 & -1 \\ 0 & -1 & 2 \end{bmatrix}$ $B = \begin{bmatrix} 3 & 0 & 0 \\ 0 & 4 & 0 \\ 0 & 0 & 3 \end{bmatrix}$

(b) $A = \begin{bmatrix} 6 & -3 & 0 \\ -3 & 6 & -3 \\ 0 & -3 & 4 \end{bmatrix}$ $B = \begin{bmatrix} 6 & 0 & 0 \\ 0 & 4 & 0 \\ 0 & 0 & 4 \end{bmatrix}$

(c) $A = \begin{bmatrix} 3 & -1 & 0 \\ -1 & 1 & -1 \\ 0 & -1 & 5 \end{bmatrix}$ $B = \begin{bmatrix} 4 & 0 & 0 \\ 0 & 1 & 0 \\ 0 & 0 & 4 \end{bmatrix}$

---

† The similarity between this procedure and the process by which the coefficients in the Fourier series of a function were determined in Sec. 6.2 is worthy of note.

In each case, verify all orthogonality relations, and, using orthogonality properties, express the vector

$$V = \begin{bmatrix} 1 \\ 2 \\ 3 \end{bmatrix}$$

as a linear combination of the characteristic vectors.

3 Find three solution vectors of the equation $(A - \lambda B)X = 0$ which are orthonormal with respect to B if

(a) $A = \begin{bmatrix} 7 & 1 & -1 \\ 1 & 4 & -1 \\ -1 & -1 & 3 \end{bmatrix}$ $B = \begin{bmatrix} 6 & 0 & 0 \\ 0 & 3 & 0 \\ 0 & 0 & 2 \end{bmatrix}$

(b) $A = \begin{bmatrix} 7 & -1 & -1 \\ -1 & 4 & 1 \\ -1 & 1 & 3 \end{bmatrix}$ $B = \begin{bmatrix} 6 & 0 & 0 \\ 0 & 3 & 0 \\ 0 & 0 & 2 \end{bmatrix}$

Are the characteristic vectors orthogonal with respect to A in the respective cases? Are they orthonormal with respect to A?

4 Prove Theorem 17.

5 In Corollary 1 of Theorem 17 why is it necessary to restrict B to be positive-definite?

6 Prove that iA is hermitian if A is skew-hermitian.

7 If A is hermitian, prove that $\bar{U}^T A U$ is also hermitian.

8 Under what conditions, if any, is it possible for every value of λ to be a characteristic value of the equation $(A - \lambda B)X = 0$?

9 Show by an example that if A and B are indefinite, the characteristic values of $(A - \lambda B)X = 0$ need not be real even though A and B are real and symmetric.

10 Show that if either A or B is nonsingular, then AB and BA have the same characteristic values. Hence prove that there are no matrices A and B, with either A or B nonsingular, such that $AB - BA = I$. (These results hold even when both A and B are singular.)

11 Show that the characteristic values of a real skew-symmetric matrix are either zero or pure imaginary.

12 Prove that if a 2×2 matrix has characteristic vectors which are orthogonal, it is symmetric.

13 Prove that if every characteristic value of a hermitian or real symmetric matrix is zero, the matrix is a null matrix. Is this true if the matrix is not symmetric?

14 Show by an example that Corollary 1 of Theorem 10 is false for symmetric matrices which are not real.

15 (a) If $(A - \lambda B)X = 0$, where B is a nonsingular matrix, show that $MX = \lambda X$ and $M^k X = \lambda^k X$, where $M = B^{-1}A$.

(b) If $C_1 = \alpha_1 X_1 + \alpha_2 X_2 + \cdots + \alpha_n X_n$ is an arbitrary linear combination of the characteristic vectors of $(A - \lambda B)X = 0$ corresponding to the distinct characteristic values $\lambda_1, \lambda_2, \ldots, \lambda_n$, show that

$$M^k C_1 = \alpha_1 \lambda_1{}^k X_1 + \alpha_2 \lambda_2{}^k X_2 + \cdots + \alpha_n \lambda_n{}^k X_n$$

16 In the notation of Exercise 15, let $C_{k+1} = M^k C_1$, and let λ_1 be the largest of the characteristic values of $(A - \lambda B)X = 0$. Prove that the ratios of corresponding components in C_{k+1} and C_k approach λ_1 as k becomes infinite.

17 Extend the results of Exercise 16 by showing that the ratios of successive components in C_{k+1} approach the ratios of the corresponding components in X_1 as k becomes infinite.

18 Using the results of Exercises 16 and 17, approximate the largest characteristic value and the corresponding characteristic vector of the system $(A - \lambda B)X = 0$, where

$$A = \begin{bmatrix} 3 & -1 & -2 \\ -1 & 2 & -2 \\ -2 & -2 & 6 \end{bmatrix} \quad \text{and} \quad B = \begin{bmatrix} 2 & 0 & 0 \\ 0 & 3 & 0 \\ 0 & 0 & 1 \end{bmatrix}$$

19   Work Exercise 18 for the system

$$3x_1 - x_2 \qquad\quad = 4\lambda x_1$$
$$-x_1 + x_2 - x_3 = \lambda x_2$$
$$- x_2 + 5x_3 = 4\lambda x_3$$

20   If A and B are symmetric matrices, explain how the results of Exercises 16 and 17 can be extended to obtain the other characteristic values and characteristic vectors of the system $(A - \lambda B)X = 0$. *Hint:* Having found the characteristic vector X_1 corresponding to the largest characteristic value λ_1, show how to determine a vector C_1 which is orthogonal to X_1 and then consider the result of carrying out the iteration procedure which is begun with such a vector.

21   (a) Find the remaining characteristic values and the corresponding vectors for the system in Exercise 18.

(b) Find the remaining characteristic values and the corresponding characteristic vectors for the system in Exercise 19.

22   (a) Show that the characteristic equation of the matrix $\begin{bmatrix} 0 & 1 \\ -c & -b \end{bmatrix}$ is the quadratic equation $\lambda^2 + b\lambda + c = 0$.

(b) Show that the polynomial equation

$$p(\lambda) = \lambda^n + a_1\lambda^{n-1} + \cdots + a_{n-1}\lambda + a_n = 0$$

is the characteristic equation of the matrix

$$\begin{bmatrix} 0 & 1 & 0 & \cdots & 0 \\ 0 & 0 & 1 & \cdots & 0 \\ \multicolumn{5}{c}{\cdots\cdots\cdots\cdots\cdots\cdots\cdots} \\ 0 & 0 & 0 & \cdots & 1 \\ -a_n & -a_{n-1} & -a_{n-2} & \cdots & -a_1 \end{bmatrix}$$

This matrix is called the **companion matrix of the polynomial** p. *Hint:* Proceed inductively, beginning with the result of part **(a)**.

23   In Exercise 22, if r_1 is a characteristic root of the companion matrix of a polynomial p, what is the corresponding characteristic vector of the companion matrix?

24   Using the results of Exercises 16 and 17, approximate the root of largest absolute value of each of the following polynomials:

(a) $x^2 - 4x + 2 = 0$ (b) $x^3 - 6x^2 + 11x - 5 = 0$

(c) $x^3 + 4x^2 + x - 5 = 0$

25   A square matrix in which the sum of the elements in each row is 1 is called a **stochastic matrix**. Show that every stochastic matrix has 1 as one of its characteristic values.

26   Show that if the characteristic roots of A are distinct and those of B are also distinct, then A and B commute; if and only if A and B have the same characteristic vectors.

27   If X_1 and X_2 are characteristic vectors of A corresponding respectively to the characteristic values λ_1 and λ_2, under what conditions, if any, is $\lambda_1 X_1 + \lambda_2 X_2$ a characteristic vector of A?

28   Let $X_1, X_2, \ldots, X_n$ be orthonormal characteristic vectors corresponding, respectively, to the characteristic values $\lambda_1, \lambda_2, \ldots, \lambda_n$ of the homogeneous equation $(A - \lambda I)X = 0$. By assuming a solution of the form $X = c_1 X_1 + c_2 X_2 + \cdots + c_n X_n$, show that

$$X = \sum_{i=1}^{n} \left(\frac{X_i^T V}{\lambda_i - \lambda} \right) X_i$$

is a solution of the nonhomogeneous equation

$$(A - \lambda I)X = V$$

where V is a given column matrix and λ is a given number distinct from each of the characteristic values of A. Under what conditions, if any, will a solution of the nonhomogeneous equation exist if λ is one of the characteristic values of A?

29 Generalize the results of Exercise 28 to the equation

$$(A - \lambda B)X = V$$

11.3 The Transformation of Matrices

In previous sections we have already encountered the idea of the transformation of matrices. For instance, in Sec. 10.4 we defined equivalent matrices as matrices A and B connected by a relation, or equivalence transformation, of the form

$$B = QAP \qquad P, Q \text{ nonsingular}$$

Again, in Sec. 11.1 (Example 2) we observed that if the variables in a quadratic form $X^T A X$ are subjected to a nonsingular linear transformation $X = PY$, then the quadratic form becomes $Y^T B Y$, where B is obtained from A by the congruence transformation

$$B = P^T A P \qquad P \text{ nonsingular}$$

In particular, we observed that if a nonsingular matrix P can be found with the property that $B = P^T A P$ is a diagonal matrix, then in terms of the new variables introduced by the substitution $X = PY$, the quadratic form $X^T A X$ becomes just a sum of squares. In this section we shall consider briefly the question of just when it is possible to transform an $n \times n$ matrix A into a diagonal matrix B by multiplying it on the right and on the left by suitable matrices P and Q.

Because an equivalence transformation is simply a composition of elementary transformations, it is clear that for any square matrix A many pairs of nonsingular matrices P and Q can be found such that QAP is diagonal. In other words, we have the following result, which is little more than a restatement of Theorem 10, Sec. 10.4, for square matrices.

THEOREM 1   Any square matrix is equivalent to a diagonal matrix.

COROLLARY 1   If a matrix A of rank r is equivalent to a diagonal matrix B, then B has exactly r nonzero diagonal elements.

A square matrix cannot in general be transformed into a diagonal matrix by a transformation more restricted than an equivalence transformation. However, in many important special cases this is possible, as the following theorems show.

THEOREM 2   A square matrix is congruent to a diagonal matrix if and only if it is symmetric.

Proof   The proof that if a matrix A is symmetric, then it is congruent to a diagonal matrix was essentially given in our discussion of the Lagrange reduction in Sec. 11.1. For if A is symmetric, then it can be regarded as the matrix of a quadratic form and the Lagrange reduction provides a linear transformation whose matrix P is a nonsingular matrix with the property that $P^T A P$ is diagonal.

On the other hand, if A is congruent to a diagonal matrix D, then there exists a nonsingular matrix P such that $A = P^TDP$. Furthermore, if this is the case, then the transpose of A is

$$A^T = (P^TDP)^T = P^TD^TP$$

However, $D^T = D$, since any diagonal matrix is obviously symmetric. Hence

$$A^T = P^TD^TP = P^TDP = A$$

Therefore, A, being equal to its transpose, is symmetric, as asserted.

From the nature of the Lagrange reduction it is evident that a symmetric matrix can be diagonalized by a congruence transformation in many ways; and among these there is always at least one which will simultaneously diagonalize a second given matrix, provided it is positive-definite. More precisely, we have the following important theorem.

THEOREM 3 Let $\lambda_1, \lambda_2, \ldots, \lambda_n$ be the characteristic values of the equation $(A - \lambda B)X = 0$, where A and B are hermitian (or real symmetric) $n \times n$ matrices and B is positive definite. Let $X_1, X_2, \ldots, X_n$ be n independent characteristic vectors corresponding to $\lambda_1, \lambda_2, \ldots, \lambda_n$; and let the X's be orthonormal with respect to B. Let M be the matrix whose columns are the characteristic vectors $X_1, X_2, \ldots, X_n$; and let D be the diagonal matrix whose diagonal elements are the characteristic values $\lambda_1, \lambda_2, \ldots, \lambda_n$. Then $\overline{M}^TBM = I$ and $\overline{M}^TAM = D$ (or $M^TBM = I$ and $M^TAM = D$).

Proof Let $\lambda_1, \lambda_2, \ldots, \lambda_n$ be the characteristic values of the equation $(A - \lambda B)X = 0$. Whether or not there are repeated roots among the λ's, we know, from Corollary 1 of Theorem 17, Sec. 11.2, that there exists a set of characteristic vectors $X_1, X_2, \ldots, X_n$ orthonormal with respect to B, that is, such that

(1)
$$\overline{X}_i^TBX_j = \begin{cases} 1 & i = j \\ 0 & i \neq j \end{cases}$$

Now, writing the modal matrix M in partitioned form, for convenience, we have

(2)
$$M = [X_1 \quad X_2 \quad \cdots \quad X_n]$$

and
$$\overline{M}^TBM = \begin{bmatrix} \overline{X}_1^T \\ \overline{X}_2^T \\ \vdots \\ \overline{X}_n^T \end{bmatrix} [BX_1 \quad BX_2 \quad \cdots \quad BX_n]$$

$$= \begin{bmatrix} \overline{X}_1^TBX_1 & \overline{X}_1^TBX_2 & \cdots & \overline{X}_1^TBX_n \\ \overline{X}_2^TBX_1 & \overline{X}_2^TBX_2 & \cdots & \overline{X}_2^TBX_n \\ \cdots & \cdots & \cdots & \cdots \\ \overline{X}_n^TBX_1 & \overline{X}_n^TBX_2 & \cdots & \overline{X}_n^TBX_n \end{bmatrix}$$

(3)
$$= I \quad \text{by (1)}$$

Also, by premultiplying Eq. (2) by A and then using the fact that for each i the X's are such that $AX_i = \lambda_i BX_i$, we have

$$AM = [AX_1 \quad AX_2 \quad \cdots \quad AX_n]$$

$$= [\lambda_1 BX_1 \quad \lambda_2 BX_2 \quad \cdots \quad \lambda_n BX_n]$$

$$= B[\lambda_1 X_1 \quad \lambda_2 X_2 \quad \cdots \quad \lambda_n X_n]$$

$$= B[X_1 \quad X_2 \quad \cdots \quad X_n] \begin{bmatrix} \lambda_1 & & & \mathbf{O} \\ & \lambda_2 & & \\ & & \ddots & \\ \mathbf{O} & & & \lambda_n \end{bmatrix}$$

$$= BMD$$

Therefore, by (3),

$$\overline{M}^T AM = \overline{M}^T(BMD) = (\overline{M}^T BM)D = D$$

which is the second assertion of the theorem.

COROLLARY 1 If A and B are hermitian (or real symmetric) matrices, if B is positive definite, and if M is a matrix whose columns are characteristic vectors of $(A - \lambda B)X = 0$ which are orthonormal with respect to B, then the substitution $X = MY$ simultaneously reduces the hermitian (or quadratic) forms $\overline{X}^T AX$ and $\overline{X}^T BX$ (or $X^T AX$ and $X^T BX$) to $\overline{Y}^T DY$ and $\overline{Y}^T IY \equiv \overline{Y}^T Y$ (or $Y^T DY$ and $Y^T IY \equiv Y^T Y$), respectively, where D is the diagonal matrix whose diagonal elements are the characteristic values which correspond respectively to the column vectors of M.

The conditions under which a square matrix can be diagonalized by a similarity transformation are contained in the next theorem.

THEOREM 4 An $n \times n$ matrix is similar to a diagonal matrix if and only if it has n independent characteristic vectors.

Proof Let A be an $n \times n$ matrix, and let us suppose first that A is similar to a diagonal matrix

$$D = \begin{bmatrix} d_{11} & & & \mathbf{O} \\ & d_{22} & & \\ & & \ddots & \\ \mathbf{O} & & & d_{nn} \end{bmatrix}$$

that is, let us suppose that there exists a matrix S with the property that

(4) $$S^{-1}AS = D$$

where D is a diagonal matrix. If, for convenience, we write S in the partitioned form

$$S = [S_1 \quad S_2 \quad \cdots \quad S_n]$$

we have, on premultiplying Eq. (4) by S,

$$AS = SD$$

or $\quad [AS_1 \quad AS_2 \quad \cdots \quad AS_n] = [S_1 \quad S_2 \quad \cdots \quad S_n] \begin{bmatrix} d_{11} & & & \mathbf{O} \\ & d_{22} & & \\ & & & \\ \mathbf{O} & & & d_{nn} \end{bmatrix}$$

$$= [d_{11}S_1 \quad d_{22}S_2 \quad \cdots \quad d_{nn}S_n]$$

Hence it follows that

$$AS_i = d_{ii}S_i = d_{ii}IS_i \qquad i = 1, 2, \ldots, n$$

which shows that $X_i = S_i$ is a characteristic vector corresponding to the characteristic value $\lambda_i = d_{ii}$ of the equation $(A - \lambda I)X = \mathbf{0}$. Thus the n columns of the transforming matrix S are characteristic vectors of the matrix A. Moreover, since the inverse of S exists, by hypothesis, it follows that $|S| \neq 0$. Hence, by Theorem 12, Sec. 10.5, the n columns of S are linearly independent. Thus the matrix A has n linearly independent characteristic vectors, and the necessity assertion of the theorem is verified.

Suppose now that A has n linearly independent characteristic vectors $X_1, X_2, \ldots,$ X_n corresponding to the (possibly repeated) characteristic values $\lambda_1, \lambda_2, \ldots, \lambda_n$. Then, by hypothesis,

$$AX_i = \lambda_i I X_i = \lambda_i X_i \qquad i = 1, 2, \ldots, n$$

Now, let S be the matrix whose columns are the characteristic vectors $X_1, X_2, \ldots,$ X_n; that is, let S be a modal matrix of A. Then, since the characteristic vectors are independent, by hypothesis, it follows, from Theorem 12, Sec. 10.5, that S^{-1} exists, and we can write

$$\begin{aligned} S^{-1}AS &= S^{-1}A[X_1 \quad X_2 \quad \cdots \quad X_n] \\ &= S^{-1}[AX_1 \quad AX_2 \quad \cdots \quad AX_n] \\ &= S^{-1}[\lambda_1 X_1 \quad \lambda_2 X_2 \quad \cdots \quad \lambda_n X_n] \\ &= S^{-1}[X_1 \quad X_2 \quad \cdots \quad X_n] \begin{bmatrix} \lambda_1 & & & \mathbf{O} \\ & \lambda_2 & & \\ & & \cdot & \\ \mathbf{O} & & & \lambda_n \end{bmatrix} \\ &= S^{-1}S \begin{bmatrix} \lambda_1 & & & \mathbf{O} \\ & \lambda_2 & & \\ & & \cdot & \\ \mathbf{O} & & & \lambda_n \end{bmatrix} \\ &= \begin{bmatrix} \lambda_1 & & & \mathbf{O} \\ & \lambda_2 & & \\ & & \cdot & \\ \mathbf{O} & & & \lambda_n \end{bmatrix} \end{aligned}$$

Hence A is similar to a diagonal matrix, and the sufficiency assertion of the theorem is also verified.

Since every hermitian and every real symmetric matrix has n linearly independent characteristic vectors (Theorem 11, Sec. 11.2), it is clear that the last theorem contains the following important special case.

COROLLARY 1   Every hermitian and every real symmetric matrix is similar to a diagonal matrix.

Using the Schmidt process, it is clear that if a matrix A has n independent characteristic vectors, it has, in fact, a set of n orthonormal characteristic vectors. Moreover, as we saw in Exercise 15, Sec. 10.3, a matrix whose columns are orthonormal is an orthogonal matrix. Hence, taking the matrix S in Theorem 4 to be a matrix whose columns are orthonormal characteristic vectors of A, we have the following results.

COROLLARY 2   Every real symmetric matrix is orthogonally similar to a diagonal matrix.

COROLLARY 3   If a matrix is orthogonally similar to a diagonal matrix, it is symmetric.

By essentially the same argument, the following companion results for hermitian matrices can be established.

COROLLARY 4   Every hermitian matrix is unitarily similar to a diagonal matrix.

COROLLARY 5   If a matrix is unitarily similar to a diagonal matrix, it is hermitian.

From Theorem 4 it is clear that not every square matrix is similar to a diagonal matrix. However, in more advanced texts† it is shown that every square matrix is similar to a triangular matrix. Furthermore, it can be shown‡ that every square matrix A is similar to an "almost diagonal" matrix C whose diagonal elements are the characteristic values of A, whose elements immediately above the principal diagonal are either 0 or 1, and whose remaining elements are all 0. This standard form of a matrix is known as the **classical** or **Jordan**§ **canonical form**.

Many of the theorems of the last two sections find their most immediate physical application in the analysis of vibrating systems, either mechanical or electrical. In particular, the orthogonality of the characteristic vectors of a matric equation makes it possible to impose initial conditions of velocity and displacement on a mechanical system with a finite number of degrees of freedom in a way that closely resembles the corresponding procedure for boundary-value problems involving continuous systems (Secs. 8.4 and 8.5). The following example illustrates these ideas.

---

† See, for instance, L. Mirsky, "Linear Algebra," p. 307, Oxford, New York, 1955.
‡ See, for instance, R. A. Rosenbaum, "Projective Geometry and Modern Algebra," pp. 316–327, Addison-Wesley, Reading, Mass., 1963.
§ Named for the French mathematician Camille Jordan (1838–1922).

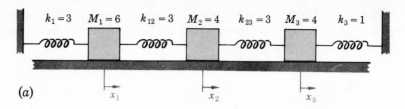

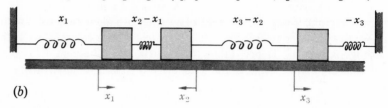

Net changes in spring lengths (x_1, x_3 shown positive; x_2 shown negative)

Figure 11.1
A three-mass system in equilibrium and in a displaced position.

The three masses shown in Fig. 11.1a are initially displaced so that

$$(x_1)_0 = 2 \qquad (x_2)_0 = -1 \qquad (x_3)_0 = 1$$

From these positions they begin to move with initial velocities

$$(v_1)_0 = 1 \qquad (v_2)_0 = 2 \qquad (v_3)_0 = 0$$

Assuming that there is no friction in the system, determine the subsequent motion of each mass.

Since friction is assumed to be negligible, the only forces acting are those transmitted to the masses by the springs directly attached to them. When the instantaneous displacements of the masses are x_1, x_2, and x_3, the lengths of the springs have changed from their unstretched, initial lengths by the respective amounts (Fig. 11.1b)

$$x_1 \qquad x_2 - x_1 \qquad x_3 - x_2 \qquad -x_3$$

Hence, the forces instantaneously exerted by the springs are, respectively,

$$3x_1 \qquad 3(x_2 - x_1) \qquad 3(x_3 - x_2) \qquad -x_3$$

plus signs indicating that the springs are in tension, minus signs that the springs are in compression. Therefore, applying Newton's law to each of the masses in turn, we obtain the three differential equations

$$6\frac{d^2x_1}{dt^2} = -3x_1 + 3(x_2 - x_1)$$

$$4\frac{d^2x_2}{dt^2} = -3(x_2 - x_1) + 3(x_3 - x_2)$$

$$4\frac{d^2x_3}{dt^2} = -3(x_3 - x_2) - x_3$$

or

$$(6D^2 + 6)x_1 - \qquad\qquad 3x_2 \qquad\qquad = 0$$

(5)

$$-3x_1 + (4D^2 + 6)x_2 - \qquad\qquad 3x_3 = 0$$

$$-3x_2 + (4D^2 + 4)x_3 = 0$$

or, in matric notation, simply

$$P(D)X = 0$$

where $\quad P(D) = \begin{bmatrix} 6D^2 + 6 & -3 & 0 \\ -3 & 4D^2 + 6 & -3 \\ 0 & -3 & 4D^2 + 4 \end{bmatrix}$ and $\quad X = \begin{bmatrix} x_1 \\ x_2 \\ x_3 \end{bmatrix}$

Since there is no dissipation of energy through friction, it is clear that each mass must vibrate around its equilibrium position with constant amplitude. Hence, as a solution we assume

$$X = A \cos \omega t = \begin{bmatrix} a_1 \\ a_2 \\ a_3 \end{bmatrix} \cos \omega t$$

that is,

$$x_1 = a_1 \cos \omega t \qquad x_2 = a_2 \cos \omega t \qquad x_3 = a_3 \cos \omega t$$

where ω is a frequency to be determined and a_1, a_2, and a_3 are the as yet unknown amplitudes through which the respective masses oscillate at this frequency. Substituting these into the differential equations (5) and dividing out the common factor $\cos \omega t$, we obtain the three algebraic equations

$$(6) \qquad \begin{aligned} (-6\omega^2 + 6)a_1 - \qquad 3a_2 \qquad\qquad &= 0 \\ -3a_1 + (-4\omega^2 + 6)a_2 - \qquad 3a_3 &= 0 \\ -3a_2 + (-4\omega^2 + 4)a_3 &= 0 \end{aligned}$$

from which to determine a_1, a_2, and a_3. This system will have a nontrivial solution if and only if the determinant of its coefficients is equal to zero. Hence we must have

$$\begin{vmatrix} -6\omega^2 + 6 & -3 & 0 \\ -3 & -4\omega^2 + 6 & -3 \\ 0 & -3 & -4\omega^2 + 4 \end{vmatrix}$$
$$= -6(4\omega^2 - 1)(\omega^2 - 1)(4\omega^2 - 9) = 0$$

Thus the system (6) has a nontrivial solution for $\omega^2 = \frac{1}{4}, 1, \frac{9}{4}$ and for no other values of ω^2. The natural frequencies of the physical system are therefore

$$\omega = \tfrac{1}{2}, 1, \tfrac{3}{2}$$

Now according to Theorem 9, Sec. 10.5, the values of a_1, a_2, and a_3 which satisfy (6) when the determinant of its coefficients is equal to zero can be read from any 2×3 submatrix of rank 2 contained in the coefficient matrix. Hence, using the matrix of the coefficients of the last two equations in (6), we have, in the three cases,

$$\omega = \tfrac{1}{2}: \quad \begin{bmatrix} -3 & 5 & -3 \\ 0 & -3 & 3 \end{bmatrix} \quad a_1 = 6 \quad a_2 = 9 \quad a_3 = 9$$

or, neglecting the irrelevant proportionality constant 3,

$$a_1 = 2 \quad a_2 = 3 \quad a_3 = 3$$

$$\omega = 1: \quad \begin{bmatrix} -3 & 2 & -3 \\ 0 & -3 & 0 \end{bmatrix} \quad a_1 = -9 \quad a_2 = 0 \quad a_3 = 9$$

or, equivalently,

$$a_1 = 1 \quad a_2 = 0 \quad a_3 = -1$$

$$\omega = \tfrac{3}{2}: \quad \begin{bmatrix} -3 & -3 & -3 \\ 0 & -3 & -5 \end{bmatrix} \quad a_1 = 6 \quad a_2 = -15 \quad a_3 = 9$$

or, equivalently,

$$a_1 = 2 \quad a_2 = -5 \quad a_3 = 3$$

Thus we have found three particular solution vectors for the system (5), namely,

$$X_1 = \begin{bmatrix} 2 \\ 3 \\ 3 \end{bmatrix} \cos \frac{t}{2} \qquad X_2 = \begin{bmatrix} 1 \\ 0 \\ -1 \end{bmatrix} \cos t \qquad X_3 = \begin{bmatrix} 2 \\ -5 \\ 3 \end{bmatrix} \cos \tfrac{3}{2}t$$

Clearly, if we had begun with the assumptions

$$x_1 = a_1 \sin \omega t \qquad x_2 = a_2 \sin \omega t \qquad x_3 = a_3 \sin \omega t$$

we would also have obtained the algebraic equations (6) and hence the same three values of ω and the same three solution vectors. Therefore we have three more particular solutions:

$$X_4 = \begin{bmatrix} 2 \\ 3 \\ 3 \end{bmatrix} \sin \frac{t}{2} \qquad X_5 = \begin{bmatrix} 1 \\ 0 \\ -1 \end{bmatrix} \sin t \qquad X_6 = \begin{bmatrix} 2 \\ -5 \\ 3 \end{bmatrix} \sin \tfrac{3}{2}t$$

and, finally, the complete solution

$$(7) \qquad X = c_1 X_1 + c_2 X_2 + c_3 X_3 + c_4 X_4 + c_5 X_5 + c_6 X_6$$

where the c's are arbitrary scalar coefficients.

To determine the values of the c's we must, of course, use the given initial conditions. The most convenient way to do this is to write the system (6) in the form

$$(8) \qquad (V - \omega^2 T)A = 0$$

where
$$V = \begin{bmatrix} 6 & -3 & 0 \\ -3 & 6 & -3 \\ 0 & -3 & 4 \end{bmatrix} \qquad T = \begin{bmatrix} 6 & 0 & 0 \\ 0 & 4 & 0 \\ 0 & 0 & 4 \end{bmatrix} \qquad A = \begin{bmatrix} a_1 \\ a_2 \\ a_3 \end{bmatrix}$$

and then recall from the last section that the solution vectors of (8), namely,

$$A_1 = \begin{bmatrix} 2 \\ 3 \\ 3 \end{bmatrix} \qquad A_2 = \begin{bmatrix} 1 \\ 0 \\ -1 \end{bmatrix} \qquad A_3 = \begin{bmatrix} 2 \\ -5 \\ 3 \end{bmatrix}$$

satisfy the orthogonality conditions

$$(9) \qquad A_i^T T A_j = 0 \qquad i \neq j$$

To take advantage of this property, we first set $t = 0$ in (7) and substitute the initial displacement vector for $X(0)$, getting

$$(10) \qquad \begin{bmatrix} 2 \\ -1 \\ 1 \end{bmatrix} = c_1 \begin{bmatrix} 2 \\ 3 \\ 3 \end{bmatrix} + c_2 \begin{bmatrix} 1 \\ 0 \\ -1 \end{bmatrix} + c_3 \begin{bmatrix} 2 \\ -5 \\ 3 \end{bmatrix}$$

Then if we multiply this equation through on the left by

$$A_1^T T \equiv [2 \quad 3 \quad 3] \begin{bmatrix} 6 & 0 & 0 \\ 0 & 4 & 0 \\ 0 & 0 & 4 \end{bmatrix} = [12 \quad 12 \quad 12]$$

the second and third terms on the right become zero because of the orthogonality condition (9), and we have simply

$$[12 \quad 12 \quad 12] \begin{bmatrix} 2 \\ -1 \\ 1 \end{bmatrix} = c_1 [12 \quad 12 \quad 12] \begin{bmatrix} 2 \\ 3 \\ 3 \end{bmatrix} \qquad \text{or} \qquad c_1 = \tfrac{1}{4}$$

Similarly, multiplying (10) on the left by

$$A_2^T T = [6 \quad 0 \quad -4] \qquad \text{and by} \qquad A_3^T T = [12 \quad -20 \quad 12]$$

in turn, we find

$$c_2 = \tfrac{4}{5} \qquad \text{and} \qquad c_3 = \tfrac{7}{20}$$

To find c_4, c_5, and c_6, we first differentiate Eq. (7), getting

$$\frac{dX}{dt} = -\tfrac{1}{2}c_1 \begin{bmatrix} 2 \\ 3 \\ 3 \end{bmatrix} \sin \frac{t}{2} - c_2 \begin{bmatrix} 1 \\ 0 \\ -1 \end{bmatrix} \sin t - \tfrac{3}{2}c_3 \begin{bmatrix} 2 \\ -5 \\ 3 \end{bmatrix} \sin \tfrac{3}{2}t$$

$$+ \tfrac{1}{2}c_4 \begin{bmatrix} 2 \\ 3 \\ 3 \end{bmatrix} \cos \frac{t}{2} + c_5 \begin{bmatrix} 1 \\ 0 \\ -1 \end{bmatrix} \cos t + \tfrac{3}{2}c_6 \begin{bmatrix} 2 \\ -5 \\ 3 \end{bmatrix} \cos \tfrac{3}{2}t$$

Then, setting $t = 0$ and substituting the initial velocity vector $\begin{bmatrix} 1 \\ 2 \\ 0 \end{bmatrix}$ for $\left. \dfrac{dX}{dt} \right|_{t=0}$, we have

$$\begin{bmatrix} 1 \\ 2 \\ 0 \end{bmatrix} = \tfrac{1}{2}c_4 \begin{bmatrix} 2 \\ 3 \\ 3 \end{bmatrix} + c_5 \begin{bmatrix} 1 \\ 0 \\ -1 \end{bmatrix} + \tfrac{3}{2}c_6 \begin{bmatrix} 2 \\ -5 \\ 3 \end{bmatrix}$$

Finally, multiplying this equation on the left by

$$A_1{}^T T = [12 \quad 12 \quad 12]$$

$$A_2{}^T T = [6 \quad\quad 0 \quad -4] \quad \text{and} \quad A_3{}^T T = [12 \quad -20 \quad\quad 12]$$

in turn, and again noting the orthogonality conditions (9), we find

$$c_4 = \tfrac{3}{4} \quad\quad c_5 = \tfrac{3}{5} \quad \text{and} \quad c_6 = -\tfrac{7}{40}$$

With the c's determined, the solution is now complete, and we have

$$X = \tfrac{1}{4}X_1 + \tfrac{4}{5}X_2 + \tfrac{7}{20}X_3 + \tfrac{3}{4}X_4 + \tfrac{3}{5}X_5 - \tfrac{7}{40}X_6$$

or, explicitly,

$$
\begin{aligned}
& x_1 = \frac{1}{2}\cos\frac{t}{2} + \frac{4}{5}\cos t + \frac{7}{10}\cos\frac{3}{2}t + \frac{3}{2}\sin\frac{t}{2} + \frac{3}{5}\sin t - \frac{7}{20}\sin\frac{3}{2}t \\
(11) \quad & x_2 = \frac{3}{4}\cos\frac{t}{2} \qquad\qquad\quad - \frac{7}{4}\cos\frac{3}{2}t + \frac{9}{4}\sin\frac{t}{2} \qquad\qquad + \frac{7}{8}\sin\frac{3}{2}t \\
& x_3 = \frac{3}{4}\cos\frac{t}{2} - \frac{4}{5}\cos t + \frac{21}{20}\cos\frac{3}{2}t + \frac{9}{4}\sin\frac{t}{2} - \frac{3}{5}\sin t - \frac{21}{40}\sin\frac{3}{2}t
\end{aligned}
$$

We have already identified the three values $\omega = \frac{1}{2}$, 1, $\frac{3}{2}$ as the natural frequencies of the system, i.e., the only frequencies at which free vibrations of the system are possible; and we have illustrated how the motion produced by an arbitrary set of initial conditions involves simultaneous vibrations at each of the natural frequencies. The vectors A_1, A_2, and A_3, associated, respectively, with the frequencies $\omega = \frac{1}{2}$, $\omega = 1$, and $\omega = \frac{3}{2}$, are called the **normal modes** of the system. Each describes the *relative* amplitudes with which the three masses would vibrate if the system were set in motion in such a way that it vibrated only at the corresponding natural frequency. The *absolute* amplitudes depend upon the c's, of course, and so are determined by the initial conditions; but at each natural frequency the *ratios* of the amplitudes with which the masses oscillate are always the same, regardless of their actual numerical values. Figure 11.2 illustrates this behavior for one full cycle of the motion at each of the three natural frequencies.

To conclude our discussion, let us now apply to this problem the results of Theorem 3. To do this, we return to the matric equation (8), namely, $(V - \omega^2 T)A = 0$, and observe that V and T are both symmetric and that T is positive-definite. Hence the hypotheses of Corollary 1, Theorem 3, are fulfilled. Therefore, if A_1^*, A_2^*, A_3^* are solution vectors of Eq. (8) which are orthonormal with respect to T, and if M is the matrix $[A_1^* \quad A_2^* \quad A_3^*]$, the substitution

$$X = MY$$

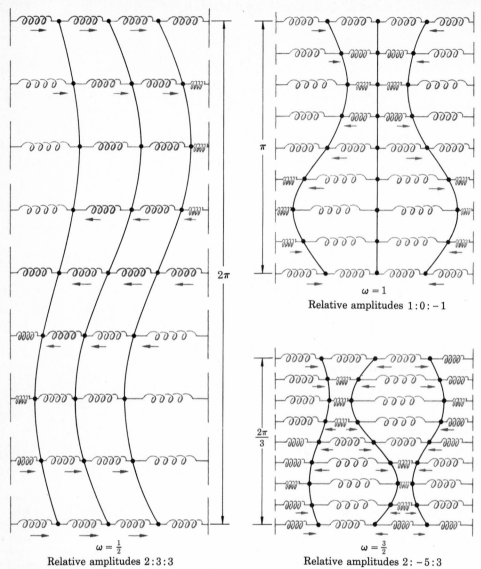

Figure 11.2
The normal modes of the system shown in Fig. 11.1.

will simultaneously reduce the quadratic forms $X^T V X$ and $X^T T X$ to the respective diagonal forms $Y^T D Y$ and $Y^T I Y \equiv Y^T Y$, where D is the diagonal matrix of characteristic values

$$\begin{bmatrix} \frac{1}{4} & 0 & 0 \\ 0 & 1 & 0 \\ 0 & 0 & \frac{9}{4} \end{bmatrix}$$

To verify this, we note first that when the solution vectors

$$A_1 = \begin{bmatrix} 2 \\ 3 \\ 3 \end{bmatrix} \qquad A_2 = \begin{bmatrix} 1 \\ 0 \\ -1 \end{bmatrix} \qquad A_3 = \begin{bmatrix} 2 \\ -5 \\ 3 \end{bmatrix}$$

are normalized with respect to T, we obtain

$$A_1^* = \frac{A_1}{\sqrt{A_1{}^T T A_1}} = \frac{1}{\sqrt{96}}\begin{bmatrix} 2 \\ 3 \\ 3 \end{bmatrix} \qquad A_2^* = \frac{A_2}{\sqrt{A_2{}^T T A_2}} = \frac{1}{\sqrt{10}}\begin{bmatrix} 1 \\ 0 \\ -1 \end{bmatrix}$$

$$A_3^* = \frac{A_3}{\sqrt{A_3{}^T T A_3}} = \frac{1}{\sqrt{160}}\begin{bmatrix} 2 \\ -5 \\ 3 \end{bmatrix}$$

Hence, the required substitution $X = MY$ is

$$\begin{bmatrix} x_1 \\ x_2 \\ x_3 \end{bmatrix} = \begin{bmatrix} \dfrac{2}{\sqrt{96}} & \dfrac{1}{\sqrt{10}} & \dfrac{2}{\sqrt{160}} \\ \dfrac{3}{\sqrt{96}} & 0 & \dfrac{-5}{\sqrt{160}} \\ \dfrac{3}{\sqrt{96}} & \dfrac{-1}{\sqrt{10}} & \dfrac{3}{\sqrt{160}} \end{bmatrix}\begin{bmatrix} y_1 \\ y_2 \\ y_3 \end{bmatrix}$$

or

$$x_1 = \frac{2y_1}{\sqrt{96}} + \frac{y_2}{\sqrt{10}} + \frac{2y_3}{\sqrt{160}}$$

(12)
$$x_2 = \frac{3y_1}{\sqrt{96}} \qquad\qquad - \frac{5y_3}{\sqrt{160}}$$

$$x_3 = \frac{3y_1}{\sqrt{96}} - \frac{y_2}{\sqrt{10}} + \frac{3y_3}{\sqrt{160}}$$

Finally, introducing these expressions into the quadratic forms

$$X^T V X = 6x_1{}^2 - 6x_1x_2 + 6x_2{}^2 - 6x_2x_3 + 4x_3{}^2$$

and
$$X^T T X = 6x_1{}^2 + 4x_2{}^2 + 4x_3{}^2$$

we obtain, respectively,

$$X^T V X = 6\left(\frac{2y_1}{\sqrt{96}} + \frac{y_2}{\sqrt{10}} + \frac{2y_3}{\sqrt{160}}\right)^2$$

$$- 6\left(\frac{2y_1}{\sqrt{96}} + \frac{y_2}{\sqrt{10}} + \frac{2y_3}{\sqrt{160}}\right)\left(\frac{3y_1}{\sqrt{96}} - \frac{5y_3}{\sqrt{160}}\right)$$

$$+ 6\left(\frac{3y_1}{\sqrt{96}} - \frac{5y_3}{\sqrt{160}}\right)^2$$

$$- 6\left(\frac{3y_1}{\sqrt{96}} - \frac{5y_3}{\sqrt{160}}\right)\left(\frac{3y_1}{\sqrt{96}} - \frac{y_2}{\sqrt{10}} + \frac{3y_3}{\sqrt{160}}\right)$$

$$+ 4\left(\frac{3y_1}{\sqrt{96}} - \frac{y_2}{\sqrt{10}} + \frac{3y_3}{\sqrt{160}}\right)^2$$

$$= \tfrac{1}{4}y_1{}^2 + y_2{}^2 + \tfrac{9}{4}y_3{}^2$$

$$X^T T X = 6\left(\frac{2y_1}{\sqrt{96}} + \frac{y_2}{\sqrt{10}} + \frac{2y_3}{\sqrt{160}}\right)^2 + 4\left(\frac{3y_1}{\sqrt{96}} - \frac{5y_3}{\sqrt{160}}\right)^2$$

$$+ 4\left(\frac{3y_1}{\sqrt{96}} - \frac{y_2}{\sqrt{10}} + \frac{3y_3}{\sqrt{160}}\right)^2$$

$$= y_1{}^2 + y_2{}^2 + y_3{}^2$$

In the present problem it is easy to identify the two quadratic forms $X^T V X$ and $X^T T X$. In fact, since the energy stored in a spring which has been stretched a distance s is $\frac{1}{2}ks^2$, it follows that the instantaneous potential energy of our system is

$$\frac{1}{2}[3x_1^2 + 3(x_2 - x_1)^2 + 3(x_3 - x_2)^2 + x_1^2]$$

$$= \frac{1}{2}(6x_1^2 - 6x_1x_2 + 6x_2^2 - 6x_2x_3 + 4x_3^2)$$

$$= \frac{1}{2}X^T V X$$

Also, the kinetic energy of a mass moving with velocity v is $\frac{1}{2}mv^2$. Hence, the instantaneous kinetic energy of our system is

$$\frac{1}{2}(6\dot{x}_1^2 + 4\dot{x}_2^2 + 4\dot{x}_3^2) = \frac{1}{2}\dot{X}^T T \dot{X}$$

From this it follows that when the system is vibrating at any one of its natural frequencies ω_i, its maximum kinetic energy is

$$\frac{\omega_i^2}{2} X^T T X$$

The new coordinates y_1, y_2, y_3 defined by (12), and in terms of which the two energy expressions appear as sums of squares, are known as the **normal coordinates** of the system.

EXERCISES

1 For each of the following matrices A, find a pair of matrices (P,Q) such that PAQ is a diagonal matrix:

(a) $\begin{bmatrix} 1 & 2 \\ 3 & 4 \end{bmatrix}$ (b) $\begin{bmatrix} 1 & -1 \\ 0 & 3 \end{bmatrix}$

(c) $\begin{bmatrix} 1 & -1 & 1 \\ 2 & 1 & 2 \\ 0 & 1 & 3 \end{bmatrix}$ (d) $\begin{bmatrix} 1 & 0 & 3 \\ 1 & -1 & 1 \\ -1 & 3 & 3 \end{bmatrix}$

2 For each of the following matrices A, find two nonsingular matrices P such that $P^T A P$ is a diagonal matrix:

(a) $\begin{bmatrix} 1 & 2 \\ 2 & 3 \end{bmatrix}$ (b) $\begin{bmatrix} 1 & -1 \\ -1 & 0 \end{bmatrix}$

(c) $\begin{bmatrix} 1 & 1 & 1 \\ 1 & 2 & 0 \\ 1 & 0 & 3 \end{bmatrix}$ (d) $\begin{bmatrix} 1 & 2 & 0 \\ 2 & 5 & 2 \\ 0 & 2 & 4 \end{bmatrix}$

3 For each of the following pairs of matrices (A,B), find a congruence transformation which will simultaneously reduce A and B to diagonal form, and carry out the diagonalization:

(a) $\begin{bmatrix} 3 & -2 \\ -2 & 4 \end{bmatrix}$ $\begin{bmatrix} 1 & 0 \\ 0 & 2 \end{bmatrix}$ (b) $\begin{bmatrix} 6 & 2 \\ 2 & 2 \end{bmatrix}$ $\begin{bmatrix} 2 & 0 \\ 0 & 1 \end{bmatrix}$

(c) $\begin{bmatrix} 4 & 3 \\ 3 & 6 \end{bmatrix}$ $\begin{bmatrix} 1 & 0 \\ 0 & 3 \end{bmatrix}$ (d) $\begin{bmatrix} 2 & 2 \\ 2 & 3 \end{bmatrix}$ $\begin{bmatrix} 1 & 1 \\ 1 & 2 \end{bmatrix}$

(e) $\begin{bmatrix} 8 & -2 & 0 \\ -2 & 3 & -1 \\ 0 & -1 & 2 \end{bmatrix}$ $\begin{bmatrix} 8 & 0 & 0 \\ 0 & 2 & 0 \\ 0 & 0 & 2 \end{bmatrix}$ (f) $\begin{bmatrix} 3 & -1 & 0 \\ -1 & 1 & -1 \\ 0 & -1 & 5 \end{bmatrix}$ $\begin{bmatrix} 4 & 0 & 0 \\ 0 & 1 & 0 \\ 0 & 0 & 4 \end{bmatrix}$

(g) $\begin{bmatrix} 6 & 0 & 2 \\ 0 & 6 & -4 \\ 2 & -4 & 6 \end{bmatrix}$ $\begin{bmatrix} 3 & 1 & 1 \\ 1 & 3 & -1 \\ 1 & -1 & 3 \end{bmatrix}$

(h) $\begin{bmatrix} 7 & -1 & 0 \\ -1 & 11 & -4 \\ 0 & -4 & 10 \end{bmatrix}$ $\begin{bmatrix} 6 & -2 & -1 \\ -2 & 10 & -5 \\ -1 & -5 & 9 \end{bmatrix}$

4 **(a)** If A and B are hermitian (or real symmetric) matrices, show that there exist congruence transformations which will simultaneously diagonalize A and B even though B is not definite.

(b) Find a congruence transformation which will simultaneously diagonalize

$$\begin{bmatrix} -2 & 1 \\ 1 & 1 \end{bmatrix} \quad \text{and} \quad \begin{bmatrix} 1 & 0 \\ 0 & -2 \end{bmatrix}.$$

(c) Find a congruence transformation which will simultaneously diagonalize

$$\begin{bmatrix} -3 & 3 \\ 3 & 0 \end{bmatrix} \quad \text{and} \quad \begin{bmatrix} -7 & 5 \\ 5 & -1 \end{bmatrix}.$$

5 Find a similarity transformation which will reduce each of the following matrices to diagonal form:

(a) $\begin{bmatrix} -3 & 2 \\ -10 & 6 \end{bmatrix}$ **(b)** $\begin{bmatrix} 0 & -2 \\ -2 & 0 \end{bmatrix}$ **(c)** $\begin{bmatrix} 2 & 1 \\ 2 & 1 \end{bmatrix}$

(d) $\begin{bmatrix} 5 & -2 & -1 \\ -1 & 4 & -1 \\ 1 & -2 & 3 \end{bmatrix}$ **(e)** $\begin{bmatrix} 2 & -3 & 3 \\ 0 & 3 & -1 \\ 0 & -1 & 3 \end{bmatrix}$ **(f)** $\begin{bmatrix} 3 & -2 & -2 \\ -1 & 2 & 0 \\ 1 & -1 & 1 \end{bmatrix}$

6 Work Example 1 with $X_0 = \begin{bmatrix} 1 \\ 2 \\ 2 \end{bmatrix}$ and $\dot{X}_0 = \begin{bmatrix} 1 \\ -1 \\ 3 \end{bmatrix}$.

7 The system shown in Fig. 11.3 begins to move with initial displacement $X_0 = \begin{bmatrix} 1 \\ 2 \end{bmatrix}$

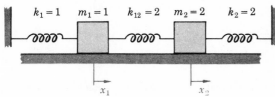

$k_1 = 1$ $m_1 = 1$ $k_{12} = 2$ $m_2 = 2$ $k_2 = 2$

x_1 x_2 **Figure 11.3**

and initial velocity $\dot{X}_0 = \begin{bmatrix} 2 \\ -1 \end{bmatrix}$. Assuming that there is no friction in the system, determine its subsequent motion.

8 The system shown in Fig. 11.4 begins to move with initial displacement $X_0 = \begin{bmatrix} 1 \\ 1 \\ -1 \end{bmatrix}$

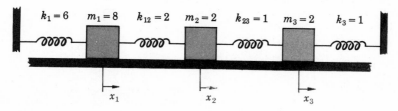

$k_1 = 6$ $m_1 = 8$ $k_{12} = 2$ $m_2 = 2$ $k_{23} = 1$ $m_3 = 2$ $k_3 = 1$

x_1 x_2 x_3

Figure 11.4

and initial velocity $\dot{X}_0 = \begin{bmatrix} 1 \\ 0 \\ 2 \end{bmatrix}$. Assuming that there is no friction in the system, determine its subsequent motion.

9 The system shown in Fig. 11.5 begins to move with initial displacement $\theta = \begin{bmatrix} 1 \\ 0 \\ 2 \end{bmatrix}$

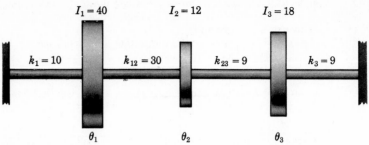

$I_1 = 40$ $I_2 = 12$ $I_3 = 18$

$k_1 = 10$ $k_{12} = 30$ $k_{23} = 9$ $k_3 = 9$

θ_1 θ_2 θ_3

Figure 11.5

and initial angular velocity $\dot{\theta} = \begin{bmatrix} 2 \\ 1 \\ 1 \end{bmatrix}$. Assuming that there is no friction in the system, determine its subsequent motion.

10 Find the natural frequencies and normal modes of the system shown in Fig. 11.6 if $k_1 = k_2 = k_3 = k_{12} = k_{23} = k_{13} = 1$ and $m_1 = m_2 = m_3 = 1$.

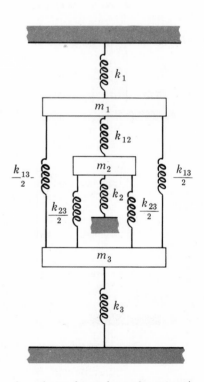

k_1

m_1

k_{12}

m_2 $\dfrac{k_{13}}{2}$

$\dfrac{k_{13}}{2}$

$\dfrac{k_{23}}{2}$ k_2 $\dfrac{k_{23}}{2}$

m_3

k_3

Figure 11.6

11 Work Exercise 10 if $k_1 = 13$, $k_2 = k_3 = k_{12} = k_{23} = k_{13} = 1$, $m_1 = 4$, $m_2 = m_3 = 1$.

12 Work Exercise 10 if $k_1 = 1$, $k_2 = 5$, $k_3 = 13$, $k_{12} = k_{23} = k_{13} = 1$, $m_1 = 4$. $m_2 = 8$, $m_3 = 16$.

13 In Exercise 10, determine the subsequent motion of the system if

$$X_0 = \begin{bmatrix} 1 \\ 2 \\ 3 \end{bmatrix} \quad \text{and} \quad \dot{X}_0 = \begin{bmatrix} 1 \\ 0 \\ 0 \end{bmatrix}$$

14　In Exercise 11, determine the subsequent motion of the system if

$$X_0 = \begin{bmatrix} 1 \\ 1 \\ 0 \end{bmatrix} \quad \text{and} \quad \dot{X}_0 = \begin{bmatrix} 0 \\ 0 \\ 1 \end{bmatrix}$$

15　In Exercise 12, determine the subsequent motion of the system if

$$X_0 = \begin{bmatrix} 1 \\ -2 \\ 1 \end{bmatrix} \quad \text{and} \quad \dot{X}_0 = \begin{bmatrix} 8 \\ -3 \\ -5 \end{bmatrix}$$

Exercises 10 to 12 show that the statement "a system with n degrees of freedom has n different natural frequencies" is not always true. Investigate this matter further by proving the following results.

16　The system shown in Fig. 11.6 will have a single natural frequency if and only if $k_{12} = k_{23} = k_{13} = 0$ and $k_1/m_1 = k_2/m_2 = k_3/m_3$, that is, if and only if it consists of three uncoupled subsystems having the same natural frequency.

17　When the system shown in Fig. 11.6 does not consist of two or more uncoupled subsystems, its frequency equation will have a simple root ω_1 and a double root ω_2, that is, the system will have exactly two natural frequencies if and only if

$$\omega_2{}^2 m_1 = k_1 + k_{12} + k_{13} + \frac{k_{12}k_{13}}{k_{23}}$$

$$\omega_2{}^2 m_2 = k_{12} + k_2 + k_{23} + \frac{k_{12}k_{23}}{k_{13}}$$

$$\omega_2{}^2 m_3 = k_{13} + k_{23} + k_3 + \frac{k_{13}k_{23}}{k_{12}}$$

18　When the system shown in Fig. 11.6 has a repeated frequency ω_2, the components d_1, d_2, d_3 of the normal mode corresponding to the repeated frequency satisfy the relation

$$\frac{d_1}{k_{23}} + \frac{d_2}{k_{13}} + \frac{d_3}{k_{12}} = 0$$

19　When the system shown in Fig. 11.6 has a repeated frequency, the components c_1, c_2, c_3 of the normal mode corresponding to the unrepeated natural frequency are proportional to

$$\frac{k_{12}k_{13}}{m_1} \qquad \frac{k_{12}k_{23}}{m_2} \qquad \frac{k_{13}k_{23}}{m_3}$$

20　When the system shown in Fig. 11.6 has a repeated frequency, the repeated frequency ω_2 and the unrepeated frequency ω_1 are connected by the relation

$$\omega_2{}^2 - \omega_1{}^2 = \frac{k_{12}k_{13}}{k_{23}m_1} + \frac{k_{12}k_{23}}{k_{13}m_2} + \frac{k_{13}k_{23}}{k_{12}m_3}$$

21　When the system shown in Fig. 11.6 is vibrating at a repeated frequency, the masses need not move in phase or with phase differences of 180°. Hence there may be no time when the three masses are simultaneously at rest.

22　In Exercise 7, find the forced response to the excitation

$$F = \begin{bmatrix} 3 \\ -4 \end{bmatrix} \sin 3t$$

23　In Exercise 8, find the forced response to the excitation

$$F = \begin{bmatrix} -16 \\ 20 \\ 16 \end{bmatrix} \cos 2t$$

11.4 Functions of a Square Matrix

In Sec. 10.2, after we had defined matric multiplication, we were able to define positive integral powers of a square matrix A and to verify that for arbitrary positive integers r and s

(1) $$A^r A^s = A^s A^r = A^{r+s}$$

Moreover, we verified in Sec. 10.3 that if A is a nonsingular matrix, it has an inverse A^{-1} such that $AA^{-1} = A^{-1}A = I$, and we defined negative integral powers of A by the relation

$$A^{-n} = (A^{-1})^n$$

Thus, after we introduced the definition $A^0 = I$, it became clear that for any nonsingular matrix A, Eq. (1) holds for all integral values of r and s. It is now natural to define polynomial functions of a square matrix and, if possible, rational fractional functions.

DEFINITION 1   A polynomial function of a square matrix A is a finite linear combination of nonnegative integral powers of A.

EXAMPLE 1

If $A = \begin{bmatrix} 1 & 2 \\ 3 & -4 \end{bmatrix}$ and $p(x) = x^2 + 5x + 4$, then

$$p(A) = A^2 + 5A + 4I = \begin{bmatrix} 7 & -6 \\ -9 & 22 \end{bmatrix} + 5\begin{bmatrix} 1 & 2 \\ 3 & -4 \end{bmatrix} + 4\begin{bmatrix} 1 & 0 \\ 0 & 1 \end{bmatrix} = \begin{bmatrix} 16 & 4 \\ 6 & 6 \end{bmatrix}$$

It is interesting to note that $p(A)$ can also be evaluated by using the factored forms of $p(x)$, namely,

$$p(x) = (x + 4)(x + 1) = (x + 1)(x + 4)$$

$$p(A) = (A + 4I)(A + I) = \left(\begin{bmatrix} 1 & 2 \\ 3 & -4 \end{bmatrix} + 4\begin{bmatrix} 1 & 0 \\ 0 & 1 \end{bmatrix}\right)\left(\begin{bmatrix} 1 & 2 \\ 3 & -4 \end{bmatrix} + \begin{bmatrix} 1 & 0 \\ 0 & 1 \end{bmatrix}\right)$$

$$= \begin{bmatrix} 5 & 2 \\ 3 & 0 \end{bmatrix}\begin{bmatrix} 2 & 2 \\ 3 & -3 \end{bmatrix} = \begin{bmatrix} 16 & 4 \\ 6 & 6 \end{bmatrix}$$

$$p(A) = (A + I)(A + 4I) = \left(\begin{bmatrix} 1 & 2 \\ 3 & -4 \end{bmatrix} + \begin{bmatrix} 1 & 0 \\ 0 & 1 \end{bmatrix}\right)\left(\begin{bmatrix} 1 & 2 \\ 3 & -4 \end{bmatrix} + 4\begin{bmatrix} 1 & 0 \\ 0 & 1 \end{bmatrix}\right)$$

$$= \begin{bmatrix} 2 & 2 \\ 3 & -3 \end{bmatrix}\begin{bmatrix} 5 & 2 \\ 3 & 0 \end{bmatrix} = \begin{bmatrix} 16 & 4 \\ 6 & 6 \end{bmatrix}$$

In this example it is of course not clear, especially in view of the noncommutative character of matric multiplication, whether the fact that $p(A)$ can be computed equally well from

$$A^2 + 5A + 4I \qquad (A + 4I)(A + I) \qquad \text{or} \qquad (A + I)(A + 4I)$$

is a result of some special property of A and $p(x)$ or is illustrative of some general principle. Actually the latter is the case. In fact, any identical relation involving sums and products of scalar polynomials is valid for the corresponding matric polynomials, as the following important, but almost obvious, theorem assures us.

THEOREM 1 Any polynomial identity between scalar polynomials implies a corresponding identity for matric polynomials.

Proof Clearly, any polynomial relation between scalar polynomials can be constructed using only the operations of addition and multiplication. For instance,

$$[f_1(x)f_2(x) + f_3(x)]f_4(x) = f_5(x)$$

is completely equivalent to the chain of relations

$$\phi(x)f_4(x) = f_5(x) \qquad \phi(x) = \Psi(x) + f_3(x) \qquad \Psi(x) = f_1(x)f_2(x)$$

Hence, to prove the theorem it is sufficient to show that for any polynomials f, g, s, p and any square matrix A

a. If $f(x) + g(x) = s(x)$, then $f(A) + g(A) = s(A)$.
b. If $f(x)g(x) = p(x)$, then $f(A)g(A) = p(A)$.

To prove the first of these, let

$$f(x) = \sum_{i=0}^{m} a_i x^i \qquad g(x) = \sum_{i=0}^{n} b_i x^i \qquad s(x) = \sum_{i=0}^{t} c_i x^i$$

where $t = \max(m,n)$, $c_i = a_i + b_i$, and the coefficients of any powers of x which are not present are understood to be zero. Then

$$f(A) + g(A) = \sum_{i=0}^{m} a_i A^i + \sum_{i=0}^{n} b_i A^i$$

$$= \sum_{i=0}^{t} (a_i + b_i)A^i = \sum_{i=0}^{t} c_i A^i = s(A)$$

as asserted.

To prove part **b**, let

$$f(x) = \sum_{i=0}^{m} a_i x^i \qquad g(x) = \sum_{j=0}^{n} b_j x^j \qquad p(x) = \sum_{k=0}^{t} c_k x^k$$

where $t = m + n$ and $c_k = \sum_{i,j} a_i b_j$, the summation extending over all values of i and j such that $i + j = k$ and, of course, $0 \le i \le m$, $0 \le j \le n$. Then, using the distributive property of matric multiplication and the associative and commutative properties of matric addition, we have

$$f(A)g(A) = \left(\sum_{i=0}^{m} a_i A^i\right)\left(\sum_{j=0}^{n} b_j A^j\right)$$

$$= \sum_{i=0}^{m}\sum_{j=0}^{n} (a_i A^i)(b_j A^j) = \sum_{i=0}^{m}\sum_{j=0}^{n} a_i b_j A^{i+j}$$

or, grouping together all terms involving the same power of A,

$$f(A)g(A) = \sum_{k=0}^{t=m+n} c_k A^k \qquad \text{where } k = i + j \text{ and } c_k = \sum_{i,j} a_i b_j$$

$$= p(A)$$

as asserted.

Since $f(x)g(x) = g(x)f(x)$, it follows from Theorem 1 that

(2) $$f(A)g(A) = g(A)f(A)$$

In other words, we have the following result.

COROLLARY 1   Any two polynomials in a matrix A commute with each other.

If $g(A)$ is a nonsingular matrix, then $g^{-1}(A)$ exists and we can premultiply and postmultiply each side of Eq. (2) by $g^{-1}(A)$, getting

$$g^{-1}(A)f(A)g(A)g^{-1}(A) = g^{-1}(A)g(A)f(A)g^{-1}(A)$$

or

(3) $$g^{-1}(A)f(A) = f(A)g^{-1}(A)$$

With this identity, we are now in a position to define rational fractional functions of a square matrix A.

DEFINITION 2   If $f(x)$ and $g(x)$ are scalar polynomials, and if A is a square matrix such that $g(A)$ is nonsingular, then either of the equal matrices $g^{-1}(A)f(A)$ and $f(A)g^{-1}(A)$ is called the quotient of $f(A)$ by $g(A)$ and is written $f(A)/g(A)$.

It is now relatively easy to prove the following extension of Theorem 1 (see Exercise 3).

THEOREM 2   Any identity between rational fractional functions of a scalar variable implies a corresponding matrix identity, provided all the matric functions are defined.

With rational functions of a square matrix now defined, it is natural to ask whether the characteristic values of a rational function of a square matrix A can be expressed in terms of the characteristic values of A. This is indeed the case, as the following chain of theorems makes clear.

THEOREM 3   If $\lambda_1, \lambda_2, \ldots, \lambda_n$ are the (possibly repeated) characteristic values of a square matrix A, and if f is any polynomial, then the determinant of $f(A)$ is given by the formula

$$|f(A)| = f(\lambda_1)f(\lambda_2)\cdots f(\lambda_n)$$

Proof   Let the factored form of the characteristic polynomial of the given matrix A be

(4) $$|A - \lambda I| = \prod_{i=1}^{n} (\lambda_i - \lambda)$$

and let the factored form of the given polynomial f be

(5) $$f(t) = c(t - r_1)(t - r_2)\cdots(t - r_k)$$

Then, since Theorem 1 assures us that identities between scalar polynomials imply corresponding matric identities, we have

$$f(A) = c(A - r_1 I)(A - r_2 I) \cdots (A - r_k I)$$

Furthermore, since the determinant of a product of square matrices is equal to the product of the determinants of the matric factors, and since the scalar factor c incorporated into any one of the matric factors reappears as the factor c^n in the determinant of that matrix, we have

$$(6) \qquad |f(A)| = c^n |A - r_1 I| \cdot |A - r_2 I| \cdots |A - r_k I| = c^n \prod_{j=1}^{k} |A - r_j I|$$

However, $|A - r_j I|$ is just the characteristic polynomial of A evaluated for $\lambda = r_j$. Hence, by (4),

$$|A - r_j I| = \prod_{i=1}^{n} (\lambda_i - r_j)$$

and therefore, substituting into (6), we have

$$|f(A)| = c^n \prod_{j=1}^{k} \prod_{i=1}^{n} (\lambda_i - r_j)$$

Next, interchanging the order in which the products are formed, by first grouping together all the factors corresponding to a given value of i and assigning a single factor c to each such group, we have

$$|f(A)| = c^n \prod_{i=1}^{n} \prod_{j=1}^{k} (\lambda_i - r_j) = \prod_{i=1}^{n} \left[c \prod_{j=1}^{k} (\lambda_i - r_j) \right]$$

Finally we observe that, with the coefficient c, the inner product in the last expression is precisely the evaluation of the factored form (5) of the given polynomial for $t = \lambda_i$. Hence

$$|f(A)| = \prod_{i=1}^{n} f(\lambda_i)$$

as asserted.

THEOREM 4   If $\lambda_1, \lambda_2, \ldots, \lambda_n$ are the characteristic values of a square matrix A, if $f = g/h$ is a rational fractional function, and if $h(A)$ is nonsingular, then the determinant of $f(A)$ is given by the formula

$$|f(A)| = f(\lambda_1) f(\lambda_2) \cdots f(\lambda_n)$$

Proof   Since, by definition, $f(A) = g(A)/h(A) = g(A)h^{-1}(A)$, and since the determinant of a product of square matrices is equal to the product of the determinants of the matric factors, we have

$$|f(A)| = |g(A)h^{-1}(A)| = |g(A)| \cdot |h^{-1}(A)|$$

Moreover, as we observed in Sec. 10.3, $|h^{-1}(A)| = 1/|h(A)|$. Therefore

$$|f(A)| = \frac{|g(A)|}{|h(A)|}$$

However, by Theorem 3, since g and h are polynomials,

$$|g(A)| = g(\lambda_1)g(\lambda_2)\cdots g(\lambda_n) \qquad \text{and} \qquad |h(A)| = h(\lambda_1)h(\lambda_2)\cdots h(\lambda_n)$$

Hence
$$|f(A)| = \frac{g(\lambda_1)g(\lambda_2)\cdots g(\lambda_n)}{h(\lambda_1)h(\lambda_2)\cdots h(\lambda_n)}$$

$$= f(\lambda_1)f(\lambda_2)\cdots f(\lambda_n)$$

as asserted.

THEOREM 5 If $\lambda_1, \lambda_2, \ldots, \lambda_n$ are the characteristic values of a square matrix A, and if $f = g/h$, where g and h are polynomials such that $|h(A)| \neq 0$, then the characteristic values of $f(A)$ are $f(\lambda_1), f(\lambda_2), \ldots, f(\lambda_n)$.

Proof Let $\phi(x)$ be the function defined by the expression

$$\phi(x) = f(x) - \lambda = \frac{g(x)}{h(x)} - \lambda = \frac{g(x) - \lambda h(x)}{h(x)}$$

Clearly, $g(x) - \lambda h(x)$ is a polynomial, and therefore $\phi(x)$ is a rational fractional function of x. Hence, by the last theorem,

$$|\phi(A)| = \phi(\lambda_1)\phi(\lambda_2)\cdots \phi(\lambda_n)$$

In other words, for all values of λ,

$$|f(A) - \lambda I| = [f(\lambda_1) - \lambda][f(\lambda_2) - \lambda]\cdots[f(\lambda_n) - \lambda]$$

The right-hand side of this identity is thus the factored form of the characteristic polynomial $|f(A) - \lambda I|$. Hence, by inspection, the roots of the characteristic equation of $f(A)$ are

$$\lambda = f(\lambda_1), f(\lambda_2), \ldots, f(\lambda_n)$$

as asserted.

COROLLARY 1 If the characteristic values of a square matrix A are $\lambda_1, \lambda_2, \ldots, \lambda_n$, then for all integral values of k if A is nonsingular and for all nonnegative integral values of k if A is singular, the characteristic values of A^k are $\lambda_1{}^k, \lambda_2{}^k, \ldots, \lambda_n{}^k$.

COROLLARY 2 If X_i is a characteristic vector corresponding to the characteristic value λ_i of a square matrix A, and if p is a polynomial, then X_i is also a characteristic vector corresponding to the characteristic value $p(\lambda_i)$ of the matrix $p(A)$.

EXAMPLE 2

As an illustration of Theorem 5, consider the matrix $A = \begin{bmatrix} 1 & -2 \\ 3 & -4 \end{bmatrix}$ and the function $\phi(x) = x/(x + 3)$. The characteristic equation of A is

$$|A - \lambda I| = \begin{vmatrix} 1 - \lambda & -2 \\ 3 & -4 - \lambda \end{vmatrix} = \lambda^2 + 3\lambda + 2 = 0$$

Hence the characteristic roots are $\lambda = -1, -2$. Therefore, according to Theorem 5, the characteristic roots of $\phi(A)$ are

$$\phi(-1) = -\tfrac{1}{2} \qquad \text{and} \qquad \phi(-2) = -2$$

To confirm this, we have, by direct calculation

$$\phi(A) = \frac{A}{A + 3I} = A(A + 3I)^{-1} = \begin{bmatrix} 1 & -2 \\ 3 & -4 \end{bmatrix} \begin{bmatrix} 4 & -2 \\ 3 & -1 \end{bmatrix}^{-1}$$

$$= \begin{bmatrix} 1 & -2 \\ 3 & -4 \end{bmatrix} \frac{1}{2} \begin{bmatrix} -1 & 2 \\ -3 & 4 \end{bmatrix}$$

$$= \frac{1}{2} \begin{bmatrix} 5 & -6 \\ 9 & -10 \end{bmatrix} = \begin{bmatrix} \frac{5}{2} & -3 \\ \frac{9}{2} & -5 \end{bmatrix}$$

The characteristic roots of $\phi(A)$ are therefore the roots of the equation

$$|\phi(A) - \lambda I| = \begin{vmatrix} \frac{5}{2} - \lambda & -3 \\ \frac{9}{2} & -5 - \lambda \end{vmatrix} = \lambda^2 + \tfrac{5}{2}\lambda + 1 = 0$$

or $-\tfrac{1}{2}$ and -2, as before.

If p is a polynomial and A is a square matrix, the evaluation of $p(A)$ is a perfectly straightforward matter. However, when A is a matrix which is similar to a diagonal matrix, the evaluation of $p(A)$ can be appreciably simplified. To establish the result upon which this simplification is based, it is convenient first to prove the following lemmas.

LEMMA 1 If $B = SAS^{-1}$, then $B^n = SA^nS^{-1}$.

Proof Clearly, the lemma is true for $n = 2$, since

$$B^2 = (SAS^{-1})(SAS^{-1}) = SA(S^{-1}S)AS^{-1} = SA^2S^{-1}$$

Assuming, then, that the lemma is true for $n = k$, we have

$$B^{k+1} = BB^k = (SAS^{-1})(SA^kS^{-1}) = SA(S^{-1}S)A^kS^{-1} = SA^{k+1}S^{-1}$$

which completes the induction and establishes the lemma.

If we now apply Lemma 1 to each term of any polynomial function of B and then use the distributive property of matric multiplication, we obtain the following result.

LEMMA 2 If $B = SAS^{-1}$, and if p is a polynomial, then $p(B) = Sp(A)S^{-1}$, that is, $p(SAS^{-1}) = Sp(A)S^{-1}$.

Furthermore, by another easy induction we can establish the following observation.

LEMMA 3 If D is the diagonal matrix

$$\begin{bmatrix} d_{11} & & \mathbf{O} \\ & d_{22} & \\ & & \cdot \\ \mathbf{O} & & d_{nn} \end{bmatrix} \quad \text{then} \quad D^k = \begin{bmatrix} d_{11}{}^k & & \mathbf{O} \\ & d_{22}{}^k & \\ & & \cdot \\ \mathbf{O} & & d_{nn}{}^k \end{bmatrix}$$

Finally, by applying Lemma 3 to each term of any polynomial function of a diagonal matrix D and then using the definition of matric addition, we have the following result.

LEMMA 4 If D is the diagonal matrix

$$\begin{bmatrix} d_{11} & & & \mathbf{O} \\ & d_{22} & & \\ & & \cdot & \\ \mathbf{O} & & & d_{nn} \end{bmatrix}$$

and p is any polynomial, then

$$p(D) = \begin{bmatrix} p(d_{11}) & & & \mathbf{O} \\ & p(d_{22}) & & \\ & & \cdot & \\ \mathbf{O} & & & p(d_{nn}) \end{bmatrix}$$

Using Lemmas 2 and 4, we can now prove the following theorem.

THEOREM 6 If A is a matrix which is similar to a diagonal matrix, that is, if

$$S^{-1}AS = D = \begin{bmatrix} \lambda_1 & & & \mathbf{O} \\ & \lambda_2 & & \\ & & \cdot & \\ \mathbf{O} & & & \lambda_n \end{bmatrix}$$

where $\lambda_1, \lambda_2, \ldots, \lambda_n$ are the characteristic values of A, then

$$p(A) = S \begin{bmatrix} p(\lambda_1) & & & \mathbf{O} \\ & p(\lambda_2) & & \\ & & \cdot & \\ \mathbf{O} & & & p(\lambda_n) \end{bmatrix} S^{-1}$$

Proof By Lemma 4,

$$p(D) = \begin{bmatrix} p(\lambda_1) & & & \mathbf{O} \\ & p(\lambda_2) & & \\ & & \cdot & \\ \mathbf{O} & & & p(\lambda_n) \end{bmatrix}$$

Also, since $S^{-1}AS = D$, it follows that $A = SDS^{-1}$. Hence, using Lemma 2, we have

$$S \begin{bmatrix} p(\lambda_1) & & & \mathbf{O} \\ & p(\lambda_2) & & \\ & & \cdot & \\ \mathbf{O} & & & p(\lambda_n) \end{bmatrix} S^{-1} = Sp(D)S^{-1} = p(SDS^{-1}) = p(A)$$

as asserted.

EXAMPLE 3

If $p(x) = x^4 - 4x^3 + 6x^2 - x - 3$ and $A = \begin{bmatrix} 0 & -2 \\ 1 & 3 \end{bmatrix}$, what is $p(A)$?

By an easy calculation we find the characteristic equation of A to be

$$|A - \lambda I| = \begin{vmatrix} -\lambda & -2 \\ 1 & 3 - \lambda \end{vmatrix} = \lambda^2 - 3\lambda + 2 = 0$$

Hence the characteristic values of A are $\lambda_1 = 1$ and $\lambda_2 = 2$; and since these are distinct, it follows from Theorem 4, Sec. 11.3, that A is similar to a diagonal matrix and Theorem 6 can be applied. Now, corresponding to λ_1 and λ_2 we have the characteristic vectors

$$X_1 = \begin{bmatrix} 2 \\ -1 \end{bmatrix} \quad \text{and} \quad X_2 = \begin{bmatrix} 1 \\ -1 \end{bmatrix}$$

and from these we can construct the modal matrix and its inverse

$$S = \begin{bmatrix} 2 & 1 \\ -1 & -1 \end{bmatrix} \quad \text{and} \quad S^{-1} = \begin{bmatrix} 1 & 1 \\ -1 & -2 \end{bmatrix}$$

According to Theorem 4, Sec. 11.3, these are matrices such that

$$S^{-1}AS = D = \begin{bmatrix} 1 & 0 \\ 0 & 2 \end{bmatrix}$$

Hence these are the matrices to be used in evaluating $p(A)$ by means of Theorem 6. Now

$$p(\lambda_1) = p(1) = -1 \quad \text{and} \quad p(\lambda_2) = p(2) = 3$$

Therefore,

$$p(A) = A^4 - 4A^3 + 6A^2 - A - 3I = S \begin{bmatrix} p(\lambda_1) & 0 \\ 0 & p(\lambda_2) \end{bmatrix} S^{-1}$$

$$= \begin{bmatrix} 2 & 1 \\ -1 & -1 \end{bmatrix} \begin{bmatrix} -1 & 0 \\ 0 & 3 \end{bmatrix} \begin{bmatrix} 1 & 1 \\ -1 & -2 \end{bmatrix}$$

$$= \begin{bmatrix} -5 & -8 \\ 4 & 7 \end{bmatrix}$$

After polynomial functions of a square matrix have been defined, it is natural to consider polynomial equations in a matric variable. In particular, now that we have developed procedures for evaluating $p(A)$, that is, solving the equation $p(A) = X$, we shall consider the problem of solving the nontrivial equation $p(X) = A$, where p is a given polynomial, A is a given square matrix, and X is a matric variable. By means of examples (see Exercise 1) it is easy to show that there are polynomial equations $p(X) = A$ which have no solution. In one important case, however, the equation $p(X) = A$ can always be solved, as the following theorem makes clear.

THEOREM 7 If A is similar to a diagonal matrix, and if p is a scalar polynomial, then the equation $p(X) = A$ is solvable for X.

Proof By hypothesis, since A is similar to a diagonal matrix D, there exists a non-singular matrix S with the property that $S^{-1}AS = D$, or $A = SDS^{-1}$, where, say,

$$D = \begin{bmatrix} d_{11} & & & \mathbf{O} \\ & d_{22} & & \\ & & \cdot & \\ \mathbf{O} & & & d_{nn} \end{bmatrix}$$

Now let r_i be any one of the roots of the equation $p(x) = d_{ii}$. Then if

$$R = \begin{bmatrix} r_1 & & & \mathbf{O} \\ & r_2 & & \\ & & \cdot & \\ \mathbf{O} & & & r_n \end{bmatrix} \quad \text{and} \quad X = SRS^{-1}$$

we have, by Lemma 2,

$$p(X) = p(SRS^{-1}) = Sp(R)S^{-1}$$

Moreover, by Lemma 4 and the fact that $p(r_i) = d_{ii}$

$$p(R) = \begin{bmatrix} p(r_1) & & & \mathbf{O} \\ & p(r_2) & & \\ & & \cdot & \\ \mathbf{O} & & & p(r_n) \end{bmatrix} = \begin{bmatrix} d_{11} & & & \mathbf{O} \\ & d_{22} & & \\ & & \cdot & \\ \mathbf{O} & & & d_{nn} \end{bmatrix} = D$$

Therefore $p(X) = Sp(R)S^{-1} = SDS^{-1} = A$, which proves that if A is similar to a diagonal matrix, then $p(X) = A$ has the solution $X = SRS^{-1}$. If the polynomial p is of degree k, the scalar equation $p(x) = d_{ii}$ has, in general, k distinct roots. Hence there are k distinct choices for each of the diagonal elements in R, and therefore $p(X) = A$ has, in general, at least k^n different solutions.

By applying the preceding theorem to the particular equation $X^2 = A$, we obtain the following corollary.

COROLLARY 1   An $n \times n$ matrix with distinct characteristic values has at least 2^n or 2^{n-1} distinct square roots, according as it is nonsingular or singular.

Proof   Let A be an $n \times n$ matrix with n distinct characteristic values $\lambda_1, \lambda_2, \ldots, \lambda_n$. It follows, then, by Theorem 4, Sec. 11.3, that there exists a nonsingular matrix S such that

$$A = S \begin{bmatrix} \lambda_1 & & & \mathbf{O} \\ & \lambda_2 & & \\ & & \cdot & \\ \mathbf{O} & & & \lambda_n \end{bmatrix} S^{-1}$$

Thus, according to the last theorem, for any choice of plus and minus signs

$$X = S \begin{bmatrix} \pm\sqrt{\lambda_1} & & & \mathbf{O} \\ & \pm\sqrt{\lambda_2} & & \\ & & \cdot & \\ \mathbf{O} & & & \pm\sqrt{\lambda_n} \end{bmatrix} S^{-1}$$

satisfies the equation $X^2 = A$. If A is nonsingular, none of the λ's is zero and there are 2^n combinations of signs each leading to a different matrix X satisfying the equation $X^2 = A$. On the other hand, if A is singular but still has distinct characteristic values, then, by Theorem 5, Sec. 11.2, one and only one of the λ's must be zero and therefore for one of the diagonal elements there is only a single choice rather than two. Hence in this case there may be no more than 2^{n-1} distinct square roots, as asserted.

EXAMPLE 4

Solve the equation $X^2 - 4X + 4I = \begin{bmatrix} 4 & 3 \\ 5 & 6 \end{bmatrix}$.

The characteristic equation of the matrix $A = \begin{bmatrix} 4 & 3 \\ 5 & 6 \end{bmatrix}$ is

$$\begin{vmatrix} 4 - \lambda & 3 \\ 5 & 6 - \lambda \end{vmatrix} = \lambda^2 - 10\lambda + 9 = 0$$

Hence the characteristic values of A are $\lambda_1 = 1$ and $\lambda_2 = 9$, and the corresponding characteristic vectors are $X_1 = \begin{bmatrix} 1 \\ -1 \end{bmatrix}$, $X_2 = \begin{bmatrix} 3 \\ 5 \end{bmatrix}$. Therefore, by Theorem 4, Sec. 11.3, A is similar to a diagonal matrix; that is,

$$S^{-1}AS = D \quad \text{or} \quad A = SAS^{-1}$$

where S is the modal matrix $\begin{bmatrix} 1 & 3 \\ -1 & 5 \end{bmatrix}$, $S^{-1} = \dfrac{1}{8} \begin{bmatrix} 5 & -3 \\ 1 & 1 \end{bmatrix}$, and $D = \begin{bmatrix} 1 & 0 \\ 0 & 9 \end{bmatrix}$. We

must now solve the equations $p(x) = d_{ii}$ for r_t $(i = 1, 2)$:

$$x^2 - 4x + 4 = d_{11} = 1 \qquad x^2 - 4x + 4 = d_{22} = 9$$
$$x = r_1 = 1, 3 \qquad\qquad x = r_2 = -1, 5$$

Pairing each possibility for r_1 with each possibility for r_2, we thus obtain four possibilities for the matrix R:

$$R_1 = \begin{bmatrix} 1 & 0 \\ 0 & -1 \end{bmatrix} \quad R_2 = \begin{bmatrix} 1 & 0 \\ 0 & 5 \end{bmatrix} \quad R_3 = \begin{bmatrix} 3 & 0 \\ 0 & -1 \end{bmatrix} \quad R_4 = \begin{bmatrix} 3 & 0 \\ 0 & 5 \end{bmatrix}$$

Then according to Theorem 7, the solutions of the given equation are

$$X_1 = SR_1S^{-1} = \begin{bmatrix} 1 & 3 \\ -1 & 5 \end{bmatrix} \begin{bmatrix} 1 & 0 \\ 0 & -1 \end{bmatrix} \frac{1}{8} \begin{bmatrix} 5 & -3 \\ 1 & 1 \end{bmatrix} = \frac{1}{4} \begin{bmatrix} 1 & -3 \\ -5 & -1 \end{bmatrix}$$

and, similarly,

$$X_2 = SR_2S^{-1} = \frac{1}{2} \begin{bmatrix} 5 & 3 \\ 5 & 7 \end{bmatrix} \qquad X_3 = SR_3S^{-1} = \frac{1}{2} \begin{bmatrix} 3 & -3 \\ -5 & 1 \end{bmatrix}$$

and $\qquad X_4 = SR_4S^{-1} = \dfrac{1}{4} \begin{bmatrix} 15 & 3 \\ 5 & 17 \end{bmatrix}$

EXERCISES

1 Prove that there is no matrix which satisfies the equation $X^2 = \begin{bmatrix} 0 & 1 \\ 0 & 0 \end{bmatrix}$.

2 Show that for particular polynomials p and particular matrices A, each of the following cases is possible:
  (a) A nonsingular, $p(A)$ nonsingular
  (b) A nonsingular, $p(A)$ singular
  (c) A singular, $p(A)$ nonsingular
  (d) A singular, $p(A)$ singular

3 Prove Theorem 2. *Hint:* Note first that it is sufficient to prove that

$$\frac{f_1(x)}{f_2(x)} + \frac{f_3(x)}{f_4(x)} = \frac{f_5(x)}{f_6(x)} \quad \text{implies} \quad \frac{f_1(A)}{f_2(A)} + \frac{f_3(A)}{f_4(A)} = \frac{f_5(A)}{f_6(A)}$$

$$\frac{f_1(x) f_3(x)}{f_2(x) f_4(x)} = \frac{f_5(x)}{f_6(x)} \quad \text{implies} \quad \frac{f_1(A) f_3(A)}{f_2(A) f_4(A)} = \frac{f_5(A)}{f_6(A)}$$

$$\frac{f_1(x)/f_2(x)}{f_3(x)/f_4(x)} = \frac{f_5(x)}{f_6(x)} \quad \text{implies} \quad \frac{f_1(A)/f_2(A)}{f_3(A)/f_4(A)} = \frac{f_5(A)}{f_6(A)}$$

Then clear of fractions in the scalar identities, use Theorem 1, and multiply the resulting identities by the appropriate inverses.

4 If A is a diagonal matrix and p is any scalar polynomial, show that $p(A)$ is also a diagonal matrix.

5 If A is a diagonal matrix and f is a rational fractional function, is $f(A)$ necessarily a diagonal matrix?.

6 Show that I_2 has infinitely many distinct square roots.

7 Prove that an $n \times n$ matrix with distinct characteristic values has no square roots other than those identified by Corollary 1, Theorem 7. *Hint:* Use the result of Exercise 17, Sec. 10.2.

8 Prove Corollary 2, Theorem 5. *Hint:* First prove the assertion for the special polynomials $p(A) = A^k$ by premultiplying the equation $AX_i = \lambda_i X_i$ by $A, A^2, \ldots$, A^{k-1}, in turn.

9 By actually constructing an infinite family of solutions, show that each of the following matric equations is satisfied by infinitely many matrices:

(a) $X^2 - 2X - 3I_2 = 0$ (b) $X^2 - 4X + 3I_2 = 0$

(c) $X^2 - 4X - 5I_2 = 0$ (d) $X^3 - 6X^2 + 11X - 6I_3 = 0$

10 Without attempting to find the solutions, show that for all·values of a and b the equation $X^2 + aX + bI_2 = 0$ has infinitely many solutions. What do you think is the generalization of this result to equations in an $n \times n$ matric variable?

11 Show that the following matric equations have no solutions:

(a) $X^2 - 2X - 3I_2 = \begin{bmatrix} -4 & 1 \\ 0 & -4 \end{bmatrix}$ (b) $X^2 - 4X + 3I_2 = \begin{bmatrix} -1 & 2 \\ 0 & -1 \end{bmatrix}$

(c) $X^2 - 4X - I_2 = \begin{bmatrix} -9 & 3 \\ 0 & -9 \end{bmatrix}$ (d) $X^2 - 4X + 3I_3 = \begin{bmatrix} 2 & 0 & 1 \\ 1 & -1 & 0 \\ 0 & 0 & -1 \end{bmatrix}$

12 If A and B commute, show that A commutes with any polynomial in B.

13 Verify each of the following identities for $X = \begin{bmatrix} 1 & 2 \\ 1 & 3 \end{bmatrix}$ and $X = \begin{bmatrix} 0 & 2 \\ 0 & -2 \end{bmatrix}$:

(a) $(X - I)^2 = X^2 - 2X + I$ (b) $X^3 - I = (X - I)(X^2 + X + I)$

(c) $\dfrac{2X}{X^2 - I} = \dfrac{I}{X - I} + \dfrac{I}{X + I}$ (d) $\dfrac{X^2}{X - 2I} = X + 2I + \dfrac{4I}{X - 2I}$

14 If $A^k = 0$ for some positive integer k, prove that every characteristic value of A is zero.

15 Solve each of the following matric equations:

(a) $X^2 - 5X + 3I = \begin{bmatrix} 1 & -4 \\ 2 & -5 \end{bmatrix}$ (b) $X^2 + 6X + 9I = \begin{bmatrix} -5 & 9 \\ -6 & 10 \end{bmatrix}$

(c) $X^3 = \begin{bmatrix} -6 & 14 \\ -7 & 15 \end{bmatrix}$ (d) $X^3 = \begin{bmatrix} 8 & -7 & -7 \\ -9 & 10 & 11 \\ 9 & -9 & -10 \end{bmatrix}$

16 If $f(x) = x/(x + 4)$, compute $f(A)$ for each of the following matrices A:

(a) $\begin{bmatrix} 1 & -4 \\ 2 & -5 \end{bmatrix}$ (b) $\begin{bmatrix} 4 & -1 \\ 6 & -1 \end{bmatrix}$ (c) $\begin{bmatrix} 2 & -1 \\ 4 & -3 \end{bmatrix}$

(d) $\begin{bmatrix} -3 & 1 & 0 \\ 1 & -4 & 0 \\ 0 & 0 & -5 \end{bmatrix}$ (e) $\begin{bmatrix} -1 & 2 & 2 \\ 2 & -1 & -2 \\ -2 & 2 & 3 \end{bmatrix}$

17 If $p(x) = x^4 - x^3 - 3x^2 + 4x + 2$, evaluate $p(A)$ for each of the following matrices A:

(a) $\begin{bmatrix} 4 & 1 \\ -3 & 0 \end{bmatrix}$ (b) $\begin{bmatrix} -1 & -2 \\ 3 & 4 \end{bmatrix}$ (c) $\begin{bmatrix} 4 & 6 \\ -3 & -5 \end{bmatrix}$

(d) $\begin{bmatrix} -4 & -9 & -3 \\ 1 & 4 & 1 \\ 3 & 3 & 2 \end{bmatrix}$ (e) $\begin{bmatrix} 2 & 1 & 1 \\ 1 & 4 & 3 \\ -1 & -1 & 0 \end{bmatrix}$

18 Verify that the characteristic values of $f(A)$ are equal to $f(\lambda_i)$ for each of the following functions and each of the given matrices:

(a) $x^2 - 2x + 3$ (b) $x^2 - 4x + 3$ (c) $x^3 + x^2 + x + 1$

(i) $\begin{bmatrix} 1 & -4 \\ 2 & -5 \end{bmatrix}$ (ii) $\begin{bmatrix} 4 & -1 \\ 6 & -1 \end{bmatrix}$ (iii) $\begin{bmatrix} 2 & 1 \\ 1 & 2 \end{bmatrix}$

19 Verify that the characteristic values of $f(A)$ are equal to $f(\lambda_i)$ for each of the following functions and each of the given matrices:

(a) $\dfrac{x}{x^2 + 1}$ (b) $\dfrac{x - 2}{x + 2}$ (c) $\dfrac{x + 1}{x^2 + x + 1}$

(i) $\begin{bmatrix} -1 & 2 \\ -1 & 2 \end{bmatrix}$ (ii) $\begin{bmatrix} 4 & 2 \\ 1 & 3 \end{bmatrix}$ (iii) $\begin{bmatrix} 5 & 2 \\ 2 & 2 \end{bmatrix}$

20 If p is a polynomial, is it possible for $p(A)$ to have a characteristic vector which is not a characteristic vector of A? Justify your answer.

11.5 The Cayley-Hamilton Theorem

Since a square null matrix, being a diagonal matrix, is obviously similar to a diagonal matrix, it follows from Theorem 7, Sec. 11.4, that the equation $p(X) = 0$ is always solvable. On the other hand, it is not immediately evident that given a square matrix A, there is always a polynomial equation with scalar coefficients, $p(X) = 0$, of which A is a solution. This is the case, however, and it is not difficult to show (see Exercise 1) that any square matrix of order n satisfies a polynomial equation whose order is at most n^2. In fact, for any square matrix there is always a polynomial equation of order n which is satisfied by A.

To prove this, it is convenient to prove first a preliminary result concerning polynomials whose coefficients are not scalars but square matrices. First, however, it is necessary to define what is meant by the value of such a polynomial, say

$$F(\lambda) = C_0 + C_1\lambda + \cdots + C_k\lambda^k$$

when a square matrix A is substituted for the scalar variable λ. Since matric multiplication is not commutative, it is clear that, in general, the various powers of A will not commute with the coefficient matrices in $F(\lambda)$. Hence, although it is true that

$$C_0 + C_1\lambda + \cdots + C_k\lambda^k = C_0 + \lambda C_1 + \cdots + \lambda^k C_k$$

the corresponding matric relation

$$C_0 + C_1 A + \cdots + C_k A^k = C_0 + AC_1 + \cdots + A^k C_k$$

is in general false. Thus it is necessary for us to assign a specific meaning to $F(A)$, and this we do by agreeing that

$$F(A) = C_0 + C_1 A + \cdots + C_k A^k$$

We have already seen (Theorem 1, Sec. 11.4) that identities relating scalar polynomials imply corresponding identities when the scalar variable is replaced by a square matrix. This is not true, however, for identical relations involving polynomials with matric coefficients. For instance, if

$$F(\lambda) = C_0 + C_1\lambda \quad \text{and} \quad G(\lambda) = D_0 + D_1\lambda$$

then for the product $P(\lambda) = F(\lambda)G(\lambda)$ we have

$$P(\lambda) = C_0D_0 + (C_0D_1 + C_1D_0)\lambda + C_1D_1\lambda^2$$

On the other hand, if we replace the scalar λ by a square matrix A, we have

$$F(A) = C_0 + C_1A \qquad \text{and} \qquad G(A) = D_0 + D_1A$$

and
$$F(A)G(A) = C_0D_0 + C_0D_1A + C_1AD_0 + C_1AD_1A$$

which is not the same as $P(A) \equiv C_0D_0 + (C_0D_1 + C_1D_0)A + C_1D_1A^2$ unless $AD_0 = D_0A$ and $AD_1 = D_1A$. However, we can prove the following theorem, which is the necessary preliminary result mentioned above.

THEOREM 1 If $F(\lambda)$ and $P(\lambda)$ are polynomials in the scalar variable λ with coefficients which are square matrices, and if $P(\lambda) = F(\lambda)(A - \lambda I)$, then $P(A) = \mathbf{0}$.

Proof In view of the fact that we have just seen that $P(\lambda) = F(\lambda)G(\lambda)$ does not imply that $P(A) = F(A)G(A)$, we cannot prove this theorem simply by substituting A for λ in the assertion of the theorem. Instead we must first multiply out the right-hand side of the given relation, express it as a polynomial in λ, and then replace λ by A. To do this, let us suppose that

$$F(\lambda) = C_0 + C_1\lambda + C_2\lambda^2 + \cdots + C_k\lambda^k$$

where $C_0, C_1, C_2, \ldots, C_k$ are $n \times n$ matrices. Then

$$\begin{aligned} P(\lambda) &= (C_0 + C_1\lambda + C_2\lambda^2 + \cdots + C_k\lambda^k)(A - \lambda I) \\ &= C_0A + C_1A\lambda + C_2A\lambda^2 + \cdots + C_kA\lambda^k \\ &\quad - C_0\lambda - C_1\lambda^2 - C_2\lambda^3 - \cdots - C_k\lambda^{k+1} \end{aligned}$$

Now substituting A for λ, we have

$$P(A) = C_0A + C_1A^2 + C_2A^3 + \cdots + C_kA^{k+1}$$
$$- C_0A - C_1A^2 - C_2A^3 - \cdots - C_kA^{k+1} = \mathbf{0}$$

as asserted.

We are now in a position to prove one of the most important results in the theory of matrices, the famous **Cayley-Hamilton theorem**.

THEOREM 2 Every square matrix satisfies its own characteristic equation.

Proof Let A be an $n \times n$ matrix whose characteristic equation is

$$|A - \lambda I| = \begin{vmatrix} a_{11} - \lambda & a_{12} & \cdots & a_{1n} \\ a_{21} & a_{22} - \lambda & \cdots & a_{2n} \\ \cdots\cdots\cdots\cdots\cdots\cdots\cdots\cdots\cdots \\ a_{n1} & a_{n2} & \cdots & a_{nn} - \lambda \end{vmatrix}$$

$$= (-1)^n[\lambda^n - \beta_1\lambda^{n-1} + \cdots + (-1)^n\beta_n] = 0$$

The adjoint of the matrix $A - \lambda I$ is clearly an $n \times n$ matrix whose elements, being the cofactors of the elements of the determinant $|A - \lambda I|$, are polynomials in λ; that is,

$$\text{adj } (A - \lambda I) = \begin{bmatrix} p_{11}(\lambda) & p_{12}(\lambda) & \cdots & p_{1n}(\lambda) \\ p_{21}(\lambda) & p_{22}(\lambda) & \cdots & p_{2n}(\lambda) \\ \cdots\cdots\cdots\cdots\cdots\cdots\cdots\cdots \\ p_{n1}(\lambda) & p_{n2}(\lambda) & \cdots & p_{nn}(\lambda) \end{bmatrix}$$

Furthermore, from the definition of matric addition, it follows that the last matrix can be written as a polynomial in λ, say $F(\lambda)$, whose coefficients are $n \times n$ matrices, the element in the ith row and jth column of the matric coefficient of λ^k being the coefficient of λ^k in $p_{ij}(\lambda)$. Now, from Corollary 1 of Theorem 1, Sec. 10.3, we have

$$[\text{adj } (A - \lambda I)](A - \lambda I) = |A - \lambda I| I$$
$$= (-1)^n [\lambda^n I - \beta_1 \lambda^{n-1} I + \cdots + (-1)^n \beta_n I]$$

that is,

$$(-1)^n [\lambda^n I - \beta_1 \lambda^{n-1} I + \cdots + (-1)^n \beta_n I] = F(\lambda)(A - \lambda I)$$

But this is a relation between polynomials in λ with matric coefficients of precisely the type covered by Theorem 1. Hence, the left-hand side must vanish when λ is replaced by the matrix A. In other words,

$$A^n - \beta_1 A^{n-1} + \cdots + (-1)^n \beta_n I = 0$$

i.e., the matrix A satisfies its own characteristic equation, as asserted.

With the Cayley-Hamilton theorem, the nth power of any square matrix A can be expressed as a linear combination of lower powers of A. Hence, by repeated applications of the Cayley-Hamilton theorem, any positive integral power of A, and therefore any polynomial in A, can be expressed as a polynomial in A of degree at most $n - 1$. Moreover, if A is nonsingular, then A^{-1} exists and in the expansion of $|A - \lambda I|$ the constant term $\beta_n = |A|$ is different from zero. Hence we can multiply the Cayley-Hamilton equation

$$A^n - \beta_1 A^{n-1} + \cdots + (-1)^n \beta_n I = 0$$

by A^{-1}, getting

$$A^{n-1} - \beta_1 A^{n-2} + \cdots + (-1)^{n-1} \beta_{n-1} I + (-1)^n \beta_n A^{-1} = 0$$

whence, solving for A^{-1}, we find

$$A^{-1} = \frac{(-1)^{n-1}}{\beta_n} [A^{n-1} - \beta_1 A^{n-2} + \cdots + (-1)^{n-1} \beta_{n-1} I]$$

In some cases this is a convenient method of obtaining the inverse of a matrix A.

EXAMPLE 1

If $A = \begin{bmatrix} -4 & 5 & 5 \\ -5 & 6 & 5 \\ -5 & 5 & 6 \end{bmatrix}$, we find by an easy calculation that

$$|A - \lambda I| = -\lambda^3 + 8\lambda^2 - 13\lambda + 6$$

Hence, by the Cayley-Hamilton theorem, it follows that

$$A^3 - 8A^2 + 13A - 6I = 0$$

as can easily be verified by direct calculation. Using this relation, we can now express higher powers of A as quadratic polynomials in A. For instance,

$$A^4 = AA^3 = A(8A^2 - 13A + 6I)$$
$$= 8A^3 - 13A^2 + 6A$$
$$= 8(8A^2 - 13A + 6I) - 13A^2 + 6A$$
$$= 51A^2 - 98A + 48I$$

and

$$A^5 = AA^4 = A(51A^2 - 98A + 48I)$$
$$= 51(8A^2 - 13A + 6I) - 98A^2 + 48A$$
$$= 310A^2 - 615A + 306I$$

Similarly, multiplying the Cayley-Hamilton equation through by A^{-1} and then solving for A^{-1}, we find

$$A^{-1} = \tfrac{1}{6}(A^2 - 8A + 13I)$$

$$= \frac{1}{6}\left(\begin{bmatrix} -34 & 35 & 35 \\ -35 & 36 & 35 \\ -35 & 35 & 36 \end{bmatrix} - 8 \begin{bmatrix} -4 & 5 & 5 \\ -5 & 6 & 5 \\ -5 & 5 & 6 \end{bmatrix} + 13 \begin{bmatrix} 1 & 0 & 0 \\ 0 & 1 & 0 \\ 0 & 0 & 1 \end{bmatrix} \right)$$

$$= \frac{1}{6}\begin{bmatrix} 11 & -5 & -5 \\ 5 & 1 & -5 \\ 5 & -5 & 1 \end{bmatrix}$$

The Cayley-Hamilton equation is not necessarily the polynomial equation of lowest degree satisfied by a given square matrix. For instance, it is easily verified that the matrix A in the last example satisfies not only the Cayley-Hamilton equation

$$A^3 - 8A^2 + 13A - 6I = 0$$

but also the simpler quadratic equation

$$A^2 - 7A + 6I = 0$$

DEFINITION 1   If A is a square matrix, any polynomial p with the property that $p(A) = 0$ is said to annihilate A.

Let us now consider the set of polynomials of minimum degree which annihilate a given square matrix A, and, for definiteness, let us assume that by multiplying them by suitable constants their leading coefficients have been made equal to 1. Clearly, all these polynomials are identical. In fact, if this is not the case, and if there are two such polynomials, f and g, then $h = f - g$ is a polynomial whose degree is lower than the degree of f and g such that

$$h(A) = f(A) - g(A) = 0 - 0 = 0$$

But, by hypothesis, f and g are polynomial annihilators of A of *minimum* degree. Hence we have a contradiction unless h is identically zero, i.e., unless f and g are the same.

We are thus justified in introducing the following definition.

DEFINITION 2   The unique polynomial with leading coefficient 1 and of minimum degree which annihilates a square matrix A is called the minimum polynomial of A.

Among the properties of **minimum polynomials**, the following are worthy of mention here.

THEOREM 3   Similar matrices have the same minimum polynomial.

Proof   Let A and B be similar matrices, so that $B = S^{-1}AS$. Then by Lemma 2, Sec. 11.4, for any polynomial p,

$$p(B) = p(S^{-1}AS) = S^{-1}p(A)S$$

From this we conclude that any polynomial which annihilates A also annihilates B, and conversely. Hence the minimum polynomials of A and B must be the same, as asserted.

THEOREM 4   The minimum polynomial of any square matrix is a divisor of any polynomial which annihilates A.

Proof   Let the minimum polynomial of a matrix A be $f(x)$, and let $\phi(x)$ be any polynomial with the property that $\phi(A) = \mathbf{0}$. Then by the division algorithm of elementary algebra,

$$\phi(x) = q(x)f(x) + r(x)$$

where the remainder polynomial $r(x)$ is either identically zero or of lower degree than the divisor polynomial $f(x)$. Then, by Theorem 1, Sec. 11.4,

$$\phi(A) = q(A)f(A) + r(A)$$

However, by hypothesis, $\phi(A) = \mathbf{0}$ and $f(A) = \mathbf{0}$. Hence $r(A) = \mathbf{0}$, and therefore $r(x)$ must be identically zero; for if it were not, it would be a polynomial which annihilated A and whose degree was less than the degree of the minimum polynomial of A, namely, $f(x)$. But if $r(x) = 0$, then the minimum polynomial $f(x)$ is a factor of $\phi(x)$, as asserted.

THEOREM 5   If the characteristic roots of a matrix A are all distinct, then, except possibly for sign, the characteristic polynomial and the minimum polynomial of A are the same.

Proof   Let A be a square matrix with distinct characteristic roots $\lambda_1, \lambda_2, \ldots, \lambda_n$, and let the characteristic polynomial of A be

$$f(\lambda) = (\lambda_1 - \lambda)(\lambda_2 - \lambda)\cdots(\lambda_n - \lambda)$$

Then, since the minimum polynomial of A, say $g(\lambda)$, must be a factor of $f(\lambda)$, it follows that if $f(\lambda)$ and $g(\lambda)$ differ in more than sign, then $g(\lambda)$ must be the product of some but not all of the factors of $f(\lambda)$. Specifically, suppose that $g(\lambda)$ does not contain the factor $\lambda_i - \lambda$. Now, by Theorem 5, Sec. 11.4, the characteristic roots of the matrix $g(A)$ are $g(\lambda_1), g(\lambda_2), \ldots, g(\lambda_n)$. However, since $g(\lambda)$ does not contain the factor $\lambda_i - \lambda$, it follows that $g(\lambda_i) \neq 0$. Hence $g(A)$ has at least one nonzero characteristic root. But if this is the case, then $g(A)$ is not a null matrix; that is, $g(A) \neq \mathbf{0}$, contrary to the hypothesis that $g(\lambda)$ is the minimum polynomial of A. This contradiction shows that $f(\lambda)$ and $g(\lambda)$ cannot differ except possibly in sign, and the theorem is established.

As an interesting application of the theory of the minimum polynomial of a matrix, we have the following result.

THEOREM 6   If A is a square matrix, and if $f(x)$ and $g(x)$ are scalar polynomials such that $g(A)$ is nonsingular, then $f(A)/g(A)$ is equal to a polynomial in A.

Proof   Since, by definition $f(A)/g(A) = f(A)g^{-1}(A)$, it is clearly sufficient to prove that $g^{-1}(A)$ is a polynomial in A. To do this, let

$$\phi(x) = x^k + c_1 x^{k-1} + \cdots + c_{k-1}x + c_k = 0$$

be the minimum polynomial of the matrix $G = g(A)$. Then

$$\phi(G) = G^k + c_1 G^{k-1} + \cdots + c_{k-1}G + c_k I = 0$$

and from this, by multiplying through by G^{-1} and transposing, we obtain

$$c_k G^{-1} = -(G^{k-1} + c_1 G^{k-2} + \cdots + c_{k-1}I)$$

Now $c_k \neq 0$, for otherwise the right-hand side of the last equation is a polynomial which annihilates G and whose degree is less than the degree, k, of the minimum polynomial of G. Hence we can divide by c_k and obtain G^{-1} as a polynomial in G. Finally, substituting $g(A)$ for G in the expression for G^{-1}, we obtain $G^{-1} = g^{-1}(A)$ as a polynomial in A, as required. It is important to note that the structure of the polynomial in A to which $g^{-1}(A)$ is equal depends upon A as well as upon $g(x)$. Hence, if $f(A)/g(A) = h(A)$, we cannot conclude that for another matrix B we necessarily have

$$\frac{f(B)}{g(B)} = h(B)$$

As we have seen, by successive applications of the Cayley-Hamilton theorem, it is possible to reduce any polynomial in an $n \times n$ matrix A to another polynomial in A whose degree is at most $n - 1$. The use of the Cayley-Hamilton theorem is not always the most convenient way of accomplishing this reduction, however, and when the characteristic values of A are all distinct, it is sometimes easier to proceed as follows.

Knowing that for any polynomial p and any $n \times n$ matrix A, the matrix $p(A)$ can be expressed as a polynomial, say $\phi(A)$, of degree at most $n - 1$, let us write

$$\phi(\lambda) = c_1[(\lambda - \lambda_2)(\lambda - \lambda_3) \cdots (\lambda - \lambda_n)]$$
$$+ c_2[(\lambda - \lambda_1)(\lambda - \lambda_3) \cdots (\lambda - \lambda_n)] + \cdots$$
$$+ c_n[(\lambda - \lambda_1)(\lambda - \lambda_2) \cdots (\lambda - \lambda_{n-1})]$$

where the ith term on the right is the product of all the factors of the characteristic polynomial of A except $\lambda - \lambda_i$. Clearly, if $c_1, c_2, \ldots, c_n$ are arbitrary and the λ's are all distinct, the right-hand side is an arbitrary polynomial of degree $n - 1$. Then

(1) $p(A) = c_1[(A - \lambda_2 I)(A - \lambda_3 I) \cdots (A - \lambda_n I)]$
$$+ c_2[(A - \lambda_1 I)(A - \lambda_3 I) \cdots (A - \lambda_n I)] + \cdots$$
$$+ c_n[(A - \lambda_1 I)(A - \lambda_2 I) \cdots (A - \lambda_{n-1} I)]$$

Now if X_k is a characteristic vector of A corresponding to the characteristic value λ_k, it follows that

(2)
$$(A - \lambda_k I)X_k = 0$$

Moreover,

$$
\begin{aligned}
(A - \lambda_j I)X_k &= [A - \lambda_k I + (\lambda_k - \lambda_j)I]X_k \\
&= (A - \lambda_k I)X_k + (\lambda_k - \lambda_j)X_k \\
&= 0 + (\lambda_k - \lambda_j)X_k = (\lambda_k - \lambda_j)X_k
\end{aligned}
$$

(3)

Hence, if we postmultiply Eq. (1) by X_k and simplify the products by successively applying Eqs. (2) and (3), we find that every product vanishes except the kth [since sooner or later each will involve the factor $(A - \lambda_k I)X_k = 0$], and we have

(4) $\qquad p(A)X_k = c_k[(\lambda_k - \lambda_1) \cdots (\lambda_k - \lambda_{k-1})(\lambda_k - \lambda_{k+1}) \cdots (\lambda_k - \lambda_n)]X_k$

Furthermore, according to Corollary 2 of Theorem 5, Sec. 11.4, X_k is a characteristic vector of the matrix $p(A)$ corresponding to the characteristic value $p(\lambda_k)$. Therefore

$$[p(A) - p(\lambda_k)I]X_k = 0 \qquad \text{or} \qquad p(A)X_k = p(\lambda_k)X_k$$

Thus Eq. (4) becomes

$$p(\lambda_k)X_k = c_k[(\lambda_k - \lambda_1) \cdots (\lambda_k - \lambda_{k-1})(\lambda_k - \lambda_{k+1}) \cdots (\lambda_k - \lambda_n)]X_k$$

which implies that

$$p(\lambda_k) = c_k \prod_{\substack{i=1 \\ i \neq k}}^{n} (\lambda_k - \lambda_i) \qquad \text{or} \qquad c_k = \frac{p(\lambda_k)}{\displaystyle\prod_{\substack{i=1 \\ i \neq k}}^{n} (\lambda_k - \lambda_i)}$$

Therefore, substituting these values for the c's into Eq. (1), we obtain the identity

(5)
$$p(A) = \sum_{k=1}^{n} \left[\frac{p(\lambda_k)}{\displaystyle\prod_{\substack{i=1 \\ i \neq k}}^{n} (\lambda_k - \lambda_i)} \prod_{\substack{i=1 \\ i \neq k}}^{n} (A - \lambda_i I) \right]$$

This important result is known as **Sylvester's identity**.[†] It may be extended to cover the case in which the characteristic values of A are not all distinct, but we shall not undertake this extension.[‡]

EXAMPLE 2

If $A = \begin{bmatrix} -15 & -14 & -40 \\ 6 & 7 & 14 \\ 5 & 4 & 14 \end{bmatrix}$, express $p(A) = A^6 - 6A^5 + 12A^4 - 12A^3 + 12A^2 - 8A + 3I$ in as simple a form as possible.

---

† Named for the English algebraist J. J. Sylvester (1814–1897).
‡ See, for instance, W. J. Duncan, R. A. Fraser, and A. R. Collar, "Elementary Matrices," pp. 78–79, Cambridge University Press, New York, 1938.

By a straightforward calculation we find for the characteristic equation of A

$$|A - \lambda I| = \begin{vmatrix} -15 - \lambda & -14 & -40 \\ 6 & 7 - \lambda & 14 \\ 5 & 4 & 14 - \lambda \end{vmatrix} = -\lambda^3 + 6\lambda^2 - 11\lambda + 6 = 0$$

Hence the characteristic values of A are $\lambda_1 = 1$, $\lambda_2 = 2$, $\lambda_3 = 3$. Therefore

$$p(\lambda_1) \equiv p(1) = 2$$
$$p(\lambda_2) \equiv p(2) = 3$$
$$p(\lambda_3) \equiv p(3) = 6$$

and, substituting into Eq. (5),

$$p(A) = \frac{2}{(1 - 2)(1 - 3)}(A - 2I)(A - 3I) + \frac{3}{(2 - 1)(2 - 3)}(A - I)(A - 3I)$$

$$+ \frac{6}{(3 - 1)(3 - 2)}(A - I)(A - 2I)$$

$$= A^2 - 2A + 3I$$

EXERCISES

1   Without using the Cayley–Hamilton theorem, prove that every $n \times n$ matrix satisfies a polynomial equation of degree at most n^2.

2   Using the Cayley–Hamilton theorem, find the inverse of each of the following matrices:

(a) $\begin{bmatrix} 2 & -4 & -4 \\ 1 & -4 & -5 \\ 1 & 4 & 5 \end{bmatrix}$ (b) $\begin{bmatrix} 2 & 1 & 1 \\ 1 & 4 & 3 \\ -1 & -1 & 0 \end{bmatrix}$ (c) $\begin{bmatrix} -4 & -9 & -3 \\ 1 & 4 & 1 \\ 3 & 3 & 2 \end{bmatrix}$

3   Find the minimum polynomial of each of the following matrices:

(a) $\begin{bmatrix} 1 & 1 & 1 \\ 1 & 1 & 1 \\ 1 & 1 & 1 \end{bmatrix}$ (b) $\begin{bmatrix} -1 & 2 & 2 \\ 2 & -1 & -2 \\ -2 & 2 & 3 \end{bmatrix}$

(c) $\begin{bmatrix} 1 & 1 & 1 \\ 1 & 1 & -1 \\ -1 & 1 & 3 \end{bmatrix}$ (d) $\begin{bmatrix} 7 & 2 & -2 \\ -6 & -1 & 2 \\ 6 & 2 & -1 \end{bmatrix}$

4   Using both the Cayley–Hamilton theorem and Sylvester's identity, evaluate $p(A) = A^5 - A^4 - 2A^3 + A^2 + A - 3I$ for each of the following matrices A:

(a) $\begin{bmatrix} 3 & -2 \\ 5 & -4 \end{bmatrix}$ (b) $\begin{bmatrix} 4 & -1 \\ 6 & -1 \end{bmatrix}$ (c) $\begin{bmatrix} -2 & -6 \\ 2 & 5 \end{bmatrix}$

(d) $\begin{bmatrix} 2 & -4 & -4 \\ 1 & -4 & -5 \\ -1 & 4 & 5 \end{bmatrix}$ (e) $\begin{bmatrix} -1 & 1 & 1 \\ 2 & -2 & -3 \\ -2 & 2 & 3 \end{bmatrix}$

5   If $f(x) = x/(x + 4)$, express $f(A)$ as a polynomial in A for each of the following matrices A:

(a) $\begin{bmatrix} 4 & 1 \\ -3 & 0 \end{bmatrix}$ (b) $\begin{bmatrix} -1 & -2 \\ 3 & 4 \end{bmatrix}$ (c) $\begin{bmatrix} 4 & 6 \\ -3 & -5 \end{bmatrix}$

(d) $\begin{bmatrix} -4 & -9 & -3 \\ 1 & 4 & 1 \\ 3 & 3 & 2 \end{bmatrix}$ (e) $\begin{bmatrix} 2 & 1 & 1 \\ 1 & 4 & 3 \\ -1 & -1 & 0 \end{bmatrix}$

11.6 Infinite Series of Matrices

In the last two sections we have considered polynomial and rational fractional functions of a square matrix. In this section we shall conclude our survey of the theory of matrices by investigating briefly, and for the most part without proof, infinite series of matrices. Once we have suitable criteria for the convergence of infinite series of matrices, it will be possible to define and study transcendental functions of square matrices, such as e^A, $\sin A$, and $\cos A$, by allowing the corresponding scalar series to have matric arguments. As our experience with scalar series suggests, we must begin with the concept of the convergence of a sequence of matrices.

DEFINITION 1   If $A_1, A_2, \ldots, A_n, \ldots$ is a sequence of $p \times q$ matrices, if $(a_{ij})_n$ is the element in the ith row and jth column of A_n, and if a_{ij} is the corresponding element in a $p \times q$ matrix A, the sequence $A_1, A_2, \ldots, A_n, \ldots$ is said to converge to the matrix A if and only if, for all values of i and j, $\lim_{n \to \infty} (a_{ij})_n = a_{ij}$.

We indicate that a sequence of matrices $\{A_n\}$ converges to a matrix by writing $A_n \to A$. A sequence of matrices which does not converge is said to **diverge** or to be **divergent**. According to Definition 1, the convergence of a sequence of $p \times q$ matrices depends on the convergence of pq scalar sequences. Hence, a simple application of familiar ideas proves the following results.

LEMMA 1   If $\{A_n\}$ and $\{B_n\}$ are two sequences of $p \times q$ matrices, and if $A_n \to A$ and $B_n \to B$, then, for all scalar constants α and β, $\alpha A_n + \beta B_n \to \alpha A + \beta B$.

LEMMA 2   If $\{A_n\}$ and $\{B_n\}$ are two sequences of suitably conformable matrices, and if $A_n \to A$ and $B_n \to B$, then $A_n B_n \to AB$.

LEMMA 3   If $\{A_n\}$ is a sequence of $p \times q$ matrices, and if $A_n \to A$, then, for suitably conformable matrices R and S, $RA_n S \to RAS$.

Proof   From the definition of matric multiplication, the element in the ith row and jth column of the product $RA_n S$ is

$$\sum_{k=1}^{p} \sum_{l=1}^{q} r_{ik}(a_{kl})_n s_{lj}$$

where, as above, $(a_{kl})_n$ denotes the element in the kth row and lth column of A_n. Since r_{ik} and s_{lj} are clearly independent of n, and since, by hypothesis, $(a_{kl})_n \to a_{kl}$, it follows that the finite sum converges to

$$\sum_{k=1}^{p} \sum_{l=1}^{q} r_{ik} a_{kl} s_{lj}$$

which is the element in the ith row and jth column of the product RAS, as asserted.

The fact that any square matrix is similar to an upper triangular matrix makes it possible to prove the following important theorem.†

---

† See, for instance, Mirsky, op. cit., p. 328.

THEOREM 1   If A is a square matrix, a necessary and sufficient condition that $A^n \to 0$ is that the absolute value of each characteristic root of A be less than 1.

We are now in a position to define the **convergence of an infinite series of matrices**.

DEFINITION 2   The series of matrices $\sum\limits_{m=0}^{\infty} c_m A_m$ is said to converge to the sum S if and only if the sequence of partial sums

$$\{S_n\} = \left\{ \sum_{m=0}^{n} c_m A_m \right\}$$

converges to S as n becomes infinite.

DEFINITION 3   A series of matrices $\sum\limits_{m=0}^{\infty} c_m A_m$ is said to converge absolutely if and only if each scalar series $\sum\limits_{m=0}^{\infty} c_m(a_{ij})_m$ is absolutely convergent.

As a criterion for the absolute convergence of a series of matrices, we have the result contained in the following theorem.

THEOREM 2   If a_m is the element of maximum absolute value in the matrix A_m, then a necessary and sufficient condition that the series $\sum\limits_{m=0}^{\infty} c_m A_m$ converge absolutely is that the scalar series $\sum\limits_{m=0}^{\infty} |c_m| \cdot |a_m|$ converge.

Proof   To prove the necessity of the condition of the theorem, let us assume that the given series $\sum\limits_{m=0}^{\infty} c_m A_m$ is absolutely convergent. Then if A_m is a $p \times q$ matrix, it follows that for all values of i and j such that $1 \le i \le p$ and $1 \le j \le q$, the series $\sum\limits_{m=0}^{\infty} c_m(a_{ij})_m$ is absolutely convergent; that is, $\sum\limits_{m=0}^{\infty} |c_m| \cdot |(a_{ij})_m|$ converges. Let $L \, (\ge 0)$ be the largest of the sums to which these pq series converge. Then, clearly, for all values of i and j and for every $n > 0$

$$\sum_{m=0}^{n} |c_m| \cdot |(a_{ij})_m| \le L$$

If we now sum the last inequality for all values of i and j, we obtain

$$\sum_{i,j} \sum_{m=0}^{n} |c_m| \cdot |(a_{ij})_m| \le \sum_{i,j} L = pqL$$

From this, by reversing the order of summation on the left and noting that $|c_m|$ is independent of i and j, we have

(1)
$$\sum_{m=0}^{n} |c_m| \sum_{i,j} |(a_{ij})_m| \le pqL$$

Now the absolute value of the element of largest absolute value in A_m surely cannot exceed the sum of the absolute values of *all* the elements in A_m; that is,

$$|a_m| \le \sum_{i,j} |(a_{ij})_m|$$

Hence, using this to underestimate the left member of (1), we have

$$\sum_{m=0}^{n} |c_m| \cdot |a_m| \leq pqL$$

Thus the series $\sum_{m=0}^{\infty} |c_m| \cdot |a_m|$ converges, since its partial sums form a bounded monotonically increasing sequence, and the necessity of the theorem is established.

To prove the sufficiency of the condition of the theorem, let us suppose that $\sum_{m=0}^{\infty} |c_m| \cdot |a_m|$ converges. Then, since $|(a_{ij})_m| \leq |a_m|$ for all i and j, it is clear that $|c_m| \cdot |(a_{ij})_m| \leq |c_m| \cdot |a_m|$ for all i and j. Hence, by an easy application of the comparison test for the convergence of scalar series, it follows that $\sum_{m=0}^{\infty} |c_m| \cdot |(a_{ij})_m|$ converges for all i and j, which proves that the given series of matrices converges absolutely, as asserted.

If a series of matrices is absolutely convergent, it follows at once from the corresponding properties of scalar series that the matric terms can be rearranged at pleasure without in any way affecting the sum of the series.

With the exception of Theorem 1, all the observations we have so far made about sequences and series of matrices apply to matrices of any shape. However, in most applications we are concerned only with series of square matrices and, in particular, with power series of such matrices. The fundamental theorem on matric power series is the following.†

THEOREM 3   If the absolute value of each characteristic root of a square matrix A is less than the radius of convergence of the scalar power series $\phi(z) = \sum_{m=0}^{\infty} c_m z^m$, then the matric power series $\phi(A) = \sum_{m=0}^{\infty} c_m A^m$ converges. If the absolute value of at least one characteristic root of A is greater than the radius of convergence of $\phi(z)$, then $\phi(A)$ diverges. In particular cases, if the absolute value of the characteristic root of A of largest absolute value is equal to the radius of convergence of $\phi(z)$, the matric series $\phi(A)$ may either converge or diverge.

COROLLARY 1   If $\phi(z)$ converges for all values of z, then $\phi(A)$ converges for all square matrices A.

COROLLARY 2   If $\phi(A) = \sum_{m=0}^{\infty} c_m A^m$ converges, and if A is similar to a diagonal matrix, i.e., if

$$S^{-1}AS = D = \begin{bmatrix} \lambda_1 & & & \mathbf{O} \\ & \lambda_2 & & \\ & & \cdot & \\ \mathbf{O} & & & \lambda_n \end{bmatrix}$$

---

† See, for instance, Mirsky, op. cit., p. 332.

then

$$\phi(A) = S \begin{bmatrix} \phi(\lambda_1) & & & \mathbf{O} \\ & \phi(\lambda_2) & & \\ & & \cdot & \\ \mathbf{O} & & & \phi(\lambda_n) \end{bmatrix} S^{-1}$$

Proof   Clearly, the partial sums of the series $\phi(A)$ are all polynomials in A. Hence, by Theorem 6, Sec. 11.4,

$$\phi_N(A) = \sum_{m=1}^{N} c_m A^m = S \begin{bmatrix} \phi_N(\lambda_1) & & & \mathbf{O} \\ & \phi_N(\lambda_2) & & \\ & & \cdot & \\ \mathbf{O} & & & \phi_N(\lambda_n) \end{bmatrix} S^{-1}$$

Now, by hypothesis, $\phi(A)$ converges. Hence, each of the scalar series $\phi(\lambda_i)$ must converge; i.e., for each i, $\phi_N(\lambda_i) \to \phi(\lambda_i)$. Therefore, as $N \to \infty$,

$$\begin{bmatrix} \phi_N(\lambda_1) & & & \mathbf{O} \\ & \phi_N(\lambda_2) & & \\ & & \cdot & \\ \mathbf{O} & & & \phi_N(\lambda_n) \end{bmatrix} \to \begin{bmatrix} \phi(\lambda_1) & & & \mathbf{O} \\ & \phi(\lambda_2) & & \\ & & \cdot & \\ \mathbf{O} & & & \phi(\lambda_n) \end{bmatrix}$$

and we have, by Lemma 3,

$$\phi(A) = \lim_{N \to \infty} \phi_N(A) = S \begin{bmatrix} \phi(\lambda_1) & & & \mathbf{O} \\ & \phi(\lambda_2) & & \\ & & \cdot & \\ \mathbf{O} & & & \phi(\lambda_n) \end{bmatrix} S^{-1}$$

The last result provides us with a useful method of evaluating the sum of certain matric power series which is sometimes preferable to using the Cayley-Hamilton theorem or Sylvester's identity.

EXAMPLE 1

What is e^A if $A = \begin{bmatrix} 0 & -2 \\ 1 & 3 \end{bmatrix}$?

The given matrix A is the one we considered in Example 3, Sec. 11.4. Hence, from that example we know that the characteristic equation of A is $\lambda^2 - 3\lambda + 2 = 0$ and that $\lambda_1 = 1$, $\lambda_2 = 2$, and

$$S^{-1}AS = \begin{bmatrix} 1 & 0 \\ 0 & 2 \end{bmatrix} \quad \text{where} \quad S = \begin{bmatrix} 2 & 1 \\ -1 & -1 \end{bmatrix} \text{ and } S^{-1} = \begin{bmatrix} 1 & 1 \\ -1 & -2 \end{bmatrix}$$

Therefore, by the last corollary,

$$e^A = S \begin{bmatrix} \phi(\lambda_1) & 0 \\ 0 & \phi(\lambda_2) \end{bmatrix} S^{-1} = \begin{bmatrix} 2 & 1 \\ -1 & -1 \end{bmatrix} \begin{bmatrix} e & 0 \\ 0 & e^2 \end{bmatrix} \begin{bmatrix} 1 & 1 \\ -1 & -2 \end{bmatrix}$$

$$= \begin{bmatrix} 2e - e^2 & 2e - 2e^2 \\ -e + e^2 & -e + 2e^2 \end{bmatrix}$$

To evaluate e^A by means of the Cayley–Hamilton theorem, we must simplify each of the powers of A in the expansion of e^A, using the relation $A^2 - 3A + 2I = 0$, or

$$A^2 = 3A - 2I$$

At first glance this would seem to be a very tedious process, since arbitrarily high powers of A are involved. However, we can shorten the work appreciably by proceeding inductively. From the fact that the characteristic equation of A is quadratic, we know that any positive integral power of A can be expressed as a linear binomial in A. Hence, if we assume

$$A^n = a_n A + b_n I \qquad n = 2, 3, 4, \ldots$$

we have

$$A^{n+1} = a_{n+1}A + b_{n+1}I = AA^n = a_n A^2 + b_n A$$
$$= a_n(3A - 2I) + b_n A = (3a_n + b_n)A - 2a_n I$$

Therefore the a's and b's satisfy the recurrence relations†

$$a_{n+1} = 3a_n + b_n \qquad \text{and} \qquad b_{n+1} = -2a_n$$

with, of course, the initial conditions $a_2 = 3$, $b_2 = -2$. From these it is easy to verify that

$$\begin{aligned} a_2 &= 3 & b_2 &= -2 \\ a_3 &= 7 & b_3 &= -6 \\ a_4 &= 15 & b_4 &= -14 \\ a_5 &= 31 & b_5 &= -30 \\ &\cdots\cdots & &\cdots\cdots \\ a_n &= 2^n - 1 & b_n &= -2^n + 2 \end{aligned}$$

Hence,

$$e^A = I + A + \frac{A^2}{2!} + \frac{A^3}{3!} + \cdots$$

$$= I + A + \frac{(2^2 - 1)A + (-2^2 + 2)I}{2!} + \frac{(2^3 - 1)A + (-2^3 + 2)I}{3!} + \cdots$$

$$= A\left(1 + \frac{2^2 - 1}{2!} + \frac{2^3 - 1}{3!} + \cdots\right) + I\left(1 + \frac{2 - 2^2}{2!} + \frac{2 - 2^3}{3!} + \cdots\right)$$

$$= A\left[\left(1 + \frac{2}{1!} + \frac{2^2}{2!} + \frac{2^3}{3!} + \cdots\right) - \left(1 + \frac{1}{1!} + \frac{1}{2!} + \frac{1}{3!} + \cdots\right)\right]$$
$$+ I\left[2\left(1 + \frac{1}{1!} + \frac{1}{2!} + \frac{1}{3!} + \cdots\right) - \left(1 + \frac{2}{1!} + \frac{2^2}{2!} + \frac{2^3}{3!} + \cdots\right)\right]$$

$$= A(e^2 - e) + I(2e - e^2)$$

$$= (e^2 - e)\begin{bmatrix} 0 & -2 \\ 1 & 3 \end{bmatrix} + (2e - e^2)\begin{bmatrix} 1 & 0 \\ 0 & 1 \end{bmatrix}$$

$$= \begin{bmatrix} 2e - e^2 & 2e - 2e^2 \\ -e + e^2 & -e + 2e^2 \end{bmatrix}$$

as before.

Finally, using Sylvester's identity, we have

$$e^A = \frac{e^{\lambda_1}}{\lambda_1 - \lambda_2}(A - \lambda_2 I) + \frac{e^{\lambda_2}}{\lambda_2 - \lambda_1}(A - \lambda_1 I)$$

$$= \frac{e}{-1}(A - 2I) + \frac{e^2}{1}(A - I)$$

$$= A(e^2 - e) + I(2e - e^2)$$

which is the first expression we obtained above when we calculated e^A by means of the Cayley-Hamilton equation.

The question of whether scalar identities such as

$$e^x e^y = e^{x+y} \qquad \cos^2 x + \sin^2 x = 1 \qquad \sin 2x = 2\sin x \cos x$$

and

$$\cos(x + y) = \cos x \cos y - \sin x \sin y$$

---

† These are examples of what are known as *difference equations*; see Sec. 4.5.

remain valid when x and y are replaced by square matrices is obviously an important one. We cannot investigate the matter here, but the applications one is likely to encounter are covered by the following theorem.†

THEOREM 4 For the elementary transcendental functions, scalar identities in a single variable remain valid when the scalar variable is replaced by a square matrix. Scalar identities in two variables remain valid when the scalar variables are replaced by square matrices only when the two matrices commute.

The familiar fact that the scalar differential equation $dx/dt = ax$ has the solution $x = x_0 e^{at}$ suggests that possibly the matric equation

$$\frac{dX}{dt} = AX$$

i.e., a system of n linear first-order constant-coefficient differential equations in the n scalar variables $x_1, x_2, \ldots, x_n$ may have a solution of the form

$$X = e^{At} X_0$$

When we use Corollary 2 of Theorem 3, and the definition of the derivative of a matrix (Exercise 21, Sec. 10.2), it is not difficult to show that if A is similar to a diagonal matrix, this is indeed the case. In fact, if A is similar to a diagonal matrix, i.e., if $A = SDS^{-1}$, where D is the diagonal matrix of the characteristic values of A, then by Corollary 2, Theorem 3,

$$e^{At} = S \begin{bmatrix} e^{\lambda_1 t} & & & \mathbf{O} \\ & e^{\lambda_2 t} & & \\ & & \cdot & \\ \mathbf{O} & & & e^{\lambda_n t} \end{bmatrix} S^{-1}$$

Hence, remembering that S, S^{-1}, and X_0 are constant matrices, and recalling the result of Exercise 22, Sec. 10.2, we have

$$\frac{dX}{dt} = \frac{d(e^{AT}X_0)}{dt} = S \frac{d}{dt} \begin{bmatrix} e^{\lambda_1 t} & & & \mathbf{O} \\ & e^{\lambda_2 t} & & \\ & & \cdot & \\ \mathbf{O} & & & e^{\lambda_n t} \end{bmatrix} S^{-1}X_0$$

$$= SD \begin{bmatrix} e^{\lambda_1 t} & & & \mathbf{O} \\ & e^{\lambda_2 t} & & \\ & & \cdot & \\ \mathbf{O} & & & e^{\lambda_n t} \end{bmatrix} S^{-1}X_0$$

$$= (SDS^{-1})S \begin{bmatrix} e^{\lambda_1 t} & & & \mathbf{O} \\ & e^{\lambda_2 t} & & \\ & & \cdot & \\ \mathbf{O} & & & e^{\lambda_n t} \end{bmatrix} S^{-1}X_0$$

$$= Ae^{At}X_0$$

$$= AX$$

which shows that the matrix $X = e^{At}X_0$ is indeed the solution of the equation $dX/dt = AX$ which assumes the value $X = X_0$ when $t = 0$, as asserted.

---

† See, for example, Mirsky, op. cit., pp. 338–341.

EXAMPLE 2

Solve the system of equations

$$\frac{dx_1}{dt} = -2x_2$$

$$\frac{dx_2}{dt} = x_1 + 3x_2$$

given that $x_1 = 1$, $x_2 = -3$, when $t = 0$, by regarding the system as a single matric equation.

In this case, the given system is equivalent to the matric equation $X' = AX$ where

$$X = \begin{bmatrix} x_1 \\ x_2 \end{bmatrix} \quad \text{and} \quad A = \begin{bmatrix} 0 & -2 \\ 1 & 3 \end{bmatrix}$$

Hence, by the preceding discussion, the solution for X is

$$X = e^{At}X_0 = \left(\exp \begin{bmatrix} 0 & -2 \\ 1 & 3 \end{bmatrix} \right) \begin{bmatrix} 1 \\ -3 \end{bmatrix}$$

Now, using the results of Example 1, noting however that $\phi(z) = e^{zt}$ rather than just e^z, we see clearly that

$$e^{At} = S \begin{bmatrix} e^{\lambda_1 t} & 0 \\ 0 & e^{\lambda_2 t} \end{bmatrix} S^{-1} = \begin{bmatrix} 2e^t - e^{2t} & 2e^t - 2e^{2t} \\ -e^t + e^{2t} & -e^t + 2e^{2t} \end{bmatrix}$$

Therefore

$$X = \begin{bmatrix} 2e^t - e^{2t} & 2e^t - 2e^{2t} \\ -e^t + e^{2t} & -e^t + 2e^{2t} \end{bmatrix} \begin{bmatrix} 1 \\ -3 \end{bmatrix}$$

or, in scalar form,

$$x_1 = (2e^t - e^{2t}) - 3(2e^t - 2e^{2t}) = -4e^t + 5e^{2t}$$

$$x_2 = (-e^t + e^{2t}) - 3(-e^t + 2e^{2t}) = 2e^t - 5e^{2t}$$

It is, of course, not difficult to check this result by solving the given system by the methods of Chap. 3 or through the use of the Laplace transform.

EXERCISES

1 Prove Lemma 1.   2 Prove Lemma 2.
3 For the matrix A of Example 1, compute e^{-A} and verify that $e^A e^{-A} = I$.
4 Using Sylvester's identity, compute e^A, $\cos A$, and $\sin A$ for each of the following matrices A:

(a) $\begin{bmatrix} 3 & -1 \\ 6 & -2 \end{bmatrix}$ (b) $\begin{bmatrix} 2 & -1 \\ 4 & -3 \end{bmatrix}$

(c) $\begin{bmatrix} 2 & 1 & 1 \\ 1 & 4 & 3 \\ -1 & -1 & 0 \end{bmatrix}$ (d) $\begin{bmatrix} -1 & 1 & 1 \\ 2 & -2 & -3 \\ -2 & 2 & 3 \end{bmatrix}$

5 Work Exercise 4 using Corollary 2.
6 Using both Sylvester's identity and the Cayley–Hamilton theorem, compute e^A for $A = \begin{bmatrix} 1 & -1 \\ -1 & 0 \end{bmatrix}$.

7 For each of the following matrices, A, verify that $\sin 2A = 2 \sin A \cos A$:

(a) $\begin{bmatrix} 3 & 2 \\ -3 & -2 \end{bmatrix}$ (b) $\begin{bmatrix} 2 & 1 \\ -3 & -2 \end{bmatrix}$

(c) $\begin{bmatrix} 1 & 1 & 1 \\ 1 & 1 & 1 \\ 1 & 1 & 1 \end{bmatrix}$ (d) $\begin{bmatrix} -1 & 2 & 2 \\ 2 & -1 & -2 \\ -2 & 2 & 3 \end{bmatrix}$

8 If $A = \begin{bmatrix} 2 & 1 \\ -2 & -1 \end{bmatrix}$ and $B = \begin{bmatrix} 1 & 0 \\ -2 & -1 \end{bmatrix}$, verify that neither of the following relations holds:

$$\sin (A \pm B) = \sin A \cos B \pm \cos A \sin B$$
$$\cos (A \pm B) = \cos A \cos B \mp \sin A \sin B$$

9 If $A = \begin{bmatrix} \frac{2}{3} & -\frac{1}{3} \\ -\frac{2}{3} & \frac{1}{3} \end{bmatrix}$ and $B = \begin{bmatrix} \frac{1}{3} & -\frac{2}{3} \\ -\frac{4}{3} & -\frac{1}{3} \end{bmatrix}$, verify that each of the following relations holds:

$$\sin (A \pm B) = \sin A \cos B \pm \cos A \sin B$$
$$\cos (A \pm B) = \cos A \cos B \mp \sin A \sin B$$

10 Determine for what values of x and y, if any, the matrices

$$A = \begin{bmatrix} x & x - 1 \\ -x & 1 - x \end{bmatrix} \quad \text{and} \quad B = \begin{bmatrix} y & y - 1 \\ -1 - y & -y \end{bmatrix}$$

commute, and verify that under these conditions

$$\sin (A \pm B) = \sin A \cos B \pm \cos A \sin B$$
$$\cos (A \pm B) = \cos A \cos B \mp \sin A \sin B$$

11 If $A = \begin{bmatrix} 1 & 0 & 2 \\ 0 & -1 & -2 \\ 2 & -2 & 0 \end{bmatrix}$, show that $\sin A = A (\sin 3)/3$. Obtain a similar expression for $\cos A$, and verify that $\cos^2 A + \sin^2 A = I$.

12 Solve the system of equations

$$\begin{aligned} dx_1/dt &= 4x_1 + 3x_2 \\ dx_2/dt &= x_1 + 6x_2 \end{aligned} \qquad x_1 = -1, x_2 = 2 \text{ when } t = 0$$

by solving the equivalent matric equation.

13 Solve the system of equations

$$\begin{aligned} dx_1/dt &= -8x_1 + 3x_2 \\ dx_2/dt &= -10x_1 + 3x_2 \end{aligned} \qquad x_1 = 1, x_2 = -1 \text{ when } t = 0$$

by solving the equivalent matric equation.

14 Solve the system of equations

$$\begin{aligned} dx_1/dt &= 5x_1 - 2x_2 \\ dx_2/dt &= 12x_1 - 5x_2 \end{aligned} \qquad x_1 = 2, x_2 = -1 \text{ when } t = 0$$

by solving the equivalent matric equation.

15 Solve the matric equation $X'' = AX$, where $A = \begin{bmatrix} 1 & 0 \\ 0 & 4 \end{bmatrix}$ given $X_0 = \begin{bmatrix} 1 \\ 1 \end{bmatrix}$ and $X_0' = \begin{bmatrix} 2 \\ -1 \end{bmatrix}$. *Hint:* By setting $x_1 = x_3$ and $x_2 = x_4$, convert the given equation into an equivalent first-order equation $X' = BX$, where X and X' are 4×1 matrices and B is a 4×4 matrix.

16 Rework Exercise 15 by first converting the given matric equation into two second-order scalar equations and then converting each of these into a matric equation involving a 2×2 matrix.

17 **(a)** If A is similar to a diagonal matrix, show that both

$$e^{At}C_1 \qquad \text{and} \qquad e^{-At}C_2$$

where C_1 and C_2 are arbitrary constant column matrices, satisfy the matric differential equation $X'' = A^2X$.

(b) If A is similar to a diagonal matrix, show that both

$$(\sin At)C_1 \qquad \text{and} \qquad (\cos At)C_2$$

where C_1 and C_2 are arbitrary constant column matrices, satisfy the matric differential equation $X'' = -A^2X$.

18 Solve each of the equations $X'' = A^2X$ and $X'' = -A^2X$ if

(a) $A = \begin{bmatrix} 0 & -2 \\ 1 & 3 \end{bmatrix}$ $X_0 = \begin{bmatrix} 0 \\ 1 \end{bmatrix}$ $X_0' = \begin{bmatrix} 1 \\ 0 \end{bmatrix}$

(b) $A = \begin{bmatrix} 3 & 1 \\ 5 & -1 \end{bmatrix}$ $X_0 = \begin{bmatrix} 1 \\ 1 \end{bmatrix}$ $X_0' = \begin{bmatrix} 0 \\ 2 \end{bmatrix}$

(c) $A = \begin{bmatrix} 4 & 6 & 6 \\ 1 & 3 & 2 \\ -1 & -4 & -3 \end{bmatrix}$ $X_0 = \begin{bmatrix} 1 \\ 0 \\ 1 \end{bmatrix}$ $X_0' = \begin{bmatrix} 2 \\ 1 \\ 0 \end{bmatrix}$

19 Solve each of the equations $X'' = BX$ and $X'' = -BX$ if

$$B = \begin{bmatrix} 4 & 3 \\ 5 & 6 \end{bmatrix} \qquad X_0 = \begin{bmatrix} 1 \\ -1 \end{bmatrix} \qquad X_0' = \begin{bmatrix} 0 \\ 1 \end{bmatrix}$$

Hint: First find the square roots of B.

20 Solve each of the equations $X'' = BX$ and $X'' = -BX$ if

$$B = \begin{bmatrix} -5 & 6 \\ -9 & 10 \end{bmatrix} \qquad X_0 = \begin{bmatrix} 3 \\ 1 \end{bmatrix} \qquad X_0' = \begin{bmatrix} 0 \\ 1 \end{bmatrix}$$

CHAPTER 12
The Calculus of Variations

12.1 Introduction

One of the fundamental problems in elementary calculus is the determination of the values of x for which a function of one variable, say $y = f(x)$, assumes its maximum and minimum values. A related problem, sometimes treated in a first course in calculus and sometimes postponed to a course in advanced calculus, is the determination of the maxima and minima of a function of more than one variable, say $F(x_1, x_2, \ldots, x_n)$. Significantly more general and more difficult than either of these problems is the determination of the function which maximizes or minimizes a quantity depending not on one or more independent *variables* but on the *functions* in a given set. For example, among all smooth curves joining $P_0 : (x_0, y_0)$ and $P_1 : (x_1, y_1)$ one might want the one of minimum length, or if P_0 and P_1 were in a vertical plane, one might want the curve along which a particle would slide from one point to the other in the least time.

In this chapter we shall first discuss the problem of finding the maxima and minima of a function of several variables subject, perhaps, to one or more side conditions or constraints. Then we shall turn our attention to the determination of necessary (though not sufficient) conditions for maximizing or minimizing a quantity which depends on the functions in a given set. Questions associated with problems of this last sort constitute the branch of mathematics known as the **calculus of variations**.

12.2 Extrema of Functions of Several Variables

If $f(x_1, \ldots, x_n)$ is a function of n variables which possesses first partial derivatives in the neighborhood of some point $P : (a_1, \ldots, a_n)$, it is shown in calculus that a necessary condition for f to have a maximum or a minimum at P is that at P each of the first partial derivatives of f be zero.† In elementary calculus, sufficient conditions for a point P to be a maximum or a minimum are usually not obtained, but with the fundamental properties of quadratic forms available, these conditions can be formulated

---

† Any point $P : (a_1, \ldots, a_n)$ whose coordinates satisfy the *necessary* conditions

$$\left.\frac{\partial f}{\partial x_1}\right|_P = \left.\frac{\partial f}{\partial x_2}\right|_P = \cdots = \left.\frac{\partial f}{\partial x_n}\right|_P = 0$$

is called a **stationary point** of f. Not all of these are extrema of f however. To determine which stationary points, if any, are maxima or minima of f we need *sufficient* conditions.

in a relatively simple way. In doing this, we shall use the representation of f in the neighborhood of P which is provided by Taylor's theorem. Hence, as we shall see, we must assume the existence around P of the partial derivatives of f of at least the third order.

Let us suppose, then, that $P:(a_1,\ldots,a_n)$ is a point at which the necessary conditions for an extremum of f are satisfied; and, using the operational form of Taylor's theorem developed in calculus, let us write

$$f(x_1,\ldots,x_n) = f(a_1,\ldots,a_n) +$$

$$\left[(x_1 - a_1)\frac{\partial}{\partial x_1} + \cdots + (x_n - a_n)\frac{\partial}{\partial x_n}\right]f(x_1,\ldots,x_n)\Bigg|_{a_1,\ldots,a_n} +$$

$$\frac{1}{2!}\left[(x_1 - a_1)\frac{\partial}{\partial x_1} + \cdots + (x_n - a_n)\frac{\partial}{\partial x_n}\right]^2 f(x_1,\ldots,x_n)\Bigg|_{a_1,\ldots,a_n} +$$

$$\frac{1}{3!}\left[(x_1 - a_1)\frac{\partial}{\partial x_1} + \cdots + (x_n - a_n)\frac{\partial}{\partial x_n}\right]^3 f(x_1,\ldots,x_n)\Bigg|_{\alpha_1,\ldots,\alpha_n}$$

where $\alpha_i = a_i + \theta(x_i - a_i)$ and $0 < \theta < 1$. By hypothesis, the necessary conditions for an extremum are satisfied at $P:(a_1,\ldots,a_n)$; that is,

$$\frac{\partial f}{\partial x_1}\Bigg|_{a_1,\ldots,a_n} = \cdots = \frac{\partial f}{\partial x_n}\Bigg|_{a_1,\ldots,a_n} = 0$$

Hence, letting $\lambda_i = x_i - a_i$, we have

(1) $f(x_1,\ldots,x_n) - f(a_1,\ldots,a_n) =$

$$\frac{1}{2!}\left[\lambda_1\frac{\partial}{\partial x_1} + \cdots + \lambda_n\frac{\partial}{\partial x_n}\right]^2 f(x_1,\ldots,x_n)\Bigg|_{a_1,\ldots,a_n} +$$

$$\frac{1}{3!}\left[\lambda_1\frac{\partial}{\partial x_1} + \cdots + \lambda_n\frac{\partial}{\partial x_n}\right]^3 f(x_1,\ldots,x_n)\Bigg|_{\alpha_1,\ldots,\alpha_n}$$

where $\alpha_i = a_i + \theta\lambda_i$ and $0 < \theta < 1$. Clearly, the first group of terms in this expression is a quadratic form in the λ's in which, specifically, the coefficient of the product $\lambda_i\lambda_j$ is

$$\frac{\partial^2 f}{\partial x_i\,\partial x_j}\Bigg|_{a_1,\ldots,a_n} \qquad i \neq j$$

$$\frac{1}{2}\frac{\partial^2 f}{\partial x_i{}^2}\Bigg|_{a_1,\ldots,a_n} \qquad i = j$$

Moreover, the terms in the second group all involve third-order products of the λ's. Hence, in general, these terms are, collectively, negligible in comparison with the second-degree terms at all points sufficiently close to, but distinct from, P.

Now $P:(a_1,\ldots,a_n)$ will be a local maximum of f if and only if the difference $f(x_1,\ldots,x_n) - f(a_1,\ldots,a_n)$ is negative for all sufficiently small values of $\lambda_i = x_i - a_i$ $(i = 1,\ldots, n)$ which are not all zero. And a sufficient condition for this to be the case is that the quadratic form in the λ's be negative-definite. Similarly, P will be a minimum of f if and only if the difference $f(x_1,\ldots,x_n) - f(a_1,\ldots,a_n)$ is positive for all sufficiently small values of the λ's which are not all zero. And a sufficient condition for this is that the quadratic form in the λ's be positive-definite. The point P will be neither a maximum nor a minimum of f if $f(x_1,\ldots,x_n) - f(a_1,\ldots,a_n)$ is sometimes positive and sometimes negative in every neighborhood of P;† and this will be the case if the quadratic form in the λ's is indefinite. Finally, if the quadratic form in the λ's is semidefinite, the quadratic form cannot change sign in the neighborhood of P but can be zero. At such points, it is no longer the dominant part of the difference $f(x_1,\ldots,x_n) - f(a_1,\ldots,a_n)$. In this case a decision about the nature of P requires a consideration of the cubic terms on the right of Eq. (1) and takes us beyond the scope of matrix theory.

EXAMPLE 1

Examine the function

$$f(x_1,x_2,x_3) = 35 - 6x_1 + 2x_3 + x_1{}^2 - 2x_1x_2 + 2x_2{}^2 + 2x_2x_3 + 3x_3{}^2$$

for maxima and minima.

To determine at what points, if any, the given function may have maxima or minima, we investigate the solutions, if any, of the three equations

$$\frac{\partial f}{\partial x_1} = -6 + 2x_1 - 2x_2 \qquad\quad = 0$$

$$\frac{\partial f}{\partial x_2} = \qquad - 2x_1 + 4x_2 + 2x_3 = 0$$

$$\frac{\partial f}{\partial x_3} = \quad 2 \qquad + 2x_2 + 6x_3 = 0$$

From these we find that the only possibility for a local extremum is the point $P:(8,5,-2)$. Clearly, $f(x_1,x_2,x_3) = 9$ and $\partial f/\partial x_1 = \partial f/\partial x_2 = \partial f/\partial x_3 = 0$ at P. Moreover, at P

$$\frac{\partial^2 f}{\partial x_1{}^2} = 2 \qquad \frac{\partial^2 f}{\partial x_1\,\partial x_2} = -2 \qquad \frac{\partial^2 f}{\partial x_1\,\partial x_3} = 0$$

$$\frac{\partial^2 f}{\partial x_2{}^2} = 4 \qquad \frac{\partial^2 f}{\partial x_2\,\partial x_3} = 2 \qquad \frac{\partial^2 f}{\partial x_3{}^3} = 6$$

Hence,

$$f(x_1,x_2,x_3) - 9 = \frac{1}{2}\,[2(x_1 - 8)^2 - 4(x_1 - 8)(x_2 - 5) + 4(x_2 - 5)^2$$

$$+ 4(x_2 - 5)(x_3 + 2) + 6(x_3 + 2)^2] + \cdots‡$$

The second-degree terms $x_1 - 8 = \lambda_1$, $x_2 - 5 = \lambda_2$, and $x_3 + 2 = \lambda_3$ in this expansion constitute a quadratic form whose matrix is

$$\begin{bmatrix} 1 & -1 & 0 \\ -1 & 2 & 1 \\ 0 & 1 & 3 \end{bmatrix}$$

By Theorem 1, Sec. 11.1, this quadratic form is positive-definite; hence, the point $(8,5,-2)$ is a local minimum of the given function.

---

† Points of this sort are said to be **saddle points** of f.

‡ Actually, since f is only a quadratic function of the x's, Taylor's expansion of f contains no terms of degree higher than the second and therefore terminates with the terms we have written explicitly.

Examine the following functions for maxima and minima:

1 $2x_1^2 + 2x_1x_2 + x_2^2 + 6x_1 + 6x_2 + 3$

2 $2x_1^2 - 2x_1x_2 + 2x_1x_3 - 4x_2x_3 + 4x_1 - 4x_2 + 4x_3 + 4$

3 $-2x_1^2 - 2x_2^2 - x_3^2 + 2x_1x_2 + 2x_2x_3 + 8x_1 - 4x_2 + 2x_3 - 3$

4 $x_1^3 - 3x_1x_2 + x_2^3$ **5** $x_1^3 + x_2^2 - 3x_1$

6 $\sin x_1 + \sin x_2 + \cos(x_1 + x_2)$

7 $x_1^2 + 2x_1x_2 + 2x_2^3 + 4x_2^2$ **8** $4x_1^3 + 2x_1^2x_2 + 6x_2x_3 + 9x_3^2$

9 $x_1^3 + 3x_1^2 + 2x_2^2 + x_3^2 - x_2x_3$

10 A long sheet of tin of width L is to be formed into a gutter of trapezoidal cross section by bending up strips of equal width along the length of the sheet. If the carrying capacity, i.e., the cross-sectional area of the gutter, is to be a maximum, determine the shape of the cross section.

12.3 Lagrange's Multipliers

In many problems, the variables in a function $f(x_1,\ldots,x_n)$ to be maximized or minimized are not all independent but are related by one or more side conditions, say

$$\phi_1(x_1,\ldots,x_n) = 0, \ \phi_2(x_1,\ldots,x_n) = 0, \ldots, \ \phi_N(x_1,\ldots,x_n) = 0 \qquad N < n$$

One way to proceed in this case is to consider the side conditions as N simultaneous equations from which, at least theoretically, N of the x's can be solved for in terms of the remaining x's. Then these N variables can be eliminated from f by substitution. Finally, the extrema of f, as a function of the remaining $n - N$ variables, can be determined by the method of the last section. This is usually referred to as the **explicit method**.

Another procedure, the **implicit method**, is based on the observation that a necessary condition that f have a maximum or minimum at a point is that at P

$$df = \frac{\partial f}{\partial x_1}\,dx_1 + \frac{\partial f}{\partial x_2}\,dx_2 + \cdots + \frac{\partial f}{\partial x_n}\,dx_n = 0$$

For if $x_1, x_2, \ldots, x_n$ are independent variables, then their differentials can be assigned arbitrarily, and taking in turn,

$$dx_i = 0 \qquad \begin{matrix} i = 1, 2, \ldots, n \\ i \neq j \end{matrix} \qquad \text{and} \qquad dx_j = 1$$

it follows that

$$\frac{\partial f}{\partial x_j} = 0 \qquad j = 1, 2, \ldots, n$$

as before. In the implicit method, the differentials of the N side conditions are calculated, giving us N equations which are linear in the n differentials $dx_1, dx_2, \ldots, dx_n$. Then these N equations are solved simultaneously for N of the differentials in terms of the others. These N differentials are then eliminated from df by substitution. Finally, the coefficient of each of the $n - N$ differentials which remain in df is equated to zero, and these $n - N$ equations, together with the N side conditions themselves, give us n equations from which all possibilities for the extrema of f can be found.

A third procedure, less obvious but often more convenient than either the explicit or implicit method, is the use of **Lagrange's multipliers**. This requires that we set up the auxiliary function

$$F(x_1,\ldots,x_n) = f(x_1,\ldots,x_n) + \lambda_1\phi_1(x_1,\ldots,x_n) + \lambda_2\phi_2(x_1,\ldots,x_n) + \cdots$$
$$+ \lambda_N\phi_N(x_1,\ldots,x_n)$$

where the ϕ's are the expressions which define the N side conditions and the λ's (Lagrange's multipliers) are unknown parameters to be determined. Then the first partial derivatives of F are calculated and set equal to zero, giving us the equations

$$\frac{\partial F}{\partial x_1} = 0, \frac{\partial F}{\partial x_2} = 0, \ldots, \frac{\partial F}{\partial x_n} = 0$$

Finally, by solving these n equations, together with the N equations of constraint,

$$\phi_1 = 0, \phi_2 = 0, \ldots, \phi_N = 0$$

we are led to all possibilities for the extrema of f. At first glance, the use of Lagrange's multipliers may seem inefficient, since it requires the solution of $n + N$ simultaneous equations, rather than n equations, as in the explicit and implicit methods. However, it is often the most convenient process, since it does not require us to decide which variables are to be considered independent and often allows us to take advantage of symmetrical occurrences of the variables.

To show the equivalence of the method of Lagrange's multipliers and the implicit method, say, we shall consider specifically the case of a function of three variables restricted by two side conditions. Other combinations can be treated in a similar fashion, and several appear in the exercises. Suppose, then, that we are to find the extrema of $f(x_1,x_2,x_3)$ subject to the independent side conditions

$$\phi_1(x_1,x_2,x_3) = 0 \quad \text{and} \quad \phi_2(x_1,x_2,x_3) = 0$$

Proceeding implicitly, we have, from the side conditions,

$$\frac{\partial \phi_1}{\partial x_1} dx_1 + \frac{\partial \phi_1}{\partial x_2} dx_2 + \frac{\partial \phi_1}{\partial x_3} dx_3 = 0$$

$$\frac{\partial \phi_2}{\partial x_1} dx_1 + \frac{\partial \phi_2}{\partial x_2} dx_2 + \frac{\partial \phi_2}{\partial x_3} dx_3 = 0$$

and from these, by Corollary 1, Theorem 9, Sec. 10.5, we obtain the expressions†

$$dx_1 = k \begin{vmatrix} \dfrac{\partial \phi_1}{\partial x_2} & \dfrac{\partial \phi_1}{\partial x_3} \\[2ex] \dfrac{\partial \phi_2}{\partial x_2} & \dfrac{\partial \phi_2}{\partial x_3} \end{vmatrix} \qquad dx_2 = -k \begin{vmatrix} \dfrac{\partial \phi_1}{\partial x_1} & \dfrac{\partial \phi_1}{\partial x_3} \\[2ex] \dfrac{\partial \phi_2}{\partial x_1} & \dfrac{\partial \phi_2}{\partial x_3} \end{vmatrix} \qquad x_3 = k \begin{vmatrix} \dfrac{\partial \phi_1}{\partial x_1} & \dfrac{\partial \phi_1}{\partial x_2} \\[2ex] \dfrac{\partial \phi_2}{\partial x_1} & \dfrac{\partial \phi_2}{\partial x_2} \end{vmatrix}$$

where k is an arbitrary proportionality constant. Substituting these into the equation

$$df = \frac{\partial f}{\partial x_1} dx_1 + \frac{\partial f}{\partial x_2} dx_2 + \frac{\partial f}{\partial x_3} dx_3 = 0$$

---

† We must assume, of course, that this system of equations in dx_1, dx_2, dx_3 is of rank 2. If this is not the case, then ϕ_1 and ϕ_2 are functionally dependent, i.e., satisfy some relation of the form $g(\phi_1,\phi_2) = 0$, contrary to the hypothesis that they are independent. For a more detailed discussion of this matter, see, for instance, Louis Brand, "Advanced Calculus," pp. 180–183, Wiley, New York, 1955.

and dividing out k, we obtain

(1)
$$\frac{\partial f}{\partial x_1}\begin{vmatrix}\dfrac{\partial \phi_1}{\partial x_2} & \dfrac{\partial \phi_1}{\partial x_3}\\[2mm] \dfrac{\partial \phi_2}{\partial x_2} & \dfrac{\partial \phi_2}{\partial x_3}\end{vmatrix} - \frac{\partial f}{\partial x_2}\begin{vmatrix}\dfrac{\partial \phi_1}{\partial x_1} & \dfrac{\partial \phi_1}{\partial x_3}\\[2mm] \dfrac{\partial \phi_2}{\partial x_1} & \dfrac{\partial \phi_2}{\partial x_3}\end{vmatrix} + \frac{\partial f}{\partial x_3}\begin{vmatrix}\dfrac{\partial \phi_1}{\partial x_1} & \dfrac{\partial \phi_1}{\partial x_2}\\[2mm] \dfrac{\partial \phi_2}{\partial x_1} & \dfrac{\partial \phi_2}{\partial x_2}\end{vmatrix} = 0$$

This, together with the conditions of constraint, $\phi_1 = \phi_2 = 0$, form a system of three equations in the unknowns x_1, x_2, x_3 from which the possible extrema of f can be found.

On the other hand, using Lagrange's multipliers, we consider the function

$$F = f + \lambda_1\phi_1 + \lambda_2\phi_2$$

getting

$$dF = \left(\frac{\partial f}{\partial x_1} + \lambda_1 \frac{\partial \phi_1}{\partial x_1} + \lambda_2 \frac{\partial \phi_2}{\partial x_1}\right) dx_1 +$$

$$\left(\frac{\partial f}{\partial x_2} + \lambda_1 \frac{\partial \phi_1}{\partial x_2} + \lambda_2 \frac{\partial \phi_2}{\partial x_2}\right) dx_2 +$$

$$\left(\frac{\partial f}{\partial x_3} + \lambda_1 \frac{\partial \phi_1}{\partial x_3} + \lambda_2 \frac{\partial \phi_2}{\partial x_3}\right) dx_3 = 0$$

as the necessary condition for an extremum. From this, since x_1, x_2, and x_3 are now regarded as *independent* variables, we infer that

$$\frac{\partial f}{\partial x_1} + \lambda_1 \frac{\partial \phi_1}{\partial x_1} + \lambda_2 \frac{\partial \phi_2}{\partial x_1} = 0$$

$$\frac{\partial f}{\partial x_2} + \lambda_1 \frac{\partial \phi_1}{\partial x_2} + \lambda_2 \frac{\partial \phi_2}{\partial x_2} = 0$$

$$\frac{\partial f}{\partial x_3} + \lambda_1 \frac{\partial \phi_1}{\partial x_3} + \lambda_2 \frac{\partial \phi_3}{\partial x_2} = 0$$

This is a system of three homogeneous linear equations in the quantities $1, \lambda_1, \lambda_2$. Since this system is to have these quantities as an obviously nontrivial solution, it follows that the determinant of the coefficients must be zero; i.e.,

(2)
$$\begin{vmatrix}\dfrac{\partial f}{\partial x_1} & \dfrac{\partial \phi_1}{\partial x_1} & \dfrac{\partial \phi_2}{\partial x_1}\\[3mm] \dfrac{\partial f}{\partial x_2} & \dfrac{\partial \phi_1}{\partial x_2} & \dfrac{\partial \phi_2}{\partial x_2}\\[3mm] \dfrac{\partial f}{\partial x_3} & \dfrac{\partial \phi_1}{\partial x_3} & \dfrac{\partial \phi_2}{\partial x_3}\end{vmatrix} = 0$$

This equation, with the side conditions $\phi_1 = \phi_2 = 0$, gives us a system of three equations in x_1, x_2, x_3 from which the possible extrema of f can be found. Finally, we note that since Eq. (1) is just the expansion of the determinant (2) according to the elements of the first column, it follows that the same conditions, and hence the same extrema, are obtained whether we use the implicit method or the method of Lagrange's multipliers.

EXAMPLE 1

Find the point on the curve $y^2 = 4x$ which is closest to the point (1,0).

Here we must minimize the function

$$s^2 \equiv f = (x - 1)^2 + y^2$$

subject to the side condition $\phi \equiv y^2 - 4x = 0$. Using the explicit method and eliminating y between the two equations, we have

$$f = (x - 1)^2 + 4x$$

$$\frac{df}{dx} = 2(x - 1) + 4 = 0$$

and $x = -1$. This is clearly absurd, however, since there is no point on the curve $y^2 = 4x$ with abscissa equal to -1.

On the other hand, using Lagrange's multipliers, we consider

$$F = f + \lambda\phi = (x - 1)^2 + y^2 + \lambda(y^2 - 4x)$$

From this we obtain

$$\frac{\partial F}{\partial x} = 2(x - 1) - 4\lambda = 0 \qquad \text{and} \qquad \frac{\partial F}{\partial y} = 2y + 2\lambda y = 0$$

From the second of these equations it follows that $y = 0$ or $\lambda = -1$. Since $\lambda = -1$ implies $x = -1$, which is obviously absurd, it must be rejected. If $y = 0$, it follows from the side condition that $x = 0$. Hence the required point is (0,0), which is clearly correct.

In this problem, the explicit method (as we applied it) failed because we made an unfortunate though natural choice of independent variable. It is true that if we had chosen y as independent variable and eliminated x by substituting from the side condition, we would have found the correct solution. However, the method of Lagrange's multipliers does not require that we be lucky enough or clever enough to guess which variable or variables to regard as independent.

EXAMPLE 2

What is the volume of the largest rectangular box which can be placed inside the ellipsoid

$$\frac{x^2}{a^2} + \frac{y^2}{b^2} + \frac{z^2}{c^2} = 1$$

so that its edges will be parallel to the coordinate axes?

Here we are to maximize the volume of the box

$$V = (2x)(2y)(2z) = 8xyz$$

subject to the condition that

$$\phi = \frac{x^2}{a^2} + \frac{y^2}{b^2} + \frac{z^2}{c^2} - 1 = 0$$

which enforces the requirement that the corners of the box be on the ellipsoid.

Using Lagrange's multipliers, we consider the function

$$F = V + \lambda\phi = 8xyz + \lambda\left(\frac{x^2}{a^2} + \frac{y^2}{b^2} + \frac{z^2}{c^2} - 1\right)$$

From this we obtain the equations

$$\frac{\partial F}{\partial x} = 8yz + \lambda\frac{2x}{a^2} = 0$$

$$\frac{\partial F}{\partial y} = 8xz + \lambda\frac{2y}{b^2} = 0$$

$$\frac{\partial F}{\partial z} = 8xy + \lambda\frac{2z}{c^2} = 0$$

Multiplying these equations by x, y, and z, respectively, we obtain

$$V = -2\lambda\frac{x^2}{a^2} \qquad V = -2\lambda\frac{y^2}{b^2} \qquad V = -2\lambda\frac{z^2}{c^2}$$

whence $x^2/a^2 = y^2/b^2 = z^2/c^2$. Substituting from these into the side condition, we find that

$$\frac{3x^2}{a^2} = 1 \qquad \text{or} \qquad x = \frac{a}{\sqrt{3}}$$

and $y = b/\sqrt{3}$, $z = c/\sqrt{3}$. Thus the maximum volume is

$$V_{\text{max}} = \frac{8\sqrt{3}}{9}\,abc$$

EXERCISES

1 Work Example 2 by:
   (a) the explicit method   (b) the implicit method
2 Find the dimensions of the box of maximum volume which can be placed so that three concurrent edges coincide with the positive coordinate axes and one corner is on the plane

$$\frac{x}{a} + \frac{y}{b} + \frac{z}{c} = 1$$

3 Find the point on the curve of intersection of $x + y + z = 1$ and $z = xy + 5$ which is closest to the origin.
4 An open-top rectangular bin is to have a specified volume V. Find the dimensions of the bin if its surface area is to be a minimum.
5 An open-top tank is to have the shape of a right circular cylinder. If its surface area is to have a specified value S, find the dimensions which make its volume a maximum.
6 A silo is to consist of a right circular cylinder with a hemispherical roof. If the silo is to have a specified volume V, find the dimensions which make its surface area a minimum.
7 Work Exercise 6 if the roof is to be a right circular cone.
8 Find the stationary values of $6x_2^2 - 6x_1x_2 + 6x_2^2 - 6x_2x_3 + 4x_3^2$ subject to the constraint $6x_1^2 + 4x_2^2 + 4x_3^2 = 1$.
9 Show that the implicit method and the method of Lagrange's multipliers lead to the same equations for the determination of the stationary values of $F(x,y,z)$ subject to the single constraint $\phi(x,y,z) = 0$.
10 Show that the implicit method and the method of Lagrange's multipliers lead to the same equations for the determination of the stationary values of $F(x,y,z,t)$ subject to the two constraints $\phi(x,y,z,t) = 0$ and $\theta(x,y,z,t) = 0$.

11 The derivation of a certain smoothing formula requires that the function $A_{-2}{}^2 + A_{-1}{}^2 + A_0{}^2 + A_1{}^2 + A_2{}^2$ be minimized subject to the conditions

$$A_{-2} + A_{-1} + A_0 + A_1 + A_2 = 1$$
$$-2A_{-2} - A_{-1} \quad\quad + A_1 + 2A_2 = 0$$
$$4A_{-2} + A_{-1} \quad\quad + A_1 + 4A_2 = 0$$

What are the values of the A's?

12 Discuss the problem of fitting a line with equation $Ax + By + C = 0$ to a set of points $(x_1, y_1), \ldots, (x_n, y_n)$ by the method of least squares if $A^2 + B^2 + C^2 = 1$.

12.4 Extremal Properties of the Characteristic Values of $(A - \lambda B)X = 0$

In this section we shall investigate certain interesting and important relations between the characteristic values of matric equations and the stationary points of corresponding quadratic forms. We begin by attempting to find the stationary values of the quadratic form $X^T A X$ subject to the constraint $X^T B X = 1$.

Using the method of Lagrange's multipliers, we first construct the function

$$F = X^T A X - \lambda(X^T B X - 1)$$

From this we obtain, at once, as conditions for extrema, the equations

$$\frac{\partial F}{\partial x_1} = 2 \sum_{j=1}^{n} a_{1j}x_j - 2\lambda \sum_{j=1}^{n} b_{1j}x_j = 2 \sum_{j=1}^{n} (a_{1j} - \lambda b_{1j})x_j = 0$$

(1)
$$\frac{\partial F}{\partial x_2} = 2 \sum_{j=1}^{n} a_{2j}x_j - 2\lambda \sum_{j=1}^{n} b_{2j}x_j = 2 \sum_{j=1}^{n} (a_{2j} - \lambda b_{2j})x_j = 0$$

$$\cdots\cdots\cdots\cdots\cdots\cdots\cdots\cdots\cdots\cdots\cdots\cdots\cdots\cdots\cdots\cdots\cdots\cdots$$

$$\frac{\partial F}{\partial x_n} = 2 \sum_{j=1}^{n} a_{nj}x_j - 2\lambda \sum_{j=1}^{n} b_{nj}x_j = 2 \sum_{j=1}^{n} (a_{nj} - \lambda b_{nj})x_j = 0$$

This system of n homogeneous linear equations in the n unknowns $x_1, x_2, \ldots, x_n$ will have a nontrivial solution (as they must, to satisfy the constraint equation) if and only if the determinant of the coefficients is equal to zero; thus

$$\begin{vmatrix} a_{11} - \lambda b_{11} & a_{12} - \lambda b_{12} & \cdots & a_{1n} - \lambda b_{1n} \\ a_{21} - \lambda b_{21} & a_{22} - \lambda b_{22} & \cdots & a_{2n} - \lambda b_{2n} \\ \cdots\cdots\cdots\cdots\cdots\cdots\cdots\cdots\cdots\cdots\cdots\cdots \\ a_{n1} - \lambda b_{n1} & a_{n2} - \lambda b_{n2} & \cdots & a_{nn} - \lambda b_{nn} \end{vmatrix} = |A - \lambda B| = 0$$

Moreover, when we have found the roots of this equation, i.e., the characteristic values of the matric equation $(A - \lambda B)X = 0$, and the corresponding characteristic vectors, if we multiply each equation in (1) by the corresponding component of any particular characteristic vector and then add the equations, we obtain

$$X_i{}^T A X_i - \lambda_i X_i{}^T B X_i = 0$$

or, since $X_i{}^T B X_i = 1$,

$$\lambda_i = X_i{}^T A X_i$$

Thus we have established the important fact that *the characteristic values of the matric equation $(A - \lambda B)X = 0$ are the stationary values of the quadratic form $X^T A X$ subject* *to the constraint* $X^T BX = 1$. It is interesting that in this case the Lagrangian multiplier λ is not just an artificial, though convenient, symbol but has an important interpretation in the problem.

A closely related problem is the determination of the stationary values of the ratio

$$\lambda = \frac{X^T AX}{X^T BX}$$

Here we have no side conditions to enforce, and, proceeding explicitly, we obtain at once the extremizing conditions

$$\frac{\partial \lambda}{\partial x_i} = \frac{(X^T BX)2 \sum\limits_{j=1}^{n} a_{ij}x_j - (X^T AX)2 \sum\limits_{j=1}^{n} b_{ij}x_j}{(X^T BX)^2} = 0 \qquad i = 1, 2, \ldots, n$$

or, multiplying by $X^T BX$ and noting that in the second term we can replace $(X^T AX)/(X^T BX)$ by λ,

$$\sum_{j=1}^{n} a_{ij}x_j - \lambda \sum_{j=1}^{n} b_{ij}x_j = 0 \qquad i = 1, 2, \ldots, n$$

This is the same system of homogeneous linear equations we obtained above; again, it will have a nontrivial solution if and only if

$$|A - \lambda B| = 0$$

This time, if the equations of condition are multiplied by the corresponding component of any characteristic vector and added, we obtain

$$X_i^T AX_i - \lambda_i X_i^T BX_i = 0$$

which implies that

$$\lambda_i = \frac{X_i^T AX_i}{X_i^T BX_i}$$

In other words, *the characteristic values of the matric equation* $(A - \lambda B)X = \mathbf{0}$ *are the stationary values of the ratio*

$$\frac{X^T AX}{X^T BX}$$

In particular, this means that *the smallest stationary value of the ratio is the smallest characteristic value and the largest stationary value of the ratio is the largest characteristic value.*

These observations indicate how information about quadratic forms can be obtained from information about the characteristic values of related matric equations. The importance of the interpretation of characteristic values as the frequencies of vibrating systems makes it natural to ask whether, conversely, the extremal properties of quadratic forms can be used to obtain information about characteristic values. This is indeed the case, as the following discussion shows.

Given the matric equation $(A - \lambda B)X = \mathbf{0}$ and the related quadratic forms $X^T AX$ and $X^T BX$, it is first convenient to apply to $X^T AX$ and $X^T BX$ the transformation described in Corollary 1, Theorem 3, Sec. 11.3.† Under this transformation the

---

† This requires us to assume that $X^T BX$ is positive-definite, a condition that is fulfilled in the applications we have in mind.

forms become, respectively, $Y^T D Y$ and $Y^T I Y$, where D is the diagonal matrix whose diagonal elements are the characteristic values of the equation $(A - \lambda B)X = \mathbf{0}$, which we assume to be named in such a way that

$$\lambda_1 \leq \lambda_2 \leq \cdots \leq \lambda_n$$

Clearly, the characteristic values of the equation $(D - \lambda I)Y = \mathbf{0}$ are the same as those of the original equation, and the sets of values which $Y^T D Y$ and $Y^T I Y$ can have are, respectively, the sets of values of $X^T A X$ and $X^T B X$.

Let us now consider the problem of finding the minimum of the ratio

(2)
$$\lambda = \frac{Y^T D Y}{Y^T I Y}$$

subject to an arbitrary linear constraint†

$$c_1 y_1 + c_2 y_2 + \cdots + c_n y_n = 0$$

Among the sets of y's satisfying this condition we can always find at least one in which y_1 and y_2 are not both zero and $y_3 = y_4 = \cdots = y_n = 0$. For such a set the ratio $Y^T D Y / Y^T I Y$ becomes simply

$$\frac{\lambda_1 y_1{}^2 + \lambda_2 y_2{}^2}{y_1{}^2 + y_2{}^2} \leq \frac{\lambda_2 y_1{}^2 + \lambda_2 y_2{}^2}{y_1{}^2 + y_2{}^2} = \lambda_2$$

Thus, for any linear constraint whatsoever, there is always at least one set of y's for which the value of the ratio $Y^T D Y / Y^T I Y$ is equal to or less than the second characteristic value λ_2. In other words, the minimum of the ratio under one linear constraint is never greater than λ_2. On the other hand, there is at least one particular constraint, namely,

$$y_1 = 0$$

for which the minimum of the ratio actually is as large as λ_2. In fact, if $y_1 = 0$, we have

$$\frac{Y^T D Y}{Y^T I Y} = \frac{\sum\limits_{j=2}^{n} \lambda_j y_j{}^2}{\sum\limits_{j=2}^{n} y_j{}^2} \geq \frac{\sum\limits_{j=2}^{n} \lambda_2 y_j{}^2}{\sum\limits_{j=2}^{n} y_j{}^2} = \lambda_2$$

and the equality sign holds for the set

$$y_1 = 0 \qquad y_2 = 1 \qquad y_3 = y_4 = \cdots = y_n = 0$$

Thus we have established the following result: *the second smallest characteristic value of the equation $(D - \lambda I)Y = \mathbf{0}$ or $(A - \lambda B)X = \mathbf{0}$ is the largest of the minimum values which the ratio $Y^T D Y / Y^T I Y$ or $X^T A X / X^T B X$ can assume under any linear constraint on the y's or x's.*

If the variables are subject to more than one linear constraint, similar results hold

---

† Nonlinear constraints can be handled by expanding them in a multiple Taylor's series of the y's and retaining only the linear terms.

for the higher characteristic values. Thus, suppose that the y's are subject to $k - 1$ arbitrary linear constraints,

$$
\begin{aligned}
c_{11}y_1 &+ c_{12}y_2 &+ \cdots + c_{1n}y_n &= 0 \\
c_{21}y_1 &+ c_{22}y_2 &+ \cdots + c_{2n}y_n &= 0 \\
&\cdots\cdots\cdots\cdots\cdots\cdots\cdots\cdots\cdots \\
c_{k-1,1}y_1 &+ c_{k-1,2}y_2 &+ \cdots + c_{k-1,n}y_n &= 0
\end{aligned}
$$

If the y's after y_k are arbitrarily set equal to zero, these equations form a system of $k - 1$ homogeneous linear equations in the k variables $y_1, y_2, \ldots, y_k$. Since such a system always has a nontrivial solution, it follows that among all sets of y's satisfying the given constraints there is always at least one in which $y_{k+1} = y_{k+2} = \cdots = y_n = 0$ and $y_1, y_2, \ldots, y_k$ are not all zero. For any such set we then have

$$
\frac{Y^TDY}{Y^TIY} = \frac{\displaystyle\sum_{j=1}^{k} \lambda_j y_j^2}{\displaystyle\sum_{j=1}^{k} y_j^2} \leq \frac{\displaystyle\sum_{j=1}^{k} \lambda_k y_j^2}{\displaystyle\sum_{j=1}^{k} y_j^2} = \lambda_k
$$

In other words, for any system of $k - 1$ linear constraints, the minimum value of the ratio Y^TDY/Y^TIY is never greater than λ_k. Moreover, there is at least one set of constraints, namely, $y_1 = y_2 = \cdots = y_{k-1} = 0$, for which the minimum value of the ratio is actually as large as λ_k. In fact, for any set of y's satisfying these particular constraints, we have

$$
\frac{Y^TDY}{Y^TIY} = \frac{\displaystyle\sum_{j=k}^{n} \lambda_j y_j^2}{\displaystyle\sum_{j=k}^{n} y_j^2} \geq \frac{\displaystyle\sum_{j=k}^{n} \lambda_k y_j^2}{\displaystyle\sum_{j=k}^{n} y_j^2} = \lambda_k
$$

and the equality sign holds for the particular set

$$
y_1 = y_2 = \cdots = y_{k-1} = 0 \qquad y_k = 1 \qquad y_{k+1} = y_{k+2} = \cdots = y_n = 0
$$

Thus we have established the following theorem.

THEOREM 1 If $\lambda_1 \leq \lambda_2 \leq \cdots \leq \lambda_n$ are the characteristic values of the equation $(D - \lambda I)Y = \mathbf{0}$ or $(A - \lambda B)X = \mathbf{0}$, then λ_k is the largest of the minimum values which the ratio Y^TDY/Y^TIY or X^TAX/X^TBX can assume under any set of $k - 1$ homogeneous linear constraints on the y's or the x's.

By an almost identical argument, beginning with the largest characteristic value, we can prove the following theorem for the characteristic values.

THEOREM 2 If $\lambda_1 \leq \lambda_2 \leq \cdots \leq \lambda_{n-k} \leq \cdots \leq \lambda_n$ are the characteristic values of the equation $(D - \lambda I)Y = \mathbf{0}$ or $(A - \lambda B)X = \mathbf{0}$, then λ_{n-k} is the smallest of the maximum values which the ratio Y^TDY/Y^TIY or X^TAX/X^TBX can assume under any set of k homogeneous linear constraints on the y's or x's.

As an important consequence of Theorems 1 and 2, we have the following result.

THEOREM 3 If the n stationary values of the ratio $\lambda = X^T A X / X^T B X$ are $\lambda_1 \leq \lambda_2 \leq \cdots \leq \lambda_k \leq \lambda_{k+1} \leq \cdots \leq \lambda_n$, and if the stationary values of λ under a single homogeneous linear constraint are $\lambda_1' \leq \lambda_2' \leq \cdots \leq \lambda_k' \leq \cdots \leq \lambda_{n-1}'$, then

$$\lambda_1 \leq \lambda_1' \leq \lambda_2 \leq \cdots \leq \lambda_k \leq \lambda_k' \leq \lambda_{k+1} \leq \cdots \leq \lambda_{n-1} \leq \lambda_{n-1}' \leq \lambda_n$$

Proof By Theorem 1, λ_{k+1} is the largest minimum of the ratio λ under *any* system of k homogeneous linear constraints. On the other hand, λ_k' is the largest minimum of λ under a subset of these k constraints consisting of $k - 1$ arbitrary constraints and the particular constraint which is given. Hence, λ_k', being a minimum under a more restricted set of constraints, is at most as large as λ_{k+1}; that is,

$$\lambda_k' \leq \lambda_{k+1}$$

Likewise, λ_k is the largest minimum of λ under any set of $k - 1$ constraints. However, as we have just observed, λ_k' is the largest minimum of λ under any set of $k - 1$ constraints and one additional specific constraint. This last constraint can serve only to increase each of the minima of which λ_k' is the largest. Hence $\lambda_k \leq \lambda_k'$, and so, finally, for each value of k,

$$\lambda_k \leq \lambda_k' \leq \lambda_{k+1}$$

as asserted.

Theorems 1 and 2 are often useful in approximating the natural frequencies of vibrating systems whose potential and kinetic energies are quadratic forms in the displacements and velocities, respectively. For if

$$\text{PE} = X^T A X \qquad \text{and} \qquad \text{KE} = \dot{X}^T B \dot{X}$$

are the instantaneous energy expressions for the system, then the substitution

$$x_i = s_i \cos \omega t$$

leads to the expressions

$$S^T A S \qquad \text{and} \qquad \omega^2 S^T B S \qquad \text{where } S = \begin{bmatrix} s_1 \\ s_2 \\ \cdot \\ s_n \end{bmatrix}$$

for the maximum values of the potential and kinetic energies. Then from the fact that during free vibrations

$$\text{PE}_{\text{max}} = \text{KE}_{\text{max}}$$

we obtain at once

$$\omega^2 = \frac{S^T A S}{S^T B S}$$

The values of ω^2, which of course are the squares of the natural frequencies of the system, can now be approximated by using the maximum-minimum properties described in Theorems 1 and 2.

EXAMPLE 1

A triple pendulum consists of three equal masses suspended by three weightless cords of equal length, as shown in Fig. 12.1. Approximate the natural frequencies of this system.

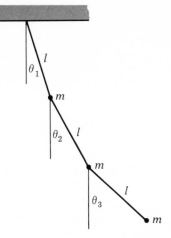

Figure 12.1
A triple pendulum.

Measuring the potential energy from the equilibrium position of the masses, we have

$$\text{PE} = mgl(1 - \cos\theta_1) + mg[l(1 - \cos\theta_1) + l(1 - \cos\theta_2)]$$
$$+ mg[l(1 - \cos\theta_1) + l(1 - \cos\theta_2) + l(1 - \cos\theta_3)]$$
$$= mgl[3(1 - \cos\theta_1) + 2(1 - \cos\theta_2) + (1 - \cos\theta_3)]$$

$$\text{KE} = \tfrac{1}{2}ml^2\{[(\cos\theta_1)\dot\theta_1]^2 + [(\sin\theta_1)\dot\theta_1]^2\}$$
$$+ \tfrac{1}{2}ml^2\{[(\cos\theta_1)\dot\theta_1 + (\cos\theta_2)\dot\theta_2]^2 + [(\sin\theta_1)\dot\theta_1 + (\sin\theta_2)\dot\theta_2]^2\}$$
$$+ \tfrac{1}{2}ml^2\{[(\cos\theta_1)\dot\theta_1 + (\cos\theta_2)\dot\theta_2 + (\cos\theta_3)\dot\theta_3]^2$$
$$+ [(\sin\theta_1)\dot\theta_1 + (\sin\theta_2)\dot\theta_2 + (\sin\theta_3)\dot\theta_3]^2\}$$
$$= \tfrac{1}{2}ml^2\{(\dot\theta_1)^2 + [(\dot\theta_1)^2 + 2\cos(\theta_1 - \theta_2)\dot\theta_1\dot\theta_2 + (\dot\theta_2)^2]$$
$$+ [(\dot\theta_1)^2 + (\dot\theta_2)^2 + (\dot\theta_3)^2 + 2\cos(\theta_1 - \theta_2)\dot\theta_1\dot\theta_2$$
$$+ 2\cos(\theta_1 - \theta_3)\dot\theta_1\dot\theta_3 + 2\cos(\theta_2 - \theta_3)\dot\theta_2\dot\theta_3]\}$$

Now, by assuming vibrations so small that we may take

$$1 - \cos\theta_i = \tfrac{1}{2}\theta_i^2 \qquad \text{and} \qquad \cos(\theta_i - \theta_j) = 1$$

these expressions become

$$\text{PE} = \tfrac{1}{2}mgl(3\theta_1^2 + 2\theta_2^2 + \theta_3^2)$$
$$\text{KE} = \tfrac{1}{2}ml^2[3(\dot\theta_1)^2 + 2(\dot\theta_2)^2 + (\dot\theta_3)^2 + 4\dot\theta_1\dot\theta_2 + 2\dot\theta_1\dot\theta_3 + 2\dot\theta_2\dot\theta_3]$$

Finally, setting $\theta_i = s_i \cos\omega t$ and equating the maximum value of the potential energy to the maximum value of the kinetic energy, we obtain

$$\omega^2 \frac{l}{g} = \frac{3s_1^2 + 2s_2^2 + s_3^2}{3s_1^2 + 2s_2^2 + s_3^2 + 4s_1 s_2 + 2s_1 s_3 + 2s_2 s_3}$$

The smallest value of $\omega^2(l/g)$ is the minimum (unconstrained) value of the ratio of these quadratic forms. Hence, making various reasonable choices for s_1, s_2, s_3, we find

s_1	s_2	s_3	$\dfrac{\omega^2 l}{g}$
1	1	1	0.428
1	1	2	0.428
2	2	3	0.420
3	3	4	0.421
3	3	5	0.422
2	1	3	0.435

It thus appears that $\omega^2 = 0.420(g/l)$ is a reasonable approximation to the square of the lowest natural frequency of the system.

To approximate the second frequency of the system, we must calculate the minimum value of the ratio for various linear constraints and select the largest of these minima. Recalling that the normal modes of a vibrating system are orthogonal with respect to the matrix of either the potential energy or the kinetic energy, we choose as our first constraint equation the relation

$$(3) \qquad [2 \ \ 2 \ \ 3] \begin{bmatrix} 3 & 0 & 0 \\ 0 & 2 & 0 \\ 0 & 0 & 1 \end{bmatrix} \begin{bmatrix} s_1 \\ s_2 \\ s_3 \end{bmatrix} = 6s_1 + 4s_2 + 3s_3 = 0$$

which expresses the fact that the vector S is orthogonal to the approximate modal vector $[2 \ 2 \ 3]$ with respect to the matrix of the potential-energy form.

We now evaluate the ratio of the two quadratic forms for various sets of values satisfying Eq. (3), getting

s_1	s_2	s_3	$\omega^2 \dfrac{l}{g}$
3	3	-10	2.38
5	3	-14	2.31
8	3	-20	2.29
4	3	-12	2.33
-3	3	2	3.77

As another constraint equation, let us try $s_1 + s_2 + s_3 = 0$. Evaluating this for the values $s_1 = 6$, $s_2 = 4$, $s_3 = -10$, we obtain $\omega^2(l/g) = 1.76$. Since this is significantly less than the (apparent) minimum (2.29) which we found from the first constraint equation, and since we are seeking the *largest* minimum, there is probably no point in working further with this constraint. Trying another, say $4s_1 + 2s_2 + 3s_3 = 0$, we find for the values $s_1 = 1$, $s_2 = 4$, $s_3 = -4$ the value $\omega^2(l/g) = 1.56$. Since this, too, is significantly less than 2.29, there is probably no point in continuing further with this constraint and we accept $\omega^2 = 2.29(g/l)$, our approximation to the second lowest frequency of the system.

To determine the highest natural frequency of the system, it is natural to begin with the constraints which express orthogonality with respect to the two (approximate) normal modes we have thus far determined. The first of these, of course, is the one we have already used, namely,

$$6s_1 + 4s_2 + 3s_3 = 0$$

The second is

$$[8 \ \ 3 \ \ -20] \begin{bmatrix} 3 & 0 & 0 \\ 0 & 2 & 0 \\ 0 & 0 & 1 \end{bmatrix} \begin{bmatrix} s_1 \\ s_2 \\ s_3 \end{bmatrix} = 24s_1 + 6s_2 - 20s_3 = 0$$

or $12s_1 + 3s_2 - 10s_3 = 0$. Now there is a unique set of s values, namely, $(49, -96, 30)$, satisfying these two constraint equations, and for these we find $\omega^2(l/g) = 5.42$. Trying several other pairs of constraints, we obtain

$$\begin{aligned} 6s_1 + 4s_2 + 3s_3 &= 0 \\ 6s_1 + \ s_2 - 6s_3 &= 0 \end{aligned} \qquad s_1 = 3 \qquad s_2 = -6 \qquad s_3 = 2 \qquad \omega^2 \frac{l}{g} = 5.42$$

$$\begin{aligned} -s_1 + \ s_2 + 3s_3 &= 0 \\ -2s_1 + \ s_2 + 4s_3 &= 0 \end{aligned} \qquad s_1 = 1 \qquad s_2 = -2 \qquad s_3 = 1 \qquad \omega^2 \frac{l}{g} = 6.00$$

$$\begin{aligned} 4s_1 + 3s_2 + \ s_3 &= 0 \\ 4s_1 + 2s_2 - \ s_3 &= 0 \end{aligned} \qquad s_1 = 5 \qquad s_2 = -8 \qquad s_3 = 4 \qquad \omega^2 \frac{l}{g} = 6.27$$

$$\begin{aligned} s_1 + \ s_2 - \ s_3 &= 0 \\ 8s_1 + 4s_2 - \ s_3 &= 0 \end{aligned} \qquad s_1 = 5 \qquad s_2 = -9 \qquad s_3 = 4 \qquad \omega^2 \frac{l}{g} = 6.17$$

On the basis of these calculations, it appears that $\omega^2 = 6.27(g/l)$ is a reasonable approximation to the square of the highest natural frequency of the system. Independent calculations show that the squares of the natural frequencies, correct to three significant figures, are

$$\omega_1{}^2 = 0.416\frac{g}{l} \qquad \omega_2{}^2 = 2.29\frac{g}{l} \qquad \omega_3{}^2 = 6.29\frac{g}{l}$$

EXERCISES

1 Using Theorem 1, approximate the characteristic values of the equation $(A - \lambda B)X = \mathbf{0}$ if

(a) $A = \begin{bmatrix} 8 & -2 & 0 \\ -2 & 3 & -1 \\ 0 & -1 & 2 \end{bmatrix}$ $\qquad B = \begin{bmatrix} 8 & 0 & 0 \\ 0 & 2 & 0 \\ 0 & 0 & 2 \end{bmatrix}$

(b) $A = \begin{bmatrix} 2 & -1 & 0 \\ -1 & 2 & -1 \\ 0 & -1 & 2 \end{bmatrix}$ $\qquad B = \begin{bmatrix} 3 & 0 & 0 \\ 0 & 4 & 0 \\ 0 & 0 & 3 \end{bmatrix}$

(c) $A = \begin{bmatrix} 1 & -1 & 0 \\ -1 & 3 & -3 \\ 0 & -3 & 6 \end{bmatrix}$ $\qquad B = \begin{bmatrix} 1 & 0 & 0 \\ 0 & 1 & 0 \\ 0 & 0 & 1 \end{bmatrix}$

2 Using Theorem 2, approximate the characteristic values of the equation $(A - \lambda B)X = \mathbf{0}$ if

(a) $A = \begin{bmatrix} 6 & -3 & 0 \\ -3 & 6 & -3 \\ 0 & -3 & 4 \end{bmatrix}$ $\qquad B = \begin{bmatrix} 6 & 0 & 0 \\ 0 & 4 & 0 \\ 0 & 0 & 4 \end{bmatrix}$

(b) $A = \begin{bmatrix} 3 & -1 & 0 \\ -1 & 1 & -1 \\ 0 & -1 & 5 \end{bmatrix}$ $\qquad B = \begin{bmatrix} 4 & 0 & 0 \\ 0 & 1 & 0 \\ 0 & 0 & 4 \end{bmatrix}$

(c) $A = \begin{bmatrix} 1 & -1 & -2 \\ -1 & 5 & -1 \\ -2 & -1 & 10 \end{bmatrix}$ $\qquad B = \begin{bmatrix} 1 & 0 & 0 \\ 0 & 1 & 0 \\ 0 & 0 & 1 \end{bmatrix}$

3 Approximate the natural frequencies of a double pendulum consisting of two equal masses suspended by cords of length l and $2l$ if:
(a) The upper cord is the cord of length l.
(b) The upper cord is the cord of length $2l$.

12.5 The Euler Equation for $\int_a^b f(x,y,y')\,dx$

Up to this point we have been concerned with maximizing and minimizing *functions*. In the remainder of this chapter we shall be concerned with maximizing and minimizing *functionals*.

DEFINITION 1 A functional is a rule which assigns to every function in a given set, or domain, of functions a unique real number.

Even though we may not have recognized them as such, we have already encountered many functionals in our past work in mathematics, among them the following:

1. For a fixed value of x and a fixed function f, the expression $f[g(x)]$ is a functional whose domain is the set of all functions g such that x is in the domain of g and $g(x)$ is in the domain of f.

2. $\int_a^b f(x)\, dx$ is a functional, since it is a rule which assigns a unique real number to each function f which is integrable on the interval $[a,b]$.

3. The Fourier coefficients

$$a_n = \frac{1}{p} \int_d^{d+2p} f(x) \cos \frac{n\pi x}{p}\, dx \qquad b_n = \frac{1}{p} \int_d^{d+2p} f(x) \sin \frac{n\pi x}{p}\, dx$$

are functionals, since they are formulas, or rules, which assign unique numerical values to each periodic function satisfying the Dirichlet conditions.

4. For each value of x, the expression

$$L(y) = a_0 y''(x) + a_1 y'(x) + a_2 y(x)$$

is a functional whose domain is the set of all functions y which are twice-differentiable at x.

5. The maximum tip deflection of a cantilever beam which is forced to vibrate by a harmonic load $w(x) \sin \omega t$ is a functional whose domain is the set of all admissible load-distribution functions $w(x)$.

6. The potential energy $V = \frac{1}{2} \int_0^l EI(x)[y''(x)]^2\, dx$ which is stored in a bent beam is a functional whose domain is the set of all admissible deflection curves $y(x)$.

7. The kinetic energy of a vibrating beam

$$T = \frac{1}{2} \int_0^l \omega^2 \rho(x) y^2(x) \sin^2 \omega t\, dx$$

at any particular time is a functional whose domain is the set of all admissible deflection curves $y(x)$.

One very important functional which we shall consider in detail is the integral

$$(1) \qquad\qquad I = \int_a^b F(x, y, y')\, dx$$

In particular, we shall attempt to find the function y in the domain of all continuously differentiable functions satisfying the end conditions $y(a) = y_1$ and $y(b) = y_2$ which maximizes or minimizes I.

Suppose that $y(x)$ is the function for which I takes on its extreme value, and let $\eta(x)$ be any continuously differentiable function which vanishes at the end points a and b. Then for any ε, the function

$$\phi(x) = y(x) + \varepsilon\eta(x)$$

satisfies the same differentiability condition and the same end conditions as y itself. Moreover, if ε is independent of x, as we shall suppose, then $\phi'(x) = y'(x) + \varepsilon\eta'(x)$. Hence for each function η, the integral

$$I(\varepsilon) = \int_a^b F(x,\, y + \varepsilon\eta,\, y' + \varepsilon\eta')\, dx$$

is a function of ε which takes on its extreme value when $\varepsilon = 0$, since

$$I(0) = \int_a^b F(x, y, y')\, dx$$

and, by hypothesis, y is the function for which I takes on its extreme value. Now if $I(\varepsilon)$ has an extremum for $\varepsilon = 0$, then

$$\frac{dI}{d\varepsilon} = 0 \quad \text{when } \varepsilon = 0$$

Hence, assuming that differentiation under the integral sign is legitimate, we have

$$\frac{dI}{d\varepsilon} = \frac{d}{d\varepsilon} \int_a^b F(x,\, y + \varepsilon\eta,\, y' + \varepsilon\eta')\,dx$$

$$= \int_a^b \frac{\partial}{\partial\varepsilon} F(x,\, y + \varepsilon\eta,\, y' + \varepsilon\eta')\,dx$$

(2) $$\qquad = \int_a^b \left(\frac{\partial F}{\partial y}\,\eta + \frac{\partial F}{\partial y'}\,\eta'\right) dx = 0$$

as a necessary condition for the desired extremum.

If we now apply integration by parts to the last term in the integral in (2), with $u = \partial F/\partial y'$ and $dv = \eta'\,dx$, we have

$$\int_a^b \frac{\partial F}{\partial y'}\,\eta'\,dx = \eta\,\frac{\partial F}{\partial y'}\Bigg|_a^b - \int_a^b \eta\,\frac{d}{dx}\left(\frac{\partial F}{\partial y'}\right) dx$$

$$= -\int_a^b \eta\,\frac{d}{dx}\left(\frac{\partial F}{\partial y'}\right) dx$$

since, by hypothesis, $\eta(a) = \eta(b) = 0$. Thus Eq. (2) reduces to

(3) $$\qquad \int_a^b \eta \left[\frac{\partial F}{\partial y} - \frac{d}{dx}\left(\frac{\partial F}{\partial y'}\right)\right] dx = 0$$

Now η is arbitrary, except for the end conditions and the requirement of differentiability. Hence, if the expression in brackets in (3) were not identically zero, η could be chosen to make the integrand in (3) positive everywhere that it was different from zero, which would contradict the vanishing of the integral. Thus our final result is that *a necessary condition for the integral $I = \int_a^b F(x,y,y')\,dx$ to have an extreme value is that y satisfy the equation*

(4) $$\qquad \frac{d}{dx}\left(\frac{\partial F}{\partial y'}\right) - \frac{\partial F}{\partial y} = 0$$

This equation is usually referred to as the **Euler equation** for the integral (1).†

Since $\partial F/\partial y'$ involves x not only explicitly but also implicitly, through y and y', we have

$$\frac{d}{dx}\left(\frac{\partial F}{\partial y'}\right) = \frac{\partial}{\partial x}\left(\frac{\partial F}{\partial y'}\right) + \frac{\partial}{\partial y}\left(\frac{\partial F}{\partial y'}\right)\frac{dy}{dx} + \frac{\partial}{\partial y'}\left(\frac{\partial F}{\partial y'}\right)\frac{dy'}{dx}$$

---

† We must be careful to note that the Euler equation (4) is not a sufficient condition, and that a function y which satisfies it does not necessarily maximize or minimize I. The fact that $\varepsilon = 0$ is a value which makes $dI/d\varepsilon = 0$ implies only that $\varepsilon = 0$ is a stationary point of $I(\varepsilon)$, at which I may have a maximum or a minimum *or a horizontal inflexion*. The corresponding solution y of the Euler equation is therefore one for which I is stationary but not necessarily a maximum or a minimum. In the elementary applications of the calculus of variations the nature of the problem will usually provide at least informal evidence that the function obtained does indeed maximize or minimize the functional under consideration, and this we shall assume.

Hence Euler's equation is equivalent to

(5)
$$F_{y'y'}\, y'' + F_{yy'}\, y' + F_{xy'} - F_y = 0$$

In general, this differential equation is of the second order in y, and the two constants of integration which appear in its solution just suffice for the imposition of the two prescribed boundary conditions.

It is easy to verify by differentiation that Eq. (5) is equivalent to

$$-\frac{1}{y'}\left[\frac{d}{dx}\,(F - F_{y'}\, y') - F_x\right] = 0$$

If F does not involve x explicitly, then $F_x \equiv \partial F/\partial x \equiv 0$ and so the last equation reduces to

$$\frac{d}{dx}\,(F - F_{y'}\, y') = 0$$

which leads at once to a first integral of Euler's equation, namely,

(6)
$$F - F_{y'}\, y' = c$$

On the other hand, if F does not involve y explicitly, then $\partial F/\partial y \equiv 0$ and the first form of Euler's equation [Eq. (4)] reduces to

$$\frac{d}{dx}\left(\frac{\partial F}{\partial y'}\right) = 0$$

which leads to the first integral

$$\frac{\partial F}{\partial y'} = c$$

In passing, we note that if the value of y is not prescribed at one of the end points of the interval $[a,b]$, then in the derivation of the Euler equation $\eta(x)$ need not vanish at that point in order for the function $\phi(x) = y(x) + \varepsilon\eta(x)$ to be in the domain of the functional (1). Hence, in the ensuing integration by parts, the integrated term

$$\left.\frac{\partial F}{\partial y'}\,\eta(x)\right|_a^b$$

will not vanish unless $\partial F/\partial y' = 0$ at the end point in question. The condition $\partial F/\partial y' = 0$ is referred to as a **natural boundary condition**.

EXAMPLE 1

What curve connecting the points $P_1:(x_1,y_1)$ and $P_2:(x_2,y_2)$ has the shortest length? Here, the answer is obvious, namely, the segment $\overline{P_1 P_2}$, but it is interesting to verify this elementary geometric fact by the calculus of variations. What we have to do, of course, is to determine the function which minimizes the integral

$$L = \int_{x=x_1}^{x=x_2} ds = \int_{x_1}^{x_2} \sqrt{1 + (y')^2}\; dx$$

For this integral, Euler's equation is

$$\frac{d}{dx}\left[\frac{\partial}{\partial y'}\sqrt{1 + (y')^2}\right] - \frac{\partial}{\partial y}\sqrt{1 + (y')^2} = 0$$

or, since the last term is obviously zero,

$$\frac{d}{dx}\left[\frac{y'}{\sqrt{1 + (y')^2}}\right] = 0$$

This implies that

$$\frac{y'}{\sqrt{1 + (y')^2}} = c$$

or, solving for y',

$$y' = \frac{c}{\sqrt{1 - c^2}} = m, \text{ say}$$

Integrating this gives

$$y = mx + b$$

The constants m and b are determined, of course, by the condition that this line must pass through P_1 and P_2.

If $x_1 = x_2$, this analysis fails, for there is no line with equation $y = mx + b$ which passes through two points with the same abscissa. However, using y as the independent variable, i.e., minimizing

$$\int_{y_1}^{y_2} \sqrt{1 + (x')^2}\, dy$$

yields the required conclusion in this case.

EXAMPLE 2

What curve joining the points $P_1 : (-a,b)$ and $P_2 : (a,b)$ generates the smallest surface area when revolved about the x axis?

In this case we are to minimize the integral

$$S = \int_{x=-a}^{x=a} 2\pi y\, ds = 2\pi \int_{-a}^{a} y\sqrt{1 + (y')^2}\, dx$$

Hence we must solve the Euler equation

$$2\pi \left\{ \left[\frac{d}{dx} \frac{yy'}{\sqrt{1 + (y')^2}}\right] - \sqrt{1 + (y')^2} \right\} = 0$$

Performing the indicated differentiation and simplifying, we obtain without difficulty the differential equation

$$yy'' - (y')^2 - 1 = 0$$

To solve this, we set

$$y'' \equiv \frac{dy'}{dx} = \frac{dy'}{dy}\frac{dy}{dx} = y'\frac{dy'}{dy}$$

getting

$$yy'\frac{dy'}{dy} - (y')^2 - 1 = 0$$

or, separating variables,

$$\frac{y'}{1 + (y')^2}\, dy' = \frac{dy}{y}$$

Integrating this, we obtain

$$\ln\left[1 + (y')^2\right] = 2\ln y - 2\ln c_1 \quad \text{ or } \quad 1 + (y')^2 = \frac{y^2}{c_1{}^2}$$

From this it follows that

$$y' = \frac{\sqrt{y^2 - c_1{}^2}}{c_1} \quad \text{ or } \quad \frac{dy}{\sqrt{y^2 - c_1{}^2}} = \frac{dx}{c_1}$$

Integrating again gives

$$\cosh^{-1}\left(\frac{y}{c_1}\right) = \frac{x}{c_1} + c_2 \qquad \text{or} \qquad y = c_1 \cosh\left(\frac{x}{c_1} + c_2\right)$$

Since P_1 and P_2 are symmetrically placed with respect to the y axis, it follows that the required curve must also be symmetric in the y axis. Hence $c_2 = 0$, and we have

$$y = c_1 \cosh\frac{x}{c_1}$$

To determine c_1 we have the equation

$$b = c_1 \cosh\frac{a}{c_1}$$

which requires some numerical procedure, such as Newton's method, for its solution.

EXERCISES

1 In Example 2, show that, depending on the relative size of a and b, the equation for c_1 may have none, one, or two solutions.

2 Find the equation of the curve which joins the points $P_1:(0,1)$ and $P_2:(2,3)$ and along which the integral

$$\int_0^2 \frac{\sqrt{1 + (y')^2}}{y}\, dx$$

is a minimum.

3 Find the natural boundary conditions for the integral

$$I = \int_0^1 [y^2 - yy' + (y')^2]\, dx$$

What is the Euler equation for I? What is the equation of the curve joining $P_1:(0,1)$ and $P_2:(1,2)$ on which I is a minimum? What is the function which minimizes I, takes on the value 1 at $x = 0$, and satisfies the natural boundary condition at $x = 1$? What is the function which minimizes I, satisfies the natural boundary condition at $x = 0$, and takes on the value 2 at $x = 1$? What is the function which minimizes I and satisfies the natural boundary condition at $x = 0$ and at $x = 1$?

4 Work Exercise 3 for the integral $\int_1^3 [x^2(y')^2 - yy']\, dx$ and the points $P_1:(1,1)$ and $P_2:(3,2)$.

5 If the cost per mile to travel in the first quadrant is equal to $1 + x$, what is the equation of the family of curves along which it is most economical to travel?

6 Determine the curve down which a particle will slide without friction from the origin to the point $P:(x_1,y_1)$ in the shortest time. *Hint:* Recall that the velocity of a particle sliding from rest down any curve is given by the formula $v^2 = 2gy$, where y is the vertical distance the particle has descended. This **brachistochrone problem** was one of the first to be solved by the methods of the calculus of variations.

7 If the first quadrant is filled with a transparent medium in which the velocity of light at any point is equal to $1 + x$, find the equation of the path along which a light ray will travel from the origin to the point $(2,3)$ in the shortest time.

8 A **geodesic** on a surface is a curve along which the distance between two points of the surface is a minimum.
(a) Find the geodesics on a right circular cylinder of radius a. *Hint:* In cylindrical coordinates the differential of arc on a cylinder is given by the formula $(ds)^2 = (a\, d\theta)^2 + (dz)^2$.
(b) Find the geodesics on a right circular cone of vertex angle α. *Hint:* In spherical coordinates, the differential of arc on a right circular cone is given by the formula $(ds)^2 = (dr)^2 + (r \sin \alpha\, d\theta)^2$.

9 The functional $I = \int_a^b F(x,y,z,y',z') \, dx$ depends on the two functions y and z. Show that necessary conditions for the determination of the functions y and z which minimize I are

$$\frac{d}{dx}\left(\frac{\partial F}{\partial y'}\right) - \frac{\partial F}{\partial y} = 0 \quad \text{and} \quad \frac{d}{dx}\left(\frac{\partial F}{\partial z'}\right) - \frac{\partial F}{\partial z} = 0$$

10 Show that the Euler equation for the integral

$$I = \int_a^b F(x,y,y',y'') \, dx$$

is

$$\frac{d^2}{dx^2}\left(\frac{\partial F}{\partial y''}\right) - \frac{d}{dx}\left(\frac{\partial F}{\partial y'}\right) + \frac{\partial F}{\partial y} = 0$$

12.6 Variations

Since a functional is just a function whose domain of definition is a set of functions, it is natural to ask whether the concept of the differential of a function can be extended to functionals. To see how this can be done, let us suppose that $F(x, y, y')$ is a functional defined on a set of functions $\{y(x)\}$, and let us develop an expression for the change in F corresponding to an assigned change in $y(x)$ *for a fixed value of x*.

If $y(x)$ is changed into the function

$$y(x) + \varepsilon\eta(x) \qquad \varepsilon \text{ independent of } x$$

we call the change $\varepsilon\eta(x)$ the **variation** of y and denote it by δy. Moreover, from the changed value of y we infer that the changed value of $y'(x)$ is

$$y'(x) + \varepsilon\eta'(x)$$

Hence we have the companion formula

$$\delta y'(x) = \varepsilon\eta'(x)$$

for the variation of $y'(x)$. Corresponding to these changes we have the change

$$\Delta F = F(x, y + \varepsilon\eta, y' + \varepsilon\eta') - F(x,y,y')$$

If we expand the first term on the right in a Maclaurin's expansion in powers of ε, we have

$$\Delta F = F(x, y, y') + \left(\frac{\partial F}{\partial y}\eta + \frac{\partial F}{\partial y'}\eta'\right)\varepsilon$$

$$+ \left(\frac{\partial^2 F}{\partial y^2}\eta^2 + 2\frac{\partial^2 F}{\partial y \, \partial y'}\eta\eta' + \frac{\partial^2 F}{\partial y'^2}\eta'^2\right)\frac{\varepsilon^2}{2!} + \cdots - F(x,y,y')$$

or, neglecting powers of ε higher than the first,

$$\Delta F \doteq \frac{\partial F}{\partial y}\eta\varepsilon + \frac{\partial F}{\partial y'}\eta'\varepsilon$$

(1)
$$= \frac{\partial F}{\partial y}\delta y + \frac{\partial F}{\partial y'}\delta y'$$

By analogy with the differential of a function, we define the last expression to be the **variation** of the functional F and denote it by δF.†

In passing, we note that in its simplest form the differential of a function is a first-order approximation to the change in the function as x varies along a particular curve, whereas the variation of a functional is a first-order approximation to the change in the functional at a particular value of x as we vary from curve to curve.

It is interesting and important to note that variations can be calculated by the same rules that apply to differentials. Specifically,

$$\delta(F_1 \pm F_2) = \delta F_1 \pm \delta F_2$$

$$\delta(F_1 F_2) = F_1 \, \delta F_2 + F_2 \, \delta F_1$$

$$\delta\left(\frac{F_1}{F_2}\right) = \frac{F_2 \, \delta F_1 - F_1 \, \delta F_2}{F_2{}^2}$$

$$\delta(F^n) = nF^{n-1} \, \delta F$$

The proof of the first of these relations is trivial. To prove the second, we have

$$\Delta(F_1 F_2) = F_1(x, y + \varepsilon\eta, y' + \varepsilon\eta')F_2(x, y + \varepsilon\eta, y' + \varepsilon\eta') - F_1 F_2$$

Hence, expanding again in terms of powers of ε, and recalling that $\varepsilon\eta = \delta y$ and $\varepsilon\eta' = \delta y'$, we obtain

$$\Delta(F_1 F_2) = \left[F_1 + \left(\frac{\partial F_1}{\partial y}\eta + \frac{\partial F_1}{\partial y'}\eta'\right)\varepsilon + \cdots\right]$$
$$\times \left[F_2 + \left(\frac{\partial F_2}{\partial y}\eta + \frac{\partial F_2}{\partial y'}\eta'\right)\varepsilon + \cdots\right] - F_1 F_2$$
$$\doteq F_1\left(\frac{\partial F_2}{\partial y}\delta y + \frac{\partial F_2}{\partial y'}\delta y'\right) + F_2\left(\frac{\partial F_1}{\partial y}\delta y + \frac{\partial F_1}{\partial y'}\delta y'\right)$$
$$= F_1 \, \delta F_2 + F_2 \, \delta F_1$$

as asserted.

From the definitive relation $\delta y(x) = \varepsilon\eta(x)$ we have at once

$$\frac{d(\delta y)}{dx} = \varepsilon\eta' = \delta y'$$

i.e., taking the variation of a functional and differentiating with respect to the independent variable are commutative operations.

We can, of course, consider functionals of more than one function, and the variations of such functionals are defined by expressions analogous to Eq. (1). For instance, for the functional $F(x,u,v,u',v')$ we have

$$(2) \qquad \delta F = \frac{\partial F}{\partial u}\delta u + \frac{\partial F}{\partial v}\delta v + \frac{\partial F}{\partial u'}\delta u' + \frac{\partial F}{\partial v'}\delta v'$$

---

† By strict analogy with the differential of a function of three variables, we might have expected the definition

$$\delta F = \frac{\partial F}{\partial x}\delta x + \frac{\partial F}{\partial y}\delta y + \frac{\partial F}{\partial y'}\delta y'$$

However, we must remember that the functional is the value of $F(x,y,y')$ at a particular value of x, that is, x is not varied in the calculation of δF, and hence $\delta x = 0$.

Equally well, we may consider variations of functionals which depend on functions of more than one variable. For example, for the functional $F(x,y,u,u_x,u_y)$, whose value depends, for fixed x and y, on the function $u(x,y)$, we have

$$\delta F = \frac{\partial F}{\partial u}\,\delta u + \frac{\partial F}{\partial u_x}\,\delta u_x + \frac{\partial F}{\partial u_y}\,\delta u_y$$

For a functional expressed as a definite integral, say the integral

$$I(y) = \int_a^b F(x,y,y')\,dx$$

which we discussed in the last section, we have, first of all,

$$\Delta I = I(y + \varepsilon\eta) - I(y)$$

If the limits of I do not depend on y, we have further

$$\Delta I = \int_a^b F(x,\,y + \varepsilon\eta,\,y' + \varepsilon\eta')\,dx - \int_a^b F(x,y,y')\,dx$$

$$= \int_a^b [F(x,\,y + \varepsilon\eta,\,y' + \varepsilon\eta') - F(x,y,y')]\,dx$$

$$= \int_a^b \Delta F(x,\,y,y')\,dx$$

The variation of I is now defined as the expression resulting when ΔF in the last integral is replaced by the first-order approximation δF; that is,

$$\delta I = \int_a^b \delta F(x,y,y')\,dx$$

In Sec. 12.3 we showed that a necessary condition for a function to have an extremum is that its *differential* vanish. We can now show, similarly, that a necessary condition for the functional I to have an extremum is that its *variation* vanish. In fact, using the results of the preceding discussion, we can write

$$\delta I = \int_a^b \delta F(x,y,y')\,dx$$

$$= \int_a^b (F_y\,\delta y + F_{y'}\,\delta y')\,dx$$

$$= \int_a^b \left[F_y\,\delta y + F_{y'}\,\frac{d(\delta y)}{dx} \right] dx$$

Now integrating the last term by parts, with $u = F_{y'}$ and $dv = [d(\delta y)/dx]dx$, we have

$$\int_a^b F_{y'}\,\frac{d(\delta y)}{dx}\,dx = F_{y'}\,\delta y \bigg|_a^b - \int_a^b \frac{d(F_{y'})}{dx}\,\delta y\,dx$$

When we assume that $\delta y \equiv \varepsilon\eta(x)$ vanishes at $x = a$ and $x = b$ because of the usual conditions on $\eta(x)$ or else that $F_{y'}$ satisfies natural boundary conditions that make it vanish at these points, it follows that the integrated portion of the last expression is equal to zero. Hence we have

$$\delta I = \int_a^b \left[F_y - \frac{d(F_{y'})}{dx} \right] \delta y \, dx$$

Since we have already seen that $d(F_{y'})/dx - F_y = 0$ is a necessary condition for an extremum of I, it follows that δI is also zero at any extremum of I. Conversely, since δy is an arbitrary variation in y, the condition $\delta I = 0$ implies that

$$\frac{d(F_{y'})}{dx} - F_y = 0$$

EXERCISES

1 Given $F(x,y,y') = (y')^2 + xy$. Compute ΔF and δF for $x = x_0$, $y = x^2$, and $\delta y = \varepsilon x^n$. Then verify Eq. (1).

2 (a) If both y and x are functions of the independent variable t, show that

$$\delta \left(\frac{dy}{dx} \right) \neq \frac{d(\delta y)}{dx}$$

by showing that the correct formula is

$$\delta \left(\frac{dy}{dx} \right) = \frac{d(\delta y)}{dx} - \frac{dy}{dx} \frac{d(\delta x)}{dx}$$

(b) Let $y = 1 + x^2$, where both y and x are functions of t. Calculate $\delta \, (dy/dx)$ and $d(\delta y)/dx$ when $x = t^3$ and $\delta x = \varepsilon t^2$ and verify the formula of part (a).

3 Derive the formula for $\delta(F_1/F_2)$.

4 Derive the formula for $\delta(F^n)$.

5 Derive Eq. (2).

12.7 The Extrema of Integrals under Constraints

Just as with functions of several variables, so with functionals extrema are often required subject to one or more constraints. Consider, in particular, the integral

$$I = \int_a^b F(x,u,v,u',v') \, dx$$

subject to the constraint $\phi(u,v) = 0$. It may, of course, be possible to proceed explicitly by using the equation $\phi(u,v)$ to eliminate either u or v from the integrand of I before the variation of I is computed and set equal to zero. On the other hand, if we compute the variation of I with both u and v present in the integrand, we obtain, as in the last section, the condition

$$\delta I = \int_a^b \left\{ \left[\frac{\partial F}{\partial u} - \frac{d}{dx} \left(\frac{\partial F}{\partial u'} \right) \right] \delta u + \left[\frac{\partial F}{\partial v} - \frac{d}{dx} \left(\frac{\partial F}{\partial v'} \right) \right] \delta v \right\} dx = 0$$

Since u and v are not independent, it follows that δu and δy are not completely arbitrary and hence we cannot conclude that their coefficients in the integrand must be zero. However, we can proceed implicitly by computing the variation of ϕ, getting

$$\delta \phi = 0 = \phi_u \, \delta u + \phi_v \, \delta v$$

and then eliminating one of the variations $(\delta u, \delta v)$, say δv, getting

$$\delta I = \int_a^b \left\{ \left[\frac{\partial F}{\partial u} - \frac{d}{dx}\left(\frac{\partial F}{\partial u'}\right) \right] \delta u + \left[\frac{\partial F}{\partial v} - \frac{d}{dx}\left(\frac{\partial F}{\partial v'}\right) \right] \left(-\frac{\phi_u \, \delta u}{\phi_v} \right) \right\} dx$$

$$= \int_a^b \left\{ \phi_v \left[\frac{\partial F}{\partial u} - \frac{d}{dx}\left(\frac{\partial F}{\partial u'}\right) \right] - \phi_u \left[\frac{\partial F}{\partial v} - \frac{d}{dx}\left(\frac{\partial F}{\partial v'}\right) \right] \right\} \frac{\delta u}{\phi_v} \, dx = 0$$

Since δu can be chosen arbitrarily, the last equation implies the necessary condition

(1) $$\phi_v \left[\frac{\partial F}{\partial u} - \frac{d}{dx}\left(\frac{\partial F}{\partial u'}\right) \right] - \phi_u \left[\frac{\partial F}{\partial v} - \frac{d}{dx}\left(\frac{\partial F}{\partial v'}\right) \right] = 0$$

Finally, we may attempt to find the extrema of I by using the method of Lagrange's multipliers. To do this, we multiply $\delta\phi$ by a Lagrangian multiplier λ, which may here be a function of x, then integrate from a to b, getting

$$\int_a^b \lambda(\phi_u \, \delta u + \phi_v \, \delta v) \, dx = 0$$

Then we add this to the equation $\delta I = 0$ and combine the integrals, getting

$$\int_a^b \left\{ \left[\frac{\partial F}{\partial u} - \frac{d}{dx}\left(\frac{\partial F}{\partial u'}\right) + \lambda\phi_u \right] \delta u + \left[\frac{\partial F}{\partial v} - \frac{d}{dx}\left(\frac{\partial F}{\partial v'}\right) + \lambda\phi_v \right] \delta v \right\} dx = 0$$

This, of course, holds for all values of λ. Hence, in particular, if we choose λ so that the coefficient of δu is equal to zero, it follows that the coefficient of δv must be zero; i.e.,

$$\frac{\partial F}{\partial v} - \frac{d}{dx}\left(\frac{\partial F}{\partial v'}\right) + \lambda\phi_v = 0$$

Similarly, if λ is chosen so that the coefficient of δv is zero, it follows that the coefficient of δu must be zero; i.e.,

$$\frac{\partial F}{\partial u} - \frac{d}{dx}\left(\frac{\partial F}{\partial u'}\right) + \lambda\phi_u = 0$$

If λ is eliminated between these two necessary conditions, the resulting equation is simply the necessary equation [Eq. (1)] obtained by the implicit method.

Sometimes the constraint is given directly as a variational condition, say $f \, \delta u + g \, \delta v = 0$. There may or may not be a function ϕ for which this is the variation, but the above procedure still applies with f and g playing, respectively, the roles of ϕ_u and ϕ_v.

In many problems the constraint is given in the form of an integral. In particular, let us attempt to determine the extrema of

$$I = \int_a^b F(x, y, y') \, dx$$

subject to the condition

$$\int_a^b G(x, y, y') \, dx = K$$

where K is a constant. Assuming (as usual) that the problem has a solution, let $y = f(x)$ be the function for which I takes on its extreme value, subject to the requirements $y(a) = y_1$, $y(b) = y_2$, and the given constraint. Let $\eta(x)$ and $\xi(x)$ be arbitrary differentiable functions of x such that

$$\eta(a) = \eta(b) = \xi(a) = \xi(b) = 0$$

and let $f(x) + \alpha\eta(x) + \beta\xi(x)$ be a family of functions which satisfy the constraint condition; i.e., functions such that

(2) $$\int_a^b G(x, f + \alpha\eta + \beta\xi, f' + \alpha\eta' + \beta\xi')\, dx - K = 0$$

This equation is in effect a relation of the form $\phi(\alpha,\beta) = 0$ and, by the implicit function theorem, will define β as a function of α in the neighborhood of $\alpha = \beta = 0$, provided

$$\frac{\partial\phi}{\partial\beta} \equiv \int_a^b (G_y\xi + G_{y'}\xi')\, dx \neq 0 \qquad \text{for } \alpha = \beta = 0$$

It can be shown that unless $G_y - d(G_{y'})/dx$ is identically zero, ξ can always be chosen so that this is the case. Assuming that this has been done, the integral

$$I = \int_a^b F(x, y + \alpha\eta + \beta\xi, y' + \alpha\eta' + \beta\xi')\, dx$$

is a function of α for which $dI/d\alpha = 0$ is a necessary condition for an extremum. Thus we must have

$$\frac{dI}{d\alpha} = \int_a^b \left[F_y \cdot \left(\eta + \xi\,\frac{d\beta}{d\alpha} \right) + F_{y'} \cdot \left(\eta' + \xi'\,\frac{d\beta}{d\alpha} \right) \right] dx$$

(3) $$= \int_a^b \left[(\eta F_y + \eta' F_{y'}) + (\xi F_y + \xi' F_{y'})\,\frac{d\beta}{d\alpha} \right] dx = 0$$

for the values $\alpha = \beta = 0$, which we know are the values for which I has its extremum.

Now from the relation $\phi(\alpha,\beta) = 0$, that is, Eq. (2), we have

$$\frac{d\beta}{d\alpha} = -\frac{\partial\phi/\partial\alpha}{\partial\phi/\partial\beta} = -\frac{\int_a^b (\eta G_y + \eta' G_{y'})\, dx}{\int_a^b (\xi G_y + \xi' G_{y'})\, dx}$$

Evaluating this for $\alpha = \beta = 0$ and substituting into Eq. (3), we obtain

$$\int_a^b \left[(\eta F_y + \eta' F_{y'}) - (\xi F_y + \xi' F_{y'})\frac{\int_a^b (\eta G_y + \eta' G_{y'})\, dx}{\int_a^b (\xi G_y + \xi' G_{y'})\, dx} \right] dx$$

$$= \int_a^b (\eta F_y + \eta' F_{y'})\, dx - \frac{\int_a^b (\xi F_y + \xi' F_{y'})\, dx}{\int_a^b (\xi G_y + \xi' G_{y'})\, dx} \int_a^b (\eta G_y + \eta' G_{y'})\, dx = 0$$

Now

$$-\frac{\int_a^b (\xi F + \xi' F_{y'})\, dx}{\int_a^b (\xi G_y + \xi' G_{y'})\, dx}$$

is a constant, say ρ, since F and G are given functions, and $\xi(x)$ has been determined. Hence the extremizing condition becomes

(4)
$$\int_a^b \left[\eta(F_y + \rho G_y) + \eta'(F_{y'} + \rho G_{y'}) \right] dx = 0$$

If the second term in the integrand in Eq. (4) is integrated by parts with $u = F_{y'} + \rho G_{y'}$, and $dv = \eta' \, dx$, it becomes

$$\eta \left[F_{y'} + \rho G_{y'} \right]_a^b - \int_a^b \eta \frac{d}{dx} (F_{y'} + \rho G_{y'}) \, dx$$

The integrated term vanishes because $\eta(a) = \eta(b) = 0$. Hence Eq. (4) becomes

$$\int_a^b \eta \left[(F_y + \rho G_y) - \frac{d}{dx} (F_{y'} + \rho G_{y'}) \right] dx = 0$$

Finally, since η is arbitrary, we conclude that

$$(F_y + \rho G_y) - \frac{d}{dx} (F_{y'} + \rho G_{y'}) = 0$$

which is precisely the Euler equation for the function $F + \rho G$.

Thus, in summary, *to find the extremum of $\int_a^b F(x,y,y') \, dx$ subject to the constraint $\int_a^b G(x,y,y') \, dx = K$, find the extremum of $\int_a^b (F + \rho G) \, dx$ subject to no constraints. The Euler equation for this integral is a second-order differential equation whose solution will contain two arbitrary constants. These, with the Lagrangian multiplier ρ, just suffice to meet the boundary conditions $y(a) = y_1$ and $y(b) = y_2$ and the condition of constraint.*

EXAMPLE 1

Find the curve of prescribed length $2L$ which joins the points $(-a,b)$ and (a,b) and has its center of gravity as low as possible.

The y coordinate of the center of gravity of the required curve is given by

$$\frac{\int_{x=-a}^{x=a} y \, ds}{\int_{x=-a}^{x=a} ds} = \frac{1}{2L} \int_{-a}^a y\sqrt{1 + (y')^2} \, dx$$

since we are given the constraint equation $\int_{x=-a}^{x=a} ds = 2L$. Hence, introducing the Lagrange multiplier in the form $\rho/2L$, we are to minimize the integral

$$\frac{1}{2L} \int_{-a}^a [y\sqrt{1 + (y')^2} + \rho\sqrt{1 + (y')^2}] \, dx$$

In this case, since the integrand does not involve x, a first integral of the Euler equation is given by Eq. (6), Sec. 12.5, and we have

$$(y + \rho)\sqrt{1 + (y')^2} - \frac{(y + \rho)(y')^2}{\sqrt{1 + (y')^2}} = c \qquad \text{or} \qquad \frac{y + \rho}{\sqrt{1 + (y')^2}} = c$$

Solved for y', this becomes

$$y' = \frac{\sqrt{(y + \rho)^2 - c^2}}{c}$$

Separating variables and integrating, we have

$$\frac{dy}{\sqrt{(y + \rho)^2 - c^2}} = \frac{dx}{c}$$

$$\cosh^{-1} \frac{y + \rho}{c} = \frac{x}{c} + k$$

$$y + \rho = c \cosh \left(\frac{x}{c} + k \right)$$

Since the required curve must obviously be symmetric in the y axis, it follows that $k = 0$, and

$$y + \rho = c \cosh \frac{x}{c}$$

Furthermore, since the curve must pass through the point (a,b), we must also have

$$\rho = c \cosh \frac{a}{c} - b$$

To satisfy the prescribed length condition requires that

$$\int_{x=-a}^{x=a} ds = \int_{-a}^{a} \sqrt{1 + \sinh^2 \frac{x}{c}} \, dx = \int_{-a}^{a} \cosh \frac{x}{c} \, dx = 2L$$

whence

$$c \sinh \frac{x}{c} \bigg|_{-a}^{a} = 2c \sinh \frac{a}{c} = 2L$$

When c has been found from this equation (by some approximate numerical procedure), the value of ρ can be found and the equation of the required curve will be determined. In particular,

$$\rho = c \cosh \frac{a}{c} - b = c \sqrt{1 + \sinh^2 \frac{a}{c}} - b = \sqrt{c^2 + L^2} - b$$

and, finally,

$$y = c \cosh \frac{x}{a} + b - \sqrt{c^2 + L^2}$$

EXERCISES

1 What is the curve of given length which joins the points $(x_1,0)$ and $(x_2,0)$ and cuts off from the first quadrant the maximum area?

2 What is the curve of minimum length which passes through the points $(x_1,0)$ and $(x_2,0)$ and cuts off from the first quadrant a given area?

3 What is the curve of given length which joins the points $(-a,b)$ and (a,b) and generates the minimum surface area when it is revolved about the x axis?

4 Determine the curve which joins the points $(0,y_1)$ and (x_2,y_2), bounds with the lines $x = 0$ and $x = x_2$ and the x axis a prescribed area, and has the average value of the square of its slope as small as possible.

5 A particle is attracted toward the origin by a force whose magnitude is proportional to the distance from the origin. Assuming a coefficient of friction μ between the particle and its path, determine the path of a given length which joins the points (x_1,y_1) and (x_2,y_2) and along which the particle will move with the minimum work done against the force of friction.

6 A particle moves without friction on the surface $\phi(x,y,z) = 0$ from the point (x_1,y_1,z_1) to the point (x_2,y_2,z_2) in the time T. If the motion takes place in such a

way that the average value of the kinetic energy of the particle is a minimum, show that the parametric equations of its path on the surface satisfy the equations

$$\frac{\ddot{x}}{\phi_x} = \frac{\ddot{y}}{\phi_y} = \frac{\ddot{z}}{\phi_z}$$

Hint: Minimize the kinetic energy $\int_0^T \frac{1}{2}m(\dot{x}^2 + \dot{y}^2 + \dot{z}^2)\,dt$ subject to the constraint $\phi(x,y,z) = 0$ by using the obvious generalization of the result of Exercise 9, Sec. 12.5.

7 Using the results of Exercise 6, find the equations of the path of a particle moving from $(0,0,1)$ to $(0,0,-1)$ on the sphere $x^2 + y^2 + z^2 = 1$ if the average kinetic energy of the particle is a minimum.

8 Work Exercise 7 for motion on the cone $x^2 + y^2 = z^2$ from $(0,1,1)$ to $(1,0,-1)$.

12.8 Sturm-Liouville Problems

In Sec. 12.4 we noted the intimate relation between the stationary values of the ratio of two quadratic forms and the characteristic values of a related matric equation. A similar relation exists between the stationary values of the ratio of two integral functionals and the characteristic values of a related Sturm-Liouville boundary-value problem.

Putting aside for the moment the question of the physical significance of the integrals, let us attempt to find the stationary values of the ratio

$$\lambda = \frac{\int_a^b [r(y')^2 - qy^2]\,dx}{\int_a^b py^2\,dx} = \frac{I_1}{I_2}$$

where p, q, and r are known functions of x, and y is an unknown function to be determined. Using the results of Sec. 12.6, we have

$$\delta\lambda = \frac{I_2\,\delta I_1 - I_1\,\delta I_2}{I_2^{\,2}} = \frac{\delta I_1 - \lambda\,\delta I_2}{I_2}$$

$$\delta I_1 = \int_a^b (2ry'\,\delta y' - 2qy\,\delta y)\,dx \qquad \text{and} \qquad \delta I_2 = \int_a^b 2py\,\delta y\,dx$$

If we now integrate the first term in δI_1 by parts, with $u = ry'$ and $dv = \delta y'\,dx = [d(\delta y)/dx]\,dx$, we obtain

$$\int_a^b 2ry'\,\delta y'\,dx = 2ry'\,\delta y\,\Big|_a^b - 2\int_a^b (ry')'\,\delta y\,dx$$

When we assume either that $\delta y = 0$ at $x = a$ and $x = b$ or that y satisfies the natural boundary condition $ry' = 0$ at a and at b, this reduces to

$$-2\int_a^b (ry')'\,\delta y\,dx$$

and we have

$$\delta\lambda = -\frac{2}{I_2}\int_a^b [(ry')' + qy + \lambda py]\,\delta y\,dx = 0$$

from which, since δy is arbitrary, we conclude that

(1) $(ry')' + (q + \lambda p)y = 0$

Thus, to find the stationary values of the ratio λ, we must solve an associated Sturm-Liouville equation with the appropriate boundary conditions.

When we investigated the corresponding algebraic problem in Sec. 12.4, we found that the characteristic values of the associated matric equation were in fact the stationary values of the ratio $\lambda = X^T A X / X^T B X$, and an analogous result is true in the present case. To verify this, let λ_k be a characteristic value of Eq. (1) with corresponding characteristic function y_k. Then for this function, the ratio λ becomes

$$\frac{\int_a^b [r(y_k')^2 - q y_k^2] \, dx}{\int_a^b p y_k^2 \, dx}$$

If we integrate the first term in the numerator by parts with $u = r y_k'$ and $dv = y_k' \, dx$, we have

$$\int_a^b r(y_k')^2 \, dx = y_k(r y_k') \Big|_a^b - \int_a^b y_k(r y_k')' \, dx$$

Under the assumed boundary conditions, the integrated term vanishes. Hence the ratio becomes

$$\frac{\int_b^a [-y_k(r y_k')' - q y_k^2] \, dx}{\int_b^a p y_k^2 \, dx}$$

Now since y_k is a solution of Eq. (1) when $\lambda = \lambda_k$, it follows that

$$-(r y_k')' = (q + \lambda_k p) y_k$$

Therefore, substituting, we have for the ratio

$$\frac{\int_b^a [y_k(q + \lambda_k p) y_k - q y_k^2] \, dx}{\int_b^a p y_k^2 \, dx} = \lambda_k$$

as asserted.

In Sec. 12.4 we obtained the ratio $\lambda = X^T A X / X^T B X$ in particular cases by equating the maximum potential energy of a system to the maximum kinetic energy of the system, and the ratio we have been considering in this section also arises in this fashion. To illustrate, let us consider a string of weight per unit length $w(x)$ stretched under tension T between two points a distance l apart and vibrating transversely in a plane. If the change of length of the string during its motion is so small that T can be assumed constant, the potential energy stored in the string by virtue of work done against T is equal to T times the change of length in the string:

$$T \int_0^l [\sqrt{1 + (y')^2} - 1] \, dx$$

If, further, the vibrations are such that $|y'| \ll 1$, then by expanding $\sqrt{1 + (y')^2}$ by the binomial expansion and retaining only the dominant term in the integrand, we have

$$\frac{T}{2} \int_0^l (y')^2 \, dx$$

as the expression for the potential energy stored in the string by virtue of its elongation. If the string is acted upon by a distributed force whose magnitude per unit length is

$f(x)y$, for example an elastic restoring force $-ky$, then there is also potential energy stored in the string by virtue of work done against this force, and its amount is

$$-\int_0^l \tfrac{1}{2} f(x) y^2 \, dx$$

Thus the potential energy instantaneously stored in the string is

$$\frac{1}{2} \int_0^l [T(y')^2 - f(x)y^2] \, dx$$

Similarly, the total instantaneous kinetic energy of the string is

$$\frac{1}{2} \int_0^l \frac{w(x)}{g} (\dot{y})^2 \, dx$$

The displacement y which we have been considering is actually a function of x and t of the form $y = X(x) \cos \omega t$. Hence, substituting into the two energy expressions and equating their maximum values, we obtain

$$\omega^2 = \frac{\int_0^l [T(X')^2 - f(x)X^2] \, dx}{\int_0^l \frac{w(x)}{g} X^2 \, dx}$$

With the obvious boundary conditions $X(0) = X(l) = 0$, this is in all respects an example of the ratio we have been considering in this section. Moreover, in this case the extremizing equation (1), namely,

$$\frac{d(TX')}{dx} + \left[f(x) + \omega^2 \frac{w(x)}{g} \right] X = 0$$

is precisely the equation for the space factor X which results when the partial differential equation governing the vibrations of a nonuniform string [Eq. (2), Sec. 8.2] is solved by the method of separation of variables.

Thus far, our discussion of the relation between extremum problems and Sturm-Liouville boundary-value problems has not revealed how the more general boundary conditions

(2)
$$a_1 X(a) - a_2 X'(a) = 0$$
$$b_1 X(b) - b_2 X'(b) = 0$$

arise. To investigate this, let us reconsider the vibrating string under the assumption that instead of being fixed, each end is restrained by a noninfinitesimal restoring force ϕ proportional to the displacement, say $\phi(0) = ay_0$ and $\phi(l) = by_l$. The potential energy stored in the system by virtue of work done against these forces is

$$\tfrac{1}{2} a y_0^2 + \tfrac{1}{2} b y_l^2$$

and hence the total potential energy of the system is now

$$\frac{1}{2} \int_0^l [T(y')^2 - f(x)y^2] \, dx + \tfrac{1}{2} a y_0^2 + \tfrac{1}{2} b y_l^2$$

To incorporate the last two terms into the integral, let $k(x)$ be an arbitrary differentiable function of x such that $k(0) = -a$ and $k(l) = b$. Then

$$\int_0^l \frac{d}{dx} [k(x) y^2] \, dx = k(x) y^2 \Big|_0^l = k(l) y^2(l) - k(0) y^2(0)$$

$$= by_l^2 + ay_0^2$$

Hence the expression for the instantaneous potential energy of the string can be rewritten in the form

$$\frac{1}{2} \int_0^l \left\{ T(y')^2 - f(x) y^2 + \frac{d}{dx} [k(x) y^2] \right\} dx$$

Again setting $y = X(x) \cos \omega t$ and equating the maximum values of the potential energy and the kinetic energy, we obtain, as the ratio to be considered,

$$\omega^2 = \frac{\displaystyle\int_0^l \left\{ T(X')^2 - f(x) X^2 + \frac{d}{dx} [k(x) X^2] \right\} dx}{\displaystyle\int_0^l \frac{w(x)}{g} X^2 \, dx}$$

Now the natural boundary condition $\partial F/\partial X' = 0$ becomes

$$2TX' + 2k(x)X = 0$$

which, evaluated at $x = 0$ and at $x = l$, yields conditions on X of the general form (2).

Not only is a variational problem of the form

$$\delta I \equiv \delta \int_a^b F(x, y, y') \, dx$$

equivalent to a related differential equation with suitable boundary conditions, but, conversely, in a large class of cases a boundary-value problem in differential equations is equivalent to a variational problem. Consider, for example, the differential equation

$$x^3 y'' + 3x^2 y' + y = x$$

or

$$(x^3 y')' + y - x = 0$$

If we multiply this by δy and integrate from a to b, we obtain

(3)
$$\int_a^b [(x^3 y')' + y - x] \, \delta y \, dx = 0$$

This, of course, is not yet a variational problem since it is not of the form

$$\int_a^b \delta F \, dx \equiv \delta \int_a^b F \, dx = 0$$

In an attempt to convert it into such a form, let us integrate the first term by parts, with $u = \delta y$ and $dv = (x^3 y')' dx$. This gives us

$$\int_a^b (x^3 y')' \, \delta y \, dx = (x^3 y') \, \delta y \Big|_a^b - \int_a^b (x^3 y') \, \delta y' \, dx$$

Therefore, since $x^3y' \, \delta y' = \delta[\frac{1}{2}x^3(y')^2]$, $y \, \delta y = \delta(\frac{1}{2}y^2)$, and $x \, \delta y = \delta(xy)$, it follows that the integral (3) can be written in the form

$$x^3y' \, \delta y \bigg|_a^b - \int_a^b \delta[\frac{1}{2}x^3(y')^2 - \frac{1}{2}y^2 + xy] \, dx = 0$$

or

$$\delta \int_a^b [\frac{1}{2}x^3(y')^2 - \frac{1}{2}y^2 + xy] \, dx - x^3y' \, \delta y \bigg|_a^b = 0$$

Thus, if y satisfies boundary conditions which make the integrated term in the last equation equal to zero, the original differential equation is equivalent to the variational problem

$$\delta \int_a^b [\frac{1}{2}x^3(y')^2 - \frac{1}{2}y^2 + xy] \, dx = 0$$

Exercise 2 makes it clear that not every linear second-order differential equation can be converted into a variational problem without preliminary changes in the form of the differential equation. Exercise 3, however, shows that after multiplication by a suitable factor, if necessary, every such equation whose solutions satisfy the appropriate boundary conditions is equivalent to a variational problem.

EXERCISES

1 Verify that finding the extremum of the ratio

$$\frac{\int_a^b [r(y')^2 - qy^2] \, dx}{\int_a^b py^2 \, dx}$$

is equivalent to finding the extremum of $\int_a^b [r(y')^2 - qy^2] \, dx$ subject to the constraint $\int_a^b py^2 \, dx = k$.

2 Can the equation $xy'' + 2y' + y - 1 = 0$ with boundary conditions $y(a) = y_1$, $y(b) = y_2$ be converted into a variational problem? Can the equation

$$x^2y'' + 2xy' + xy - x = 0$$

with the same boundary conditions be converted into a variational problem?

3 Show that a second-order differential equation

$$py'' + qy' + ry - \lambda y = 0$$

can always be put into a form which can be converted into a variational problem by multiplying it by $(1/p) \exp [\int (q/p) \, dx]$ provided suitable boundary conditions are assumed. What are these boundary conditions? What is the equivalent variational problem?

4 By multiplying the first equation by δy and the second equation by δz, then adding and integrating, show that the system

$$(p_1y')' + (p_2z')' + q_1y + q_2z = 0$$
$$(p_2y')' + (p_3z')' + q_2y + q_3z = 0$$

can be converted into a variational problem if appropriate boundary conditions are satisfied at $x = a$ and $x = b$. What is the variational problem, and what are the necessary boundary conditions?

5 Show that if suitable boundary conditions are satisfied at $x = a$ and $x = b$, the fourth-order equation

$$(py'')'' + (qy')' + ry + \lambda y = 0$$

can be converted into a variational problem. What is the variational problem, and what are the necessary boundary conditions?

6 Convert the boundary-value problem $xy'' + 2y' + y - 1 = 0$, $y'(0) = 0$, $y(1) = 0$, into a variational problem and find the function of the family $y = a(1 - x^2)$ which best approximates the minimizing function.

7 Work Exercise 6 using the most general quadratic function which satisfies the boundary conditions $y(0) = 0$, $y'(1) = 0$.

8 Convert the boundary-value problem $y'' + \lambda^2 y = 0$, $y(0) = y(\pi) = 0$, into a variational problem and approximate the smallest characteristic value by finding the value of λ for which the variational problem has a nontrivial solution of the form $y = ax(\pi - x)$. How does this approximate value of λ_1 compare with the first characteristic value of the original boundary-value problem?

9 Work Exercise 8 using the most general quadratic function which satisfies the boundary conditions $y(0) = y'(\pi) = 0$.

10 In Exercise 8, approximate the first two characteristic values by finding the values of λ for which the variational problem has a nontrivial solution of the form $y = ax(\pi - x) + bx^2(\pi - x)$.†

12.9 Hamilton's Principle and Lagrange's Equation

Although Newton's law, $F = ma$, suffices for the formulation of many problems in dynamics, there are refinements and extensions which often provide more effective methods of attack. In this section we shall undertake a brief investigation of two of these, Hamilton's principle and Lagrange's equation(s).

Let us consider first a mass particle moving in a force field $\mathbf{F}$. Let $\mathbf{r}(t)$ be the vector‡ from the origin to the instantaneous position of the particle. Then according to Newton's law in vector form, the actual path of the particle is described by the equation

$$m \frac{d^2\mathbf{r}}{dt^2} - \mathbf{F} = 0$$

Now consider any other path joining the points where the particle is located at $t = t_1$ and at $t = t_2$. Such a path is, of course, described by the vector function $\mathbf{r} + \delta\mathbf{r}$, where

$$\delta\mathbf{r}|_{t_1} = \delta\mathbf{r}|_{t_2} = 0$$

(Fig. 12.2). If we now form the scalar product of (the vector) $\delta\mathbf{r}$ and the terms in (the vector form of) Newton's law, and integrate from t_1 to t_2, we obtain

(1)
$$\int_{t_1}^{t_2} (m\ddot{\mathbf{r}} \cdot \delta\mathbf{r} - \mathbf{F} \cdot \delta\mathbf{r}) \, dt = 0$$

Applying integration by parts to the first term in Eq. (1), with $\mathbf{u} = \delta\mathbf{r}$ and $dv = \ddot{\mathbf{r}} \, dt$, we obtain

$$m\dot{\mathbf{r}} \cdot \delta\mathbf{r} \Big|_{t_1}^{t_2} - m \int_{t_1}^{t_2} \dot{\mathbf{r}} \cdot \delta\dot{\mathbf{r}} \, dt$$

---

† Exercises 8 to 10 illustrate the so-called **Rayleigh-Ritz method** of approximating characteristic values by finding solutions of a variational problem in a restricted family of functions satisfying the boundary conditions of the actual problem. The method is named for the English mathematical physicist Lord Rayleigh (1842–1919) and the Swiss mathematical physicist Walther Ritz (1878–1909).

‡ The relatively few concepts from vector analysis which we need in this section can be found in Secs. 13.1 to 13.3.

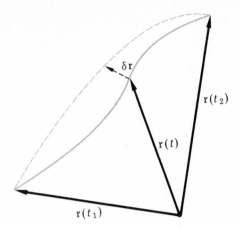

Figure 12.2
The actual path of a moving particle and
a possible variation of that path.

The integrated term vanishes because of the properties of $\delta \mathbf{r}$. Moreover,

$$m\dot{\mathbf{r}} \cdot \delta \dot{\mathbf{r}} = \frac{m}{2} \delta(\dot{\mathbf{r}} \cdot \dot{\mathbf{r}}) = \delta \left(\frac{m}{2} v^2 \right) = \delta T$$

where T is the kinetic energy of the moving particle. Hence Eq. (1) can be rewritten

(2)
$$\int_{t_1}^{t_2} (\delta T + \mathbf{F} \cdot \delta \mathbf{r}) \, dt = 0$$

This is **Hamilton's principle**† in its general form, as applied to the motion of a single mass particle in a force field that can be either conservative or nonconservative.

If $\mathbf{F}$ is conservative, Hamilton's principle assumes an even simpler form. For if $\mathbf{F}$ is conservative, then there exists a scalar function $\phi(x, y, z)$ such that $\mathbf{F} \cdot \delta \mathbf{r} = d\phi$ or, equivalently, $\mathbf{F} = \nabla \phi$.‡ The function ϕ is called the **potential function**, and $-\phi$ is (to within an additive constant) the potential energy of the particle in the field. Now

$$\mathbf{F} = \nabla \phi = \frac{\partial \phi}{\partial x} \mathbf{i} + \frac{\partial \phi}{\partial y} \mathbf{j} + \frac{\partial \phi}{\partial z} \mathbf{k}$$

and
$$\delta \mathbf{r} = \delta x \, \mathbf{i} + \delta y \, \mathbf{j} + \delta z \, \mathbf{k}$$

Hence

$$\mathbf{F} \cdot \delta \mathbf{r} = \frac{\partial \phi}{\partial x} \delta x + \frac{\partial \phi}{\partial y} \delta y + \frac{\partial \phi}{\partial z} \delta z = \delta \phi$$

and therefore the equation $\int_{t_1}^{t_2} (\delta T + \mathbf{F} \cdot \delta \mathbf{r}) \, dt = 0$ can be rewritten

$$\int_{t_1}^{t_2} (\delta T + \delta \phi) \, dt = 0 \qquad \text{or} \qquad \delta \int_{t_1}^{t_2} (T + \phi) \, dt = 0$$

or, finally, since $\phi = -V$, where V is the potential energy of the system,

(3)
$$\delta \int_{t_1}^{t_2} (T - V) \, dt = 0$$

---

† Named for the Irish mathematician W. R. Hamilton (1805–1865).
‡ See Theorem 6, Sec. 13.5.

This is Hamilton's principle for a single mass particle in a conservative field. The function $T - V$ is usually referred to as the **Lagrangian function** or the **kinetic potential**. Hamilton's principle can, of course, be extended to a system of discrete particles by summation and to continuous systems by integration.

In many elementary problems, a dynamical system is described in terms of coordinates which are distances. This is not necessarily the case, however. For instance, problems are often formulated in terms of polar coordinates, one of which is a distance and one of which is an angle. In more advanced applications, **generalized coordinates**, such as angles and areas as well as distances, are common, and the expressions for F, T, and V in the various forms of Hamilton's principle will often involve coordinates of several different kinds.

In general, the kinetic energy T of a system will depend not only on the **generalized velocities**, i.e., the time derivatives of the generalized coordinates, but also the generalized coordinates themselves. Similarly, the potential energy V may involve the generalized velocities as well as the generalized coordinates. However, in a conservative system, in which, by definition, the potential energy depends only on position, V will not involve the generalized velocities.

In a conservative system, the work done by the various forces when the generalized coordinates $\{q_i\}$ are given small changes $\{\delta q_i\}$ is

$$\delta\phi = -\delta V = -\left(\frac{\partial V}{\partial q_1}\delta q_1 + \frac{\partial V}{\partial q_2}\delta q_2 + \cdots + \frac{\partial V}{\partial q_n}\delta q_n\right)$$

$$= Q_1\,\delta q_1 + Q_2\,\delta q_2 + \cdots + Q_n\,\delta q_n$$

where we have introduced the conventional symbol Q_i for $-\partial V/\partial q_i$. The typical term in this expression, $Q_i\,\delta q_i$, is the work done in a displacement in which δq_i is different from zero but all other δq's are equal to zero. Since q_i is not necessarily a distance, Q_i is not necessarily a force. Nonetheless, the Q's are referred to as **generalized forces**.

The Euler equations for the integral (3), namely,

(4) $$\frac{d}{dt}\left[\frac{\partial(T - V)}{\partial\dot{q}_i}\right] - \frac{\partial(T - V)}{\partial q_i} = 0 \qquad i = 1, 2, \ldots, n$$

are known as **Lagrange's equations**. In a conservative system, since the potential energy V is independent of the generalized velocities, Eq. (4) can be simplified to

(4a) $$\frac{d}{dt}\left(\frac{\partial T}{\partial\dot{q}_i}\right) - \frac{\partial T}{\partial q_i} + \frac{\partial V}{\partial q_i} = 0 \qquad i = 1, 2, \ldots, n$$

Moreover, in a conservative system $\partial V/\partial q_i = -Q_i$, and so Eq. (4a) can also be written

(5) $$\frac{d}{dt}\left(\frac{\partial T}{\partial\dot{q}_i}\right) - \frac{\partial T}{\partial q_i} = Q_i \qquad i = 1, 2, \ldots, n$$

Although we shall not prove the fact, this equation is also correct for nonconservative systems, the only difference being that in a nonconservative system the generalized forces cannot be derived from a potential function.

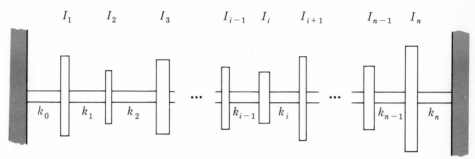

Figure 12.3
A system of elastically coupled disks.

EXAMPLE 1

Using Lagrange's equations, derive the system of differential equations describing the torsional vibrations of the system of elastically coupled disks shown in Fig. 12.3.

Taking as coordinates the angles of twist θ_i of the respective disks, we have at once

$$T = \tfrac{1}{2}(I_1\dot\theta_1{}^2 + I_2\dot\theta_2{}^2 + \cdots + I_i\dot\theta_i{}^2 + \cdots + I_n\dot\theta_n{}^2)$$

Also, since the potential energy stored in a twisted shaft is

$$\tfrac{1}{2}\ \text{modulus} \times (\text{angle of twist})^2$$

we have

$$V = \tfrac{1}{2}[k_0\theta_1{}^2 + k_1(\theta_1 - \theta_2)^2 + \cdots + k_{i-1}(\theta_{i-1} - \theta_i)^2$$
$$+ k_i(\theta_i - \theta_{i+1})^2 + \cdots + k_{n-1}(\theta_{n-1} - \theta_n)^2 + k_n\theta_n{}^2]$$

Hence, Lagrange's equations are

$$I_1\ddot\theta_1 + k_0\theta_1 + k_1(\theta_1 - \theta_2) = 0$$
$$I_2\ddot\theta_2 + k_1(\theta_1 - \theta_2)(-1) + k_2(\theta_2 - \theta_3) = 0$$
$$\dots\dots\dots\dots\dots\dots\dots\dots\dots\dots\dots\dots\dots\dots\dots\dots$$
$$I_i\ddot\theta_i + k_{i-1}(\theta_{i-1} - \theta_i)(-1) + k_i(\theta_i - \theta_{i+1}) = 0$$
$$\dots\dots\dots\dots\dots\dots\dots\dots\dots\dots\dots\dots\dots\dots\dots\dots$$
$$I_n\ddot\theta_n + k_{n-1}(\theta_{n-1} - \theta_n) + k_n\theta_n = 0$$

These equations are easy to set up by elementary methods based on Newton's law in torsional form, but the use of Lagrange's equations eliminates the need to check the signs of the various torques, which is sometimes a troublesome detail.

EXERCISES

1 Use Lagrange's equation to set up the differential equation of a freely falling body.
2 Use Lagrange's equation to set up the differential equation of a spring-suspended mass.
3 A mass m hangs from a spring of modulus k. If the point of suspension moves vertically according to the law $s = a \sin \omega t$, use Lagrange's equation to set up the differential equation describing the motion of the weight, and solve the equation if the point of suspension begins its motion when the mass is at rest in its equilibrium position.
4 A pendulum consists of a mass m suspended from a rigid but weightless rod of length l. If the point of suspension of the pendulum moves horizontally according to the law $s = a \sin \omega t$, set up the expressions for T and V in terms of s and the inclination angle θ of the pendulum. Then use Lagrange's equation to obtain the differential equation describing the motion of the pendulum. What appears to be the limiting behavior of the pendulum as $\omega \to \infty$?

5   In Example 1, Sec. 11.3, use Lagrange's equations to obtain the differential equations describing the motion of the system in terms of the normal coordinates y_1, y_2, y_3.

6   A particle of mass m slides without friction on a straight rod which passes through the origin and rotates with constant velocity about the z axis in such a way that it describes a right circular cone of vertex angle α. Use Lagrange's equation to set up the equation of motion of the particle, and find the solution corresponding to the initial conditions $r = r_0, \dot{r}_0 = 0$, where r is the distance of the particle from the vertex of the cone.

7   The point of suspension of a simple pendulum can slide without friction along a horizontal line against elastic restoring forces, as shown in Fig. 12.4a. Letting x

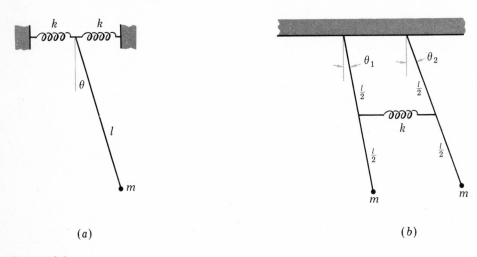

(a)   (b)

Figure 12.4

be the horizontal displacement of the point of suspension and θ be the angle the pendulum makes with the vertical, determine the kinetic and potential energies of the system. Using Lagrange's equations, obtain the equations of motion of the system. Under the assumption of small vibrations, what are the natural frequencies of the system?

8   Assuming that the spring remains horizontal, use Lagrange's equations to set up the differential equations describing the motion of the coupled pendulums shown in Fig. 12.4b. Assuming small vibrations, what are the natural frequencies of the system?

9   A particle moves without friction in the xy plane under such conditions that its potential energy depends only on its distance from the origin. Using Lagrange's equations, set up the differential equations describing its motion, and deduce Kepler's law: *the radius vector from the origin to the particle sweeps out area at a constant rate.*

CHAPTER 13
Vector Analysis

13.1 The Algebra of Vectors

In Sec. 10.2, in our discussion of determinants and matrices, we introduced the concept of a vector as an ordered set of n quantities, say $(a_1, a_2, \ldots, a_n)$. In the present chapter we shall undertake the study of what is known as **vector analysis**, using the more traditional (and limited) definition of a **vector** as a quantity, e.g., force, velocity, or acceleration, which possesses both magnitude and direction. Although this approach has been all but abandoned in pure mathematics because it is unnecessarily restricted, it is still the usual approach in physics and engineering and the results to which it leads are extremely useful in these fields.

Almost any physical discussion will involve, in addition to vector quantities, other quantities, e.g., volume, mass, and work, which possess only magnitude and are known as **scalars**. To distinguish vectors from scalars we shall consistently write the former in boldface type **V**. This is a common notation; in writing by hand a vector quantity is indicated by putting an arrow above the symbol.

A scalar quantity can be adequately represented by a mark on a fixed scale. To represent a vector quantity, however, we must use a directed line segment whose direction is the same as the direction of the vector and whose length is equal (on some convenient scale) to the magnitude of the vector. For convenience, we shall often refer to the representative line segment as though it were the vector itself. The magnitude or length of a vector **A** is called the **absolute value** of the vector and is indicated either by enclosing the symbol for the vector between ordinary absolute-value bars or simply by setting the symbol for the vector in ordinary rather than boldface type. Thus,

$$A = |\mathbf{A}|$$

represents the magnitude, or absolute value, of the vector **A**. Regardless of its direction, a vector whose length, or absolute value, is unity is called a **unit vector**. A vector is said to be **zero** if and only if its absolute value is zero.† The direction of a zero vector is undefined.

Two vectors whose magnitudes, or lengths, are equal and whose directions are the same are said to be **equal**, regardless of the points in space from which they may be

---

† We shall denote zero vectors, as we did zero matrices in Chaps. 10 and 11, by a zero set in boldface type, thus **0**.

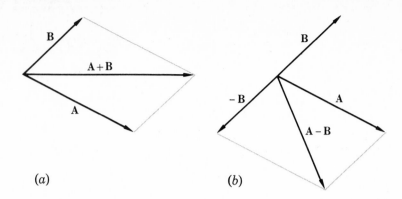

(a) (b)

Figure 13.1
The addition and subtraction of vectors.

drawn.† If two vectors have the same length but are oppositely directed, either is said to be the **negative** of the other.

The **sum** of two vectors **A** and **B** is defined by the familiar parallelogram law; i.e., if **A** and **B** are drawn from the same point, or origin, and if the parallelogram having **A** and **B** as adjacent sides is constructed, then the sum **A** + **B** is the vector represented by the diagonal of the parallelogram which passes through the common origin of **A** and **B** (Fig. 13.1a). From this definition it is evident that

$$\mathbf{A} + \mathbf{B} = \mathbf{B} + \mathbf{A}$$

i.e., that *vector addition is commutative*, and that

$$\mathbf{A} + (\mathbf{B} + \mathbf{C}) = (\mathbf{A} + \mathbf{B}) + \mathbf{C}$$

i.e., that *vector addition is associative*. By the **difference** of vectors **A** and **B**, we mean the sum of the first and the negative of the second; i.e.,

$$\mathbf{A} - \mathbf{B} = \mathbf{A} + (-\mathbf{B})$$

(Fig. 13.1b). By the product of a scalar a and a vector **A** we mean the vector $a\mathbf{A} = \mathbf{A}a$ whose length is equal to the product of $|a|$ and the magnitude of **A** and whose direction is the same as the direction of **A** if a is positive and opposite to it if a is negative.

In addition to the product of a scalar and a vector, two other types of products are defined in vector analysis. The first of these is the **scalar** or **dot** or **inner product**, indicated by placing a dot between the two factors. By definition, this is a *scalar* equal to the product of the absolute values of the two vector factors and the cosine of the angle between their (positive) directions; i.e.,

(1) $$\mathbf{A} \cdot \mathbf{B} = |\mathbf{A}|\,|\mathbf{B}|\cos\theta = AB\cos\theta$$

---

† In other words, a vector quantity can be represented equally well by any of infinitely many equivalent line segments, all having the same length and the same direction. It is therefore customary to say that *a vector can be moved parallel to itself without change*. In some applications, however, e.g., in dealing with forces whose points of application or lines of action cannot be shifted, it is necessary to think of a vector as fixed or at least limited in position. Such vectors are usually said to be **bound**, in contrast to unrestricted vectors, which are said to be **free**.

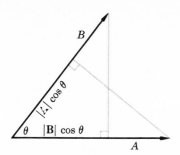

Figure 13.2
The geometrical interpretation of the scalar product.

Since $|\mathbf{A}|\cos\theta$ is just the projection of the vector $\mathbf{A}$ in the direction of $\mathbf{B}$, and since $|\mathbf{B}|\cos\theta$ is the projection of the vector $\mathbf{B}$ in the direction of $\mathbf{A}$, it follows that *the dot product of two vectors is equal to the length of either of them multiplied by the projection of the other upon it* (Fig. 13.2). Two particular cases of this are worthy of note. If one of the vectors, say $\mathbf{A}$, is of unit length, then $\mathbf{A}\cdot\mathbf{B}$ becomes simply

$$|\mathbf{B}|\cos\theta = B\cos\theta$$

which is just the projection, or component, of $\mathbf{B}$ in the direction of the unit vector $\mathbf{A}$. On the other hand, if $\mathbf{B} = \mathbf{A}$, then, obviously,

$$(2) \qquad \mathbf{A}\cdot\mathbf{A} = |\mathbf{A}|^2 = A^2$$

From the relation between dot products and projections it is easy to show that *dot multiplication is distributive over addition*; i.e.,

$$(3) \qquad \mathbf{A}\cdot(\mathbf{B} + \mathbf{C}) = \mathbf{A}\cdot\mathbf{B} + \mathbf{A}\cdot\mathbf{C}$$

Moreover, from the definitive relation (1) it is clear that *dot multiplication is commutative*; i.e.,

$$(4) \qquad \mathbf{A}\cdot\mathbf{B} = \mathbf{B}\cdot\mathbf{A}$$

However, if the dot product of two vectors is zero, it does not follow that one or the other of the factors is zero, for there is a third possibility, namely, $\cos\theta = 0$. Thus, *if $\mathbf{A}\cdot\mathbf{B} = 0$, then either at least one of the vectors $(\mathbf{A},\mathbf{B})$ is zero or $\mathbf{A}$ and $\mathbf{B}$ are perpendicular.*

The third type of product with which we shall deal is the **vector**, or **cross, product**, indicated by putting a cross between the factors.† If $\mathbf{A}$ and $\mathbf{B}$ are the factors, then by definition $\mathbf{A} \times \mathbf{B}$ is a *vector* $\mathbf{V}$ whose absolute value is the product of the absolute values of $\mathbf{A}$, $\mathbf{B}$ and the sine of the angle between them and whose direction is perpendicular to the plane determined by $\mathbf{A}$ and $\mathbf{B}$ and so sensed that a right-handed screw turned from $\mathbf{A}$ toward $\mathbf{B}$ through the smaller of the angles determined by these vectors would advance in the direction of $\mathbf{V}$ (Fig. 13.3*a*). Since $|\mathbf{B}|\,|\sin\theta|$ is the

---

† Meaning has also been given to the symbol **AB**, and in fact under the name **dyad** such combinations have been extensively studied, for instance, by J. W. Gibbs and E. B. Wilson, "Vector Analysis," Yale University Press, New Haven, Conn., 1929. We shall not consider them in our work, however, since they are actually special cases of what are known as *tensors*, which we shall consider from a somewhat different point of view in the next chapter. For us, the only possible product combinations of two vectors will be the dot and cross products themselves.

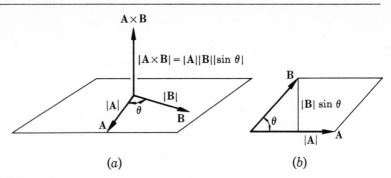

Figure 13.3
The geometrical interpretation of the vector product.

projection of **B** in the direction perpendicular to **A** or, in other words, is the altitude of the parallelogram determined when **A** and **B** are drawn from a common point, it follows that the magnitude of **A** × **B**, namely, $|\mathbf{A}|(|\mathbf{B}| \ |\sin \theta|)$, is equal to the area of this parallelogram (Fig. 13.3*b*).

From the relation between cross products and areas it is easy to show that *cross multiplication is distributive over addition*; i.e.,

$$(5) \qquad \mathbf{A} \times (\mathbf{B} + \mathbf{C}) = \mathbf{A} \times \mathbf{B} + \mathbf{A} \times \mathbf{C}$$

However, since the direction of **A** × **B** is determined by the right-hand rule, it is clear that interchanging **A** and **B** reverses the direction, or sign, of their product. Hence *cross multiplication is not commutative*, and we have, in fact,

$$(6) \qquad \mathbf{A} \times \mathbf{B} = -\mathbf{B} \times \mathbf{A}$$

Multiplication in which products obey this rule is sometimes said to be **anticommutative**.

From the foregoing it is clear that we must be careful to preserve the proper order of factors in any expression involving vector multiplication. Moreover, if **A** × **B** = **0**, we cannot conclude that either **A** or **B** is zero, for this product will also vanish if $\sin \theta = 0$. Hence *if* **A** × **B** = **0**, *then either at least one of the vectors* **A**, **B** *is zero or* **A** *and* **B** *are parallel.*

It is often convenient to be able to refer vector expressions to a cartesian frame of reference. To provide for this we define **i**, **j**, and **k** to be vectors of unit length directed, respectively, along the positive *x*, *y*, and *z* axes of a right-handed rectangular coordinate system. Then *x***i**, *y***j**, and *z***k** represent vectors of lengths *x*, *y*, and *z* whose directions are those of the respective axes; and from the definition of vector addition it is evident that the vector joining the origin to a general point $P:(x,y,z)$ (Fig. 13.4) can be written

$$(7) \qquad \mathbf{R} = x\mathbf{i} + y\mathbf{j} + z\mathbf{k}$$

In more general terms, any vector whose components along the coordinate axes are, respectively, a_1, a_2, and a_3 can be written

$$\mathbf{A} = a_1\mathbf{i} + a_2\mathbf{j} + a_3\mathbf{k}$$

If, further,

$$\mathbf{B} = b_1\mathbf{i} + b_2\mathbf{j} + b_3\mathbf{k}$$

then

$$\mathbf{A} \pm \mathbf{B} = (a_1 \pm b_1)\mathbf{i} + (a_2 \pm b_2)\mathbf{j} + (a_3 \pm b_3)\mathbf{k}$$

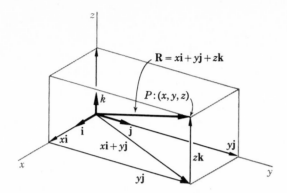

Figure 13.4
The representation of a vector as a linear combination of the unit vectors **i**, **j**, **k**.

Clearly, *two vectors will be equal if and only if their respective components are equal.* Hence any vector equation implies three scalar equations.

Since the dot product of perpendicular vectors is zero, it follows that

$$(8) \qquad\qquad \mathbf{i}\cdot\mathbf{j} = \mathbf{j}\cdot\mathbf{k} = \mathbf{k}\cdot\mathbf{i} = 0$$

Moreover, applying (2) to the unit vectors **i**, **j**, **k**, we have

$$(9) \qquad\qquad \mathbf{i}\cdot\mathbf{i} = \mathbf{j}\cdot\mathbf{j} = \mathbf{k}\cdot\mathbf{k} = 1$$

Hence, if we write

$$\mathbf{A}\cdot\mathbf{B} = (a_1\mathbf{i} + a_2\mathbf{j} + a_3\mathbf{k})\cdot(b_1\mathbf{i} + b_2\mathbf{j} + b_3\mathbf{k})$$

and use the fact that dot multiplication is distributive over addition [Eq. (3)] to expand and simplify, we obtain the important result

$$(10) \qquad\qquad \mathbf{A}\cdot\mathbf{B} = a_1b_1 + a_2b_2 + a_3b_3$$

In particular, taking $\mathbf{B} = \mathbf{A}$, we have

$$\mathbf{A}\cdot\mathbf{A} = |A|^2 = a_1^2 + a_2^2 + a_3^2$$

or

$$(11) \qquad\qquad \mathbf{A} = \sqrt{a_1^2 + a_2^2 + a_3^2}$$

On the other hand, if we write $\mathbf{A}\cdot\mathbf{B} = |A|\,|B|\cos\theta$ and then solve for $\cos\theta$, using (10) and (11), we obtain the useful formula

$$(12) \qquad\qquad \cos\theta = \frac{a_1b_1 + a_2b_2 + a_3b_3}{\sqrt{a_1^2 + a_2^2 + a_3^2}\,\sqrt{b_1^2 + b_2^2 + b_3^2}}$$

a result familiar from analytic geometry, where the a's and b's were introduced not as the components of two vectors but as the direction numbers of two straight lines.

For the cross products of the unit vectors **i**, **j**, **k** we find at once

$$(13) \qquad \begin{aligned} \mathbf{i}\times\mathbf{i} &= \mathbf{j}\times\mathbf{j} = \mathbf{k}\times\mathbf{k} = 0 \\ \mathbf{i}\times\mathbf{j} &= -\mathbf{j}\times\mathbf{i} = \mathbf{k} \\ \mathbf{j}\times\mathbf{k} &= -\mathbf{k}\times\mathbf{j} = \mathbf{i} \\ \mathbf{k}\times\mathbf{i} &= -\mathbf{i}\times\mathbf{k} = \mathbf{j} \end{aligned}$$

Hence, using (13) and the fact that cross multiplication is distributive over addition, we obtain for

$$\mathbf{A} \times \mathbf{B} = (a_1\mathbf{i} + a_2\mathbf{j} + a_3\mathbf{k}) \times (b_1\mathbf{i} + b_2\mathbf{j} + b_3\mathbf{k})$$

the expression

(14) $$\mathbf{A} \times \mathbf{B} = (a_2b_3 - a_3b_2)\mathbf{i} - (a_1b_3 - a_3b_1)\mathbf{j} + (a_1b_2 - a_2b_1)\mathbf{k}$$

which is precisely the expanded form of the determinant

(15) $$\mathbf{A} \times \mathbf{B} = \begin{vmatrix} \mathbf{i} & \mathbf{j} & \mathbf{k} \\ a_1 & a_2 & a_3 \\ b_1 & b_2 & b_3 \end{vmatrix}$$

The anticommutative character of vector multiplication thus corresponds to the fact that interchanging two rows of a determinant changes the sign of the determinant.

EXAMPLE 1

Using vector methods, derive the law of cosines.

To do this, let directions be assigned to the sides of the given triangle as in Fig. 13.5. Then $\mathbf{C} = \mathbf{A} - \mathbf{B}$, and we have

$$\mathbf{C} \cdot \mathbf{C} = (\mathbf{A} - \mathbf{B}) \cdot (\mathbf{A} - \mathbf{B}) = \mathbf{A} \cdot \mathbf{A} - 2\mathbf{A} \cdot \mathbf{B} + \mathbf{B} \cdot \mathbf{B}$$

or, using (1) and (2),

$$C^2 = A^2 + B^2 - 2AB \cos \theta$$

which is the law of cosines.

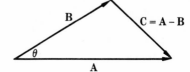

Figure 13.5
The triangle used in the vector derivation
of the law of cosines.

EXAMPLE 2

If (x,y,z) and (x',y',z') are two right-handed coordinate systems having a common origin, obtain by vector methods the transformation equations connecting the two sets of coordinates.

To do this, let $\mathbf{i}, \mathbf{j}, \mathbf{k}$ and $\mathbf{i}', \mathbf{j}', \mathbf{k}'$ be unit vectors in the directions of the respective axes (Fig. 13.6), and let P be a general point in space having coordinates (x,y,z) and (x',y',z') in the respective systems. Now the coordinates (x',y',z') are simply the components of the vector OP along the x', y', z' axes. Hence, if we write

$$\mathbf{R} = OP = x\mathbf{i} + y\mathbf{j} + z\mathbf{k}$$

and observe that the dot products of this vector with the unit vectors $\mathbf{i}'$, $\mathbf{j}'$, and $\mathbf{k}'$ are its components in these directions, we find the required transformation formulas to be

$$x' = \mathbf{R} \cdot \mathbf{i}' = (x\mathbf{i} + y\mathbf{j} + z\mathbf{k}) \cdot \mathbf{i}' = x(\mathbf{i} \cdot \mathbf{i}') + y(\mathbf{j} \cdot \mathbf{i}') + z(\mathbf{k} \cdot \mathbf{i}')$$
$$y' = \mathbf{R} \cdot \mathbf{j}' = (x\mathbf{i} + y\mathbf{j} + z\mathbf{k}) \cdot \mathbf{j}' = x(\mathbf{i} \cdot \mathbf{j}') + y(\mathbf{j} \cdot \mathbf{j}') + z(\mathbf{k} \cdot \mathbf{j}')$$
$$z' = \mathbf{R} \cdot \mathbf{k}' = (x\mathbf{i} + y\mathbf{j} + z\mathbf{k}) \cdot \mathbf{k}' = x(\mathbf{i} \cdot \mathbf{k}') + y(\mathbf{j} \cdot \mathbf{k}') + z(\mathbf{k} \cdot \mathbf{k}')$$

From (1), the products $\mathbf{i} \cdot \mathbf{i}', \mathbf{j} \cdot \mathbf{i}', \ldots, \mathbf{k} \cdot \mathbf{k}'$ are just the cosines of the angles between the various axes of the two systems and are known from the data of the problem.

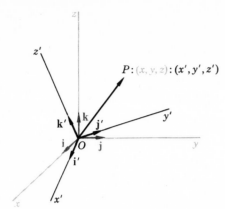

Figure 13.6
Two rectangular coordinate systems
with the same origin.

Products involving three rather than two vectors offer the following possibilities:

$$(\mathbf{A} \cdot \mathbf{B})\mathbf{C} \qquad \mathbf{A} \cdot (\mathbf{B} \times \mathbf{C}) \qquad \mathbf{A} \times (\mathbf{B} \times \mathbf{C})$$

The first of these can be dismissed with a word. In fact $\mathbf{A} \cdot \mathbf{B}$ is just a scalar, and thus $(\mathbf{A} \cdot \mathbf{B})\mathbf{C}$ is simply a vector whose length is $|\mathbf{A} \cdot \mathbf{B}|$ times the length of $\mathbf{C}$ and whose direction is the same as that of $\mathbf{C}$ or opposite to it, according to $\mathbf{A} \cdot \mathbf{B}$ is positive or negative.

For the product $\mathbf{A} \cdot (\mathbf{B} \times \mathbf{C})$, which is known as a **scalar triple product**, we observe first that the parentheses enclosing the vector product $\mathbf{B} \times \mathbf{C}$ are superfluous. There is, in fact, only one alternative interpretation, namely, $(\mathbf{A} \cdot \mathbf{B}) \times \mathbf{C}$ and this is meaningless, since both factors in a cross product must be vectors, whereas $\mathbf{A} \cdot \mathbf{B}$ is a scalar. Thus, no meaning but the intended one can be attached to the expression $\mathbf{A} \cdot \mathbf{B} \times \mathbf{C}$, and hence it is customary to omit the parentheses.

Geometrically, *the scalar triple product* $\mathbf{A} \cdot \mathbf{B} \times \mathbf{C}$ *represents the volume of the parallelepiped having the vectors* $\mathbf{A}$, $\mathbf{B}$, *and* $\mathbf{C}$ *as concurrent edges*. For, if we regard the parallelogram having $\mathbf{B}$ and $\mathbf{C}$ as adjacent sides as the base of this figure, then $\mathbf{B} \times \mathbf{C}$ is a vector whose direction is perpendicular to the base and whose magnitude is equal to the area of the base. Moreover, the altitude of the parallelepiped is the projection of $\mathbf{A}$ on $\mathbf{B} \times \mathbf{C}$ (Fig. 13.7a). Hence, $\mathbf{A} \cdot \mathbf{B} \times \mathbf{C}$, whose value is just the magnitude of $\mathbf{B} \times \mathbf{C}$ multiplied by the projection of $\mathbf{A}$ on $\mathbf{B} \times \mathbf{C}$, is numerically equal to the volume of the parallelepiped. If θ is less than $\pi/2$, that is, if $\mathbf{A}$ and $\mathbf{B} \times \mathbf{C}$ lie on the same side of the plane of $\mathbf{B}$ and $\mathbf{C}$, then $\cos \theta$ is positive and so is the scalar

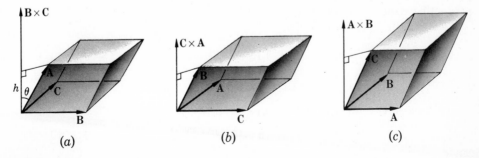

(a) (b) (c)

Figure 13.7
The geometrical interpretation of the scalar triple product.

Figure 13.8
Cyclic and anticyclic permutations.

triple product. In particular, changing the order of the factors **B** and **C** gives the product **C** × **B**, whose direction, of course, is opposite to that of **B** × **C**; hence

$$(16) \qquad \mathbf{A} \cdot \mathbf{B} \times \mathbf{C} = -\mathbf{A} \cdot \mathbf{C} \times \mathbf{B}$$

Since the volume of a parallelepiped is independent of the face chosen as its base, it follows by applying the preceding argument to Fig. 13.7b and c, that **B** · **C** × **A** and **C** · **A** × **B** also give the volume of the same parallelepiped. From this fact, together with (16), we therefore find that

$$(17) \qquad \mathbf{A} \cdot \mathbf{B} \times \mathbf{C} = \mathbf{B} \cdot \mathbf{C} \times \mathbf{A} = \mathbf{C} \cdot \mathbf{A} \times \mathbf{B} = -\mathbf{A} \cdot \mathbf{C} \times \mathbf{B}$$

$$= -\mathbf{B} \cdot \mathbf{A} \times \mathbf{C} = -\mathbf{C} \cdot \mathbf{B} \times \mathbf{A}$$

The first three of these arrangements can be obtained by starting anywhere on the circle in Fig. 13.8 and reading the letters in the counterclockwise direction. For this reason they are said to be **cyclic permutations** of one another. Similarly, the last three arrangements are cyclic permutations of one another which can be obtained by reading the letters in Fig. 13.8 in the clockwise direction. Thus (17) asserts that *any cyclic permutation of the factors in a scalar triple product leaves the value of the product unchanged, whereas any permutation which reverses the original cyclic order changes the sign of the product.*

Furthermore, since the order of the factors in a dot product is immaterial, we find, by considering the first and third members of (17), that $\mathbf{A} \cdot \mathbf{B} \times \mathbf{C} = \mathbf{C} \cdot \mathbf{A} \times \mathbf{B} = \mathbf{A} \times \mathbf{B} \cdot \mathbf{C}$, which shows that *in any scalar triple product the dot and the cross can be interchanged without altering the value of the product.* For this reason it is customary to omit these symbols and write a scalar triple product simply as [**ABC**].

If the vectors **A**, **B**, **C** all lie in the same plane or are parallel to the same plane, they necessarily form a parallelepiped of zero volume, and conversely. Hence [**ABC**] = 0 is a necessary and sufficient condition that three vectors **A**, **B**, and **C** be parallel to one and the same plane. In particular, if two factors of a scalar triple product have the same direction, the product is zero.

Algebraically, if we write

$$\mathbf{A} = a_1\mathbf{i} + a_2\mathbf{j} + a_3\mathbf{k} \qquad \mathbf{B} = b_1\mathbf{i} + b_2\mathbf{j} + b_3\mathbf{k} \qquad \mathbf{C} = c_1\mathbf{i} + c_2\mathbf{j} + c_3\mathbf{k}$$

we have

$$\mathbf{A} \cdot \mathbf{B} \times \mathbf{C} = (a_1\mathbf{i} + a_2\mathbf{j} + a_3\mathbf{k}) \cdot$$

$$[(b_2c_3 - b_3c_2)\mathbf{i} - (b_1c_3 - b_3c_1)\mathbf{j} + (b_1c_2 - b_2c_1)\mathbf{k}]$$

$$= a_1(b_2c_3 - b_3c_2) - a_2(b_1c_3 - b_3c_1) + a_3(b_1c_2 - b_2c_1)$$

which is just the expanded form of the determinant

(18)
$$[\mathbf{ABC}] = \begin{vmatrix} a_1 & a_2 & a_3 \\ b_1 & b_2 & b_3 \\ c_1 & c_2 & c_3 \end{vmatrix}$$

The relations in (17) are thus equivalent to the familiar fact that interchanging any two rows in a determinant changes the sign of the determinant.

EXAMPLE 3

If **A**, **B**, and **C** are three vectors which are not parallel to the same plane, show that any vector **V** can be expressed as a linear combination of **A**, **B**, and **C**.

If we write

(19)
$$\mathbf{V} = a\mathbf{A} + b\mathbf{B} + c\mathbf{C}$$

where a, b, and c are scalar constants to be determined, and form the cross product of each member with the vector **B**, we obtain

$$\mathbf{V} \times \mathbf{B} = a\mathbf{A} \times \mathbf{B} + b\mathbf{B} \times \mathbf{B} + c\mathbf{C} \times \mathbf{B} = a\mathbf{A} \times \mathbf{B} + c\mathbf{C} \times \mathbf{B}$$

where the term $\mathbf{B} \times \mathbf{B}$ vanishes because its factors are identical. Now if we form the dot product of the last result with the vector **C**, we have

$$\mathbf{V} \times \mathbf{B} \cdot \mathbf{C} = a\mathbf{A} \times \mathbf{B} \cdot \mathbf{C} + c\mathbf{C} \times \mathbf{B} \cdot \mathbf{C} = a\mathbf{A} \times \mathbf{B} \cdot \mathbf{C}$$

where the term $\mathbf{C} \times \mathbf{B} \cdot \mathbf{C}$ vanishes because it is a scalar triple product with two factors identical. By hypothesis, **A**, **B**, and **C** are not parallel to the same plane. Hence $\mathbf{A} \times \mathbf{B} \cdot \mathbf{C}$ is different from zero, and we can solve for a, getting

$$a = \frac{[\mathbf{VBC}]}{[\mathbf{ABC}]}$$

In the same way we can obtain the remaining constants in the required linear combination:

$$b = \frac{[\mathbf{AVC}]}{[\mathbf{ABC}]} \qquad c = \frac{[\mathbf{ABV}]}{[\mathbf{ABC}]}$$

Thus, under the conditions of the problem,

(20)
$$\mathbf{V} = \frac{[\mathbf{VBC}]}{[\mathbf{ABC}]}\mathbf{A} + \frac{[\mathbf{AVC}]}{[\mathbf{ABC}]}\mathbf{B} + \frac{[\mathbf{ABV}]}{[\mathbf{ABC}]}\mathbf{C}$$

The following special case of this result is often useful. If **V** is any vector parallel to the plane determined by **A** and **B**, then $[\mathbf{ABV}] = 0$ and the last term in the expansion (20) is zero. Hence it follows that *if* **A** *and* **B** *are vectors which are not parallel to the same line and if* **V** *is any vector parallel to the plane determined by* **A** *and* **B**, *then* **V** *can be expressed as a linear combination of* **A** *and* **B**.

To express the vector triple product $\mathbf{A} \times (\mathbf{B} \times \mathbf{C})$ in a simpler expanded form, let us consider first the general case in which neither **A**, **B**, nor **C** is a zero vector and **B** and **C** are not parallel. Now, from the definition of a cross product, it is clear that $\mathbf{A} \times (\mathbf{B} \times \mathbf{C})$ is a vector perpendicular to **A** and to $\mathbf{B} \times \mathbf{C}$. But $\mathbf{B} \times \mathbf{C}$ is itself perpendicular to the plane of **B** and **C**, and thus any vector such as $\mathbf{A} \times (\mathbf{B} \times \mathbf{C})$ which is perpendicular to $\mathbf{B} \times \mathbf{C}$ must lie in the plane of **B** and **C** (Fig. 13.9). Hence, by Example 3, the vector $\mathbf{A} \times (\mathbf{B} \times \mathbf{C})$ must be expressible as a linear combination of **B** and **C**; that is,

$$\mathbf{A} \times (\mathbf{B} \times \mathbf{C}) = \lambda\mathbf{B} + \mu\mathbf{C}$$

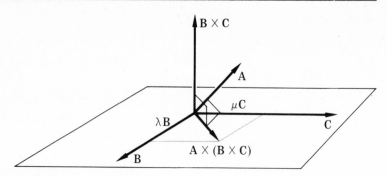

Figure 13.9
The geometrical interpretation of the vector triple product.

To find λ and μ, we first use the fact that $\mathbf{A} \times (\mathbf{B} \times \mathbf{C})$ is also perpendicular to $\mathbf{A}$ and hence that its dot product with $\mathbf{A}$ must be zero:

$$\mathbf{A} \cdot [\mathbf{A} \times (\mathbf{B} \times \mathbf{C})] = \mathbf{A} \cdot (\lambda \mathbf{B} + \mu \mathbf{C}) = \lambda(\mathbf{A} \cdot \mathbf{B}) + \mu(\mathbf{A} \cdot \mathbf{C}) = 0$$

Thus there must exist a value of v such that

$$\lambda = v(\mathbf{A} \cdot \mathbf{C}) \qquad \mu = -v(\mathbf{A} \cdot \mathbf{B})$$

and

(21) $$\mathbf{A} \times (\mathbf{B} \times \mathbf{C}) = v[(\mathbf{A} \cdot \mathbf{C})\mathbf{B} - (\mathbf{A} \cdot \mathbf{B})\mathbf{C}]$$

To find v, it is convenient to consider first the special case in which $\mathbf{A} = \mathbf{B}$:

(22) $$\mathbf{B} \times (\mathbf{B} \times \mathbf{C}) = v_1[(\mathbf{B} \cdot \mathbf{C})\mathbf{B} - (\mathbf{B} \cdot \mathbf{B})\mathbf{C}]$$

Let θ be the angle between $\mathbf{B}$ and $\mathbf{C}$, and form the dot product of $\mathbf{C}$ with each side of the last identity. Then, interchanging the first two factors in the scalar triple product on the left and applying Eqs. (1) and (2) to the resulting dot products on the right, we have

$$[\mathbf{B} \times (\mathbf{B} \times \mathbf{C})] \cdot \mathbf{C} = -(\mathbf{B} \times \mathbf{C}) \cdot (\mathbf{B} \times \mathbf{C}) = v_1[(\mathbf{B} \cdot \mathbf{C})(\mathbf{B} \cdot \mathbf{C}) - (\mathbf{B} \cdot \mathbf{B})(\mathbf{C} \cdot \mathbf{C})]$$

$$- |\mathbf{B} \times \mathbf{C}|^2 = v_1(B^2 C^2 \cos^2 \theta - B^2 C^2)$$

$$- B^2 C^2 \sin^2 \theta = v_1 B^2 C^2 (\cos^2 \theta - 1)$$

$$= -v_1 B^2 C^2 \sin^2 \theta$$

Hence, $v_1 = 1$, and (22) becomes specifically

(23) $$\mathbf{B} \times (\mathbf{B} \times \mathbf{C}) = (\mathbf{B} \cdot \mathbf{C})\mathbf{B} - (\mathbf{B} \cdot \mathbf{B})\mathbf{C}$$

We now return to Eq. (21) and form the dot product of both members with $\mathbf{B}$:

$$\mathbf{A} \times (\mathbf{B} \times \mathbf{C}) \cdot \mathbf{B} = v[(\mathbf{A} \cdot \mathbf{C})(\mathbf{B} \cdot \mathbf{B}) - (\mathbf{A} \cdot \mathbf{B})(\mathbf{C} \cdot \mathbf{B})]$$

Now, by interchanging the last two factors in the scalar triple product on the left, we have

$$-\mathbf{A} \cdot \mathbf{B} \times (\mathbf{B} \times \mathbf{C}) = v[(\mathbf{A} \cdot \mathbf{C})(\mathbf{B} \cdot \mathbf{B}) - (\mathbf{A} \cdot \mathbf{B})(\mathbf{C} \cdot \mathbf{B})]$$

or, applying Eq. (23) to the left-hand side,

$$-\mathbf{A}\cdot[(\mathbf{B}\cdot\mathbf{C})\mathbf{B} - (\mathbf{B}\cdot\mathbf{B})\mathbf{C}] = v[(\mathbf{A}\cdot\mathbf{C})(\mathbf{B}\cdot\mathbf{B}) - (\mathbf{A}\cdot\mathbf{B})(\mathbf{C}\cdot\mathbf{B})]$$

which will be true if and only if $v = 1$.† Hence, in general, from Eq. (21),

$$(24) \qquad\qquad \mathbf{A} \times (\mathbf{B} \times \mathbf{C}) = (\mathbf{A}\cdot\mathbf{C})\mathbf{B} - (\mathbf{A}\cdot\mathbf{B})\mathbf{C}$$

Moreover, if either $\mathbf{A}$, $\mathbf{B}$, or $\mathbf{C}$ is zero, or if $\mathbf{B}$ and $\mathbf{C}$ have the same direction, it is evident, by inspection, that Eq. (24) still holds. Hence the restrictions we imposed upon $\mathbf{A}$, $\mathbf{B}$, and $\mathbf{C}$ at the beginning of our discussion can be eliminated, and Eq. (24) is correct in all cases.

By a straightforward application of Eq. (24) we find that

$$(\mathbf{A} \times \mathbf{B}) \times \mathbf{C} = -\mathbf{C} \times (\mathbf{A} \times \mathbf{B}) = -(\mathbf{C}\cdot\mathbf{B})\mathbf{A} + (\mathbf{C}\cdot\mathbf{A})\mathbf{B}$$

which is *not* equal to $\mathbf{A} \times (\mathbf{B} \times \mathbf{C})$. Hence the position of the parentheses in a vector triple product is significant.

With a knowledge of scalar and vector triple products, products involving more than three vectors can be expanded without difficulty. For instance

$$(\mathbf{A} \times \mathbf{B})\cdot(\mathbf{C} \times \mathbf{D})$$

can be regarded as the scalar triple product of the vectors $\mathbf{A}$, $\mathbf{B}$, and $\mathbf{C} \times \mathbf{D}$. This allows us to write

$$\begin{aligned}
\mathbf{A} \times \mathbf{B}\cdot(\mathbf{C} \times \mathbf{D}) &= \mathbf{A}\cdot[\mathbf{B} \times (\mathbf{C} \times \mathbf{D})] \\
&= \mathbf{A}\cdot[(\mathbf{B}\cdot\mathbf{D})\mathbf{C} - (\mathbf{B}\cdot\mathbf{C})\mathbf{D}] \\
&= (\mathbf{A}\cdot\mathbf{C})(\mathbf{B}\cdot\mathbf{D}) - (\mathbf{A}\cdot\mathbf{D})(\mathbf{B}\cdot\mathbf{C})
\end{aligned}$$

This result is sometimes referred to as **Lagrange's identity**.

Similarly, $(\mathbf{A} \times \mathbf{B}) \times (\mathbf{C} \times \mathbf{D})$ can be thought of as the vector triple product of $\mathbf{A} \times \mathbf{B}$, $\mathbf{C}$, and $\mathbf{D}$ or of $\mathbf{A}$, $\mathbf{B}$, and $\mathbf{C} \times \mathbf{D}$. Taking the former point of view and applying (24), we find

$$\begin{aligned}
(\mathbf{A} \times \mathbf{B}) \times (\mathbf{C} \times \mathbf{D}) &= (\mathbf{A} \times \mathbf{B}\cdot\mathbf{D})\mathbf{C} - (\mathbf{A} \times \mathbf{B}\cdot\mathbf{C})\mathbf{D} \\
&= [ABD]\mathbf{C} - [ABC]\mathbf{D}
\end{aligned}$$

which is a vector in the plane of $\mathbf{C}$ and $\mathbf{D}$. From the latter point of view,

$$\begin{aligned}
(\mathbf{A} \times \mathbf{B}) \times (\mathbf{C} \times \mathbf{D}) &= -(\mathbf{C} \times \mathbf{D}) \times (\mathbf{A} \times \mathbf{B}) \\
&= -(\mathbf{C} \times \mathbf{D}\cdot\mathbf{B})\mathbf{A} + (\mathbf{C} \times \mathbf{D}\cdot\mathbf{A})\mathbf{B} \\
&= [CDA]\mathbf{B} - [CDB]\mathbf{A}
\end{aligned}$$

which is a vector in the plane of $\mathbf{A}$ and $\mathbf{B}$. These two results together show that $(\mathbf{A} \times \mathbf{B}) \times (\mathbf{C} \times \mathbf{D})$ is directed along the line of intersection of the plane of $\mathbf{A}$ and $\mathbf{B}$ and the plane of $\mathbf{C}$ and $\mathbf{D}$.

---

† Unless, of course, $(\mathbf{A}\cdot\mathbf{C})(\mathbf{B}\cdot\mathbf{B}) - (\mathbf{A}\cdot\mathbf{B})(\mathbf{C}\cdot\mathbf{B}) = 0$, in which case the value of v is irrelevant.

1 For each of the following sets of vectors

(a) $A = 2i - 2j + k$
    $B = i + 8j - 4k$
    $C = 12i - 4j - 3k$

(b) $A = 2i - 3j + 6k$
    $B = 10i + 2j + 11k$
    $C = 2i - 9j - 6k$

(c) $A = 10i + 10j + 5k$
    $B = 5i - 2j - 14k$
    $C = 4i + 7j - 4k$

what are the lengths of **A**, **B**, and **C**? What is $A \cdot B$? $A \times C$? The projection of **C** on **B**? The angle between **A** and **B**? [**ABC**]? $A \times (B \times C)$? $(A \times B) \times C$? The volume of the parallelepiped having $A + C$, $A - C$, and **B** as concurrent edges? The volume of the parallelepiped having $A + C$, $A - C$, and **C** as concurrent edges?

2 If **A**, **B**, and **C** are any three vectors, prove that

$$A \times (B \times C) + B \times (C \times A) + C \times (A \times B) = 0$$

3 Prove that $(A \times B) \cdot (C \times D) + (B \times C) \cdot (A \times D) + (C \times A) \cdot (B \times D) = 0$.

4 If the plane determined by **A** and **B** is perpendicular to the plane determined by **C** and **D**, show that $(A \times B) \cdot (C \times D) = 0$.

5 Show that the volume of the parallelepiped having $A + B$, $B + C$, and $C + A$ as concurrent edges is twice the volume of the parallelepiped having **A**, **B**, and **C** as concurrent edges.

6 If **A** is a given vector and $A \cdot X = A \cdot Y$, can we conclude that $X = Y$?

7 Are two vectors equal if they have equal components in a given direction? In two given direction? In three given directions? In an arbitrary direction?

8 Find the unit vector which is perpendicular to both $i - 2j + k$ and $3i + j - 2k$.

9 Find the unit vector which is parallel to the plane of $i + j - 2k$ and $3i - 2j + k$ and perpendicular to $2i + 2j - k$.

10 Show that if four vectors **A**, **B**, **C**, **D** are coplanar, then $(A \times B) \times (C \times D) = 0$. Is the converse true?

11 Show that if $A + B + C = 0$, then $A \times B = B \times C = C \times A$. Is the converse true?

12 Prove that two nonzero vectors are linearly dependent if and only if they are parallel.

13 Prove that three vectors are linearly dependent if and only if they are parallel to the same plane.

14 Prove that four vectors are always linearly dependent. *Hint:* Expand $(A \times B) \times (C \times D)$ in two different ways and equate the results.

15 Prove that for all values of the a's and b's

$$(a_1 b_1 + a_2 b_2 + a_3 b_3)^2 \leq (a_1^2 + a_2^2 + a_3^2)(b_1^2 + b_2^2 + b_3^2)$$

This is the special case $n = 3$ of **Cauchy's inequality**,

$$\left(\sum_{i=1}^{n} a_i b_i \right)^2 \leq \left(\sum_{i=1}^{n} a_i^2 \right) \left(\sum_{i=1}^{n} b_i^2 \right)$$

16 By considering the dot product of the two vectors

$$A = a_1 i + a_2 j \quad \text{and} \quad B = b_1 i + b_2 j$$

derive the formula for the cosine of the difference of two angles.

17 By considering the cross product of the two vectors in Exercise 16, derive the formula for the sine of the difference of two angles.

18 Show that if $A = a_1 i + a_2 j + a_3 k$ is a constant vector drawn from the origin, the locus of the end points of the vectors $R = xi + yj + zk$ which satisfy the equation $(R - A) \cdot A = 0$ is a plane perpendicular to **A** at its end point. What is the locus of the end points of the vectors which satisfy the equation $(R - A) \cdot R = 0$? The equation $(R - A) \cdot (R - A) = 0$?

19 Three noncollinear points L, M, and N lie in a plane p. Prove that if **L**, **M**, and **N** are the vectors to these points from an arbitrary origin, then the vector $(\mathbf{L} \times \mathbf{M}) + (\mathbf{M} \times \mathbf{N}) + (\mathbf{N} \times \mathbf{L})$ is perpendicular to p.

20 Carry through in detail the geometrical proof that dot multiplication is distributive over addition.

21 Carry through in detail the geometrical proof that cross multiplication is distributive over addition.

22 Prove that $(\mathbf{A} \times \mathbf{B}) \cdot (\mathbf{B} \times \mathbf{C}) \times (\mathbf{C} \times \mathbf{A}) = [ABC]^2$.

23 If **A**, **B**, and **C** are any three independent vectors, the vectors

$$\mathbf{U} = \frac{\mathbf{B} \times \mathbf{C}}{[ABC]} \qquad \mathbf{V} = \frac{\mathbf{C} \times \mathbf{A}}{[ABC]} \qquad \mathbf{W} = \frac{\mathbf{A} \times \mathbf{B}}{[ABC]}$$

are said to form a set **reciprocal** to the set **A**, **B**, **C**. Show that

$$\mathbf{A} \cdot \mathbf{U} = \mathbf{B} \cdot \mathbf{V} = \mathbf{C} \cdot \mathbf{W} = 1 \qquad \text{and} \qquad [UVW] = \frac{1}{[ABC]}$$

If $\mathbf{A} = \mathbf{i} + 2\mathbf{j} - 2\mathbf{k}$, $\mathbf{B} = \mathbf{i} + 8\mathbf{j} + 4\mathbf{k}$, and $\mathbf{C} = 12\mathbf{i} - 4\mathbf{j} + 3\mathbf{k}$, express the vector $\mathbf{i} + 2\mathbf{j} + 3\mathbf{k}$ as a linear combination of **A**, **B**, and **C** and also as a linear combination of the vectors **U**, **V**, and **W** of the set reciprocal to **A**, **B**, and **C**.

24 Show that if $\mathbf{A} = a_1\mathbf{i} + a_2\mathbf{j} + a_3\mathbf{k}$, $\mathbf{B} = b_1\mathbf{i} + b_2\mathbf{j} + b_3\mathbf{k}$, $\mathbf{C} = c_1\mathbf{i} + c_2\mathbf{j} + c_3\mathbf{k}$, and $\mathbf{D} = d_1\mathbf{i} + d_2\mathbf{j} + d_3\mathbf{k}$, then the system of equations

$$a_1 x + b_1 y + c_1 z = d_1$$
$$a_2 x + b_2 y + c_3 z = d_2$$
$$a_3 x + b_3 y + c_3 z = d_3$$

is equivalent to the single vector equation $x\mathbf{A} + y\mathbf{B} + z\mathbf{C} = \mathbf{D}$. Assuming that $[ABC] \neq 0$, solve this vector equation for x, y, and z, and show that the result is equivalent to that obtained from the algebraic form of the system by using Cramer's rule (Theorem 8, Sec. 10.3).

25 In mechanics the **moment** M **of a force F about a point** O is defined as the magnitude of **F** times the perpendicular distance from the point O to the line of action of **F**. If the **vector moment M** is defined as the vector whose magnitude is M and whose direction is perpendicular to the plane of O and **F**, show that $\mathbf{M} = \mathbf{R} \times \mathbf{F}$, where **R** is the vector from O to any point on the line of action of **F**. Would $\mathbf{M} = \mathbf{F} \times \mathbf{R}$ be an equally acceptable definition? Explain.

26 Using vector methods, show that the segment joining the midpoints of two sides of any triangle is parallel to the third side and half as long. *Hint:* Let $\triangle A_1 A_2 A_3$ be an arbitrary triangle, let A_4 be the midpoint of $A_1 A_3$, let A_5 be the midpoint of $A_2 A_3$, and let $A_i A_j$ denote the vector whose initial point is A_i and whose terminal point is A_j (Fig. 13.10). Then verify that $A_1 A_2 = A_1 A_3 + A_3 A_2$, $A_1 A_4 = \frac{1}{2} A_1 A_3$,

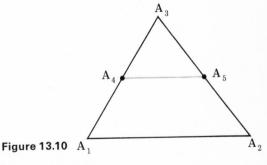

Figure 13.10

$A_2 A_5 = \frac{1}{2} A_2 A_3$, and $A_1 A_4 + A_4 A_5 = A_1 A_2 + A_2 A_5$. Finally, combine these relations to obtain the conclusion that

$$A_4 A_5 = \frac{1}{2} A_1 A_2$$

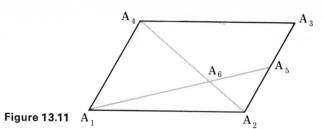

Figure 13.11 A_1

27   If $A_1A_2A_3A_4$ is an arbitrary parallelogram (Fig. 13.11), prove by vector methods that the line joining A_1 to the midpoint of A_2A_3 intersects the diagonal A_2A_4 in one of the trisection points of the diagonal. *Hint:* Let A_5 be the midpoint of A_2A_3, let A_6 be the intersection of A_1A_5 and A_2A_4, let $A_1A_6 = xA_1A_5$, and let $A_2A_6 = yA_2A_4$. Then by obtaining two different representations of $\mathbf{A_1A_6}$, show that

$$(1 - y)\mathbf{A_1A_2} + y\mathbf{A_1A_4} = x(\mathbf{A_1A_2} + \tfrac{1}{2}\mathbf{A_1A_4})$$

and finally show that this implies that $x = \tfrac{2}{3}$ and $y = \tfrac{1}{3}$.

28   Show by vector methods that the diagonals of a parallelogram bisect each other.

29   Show by vector methods that the medians of a triangle meet in a point of trisection of each median.

30   Using vector methods, show that the altitudes of an arbitrary triangle are concurrent.

13.2   Vector Functions of One Variable

If t is a scalar variable, and if to each value of t in some interval there corresponds a value of a vector $\mathbf{V}$, we say that $\mathbf{V}$ is a **vector function** of t. Since the component of a vector in any direction is known whenever the vector itself is known, it follows that if $\mathbf{V}$ is a function of t, so too are its components in the directions of the unit vectors $\mathbf{i}$, $\mathbf{j}$, and $\mathbf{k}$. Hence we can write

(1) $$\mathbf{V}(t) = V_1(t)\mathbf{i} + V_2(t)\mathbf{j} + V_3(t)\mathbf{k}$$

In particular, we say that $\mathbf{V}(t)$ is continuous if and only if the three scalar functions $V_1(t)$, $V_2(t)$, and $V_3(t)$ are continuous.

If the independent variable t of a vector function $\mathbf{V}(t)$ changes by an amount Δt, the function will in general change both in magnitude and in direction. Specifically, corresponding to the scalar increment Δt we have the vector increment

$$\Delta \mathbf{V} = \mathbf{V}(t + \Delta t) - \mathbf{V}(t)$$

$$= [V_1(t + \Delta t)\mathbf{i} + V_2(t + \Delta t)\mathbf{j} + V_3(t + \Delta t)\mathbf{k}] - [V_1(t)\mathbf{i} + V_2(t)\mathbf{j} + V_3(t)\mathbf{k}]$$

(2) $$= \Delta V_1\mathbf{i} + \Delta V_2\mathbf{j} + \Delta V_3\mathbf{k}$$

By the **derivative of a vector function** $\mathbf{V}(t)$ we mean, as usual,

$$\frac{d\mathbf{V}}{dt} = \lim_{\Delta t \to 0} \frac{\mathbf{V}(t + \Delta t) - \mathbf{V}(t)}{\Delta t} = \lim_{\Delta t \to 0} \frac{\Delta \mathbf{V}}{\Delta t}$$

or, using (2),

$$\frac{d\mathbf{V}}{dt} = \lim_{\Delta t \to 0} \frac{\Delta V_1}{\Delta t}\mathbf{i} + \lim_{\Delta t \to 0} \frac{\Delta V_2}{\Delta t}\mathbf{j} + \lim_{\Delta t \to 0} \frac{\Delta V_3}{\Delta t}\mathbf{k}$$

(3) $$= \frac{dV_1}{dt}\mathbf{i} + \frac{dV_2}{dt}\mathbf{j} + \frac{dV_3}{dt}\mathbf{k}$$

From (3) we are motivated to define the **differential of a vector function** $\mathbf{V}(t)$ to be

(4) $$dV = dV_1\,\mathbf{i} + dV_2\,\mathbf{j} + dV_3\,\mathbf{k}$$

In particular, for the very important vector

(5) $$\mathbf{R} = x\mathbf{i} + y\mathbf{j} + z\mathbf{k}$$

drawn from the origin to the point (x,y,z), we have

(6) $$d\mathbf{R} = dx\,\mathbf{i} + dy\,\mathbf{j} + dz\,\mathbf{k}$$

From the definition of the derivative of a vector function of one variable it follows that sums, differences, and products of vectors can be differentiated by formulas just like those of ordinary calculus, provided that the proper order of factors is maintained wherever the order is significant. Specifically, we have

(7) $$\frac{d(\mathbf{U} \pm \mathbf{V})}{dt} = \frac{d\mathbf{U}}{dt} \pm \frac{d\mathbf{V}}{dt}$$

(8) $$\frac{d(\phi\mathbf{V})}{dt} = \frac{d\phi}{dt}\mathbf{V} + \phi\frac{d\mathbf{V}}{dt}$$

(9) $$\frac{d(\mathbf{U}\cdot\mathbf{V})}{dt} = \frac{d\mathbf{U}}{dt}\cdot\mathbf{V} + \mathbf{U}\cdot\frac{d\mathbf{V}}{dt}$$

(10) $$\frac{d(\mathbf{U} \times \mathbf{V})}{dt} = \frac{d\mathbf{U}}{dt} \times \mathbf{V} + \mathbf{U} \times \frac{d\mathbf{V}}{dt}$$

(11) $$\frac{d[\mathbf{UVW}]}{dt} = \left[\frac{d\mathbf{U}}{dt}\mathbf{VW}\right] + \left[\mathbf{U}\frac{d\mathbf{V}}{dt}\mathbf{W}\right] + \left[\mathbf{UV}\frac{d\mathbf{W}}{dt}\right]$$

(12) $$\frac{d[\mathbf{U} \times (\mathbf{V} \times \mathbf{W})]}{dt} = \frac{d\mathbf{U}}{dt} \times (\mathbf{V} \times \mathbf{W}) + \mathbf{U} \times \left(\frac{d\mathbf{V}}{dt} \times \mathbf{W}\right) + \mathbf{U} \times \left(\mathbf{V} \times \frac{d\mathbf{W}}{dt}\right)$$

The simplest example of a vector function of one variable is the set of vectors drawn from the origin to the points of a curve C on which the scalar variable t is a parameter. In fact, a general point on C is associated with a unique value of the parameter, say $t = t_1$, and determines with the origin a unique vector $\mathbf{V}(t_1)$ (Fig. 13.12a). This correspondence between the values of t and the vectors $\mathbf{V}(t)$ is clearly a vector function of t according to our definition. Conversely, if the values of a

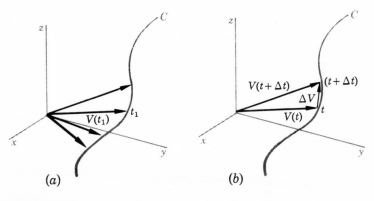

(a) (b)

Figure 13.12
The geometrical interpretation of a vector function of one variable.

continuous vector function $V(t)$ are drawn from a common origin, their end points will define a curve C whose points will be in correspondence with the values of the scalar variable t.

This point of view leads to an important geometric interpretation of the derivative dV/dt. For, since Δt is just a scalar, the quotient $\Delta V/\Delta t$ is a well-defined vector having the same direction as ΔV itself. Moreover, as Fig. 13.12b shows, the direction of ΔV is that of an infinitesimal chord of the curve C. Therefore, as Δt approaches 0, the direction of ΔV, and hence the direction of $\Delta V/\Delta t$, approaches the direction of the tangent to C. That is, dV/dt is a vector tangent to the curve C which is the locus of the end points of the vector $V(t)$.† In particular, if the scalar variable t is taken to be the arc length s of C, measured from some reference point on C, we have

$$\left|\frac{dV}{ds}\right| = \lim_{\Delta s \to 0} \frac{|\Delta V|}{|\Delta s|} = \lim_{\Delta s \to 0} \frac{\text{infinitesimal chord of } C}{\text{infinitesimal arc of } C} = 1$$

Hence, *if s is the arc length of the curve C defined by the end points of the vectors* $V(s)$, *then dV/ds is a unit tangent to C.*

EXAMPLE 1

At what point or points is the tangent to the curve $x = t^3$, $y = 5t^2$, $z = 10t$ perpendicular to the tangent at the point where $t = 1$?

From our earlier discussion it is clear that the given curve is equivalent to the vector function

$$V(t) = t^3 \mathbf{i} + 5t^2 \mathbf{j} + 10t \mathbf{k}$$

Moreover, the tangent to this curve at a general point t is

$$\frac{dV}{dt} = 3t^2 \mathbf{i} + 10t \mathbf{j} + 10\mathbf{k}$$

and, in particular, at $t = 1$ the tangent is

$$3\mathbf{i} + 10\mathbf{j} + 10\mathbf{k}$$

Using the fact that two nonzero vectors are perpendicular if and only if their dot product vanishes, it follows that the tangent at a general point t will be perpendicular to the tangent at the point $t = 1$ if and only if

$$3(3t^2) + 10(10t) + 10(10) \equiv 9t^2 + 100t + 100 = 0$$

This condition holds for the two values $t = -\frac{10}{9}, -10$. Hence, evaluating the x, y, and z coordinates of the points with these parameters, it follows that the tangent at $(-\frac{1000}{729}, \frac{500}{81}, -\frac{100}{9})$ and the tangent at $(-1{,}000, 500, -100)$ are both perpendicular to the tangent at $t = 1$ and that these are the only points with this property.

EXAMPLE 2

From the point of view of vector analysis discuss the problem of determining the velocity and acceleration of a particle moving along a curve C.

To do this, let us suppose that the path C, which is the locus of the instantaneous positions of the moving particle, is defined by the vector function $P(t)$, where t is the time. In other words, $P(t)$ is the vector drawn from the origin to the position of the moving particle at the general time t.

Now let s be the arc length of C. Then by the chain rule, we can write

$$(13) \qquad \frac{dP}{dt} = \frac{dP}{ds}\frac{ds}{dt}$$

---

† Unless, of course, $dV/dt = 0$, as it may at a singular point of the curve C.

Since ds/dt is the speed v of the moving particle, and since $d\mathbf{P}/ds$ is a unit vector tangent to the path of the particle, it follows from (13) that the vector

(14)
$$\mathbf{v} = \frac{d\mathbf{P}}{dt}$$

agrees both in magnitude and in direction with the velocity of the particle and thus can properly be called its **vector velocity**.

Moreover, if we define the **vector acceleration** of the particle as the time derivative of its vector velocity and for convenience denote the general unit vector tangent to C, namely, $d\mathbf{P}/ds$, by the symbol $\mathbf{T}$, so that (14) becomes

$$\mathbf{v} = v\mathbf{T}$$

we can write

$$\mathbf{a} = \frac{d\mathbf{v}}{dt} = \frac{d(v\mathbf{T})}{dt} = \frac{dv}{dt}\mathbf{T} + v\frac{d\mathbf{T}}{dt} = \frac{dv}{dt}\mathbf{T} + v\frac{d\mathbf{T}}{ds}\frac{ds}{dt}$$

(15)
$$= \frac{dv}{dt}\mathbf{T} + v^2\frac{d\mathbf{T}}{ds}$$

In the first term on the right in (15), the scalar quantity dv/dt is the rate of change of the tangential speed v. Therefore, since $\mathbf{T}$ is by definition a unit vector tangent to C, the product $(dv/dt)\,\mathbf{T}$ is in magnitude and direction just the **tangential acceleration** of the moving particle.

To interpret the second term on the right in (15), we observe that since $\mathbf{T}$ is a unit vector, it can vary only in direction. Hence, if the various values of $\mathbf{T}$ are drawn from a common origin, the locus of their end points will be a curve Γ on a sphere of unit radius. Now the length of the increment $\Delta\mathbf{T}$ corresponding to two adjacent points P_1' and P_2' on the curve Γ (Fig. 13.13b) is approximately the length of the arc $P_1'P_2'$, which in turn is equal to $\Delta\theta$, where $\Delta\theta$ is the angle between the tangents to C at the points P_1 and P_2, a distance Δs apart (Fig. 13.13b). Hence

$$\left|\frac{d\mathbf{T}}{ds}\right| = \lim_{\Delta s \to 0}\left|\frac{\Delta\mathbf{T}}{\Delta s}\right| = \lim_{\Delta s \to 0}\frac{\text{angle between tangents to } C \text{ at } P_1 \text{ and } P_2}{\text{arc length along } C \text{ between } P_1 \text{ and } P_2}$$

$$= \text{curvature of } C \text{ at } P_1$$

Moreover, from Fig. 13.13b it is evident that the limiting position of ΔT as Δs (and hence $\Delta\theta$) approaches 0 is perpendicular to $\mathbf{T}$ in the plane which $\mathbf{T}$ and $\mathbf{T} + \Delta\mathbf{T}$

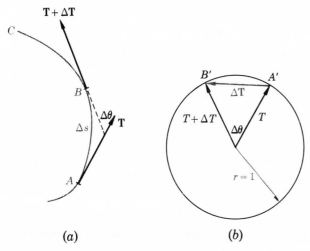

(a)   (b)

Figure 13.13
The unit tangents to a space curve plotted from a common origin.

determine in the limit. If C is a plane curve, this, of course, is the unique plane in which C lies. If C is a twisted curve, this plane, which is known as the **osculating plane**, will vary from point to point along C. Hence, to summarize, *at any point P on the curve C, $d\mathbf{T}/ds$ is a vector which is perpendicular to C in the osculating plane of C at P and whose magnitude is equal to the curvature K of C at P.* If, finally, we let $\mathbf{N}$ denote a unit normal drawn toward the concave side of C in the osculating plane and define, as usual, the radius of curvature of C to be

$$\rho = \frac{1}{K}$$

we can write Eq. (15) in the form

(16)
$$a = \frac{dv}{dt}\mathbf{T} + \frac{v^2}{\rho}\mathbf{N}$$

which shows that at any point in its path, *the vector acceleration of a moving particle is the sum of a component of magnitude dv/dt along the tangent to the path and a component of magnitude v^2/ρ normal to the path in the osculating plane to the path.*

EXERCISES

1   If $\mathbf{P} = \mathbf{A}\cos \omega t + \mathbf{B}\sin \omega t$, where $\mathbf{A}$ and $\mathbf{B}$ are arbitrary constant vectors, show that $\mathbf{P} \times d\mathbf{P}/dt$ is a constant and that $d^2\mathbf{P}/dt^2 + \omega^2\mathbf{P} = \mathbf{0}$.

2   If $\mathbf{P}$ is any vector, show that

$$\frac{d}{dt}\left(\mathbf{P} \times \frac{d\mathbf{P}}{dt}\right) = \mathbf{P} \times \frac{d^2\mathbf{P}}{dt^2}$$

3   Find the derivative of:

(a) $\mathbf{V} \cdot \dfrac{d\mathbf{V}}{dt} \times \dfrac{d^2\mathbf{V}}{dt^2}$
(b) $\mathbf{V} \times \left(\dfrac{d\mathbf{V}}{dt} \times \dfrac{d^2\mathbf{V}}{dt^2}\right)$

4   If $\mathbf{V}$ is an arbitrary vector function of t, is $|d\mathbf{V}| = d|\mathbf{V}|$?

5   If $\mathbf{V}$ is an arbitrary vector function of t, show that $\mathbf{V} \cdot d\mathbf{V}/dt = V(dV/dt)$.

6   What is the angle between the tangents to the curve $x = t$, $y = t^2$, $z = t^3$ at the points where $t = 1$ and $t = -1$?

7   Show that there are no pairs of points on the curve $x = t$, $y = t^2$, $z = t^3$ at which the tangents are parallel. Are there such pairs of points on the curve $x = 3t^4 - 6t^2 + 12t$, $y = 4t^3 - 6t^2$, $z = 12t$? On the curve $x = 15t$, $y = 5t^3$, $z = 15t + 3t^5$?

8   If $\mathbf{R} = t^2\mathbf{i} - t^3\mathbf{j} + t^4\mathbf{k}$ is the vector from the origin to a moving particle, find the velocity of the particle when $t = 1$. What is the component of this velocity in the direction of the vector $8\mathbf{i} - \mathbf{j} + 4\mathbf{k}$? What is the vector acceleration of the particle? What are the tangential and normal components of this acceleration?

9   If a particle starts to move from rest at the point $(0,1,2)$ with component accelerations $a_x = 1 + t$, $a_y = t^3$, $a_z = 2t - t^2$, find the vector from the origin to the instantaneous position of the particle.

10   If $\mathbf{R}_1, \mathbf{R}_2, \ldots, \mathbf{R}_n$ are the vectors from the origin to the respective mass particles $m_1, m_2, \ldots, m_n$, the end point of the vector

$$\mathbf{C} = \frac{\displaystyle\sum_{i=1}^{n} m_i\mathbf{R}_i}{\displaystyle\sum_{i=1}^{n} m_i}$$

is called the **center of gravity** of the system of particles. Show that for any vector $\mathbf{R}$,

$$\sum_{i=1}^{n} m_i(\mathbf{R} - \mathbf{R}_i) \cdot (\mathbf{R} - \mathbf{R}_i) = m(\mathbf{R} - \mathbf{C}) \cdot (\mathbf{R} - \mathbf{C}) + \sum_{i=1}^{n} m_i(\mathbf{C} - \mathbf{R}_i) \cdot (\mathbf{C} - \mathbf{R}_i)$$

where m is the total mass of all the particles.

11 If a particle moves under the influence of a force which is always directed toward the origin, show that $\mathbf{R} \times d^2\mathbf{R}/dt^2 = 0$, where $\mathbf{R}$ is the vector from the origin to the particle. *Hint:* Newton's law, i.e., mass × acceleration = force, remains correct when the acceleration and the force are interpreted as vector quantities.

12 If $\mathbf{R}(t)$ is the vector from the origin to the instantaneous position of a particle moving along a curve C, show that $\mathbf{R} \times d\mathbf{R}$ is equal to twice the area of the sector defined by the two vectors $\mathbf{R}(t)$ and $\mathbf{R}(t + \Delta t) \equiv \mathbf{R} + d\mathbf{R}$ and the arc of C which they intercept. Hence show that:
(a) If $\mathbf{R} \times d\mathbf{R} = 0$, the vector $\mathbf{R}$ has a constant direction.
(b) If $\mathbf{R} \times d^2\mathbf{R}/dt^2 = 0$, the particle moves so that the radius vector sweeps out equal area in equal times. This is a generalization of one of the laws of planetary motion discovered by Johannes Kepler (1571–1630).

13 Show that

$$d\left(\frac{\mathbf{u}(t)}{|\mathbf{u}(t)|}\right) = \frac{\dot{\mathbf{u}}(\mathbf{u} \cdot \mathbf{u}) - \mathbf{u}(\mathbf{u} \cdot \dot{\mathbf{u}})}{(\mathbf{u} \cdot \mathbf{u})^{3/2}}$$

14 While a disk is rotating in a horizontal plane with constant angular velocity ω, a particle moves radially outward from the center of the disk, its position vector in space being $\mathbf{R}(t) = t\mathbf{b}$, where $\mathbf{b}$ is a unit vector rotating with the disk. By noting first that $\mathbf{b}$ can be expressed in the form

$$\mathbf{b} = \cos \omega t \, \mathbf{i} + \sin \omega t \, \mathbf{j}$$

where $\mathbf{i}$ and $\mathbf{j}$ are mutually perpendicular vectors which are fixed in space, show that the velocity and acceleration of the particle are, respectively,

$$\dot{\mathbf{R}}(t) = \mathbf{b} + t\dot{\mathbf{b}} \qquad \text{and} \qquad \ddot{\mathbf{R}}(t) = 2\dot{\mathbf{b}} + t\ddot{\mathbf{b}} = 2\dot{\mathbf{b}} - \omega^2 t\mathbf{b}$$

[The component of acceleration corresponding to the term $2\dot{\mathbf{b}}$ is called the **Coriolis acceleration**, after the French physicist Gaspard Gustave de Coriolis (1792–1843).]

15 In Exercise 14, if the position vector of the particle in space is $\mathbf{R}(t) = x(t)\mathbf{i} + y(t)\mathbf{j}$, where $\mathbf{i}$ and $\mathbf{j}$ are mutually perpendicular unit vectors fixed with respect to the disk, show that $d\mathbf{i}/dt = \omega\mathbf{j}$ and $d\mathbf{j}/dt = -\omega\mathbf{i}$. Then show that

$$\dot{\mathbf{R}}(t) = (\dot{x}\mathbf{i} + \dot{y}\mathbf{j}) + (-\omega y\mathbf{i} + \omega x\mathbf{j})$$

$$\ddot{\mathbf{R}}(t) = (\ddot{x}\mathbf{i} + \ddot{y}\mathbf{j}) + 2\omega(-\dot{y}\mathbf{i} + \dot{x}\mathbf{j}) - \omega^2(x\mathbf{i} + y\mathbf{j})$$

Verify that the relative velocity vector, $\dot{x}\mathbf{i} + \dot{y}\mathbf{j}$, and the Coriolis acceleration, namely, $2\omega(-\dot{y}\mathbf{i} + \dot{x}\mathbf{j})$, are perpendicular.

16 If $\mathbf{T}$ is a unit vector tangent to a curve C, and if $\mathbf{N}$ is the unit normal to C in the osculating plane, the vector $\mathbf{B} = \mathbf{T} \times \mathbf{N}$ is called the **binormal** to C at the point where $\mathbf{T}$ and $\mathbf{N}$ are drawn. Using the fact that $d\mathbf{T}/ds = \mathbf{N}/\rho$, show that

$$\frac{d\mathbf{B}}{ds} = \mathbf{T} \times \frac{d\mathbf{N}}{ds}$$

and hence that $d\mathbf{B}/ds$ has the same direction as $\mathbf{N}$. (The absolute value of $d\mathbf{B}/ds$ is called the **torsion** of the curve C and measures the rate at which the osculating plane turns as we move along C.)

17 What is the equation of the osculating plane to the space curve $x = t^4$, $y = t^2$, $z = t^3$ at the point $P_1:(x_1, y_1, z_1)$? *Hint:* Let $P:(x, y, z)$ be a general point in the osculating plane, and impose the condition that the vector joining P to P_1 be coplanar with the vectors $\mathbf{T}$ and $d\mathbf{T}/ds$ at P_1.

18 What is the equation of the osculating plane to the curve $x = t$, $y = t^2$, $z = t^3$ at the point whose parameter is $t = 1$? What is the equation of the tangent to this curve at the point $t = 1$? What is the equation of the normal at $t = 1$? What is the equation of the binormal at $t = 1$?

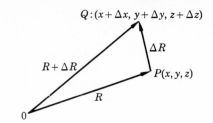

Figure 13.14
The coordinate vectors of two
neighboring points.

13.3 The Operator ∇

Let $\phi(x,y,z)$ be a scalar function of position possessing first partial derivatives with respect to x, y, and z throughout some region of space, and let $\mathbf{R} = x\mathbf{i} + y\mathbf{j} + z\mathbf{k}$ be the vector drawn from the origin to a general point $P:(x,y,z)$. If we move from P to a neighboring point $Q:(x + \Delta x, y + \Delta y, z + \Delta z)$ (Fig. 13.14), the function ϕ will change by an amount $\Delta\phi$ whose exact value, as derived in calculus, is

$$(1) \qquad \Delta\phi = \frac{\partial\phi}{\partial x}\,\Delta x + \frac{\partial\phi}{\partial y}\,\Delta y + \frac{\partial\phi}{\partial z}\,\Delta z + \varepsilon_1\,\Delta x + \varepsilon_2\,\Delta y + \varepsilon_3\,\Delta z$$

where ε_1, ε_2, ε_3 are quantities which approach zero as Q approaches P, that is, as Δx, Δy, and Δz approach zero. If we divide the change $\Delta\phi$ by the distance $\Delta s = |\Delta\mathbf{R}|$ between P and Q, we obtain a measure of the rate at which ϕ changes when we move from P to Q:

$$(2) \qquad \frac{\Delta\phi}{\Delta s} = \frac{\partial\phi}{\partial x}\frac{\Delta x}{\Delta s} + \frac{\partial\phi}{\partial y}\frac{\Delta y}{\Delta s} + \frac{\partial\phi}{\partial z}\frac{\Delta z}{\Delta s} + \varepsilon_1\frac{\Delta x}{\Delta s} + \varepsilon_2\frac{\Delta y}{\Delta s} + \varepsilon_3\frac{\Delta z}{\Delta s}$$

For instance, if $\phi(x,y,z)$ is the temperature at the general point $P:(x,y,z)$, then $\Delta\phi/\Delta s$ is the average rate of change of temperature in the direction in which Δs is measured. The limiting value of $\Delta\phi/\Delta s$ as Q approaches P along the segment PQ is called the **derivative of** ϕ **in the direction** PQ or simply the **directional derivative of** ϕ. Clearly, in the limit the last three terms in (2) become zero, and we have explicitly

$$(3) \qquad \frac{d\phi}{ds} = \frac{\partial\phi}{\partial x}\frac{dx}{ds} + \frac{\partial\phi}{\partial y}\frac{dy}{ds} + \frac{\partial\phi}{\partial z}\frac{dz}{ds}$$

The first factor in each product on the right in (3) depends only on ϕ and the coordinates of the point at which the derivatives of ϕ are evaluated. The second factor in each product is independent of ϕ and depends only on the direction in which the derivative is being computed. This observation suggests that $d\phi/ds$ can be thought of as the dot product of two vectors, one depending only on ϕ and the coordinates of P, the other depending only on the direction of ds; and in fact we can write

$$
\frac{d\phi}{ds} = \left(\frac{\partial\phi}{\partial x}\mathbf{i} + \frac{\partial\phi}{\partial y}\mathbf{j} + \frac{\partial\phi}{\partial z}\mathbf{k}\right)\cdot\left(\frac{dx}{ds}\mathbf{i} + \frac{dy}{ds}\mathbf{j} + \frac{dz}{ds}\mathbf{k}\right)
$$

$$(4) \qquad = \left(\frac{\partial\phi}{\partial x}\mathbf{i} + \frac{\partial\phi}{\partial y}\mathbf{j} + \frac{\partial\phi}{\partial z}\mathbf{k}\right)\cdot\frac{d\mathbf{R}}{ds}$$

The vector function

$$\frac{\partial\phi}{\partial x}\mathbf{i} + \frac{\partial\phi}{\partial y}\mathbf{j} + \frac{\partial\phi}{\partial z}\mathbf{k}$$

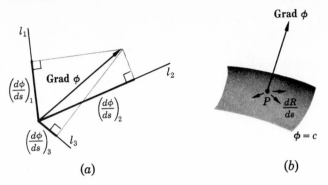

Figure 13.15
The geometrical interpretation of the gradient.

is known as the **gradient of** ϕ or simply **grad** ϕ, and in this notation (4) can be rewritten in the form

$$(4a) \qquad \frac{d\phi}{ds} = (\text{grad } \phi) \cdot \frac{d\mathbf{R}}{ds}$$

To determine the significance of **grad** ϕ, we observe first that since Δs is by definition just the length of $\Delta \mathbf{R}$, it follows that $d\mathbf{R}/ds$ is a unit vector. Hence the dot product $(\text{grad } \phi) \cdot d\mathbf{R}/ds$ is just the projection of **grad** ϕ in the direction of $d\mathbf{R}/ds$. Thus, according to (4a), **grad** ϕ *has the property that its projection in any direction is equal to the derivative of ϕ in that direction* (Fig. 13.15). Since the maximum projection of a vector is the vector itself, it is clear that **grad** ϕ *extends in the direction of the greatest rate of change of ϕ and has that rate of change for its length.*

If we set $\phi(x,y,z) = c$, we obtain, as c takes on different values, a family of surfaces known as the **level surfaces**† of ϕ, and on the assumption that ϕ is single-valued, one and only one level surface passes through any given point P. If we now consider the level surface which passes through P and fix our attention on neighboring points Q which lie on this surface, we have $d\phi/ds = 0$, since $\Delta\phi = 0$, because by definition ϕ has the same value at all points of a level surface. Hence, by (4a),

$$(5) \qquad (\text{grad } \phi) \cdot \frac{d\mathbf{R}}{ds} = 0$$

for any vector $d\mathbf{R}/ds$ which has the limiting direction of a secant PQ of the level surface. Clearly such vectors are all tangent to $\phi = c$ at the point P; hence, from the vanishing of the dot product in (5) it follows that **grad** ϕ is perpendicular to every tangent to the level surface at P. In other words, *the gradient of ϕ at any point P is perpendicular to the level surface of ϕ which passes through that point* (Fig. 13.15b). Evidently, **grad** ϕ is related to the level surfaces of ϕ in a way which is independent

---

† This name, which is used regardless of the number of independent variables, is suggested by the analogy between the general case and the two-dimensional topographic interpretation in which $\phi(x,y)$ is the elevation at the point (x,y) and the loci $\phi(x,y) = c$ are the contour lines, i.e., curves whose points are all at the same *level* or whose elevation above (or below) the xy plane is a constant.

of the particular coordinate system used to describe ϕ. In other words, **grad** ϕ depends only on the intrinsic properties of ϕ. It follows, therefore, that in the expression

$$\mathbf{grad}\ \phi = \frac{\partial \phi}{\partial x}\mathbf{i} + \frac{\partial \phi}{\partial y}\mathbf{j} + \frac{\partial \phi}{\partial z}\mathbf{k}$$

i, j, and **k** can be replaced by any other set of mutually perpendicular unit vectors provided that $\partial \phi/\partial x$, $\partial \phi/\partial y$, $\partial \phi/\partial z$ are replaced by the directional derivatives of ϕ along the new axes.

The gradient of a function is frequently written in operational form as

$$\mathbf{grad}\ \phi = \left(\mathbf{i}\frac{\partial}{\partial x} + \mathbf{j}\frac{\partial}{\partial y} + \mathbf{k}\frac{\partial}{\partial z}\right)\phi$$

The operational "vector" thus defined is usually denoted by the symbol ∇ (read "del"), i.e.,

(6)
$$\nabla \equiv \mathbf{i}\frac{\partial}{\partial x} + \mathbf{j}\frac{\partial}{\partial y} + \mathbf{k}\frac{\partial}{\partial z}$$

In this notation our earlier results can be written

(7)
$$\mathbf{grad}\ \phi = \nabla\phi$$

(8)
$$\frac{d\phi}{ds} = \nabla\phi \cdot \frac{d\mathbf{R}}{ds}$$

(9)
$$d\phi = \nabla\phi \cdot d\mathbf{R}$$

Also, if ϕ is a function of a single variable u which, in turn, is a function of x, y, and z, then

$$\nabla\phi = \frac{\partial \phi}{\partial x}\mathbf{i} + \frac{\partial \phi}{\partial y}\mathbf{j} + \frac{\partial \phi}{\partial z}\mathbf{k}$$

$$= \frac{d\phi}{du}\frac{\partial u}{\partial x}\mathbf{i} + \frac{d\phi}{du}\frac{\partial u}{\partial y}\mathbf{j} + \frac{d\phi}{du}\frac{\partial u}{\partial z}\mathbf{k}$$

$$= \frac{d\phi}{du}\left(\frac{\partial u}{\partial x}\mathbf{i} + \frac{\partial u}{\partial y}\mathbf{j} + \frac{\partial u}{\partial z}\mathbf{k}\right)$$

(10)
$$= \frac{d\phi}{du}\nabla u$$

EXAMPLE 1

What is the directional derivative of the function $\phi(x,y,z) = xy^2 + yz^3$ at the point $(2,-1,1)$ in the direction of the vector $\mathbf{i} + 2\mathbf{j} + 2\mathbf{k}$?

Our first step must be to find the gradient of ϕ at the point $(2,-1,1)$. This is

$$\nabla\phi = \frac{\partial(xy^2 + yz^3)}{\partial x}\mathbf{i} + \frac{\partial(xy^2 + yz^3)}{\partial y}\mathbf{j} + \frac{\partial(xy^2 + yz^3)}{\partial z}\mathbf{k}\bigg|_{2,-1,1}$$

$$= y^2\mathbf{i} + (2xy + z^3)\mathbf{j} + 3yz^2\mathbf{k}\bigg|_{2,-1,1}$$

$$= \mathbf{i} - 3\mathbf{j} - 3\mathbf{k}$$

The projection of this in the direction of the given vector will be the required directional derivative. Since this projection can be found at once as the dot product of $\nabla\phi$ and a unit vector in the given direction, we next reduce $\mathbf{i} + 2\mathbf{j} + 2\mathbf{k}$ to a unit vector by dividing it by its magnitude, getting

$$\frac{\mathbf{i} + 2\mathbf{j} + 2\mathbf{k}}{\sqrt{1 + 4 + 4}} = \tfrac{1}{3}\mathbf{i} + \tfrac{2}{3}\mathbf{j} + \tfrac{2}{3}\mathbf{k}$$

The answer to our problem is therefore

$$\nabla\phi \cdot (\tfrac{1}{3}\mathbf{i} + \tfrac{2}{3}\mathbf{j} + \tfrac{2}{3}\mathbf{k}) = (\mathbf{i} - 3\mathbf{j} - 3\mathbf{k}) \cdot (\tfrac{1}{3}\mathbf{i} + \tfrac{2}{3}\mathbf{j} + \tfrac{2}{3}\mathbf{k}) = -\tfrac{11}{3}$$

The negative sign, of course, indicates that ϕ decreases in the given direction.

EXAMPLE 2

What is the unit normal to the surface $xy^3z^2 = 4$ at the point $(-1,-1,2)$?

Let us regard the given surface as a particular level surface of the function $\phi = xy^3z^2$. Then the gradient of this function at the point $(-1,-1,2)$ will be perpendicular to the level surface through $(-1,-1,2)$, which is the given surface. When this gradient has been found, the unit normal can be obtained at once by dividing the gradient by its magnitude:

$$\nabla\phi = \frac{\partial(xy^3z^2)}{\partial x}\mathbf{i} + \frac{\partial(xy^3z^2)}{\partial y}\mathbf{j} + \frac{\partial(xy^3z^2)}{\partial z}\mathbf{k}\bigg|_{-1,-1,2}$$

$$= -4\mathbf{i} - 12\mathbf{j} + 4\mathbf{k}$$

$$|\nabla\phi| = \sqrt{16 + 144 + 16} = 4\sqrt{11}$$

$$\frac{\nabla\phi}{|\nabla\phi|} = \frac{-4\mathbf{i} - 12\mathbf{j} + 4\mathbf{k}}{4\sqrt{11}} = -\frac{1}{\sqrt{11}}\mathbf{i} - \frac{3}{\sqrt{11}}\mathbf{j} + \frac{1}{\sqrt{11}}\mathbf{k}$$

It may be necessary to reverse the direction of this result by multiplying it by -1, depending on which side of the surface we wish the normal to extend.

The vector character of the operator ∇ suggests that we also consider dot and cross products in which it appears as one factor. If $\mathbf{F} = F_1\mathbf{i} + F_2\mathbf{j} + F_3\mathbf{k}$ is a vector whose components are differentiable functions of x, y, and z, this leads to the combinations

$$\nabla \cdot \mathbf{F} = \left(\mathbf{i}\frac{\partial}{\partial x} + \mathbf{j}\frac{\partial}{\partial y} + \mathbf{k}\frac{\partial}{\partial z}\right) \cdot (F_1\mathbf{i} + F_2\mathbf{j} + F_3\mathbf{k})$$

(11)
$$= \frac{\partial F_1}{\partial x} + \frac{\partial F_2}{\partial y} + \frac{\partial F_3}{\partial z}$$

which is known as the **divergence** of the vector $\mathbf{F}$, and

$$\nabla \times \mathbf{F} = \left(\mathbf{i}\frac{\partial}{\partial x} + \mathbf{j}\frac{\partial}{\partial y} + \mathbf{k}\frac{\partial}{\partial z}\right) \cdot (F_1\mathbf{i} + F_2\mathbf{j} + F_3\mathbf{k})$$

$$= \mathbf{i}\left(\frac{\partial F_3}{\partial y} - \frac{\partial F_2}{\partial z}\right) - \mathbf{j}\left(\frac{\partial F_3}{\partial x} - \frac{\partial F_1}{\partial z}\right) + \mathbf{k}\left(\frac{\partial F_2}{\partial x} - \frac{\partial F_1}{\partial y}\right)$$

(12)
$$= \begin{vmatrix} \mathbf{i} & \mathbf{j} & \mathbf{k} \\ \dfrac{\partial}{\partial x} & \dfrac{\partial}{\partial y} & \dfrac{\partial}{\partial z} \\ F_1 & F_2 & F_3 \end{vmatrix}$$

which is known as the **curl** of F.

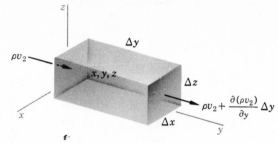

Figure 13.16
A typical volume element in a region filled with a moving fluid.

Both the divergence and the curl admit of physical interpretations which justify their names. For instance, to illustrate the significance of the divergence, consider a region of space filled with a moving fluid, and let

$$\mathbf{v} = v_1\mathbf{i} + v_2\mathbf{j} + v_3\mathbf{k}$$

be a vector function representing at each point the instantaneous velocity of the particle of fluid. If we fix our attention on an infinitesimal volume (Fig. 13.16) in the region occupied by the fluid, there will be flow through each of its faces, and as a result the amount of fluid within the element may vary. To measure this variation, let us compute the loss of fluid from the element in the time Δt.

The volume of fluid which passes through one face of the element ΔV in time Δt is approximately equal to the component of the fluid velocity normal to the face times the area of the face times Δt, and the corresponding mass flow is, of course, the product of this volume and the density of the fluid ρ. Hence, computing the loss of fluid through each face in turn (remembering that since the fluid is not assumed to be incompressible, the density as well as the velocity may vary from point to point), we have

Right face:
$$\left[\rho v_2 + \frac{\partial(\rho v_2)}{\partial y}\, \Delta y \right] \Delta x\, \Delta z\, \Delta t$$

Left face:
$$- \rho v_2\, \Delta x\, \Delta z\, \Delta t$$

Front face:
$$\left[\rho v_1 + \frac{\partial(\rho v_1)}{\partial x}\, \Delta x \right] \Delta y\, \Delta z\, \Delta t$$

Rear face:
$$- \rho v_1\, \Delta y\, \Delta z\, \Delta t$$

Top face:
$$\left[\rho v_3 + \frac{\partial(\rho v_3)}{\partial z}\, \Delta z \right] \Delta x\, \Delta y\, \Delta t$$

Bottom face:
$$- \rho v_3\, \Delta x\, \Delta y\, \Delta t$$

If we add these and convert the resulting estimate of the absolute loss of fluid from ΔV in the interval Δt into the loss per unit volume per unit time by dividing by $\Delta V\, \Delta t \equiv \Delta x\, \Delta y\, \Delta z\, \Delta t$, we obtain in the limit

$$\text{Rate of loss per unit volume} = \frac{\partial(\rho v_1)}{\partial x} + \frac{\partial(\rho v_2)}{\partial y} + \frac{\partial(\rho v_3)}{\partial z}$$

which is precisely the divergence of the vector $\rho \mathbf{v}$. Thus fluid mechanics affords one possible interpretation of the divergence as the rate of loss of fluid per unit volume. If the fluid is incompressible, there can be neither gain nor loss of fluid in a general element. Hence, since the density ρ is constant for an incompressible fluid, we must have

$$\nabla \cdot \rho\mathbf{v} = \rho\nabla \cdot \mathbf{v} = 0 \quad \text{or} \quad \nabla \cdot \mathbf{v} = 0$$

which is known as the **equation of continuity** for incompressible fluids. However, if ΔV encloses a source of fluid, there is a net loss of fluid through the surface of ΔV equal to the amount *diverging* from the source. Similar results, of course, hold for such things as electric and magnetic flux, which exhibit many of the properties of incompressible fluids.

To find a possible interpretation of the curl, let us consider a body rotating with uniform angular speed ω about an axis l. Let us define the **vector angular velocity Ω** to be a vector of length ω extending along l in the direction in which a right-handed screw would advance if subject to the same rotation as the body. Finally, let $\mathbf{R}$ be the vector drawn from any point O on the axis l to an arbitrary point P in the body.

From Fig. 13.17 it is evident that the radius at which P rotates is $|\mathbf{R}| \cdot |\sin \theta|$. Hence, the linear speed of P is

$$|\mathbf{v}| = \omega|\mathbf{R}| \cdot |\sin \theta| = |\Omega| \cdot |\mathbf{R}| \cdot |\sin \theta| = |\Omega \times \mathbf{R}|$$

Moreover, the vector velocity $\mathbf{v}$ is directed perpendicular to the plane of Ω and $\mathbf{R}$, so that Ω, $\mathbf{R}$, and $\mathbf{v}$ form a right-handed system. Hence, the cross product $\Omega \times \mathbf{R}$ gives not only the magnitude of $\mathbf{v}$ but the direction as well.

If we now take the point O as the origin of coordinates, we can write

$$\mathbf{R} = x\mathbf{i} + y\mathbf{j} + z\mathbf{k} \quad \text{and} \quad \Omega = \Omega_1\mathbf{i} + \Omega_2\mathbf{j} + \Omega_3\mathbf{k}$$

Hence, the equation $\mathbf{v} = \Omega \times \mathbf{R}$ can be written at length in the form

$$\mathbf{v} = (\Omega_2 z - \Omega_3 y)\mathbf{i} - (\Omega_1 z - \Omega_3 x)\mathbf{j} + (\Omega_1 y - \Omega_2 x)\mathbf{k}$$

If we take the curl of $\mathbf{v}$, we therefore have

$$\nabla \times \mathbf{v} = \begin{vmatrix} \mathbf{i} & \mathbf{j} & \mathbf{k} \\ \dfrac{\partial}{\partial x} & \dfrac{\partial}{\partial y} & \dfrac{\partial}{\partial z} \\ \Omega_2 z - \Omega_3 y & -(\Omega_1 z - \Omega_3 x) & \Omega_1 y - \Omega_2 x \end{vmatrix}$$

Expanding this, remembering that Ω is a constant vector, we find

$$\nabla \times \mathbf{v} = 2\Omega_1\mathbf{i} + 2\Omega_2\mathbf{j} + 2\Omega_3\mathbf{k} = 2\Omega$$

$$\Omega = \tfrac{1}{2}\nabla \times \mathbf{v}$$

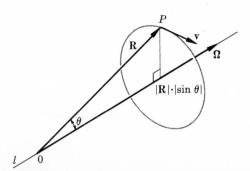

Figure 13.17
A physical interpretation of the curl.

The angular velocity of a uniformly rotating body is thus equal to one-half the curl of the linear velocity of any point of the body. The aptness of the name *curl* in this connection is apparent.

The results of applying the operator $\mathbf{V}$ to various combinations of scalar and vector functions can be found by the following formulas:†

(13) $$\mathbf{V} \cdot \phi\mathbf{v} = \phi\mathbf{V} \cdot \mathbf{v} + \mathbf{v} \cdot \mathbf{V}\phi$$

(14) $$\mathbf{V} \times \phi\mathbf{v} = \phi\mathbf{V} \times \mathbf{v} + (\mathbf{V}\phi) \times \mathbf{v}$$

(15) $$\mathbf{V} \cdot (\mathbf{u} \times \mathbf{v}) = \mathbf{v} \cdot \mathbf{V} \times \mathbf{u} - \mathbf{u} \cdot \mathbf{V} \times \mathbf{v}$$

(16) $$\mathbf{V} \times (\mathbf{u} \times \mathbf{v}) = \mathbf{v} \cdot \mathbf{V}\mathbf{u} - \mathbf{u} \cdot \mathbf{V}\mathbf{v} + \mathbf{u}\mathbf{V} \cdot \mathbf{v} - \mathbf{v}\mathbf{V} \cdot \mathbf{u}$$

(17) $$\mathbf{V}(\mathbf{u} \cdot \mathbf{v}) = \mathbf{u} \cdot \mathbf{V}\mathbf{v} + \mathbf{v} \cdot \mathbf{V}\mathbf{u} + \mathbf{u} \times (\mathbf{V} \times \mathbf{v}) + \mathbf{v} \times (\mathbf{V} \times \mathbf{u})$$

(18) $$\mathbf{V} \times \mathbf{V}\phi = \mathbf{0}$$

(19) $$\mathbf{V} \cdot \mathbf{V} \times \mathbf{v} = 0$$

(20) $$\mathbf{V} \times (\mathbf{V} \times \mathbf{v}) = \mathbf{V}(\mathbf{V} \cdot \mathbf{v}) - \mathbf{V} \cdot \mathbf{V}\mathbf{v} = \mathbf{V}(\mathbf{V} \cdot \mathbf{v}) - \mathbf{V}^2\mathbf{v}$$

These identities can all be verified by direct expansion. For instance, to prove (13), we have

$$\mathbf{V} \cdot (\phi\mathbf{v}) = \mathbf{V} \cdot [\phi(v_1\mathbf{i} + v_2\mathbf{j} + v_3\mathbf{k})]$$

$$= \frac{\partial(\phi v_1)}{\partial x} + \frac{\partial(\phi v_2)}{\partial y} + \frac{\partial(\phi v_3)}{\partial z}$$

$$= \left(\phi \frac{\partial v_1}{\partial x} + v_1 \frac{\partial \phi}{\partial x}\right) + \left(\phi \frac{\partial v_2}{\partial y} + v_2 \frac{\partial \phi}{\partial y}\right) + \left(\phi \frac{\partial v_3}{\partial z} + v_3 \frac{\partial \phi}{\partial z}\right)$$

which, on regrouping, is simply $\phi\mathbf{V} \cdot \mathbf{v} + \mathbf{v} \cdot \mathbf{V}\phi$ as asserted.

In general, however, it is easier to establish formulas like those in the above list by treating $\mathbf{V}$ as a vector, manipulating the expressions according to the appropriate formulas from vector algebra, and finally giving $\mathbf{V}$ its operational meaning. Since $\mathbf{V}$ is a linear combination of scalar differential operators which obey the usual product rule of differentiation, i.e., act on the factors in a product one at a time, it is clear that $\mathbf{V}$ itself has this property. In other words, we can apply $\mathbf{V}$ to products of various sorts by assuming that each of the factors, in turn, is the only one which is variable and adding the partial results so obtained. As a notation to aid us in determining these partial results, it is helpful to attach to $\mathbf{V}$, whenever it is followed by more than one factor, a subscript indicating the one factor upon which it is currently allowed to operate.

To prove (14), using the second, more formal, procedure, we suppose first that the scalar function ϕ is a constant; i.e., we let $\mathbf{V}$ operate only on the vector $\mathbf{v}$. Then we can write

$$\mathbf{V}_v \times (\phi\mathbf{v}) = \phi \mathbf{V} \times \mathbf{v}$$

---

† We must remember, however, that these results are correct only for the cartesian form of the operator $\mathbf{V}$ given by Eq. (6). Different formulas arise when $\mathbf{V}$ is expressed in terms of more general coordinate systems.

where the subscript has been omitted from the right-hand side, since it is always completely clear what ∇ operates on when it is followed by just one factor. Similarly, if we regard $\mathbf{v}$ as constant and ϕ as variable, we have

$$\nabla_\phi \times (\phi\mathbf{v}) = (\nabla\phi) \times \mathbf{v}$$

where the parentheses now restrict the effect of ∇ to the factor ϕ alone and so make a subscript on ∇ unnecessary. Finally, adding our two partial results, we have

$$\nabla_v \times (\phi\mathbf{v}) + \nabla_\phi \times (\phi\mathbf{v}) \equiv \nabla \times (\phi\mathbf{v}) = \phi\nabla \times \mathbf{v} + (\nabla\phi) \times \mathbf{v}$$

To prove (15), we have, from the cyclic properties of scalar triple products,

$$\nabla_u \cdot (\mathbf{u} \times \mathbf{v}) = \mathbf{v} \cdot \nabla \times \mathbf{u} \qquad \text{and} \qquad \nabla_v \cdot (\mathbf{u} \times \mathbf{v}) = -\mathbf{u} \cdot \nabla \times \mathbf{v}$$

Hence, adding these two partial results, we find

$$\nabla_u \cdot (\mathbf{u} \times \mathbf{v}) + \nabla_v \cdot (\mathbf{u} \times \mathbf{v}) \equiv \nabla \cdot (\mathbf{u} \times \mathbf{v}) = \mathbf{v} \cdot \nabla \times \mathbf{u} - \mathbf{u} \cdot \nabla \times \mathbf{v}$$

To prove (16), we have

$$\nabla_u \times (\mathbf{u} \times \mathbf{v}) = (\nabla_u \cdot \mathbf{v})\mathbf{u} - (\nabla_u \cdot \mathbf{u})\mathbf{v} = \mathbf{v} \cdot \nabla\mathbf{u} - \mathbf{v}\nabla \cdot \mathbf{u}$$

and
$$\nabla_v \times (\mathbf{u} \times \mathbf{v}) = (\nabla_v \cdot \mathbf{v})\mathbf{u} - (\nabla_v \cdot \mathbf{u})\mathbf{v} = \mathbf{u}\nabla \cdot \mathbf{v} - \mathbf{u} \cdot \nabla\mathbf{v}$$

Adding gives

$$\nabla_u \times (\mathbf{u} \times \mathbf{v}) + \nabla_v \times (\mathbf{u} \times \mathbf{v}) \equiv \nabla \times (\mathbf{u} \times \mathbf{v})$$
$$= \mathbf{v} \cdot \nabla\mathbf{u} - \mathbf{u} \cdot \nabla\mathbf{v} + \mathbf{u}\nabla \cdot \mathbf{v} - \mathbf{v}\nabla \cdot \mathbf{u}$$

To prove (17), we note that

$$\mathbf{u} \times (\nabla \times \mathbf{v}) \equiv \mathbf{u} \times (\nabla_v \times \mathbf{v}) = (\mathbf{u} \cdot \mathbf{v})\nabla_v - (\mathbf{u} \cdot \nabla)\mathbf{v}$$
$$= \nabla_v(\mathbf{u} \cdot \mathbf{v}) - \mathbf{u} \cdot \nabla\mathbf{v}$$

and
$$\mathbf{v} \times (\nabla \times \mathbf{u}) \equiv \mathbf{v} \times (\nabla_u \times \mathbf{u}) = (\mathbf{v} \cdot \mathbf{u})\nabla_u - (\mathbf{v} \cdot \nabla)\mathbf{u}$$
$$= \nabla_u(\mathbf{u} \cdot \mathbf{v}) - \mathbf{v} \cdot \nabla\mathbf{u}$$

Hence, transposing and adding, we find

$$\nabla_u(\mathbf{u} \cdot \mathbf{v}) + \nabla_v(\mathbf{u} \cdot \mathbf{v}) \equiv \nabla(\mathbf{u} \cdot \mathbf{v})$$
$$= \mathbf{u} \times (\nabla \times \mathbf{v}) + \mathbf{v} \times (\nabla \times \mathbf{u}) + \mathbf{u} \cdot \nabla\mathbf{v} + \mathbf{v} \cdot \nabla\mathbf{u}$$

The fact that the operational coefficient $\nabla \times \nabla$ in formula (18) appears as the cross product of two identical factors suggests that $\nabla \times \nabla\phi$ is indeed zero, and it is easy to verify that this is the case (see Exercise 8). Similarly, the fact that the left member of formula (19) appears as a scalar triple product with two identical factors suggests the truth of the formula, and again it is not difficult to prove that $\nabla \cdot \nabla \times \mathbf{v}$ is always equal to zero (see Exercise 9). It should be noted, however, that the structure of these formulas is not sufficient to establish their truth, since the formal scalar triple product $\nabla \times \mathbf{v} \cdot \mathbf{v}$ is not zero in general, even though it has two identical factors (see Exercise 11).

Finally, it is easy to establish formula (20) by direct expansion (see Exercise 10), or it can be verified by applying the usual rule for expanding a vector triple product:

$$\mathbf{V} \times (\mathbf{V} \times \mathbf{v}) = (\mathbf{V} \cdot \mathbf{v})\mathbf{V} - (\mathbf{V} \cdot \mathbf{V})\mathbf{v} = \mathbf{V}(\mathbf{V} \cdot \mathbf{v}) - \mathbf{V}^2\mathbf{v}$$

where the conventional symbol $\mathbf{V}^2$ has been substituted for the second-order operator

$$\mathbf{V} \cdot \mathbf{V} = \left(\mathbf{i}\frac{\partial}{\partial x} + \mathbf{j}\frac{\partial}{\partial y} + \mathbf{k}\frac{\partial}{\partial z} \right) \cdot \left(\mathbf{i}\frac{\partial}{\partial x} + \mathbf{j}\frac{\partial}{\partial y} + \mathbf{k}\frac{\partial}{\partial z} \right)$$

$$= \frac{\partial^2}{\partial x^2} + \frac{\partial^2}{\partial y^2} + \frac{\partial^2}{\partial z^2}$$

EXERCISES

1 Compute the divergence and curl of each of the following vectors:
   (a) $y^2\mathbf{i} + 2x^2z\mathbf{j} - xyz\mathbf{k}$ (b) $xyz\mathbf{i} + 3x^2y\mathbf{j} + (xz^2 - y^2z)\mathbf{k}$
2 Compute the gradient of each of the following functions:
   (a) $x^2 + 2yz$ (b) $x \sin yz$ (c) $x^3 + y^3 - 3xyz$
3 What is the directional derivative of the function $2xy + z^2$ at the point $(1,-1,3)$ in the direction of the vector $\mathbf{i} + 2\mathbf{j} + 2\mathbf{k}$?
4 What is the unit normal to the surface $z = x^2 + y^2$ at the point $(1,-2,5)$?
5 What is the angle between the normals to the surface $xy = z^2$ at the points $(1,4,2)$ and $(-3,-3,3)$?
6 Verify Eq. (14) by direct expansion.
7 Verify Eq. (15) by direct expansion.
8 Verify Eq. (18) by direct expansion.
9 Verify Eq. (19) by direct expansion.
10 Verify Eq. (20) by direct expansion.
11 Show by an example that in general $\mathbf{V} \times \mathbf{v} \cdot \mathbf{v} \neq 0$.
12 Prove that the curl of any vector whose direction is constant is perpendicular to that direction.
13 What is the generalization of Eq. (10) to the case in which ϕ is a function of u, v, and w, where u, v, and w are each functions of x, y, and z?

In the following exercises, $\mathbf{R} = x\mathbf{i} + y\mathbf{j} + z\mathbf{k}$, as usual, and $r = |\mathbf{R}| = \sqrt{x^2 + y^2 + z^2}$.

14 Prove that $\mathbf{V} \times \mathbf{R} = 0$. What is $\mathbf{V} \cdot \mathbf{R}$?
15 What is $(\mathbf{A} \cdot \mathbf{V})\mathbf{R}$? If $\mathbf{A}$ is an arbitrary constant vector, prove that $\mathbf{V}(\mathbf{A} \cdot \mathbf{R}) = \mathbf{A}$.
16 Prove that $\mathbf{V}r^n = nr^{n-2}\mathbf{R}$.
17 For what values of n is $\nabla^2 r^n = 0$?
18 Prove that the curl of $f(r)\mathbf{R}$ is identically zero.
19 Determine n so that $\mathbf{V} \cdot (r^n\mathbf{R})$ will vanish identically.
20 Prove that $(\mathbf{A} \times \mathbf{V}) \times \mathbf{R} = -2\mathbf{A}$. What is $(\mathbf{A} \times \mathbf{V}) \cdot \mathbf{R}$?
21 Prove that $\mathbf{V} \cdot [(1/r)(\mathbf{A} \times \mathbf{R})] = 0$ for any constant vector $\mathbf{A}$.
22 Prove that

$$\mathbf{V} \times \left[\frac{1}{r}(\mathbf{A} \times \mathbf{R}) \right] = \frac{1}{r}\mathbf{A} + \frac{\mathbf{A} \cdot \mathbf{R}}{r^3}\mathbf{R}$$

   for any constant vector $\mathbf{A}$.
23 Prove that

$$\mathbf{V}\phi_1 \times \mathbf{V}\phi_2 = \mathbf{V} \times (\phi_1 \mathbf{V}\phi_2) = -\mathbf{V} \times (\phi_2 \mathbf{V}\phi_1)$$

24 If $u = x + y + z$, $v = x + y$, and $w = -2xz - 2yz - z^2$, show that $[\mathbf{V}u \; \mathbf{V}v \; \mathbf{V}w] = 0$.
25 If three functions u, v, and w are connected by a relation $f(u,v,w) = 0$, prove that $[\mathbf{V}u \; \mathbf{V}v \; \mathbf{V}w] = 0$. *Hint:* Consider the dot product of $\mathbf{V}f$ and $\mathbf{V}u \times \mathbf{V}v$.
26 If $\mathbf{V}_1$ and $\mathbf{V}_2$ are the vectors which join the fixed points $P_1(x_1,y_1,z_1)$ and $P_2:(x_2,y_2,z_2)$ to the variable point $P:(x,y,z)$, prove that the gradient of $\mathbf{V}_1 \cdot \mathbf{V}_2$ is $\mathbf{V}_1 + \mathbf{V}_2$. What is $\mathbf{V} \cdot (\mathbf{V}_1 \times \mathbf{V}_2)$? What is $\mathbf{V} \times (\mathbf{V}_1 \times \mathbf{V}_2)$?

13.4 Line, Surface, and Volume Integrals

In the rest of our work in vector analysis and in much of the work ahead in the chapters on complex variables, a simple extension of the familiar process of integration known as line integration will be of fundamental importance. Although in vector analysis we are usually concerned with line integrals taken along space curves, it is convenient to begin our discussion with a consideration of line integration along plane curves, since the applications of line integration in our study of complex variables will be exclusively in two dimensions. In both the two- and the three-dimensional case, our work will involve only continuous curves which are **sectionally smooth**; i.e., curves which are continuous and consist of a finite number of arcs on each of which the tangent changes continuously. Clearly, such curves can have at most a finite number of "corners" where the direction of the tangent changes abruptly. Moreover, as we learned in calculus, the length of such a curve between any two of its points is finite.

Let $F(x,y)$† be a function of x and y, and let C be a continuous, sectionally smooth curve joining the points A and B. Furthermore, let the arc of C between A and B be divided into n segments Δs_i whose projections on the x and y axes are, respectively, Δx_i and Δy_i, and let (ξ_i, η_i) be the coordinates of an arbitrary point in the segment Δs_i (Fig. 13.18).

If we evaluate the given function $F(x,y)$ at each of the points (ξ_i, η_i) and form the products

$$F(\xi_i, \eta_i)\, \Delta x_i \qquad F(\xi_i, \eta_i)\, \Delta y_i \qquad F(\xi_i, \eta_i)\, \Delta s_i$$

and then sum over all the subdivisions of the arc AB, we have the three sums

$$\sum_{i=1}^{n} F(\xi_i, \eta_i)\, \Delta x_i \qquad \sum_{i=1}^{n} F(\xi_i, \eta_i)\, \Delta y_i \qquad \sum_{i=1}^{n} F(\xi_i, \eta_i)\, \Delta s_i$$

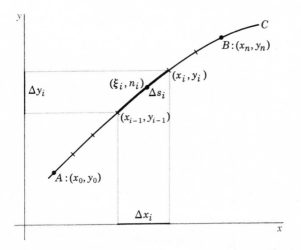

Figure 13.18
The subdivision of an arc preparatory to defining a line integral.

---

† $F(x,y)$ bears no relation to the equation of C and is merely a function defined throughout some region in the xy plane containing the portion of the curve C under consideration.

The limits of these sums, as n becomes infinite in such a way that the length of each Δs_i approaches zero, are known as **line integrals** and are written, respectively,

$$\int_C F(x,y)\,dx \qquad \int_C F(x,y)\,dy \qquad \int_C F(x,y)\,ds$$

It can be shown† that the continuity of $F(x,y)$ is a sufficient condition for the existence of the limits which define these integrals.

In these definitions, Δx_i and Δy_i are signed quantities, whereas Δs_i is intrinsically positive. Thus the following properties of ordinary definite integrals

    a. $\int_A^B c\phi(t)\,dt = c\int_A^B \phi(t)$ c a constant

    b. $\int_A^B [\phi_1(t) \pm \phi_2(t)]\,dt = \int_A^B \phi_1(t)\,dt \pm \int_A^B \phi_2(t)\,dt$

    c. $\int_A^B \phi(t)\,dt = -\int_B^A \phi(t)\,dt$

    d. $\int_A^P \phi(t)\,dt + \int_P^B \phi(t)\,dt = \int_A^B \phi(t)\,dt$

are equally valid for line integrals of the first two types, provided that throughout each formula the curve joining A and B remains the same. On the other hand, line integrals of the third type, although they do have properties **a** and **b**, do not have property **c**, since, in fact,

$$\int_A^B F(x,y)\,ds = \int_B^A F(x,y)\,ds$$

Moreover, property **d** holds for these integrals if and only if P is between A and B on the path of integration. In general, we shall be more interested in integrals of the first two types than in those of the third.

In many problems the path of integration C will consist of one or more simple closed curves‡ forming the boundary of a region R. Clearly, integration can be performed in either of two directions around a closed curve, and it is important that we be able to distinguish between them. This is done by defining the **positive direction** around a closed curve as the direction in which an observer would move if he traversed the curve in such a way that the area of R was always on his left. According to this definition, if R is the interior of a simple closed curve C, then the positive direction around C is the counterclockwise direction. If R is the region exterior to a simple closed curve C, then the positive direction around C is clockwise. If R is the region interior to a simple closed curve C_1 and exterior to a second simple closed curve C_2, then the positive direction of traversing the entire boundary of R, namely $C = C_1 + C_2$, is counterclockwise around C_1 and clockwise around C_2.

Much of the initial strangeness of line integrals will disappear if we observe that the ordinary definite integrals of elementary calculus are just line integrals in which the curve C is the x axis and the integrand is a function of x alone. Moreover, the evaluation of line integrals can be reduced to the evaluation of ordinary definite integrals, as the following example shows.

---

† See, for instance, D. V. Widder, "Advanced Calculus," p. 187, Prentice-Hall, Englewood Cliffs, N.J., 1947.

‡ For our purposes it is sufficient to define a **simple closed curve** as a closed, sectionally smooth curve which does not cross itself. That this is not the whole story, however, can be inferred from G. T. Whyburn, What Is a Curve?, *Am. Math. Mon.*, vol. 49, pp. 493–497, October 1942.

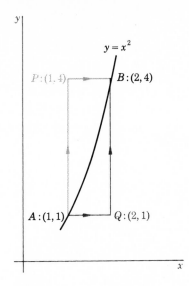

Figure 13.19
Possible paths for line integration
from $A:(1,1)$ to $B:(2,4)$.

EXAMPLE 1

What is the value of $\int_B^A [1/(x + y)]\, dx$ along each of the paths shown in Fig. 13.19?

Before this integral can be evaluated, y must be expressed in terms of x. To do this, we recall from the definition of a line integral that the integrand is always to be evaluated *along the path of integration*. Along the parabolic arc joining A and B, we have $y = x^2$, and making this substitution in the given line integral yields the ordinary definite integral

$$\int_1^2 \frac{dx}{x + x^2} = \int_1^2 \left(\frac{1}{x} - \frac{1}{1 + x}\right) dx = \left[\ln x - \ln (1 + x)\right]_1^2 = \ln \tfrac{4}{3}$$

Similarly, along the straight-line path from A to B, we have $y = 3x - 2$, and making this substitution in the integrand of the given integral, we obtain the ordinary definite integral

$$\int_1^2 \frac{dx}{x + (3x - 2)} = \frac{1}{4}\left[\ln (4x - 2)\right]_1^2 = \tfrac{1}{4}(\ln 6 - \ln 2) = \tfrac{1}{4} \ln 3$$

To compute the line integral along the path APB, we must perform two integrations, one along AP and one along PB, since the relation expressing y in terms of x is different on these two segments. Along AP the integral is obviously zero, since x remains constant and therefore in the sum leading to the integral each Δx_i is zero. Along PB, on which $y = 4$, we have the integral

$$\int_1^2 \frac{dx}{x + 4} = \left[\ln (x + 4)\right]_1^2 = \ln \tfrac{6}{5}$$

which is thus the value of the integral along the entire path APB.

Along the path AQB we again have two integrations to perform. Along AQ, on which $y = 1$, we have the integral

$$\int_1^2 \frac{dx}{x + 1} = \left[\ln (x + 1)\right]_1^2 = \ln \tfrac{3}{2}$$

Along the vertical segment QB the integral is again zero. Hence for the entire path AQB the value of the given integral is $\ln \tfrac{3}{2}$.

This example not only illustrates the computational details of line integration but also shows that in general a line integral depends not only on the end points of the integration but also upon the particular path which joins them.

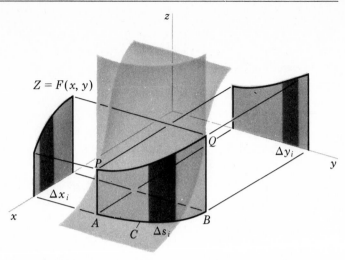

Figure 13.20
The interpretation of a line integral as an area.

It is possible, as in ordinary integration, to interpret a line integral as an area. For if we think of the integrand function $F(x,y)$ as defining a surface extending above some region in the xy plane, then the vertical cylindrical surface standing on the arc AB as base, or directrix, will cut the surface $z = F(x,y)$ in some curve, such as the arc PQ shown in Fig. 13.20. This curve is clearly the upper boundary of the portion $ABQP$ of the cylindrical surface which lies above the xy plane, below the surface $z = F(x,y)$, and between the generators AP and BQ. Moreover, the product $F(\xi_i,\eta_i) \, \Delta s_i$ is approximately the area of the vertical strip of this portion of the surface which stands above the infinitesimal base Δs_i. Hence the sum

$$\sum_{i=1}^{n} F(\xi_i,\eta_i) \, \Delta s_i$$

is approximately equal to the curved area $ABQP$, and, in the limit, the integral

$$\int_C F(x,y) \, ds$$

gives this area exactly.

In a similar fashion, the product $F(\xi_i,\eta_i) \, \Delta x_i$ is approximately the area of the projection on the xz plane of the vertical strip standing on Δs_i; the sum

$$\sum_{i=1}^{n} F(\xi_i,\eta_i) \, \Delta x_i$$

represents approximately the area of the projection on the xz plane of the entire curved area $ABQP$; and in the limit, the integral

$$\int_C F(x,y) \, dx$$

gives the projected area exactly. In the same way, the integral

$$\int_C F(x,y) \, dy$$

represents the area of the projection of $ABQP$ on the yz plane.

Although this geometrical interpretation of line integrals as areas is vivid and easily grasped, it obscures the fact that almost invariably in applications the function $F(x,y)$ describes some physical property of the plane of integration and is actually unrelated to any other region of space.

EXAMPLE 2

If a particle is attracted toward the origin by a force whose magnitude is proportional to the distance r of the particle from the origin, how much work is done when the particle is moved from the point $(0,1)$ to the point $(1,2)$ along the path $y = 1 + x^2$ assuming a coefficient of friction μ between the particle and the path?

Let θ be the angle which the tangent to the curve at a general point $P:(x,y)$ makes with the x axis; let ϕ be the angle which the radius vector to P makes with the x axis; and let α be the angle between the tangent and the radius vector at P (Fig. 13.21). In moving the particle an infinitesimal distance Δs along the path, work must be done against two forces, namely, the tangential component of the central force

$$F_t = F \cos \alpha = kr \cos \alpha$$

and the frictional force

$$F_f = \mu F_n = \mu F \sin \alpha = \mu kr \sin \alpha$$

arising from the component of the central force which is perpendicular to the path and which acts to press the particle against the path. The infinitesimal amount of work done against these forces in moving a distance Δs along the curve is approximately

$$\Delta W = F_t \, \Delta s + F_f \, \Delta s = (kr \cos \alpha + \mu kr \sin \alpha) \, \Delta s$$

From the exterior-angle theorem of plane geometry, $\alpha = \phi - \theta$. Hence, summing and passing to the limit as $\Delta s \to 0$, we have

$$W = k \int_{0,1}^{1,2} r \cos (\phi - \theta) \, ds + \mu k \int_{0,1}^{1,2} r \sin (\phi - \theta) \, ds$$

$$= k \int_{0,1}^{1,2} r(\cos \phi \cos \theta + \sin \phi \sin \theta) \, ds$$

$$+ \mu k \int_{0,1}^{1,2} r(\sin \phi \cos \theta - \cos \phi \sin \theta) \, ds$$

Now

$$r \cos \phi = x \qquad r \sin \phi = y$$

and

$$\cos \theta \, ds = dx \qquad \sin \theta \, ds = dy$$

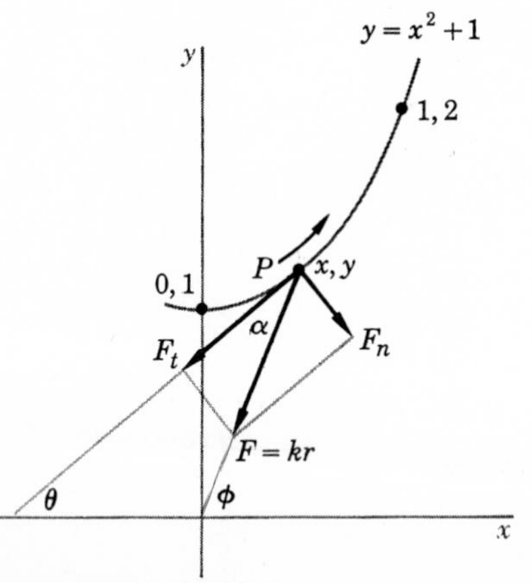

Figure 13.21
The resolution of a central force into tangential and normal components along a curve.

Therefore, substituting these into the last expression for W, we have

$$W = k \int_{0,1}^{1,2} (x\,dx + y\,dy) + \mu k \int_{0,1}^{1,2} (y\,dx - x\,dy)$$

The first of these integrals can be written very simply as

$$\frac{k}{2} \int_{0,1}^{1,2} d(x^2 + y^2)$$

which, regardless of the relation between x and y, that is, *independent of the path*, can be integrated at once, giving

$$\frac{k}{2}(x^2 + y^2)\Big|_{0,1}^{1,2} = 2k$$

The second integral in the expression for W is not an exact differential and thus cannot be integrated until y is expressed in terms of x, or vice versa; therefore, as usual, due account must be taken of the path of integration. Now, along the path, we have $y = x^2 + 1$ and $x = \sqrt{y - 1}$. Hence

$$\mu k \int_{0,1}^{1,2} (y\,dx - x\,dy) = \mu k \int_0^1 (x^2 + 1)\,dx - \mu k \int_1^2 \sqrt{y - 1}\,dy$$

$$= \mu k \left[\frac{x^3}{3} + x\right]_0^1 - \mu k \left[\frac{2(y - 1)^{3/2}}{3}\right]_1^2 = \frac{2\mu k}{3}$$

The total amount of work done in the course of the motion is therefore

$$2k + \frac{2\mu k}{3}$$

The first term represents recoverable work stored as potential energy in the system; the second term represents irrecoverable work dissipated as heat through friction.

The extension of line integration to paths in three dimensions is easily accomplished. Let $F(x,y,z)$ be a continuous function of x, y, and z, and let C be a continuous, sectionally smooth curve joining the points A and B. Furthermore, let the arc of C between A and B be divided in an arbitrary manner into n subintervals Δs_i whose projections on the coordinate axes are Δx_i, Δy_i, and Δz_i, and let an arbitrary point $P_i:(\xi_i,\eta_i,\zeta_i)$ be chosen in each Δs_i. We now evaluate $F(x,y,z)$ at each of the points P_i and form the sums

$$\sum_{i=1}^n F(\xi_i,\eta_i,\zeta_i)\,\Delta x_i \qquad \sum_{i=1}^n F(\xi_i,\eta_i,\zeta_i)\,\Delta y_i$$

$$\sum_{i=1}^n F(\xi_i,\eta_i,\zeta_i)\,\Delta z_i \qquad \sum_{i=1}^n F(\xi_i,\eta_i,\zeta_i)\,\Delta s_i$$

The limits of these sums as n becomes infinite in such a way that the length of each Δs_i approaches zero define the respective line integrals:

$$\int_C F(x,y,z)\,dx \qquad \int_C F(x,y,z)\,dy \qquad \int_C F(x,y,z)\,dz \qquad \int_C F(x,y,z)\,ds$$

Because of the difficulty of defining a space curve C as the intersection of several surfaces, it is customary to use a parametric representation for C. Hence, line integrals in three dimensions are ordinarily evaluated by integrating in terms of the parameter on C after the variables in the integrand have been replaced by their expressions in terms of the parameter.

EXAMPLE 3

What is $\int_C (xy + z^2)\, ds$, where C is the arc of the helix

$$x = \cos t \qquad y = \sin t \qquad z = t$$

which joins the points $(1,0,0)$ and $(-1,0,\pi)$?

Since $(ds)^2 = (dx)^2 + (dy)^2 + (dz)^2$, and since

$$dx = -\sin t\, dt \qquad dy = \cos t\, dt \qquad dz = dt$$

we have at once that

$$ds = \sqrt{\sin^2 t + \cos^2 t + 1}\,|dt| = \sqrt{2}\,|dt|$$

Furthermore, it is clear that the point $(1,0,0)$ corresponds to the parametric value $t = 0$ and that the point $(-1,0,\pi)$ corresponds to the parametric value $t = \pi$. Hence, when the integrand is expressed in terms of the parameter t, the required integral becomes

$$\int_0^\pi (\cos t \sin t + t^2)\sqrt{2}\, dt = \sqrt{2}\left[\frac{\cos^2 t}{2} + \frac{t^3}{3}\right]_0^\pi = \frac{\sqrt{2}\,\pi^3}{3}$$

The concept of a line integral generalizes at once to **surface** and **volume integrals**. To describe the former, let $F(x,y,z)$ be a continuous function of x, y, and z, and let S be a given regular† surface or portion of a regular surface in the region where $F(x,y,z)$ is defined. Let S be subdivided in an arbitrary manner into n elements ΔS_i (Fig. 13.22), and in each element let an arbitrary point $P_i:(\xi_i,\eta_i,\zeta_i)$ be chosen. Finally, let $F(x,y,z)$ be evaluated at each of the points P_i. Then the limit of the sum

$$\sum_{i=1}^{n} F(\xi_i,\eta_i,\zeta_i)\, \Delta S_i$$

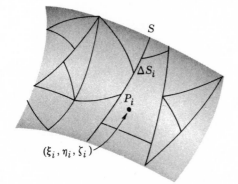

Figure 13.22
The subdivision of a surface preparatory to defining a surface integral.

---

† A surface is said to be **smooth** if at each of its points there exists a tangent plane which varies continuously as the point varies continuously on the surface. A smooth surface is said to be **orientable** if it is two-sided; i.e., if it is possible at each point to identify consistently a unique direction normal to the surface. A surface which can be subdivided by a finite number of sectionally smooth curves into pieces which are all orientable (and therefore smooth) is said to be **regular**. For a discussion of smooth surfaces which are not regular, i.e., smooth one-sided surfaces, see, for instance, Richard Courant and Herbert Robbins, "What Is Mathematics?," pp. 259–264, Oxford, New York, 1951.

as n becomes infinite in such a way that not only the area of each ΔS_i but also its maximum chord approaches zero is the **surface integral**

$$\iint_S F(x,y,z)\,dS$$

Similarly, given a function $F(x,y,z)$ and a region of space V, we can subdivide V into arbitrary subregions ΔV_i, then evaluate $F(x,y,z)$ at an arbitrary point $P_i:(\xi_i,\eta_i,\zeta_i)$ in each ΔV_i and form the sum

$$\sum_{i=1}^{n} F(\xi_i,\eta_i,\zeta_i)\,\Delta V_i$$

The limit of this sum as n becomes infinite in such a way that not only the volume of each ΔV_i but also its maximum chord approaches zero, is the **volume integral**

$$\iiint_V F(x,y,z)\,dV$$

EXAMPLE 4

What is the integral of the function $x^2 z$ taken over the entire surface of the right circular cylinder of height h which stands on the circle $x^2 + y^2 = a^2$? What is the integral of the given function taken throughout the volume of the cylinder?

To answer the first question, we must perform three integrations; i.e., we must integrate separately over the curved surface, the lower base, and the upper base of the cylinder. In each case, of course, we must employ a subdivision of the appropriate portion of the surface which will lead, if possible, to integrals that can conveniently be evaluated. This is most easily done by using cylindrical coordinates, as shown in Fig. 13.23. Then, on the curved surface, say S_1, we have

$$dS_1 = a\,d\theta\,dz \qquad x = a\cos\theta \qquad z = z$$

Figure 13.23
A typical volume element in cylindrical coordinates.

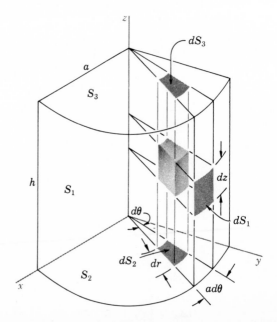

and the integral over this portion of the entire surface becomes

$$\iint_{S_1} x^2 z \, dS_1 = \int_0^h \int_0^{2\pi} (a \cos \theta)^2 z (a \, d\theta \, dz) = a^3 \int_0^h z \int_0^{2\pi} \cos^2 \theta \, d\theta \, dz$$

$$= a^3 \int_0^h z \left[\frac{\theta}{2} + \frac{\sin 2\theta}{4} \right]_0^{2\pi} dz = \pi a^3 \int_0^h z \, dz = \frac{\pi a^3 h^2}{2}$$

On the lower base, say S_2, we have

$$dS_2 = r \, dr \, d\theta \qquad x = r \cos \theta \qquad z = 0$$

However, because of the factor z, the integrand vanishes identically on S_2, and without further calculations we have

$$\iint_{S_2} x^2 z \, dS_2 = 0$$

On the upper base, say S_3, we have $dS_3 = r \, dr \, d\theta$, $x = r \cos \theta$, and $z = h$. Hence

$$\iint_{S_3} x^2 z \, dz = \int_0^{2\pi} \int_0^a (r \cos \theta)^2 h (r \, dr \, d\theta) = h \int_0^{2\pi} \cos^2 \theta \int_0^a r^3 \, dr \, d\theta$$

$$= h \int_0^{2\pi} \cos^2 \theta \left[\frac{r^4}{4} \right]_0^a d\theta = \frac{a^4 h}{4} \left[\frac{\theta}{2} + \frac{\sin 2\theta}{4} \right]_0^{2\pi} = \frac{\pi a^4 h}{4}$$

The integral over the entire surface S is, of course, the sum of the integrals over S_1, S_2 and S_3, that is

$$\iint_S x^2 z \, dS = \frac{\pi a^3 h^2}{2} + 0 + \frac{\pi a^4 h}{4} = \frac{\pi a^3 h (2h + a)}{4}$$

In computing the required volume integral, it is also convenient to use cylindrical coordinates. Doing this, we have $dV = r \, dr \, d\theta \, dz$, $x = r \cos \theta$, $z = z$, and the required integral becomes

$$\iiint_V x^2 z \, dV = \int_0^h \int_0^{2\pi} \int_0^a (r \cos \theta)^2 z (r \, dr \, d\theta \, dz) = \frac{a^4}{4} \int_0^h \int_0^{2\pi} z \cos^2 \theta \, d\theta \, dz$$

$$= \frac{\pi a^4}{4} \int_0^h z \, dz = \frac{\pi a^4 h^2}{8}$$

For the most part, our interest in line, surface, and volume integrals will be theoretical rather than computational; i.e., we shall use them far more often in derivations than in numerical calculations. Fundamental among the theorems we shall need for this purpose is **Green's lemma,**[†] which relates the line integral of a function taken around the boundary of a plane region to the surface integral of an associated function taken over the region itself.

THEOREM 1 If R is a plane region bounded by a finite number of simple closed curves, and if $U(x,y)$, $V(x,y)$, $\partial U/\partial y$, and $\partial V/\partial x$ are continuous at all points of R and its boundary C, then

$$\int_C U \, dx + V \, dy = \iint_R \left(\frac{\partial V}{\partial x} - \frac{\partial U}{\partial y} \right) dx \, dy$$

provided the line integral is taken in the positive direction around C.

---

† Named for the English mathematical physicist George Green (1793–1841).

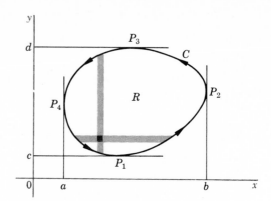

Figure 13.24
A plane region and its boundary.

Proof   Let us first suppose that the boundary of R is a single simple closed curve C with the property that any line parallel to either of the coordinate axes cuts it in at most two points, and let us draw the horizontal and vertical lines which circumscribe C (Fig. 13.24). Then the arcs $P_4P_1P_2$ and $P_4P_3P_2$ define single-valued functions of x, which we shall call $f_1(x)$ and $f_2(x)$, respectively. Similarly, the arcs $P_1P_4P_3$ and $P_1P_2P_3$ define single-valued functions of y which we shall call $g_1(y)$ and $g_2(y)$, respectively. Now consider

$$I_1 = \iint_R \frac{\partial V}{\partial x}\, dx\, dy$$

To carry out this integration over R, it is sufficient to integrate with respect to x from the arc $P_1P_4P_3$ to the arc $P_1P_2P_3$ and then to integrate with respect to y from c to d. Hence

$$I_1 = \int_c^d \int_{g_1(y)}^{g_2(y)} \frac{\partial V}{\partial x}\, dx\, dy$$

The inner integration can easily be performed, and we find

$$I_1 = \int_c^d V(x,y)\Big|_{g_1(y)}^{g_2(y)} dy = \int_c^d V[g_2(y),y]\, dy - \int_c^d V[g_1(y),y]\, dy$$

$$= \int_c^d V[g_2(y),y]\, dy + \int_d^c V[g_1(y),y]\, dy$$

Obviously, the integrand of the first of these integrals is just $V(x,y)$ evaluated for $x = g_2(y)$. Hence this integral is precisely the line integral

$$\int_c^d V(x,y)\, dy$$

taken along the path $x = g_2(y)$ from P_1 (where $y = c$) to P_3 (where $y = d$). Similarly, the second integral is just the same line integral taken this time along the path $x = g_1(y)$ in the direction from P_3 through P_4 to P_1. Together they constitute the line integral of $V(x,y)$ around the entire closed curve C in the positive direction; hence,

(1)
$$\iint_R \frac{\partial V}{\partial x}\, dx\, dy = \int_C V(x,y)\, dy$$

In the same way, if we consider

$$I_2 = \iint_R \frac{\partial U}{\partial y}\, dx\, dy = \iint_R \frac{\partial U}{\partial y}\, dy\, dx$$

we can write more specifically

$$I_2 = \int_a^b \int_{f_1(x)}^{f_2(x)} \frac{\partial U}{\partial y}\, dy\, dx$$

Then, performing the inner integration, we have

$$I_2 = \int_a^b U(x,y)\Big|_{f_1(x)}^{f_2(x)} dx = \int_a^b U[x, f_2(x)]\, dx - \int_a^b U[x, f_1(x)]\, dx$$

$$= -\int_b^a U[x, f_2(x)]\, dx - \int_a^b U[x, f_1(x)]\, dx$$

The first of these integrals is just the negative of the line integral of $U(x,y)$ along the path $y = f_2(x)$ in the direction from P_2 through P_3 to P_4. The second is the negative of the line integral of $U(x,y)$ along $y = f_1(x)$ from P_4 through P_1 to P_2. Together they constitute the negative of the line integral of $U(x,y)$ entirely around the closed curve C in the same direction (the positive direction) in which we integrated in (1), i.e.,

$$(2) \qquad \iint_R \frac{\partial U}{\partial y}\, dx\, dy = -\int_C U(x,y)\, dx$$

Finally, if we subtract (2) from (1) and combine the integrals on each side, we obtain

$$(3) \qquad \int_C U\, dx + V\, dy = \iint_R \left(\frac{\partial V}{\partial x} - \frac{\partial U}{\partial y} \right) dx\, dy$$

which establishes Green's lemma for the special regions we have thus far been considering.

It is a simple matter, now, to extend Green's lemma to regions whose boundaries do not satisfy the condition that every line parallel to either of the coordinate axes cut them in at most two points. For if this is not the case, the region R can be divided into subregions R_i whose boundaries C_i do have this property (Fig. 13.25). Then

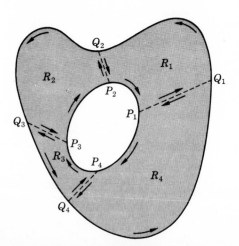

Figure 13.25
A plane region R subdivided into simpler regions R_1, R_2, R_3, R_4.

Eq. (3) can be applied to each subregion, following which the addition of these results yields Green's lemma for the general region R itself. For instance, for the region shown in Fig. 13.25, when Green's lemma is applied to each subregion, we have

$$\int_{C_1} U\,dx + V\,dy = \iint_{R_1} \left(\frac{\partial V}{\partial x} - \frac{\partial U}{\partial y}\right) dx\,dy$$

$$\int_{C_2} U\,dx + V\,dy = \iint_{R_2} \left(\frac{\partial V}{\partial x} - \frac{\partial U}{\partial y}\right) dx\,dy$$

$$\int_{C_3} U\,dx + V\,dy = \iint_{R_3} \left(\frac{\partial V}{\partial x} - \frac{\partial U}{\partial y}\right) dx\,dy$$

$$\int_{C_4} U\,dx + V\,dy = \iint_{R_4} \left(\frac{\partial V}{\partial x} - \frac{\partial U}{\partial y}\right) dx\,dy$$

When these results are added, the four integrals on the right combine to give exactly

$$\iint_R \left(\frac{\partial V}{\partial x} - \frac{\partial U}{\partial y}\right) dx\,dy$$

since $R_1 + R_2 + R_3 + R_4 = R$. Moreover, the four line integrals on the left combine to give the line integral around the two curves which form the boundary of R *plus* a set of line integrals taken along the auxiliary boundary arcs P_iQ_i. Because U and V are continuous throughout R, these integrals cancel in pairs, since each of the arcs P_iQ_i is traversed twice, once in one direction and once in the opposite direction. Hence we are left with

$$\int_C U\,dx + V\,dy = \iint_R \left(\frac{\partial V}{\partial x} - \frac{\partial U}{\partial y}\right) dx\,dy$$

which is the assertion of Green's lemma.

EXERCISES

1 Compute the value of the line integral $\int_C \dfrac{dy}{x+y}$ along each of the paths shown in Fig. 13.19.

2 Evaluate $\int_{0,1}^{2,3} (2xy - 1)\,dx + (x^2 + 1)\,dy$ along:

   (a) $y = x + 1$ (b) $y = \dfrac{x^2}{2} + 1$

3 Evaluate $\int_{-1,0}^{1,0} y(1 + x)\,dy$ along:
   (a) the x axis   (b) $y = 1 - x^2$

4 Evaluate $\int_{0,0}^{1,1} x\,ds$ along:

   (a) $y = x$ (b) $y = x^2$ (c) $y = x^{3/2}$

5 Evaluate $\int x^2 y^2\,ds$ around the circle $x^2 + y^2 = 1$. *Hint:* Use polar coordinates.

6 Along what curve of the family $y = kx(1 - x)$ does the integral $\int_{0,0}^{1,0} y(x - y)\,dx$ attain its largest value?

7 Along what curve of the family $y = x^n$ does the integral $\int_{0,0}^{1,1} (25xy - 8y^2)\,dx$ attain its largest value?

8 What is the value of the line integral $\int_C 3x^2y^2\, dx + 2x^3y\, dy$ in the positive direction around C, where C is the ellipse $x^2 + 4y^2 = 4$?

9 Evaluate $\iint_S (x + y)z\, dS$, where S is the surface of the cube whose vertices are $(0,0,0)$, $(1,0,0)$, $(1,1,0)$, $(0,1,0)$, $(0,0,1)$, $(1,0,1)$, $(1,1,1)$, $(0,1,1)$.

10 Evaluate $\iint_S (x + y + z)\, dS$, where S is the portion of the surface of the sphere $x^2 + y^2 + z^2 = a^2$ which lies in the first octant. *Hint:* Use spherical coordinates.

11 Evaluate $\iiint_V x^2z\, dV$, where V is the volume under the surface $x^2 + y^2 + z^2 = a^2$ and above the xy plane.

12 Verify Green's lemma for the integral $\int (x^2 + y)\, dx - xy^2\, dy$ taken around the boundary of the square whose vertices are $(0,0)$, $(1,0)$, $(1,1)$, and $(0,1)$.

13 Verify Green's lemma for the integral $\int (x - y)\, dx + (x + y)\, dy$ taken around the boundary of the finite area in the first quadrant between the curves $y = x^2$ and $y^2 = x$.

14 Verify Green's lemma for the integral $\int (x - 2y)\, dx + x\, dy$ taken around the circle $x^2 + y^2 = a^2$.

15 If a particle is attracted toward the origin by a force proportional to the nth power of the distance from the origin, show that the work done against this force in moving the particle from the point (x_0, y_0) to the point (x_1, y_1) is independent of the path along which the particle is moved. What is the amount of work done?

16 A particle is attracted toward the origin by a force proportional to the cube of the distance from the origin. How much work is done in moving the particle from the origin to the point $(1,1)$ if in each case the coefficient of friction between the particle and the path is μ and if motion takes place:
(a) Along the path $y = x$ (b) Along the path $y = x^2$
(c) Along the x axis to $(1,0)$ and then vertically to $(1,1)$
(d) Along the y axis to $(0,1)$ and then horizontally to $(1,1)$

17 Using Green's lemma, show that the area bounded by any simple closed curve C is given by the formula $A = \dfrac{1}{2} \int_C x\, dy - y\, dx$. Is this formula correct for regions bounded by more than one simple closed curve?

18 Let R be the region bounded by a simple closed curve C, let the area of R be A, and let the coordinates of the center of gravity of R be $(\bar{x}, \bar{y})$. Use Green's lemma to show that

$$\frac{1}{2} \int_C x^2\, dy = A\bar{x} \qquad \text{and} \qquad \int_C xy\, dy = A\bar{y}$$

19 If I_x and I_y are, respectively, the moments of inertia about the x axis and y axis of the area bounded by a simple closed curve C, show that

$$I_x = \int_C xy^2\, dy \qquad \text{and} \qquad I_y = -\int_C x^2y\, dx$$

Obtain a similar formula for the polar moment of inertia about the origin of area bounded by C.

20 Discuss the extension of Green's lemma to regions whose boundaries contain segments which are parallel to one or the other of the coordinate axes.

21 If U, V, $\partial U/\partial y$, and $\partial V/\partial x$ are continuous, and if $\partial U/\partial y = \partial V/\partial x$ at all points in the interior of a simple closed curve C, show that $\int_\Gamma U\, dx + V\, dy = 0$ for any simple closed curve Γ which lies entirely within C.

22 Show that Green's lemma fails to hold for the functions

$$U = -\frac{y}{x^2 + y^2} \quad \text{and} \quad V = \frac{x}{x^2 + y^2}$$

if R is the interior of the circle C: $x^2 + y^2 = 1$. Explain.

23 Using Green's lemma, establish the formula

$$\iint_R \left(\frac{\partial^2 F}{\partial x^2} + \frac{\partial^2 F}{\partial y^2} \right) dx \, dy = \int_C \frac{dF}{dn} \, ds$$

where R is the region bounded by the simple closed curve C and dF/dn is the directional derivative of F in the direction of the outer normal to C.

24 By setting $U = f \dfrac{\partial g}{\partial x}$ and $V = f \dfrac{\partial g}{\partial y}$ in Green's lemma, show that

$$\iint_R \left(\frac{\partial f}{\partial x} \frac{\partial g}{\partial y} - \frac{\partial f}{\partial y} \frac{\partial g}{\partial x} \right) dx \, dy = \int_C f \, dg$$

where R is the region bounded by the simple closed curve C. What is $\displaystyle\int_C g \, df$?

25 By setting $U = f \dfrac{\partial g}{\partial y}$ and $V = -f \dfrac{\partial g}{\partial x}$ in Green's lemma, show that

$$\iint_R f \cdot \left(\frac{\partial^2 g}{\partial x^2} + \frac{\partial^2 g}{\partial y^2} \right) dx \, dy + \iint_R \left(\frac{\partial f}{\partial x} \frac{\partial g}{\partial x} + \frac{\partial f}{\partial y} \frac{\partial g}{\partial y} \right) dx \, dy = \int_C f \frac{dg}{dn} \, ds$$

where R is the region bounded by the simple closed curve C and dg/dn is the directional derivative of g in the direction of the outer normal to C.

13.5 Integral Theorems

Most of the integrals we encounter in vector analysis are scalar quantities. For instance, given a vector function $\mathbf{F}(x,y,z)$, we are often interested in the integral of its tangential component along a curve C or in the integral of its normal component over a surface S. In the first case, if $\mathbf{R}$ is the vector from the origin to a general point of C, so that $d\mathbf{R}/ds \equiv \mathbf{T}$ is the unit vector tangent to C at a general point, then $\mathbf{F} \cdot \mathbf{T}$ is the tangential component of $\mathbf{F}$ and

$$\int_C \mathbf{F} \cdot \mathbf{T} \, ds = \int_C \mathbf{F} \cdot \frac{d\mathbf{R}}{ds} \, ds$$

(1)
$$= \int_C \mathbf{F} \cdot d\mathbf{R}$$

is the integral of this component along the curve C. In the second case, if $\mathbf{N}$ is the unit normal to S at a general point, then $\mathbf{F} \cdot \mathbf{N}$ is the normal component of $\mathbf{F}$ and

(2)
$$\iint_S \mathbf{F} \cdot \mathbf{N} \, dS\dagger$$

---

† Some writers denote the differential vector $\mathbf{N} \, dS$ by $d\mathbf{S}$ or $d\mathbf{A}$.

is the integral of this component over the surface S. Other scalar integrals of frequent occurrence are the surface integral of the normal component of the curl of $\mathbf{F}$

$$(3) \qquad \iint_S (\nabla \times \mathbf{F}) \cdot \mathbf{N} \, dS$$

and the volume integral of the divergence of $\mathbf{F}$

$$(4) \qquad \iiint_V \nabla \cdot \mathbf{F} \, dV$$

Fundamental in many of the applications of vector analysis is the so-called **divergence theorem**, which asserts the equality of the integrals (2) and (4) when V is the volume bounded by the closed regular surface S.

THEOREM 1   If $F(x,y,z)$ and $\nabla \cdot \mathbf{F}$ are continuous over the closed regular surface S and its interior V, and if $\mathbf{N}$ is the unit vector perpendicular to S at a general point and extending outward from S, then

$$\iint_S \mathbf{N} \cdot \mathbf{F} \, dS = \iiint_V \nabla \cdot \mathbf{F} \, dV$$

Proof   To prove this theorem, we shall first suppose that S is a closed surface such that no line parallel to one of the coordinate axes cuts it in more than two points. Now if $\mathbf{F} = u\mathbf{i} + v\mathbf{j} + w\mathbf{k}$, the assertion of the theorem can be written at length in the form

$$\iint_S \mathbf{N} \cdot (u\mathbf{i} + v\mathbf{j} + w\mathbf{k}) \, dS = \iiint_V \left(\frac{\partial u}{\partial x} + \frac{\partial v}{\partial y} + \frac{\partial w}{\partial z} \right) dV$$

or

$$(5) \qquad \iint_S \mathbf{N} \cdot i u \, dS + \iint_S \mathbf{N} \cdot j v \, dS + \iint_S \mathbf{N} \cdot k w \, dS$$

$$= \iiint_V \frac{\partial u}{\partial x} \, dV + \iiint_V \frac{\partial v}{\partial y} \, dV + \iiint_V \frac{\partial w}{\partial z} \, dV$$

We shall establish (5) by proving that respective integrals on each side are equal. To do this, let us consider first the integral

$$\iiint_V \frac{\partial w}{\partial z} \, dV$$

Under our assumption that no line parallel to one of the coordinate axes meets S in more than two points, it follows, in particular, that S is a double-valued surface over its projection on the xy plane and hence can be thought of as consisting of a lower half, say S_1, and an upper half, say S_2. Then if we take $dV = dx \, dy \, dz$ and perform the z integration first, we have

$$(6) \qquad \int \int \int_{z \text{ on } S_1}^{z \text{ on } S_2} \frac{\partial w}{\partial z} \, dz \, dx \, dy = \iint (w|_{\text{on } S_2} - w|_{\text{on } S_1}) \, dx \, dy$$

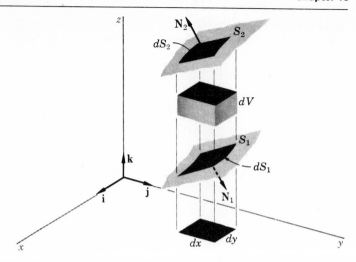

Figure 13.26
Integration in the z direction from S_1 to S_2 in the proof of the divergence theorem.

where, of course, x and y range over the area in the xy plane which is the projection of S. Moreover, the elements dS_1 and dS_2 can be defined so that they have $dx\,dy$ as their common projection on the xy plane (Fig. 13.26). Now since $\mathbf{k}$, $\mathbf{N_1}$, and $\mathbf{N_2}$ are all unit vectors, it follows that $\mathbf{k} \cdot \mathbf{N_1}$ and $\mathbf{k} \cdot \mathbf{N_2}$ are, respectively, the cosines of the angles between the normal to the xy plane, namely, $\mathbf{k}$, and the outer normals to dS_1 and dS_2; that is, they are numerically the cosines of the angles through which dS_1 and dS_2 are projected onto the element $dx\,dy$. Hence

$$dx\,dy = -\mathbf{k} \cdot \mathbf{N_1}\,dS_1 = \mathbf{k} \cdot \mathbf{N_2}\,dS_2$$

where the minus sign is necessary in the first equality because the outer normal $\mathbf{N_1}$ to dS_1 makes an angle of more than 90° with the direction of $\mathbf{k}$ and thus $\mathbf{k} \cdot \mathbf{N_1}$ is negative, whereas both $dx\,dy$ and dS_1 are clearly positive. Therefore, substituting for $dx\,dy$ in the right-hand side of (6), i.e., transferring the integration from the common projection of S_1 and S_2 back onto S_1 and S_2 themselves, we have

$$\iiint_V \frac{\partial w}{\partial z}\,dV = \iint w|_{\text{on } S_2}\,dx\,dy - \iint w|_{\text{on } S_1}\,dx\,dy$$

$$= \iint w|_{\text{on } S_2}\,\mathbf{k} \cdot \mathbf{N_2}\,dS_2 + \iint w|_{\text{on } S_1}\,\mathbf{k} \cdot \mathbf{N_1}\,dS_1$$

$$= \iint_{S_2} w\mathbf{k} \cdot \mathbf{N}\,dS + \iint_{S_1} w\mathbf{k} \cdot \mathbf{N}\,dS$$

where the subscripts have been dropped from the integrands as superfluous, since the ranges of integration are now explicitly indicated. Finally, since S_1 and S_2 together make up the entire closed surface S, we can combine the last two integrals, getting

$$\iiint_V \frac{\partial w}{\partial z}\,dV = \iint_S w\mathbf{k} \cdot \mathbf{N}\,dS$$

Similarly we can show that

$$\iiint_V \frac{\partial u}{\partial x}\, dV = \iint_S u\mathbf{i} \cdot \mathbf{N}\, dS$$

$$\iiint_V \frac{\partial v}{\partial y}\, dV = \iint_S v\mathbf{j} \cdot \mathbf{N}\, dS$$

Adding the last three equations, we obtain the expanded form (5) of the divergence theorem, under the assumption that S is exactly two-valued over its projections on each of the coordinate planes.

On the other hand, if S does not have this property, then (very much as in our extension of Green's lemma to more general regions in the last section) we can partition its interior V into subregions V_i whose boundaries S_i do have this property. Now, applying our limited result to each of these regions, we obtain a set of equations of the form

$$\iint_{S_i} \mathbf{N} \cdot \mathbf{F}\, dS = \iiint_{V_i} \nabla \cdot \mathbf{F}\, dV$$

If these are added, the sum of the volume integrals is, of course, just the integral of $\nabla \cdot \mathbf{F}$ throughout the entire volume V. The sum of the surface integrals is equal to the integral of $\mathbf{N} \cdot \mathbf{F}$ over the original surface S plus a set of integrals over the auxiliary boundary surfaces which were introduced when V was subdivided. These cancel in pairs, however, since the integration extends twice over each interface, with integrands which are identical except for the oppositely directed unit normals they contain as factors. Thus, our proof can be extended to volumes bounded by general closed regular surfaces, and Theorem 1 is established.

EXAMPLE 1

Prove that

$$\iint_S \mathbf{N} \times \mathbf{F}\, dS = \iiint_V \nabla \times \mathbf{F}\, dV$$

To show this, let us apply the divergence theorem to the vector $\mathbf{F} \times \mathbf{C}$, where $\mathbf{C}$ is an arbitrary constant vector. Then

$$\iint_S \mathbf{N} \cdot (\mathbf{F} \times \mathbf{C})\, dS = \iiint_V \nabla \cdot (\mathbf{F} \times \mathbf{C})\, dV$$

Now taking advantage of the fact that $\mathbf{C}$ is a constant vector and that a cyclic permutation of the elements of a scalar triple product leaves the product unchanged, we can write

$$\iint_S \mathbf{C} \cdot \mathbf{N} \times \mathbf{F}\, dS = \iiint_V \mathbf{C} \cdot \nabla \times \mathbf{F}\, dV$$

or, removing the constant vector $\mathbf{C}$ from each integral,

$$\mathbf{C} \cdot \iint_S \mathbf{N} \times \mathbf{F}\, dS = \mathbf{C} \cdot \iiint_V \nabla \times \mathbf{F}\, dV$$

Since $\mathbf{C}$ is an arbitrary vector, this equation asserts that the vectors

$$\iint_S \mathbf{N} \times \mathbf{F}\, dS \qquad \text{and} \qquad \iiint_V \nabla \times \mathbf{F}\, dV$$

have equal projections in all directions and hence must be equal, as asserted.

Various important theorems stem from the divergence theorem. For instance, if u and v are two sufficiently differentiable scalar point functions, and if we set

$$\mathbf{F} = u\,\nabla v$$

then, by Eq. (13), Sec. 13.3,

$$\nabla \cdot \mathbf{F} = \nabla \cdot (u\,\nabla v) = u\nabla \cdot \nabla v + \nabla u \cdot \nabla v = \nabla u \cdot \nabla v + u\,\nabla^2 v$$

Hence, applying the divergence theorem to the vector $\mathbf{F} = u\,\nabla v$, we have

$$(7) \qquad \iiint_V (\nabla u \cdot \nabla v + u\,\nabla^2 v)\, dV = \iint_S \mathbf{N} \cdot u\,\nabla v\, dS$$

Similarly, if we interchange the roles of u and v in (7), we obtain

$$(8) \qquad \iiint_V (\nabla v \cdot \nabla u + v\,\nabla^2 u)\, dV = \iint_S \mathbf{N} \cdot v\,\nabla u\, dS$$

Finally, if we subtract (8) from (7), we obtain **Green's theorem.†**

THEOREM 2 If V is the volume bounded by a closed regular surface S, and if $u(x,y,z)$ and $v(x,y,z)$ are scalar functions having continuous second partial derivatives, then

$$\iiint_V (u\,\nabla^2 v - v\,\nabla^2 u)\, dV = \iint_S \mathbf{N} \cdot (u\,\nabla v - v\,\nabla u)\, dS$$

where $\mathbf{N}$ is the outer normal to the surface S which bounds V.

Another result of some importance can be obtained by applying the divergence theorem to the function $\mathbf{F} = \mathbf{R}/r^3$, where, as usual,

$$\mathbf{R} = x\mathbf{i} + y\mathbf{j} + z\mathbf{k} \qquad \text{and} \qquad r = |\mathbf{R}| = \sqrt{x^2 + y^2 + z^2}$$

Thus, substituting into the divergence theorem, we have

$$(9) \qquad \iint_S \mathbf{N} \cdot \frac{\mathbf{R}}{r^3}\, dS = \iiint_V \nabla \cdot \frac{\mathbf{R}}{r^3}\, dV$$

Now, by Eq. (13), Sec. 12.3, and Exercise 17, Sec. 12.3,

$$\nabla \cdot \frac{\mathbf{R}}{r^3} = \frac{1}{r^3}\nabla \cdot \mathbf{R} + \mathbf{R} \cdot \nabla \frac{1}{r^3} = \frac{3}{r^3} + \mathbf{R} \cdot \frac{d(1/r^3)}{dr}\,\nabla r$$

$$= \frac{3}{r^3} + \mathbf{R} \cdot \left(-\frac{3}{r^4}\frac{\mathbf{R}}{r}\right) = \frac{3}{r^3} - 3\frac{\mathbf{R} \cdot \mathbf{R}}{r^5} = 0$$

Hence, we conclude from (9) that

$$(10) \qquad \iint_S \mathbf{N} \cdot \frac{\mathbf{R}}{r^3}\, dS = 0$$

provided, of course, that r is different from zero at all points on and within S, that is, provided that the origin from which $\mathbf{R}$ is drawn does not lie on S or within the volume enclosed by S.

---

† This should not be confused with *Green's lemma*, Theorem 1, Sec. 13.4.

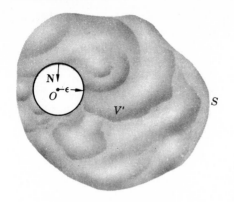

Figure 13.27
A singular point excluded from a three-dimensional region by an auxiliary spherical boundary.

Since the divergence theorem requires that the function to which it is applied have continuous first partial derivatives throughout the volume of integration, it cannot be applied to $\mathbf{R}/r^3$ if the origin of $\mathbf{R}$ is within S. In this case we therefore modify the region of integration by constructing a sphere S' of radius ε having the origin O as center (Fig. 13.27). In the region V' between S and S' the function $\mathbf{R}/r^3$ satisfies the conditions of the divergence theorem, and thus Eq. (10) can properly be applied, giving

$$(11) \quad \iint_{S+S'} \mathbf{N} \cdot \frac{\mathbf{R}}{r^3} \, dS = 0 \quad \text{or} \quad \iint_{S} \mathbf{N} \cdot \frac{\mathbf{R}}{r^3} \, dS + \iint_{S'} \mathbf{N} \cdot \frac{\mathbf{R}}{r^3} \, dS = 0$$

At any point of S', the direction of the normal which extends outward from the volume V' is opposite to $\mathbf{R}$. Hence, the unit outer normal to S' is $\mathbf{N} = -\mathbf{R}/\varepsilon$, since on S' the length of the radius vector $\mathbf{R}$ is $r = \varepsilon$. Therefore, in the last integral,

$$\mathbf{N} \cdot \mathbf{R} = -\frac{\mathbf{R}}{\varepsilon} \cdot \mathbf{R} = -\frac{\varepsilon^2}{\varepsilon} = -\varepsilon$$

and Eq. (11) becomes

$$\iint_{S} \mathbf{N} \cdot \frac{\mathbf{R}}{r^3} \, dS + \iint_{S'} \frac{-\varepsilon}{\varepsilon^3} \, dS = 0$$

or

$$\iint_{S} \mathbf{N} \cdot \frac{\mathbf{R}}{r^3} \, dS = \frac{1}{\varepsilon^2} \iint_{S'} dS = \frac{4\pi\varepsilon^2}{\varepsilon^2} = 4\pi$$

This result, coupled with Eq. (10), gives us **Gauss' theorem,**

THEOREM 3   If S is a closed regular surface, then

$$\iint_{S} \mathbf{N} \cdot \frac{\mathbf{R}}{r^3} \, dS = \begin{cases} 0 & O \text{ outside } S \\ 4\pi & O \text{ inside } S \end{cases}$$

Another integral formula of great importance in vector analysis is **Stokes' theorem.**†

---

† Named for the English mathematical physicist G. G. Stokes (1819–1903).

THEOREM 4 If S is the portion of a regular surface bounded by the closed curve C, and if $F(x,y,z)$ is a vector function possessing continuous first partial derivatives, then

$$\int_C \mathbf{F} \cdot d\mathbf{R} = \iint_S \mathbf{N} \cdot \mathbf{V} \times \mathbf{F} \, dS$$

provided the direction of integration around C is positive with respect to the side of S on which the unit normal N is drawn.

Proof To prove this theorem we suppose first that S has the property that it is single-valued above its projections on each of the coordinate planes. If we write $\mathbf{F} = u\mathbf{i} + v\mathbf{j} + w\mathbf{k}$, Stokes' theorem becomes

$$\int_C u \, dx + v \, dy + w \, dz = \iint_S \mathbf{N} \cdot \mathbf{V} \times (u\mathbf{i} + v\mathbf{j} + w\mathbf{k}) \, dS$$

(12)
$$= \iint_S \mathbf{N} \cdot \mathbf{V} \times u\mathbf{i} \, dS + \iint_S \mathbf{N} \cdot \mathbf{V} \times v\mathbf{j} \, dS$$

$$+ \iint_S \mathbf{N} \cdot \mathbf{V} \times w\mathbf{k} \, ds$$

and, to establish the theorem it is sufficient to show that respective integrals on the two sides of the last equation are equal. We consider first the integral

$$\iint_S \mathbf{N} \cdot \mathbf{V} \times u\mathbf{i} \, dS$$

taken over the *closed* surface consisting of S, its projection on the xy plane, say S', and the cylindrical surface, say S'', which projects S onto S' (Fig. 13.28a). If we apply

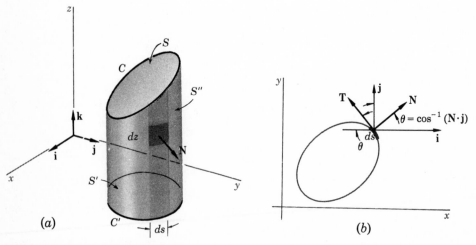

Figure 13.28
The closed surface $S + S' + S''$ employed in the proof of Stokes' theorem.

the divergence theorem to the vector $\mathbf{V} \times u\mathbf{i}$ over this surface and the volume it encloses and note from Eq. (19), Sec. 13.3, that $\mathbf{V} \cdot (\mathbf{V} \times u\mathbf{i}) = 0$, we obtain

$$\iint_S \mathbf{N} \cdot \mathbf{V} \times u\mathbf{i} \, dS + \iint_{S'} \mathbf{N} \cdot \mathbf{V} \times u\mathbf{i} \, dS + \iint_{S''} \mathbf{N} \cdot \mathbf{V} \times u\mathbf{i} \, dS$$

$$= \iiint_V \mathbf{V} \cdot (\mathbf{V} \times u\mathbf{i}) \, dV = 0$$

or

$$(13) \qquad \iint_S \mathbf{N} \cdot \mathbf{V} \times u\mathbf{i} \, dS = - \iint_{S'} \mathbf{N} \cdot \mathbf{V} \times u\mathbf{i} \, dS - \iint_{S''} \mathbf{N} \cdot \mathbf{V} \times u\mathbf{i} \, dS$$

Now

$$\mathbf{V} \times u\mathbf{i} = \begin{vmatrix} \mathbf{i} & \mathbf{j} & \mathbf{k} \\ \dfrac{\partial}{\partial x} & \dfrac{\partial}{\partial y} & \dfrac{\partial}{\partial z} \\ u & 0 & 0 \end{vmatrix} = \mathbf{j} \dfrac{\partial u}{\partial z} - \mathbf{k} \dfrac{\partial u}{\partial y}$$

Moreover, on S' the outer normal $\mathbf{N}$ is clearly equal to $-\mathbf{k}$. Hence, on S' we have

$$\mathbf{N} \cdot \mathbf{V} \times u\mathbf{i} = -\mathbf{k} \cdot \left(\mathbf{j} \dfrac{\partial u}{\partial z} - \mathbf{k} \dfrac{\partial u}{\partial y} \right) = \dfrac{\partial u}{\partial y}$$

and

$$\iint_{S'} \mathbf{N} \cdot \mathbf{V} \times u\mathbf{i} \, dS = \iint_{S'} \dfrac{\partial u}{\partial y} \, dS$$

If we now apply Green's lemma (Theorem 1, Sec. 13.4) to the integral on the right, the last equation becomes

$$(14) \qquad \iint_{S'} \mathbf{N} \cdot \mathbf{V} \times u\mathbf{i} \, dS = - \int_{C'} u \, dx$$

where C' is the boundary of the region S' which is the projection of S.

To investigate the second integral on the right in Eq. (13), we note first that since S'' is a cylindrical surface whose generators are parallel to the z axis, the normals to S'' are all perpendicular to the vector $\mathbf{k}$. Therefore, on S'' we have

$$\mathbf{N} \cdot \mathbf{V} \times u\mathbf{i} = \mathbf{N} \cdot \left(\mathbf{j} \dfrac{\partial u}{\partial z} - \mathbf{k} \dfrac{\partial u}{\partial y} \right) = \mathbf{N} \cdot \mathbf{j} \dfrac{\partial u}{\partial z}$$

Next, taking $dS = dz \, ds$ (Fig. 13.28a), and noting that $\mathbf{N} \cdot \mathbf{j}$ is independent of z, we have

$$(15) \qquad \iint_{S''} \mathbf{N} \cdot \mathbf{V} \times u\mathbf{i} \, dS = \int_{C'} \mathbf{N} \cdot \mathbf{j} \int_{z \text{ on } S'}^{z \text{ on } S} \dfrac{\partial u}{\partial z} \, dz \, ds = \int_{C'} (u|_S - u|_{S'}) \, \mathbf{N} \cdot \mathbf{j} \, ds$$

Now $\mathbf{N} \cdot \mathbf{j}$ is equal to the cosine of the angle θ between the normal $\mathbf{N}$ and the positive y axis, and this is numerically equal but opposite in sign to the cosine of the angle between the directed tangent to C' and the positive x axis (Fig. 13.28b). Hence $\mathbf{N} \cdot \mathbf{j} \, ds = -dx$, and Eq. (15) becomes

$$(16) \qquad \iint_{S''} \mathbf{N} \cdot \mathbf{V} \times u\mathbf{i} \, dS = - \int_{C'} u|_S \, dx + \int_{C'} u|_{S'} \, dx$$

In the first integral on the right in (16), the integrand, being evaluated at points of S which are directly above the curve C', is actually evaluated along the curve C. Moreover, because C' is the projection of C in the z direction, the variation of x around C' is exactly the same as the variation of x around C. Hence, in this integral we can properly replace the indicated path of integration C' by the curve C, getting

(17)
$$\iint_{S_n} \mathbf{N} \cdot \nabla \times u\mathbf{i} \, dS = -\int_C u \, dx + \int_{C'} u \, dx$$

Therefore, substituting from (14) and (17) into (13), we have

$$\iint_S \mathbf{N} \cdot \nabla \times u\mathbf{i} \, dS = -\left(-\int_{C'} u \, dx\right) - \left(-\int_C u \, dx + \int_{C'} u \, dx\right)$$

(18)
$$= \int_C u \, dx$$

In precisely the same way we can show that

(19)
$$\iint_S \mathbf{N} \cdot \nabla \times v\mathbf{j} \, dS = \int_C v \, dy$$

(20)
$$\iint_S \mathbf{N} \cdot \nabla \times w\mathbf{k} \, dS = \int_C w \, dz$$

Finally by adding (18), (19), and (20) we obtain Eq. (12).

It is now a simple matter to extend Eq. (12) to surfaces which are not single-valued above their projections on the coordinate planes. For if this is not the case, we can always divide S into subregions S_i which do have this property and then apply Eq. (12) to each S_i and its boundary C_i, getting the set of equations

$$\int_{C_1} \mathbf{F} \cdot d\mathbf{R} = \iint_{S_1} \mathbf{N} \cdot \nabla \times \mathbf{F} \, dS$$
$$\cdots\cdots\cdots\cdots\cdots\cdots\cdots\cdots\cdots$$
$$\int_{C_n} \mathbf{F} \cdot d\mathbf{R} = \iint_{S_n} \mathbf{N} \cdot \nabla \times \mathbf{F} \, dS$$

When these are added, the surface integrals combine to give precisely the surface integral over S itself, since $S_1 + \cdots + S_n = S$. At the same time, the line integrals combine to give the line integral around the actual boundary of S plus the line integral along all the auxiliary boundary arcs taken twice in opposite directions (Fig. 13.29).

Figure 13.29
A portion of a surface S subdivided into simpler regions $S_1, S_2, \ldots, S_n$.

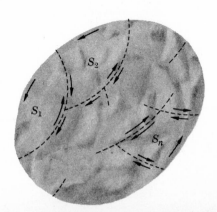

Since the latter cancel identically, the line integral around C itself is all that remains and the theorem follows in the general case.

If A and B are two arbitrary points in space, it is often important to know whether the line integral

(21)
$$\int_A^B \mathbf{F} \cdot d\mathbf{R}$$

is independent of the path along which the integral is calculated. As a first step in establishing criteria for this, we observe that if the integral (21) is independent of the path, then

$$\int \mathbf{F} \cdot d\mathbf{R}$$

taken around any closed path is zero. For let C be any simple closed curve, and let A and B be any two points on C (Fig. 13.30). Then since the integral is independent of the path, by hypothesis, we have

$$\int_{APB} \mathbf{F} \cdot d\mathbf{R} = \int_{AQB} \mathbf{F} \cdot d\mathbf{R}$$

If we reverse the direction of integration in the integral on the right, we have

$$\int_{APB} \mathbf{F} \cdot d\mathbf{R} = -\int_{BQA} \mathbf{F} \cdot d\mathbf{R}$$

or, transposing,

$$\int_{APB} \mathbf{F} \cdot d\mathbf{R} + \int_{BQA} \mathbf{F} \cdot d\mathbf{R} = \int_C \mathbf{F} \cdot d\mathbf{R} = 0$$

as asserted.

Conversely, if $\int \mathbf{F} \cdot d\mathbf{R}$ is zero around every closed curve in a region, then the integral (21) is independent of the path. For if APB and AQB are any two paths joining A and B (Fig. 13.30), we have, by hypothesis,

$$\int_{APB} \mathbf{F} \cdot d\mathbf{R} + \int_{BQA} \mathbf{F} \cdot d\mathbf{R} = 0$$

whence, reversing the direction of integration along BQA and transposing, we have

$$\int_{APB} \mathbf{F} \cdot d\mathbf{R} = \int_{AQB} \mathbf{F} \cdot d\mathbf{R}$$

as asserted.

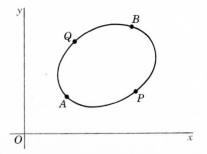

Figure 13.30
Two paths from A to B forming a simple closed curve.

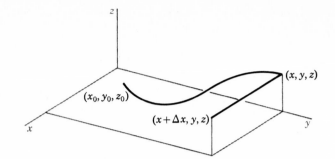

Figure 13.31
A convenient path of
integration from (x_0, y_0, z_0)
through (x, y, z) to
$(x + \Delta x, y, z)$.

Now if the integral (21) is independent of the path, then when we integrate from a fixed point $P_0:(x_0, y_0, z_0)$ to a variable point $P:(x, y, z)$, the result is a function only of the coordinates x, y, z of the variable end point. That is, if $\mathbf{F} = u\mathbf{i} + v\mathbf{j} + z\mathbf{k}$, we can appropriately write

$$\int_{P_0}^{P} \mathbf{F} \cdot d\mathbf{R} = \int_{P_0}^{P} u\, dx + v\, dy + w\, dz = \phi(x, y, z)$$

In what follows it will be necessary to know the partial derivatives of the function ϕ defined by the last equation. To obtain these, it is convenient to go back to the fundamental definition of a derivative and write, for the x partial derivative, for instance,

$$\frac{\partial \phi}{\partial x} = \lim_{\Delta x \to 0} \frac{\phi(x + \Delta x, y, z) - \phi(x, y, z)}{\Delta x}$$

$$= \lim_{\Delta x \to 0} \frac{1}{\Delta x} \left(\int_{x_0, y_0, z_0}^{x + \Delta x, y, z} u\, dx + v\, dy + w\, dz - \int_{x_0, y_0, z_0}^{x, y, z} u\, dx + v\, dy + w\, dz \right)$$

Since, by hypothesis, these integrals are independent of the path, we can use any paths we find convenient. In particular, in the integral from (x_0, y_0, z_0) to $(x + \Delta x, y, z)$ we shall let the path of integration consist of any smooth curve joining (x_0, y_0, z_0) to (x, y, z) plus the segment of the straight line joining (x, y, z) to $(x + \Delta x, y, z)$ (Fig. 13.31). Then

$$\frac{\partial \phi}{\partial x} = \lim_{\Delta x \to 0} \frac{1}{\Delta x} \left[\left(\int_{x_0, y_0, z_0}^{x, y, z} u\, dx + v\, dy + w\, dz + \int_{x, y, z}^{x + \Delta x, y, z} u\, dx + v\, dy + w\, dz \right) \right.$$

$$\left. - \int_{x_0, y_0, z_0}^{x, y, z} u\, dx + v\, dy + w\, dz \right]$$

$$= \lim_{\Delta x \to 0} \frac{1}{\Delta x} \int_{x, y, z}^{x + \Delta x, y, z} u\, dx + v\, dy + w\, dz$$

Along the path of integration in the last integral we have

$$dy \equiv 0 \qquad \text{and} \qquad dz \equiv 0$$

Hence,

$$\frac{\partial \phi}{\partial x} = \lim_{\Delta x \to 0} \frac{1}{\Delta x} \int_{x}^{x + \Delta x} u\, dx$$

Since u is assumed to be continuous, the law of the mean for integrals can be applied to the last expression and we have

$$\frac{\partial \phi}{\partial x} = \lim_{\Delta x \to 0} \frac{1}{\Delta x} \left[u(x + \theta \, \Delta x, y, z) \, \Delta x \right] \qquad 0 < \theta < 1$$

$$= u(x, y, z)$$

In the same way, the partial derivatives with respect to y and z can be determined, and we have the following theorem.

THEOREM 5 If $\mathbf{F} = u\mathbf{i} + v\mathbf{j} + w\mathbf{k}$ is a continuous vector function of x, y, and z with the property that

$$\int \mathbf{F} \cdot d\mathbf{R} = \int u \, dx + v \, dy + w \, dz$$

is independent of the path, then the partial derivatives of the function

$$\phi(x, y, z) \equiv \int_{P_0}^{P} \mathbf{F} \cdot d\mathbf{R} = \int_{P_0}^{P} u \, dx + v \, dy + w \, dz$$

are

$$\frac{\partial \phi}{\partial x} = u \qquad \frac{\partial \phi}{\partial y} = v \qquad \frac{\partial \phi}{\partial z} = w$$

We are now in a position to show that if $\mathbf{F} = u\mathbf{i} + v\mathbf{j} + w\mathbf{k}$ is a continuous vector function, and if $\int \mathbf{F} \cdot d\mathbf{R}$ is independent of the path, then $\mathbf{F}$ is the gradient of some scalar function ϕ. In fact, if we define

$$\phi(x, y, z) = \int_{P_0}^{P} \mathbf{F} \cdot d\mathbf{R} = \int_{P_0}^{P} u \, dx + v \, dy + w \, dz$$

we have, by Theorem 5,

$$\nabla \phi \equiv \frac{\partial \phi}{\partial x} \mathbf{i} + \frac{\partial \phi}{\partial y} \mathbf{j} + \frac{\partial \phi}{\partial z} \mathbf{k} = u\mathbf{i} + v\mathbf{j} + w\mathbf{k} = \mathbf{F}$$

as asserted.

Before we can state a correct converse of the last result, we must distinguish between two types of regions in space. On the one hand, a region V may have the property that every simple closed curve within it can be continuously contracted into a point without having to leave the region at any stage. Regions of this type are said to be **simply connected**; as examples, we have the interior of a sphere, the exterior of a sphere, and the space between two concentric spheres. On the other hand, a region V may contain simple closed curves which cannot be continuously contracted to a point without having to leave the region at some stage. Such regions are said to be **multiply connected**; as an example we have the space between two infinitely long coaxial cylinders, within which it is clearly impossible to shrink into a single point any closed curve encircling the inner cylindrical boundary. Both the interior and the exterior of a torus are also examples of multiply connected regions.†

---

† The distinction between simply connected and multiply connected regions applies equally well in the plane, of course, and in our study of functions of a complex variable it will often be an important distinction.

Now suppose that throughout a simply connected region V, the vector function $\mathbf{F}$ is the gradient of a scalar function ϕ. Let C be an arbitrary closed curve in V, and let S be any regular surface having C as its boundary curve and lying entirely in V. Since V is simply connected, it is clear that such a surface can always be found. Now, by Stokes' theorem,

$$\int_C \mathbf{F} \cdot d\mathbf{R} = \iint_S \mathbf{N} \cdot \mathbf{V} \times \mathbf{F} \, dS = \iint_S \mathbf{N} \cdot \mathbf{V} \times \mathbf{V}\phi \, dS = 0$$

since, by Eq. (18), Sec. 13.3, the curl of any gradient is identically zero. Thus $\int_C \mathbf{F} \cdot d\mathbf{R} = 0$ for every closed curve C in V, and hence, by one of our earlier observations, $\int_A^B \mathbf{F} \cdot d\mathbf{R}$ is independent of the path in the *simply connected* region V.

If V is multiply connected, the preceding argument breaks down because if the curve C encircles an inner boundary of V, then every surface S spanning C will perforce contain points which are not in V and where, for all we know, ϕ may not exist. If this is the case, Stokes' theorem cannot be applied and the conclusion that $\int \mathbf{F} \cdot d\mathbf{R}$ is independent of the path cannot be inferred. It is easy to show by an example that this can actually happen (see Exercise 30).

Finally, we note that if $\mathbf{F}$ is the gradient of a function ϕ, then $\mathbf{F} \cdot d\mathbf{R} = \mathbf{V}\phi \cdot d\mathbf{R} = d\phi$; that is, $\mathbf{F} \cdot d\mathbf{R}$ is an exact differential.

The results of the preceding discussion can now be summarized in the following theorem.

THEOREM 6   If $\mathbf{F} = u\mathbf{i} + v\mathbf{j} + w\mathbf{k}$ is a vector function of x, y, and z possessing continuous first partial derivatives at all points of a simply connected region V, then the following statements are all equivalent; i.e., any one of them implies each of the others:

a. $\int_A^B \mathbf{F} \cdot d\mathbf{R} \equiv \int_A^B u \, dx + v \, dy + w \, dz$ is independent of the path between A and B.
b. $\int \mathbf{F} \cdot d\mathbf{R} \equiv \int u \, dx + v \, dy + w \, dz$ is zero around every closed curve in V.
c. $\mathbf{F}$ is the gradient of the scalar point function

$$\phi(x,y,z) = \int_{P_0}^P \mathbf{F} \cdot d\mathbf{R} \equiv \int_{P_0}^P u \, dx + v \, dy + w \, dz$$

d. The curl of $\mathbf{F}$ vanishes identically.
e. $\mathbf{F} \cdot d\mathbf{R} \equiv u \, dx + v \, dy + w \, dz$ is an exact differential.

EXERCISES

1   If $\mathbf{F} = 2y\mathbf{i} + x\mathbf{j} + z^2\mathbf{k}$, evaluate $\int_{0,0,0}^{1,1,1} \mathbf{F} \cdot d\mathbf{R}$ along:
   (a) The rectilinear path from $(0,0,0)$ to $(1,0,0)$ to $(1,1,0)$ to $(1,1,1)$
   (b) The rectilinear path from $(0,0,0)$ to $(1,1,0)$ to $(1,1,1)$
   (c) The straight line joining $(0,0,0)$ to $(1,1,1)$
   (d) The curve which is the intersection of the paraboloid $x^2 + y^2 = 2z$ and the plane $x = y$

2   If $\mathbf{F} = x\mathbf{i} + y\mathbf{j} + 2\mathbf{k}$, evaluate $\iint_S \mathbf{F} \cdot \mathbf{N} \, dS$ over:
   (a) The surface of the cube whose vertices are $(0,0,0)$, $(1,0,0)$, $(1,1,0)$, $(0,1,0)$, $(0,0,1)$, $(1,0,1)$, $(1,1,1)$, $(0,1,1)$
   (b) The portion of the plane $x + 2y + 3z = 6$ which lies in the first octant
   (c) The entire surface of the sphere $x^2 + y^2 + z^2 = 1$
   (d) The portion of the cone $x^2 + y^2 - (1 - z)^2 = 0$ between the plane $z = 0$ and the plane $z = 1$

3 If $\mathbf{F} = y\mathbf{i} + x\mathbf{j} + z^2\mathbf{k}$, evaluate $\iiint_V \boldsymbol{\nabla} \cdot \mathbf{F}\, dV$ throughout:

   (a) The volume bounded by the cube whose vertices are $(0,0,0)$, $(1,0,0)$, $(1,1,0)$ $(0,1,0)$, $(0,0,1)$, $(1,0,1)$, $(1,1,1)$, $(0,1,1)$

   (b) The volume cut off from the first octant by the plane $x + 2y + 3z = 6$

   (c) The upper half of the volume within the sphere $x^2 + y^2 + z^2 = 1$

   (d) The volume under the paraboloid $z = 1 - x^2 - y^2$ and above the plane $z = 0$

4 Write the divergence theorem in cartesian form.

5 Write Green's theorem in cartesian form.

6 Write Gauss' theorem in cartesian form.

7 Write Stokes' theorem in cartesian form.

8 If S is a closed surface, what is $\iint_S \mathbf{N} \cdot \boldsymbol{\nabla} \times \mathbf{F}\, dS$?

9 If $\mathbf{T}$ is the variable unit tangent to a curve C, what is $\int_C \mathbf{T} \cdot d\mathbf{R}$? Can Stokes' theorem be used to evaluate this integral?

10 If $\mathbf{A}$ is a constant vector and C is a closed curve, show that $\int_C \mathbf{A} \cdot d\mathbf{R} = 0$. What is $\int_C d\mathbf{R}$?

11 If C is a closed curve, show that $\int_C \mathbf{R} \cdot d\mathbf{R} = 0$.

12 If C is a closed curve, show that $\int_C (u\, \boldsymbol{\nabla}v + v\, \boldsymbol{\nabla}u) \cdot d\mathbf{R} = 0$.

13 If S is a closed surface, show that $\iint_S \mathbf{N} \cdot \mathbf{R}\, dS = 3V$, where V is the volume enclosed by S.

14 If S is an arbitrary closed surface and $\iint_S \mathbf{N} \cdot \mathbf{F}\, dS = 0$, can we conclude that $\mathbf{F} \equiv 0$? Can we if S is an arbitrary open surface?

15 By applying the divergence theorem to the vector $\phi\mathbf{A}$, where $\mathbf{A}$ is an arbitrary constant vector, show that $\iint_S \phi\mathbf{N}\, dS = \iiint_V \boldsymbol{\nabla}\phi\, dV$. What is $\iint_S \mathbf{N}\, dS$?

16 By applying Stokes' theorem to the vector $\phi\mathbf{A}$, where $\mathbf{A}$ is an arbitrary constant vector, show that $\int_C \phi\, d\mathbf{R} = \iint_S \mathbf{N} \times \boldsymbol{\nabla}\phi\, dS$.

17 If S is an open surface, what is $\iint_S \mathbf{N} \times \mathbf{R}\, dS$? *Hint:* Use the result of Exercise 16.

18 By applying Stokes' theorem to the vector $\mathbf{F} \times \mathbf{A}$, where $\mathbf{A}$ is an arbitrary constant vector, show that $\int_C d\mathbf{R} \times \mathbf{F} = \iint_S (\mathbf{N} \times \boldsymbol{\nabla}) \times \mathbf{F}\, dS$. What is $\int_C d\mathbf{R} \times \mathbf{R}$?

19 Verify the divergence theorem for the function $2xz\mathbf{i} + yz\mathbf{j} + z^2\mathbf{k}$ over the upper half of the sphere $x^2 + y^2 + z^2 = a^2$.

20 Verify the divergence theorem for the function $y\mathbf{i} + x\mathbf{j} + z^2\mathbf{k}$ over the cylindrical region bounded by $x^2 + y^2 = a^2$, $z = 0$, and $z = a$.

21 Verify the divergence theorem for the function $x^2\mathbf{i} + z\mathbf{j} + yz\mathbf{k}$ over the cube whose vertices are $(0,0,0)$, $(1,0,0)$, $(1,1,0)$, $(0,1,0)$, $(0,0,1)$, $(1,0,1)$, $(1,1,1)$, and $(0,1,1)$.

22 Verify Stokes' theorem for the function $xy\mathbf{i} + yz\mathbf{j} + z^2\mathbf{k}$ over the cube described in Exercise 21 if the face of the cube in the xy plane is missing.

23 What is the surface integral of the normal component of the curl of the vector $(x + y)\mathbf{i} + (y - x)\mathbf{j} + z^3\mathbf{k}$ over the upper half of the sphere $x^2 + y^2 + z^2 = 1$?

24 If at each point of a surface S the vector $\mathbf{F}(x,y,z)$ is perpendicular to S, prove that the curl of $\mathbf{F}$ either vanishes identically or is everywhere tangent to S. *Hint:* Apply Stokes' theorem to $\mathbf{F}$ over the portion of S bounded by an arbitrary closed curve on S.

25 If at each point of a closed surface S the vector $\mathbf{F}(x,y,z)$ is perpendicular to S, prove that $\iiint_V \boldsymbol{\nabla} \times \mathbf{F}\, dV = \mathbf{0}$. *Hint:* Use the result of Example 1.

26 If $\mathbf{A}$ is an arbitrary constant vector, show that $\iint_S \mathbf{N} \times (\mathbf{A} \times \mathbf{R})\, dS = 2VA$, where V is the volume bounded by the closed surface S. *Hint:* Use the result of Example 1.

27 Show that

$$\iiint_V \left(\frac{\partial^2 \phi}{\partial x^2} + \frac{\partial^2 \phi}{\partial y^2} + \frac{\partial^2 \phi}{\partial z^2}\right) dV = \iint_S \frac{d\phi}{dn}\, dS$$

where $d\phi/dn$ is the directional derivative of ϕ in the direction of the outer normal to the closed surface S which bounds the volume V.

28 If $\phi(x,y,z)$ is a solution of Laplace's equation, show that

$$\iiint_V \left[\left(\frac{\partial \phi}{\partial x}\right)^2 + \left(\frac{\partial \phi}{\partial y}\right)^2 + \left(\frac{\partial \phi}{\partial z}\right)^2\right] dV = \iint_S \phi \frac{d\phi}{dn}\, dS$$

where $d\phi/dn$ is the directional derivative of ϕ in the direction of the outer normal to the closed surface S. Hence show also that

$$\iint_S \phi \frac{d\phi}{dn}\, dS > 0$$

if ϕ is a solution of Laplace's equation.

29 Extend Gauss' theorem to the case in which O lies *on* the surface S.

30 Show that although the function

$$\mathbf{F} = \frac{-y}{x^2 + y^2}\,\mathbf{i} + \frac{x}{x^2 + y^2}\,\mathbf{j} + \mathbf{k}$$

is continuous and equal to the gradient of

$$\phi(x,y,z) = \tan^{-1}\frac{y}{x} + z$$

at all points of the region between the two cylinders

$$x^2 + y^2 = \tfrac{1}{4} \quad \text{and} \quad x^2 + y^2 = 4$$

the integral $\int \mathbf{F} \cdot d\mathbf{R}$ is not independent of the path in this region. *Hint:* Take A to be $(-1,0,0)$ and B to be $(1,0,0)$, and compute $\displaystyle\int_A^B \mathbf{F} \cdot d\mathbf{R}$ along the upper and lower arcs of the circle $x^2 + y^2 = 1$, $z = 0$.

13.6 Further Applications

One of the most important uses of vector analysis is in the concise formulation of physical laws and the derivation of other results from those laws. As a first example of this sort, we shall develop the concept of *potential* and obtain the partial differential equation satisfied by the gravitational potential.

To do this, let us suppose that we have a **field of force** of some kind, or, in other words, let us consider a region of space in which at every point a force vector $\mathbf{F}$ is defined. The field might, for instance, be **gravitational**, in which case $\mathbf{F}(x,y,z)$ would be the force acting on a unit mass at the general point $P{:}(x,y,z)$ because of the attraction of other masses present in the region. On the other hand, the field might be **electrostatic**, in which case $\mathbf{F}(x,y,z)$ would be the force acting on a unit charge at the general point $P{:}(x,y,z)$ because of the attraction or repulsion of other charges present in the region. Or the field might be **magnetic**, in which case $\mathbf{F}(x,y,z)$ would be the force acting on a unit magnetic pole situated at the point $P{:}(x,y,z)$. In any case, the force $\mathbf{F}$ experienced by a unit test body of the appropriate nature is called the **field intensity**.

The amount of work that must be done when a unit test body is moved along an arbitrary curve in the force field defined by a vector function $\mathbf{F}$ is the line integral of the tangential component of $\mathbf{F}$; that is,

$$W = \int_C \mathbf{F} \cdot d\mathbf{R}$$

If there is no dissipation of energy through friction or similar effects, then, according to the law of the conservation of energy, this integral must be zero around every closed path, and hence, by Theorem 6, Sec. 13.5, it must be independent of the path between any given points A and B. Fields for which this is the case are said to be **conservative**. Furthermore, according to Theorem 6, Sec. 13.5, it is clear that in a conservative field the force vector $\mathbf{F}$ is the gradient of the scalar function

$$\phi(x,y,z) = \int_{P_0}^{P} \mathbf{F} \cdot d\mathbf{R}$$

The function ϕ is called the **potential function**† of the field. In most problems, the masses or charges which produce $\mathbf{F}$ are given, and it is required to find $\mathbf{F}$ itself. Since $\mathbf{F} = \nabla\phi$, it is clear that knowing ϕ is equivalent to knowing $\mathbf{F}$, and hence the determination of ϕ is of prime importance in most field problems.

Assuming, for definiteness, that we are dealing with a gravitational field, let $\mathbf{F}$ be the field intensity at a general point $P:(x,y,z)$, and let $\Delta\mathbf{F}$ be the contribution to $\mathbf{F}$ due to the infinitesimal mass Δm_1 in an infinitesimal volume $\Delta V_1 = \Delta x_1 \, \Delta y_1 \, \Delta z_1$ enclosing the point $P_1:(x_1,y_1,z_1)$. According to Newton's law of universal gravitation, $\Delta\mathbf{F}$ is a vector whose magnitude is

$$\Delta F = k \frac{1 \cdot \Delta m_1}{r^2}$$

where
$$r^2 = (x - x_1)^2 + (y - y_1)^2 + (z - z_1)^2$$

and whose direction is opposite to that of the vector

$$\mathbf{R} = (x - x_1)\mathbf{i} + (y - y_1)\mathbf{j} + (z - z_1)\mathbf{k}$$

extending from P_1 to P (Fig. 13.32). In other words, if units are so chosen that the

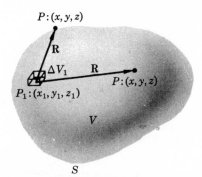

$P:(x,y,z)$

$\mathbf{R}$

ΔV_1 $\mathbf{R}$

$P:(x,y,z)$

$P_1:(x_1,y_1,z_1)$

V

S

Figure 13.32
Calculating the potential at a point P due to the material in a volume element ΔV_1.

---

† Many writers define the potential to be $\int_P^{P_0} \mathbf{F} \cdot d\mathbf{R}$, in which case $\mathbf{F} = -\nabla\phi$. In particular, P_0 is often taken to be infinitely distant, so that $\phi = \int_P^{\infty} \mathbf{F} \cdot d\mathbf{R}$.

constant in Newton's law is equal to unity, then the field intensity at P due to the infinitesimal mass Δm_1 at P_1 is

(1)
$$\Delta \mathbf{F} = -\frac{\Delta m_1}{r^2} \cdot \frac{\mathbf{R}}{r} = -\rho(x_1,y_1,z_1) \, \Delta V_1 \frac{\mathbf{R}}{r^3}$$

where $\rho(x_1,y_1,z_1)$ is the density of the material at the point P_1.

Now let S be an arbitrary closed regular surface bounding a volume V, and let I denote the integral over S of the normal component of the force due to all the attracting material in the field. By hypothesis, $\mathbf{F} = \mathbf{V}\phi$; hence

(2)
$$I = \iint_S \mathbf{N} \cdot \mathbf{F} \, dS = \iint_S \mathbf{N} \cdot \mathbf{V}\phi \, dS$$

However, I can also be computed by first determining the part of it, ΔI, due to the material in an arbitrary volume element ΔV_1 and then taking into account all such infinitesimal contributions by integrating over the entire field. From this point of view we have, from (1) and (2),

$$\Delta I = \iint_S \mathbf{N} \cdot \Delta \mathbf{F} \, dS = -\iint_S [\rho(x_1,y_1,z_1) \, \Delta V_1] \mathbf{N} \cdot \frac{\mathbf{R}}{r^3} \, dS$$

$$= -\rho(x_1,y_1,z_1) \, \Delta V_1 \iint_S \mathbf{N} \cdot \frac{\mathbf{R}}{r^3} \, dS$$

since x_1, y_1, z_1 are constant with respect to the x,y,z integration over S. The last integral can, of course, be evaluated by Gauss' theorem (Theorem 3, Sec. 13.5). Specifically, if the origin of $\mathbf{R}$, namely, the point $P_1{:}(x_1,y_1,z_1)$, is within S, the value of the integral is 4π; otherwise the value of the integral is 0. Hence

$$\Delta I = \begin{cases} -4\pi\rho(x_1,y_1,z_1) \, \Delta V_1 & \Delta V_1 \text{ within } S \\ 0 & \Delta V_1 \text{ outside } S \end{cases}$$

and, therefore, in computing I it is necessary to integrate only over the volume V bounded by S. Doing this, we find

$$I = \int dI = -4\pi \iiint_V \rho(x_1,y_1,z_1) \, dV_1$$

or, since x_1, y_1, z_1 are just dummy variables,

(3)
$$I = -4\pi \iiint_V \rho(x,y,z) \, dV$$

Equating the two expressions (2) and (3) which we now have for I, we get

$$\iint_S \mathbf{N} \cdot \mathbf{V}\phi \, dS = -4\pi \iiint_V \rho(x,y,z) \, dV$$

If we now apply the divergence theorem to the integral on the left, we have

$$\iiint_V \mathbf{V} \cdot (\mathbf{V}\phi) \, dV = -4\pi \iiint_V \rho(x,y,z) \, dV$$

or

$$\iiint_V [\nabla^2\phi + 4\pi\rho(x,y,z)] \, dV = 0$$

Since this holds for an arbitrary volume V, it follows that the integrand must vanish identically,† and therefore that

(4)
$$\nabla^2 \phi = -4\pi\rho(x,y,z)$$

This is **Poisson's equation**,‡ and we have thus shown that *in regions occupied by matter, the gravitational potential satisfies Poisson's equation.* In empty space $\rho(x,y,z) = 0$, and thus *in empty space the gravitational potential satisfies Laplace's equation*

(5)
$$\nabla^2 \phi = 0$$

Results similar to these hold for the electrostatic and magnetic potentials.

As a second example of the use of vector analysis in formulating physical laws in mathematical terms, we shall now derive *Maxwell's equations*§ for electric and magnetic fields. To do this we shall have to work with the vector quantities

$\mathbf{E}$ = electric intensity
$\mathbf{H}$ = magnetic intensity
$\mathbf{D} = \varepsilon\mathbf{E}$ = electric flux density
$\mathbf{B} = \mu\mathbf{H}$ = magnetic flux density
$\mathbf{J}$ = current density

and the scalars

ε = permittivity
μ = permeability
σ = conductivity
Q = charge density
$q = \iiint_V Q \, dV$ = total charge within volume V
$\phi = \iint_S \mathbf{N} \cdot \mathbf{B} \, dS$ = total magnetic flux passing through surface S
$i = \iint_S \mathbf{N} \cdot \mathbf{J} \, dS$ = total current flowing through surface S

These quantities are connected by a number of equations expressing relations discovered experimentally in the early years of the nineteenth century, chiefly by Michael Faraday (1791–1867). In particular, we have

a. Faraday's law,

(6)
$$\int_C \mathbf{E} \cdot d\mathbf{R} = -\frac{\partial \phi}{\partial t}$$

which asserts that the integral of the tangential component of the electric intensity vector around any closed curve C is equal but opposite in sign to the rate of change of the magnetic flux passing through any surface spanning C

---

† Suppose that this is not the case, and let P_0 be a point at which the integrand does not vanish. Then, if $\rho(x,y,z)$ and $\nabla^2\phi$ are continuous (as we have implicitly assumed), it follows that throughout some sufficiently small three-dimensional region V_0 enclosing P_0 the integrand has everywhere the same sign it has at P_0. Integrating over V_0, we then obtain an integral which is not equal to zero, contrary to the fact that the integral has been shown to be zero for every volume V.
‡ Named for the French mathematical physicist Simeon Denis Poisson (1781–1840).
§ Named for the Scottish mathematical physicist James Clerk Maxwell (1831–1879).

b. Ampere's law,

(7)
$$\int_C \mathbf{H} \cdot d\mathbf{R} = i$$

which asserts that the integral of the tangential component of the magnetic intensity vector around any closed curve C is equal to the current flowing through any surface spanning C

c. Gauss' law for electric fields,

(8)
$$\iint_S \mathbf{N} \cdot \mathbf{D} \, dS = q$$

which asserts that the integral of the normal component of the electric flux density over any closed surface S is equal to the total electric charge enclosed by S

d. Gauss' law for magnetic fields,

(9)
$$\iint_S \mathbf{N} \cdot \mathbf{B} \, dS = 0$$

which asserts that the total magnetic flux passing through any closed surface S is zero

If we now apply Stokes' theorem to Faraday's law (6), we have

$$\iint_S \mathbf{N} \cdot \mathbf{V} \times \mathbf{E} \, dS = -\frac{\partial \phi}{\partial t}$$

and, substituting for ϕ from its definition in terms of $\mathbf{B}$,

$$\iint_S \mathbf{N} \cdot \mathbf{V} \times \mathbf{E} \, dS = -\frac{\partial}{\partial t} \left(\iint_S \mathbf{N} \cdot \mathbf{B} \, dS \right) = -\iint_S \mathbf{N} \cdot \frac{\partial \mathbf{B}}{\partial t} \, dS$$

Since S is an arbitrary surface spanning the arbitrary closed curve C, the last equation can hold only if

(10)
$$\mathbf{V} \times \mathbf{E} = -\frac{\partial \mathbf{B}}{\partial t}$$

Similarly, by applying Stokes' theorem to Ampere's law (7), we obtain

$$\iint_S \mathbf{N} \cdot \mathbf{V} \times \mathbf{H} \, dS = i = \iint_S \mathbf{N} \cdot \mathbf{J} \, dS$$

and again, since S is an arbitrary open surface, we conclude that the vectors being integrated over S must be identical; i.e.,

(11)
$$\mathbf{V} \times \mathbf{H} = \mathbf{J}$$

As Maxwell was the first to realize, the current density $\mathbf{J}$ consists of two parts, namely, a conduction current density

$$\mathbf{J}_c = \sigma \mathbf{E}$$

due to the flow of electric charges, and a displacement current density

$$\mathbf{J}_d = \frac{\partial \mathbf{D}}{\partial t} = \varepsilon \frac{\partial \mathbf{E}}{\partial t}$$

due to the time variation of the electric field. Thus,

$$\mathbf{J} = \sigma\mathbf{E} + \varepsilon\,\frac{\partial\mathbf{E}}{\partial t}$$

and (11) becomes

(12) $$\mathbf{V} \times \mathbf{H} = \sigma\mathbf{E} + \varepsilon\,\frac{\partial\mathbf{E}}{\partial t}$$

Next we apply the divergence theorem to the first of Gauss' laws, (8), getting

$$\iiint_V \mathbf{V}\cdot\mathbf{D}\,dV = q = \iiint_V Q\,dV$$

whence, since V is arbitrary, it follows that

(13) $$\mathbf{V}\cdot\mathbf{D} = Q$$

In the same way, by applying the divergence theorem to Gauss' second law, (9), we find that

$$\iiint_V \mathbf{V}\cdot\mathbf{B}\,dV = 0$$

and therefore, since V is arbitrary,

(14) $$\mathbf{V}\cdot\mathbf{B} = 0$$

If we take the curl of Eq. (10), we obtain

$$\mathbf{V} \times (\mathbf{V} \times \mathbf{E}) = -\mathbf{V} \times \frac{\partial\mathbf{B}}{\partial t} = -\frac{\partial}{\partial t}(\mathbf{V} \times \mathbf{B}) = -\mu\frac{\partial}{\partial t}(\mathbf{V} \times \mathbf{H})$$

If we expand the term $\mathbf{V} \times (\mathbf{V} \times \mathbf{E})$ by means of Eq. (20), Sec. 13.5, the last equation becomes

$$\mathbf{V}(\mathbf{V}\cdot\mathbf{E}) - \mathbf{V}^2\mathbf{E} = -\mu\frac{\partial}{\partial t}(\mathbf{V} \times \mathbf{H})$$

and, substituting for $\mathbf{V} \times \mathbf{H}$ from (12),

(15) $$\mathbf{V}(\mathbf{V}\cdot\mathbf{E}) - \mathbf{V}^2\mathbf{E} = -\mu\frac{\partial}{\partial t}\left(\sigma\mathbf{E} + \varepsilon\,\frac{\partial\mathbf{E}}{\partial t}\right)$$

If the space charge density Q is zero, as it is to a high degree of approximation in both good dielectrics and good conductors, then from (13) and the relation $\mathbf{D} = \varepsilon\mathbf{E}$ we see that

$$\mathbf{V}\cdot\mathbf{E} = 0$$

Therefore Eq. (15) reduces to

$$\mathbf{V}^2\mathbf{E} = \mu\varepsilon\,\frac{\partial^2\mathbf{E}}{\partial t^2} + \mu\sigma\,\frac{\partial\mathbf{E}}{\partial t}$$

which is **Maxwell's equation for the electric intensity vector E.**
Similarly, if we take the curl of Eq. (12), we obtain

$$\mathbf{V} \times (\mathbf{V} \times \mathbf{H}) = \mathbf{V} \times \left(\sigma\mathbf{E} + \varepsilon\,\frac{\partial\mathbf{E}}{\partial t}\right)$$

and, expanding the left-hand side,

$$\mathbf{V}(\mathbf{V} \cdot \mathbf{H}) - \nabla^2 \mathbf{H} = \sigma \mathbf{V} \times \mathbf{E} + \varepsilon \mathbf{V} \times \frac{\partial \mathbf{E}}{\partial t}$$

$$= \sigma \mathbf{V} \times \mathbf{E} + \varepsilon \frac{\partial}{\partial t} (\mathbf{V} \times \mathbf{E})$$

Now, substituting for $\mathbf{V} \times \mathbf{E}$ from (10), we have

$$\mathbf{V}(\mathbf{V} \cdot \mathbf{H}) - \nabla^2 \mathbf{H} = \sigma \left(-\frac{\partial \mathbf{B}}{\partial t} \right) + \varepsilon \left(-\frac{\partial^2 \mathbf{B}}{\partial t^2} \right)$$

But $\mathbf{B} = \mu \mathbf{H}$, by definition. Hence (14) implies that $\mathbf{V} \cdot \mathbf{H} = 0$, and therefore the last equation reduces to

$$\nabla^2 \mathbf{H} = \mu \varepsilon \frac{\partial^2 \mathbf{H}}{\partial t^2} + \mu \sigma \frac{\partial \mathbf{H}}{\partial t}$$

which is **Maxwell's equation for the magnetic intensity vector H.**

For a perfect dielectric, $\sigma = 0$. Hence in this case Maxwell's equations reduce to the three-dimensional wave equations

$$\nabla^2 \mathbf{E} = \mu \varepsilon \frac{\partial^2 \mathbf{E}}{\partial t^2} \qquad \text{and} \qquad \nabla^2 \mathbf{H} = \mu \varepsilon \frac{\partial^2 \mathbf{H}}{\partial t^2}$$

On the other hand, in a good conductor the terms arising from the displacement current, i.e., the terms containing the second time derivatives, are negligible, and Maxwell's equations reduce to

$$\nabla^2 \mathbf{E} = \mu \sigma \frac{\partial \mathbf{E}}{\partial t} \qquad \text{and} \qquad \nabla^2 \mathbf{H} = \mu \sigma \frac{\partial \mathbf{H}}{\partial t}$$

which are examples of the three-dimensional heat equation.

As a final application of the methods of vector analysis, we shall investigate the question whether a solution of the heat equation satisfying prescribed boundary and initial conditions over a given region is necessarily unique. In our discussion of boundary-value problems in Chap. 8 we proceeded on the assumption that this was the case. Nevertheless, examples have been given[†] of solutions of the one-dimensional heat equation

$$a^2 \frac{\partial u}{\partial t} = \frac{\partial^2 u}{\partial x^2}$$

which possess derivatives of all orders for all values of x and t, satisfy identical initial conditions everywhere on the entire x axis, and yet are different! Confronted with such a clear-cut failure of intuition, we must regard the uniqueness question as of more than academic interest and any positive result as having important practical significance.

Let us suppose, then, that we are to solve the three-dimensional heat equation

$$a^2 \frac{\partial u}{\partial t} = \nabla^2 u$$

---

† See, for instance, P. C. Rosenbloom and D. V. Widder, A Temperature Function Which Vanishes Identically, *Am. Math. Mon.*, vol. 65, p. 607, October 1958.

throughout a region V bounded by the closed surface S, subject to the boundary condition

$$u = f(x,y,z,t) \qquad \text{on } S$$

and the initial condition

$$u(x,y,z,0) = g(x,y,z) \qquad \text{throughout } V$$

Furthermore, let us suppose that we have two solutions of this problem, u_1 and u_2, each of which, with its derivatives through the second, is continuous in V.

If we define a new function

$$w(x,y,z,t) = u_2(x,y,z,t) - u_1(x,y,z,t)$$

it is clear from the linearity of the heat equation that w also satisfies this equation. Moreover, w obviously assumes boundary and initial conditions which are identically zero. Finally, w is continuous and differentiable, since it is the difference of two functions with these properties.

Now consider the volume integral

$$(16) \qquad J(t) = \frac{1}{2} \iiint_V w^2(x,y,z,t)\, dV \qquad t \geq 0$$

Clearly, $J(t)$ is a continuous function of t which is always equal to or greater than zero, since its integrand is everywhere nonnegative. Also, since $w = 0$ when $t = 0$, it follows that $J(0) = 0$. Now, differentiating with respect to t inside the integral sign, we have

$$J'(t) = \frac{1}{2} \iiint_V 2w\, \frac{\partial w}{\partial t}\, dV$$

and thus, since w satisfies the heat equation, on substituting for $\partial w/\partial t$ we have

$$(17) \qquad J'(t) = \frac{1}{a^2} \iiint_V w\, \nabla^2 w\, dV$$

According to Eq. (7), Sec. 13.5, with both u and v in that formula taken to be the function w of the present problem, we have

$$(18) \qquad \iiint_V (w\, \nabla^2 w + \nabla w \cdot \nabla w)\, dV = \iint_S \mathbf{N} \cdot w\, \nabla w\, dS$$

Since the function w vanishes identically on S, the integral on the right side of (18) is zero, and we have

$$\iiint_V w\, \nabla^2 w\, dV = -\iiint_V \nabla w \cdot \nabla w\, dV$$

Hence, substituting into (17),

$$J'(t) = -\frac{1}{a^2} \iiint_V \nabla w \cdot \nabla w\, dV$$

$$= -\frac{1}{a^2} \iiint_V \left[\left(\frac{\partial w}{\partial x}\right)^2 + \left(\frac{\partial w}{\partial y}\right)^2 + \left(\frac{\partial w}{\partial z}\right)^2 \right] dV$$

which shows that

$$J'(t) \leq 0 \qquad \text{for } t \geq 0$$

Now, by the law of the mean,

$$\frac{J(t) - J(0)}{t} = J'(t_1) \qquad 0 < t_1 < t$$

or

$$J(t) = J(0) + tJ'(t_1) \qquad 0 < t_1 < t$$

But we have already verified that $J(0) = 0$. Hence the last equation reduces to

$$J(t) = tJ'(t_1)$$

which shows that

(19) $$J(t) \leq 0 \qquad \text{for } t \geq 0$$

since we have just proved that $J'(t)$ is nonpositive for $t \geq 0$. However, as we observed earlier, the definition of $J(t)$ shows that

(20) $$J(t) \geq 0 \qquad \text{for } t \geq 0$$

The only way in which the inequalities (19) and (20) can simultaneously be fulfilled is for $J(t)$ to be identically zero. But this is possible if and only if the integrand of $J(t)$ vanishes identically. Hence

$$w(x,y,z,t) \equiv u_2(x,y,z,t) - u_1(x,y,z,t) \equiv 0$$

or

$$u_2(x,y,z,t) \equiv u_1(x,y,z,t)$$

Thus, *in bounded regions, twice differentiable solutions of the heat equation satisfying prescribed surface and initial temperature conditions are unique.*

EXERCISES

1 What is the potential function for a central-force field in which the attraction on a particle varies directly as the square of the distance from the origin? Inversely as the distance from the origin?

2 What is the potential function of the force field due to uniform rotation about the z axis?

3 What is the potential function for the gravitational field of a uniform circular disk at any point on the axis of the disk?

4 What is the potential function for the gravitational field of a uniform sphere of radius a and mass M? Show that the attraction of the sphere at a point P a distance r from the center of the sphere is

$$\mathbf{F} = \begin{cases} -\dfrac{M\mathbf{R}}{a^3} & r \leq a \\[2ex] -\dfrac{M\mathbf{R}}{r^3} & r \geq a \end{cases}$$

5 Show that the electrostatic field intensity at a point P due to a set of charges q_i is equal to

$$\mathbf{E} = -\sum_{i=1}^{n} \frac{q_i}{r_i^3} \mathbf{R}_i$$

where $\mathbf{R}_i$ is the vector from the point P to the point P_i where the charge q_i is located. Verify that $\mathbf{V} \cdot \mathbf{E} = 0$ in this case.

6 Show that the work done in bringing a charge of strength q from infinity to a point at a distance of r_0 from a fixed charge q_0 is qq_0/r_0. Using this result, determine the total energy in the electrostatic field defined by the fixed charges $q_1, q_2, \ldots, q_n$ whose mutual distances are $\{r_{ij}.\}$

7   If a **conductor** is defined to be a body in whose interior the electric field is everywhere zero, show that any charge on a conductor must be located entirely on its surface.

8   Let V_1 and V_2 be two regions with respective dielectric constants ε_1 and ε_2, and let S be the surface of discontinuity which separates them. By applying Gauss' law for electric fields to a closed cylindrical surface of infinitesimal height whose bases are parallel to S in the respective media, show that if there are no charges on S, the normal component of the electric flux density is continuous across S. Similarly, by applying Faraday's law to a rectangle of negligible width whose longer sides are parallel to S in the respective media, prove that if the field is conservative, the tangential component of the electric intensity is continuous across S.

9   What is the electric field in the empty space between the perfectly conducting, infinite planes $y = 0$ and $y = l$ if

$$\mathbf{E}\Big|_{t=0} = \mathbf{i} + \mathbf{k} \quad \text{and} \quad \frac{\partial \mathbf{E}}{\partial t}\Big|_{t=0} = \mathbf{i} - \mathbf{k}?$$

Hint: From the nature of the region of the problem and the initial conditions, it is clear that the field has no component in the y direction and that E_x and E_z are functions only of y.

10   Prove that a solution of the heat equation, possessing continuous second partial derivatives, which takes on prescribed initial values throughout a region V and whose normal derivative takes on prescribed values on the surface S which encloses V is unique.

11   Prove that a solution $u(x,y,z,t)$ of the heat equation possessing continuous second partial derivatives which takes on prescribed initial values throughout a region V and for which the expression

$$u(x,y,z,t) + h^2\mathbf{N} \cdot \nabla u(x,y,z,t)$$

takes on prescribed values on the surface S which encloses V is unique.

12   Prove that a vector function is uniquely determined at all points of a region V if its curl and divergence throughout V and its normal component over S are known. *Hint:* Assume that there are two such functions, verify that their difference must be the gradient of some scalar function w, and then use Eq. (7), Sec. 13.5, with u and v each taken equal to w

CHAPTER 14
Tensor Analysis

14.1 Introduction

In Chap. 10 we introduced the concept of a vector as either a $1 \times n$ or an $n \times 1$ matrix, i.e., as an ordered set of n quantities. In the last chapter we took a somewhat less abstract point of view and regarded a vector as a quantity which could be represented by a directed line segment. Using this interpretation, we then developed the algebra and calculus of vectors, working implicitly (and sometimes explicitly) in a rectangular coordinate system. In spite of this, however, it should be clear that we were dealing with quantities which are independent of any particular coordinate system. For example, although the description of a point, i.e., its coordinates, may change from one coordinate system to another, the point is recognizably the same in all coordinate systems. Similarly, although the formula by which it is computed may change, the length of a particular vector must be the same in all coordinate systems.

In this chapter we shall pursue further this idea of invariance and adopt as our fundamental idea of a vector the concept of a quantity invariant under any transformation of coordinates. This will lead us to the idea of the covariant and contravariant representation of vectors and thence to the highly important concept of a *tensor*. Although we cannot undertake a detailed discussion of tensor analysis, we shall indicate some of its principal features and illustrate the remarkable economy of the tensor notation.

14.2 Oblique Coordinates

Because of the need to distinguish between what we shall soon refer to as *covariant* and *contravariant vectors*, it is necessary that our notation employ indices not only in the familiar subscript position but in the superscript position as well. In tensor analysis, this requirement takes precedence over the usual exponential symbolism, and henceforth when we write, say,

$$\zeta^\alpha$$

α will be a distinguishing index, like subscripts heretofore, and *not* an exponent. If and when we wish to indicate the αth power of a quantity ξ, we shall always use parentheses and write

$$(\xi)^\alpha$$

With this convention in mind, and as a relatively simple example of the generalized coordinates we shall subsequently investigate, let us consider a system of coordinates

$(\bar{x}^1,\bar{x}^2,\bar{x}^3)$ connected with a system of rectangular coordinates (x^1,x^2,x^3) by the equations

$$\bar{x}^1 = a_{11}x^1 + a_{12}x^2 + a_{13}x^3$$
$$\bar{x}^2 = a_{21}x^1 + a_{22}x^2 + a_{23}x^3$$
$$\bar{x}^3 = a_{31}x^1 + a_{32}x^2 + a_{33}x^3$$

or, in matric form,

(1) $$\bar{X} = AX$$

and

(2) $$X = A^{-1}\bar{X}$$

where, as usual,

$$X = \begin{bmatrix} x^1 \\ x^2 \\ x^3 \end{bmatrix} \qquad \bar{X}\dagger = \begin{bmatrix} \bar{x}^1 \\ \bar{x}^2 \\ \bar{x}^3 \end{bmatrix} \qquad A = \begin{bmatrix} a_{11} & a_{12} & a_{13} \\ a_{21} & a_{22} & a_{23} \\ a_{31} & a_{32} & a_{33} \end{bmatrix}$$

and A^{-1} is the inverse of the matrix A.

The locus of points for which $\bar{x}^1 = 0$ is, of course, the plane

$$\pi_1: \quad a_{11}x^1 + a_{12}x^2 + a_{13}x^3 = 0$$

Similarly, the locus of points for which $\bar{x}^2 = 0$ is the plane

$$\pi_2: \quad a_{21}x^1 + a_{22}x^2 + a_{23}x^3 = 0$$

and the locus of the points for which $\bar{x}^3 = 0$ is the plane

$$\pi_3: \quad a_{31}x^1 + a_{32}x^2 + a_{33}x^3 = 0$$

Clearly, on the line of intersection of π_2 and π_3, both $\bar{x}^2$ and $\bar{x}^3$ are zero, and $\bar{x}^1$ alone varies. This line can therefore be thought of as the $\bar{x}^1$ axis. In the same fashion, we can identify the line of intersection of π_1 and π_3 as the $\bar{x}^2$ axis and the line of intersection of π_1 and π_2 as the $\bar{x}^3$ axis (Fig. 14.1a). Obviously, the $\bar{x}^1$, $\bar{x}^2$, $\bar{x}^3$ axes are concurrent. Moreover, since $|A| \neq 0$, these lines are distinct and noncoplanar. In general, however, they will not be mutually perpendicular, and for this reason they are said to be the axes of an **oblique coordinate system**.

Since the $\bar{x}^1$, $\bar{x}^2$, and $\bar{x}^3$ axes are noncoplanar, any vector can be expressed as a linear combination of arbitrary reference vectors along the three oblique axes. By analogy with the unit vectors $\mathbf{i}$, $\mathbf{j}$, $\mathbf{k}$ (or $\mathbf{e}_1$, $\mathbf{e}_2$, $\mathbf{e}_3$, as we shall now denote them) it might seem natural to choose vectors of unit length for this purpose. However, because the oblique coordinates $\bar{x}^1$, $\bar{x}^2$, $\bar{x}^3$ are *not* distance measures along the coordinate axes, as x^1, x^2, x^3 are along the axes of a rectangular coordinate system, it turns out to be more convenient to take the new reference vectors $\bar{\mathbf{e}}_1$, $\bar{\mathbf{e}}_2$, $\bar{\mathbf{e}}_3$ to be, respectively, the vectors from the origin to the points whose oblique coordinates are $(1,0,0)$, $(0,1,0)$, and $(0,0,1)$ (Fig. 14.1b).

---

† We shall use overbars to denote not only the new coordinates themselves but also all quantities referred to the new coordinate system. Thus, if P is the name of a point described in the original (rectangular) coordinate system by the coordinates (p^1,p^2,p^3), then $\bar{P}$ is the name we shall use for this same point thought of as described by the new coordinates $(\bar{p}^1,\bar{p}^2,\bar{p}^3)$ determined by Eq. (1).

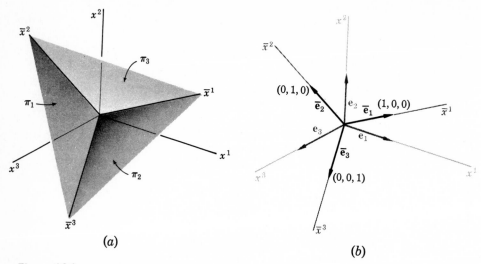

Figure 14.1
A rectangular and an oblique coordinate system with their related reference vectors.

To determine the lengths of the reference vectors $\bar{e}_1$, $\bar{e}_2$, $\bar{e}_3$ and to obtain the formula for measuring distances in general in oblique coordinates, let us consider the vector $\bar{\mathbf{V}}$† extending from the point $\bar{P}$ whose matrix of oblique coordinates is

$$\bar{V}_P = \begin{bmatrix} \bar{p}^1 \\ \bar{p}^2 \\ \bar{p}^3 \end{bmatrix}$$

to the point $\bar{Q}$ whose matrix of oblique coordinates is

$$\bar{V}_Q = \begin{bmatrix} \bar{q}^1 \\ \bar{q}^2 \\ \bar{q}^3 \end{bmatrix}$$

From Eq. (2) it follows that the rectangular coordinates of $\bar{P}\,(=P)$ and $\bar{Q}\,(=Q)$ are defined, respectively, by the matrices

$$V_P = A^{-1}\bar{V}_P \qquad \text{and} \qquad V_Q = A^{-1}\bar{V}_Q$$

Hence, in rectangular coordinates, the vector $\bar{\mathbf{V}} = \bar{\mathbf{V}}_Q - \bar{\mathbf{V}}_P$ (Fig. 14.2a), defined in oblique coordinates by the matrix of components $\bar{V} = \bar{V}_Q - \bar{V}_P$, becomes the vector $\mathbf{V} = \mathbf{V}_Q - \mathbf{V}_P$ (Fig. 14.2b) defined by the matrix

$$V = V_Q - V_P = A^{-1}\bar{V}_Q - A^{-1}\bar{V}_P = A^{-1}(\bar{V}_Q - \bar{V}_P) = A^{-1}\bar{V}$$

In rectangular coordinates, the square of the length of a vector $\mathbf{V}$ whose matrix of components is V is given by the formula

$$\mathbf{V} \cdot \mathbf{V} = V^T I V = V^T G V$$

---

† In this chapter we shall use boldface symbols to denote vectors only when we are considering them as directed line segments, as we did in the last chapter. In particular, if $\mathbf{V}$ is a vector considered in the geometric sense, we shall use the symbol V to denote not the length of $\mathbf{V}$ but rather the matrix of the components of $\mathbf{V}$ along the appropriate set of axes.

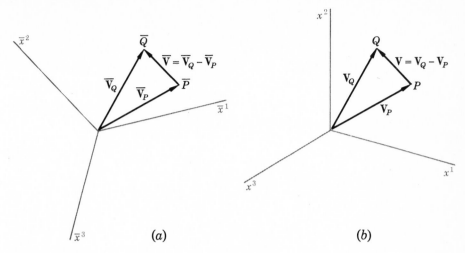

Figure 14.2
A vector **V** represented in each of two coordinate systems.

where, for later convenience, we have introduced G as another name for the matrix which is I in this case but not in general. Therefore, since we require the length of a given vector to be the same in all coordinate systems, we define the scalar product of a vector with itself in oblique coordinates by the condition that

$$\bar{V} \cdot \bar{V} = V \cdot V = (A^{-1}\bar{V})^T I (A^{-1}\bar{V}) = \bar{V}^T (A^{-1})^T I A^{-1} \bar{V}$$
$$= \bar{V}^T [(A^{-1})^T I A^{-1}] \bar{V}$$
(3) $$= \bar{V}^T \bar{G} \bar{V}$$

Similarly, for distinct vectors $\bar{U}$ and $\bar{V}$, we define

$$\bar{U} \cdot \bar{V} = U \cdot V = (A^{-1}\bar{U})^T I (A^{-1}\bar{V}) = \bar{U}^T (A^{-1})^T I A^{-1} \bar{V}$$
$$= \bar{U}^T [(A^{-1})^T I A^{-1}] \bar{V}$$
(4) $$= \bar{U}^T \bar{G} \bar{V}$$

Thus, *the metrical properties of space, which in rectangular coordinates are determined by the identity matrix $I = G$, are in oblique coordinates determined by the matrix $(A^{-1})^T I A^{-1} = \bar{G}$, where A is the matrix of the transformation $\bar{X} = AX$ from rectangular to oblique coordinates.*

When we denote $\bar{g}_{ij}$ the element in the ith row and jth column of the matrix $\bar{G} = (A^{-1})^T A^{-1}$, it is clear from Eqs. (3) and (4) that for the reference vectors $\bar{e}_1, \bar{e}_2, \bar{e}_3$ defined by the matrices

$$\bar{e}_1 = \begin{bmatrix} 1 \\ 0 \\ 0 \end{bmatrix} \quad \bar{e}_2 = \begin{bmatrix} 0 \\ 1 \\ 0 \end{bmatrix} \quad \bar{e}_3 = \begin{bmatrix} 0 \\ 0 \\ 1 \end{bmatrix}$$

we have

(5) $$\bar{e}_i \cdot \bar{e}_j = \bar{g}_{ij}$$

In particular, the lengths of $\bar{e}_1, \bar{e}_2,$ and $\bar{e}_3$ are, respectively,

$$|\bar{e}_1| = \sqrt{\bar{g}_{11}} \quad |\bar{e}_2| = \sqrt{\bar{g}_{22}} \quad |\bar{e}_3| = \sqrt{\bar{g}_{33}}$$

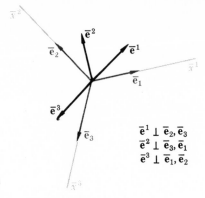

Figure 14.3
The base vectors $\bar{\mathbf{e}}_1$, $\bar{\mathbf{e}}_2$, $\bar{\mathbf{e}}_3$ and the reciprocal base
vectors $\bar{\mathbf{e}}^1$, $\bar{\mathbf{e}}^2$, $\bar{\mathbf{e}}^3$ in an oblique coordinate system.

As a consequence of these formulas, it follows that *the length of $\bar{\mathbf{e}}_i$ is such that if $\mathbf{R}$
is the vector extending along the $\bar{x}^i$ axis from the origin to the point for which $\bar{x}^i = \bar{a}^i$,
then the length of $\bar{\mathbf{R}}$ is equal to the absolute value of the coordinate $\bar{a}^i$ times the length
of $\bar{\mathbf{e}}_i$ that is,* $|\bar{\mathbf{R}}| = |\bar{a}^i| \cdot |\bar{\mathbf{e}}_i|$.

In a rectangular coordinate system a unique set of directions for the reference
vectors is clearly identified by the axes of the system. In oblique coordinates this is
not the case; for although the oblique axes certainly define a set of directions in which
reference vectors can naturally be chosen, there is another set distinct from the first
which is also intrinsic in the system, namely, the directions perpendicular to the
coordinate planes π_1, π_2, π_3. As base vectors in these directions it is customary to
take vectors $\bar{\mathbf{e}}^1$, $\bar{\mathbf{e}}^2$, $\bar{\mathbf{e}}^3$ defined by the conditions

(6)
$$\bar{\mathbf{e}}^i \cdot \bar{\mathbf{e}}_j = \begin{cases} 0 & i \neq j \\ 1 & i = j \end{cases}$$

For $i \neq j$ these relations fix the directions of the new reference vectors, and for
$i = j$ they determine their lengths and sense. The vectors $\bar{\mathbf{e}}^1$, $\bar{\mathbf{e}}^2$, $\bar{\mathbf{e}}^3$ are said to form a
set **reciprocal** to the set $\bar{\mathbf{e}}_1$, $\bar{\mathbf{e}}_2$, $\bar{\mathbf{e}}_3$, and vice versa† (Fig. 14.3). From their definition
it is clear that $\bar{\mathbf{e}}^1$, $\bar{\mathbf{e}}^2$, $\bar{\mathbf{e}}^3$ are noncoplanar and hence can be used as a basis for the
representation of any vector. Thus, when we use oblique coordinates, any vector $\bar{\mathbf{V}}$
has two different but equally natural representations: It can be expressed as a linear
combination of the base vectors $\bar{\mathbf{e}}_1$, $\bar{\mathbf{e}}_2$, $\bar{\mathbf{e}}_3$, or it can be expressed as a linear combination of the vectors of the reciprocal set $\bar{\mathbf{e}}^1$, $\bar{\mathbf{e}}^2$, $\bar{\mathbf{e}}^3$.

In particular, the vectors in each of the sets $\bar{\mathbf{e}}_1$, $\bar{\mathbf{e}}_2$, $\bar{\mathbf{e}}_3$ and $\bar{\mathbf{e}}^1$, $\bar{\mathbf{e}}^2$, $\bar{\mathbf{e}}^3$ must be
expressible as linear combinations of the vectors in the other set. Specifically, if we
write

$$\bar{\mathbf{e}}_1 = \mu_{11}\bar{\mathbf{e}}^1 + \mu_{12}\bar{\mathbf{e}}^2 + \mu_{13}\bar{\mathbf{e}}^3$$
$$\bar{\mathbf{e}}_2 = \mu_{21}\bar{\mathbf{e}}^1 + \mu_{22}\bar{\mathbf{e}}^2 + \mu_{23}\bar{\mathbf{e}}^3$$
$$\bar{\mathbf{e}}_3 = \mu_{31}\bar{\mathbf{e}}^1 + \mu_{32}\bar{\mathbf{e}}^2 + \mu_{33}\bar{\mathbf{e}}^3$$

---

† It is evident that in rectangular coordinates the set of base vectors and the set of reciprocal
vectors are the same; that is, $\mathbf{i} = \mathbf{e}_1 = \mathbf{e}^1$, $\mathbf{j} = \mathbf{e}_2 = \mathbf{e}^2$, $\mathbf{k} = \mathbf{e}_3 = \mathbf{e}^3$. It is for this reason
that the concept of reciprocal sets of vectors was not introduced in the last chapter (except
in Exercise 23, Sec. 13.1).

and then form the scalar product of each side of the ith equation with $\bar{\mathbf{e}}_j$, we obtain

$$\bar{\mathbf{e}}_i \cdot \bar{\mathbf{e}}_j = \mu_{i1}\bar{\mathbf{e}}^1 \cdot \bar{\mathbf{e}}_j + \mu_{i2}\bar{\mathbf{e}}^2 \cdot \bar{\mathbf{e}}_j + \mu_{i3}\bar{\mathbf{e}}^3 \cdot \bar{\mathbf{e}}_j$$

Hence, using Eqs. (5) and (6), we find

$$\bar{g}_{ij} = \mu_{ij}$$

and, therefore,

(7)
$$\begin{aligned}
\bar{\mathbf{e}}_1 &= \bar{g}_{11}\bar{\mathbf{e}}^1 + \bar{g}_{12}\bar{\mathbf{e}}^2 + \bar{g}_{13}\bar{\mathbf{e}}^3 \\
\bar{\mathbf{e}}_2 &= \bar{g}_{21}\bar{\mathbf{e}}^1 + \bar{g}_{22}\bar{\mathbf{e}}^2 + \bar{g}_{23}\bar{\mathbf{e}}^3 \\
\bar{\mathbf{e}}_3 &= \bar{g}_{31}\bar{\mathbf{e}}^1 + \bar{g}_{32}\bar{\mathbf{e}}^2 + \bar{g}_{33}\bar{\mathbf{e}}^3
\end{aligned}$$

If we define the matrices

$$\bar{V}_e = \begin{bmatrix} \bar{\mathbf{e}}_1 \\ \bar{\mathbf{e}}_2 \\ \bar{\mathbf{e}}_3 \end{bmatrix} \quad \text{and} \quad \bar{V}^e = \begin{bmatrix} \bar{\mathbf{e}}^1 \\ \bar{\mathbf{e}}^2 \\ \bar{\mathbf{e}}^3 \end{bmatrix}$$

Eq. (7) can be written more compactly in the form

(8)
$$\bar{V}_e = \bar{G}\bar{V}^e$$

from which it follows that

(9)
$$\bar{V}^e = \bar{G}^{-1}\bar{V}_e$$

or

(10)
$$\begin{aligned}
\bar{\mathbf{e}}^1 &= \bar{g}^{11}\bar{\mathbf{e}}_1 + \bar{g}^{12}\bar{\mathbf{e}}_2 + \bar{g}^{13}\bar{\mathbf{e}}_3 \\
\bar{\mathbf{e}}^2 &= \bar{g}^{21}\bar{\mathbf{e}}_1 + \bar{g}^{22}\bar{\mathbf{e}}_2 + \bar{g}^{23}\bar{\mathbf{e}}_3 \\
\bar{\mathbf{e}}^3 &= \bar{g}^{31}\bar{\mathbf{e}}_1 + \bar{g}^{32}\bar{\mathbf{e}}_2 + \bar{g}^{33}\bar{\mathbf{e}}_3
\end{aligned}$$

where $\bar{g}^{ij}$ is the element in the ith row and jth column of $\bar{G}^{-1}$; that is,

$$\bar{g}^{ij} = \frac{\bar{G}_{ji}}{|\bar{G}|}$$

Of course, since $\bar{G} = [\bar{g}_{ij}] = (A^{-1})^T A^{-1}$ is symmetric, so is its inverse $\bar{G}^{-1} = [\bar{g}^{ij}] = AA^T$. From (10) and (6) it follows immediately that

(11)
$$\bar{\mathbf{e}}^i \cdot \bar{\mathbf{e}}^j = \bar{g}^{ij}$$

and hence, more generally, that

$$\bar{\mathbf{V}} \cdot \bar{\mathbf{V}} = (\bar{v}_1\bar{\mathbf{e}}^1 + \bar{v}_2\bar{\mathbf{e}}^2 + \bar{v}_3\bar{\mathbf{e}}^3) \cdot (\bar{v}_1\bar{\mathbf{e}}^1 + \bar{v}_2\bar{\mathbf{e}}^2 + \bar{v}_3\bar{\mathbf{e}}^3) = \bar{V}^T \bar{G}^{-1} \bar{V}$$

Thus, *in oblique coordinates the metrical properties of space, which are determined by the matrix* $\bar{G} = [\bar{g}_{ij}] = (A^{-1})^T A^{-1}$ *if vectors are represented in terms of the base vectors* $\bar{\mathbf{e}}_1$, $\bar{\mathbf{e}}_2$, $\bar{\mathbf{e}}_3$, *are determined equally well by the inverse matrix* $\bar{G}^{-1} = [\bar{g}^{ij}] = AA^T$ *if vectors are represented in terms of the reciprocal base vectors* $\bar{\mathbf{e}}^1$, $\bar{\mathbf{e}}^2$, $\bar{\mathbf{e}}^3$.

It is also instructive to consider the representation of the vectors $\mathbf{i} = \mathbf{e}_1 = \mathbf{e}^1$, $\mathbf{j} = \mathbf{e}_2 = \mathbf{e}^2$, $\mathbf{k} = \mathbf{e}_3 = \mathbf{e}^3$ in terms of the vectors $\bar{\mathbf{e}}_1$, $\bar{\mathbf{e}}_2$, $\bar{\mathbf{e}}_3$ and $\bar{\mathbf{e}}^1$, $\bar{\mathbf{e}}^2$, $\bar{\mathbf{e}}^3$, and vice versa. Specifically, since $\bar{\mathbf{e}}_1$, $\bar{\mathbf{e}}_2$, $\bar{\mathbf{e}}_3$ are, respectively, the vectors from the origin $\bar{O}$ ($= O$) to the points whose oblique coordinates are $(1,0,0)$, $(0,1,0)$, and $(0,0,1)$, and since, from the transformation equation $X = A^{-1}\bar{X}$ [Eq. (2)], these points have rectangular coordinates (a^{11},a^{21},a^{31}), (a^{12},a^{22},a^{32}), and (a^{13},a^{23},a^{33}), where

$a^{ij} = A_{ji}/|A|$ is the element in the ith row and jth column of the matrix A^{-1}, it follows from the definition of vector addition that

$$\bar{\mathbf{e}}_1 = a^{11}\mathbf{i} + a^{21}\mathbf{j} + a^{31}\mathbf{k} = a^{11}\mathbf{e}_1 + a^{21}\mathbf{e}_2 + a^{31}\mathbf{e}_3$$

(12)
$$\bar{\mathbf{e}}_2 = a^{12}\mathbf{i} + a^{22}\mathbf{j} + a^{32}\mathbf{k} = a^{12}\mathbf{e}_1 + a^{22}\mathbf{e}_2 + a^{32}\mathbf{e}_3$$

$$\bar{\mathbf{e}}_3 = a^{13}\mathbf{i} + a^{23}\mathbf{j} + a^{33}\mathbf{k} = a^{13}\mathbf{e}_1 + a^{23}\mathbf{e}_2 + a^{33}\mathbf{e}_3$$

Thus, introducing the matrices

$$\bar{V}_e = \begin{bmatrix} \bar{\mathbf{e}}_1 \\ \bar{\mathbf{e}}_2 \\ \bar{\mathbf{e}}_3 \end{bmatrix} \quad \text{and} \quad V_e = \begin{bmatrix} \mathbf{i} \\ \mathbf{j} \\ \mathbf{k} \end{bmatrix} = \begin{bmatrix} \mathbf{e}_1 \\ \mathbf{e}_2 \\ \mathbf{e}_3 \end{bmatrix} = \begin{bmatrix} \mathbf{e}^1 \\ \mathbf{e}^2 \\ \mathbf{e}^3 \end{bmatrix} = V^e$$

we have

(13)
$$\bar{V}_e = (A^{-1})^T V_e = (A^T)^{-1} V_e$$

Either in the same fashion or directly from (13), we obtain

(14)
$$V_e = A^T \bar{V}_e$$

that is,

$$\mathbf{e}_1 = a_{11}\bar{\mathbf{e}}_1 + a_{21}\bar{\mathbf{e}}_2 + a_{31}\bar{\mathbf{e}}_3$$

(15)
$$\mathbf{e}_2 = a_{12}\bar{\mathbf{e}}_1 + a_{22}\bar{\mathbf{e}}_2 + a_{32}\bar{\mathbf{e}}_3$$

$$\mathbf{e}_3 = a_{13}\bar{\mathbf{e}}_1 + a_{23}\bar{\mathbf{e}}_2 + a_{33}\bar{\mathbf{e}}_3$$

as the equations expressing $\mathbf{i} = \mathbf{e}_1$, $\mathbf{j} = \mathbf{e}_2$, $\mathbf{k} = \mathbf{e}_3$ in terms of the base vectors $\bar{\mathbf{e}}_1$, $\bar{\mathbf{e}}_2$, $\bar{\mathbf{e}}_3$ of the oblique system.

To obtain the equations relating the reciprocal vectors $\bar{\mathbf{e}}^1$, $\bar{\mathbf{e}}^2$, $\bar{\mathbf{e}}^3$ in the oblique system to the vectors $\mathbf{i} = \mathbf{e}^1$, $\mathbf{j} = \mathbf{e}^2$, $\mathbf{k} = \mathbf{e}^3$ of the original rectangular system, we begin with the relation (9), that is, $\bar{V}^e = \bar{G}^{-1}\bar{V}_e$. From this, using (13) and the fact that $\bar{G}^{-1} = AA^T$ and $V_e = V^e$, we have

(16)
$$\bar{V}^e = (AA^T)(A^T)^{-1} V^e = A V^e$$

that is,

$$\bar{\mathbf{e}}^1 = a_{11}\mathbf{e}^1 + a_{12}\mathbf{e}^2 + a_{13}\mathbf{e}^3$$

(17)
$$\bar{\mathbf{e}}^2 = a_{21}\mathbf{e}^1 + a_{22}\mathbf{e}^2 + a_{23}\mathbf{e}^3$$

$$\bar{\mathbf{e}}^3 = a_{31}\mathbf{e}^1 + a_{32}\mathbf{e}^2 + a_{33}\mathbf{e}^3$$

Solving (16) for V^e, we have, of course,

(18)
$$V^e = A^{-1}\bar{V}^e$$

or

$$\mathbf{e}^1 = a^{11}\bar{\mathbf{e}}^1 + a^{12}\bar{\mathbf{e}}^2 + a^{13}\bar{\mathbf{e}}^3$$

(19)
$$\mathbf{e}^2 = a^{21}\bar{\mathbf{e}}^1 + a^{22}\bar{\mathbf{e}}^2 + a^{23}\bar{\mathbf{e}}^3$$

$$\mathbf{e}^3 = a^{31}\bar{\mathbf{e}}^1 + a^{32}\bar{\mathbf{e}}^2 + a^{33}\bar{\mathbf{e}}^3$$

Suppose now that we have a vector

$$\mathbf{V} = \mathbf{V}^r = u\mathbf{i} + v\mathbf{j} + z\mathbf{k} \equiv v^1\mathbf{e}_1 + v^2\mathbf{e}_2 + v^3\mathbf{e}_3 \equiv v_1\mathbf{e}^1 + v_2\mathbf{e}^2 + v_3\mathbf{e}^3 = \mathbf{V}_r$$

where, since **V** is given in a rectangular coordinate system, $\mathbf{e}_i = \mathbf{e}^i$ and therefore $v^i = v_i$. If we express $\mathbf{V} = \mathbf{V}^r$ in terms of the base vectors $\bar{\mathbf{e}}_1$, $\bar{\mathbf{e}}_2$, $\bar{\mathbf{e}}_3$ of the oblique system by means of (15), we obtain, after collecting terms,

$$\bar{\mathbf{V}}^r = (v^1 a_{11} + v^2 a_{12} + v^3 a_{13})\bar{\mathbf{e}}_1 + (v^1 a_{21} + v^2 a_{22} + v^3 a_{23})\bar{\mathbf{e}}_2$$
$$+ (v^1 a_{31} + v^2 a_{32} + v^3 a_{33})\bar{\mathbf{e}}_3$$

(20)
$$= \bar{v}^1\bar{\mathbf{e}}_1 + \bar{v}^2\bar{\mathbf{e}}_2 + \bar{v}^3\bar{\mathbf{e}}_3$$

Thus, when **V** is transformed from its representation in terms of the base vectors $\mathbf{e}_1$, $\mathbf{e}_2$, $\mathbf{e}_3$ to the corresponding representation in terms of the base vectors $\bar{\mathbf{e}}_1$, $\bar{\mathbf{e}}_2$, $\bar{\mathbf{e}}_3$, the components of $\mathbf{V} = \mathbf{V}^r$ transform according to the law

(21)
$$\bar{v}^i = a_{i1}v^1 + a_{i2}v^2 + a_{i3}v^3$$

or

(22)
$$\bar{V}^r = AV^r$$

Similarly, if we express $\mathbf{V} = \mathbf{V}_r$ in terms of the reciprocal base vectors $\bar{\mathbf{e}}^1$, $\bar{\mathbf{e}}^2$, $\bar{\mathbf{e}}^3$ by means of (19), we obtain the representation

$$\mathbf{V}_r = (v_1 a^{11} + v_2 a^{21} + v_3 a^{31})\bar{\mathbf{e}}^1 + (v_1 a^{12} + v_2 a^{22} + v_3 a^{32})\bar{\mathbf{e}}^2$$
$$+ (v_1 a^{13} + v_2 a^{23} + v_3 a^{33})\bar{\mathbf{e}}^3$$

(23)
$$= \bar{v}_1\bar{\mathbf{e}}^1 + \bar{v}_2\bar{\mathbf{e}}^2 + \bar{v}_3\bar{\mathbf{e}}^3$$

Thus, when **V** is transformed from its representation in terms of the base vectors $\mathbf{e}^1$, $\mathbf{e}^2$, $\mathbf{e}^3$ to its corresponding representation in terms of the reciprocal base vectors $\bar{\mathbf{e}}^1$, $\bar{\mathbf{e}}^2$, $\bar{\mathbf{e}}^3$, its components transform according to the law

(24)
$$\bar{v}_i = a^{1i}v_1 + a^{2i}v_2 + a^{3i}v_3$$

or

(25)
$$\bar{V}_r = (A^{-1})^T V_r$$

Equations (24) and (25) have exactly the same form as Eqs. (12) and (13) for the transformation of the base vectors $\mathbf{e}_1$, $\mathbf{e}_2$, $\mathbf{e}_3$, and for this reason, the representation of **V** in terms of the reciprocal base vectors is called the **covariant**† representation of **V**. On the other hand, Eqs. (21) and (22) have the form of Eqs. (16) and (17) for the transformation of the reciprocal base vectors $\mathbf{e}^1$, $\mathbf{e}^2$, $\mathbf{e}^3$, and for this reason, the representation of **V** in terms of the base vectors themselves is called the **contravariant**‡ representation of **V**.

From the expanded form of Eq. (1) it is clear that

$$a_{ij} = \frac{\partial \bar{x}^i}{\partial x^j}$$

and from the expanded form of Eq. (2) it is clear that

$$a^{ij} = \frac{\partial x^i}{\partial \bar{x}^j}$$

---

† Co- = *with* or *alike*.
‡ Contra- = *against* or *opposite to*.

There is no particular reason for introducing this notation in the study of oblique coordinates, but it may be helpful as a preparation for the work of the next section on generalized coordinates to rewrite some of the important formulas of this section in terms of the partial derivatives of the transformation equations.

For the matrix of the transformation and its inverse we have, respectively,

$$A = [a_{ij}] = \left[\frac{\partial \bar{x}^i}{\partial x^j}\right] \quad \text{and} \quad A^{-1} = [a^{ij}] = \left[\frac{\partial x^i}{\partial \bar{x}^j}\right]$$

For the general element of the matrix $\bar{G} = (A^{-1})^T A^{-1}$, which in oblique coordinates defines the metrical properties of space, we have

$$\bar{g}_{ij} = \sum_k a^{ki} a^{kj} = \sum_k \frac{\partial x^k}{\partial \bar{x}^i} \frac{\partial x^k}{\partial \bar{x}^j}$$

This formula, as it stands, does not hold for the generalized coordinate systems we shall study in the next section. To obtain the appropriate general form we observe that in rectangular coordinates the matrix G is just the identity, and therefore g_{kl} is 1 if $k = l$ and 0 if $k \neq l$. Hence, if the factor g_{kl} is inserted in the expression being summed and the summation extended over l as well as k, the effect in this case is simply to include a number of terms each of which is equal to zero. On the other hand, the formula for $\bar{g}_{ij}$ as thus modified, namely,

$$(26) \qquad \bar{g}_{ij} = \sum_{k,l} g_{kl} \frac{\partial x^k}{\partial \bar{x}^i} \frac{\partial x^l}{\partial \bar{x}^j}$$

is now correct for more general coordinate systems in which g_{kl} is not equal to 0 when k and l are different.

Likewise, for the matrix $\bar{G}^{-1} = \bar{g}^{ij} = A A^T$, we have

$$\bar{g}^{ij} = \sum_k a_{ik} a_{jk} = \sum_k \frac{\partial \bar{x}^i}{\partial x^k} \frac{\partial \bar{x}^j}{\partial x^k}$$

or, inserting the factor $g^{kl} \equiv g_{kl}$,

$$(27) \qquad \bar{g}^{ij} = \sum_{k,l} g^{kl} \frac{\partial \bar{x}^i}{\partial x^k} \frac{\partial \bar{x}^j}{\partial x^l}$$

For the relations connecting the base vectors $\bar{e}_1$, $\bar{e}_2$, $\bar{e}_3$ and the vectors e_1, e_2, e_3, we have from (12) and (15)

$$(28) \qquad \bar{e}_i = \sum_k a^{ki} e_k = \sum_k e_k \frac{\partial x^k}{\partial \bar{x}^i}$$

and

$$(29) \qquad e_k = \sum_i a_{ik} \bar{e}_i = \sum_i \bar{e}_i \frac{\partial \bar{x}^i}{\partial x^k}$$

For the relations connecting the reciprocal base vectors $\bar{e}^1$, $\bar{e}^2$, $\bar{e}^3$ and the vectors e^1, e^2, e^3, we have from (17) and (19)

$$(30) \qquad \bar{e}^i = \sum_k a_{ik} e^k = \sum_k e^k \frac{\partial \bar{x}^i}{\partial x^k}$$

and

$$(31) \qquad e^k = \sum_i a^{ki} \bar{e}^i = \sum_i \bar{e}^i \frac{\partial x^k}{\partial \bar{x}^i}$$

For the components of a vector represented covariantly, we have from the law of transformation (24)

$$(32) \qquad \bar{v}_i = \sum_k a^{ki} v_k = \sum_k v_k \frac{\partial x^k}{\partial \bar{x}^i}$$

For the components of a vector represented contravariantly, we have from the law of transformation (21)

$$(33) \qquad \bar{v}^i = \sum_k a_{ik} v^k = \sum_k v^k \frac{\partial \bar{x}^i}{\partial x^k}$$

If we have a general transformation of coordinates, say

$$\bar{x}^i = \bar{x}^i(x^1, x^2, x^3) \qquad i = 1, 2, 3$$

then any vector whose components transform according to the law (32) is called a **covariant vector** and any vector whose components transform according to the law (33) is called a **contravariant vector**. In rectangular coordinates, as we pointed out earlier, the base set e_1, e_2, e_3 and the reciprocal set e^1, e^2, e^3 are identical. Hence, there is no distinction between covariant and contravariant vectors, and no need to introduce the two concepts, in elementary vector analysis.

EXERCISES

1   Prove that, for any nonsingular matrix A, the product $\bar{G} = (A^{-1})^T A^{-1}$ is symmetric.

2   What is the condition that the set of base vectors $\bar{e}_1$, $\bar{e}_2$, $\bar{e}_3$ and the set of reciprocal vectors $\bar{e}^1$, $\bar{e}^2$, $\bar{e}^3$ be the same?

3   (a) Let x^1, x^2, x^3 and $\bar{x}^1, \bar{x}^2, \bar{x}^3$ be, respectively, rectangular and oblique coordinates connected by the transformation equations

$$\begin{aligned}
\bar{x}^1 &= 2x^1 && + x^3 \\
\bar{x}^2 &= x^1 + 2x^2 + 3x^3 \\
\bar{x}^3 &= x^1 + x^2 + x^3
\end{aligned}$$

Working directly from their definitions, determine the rectangular representation of the base vectors $\bar{e}_1$, $\bar{e}_2$, $\bar{e}_3$ and the reciprocal vectors $\bar{e}^1$, $\bar{e}^2$, $\bar{e}^3$. Thence verify that Eqs. (12) and (17) are satisfied.

(b) Work part (a) if the matrix of the transformation to oblique coordinates is

$$A = \begin{bmatrix} 1 & 2 & -1 \\ 2 & 1 & 3 \\ 1 & 1 & 1 \end{bmatrix}$$

4   (a) In part (a) of Exercise 3, what is the distance from the origin to the point whose oblique coordinates are $(1,1,1)$? What is the distance between the points whose oblique coordinates are $(1,1,1)$ and $(1,2,3)$?

(b) In part (b) of Exercise 3, what is the distance from the origin to the points whose oblique coordinates are $(1,0,-1)$ and $(2,1,1)$?

5   (a) If U and V are two vectors represented contravariantly in an oblique coordinate system connected with a rectangular coordinate system by the transformation $\bar{X} = AX$, show that the angle between U and V is given by the formula

$$\cos \theta = \frac{\bar{U}^T \bar{G} \bar{V}}{\sqrt{\bar{U}^T \bar{G} \bar{U}} \sqrt{\bar{V}^T \bar{G} \bar{V}}}$$

(b) What is the angle between two vectors U and V represented covariantly?

6   Prove that $\bar{G}$ is positive-definite.

14.3 Generalized Coordinates

Let x^1, x^2, x^3 be three independent, single-valued, differentiable scalar point functions such that to every point of some region R of three-dimensional euclidean space there corresponds a unique triple of values (x^1, x^2, x^3) and such that to every triple of values (x^1, x^2, x^3) within ranges determined by the nature of R there corresponds a unique point of R. Then x^1, x^2, x^3 are called **generalized coordinates** in R, and the correspondence between the points of R and the number triples (x^1, x^2, x^3) is called a **generalized coordinate system** for R. Rectangular, cylindrical, spherical, and now oblique coordinates, are familiar examples of generalized coordinates.

Through each point P of R there passes a unique surface S^1 on which x^1 is constant, a unique surface S^2 on which x^2 is constant, and a unique surface S^3 on which x^3 is constant. These surfaces intersect by pairs in curves, called **parametric curves**, which pass through P and on which one and only one of the generalized coordinates varies. In general, the tangents to the three parametric curves which pass through a point will be noncoplanar, and we shall suppose this to be the case throughout R.

In all but rectangular and oblique coordinate systems, the tangents to the parametric curves will vary in direction from point to point, and no one set of directions is singled out as any more natural than any other for the directions of a set of base vectors for R. However, *at each point*, vectors along the tangents to the parametric curves through that point provide a natural basis for the representation of vectors extending from that point as origin (Fig. 14.4), and our development will be based on this concept of local base vectors and, of course, the related concept of local reciprocal base vectors.

The **local base vectors** $\mathbf{e}_1$, $\mathbf{e}_2$, $\mathbf{e}_3$ at any point P we define to have, respectively, the directions of the tangents to the x^1, x^2, x^3 parametric curves at P, and to have lengths $|\mathbf{e}_i| = \sqrt{\mathbf{e}_i \cdot \mathbf{e}_i}$ such that if ds is the infinitesimal distance along the x^i parametric curve corresponding to the infinitesimal change dx^i in x^i, then

$$(1) \qquad\qquad ds = |\mathbf{e}_i\, dx^i| = \sqrt{\mathbf{e}_i \cdot \mathbf{e}_i}\, |dx^i|$$

Figure 14.4
The parametric curves and the local base vectors at a point P in a generalized coordinate system.

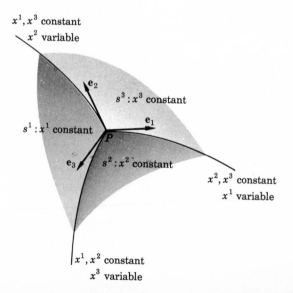

At P we define the local reciprocal base vectors $\mathbf{e}^1$, $\mathbf{e}^2$, $\mathbf{e}^3$ precisely as we did in oblique coordinates, namely, by the conditions

$$\mathbf{e}^i \cdot \mathbf{e}_j = \begin{cases} 0 & i \neq j \\ 1 & i = j \end{cases}$$

where, as usual, $\mathbf{e}^i \cdot \mathbf{e}_j = |\mathbf{e}^i|\,|\mathbf{e}_j| \cos(\mathbf{e}^i, \mathbf{e}_j)$.

Since our definitions for the local base vectors and the corresponding reciprocal vectors involve the notion of length, we must, of course, have some method of measuring distances. To do this, we assume the existence throughout R of a symmetric, positive-definite matrix

$$G = [g_{ij}]$$

whose elements are functions of the generalized coordinates and which has the property that if

$$dX = \begin{bmatrix} dx^1 \\ dx^2 \\ dx^3 \end{bmatrix}$$

then the distance ds from $P{:}(x^1, x^2, x^3)$ to $Q{:}(x^1 + dx^1, x^2 + dx^2, x^3 + dx^3)$ is given by the formula

$$(2) \qquad (ds)^2 = (dX)^T G\, dX = \sum_{i,j} g_{ij}\, dx^i\, dx^j$$

Thus, if $x^1 = x^1(t)$, $x^2 = x^2(t)$, $x^3 = x^3(t)$ are the parametric equations of a curve, then the length of the curve between the points P_1 and P_2 at which t has the values t_1 and t_2, respectively, is

$$\int_{P_1}^{P_2} ds = \int_{P_1}^{P_2} \sqrt{\sum_{i,j} g_{ij}\, dx^i\, dx^j} = \int_{t_1}^{t_2} \sqrt{\sum_{i,j} g_{ij} \frac{dx^i}{dt} \frac{dx^j}{dt}}\, dt$$

In particular, for the length of an arbitrary infinitesimal vector $\mathbf{e}_1\, dx^1 + \mathbf{e}_2\, dx^2 + \mathbf{e}_3\, dx^3$ we have

$$(ds)^2 = (\mathbf{e}_1\, dx^1 + \mathbf{e}_2\, dx^2 + \mathbf{e}_3\, dx^3) \cdot (\mathbf{e}_1\, dx^1 + \mathbf{e}_2\, dx^2 + \mathbf{e}_3\, dx^3)$$
$$= \sum_{i,j} \mathbf{e}_i \cdot \mathbf{e}_j\, dx^i\, dx^j = \sum_{i,j} g_{ij}\, dx^i\, dx^j$$

where the last equality follows from Eq. (2). Since the differentials of the coordinate variables are independent and arbitrary, it follows that the coefficients of corresponding terms in the last two sums must be equal. Therefore,

$$(3) \qquad \mathbf{e}_i \cdot \mathbf{e}_j = g_{ij}$$

In particular,

$$(4) \qquad |\mathbf{e}_i| = \sqrt{\mathbf{e}_i \cdot \mathbf{e}_i} = \sqrt{g_{ii}}$$

Using Eqs. (3) and (4), we can determine without integration the length of a noninfinitesimal vector $\mathbf{V} = v^1\mathbf{e}_1 + v^2\mathbf{e}_2 + v^3\mathbf{e}_3$ expressed in terms of the base vectors at a point P. In fact,

$$|\mathbf{V}|^2 = \mathbf{V} \cdot \mathbf{V} = (v^1\mathbf{e}_1 + v^2\mathbf{e}_2 + v^3\mathbf{e}_3) \cdot (v^1\mathbf{e}_1 + v^2\mathbf{e}_2 + v^3\mathbf{e}_3)$$
$$= \sum_{i,j} \mathbf{e}_i \cdot \mathbf{e}_j v^i v^j = \sum_{i,j} g_{ij} v^i v^j = V^T G V$$

Hence,

(5)
$$|V| = \sqrt{V^T G V}$$

where, of course, the elements g_{ij} of G are to be evaluated at the point P at which e_1, e_2, e_3 are the base vectors.

From (3) we also draw the important conclusion that *a necessary and sufficient condition that the parametric curves be orthogonal at every point of R is that* $g_{ij} = 0$ *for* $i \neq j$ *at all points of R.*

By exactly the same reasoning that we used to derive Eqs. (7) and (10) in the last section, we can now prove that for the local base vectors and the local reciprocal base vectors we have the following relations:

(6)
$$e_i = \sum_k g_{ik} e^k$$

(7)
$$e^i = \sum_k g^{ik} e_k$$

where, as in the last section, g^{ik} is the element in the ith row and kth column of the matrix G^{-1} which is the inverse of $G = [g_{ik}]$. Furthermore, by forming the scalar product of Eq. (7) with e^j and using the definitive relation $e^j \cdot e_k = \delta_k{}^j$, where $\delta_k{}^j$ is the Kronecker delta,† we obtain the following companion result to Eq. (3):

(8)
$$e^i \cdot e^j = g^{ij}$$

Equation (7) is not the only formula which can be used to express the local reciprocal vectors in terms of the local base vectors. Specifically, with cross products defined just as they were in Sec. 13.1, it is easy to verify that e^1, e^2, e^3 are given in terms of e_1, e_2, e_3 by the formulas

(9)
$$e^1 = \frac{e_2 \times e_3}{[e_1 e_2 e_3]} \qquad e^2 = \frac{e_3 \times e_1}{[e_1 e_2 e_3]} \qquad e^3 = \frac{e_1 \times e_2}{[e_1 e_2 e_3]}$$

Hence, using the result of Exercise 22, Sec. 13.1, we have

$$[e^1 e^2 e^3] = \frac{e_2 \times e_3}{[e_1 e_2 e_3]} \cdot \frac{e_3 \times e_1}{[e_1 e_2 e_3]} \times \frac{e_1 \times e_2}{[e_1 e_2 e_3]} = \frac{[e_1 e_2 e_3]^2}{[e_1 e_2 e_3]^3}$$

(10)
$$= \frac{1}{[e_1 e_2 e_3]}$$

Moreover, by using Eq. (6) in conjunction with Eq. (10), the numerical value of $[e_1 e_2 e_3]$ (and hence of $[e^1 e^2 e^3]$) can easily be found. For, by (6),

$$[e_1 e_2 e_3] = \left(\sum_i g_{1i} e^i \right) \cdot \left(\sum_j g_{2j} e^j \right) \times \left(\sum_k g_{3k} e^k \right)$$

(11)
$$= \sum_{i,j,k} g_{1i} g_{2j} g_{3k} [e^i e^j e^k]$$

Now of the $3^3 = 27$ terms which arise as i, j, and k range independently over the numbers 1, 2, 3, twenty-one are zero, because the scalar triple product $[e^i e^j e^k]$

---

† Defined at the end of Sec. 10.1.

contains at least one repeated factor. Of the remaining six terms in the last sum there are three, corresponding to the sets of values $(i,j,k) = (1,2,3), (2,3,1), (3,1,2)$, in which

$$[e^i e^j e^k] = [e^1 e^2 e^3] = \frac{1}{[e_1 e_2 e_3]}$$

In the remaining three terms, corresponding to the sets of values $(i,j,k) = (1,3,2)$, $(3,2,1), (2,1,3)$, the factor $[e^i e^j e^k]$ is equal to

$$-[e^1 e^2 e^3] = -\frac{1}{[e_1 e_2 e_3]}$$

Hence, factoring $1/[e_1 e_2 e_3]$ from the sum in (11) and cross-multiplying, we have

$$[e_1 e_2 e_3]^2 = g_{11} g_{22} g_{33} + g_{12} g_{23} g_{31} + g_{13} g_{21} g_{32}$$
$$- g_{11} g_{23} g_{32} - g_{13} g_{22} g_{31} - g_{12} g_{21} g_{33}$$

Since the sum on the right in the last equation is precisely the expansion of the determinant of the matrix $G = [g_{ij}]$, we have thus established the useful result

$$(12) \qquad\qquad [e_1 e_2 e_3]^2 = |G|$$

where G is the matrix which defines the metrical properties of space.

We now turn our attention to transformations from one set of generalized coordinates to another. In particular, we are interested in the laws of transformation for the fundamental matrices G and G^{-1}, the local base vectors e_1, e_2, e_3, the local reciprocal vectors e^1, e^2, e^3, and vectors expressed in terms of these reference vectors, which are induced by a transformation of coordinates.

Let us suppose, then, that we have two systems of coordinates (x^1, x^2, x^3) and $(\bar{x}^1, \bar{x}^2, \bar{x}^3)$ connected by transformation equations of the form

$$\bar{x}^1 = \bar{x}^1(x^1, x^2, x^3)$$
$$\bar{x}^2 = \bar{x}^2(x^1, x^2, x^3)$$
$$\bar{x}^3 = \bar{x}^3(x^1, x^2, x^3)$$

or, simply,

$$(13) \qquad\qquad T: \quad \bar{x}^i = \bar{x}^i(x^1, x^2, x^3) \qquad i = 1, 2, 3$$

In particular cases, Eqs. (13) might be the equations connecting a rectangular and an oblique coordinate system, as in the last section, a rectangular and a cylindrical coordinate system, a rectangular and a spherical coordinate system, or a cylindrical and a spherical coordinate system.

Naturally, we wish a point with coordinates (x^1, x^2, x^3) in the x system to have a unique set of coordinates $(\bar{x}^1, \bar{x}^2, \bar{x}^3)$ in the $\bar{x}$ system. Hence we require that throughout the region R with which we are concerned, the $\bar{x}^i$'s be single-valued functions of the x^i's. Moreover, we wish the point with coordinates $(\bar{x}^1, \bar{x}^2, \bar{x}^3)$ to have a unique set of x coordinates. Hence we also require that Eqs. (13) be solvable for x^1, x^2, x^3 as single-valued functions of $\bar{x}^1, \bar{x}^2, \bar{x}^3$, say

$$(14) \qquad\qquad T^{-1}: \quad x^i = x^i(\bar{x}^1, \bar{x}^2, \bar{x}^3) \qquad i = 1, 2, 3$$

In advanced calculus it is shown† that if the first partial derivatives of the

---

† See, for instance, R. C. Buck, "Advanced Calculus," p. 215, McGraw-Hill, New York, 1956.

coordinate functions $\bar{x}^i$ in T are continuous and if, throughout R, the so-called **jacobian determinant**†

$$(15) \qquad |J| = \frac{\partial(\bar{x}^1, \bar{x}^2, \bar{x}^3)}{\partial(x^1, x^2, x^3)} = \begin{vmatrix} \dfrac{\partial \bar{x}^1}{\partial x^1} & \dfrac{\partial \bar{x}^1}{\partial x^2} & \dfrac{\partial \bar{x}^1}{\partial x^3} \\[2mm] \dfrac{\partial \bar{x}^2}{\partial x^1} & \dfrac{\partial \bar{x}^2}{\partial x^2} & \dfrac{\partial \bar{x}^2}{\partial x^3} \\[2mm] \dfrac{\partial \bar{x}^3}{\partial x^1} & \dfrac{\partial \bar{x}^3}{\partial x^2} & \dfrac{\partial \bar{x}^3}{\partial x^3} \end{vmatrix}$$

is different from zero, as we shall suppose, then around any interior point of R there exists a neighborhood in which T has a single-valued inverse (14). Naturally, since the equations of the inverse transformation (14) are to be uniquely solvable for x^1, x^2, x^3, the jacobian determinant of the inverse transformation, namely,

$$(16) \qquad |\bar{J}| = \left| \frac{\partial(x^1, x^2, x^3)}{\partial(\bar{x}^1, \bar{x}^2, \bar{x}^3)} \right| = \begin{vmatrix} \dfrac{\partial x^1}{\partial \bar{x}^1} & \dfrac{\partial x^1}{\partial \bar{x}^2} & \dfrac{\partial x^1}{\partial \bar{x}^3} \\[2mm] \dfrac{\partial x^2}{\partial \bar{x}^1} & \dfrac{\partial x^2}{\partial \bar{x}^2} & \dfrac{\partial x^2}{\partial \bar{x}^3} \\[2mm] \dfrac{\partial x^3}{\partial \bar{x}^1} & \dfrac{\partial x^3}{\partial \bar{x}^2} & \dfrac{\partial x^3}{\partial \bar{x}^3} \end{vmatrix}$$

must also be different from zero throughout R.

We have now reached the point where it is necessary to introduce the so-called **Einstein summation convention**. Just as the summation symbol $\sum$ effects a great notational economy when it is used instead of writing a sum of terms at length, so this convention replaces the symbol $\sum$ with a notation still shorter and much more suggestive. Briefly, the convention is this. *If any term contains the same letter twice as a distinguishing index, it is understood that the term is to be summed for all values of the repeated index.* For example, using the summation convention with the understanding that the range of our indices is 1 to 3, we can write the differential of $\bar{x}^i$ in the equivalent forms

$$d\bar{x}^i = \frac{\partial \bar{x}^i}{\partial x^1} dx^1 + \frac{\partial \bar{x}^i}{\partial x^2} dx^2 + \frac{\partial \bar{x}^i}{\partial x^3} dx^3 = \sum_{j=1}^{3} \frac{\partial \bar{x}^i}{\partial x^j} dx^j = \frac{\partial \bar{x}^i}{\partial x^j} dx^j$$

In the last expression, the index i identifies the particular variable $\bar{x}^i$ whose differential is being considered and cannot be changed without changing the meaning of the expression. On the other hand, the index j merely indicates that summation over a certain range is to be carried out and, like the variable of integration in a definite integral, can be changed at pleasure to any other letter, except i, of course. Thus we can write equally well

$$d\bar{x}^i = \frac{\partial \bar{x}^i}{\partial x^j} dx^j = \frac{\partial \bar{x}^i}{\partial x^k} dx^k = \frac{\partial \bar{x}^i}{\partial x^l} dx^l = \cdots$$

---

† Named for the German mathematician K. G. J. Jacobi (1804–1851). We shall frequently refer to the **jacobian matrix** J simply as the **jacobian** of the transformation T.

An index which can be arbitrarily replaced in this fashion by another is usually called a **dummy index** or an **umbral index**.

The summation convention also permits more than one pair of repeated indices in a term to be summed. For instance, applying the convention first to the repeated index i and then to the repeated index j, we have

$$
\begin{aligned}
g_{ij}\, dx^i\, dx^j &= g_{1j}\, dx^1\, dx^j + g_{2j}\, dx^2\, dx^j + g_{3j}\, dx^3\, dx^j \\
&= (g_{11}\, dx^1\, dx^1 + g_{12}\, dx^1\, dx^2 + g_{13}\, dx^1\, dx^3) \\
&\quad + (g_{21}\, dx^2\, dx^1 + g_{22}\, dx^2\, dx^2 + g_{23}\, dx^2\, dx^3) \\
&\quad + (g_{31}\, dx^3\, dx^1 + g_{32}\, dx^3\, dx^2 + g_{33}\, dx^3\, dx^3) = \sum_{i,j} g_{ij}\, dx^i\, dx^j
\end{aligned}
$$

It should be noted that

$$
g_{ij}\, dx^i\, dx^j \neq g_{ii}\, dx^i\, dx^i
$$

since the latter is equal to the simpler sum

$$
g_{11}\, dx^1\, dx^1 + g_{22}\, dx^2\, dx^2 + g_{33}\, dx^3\, dx^3
$$

Hence, unless the more restricted meaning is intended, the same index cannot be used a second time in the same term as a dummy index.

Before resuming our discussion of coordinate transformations, it will be helpful to introduce several simple lemmas.

LEMMA 1   If (x^1,x^2,x^3) and $(\bar{x}^1,\bar{x}^2,\bar{x}^3)$ are coordinates connected by a transformation

$$
\bar{x}^i = \bar{x}^i(x^1,x^2,x^3) \qquad i = 1, 2, 3
$$

then

$$
\frac{\partial \bar{x}^i}{\partial x^\alpha}\frac{\partial x^\alpha}{\partial \bar{x}^j} = \delta^i_j
$$

Proof   By hypothesis, $\bar{x}^i$ is a differentiable function of x^1, x^2, x^3, which in turn are differentiable functions of $\bar{x}^1$, $\bar{x}^2$, $\bar{x}^3$. Hence by the chain rule of partial differentiation,

$$
\frac{\partial \bar{x}^i}{\partial \bar{x}^i} = 1 = \frac{\partial \bar{x}^i}{\partial x^1}\frac{\partial x^1}{\partial \bar{x}^i} + \frac{\partial \bar{x}^i}{\partial x^2}\frac{\partial x^2}{\partial \bar{x}^i} + \frac{\partial \bar{x}^i}{\partial x^3}\frac{\partial x^3}{\partial \bar{x}^i} = \frac{\partial \bar{x}^i}{\partial x^\alpha}\frac{\partial x^\alpha}{\partial \bar{x}^i} \qquad i \text{ not summed}
$$

and, since the $\bar{x}^i$'s are independent,

$$
\frac{\partial \bar{x}^i}{\partial \bar{x}^j} = 0 = \frac{\partial \bar{x}^i}{\partial x^1}\frac{\partial x^1}{\partial \bar{x}^j} + \frac{\partial \bar{x}^i}{\partial x^2}\frac{\partial x^2}{\partial \bar{x}^j} + \frac{\partial \bar{x}^i}{\partial x^3}\frac{\partial x^3}{\partial \bar{x}^j} = \frac{\partial \bar{x}^i}{\partial x^\alpha}\frac{\partial x^\alpha}{\partial \bar{x}^j} \qquad i \neq j
$$

These two relations together establish the assertion of the lemma. Of course by an identical proof it follows that

$$
\frac{\partial x^i}{\partial \bar{x}^\alpha}\frac{\partial \bar{x}^\alpha}{\partial x^j} = \delta^i_j
$$

LEMMA 2   If ϕ^i and $\bar{\phi}^i$ $(i = 1, 2, 3)$ are, respectively, functions of x^1, x^2, x^3 and $\bar{x}^1$, $\bar{x}^2$, $\bar{x}^3$, then

$$
\bar{\phi}^i = \phi^\alpha \frac{\partial \bar{x}^i}{\partial x^\alpha} \qquad \text{implies} \qquad \bar{\phi}^i \frac{\partial x^\alpha}{\partial \bar{x}^i} = \phi^\alpha
$$

Proof   To provide us with further insight into the efficiency of the summation convention, let us first prove this lemma using the more familiar $\sum$ notation. We are given the relation

$$\bar{\phi}^i = \sum_{\alpha=1}^{3} \phi^\alpha \frac{\partial \bar{x}^i}{\partial x^\alpha} = \sum_{\beta=1}^{3} \phi^\beta \frac{\partial \bar{x}^i}{\partial x^\beta}$$

If we now multiply both sides of this equation in its second form by $\partial x^\alpha / \partial \bar{x}^i$ and then sum over i, we have

$$\sum_{i=1}^{3} \bar{\phi}^i \frac{\partial x^\alpha}{\partial \bar{x}^i} = \sum_{i=1}^{3} \left(\frac{\partial x^\alpha}{\partial \bar{x}^i} \sum_{\beta=1}^{3} \phi^\beta \frac{\partial \bar{x}^i}{\partial x^\beta} \right)$$

or, interchanging the order of summation on the right,

$$\sum_{i=1}^{3} \bar{\phi}^i \frac{\partial x^\alpha}{\partial \bar{x}^i} = \sum_{\beta=1}^{3} \left(\phi^\beta \sum_{i=1}^{3} \frac{\partial x^\alpha}{\partial \bar{x}^i} \frac{\partial \bar{x}^i}{\partial x^\beta} \right)$$

Now, by Lemma 1, the inner sum on the right is equal to $\delta_\beta{}^\alpha$. Hence, the right-hand side reduces to

$$\sum_{\beta=1}^{3} \delta_\beta{}^\alpha \phi^\beta$$

which is equal to zero unless $\beta = \alpha$. Therefore, finally,

$$\sum_{i=1}^{3} \bar{\phi}^i \frac{\partial x^\alpha}{\partial \bar{x}^i} \equiv \bar{\phi}^i \frac{\partial x^\alpha}{\partial \bar{x}^i} = \phi^\alpha$$

as asserted.

Using the summation convention, our proof would have proceeded as follows. Introducing the dummy index β in place of α, we begin with

$$\bar{\phi}^i = \phi^\beta \frac{\partial \bar{x}^i}{\partial x^\beta}$$

Now, multiplying both sides by $\partial x^\alpha / \partial \bar{x}^i$ and using Lemma 1, we have

$$\bar{\phi}^i \frac{\partial x^\alpha}{\partial \bar{x}^i} = \phi^\beta \frac{\partial x^\alpha}{\partial \bar{x}^i} \frac{\partial \bar{x}^i}{\partial x^\beta} = \phi^\beta \delta_\beta{}^\alpha = \phi^\alpha \quad .$$

The converse assertion is, of course, established in exactly the same fashion.

LEMMA 3   If ϕ^{ij} and $\bar{\phi}^{ij}$ ($i, j = 1, 2, 3$) are, respectively, functions of x^1, x^2, x^3 and $\bar{x}^1, \bar{x}^2, \bar{x}^3$, then any one of the relations

$$\bar{\phi}^{ij} = \phi^{\alpha\beta} \frac{\partial \bar{x}^i}{\partial x^\alpha} \frac{\partial \bar{x}^j}{\partial x^\beta}$$

$$\bar{\phi}^{ij} \frac{\partial x^\beta}{\partial \bar{x}^j} = \phi^{\alpha\beta} \frac{\partial \bar{x}^i}{\partial x^\alpha}$$

$$\bar{\phi}^{ij} \frac{\partial x^\alpha}{\partial \bar{x}^i} = \phi^{\alpha\beta} \frac{\partial \bar{x}^j}{\partial x^\beta}$$

$$\bar{\phi}^{ij} \frac{\partial x^\alpha}{\partial \bar{x}^i} \frac{\partial x^\beta}{\partial \bar{x}^j} = \phi^{\alpha\beta}$$

implies each of the others.

Proof Because of the near identity of the arguments, it will be sufficient to establish just one of the assertions of the lemma, say the assertion that

$$\bar{\phi}^{ij} = \phi^{\alpha\beta} \frac{\partial \bar{x}^i}{\partial x^\alpha} \frac{\partial \bar{x}^j}{\partial x^\beta} \quad \text{implies} \quad \bar{\phi}^{ij} \frac{\partial x^\alpha}{\partial \bar{x}^i} \frac{\partial x^\beta}{\partial \bar{x}^j} = \phi^{\alpha\beta}$$

To do this, let us write the first relation using a and b in place of α and β, and then let us multiply both sides by $(\partial x^\alpha/\partial \bar{x}^i)(\partial x^\beta/\partial \bar{x}^j)$ and sum over i and j. This gives us, by Lemma 1,

$$\bar{\phi}^{ij} \frac{\partial x^\alpha}{\partial \bar{x}^i} \frac{\partial x^\beta}{\partial \bar{x}^j} = \phi^{ab} \left(\frac{\partial x^\alpha}{\partial \bar{x}^i} \frac{\partial \bar{x}^i}{\partial x^a} \right) \left(\frac{\partial x^\beta}{\partial \bar{x}^j} \frac{\partial \bar{x}^j}{\partial x^b} \right)$$

$$= \phi^{ab} \, \delta_a^{\ \alpha} \, \delta_b^{\ \beta}$$

$$= \phi^{\alpha\beta}$$

as asserted.

In Secs. 10.2 and 10.3, when we considered linear transformations such as

$$T_1: \quad Y = AX \quad \text{and} \quad T_2: \quad Z = BY \quad A, B \text{ nonsingular}$$

we observed that the matrices of the inverse transformations T_1^{-1} and T_2^{-1} are A^{-1} and B^{-1}, respectively, and that the matrix of the transformation resulting when T_1 is followed by T_2 is BA. Since linear transformations are obviously special cases of the transformation (13), it is natural to ask whether general coordinate transformations have comparable properties. The answer is yes, and in fact we have the following theorems, the proof of the second of which we leave as an exercise.

THEOREM 1 If $T: \bar{x}^\alpha = \bar{x}^\alpha(x^1, x^2, x^3)$ is a transformation with jacobian J, then the jacobian $\bar{J}$ of the inverse transformation $T^{-1}: x^\alpha = x^\alpha(\bar{x}^1, \bar{x}^2, \bar{x}^3)$ is J^{-1}.

Proof By definition, the jacobian of the direct transformation T is $J = [\partial \bar{x}^i/\partial x^k]$, and the jacobian of the inverse transformation T^{-1} is $\bar{J} = [\partial x^k/\partial \bar{x}^j]$. From the definition of matric multiplication, the element in the ith row and jth column of the product $J\bar{J}$ is $(\partial \bar{x}^i/\partial x^k)(\partial x^k/\partial \bar{x}^j)$, and, by Lemma 1, this sum is equal to δ_j^i. Thus

$$J\bar{J} = [\delta_j^i] = I$$

Hence $\bar{J} = J^{-1}$; that is, the matrix $\bar{J}$ of the inverse transformation T^{-1} is the inverse of the matrix J of the direct transformation T, as asserted.

COROLLARY 1 If J is the jacobian of the transformation $T: \bar{x}^\alpha = \bar{x}^\alpha(x^1, x^2, x^3)$, then $\partial x^i/\partial \bar{x}^j$ is equal to $1/|J|$ times the cofactor of $\partial \bar{x}^j/\partial x^i$ in $|J|$.

THEOREM 2 If $T_1: \bar{x}^\alpha = \bar{x}^\alpha(x^1, x^2, x^3)$ is a transformation with jacobian J_1, and if $T_2: \bar{\bar{x}}^\beta = \bar{\bar{x}}^\beta(\bar{x}^1, \bar{x}^2, \bar{x}^3)$ is a transformation with jacobian J_2, then the jacobian of the transformation $T_2 T_1$ is $J_2 J_1$.

Let us now determine how the fundamental differential quadratic form $(ds)^2 = g_{ij}\,dx^i\,dx^j$ transforms when the coordinates x^1, x^2, x^3 are transformed into the coordinates $\bar{x}^1$, $\bar{x}^2$, $\bar{x}^3$ by means of Eqs. (13). For dx^i and dx^j we have, of course,

$$dx^i = \frac{\partial x^i}{\partial \bar{x}^\alpha}\,d\bar{x}^\alpha \qquad \text{and} \qquad dx^j = \frac{\partial x^j}{\partial \bar{x}^\beta}\,d\bar{x}^\beta$$

Hence $(ds)^2$ becomes the quadratic form

$$g_{ij}\frac{\partial x^i}{\partial \bar{x}^\alpha}\frac{\partial x^j}{\partial \bar{x}^\beta}\,d\bar{x}^\alpha\,d\bar{x}^\beta$$

Therefore, if we write the quadratic form after transformation as

$$\bar{g}_{\alpha\beta}\,d\bar{x}^\alpha\,d\bar{x}^\beta$$

it follows that the coefficients $\bar{g}_{\alpha\beta}$ transform according to the law

(17) $$\bar{g}_{\alpha\beta} = g_{ij}\frac{\partial x^i}{\partial \bar{x}^\alpha}\frac{\partial x^j}{\partial \bar{x}^\beta}$$

as we verified in the particular case of a transformation from rectangular coordinates to oblique coordinates in the last section [Eq. (26), Sec. 14.2]. Of course, in considering the transformation from $\bar{x}$ coordinates to x coordinates, an argument identical with the one we have just given provides us with the companion formula

(18) $$g_{ij} = \bar{g}_{\alpha\beta}\frac{\partial \bar{x}^\alpha}{\partial x^i}\frac{\partial \bar{x}^\beta}{\partial x^j}$$

which also follows from an obvious modification of Lemma 3.

Formula (18) also leads to the following interesting conclusion. From the rule for the multiplication of matrices, the element in the ith row and jth column of the product $J^T\bar{G}J$ is

$$\frac{\partial \bar{x}^\alpha}{\partial x^i}\,\bar{g}_{\alpha\beta}\,\frac{\partial \bar{x}^\beta}{\partial x^j}$$

However, by (18), this is the element g_{ij} in the ith row and jth column of G. Hence, taking determinants, we have

$$\left|\frac{\partial \bar{x}^\alpha}{\partial x^i}\,\bar{g}_{\alpha\beta}\,\frac{\partial \bar{x}^\beta}{\partial x^j}\right| = |g_{ij}| \qquad \text{or} \qquad |J^T\bar{G}J| = |G|$$

or, finally,

(19) $$|J|^2\,|\bar{G}| = |G|$$

EXAMPLE 1

Obtain the formula for the differential of arc length in spherical coordinates.

Since we know the formula for the differential of arc length in rectangular coordinates, namely,

(20) $$(ds)^2 = (dx)^2 + (dy)^2 + (dz)^2$$

we can obtain the corresponding formula in spherical coordinates by transformation from rectangular coordinates. To do this, let x^1, x^2, x^3 denote, respectively, the rec-

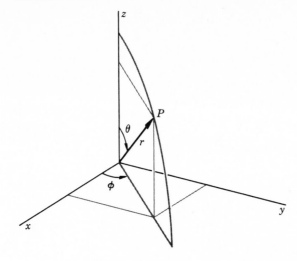

Figure 14.5
The relation between rectangular
and spherical coordinates.

tangular coordinates x, y, z, and let $\bar{x}^1$, $\bar{x}^2$, $\bar{x}^3$ denote, respectively, the spherical coordinates r, θ, ϕ (Fig. 14.5). Then, as usual, we have

$$x^1 = \bar{x}^1 \sin \bar{x}^2 \cos \bar{x}^3$$
$$T^{-1}: \quad x^2 = \bar{x}^1 \sin \bar{x}^2 \sin \bar{x}^3$$
$$x^3 = \bar{x}^1 \cos \bar{x}^2$$

and from these

$$\frac{\partial x^1}{\partial \bar{x}^1} = \sin \bar{x}^2 \cos \bar{x}^3 \qquad \frac{\partial x^1}{\partial \bar{x}^2} = \bar{x}^1 \cos \bar{x}^2 \cos \bar{x}^3 \qquad \frac{\partial x^1}{\partial \bar{x}^3} = -\bar{x}^1 \sin \bar{x}^2 \sin \bar{x}^3$$

$$\frac{\partial x^2}{\partial \bar{x}^1} = \sin \bar{x}^2 \sin \bar{x}^3 \qquad \frac{\partial x^2}{\partial \bar{x}^2} = \bar{x}^1 \cos \bar{x}^2 \sin \bar{x}^3 \qquad \frac{\partial x^2}{\partial \bar{x}^3} = \bar{x}^1 \sin \bar{x}^2 \cos \bar{x}^3$$

$$\frac{\partial x^3}{\partial \bar{x}^1} = \cos \bar{x}^2 \qquad \frac{\partial x^3}{\partial \bar{x}^2} = -\bar{x}^1 \sin \bar{x}^2 \qquad \frac{\partial x^3}{\partial \bar{x}^3} = 0$$

Hence, substituting into Eq. (17) and noting from Eq. (20) that $g_{ij} = \delta_j{}^i$, we have

$$\bar{g}_{11} = (\sin \bar{x}^2 \cos \bar{x}^3)^2 + (\sin \bar{x}^2 \sin \bar{x}^3)^2 + (\cos \bar{x}^2)^2$$
$$= 1$$
$$\bar{g}_{22} = (\bar{x}^1 \cos \bar{x}^2 \cos \bar{x}^3)^2 + (\bar{x}^1 \cos \bar{x}^2 \sin \bar{x}^3)^2 + (-\bar{x}^1 \sin \bar{x}^2)^2$$
$$= (\bar{x}^1)^2$$
$$\bar{g}_{33} = (-\bar{x}^1 \sin \bar{x}^2 \sin \bar{x}^3)^2 + (\bar{x}^1 \sin \bar{x}^2 \cos \bar{x}^3)^2$$
$$= (\bar{x}^1 \sin \bar{x}^2)^2$$
$$\bar{g}_{12} = (\sin \bar{x}^2 \cos \bar{x}^3)(\bar{x}^1 \cos \bar{x}^2 \cos \bar{x}^3) + (\sin \bar{x}^2 \sin \bar{x}^3)(\bar{x}^1 \cos \bar{x}^2 \sin \bar{x}^3)$$
$$\quad + (\cos \bar{x}^2)(-\bar{x}^1 \sin \bar{x}^2)$$
$$= 0$$
$$\bar{g}_{13} = (\sin \bar{x}^2 \cos \bar{x}^3)(-\bar{x}^1 \sin \bar{x}^2 \sin \bar{x}^3) + (\sin \bar{x}^2 \sin \bar{x}^3)(\bar{x}^1 \sin \bar{x}^2 \cos \bar{x}^3)$$
$$= 0$$
$$\bar{g}_{23} = (\bar{x}^1 \cos \bar{x}^2 \cos \bar{x}^3)(-\bar{x}^1 \sin \bar{x}^2 \sin \bar{x}^3) + (\bar{x}^1 \cos \bar{x}^2 \sin \bar{x}^3)(\bar{x}^1 \sin \bar{x}^2 \cos \bar{x}^3)$$
$$= 0$$

and, finally,

$$(ds)^2 = \bar{g}_{ij}\, d\bar{x}^i\, d\bar{x}^j$$
$$= (d\bar{x}^1)^2 + (\bar{x}^1)^2 (d\bar{x}^2)^2 + (\bar{x}^1 \sin \bar{x}^2)^2 (d\bar{x}^3)^2$$
$$= (dr)^2 + r^2 (d\theta)^2 + (r \sin \theta)^2 (d\phi)^2$$

When the coordinates x^1, x^2, x^3 are replaced by the coordinates $\bar{x}^1$, $\bar{x}^2$, $\bar{x}^3$, there is, of course, a new set of parametric curves passing through an arbitrary point P and hence a new set of base vectors $\bar{e}_1$, $\bar{e}_2$, $\bar{e}_3$ and a new set of reciprocal base vectors $\bar{e}^1$, $\bar{e}^2$, $\bar{e}^3$. To obtain the relations between the old and the new base vectors, let us consider an arbitrary infinitesimal displacement ds expressed in terms of each system of coordinates:

(21) $$ds = dx^\alpha\, e_\alpha = d\bar{x}^i\, \bar{e}_i$$

Now from the transformation equations we have

$$d\bar{x}^i = \frac{\partial \bar{x}^i}{\partial x^\alpha}\, dx^\alpha$$

and hence, from (21), we can write

$$dx^\alpha\, e_\alpha = \frac{\partial \bar{x}^i}{\partial x^\alpha}\, dx^\alpha\, \bar{e}_i$$

Since the differentials are arbitrary, it follows that the coefficients of corresponding differentials on each side of the last equation must be equal. Thus

(22) $$e_\alpha = \frac{\partial \bar{x}^i}{\partial x^\alpha}\, \bar{e}_i$$

Similarly, or by Lemma 2,

(23) $$\bar{e}_i = \frac{\partial x^\alpha}{\partial \bar{x}^i}\, e_\alpha$$

Formulas (29) and (28) of Sec. 14.2 are, respectively, special cases of these relations.

Knowing from Eq. (6) how the local base vectors are expressed in terms of the local reciprocal base vectors in any coordinate system, and knowing from Eq. (23) how the local base vectors transform, we can now determine how the local reciprocal base vectors transform. For, beginning with the relation (6) for the new coordinate system, namely,

$$\bar{e}_i = \bar{g}_{ij}\bar{e}^j$$

and substituting from Eqs. (23) and (17), we have

$$\frac{\partial x^\alpha}{\partial \bar{x}^i}\, e_\alpha = g_{\alpha\beta}\, \frac{\partial x^\alpha}{\partial \bar{x}^i}\, \frac{\partial x^\beta}{\partial \bar{x}^j}\, \bar{e}^j$$

Multiplying this equation by $\partial \bar{x}^i / \partial x^\gamma$ and summing each side with respect to i, we obtain

$$\left(\frac{\partial x^\alpha}{\partial \bar{x}^i}\, \frac{\partial \bar{x}^i}{\partial x^\gamma}\right) e_\alpha = g_{\alpha\beta}\left(\frac{\partial x^\alpha}{\partial \bar{x}^i}\, \frac{\partial \bar{x}^i}{\partial x^\gamma}\right)\frac{\partial x^\beta}{\partial \bar{x}^j}\, \bar{e}^j$$

or, using Lemma 1, and noting that $\delta_\gamma{}^\alpha = 0$ unless $\alpha = \gamma$,

$$\delta_\gamma{}^\alpha e_\alpha = g_{\alpha\beta}\, \delta_\gamma{}^\alpha \frac{\partial x^\beta}{\partial \bar{x}^j}\, \bar{e}^j \qquad \text{and} \qquad e_\gamma = g_{\gamma\beta}\, \frac{\partial x^\beta}{\partial \bar{x}^j}\, \bar{e}^j$$

If we multiply the last equation by $g^{\lambda\gamma}$ and sum with respect to γ, making use of the fact that $[g^{ij}]$ is the inverse of $[g_{ij}]$ and hence that $g^{\lambda\gamma}g_{\gamma\beta} = \delta_\beta{}^\lambda$, we have

$$g^{\lambda\gamma}\mathbf{e}_\gamma = g^{\lambda\gamma}g_{\gamma\beta}\frac{\partial x^\beta}{\partial \bar{x}^j}\bar{\mathbf{e}}^j = \delta_\beta{}^\lambda \frac{\partial x^\beta}{\partial \bar{x}^j}\bar{\mathbf{e}}^j$$

and, using Eq. (7),

(24)
$$\mathbf{e}^\lambda = \frac{\partial x^\lambda}{\partial \bar{x}^j}\bar{\mathbf{e}}^j$$

Similarly, or by using Lemma 2,

(25)
$$\bar{\mathbf{e}}^j = \frac{\partial \bar{x}^j}{\partial x^\lambda}\mathbf{e}^\lambda$$

Equations (31) and (30) of Sec. 14.2 are, respectively, special cases of these relations.

From Eq. (8) applied to the new coordinate system, we have

$$\bar{\mathbf{e}}^i \cdot \bar{\mathbf{e}}^j = \bar{g}^{ij}$$

Hence, using (25), we have

$$\left(\mathbf{e}^\alpha \frac{\partial \bar{x}^i}{\partial x^\alpha}\right) \cdot \left(\mathbf{e}^\beta \frac{\partial \bar{x}^j}{\partial x^\beta}\right) = \bar{g}^{ij}$$

or, since $\mathbf{e}^\alpha \cdot \mathbf{e}^\beta = g^{\alpha\beta}$,

(26)
$$\bar{g}^{ij} = g^{\alpha\beta}\frac{\partial \bar{x}^i}{\partial x^\alpha}\frac{\partial \bar{x}^j}{\partial x^\beta}$$

which is the law of transformation for the g^{ij}'s. Equation (27), Sec. 13.2, is a special case of this result. Similarly, or by Lemma 3,

$$g^{\alpha\beta} = \bar{g}^{ij}\frac{\partial x^\alpha}{\partial \bar{x}^i}\frac{\partial x^\beta}{\partial \bar{x}^j}$$

When a vector represented contravariantly, i.e., a vector expressed in terms of the local base vectors $\mathbf{e}_1$, $\mathbf{e}_2$, $\mathbf{e}_3$, say

$$\mathbf{V} = v^1\mathbf{e}_1 + v^2\mathbf{e}_2 + v^3\mathbf{e}_3 = v^\alpha\mathbf{e}_\alpha$$

is expressed in terms of the corresponding local base vectors $\bar{\mathbf{e}}_1$, $\bar{\mathbf{e}}_2$, $\bar{\mathbf{e}}_3$ of a new coordinate system, we have, using (22), the new representation

$$\bar{\mathbf{V}} = v^\alpha \left(\frac{\partial \bar{x}^i}{\partial x^\alpha}\bar{\mathbf{e}}_i\right) = \left(v^\alpha\frac{\partial \bar{x}^i}{\partial x^\alpha}\right)\bar{\mathbf{e}}_i = \bar{v}^i\bar{\mathbf{e}}_i$$

Hence, the components of a contravariant vector transform according to the law

(27)
$$\bar{v}^i = v^\alpha \frac{\partial \bar{x}^i}{\partial x^\alpha}$$

Similarly, for a vector represented covariantly, i.e., a vector expressed in terms of the local reciprocal base vectors, say

$$\mathbf{V} = v_1\mathbf{e}^1 + v_2\mathbf{e}^2 + v_3\mathbf{e}^3 = v_\alpha\mathbf{e}^\alpha$$

we have, using (24), the new representation

$$\bar{\mathbf{V}} = v_\alpha \left(\bar{\mathbf{e}}^i \frac{\partial x^\alpha}{\partial \bar{x}^i}\right) = \left(v_\alpha \frac{\partial x^\alpha}{\partial \bar{x}^i}\right)\bar{\mathbf{e}}^i = \bar{v}_i\bar{\mathbf{e}}^i$$

Hence the components of a covariant vector transform according to the law

(28)
$$\bar{v}_i = v_\alpha \frac{\partial x^\alpha}{\partial \bar{x}^i}$$

Equations (33) and (32), Sec. 14.2, are special cases of Eqs. (27) and (28), respectively.

EXERCISES

1   If the range of each index is 3, write out each of the following sums:
   (a) $f(x_i)\,\Delta x_i$ (b) $a_{ij}x_i x_j$ (c) $a_{ij}x_i y_j$
   (d) $a_{ii}x_i x_i$ (e) $(a_i x^i)^2$

   (f) $\dfrac{\partial x^i}{\partial y^j}\dfrac{\partial y^j}{\partial x^k}z^k$ (g) $\dfrac{\partial x^i}{\partial y^j}\dfrac{\partial y^j}{\partial x^k}z_i$

2   If the range of each index is n, show that:
   (a) $\delta_j{}^i \delta_k{}^j = \delta_k{}^i$ (b) $\delta_i{}^i = \delta_j{}^j = n$
   (c) $\delta_j{}^k A_j = A_k$ (d) $\delta_k{}^i A^i A_k = A^k A_k$
   (e) $\delta_j{}^i A_{ijk}\delta_i{}^k = A_{iil}$ (f) $\delta_j{}^i A_{ik}A_{jl}\delta_k{}^l = A_{ik}A_{ik}$
3   Write out the proofs of Theorems 2 and 7, Sec. 10.2, using the summation convention rather than the Σ notation.
4. Write out the proofs of the remaining assertions of Lemma 3.
5   Prove the following lemma. If ϕ_{ij} and $\bar{\phi}_{ij}$ $(i, j = 1, 2, 3)$ are, respectively, functions of x^1, x^2, x^3 and $\bar{x}^1$, $\bar{x}^2$, $\bar{x}^3$, then any one of the relations

$$\bar{\phi}_{ij} = \phi_{\alpha\beta}\frac{\partial x^\alpha}{\partial \bar{x}^i}\frac{\partial x^\beta}{\partial \bar{x}^j}$$

$$\bar{\phi}_{ij}\frac{\partial \bar{x}^j}{\partial x^\beta} = \phi_{\alpha\beta}\frac{\partial x^\alpha}{\partial \bar{x}^i}$$

$$\bar{\phi}_{ij}\frac{\partial \bar{x}^i}{\partial x^\alpha} = \phi_{\alpha\beta}\frac{\partial x^\beta}{\partial \bar{x}^j}$$

$$\bar{\phi}_{ij}\frac{\partial \bar{x}^i}{\partial x^\alpha}\frac{\partial \bar{x}^j}{\partial x^\beta} = \phi_{\alpha\beta}$$

implies each or the others.
6   Verify Lemma 1 if:
   (a) The x's are rectangular coordinates, and the $\bar{x}$'s are cylindrical coordinates.
   (b) The x's are rectangular coordinates, and the $\bar{x}$'s are oblique coordinates.
   (c) The x's are cylindrical coordinates, and the $\bar{x}$'s are spherical coordinates.
7   If $(\bar{x}^1, \bar{x}^2, \bar{x}^3)$ are the cylindrical coordinates (r, θ, z), and if (x^1, x^2, x^3) are the cartesian coordinates (x, y, z), verify that $\phi^1 = x^1$, $\phi^2 = x^2$, $\phi^3 = x^3$ and $\phi^1 = \bar{x}^1$, $\phi^2 = 0$, $\phi^3 = \bar{x}^3$ satisfy the relations described in Lemma 1.
8   If $(\bar{x}^1, \bar{x}^2, \bar{x}^3)$ and (x^1, x^2, x^3) are, respectively, cylindrical coordinates and rectangular coordinates, and if $\phi^1 = x^2 x^3$, $\phi^2 = x^1 x^3$, $\phi^3 = x^1 x^2$, use the first relation in Lemma 2 to calculate $\bar{\phi}^1$, $\bar{\phi}^2$, $\bar{\phi}^3$ and then verify that the second relation in the lemma is satisfied.
9   If x^1, x^2, x^3 are functions of the cartesian coordinates (x, y, z) defined by the equations $x^1 = (x)^2 + (y)^2 - 2z$, $x^2 = (x)^2 - (y)^2 - 2z$, $x^3 = (x)^2 + (y)^2 + (z)^2 - 2z$, determine at what points, if any, the tangents to the parametric curves defined by these three functions are coplanar. Describe at least two regions in which these equations have a single-valued inverse and find the inverse in each region.
10 If x^1, x^2, x^3 are three functions of x, y, z, show that the tangents to the parametric curves defined at a point P by these functions are noncoplanar if and only if the jacobian determinant $|\partial(x^1, x^2, x^3)/\partial(x, y, z)|$ is different from zero at P.

11 (a) What is the differential of arc in cylindrical coordinates?
    (b) What are G and G^{-1} in cylindrical coordinates?

Each of the following problems refers to a cylindrical coordinate system.

12 (a) What are the lengths of the local base vectors at $(2,0,0)$? At $(2,0,1)$? At $(2,\pi/3,0)$? At $(2,\pi/3,1)$?
    (b) What are the lengths of the local reciprocal base vectors at each of the points in part (a)?

13 If e_1, e_2, e_3 are the base vectors at the point $(2,\pi/3,1)$, and if

$$U = 2e_1 + 3e_2 + e_3 \qquad \text{and} \qquad V = e_1 - e_2 + 2e_3$$

what is the length of U? The length of V? The angle between U and V?

14 Let V be the vector extending from the point $(2,0,1)$ to the point $(2,\pi/3,1)$. Express V in terms of the base vectors at $(2,0,1)$ and also in terms of the base vectors at $(2,\pi/3,1)$. Find the length of V using each of these representations.

15 Work Exercise 14 using the reciprocal base vectors at $(2,0,1)$ and at $(2,\pi/3,1)$.

14.4 Tensors

In the last section, without referring to them by name we had already started working with tensors. Now, with the experience we have gained from our discussion of coordinate transformations in three dimensions, we can make the matter explicit.

Let $(x^1,x^2,\ldots,x^n)$ and $(\bar{x}^1,\bar{x}^2,\ldots,\bar{x}^n)$ be generalized coordinates in n dimensions, and let the two systems of coordinates be related by the transformation equations

(1)
$$T: \quad \bar{x}^i = \bar{x}^i(x^1,x^2,\ldots,x^n)$$
$$T^{-1}: \quad x^i = x^i(\bar{x}^1,\bar{x}^2,\ldots,\bar{x}^n)$$

Once we pass beyond the three-dimensional space of experience, geometric intuition is of little help to us. However, it can be shown that in n dimensions, just as in three dimensions, there are n parametric curves passing through an arbitrary point and on each of these curves one and only one of the generalized coordinates varies. Moreover, if the jacobian determinant of the transformation (1) is different from zero at a point, then at that point the tangents to the parametric curves can be shown to be linearly independent. Hence, if local base vectors e_i $(i = 1, 2,\ldots,n)$ are defined at an arbitrary point P by the conditions that

$$ds\dagger = |e_i \, dx^i| = \sqrt{e_i \cdot e_i} \, |dx^i| \qquad i \text{ not summed}$$

then any vector originating at P can be expressed as a linear combination of these vectors. Furthermore, a set of independent local reciprocal base vectors e^i $(i = 1, 2, \ldots, n)$ can be defined at P by the same conditions we used in three dimensions, namely,

$$e^i \cdot e_j = \delta_j{}^i$$

and any vector originating at P can also be expressed as a linear combination of the reciprocal base vectors at P. In fact, all the results of the last section, with the exception of those involving scalar triple products, are equally correct in n dimensions,

---

† Just as in three dimensions, so in n dimensions we assume that the metrical properties of space are defined by a positive-definite differential quadratic form $(ds)^2 = g_{ij} \, dx^i \, dx^j$ $(i, j = 1, 2, \ldots, n)$ whose matrix $G = [g_{ij}]$ is, of course, nonsingular.

provided only that the summation convention is understood to cover the range 1 to n instead of the range 1 to 3.

By a **scalar**, or **tensor of rank zero**, we mean a quantity S whose descriptions in the two coordinate systems are connected by the relation

$$(2) \qquad \bar{S}(\bar{x}^1, \bar{x}^2, \ldots, \bar{x}^n) = S(x^1, x^2, \ldots, x^n)$$

By a **contravariant vector**, or a **contravariant tensor of rank 1**, we mean a set of n quantities ξ^i, called **components**, whose descriptions in the two coordinate systems are connected by the relations

$$(3) \qquad \bar{\xi}^i(\bar{x}^1, \bar{x}^2, \ldots, \bar{x}^n) = \xi^\alpha(x^1, x^2, \ldots, x^n) \frac{\partial \bar{x}^i}{\partial x^\alpha} \qquad i = 1, 2, \ldots, n$$

Since $d\bar{x}^i = (\partial \bar{x}^i / \partial x^\alpha)\, dx^\alpha$, it follows that the differentials of the coordinate variables are the components of a contravariant tensor of rank 1.

By a **covariant vector**, or **covariant tensor of rank 1**, we mean a set of n quantities ξ_i, also called **components**, whose descriptions in the two coordinate systems are connected by the relations

$$(4) \qquad \bar{\xi}_i(\bar{x}^1, \bar{x}^2, \ldots, \bar{x}^n) = \zeta_\alpha(x^1, x^2, \ldots, x^n) \frac{\partial x^\alpha}{\partial \bar{x}^i} \qquad i = 1, 2, \ldots, n$$

If f is a scalar point function, for which, therefore

$$\bar{f}(\bar{x}^1, \bar{x}^2, \ldots, \bar{x}^n) = f(x^1, x^2, \ldots, x^n)$$

then

$$\frac{\partial \bar{f}}{\partial \bar{x}^i} = \frac{\partial f}{\partial x^\alpha} \frac{\partial x^\alpha}{\partial \bar{x}^i}$$

Hence, the n quantities $\partial f / \partial x^i$ are the components of a covariant tensor of rank 1, which we recognize as the gradient of f.

A **contravariant tensor of rank 2** is a set of n^2 quantities ξ^{ij} whose descriptions in the two coordinate systems are connected by the relations

$$(5) \qquad \bar{\xi}^{ij} = \xi^{\alpha\beta} \frac{\partial \bar{x}^i}{\partial x^\alpha} \frac{\partial \bar{x}^j}{\partial x^\beta} \qquad i, j = 1, 2, \ldots, n$$

From Eq. (26), Sec. 14.3, it is clear that the elements g^{ij} of the matrix G^{-1} form a contravariant tensor of rank 2. Indices which identify the components of a contravariant tensor, i.e., indices in the superscript position on the components of a tensor, are called **contravariant indices**.

A **covariant tensor of rank 2** is a set of n^2 quantities ξ_{ij} whose representations in the two coordinate systems are connected by the relations

$$(6) \qquad \bar{\xi}_{ij} = \xi_{\alpha\beta} \frac{\partial x^\alpha}{\partial \bar{x}^i} \frac{\partial x^\beta}{\partial \bar{x}^j} \qquad i, j = 1, 2, \ldots, n$$

From Eq. (17), Sec. 14.3, it is clear that the elements g_{ij} of the fundamental matrix G form a covariant tensor of rank 2. This tensor is often called the **fundamental metric tensor**. Indices which identify the components of a covariant tensor, i.e., indices in the subscript position on the components of a tensor, are called **covariant indices**.

A **mixed tensor of rank 2** is a set of n^2 quantities $\xi_j{}^i$ whose descriptions in the two coordinate systems are connected by the relations

(7)
$$\bar{\xi}_j{}^i = \xi_\beta{}^\alpha \frac{\partial \bar{x}^i}{\partial x^\alpha} \frac{\partial x^\beta}{\partial \bar{x}^j} \qquad i, j = 1, 2, \ldots, n$$

Although we shall leave the proof of this fact as an exercise, $\delta_j{}^i$ is an example of a mixed tensor of rank 2. It should be noted that the components of a mixed tensor of rank 2 are identified by one upper, or contravariant, index and one lower, or covariant, index.

A tensor ξ^{ij} (or ξ_{ij}) such that $\xi^{ij} = \xi^{ji}$ (or $\xi_{ij} = \xi_{ji}$) for all values of i and j is said to be **symmetric**. A tensor ξ^{ij} (or ξ_{ij}) such that $\xi^{ij} = -\xi^{ji}$ (or $\xi_{ij} = -\xi_{ji}$) is said to be **skew-symmetric** or **alternating**.

Clearly, the concept of a tensor can be generalized to include tensors of arbitrary rank r with any number k ($0 \le k \le r$) of covariant indices and $r - k$ contravariant indices. For instance, a set of n^5 quantities ξ_{uvw}^{ij} whose descriptions in the two coordinate systems are connected by the relations

$$\bar{\xi}_{uvw}^{ij} = \xi_{\delta\gamma\varepsilon}^{\alpha\beta} \frac{\partial \bar{x}^i}{\partial x^\alpha} \frac{\partial \bar{x}^j}{\partial x^\beta} \frac{\partial x^\delta}{\partial \bar{x}^u} \frac{\partial x^\gamma}{\partial \bar{x}^v} \frac{\partial x^\varepsilon}{\partial \bar{x}^w}$$

constitutes a mixed tensor of rank 5 with two contravariant indices, i and j, and three covariant indices, u, v, and w.

From the definition of a tensor as a set of quantities which transform in a prescribed way, it is clear that a tensor can be constructed by specifying its components in one coordinate system arbitrarily and then letting the appropriate transformation laws define its components in all other coordinate systems.

The algebra of tensors is based primarily upon the following observations.

Two tensors are equal if and only if they have the same rank and the same number of indices of each type and have their corresponding components equal in one coordinate system and hence in all. In particular, *if the components of a tensor are all zero in one coordinate system, they are zero in every coordinate system.*

If T_1 and T_2 are tensors of the same type, then the set of quantities obtained by adding respective components of T_1 and T_2 is a tensor $T_1 + T_2$ of the same type as T_1 and T_2.

If T_1 is a tensor of rank r_1 with c_1 contravariant and γ_1 covariant indices, and if T_2 is a tensor of rank r_2 with c_2 contravariant and γ_2 covariant indices, then the set of quantities obtained by multiplying each component of T_1 by each component of T_2 is a tensor $T_1 T_2$, called the **outer product** of T_1 and T_2, of rank $r_1 + r_2$ with $c_1 + c_2$ contravariant indices and $\gamma_1 + \gamma_2$ covariant indices. For instance, if T_1 is the tensor ξ^{ij} and T_2 is the tensor $\xi_l{}^k$, then the general term $\xi^{ij}\xi_l{}^k$ of the outer product $T_1 T_2$ transforms according to the law

$$\bar{\xi}^{ij}\bar{\xi}_l{}^k = \left(\xi^{\alpha\beta} \frac{\partial \bar{x}^i}{\partial x^\alpha} \frac{\partial \bar{x}^j}{\partial x^\beta} \right) \left(\xi_\delta{}^\gamma \frac{\partial \bar{x}^k}{\partial x^\gamma} \frac{\partial x^\delta}{\partial \bar{x}^l} \right) = \xi^{\alpha\beta}\xi_\delta{}^\gamma \frac{\partial \bar{x}^i}{\partial x^\alpha} \frac{\partial \bar{x}^j}{\partial x^\beta} \frac{\partial \bar{x}^k}{\partial x^\gamma} \frac{\partial x^\delta}{\partial \bar{x}^l}$$

which shows that $T_1 T_2 = \xi^{ij}\xi_l{}^k$ is a tensor, say η_l^{ijk}, of rank 4 with three contravariant and one covariant indices.

If, in a tensor of any type, a contravariant index is summed against a covariant index by simply setting one index equal to the other and thereby invoking the summation convention, the resulting set of quantities is a tensor with one less contravariant index and one less covariant index. For example, since the tensor $\xi_k{}^{ij}$ transforms according to the law

$$\xi_k{}^{ij} = \xi_\gamma{}^{\alpha\beta} \frac{\partial \bar{x}^i}{\partial x^\alpha} \frac{\partial \bar{x}^j}{\partial x^\beta} \frac{\partial x^\gamma}{\partial \bar{x}^k}$$

we have, setting $j = k$,

$$\xi_j{}^{ij} = \xi_\gamma{}^{\alpha\beta} \frac{\partial \bar{x}^i}{\partial x^\alpha} \frac{\partial \bar{x}^j}{\partial x^\beta} \frac{\partial x^\gamma}{\partial \bar{x}^j}$$

$$= \xi_\gamma{}^{\alpha\beta} \frac{\partial \bar{x}^i}{\partial x^\alpha} \delta_\beta{}^\gamma \qquad \text{by Lemma 1, Sec. 14.3}$$

$$= \xi_\gamma{}^{\alpha\gamma} \frac{\partial \bar{x}^i}{\partial x^\alpha}$$

since only when $\beta = \gamma$ is $\delta_\beta{}^\gamma \neq 0$. Hence, $\xi_j{}^{ij}$ transforms as a contravariant tensor of rank 1; that is, $\xi_j{}^{ij}$ is a contravariant vector, say η^i. This process of obtaining one tensor from another is known as **contraction**. Obviously, the process of contraction can be repeated as long as there are indices of each type. When the process of contraction is applied to the outer product of two tensors in such a way that at each stage one of the two indices involved in the contraction belongs to the first factor and the other to the second, the resulting tensor is said to be the **inner product** of the two tensors with respect to the given set of indices.

The converse of the last observation is also important: *a set of n^r quantities is a tensor provided an inner product of the set and an arbitrary tensor is also a tensor*. The proof of this assertion should be sufficiently clear from the argument for the special case $r = 2$. Suppose, then, that $\xi_\beta{}^\alpha$ is a set of n^2 quantities such that for an arbitrary tensor of the second rank, say $\eta_\delta{}^\gamma$, we have

(8) $$\xi_\beta{}^\alpha \eta_\delta{}^\beta = \zeta_\delta{}^\alpha$$

where $\zeta_\delta{}^\alpha$ is a tensor. Under an arbitrary transformation of coordinates we have, of course,

(9) $$\bar{\xi}_b{}^a \bar{\eta}_d{}^b = \zeta_d{}^a$$

Now, since $\eta_\delta{}^\beta$ and $\zeta_\delta{}^\alpha$ are tensors of the indicated type, we have, by definition,

$$\bar{\eta}_d{}^b = \eta_\delta{}^\beta \frac{\partial \bar{x}^b}{\partial x^\beta} \frac{\partial x^\delta}{\partial \bar{x}^d} \qquad \text{and} \qquad \zeta_d{}^a = \zeta_\delta{}^\alpha \frac{\partial \bar{x}^a}{\partial x^\alpha} \frac{\partial x^\delta}{\partial \bar{x}^d}$$

Hence, substituting into (9) and then using (8), we have

$$\bar{\xi}_b{}^a \eta_\delta{}^\beta \frac{\partial \bar{x}^b}{\partial x^\beta} \frac{\partial x^\delta}{\partial \bar{x}^d} = \zeta_\delta{}^\alpha \frac{\partial \bar{x}^a}{\partial x^\alpha} \frac{\partial x^\delta}{\partial \bar{x}^d} = \xi_\beta{}^\alpha \eta_\delta{}^\beta \frac{\partial \bar{x}^a}{\partial x^\alpha} \frac{\partial x^\delta}{\partial \bar{x}^d}$$

From this, by transposing and collecting terms, we obtain

$$\frac{\partial x^\delta}{\partial \bar{x}^d} \left(\bar{\xi}_b{}^a \frac{\partial \bar{x}^b}{\partial x^\beta} - \xi_\beta{}^\alpha \frac{\partial \bar{x}^a}{\partial x^\alpha} \right) \eta_\delta{}^\beta = 0$$

If we now form the inner product of the expression on the left with $\partial \bar{x}^d / \partial x^\varepsilon$ and recall from Lemma 1, Sec. 14.3, that

$$\frac{\partial x^\delta}{\partial \bar{x}^d} \frac{\partial \bar{x}^d}{\partial x^\varepsilon} = \delta_\varepsilon^{\ \delta}$$

we have

$$\delta_\varepsilon^{\ \delta} \left(\xi_b^{\ a} \frac{\partial \bar{x}^b}{\partial x^\beta} - \xi_\beta^{\ a} \frac{\partial \bar{x}^a}{\partial x^\alpha} \right) \eta_\delta^{\ \beta} = 0 \quad \text{or} \quad \left(\xi_b^{\ a} \frac{\partial \bar{x}^b}{\partial x^\beta} - \xi_\beta^{\ a} \frac{\partial \bar{x}^a}{\partial x^\alpha} \right) \eta_\varepsilon^{\ \beta} = 0$$

Now, since $\eta_\varepsilon^{\ \beta}$ is completely arbitrary, it may be chosen, in turn, to be a tensor each of whose components except one is equal to zero. Hence it follows that the expression in parentheses in the last equation must be identically zero. Therefore,

$$\xi_b^{\ a} \frac{\partial \bar{x}^b}{\partial x^\beta} = \xi_\beta^{\ a} \frac{\partial \bar{x}^a}{\partial x^\alpha}$$

Finally, if we form the inner product of each member of this equation with $\partial x^\beta / \partial \bar{x}^e$ and again use Lemma 1, Sec. 14.3, we have

$$\xi_b^{\ a} \delta_e^{\ b} = \xi_\beta^{\ a} \frac{\partial \bar{x}^a}{\partial x^\alpha} \frac{\partial x^\beta}{\partial \bar{x}^e}$$

or

$$\xi_e^{\ a} = \xi_\beta^{\ a} \frac{\partial \bar{x}^a}{\partial x^\alpha} \frac{\partial x^\beta}{\partial \bar{x}^e} \equiv \xi_\varepsilon^{\ a} \frac{\partial \bar{x}^a}{\partial x^\alpha} \frac{\partial x^\varepsilon}{\partial \bar{x}^e}$$

which is precisely the law of transformation for a mixed tensor of rank 2. In other words, $\xi_\varepsilon^{\ \alpha}$ is a tensor, as asserted. The property we have confirmed in this particular case is often referred to as the **quotient law** for tensors.

The preceding observation is frequently the most effective means of proving that a set of quantities is a tensor. For instance, by its use we can establish the following interesting result. *If the elements of a nonsingular matrix* $[f_{ij}]$ *are the components of a covariant tensor of rank 2, then the elements of the inverse matrix* $[f^{ij}]$ *are the components of a contravariant tensor of rank 2.* To prove this, let ξ^α be any contravariant vector. Then, by the process of contraction,

$$\eta_i = f_{i\alpha} \xi^\alpha$$

is a covariant vector. Moreover, since $[f_{ij}]$ is nonsingular and ξ^α is arbitrary, η_i is also arbitrary; i.e., any covariant vector η_i can be obtained in this fashion from a suitable contravariant vector ξ^α. If we now form the inner product of each member of the last equation with $f^{\beta i}$, we have

$$f^{\beta i} \eta_i = f^{\beta i} f_{i\alpha} \xi^\alpha$$

However, since $[f_{ij}]$ and $[f^{ij}]$ are inverses, it follows that

$$f^{\beta i} f_{i\alpha} = \delta_\alpha^{\ \beta}$$

Hence,

$$f^{\beta i} \eta_i = \delta_\alpha^{\ \beta} \xi^\alpha = \xi^\beta$$

Since η_i is arbitrary, it follows from the quotient law that f^{ij} is a tensor, and in fact a contravariant tensor, as asserted.

EXERCISES

1   Verify that $\delta_j{}^i$ is a mixed tensor of rank 2.

2   Is $\delta_1{}^i$ a tensor? Is $\delta_j{}^2$ a tensor?

3   Verify that if T_1 and T_2 are tensors of the same type, then $T_1 \pm T_2$ is also a tensor of the same type.

4   Verify that if T is a tensor and ϕ is a scalar, then the set of quantities obtained by multiplying each component of T by ϕ is a tensor of the same type as T.

5   Is a tensor obtained if two covariant indices in a tensor are summed against each other? Is a tensor obtained if two contravariant indices in a tensor are summed against each other?

6   If the elements of a nonsingular matrix $[f_j{}^i]$ are the components of a mixed tensor of rank 2, do the elements of the inverse matrix form a tensor?

7   (a) Let $\xi_\beta{}^\alpha$ be a set of n^2 quantities, and let $\eta^{\beta\gamma}$ be an arbitrary contravariant tensor of rank 2. Show that $\xi_\beta{}^\alpha$ is a tensor if the product $\xi_\beta{}^\alpha \eta^{\beta\gamma}$ is a tensor $\zeta^{\alpha\gamma}$.
    (b) Show that $\xi_\beta{}^\alpha$ is a tensor if its inner product with an arbitrary covariant tensor is also a tensor.

8   (a) Show that $\xi_{\alpha\beta}$ is a tensor if its inner product with an arbitrary mixed tensor $\eta_\gamma{}^\beta$ is a tensor.
    (b) Show that $\xi^{\alpha\beta}$ is a tensor if the product $\xi^{\alpha\beta}\eta_{\beta\gamma}$ is a tensor, where $\eta_{\beta\gamma}$ is an arbitrary covariant tensor of rank 2.

9   If $\eta_{\alpha\beta}{}^\delta$ is an arbitrary tensor of the indicated type, show that the n^3 quantities $\xi_\gamma{}^{\alpha\beta}$ form a tensor if the product $\xi_\gamma{}^{\alpha\beta}\eta_{\alpha\beta}{}^\delta$ is a tensor $\zeta_\gamma{}^\delta$.

10 Show how the contravariant representation of a vector can be obtained from its covariant representation, and vice versa.

14.5 Divergence and Curl

We have already seen (Sec. 14.4) that if f is a scalar point function, then $\partial f / \partial x^i$ is a covariant vector, which we recognized as the gradient of f. We now turn our attention to the determination of the divergence and curl of vectors in generalized coordinates.

In rectangular coordinates, the divergence of a vector $\xi^1\mathbf{i} + \xi^2\mathbf{j} + \xi^3\mathbf{k}$ is the scalar quantity

$$\frac{\partial \xi^1}{\partial x^1} + \frac{\partial \xi^2}{\partial x^2} + \frac{\partial \xi^3}{\partial x^3} = \frac{\partial \xi^a}{\partial x^a}$$

Corresponding to this, in generalized coordinates the divergence of a contravariant vector ξ^a is given by the formula

(1)
$$\frac{1}{\sqrt{|G|}} \frac{\partial}{\partial x^a} (\sqrt{|G|}\ \xi^a)$$

where G is the metric tensor of the space. In rectangular coordinates, as we observed in Sec. 14.2, we have

$$G = \begin{bmatrix} 1 & 0 & 0 \\ 0 & 1 & 0 \\ 0 & 0 & 1 \end{bmatrix} \quad \text{and} \quad |G| = 1$$

Hence the familiar expression for the divergence of ξ^a in rectangular coordinates is a special case of the general formula (1). However, before this can be accepted as a definition of the divergence in an arbitrary coordinate system, we must prove that it is a scalar invariant, i.e., we must prove that it is the same in all coordinate systems.

To do this, let us consider the given expression in a second coordinate system and perform the indicated differentiation:

$$\frac{1}{\sqrt{|\bar{G}|}} \frac{\partial}{\partial \bar{x}^i} (\sqrt{|\bar{G}|}\ \bar{\xi}^i) = \frac{1}{\sqrt{|\bar{G}|}} \left(\frac{1}{2\sqrt{|\bar{G}|}} \frac{\partial |\bar{G}|}{\partial \bar{x}^i} \bar{\xi}^i + \sqrt{|\bar{G}|}\ \frac{\partial \bar{\xi}^i}{\partial \bar{x}^i} \right)$$

$$(2) \qquad\qquad\qquad = \frac{1}{2|\bar{G}|} \frac{\partial |\bar{G}|}{\partial x^a} \frac{\partial x^a}{\partial \bar{x}^i} \bar{\xi}^i + \frac{\partial \bar{\xi}^i}{\partial \bar{x}^i}$$

By hypothesis, ξ^a is a contravariant vector. Hence

$$(3) \qquad\qquad\qquad \bar{\xi}^i = \xi^a \frac{\partial \bar{x}^i}{\partial x^a}$$

and

$$(4) \qquad\qquad\qquad \bar{\xi}^i \frac{\partial x^a}{\partial \bar{x}^i} = \xi^a$$

and therefore Eq. (2) can be written

$$(5) \qquad \frac{1}{\sqrt{|\bar{G}|}} \frac{\partial}{\partial \bar{x}^i} (\sqrt{|\bar{G}|}\ \bar{\xi}^i) = \left(\frac{1}{2|\bar{G}|} \frac{\partial |\bar{G}|}{\partial x^a} \right) \xi^a + \frac{\partial}{\partial \bar{x}^i} \left(\xi^a \frac{\partial \bar{x}^i}{\partial x^a} \right)$$

Also, from Eq. (19), Sec. 14.3, $|\bar{G}| = |G| \cdot |J|^{-2}$, where J is the jacobian of the transformation. Hence, differentiating this equation with respect to x^a and dividing by $|\bar{G}|$, we have

$$(6) \qquad\qquad \frac{1}{|\bar{G}|} \frac{\partial |\bar{G}|}{\partial x^a} = \frac{1}{|G|} \frac{\partial |G|}{\partial x^a} - \frac{2}{|J|} \frac{\partial |J|}{\partial x^a}$$

Thus, substituting from (6) into (5), performing the indicated differentiation, and rearranging, we obtain

$$\frac{1}{\sqrt{|\bar{G}|}} \frac{\partial}{\partial \bar{x}^i} (\sqrt{|\bar{G}|}\ \bar{\xi}^i) = \frac{1}{2} \left(\frac{1}{|G|} \frac{\partial |G|}{\partial x^a} - \frac{2}{|J|} \frac{\partial |J|}{\partial x^a} \right) \xi^a + \frac{\partial}{\partial \bar{x}^i} \left(\xi^a \frac{\partial \bar{x}^i}{\partial x^a} \right)$$

$$= \frac{1}{2|G|} \frac{\partial |G|}{\partial x^a} \xi^a - \frac{1}{|J|} \frac{\partial |J|}{\partial x^a} \xi^a + \frac{\partial \xi^a}{\partial \bar{x}^i} \frac{\partial \bar{x}^i}{\partial x^a} + \xi^a \frac{\partial^2 \bar{x}^i}{\partial x^b\, \partial x^a\, \partial \bar{x}^i} \frac{\partial x^b}{}$$

$$(7) \qquad = \left(\frac{1}{2|G|} \frac{\partial |G|}{\partial x^a} \xi^a + \frac{\partial \xi^a}{\partial x^a} \right) + \left(\frac{\partial^2 \bar{x}^i}{\partial x^a\, \partial x^b} \frac{\partial x^b}{\partial \bar{x}^i} - \frac{1}{|J|} \frac{\partial |J|}{\partial x^a} \right) \xi^a$$

Since the first quantity in parentheses in Eq. (7) is precisely

$$\frac{1}{\sqrt{|G|}} \frac{\partial}{\partial x^a} (\sqrt{|G|}\ \xi^a)$$

our proof will be complete if we can show that the second quantity in parentheses is zero. To do this, we recall from the rule for differentiating a determinant that the derivative of $|J|$ is equal to the sum of n determinants, the ith one of which is identical with $|J|$ except that the ith row consists of the derivatives of the elements of the ith row of $|J|$. Therefore, if we denote by J_i^b the cofactor of the element in the ith row and bth column of $|J|$, then, in expanded form,

$$\frac{\partial |J|}{\partial x^a} = \frac{\partial^2 \bar{x}^i}{\partial x^a\, \partial x^b} J_i^b$$

However, by Corollary 1 of Theorem 1, Sec. 14.3,

$$J_i^b = \frac{\partial x^b}{\partial \bar{x}^i} |J|$$

Hence,

$$\frac{1}{|J|} \frac{\partial |J|}{\partial x^a} = \frac{\partial^2 \bar{x}^i}{\partial x^a \, \partial x^b} \frac{\partial x^b}{\partial \bar{x}^i}$$

and thus the second quantity in parentheses in Eq. (7) is indeed zero, and the scalar

$$\frac{1}{\sqrt{|G|}} \frac{\partial}{\partial x^a} (\sqrt{|G|} \, \xi^a)$$

is invariant and hence equal to the divergence in every coordinate system.

If we use the covariant representation ξ_a instead of the contravariant representation ξ^a, then, since

$$\xi^a = g^{ab}\xi_b$$

(see Exercise 10, Sec. 14.4), we have for the divergence

(8)
$$\frac{1}{\sqrt{|G|}} \frac{\partial}{\partial x^a} (\sqrt{|G|} \, g^{ab}\xi_b)$$

If ξ_a is the gradient of a scalar point function, i.e., if ξ_a is the covariant vector $\partial f/\partial x^b$, then its divergence is called the **laplacian** of f. In other words, in generalized coordinates,

(9)
$$\nabla^2 f = \frac{1}{\sqrt{|G|}} \frac{\partial}{\partial x^a} \left(\sqrt{|G|} \, g^{ab} \frac{\partial f}{\partial x^b} \right)$$

EXAMPLE 1

Obtain the expression for $\nabla^2 f$ in cylindrical coordinates.

By direct calculation, as in Example 1, Sec. 14.3, or by observing from a figure that

$$(ds)^2 = (dr)^2 + (r\,d\theta)^2 + (dz)^2$$

we find that in cylindrical coordinates

$$G = [g_{ij}] = \begin{bmatrix} 1 & 0 & 0 \\ 0 & r^2 & 0 \\ 0 & 0 & 1 \end{bmatrix} \quad \text{and} \quad G^{-1} = [g^{ij}] = \begin{bmatrix} 1 & 0 & 0 \\ 0 & \dfrac{1}{r^2} & 0 \\ 0 & 0 & 1 \end{bmatrix}$$

Therefore, from (9),

$$\begin{aligned}
\nabla^2 \phi &= \frac{1}{r} \left[\frac{\partial}{\partial r}\left(r\,\frac{\partial \phi}{\partial r} \right) + \frac{\partial}{\partial \theta}\left(r\,\frac{1}{r^2}\frac{\partial \phi}{\partial \theta} \right) + \frac{\partial}{\partial z}\left(r\,\frac{\partial \phi}{\partial z} \right) \right] \\
&= \frac{1}{r} \left[\left(r\,\frac{\partial^2 \phi}{\partial r^2} + \frac{\partial \phi}{\partial r} \right) + \frac{1}{r}\frac{\partial^2 \phi}{\partial \theta^2} + r\,\frac{\partial^2 \phi}{\partial z^2} \right] \\
&= \frac{\partial^2 \phi}{\partial r^2} + \frac{1}{r^2}\frac{\partial^2 \phi}{\partial \theta^2} + \frac{\partial^2 \phi}{\partial z^2} + \frac{1}{r}\frac{\partial \phi}{\partial r}
\end{aligned}$$

Consider, now, an arbitrary covariant vector ξ_a. From its law of transformation

$$\bar{\xi}_\alpha = \xi_a \frac{\partial x^a}{\partial \bar{x}^\alpha}$$

we have, by differentiation,

$$\frac{\partial \bar{\xi}_\alpha}{\partial \bar{x}^\beta} = \frac{\partial \xi_a}{\partial x^b} \frac{\partial x^b}{\partial \bar{x}^\beta} \frac{\partial x^a}{\partial \bar{x}^\alpha} + \xi_a \frac{\partial^2 x^a}{\partial \bar{x}^\beta \partial \bar{x}^\alpha}$$

Similarly, of course,

$$\frac{\partial \bar{\xi}_\beta}{\partial \bar{x}^\alpha} = \frac{\partial \xi_b}{\partial x^a} \frac{\partial x^a}{\partial \bar{x}^\alpha} \frac{\partial x^b}{\partial \bar{x}^\beta} + \xi_b \frac{\partial^2 x^b}{\partial \bar{x}^\alpha \partial \bar{x}^\beta}$$

Hence, subtracting the last two equations, we have

$$\frac{\partial \bar{\xi}_\alpha}{\partial \bar{x}^\beta} - \frac{\partial \bar{\xi}_\beta}{\partial \bar{x}^\alpha} = \left(\frac{\partial \xi_a}{\partial x^b} - \frac{\partial \xi_b}{\partial x^a} \right) \frac{\partial x^a}{\partial \bar{x}^\alpha} \frac{\partial x^b}{\partial \bar{x}^\beta}$$

where the other terms cancel, since the order of partial differentiation is immaterial and since a and b are just dummy indices. From the law of transformation defined by the last equation, it follows that

$$\frac{\partial \xi_a}{\partial x^b} - \frac{\partial \xi_b}{\partial x^a}$$

is a covariant tensor of the second rank. Clearly, it is a generalization of the familiar notion of the curl of a vector.

More specifically, in three dimensions, let ξ_{ab} be an arbitrary alternating covariant tensor of the second rank, for which, by definition, $\xi_{ab} = -\xi_{ba}$. From ξ_{ab} we can construct the expression

$$\tfrac{1}{2}\xi_{ab}\mathbf{e}^a \times \mathbf{e}^b \equiv \xi_{12}\mathbf{e}^1 \times \mathbf{e}^2 + \xi_{23}\mathbf{e}^2 \times \mathbf{e}^3 + \xi_{31}\mathbf{e}^3 \times \mathbf{e}^1$$

Moreover, if we use the fact (see Exercise 1, below) that

$$\mathbf{e}^i \times \mathbf{e}^j = \frac{\mathbf{e}_k}{[\mathbf{e}_1\mathbf{e}_2\mathbf{e}_3]} \qquad i, j, k \text{ any cyclic permutation of } 1, 2, 3$$

we can write $\tfrac{1}{2}\xi_a{}^b\mathbf{e}^a \times \mathbf{e}^b$ in the form

$$\xi_{12} \frac{\mathbf{e}_3}{[\mathbf{e}_1\mathbf{e}_2\mathbf{e}_3]} + \xi_{23} \frac{\mathbf{e}_1}{[\mathbf{e}_1\mathbf{e}_2\mathbf{e}_3]} + \xi_{31} \frac{\mathbf{e}_2}{[\mathbf{e}_1\mathbf{e}_2\mathbf{e}_3]}$$

Since this expression is a linear combination of the local base vectors $\mathbf{e}_1, \mathbf{e}_2, \mathbf{e}_3$, it is the representation of a contravariant vector; that is $\tfrac{1}{2}\xi_a{}^b\mathbf{e}^a \times \mathbf{e}^b$ is a contravariant tensor of rank 1. Finally, if we recall from Eq. (12), Sec. 14.3, that

$$[\mathbf{e}_1\mathbf{e}_2\mathbf{e}_3]^2 = |G|$$

we can write this tensor in the form

(10) $$\frac{\xi_{23}}{-\sqrt{|G|}} \mathbf{e}_1 + \frac{\xi_{31}}{-\sqrt{|G|}} \mathbf{e}_2 + \frac{\xi_{12}}{-\sqrt{|G|}} \mathbf{e}_3$$

If $\xi_{ab} = \partial \xi_a/\partial x^b - \partial \xi_b/\partial x^a$, where ξ_a is a covariant vector, then the expression (10), with the negative square root used, as indicated, is precisely the curl of ξ_a, as we defined it in rectangular coordinates in Sec. 13.3.

1   Using Eqs. (9) and (10), Sec. 14.3, or otherwise, show that

$$\mathbf{e}^2 \times \mathbf{e}^3 = \frac{\mathbf{e}_1}{[\mathbf{e}_1\mathbf{e}_2\mathbf{e}_3]} \qquad \mathbf{e}^3 \times \mathbf{e}^1 = \frac{\mathbf{e}_2}{[\mathbf{e}_1\mathbf{e}_2\mathbf{e}_3]} \qquad \mathbf{e}^1 \times \mathbf{e}^2 = \frac{\mathbf{e}_3}{[\mathbf{e}_1\mathbf{e}_2\mathbf{e}_3]}$$

2   (a) What is the divergence of a contravariant vector in cylindrical coordinates?
    (b) What is the divergence of a covariant vector in cylindrical coordinates?
3   (a) What is the divergence of a contravariant vector in spherical coordinates?
    (b) What is the divergence of a covariant vector in spherical coordinates?
4   Obtain the expression for $\nabla^2 f$ in spherical coordinates.
5   If ξ_a is the gradient of a scalar function f, show that the curl of ξ_a vanishes
    identically.

14.6 Covariant Differentiation

Since the components of a tensor are functions of the generalized coordinates, it is obvious that they can be differentiated with respect to the coordinate variables. However, the quantities thus obtained are of no intrinsic interest since they are not the components of a tensor. For instance, if ξ^d is a contravariant vector, and if we differentiate the transformation equation

$$\bar{\xi}^\delta = \xi^d \frac{\partial \bar{x}^\delta}{\partial x^d}$$

partially with respect to $\bar{x}^\beta$, we have

(1)
$$\frac{\partial \bar{\xi}^\delta}{\partial \bar{x}^\beta} = \left(\frac{\partial \xi^d}{\partial x^b} \frac{\partial x^b}{\partial \bar{x}^\beta} \right) \frac{\partial \bar{x}^\delta}{\partial x^d} + \xi^d \left(\frac{\partial^2 \bar{x}^\delta}{\partial x^b \, \partial x^d} \frac{\partial x^b}{\partial \bar{x}^\beta} \right)$$

Clearly, if the second term on the right were not present, $\partial \xi^d / \partial x^b$ would be a mixed tensor of rank 2, since the first term on the right in (1) is precisely what is given by the law of transformation for a tensor with one covariant and one contravariant index. It is also interesting to note that if the equations connecting the two sets of generalized coordinates are linear, as they are for transformations between rectangular and oblique coordinates, then the second term in (1) *is* missing. These observations raise the important question of whether or not it is possible to add "correction" terms $C_b{}^d$ to the partial derivatives $\partial \xi^d / \partial x^b$ so that

$$\frac{\partial \xi^d}{\partial x^b} + C_b{}^d$$

will be a mixed tensor of rank 2. This is indeed possible, and although we cannot go deeply into the matter, we shall determine the appropriate correction terms and define the so-called *covariant derivative*.

Since ξ^d enters linearly in the second term in (1), it is almost obvious that the terms to be added to $\partial \xi^d / \partial x^b$ to eliminate the second sum should be linear in the ξ's, say,

$$C_b{}^d = \Gamma_{ab}{}^d \, \xi^a$$

To determine the coefficient function $\Gamma_{ab}{}^d$, we consider first the metric tensor g_{ab}. Beginning with its law of transformation, namely,

$$\bar{g}_{\alpha\beta} = g_{ab} \frac{\partial x^a}{\partial \bar{x}^\alpha} \frac{\partial x^b}{\partial \bar{x}^\beta}$$

and differentiating each side with respect to $\bar{x}^\gamma$, we obtain

(2) $\quad \dfrac{\partial \bar{g}_{\alpha\beta}}{\partial \bar{x}^\gamma} = \dfrac{\partial g_{ab}}{\partial x^c} \dfrac{\partial x^c}{\partial \bar{x}^\gamma} \dfrac{\partial x^a}{\partial \bar{x}^\alpha} \dfrac{\partial x^b}{\partial \bar{x}^\beta} + \left[g_{ab} \dfrac{\partial^2 x^a}{\partial \bar{x}^\gamma \partial \bar{x}^\alpha} \dfrac{\partial x^b}{\partial \bar{x}^\beta} \right] + \left\{ g_{ab} \dfrac{\partial x^a}{\partial \bar{x}^\alpha} \dfrac{\partial^2 x^b}{\partial \bar{x}^\gamma \partial \bar{x}^\beta} \right\}$

From this, by first interchanging β and γ and then interchanging γ and α, and making the corresponding permutations of the dummy indices a, b, c in the first term, we obtain, respectively,

(3) $\quad \dfrac{\partial \bar{g}_{\alpha\gamma}}{\partial \bar{x}^\beta} = \dfrac{\partial g_{ac}}{\partial x^b} \dfrac{\partial x^b}{\partial \bar{x}^\beta} \dfrac{\partial x^a}{\partial \bar{x}^\alpha} \dfrac{\partial x^c}{\partial \bar{x}^\gamma} + g_{ab} \dfrac{\partial^2 x^a}{\partial \bar{x}^\beta \partial \bar{x}^\alpha} \dfrac{\partial x^b}{\partial \bar{x}^\gamma} + \left\{ g_{ab} \dfrac{\partial x^a}{\partial \bar{x}^\alpha} \dfrac{\partial^2 x^b}{\partial \bar{x}^\beta \partial \bar{x}^\gamma} \right\}$

(4) $\quad \dfrac{\partial \bar{g}_{\gamma\beta}}{\partial \bar{x}^\alpha} = \dfrac{\partial g_{cb}}{\partial x^a} \dfrac{\partial x^a}{\partial \bar{x}^\alpha} \dfrac{\partial x^c}{\partial \bar{x}^\gamma} \dfrac{\partial x^b}{\partial \bar{x}^\beta} + \left[g_{ab} \dfrac{\partial^2 x^a}{\partial \bar{x}^\alpha \partial \bar{x}^\gamma} \dfrac{\partial x^b}{\partial \bar{x}^\beta} \right] + g_{ab} \dfrac{\partial x^a}{\partial \bar{x}^\gamma} \dfrac{\partial^2 x^b}{\partial \bar{x}^\alpha \partial \bar{x}^\beta}$

Now, subtracting (2) from the sum of (3) and (4), noting that the quantities in brackets and the quantities in braces cancel respectively, we obtain

$$\dfrac{\partial \bar{g}_{\gamma\beta}}{\partial \bar{x}^\alpha} + \dfrac{\partial \bar{g}_{\alpha\gamma}}{\partial \bar{x}^\beta} - \dfrac{\partial \bar{g}_{\alpha\beta}}{\partial \bar{x}^\gamma} = \left(\dfrac{\partial g_{cb}}{\partial x^a} + \dfrac{\partial g_{ac}}{\partial x^b} - \dfrac{\partial g_{ab}}{\partial x^c} \right) \dfrac{\partial x^a}{\partial \bar{x}^\alpha} \dfrac{\partial x^b}{\partial \bar{x}^\beta} \dfrac{\partial x^c}{\partial \bar{x}^\gamma}$$

$$+ g_{ab} \dfrac{\partial^2 x^a}{\partial \bar{x}^\beta \partial \bar{x}^\alpha} \dfrac{\partial x^b}{\partial \bar{x}^\gamma} + g_{ab} \dfrac{\partial x^a}{\partial \bar{x}^\gamma} \dfrac{\partial^2 x^b}{\partial \bar{x}^\alpha \partial \bar{x}^\beta}$$

Finally, interchanging the dummy indices a and b in the last term and recalling that $g_{ba} = g_{ab}$, we have

(5) $\quad \dfrac{\partial \bar{g}_{\gamma\beta}}{\partial \bar{x}^\alpha} + \dfrac{\partial \bar{g}_{\alpha\gamma}}{\partial \bar{x}^\beta} - \dfrac{\partial \bar{g}_{\alpha\beta}}{\partial \bar{x}^\gamma} = \left(\dfrac{\partial g_{cb}}{\partial x^a} + \dfrac{\partial g_{ac}}{\partial x^b} - \dfrac{\partial g_{ab}}{\partial x^c} \right) \dfrac{\partial x^a}{\partial \bar{x}^\alpha} \dfrac{\partial x^b}{\partial \bar{x}^\beta} \dfrac{\partial x^c}{\partial \bar{x}^\gamma} + 2 g_{ab} \dfrac{\partial^2 x^a}{\partial \bar{x}^\alpha \partial \bar{x}^\beta} \dfrac{\partial x^b}{\partial \bar{x}^\gamma}$

The quantities

(6) $\qquad\qquad\qquad \Gamma_{c,ab} = \dfrac{1}{2} \left(\dfrac{\partial g_{cb}}{\partial x^a} + \dfrac{\partial g_{ac}}{\partial x^b} - \dfrac{\partial g_{ab}}{\partial x^c} \right)$

whose law of transformation is given by Eq. (5), are known as **Christoffel symbols of the first kind**.† Incidentally, because of the second term on the right in the transformation equation (5), it is clear that $\Gamma_{c,ab}$ is *not* a tensor.

The **Christoffel symbols of the second kind** are, by definition, the quantities

(7) $\qquad\qquad\qquad\qquad \Gamma_{ab}{}^d = g^{dc} \Gamma_{c,ab}$

To obtain their law of transformation, we begin by recalling from Eq. (26), Sec. 14.3, that

$$\bar{g}^{\delta\gamma} = g^{di} \dfrac{\partial \bar{x}^\delta}{\partial x^d} \dfrac{\partial \bar{x}^\gamma}{\partial x^i}$$

---

† Named for the Swiss mathematician E. B. Christoffel (1829–1900).

Hence,

$$
\begin{aligned}
\bar{\Gamma}_{\alpha\beta}{}^{\delta} &= \bar{g}^{\delta\gamma}\bar{\Gamma}_{\gamma,\alpha\beta} = \tfrac{1}{2}\bar{g}^{\delta\gamma}\left(\frac{\partial\bar{g}_{\gamma\beta}}{\partial\bar{x}^{\alpha}} + \frac{\partial\bar{g}_{\alpha\gamma}}{\partial\bar{x}^{\beta}} - \frac{\partial\bar{g}_{\alpha\beta}}{\partial\bar{x}^{\gamma}}\right) \\
&= \frac{1}{2}\left(g^{di}\frac{\partial\bar{x}^{\delta}}{\partial x^{d}}\frac{\partial\bar{x}^{\gamma}}{\partial x^{i}}\right)\left[\left(\frac{\partial g_{cb}}{\partial x^{a}} + \frac{\partial g_{ac}}{\partial x^{b}} - \frac{\partial g_{ab}}{\partial x^{c}}\right)\frac{\partial x^{a}}{\partial\bar{x}^{\alpha}}\frac{\partial x^{b}}{\partial\bar{x}^{\beta}}\frac{\partial x^{c}}{\partial\bar{x}^{\gamma}} + 2g_{ab}\frac{\partial^{2}x^{a}}{\partial\bar{x}^{\alpha}\partial\bar{x}^{\beta}}\frac{\partial x^{b}}{\partial\bar{x}^{\gamma}}\right] \\
&= \tfrac{1}{2}g^{di}\left(\frac{\partial g_{cb}}{\partial x^{a}} + \frac{\partial g_{ac}}{\partial x^{b}} - \frac{\partial g_{ab}}{\partial x^{c}}\right)\frac{\partial x^{a}}{\partial\bar{x}^{\alpha}}\frac{\partial x^{b}}{\partial\bar{x}^{\beta}}\frac{\partial\bar{x}^{\delta}}{\partial x^{d}}\left[\frac{\partial x^{c}}{\partial\bar{x}^{\gamma}}\frac{\partial\bar{x}^{\gamma}}{\partial x^{i}}\right] \\
&\quad + g^{di}g_{ab}\frac{\partial^{2}x^{a}}{\partial\bar{x}^{\alpha}\partial\bar{x}^{\beta}}\frac{\partial\bar{x}^{\delta}}{\partial x^{d}}\left[\frac{\partial x^{b}}{\partial\bar{x}^{\gamma}}\frac{\partial\bar{x}^{\gamma}}{\partial x^{i}}\right]
\end{aligned}
$$

Now, by Lemma 1, Sec. 14.3, the bracketed terms become

$$
\frac{\partial x^{c}}{\partial\bar{x}^{\gamma}}\frac{\partial\bar{x}^{\gamma}}{\partial x^{i}} = \delta_{i}{}^{c} \qquad \text{and} \qquad \frac{\partial x^{b}}{\partial\bar{x}^{\gamma}}\frac{\partial\bar{x}^{\gamma}}{\partial x^{i}} = \delta_{i}{}^{b}
$$

Therefore, the last equation simplifies to

$$
\bar{\Gamma}_{\alpha\beta}{}^{\delta} = \tfrac{1}{2}g^{dc}\left(\frac{\partial g_{cb}}{\partial x^{a}} + \frac{\partial g_{ac}}{\partial x^{b}} - \frac{\partial g_{ab}}{\partial x^{c}}\right)\frac{\partial x^{a}}{\partial\bar{x}^{\alpha}}\frac{\partial x^{b}}{\partial\bar{x}^{\beta}}\frac{\partial\bar{x}^{\delta}}{\partial x^{d}} + g^{db}g_{ab}\frac{\partial^{2}x^{a}}{\partial\bar{x}^{\alpha}\partial\bar{x}^{\beta}}\frac{\partial\bar{x}^{\delta}}{\partial x^{d}}
$$

Furthermore, since $[g^{ij}]$ and $[g_{ij}]$ are inverse matrices, it follows that

$$
g^{db}g_{ab} = g^{db}g_{ba} = \delta_{a}{}^{d}
$$

Hence, the last term in the preceding equation reduces to

$$
\frac{\partial^{2}x^{a}}{\partial\bar{x}^{\alpha}\partial\bar{x}^{\beta}}\frac{\partial\bar{x}^{\delta}}{\partial x^{a}}
$$

and we have, finally, the law of transformation

$$
\begin{aligned}
\bar{\Gamma}_{\alpha\beta}{}^{\delta} &= \tfrac{1}{2}g^{dc}\left(\frac{\partial g_{cb}}{\partial x^{a}} + \frac{\partial g_{ac}}{\partial x^{b}} - \frac{\partial g_{ab}}{\partial x^{c}}\right)\frac{\partial x^{a}}{\partial\bar{x}^{\alpha}}\frac{\partial x^{b}}{\partial\bar{x}^{\beta}}\frac{\partial\bar{x}^{\delta}}{\partial x^{d}} + \frac{\partial^{2}x^{a}}{\partial\bar{x}^{\alpha}\partial\bar{x}^{\beta}}\frac{\partial\bar{x}^{\delta}}{\partial x^{a}} \\
&= g^{dc}\Gamma_{c,ab}\frac{\partial x^{a}}{\partial\bar{x}^{\alpha}}\frac{\partial x^{b}}{\partial\bar{x}^{\beta}}\frac{\partial\bar{x}^{\delta}}{\partial x^{d}} + \frac{\partial^{2}x^{a}}{\partial\bar{x}^{\alpha}\partial\bar{x}^{\beta}}\frac{\partial\bar{x}^{\delta}}{\partial x^{a}} \\
&= \Gamma_{ab}{}^{d}\frac{\partial x^{a}}{\partial\bar{x}^{\alpha}}\frac{\partial x^{b}}{\partial\bar{x}^{\beta}}\frac{\partial\bar{x}^{\delta}}{\partial x^{d}} + \frac{\partial^{2}x^{a}}{\partial\bar{x}^{\alpha}\partial\bar{x}^{\beta}}\frac{\partial\bar{x}^{\delta}}{\partial x^{a}}
\end{aligned}
$$

(8)

Because of the second term on the right in (8), it is clear that $\Gamma_{ab}{}^{d}$, like $\Gamma_{c,ab}$, is *not* a tensor.

We can now establish the fundamental result that $\partial\xi^{d}/\partial x^{b} + \Gamma_{ab}{}^{d}\xi^{a}$ is a mixed tensor of rank 2. In fact, knowing the law of transformation for $\partial\xi^{d}/\partial x^{b}$, namely, Eq. (1), and the law of transformation for $\Gamma_{ab}{}^{d}$, namely, Eq. (8), we have

$$
\frac{\partial\bar{\xi}^{\delta}}{\partial\bar{x}^{\beta}} + \bar{\Gamma}_{\alpha\beta}{}^{\delta}\bar{\xi}^{\alpha} = \frac{\partial\xi^{d}}{\partial x^{b}}\frac{\partial x^{b}}{\partial\bar{x}^{\beta}}\frac{\partial\bar{x}^{\delta}}{\partial x^{d}} + \xi^{d}\frac{\partial^{2}\bar{x}^{\delta}}{\partial x^{b}\partial x^{d}}\frac{\partial x^{b}}{\partial\bar{x}^{\beta}}
$$

$$
+ \left(\Gamma_{ab}{}^{d}\frac{\partial x^{a}}{\partial\bar{x}^{\alpha}}\frac{\partial x^{b}}{\partial\bar{x}^{\beta}}\frac{\partial\bar{x}^{\delta}}{\partial x^{d}} + \frac{\partial^{2}x^{a}}{\partial\bar{x}^{\alpha}\partial\bar{x}^{\beta}}\frac{\partial\bar{x}^{\delta}}{\partial x^{a}}\right)\xi^{i}\frac{\partial\bar{x}^{\alpha}}{\partial x^{i}}
$$

or, replacing the dummy index d by i in the second term, replacing the dummy index a by b in the fourth term, and observing that in the third term $(\partial x^a/\partial \bar{x}^\alpha)(\partial \bar{x}^\alpha/\partial x^i) = \delta_i{}^a$,

$$\frac{\partial \bar{\xi}^\delta}{\partial \bar{x}^\beta} + \bar{\Gamma}_{\alpha\beta}{}^\delta \, \bar{\xi}^\alpha = \frac{\partial \xi^d}{\partial x^b} \frac{\partial x^b}{\partial \bar{x}^\beta} \frac{\partial \bar{x}^\delta}{\partial x^d} + \xi^i \frac{\partial^2 \bar{x}^\delta}{\partial x^b \partial x^i} \frac{\partial x^b}{\partial \bar{x}^\beta}$$

$$+ \, \xi^i \Gamma_{ab}{}^d \, \delta_i{}^a \frac{\partial x^b}{\partial \bar{x}^\beta} \frac{\partial \bar{x}^\delta}{\partial x^d} + \xi^i \frac{\partial^2 x^b}{\partial \bar{x}^\alpha \partial \bar{x}^\beta} \frac{\partial \bar{x}^\delta}{\partial x^b} \frac{\partial \bar{x}^\alpha}{\partial x^i}$$

(9)
$$= \left(\frac{\partial \xi^d}{\partial x^b} + \xi^a \Gamma_{ab}{}^d \right) \frac{\partial x^b}{\partial \bar{x}^\beta} \frac{\partial \bar{x}^\delta}{\partial x^d} + \xi^i \left[\frac{\partial^2 \bar{x}^\delta}{\partial x^b \partial x^i} \frac{\partial x^b}{\partial \bar{x}^\beta} + \frac{\partial^2 x^b}{\partial \bar{x}^\alpha \partial \bar{x}^\beta} \frac{\partial \bar{x}^\delta}{\partial x^b} \frac{\partial \bar{x}^\alpha}{\partial x^i} \right]$$

Now, $(\partial \bar{x}^\delta/\partial x^b)(\partial x^b/\partial \bar{x}^\beta) = \delta_\beta{}^\delta$. Hence, differentiating with respect to x^i, we have

$$\frac{\partial^2 \bar{x}^\delta}{\partial x^i \partial x^b} \frac{\partial x^b}{\partial \bar{x}^\beta} + \frac{\partial \bar{x}^\delta}{\partial x^b} \frac{\partial^2 x^b}{\partial \bar{x}^\alpha \partial \bar{x}^\beta} \frac{\partial \bar{x}^\alpha}{\partial x^i} = 0$$

Therefore, the expression in brackets in Eq. (9) is equal to zero, and we have

$$\frac{\partial \bar{\xi}^\delta}{\partial \bar{x}^\beta} + \bar{\Gamma}_{\alpha\beta}{}^\delta \, \bar{\xi}^\alpha = \left(\frac{\partial \xi^d}{\partial x^b} + \Gamma_{ab}{}^d \, \xi^a \right) \frac{\partial x^b}{\partial \bar{x}^\beta} \frac{\partial \bar{x}^\delta}{\partial x^d}$$

which proves that

(10)
$$\frac{\partial \xi^d}{\partial x^b} + \Gamma_{ab}{}^d \, \xi^a$$

is a mixed tensor of the second rank, as asserted.

The expression (10) is called the **covariant derivative of the contravariant vector** ξ^d and is frequently denoted by $D\xi^d/\partial x^b$.

In very much the same way it can be shown that, if ξ_d is a covariant vector, then

(11)
$$\frac{\partial \xi_d}{\partial x^b} - \Gamma_{db}{}^a \, \xi_a$$

is a mixed tensor of rank 2. This expression is known as the **covariant derivative of the covariant vector** ξ_d and is denoted by $D\xi_d/\partial x^b$.

It can also be shown that any tensor has a covariant derivative, in which a term like the second term in (10) enters for each contravariant index in the tensor and a term like the second term in (11) enters for each covariant index. Thus, for tensors of the second rank, we have the formulas

(12)
$$\frac{D\xi^{de}}{\partial x^b} = \frac{\partial \xi^{de}}{\partial x^b} + \Gamma_{ib}{}^d \, \xi^{ie} + \Gamma_{ib}{}^e \, \xi^{di}$$

(13)
$$\frac{D\xi_e{}^d}{\partial x^b} = \frac{\partial \xi_e{}^d}{\partial x^b} + \Gamma_{ib}{}^d \, \xi_e{}^i - \Gamma_{eb}{}^i \, \xi_i{}^d$$

(14)
$$\frac{D\xi_{de}}{\partial x^b} = \frac{\partial \xi_{de}}{\partial x^b} - \Gamma_{db}{}^i \, \xi_{ie} - \Gamma_{eb}{}^i \, \xi_{di}$$

EXERCISES

1   (a) Show that $\Gamma_{d,ab} = \Gamma_{d,ba}$
    (b) Show that $\Gamma_{ab}{}^{d} = \Gamma_{ba}{}^{d}$
    (c) Show that $\Gamma_{d,ab} + \Gamma_{a,bd} = \dfrac{\partial g_{ad}}{\partial x^b}$

    (d) Show that a necessary and sufficient condition that the Christoffel symbols all be zero is that the g_{ij}'s be constants.
2   (a) Calculate the Christoffel symbols for a cylindrical coordinate system.
    (b) Calculate the Christoffel symbols for a spherical coordinate system.
3   If ϕ is a scalar function and ξ^d is a contravariant vector, show that

$$\frac{D(\phi\xi^d)}{\partial x^b} = \frac{\partial \phi}{\partial x^b}\,\xi^d + \phi\,\frac{D\xi^d}{\partial x^b}$$

4   Prove that $Dg_{ij}/\partial x^k = 0$.
5   Prove that

$$\frac{D(\xi^a\eta_b)}{\partial x^c} = \frac{D\xi^a}{\partial x^c}\,\eta_b + \xi^a\,\frac{D\eta_b}{\partial x^c}$$

CHAPTER 15
Analytic Functions of a Complex Variable

15.1 Introduction

In our work up to this point we have frequently found the use of complex numbers necessary or at least convenient. For instance, we encountered them in the solution of linear differential equations with constant coefficients in Chap. 2. In Chap. 5 they appeared in the complex impedance, which we found useful in the determination of the steady-state behavior of electric circuits. Then in Chap. 6 their use led to the important complex exponential form of Fourier series and ultimately to the inversion integral of Laplace transform theory. Finally, in Chap. 9, we found that certain physical problems require the consideration of Bessel functions of complex arguments.

None of these applications, with the exception of the inversion integral, for which fortunately we had no immediate need, required any knowledge of the properties of complex numbers or functions of a complex variable beyond what is ordinarily acquired in courses in college algebra and calculus. There are, however, large areas of applied mathematics in which familiarity with the theory of functions of a complex variable beyond this minimum is indispensable. In this and the next three chapters we shall develop the major features of this theory and illustrate some of its more striking applications.

15.2 Algebraic Preliminaries

By a **complex number** we mean a number of the form $z = x + iy$, where x and y are real numbers and i is the so-called **imaginary unit** whose existence is postulated such that $i^2 = -1$. The real number x is called the **real component** or **real part** of z. The real number y is called the **imaginary component** or **imaginary part** of z. The real and imaginary parts of a complex number or expression z are often denoted by the respective symbols

$$\mathscr{R}(z) \quad \text{and} \quad \mathscr{I}(z)$$

It is important to keep in mind that $\mathscr{I}(z)$, as here defined, is a real quantity.

Two complex numbers $a + ib$ and $c + id$ are said to be **equal** if and only if the real and imaginary parts of the first are, respectively, equal to the real and imaginary parts of the second. In particular, the vanishing of a complex number implies not one but two conditions, namely, that both the real part and the imaginary part of the given number are zero.

EXAMPLE 1

If $(x + y + 2) + (x^2 + y)i = 0$, then

$$x + y + 2 = 0 \quad \text{and} \quad x^2 + y = 0$$

From this pair of simultaneous equations it follows necessarily that $x = 2$ and $y = -4$ or $x = -1$ and $y = -1$.

If $z = x + iy$, then the **negative** of z is the complex number

$$-z = -x - iy$$

If two complex numbers differ only in the sign of their imaginary parts, either one is said to be the **conjugate** of the other. The conjugate of a complex number z is usually written $\bar{z}$ or z^*.

Addition, subtraction, and **multiplication** of complex numbers follow the familiar rules for real quantities, with the additional provision that in multiplication, all powers of i are to be reduced as far as possible by applying the definitive property of i and its obvious extensions:

$$i^2 = -1$$
$$i^3 = i^2 i = -i$$
$$i^4 = i^2 i^2 = 1$$
$$i^5 = i^4 i = i$$
$$\cdots\cdots\cdots\cdots$$

Thus

$$(a + ib) \pm (c + id) = (a \pm c) + (b \pm d)i$$

and

$$(a + ib)(c + id) = (ac - bd) + (bc + ad)i$$

Division of complex numbers is defined as the inverse of multiplication; that is, $(a + ib)/(c + id)$ is the complex number $z = x + iy$ which satisfies the equation

$$(c + id)(x + iy) = a + ib$$

Performing the indicated multiplication, we find

$$(cx - dy) + (dx + cy)i = a + ib$$

which implies that

$$cx - dy = a \quad \text{and} \quad dx + cy = b$$

Solving these for x and y, we obtain

$$x = \frac{ac + bd}{c^2 + d^2} \quad \text{and} \quad y = \frac{bc - ad}{c^2 + d^2}$$

Hence,

$$\frac{a + ib}{c + id} = \frac{ac + bd}{c^2 + d^2} + \frac{bc - ad}{c^2 + d^2}i$$

provided, of course, that $c^2 + d^2 \neq 0$, that is, provided that the divisor $c + id$ is not equal to zero. In practice, the quotient of two complex numbers is usually found by multiplying both numerator and denominator by the conjugate of the denominator:

$$\frac{a + ib}{c + id} = \frac{a + ib}{c + id}\frac{c - id}{c - id} = \frac{ac + bd}{c^2 + d^2} + \frac{bc - ad}{c^2 + d^2}i \quad c + id \neq 0$$

Conjugate complex numbers have various simple though important properties. For instance, if $z = x + iy$, then

(1) $$z\bar{z} = (x + iy)(x - iy) = x^2 + y^2$$

which is a purely real quantity. This is the basis for the use of conjugates in division. Also,

$$z + \bar{z} = (x + iy) + (x - iy) = 2x = 2\mathcal{R}(z)$$

or

(2) $$\mathcal{R}(z) = \frac{z + \bar{z}}{2}$$

and

$$z - \bar{z} = (x + iy) - (x - iy) = 2iy = 2i\mathcal{I}(z)$$

or

(3) $$\mathcal{I}(z) = \frac{z - \bar{z}}{2i}$$

In taking the conjugate of a complicated expression, the following results are useful:

(4) $$\overline{z_1 \pm z_2} = \bar{z}_1 \pm \bar{z}_2$$

(5) $$\overline{z_1 z_2} = \bar{z}_1 \bar{z}_2$$

(6) $$\overline{\left(\frac{z_1}{z_2}\right)} = \frac{\bar{z}_1}{\bar{z}_2} \qquad z_2 \neq 0$$

The proofs of these all follow immediately from the four laws of operation and the definition of conjugates, and we shall leave them as exercises.

EXERCISES

1 Prove that if a number is equal to its conjugate, it is necessarily real.
2 Prove that any number is equal to the conjugate of its conjugate.
3 Prove that if both $z_1 + z_2$ and $z_1 z_2$ are real, then either z_1 and z_2 are both real or $z_1 = \bar{z}_2$.
4 Prove that if the product of two complex numbers is zero, at least one of the numbers must be zero.
5 Prove Eq. (4).   6 Prove Eq. (5).   7 Prove Eq. (6).
8 Verify that $z = (1 + i\sqrt{3})/2$ satisfies the equation $z^2 - z + 1 = 0$.
9 What is $\mathcal{R}(z^3 - 2z)$? What is $\mathcal{I}(z^3 - 2z)$?
10 Verify that, for all combinations of signs, $z = (\pm 1 \pm i)/\sqrt{2}$ satisfies the equation $z^4 + 1 = 0$.

Reduce each of the following expressions to the form $a + ib$:

11 $(1 - i)^2 + (2 + i)^2$

12 $(1 - 2i)(3 + 2i)^2$

13 $i(2 + 3i)^4$

14 $\dfrac{1 + i}{1 - i} - \dfrac{1 - i}{1 + i}$

15 $\dfrac{1 + i}{(3 - i)(1 - 2i)}$

16 $\dfrac{(1 + i)^3}{(2 + i)(1 + 2i)}$

17 Solve the equation $(x^2 y - 2) + (x + 2xy - 5)i = 0$ for x and y.
18 If $F(z)$ is a polynomial in z with real coefficients and $F(2 + 3i) = 1 - i$, what is $F(2 - 3i)$? Is $F(a - ib)$ determined by a knowledge of $F(a + ib)$ if the coefficients of $F(z)$ are not all real?

19 If $B\bar{B} > (A + \bar{A})(C + \bar{C})$, show that the equation

$$(A + \bar{A})z\bar{z} + Bz + \bar{B}\bar{z} + (C + \bar{C}) = 0$$

represents a real circle, and find its center and radius.

20 A number system is said to be **ordered** if it contains a subset of numbers P, called **positive numbers**, with the following properties:

(a) For any two numbers p_1 and p_2 in P, both the sum $p_1 + p_2$ and the product $p_1 p_2$ are also in P, that is, are positive numbers.

(b) For any number p in the system, exactly one of the following possibilities holds: p is in P, $p = 0$, or $-p$ is in P.

If a number system is ordered, it is possible to define the relation *greater than* or $>$, by saying that $a > b$ means that $a - b$ is a number in P, that is, is positive. Prove that the complex numbers cannot be ordered and that it is therefore meaningless to say that one complex number is greater than another. *Hint:* Note first that $i \neq 0$, and then show that no matter how the subset P is defined, both the assumption that i is in P and the assumption that $-i$ is in P lead at once to contradictions.

15.3 The Geometric Representation of Complex Numbers

A complex number is represented geometrically either by the point P whose abscissa and ordinate are, respectively, the real and the imaginary components of the given number or by the directed line segment, or vector, which joins the origin to this point. When used in this fashion for representing complex numbers, the cartesian plane is referred to as the **argand diagram**† or the **complex plane** or simply the z **plane**.

The vector OP which represents the complex number $x + iy$ possesses two important attributes besides its components x and y. These are its length

(1)
$$r = \sqrt{x^2 + y^2}$$

and its direction angle

(2)
$$\theta = \tan^{-1} \frac{y}{x}‡$$

Since (Fig. 15.1) $x = r \cos \theta$ and $y = r \sin \theta$, it is evident that $x + iy$ can be written in the equivalent form

(3)
$$z = r \cos \theta + ir \sin \theta = r(\cos \theta + i \sin \theta)$$

Figure 15.1
The modulus r, the amplitude θ, and the components x and y of the complex number $z = x + iy$.

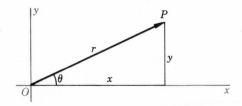

---

† Named for the French mathematician J. R. Argand (1768–1822), although the Norwegian Caspar Wessel (1745–1818) published a discussion of this method of representation 9 years before Argand did.

‡ Actually, $\tan^{-1} (y/x)$ defines two sets of angles in opposite quadrants, the angles of one set being angles of z, the others not. Hence, in using the formula $\theta = \tan^{-1} (y/x)$ one must be careful to select the angles in the proper quadrant, as determined by the signs of x and y.

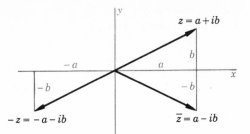

Figure 15.2
The relation between z, $-z$, and $\bar{z}$.

This is known as the **polar** or **trigonometric form** of a complex number and is some-times abbreviated to

$$r \text{ cis } \theta$$

in which only the initial letters of *cosine* and *sine* are retained. The length r is called the **absolute value** or **modulus** of z (written mod z). The angle θ is called the **amplitude** or **argument** of z (written arg z).

The various combinations of complex numbers we have thus far discussed can easily be interpreted geometrically. For instance, Fig. 15.2 shows that the negative of a complex number is the reflection of that number† in the origin, while the conjugate of a complex number is the reflection of that number in the real axis. The geometrical addition of complex numbers is shown in Fig. 15.3a. By drawing one complex number from the terminus of the other and completing the triangle thus formed, a third complex number is determined whose components are precisely those of the sum $z_1 + z_2$. Figure 15.3b shows the construction for the difference of two complex numbers, i.e., for the sum $z_1 + (-z_2)$. Evidently, $z_1 - z_2$ is identical in length and direction with the vector drawn from the end of z_2 to the end of z_1.

Both the sum and the difference of two complex numbers can be described in terms of the parallelogram having the given numbers for adjacent sides; for the sum is

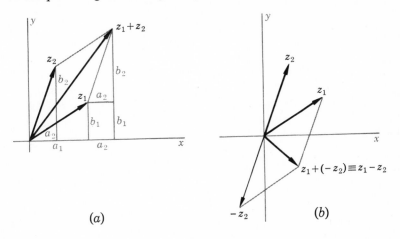

(a)   (b)

Figure 15.3
The sum and difference of the complex numbers $z_1 = a_1 + ib_1$ and $z_2 = a_2 + ib_2$.

† For conciseness of expression, we shall often speak of a complex number and its geometric image as though they were the same thing.

simply the diagonal of the parallelogram which passes through the common origin of the two vectors, and the difference is just the other diagonal, properly directed. Much of the usefulness of complex numbers in elementary engineering applications stems from the fact that they add according to the parallelogram law. Since this is the experimentally established law for the addition of such things as forces, velocities, currents, and voltages, it is evident that in two dimensions complex numbers, like ordinary vectors in three dimensions, can conveniently be used to represent such quantities.

Although we shall have no occasion to use it, a graphical process for multiplying and dividing complex numbers can also be devised. It is based upon the following exceedingly important considerations. If we have two complex numbers given in polar form, their product can be written

$$z_1 z_2 = [r_1(\cos \theta_1 + i \sin \theta_1)][r_2(\cos \theta_2 + i \sin \theta_2)]$$
$$= r_1 r_2 [(\cos \theta_1 \cos \theta_2 - \sin \theta_1 \sin \theta_2)$$
$$+ i(\sin \theta_1 \cos \theta_2 + \cos \theta_1 \sin \theta_2)]$$

(4) $$= r_1 r_2 [\cos(\theta_1 + \theta_2) + i \sin(\theta_1 + \theta_2)]$$

and their quotient can be written

$$\frac{z_1}{z_2} = \frac{r_1(\cos \theta_1 + i \sin \theta_1)}{r_2(\cos \theta_2 + i \sin \theta_2)}$$

$$= \frac{r_1(\cos \theta_1 + i \sin \theta_1)(\cos \theta_2 - i \sin \theta_2)}{r_2(\cos \theta_2 + i \sin \theta_2)(\cos \theta_2 - i \sin \theta_2)}$$

$$= \frac{r_1}{r_2} \frac{(\cos \theta_1 \cos \theta_2 + \sin \theta_1 \sin \theta_2) + i(\sin \theta_1 \cos \theta_2 - \cos \theta_1 \sin \theta_2)}{\cos^2 \theta_2 + \sin^2 \theta_2}$$

(5) $$= \frac{r_1}{r_2} [\cos (\theta_1 - \theta_2) + i \sin (\theta_1 - \theta_2)] \qquad r_2 \neq 0$$

In words, then, *the product of two complex numbers is a complex number whose absolute value is the product of the absolute values of the two factors and whose amplitude is the sum of the amplitudes of the two factors*, and *the quotient of two complex numbers is a complex number whose absolute value is the quotient of the absolute values of the numbers and whose amplitude is the difference of their amplitudes*. The behavior of the angles of complex numbers when the numbers are multiplied or divided is concisely expressed by the formulas

(6) $$\arg z_1 z_2 = \arg z_1 + \arg z_2$$

(7) $$\arg \frac{z_1}{z_2} = \arg z_1 - \arg z_2$$

In Sec. 15.7, when we have succeeded in writing a general complex number as an exponential, the reason for the striking resemblance of these results to the corresponding logarithmic formulas will become apparent.

The extension of these ideas to products of more than two factors is obvious, and we can write at once

$$z_1 z_2 \cdots z_n = r_1 r_2 \cdots r_n [\cos (\theta_1 + \theta_2 + \cdots + \theta_n) + i \sin (\theta_1 + \theta_2 + \cdots + \theta_n)]$$

In particular, if all the z's are the same, we have the important result

$$(8) \qquad z^n = r^n(\cos n\theta + i \sin n\theta)$$

If $r = 1$, this is known as **Demoivre's theorem.**† Since the law of division in polar form (5) gives for the numbers 1 and z the quotient

$$\frac{1}{z} = \frac{1}{r}\left[\cos(0 - \theta) + i \sin(0 - \theta)\right] = \frac{1}{r}\left[\cos(-\theta) + i \sin(-\theta)\right]$$

which is just the content of Eq. (8) for $n = -1$, it is clear that Eq. (8) is valid for all integral values of n, both positive and negative.

The extension of Eq. (8) to roots of integral order is an easy matter. In fact, an nth root of $z = r(\cos \theta + i \sin \theta)$ is defined to be any number $w = R(\cos \phi + i \sin \phi)$ such that $w^n = z$, that is,

$$R^n(\cos n\phi + i \sin n\phi) = r(\cos \theta + i \sin \theta)$$

Since two complex numbers which are equal must have the same modulus, it follows that

$$R^n = r \qquad \text{or} \qquad R = r^{1/n}$$

It should be noted that only real numbers are involved in the determination of R, since $r^{1/n}$ is the *real* nth root of the positive quantity r and can always be found by an ordinary logarithmic calculation. Furthermore, the angles of equal complex numbers must either be equal or differ by an integral multiple of 2π. Hence

$$n\phi = \theta + 2k\pi \qquad \text{or} \qquad \phi = \frac{\theta + 2k\pi}{n}$$

For $k = 0, 1, \ldots, n - 1$, these values of ϕ define n distinct angles which identify n different complex numbers. After this, as k increases through the values $n, n + 1, n + 2, \ldots$, the same angles are repeated, again and again, each time with an irrelevant increment of 2π in their measures. Thus, *there are exactly n distinct values of $w = z^{1/n}$*:

$$(9) \quad w = z^{1/n} = r^{1/n}\left(\cos \frac{\theta + 2k\pi}{n} + i \sin \frac{\theta + 2k\pi}{n}\right) \qquad k = 0, 1, \ldots, n - 1$$

In the complex plane these are represented by radii of the circle with center at the origin and radius $r^{1/n}$, spaced at equal angular intervals of $2\pi/n$, beginning with the radius whose angle is θ/n.

With integral powers and roots defined, the general rational power of a complex number can be defined at once. In fact

$$z^{p/q} = (z^{1/q})^p = \left[r^{1/q}\left(\cos \frac{\theta + 2k\pi}{q} + i \sin \frac{\theta + 2k\pi}{q}\right)\right]^p$$

$$(10) \qquad = r^{p/q}\left[\cos \frac{p}{q}(\theta + 2k\pi) + i \sin \frac{p}{q}(\theta + 2k\pi)\right] \qquad k = 0, 1, \ldots, n -$$

The definition of z^α when α is not a rational number, however, must be postponed until Sec. 15.7.

---

† Named for the French mathematician Abraham Demoivre (1667–1754) although an equivalent form had been obtained earlier by the English mathematician Roger Cotes (1682–1716).

EXAMPLE 1

Find the fourth roots of $-8i$.

To do this, we first note that the angle, or amplitude, of $-8i$ is $3\pi/2$ and the length, or modulus, is 8. Hence, in standard polar form,

$$-8i = 8\left(\cos\frac{3\pi}{2} + i\sin\frac{3\pi}{2}\right)$$

From this, by applying Eq. (9), we find that the four fourth roots of $-8i$ are given by the expression

$$8^{1/4}\left[\cos\frac{1}{4}\left(\frac{3\pi}{2} + 2k\pi\right) + i\sin\frac{1}{4}\left(\frac{3\pi}{2} + 2k\pi\right)\right] \qquad k = 0, 1, 2, 3$$

or, explicitly,

$$r_1 = 8^{1/4}\left(\cos\frac{3\pi}{8} + i\sin\frac{3\pi}{8}\right) \qquad k = 0$$

$$r_2 = 8^{1/4}\left(\cos\frac{7\pi}{8} + i\sin\frac{7\pi}{8}\right) \qquad k = 1$$

$$r_3 = 8^{1/4}\left(\cos\frac{11\pi}{8} + i\sin\frac{11\pi}{8}\right) \qquad k = 2$$

$$r_4 = 8^{1/4}\left(\cos\frac{15\pi}{8} + i\sin\frac{15\pi}{8}\right) \qquad k = 3$$

The coefficient $8^{1/4}$ is, of course, the *real* fourth root of 8, the value of which is found by a simple logarithmic calculation to be 1.682.

EXAMPLE 2

Using Demoivre's theorem and the binomial expansion, express $\cos 4\theta$ and $\sin 4\theta$ in terms of powers of $\cos\theta$ and $\sin\theta$.

To do this we consider $(\cos\theta + i\sin\theta)^4$ and expand it first by Demoivre's theorem and then by the binomial expansion. This gives the identity

$$(\cos 4\theta + i\sin 4\theta) = \cos^4\theta + 4i\cos^3\theta\sin\theta + 6i^2\cos^2\theta\sin^2\theta$$
$$+ 4i^3\cos\theta\sin^3\theta + i^4\sin^4\theta$$
$$= (\cos^4\theta - 6\cos^2\theta\sin^2\theta + \sin^4\theta)$$
$$+ i(4\cos^3\theta\sin\theta - 4\cos\theta\sin^3\theta)$$

Equating real and imaginary parts of these equal complex expressions, we obtain the required formulas:

$$\cos 4\theta = \cos^4\theta - 6\cos^2\theta\sin^2\theta + \sin^4\theta$$
$$\sin 4\theta = 4(\cos^3\theta\sin\theta - \cos\theta\sin^3\theta)$$

EXERCISES

1 Show that multiplying a complex number by i rotates it through 90° without changing its length. Find the effect of multiplying a complex number by:
   (a) $-i$ (b) $\sqrt{i}$

2 A square lies entirely in the second quadrant. If one of its sides joins the points $z_1 = -3$ and $z_2 = 2i$, find the coordinates of the other two vertices.

3 Find all the distinct fourth roots of -1.

4 Find all the distinct fifth roots of 32.

5 Find all the distinct cube roots of i.

6 Express the complex number $8 - 8\sqrt{3}\,i$ in polar form and find its distinct fourth roots.

7 Find the distinct cube roots of $1 + i$ and reduce each to the form $a + ib$, where a and b are decimal fractions.

8 Find all the distinct values of $(1 - i)^{5/4}$.

9 Find all the distinct values of $(-1 - i)^{4/5}$.

10 Using Demoivre's theorem, express $\cos 5\theta$ and $\sin 5\theta$ in terms of powers of $\cos \theta$ and $\sin \theta$.

11 Show that if n is an integer, both $\cos n\theta$ and $(\sin n\theta)/(\sin \theta)$ can be expressed as polynomials in $\cos \theta$.

12 If z_1 and z_2 are complex numbers, what point is represented by the number $(z_1 + z_2)/2$? What is the locus of the points $\lambda z_1 + \mu z_2$, where λ and μ are real parameters and $\lambda + \mu = 1$?

13 Show that the centroid of a system of three particles of equal mass situated at the points z_1, z_2, z_3, respectively, is the point $(z_1 + z_2 + z_3)/3$. Where is the centroid of a system of three masses m_1, m_2, m_3 situated respectively at the points z_1, z_2, z_3?

14 Show that three points z_1, z_2, z_3 in the complex plane are collinear if and only if there exist real numbers p, q, r, not all zero, such that $p + q + r = 0$ and $pz_1 + qz_2 + rz_3 = 0$.

15 Show that the triangle whose vertices are the points z_1, z_2, z_3 and the triangle whose vertices are the points z_4, z_5, z_6 are similar if and only if

$$\begin{vmatrix} z_1 & z_4 & 1 \\ z_2 & z_5 & 1 \\ z_3 & z_6 & 1 \end{vmatrix} = 0$$

16 Show that if the points z_1, z_2, z_3 are the vertices of an equilateral triangle, then

$$z_1{}^2 + z_2{}^2 + z_3{}^2 = z_1 z_2 + z_2 z_3 + z_3 z_1$$

Is the converse true?

17 Using the polar form of the multiplication law, devise a geometrical construction for the product of two complex numbers.

18 Devise a geometrical construction for the quotient of two complex numbers.

15.4 Absolute Values

We have already defined the absolute value of a complex number z to be the length of the vector which represents z; that is,

$$|z| = \sqrt{x^2 + y^2} = \sqrt{\mathscr{R}^2(z) + \mathscr{I}^2(z)}$$

From this it is evident that *a complex number is zero if and only if its absolute value is zero*. Since both $\mathscr{R}^2(z)$ and $\mathscr{I}^2(z)$ are nonnegative real numbers, it is also clear, by dropping first one and then the other of these quantities from the last equation, that†

(1) $$|z| \geq \mathscr{R}(z)$$

(2) $$|z| \geq \mathscr{I}(z)$$

---

† We must always keep in mind the fact that the complex numbers cannot be ordered and that *greater than* and *less than* have meaning only when applied to real numbers (see Exercise 20, Sec. 15.2).

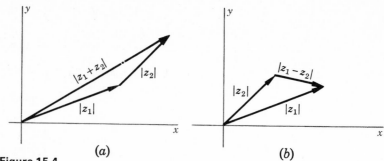

Figure 15.4
The triangle inequality applied to complex numbers.

Moreover, from the definition of conjugate complex numbers, it follows that

(3) $$|z| = |\bar{z}|$$

and

(4) $$|z\bar{z}| = |z|^2$$

Also, from Eqs. (4) and (5), Sec. 15.3, for the products and quotients of complex numbers expressed in polar form, it is clear that

(5) $$|z_1 z_2| = |z_1| \cdot |z_2|$$

and

(6) $$\left|\frac{z_1}{z_2}\right| = \frac{|z_1|}{|z_2|}$$

Since the length of any side of a triangle must be equal to or less than the sum of the lengths of the other two sides, it follows from the geometric addition of complex numbers (Fig. 15.4a) that

(7) $$|z_1 + z_2| \leq |z_1| + |z_2|$$

This can readily be extended to three terms, for

$$\begin{aligned} |z_1 + z_2 + z_3| &= |z_1 + (z_2 + z_3)| \\ &\leq |z_1| + |z_2 + z_3| \\ &\leq |z_1| + |z_2| + |z_3| \end{aligned}$$

The important extension to n terms is obvious:

(8) $$\left|\sum_{k=1}^{n} z_k\right| \leq \sum_{k=1}^{n} |z_k|$$

It is also geometrically evident that the length of any side of a triangle must be at least as great as the difference of the lengths of the other two sides (Fig. 15.4b). Hence

(9) $$|z_1 - z_2| \geq \left||z_1| - |z_2|\right| \geq 0$$

If it happens that $|z_1|$ is equal to or greater than $|z_2|$, the outer absolute-value signs on the right-hand side are, of course, unnecessary.

EXAMPLE 1

Describe the region in the z plane defined by the inequality $\mathscr{R}(z) > 1$.

If the real part of z is greater than 1, the image of z must be a point to the right of the line $x = 1$. Hence, the given inequality defines the set of all points in the half plane to the right of this line. Since the equality sign is not included in the definition of the region, points actually on the line $x = 1$ do not belong to the region.

EXAMPLE 2

What region in the z plane is defined by the inequality $|z - z_0| \leq 9$?

In words, the given inequality asserts that the distance between the image point of z and the fixed point which is the image of z_0 is equal to or less than 9. This clearly defines the set of all points within and on the circumference of the circle of radius 9 which has the image of z_0 as its center (Fig. 15.5). In the work that lies ahead, we shall frequently have to consider regions of this type.

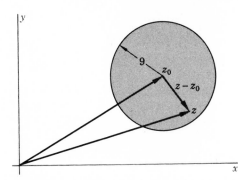

Figure 15.5
The circular region $|z - z_0| \leq 9$.

EXAMPLE 3

If $w = (z + i)/(iz + 1)$, show that the restriction $\mathscr{I}(z) \leq 0$ implies the restriction $|w| \leq 1$.

Since we are asked to establish a certain property of $|w|$, our first step is to compute this quantity. This can be done in various ways, but it is probably most convenient to construct the product

$$w\bar{w} = |w|^2 = \frac{z + i}{iz + 1} \overline{\left(\frac{z + i}{iz + 1}\right)}$$

Since the conjugate of a quotient is the quotient of the conjugates, this can be written as

$$|w|^2 = \frac{z + i}{iz + 1} \frac{\overline{z + i}}{\overline{iz + 1}}$$

Moreover, the conjugate of a sum is the sum of the conjugates; hence we have further

$$|w|^2 = \frac{z + i}{iz + 1} \frac{\bar{z} + \bar{i}}{\overline{iz} + 1}$$

Finally, since $\bar{i} = -i$ and $\overline{iz} = \bar{i}\bar{z} = -i\bar{z}$, we have

$$|w|^2 = \frac{z + i}{iz + 1} \frac{\bar{z} - i}{-i\bar{z} + 1} = \frac{(z\bar{z} + 1) - i(z - \bar{z})}{(z\bar{z} + 1) + i(z - \bar{z})} = \frac{z\bar{z} + 1 + 2\mathscr{I}(z)}{z\bar{z} + 1 - 2\mathscr{I}(z)}$$

Since $z\bar{z} + 1$ is a positive quantity, it is clear that if $\mathscr{I}(z) \leq 0$, as given, then the numerator of the last fraction is equal to or less than the denominator. Thus $|w|^2$, and hence $|w|$, is at most equal to 1 under the given conditions.

Since the restriction $\mathscr{I}(z) \leq 0$ implies that z lies on or below the real axis in the plane in which z is plotted, and since $|w| \leq 1$ implies that w lies on or inside the unit circle in the plane in which w is plotted, it follows that the given relation

$$w = \frac{z + i}{iz + 1}$$

can be thought of as a transformation, or mapping, which sends the lower half of the z plane, point by point, into the region consisting of the unit circle and its interior in the w plane. Mappings of this sort are of considerable importance in applied mathematics, and in Chap. 18 we shall examine their properties in more detail.

EXERCISES

1   If a and b are real, show that $|(a + ib)/(b + ia)| = 1$. Is this true if a and b are not real?

2   Determine $|z|$, $\mathscr{R}(z)$, and $\mathscr{I}(z)$ if

$$z = \frac{(3 + 4i)(12 - 5i)}{1 + i}$$

3   Under what conditions, if any, does $|z_1 + z_2| = |z_1| + |z_2|$?

4   Show that $|x + iy| \geq (|x| + |y|)/\sqrt{2}$. Under what conditions, if any, will the equality sign hold?

5   Show that $|z_1 - z_2|^2 + |z_1 + z_2|^2 = 2|z_1|^2 + 2|z_2|^2$.

6   Show that the locus of points for which $|(z - 1)/(z + 1)| = k$, where k is a positive constant different from 1, is a circle. What is the locus if $k = 1$? If $k = 0$? If $k < 0$?

7   What region in the z plane is defined by the inequalities $0 < \mathscr{R}(z) \leq \mathscr{I}(z)$?

8   What region in the z plane is defined by the inequality $|z - 1| \leq \mathscr{R}(z)$?

9   If $w = [i(1 - z)]/(1 + z)$, prove that $|z| < 1$ implies $\mathscr{I}(w) > 0$.

10   Without using any properties of the polar representation of complex numbers, prove that $|z_1 z_2| = |z_1| \cdot |z_2|$.

11   Prove algebraically that $|z_1 + z_2| \leq |z_1| + |z_2|$. *Hint:* Consider the identity $|z_1 + z_2|^2 = (z_1 + z_2)(\overline{z_1} + \overline{z_2})$ and at the appropriate points use Eq. (2), Sec. 15.2, and formula (1) of this section.

12   Prove algebraically that $|z_1 - z_2| \geq ||z_1| - |z_2|| \geq 0$.

13   Prove that if $z + 1/z$ is real, then either z is real or the absolute value of z is 1.

14   If $z_1, z_2, \ldots, z_n$ and $w_1, w_2, \ldots, w_n$ are complex numbers, prove that

$$\left| \sum_{k=1}^{n} z_k w_k \right|^2 \leq \sum_{k=1}^{n} |z_k|^2 \sum_{k=1}^{n} |w_k|^2$$

This result is sometimes known as **Cauchy's inequality.** *Hint:* Consider the discriminant of the quadratic equation in λ

$$\sum_{k=1}^{n} (|z_k|\lambda - |w_k|)^2 = 0$$

15   Consider the set of all complex numbers $z = x + iy$, the set of all matrices of the form $M = \begin{bmatrix} x & y \\ -y & x \end{bmatrix}$, and the one-to-one correspondence between these two sets defined by the association $x + iy \leftrightarrow \begin{bmatrix} x & y \\ -y & x \end{bmatrix}$. Using the usual definitions for the sum and product of two complex numbers and the sum and product of two matrices, show that if $z_j \leftrightarrow M_j$ and $z_k \leftrightarrow M_k$, then $z_j + z_k \leftrightarrow M_j + M_k$ and $z_j z_k \leftrightarrow M_j M_k$. If $z_k \leftrightarrow M_k$, show that the determinant of M_k is equal to the square of the absolute value of z_k. What are the characteristic values of M_k? (A one-to-one relation-preserving correspondence, like this one, is called an **isomorphism.**)

15.5 Functions of a Complex Variable

If $z = x + iy$ and $w = u + iv$ are two complex variables, and if for each value of z in some portion of the complex plane, one or more values of w are defined, then w is said to be a **function**† of z, and we write

$$w = f(z)$$

If $w = f(z)$, that is, if

$$u + iv = f(x + iy)$$

it follows that the real numbers u and v are themselves determined by the real numbers x and y. Hence, the assertion that w is a function of $z = x + iy$ can also be written

$$\text{(1)} \qquad\qquad w = u(x,y) + iv(x,y)$$

where $u(x,y)$ and $v(x,y)$ are real-valued functions of the real variables x and y. Clearly, whenever a value of z is given, values of x and y are thereby provided, and thus one or more values of w are determined by (1). For example, if

$$w = f(z) = (x^2 - y) + i(x + y^2)$$

and if $z = 1 + 2i$, then $x = 1$ and $y = 2$, and thus

$$f(1 + 2i) = (1^2 - 2) + i(1 + 2^2) = -1 + 5i$$

If w is defined as a function of z in the form (1), it may be possible by suitable manipulations to rearrange the expression $u(x,y) + iv(x,y)$ so that x and y occur only in the binomial combination $x + iy$. For instance,

$$w = (x^2 - y^2) + 2ixy$$

is immediately recognizable as

$$w = (x + iy)^2 = z^2$$

and

$$w = \frac{x}{x^2 + y^2} - i\,\frac{y}{x^2 + y^2}$$

is nothing but the standard complex form of

$$w = \frac{1}{x + iy} = \frac{1}{z}$$

On the other hand, it may be impossible to express w in a form involving only the explicit combination $x + iy$ without using such "artificial" expressions as $\mathcal{R}(z) \equiv x$ and $\mathcal{I}(z) \equiv y$, with which, of course, any formula in x and y can be written in terms of z. For instance, unless we resort to these artificial functions, no rearrangement of the formula

$$w = 7x + 3iy = 4\mathcal{R}(z) + 3z = 7z - 4i\mathcal{I}(z) = 5z + 2\bar{z}$$

---

† The best modern usage restricts the word *function* to the case in which to each value of the independent variable there corresponds a single value of the dependent variable. However, it is convenient, and not uncommon, in complex-variable theory to define a function as we have just done and to speak of both *single-valued functions* and *multiple-valued functions*.

can reduce w to explicit dependence on z alone. In our work, and in fact in most applications of complex-variable theory, the only functions of real interest will be those which can be written in terms of z alone, without recourse to $\bar{z}$, $\mathscr{R}(z)$, $\mathscr{I}(z)$, and similar expressions.

Frequently our interest in a function will be restricted to its behavior at the points of some specified part of the z plane. However, before we can undertake discussions of this sort, we must define and explain some of the simpler properties of the sets of points we intend to consider.

By a **neighborhood** of a point z_0 we mean any set consisting of all the points which satisfy an inequality of the form

$$|z - z_0| < \varepsilon \qquad \varepsilon > 0$$

Geometrically speaking, a neighborhood of z_0 thus consists of all the points within but not on a circle having z_0 as center. A point z_0 belonging to a set S is said to be an **interior point** of S if there exists at least one neighborhood of z_0 whose points all belong to S. A set each of whose points is an interior point is said to be **open**. A point z_0 not belonging to a set S is said to be **exterior** to S or an **exterior point** of S if there exists at least one neighborhood of z_0 none of whose points belongs to S. Intermediate between points interior to S and points exterior to S are the boundary points of S. A point z_0 is said to be a **boundary point** of a set S if every neighborhood of z_0 contains both points belonging to S and points not belonging to S. The boundary points of a set may or may not belong to the set, depending upon its definition.

A point z_0 is said to be a **limit point** of a set if every neighborhood of the point contains at least one point of the set distinct from z_0. A set which contains all its limit points is said to be **closed**. Obviously, a set can be defined to contain some but not all of its limit points; hence it is clear that a set may be neither open nor closed.

If a set S has the property that every pair of its points can be joined by a polygonal line whose points all belong to the set, it is said to be **connected**. An open connected set is said to be a **region** or a **domain**. A set consisting of a region together with all its limit points is called a **closed region**. A connected set S with the property that every simple closed curve† which can be drawn in its interior contains only points of S is said to be **simply connected**. If it is possible to draw in S at least one simple closed curve whose interior contains one or more points not belonging to S, then S is said to be **multiply connected**.‡ If there exists a circle with center at the origin enclosing all the points of a set S, that is, if there exists a number d such that

$$|z| < d \qquad \text{for all } z \text{ in } S$$

then S is said to be **bounded**. A set which is not bounded is said to be **unbounded**. The set consisting of the points between two concentric circles is said to be an **annular region** or an **annulus**.

---

† See the footnote to Theorem 1, Sec. 13.4.
‡ In two dimensions the definitions of simply connected sets and multiply connected sets given at the end of Sec. 13.5 are clearly equivalent to those of the present section.

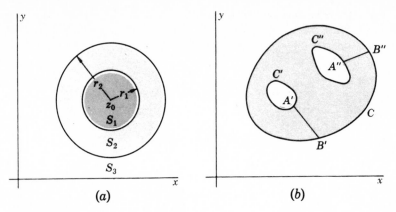

Figure 15.6
Typical regions in the complex plane.

The preceding ideas are illustrated in Fig. 15.6a, where the three sets

$$S_1: \qquad |z - z_0| < r_1$$
$$S_2: \quad r_1 \le |z - z_0| < r_2$$
$$S_3: \quad r_2 \le |z - z_0|$$

are shown. The set S_1 consists of all points interior to the circle $|z - z_0| = r_1$. It is bounded and simply connected. Since points on the boundary circle $|z - z_0| = r_1$ are not included in the definition of S_1, the set is open and is therefore a domain. In particular, it is a neighborhood of z_0. The set S_2 consists of all the points in the annulus between the circles $|z - z_0| = r_1$ and $|z - z_0| = r_2$ plus the points on the inner boundary of the annulus but not those on the outer boundary. Since S_2 thus contains some but not all of its boundary points, it is neither open nor closed and is therefore neither a domain nor a closed region. Clearly, there are closed curves in S_2, namely, any curve encircling the inner boundary, which enclose points not belonging to S_2, namely, the points of S_1. Hence S_2 is multiply connected. Obviously, S_2 is bounded. The set S_3 consists of all points on and outside the circle $|z - z_0| = r_2$. It is therefore unbounded, closed, and multiply connected.

Because simply connected regions are in many respects easier to work with than multiply connected regions, it is often desirable to be able to reduce the latter to the former. This can always be done by modifying the given multiply connected region through the introduction of auxiliary boundary arcs, or **crosscuts**, joining boundary curves that were originally disconnected. The effectiveness of this technique is illustrated in Fig. 15.6b, which shows a closed region originally multiply connected with one outer boundary curve C and two inner boundary curves C' and C''. The introduction of the auxiliary boundary arcs $A'B'$ and $A''B''$ clearly makes it impossible to draw closed curves which lie entirely in the *interior* of the modified region and at the same time encircle either of the inner boundaries C' and C''. The modified region is therefore simply connected, as desired.

It will often be necessary for us to consider the **limit of a function of** z as z approaches some particular value z_0. The basis for this is the following definition.

DEFINITION 1 If $f(z)$ is a single-valued function of z and w_0 is a complex constant, and if for every $\varepsilon > 0$ there exists a positive number $\delta(\varepsilon)$ such that $|f(z) - w_0| < \varepsilon$ for all z such that $0 < |z - z_0| < \delta$, then w_0 is said to be the limit of $f(z)$ as z approaches z_0.

In less technical terms, w_0 is the limit of $f(z)$ as z approaches z_0 provided that $f(z)$ can be kept arbitrarily close to w_0 by keeping z sufficiently close to, but distinct from, z_0.

EXAMPLE 1

If

$$f(z) = \frac{(x + y)^2}{x^2 + y^2}$$

show that

$$\lim_{x \to 0} [\lim_{y \to 0} f(z)] = 1 \quad \text{and} \quad \lim_{y \to 0} [\lim_{x \to 0} f(z)] = 1$$

but that $\lim_{z \to 0} f(z)$ does not exist.

Clearly,

$$\lim_{x \to 0} [\lim_{y \to 0} f(z)] = \lim_{x \to 0} \left[\lim_{y \to 0} \frac{(x + y)^2}{x^2 + y^2} \right] = \lim_{x \to 0} (1) = 1$$

and

$$\lim_{y \to 0} [\lim_{x \to 0} f(z)] = \lim_{y \to 0} \left[\lim_{x \to 0} \frac{(x + y)^2}{x^2 + y^2} \right] = \lim_{y \to 0} (1) = 1$$

as asserted. On the other hand, for $\lim_{z \to 0} f(z)$ to exist, it is necessary that $f(z)$ approach the same value along all paths leading to the origin, and this is not the case; for along the paths $y = mx$ we have

$$\lim_{z \to 0} f(z) = \lim_{z \to 0} \frac{(x + y)^2}{x^2 + y^2} = \lim_{x \to 0} \frac{(1 + m)^2}{1 + m^2} = \frac{(1 + m)^2}{1 + m^2}$$

The limiting value here clearly depends on m; that is, $f(z)$ approaches different values along different radial lines, and hence no limit exists.

Closely associated with the concept of a limit is the concept of **continuity**.

DEFINITION 2 The function $f(z)$ is continuous at the point z_0 provided that $\lim_{z \to z_0} f(z) = f(z_0)$.

In other words, for a function to be continuous at a point z_0, the function must have both a value at that point and a limit as z approaches that point and the two must be equal. If $f(z)$ is continuous at every point of a region, it is said to be **continuous throughout that region**.

In addition to the fundamental theorems on limits that we encountered in calculus, there are various theorems on continuous functions which we shall need from time to time. For the most part these appear almost self-evident, although their proofs are by no means trivial. We shall merely list them here and refer to standard texts on advanced calculus for their proofs.†

---

† See, for instance, A. E. Taylor, "Advanced Calculus," pp. 494–503, Ginn, Boston, 1955.

THEOREM 1 Sums, differences, and products of continuous functions, and quotients of continuous functions provided the divisor functions are different from zero, are continuous.

THEOREM 2 A continuous function of a continuous function is a continuous function.

THEOREM 3 A necessary and sufficient condition that

$$f(z) = u(x,y) + iv(x,y)$$

be continuous is that the real functions $u(x,y)$ and $v(x,y)$ be continuous.

THEOREM 4 If $f(z)$ is continuous at a point z_0, and if $f(z_0) \neq 0$, then there exists a neighborhood of z_0 throughout which $f(z)$ is different from 0.

THEOREM 5 If $f(z)$ is continuous over a bounded, closed region R, then there exists a positive constant M such that $|f(z)| < M$ for all values of z in R.

EXERCISES

1   If $f(z) = xy + i(x^2 - y^2)$, what is $f(-1 + 2i)$?

2   If $f(z) = z + (\bar{z})^2 + \mathscr{I}(z\bar{z})$, what is $f(2 + i)$?

3   Express $(2xy + 2x - 1) - i(x^2 - y^2 - 2y)$ as a polynomial in the binomial argument $z = x + iy$.

4   Express $x^2 + iy^2$ in terms of z and $\bar{z}$.

5   Describe each of the following sets of points, telling whether it is bounded or unbounded, open or closed, and simply or multiply connected:

    (a) $\mathscr{I}(z) > 0$ (b) $2 \leq |z| \leq 3$ (c) $|z - 1| > 4$

    (d) $0 \leq \mathscr{R}(z) \leq 1$ (e) $|z^2 - 1| \leq \frac{3}{4}$ (f) $0 \leq \mathscr{I}(z) < \mathscr{R}(z)$

6   Using Definition 1, show that:

    (a) $\lim\limits_{z \to 2i} \dfrac{z^2 + 4}{z - 2i} = 4i$ (b) $\lim\limits_{z \to 1} \dfrac{2z}{z + i} = 1 - i$

7   Determine at what points, if any, each of the following functions fails to be continuous and explain why:

    (a) $f(z) = \dfrac{z}{z^2 + 1}$ (b) $f(z) = \dfrac{z^2 - 1}{z - 1}$ (c) $f(z) = (x + y^2) + ixy$

    (d) $f(z) = \begin{cases} \dfrac{z^2 + 3iz - 2}{z + i} & z \neq -i \\ i & z = -i \end{cases}$ (e) $f(z) = \begin{cases} z^2 + iz + 2 & z \neq i \\ i & z = i \end{cases}$

8   Show that $\lim\limits_{z \to 0} [xy/(x^2 + y^2)]$ does not exist.

9   Show that $\lim\limits_{z \to 0} [x^2 y/(x^4 + y^2)]$ does not exist even though this function approaches the same limit along every straight line through the origin.

10   If

$$f(z) = \begin{cases} x \sin \dfrac{1}{y} & y \neq 0 \\ 0 & y = 0 \end{cases}$$

show that $\lim\limits_{y \to 0}[\lim\limits_{x \to 0} f(z)]$ and $\lim\limits_{z \to 0} f(z)$ exist and are equal, but that $\lim\limits_{x \to 0}[\lim\limits_{y \to 0} f(z)]$ does not exist.

11 Show that every neighborhood of a limit point of a set S contains infinitely many points of S.

12 Prove Theorem 2.

15.6 Analytic Functions

The **derivative of a function of a complex variable** $w = f(z)$ is defined to be

(1)
$$\frac{dw}{dz} = w' = f'(z) = \lim_{\Delta z \to 0} \frac{f(z + \Delta z) - f(z)}{\Delta z}$$

This definition is formally identical with that for the derivative of a function of a real variable. Moreover, since the general theory of limits is phrased in terms of absolute values, it is valid for complex variables as well as for real variables. Hence it is clear that formulas for the differentiation of functions of a real variable will have identical counterparts in the field of complex numbers when the corresponding functions of a complex variable are suitably defined. In particular, such familiar formulas as

$$\frac{d(w_1 \pm w_2)}{dz} = \frac{dw_1}{dz} \pm \frac{dw_2}{dz}$$

$$\frac{d(w_1 w_2)}{dz} = w_1 \frac{dw_2}{dz} + w_2 \frac{dw_1}{dz}$$

$$\frac{d(w_1/w_2)}{dz} = \frac{w_2(dw_1/dz) - w_1(dw_2/dz)}{w_2{}^2} \qquad w_2 \neq 0$$

$$\frac{d(w^n)}{dz} = nw^{n-1} \frac{dw}{dz}$$

are valid when w_1, w_2, and w are differentiable functions of a complex variable z. However, $\Delta z = \Delta x + i\,\Delta y$ is itself a complex variable, and the question of just how it is to approach zero involves difficulties which have no counterpart in the differentiation of functions of a real variable.

In Fig. 15.7, it is clear that Δz can approach zero in infinitely many ways; i.e., a point $Q: z + \Delta z$ can approach the point $P: z$ along infinitely many different paths. In particular, Q can approach P along the line AP on which Δx is zero or along the line BP on which Δy is zero. Clearly, *for the derivative of $f(z)$ to exist, it is necessary*

Figure 15.7
Various ways in which Δz can approach zero.

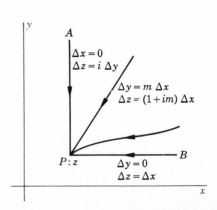

that the limit of the difference quotient (1) *be the same no matter how* Δz *approaches zero.* How severe a restriction this is can be seen by considering the simple function

$$w = f(z) = \bar{z} = x - iy$$

Giving to z the increment $\Delta z = \Delta x + i\,\Delta y$ means that x changes by the amount Δx and y changes by the amount Δy. Hence, for the given function $x - iy$, we have

$$\frac{f(z + \Delta z) - f(z)}{\Delta z} = \frac{[(x + \Delta x) - i(y + \Delta y)] - (x - iy)}{\Delta x + i\,\Delta y} = \frac{\Delta x - i\,\Delta y}{\Delta x + i\,\Delta y}$$

Now if Δz is real, so that $\Delta y = 0$, we have

$$\lim_{\Delta z \to 0} \frac{\Delta x - i\,\Delta y}{\Delta x + i\,\Delta y} = \lim_{\Delta x \to 0} \frac{\Delta x}{\Delta x} = 1$$

On the other hand, if Δz is imaginary, so that $\Delta x = 0$, we have

$$\lim_{\Delta z \to 0} \frac{\Delta x - i\,\Delta y}{\Delta x + i\,\Delta y} = \lim_{\Delta y \to 0} \frac{-i\,\Delta y}{i\,\Delta y} = -1$$

More generally, if we let Δz approach zero in such a way that $\Delta y = m\,\Delta x$, we have

$$\lim_{\Delta z \to 0} \frac{\Delta x - i\,\Delta y}{\Delta x + i\,\Delta y} = \lim_{\Delta x \to 0} \frac{\Delta x - im\,\Delta x}{\Delta x + im\,\Delta x} = \frac{1 - im}{1 + im} = \frac{(1 - m^2) - 2im}{1 + m^2}$$

Thus there are infinitely many complex values which the difference quotient for $f(z) = x - iy$ can be made to approach by suitably choosing the manner in which Δz is to approach zero. It is therefore clear that $\bar{z} = x - iy$ has no derivative.

That a function as simple as $f(z) = x - iy$ should have no derivative seems at first glance a discouraging state of affairs. However, there are many functions of z which do have derivatives, and in applications it is these functions which are of importance. Our immediate task is to identify these functions by obtaining conditions for the existence of the derivative of a function of a complex variable.

To do this, consider

$$w = f(z) = u(x,y) + iv(x,y)$$

By definition

$$\frac{dw}{dz} = \lim_{\Delta z \to 0} \frac{\Delta w}{\Delta z}$$

$$(2) \qquad = \lim_{\substack{\Delta x \to 0 \\ \Delta y \to 0}} \frac{[u(x + \Delta x, y + \Delta y) + iv(x + \Delta x, y + \Delta y)] - [u(x,y) + iv(x,y)]}{\Delta x + i\,\Delta y}$$

Now if Δz is real, i.e., if $\Delta y = 0$, we obtain

$$\frac{dw}{dz} = \lim_{\Delta x \to 0} \frac{[u(x + \Delta x, y) + iv(x + \Delta x, y)] - [u(x,y) + iv(x,y)]}{\Delta x}$$

$$= \lim_{\Delta x \to 0} \left[\frac{u(x + \Delta x, y) - u(x,y)}{\Delta x} + i\,\frac{v(x + \Delta x, y) - v(x,y)}{\Delta x} \right]$$

The two difference quotients which appear in the last expression are precisely those whose limits define the partial derivatives of u and v with respect to x. Hence it appears that

$$(3) \qquad \frac{dw}{dz} = \frac{\partial u}{\partial x} + i \frac{\partial v}{\partial x}$$

On the other hand, if Δz is imaginary, i.e., if $\Delta x = 0$, we find from (2) that

$$\frac{dw}{dz} = \lim_{\Delta y \to 0} \frac{[u(x, y + \Delta y) + iv(x, y + \Delta y)] - [u(x,y) + iv(x,y)]}{i \, \Delta y}$$

$$= \lim_{\Delta y \to 0} \left[\frac{u(x, y + \Delta y) - u(x,y)}{i \, \Delta y} + i \frac{v(x, y + \Delta y) - v(x,y)}{i \, \Delta y} \right]$$

$$= \frac{1}{i} \frac{\partial u}{\partial y} + \frac{\partial v}{\partial y}$$

or, finally,

$$(4) \qquad \frac{dw}{dz} = \frac{\partial v}{\partial y} - i \frac{\partial u}{\partial y}$$

Obviously, if the derivative dw/dz is to exist, it is necessary that the two expressions we have just derived for it should be the same. Hence, from (3) and (4),

$$\frac{\partial u}{\partial x} + i \frac{\partial v}{\partial x} = \frac{\partial v}{\partial y} - i \frac{\partial u}{\partial y}$$

which requires that

$$(5a) \qquad \frac{\partial u}{\partial x} = \frac{\partial v}{\partial y}$$

$$(5b) \qquad \frac{\partial u}{\partial y} = -\frac{\partial v}{\partial x}$$

These two extremely important conditions, which are known as the **Cauchy-Riemann equations,**[†] have arisen here from a consideration of only two of the infinitely many ways in which Δz can approach zero. It is therefore natural to expect that severe additional conditions will be necessary to ensure that along these other paths $\Delta w/\Delta z$ will also approach the same limit dw/dz. This is not the case, however, and it can be proved without great difficulty[‡] that if u and v, together with their first partial derivatives u_x, u_y, v_x, v_y are continuous in some neighborhood of the point z_0, then the Cauchy-Riemann equations are not only necessary but also sufficient conditions for the existence of the derivative of $w = u(x,y) + iv(x,y)$ at $z = z_0$.

If $w = f(z)$ possesses a derivative at $z = z_0$ and at every point in some neighborhood of z_0, then $f(z)$ is said to be **analytic** at z_0 and z_0 is called a **regular point** of the function. If $f(z)$ is not analytic at z_0, but if every neighborhood of z_0 contains points at which $f(z)$ is analytic, then z_0 is called a **singular point** of $f(z)$. A function

---

† After the French mathematician Augustin Louis Cauchy (1789–1857), and the German mathematician Georg Friedrich Bernhard Riemann (1826–1866).
‡ See, for instance, Einar Hille, "Analytic Function Theory," vol. 1, pp. 78–80, Ginn, Boston, 1959.

analytic at every point of a region R we shall call **analytic in** R. Although most writers use this term, a few substitute such adjectives as **regular** and **holomorphic**. As a summary of our discussion we have the following theorem.

THEOREM 1 If u and v are real single-valued functions of x and y which, with their four first partial derivatives, are continuous throughout a region R, then the Cauchy-Riemann equations

$$\frac{\partial u}{\partial x} = \frac{\partial v}{\partial y} \quad \text{and} \quad \frac{\partial u}{\partial y} = -\frac{\partial v}{\partial x}$$

are both necessary and sufficient conditions that $f(z) = u(x,y) + iv(x,y)$ be analytic in R. In this case, the derivative of $f(z)$ is given by either

$$f'(z) = \frac{\partial u}{\partial x} + i\frac{\partial v}{\partial x} \quad \text{or} \quad f'(z) = \frac{\partial v}{\partial y} - i\frac{\partial u}{\partial y}$$

EXAMPLE 1

For $f(z) = \bar{z} = x - iy$, we have $u = x$ and $v = -y$. In this case

$$\frac{\partial u}{\partial x} = 1 \qquad \frac{\partial u}{\partial y} = 0 \qquad \frac{\partial v}{\partial x} = 0 \qquad \frac{\partial v}{\partial y} = -1$$

and although the second of the Cauchy-Riemann equations is satisfied everywhere, the first is nowhere satisfied. Hence, there is no point in the z plane where $f'(z)$ exists, which, of course, confirms our earlier investigation of this function.

EXAMPLE 2

For $f(z) = z\bar{z} = x^2 + y^2$, we have $u = x^2 + y^2$ and $v = 0$. In this case the four first partial derivatives

$$\frac{\partial u}{\partial x} = 2x \qquad \frac{\partial u}{\partial y} = 2y \qquad \frac{\partial v}{\partial x} = 0 \qquad \frac{\partial v}{\partial y} = 0$$

are continuous everywhere. However, the Cauchy-Riemann equations, which in this case are, respectively,

$$2x = 0 \quad \text{and} \quad 2y = 0$$

are satisfied only at the origin. Hence $z = 0$ is the only point at which $f'(z)$ exists, and therefore $f(z) = z\bar{z}$ is nowhere analytic.

EXAMPLE 3

For $f(z) = z^2 = (x^2 - y^2) + 2ixy$, we have

$$\frac{\partial u}{\partial x} = 2x \qquad \frac{\partial u}{\partial y} = -2y \qquad \frac{\partial v}{\partial x} = 2y \qquad \frac{\partial v}{\partial y} = 2x$$

and, clearly, the Cauchy-Riemann equations are identically satisfied. Moreover, the first partial derivatives of u and v are everywhere continuous. Hence the derivative $f'(z)$ exists at all points of the z plane, and its value from either (3) or (4) is

$$f'(z) = 2x + 2iy = 2z$$

This, of course, is exactly what formal differentiation of z^2 according to the power rule would give.

Analytic functions have a great many important properties, many of which we shall investigate in later sections. At this point we note only the following.

PROPERTY 1   If both the real part and the imaginary part of an analytic function have continuous second partial derivatives, then they satisfy Laplace's equation

$$\frac{\partial^2 \phi}{\partial x^2} + \frac{\partial^2 \phi}{\partial y^2} = 0$$

Proof   Let $w = u(x,y) + iv(x,y)$ be an analytic function of z. Then u and v must satisfy the Cauchy-Riemann equations, namely,

$$\frac{\partial u}{\partial x} = \frac{\partial v}{\partial y} \quad \text{and} \quad \frac{\partial u}{\partial y} = -\frac{\partial v}{\partial x}$$

If we differentiate the first of these with respect to x and the second with respect to y and add the results, we obtain our first assertion:

$$\frac{\partial^2 u}{\partial x^2} = \frac{\partial^2 v}{\partial x\, \partial y}$$

$$\frac{\partial^2 u}{\partial y^2} = -\frac{\partial^2 v}{\partial y\, \partial x}$$

$$\overline{\qquad\qquad\qquad\qquad}$$

$$\frac{\partial^2 u}{\partial x^2} + \frac{\partial^2 u}{\partial y^2} = 0$$

In exactly the same way we can establish our second assertion, namely, that v also satisfies Laplace's equation.

The existence of the second partial derivatives and their continuity, which makes the order of differentiation in the cross partial derivatives immaterial, must here be assumed. Later we shall show that an analytic function possesses not only a first derivative but derivatives of *all* orders, which implies the existence and continuity of all the partial derivatives of u and v.

A function which possesses continuous second partial derivatives and satisfies Laplace's equation is usually called a **harmonic function**. Two harmonic functions u and v so related that $u + iv$ is an analytic function are called **conjugate harmonic functions**.† This use of the word *conjugate* must not be confused with its use in describing $\bar{z}$, the complex number conjugate to z.

PROPERTY 2   If $w = u(x,y) + iv(x,y)$ is an analytic function of z, then the curves of the family $u(x,y) = c$ are orthogonal trajectories of the curves of the family $v(x,y) = k$, and vice versa.

---

† The order in the pair (u,v) is important, as Exercise 6 makes clear.

Proof To prove this, we compute the slope of the general curve of each family by implicit differentiation, getting for the curves $u(x, y) = c$ the expression

(6)
$$\frac{dy}{dx} = -\frac{\partial u/\partial x}{\partial u/\partial y}$$

and for the curves $v(x, y) = k$ the expression

(7)
$$\frac{dy}{dx} = -\frac{\partial v/\partial x}{\partial v/\partial y}$$

By hypothesis, $w = u + iv$ is an analytic function. Hence it follows from Theorem 1 that u and v satisfy the Cauchy-Riemann equations. When these equations are used, expression (7) for the slope of the general curve of the family $v(x, y) = k$ can be rewritten

$$\frac{dy}{dx} = \frac{\partial u/\partial y}{\partial u/\partial x}$$

which, at any common point, is just the negative reciprocal of the slope of the general curve of the family $u(x, y) = c$, as given by Eq. (6). This suffices to prove that the two families of curves are orthogonal trajectories, as asserted.

PROPERTY 3 If in any analytic function $w = u(x, y) + iv(x, y)$, the variables x and y are replaced by their equivalents in terms of z and $\bar{z}$, namely,

$$x = \frac{z + \bar{z}}{2} \qquad \text{and} \qquad y = \frac{z - \bar{z}}{2i}$$

then w will appear as a function of z alone.

Proof Although z and $\bar{z}$ are clearly dependent, since either is determined when the other is given, we can regard w, by virtue of the given substitutions, as formally a function of two new independent variables z and $\bar{z}$. To show that w depends only on z and does not involve $\bar{z}$, it is sufficient to compute $\partial w/\partial \bar{z}$ and verify that it is identically zero. By the chain rule for partial derivatives,

$$\frac{\partial w}{\partial \bar{z}} = \frac{\partial(u + iv)}{\partial \bar{z}} = \frac{\partial u}{\partial \bar{z}} + i\frac{\partial v}{\partial \bar{z}} = \left(\frac{\partial u}{\partial x}\frac{\partial x}{\partial \bar{z}} + \frac{\partial u}{\partial y}\frac{\partial y}{\partial \bar{z}}\right) + i\left(\frac{\partial v}{\partial x}\frac{\partial x}{\partial \bar{z}} + \frac{\partial v}{\partial y}\frac{\partial y}{\partial \bar{z}}\right)$$

Moreover, from the equations expressing x and y in terms of z and $\bar{z}$, we have

$$\frac{\partial x}{\partial \bar{z}} = \frac{1}{2} \qquad \text{and} \qquad \frac{\partial y}{\partial \bar{z}} = -\frac{1}{2i} = \frac{i}{2}$$

Hence we can write

$$\frac{\partial w}{\partial \bar{z}} = \left(\frac{1}{2}\frac{\partial u}{\partial x} + \frac{i}{2}\frac{\partial u}{\partial y}\right) + i\left(\frac{1}{2}\frac{\partial v}{\partial x} + \frac{i}{2}\frac{\partial v}{\partial y}\right) = \frac{1}{2}\left(\frac{\partial u}{\partial x} - \frac{\partial v}{\partial y}\right) + \frac{i}{2}\left(\frac{\partial u}{\partial y} + \frac{\partial v}{\partial x}\right)$$

Since w, by hypothesis, is an analytic function, u and v satisfy the Cauchy-Riemann equations. Hence each of the quantities in parentheses in the last expression is equal to zero. Thus $\partial w/\partial \bar{z} \equiv 0$; and therefore w is independent of $\bar{z}$; that is, w depends on x and y only through the combination $z = x + iy$, as asserted.

1   Show that at no point in the z plane does the derivative of $f(z) = \mathcal{R}(z) = x$ exist. Does this contradict the fact that according to the rules of calculus $dx/dx = 1$?

2   At what points does $f(z) = (z - 2)/[(z + 1)(z^2 + 1)]$ fail to be analytic?

3   Where are the Cauchy-Riemann equations satisfied by the function $f(z) = xy^2 + ix^2y$? Where does $f'(z)$ exist? Where is $f(z)$ analytic?

4   Verify by direct substitution that $\mathcal{R}(z^3)$ and $\mathcal{I}(z^3)$ satisfy Laplace's equation.

5   If $u + iv$ is an analytic function, under what conditions, if any, will $v + iu$ be analytic?

6   If u and v are conjugate harmonic functions, show that v and $-u$ as well as $-v$ and u are also conjugate harmonic functions but that v and u are not.

7   Does there exist an analytic function for which $u = x^2 + y$?

8   Show that $u = e^x \cos y$ is a harmonic function and determine v so that $u + iv$ is an analytic function. *Hint:* Use the first of the Cauchy-Riemann equations to determine $\partial v/\partial y$, then integrate this expression with respect to y, and finally use the second Cauchy-Riemann equation to determine the arbitrary function of x introduced by this integration.

9   If u and v are harmonic in a region R, show that

$$\left(\frac{\partial u}{\partial y} - \frac{\partial v}{\partial x}\right) + i\left(\frac{\partial u}{\partial x} + \frac{\partial v}{\partial y}\right)$$

is analytic in R.

10   Is the converse of Property 2 true, i.e., if $u(x,y) = c$ and $v(x,y) = k$ are orthogonal trajectories, is $u + iv$ necessarily an analytic function?

11   Show that the various values approached by the difference quotient of $f(z) = \bar{z}$ as $\Delta z \to 0$ along the lines $y = mx$ all lie on a circle.

12   Prove that if $f'(z) \equiv 0$, then $f(z)$ is a constant.

13   If both $f(z)$ and $\overline{f(z)}$ are analytic functions, show that $f(z)$ is a constant.

14   If $f(z)$ is an analytic function for which $u^2 + v^2$ is a constant, show that $f(z)$ is a constant.

15   Prove L'Hospital's rule for analytic functions: if $f(z)$ and $g(z)$ are analytic functions in a region containing z_0, if $f(z_0) = g(z_0) = 0$, and if $g'(z_0) \neq 0$, then

$$\lim_{z \to z_0} \frac{f(z)}{g(z)} = \frac{f'(z_0)}{g'(z_0)}$$

16   If $f(z)$ is an analytic function, show that

(a) $\left[\dfrac{\partial}{\partial x}|f(z)|\right]^2 + \left[\dfrac{\partial}{\partial y}|f(z)|\right]^2 = |f'(z)|^2$

(b) $\left(\dfrac{\partial^2}{\partial x^2} + \dfrac{\partial^2}{\partial y^2}\right)|f(z)|^2 = 4|f'(z)|^2$

17   If the analytic function $f(z) = u(x,y) + iv(x,y)$ is expressed in terms of the polar coordinates r and θ, show that

$$\frac{\partial u}{\partial r} = \frac{1}{r}\frac{\partial v}{\partial \theta} \qquad \text{and} \qquad \frac{\partial v}{\partial r} = -\frac{1}{r}\frac{\partial u}{\partial \theta}$$

18   If the analytic function $f(z) = u(x,y) + iv(x,y)$ is expressed in terms of the polar coordinates r and θ, show that

$$f'(z) = (\cos\theta - i\sin\theta)\frac{\partial f}{\partial r} = -\frac{\sin\theta + i\cos\theta}{r}\frac{\partial f}{\partial \theta}$$

19   Using the results of Exercise 17, show that when an analytic function is expressed in terms of polar coordinates, both its real part and its imaginary part satisfy Laplace's equation in polar coordinates,

$$\frac{\partial^2 \phi}{\partial r^2} + \frac{1}{r}\frac{\partial \phi}{\partial r} + \frac{1}{r^2}\frac{\partial^2 \phi}{\partial \theta^2} = 0$$

20 Show that at the origin the function

$$f(z) = \begin{cases} \dfrac{x^3 - y^3}{x^2 + y^2} + i\,\dfrac{x^3 + y^3}{x^2 + y^2} & z \neq 0 \\ 0 & z = 0 \end{cases}$$

satisfies the Cauchy-Riemann equations but does not have a derivative. Explain. *Hint:* Compute the first partial derivatives of u and v at the origin by determining the limits of the appropriate difference quotients.

15.7 The Elementary Functions of z

The exponential function e^z is of fundamental importance, not only for its own sake but also as a basis for defining all the other elementary transcendental functions. In its definition we seek to preserve as many of the familiar properties of the real exponential function e^x as possible. Specifically, we desire that

a. e^z shall be single-valued and analytic.
b. $de^z/dz = e^z$.
c. e^z shall reduce to e^x when $\mathcal{I}(z) = 0$.

If we let

(1) $$e^z = u + iv$$

and recall from Eq. (3), Sec. 15.6, that the derivative of an analytic function can be written in the form

$$f'(z) = \frac{\partial u}{\partial x} + i\,\frac{\partial v}{\partial x}$$

then, to satisfy condition **b**, we must have

$$\frac{\partial u}{\partial x} + i\,\frac{\partial v}{\partial x} = u + iv$$

Hence, equating real and imaginary parts, we must have

(2) $$\frac{\partial u}{\partial x} = u$$

(3) $$\frac{\partial v}{\partial x} = v$$

Now Eq. (2) will be satisfied if

(4) $$u = e^x \phi(y)$$

where $\phi(y)$ is any function of y. Furthermore, since e^z is to be analytic (condition **a**), u and v must satisfy the Cauchy-Riemann equations. Hence, by using the second of these equations Eq. (3) can be written

(5) $$-\frac{\partial u}{\partial y} = v$$

Differentiating this with respect to y, we obtain

$$\frac{\partial^2 u}{\partial y^2} = -\frac{\partial v}{\partial y}$$

or, replacing $\partial v/\partial y$ with $\partial u/\partial x$ according to the first of the Cauchy-Riemann equations,

$$\frac{\partial^2 u}{\partial y^2} = -\frac{\partial u}{\partial x}$$

Finally, when we use (2), this becomes

$$\frac{\partial^2 u}{\partial y^2} = -u$$

which, on substituting $u = e^x \phi(y)$ from (4), reduces to

$$e^x \phi''(y) = -e^x \phi(y) \qquad \text{or} \qquad \phi''(y) = -\phi(y)$$

This is a simple linear differential equation whose solution can be written down at once:

$$\phi(y) = A \cos y + B \sin y$$

Hence, from (4),

$$u = e^x \phi(y) = e^x (A \cos y + B \sin y)$$

and, from (5),

$$v = -\frac{\partial u}{\partial y} = -e^x(-A \sin y + B \cos y)$$

Therefore, from (1),

$$e^z = u + iv = e^x[(A \cos y + B \sin y) + i(A \sin y - B \cos y)]$$

Finally, if this is to reduce to e^x when $y = 0$, as required by condition **c**, we must have

$$e^x = e^x(A - iB)$$

which will be true if and only if $A = 1$ and $B = 0$.

Thus we have been led inevitably to the conclusion that *if* there is a function of z satisfying the conditions **a, b,** and **c,** *then* it must be

(6) $$e^z = e^{x+iy} = e^x(\cos y + i \sin y)$$

That this expression does indeed meet our requirements can be checked immediately; hence we adopt it as the definition of e^z.

It is important to note that the right-hand side of (6) is in standard polar form. Hence,

$$\text{mod } e^z = |e^z| = e^x \qquad \text{and} \qquad \text{arg } e^z = y$$

The possibility of writing any complex number in exponential form is now apparent, for, applying (6), with $x = 0$ and $y = 0$, we have

(7) $$\cos \theta + i \sin \theta = e^{i\theta}$$

and thus

(8) $$r(\cos \theta + i \sin \theta) = re^{i\theta}$$

The fact that the angle, or argument, of a complex number is actually an exponent explains why the angles of complex numbers are added when the numbers are multiplied and subtracted when the numbers are divided, as we discovered in Sec. 15.3.

From the relation

$$e^{i\theta} = \cos\theta + i\sin\theta$$

and its obvious companion

$$e^{-i\theta} = \cos(-\theta) + i\sin(-\theta) = \cos\theta - i\sin\theta$$

we obtain, by addition and subtraction, the so-called **Euler formulas**

$$\cos\theta = \frac{e^{i\theta} + e^{-i\theta}}{2} \quad \text{and} \quad \sin\theta = \frac{e^{i\theta} - e^{-i\theta}}{2i}$$

On the basis of these equations, we extend the definitions of the sine and cosine into the complex domain by the formulas

$$(9) \qquad\qquad \cos z = \frac{e^{iz} + e^{-iz}}{2}$$

$$(10) \qquad\qquad \sin z = \frac{e^{iz} - e^{-iz}}{2i}$$

From these definitions it is easy to establish the validity of such familiar formulas as

$$\cos^2 z + \sin^2 z = 1$$

$$\cos(z_1 \pm z_2) = \cos z_1 \cos z_2 \mp \sin z_1 \sin z_2$$

$$\sin(z_1 \pm z_2) = \sin z_1 \cos z_2 \pm \cos z_1 \sin z_2$$

$$\frac{d(\cos z)}{dz} = -\sin z$$

$$\frac{d(\sin z)}{dz} = \cos z$$

If we expand the exponentials in (9), we find

$$\cos z = \frac{e^{i(x+iy)} + e^{-i(x+iy)}}{2}$$

$$= \frac{e^{-y}e^{ix} + e^{y}e^{-ix}}{2}$$

$$= \frac{e^{-y}(\cos x + i\sin x) + e^{y}(\cos x - i\sin x)}{2}$$

$$= \cos x\, \frac{e^{y} + e^{-y}}{2} - i\sin x\, \frac{e^{y} - e^{-y}}{2}$$

or, using the usual definitions of the hyperbolic functions of real variables,

$$(11) \qquad \cos z = \cos(x + iy) = \cos x \cosh y - i\sin x \sinh y$$

Similarly, it is easy to show that

$$(12) \qquad \sin z = \sin(x + iy) = \sin x \cosh y + i\cos x \sinh y$$

In particular, taking $x = 0$ in (11) and (12), we find

(13) $$\cos iy = \cosh y$$

(14) $$\sin iy = i \sinh y$$

The remaining trigonometric functions of z are defined in terms of $\cos z$ and $\sin z$ by means of the usual identities.

EXAMPLE 1

What is $\cos (1 + 2i)$?
By direct use of (11), we have

$$\cos (1 + 2i) = \cos 1 \cosh 2 - i \sin 1 \sinh 2$$
$$= (0.5403)(3.7622) - i(0.8415)(3.6269)$$
$$= 2.033 - 3.052i$$

EXAMPLE 2

Prove that the only values of z for which $\sin z = 0$ are the real values $z = 0, \pm \pi, \pm 2\pi, \ldots$.
From (12), $\sin z = \sin x \cosh y + i \cos x \sinh y$. Hence, if $\sin z$ is to vanish, it is necessary that simultaneously

$$\sin x \cosh y = 0 \quad \text{and} \quad \cos x \sinh y = 0$$

Since y is a real number, it follows from the familiar properties of the hyperbolic cosine that $\cosh y \geq 1$. Hence, the first of these equations can hold only if $\sin x = 0$; that is, only if

$$x = 0, \pm \pi, \pm 2\pi, \ldots$$

But for these values of x, $\cos x$ is either 1 or -1 and therefore cannot vanish. Thus for the second equation to hold, it is necessary that $\sinh y = 0$. Since y is real, the familiar properties of the hyperbolic sine can be invoked, leading to the conclusion that

$$y = 0$$

Hence, the only values of z for which $\sin z = 0$ are of the form

$$z = n\pi + 0i = n\pi \quad n = 0, \pm 1, \pm 2, \ldots$$

The hyperbolic functions of z we define simply by extending the familiar definitions into the complex-number field:

(15) $$\cosh z = \frac{e^z + e^{-z}}{2}$$

(16) $$\sinh z = \frac{e^z - e^{-z}}{2}$$

By expanding the exponentials and regrouping, as we did in deriving (11), we obtain without difficulty the formulas

(17) $$\cosh z = \cosh x \cos y + i \sinh x \sin y$$

(18) $$\sinh z = \sinh x \cos y + i \cosh x \sin y$$

In particular, setting $x = 0$, we find

(19) $$\cosh iy = \cos y$$

(20) $$\sinh iy = i \sin y$$

The remaining hyperbolic functions are defined in terms of cosh z and sinh z via the usual identities.

The logarithm of z we define implicitly as the function $w = \ln z$ which satisfies the equation

(21) $$e^w = z$$

If we let $w = u + iv$ and $z = re^{i\theta}$, Eq. (21) becomes

$$e^{u+iv} = e^u e^{iv} = re^{i\theta}$$

Hence $e^u = r$, or $u = \ln r$, and $v = \theta$. Thus

$$w = u + iv = \ln r + i\theta$$

(22) $$= \ln |z| + i \arg z$$

If we let θ_1 be the **principal argument** of z, that is, the particular argument of z which lies in the interval $-\pi < \theta \le \pi$, Eq. (22) can be written

(22a) $$\ln z = \ln |z| + i(\theta_1 + 2n\pi) \qquad n = 0, \pm 1, \pm 2, \ldots$$

which shows that the logarithmic function is infinitely many-valued. For any particular value of n a unique branch of the function is determined, and the logarithm becomes effectively single-valued. If $n = 0$, the resulting branch of the logarithmic function is called the **principal value**.

For every n, the corresponding branch of ln z is obviously discontinuous at $z = 0$. Moreover, for each n the corresponding branch is also discontinuous at every point of the negative real axis. To verify this, we note that if $n = n_0$, the corresponding branch of ln z is, by Eq. (22a),

$$\ln |z| + i \arg z \qquad \text{where } (2n_0 - 1)\pi < \arg z \le (2n_0 + 1)\pi$$

Hence, if P is an arbitrary point on the negative real axis, the limit of arg z as z approaches P through the second quadrant is $(2n_0 + 1)\pi$ while the limit of arg z as z approaches P through the third quadrant is $(2n_0 - 1)\pi$. Since these two values are different, it follows that on any particular branch, ln z does not approach a limit as z approaches an arbitrary point on the negative real axis and therefore is discontinuous at every such point.

At all points except the points on the nonpositive real axis, each branch of ln z is continuous and analytic. In fact, from the definition

$$\ln z = \ln |z| + i \arg z = \tfrac{1}{2} \ln (x^2 + y^2) + i \tan^{-1} \frac{y}{x}$$

it is easy to verify that the Cauchy-Riemann equations are satisfied everywhere except at the origin. Moreover, from the preceding discussion, it is clear that

$$u = \tfrac{1}{2} \ln (x^2 + y^2) \qquad \text{and} \qquad v = \tan^{-1} \frac{y}{x}$$

are continuous except on the nonpositive real axis. Hence, by Theorem 1, Sec. 15.6, it follows that everywhere except on the nonpositive real axis

$$\frac{d(\ln z)}{dz} = \frac{\partial u}{\partial x} + i\,\frac{\partial v}{\partial x}$$

$$= \frac{x}{x^2 + y^2} - i\,\frac{y}{x^2 + y^2} = \frac{x - iy}{x^2 + y^2} = \frac{\bar{z}}{z\bar{z}} = \frac{1}{z}$$

as expected.

The familiar laws for the logarithms of real quantities all hold for the logarithms of complex quantities in the following sense. If a suitable choice is made among the infinite number of possible values of $\ln z_1 z_2$, $\ln (z_1/z_2)$, and $\ln z^n$, then

$$\ln z_1 z_2 = \ln z_1 + \ln z_2$$

$$\ln \frac{z_1}{z_2} = \ln z_1 - \ln z_2$$

$$\ln z^m = m \ln z$$

For example, to show that $\ln z_1 z_2 = \ln z_1 + \ln z_2$, let $z_1 = r_1 e^{i\theta_1}$ and $z_2 = r_2 e^{i\theta_2}$, where θ_1 and θ_2 are the principal arguments of z_1 and z_2, respectively. Then

$$\begin{aligned}
\ln z_1 + \ln z_2 &= \left[\ln r_1 + i(\theta_1 + 2n_1\pi)\right] + \left[\ln r_2 + i(\theta_2 + 2n_2\pi)\right] \\
&= (\ln r_1 + \ln r_2) + i[(\theta_1 + \theta_2) + 2(n_1 + n_2)\pi] \\
&= \ln r_1 r_2 + i[(\theta_1 + \theta_2) + 2n_3\pi] \\
&= \ln |z_1 z_2| + i \arg z_1 z_2 \\
&= \ln z_1 z_2
\end{aligned}$$

since $\theta_1 + \theta_2 + 2(n_1 + n_2)\pi$ is *one* of the arguments of $z_1 z_2$.

However, the familiar laws of logarithms are not necessarily true if we restrict ourselves to a particular branch of $\ln z$. For instance, in terms of principal values,

$$\ln i = i\,\frac{\pi}{2} \qquad \ln(-1 + i) = \ln \sqrt{2} + i\,\frac{3\pi}{4}$$

and
$$\ln\left[i(-1 + i)\right] = \ln(-1 - i) = \ln \sqrt{2} - i\,\frac{3\pi}{4}$$

whereas

$$\ln i + \ln(-1 + i) = i\,\frac{\pi}{2} + \left(\ln \sqrt{2} + i\,\frac{3\pi}{4}\right) = \ln \sqrt{2} + i\,\frac{5\pi}{4}$$

Clearly, the principal value of $\ln[i(-1 + i)]$ differs from the sum of the principal values of $\ln i$ and $\ln(-1 + i)$ by $2\pi i$. For principal values, the proper generalizations of the familiar laws of logarithms are contained in the following theorem, whose proof we shall leave as an exercise.

THEOREM 1   The principal values of $\ln z$ satisfy the following relations:

$$\ln z_1 z_2 = \begin{cases} \ln z_1 + \ln z_2 - 2i\pi & \pi < \arg z_1 + \arg z_2 \le 2\pi \\ \ln z_1 + \ln z_2 & -\pi < \arg z_1 + \arg z_2 \le \pi \\ \ln z_1 + \ln z_2 + 2i\pi & -2\pi < \arg z_1 + \arg z_2 \le -\pi \end{cases}$$

$$\ln \frac{z_1}{z_2} = \begin{cases} \ln z_1 - \ln z_2 - 2i\pi & \pi < \arg z_1 - \arg z_2 \le 2\pi \\ \ln z_1 - \ln z_2 & -\pi < \arg z_1 - \arg z_2 \le \pi \\ \ln z_1 - \ln z_2 + 2i\pi & -2\pi < \arg z_1 - \arg z_2 \le -\pi \end{cases}$$

$$\ln z^m = m \ln z - 2ki\pi \qquad\qquad\qquad\qquad m \text{ an integer}$$

where k is the unique integer such that $(m/2\pi) \arg z - \frac{1}{2} \le k < (m/2\pi) \arg z + \frac{1}{2}$.

General powers of z are defined by the formula

(23) $$z^\alpha = \exp(\alpha \ln z)$$

which generalizes a familiar result for real variables which we frequently found useful in solving linear first-order differential equations. Since $\ln z$ is infinitely many-valued, so too is z^α, in general. Specifically,

$$z^\alpha = \exp(\alpha \ln z) = \exp\{\alpha[\ln|z| + i(\theta_1 + 2n\pi)]\} = \exp(\alpha \ln|z|)e^{\alpha\theta_1 i}e^{2n\alpha\pi i}$$

The last factor in this product clearly involves infinitely many different values unless α is a rational number, say p/q, in which case, as we saw in our discussion of Demoivre's theorem in Sec. 15.3, there are only q distinct values.†

EXAMPLE 3

What is the principal value of $(1 + i)^{2-i}$?
     By definition,

$$(1 + i)^{2-i} = \exp[(2 - i)\ln(1 + i)]$$

$$= \exp\left\{(2 - i)\left[\ln\sqrt{2} + i\left(\frac{\pi}{4} + 2n\pi\right)\right]\right\}$$

The principal value of this, obtained by taking $n = 0$, is

$$\exp\left[(2 - i)\left(\ln\sqrt{2} + i\frac{\pi}{4}\right)\right] = \exp\left[\left(2\ln\sqrt{2} + \frac{\pi}{4}\right) + i\left(-\ln\sqrt{2} + \frac{\pi}{2}\right)\right]$$

$$= \exp\left(\ln 2 + \frac{\pi}{4}\right)\left[\cos\left(\frac{\pi}{2} - \ln\sqrt{2}\right)\right.$$

$$\left. + i\sin\left(\frac{\pi}{2} - \ln\sqrt{2}\right)\right]$$

$$= \exp\left(\ln 2 + \frac{\pi}{4}\right)[\sin(\ln\sqrt{2}) + i\cos(\ln\sqrt{2})]$$

$$= e^{1.4785}(\sin 0.3466 + i\cos 0.3466)$$

$$= 1.490 + 4.126i$$

---

† However, in the particular case $z = e$, the expression $z^\alpha = e^\alpha$ is single-valued for all values of α, rational or not, since $\exp(\alpha_r + i\alpha_i)$ was defined simply as $\exp(\alpha_r)(\cos\alpha_i + i\sin\alpha_i)$, which is clearly a unique complex number.

The inverse trigonometric and hyperbolic functions we define implicitly. For instance,

$$w = \cos^{-1} z$$

we define as the value or values of w which satisfy the equation

$$z = \cos w = \frac{e^{iw} + e^{-iw}}{2}$$

From this, by obvious steps, we obtain successively

$$e^{2iw} - 2ze^{iw} + 1 = 0$$

$$e^{iw} = z \pm \sqrt{z^2 - 1}$$

and finally, by taking logarithms and solving for w,

$$(24) \qquad\qquad w = \cos^{-1} z = -i \ln (z \pm \sqrt{z^2 - 1})$$

Since the logarithm is infinitely many-valued, so too is $\cos^{-1} z$.

Similarly, we can obtain the formulas

$$(25) \qquad\qquad \sin^{-1} z = -i \ln (iz \pm \sqrt{1 - z^2})$$

$$(26) \qquad\qquad \tan^{-1} z = \frac{i}{2} \ln \frac{i + z}{i - z}$$

$$(27) \qquad\qquad \cosh^{-1} z = \ln (z \pm \sqrt{z^2 - 1})$$

$$(28) \qquad\qquad \sinh^{-1} z = \ln (z \pm \sqrt{z^2 + 1})$$

$$(29) \qquad\qquad \tanh^{-1} z = \tfrac{1}{2} \ln \frac{1 + z}{1 - z}$$

From these, after their principal values have been suitably defined by choosing the positive square root and the principal value of the logarithm in each case, the usual differentiation formulas can be obtained without difficulty.

EXERCISES

1 Prove that $e^{z_1} e^{z_2} = e^{z_1 + z_2}$.
2 Prove that $\cos^2 z + \sin^2 z = 1$.
3 Prove that $\cos (z_1 \pm z_2) = \cos z_1 \cos z_2 \mp \sin z_1 \sin z_2$.
4 Prove that $\sin (z_1 \pm z_2) = \sin z_1 \cos z_2 \pm \cos z_1 \sin z_2$.
5 Prove that $d(\cos z)/dz = -\sin z$.
6 Prove that $d(\sin z)/dz = \cos z$.
7 Express each of the following in the form $a + ib$, where a and b are decimal fractions, but giving only principal values:
    (a) $\sin (2 - i)$ **(b)** $\cosh (1 + i)$ **(c)** $\sinh (2 + 3i)$
    (d) $\cos^{-1} 2$ **(e)** $\tan i$ **(f)** $\tanh^{-1} 2$
    (g) $\ln (-3 + 4i)$ **(h)** 2^i **(i)** $(1 - i)^{2 - 3i}$
8 Show that all the values of $(1 + i)^{1 - i}$ have the same amplitude.
9 Prove that there is no value of z for which $e^z = 0$.
10 If $g(x,y)$ is a real function of x and y, what is $|e^{g(x,y)i}|$?
11 Prove that $\overline{e^z} = e^{\bar{z}}$.   12 Prove that $\overline{\cos z} = \cos \bar{z}$.
13 Is $\overline{\ln z} = \ln \bar{z}$?   14 Is $\overline{\sin z} = \sin \bar{z}$?
15 Prove that the only zeros of $\cos z$ are the values $\pm \pi/2, \pm 3\pi/2, \pm 5\pi/2, \ldots$.

16 Find all solutions of the equation $\sin z = 3$.

17 Find all solutions of the equation $\cosh z = -2$.

18 Find all solutions of the equation $e^z = -2$.

19 By inspection, $e^0 > 0$ and $e^{i\pi} < 0$, yet by Exercise 9 there is no value of z for which $e^z = 0$ even though e^z is everywhere continuous. Explain.

20 Show that Rolle's theorem fails to hold for the function $e^{iz} - 1$ even though the conditions of the theorem appear to be satisfied with respect to the two values $z = 0$ and $z = 2\pi$. Explain.

21 (a) Show that $\dfrac{d(\cos^{-1} z)}{dz} = -\dfrac{1}{\sqrt{1 - z^2}}$.

   (b) Show that $\dfrac{d(\sin^{-1} z)}{dz} = \dfrac{1}{\sqrt{1 - z^2}}$.

   (c) What is $\dfrac{d(\tan^{-1} z)}{dz}$?

22 (a) Show that $\dfrac{d(\cosh^{-1} z)}{dz} = \dfrac{1}{\sqrt{z^2 - 1}}$.

   (b) Show that $\dfrac{d(\sinh^{-1} z)}{dz} = \dfrac{1}{\sqrt{1 + z^2}}$.

   (c) What is $\dfrac{d(\tanh^{-1} z)}{dz}$?

23 Show that $|\sin z|^2 = \sin^2 x + \sinh^2 y$ and that $|\cos z|^2 = \cos^2 x + \sinh^2 y$. What is $|\sinh z|^2$? What is $|\cosh z|^2$?

24 If $z = x + iy$, show that $|\sinh y| \leq |\sin z| \leq |\cosh y|$.

25 If $z = x + iy$, show that $|\sinh y| \leq |\cos z| \leq |\cosh y|$.

26 (a) If $|z| \leq 1$, show that $|\sin z| \leq \frac{6}{5}|z|$.

   (b) Obtain an upper bound for $|\cos z|$ given that $|z| \leq 1$.

27 Show that $\sin \bar{z}$ and $\cos \bar{z}$ are not analytic functions of z.

28 Prove that

$$\tan z = \frac{\sin 2x + i \sinh 2y}{\cos 2x + \cosh 2y}$$

29 If $w = u + iv$ is the principal value of $\ln z$, plot the curves of the families $u = c$ and $v = k$.

30 If $w = \ln z$, what is the net change in w as z varies continuously in the counterclockwise direction around the curve C and returns to its initial value if:

   (a) C is the curve $|z| = 1$?   (b) C is the curve $|z - 1| = 2$?

   (c) C is the curve $|z - 2| = 1$?

31 Prove that there are values for the various logarithms such that:

   (a) $\ln (z_1/z_2) = \ln z_1 - \ln z_2$ (b) $\ln z^m = m \ln z$.

32 Prove Theorem 1.

15.8 Integration in the Complex Plane

Line integrals in the complex plane are defined as follows. Let $f(z) = u(x,y) + iv(x,y)$ be any continuous function of z, analytic or not, and let C be a sectionally smooth arc joining the points A and B. Divide C into n subintervals Δs_k by the points z_k ($k = 1, 2, \ldots, n - 1$), and let $\Delta z_k = z_k - z_{k-1}$ be the infinitesimal chord determined by Δs_k. Finally, in each subinterval on C, choose an arbitrary point $\zeta_k = \xi_k + i\eta_k$ (Fig. 15.8). Then, if it exists, the limit of the sum

(1)
$$\sum_{k=1}^{n} f(\zeta_k)\, \Delta z_k$$

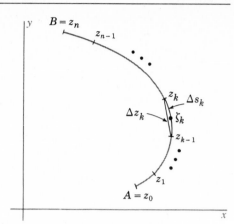

Figure 15.8
The subdivision of an arc preparatory to
defining a line integral in the complex plane.

as n becomes infinite in such a way that the length of each chord Δz_k approaches zero
is called the **line integral** of $f(z)$ along C:

$$(2) \qquad \int_C f(z)\, dz = \lim_{n \to \infty} \sum_{k=1}^{n} f(\zeta_k)\, \Delta z_k$$

In the special case in which A and B coincide and the path of integration forms a
closed curve, the integral in (2) is often called a **contour integral** and is sometimes
denoted

$$\oint f(z)\, dz$$

In working with complex line integrals it is frequently necessary to establish bounds
on their absolute values. To do this, let us return to the definitive sum (1) and apply
to it the fundamental fact that the absolute value of a sum of complex numbers is
less than or equal to the sum of their absolute values [Eq. (8), Sec. 15.4]. Then

$$\left| \sum_{k=1}^{n} f(\zeta_k)\, \Delta z_k \right| \le \sum_{k=1}^{n} |f(\zeta_k)\, \Delta z_k| = \sum_{k=1}^{n} |f(\zeta_k)|\, |\Delta z_k|$$

the last equality following from the fact that the absolute value of a product is equal
to the product of the absolute values [Eq. (5), Sec. 15.4]. As $n \to \infty$, this yields a
corresponding inequality for the integrals which are the limits of the respective sums:

$$(3) \qquad \left| \int_C f(z)\, dz \right| \le \int_C |f(z)|\, |dz|$$

The integral on the right is the real line integral

$$\int_C \sqrt{u^2 + v^2}\, \sqrt{(dx)^2 + (dy)^2} = \int_C \sqrt{u^2 + v^2}\, ds$$

where ds is the differential of arc length on C, which of course exists since C is assumed
to be sectionally smooth. In particular, if $f(z) \equiv 1$, we have the simple but important
result

$$(4) \qquad \int_C |dz| = \int_C ds = L$$

where L is the length of the path of integration. Since $f(z)$ is assumed to be continuous on the path of integration, including the end points A and B, it follows that $f(z)$ is a bounded function of z on the path of integration; in other words, there exists a constant M such that $|f(z)| \leq M$ for all values of z on C. Hence we have, from (3),

$$\left| \int_C f(z)\, dz \right| \leq \int_C |f(z)|\, |dz| \leq \int_C M|dz| = M \int_C |dz|$$

Therefore, using (4), we obtain the important inequality

$$(5) \qquad\qquad \left| \int_C f(z)\, dz \right| \leq ML$$

where M is any bound for $|f(z)|$ on the path of integration and L is the length of the path of integration.

Complex line integrals can readily be expressed in terms of real line integrals. For the sum (1) can be written

$$\sum_{k=1}^{n} [u(\xi_k,\eta_k) + iv(\xi_k,\eta_k)](\Delta x_k + i\,\Delta y_k) = \sum_{k=1}^{n} [u(\xi_k,\eta_k)\,\Delta x_k - v(\xi_k,\eta_k)\,\Delta y_k]$$

$$+ i \sum_{k=1}^{n} [v(\xi_k,\eta_k)\,\Delta x_k + u(\xi_k,\eta_k)\,\Delta y_k]$$

and, in the limit, the last expression yields the relation

$$(6) \qquad \int_C f(z)\, dz = \int_C u\, dx - v\, dy + i \int_C v\, dx + u\, dy$$

$$= \int_C (u + iv)(dx + i\, dy)$$

From (6) and the known properties of real line integrals (Sec. 13.4) or directly from the definition (2), it is easy to see that when the same path of integration is used in each integral, we have

$$(7) \qquad\qquad \int_A^B f(z)\, dz = - \int_B^A f(z)\, dz$$

$$(8) \qquad\qquad \int_A^B kf(z)\, dz = k \int_A^B f(z)\, dz$$

$$(9) \qquad \int_A^B [f(z) \pm g(z)]\, dz = \int_A^B f(z)\, dz \pm \int_A^B g(z)\, dz$$

and if D is a third point on the arc AB,

$$(10) \qquad\qquad \int_A^B f(z)\, dz = \int_A^D f(z)\, dz + \int_D^B f(z)\, dz$$

EXAMPLE 1

If C is a circle of radius r and center z_0, and if n is an integer, what is the value of

$$\int_C \frac{dz}{(z - z_0)^{n+1}}$$

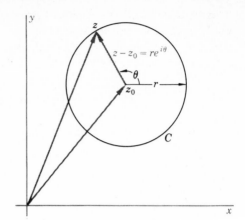

Figure 15.9
The circle $z - z_0 = re^{i\theta}$.

For convenience, let us make the substitution $z - z_0 = re^{i\theta}$, noting that θ varies from 0 to 2π as z varies around the circle in the counterclockwise direction (Fig. 15.9). Then $dz = rie^{i\theta}\, d\theta$, and the integral becomes

$$\int_0^{2\pi} \frac{rie^{i\theta}\, d\theta}{r^{n+1}e^{i(n+1)\theta}} = \frac{i}{r^n} \int_0^{2\pi} e^{-in\theta}\, d\theta$$

If $n = 0$, this reduces to

$$i \int_0^{2\pi} d\theta = 2\pi i$$

On the other hand, if $n \neq 0$, we have

$$\frac{i}{r^n} \int_0^{2\pi} (\cos n\theta - i \sin n\theta)\, d\theta = 0$$

This is an important result to which we shall have occasion to refer from time to time.

The form of the real line integrals in (6) suggests that Green's lemma (Theorem 1, Sec. 13.4) and the related results in Theorems 5 and 6, Sec. 13.4, may be useful in studying line integration in the complex plane, and this is indeed the case. Hence, for ease of reference, we repeat this important material, appropriately specialized to the two-dimensional applications we now have in mind.†

THEOREM 1   If R is a region, either simply or multiply connected, whose boundary C is sectionally smooth, and if $P(x,y)$, $Q(x,y)$, $\partial P/\partial y$, and $\partial Q/\partial x$ are continuous in and on the boundary of R, then

$$\int_C P\, dx + Q\, dy = \iint_R \left(\frac{\partial Q}{\partial x} - \frac{\partial P}{\partial y} \right) dx\, dy$$

where the integration is taken around C in the positive direction with respect to the interior of R.

---

† To avoid confusion with u and v in the standard notation for a function of a complex variable, namely $f(z) = u + iv$, we here use P and Q in place of the symbols U and V we used in Chap. 13.

THEOREM 2   In any region where $\int P\, dx + Q\, dy$ is independent of the path, the partial derivatives of the function

$$\phi(x,y) = \int_{a,b}^{x,y} P(x,y)\, dx + Q(x,y)\, dy$$

are

$$\frac{\partial \phi}{\partial x} = P(x,y) \qquad \text{and} \qquad \frac{\partial \phi}{\partial y} = Q(x,y)$$

THEOREM 3   If $\partial Q/\partial x = \partial P/\partial y$ at all points of a simply connected region R, then in R the integral

$$\int P(x,y)\, dx + Q(x,y)\, dy$$

is independent of the path, and conversely.

As a first application of Green's lemma, we have Cauchy's theorem, perhaps the most fundamental and far-reaching result in the theory of analytic functions.

THEOREM 4   If R is a region, either simply or multiply connected, whose boundary C is sectionally smooth, and if $f(z)$ is analytic and $f'(z)$ is continuous within and on the boundary of R, then

$$\int_C f(z)\, dz = 0$$

Proof   We begin by recalling from Eq. (6) that

$$\int_C f(z)\, dz = \int_C u\, dx - v\, dy + i \int_C v\, dx + u\, dy$$

Now the hypothesis that $f'(z)$ is continuous means that the partial derivatives $\partial u/\partial x$, $\partial u/\partial y$, $\partial v/\partial x$, $\partial v/\partial y$ exist and are continuous throughout R. Hence, Green's lemma can be applied to each of the line integrals on the right in the last expression, giving

$$\int_C f(z)\, dz = \iint_R \left(-\frac{\partial v}{\partial x} - \frac{\partial u}{\partial y} \right) dx\, dy + i \iint_R \left(\frac{\partial u}{\partial x} - \frac{\partial v}{\partial y} \right) dx\, dy$$

However, u and v necessarily satisfy the Cauchy-Riemann equations, since, by hypothesis, $f(z)$ is analytic. Therefore, the integrand of each of the double integrals vanishes identically in R, leaving

$$\int_C f(z)\, dz = 0$$

as asserted.

The last theorem can be proved without making use of the hypothesis that $f'(z)$ is continuous.† The French mathematician Edouard Goursat (1858–1936) was the

---

† See, for example, E. G. Phillips, "Functions of a Complex Variable," pp. 89–92, Interscience, New York, 1945.

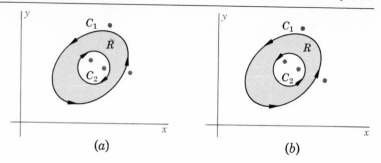

Figure 15.10
Contours which can be continuously deformed into each other.

first to do this, and in his honor the more general form of the result is usually referred to as the **Cauchy-Goursat theorem**.

In particular, if $f(z)$ is analytic in and on the boundary of the region R between two simple closed curves, we have, from the Cauchy-Goursat theorem,

$$\int_{C_1} f(z)\, dz + \int_{C_2} f(z)\, dz = 0$$

provided that each curve is traversed in the positive direction, as shown in Fig. 15.10a. On the other hand, if we reverse the direction of integration around the inner curve C_2 and transpose the resulting integral, we obtain

$$\int_{C_1} f(z)\, dz = \int_{C_2} f(z)\, dz$$

each integration now being performed in the counterclockwise sense, as shown in Fig. 15.10b. Since there may be points in the interior of C_2 (which, of course, is not a part of R) where $f(z)$ is not analytic, we cannot assert that either of these integrals is zero. However, we have shown that they both have the same value. This result can be summarized in the highly important **principle of the deformation of contours**.

THEOREM 5 The line integral of an analytic function around any simple closed curve is equal to the line integral of the same function around any other simple closed curve into which the first can be continuously deformed without passing through a point where $f(z)$ is nonanalytic.

If $f(z)$ is analytic throughout a simply connected region R, then according to the Cauchy-Goursat theorem,

$$\int_C f(z)\, dz = 0$$

for every simple closed curve in R. But, as we saw in the discussion which led to Theorem 6, Sec. 13.5, this implies that the line integral of $f(z)$ between any two points A and B in R is independent of the path. On the other hand, in multiply connected regions this observation is not necessarily true, since two different paths joining A and B may form a closed path encircling one of the inner boundaries of R and there is no assurance that the integral of $f(z)$ around such a path is zero. Thus, summarizing, we have the following theorem.

THEOREM 6   In any simply connected region in which $f(z)$ is analytic, the integral $\int f(z)\, dz$ is independent of the path.

Using Theorems 2 and 3, we can establish the following interesting result.

THEOREM 7   If $u(x,y)$ is a solution of Laplace's equation in a region R, then in R there exists an analytic function having u as its real part, namely, $f(z) = u + iv$, where

$$v(x, y) = \int_{a,b}^{x,y} -\frac{\partial u}{\partial y}\, dx + \frac{\partial u}{\partial x}\, dy$$

and the path of integration from (a,b) to (x,y) lies entirely in R.

Proof   Suppose first that R is simply connected. Then in R the integral defining v is independent of the path between the arbitrary fixed point (a,b) and the variable point (x,y), since the condition for independence provided by Theorem 3 is in this case

$$\frac{\partial(\partial u/\partial x)}{\partial x} = \frac{\partial(-\partial u/\partial y)}{\partial y} \qquad \text{or} \qquad \frac{\partial^2 u}{\partial x^2} = -\frac{\partial^2 u}{\partial y^2}$$

which is true because of the hypothesis that u satisfies Laplace's equation. Theorem 2 can therefore be applied to the integral which defines v, and we have

$$\frac{\partial v}{\partial x} = -\frac{\partial u}{\partial y} \qquad \text{and} \qquad \frac{\partial v}{\partial y} = \frac{\partial u}{\partial x}$$

These are precisely the Cauchy-Riemann equations, which, if the derivatives are continuous, are the conditions that $f(z) = u + iv$ be an analytic function. But $\partial u/\partial x$ and $\partial u/\partial y$, and hence $\partial v/\partial y$ and $-\partial v/\partial x$, to which these are respectively equal, must be continuous, since the second partial derivatives $\partial^2 u/\partial x^2$ and $\partial^2 u/\partial y^2$ are known to exist. Hence, if R is simply connected, $f(z) = u + iv$ is analytic, as asserted.

On the other hand, if R is multiply connected, then, by the principle of the deformation of contours, the possible values of v differ at most by constants independent of the end points. And, clearly, a constant added to v will not affect the analyticity of $u + iv$. This completes the proof of the theorem.

One of the most important consequences of Cauchy's theorem is **Cauchy's integral formula**.

THEOREM 8   If $f(z)$ is analytic within and on the boundary C of a simply connected region R whose boundary C is sectionally smooth, and if z_0 is any point in the interior of R, then

$$f(z_0) = \frac{1}{2\pi i} \int_C \frac{f(z)}{z - z_0}\, dz$$

the integration around C being taken in the positive sense.

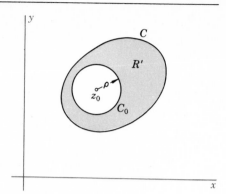

Figure 15.11
The circle C_0 used in the proof of Cauchy's integral formula.

Proof Let C_0 be a circle with center at z_0 and arbitrary radius small enough for C_0 to lie entirely in R (Fig. 15.11). Now by hypothesis, $f(z)$ is analytic everywhere within R. Hence the function $f(z)/(z - z_0)$ is analytic everywhere within R except at the one point $z = z_0$. In particular, it is analytic everywhere in the region R' between C and C_0. Hence, by Theorem 5, the integral of this function around C is equal to its integral around C_0. That is,

$$\int_C \frac{f(z)}{z - z_0}\, dz = \int_{C_0} \frac{f(z)}{z - z_0}\, dz = \int_{C_0} \frac{f(z_0) + [f(z) - f(z_0)]}{z - z_0}\, dz$$

(11)
$$= f(z_0) \int_{C_0} \frac{dz}{z - z_0} + \int_{C_0} \frac{f(z) - f(z_0)}{z - z_0}\, dz$$

By Example 1, the first integral on the right is equal to $2\pi i$. Hence, the assertion of the theorem will be established if we can show that the last integral vanishes. To do this, we observe that

(12)
$$\left| \int_{C_0} \frac{f(z) - f(z_0)}{z - z_0}\, dz \right| \leq \int_{C_0} \frac{|f(z) - f(z_0)|}{|z - z_0|}\, |dz|$$

On C_0 we have $|z - z_0| = \rho$. Moreover, since $f(z)$ is analytic and hence continuous, it follows that, for any $\varepsilon > 0$, there exists a δ such that

$$|f(z) - f(z_0)| < \varepsilon \qquad \text{provided } |z - z_0| \equiv \rho < \delta$$

Choosing the radius ρ to be less than δ and inserting these estimates in the right member of (12), we therefore have

$$\left| \int_{C_0} \frac{f(z) - f(z_0)}{z - z_0}\, dz \right| < \int_{C_0} \frac{\varepsilon}{\rho}\, |dz| = \frac{\varepsilon}{\rho} \int_{C_0} |dz| = \frac{\varepsilon}{\rho}\, 2\pi\rho = 2\pi\varepsilon$$

Since the integral on the left is independent of ε, yet cannot exceed $2\pi\varepsilon$, which can be made arbitrarily small, it follows that the absolute value of the integral, and hence the integral itself, is zero. Thus, (11) reduces to

$$\int_{C_0} \frac{f(z)}{z - z_0}\, dz = f(z_0)2\pi i + 0$$

whence,
$$f(z_0) = \frac{1}{2\pi i} \int_{C_0} \frac{f(z)}{z - z_0}\, dz$$

as asserted. Cauchy's integral formula is also true for multiply connected regions, but we shall leave as an exercise the easy modification of our proof required to establish this fact.

EXAMPLE 2

Find the values of

$$\int_C \frac{e^z}{z^2 + 1} \, dz$$

if C is a circle of unit radius with center at **(a)** $z = i$ and **(b)** $z = -i$.

In **(a)** we think of the integral as written in the form

$$\int_C \frac{e^z}{z + i} \frac{dz}{z - i}$$

and identify z_0 as i and $f(z)$ as $e^z/(z + i)$. The function $f(z)$ is analytic everywhere within and on the given circle of unit radius around $z = i$. (In fact, it is analytic everywhere except at $z = -i$.) Therefore we can apply Cauchy's integral formula, getting

$$\int_C \frac{e^z}{z + i} \frac{dz}{z - i} = 2\pi i f(z_0) = 2\pi i f(i) = 2\pi i \frac{e^i}{2i} = \pi(\cos 1 + i \sin 1)$$

In **(b)** we identify z_0 as $-i$ and $f(z)$ as $e^z/(z - i)$. Then Cauchy's integral formula gives immediately

$$\int_C \frac{e^z}{z - i} \frac{dz}{z + i} = 2\pi i f(z_0) = 2\pi i f(-i) = 2\pi i \frac{e^{-i}}{-2i} = -\pi(\cos 1 - i \sin 1)$$

From Cauchy's integral formula, which expresses the value of an analytic function at an interior point of a region R in terms of its values on the boundary of the region, we can readily obtain an expression for the derivative of a function at an interior point of R in terms of the boundary values of the function. In fact, from the definition of a derivative and Cauchy's integral formula, we have

$$f'(z_0) = \lim_{\Delta z_0 \to 0} \frac{f(z_0 + \Delta z_0) - f(z_0)}{\Delta z_0}$$

$$= \lim_{\Delta z_0 \to 0} \frac{1}{\Delta z_0} \left[\frac{1}{2\pi i} \int_C \frac{f(z) \, dz}{z - (z_0 + \Delta z_0)} - \frac{1}{2\pi i} \int_C \frac{f(z) \, dz}{z - z_0} \right]$$

$$= \lim_{\Delta z_0 \to 0} \frac{1}{\Delta z_0} \left\{ \frac{1}{2\pi i} \int_C f(z) \left[\frac{1}{z - (z_0 + \Delta z_0)} - \frac{1}{z - z_0} \right] dz \right\}$$

$$= \lim_{\Delta z_0 \to 0} \frac{1}{2\pi i} \int_C \frac{f(z) \, dz}{(z - z_0 - \Delta z_0)(z - z_0)}$$

Assuming that the limit of the integral is equal to the integral of the limit in the last expression, and letting $\Delta z_0 \to 0$ in the integrand, we obtain the desired result:

$$f'(z_0) = \frac{1}{2\pi i} \int_C \frac{f(z) \, dz}{(z - z_0)^2}$$

That the interchange of the operations of integration and taking the limit is legitimate can be established by adding and subtracting the desired final integral, namely

$$\frac{1}{2\pi i} \int_C \frac{f(z) \, dz}{(z - z_0)^2}$$

and then showing that the absolute value of the difference

(13)
$$\frac{1}{2\pi i} \int_C \frac{f(z)\, dz}{(z - z_0 - \Delta z_0)(z - z_0)} - \frac{1}{2\pi i} \int \frac{f(z)\, dz}{(z - z_0)^2}$$

approaches zero as $\Delta z_0 \to 0$.

Continuing in the same way, we obtain the additional formulas

$$f''(z_0) = \frac{2!}{2\pi i} \int_C \frac{f(z)\, dz}{(z - z_0)^3}$$

$$f'''(z_0) = \frac{3!}{2\pi i} \int_C \frac{f(z)\, dz}{(z - z_0)^4}$$

$$\cdots\cdots\cdots\cdots\cdots\cdots\cdots\cdots$$

These results could also have been obtained formally by repeated differentiation of Cauchy's integral formula with respect to the parameter z_0.

From the preceding discussion we conclude not only that an analytic function possesses derivatives of all orders but also that each derivative is itself analytic since it too possesses a derivative. This completes the proof of the following theorem.

THEOREM 9   If $f(z)$ is analytic throughout a closed simply connected region R, then at any interior point z_0 of R the derivatives of $f(z)$ of all orders exist and are analytic. Moreover,

$$f^{(n)}(z_0) = \frac{n!}{2\pi i} \int_C \frac{f(z)\, dz}{(z - z_0)^{n+1}}$$

where C is the boundary of R.

It is interesting to note that functions of a real variable do not in general possess the derivative properties described by Theorem 9, for at particular points a function of a real variable may possess one or more derivatives without the derivatives of all orders existing. For instance, at the origin the function $x^{7/3}$ possesses a first and a second derivative but no derivative of higher order.

Using Theorem 9, we can now prove the converse of Cauchy's theorem, which is known as **Morera's theorem.†**

THEOREM 10   If $f(z)$ is continuous in a region R, and if $\int_C f(z)\, dz = 0$ for every simple closed curve C which can be drawn in R, then $f(z)$ is analytic in R.

Proof   To prove this, we observe, as in the proof of Theorem 6, Sec. 13.5, that if the line integral of $f(z)$ around every closed curve in R is zero, then the line integral of $f(z)$ between a fixed point z_0 and a variable point z in R is independent of the path and hence is a function of z alone, say

$$F(z) = \int_{z_0}^z f(z)\, dz$$

If we let $f(z) = u + iv$ and $F(z) = U + iV$, this can be written

$$F(z) = U + iV = \int_{x_0, y_0}^{x, y} u\, dx - v\, dy + i \int_{x_0, y_0}^{x, y} v\, dx + u\, dy$$

---

† Named for the Italian mathematician Giacinto Morera (1856–1909).

or, equating real and imaginary parts,

$$U = \int_{x_0, y_0}^{x, y} u\, dx - v\, dy \quad \text{and} \quad V = \int_{x_0, y_0}^{x, y} v\, dx + u\, dy$$

By Theorem 2, each of these integrals can be differentiated partially with respect to x and y, and we find

$$\frac{\partial U}{\partial x} = u \qquad \frac{\partial U}{\partial y} = -v \qquad \frac{\partial V}{\partial x} = v \qquad \frac{\partial V}{\partial y} = u$$

From these it is obvious that

$$\frac{\partial U}{\partial x} = \frac{\partial V}{\partial y} \quad \text{and} \quad \frac{\partial U}{\partial y} = -\frac{\partial V}{\partial x}$$

or, in other words, that U and V satisfy the Cauchy-Riemann equations. Moreover, since u and v are continuous, because of the hypothesis that $f(z) = u + iv$ is continuous, it follows that $\partial U/\partial x$, $\partial U/\partial y$, $\partial V/\partial x$, $\partial V/\partial y$ are continuous. Hence, $F(z) = U + iV$ is an analytic function whose derivative, in fact, is

$$F'(z) = \frac{\partial U}{\partial x} + i\frac{\partial V}{\partial x} = u + iv = f(z)$$

Being the derivative of an analytic function, $f(z)$ is therefore analytic, by Theorem 9, as asserted.

Beginning with the formula for $f^{(n)}(z_0)$ provided by Theorem 9, we can now establish **Cauchy's inequality**.

THEOREM 11 If $f(z)$ is analytic within and on a circle C of radius r with center at z_0, then

$$|f^{(n)}(z_0)| \leq \frac{n!\, M}{r^n}$$

where M is the maximum value of $f(z)$ on C.

Proof From Theorem 9, we have

$$
\begin{aligned}
|f^{(n)}(z_0)| &= \left| \frac{n!}{2\pi i} \int_C \frac{f(z)\, dz}{(z - z_0)^{n+1}} \right| \\
&\leq \frac{n!}{2\pi} \int_C \frac{|f(z)|\, |dz|}{|z - z_0|^{n+1}} \\
&\leq \frac{n!}{2\pi} \frac{M}{r^{n+1}} \int_C |dz| \\
&= \frac{n!}{2\pi} \frac{M}{r^{n+1}} 2\pi r \\
&= \frac{n!\, M}{r^n}
\end{aligned}
$$

as asserted.

For the special case $n = 0$, Cauchy's inequality becomes

$$|f(z_0)| \leq M$$

which shows that on every circle around z_0, no matter how small, $|f(z)|$ has a maximum value M which is at least as great as $f(z_0)$. In other words, we have the following result, usually referred to as the **maximum modulus theorem.**

THEOREM 12   The absolute value of a nonconstant function $f(z)$ cannot have a maximum at any point where the function is analytic.

EXERCISES

1 Evaluate $\displaystyle\int_0^{3+i} z^2 \, dz$:

   (a) Along the line $y = x/3$.
   (b) Along the real axis to 3 and then vertically to $3 + i$.
   (c) Along the imaginary axis to i and then horizontally to $3 + i$.

2 Evaluate $\displaystyle\int_0^{3+i} (\bar{z})^2 \, dz$ along each of the paths used in Exercise 1.

3 Evaluate $\displaystyle\int_0^{1+i} (x^2 + iy) \, dz$ along the paths $y = x$ and $y = x^2$.

4 Obtain an upper bound for the absolute value of the integral $\displaystyle\int_0^{1+i} e^{-z^2} \, dz$:

   (a) Along $y = x$.   (b) Along $y = x^2$.
   (c) Along the real axis to 1 and then vertically to $1 + i$.

5 Obtain an upper bound for the absolute value of the integral

$$\frac{1}{2\pi i} \int \frac{e^{2z}}{z^2 + 1} \, dz$$

   taken around the circle $|z| = 3$. What is the value of this integral if the path of integration is the circle $|z| = \frac{1}{2}$?

6 What is the value of

$$\int_C \frac{3z^2 + 7z + 1}{z + 1} \, dz$$

   (a) If C is the circle $|z + 1| = 1$?
   (b) If C is the circle $|z + i| = 1$?
   (c) If C is the ellipse $x^2 + 2y^2 = 8$?

7 What is the value of

$$\int_C \frac{z + 4}{z^2 + 2z + 5} \, dz$$

   (a) If C is the circle $|z| = 1$?
   (b) If C is the circle $|z + 1 - i| = 2$?
   (c) If C is the circle $|z + 1 + i| = 2$?

8 What is the value of

$$\int \frac{e^z}{(z + 1)^2} \, dz$$

   around the circle $|z - 1| = 3$?

9 What is the value of

$$\int \frac{z + 1}{z^3 - 2z^2} \, dz$$

   (a) Around the circle $|z| = 1$?
   (b) Around the circle $|z - 2 - i| = 2$?
   (c) Around the circle $|z - 1 - 2i| = 2$?

10 Show that Cauchy's integral formula is valid in multiply connected regions.

11 Complete the proof of Theorem 9 by showing that the absolute value of the difference (13) approaches zero as Δz_0 approaches zero.

12 Prove that if $f(z)$ is analytic inside and on a circle C with center at z_0, then the average value of $f(z)$ on C is $f(z_0)$. This result is sometimes called **Gauss' mean-value theorem**. *Hint:* Make the substitution $z = z_0 + re^{i\theta}$ in Cauchy's integral formula, where r is the radius of C.

13 Prove that if $f(z)$ is analytic inside and on a simple closed curve C, and if $f(z) \neq 0$ inside C, then $|f(z)|$ must assume its minimum value on C. This result is sometimes known as the **minimum-modulus theorem**. *Hint:* Apply the maximum-modulus theorem to $1/f(z)$.

14 Using Theorem 9, show that

$$\frac{x^n}{n!} = \frac{1}{2\pi i} \int_C \frac{e^{xz}}{z^{n+1}} \, dz$$

where C is any simple closed curve enclosing the origin.

15 Observing that the result of Exercise 14 can be written

$$\left(\frac{x^n}{n!}\right)^2 = \frac{1}{2\pi i} \int_C \frac{x^n e^{xz}}{n! z^{n+1}} \, dz$$

show that

$$\sum_{n=0}^{\infty} \left(\frac{x^n}{n!}\right)^2 \equiv I_0(2x) = \frac{1}{2\pi} \int_0^{2\pi} \exp(2x \cos \theta) \, d\theta$$

where I_0 is the modified Bessel function of order 0. *Hint:* Take C to be the circle $|z| = 1$ and then put $z = e^{i\theta}$.

16 Proceeding essentially as in Exercise 15, show that

$$J_0(2x) = \frac{1}{2\pi} \int_0^{2\pi} \exp(2ix \sin \theta) \, d\theta$$

where J_0 is the Bessel function of order 0.

17 Proceeding as in Exercises 15 and 16, obtain integral formulas for $I_k(2x)$ and $J_k(2x)$, where k is an integer.

18 If $u(x,y)$ is harmonic, i.e., satisfies Laplace's equation, in the closed region R bounded by a simple C, prove that the maximum value of $u(x,y)$ in R always occurs on C and not in the interior of R. *Hint:* Apply the maximum-modulus theorem to the function $e^{f(z)}$, where $f(z)$ is the analytic function having $u(x,y)$ as its real part.

19 (a) Taking C to be the circle defined by $z = Re^{i\theta}$, and letting $z_0 = re^{i\theta}$, where $r < R$, show that Cauchy's integral formula becomes

$$f(re^{i\phi}) = \frac{1}{2\pi i} \int_0^{2\pi} \frac{f(Re^{i\theta})}{Re^{i\theta} - re^{i\phi}} (iRe^{i\theta} \, d\theta)$$

(b) Show that

$$\frac{1}{2\pi i} \int_0^{2\pi} \frac{f(Re^{i\theta})}{Re^{i\theta} - (R^2/r)e^{i\phi}} (iRe^{i\theta} \, d\theta) = 0$$

20 By subtracting the two integrals in Exercise 19 and equating real parts in the resulting equation, obtain **Poisson's formula**:

$$u(r, \phi) = \frac{1}{2\pi} \int_0^{2\pi} \frac{(R^2 - r^2)u(R,\theta)}{R^2 - 2Rr \cos(\theta - \phi) + r^2} \, d\theta$$

CHAPTER 16

Infinite Series
in the Complex Plane

16.1 Series of Complex Terms

Most of the definitions and theorems relating to infinite series of real terms can be applied with little or no change to series whose terms are complex. To restate these briefly, let

(1)
$$f_1(z) + f_2(z) + \cdots + f_n(z) + \cdots$$

be a series whose terms are functions of a complex variable z. Then the **partial sums** of this series are defined to be the finite sums

$$S_1(z) = f_1(z)$$
$$S_2(z) = f_1(z) + f_2(z)$$
$$\cdots\cdots\cdots\cdots\cdots\cdots\cdots$$
$$S_n(z) = f_1(z) + f_2(z) + \cdots + f_n(z)$$

The series (1) is said to converge to the **sum** $S(z)$ in a region R provided that for all z in R the limit of the nth partial sum $S_n(z)$ as n becomes infinite is $S(z)$.

According to the technical definition of a limit, this requires that for any $\varepsilon > 0$ there should exist an integer N, depending in general on ε and also on the particular value of z under consideration, such that

$$|S(z) - S_n(z)| < \varepsilon \qquad \text{for all } n > N$$

The difference $S(z) - S_n(z)$ is evidently just the remainder after the first n terms in $S(z)$, say $R_n(z)$. Thus the definition of convergence requires that the limit of $|R_n(z)|$ as n becomes infinite be zero. A series which has a sum, as just defined, is said to be **convergent**, and the set of all values of z for which it converges is called the **region of convergence** of the series. A series which is not convergent is said to be **divergent**.

If the absolute values of the terms in (1) form a convergent series

$$|f_1(z)| + |f_2(z)| + \cdots + |f_n(z)| + \cdots$$

then (1) is said to be **absolutely convergent**. If the series (1) converges but is not absolutely convergent, it is said to be **conditionally convergent**. Absolute convergence is an important property because it is a sufficient (though not a necessary) condition for ordinary convergence. Moreover, the terms of an absolutely convergent series can be rearranged in any manner whatsoever without affecting the sum of the series, whereas rearranging the terms of a conditionally convergent series may alter the sum of the series or even cause the series to diverge.

From the definition of convergence it is easy to prove the following theorem.

THEOREM 1   A necessary and sufficient condition that the series of complex terms

$$f_1(z) + f_2(z) + \cdots + f_n(z) + \cdots$$

should converge is that the series of the real parts and the series of the imaginary parts of these terms should each converge. Moreover, if

$$\sum_{n=1}^{\infty} \mathscr{R}(f_n) \quad \text{and} \quad \sum_{n=1}^{\infty} \mathscr{I}(f_n)$$

converge to the respective functions $R(z)$ and $I(z)$, then the given series converges to $R(z) + iI(z)$.

Of all the tests for the convergence of infinite series, the most useful is probably the familiar **ratio test**, which applies to series whose terms are complex as well as to series whose terms are real.

THEOREM 2   For the series $f_1(z) + f_2(z) + \cdots + f_n(z) + \cdots$ let

$$\lim_{n \to \infty} \left| \frac{f_{n+1}(z)}{f_n(z)} \right| = |r(z)|$$

Then the given series converges absolutely for those values of z for which $0 \le |r(z)| < 1$ and diverges for those values of z for which $|r(z)| > 1$. The values of z for which $|r(z)| = 1$ form the boundary of the region of convergence of the series, and at these points the ratio test provides no information about the convergence or divergence of the series.

EXAMPLE 1

Find the region of convergence of the series

$$1 + \frac{1}{2^2}\frac{z+1}{z-1} + \frac{1}{3^2}\left(\frac{z+1}{z-1}\right)^2 + \frac{1}{4^2}\left(\frac{z+1}{z-1}\right)^3 + \cdots$$

Applying the ratio test, we find

$$\left| \frac{f_{n+1}(z)}{f_n(z)} \right| = \left| \frac{\dfrac{1}{(n+1)^2}\left(\dfrac{z+1}{z-1}\right)^n}{\dfrac{1}{n^2}\left(\dfrac{z+1}{z-1}\right)^{n-1}} \right| = \left| \frac{n^2}{(n+1)^2}\frac{z+1}{z-1} \right|$$

As n becomes infinite, this ratio approaches $|(z+1)/(z-1)|$. Hence the values of z for which the series surely converges are those in the region defined by the inequality

$$\left| \frac{z+1}{z-1} \right| < 1$$

that is, by $|z + 1| < |z - 1|$

Since $|z + 1|$ is just the distance from z to the point -1 and $|z - 1|$ is just the distance from z to the point 1, z is restricted to be nearer to the point -1 than to the point 1. In other words, z must lie to the left of the perpendicular bisector of the segment joining -1 and 1; that is, z must lie in the left half of the complex plane. The boundary cases for which the test fails are the values of z which are equidistant from -1 and 1, that is, the values of z on the imaginary axis. But for these points, the related series of absolute values is the convergent real series

$$1 + \frac{1}{2^2} + \frac{1}{3^2} + \frac{1}{4^2} + \cdots$$

Hence, for all values of z on the imaginary axis, the given series, being absolutely convergent, is convergent. Thus these points also belong to the region of convergence.

The sum or difference of two convergent series can be found by term-by-term addition or subtraction of the series. If two series converge absolutely, their product can be found by multiplying the two series together as though they were polynomials. To establish conditions under which series can legitimately be integrated or differentiated term by term, however, the concept of **uniform convergence** is required.

DEFINITION 1   A series of functions is said to converge uniformly to the function $S(z)$ in a region R, either open or closed, if corresponding to an arbitrary $\varepsilon > 0$ there exists a positive integer N, depending on ε but not on z, such that for every value of z in R

$$|S(z) - S_n(z)| < \varepsilon \qquad \text{for all } n > N$$

In other words, if a series converges uniformly in a region R, then corresponding to any $\varepsilon > 0$ there exists an integer N such that *everywhere* in R the sum of the series $S(z)$ can be approximated with an error less than ε by using *no more than N terms* of the series. It may well be that fewer than N terms will suffice at most of the points of the region, but *nowhere* will more than N be required. This is in sharp contrast to ordinary convergence; for in the neighborhood of certain points in a region of ordinary convergence it may be that no limit can be set on the number of terms required to secure a prescribed degree of accuracy.

EXAMPLE 2

Discuss the convergence of the series

$$z^2 + \frac{z^2}{1 + z^2} + \frac{z^2}{(1 + z^2)^2} + \frac{z^2}{(1 + z^2)^3} + \cdots + \frac{z^2}{(1 + z^2)^{n-1}} + \cdots$$

in the 90° sector bounded by the right halves of the lines $y = \pm x$ (Fig. 16.1a).

The given series is a geometric progression which will converge for all values of z for which the absolute value of the common ratio, namely,

$$|r| = \left| \frac{1}{1 + z^2} \right|$$

is less than 1. Now the angle of z is restricted, by hypothesis, to be between $-\pi/4$ and $\pi/4$. Hence the angle of z^2 must be between $-\pi/2$ and $\pi/2$. Therefore (Fig. 16.1b) the vectors representing the numbers 1, z^2, and $1 + z^2$ are the sides of a triangle in which the side $1 + z^2$ is opposite the largest angle. Thus, for every z in the given region D, we have

$$|1 + z^2| \geq 1 \qquad \text{and} \qquad \frac{1}{|1 + z^2|} \leq 1$$

and the equality sign holds only for the value $z = 0$. Thus the given series converges for all values of z in D, and its sum is

$$S(z) = \begin{cases} \dfrac{a}{1 - r} = \dfrac{z^2}{1 - 1/(1 + z^2)} = 1 + z^2 & z \neq 0 \\ 0 + 0 + 0 + 0 + \cdots = 0 & z = 0 \end{cases}$$

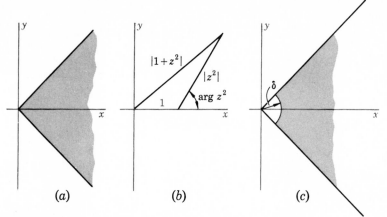

Figure 16.1
A 90° sector before and after modification to exclude its vertex.

Now let an arbitrary $\varepsilon > 0$ be given, and let us attempt to determine how many terms of the series must be taken in order that

$$|S(z) - S_n(z)| < \varepsilon$$

This difference, i.e., the remainder after n terms of the series, is just the geometric progression

$$\frac{z^2}{(1 + z^2)^n} + \frac{z^2}{(1 + z^2)^{n+1}} + \frac{z^2}{(1 + z^2)^{n+2}} + \cdots$$

whose sum is

$$R_n(z) = \begin{cases} \dfrac{1}{(1 + z^2)^{n-1}} & z \neq 0 \\ 0 & z = 0 \end{cases}$$

Hence, to show that the given series converges *uniformly* in D, we must find, if possible, a value of N such that

$$(2) \qquad |R_n(z)| = \frac{1}{|1 + z^2|^{n-1}} < \varepsilon \qquad \begin{array}{l} \text{for all } n > N \\ \text{and all } z \text{ in } D \end{array}$$

Now $|1 + z^2| \leq 1 + |z^2| = 1 + |z|^2$. Hence, underestimating $R_n(z)$ by overestimating its denominator, we have

$$|R_n(z)| = \frac{1}{|1 + z^2|^{n-1}} \geq \frac{1}{(1 + |z|^2)^{n-1}}$$

From this inequality we observe that if it should be impossible to find an integer N such that

$$(3) \qquad \frac{1}{(1 + |z|^2)^{n-1}} < \varepsilon \qquad \begin{array}{l} \text{for all } n > N \\ \text{and all } z \text{ in } D \end{array}$$

then surely it will be impossible to find an integer N which will suffice to keep

$$|R_n(z)| < \varepsilon$$

everywhere in D. And this is indeed the case, for if we attempt to solve the inequality (3) for n, we find, by obvious steps,

$$(1 + |z|^2)^{n-1} > \frac{1}{\varepsilon}$$

$$(n - 1) \ln (1 + |z|^2) > \ln \frac{1}{\varepsilon} = -\ln \varepsilon$$

$$n > 1 - \frac{\ln \varepsilon}{\ln (1 + |z|^2)}$$

But for values of z within the sector of the problem and sufficiently close to the origin, $\ln (1 + |z|^2)$ can be made arbitrarily close to $\ln 1$, that is, zero. Hence n is unbounded, and there exists no integer N for which (3) holds. Since $|R_n(z)|$ is larger than the fraction in (3), it is clear that the fundamental requirement of uniform convergence (2) cannot be fulfilled. Thus the convergence of the given series in the original region D is nonuniform.

On the other hand, if we restrict z to the infinite region D' bounded by the given rays and a circular arc of small but fixed radius δ, as shown in Fig. 16.1c, the series does converge uniformly. In fact, the law of cosines applied in Fig. 16.1b gives

$$|1 + z^2|^2 = 1 + |z^2|^2 + 2|z^2| \cos (\arg z^2)$$

Reducing the right-hand side by dropping the last term, which is surely nonnegative since $-\pi/2 \le \arg z^2 \le \pi/2$, gives

$$|1 + z^2|^2 \ge 1 + |z^2|^2 = 1 + |z|^4$$

Hence, overestimating $|R_n(z)|$ by underestimating its denominator, we can write

$$|R_n(z)| = \frac{1}{|1 + z^2|^{n-1}} \le \frac{1}{(1 + |z|^4)^{(n-1)/2}}$$

From this it is clear that if we can find an integer N such that

$$(4) \qquad \frac{1}{(1 + |z|^4)^{(n-1)/2}} < \varepsilon \qquad \begin{array}{l} \text{for all } n > N \\ \text{and all } z \text{ in } D' \end{array}$$

then surely for the same N we shall have

$$(5) \qquad |R_n(z)| < \varepsilon \qquad \begin{array}{l} \text{for all } n > N \\ \text{and all } z \text{ in } D' \end{array}$$

Hence we attempt to solve the inequality in (4) for n:

$$(1 + |z|^4)^{(n-1)/2} > \frac{1}{\varepsilon}$$

$$\frac{n - 1}{2} \ln (1 + |z|^4) > \ln \frac{1}{\varepsilon} = -\ln \varepsilon$$

$$n > 1 - \frac{2 \ln \varepsilon}{\ln (1 + |z|^4)}$$

The most unfavorable case, i.e., the largest possible value of the fraction on the right, occurs when $|z|$ is as small as possible. But in the modified region D' which we are now considering, the smallest possible value of $|z|$ is δ, which yields

$$n > 1 - \frac{2 \ln \varepsilon}{\ln (1 + \delta^4)}$$

If we choose N to be the first integer equal to or greater than the expression on the right, then (4) will surely hold. But, as we observed above, if (4) is satisfied, so too is (5), and hence, in the modified region D' the given series does converge uniformly.

Usually, uniform convergence is established not by a direct application of the definition, as in Example 2, but by the so-called **Weierstrass M test.**†

THEOREM 3   If a sequence of positive constants M_n exists such that $|f_n(z)| \leq M_n$ for all positive integers n and for all values of z in a given region D, and if the series

$$M_1 + M_2 + M_3 + \cdots + M_n + \cdots$$

is convergent, then the series

$$f_1(z) + f_2(z) + f_3(z) + \cdots + f_n(z) + \cdots$$

converges uniformly in D.

Proof   To prove this, we must show that for any $\varepsilon > 0$ there exists an integer N, independent of z, such that for all values of z in D the absolute value of the remainder after n terms of the series of the f's is less than ε whenever n exceeds N. To do this, we note that

$$|R_n(z)| = |f_{n+1}(z) + f_{n+2}(z) + \cdots |$$
$$\leq |f_{n+1}(z)| + |f_{n+2}(z)| + \cdots$$
(6) $$\leq M_{n+1} + M_{n+2} + \cdots$$

The last expression is just the remainder after n terms of the series of the M's. Since this series is convergent, by hypothesis, it follows that for every $\varepsilon > 0$ there exists an N such that this remainder is less than ε for all $n > N$. This value of N, arising as it does from a series of constants, is obviously independent of z. Moreover, from the inequality (6) it is clear that whenever n exceeds this N, $|R_n(z)| < \varepsilon$ for all values of z in D. Hence the series of the f's is uniformly convergent, as asserted. Incidentally, this theorem implies a comparison test which proves that the series of the f's is also absolutely convergent.

The Weierstrass M test is merely a sufficient test; i.e., there exist uniformly convergent series whose terms cannot be dominated by the respective terms of any convergent series of positive constants.‡ The M test suffices for almost all applications, however.

One useful property of uniformly convergent series is contained in the following theorem.

THEOREM 4   If the terms of a uniformly convergent series are multiplied by any bounded function of z, the resulting series will also converge uniformly.

Proof   Let D be the region of uniform convergence of the series

$$f_1(z) + f_2(z) + \cdots + f_n(z) + \cdots$$

and let g be a function such that

$$|g(z)| < M$$

---

† Karl Weierstrass (1815–1897), a German mathematician, is often called the "father of modern rigor."
‡ One example of such a series will be found in Exercise 6.

for all values of z in D. Now since the series of the f's converges uniformly, it follows that corresponding to the infinitesimal ε/M there exists an integer N such that

$$|R_n(z)| = |f_{n+1}(z) + f_{n+2}(z) + \cdots| < \frac{\varepsilon}{M} \qquad \begin{array}{l} \text{for all } n > N \\ \text{and all } z \text{ in } D \end{array}$$

Hence,

$$
\begin{aligned}
|g(z)f_{n+1}(z) + g(z)f_{n+2}(z) + \cdots| &= |g(z)| \cdot |f_{n+1}(z) + f_{n+2}(z) + \cdots| \\
&\le M|f_{n+1}(z) + f_{n+2}(z) + \cdots| \\
&\le M\frac{\varepsilon}{M} \\
&= \varepsilon \qquad \begin{array}{l} \text{for all } n > N \\ \text{and all } z \text{ in } D \end{array}
\end{aligned}
$$

But this is precisely the condition that the product series

$$g(z)f_1(z) + g(z)f_2(z) + \cdots + g(z)f_n(z) + \cdots$$

is uniformly convergent.

One important consequence of uniform convergence is embodied in the following theorem.

THEOREM 5   The sum of a uniformly convergent series of continuous functions is a continuous function.

Proof   Let

$$f(z) = f_1(z) + f_2(z) + \cdots + f_n(z) + \cdots = S_n(z) + R_n(z)$$

be a uniformly convergent series in which each term is a continuous function of z, and let $\varepsilon/3$ be an arbitrary infinitesimal. Then since the series converges uniformly, there exists an integer N, depending only on ε, such that

$$|R_n(z_0)| < \frac{\varepsilon}{3} \qquad \text{for all } n > N$$

and any value of z_0 in the region of uniform convergence. Likewise, if Δz_1 is an increment such that $z_0 + \Delta z$ is in the region of uniform convergence whenever $|\Delta z| < |\Delta z_1|$, we also have

$$|R_n(z_0 + \Delta z)| < \frac{\varepsilon}{3} \qquad \begin{array}{l} \text{for any } n > N \\ \text{and any } \Delta z \text{ such that } |\Delta z| < |\Delta z_1| \end{array}$$

Moreover, since each term of the given series is a continuous function, and since any *finite* sum of continuous functions is necessarily continuous, it follows that $S_n(z)$ is continuous and hence for any value z_0 there exists an increment $\Delta_2 z$ such that

$$|S_n(z_0 + \Delta z) - S_n(z_0)| < \frac{\varepsilon}{3} \qquad \text{for all } \Delta z\text{'s such that } |\Delta z| < |\Delta_2 z|$$

Now, choosing $n > N$, we can write

$$|f(z_0 + \Delta z) - f(z_0)| = |[S_n(z_0 + \Delta z) + R_n(z_0 + \Delta z)] - [S_n(z_0) + R_n(z_0)]|$$
$$\leq |S_n(z_0 + \Delta z) - S_n(z_0)| + |R_n(z_0 + \Delta z)| + |R_n(z_0)|$$

Hence, for all values of Δz whose absolute values are less than the smaller of the quantities $|\Delta_1 z|$ and $|\Delta_2 z|$ it follows that

$$|f(z_0 + \Delta z) - f(z_0)| < \frac{\varepsilon}{3} + \frac{\varepsilon}{3} + \frac{\varepsilon}{3} = \varepsilon$$

which is precisely what we mean by saying that $f(z)$ is continuous at $z = z_0$. Since z_0 was an arbitrary point of the region of uniform convergence, the assertion of the theorem is established.

Theorem 5 makes no assertion about the sum of a series of continuous functions if the convergence is nonuniform. However, specific examples make it clear that in such cases the sum need not be continuous. For instance, Example 2, in which we found the sum of the series

$$z^2 + \frac{z^2}{1 + z^2} + \frac{z^2}{(1 + z^2)^2} + \frac{z^2}{(1 + z^2)^3} + \cdots$$

to be

$$f(z) = \begin{cases} 1 + z^2 & z \neq 0 \\ 0 & z = 0 \end{cases}$$

shows that the limit of a sum of continuous functions may be discontinuous if the convergence is nonuniform. In fact, in the neighborhood of $z = 0$, where the convergence is nonuniform, the sum jumps abruptly from $1 + z^2$ to 0, even though every term of the series is a continuous function of z for all values of z except $z = \pm i$.

One of the most important properties of uniformly convergent series is given by the following theorem.

THEOREM 6   The integral of the sum of a uniformly convergent series of continuous functions along any curve C lying entirely in the region of uniform convergence can be found by term-by-term integration of the series. Moreover, if each term of the series is analytic, so too is the sum.

Proof   Let the given series be

$$f(z) = f_1(z) + f_2(z) + \cdots + f_n(z) + \cdots$$

Then, to establish the theorem, we must show that

$$\int_C f(z)\, dz = \int_C f_1(z)\, dz + \int_C f_2(z)\, dz + \cdots + \int_C f_n(z)\, dz + \cdots$$

which, in accordance with the usual definition of convergence, requires that we prove the existence, for every $\varepsilon > 0$, of an integer N such that

$$\left| \int_C f(z)\, dz - \sum_{i=1}^{n} \int_C f_i(z)\, dz \right| < \varepsilon \qquad \text{for all } n > N$$

Now for any *finite* sum it is true that the integral of a sum is equal to the sum of the integrals. Hence, the left member of the last inequality can be written

$$\left| \int_C f(z)\, dz - \int_C \sum_{i=1}^{n} f_i(z)\, dz \right| = \left| \int_C \left[f(z) - \sum_{i=1}^{n} f_i(z) \right] dz \right| = \left| \int_C R_n(z)\, dz \right|$$

Let L be the length of the path of integration. Then, from the uniform convergence of the given series, we know that there exists an integer N such that

$$|R_n(z)| < \frac{\varepsilon}{L} \qquad \text{for all } n > N$$

and for all z's in the region of uniform convergence, in particular for all values of z on the path of integration C. If $n > N$, we can therefore write

$$\left| \int_C f(z)\, dz - \sum_{i=1}^{n} \int_C f_i(z)\, dz \right| = \left| \int_C R_n(z)\, dz \right|$$

$$\le \int_C |R_n(z)| \cdot |dz| < \frac{\varepsilon}{L} \int_C |dz| = \frac{\varepsilon}{L} L = \varepsilon$$

which establishes the first part of the theorem.

To establish the second part, we suppose that the region of uniform convergence D is either simply connected or has been made simply connected by suitable crosscuts. Then if each term f_i is analytic in D, it follows from Cauchy's theorem that the integral of each term around any simple closed curve in D (or its simply connected modification) is zero. Hence, the integral of the sum $f(z)$ around any closed curve is zero, and thus, by Morera's theorem, $f(z)$ is analytic. This completes the proof of the theorem.

The companion result on the term-by-term differentiation of series is contained in the following theorem.

THEOREM 7   If $f(z)$ is the sum of a uniformly convergent series of analytic functions, then the derivative of $f(z)$ at any interior point of the region of uniform convergence can be found by term-by-term differentiation of the series.

Proof   Let z be a general point of the region of uniform convergence D of the given series, and let C be a simple closed curve drawn around z in D. If we write the given series as

$$f(t) = f_1(t) + f_2(t) + \cdots + f_n(t) + \cdots$$

where t is a variable ranging over the values of z on C, we can multiply by the bounded function

$$\frac{1}{2\pi i(t - z)^2}$$

and, by Theorem 4, the resulting series

$$\frac{f(t)}{2\pi i(t - z)^2} = \frac{f_1(t)}{2\pi i(t - z)^2} + \frac{f_2(t)}{2\pi i(t - z)^2} + \cdots + \frac{f_n(t)}{2\pi i(t - z)^2} + \cdots$$

will also converge uniformly. By Theorem 6, it can therefore be integrated term by term around C, giving

$$\frac{1}{2\pi i} \int_C \frac{f(t)\, dt}{(t - z)^2} = \frac{1}{2\pi i} \int_C \frac{f_1(t)\, dt}{(t - z)^2} + \frac{1}{2\pi i} \int_C \frac{f_2(t)\, dt}{(t - z)^2} + \cdots$$

$$+ \frac{1}{2\pi i} \int_C \frac{f_n(t)\, dt}{(t - z)^2} + \cdots$$

But these integrals, by the first generalization of Cauchy's integral formula (Theorem 9, Sec. 15.8) are precisely the derivatives of the respective terms of the given series at the point z. Hence

$$f'(z) = f_1'(z) + f_2'(z) + \cdots + f_n'(z) + \cdots$$

which establishes the theorem.

It is interesting and important to note that Theorem 7 does not apply to series of functions of the real variable x. To justify term-by-term differentiation of such series, we require not uniform convergence of the original series but uniform convergence of the series resulting from the term-by-term differentiation. More precisely, we have the following theorem, which is proved in most texts on advanced calculus.†

THEOREM 8 If $f(x) = f_1(x) + f_2(x) + \cdots + f_n(x) + \cdots$ is a convergent series of functions of the real variable x, each of which possesses a continuous first derivative, then $f'(x)$ can be found by term-by-term differentiation provided the series of the derivatives is uniformly convergent.

EXERCISES

1 Find the region of convergence and the sum of each of the following series:

(a) $1 + (z - i) + (z - i)^2 + (z - i)^3 + \cdots$

(b) $z(1 - z) + z^2(1 - z) + z^3(1 - z) + \cdots$

(c) $\dfrac{1}{2}\dfrac{z + 1}{z - 1} + \dfrac{1}{2^2}\left(\dfrac{z + 1}{z - 1}\right)^2 + \dfrac{1}{2^3}\left(\dfrac{z + 1}{z - 1}\right)^3 + \cdots$

(d) $\dfrac{1}{2(z + i)} + \dfrac{1}{2^2(z + i)^2} + \dfrac{1}{2^3(z + i)^3} + \cdots$

2 Find the region of convergence of each of the following series:

(a) $1 + \dfrac{1}{2^2}\dfrac{\mathscr{R}(z)}{z + 1} + \dfrac{1}{3^2}\left[\dfrac{\mathscr{R}(z)}{z + 1}\right]^2 + \dfrac{1}{4^2}\left[\dfrac{\mathscr{R}(z)}{z + 1}\right]^3 + \cdots$

(b) $1 + \dfrac{1}{2^2}\dfrac{\mathscr{I}(z)}{z + 1} + \dfrac{1}{3^2}\left[\dfrac{\mathscr{I}(z)}{z + 1}\right]^2 + \dfrac{1}{4^2}\left[\dfrac{\mathscr{I}(z)}{z + 1}\right]^3 + \cdots$

3 Show that the entire region of convergence of the series of Example 2 consists of the exterior of the lemniscate $(x^2 - y^2 + 1)^2 + 4x^2y^2 = 1$ together with the origin.

4 Find the region of convergence of the series

$$\frac{4 - z^2}{1^2} + \frac{(4 - z^2)^2}{2^2} + \frac{(4 - z^2)^3}{3^2} + \cdots$$

† See, for instance, A. E. Taylor, "Advanced Calculus," p. 602, Ginn, Boston, 1955.

5 **(a)** Show that the series in Exercise 1*b* converges uniformly for $|z| \leq \rho < 1$ but does not converge uniformly for $|z| \leq 1$.

(b) Show that the series $x + x(1 - x) + x(1 - x)^2 + x(1 - x)^3 + \cdots$ converges for $0 \leq x < 2$ but that the convergence is not uniform in any subinterval which contains the origin. Is the convergence uniform for $0 < \rho_1 \leq x \leq \rho_2 < 2$?

6 Show that the series

$$\frac{1}{1 + x^2} - \frac{1}{2 + x^2} + \frac{1}{3 + x^2} - \frac{1}{4 + x^2} + \cdots$$

converges uniformly over any interval of the x axis but that this cannot be established by the Weierstrass M test.

7 **(a)** Determine the region of convergence and the sum of the series

$$\frac{z}{(0z + 1)(z + 1)} + \frac{z}{(z + 1)(2z + 1)} + \frac{z}{(2z + 1)(3z + 1)} + \frac{z}{(3z + 1)(4z + 1)} + \cdots$$

Show that for $|z| \geq \rho > 1$ the convergence is uniform. *Hint:* Express each term as a sum of partial fractions.

(b) Determine the region of convergence and the sum of the series

$$\frac{1}{z(z + 1)} + \frac{1}{(z + 1)(z + 2)} + \frac{1}{(z + 2)(z + 3)} + \frac{1}{(z + 3)(z + 4)} + \cdots$$

In what region, if any, does the series converge uniformly?

8 What is the region of convergence of the series $\sum\limits_{n=1}^{\infty} \dfrac{e^{inz}}{n^{3/2}}$? Where does the series converge uniformly?

9 Show by an example that the sum of a nonuniformly convergent series of continuous functions may be continuous.

10 Prove Theorem 1.

16.2 Taylor's Expansion

Very often the series with which one has to deal in applications are those which are studied formally in elementary calculus under the name of *Taylor's series*. Their systematic study begins with **Taylor's theorem.**†

THEOREM 1 If $f(z)$ is analytic throughout the region bounded by a simple closed curve C, and if both z and a are interior to C, then

$$f(z) = f(a) + f'(a)(z - a) + f''(a)\frac{(z - a)^2}{2!} + \cdots + f^{(n-1)}(a)\frac{(z - a)^{n-1}}{(n - 1)!} + R_n$$

where
$$R_n = \frac{(z - a)^n}{2\pi i} \int_C \frac{f(t)\,dt}{(t - a)^n(t - z)}$$

Proof We first note that after adding and subtracting a in the denominator of the integrand, Cauchy's integral formula can be written in the form

$$f(z) = \frac{1}{2\pi i} \int_C \frac{f(t)\,dt}{t - z} = \frac{1}{2\pi i} \int_C \frac{f(t)}{t - a} \frac{1}{1 - (z - a)/(t - a)}\,dt$$

---

† Named for the English mathematician Brook Taylor (1685–1731).

Then from this, by applying the identity

$$\frac{1}{1-u} = 1 + u + u^2 + u^3 + \cdots + u^{n-1} + \frac{u^n}{1-u}$$

to the factor $1/[1 - (z - a)/(t - a)]$ in the last integral, we have

$$f(z) = \frac{1}{2\pi i}\int_c \frac{f(t)}{t-a}\left[1 + \frac{z-a}{t-a} + \left(\frac{z-a}{t-a}\right)^2 + \cdots\right.$$

$$\left. + \left(\frac{z-a}{t-a}\right)^{n-1} + \frac{(z-a)^n/(t-a)^n}{1-(z-a)/(t-a)}\right]dt$$

$$= \frac{1}{2\pi i}\int_c \frac{f(t)\,dt}{t-a} + \frac{z-a}{2\pi i}\int_c \frac{f(t)\,dt}{(t-a)^2} + \cdots$$

$$+ \frac{(z-a)^{n-1}}{2\pi i}\int_c \frac{f(t)\,dt}{(t-a)^n} + \frac{(z-a)^n}{2\pi i}\int_c \frac{f(t)\,dt}{(t-a)^n(t-z)}$$

From the generalizations (Theorem 9, Sec. 15.8) of Cauchy's integral formula it is evident that, except for the necessary factorials, the first n integrals in the last expression are precisely the corresponding derivatives of $f(z)$ evaluated at the point $z = a$. Hence,

$$f(z) = f(a) + f'(a)(z - a) + \cdots + f^{(n-1)}(a)\frac{(z-a)^{n-1}}{(n-1)!}$$

$$+ \frac{(z-a)^n}{2\pi i}\int_c \frac{f(t)\,dt}{(t-a)^n(t-z)}$$

which establishes the theorem.

By **Taylor's series** we mean the infinite expansion suggested by the last theorem, namely,

$$f(z) \sim f(a) + f'(a)(z - a) + f''(a)\frac{(z-a)^2}{2!} + \cdots + f^{(n-1)}(a)\frac{(z-a)^{n-1}}{(n-1)!} + \cdots$$

To show that this series actually converges to $f(z)$, we must show, as usual, that the absolute value of the difference between $f(z)$ and the sum of the first n terms of the series approaches zero as n becomes infinite. From Taylor's theorem it is evident that this difference is

$$R_n(z) = \frac{(z-a)^n}{2\pi i}\int_c \frac{f(t)\,dt}{(t-a)^n(t-z)}$$

Accordingly, we must determine the values of z for which the absolute value of this integral approaches zero as n becomes infinite.

To do this, let C_1 and C_2 be two circles of radii r_1 and r_2 having their centers at the point a and lying entirely in the interior of C (Fig. 16.2). Since $f(z)$ is analytic throughout the interior of C, the entire integrand of $R_n(z)$ is analytic in the region between C and C_2 provided that z, like a, lies in the interior of C_2. Under these conditions, the integral around C can be replaced by the integral around C_2. If, in

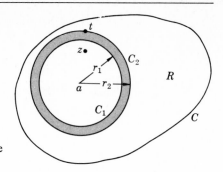

Figure 16.2
The circles C_1 and C_2 used in the proof of the
convergence of Taylor's series.

addition, z is interior to C_1, then for all values of t on C_2 (the values of t which are
now involved in the integration) we have

$$|t - a| = r_2$$
$$|z - a| < r_1$$
$$|t - z| > r_2 - r_1$$

and
$$|f(t)| \leq M$$

where M is the maximum of $|f(z)|$ on C_2. Hence, overestimating factors in the
numerator and underestimating factors in the denominator, we have

$$
\begin{aligned}
|R_n(z)| &= \left| \frac{(z - a)^n}{2\pi i} \int_c \frac{f(t)\, dt}{(t - a)^n (t - z)} \right| \\
&\leq \frac{|z - a|^n}{|2\pi i|} \int_c \frac{|f(t)|\, |dt|}{|t - a|^n |t - z|} \\
&< \frac{r_1^n}{2\pi} \int_c \frac{M\, |dt|}{r_2^n (r_2 - r_1)} \\
&= \frac{r_1^n M}{2\pi r_2^n (r_2 - r_1)}\, 2\pi r_2 \\
&= M \left(\frac{r_1}{r_2} \right)^n \frac{r_2}{r_2 - r_1}
\end{aligned}
$$

Since $0 < r_1 < r_2$, the fraction $(r_1/r_2)^n$ approaches zero as n becomes infinite and
therefore the limit of $|R_n(z)|$ is zero. Thus we have established the following important
theorem.

THEOREM 2   Taylor's series

$$f(z) = f(a) + f'(a)(z - a) + f''(a) \frac{(z - a)^2}{2!} + f'''(a) \frac{(z - a)^3}{3!} + \cdots$$

is a valid representation of $f(z)$ at all points in the interior of any circle having its
center at a and within which $f(z)$ is analytic.

   If we let Γ be the largest circle with center at $z = a$ and within which $f(z)$ is
everywhere analytic, Theorem 2 guarantees that the Taylor's series of $f(z)$ converges
to $f(z)$ at all points in the interior of Γ. However, it does not provide any information

about the behavior of the series outside Γ. Actually, the Taylor's series of $f(z)$ converges to $f(z)$ only within and possibly on the circle Γ and either diverges or else converges to a limit other than $f(z)$ everywhere outside Γ. The next two theorems make these observations more precise.

THEOREM 3   If the power series

$$c_0 + c_1(z - a) + c_2(z - a)^2 + c_3(z - a)^3 + \cdots$$

converges for $z = z_1$, then it converges absolutely for all values of z such that $|z - a| < |z_1 - a|$ and converges uniformly for all values of z such that $|z - a| \leq r < |z_1 - a|$. Moreover, the sum to which it converges is analytic.

Proof   Since the given series converges when $z = z_1$, it follows that the terms of the series are bounded for this value of z. In other words, there exists a positive constant M such that

$$|c_n(z_1 - a)^n| = |c_n| \cdot |z_1 - a|^n \leq M \qquad \text{for } n = 0, 1, 2, \ldots$$

Now let z_0 be any value of z such that

$$|z_0 - a| < |z_1 - a|$$

i.e., let z_0 be any point which is nearer to a than z_1 is. Then, for the general term of the series when $z = z_0$, we have

$$|c_n(z_0 - a)^n| = |c_n| \cdot |z_0 - a|^n = |c_n| \cdot |z_1 - a|^n \cdot \left|\frac{z_0 - a}{z_1 - a}\right|^n \leq M \left|\frac{z_0 - a}{z_1 - a}\right|^n$$

If we set

$$(1) \qquad\qquad \left|\frac{z_0 - a}{z_1 - a}\right| = k$$

where k is obviously less than 1, this shows that the absolute values of the terms of the series

$$(2) \qquad\qquad c_0 + c_1(z - a) + c_2(z - a)^2 + c_3(z - a)^3 + \cdots$$

are dominated, respectively, by the terms of the series of positive constants

$$(3) \qquad\qquad M + Mk + Mk^2 + Mk^3 + \cdots$$

This is a geometric series whose common ratio k is numerically less than 1. It therefore converges, and provides a comparison test which establishes the absolute convergence of the given series (2).

Unfortunately, the series (3) does not provide a test series which can be used in applying the Weierstrass M test to the series (2), because it is clear from the definition of k in Eq. (1) that the terms of the series (3) depend on z_0. However, for values of z_0 such that

$$(4) \qquad\qquad |z_0 - a| \leq r < |z_1 - a|$$

we have

$$k = \left|\frac{z_0 - a}{z_1 - a}\right| \leq \frac{r}{|z_1 - a|} = \lambda, \text{ say}$$

and λ is clearly a positive constant less than 1 which *is* independent of the general point z_0. Hence, for all values of z_0 satisfying the condition (4), the series (2) is dominated term by term by the convergent geometric series of positive constants

$$M + M\lambda + M\lambda^2 + M\lambda^3 + \cdots$$

and therefore, by Theorem 3, Sec. 16.1, the series (2) is uniformly convergent.

Finally, since each term $c_n(z - a)^n$ of the given series is an analytic function, and since any point in the interior of the circle $|z - a| = |z_1 - a|$ can be included within a circle of the form $|z - a| = r < |z_1 - a|$, it follows from the second part of Theorem 6, Sec. 16.1, that within the circle $|z - a| = |z_1 - a|$ the function to which the series converges is analytic.

Now let α be the singular point of $f(z)$ which is nearest to the center of the expansion $z = a$, and suppose that the Taylor's series of $f(z)$ converges for some value $z = z_1$ farther from a than α is. By the last theorem, the series must converge at all points nearer to a than z_1 is, and, moreover, the sum to which it converges must be analytic at every such point. Therefore the series converges to a function which is analytic at α, in apparent contradiction of the assumption that α is a point where $f(z)$ is not analytic. Thus we must either abandon the supposition that the series converges at points farther from a than the distance $|\alpha - a|$, or else we must accept the fact that the function to which the series converges in the neighborhood of α is different from $f(z)$. Each of these alternatives is possible,† but in either case we have established the following theorem.

THEOREM 4   It is impossible for the Taylor's series of a function $f(z)$ to converge to $f(z)$ outside the circle whose center is the point of expansion $z = a$ and whose radius is the distance from a to the nearest singularity of $f(z)$.

The largest circle which can be drawn around the point of expansion $z = a$ such that the Taylor's series of $f(z)$ converges everywhere in its interior is called the **circle of convergence** of the series, and the radius of this circle is called the **radius of convergence** of the series. In general, the radius of convergence of the Taylor's series of $f(z)$ around the point $z = a$ is equal to the distance from a to the nearest singularity of $f(z)$, and in any event it is at least as large as this distance. In particular, if the nearest singular point α has the property that $f(z)$ becomes infinite as z approaches α, then the radius of convergence is equal to $|\alpha - a|$. Of course, this entire discussion applies without change to the case $a = 0$, which is usually called **Maclaurin's series**.‡

The notion of the circle of convergence is often useful in determining the interval

---

† The usual situation is that the series diverges for all values of z such that $|z - a| > |\alpha - a|$. It is possible, however, that for values of z such that $|z - a| > |\alpha - a|$ the Taylor's series of $f(z)$ converges to an analytic function which is different from $f(z)$. Examples of this behavior will be found in Exercises 11 and 12.

‡ Named for the Scottish mathematician Colin Maclaurin (1698–1746), although another Scottish mathematician, James Stirling (1692–1770), anticipated by 25 years Maclaurin's use of this result.

of convergence of a series arising as the expansion of a function of a real variable. To illustrate, consider

$$f(z) = \frac{1}{1 + z^2} = 1 - z^2 + z^4 - z^6 + \cdots$$

This will converge throughout the largest circle around the origin in which $f(z)$ is analytic. Now, by inspection, $f(z)$ becomes infinite as z approaches $\pm i$, and even though one may be concerned only with real values of z [for which $1/(1 + x^2)$ is everywhere infinitely differentiable], these singularities in the complex plane set an inescapable limit to the interval of convergence on the x axis. We can, in fact, have convergence around $x = a$ on the real axis only over the horizontal diameter of the circle of convergence around the point $z = a$ in the complex plane.

As an application of Taylor's expansion, we shall conclude this section by establishing the simple but important result known as the **theorem of Liouville.**†

THEOREM 5 If $f(z)$ is bounded and analytic for all values of z, then $f(z)$ is a constant.

Proof To prove this, we observe first that since $f(z)$ is everywhere analytic, it possesses a power-series expansion about the origin

$$f(z) = f(0) + f'(0)z + \cdots + \frac{f^{(n)}(0)}{n!} z^n + \cdots$$

which converges and represents it for all values of z. Now if C is an arbitrary circle having the origin as center, it follows from Cauchy's inequality (Theorem 11, Sec. 15.8) that

$$|f^{(n)}(0)| \le \frac{n! \, M_C}{r^n}$$

where r is the radius of C and M_C is the maximum value of $|f(z)|$ on C. Hence, for the coefficient of z^n in the expansion of $f(z)$, we have

$$\left| \frac{f^{(n)}(0)}{n!} \right| \le \frac{M_C}{r^n} \le \frac{M}{r^n}$$

where M, the bound on $|f(z)|$ for *all* values of z, which exists by hypothesis, is independent of r. Since r can be taken arbitrarily large, it follows, therefore, that the coefficient of z^n is zero for $n = 1, 2, 3, \ldots$. In other words, for all values of z,

$$f(z) = f(0)$$

which proves the theorem.

A function which is analytic for all values of z is called an **entire function** or an **integral function**, and Liouville's theorem thus states that *any entire function which is bounded for all values of z is a constant.*

---

† Named for the French mathematician Joseph Liouville (1809–1882) but actually due to Cauchy.

EXERCISES

1   Expand $f(z) = (z - 1)/(z + 1)$ in a Taylor's series (a) about the point $z = 0$ and
    (b) about the point $z = 1$. Determine the radius of convergence of each series.

2   Expand $f(z) = [1/(z + 1)(z + 2)]$ in a Taylor's series (a) about the point $z = 0$
    and (b) about the point $z = 2$. Determine the radius of convergence of each series.

3   Expand $f(z) = \cosh z$ in a Taylor's series about the point $z = i\pi$. What is the
    radius of convergence of this series?

4   Obtain the first three nonzero terms in the Maclaurin expansion of (a) $\tan z$ and
    (b) $\sec z$. What is the radius of convergence of each series?

5   Without obtaining the series, determine the radius of convergence of each of the
    following expansions:

    (a) $\text{Tan}^{-1} z$ around $z = 1$ (b) $\dfrac{1}{e^z - 1}$ around $z = 4i$

    (c) $\dfrac{x}{x^2 + 2x + 10}$ around $x = 0$ (d) $\dfrac{1}{x^2 - 9x + 10}$ around $x = -1$

6   If the radius of convergence of the power series $\sum\limits_{n=0}^{\infty} c_n(z - a)^n$ is R, prove that
    the radius of convergence of the termwise derivative of this series is also R.

7   If $\sum\limits_{n=0}^{\infty} c_n(z - a)^n$ converges to $f(z)$, show that in the interior of the circle of
    convergence the derivative of $f(z)$ can be found by termwise differentiation of the
    series and the integral of $f(z)$ can be found by termwise integration of the series.

8   Prove that if a function is analytic at a point $z = a$, its power-series expansion in
    the neighborhood of that point is unique.

9   Show that if the values of an analytic function $f(z)$ are known at the points of an
    arbitrarily short arc having $z = a$ as one end point, then the Taylor's expansion of
    $f(z)$ around $z = a$ is completely determined.

10   Prove that if $\sum\limits_{n=0}^{\infty} c_n z^n$ converges absolutely at one point on its circle of con-
    vergence, then it converges absolutely and uniformly in the closed region bounded
    by its circle of convergence.

11   Find the Maclaurin expansion of

$$f(z) = \int_c \frac{dt}{t - z} \qquad |z| \neq 1$$

    if C is the circle $|t| = 1$. Show that this expansion converges for all values of z
    but represents $f(z)$ only in the interior of C. *Hint:* Recall Cauchy's integral formulas
    and the results of Example 1, Sec. 15.8.

12   Find the Taylor's series of the principal value of $\ln z$ around the point $z = -1 + i$.
    Show that the radius of convergence of this series is $\sqrt{2}$ but that the series represents
    the principal value of $\ln z$ only within a circle of radius 1 about the point
    $z = -1 + i$.

13   Prove that every polynomial equation $P(z) = 0$ has at least one root. *Hint:*
    Assume the contrary and apply the theorem of Liouville to the function

$$f(z) = \frac{1}{P(z)}$$

14   (a) If z_0 is a point within the circle Γ where the Taylor's series of a function $f(z)$
    converges to $f(z)$, then by Exercise 7, $f(z_0)$ and all the derivatives of $f(z)$ at $z = z_0$
    can be computed. Hence, by using these values, the expansion of $f(z)$ around
    $z = z_0$ can be constructed. Show that in general this new series will converge in
    the interior of a circle Γ' which lies partly inside and partly outside Γ. [When this
    is the case, the second series is said to be an **analytic continuation** of the first, since

it provides a series representation of $f(z)$ beyond that provided by the first series].

(b) Show that the series

$$\sum_{n=0}^{\infty} \frac{z^n}{2^{n+1}} \quad \text{and} \quad \sum_{n=0}^{\infty} \frac{(z-i)^n}{(2-i)^{n+1}}$$

are analytic continuations of each other and determine the region common to their respective circles of convergence.

15 Show that no analytic continuation beyond $|z| = 1$ is possible for the function defined by the series

$$1 + z + z^2 + z^4 + z^8 + \cdots = 1 + \sum_{n=0}^{\infty} z^{2^n}$$

Hint: Note first that $z = 1$ is a singular point of the given function. Then note that

$$f(z) = z + f(z^2)$$
$$= z + z^2 + f(z^4)$$
$$= z + z^2 + z^4 + f(z^8)$$
$$\cdots\cdots\cdots\cdots\cdots\cdots\cdots\cdots$$

so that not only the value $z = 1$ but also the values of z such that $z^2 = 1$, $z^4 = 1$, $z^8 = 1, \ldots$ are singular points of $f(z)$. Finally, consider the distribution of these singularities around the circumference of the circle of convergence $|z| = 1$. A curve, such as this, beyond which a function cannot be extended by analytic continuation is called a **natural boundary** of the function.

16.3 Laurent's Expansion

In many applications it is necessary to expand functions around points at which or in the neighborhood of which the functions are not analytic. The method of Taylor's series is obviously inapplicable in such cases, and a new type of series known as **Laurent's expansion**† is required. This furnishes us with a representation which is valid in the annular ring bounded by two concentric circles, provided that the function being expanded is analytic everywhere on and between the two circles. As with Taylor's series, the function may have singular points outside the larger circle, and, as the essentially new feature, it may also have singular points within the inner circle. The price we pay for this is that negative as well as positive powers of $z - a$ now appear in the expansion and that the coefficients, even of the positive powers of $z - a$, cannot be expressed in terms of the evaluated derivatives of the function. The precise result is given by the following theorem.

THEOREM 1 If $f(z)$ is analytic throughout the closed region R bounded by two concentric circles, then at any point in the annular ring bounded by the circles, $f(z)$ can be represented by the series

$$f(z) = \sum_{n=-\infty}^{\infty} a_n (z - a)^n$$

† Named for the French mathematician Hermann Laurent (1841–1908).

where a is the common center of the circles and

$$a_n = \frac{1}{2\pi i} \int_C \frac{f(t)\,dt}{(t-a)^{n+1}}$$

each integral being taken in the counterclockwise sense around any curve C lying in the annulus and encircling its inner boundary.

Proof Let z be an arbitrary point of the given annulus. Then, according to Cauchy's integral formula, we can write

$$f(z) = \frac{1}{2\pi i} \int_{C_2+C_1} \frac{f(t)\,dt}{t-z}$$

$$= \frac{1}{2\pi i} \int_{C_2} \frac{f(t)\,dt}{t-z} + \frac{1}{2\pi i} \int_{C_1} \frac{f(t)\,dt}{t-z}$$

where C_2 is traversed in the counterclockwise direction and C_1 is traversed in the clockwise direction, in order that the entire integration shall be in the positive direction (Fig. 16.3). Reversing the sign of the integral around C_1 and also changing the direction of integration from clockwise to counterclockwise, we can write

$$f(z) = \frac{1}{2\pi i} \int_{C_2} \frac{f(t)\,dt}{t-z} - \frac{1}{2\pi i} \int_{C_1} \frac{f(t)\,dt}{t-z}$$

$$= \frac{1}{2\pi i} \int_{C_2} \frac{f(t)}{t-a} \frac{1}{1-(z-a)/(t-a)}\,dt$$

$$+ \frac{1}{2\pi i} \int_{C_1} \frac{f(t)}{z-a} \frac{1}{1-(t-a)/(z-a)}\,dt$$

Now, in each of these integrals let us apply the identity

$$\frac{1}{1-u} = 1 + u + u^2 + \cdots + \frac{u^n}{1-u}$$

Figure 16.3
The circles C_1 and C_2 used in the derivation of Laurent's expansion.

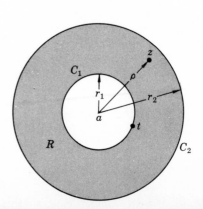

to the last factor. Then

$$f(z) = \frac{1}{2\pi i} \int_{C_2} \frac{f(t)}{t - a} \left[1 + \frac{z - a}{t - a} + \cdots + \left(\frac{z - a}{t - a}\right)^{n-1} + \frac{(z - a)^n/(t - a)^n}{1 - (z - a)/(t - a)} \right] dt$$

$$+ \frac{1}{2\pi i} \int_{C_1} \frac{f(t)}{z - a} \left[1 + \frac{t - a}{z - a} + \cdots + \left(\frac{t - a}{z - a}\right)^{n-1} \right.$$

$$\left. + \frac{(t - a)^n/(z - a)^n}{1 - (t - a)/(z - a)} \right] dt$$

$$= \frac{1}{2\pi i} \int_{C_2} \frac{f(t)\, dt}{t - a} + \frac{z - a}{2\pi i} \int_{C_2} \frac{f(t)\, dt}{(t - a)^2} + \cdots$$

$$+ \frac{(z - a)^{n-1}}{2\pi i} \int_{C_2} \frac{f(t)\, dt}{(t - a)^n} + R_{n2}$$

$$+ \frac{1}{2\pi i(z - a)} \int_{C_1} f(t)\, dt + \frac{1}{2\pi i(z - a)^2} \int_{C_1} (t - a)f(t)\, dt + \cdots$$

$$+ \frac{1}{2\pi i(z - a)^n} \int_{C_1} (t - a)^{n-1}f(t)\, dt + R_{n1}$$

where

$$R_{n2} = \frac{(z - a)^n}{2\pi i} \int_{C_2} \frac{f(t)\, dt}{(t - a)^n(t - z)}$$

$$R_{n1} = \frac{1}{2\pi i(z - a)^n} \int_{C_1} \frac{(t - a)^n f(t)\, dt}{z - t}$$

The truth of the theorem will be established if we can show that

$$\lim_{n \to \infty} R_{n2} = 0 \quad \text{and} \quad \lim_{n \to \infty} R_{n1} = 0 \qquad .$$

The proof of the first of these equations we can pass over without comment, because it was given in complete detail in the derivation of Taylor's series in Sec. 16.2. To prove the second, we note that for values of t on C_1 (Fig. 16.3),

$$|t - a| = r_1$$
$$|z - a| = \rho \quad \text{say, where } \rho > r_1$$
$$|z - t| = |(z - a) - (t - a)| \geq \rho - r_1$$

and

$$|f(t)| \leq M$$

where M is the maximum of $|f(z)|$ on C_1. Thus

$$|R_{n1}| = \left| \frac{1}{2\pi i(z - a)^n} \int_{C_1} \frac{(t - a)^n f(t)\, dt}{z - t} \right|$$

$$\leq \frac{1}{|2\pi i|\, |z - a|^n} \int_{C_1} \frac{|t - a|^n |f(t)|\, |dt|}{|z - t|}$$

$$\leq \frac{r_1^n M}{2\pi\rho^n(\rho - r_1)} \int_{C_1} |dt|$$

$$= \frac{M}{2\pi} \left(\frac{r_1}{\rho}\right)^n \frac{2\pi r_1}{\rho - r_1}$$

$$= M \left(\frac{r_1}{\rho}\right)^n \frac{r_1}{\rho - r_1}$$

Since $0 < r_1/\rho < 1$, the last expression approaches zero as n becomes infinite. Hence, $\lim\limits_{n \to \infty} R_{n1} = 0$; and thus we have

$$f(z) = \frac{1}{2\pi i} \int_{C_2} \frac{f(t)\, dt}{t - a} + \left[\frac{1}{2\pi i} \int_{C_2} \frac{f(t)\, dt}{(t - a)^2} \right] (z - a)$$

$$+ \left[\frac{1}{2\pi i} \int_{C_2} \frac{f(t)\, dt}{(t - a)^3} \right] (z - a)^2 + \cdots$$

$$+ \left[\frac{1}{2\pi i} \int_{C_1} f(t)\, dt \right] \frac{1}{z - a} + \left[\frac{1}{2\pi i} \int_{C_1} (t - a)f(t)\, dt \right] \frac{1}{(z - a)^2} + \cdots$$

Since $f(z)$ is analytic throughout the region between C_1 and C_2, the paths of integration C_1 and C_2 can be replaced by any other curve C within this region and encircling C_1. The resulting integrals are precisely the coefficients a_n described by the theorem; hence, our proof is complete.

It should be noted that the coefficients of the positive powers of $z - a$ in Laurent's expansion, although identical in form with the integrals of Theorem 9, Sec. 15.8, *cannot* be replaced by the derivative expressions

$$\frac{f^{(n)}(a)}{n!}$$

as they were in the derivation of Taylor's series, since $f(z)$ is not analytic throughout the entire interior of C_2 (or C) and hence Cauchy's generalized integral formula cannot be applied. Specifically, $f(z)$ may have many points of nonanalyticity within C_1 and therefore within C_2 (or C).

In many instances the Laurent expansion of a function is found not through the use of the last theorem but by algebraic manipulations suggested by the nature of the function. In particular, in dealing with the quotients of polynomials, it is often advantageous to express them in terms of partial fractions and then expand the various denominators in series of the appropriate form through the use of the binomial expansion, which we list here for reference.

THEOREM 2   The expansion

$$(s + t)^n = s^n + ns^{n-1}t + \frac{n(n-1)}{2!} s^{n-2}t^2 + \frac{n(n-1)(n-2)}{3!} s^{n-3}t^3 + \cdots$$

is valid for all values of n if $|s| > |t|$. If $|s| \le |t|$, the expansion is valid only if n is a nonnegative integer.

That such procedures are correct follows from the fact that *the Laurent expansion of a function over a given annulus is unique.* In other words, if an expansion of the Laurent type is found by any process, it must be *the* Laurent expansion.

EXAMPLE 1

Find the Laurent expansion of the function

$$f(z) = \frac{7z - 2}{(z + 1)z(z - 2)}$$

in the annulus $1 < |z + 1| < 3$.

As a preliminary step it is convenient to apply the method of partial fractions to $f(z)$ and express it in the form

$$f(z) = \frac{-3}{z + 1} + \frac{1}{z} + \frac{2}{z - 2}$$

Now, after suitable rearrangement, these fractions can be expanded into infinite series by means of Theorem 2 and added to give the required expansion of $f(z)$.

To do this, we observe that since the center of the given annulus is $z = -1$, the series we are seeking must be one involving powers of $z + 1$. Hence, we modify the second and third terms in the partial-fraction representation of $f(z)$ so that z will appear in the combination $z + 1$. This gives us the equivalent representation

$$f(z) = \frac{-3}{z + 1} + \frac{1}{(z + 1) - 1} + \frac{2}{(z + 1) - 3}$$

$$= -3(z + 1)^{-1} + [(z + 1) - 1]^{-1} + 2[(z + 1) - 3]^{-1}$$

However, according to Theorem 2, the series for $[(z + 1) - 3]^{-1}$ will converge only where $|z + 1| > 3$, whereas we require an expansion valid for $|z + 1| < 3$. Hence we rewrite this term in the other order, $[-3 + (z + 1)]^{-1}$, before expanding it. Now we can apply Theorem 2, obtaining

$$f(z) = -3(z + 1)^{-1} + [(z + 1) - 1]^{-1} + 2[-3 + (z + 1)]^{-1}$$

$$= -3(z + 1)^{-1} + [(z + 1)^{-1} + (z + 1)^{-2} + (z + 1)^{-3} + \cdots]$$

$$+ 2\left[-\frac{1}{3} - \frac{z + 1}{9} - \frac{(z + 1)^2}{27} - \frac{(z + 1)^3}{81} - \cdots \right]$$

$$= \cdots + (z + 1)^{-3} + (z + 1)^{-2} - 2(z + 1)^{-1} - \frac{2}{3}$$

$$- \frac{2}{9}(z + 1) - \frac{2}{27}(z + 1)^2 - \frac{2}{81}(z + 1)^3 + \cdots \qquad 1 < |z + 1| < 3$$

It is important to note that $f(z)$ has two other Laurent expansions around the point $z = -1$. One is valid in the annular region between a circle of arbitrarily small radius around $z = -1$ and a circle of unit radius around $z = -1$. The other is valid in the region exterior to a circle of radius 3 around $z = -1$ (Fig. 16.4). Each of these can be found, as above, by suitably rearranging the terms in the partial-fraction representation of $f(z)$ and then expanding these terms by means of Theorem 2. Thus in the innermost region we have

$$f(z) = -3(z + 1)^{-1} + [-1 + (z + 1)]^{-1} + 2[-3 + (z + 1)]^{-1}$$

$$= -3(z + 1)^{-1} + [-1 - (z + 1) - (z + 1)^2 - (z + 1)^3 - \cdots]$$

$$+ 2\left[-\frac{1}{3} - \frac{z + 1}{9} - \frac{(z + 1)^2}{27} - \frac{(z + 1)^3}{81} - \cdots \right]$$

$$= -3(z + 1)^{-1} - \frac{5}{3} - \frac{11}{9}(z + 1) - \frac{29}{27}(z + 1)^2$$

$$- \frac{83}{81}(z + 1)^3 - \cdots \qquad 0 < |z + 1| < 1$$

Similarly, in the outermost region we have

$$f(z) = -3(z + 1)^{-1} + [(z + 1) - 1]^{-1} + 2[(z + 1) - 3]^{-1}$$

$$= -3(z + 1)^{-1} + [(z + 1)^{-1} + (z + 1)^{-2} + (z + 1)^{-3} + \cdots]$$

$$+ 2[(z + 1)^{-1} + 3(z + 1)^{-2} + 9(z + 1)^{-3} + \cdots]$$

$$= \cdots + 19(z + 1)^{-3} + 7(z + 1)^{-2} \qquad |z + 1| > 3$$

Incidentally, the fact that we have obtained these Laurent expansions without using the general theory means that we can evaluate the integrals in the coefficient formulas by comparing them with the numerical values of the coefficients we have found by independent means. For instance, in the first expansion the coefficient of $(z + 1)^{-1}$

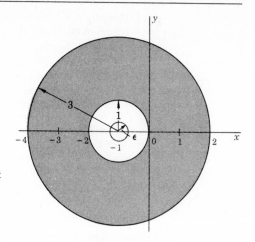

Figure 16.4
The regions of validity of the three Laurent expansions of $\dfrac{7z - 2}{(z + 1)z(z - 2)}$ around $z = -1$.

is -2. On the other hand, according to the theory of Laurent's expansion, the co-efficient of this term is

$$a_{-1} = \frac{1}{2\pi i} \int_C f(z)\, dz = \frac{1}{2\pi i} \int_C \frac{7z - 2}{(z + 1)z(z - 2)}\, dz$$

where C is any closed curve lying in the interior of the circle $|z + 1| = 3$ and enclosing the circle $|z + 1| = 1$. Thus, although we have done nothing resembling an integration, we have nonetheless shown that

$$\frac{1}{2\pi i} \int_C \frac{7z - 2}{(z + 1)z(z - 2)}\, dz = -2 \quad \text{or} \quad \int_C \frac{7z - 2}{(z + 1)z(z - 2)}\, dz = -4\pi i$$

a result, incidentally, which could not have been obtained by a direct application of Cauchy's integral formula, as in Example 2, Sec. 15.8.

EXERCISES

1   Expand $f(z) = 1/(z - 1)(z - 2)$:
   (a) For $|z| < 1$ (b) For $1 < |z| < 2$ (c) For $2 < |z|$
   (d) For $0 < |z - 1| < 1$ (e) For $|z - 1| > 1$
   (f) For $0 < |z - 2| < 1$ (g) For $|z - 2| > 1$

2   Obtain two distinct Laurent expansions for $f(z) = (3z + 1)/(z^2 - 1)$ around $z = 1$, and tell where each converges.

3   Expand $f(z) = 1/z^2(z - i)$ in two different Laurent expansions around $z = i$, and tell where each converges.

4   Construct all the Laurent expansions of $f(z) = 1/z(z - 1)(z - 2)$ around $z = -1$, and tell where each converges.

5   Find the value of $\displaystyle\int_C f(z)\, dz$ if C is the circle $|z| = 3$ and $f(z)$ is:

   (a) $\dfrac{1}{z(z + 2)}$ (b) $\dfrac{z + 2}{z(z + 1)}$ (c) $\dfrac{1}{(z + 1)^2}$

   (d) $\dfrac{1}{z(z + 1)^2}$ (e) $\dfrac{z}{(z + 1)(z + 2)}$ (f) $\dfrac{1}{z(z + 1)(z + 4)}$

6   If k is a real number such that $k^2 < 1$, prove that

$$\sum_{n=0}^{\infty} k^n \sin (n + 1)\theta = \frac{\sin \theta}{1 - 2k \cos \theta + k^2}$$

$$\sum_{n=0}^{\infty} k^n \cos (n + 1)\theta = \frac{\cos \theta - k}{1 - 2k \cos \theta + k^2}$$

Hint: Expand $(z - k)^{-1}$ for $|z| > k$, set $z = e^{i\theta}$, and equate real and imaginary components in the resulting expression.

7 Criticize the following argument: Since (by long division, for instance)

$$\frac{z}{1-z} = z + z^2 + z^3 + z^4 + \cdots$$

and

$$\frac{z}{z-1} = 1 + \frac{1}{z} + \frac{1}{z^2} + \frac{1}{z^3} + \cdots$$

and since

$$\frac{z}{1-z} + \frac{z}{z-1} = 0$$

therefore, by adding these two series we obtain

$$\cdots + \frac{1}{z^3} + \frac{1}{z^2} + \frac{1}{z} + 1 + z + z^2 + z^3 + z^4 + \cdots = 0$$

8 Criticize the following argument: the series

$$\frac{1}{z} + 1 + z + z^2 + z^3 + z^4 + \cdots$$

converges to the sum $S(z) = 1/z(1-z)$ for all values of z such that $|z| < 1$, *including $z = 0$*, since

$$|S(z) - S_n(z)| = \left| \frac{1}{z(1-z)} - \left(\frac{1}{z} + 1 + z + \cdots + z^{n-2} \right) \right|$$

$$= \left| \frac{1}{z} + \frac{1}{1-z} - \frac{1}{z} - 1 - z - \cdots - z^{n-2} \right|$$

$$= \left| \frac{1}{1-z} - 1 - z - \cdots - z^{n-2} \right|$$

$$= \left| \frac{z^{n-1}}{1-z} \right|$$

and this expression clearly approaches 0 as n becomes infinite for *all* values of z such that $|z| < 1$.

9 **(a)** Show that the Laurent expansion of $f(z) = \sinh[z + (1/z)]$ in powers of z is

$$\sum_{n=-\infty}^{\infty} a_n z^n \quad \text{where} \quad a_n = \frac{1}{2\pi} \int_0^{2\pi} \cos n\theta \sinh(2\cos\theta)\, d\theta$$

Hint: In the formula for a provided by Theorem 1, take the curve C to be the circle $|z| = 1$. On this circle let the variable of integration t be taken in the form $t = e^{i\theta}$. Finally, verify that the imaginary part of the integral for a_n is equal to zero.

(b) Show that the coefficients in the Laurent expansion of $f(z) = \sin[z + (1/z)]$ in powers of z are given by the formula

$$a_n = \frac{1}{2\pi} \int_0^{2\pi} \cos n\theta \sin(2\cos\theta)\, d\theta$$

(c) What are the coefficients in the Laurent expansion of $f(z) = \cosh[z + (1/z)]$ in powers of z?

(d) What are the coefficients in the Laurent expansion of $f(z) = \cos[z + (1/z)]$ in powers of z?

10 Show that the Laurent expansion of the function

$$f(z) = \exp\left[\frac{u}{2} \left(z - \frac{1}{z} \right) \right] \qquad |z| > 0$$

is

$$\sum_{n=-\infty}^{\infty} a_n z^n \quad \text{where} \quad a_n = \frac{1}{2\pi} \int_0^{2\pi} \cos(n\theta - u\sin\theta)\, d\theta$$

and verify that a_n is the Bessel function $J_n(u)$. *Hint:* In the integral for a_n given by Theorem 1, let $z = e^{i\theta}$; then recall Eq. (12), Sec. 9.5.

11   If $F(\theta)$ is the function of θ to which $f(z)$ reduces on the circle $|z| = 1$, show that

$$F(\theta) = \sum_{n=-\infty}^{\infty} a_n e^{in\theta} \qquad \text{where} \qquad a_n = \frac{1}{2\pi} \int_0^{2\pi} F(\theta) e^{-in\theta} \, d\theta$$

12   Using the result of Exercise 11, show that

$$F(\theta) = \frac{1}{2\pi} \int_0^{2\pi} F(\phi) \, d\phi + \frac{1}{\pi} \sum_{n=1}^{\infty} \int_0^{2\pi} F(\phi) \cos n(\theta - \phi) \, d\phi$$

CHAPTER 17

The Theory of Residues

17.1 The Residue Theorem

In Sec. 15.6 we defined a singular point of a function $f(z)$ as a point .where $f(z)$ is not analytic but in every neighborhood of which there are points where $f(z)$ is analytic. If $z = a$ is a singular point of the function $f(z)$ but there exists a neighborhood of a in which there are no other singular points of $f(z)$, then $z = a$ is called an **isolated singular point**. Clearly, if $z = a$ is an isolated singularity of $f(z)$, then $f(z)$ will possess a Laurent expansion around $z = a$ which will represent $f(z)$ in the interior of an annulus whose outer radius is the distance from a to the nearest of the other singular points of $f(z)$ and whose inner radius can be taken arbitrarily small.

If the Laurent expansion of $f(z)$ in the neighborhood of an isolated singular point $z = a$ contains only a finite number of negative powers of $z - a$, then $z = a$ is called a **pole** of $f(z)$. If $(z - a)^{-m}$ is the highest negative power in the expansion, the pole is said to be of **order** m and the sum of all terms containing negative powers, namely,

$$\frac{a_{-m}}{(z - a)^m} + \cdots + \frac{a_{-2}}{(z - a)^2} + \frac{a_{-1}}{z - a}$$

is called the **principal part** of $f(z)$ at $z = a$. If the Laurent expansion of $f(z)$ in the neighborhood of an isolated singular point $z = a$ contains infinitely many negative power of $z - a$, then $z = a$ is called an **essential singularity** of $f(z)$. For instance, since

$$\frac{1}{z(z - 1)^2} = \frac{[1 + (z - 1)]^{-1}}{(z - 1)^2}$$

$$= \frac{1}{(z - 1)^2} - \frac{1}{z - 1} + 1 - (z - 1) + \cdots \qquad 0 < |z - 1| < 1$$

this function has a pole of order 2 at $z = 1$ and its principal part there is

$$\frac{1}{(z - 1)^2} - \frac{1}{z - 1} \text{†}$$

---

† It should be noted that although we can also write

$$\frac{1}{z(z - 1)^2} = \frac{[(z - 1) + 1]^{-1}}{(z - 1)^2} = \cdots + \frac{1}{(z - 1)^5} - \frac{1}{(z - 1)^4} + \frac{1}{(z - 1)^3} \qquad |z - 1| > 1$$

the fact that this expansion contains infinitely many negative powers of $z - 1$ does not contradict our observation that $1/z(z - 1)^2$ has a pole of order 2 at $z = 1$. For this series is valid only *outside* the circle $|z - 1| = 1$, whereas the presence of poles and essential singularities is determined by the structure of the particular Laurent expansion which is valid in the *innermost* annulus, or deleted neighborhood, of the point in question.

On the other hand, since $e^{1/z}$ is represented for all values of z except $z = 0$ by the series

$$e^{1/z} = 1 + \frac{1}{z} + \frac{1}{2!\, z^2} + \frac{1}{3!\, z^3} + \frac{1}{4!\, z^4} + \cdots$$

it has an essential singularity at the origin.

In passing, we note that if the terms in the expansion of $f(z)$ around a pole of order m, say $z = a$, are put over a common denominator, $f(z)$ will contain the factor $1/(z - a)^m$. Conversely, if a function $f(z)$ is expressed as a fraction in lowest terms, then the presence of a factor of the form $(z - a)^m$ in the denominator implies that $f(z)$ has a pole of order m at $z = a$. In most applications this is how the poles of a function are found.

As we suggested at the end of the last chapter, the coefficient a_{-1} of the term $(z - a)^{-1}$ in the Laurent expansion of a function $f(z)$ is of great importance because of its connection with the integral of the function, through the formula

$$a_{-1} = \frac{1}{2\pi i} \int_C f(z)\, dz$$

In particular, the coefficient of $(z - a)^{-1}$ in the expansion of $f(z)$ *in the neighborhood of an isolated singular point* is called the **residue** of $f(z)$ at that point.

Consider a simple closed curve C containing in its interior a number of isolated singularities of a function $f(z)$. If around each singular point we draw a circle so small that it encloses no other singular points (Fig. 17.1), these circles, together with the curve C, form the boundary of a multiply connected region in which $f(z)$ is everywhere analytic and to which Cauchy's theorem can therefore be applied. This gives

$$\frac{1}{2\pi i} \int_C f(z)\, dz + \frac{1}{2\pi i} \int_{C_1} f(z)\, dz + \cdots + \frac{1}{2\pi i} \int_{C_n} f(z)\, dz = 0$$

If we reverse the direction of integration around each of the circles and change the sign of each integral to compensate, this can be written

$$\frac{1}{2\pi i} \int_C f(z)\, dz = \frac{1}{2\pi i} \int_{C_1} f(z)\, dz + \cdots + \frac{1}{2\pi i} \int_{C_n} f(z)\, dz$$

Figure 17.1
The circles $C_1, C_2, \ldots, C_n$ enclosing, respectively, the singular points $z_1, z_2, \ldots, z_n$ within a simple closed curve.

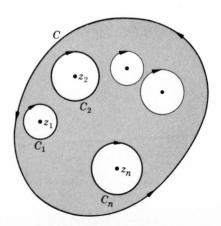

where all the integrals are now taken in the counterclockwise sense. But the integrals on the right are, by definition, just the residues of $f(z)$ at the various isolated singularities within C. Hence we have established the important **residue theorem**.

THEOREM 1   If C is a closed curve, and if $f(z)$ is analytic within and on C except at a finite number of singular points in the interior of C, then

$$\int_C f(z)\,dz = 2\pi i(r_1 + r_2 + \cdots + r_n)$$

where $r_1, r_2, \ldots, r_n$ are the residues of $f(z)$ at its singular points within C.

EXAMPLE 1

What is the integral of

$$f(z) = \frac{-3z + 4}{z(z - 1)(z - 2)}$$

around the circle $|z| = \frac{3}{2}$?

In this case, although there are three singular points of the function, namely, the three first-order poles at $z = 0$, $z = 1$, and $z = 2$, only $z = 0$ and $z = 1$ lie within the path of integration. Hence the core of the problem is to find the residues of $f(z)$ at these two points.

To do this, it is natural to begin by constructing the partial-fraction representation of $f(z)$, namely,

$$f(z) = \frac{2}{z} - \frac{1}{z - 1} - \frac{1}{z - 2}$$

Then in the neighborhood of $z = 0$ we can write

$$f(z) = \frac{2}{z} + (1 - z)^{-1} + (2 - z)^{-1}$$

$$= \frac{2}{z} + (1 + z + z^2 + \cdots) + \left(\frac{1}{2} + \frac{z}{4} + \frac{z^2}{8} + \cdots\right)^{\dagger}$$

$$= \frac{2}{z} + \frac{3}{2} + \frac{5}{4}z + \frac{9}{8}z^2 + \cdots$$

Thus the residue of $f(z)$ at $z = 0$, that is, the coefficient of the term $1/z$ in the last expansion, is 2. Also, in the neighborhood of $z = 1$, we have

$$f(z) = 2[1 + (z - 1)]^{-1} - \frac{1}{z - 1} + [1 - (z - 1)]^{-1}$$

$$= 2[1 - (z - 1) + (z - 1)^2 - \cdots]$$

$$- \frac{1}{z - 1} + [1 + (z - 1) + (z - 1)^2 + \cdots]$$

$$= \frac{-1}{z - 1} + 3 - (z - 1) + 3(z - 1)^2 - \cdots$$

---

† Since $1/(z - 1)$ and $1/(z - 2)$ are both analytic in the neighborhood of $z = 0$, it is evident in advance that their expansions around $z = 0$ will be not Laurent but Taylor's series and hence will contain no negative powers of z. Thus neither of these terms can contribute to the residue of $f(z)$ at $z = 0$, and so it is actually unnecessary to obtain their expansions. The same thing is true of the terms $2/z$ and $1/(z - 2)$ around $z = 1$.

Hence the residue of $f(z)$ at $z = 1$ is -1. Therefore, according to the residue theorem,

$$\int_c \frac{-3z + 4}{z(z - 1)(z - 2)}\, dz = 2\pi i[2 + (-1)] = 2\pi i$$

Since the determination of residues by the use of series expansions in the manner just illustrated is often tedious and sometimes very difficult, it is desirable to have a simpler alternative procedure. Such a process is provided by the following considerations. Suppose first that $f(z)$ has a simple, or first-order, pole at $z = a$. It follows then that we can write

$$f(z) = \frac{a_{-1}}{z - a} + a_0 + a_1(z - a) + \cdots$$

If we multiply this identity by $z - a$, we get

$$(z - a)f(z) = a_{-1} + a_0(z - a) + a_1(z - a)^2 + \cdots$$

Now if we let z approach a, we obtain for the residue

$$a_{-1} = \lim_{z \to a} (z - a)f(z)$$

If $f(z)$ has a second-order pole at $z = a$, then

$$f(z) = \frac{a_{-2}}{(z - a)^2} + \frac{a_{-1}}{z - a} + a_0 + a_1(z - a) + a_2(z - a)^2 + \cdots$$

To obtain the residue a_{-1} we must first multiply this identity by $(z - a)^2$, getting

$$(z - a)^2 f(z) = a_{-2} + a_{-1}(z - a) + a_0(z - a)^2 + a_1(z - a)^3 + a_2(z - a)^4 + \cdots$$

and then differentiate with respect to z before we let z approach a. The result this time is

$$a_{-1} = \lim_{z \to a} \frac{d}{dz}\left[(z - a)^2 f(z)\right]$$

The same procedure can be extended to poles of higher order, leading to the formula contained in the following theorem.

THEOREM 2   If $f(z)$ has a pole of order m at $z = a$, then the residue of $f(z)$ at $z = a$ is

$$a_{-1} = \frac{1}{(m - 1)!} \lim_{z \to a} \frac{d^{m-1}}{dz^{m-1}}\left[(z - a)^m f(z)\right]$$

In many problems the order of the pole at $z = a$ will not be known in advance. In such cases it is still possible to apply Theorem 2 by taking $m = 1, 2, 3, \ldots$ in turn, until for the *first* time a finite limit is obtained for a_{-1}. The value of m for which this occurs is the order of the pole, and the value of a_{-1} thus determined is the residue. If $f(z)$ has an essential singularity at $z = a$, however, this process fails and the residue cannot be determined by means of Theorem 2.

EXAMPLE 2

What is the residue of

$$f(z) = \frac{1 + z}{1 - \cos z}$$

at the origin?

Here the order of the pole is unknown; so it appears that we may have to proceed tentatively, trying $m = 1, 2, 3, \ldots$ in turn, until for the first time we obtain a finite value for the residue a_{-1}. However, if we replace $\cos z$ by its Maclaurin expansion, we obtain for $f(z)$ the expression

$$\frac{1 + z}{1 - (1 - z^2/2 + z^4/24 - \cdots)} = \frac{2(1 + z)}{z^2(1 - z^2/12 + \cdots)}$$

and the factor z^2 in the denominator now identifies the pole as of the second order. Hence, applying Theorem 2 with $m = 2$, we have

$$
\begin{aligned}
a_{-1} &= \lim_{z \to 0} \frac{d}{dz} \left[z^2 \frac{2(1 + z)}{z^2(1 - z^2/12 + \cdots)} \right] \\
&= \lim_{z \to 0} 2 \frac{(1 - z^2/12 + \cdots) - (1 + z)(-z/6 + \cdots)}{(1 - z^2/12 + \cdots)^2} \\
&= 2
\end{aligned}
$$

The expansion of a rational fractional function $p(x)/q(x)$ in terms of partial fractions is a well-known procedure which we used frequently in our work with the Laplace transformation. Using the residue theorem, we can now establish a result, known as the **Mittag-Leffler expansion theorem**,† which generalizes this type of expansion to a much wider class of functions.

THEOREM 3   If $f(z)$ is a function which is analytic at $z = 0$ and whose only singularities are first-order poles at the points $a_1, a_2, a_3, \ldots$, where the residues are, respectively, $b_1, b_2, b_3, \ldots$; if $\{C : |z| = R_N\}$ is a set of circles none of which passes through any singularity of $f(z)$ and whose radii $\{R_N\}$ become infinite as $N \to \infty$; and if on each of these circles $|f(z)|$ is bounded by a constant M which is independent of N, then

$$f(z) = f(0) + \sum_{k=1}^{\infty} b_k \left(\frac{1}{z - a_k} + \frac{1}{a_k} \right)$$

Proof   Suppose $f(z)$ has first-order poles, with corresponding residues $b_1, b_2, b_3, \ldots$ at $z = a_1, a_2, a_3, \ldots$, and suppose that $z = t$ is not one of these points. Then the function

$$\frac{f(z)}{z - t}$$

has first-order poles at $z = a_1, a_2, a_3, \ldots$ and at $z = t$. At $z = a_k$ the residue of $f(z)/(z - t)$ is

$$\lim_{z \to a_k} (z - a_k) \frac{f(t)}{z - t} = \frac{b_k}{a_k - t}$$

† Named for the Swedish mathematician G. M. Mittag-Leffler (1846–1927).

and at $z = t$ the residue is

$$\lim_{z \to t} (z - t) \frac{f(z)}{z - t} = f(t)$$

Hence, applying the residue theorem to the region bounded by C_N, assumed large enough to include the point $z = t$, we have

(1)
$$\frac{1}{2\pi i} \int_{C_N} \frac{f(z)}{z - t} \, dz = f(t) + \sum \frac{b_k}{a_k - t}$$

where the summation extends over all the poles of $f(z)$ within C_N. If we set $t = 0$ in this equation, we obtain

$$\frac{1}{2\pi i} \int_{C_N} \frac{f(z)}{z} \, dz = f(0) + \sum \frac{b_k}{a_k}$$

and, subtracting this from (1), we have

$$f(t) - f(0) + \sum b_k \left(\frac{1}{a_k - t} - \frac{1}{a_k} \right) = \frac{1}{2\pi i} \int_{C_N} f(z) \left(\frac{1}{z - t} - \frac{1}{z} \right) dz$$

or
$$f(t) - f(0) - \sum b_k \left(\frac{1}{t - a_k} + \frac{1}{a_k} \right) = \frac{t}{2\pi i} \int_{C_N} f(z) \frac{1}{z(z - t)} \, dz$$

The assertion of the theorem will be established if we can show that the limit of the last integral as $N \to \infty$ is zero.

Now for values of z on C_N, we have

$$|z - t| \geq |z| - |t| = R_N - |t|$$

Hence, remembering that, by hypothesis, $|f(z)|$ is bounded on C_N by a number M which is independent of N, we find

$$\left| \frac{t}{2\pi i} \int_{C_N} \frac{f(z)}{(z - t)z} \, dz \right| \leq \frac{|t|}{2\pi} \int_{C_N} \frac{|f(z)| \, |dz|}{|z - t| \, |z|} \leq \frac{|t|}{2\pi} \frac{M}{(R_N - |t|)R_N} \int_{C_N} |dz|$$

$$= \frac{|t|M}{R_N - |t|}$$

Since M is a constant, this fraction does approach zero as N, and hence R_N, becomes infinite. Thus our proof is complete.

EXERCISES

1　Find the residue of $f(z) = z/(z^2 + 1)$ (a) at $z = i$ and (b) at $z = -i$.
2　Find the residue of $f(z) = (z + 1)/[z^2(z - 2)]$ (a) at $z = 0$ and (b) at $z = 2$.
3　Find the residue of $f(z) = z/(z^2 + 2z + 5)$ at each of its poles.
4　What is the residue of $f(z) = 1/(z + 1)^3$ at $z = -1$?
5　What is the residue of $f(z) = \tan z$ at $z = \pi/2$?
6　What is the residue of $f(z) = z/(\cosh z - \cos z)$ at $z = 0$?
7　What is the residue of $f(z) = 1/(z - \sin z)$ at $z = 0$?
8　What is the residue of $f(z) = 1/(e^z - 1)$ at $z = 0$?

9   If C is the circle $|z| = 4$, evaluate $\displaystyle\int_C f(z)\,dz$ for each of the following functions:

(a) $\dfrac{z}{z^2 - 1}$

(b) $\dfrac{z + 1}{z^2(z + 2)}$

(c) $\dfrac{1}{z(z - 2)^3}$

(d) $\dfrac{z^2}{(z^2 + 3z + 2)^2}$

(e) $\dfrac{1}{z^2 + z + 1}$

(f) $\dfrac{1}{z(z^2 + 6z + 4)}$

10 If C is the circle $|z| = 2$, evaluate $\displaystyle\int_C f(z)\,dz$ for each of the following functions:

(a) $\tan z$

(b) $\dfrac{1}{z \sin z}$

(c) $\dfrac{1}{z^2 \sin z}$

(d) $\dfrac{e^{-z}}{z^2}$

(e) $ze^{1/z}$

(f) $\dfrac{z}{\cos z}$

11 Apply Theorem 3 to each of the following functions, and verify that the result is equivalent to the ordinary partial-fraction expansion:

(a) $\dfrac{2z + 1}{z^3 + 6z^2 + 11z + 6}$

(b) $\dfrac{1}{(z + 1)(z^2 + 1)}$

(c) $\dfrac{z}{(z^2 + 1)(z^2 + 2z + 2)}$

12 (a) Show that

$$f(z) = \begin{cases} \cot z - \dfrac{1}{z} & z \neq 0 \\ 0 & z = 0 \end{cases}$$

is uniformly bounded on the sequence of circles $\{C_N\colon |z| = (2N + 1)(\pi/2)\}$. *Hint:* Show first that

$$\left| \cot z - \frac{1}{z} \right| = \left| \frac{z \cos z - \sin z}{z \sin z} \right| \leq \frac{|z| + 1}{|z|}\,\frac{\cosh y}{\sqrt{\sin^2 x + \sinh^2 y}}$$

and then consider the portions of C_N for which $|x| < R_N - 1/R_N$ and $|x| \geq R_N - 1/R_N$.
(b) Using Theorem 3 and the result of part (a), show that

$$\cot z = \frac{1}{z} + \sum_{k=1}^{\infty} \left(\frac{1}{z - n\pi} + \frac{1}{n\pi} \right)$$

13 (a) Show that $\cot \pi z$ is uniformly bounded on the squares of the family shown in Fig. 17.2. *Hint:* Consider separately the portions of C_N for which $|y| > \frac{1}{2}$ and $|y| \leq \frac{1}{2}$, and in the second case show first that if $z = \pm(N + \frac{1}{2}) + iy$, then $|\cot \pi z| = |\tanh \pi y|$.
(b) Use the result of part (a) to show that if $f(z)$ is a function which has only a finite number of poles, none of which occurs at a point of the set $z = \pm n$, and if on each square of the sequence described in part (a) $|f(z)| < M/|z|^k$, where M and k are independent of N and $k > 1$, then

$$\sum_{n=-\infty}^{\infty} f(n) = -\sum \text{ residues of } \pi \cot \pi z\, f(z) \text{ at poles of } f(z)$$

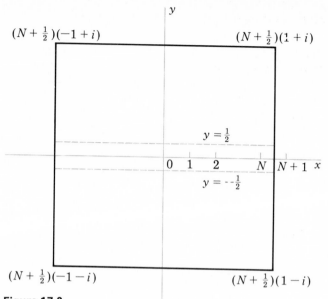

Figure 17.2

14 Show that under the conditions on $f(z)$ described in part **(b)** of Exercise 13,

$$\sum_{n=-\infty}^{\infty} (-1)^n f(n) = -\sum \text{ residues of } \pi \csc \pi z \; f(z) \text{ at poles of } f(z)$$

15 Using the result of part **(b)** of Exercise 13, find the sum of each of the following series:

(a) $\displaystyle\sum_{n=-\infty}^{\infty} \frac{1}{n^2 + a^2}$

(b) $\displaystyle\sum_{n=-\infty}^{\infty} \frac{1}{(n + a)^2 + b^2}$

(c) $\displaystyle\sum_{n=-\infty}^{\infty} \frac{1}{(n + a)^3}$

(d) $\displaystyle\sum_{n=-\infty}^{\infty} \frac{1}{n^4 + 4a^4}$

16 Using the result of Exercise 14, find the sum of each of the following series:

(a) $\displaystyle\sum_{n=-\infty}^{\infty} (-1)^n \frac{1}{n^2 + a^2}$

(b) $\displaystyle\sum_{n=-\infty}^{\infty} (-1)^n \frac{1}{(n + a)^2}$

(c) $\displaystyle\sum_{n=-\infty}^{\infty} (-1)^n \frac{n}{n^3 + 8a^3}$

17.2 The Evaluation of Real Definite Integrals

There are several large and important classes of real definite integrals whose evaluation by the theory of residues can be made a routine matter. The results in question are contained in the next three theorems.

THEOREM 1 If $R(\cos \theta, \sin \theta)$ is a rational function of $\cos \theta$ and $\sin \theta$ which is finite on the closed interval $0 \leq \theta \leq 2\pi$, and if $f(z)$ is the function obtained from R by the substitutions

$$\cos \theta = \frac{z + z^{-1}}{2} \quad \text{and} \quad \sin \theta = \frac{z - z^{-1}}{2i}$$

then $\int_0^{2\pi} R(\cos\theta, \sin\theta)\, d\theta$ is equal to $2\pi i$ times the sum of the residues of the function $f(z)/iz$ at such of its poles as lie within the unit circle $|z| = 1$.

Proof   As a first step, let us transform the given integral by means of the substitution $z = e^{i\theta}$, according to which

$$\cos\theta = \frac{e^{i\theta} + e^{-i\theta}}{2} = \frac{z + z^{-1}}{2} \qquad \sin\theta = \frac{e^{i\theta} - e^{-i\theta}}{2i} = \frac{z - z^{-1}}{2i} \qquad d\theta = \frac{dz}{iz}$$

Under this transformation the original integrand becomes a rational function of z, which we call $f(z)$. Furthermore, as θ ranges from 0 to 2π, the relation $z = e^{i\theta}$ shows that z ranges around the unit circle $|z| = 1$. Hence, the transformed integral is

$$\int_C f(z)\,\frac{dz}{iz}$$

where C is the unit circle. By the residue theorem, the value of this integral is $2\pi i$ times the sum of the residues at those poles of its integrand, namely, $f(z)/iz$, which lie within the unit circle. Since this integral is equal to the original one, the theorem is established.

EXAMPLE 1

Evaluate

$$\int_0^{2\pi} \frac{\cos 2\theta\, d\theta}{1 - 2p\cos\theta + p^2} \qquad -1 < p < 1$$

We note first that by adding and subtracting $2p$, the denominator of the integrand can be written in either of two equivalent forms:

$$1 - 2p\cos\theta + p^2 = 1 - 2p + p^2 + 2p - 2p\cos\theta = (1 - p)^2 + 2p(1 - \cos\theta)$$

$$= 1 + 2p + p^2 - 2p - 2p\cos\theta = (1 + p)^2 - 2p(1 + \cos\theta)$$

From the first of these it is clear that if $0 \le p < 1$, the denominator is different from zero for all values of θ; and from the second it is clear that if $-1 < p \le 0$, the denominator is also different from zero for all values of θ. Hence if $-1 < p < 1$, the integrand is finite on the closed interval $0 \le \theta \le 2\pi$ and Theorem 1 can be applied. Now

$$\cos 2\theta = \frac{e^{2i\theta} + e^{-2i\theta}}{2} = \frac{z^2 + z^{-2}}{2}$$

and thus the given integral becomes

$$\int_C \frac{z^2 - z^{-2}}{2} \frac{1}{1 - 2p(z + z^{-1})/2 + p^2} \frac{dz}{iz} = \int_C \frac{z^4 + 1}{2z^2} \frac{z}{z - pz^2 - p + p^2 z} \frac{dz}{iz}$$

$$= \int_C \frac{(1 + z^4)\, dz}{2iz^2(1 - pz)(z - p)}$$

Of the three poles of the integrand, only the first-order pole at $z = p$ and the second-order pole at $z = 0$ lie within the unit circle C. For the residue at the former we have

$$\lim_{z \to p} (z - p) \frac{1 + z^4}{2iz^2(1 - pz)(z - p)} = \frac{1 + p^4}{2ip^2(1 - p^2)}$$

For the residue at the second-order pole $z = 0$, we have

$$\lim_{z \to 0} \frac{d}{dz} \left[z^2 \frac{1 + z^4}{2iz^2(z - pz^2 - p + p^2z)} \right]$$

$$= \lim_{z \to 0} \frac{(z - pz^2 - p + p^2z)(4z^3) - (1 + z^4)(1 - 2pz + p^2)}{2i(z - pz^2 - p + p^2z)^2}$$

$$= -\frac{1 + p^2}{2ip^2}$$

By Theorem 1, the value of the integral is therefore

$$2\pi i \left[\frac{1 + p^4}{2ip^2(1 - p^2)} - \frac{1 + p^2}{2ip^2} \right] = \frac{2\pi p^2}{1 - p^2}$$

THEOREM 2 If $Q(z)$ is a function which is analytic in the upper half of the z plane except at a finite number of poles none of which lies on the real axis, and if $zQ(z)$ converges uniformly to zero when $z \to \infty$ through values for which $0 \le \arg z \le \pi$, then $\int_{-\infty}^{\infty} Q(x)\, dx$ is equal to $2\pi i$ times the sum of the residues at the poles of $Q(z)$ which lie in the upper half plane.

Proof Consider a semicircular contour with center at $z = 0$ and with radius R large enough to include all the poles of $Q(z)$ which lie in the upper half plane (Fig. 17.3). Then, by the residue theorem,

$$\int_{C_1 + C_2} Q(z)\, dz = 2\pi i \sum \text{residues of } Q(z) \text{ at all poles within } C_1 + C_2$$

or, noting that z is real along C_1,

$$\int_{-R}^{R} Q(x)\, dx + \int_{C_2} Q(z)\, dz = 2\pi i \sum \text{residues}$$

Hence,

(1) $$\left| \int_{-R}^{R} Q(x)\, dx - 2\pi i \sum \text{residues} \right| = \left| -\int_{C_2} Q(z)\, dz \right|$$

and to prove the theorem we must show that the limit of the right-hand side of this equation as $R \to \infty$ is zero. To do this, put $z = Re^{i\theta}$ and $dz = Rie^{i\theta}\, d\theta = iz\, d\theta$ in the integral. Then

$$\left| -\int_{C_2} Q(z)\, dz \right| = \left| -\int_0^{\pi} Q(z)iz\, d\theta \right| \le \int_0^{\pi} |zQ(z)|\, |d\theta|$$

Figure 17.3
A semicircular contour enclosing all the poles of
a function which lie in the upper half plane.

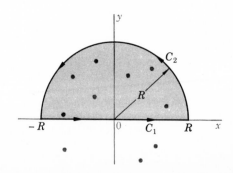

Now the hypothesis that $|zQ(z)|$ converges *uniformly* to zero as $z \to \infty$ and $0 \le \arg z \le \pi$ means that given any small positive quantity, say ε/π, there exists a radius R_0, depending on ε, of course, *but not on* θ, such that

$$|zQ(z)| < \frac{\varepsilon}{\pi}$$

whenever $R > R_0$ and $0 \le \theta \le \pi$. Thus for any semicircular arc C_2 whose radius is greater than R_0, we have

$$\int_0^\pi |zQ(z)\, d\theta| < \frac{\varepsilon}{\pi} \int_0^\pi |d\theta| = \varepsilon$$

This, coupled with (1), proves that

$$\lim_{R \to \infty} \int_{-R}^R Q(x)\, dx = 2\pi i \sum \text{residues}$$

Since the limit on the left is what is meant by $\int_{-\infty}^\infty Q(x)\, dx$,† the theorem is established.

In particular, the quotient of two polynomials $p(x)/q(x)$ automatically satisfies all the hypotheses of the last theorem whenever the degree of the denominator exceeds the degree of the numerator by at least 2. Hence, we have the following important corollary.

COROLLARY 1   If $p(x)$ and $q(x)$ are real polynomials such that the degree of $q(x)$ is at least 2 more than the degree of $p(x)$, and if $q(x) = 0$ has no real roots, then

$$\int_{-\infty}^\infty \frac{p(x)}{q(x)}\, dx = 2\pi i \sum \text{residues of } p(z)/q(z) \text{ at its poles in the upper half plane}$$

EXAMPLE 2

Evaluate

$$\int_{-\infty}^\infty \frac{x^2}{(x^2 + a^2)(x^2 + b^2)}\, dx$$

This is an integral to which Corollary 1 can surely be applied. The only poles of

$$\frac{z^2}{(z^2 + a^2)(z^2 + b^2)}$$

are at $z = \pm ai, \pm bi$. Of these, only $z = ai$ and $z = bi$ lie in the upper half plane. At $z = ai$ the residue is

$$\lim_{z \to ai} (z - ai) \frac{z^2}{(z - ai)(z + ai)(z^2 + b^2)} = \frac{-a^2}{2ai(-a^2 + b^2)} = \frac{a}{2i(a^2 - b^2)}$$

---

† Actually, $\displaystyle\lim_{R \to \infty} \int_{-R}^R Q(x)\, dx$ is only the principal value of the integral $\int_{-\infty}^\infty Q(x)\, dx$, whose general definition is

$$\lim_{R \to \infty} \int_{-R}^0 Q(x)\, dx + \lim_{S \to \infty} \int_0^S Q(x)\, dx$$

where R and S become infinite independently of each other. As the simple function $Q(x) = x$ shows, the principal value of an integral may exist when the integral itself is undefined. However, under the relatively stringent conditions of Theorem 2, the existence of the principal value implies the existence of the integral under its general definition.

From symmetry, the residue at $z = bi$ is obviously $b/2i(b^2 - a^2)$. Hence the value of the integral is

$$2\pi i \left[\frac{a}{2i(a^2 - b^2)} + \frac{b}{2i(b^2 - a^2)} \right] = \frac{\pi}{a + b}$$

If $Q(z)$ satisfies all the hypotheses of Theorem 2, then so does $e^{imz}Q(z)$, provided $m > 0$. For e^{imz} is analytic everywhere, and under the assumption that $m > 0$ its absolute value is

$$|e^{imz}| = |e^{im(x + iy)}| = |e^{imx}e^{-my}| = e^{-my}$$

which is less than or equal to 1 for all values of y in the upper half plane. Therefore

$$|e^{imz}zQ(z)| \leq |zQ(z)|$$

and thus, if the latter converges uniformly to zero when z and $0 \leq \arg z \leq \pi$, so will the former. Hence, the conclusions of Theorem 2 can be applied equally well to $e^{imz}Q(z)$, and we can write

(2) $$\int_{-\infty}^{\infty} e^{imx}Q(x) \, dx = 2\pi i \sum \text{residues of } e^{imz}Q(z) \text{ at its poles in upper half plane}$$

Separating the integral in (2) into its real and its imaginary parts and equating these to the corresponding parts of the expression on the right, we obtain the following useful result.

COROLLARY 2 If $Q(z)$ is analytic in the upper half of the z plane except at a finite number of poles none of which lies on the real axis, and if $zQ(z)$ converges uniformly to zero when z becomes infinite through the upper half plane, then

$$\int_{-\infty}^{\infty} \cos mx \, Q(x) \, dx = -2\pi \sum \text{imaginary parts of residues of } e^{imz}Q(z)$$

at its poles in the upper half plane

$$\int_{-\infty}^{\infty} \sin mx \, Q(x) \, dx = 2\pi \sum \text{real parts of residues of } e^{imz}Q(z)$$

at its poles in the upper half plane

EXAMPLE 3

Evaluate

$$\int_{-\infty}^{\infty} \frac{\cos mx}{1 + x^2} \, dx$$

To do this, we consider the related function $e^{imz}/(1 + z^2)$. The only pole of this function in the upper half plane is $z = i$, and the residue there is

$$\lim_{z \to i} (z - i) \frac{e^{imz}}{(z - i)(z + i)} = \frac{e^{-m}}{2i} = -\frac{ie^{-m}}{2}$$

Hence, by Corollary 2,

$$\int_{-\infty}^{\infty} \frac{\cos mx}{1 + x^2} \, dx = -2\pi \mathscr{I} \left(-\frac{ie^{-m}}{2} \right) = \pi e^{-m}$$

Incidentally, the fact that the residue at $z = i$ is a pure imaginary confirms the observation, obvious from symmetry, that

$$\int_{-\infty}^{\infty} \frac{\sin mx}{1 + x^2}\, dx = 0$$

As a final result on the evaluation of real definite integrals by the method of residues, we have the following theorem, whose proof we omit because of its relative intricacy.[†]

THEOREM 3 If $Q(z)$ is analytic everywhere in the z plane except at a finite number of poles none of which lies on the positive half of the real axis, and if $|z\ Q(z)|$ converges uniformly to zero when $z \to 0$ and when $z \to \infty$, then

$$\int_0^{\infty} x^{a-1} Q(x)\, dx = \frac{\pi}{\sin a\pi} \sum \text{residues of } (-z)^{a-1} Q(z) \text{ at all its poles}$$

provided that arg z is taken in the interval $(-\pi, \pi)$.

EXAMPLE 4

Evaluate

$$\int_0^{\infty} \frac{x^{a-1}}{1 + x^2}\, dx \qquad 0 < a < 2$$

For a within the specified range, the conditions of Theorem 3 are fulfilled; hence the given integral is equal to $\pi/(\sin a\pi)$ times the sum of the residues of $(-z)^{a-1}/(1 + z^2)$ at $z = \pm i$. At $z = i$, we have for the residue

$$\lim_{z \to i} (z - i) \frac{(-z)^{a-1}}{(z - i)(z + i)} = \frac{(-i)^{a-1}}{2i} = \frac{(e^{-i\pi/2})^{a-1}}{2i} = \frac{e^{-i\pi(a-1)/2}}{2i}$$

At $z = -i$, we find the residue to be

$$\lim_{z \to -i} (z + i) \frac{(-z)^{a-1}}{(z + i)(z - i)} = \frac{i^{a-1}}{-2i} = \frac{(e^{i\pi/2})^{a-1}}{-2i} = \frac{e^{i\pi(a-1)/2}}{-2i}$$

The value of the integral is therefore

$$\frac{\pi}{\sin a\pi} \frac{e^{i\pi(a-1)/2} - e^{-i\pi(a-1)/2}}{-2i} = -\frac{\pi}{\sin a\pi} \sin \frac{(a-1)\pi}{2}$$

$$= \frac{\pi}{\sin a\pi} \cos \frac{a\pi}{2} = \frac{\pi}{2 \sin (a\pi/2)}$$

For definite integrals not covered by the theorems of this section, evaluation by the method of residues, when possible at all, usually requires considerable ingenuity in selecting the appropriate contour and in eliminating the integrals over all but the desired portion of the contour. Several examples of this sort will be found, with hints, in the exercises.

[†] See, for instance, E. T. Whittaker and G. N. Watson, "Modern Analysis," p. 117, Macmillan New York, 1943.

EXERCISES

Evaluate the following integrals by the method of residues:

1 $\displaystyle\int_0^{2\pi} \frac{d\theta}{1 - 2p \sin \theta + p^2}$ $-1 < p < 1$

2 $\displaystyle\int_0^{2\pi} \frac{d\theta}{(a + b \cos \theta)^2}$ $0 < b < a$ **3** $\displaystyle\int_0^{2\pi} \frac{d\theta}{\cos \theta + 2 \sin \theta + 3}$

4 $\displaystyle\int_0^{2\pi} \frac{d\theta}{2 \cos \theta + 3 \sin \theta + 7}$ **5** $\displaystyle\int_0^{2\pi} \frac{\sin^2 \theta \, d\theta}{a + b \cos \theta}$ $0 < b < a$

6 $\displaystyle\int_0^{\pi} \frac{\cos 2\theta \, d\theta}{5 + 4 \cos \theta}$ **7** $\displaystyle\int_{-\infty}^{\infty} \frac{dx}{x^4 + a^4}$

8 $\displaystyle\int_{-\infty}^{\infty} \frac{dx}{(1 + x^2)^3}$ **9** $\displaystyle\int_{-\infty}^{\infty} \frac{x^2 \, dx}{1 + x^6}$

10 $\displaystyle\int_{-\infty}^{\infty} \frac{x^2 \, dx}{(1 + x^4)^2}$ **11** $\displaystyle\int_0^{\infty} \frac{dx}{(a^2 + x^2)^2}$

12 $\displaystyle\int_0^{\infty} \frac{dx}{1 + x^6}$ **13** $\displaystyle\int_{-\infty}^{\infty} \frac{\cos mx}{(x - a)^2 + b^2} \, dx$

14 $\displaystyle\int_{-\infty}^{\infty} \frac{\sin mx}{(x - a)^2 + b^2} \, dx$ **15** $\displaystyle\int_0^{\infty} \frac{\cos mx}{(a^2 + x^2)^2} \, dx$

16 $\displaystyle\int_0^{\infty} \frac{\cos mx}{1 + x^4} \, dx$ **17** $\displaystyle\int_{-\infty}^{\infty} \frac{\cos mx}{(x^2 + a^2)(x^2 + b^2)} \, dx$

18 $\displaystyle\int_{-\infty}^{\infty} \frac{x \sin mx}{(x^2 + a^2)(x^2 + b^2)} \, dx$ **19** $\displaystyle\int_{-\infty}^{\infty} \frac{x \sin mx}{1 + x^4} \, dx$

20 $\displaystyle\int_0^{\infty} \frac{x^{a-1}}{(x + b)(x + c)} \, dx$ $\begin{array}{l} 0 < a < 2 \\ 0 < b, c \end{array}$

21 $\displaystyle\int_0^{\infty} \frac{x^{a-1}}{(x - b)^2 + c^2} \, dx$ $0 < a < 2$

22 $\displaystyle\int_0^{\infty} \frac{x^{a-1}}{(x + b)(x + c)(x + d)} \, dx$ $\begin{array}{l} 0 < a < 3 \\ 0 < b, c, d \end{array}$

23 $\displaystyle\int_0^{\infty} \frac{x^{a-1}}{1 + x^3} \, dx$ $0 < a < 3$ **24** $\displaystyle\int_0^{\infty} \frac{x^{a-1}}{1 + x^4} \, dx$ $0 < a < 4$

25 Show that $\Gamma(a)\Gamma(1 - a) = \pi/(\sin a\pi)$ if $0 < a < 1$. *Hint:* Consider the integral $\displaystyle\int_0^{\infty} \frac{y^{a-1}}{1 + y} \, dy$, and evaluate it first by the method of residues and then by making the substitution $y = x/(1 - x)$ and expressing it in terms of gamma functions.

26 Show that

$$\int_0^{\infty} \frac{\sin x}{x} \, dx = \frac{\pi}{2}$$

Hint: Integrate e^{iz}/z around the contour shown in Fig. 17.4, and let $r \to 0$ and $R \to \infty$.

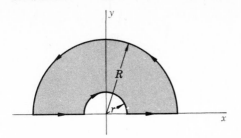

Figure 17.4

27 Show that

$$\int_0^\infty \frac{\cos x}{\sqrt{x}}\,dx = \int_0^\infty \frac{\sin x}{\sqrt{x}}\,dx = \sqrt{\frac{\pi}{2}}$$

Hint: Integrate $e^{iz}/\sqrt{z}$ around the contour shown in Fig. 17.5, let $r \to 0$ and $R \to \infty$, and recall (Exercise 14, Sec. 7.3) that

$$\int_0^\infty e^{-x^2}\,dx = \frac{\sqrt{\pi}}{2}$$

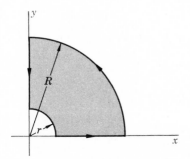

Figure 17.5

28 Show that

$$\int_{-\infty}^\infty \frac{\cos mx}{e^x + e^{-x}}\,dx = \frac{\pi}{e^{m\pi/2} + e^{-m\pi/2}}$$

Hint: Integrate the function $e^{imz}/(e^z + e^{-z})$ around the contour shown in Fig. 17.6, and let $R \to \infty$.

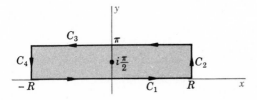

Figure 17.6

29 If $f(z)$ has a number of first-order poles on the real axis but otherwise satisfies all the conditions of Theorem 2, show that the principal value of $\displaystyle\int_{-\infty}^\infty e^{imx} f(x)\,dx$ is equal to $2\pi i$ times the sum of the residues of $e^{imz} f(z)$ at its poles in the upper half plane plus $i\pi$ times the sum of the residues of $e^{imz} f(z)$ at its poles on the real axis. *Hint:* Use a contour like that shown in Fig. 17.4, suitably indented around each of the poles of $f(z)$ which lies on the real axis.

30 If n is a positive integer, show that

$$\int_0^{2\pi} \exp(\cos\theta)\cos(n\theta - \sin\theta)\,d\theta = \frac{2\pi}{n!}$$

31 Find the Fourier expansion of the periodic function

$$\frac{1}{a + b\cos\theta} \qquad 0 < b < a$$

Discuss from the point of view of Theorem 3, Sec. 6.3, the limiting behavior of the Fourier coefficients of this function as $n \to \infty$.

32 Show that

$$\int_0^\infty \frac{\ln(x^2 + 1)}{1 + x^2}\,dx = \pi \ln 2$$

Hint: Consider the integral of the function $[\ln(z + i)]/(z^2 + 1)$ around the contour shown in Fig. 17.7.

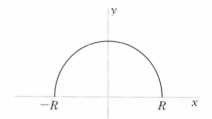

Figure 17.7

33 Show that

$$\int_0^\infty \frac{(\ln x)^2}{1 + x^2}\,dx = \frac{\pi^3}{8} \qquad \text{and} \qquad \int_0^\infty \frac{\ln x}{1 + x^2}\,dx = 0$$

Hint: Integrate the function $(\ln z)^2/(1 + z^2)$ around the contour shown in Fig. 17.4.

34 Find the values of the integrals

$$\int_0^\infty \frac{(\ln x)^2}{a^2 + x^2}\,dx \qquad \text{and} \qquad \int_0^\infty \frac{\ln x}{a^2 + x^2}\,dx$$

17.3 The Complex Inversion Integral

We are now in a position to appreciate more fully the complex inversion integral of Laplace transform theory. In Sec. 6.8 we defined the Laplace transform of a function $f(t)$ to be

(1)
$$\mathcal{L}\{f(t)\} = \int_0^\infty f(t)e^{-st}\,dt$$

and we showed that conversely

(2)
$$f(t) = \frac{1}{2\pi i}\int_{a-i\infty}^{a+i\infty} \mathcal{L}\{f(t)\}e^{st}\,ds$$

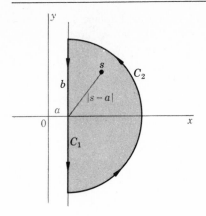

Figure 17.8
The contour used to obtain the complex inversion
integral.

s being a complex variable. It is interesting now to reconsider the derivation of (2) in the light of complex-variable theory and to investigate how this formula can be applied to the determination of a function when its transform is known.

In the complex plane let $\phi(z)$ be a function of z which is analytic on the line $x = a$ and in the entire half plane R to the right of this line. Moreover, let $|\phi(z)|$ approach zero uniformly as z becomes infinite through this half plane. Then if s is any point in the half plane R, we can choose a semicircular contour $C = C_1 + C_2$, as shown in Fig. 17.8, and apply Cauchy's integral formula, getting

$$(3) \qquad \phi(s) = \frac{1}{2\pi i} \int_C \frac{\phi(z)}{z - s} \, dz = \frac{1}{2\pi i} \int_{a+ib}^{a-ib} \frac{\phi(z)}{z - s} \, dz + \frac{1}{2\pi i} \int_{C_2} \frac{\phi(z)}{z - s} \, dz$$

Now the minimum value of $|z - s|$ occurs when z is collinear with a and s. Hence, for all values of z on C_2 and for b sufficiently large we have

$$|z - s| \geq b - |s - a| \geq b - |s| - |a|$$

whether a is positive, as shown in Fig. 17.8, or negative. Thus, letting M denote the maximum value of $|\phi(z)|$ on C_2, we have

$$\left| \int_{C_2} \frac{\phi(z)}{z - s} \, dz \right| \leq \int_{C_2} \frac{|\phi(z)|}{|z - s|} \, |dz| \leq \frac{M}{b - |s| - |a|} \int_{C_2} |dz| = \frac{\pi b M}{b - |s| - |a|}$$

As b becomes infinite, the fraction

$$\frac{b}{b - |s| - |a|}$$

approaches 1 and at the same time M approaches zero since, by hypothesis, $|\phi(z)|$ converges uniformly to zero as z becomes infinite through the right half plane R. Hence,

$$\lim_{b \to \infty} \int_{C_2} \frac{\phi(z)}{z - s} \, dz = 0$$

and in the limit we have, from (3),

$$\phi(s) = \lim_{b \to \infty} \frac{1}{2\pi i} \int_{a+ib}^{a-ib} \frac{\phi(z)}{z - s} \, ds = \frac{1}{2\pi i} \int_{a-i\infty}^{a+i\infty} \frac{\phi(z)}{s - z} \, dz$$

Let us now attempt to determine the function of t whose Laplace transform is $\phi(s)$. Taking the inverse of $\phi(s)$ as defined by the last expression, we have

$$\mathscr{L}^{-1}\{\phi(s)\} = f(t) = \mathscr{L}^{-1}\left\{\frac{1}{2\pi i}\int_{a-i\infty}^{a+i\infty}\frac{\phi(z)}{s-z}\,dz\right\}$$

Assuming that the operations of integrating along the vertical line $x = a$ and applying the inverse Laplace transformation can be interchanged, the last equation can be written

$$f(t) = \frac{1}{2\pi i}\int_{a-i\infty}^{a+i\infty}\mathscr{L}^{-1}\left\{\frac{\phi(z)}{s-z}\right\}dz$$

or, since the operator $\mathscr{L}^{-1}$ refers only to the variable s,

$$f(t) = \frac{1}{2\pi i}\int_{a-i\infty}^{a+i\infty}\phi(z)\mathscr{L}^{-1}\left\{\frac{1}{s-z}\right\}dz$$

Now the specific result

$$\mathscr{L}^{-1}\left\{\frac{1}{s-z}\right\} = e^{zt}$$

is known to us through independent reasoning. Hence we have finally

$$f(t) = \frac{1}{2\pi i}\int_{a-i\infty}^{a+i\infty}\phi(z)e^{tz}\,dz$$

which, except that the variable of integration is z instead of s, is exactly Eq. (2). From this result it is clear that *the inversion integral is a line integral in the complex plane taken along a vertical line to the right of all singularities of the transform $\phi(s)$ or along any other path into which this can legitimately be deformed.*

In the usual applications, the evaluation of the complex inversion integral is accomplished by the method of residues, using a semicircular contour whose diameter is the segment joining the points $a - ib$ and $a + ib$ and whose radius b is large enough to ensure that all the poles of the transform are within the contour (Fig. 17.9). Specifically, we have the following result.

THEOREM 1 If the Laplace transform $\phi(s)$ is an analytic function of s except at a finite number of poles each of which lies to the left of the vertical line $\mathscr{R}(s) = a$, and if $s\phi(s)$ is bounded as s becomes infinite through the half plane $\mathscr{R}(z) \le a$, then

$$\mathscr{L}^{-1}\{\phi(s)\} = \sum \text{ residues of } \phi(s)e^{st} \text{ at each of its poles}$$

Proof Using the contour shown in Fig. 17.9, we have by the residue theorem,

$$\frac{1}{2\pi i}\int_{a-ib}^{a+ib}\phi(s)e^{st}\,ds + \frac{1}{2\pi i}\int_{C_2}\phi(s)e^{st}\,ds = \sum \text{ residues of } \phi(s)e^{st}$$

Hence,

(4) $$\left|\frac{1}{2\pi i}\int_{a-ib}^{a+ib}\phi(s)e^{st}\,ds - \sum \text{ residues of } \phi(s)e^{st}\right| = \left|-\frac{1}{2\pi i}\int_{C_2}\phi(s)e^{st}\,ds\right|$$

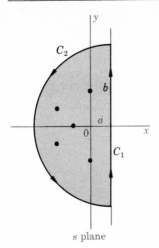

Figure 17.9
A typical contour used in the evaluation of the complex
inversion integral.

s plane

Now along C_2 we have

$$s = a + be^{i\theta} = a + b(\cos\theta + i\sin\theta) \qquad \frac{\pi}{2} \le \theta \le \frac{3\pi}{2}$$

and

$$ds = ibe^{i\theta}\, d\theta = i(s - a)\, d\theta$$

and, for sufficiently large s,

$$|s - a| \le |s| + |a| < 2|s| \qquad \text{and} \qquad |s\phi(s)| < M$$

Therefore

$$\left| -\frac{1}{2\pi i}\int_{C_2} \phi(s)e^{st}\, ds \right| \le \frac{1}{2\pi}\int_{C_2} |\phi(s)|\, |e^{st}|\, |ds|$$

$$= \frac{1}{2\pi}\int_{\pi/2}^{3\pi/2} |\phi(s)|\, |\exp\{t[a + b(\cos\theta + i\sin\theta)]\}|\, |ibe^{i\theta}\, d\theta|$$

$$= \frac{1}{2\pi}\int_{\pi/2}^{3\pi/2} |\phi(s)|\, |s - a|\, \exp[t(a + b\cos\theta)]\, d\theta$$

$$\le \frac{1}{2\pi}\int_{\pi/2}^{3\pi/2} 2|s\phi(s)|e^{at}\exp(bt\cos\theta)\, d\theta$$

$$\le \frac{1}{\pi} Me^{at}\int_{\pi/2}^{3\pi/2} \exp(bt\cos\theta)\, d\theta$$

If we now set $\theta = \pi/2 + \alpha$ and then take advantage of the symmetry of the resulting
integrand, the last integral becomes

$$\frac{M}{\pi}e^{at}\int_0^{\pi} \exp(-bt\sin\alpha)\, d\alpha = \frac{2M}{\pi}e^{at}\int_0^{\pi/2} \exp(-bt\sin\alpha)\, d\alpha$$

Now it is evident from Fig. 17.10 that

$$\sin\alpha \ge \frac{2\alpha}{\pi} \qquad \text{for } 0 \le \alpha \le \frac{\pi}{2}$$

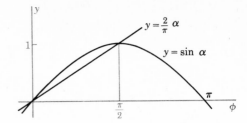

Figure 17.10
Plot showing that $\sin \alpha \geq 2\alpha/\pi \geq 0$ for $0 \leq \alpha \leq \pi/2$.

Hence the last integral is overestimated if we replace $\sin \alpha$ in the exponent by the smaller *positive* quantity $2\alpha/\pi$. Doing this and then performing the integration, we have

$$\left| \frac{1}{2\pi i} \int_{C_2} \phi(s)e^{st}\, ds \right| \leq \frac{2M}{\pi} e^{at} \left[\frac{e^{-2bt\alpha/\pi}}{-2bt/\pi} \right]_0^{\pi/2}$$

$$= \frac{2M}{\pi} e^{at} \left[-\frac{\pi}{2bt} (e^{-bt} - 1) \right]$$

For $t \geq 0$, the last expression clearly approaches zero as b becomes infinite. Hence, returning to Eq. (4), it is clear that

$$\lim_{b \to \infty} \frac{1}{2\pi i} \int_{a-ib}^{a+ib} \phi(s)e^{st}\, ds \equiv \frac{1}{2\pi i} \int_{a-i\infty}^{a+i\infty} \phi(s)e^{st}\, ds$$

$$= \mathscr{L}^{-1}\{\phi(s)\}$$

$$= \sum \text{residues of } \phi(s)e^{st}$$

as asserted.

The proof of the last theorem breaks down if $\phi(s)$ has infinitely many poles, because then as $b \to \infty$ there will always be semicircles C_2 on which $|s\phi(s)|$ is not bounded. However, to prove the theorem it is not necessary to know that $|s\phi(s)|$ is bounded for *all* values of s to the left of the line $\mathscr{R}(s) = a$ and outside some sufficiently large semicircle. It is sufficient, in fact, that there should exist a sequence of semicircles $\{(C_2)_N\}$ whose radii $\{b_N\}$ become infinite as N becomes infinite and a number M, independent of N, such that $|s\phi(s)| < M$ on every semicircle in the sequence. If $\phi(s)$ is a function with infinitely many poles which is uniformly bounded on such a sequence of semicircles, it can be shown that Theorem 1 is still valid.†

EXAMPLE 1

What is $\mathscr{L}^{-1}\{1/[(s + a)^2 + b^2]\}$?

Using Theorem 1, we have only to compute the residues of

$$\frac{e^{st}}{(s + a)^2 + b^2}$$

at its two first-order poles $-a \pm ib$. At $s = -a + ib$, we have for the residue

$$\lim_{s \to -a+ib} \frac{[s - (-a + ib)]e^{st}}{[s - (-a + ib)][s - (-a - ib)]} = \frac{e^{(-a+ib)t}}{2ib}$$

---

† This is essentially the hypothesis we needed to prove the Mittag-Leffler expansion theorem in the last section. For a more detailed discussion of this point, see, for instance, R. V. Churchill, "Operational Mathematics," 2d ed., pp. 190–193, McGraw-Hill, New York, 1958.

and at $s = -a - ib$ we have for the residue

$$\lim_{s \to -a-ib} \frac{[s - (-a - ib)]e^{st}}{[s - (-a + ib)][s - (-a - ib)]} = \frac{e^{(-a-ib)t}}{-2ib}$$

Hence, by Theorem 1,

$$f(t) = \mathscr{L}^{-1}\{\phi(s)\} = \frac{e^{(-a+ib)t}}{2ib} + \frac{e^{(-a-ib)t}}{-2ib}$$

$$= e^{-at} \frac{e^{ibt} - e^{-ibt}}{2ib}$$

$$= \frac{e^{-at} \sin bt}{b}$$

This example, of course, has been merely a new approach to a result with which we were already familiar. However, in more difficult applications, the use of the complex inversion integral and contour integration is often either the only way or the best way of finding a function when its transform is known.

EXAMPLE 2

Find $\mathscr{L}^{-1}\{1/(s \cosh as)\}$.

Obviously, in this case the function $\phi(s)$ has a first-order pole at $s = 0$. Moreover, since $\cosh as = \cos ias$ [Eq. (13), Sec. 15.7], it follows that $\phi(s)$ has infinitely many other first-order poles, namely, the points where

$$ias = \pm \frac{(2n - 1)\pi}{2} \quad \text{or} \quad s = \pm \frac{(2n - 1)\pi}{2ia} \quad n = 0, 1, 2, \ldots$$

However, if we set $s = \sigma + i\omega$, we have, by Eq. (17), Sec. 14.7,

$$|s\phi(s)| = \left|\frac{1}{\cosh as}\right| = \frac{1}{(\cosh^2 a\sigma \cos^2 a\omega + \sinh^2 a\sigma \sin^2 a\omega)^{1/2}}$$

$$= \frac{1}{(\cosh^2 a\sigma - \sin^2 a\omega)^{1/2}}$$

and it is not difficult to show that this expression has a bound independent of N on the semicircles $\{(C_2)_N\}$ determined by the center a and the sequence of points $s = \{N\pi i/a\}$. Hence the inverse of $\phi(s)$ is simply the sum of the residues of

$$\frac{e^{st}}{s \cosh as}$$

at each of its poles, i.e., at the poles of $\phi(s)$.

At $s = 0$ the residue is

$$\lim_{s \to 0} \frac{se^{st}}{s \cosh as} = 1$$

and at $s = (2n - 1)\pi/2ia$ [using L'Hospital's rule and Eq. (20), Sec. 15.7, to evaluate the indeterminacy], the residue is

$$\lim_{s \to (2n-1)\pi/2ia} \frac{[s - (2n - 1)\pi/2ia]e^{st}}{s \cosh as} = \frac{e^{(2n-1)\pi t/2ia}}{[(2n - 1)\pi/2ia] a \sinh [(2n - 1)\pi/2i]}$$

$$= \frac{2(-1)^n e^{(2n-1)\pi t/2ia}}{(2n - 1)\pi}$$

Similarly, at $s = -[(2n - 1)\pi/2ia]$, the residue is

$$\frac{2(-1)^n e^{-(2n-1)\pi t/2ia}}{(2n - 1)\pi}$$

Hence, pairing the terms which correspond to the same value of n,

$$f(t) = \mathcal{L}^{-1}\{\phi(s)\} = 1 + \frac{2}{\pi}\sum_{n=1}^{\infty}\frac{(-1)^n}{2n-1}[e^{(2n-1)\pi t/2ia} + e^{-(2n-1)\pi t/2ia}]$$

$$= 1 + \frac{4}{\pi}\sum_{n=1}^{\infty}\frac{(-1)^n}{2n-1}\cos\frac{(2n-1)\pi t}{2a}$$

EXERCISES

Using the complex inversion integral, find the inverses of the following Laplace transforms. In each case discuss the resemblance of the method of residues to the use of the Heaviside expansion theorems (Sec. 7.5).

1 $\dfrac{1}{(s+1)(s+3)}$ **2** $\dfrac{1}{(s+2)^2}$ **3** $\dfrac{1}{s^2+4}$

4 $\dfrac{s}{s^2+4s+13}$ **5** $\dfrac{1}{s(s^2+1)}$ **6** $\dfrac{s}{s^3+1}$

7 $\dfrac{s}{(s^2+4)^2}$ **8** $\dfrac{1}{(s^2+9)(s^2+4)}$ **9** $\dfrac{s+1}{(s+2)^2(s+3)}$

10 $\dfrac{1}{(s^2+2s+5)^2}$

11 Complete Example 2 by showing that

$$\frac{1}{(\cosh^2 a\sigma - \sin^2 a\omega)^{1/2}}$$

is uniformly bounded on the semicircles $(C_2)_N$ described in the example.

12 Complete the solution of Exercise 8, Sec. 8.7, by finding the angular displacement at a general point x.

Find the inverse of each of the following transforms:

13 $\dfrac{1}{s\sinh as}$ **14** $\dfrac{1}{(s+b)\cosh as}$ **15** $\dfrac{\sinh x\sqrt{s}}{s\sinh\sqrt{s}}$

16 $[I_0(r\sqrt{s})/sI_0(\sqrt{s})]$, where I_0 is the modified Bessel function of the first kind.

17.4 Stability Criteria

In the analysis of many physical systems a complete description of the behavior of the system is unnecessary, and all that is required is a knowledge of whether the system is stable, i.e., whether its response to a bounded excitation remains bounded or becomes infinite as $t \to \infty$. As we shall see in this section, this question can be answered by analyzing the Laplace transform of the response without actually determining the response itself.

We begin by supposing that by methods like those described in Chap. 7 we have obtained the Laplace transform of the response of the system $\mathcal{L}\{y(t)\} \equiv \phi(s)$ and that $\phi(s)$ is a rational function; i.e.,

$$\phi(s) = \frac{P(s)}{Q(s)}$$

where P and Q are real polynomials in the complex variable $s = a + i\omega$. We know from algebra that any real polynomial, such as $Q(s)$, can always be factored into real linear and quadratic factors that may or may not be repeated. Moreover, we know from the Heaviside theorems (Sec. 7.5) that the form of the inverse $y(t) = \mathscr{L}^{-1}\{\phi(s)\}$ is determined completely and solely by the factors of $Q(s)$ and that the only terms that can possibly occur in it are those given in Table 17.1. Clearly, terms of the forms 1 and 2 are stable in all cases, for although they do not approach zero as $t \rightarrow \infty$, they do remain finite. Terms of the forms 3, 4, 7, and 8 are stable if and only if a is negative, in which case they not only remain finite but in fact approach zero as $t \rightarrow \infty$. Terms of the forms 5 and 6 are unstable in all cases, since the factor t means that each becomes unbounded as $t \rightarrow \infty$. Translating these observations into conditions on the roots of the polynomial equation $Q(s) = 0$, we see that the response $y(t)$ will be stable if and only if the following conditions are met:

a. Every unrepeated real root is nonpositive.
b. Every repeated real root is negative.
c. Every pure imaginary root is unrepeated.
d. Every general complex root has negative real part.

Geometrically speaking, these conditions can be described as follows.

THEOREM 1   In order for the function

$$y(t) = \mathscr{L}^{-1}\left\{\frac{P(s)}{Q(s)}\right\}$$

to be stable it is necessary and sufficient that the equation $Q(s) = 0$ have no roots to the right of the imaginary axis in the complex s plane and that any roots on the imaginary axis in the s plane be unrepeated.

Various methods are available for determining whether the roots of a polynomial equation all have nonpositive real parts.† In general, however, these are more

Table 17.1

Factor	Term
From unrepeated factors	
1. s	1
2. $s^2 + b^2$	$\cos bt$, $\sin bt$
3. $s - a$	e^{at}
4. $(s - a)^2 + b^2$	$e^{at} \cos bt$, $e^{at} \sin bt$
From repeated factors	
5. s^n, $n > 1$	t^k, $0 \le k \le n - 1$
6. $(s^2 + b^2)^n$, $n > 1$	$t^k \cos bt$, $t^k \sin bt$, $0 \le k \le n - 1$
7. $(s - a)^n$, $n > 1$	$t^k e^{at}$, $0 \le k \le n - 1$
8. $[(s - a)^2 + b^2]^n$, $n > 1$	$t^k e^{at} \cos bt$, $t^k e^{at} \sin bt$, $0 \le k \le n - 1$

---

† See, for instance, A. Bronwell, "Advanced Mathematics in Physics and Engineering," pp. 386–413, McGraw-Hill, New York, 1953, and E. A. Guilleman, "The Mathematics of Circuit Analysis," pp. 395–409, Wiley, New York, 1953.

conveniently formulated as methods for determining whether the roots all have real parts that are strictly negative, and most though not all of our results will be of this nature. This is not a serious disadvantage, because in practice zero roots and pure imaginary roots, i.e., roots whose real parts are zero, if they occur at all, are usually easily recognizable.

A preliminary result of considerable importance is contained in the following theorem.

THEOREM 2   If the real part of each root of the polynomial equation $Q(s) = 0$ is less than or equal to zero, then the coefficients in $Q(s)$ all have the same sign.

Proof   We observe first that it is no specialization to interpret the condition of the theorem as asserting that all coefficients in $Q(s)$ are positive. For the case in which all coefficients are negative can be converted into the case in which all coefficients are positive, and vice versa, simply by multiplying $Q(s) = 0$ by -1, which, of course, in no way alters the roots of this equation. Now if every root of $Q(s) = 0$ has a nonpositive real part, then the only possible factors of $Q(s)$ are of the form

$$s + a_i \quad \text{and} \quad (s + a_j)^2 + b_j^2 \quad \text{where } a_i, a_j \geq 0$$

Since these factors contain only nonnegative terms, and since $Q(s)$ is simply the product of a finite number of these factors, it is clear that every nonzero coefficient in $Q(s)$ must be positive, as asserted.

It is also clear from the preceding argument that if every a is positive, so that all roots of $Q(s) = 0$ have real parts that are strictly negative, then there can be no zero coefficients in $Q(s)$; that is, all terms must be present. Hence, restating this observation contrapositively, we have the following corollary.

COROLLARY 1   If one or more terms are missing from $Q(s)$, then the equation $Q(s) = 0$ has at least one root whose real part is nonnegative.

The condition of Theorem 2 is only a necessary and not a sufficient one; i.e., it *cannot* be asserted, conversely, that if the coefficients in $Q(s)$ all have the same sign, then the real part of each root of $Q(s) = 0$ is nonpositive. For instance,

$$s^4 + s^3 + s^2 + 11s + 10$$

contains only terms with positive coefficients; yet the roots of the equation

$$s^4 + s^3 + s^2 + 11s + 10 = 0$$

are $s = -1, -2, 1 \pm 2i$

and the two complex roots have positive real parts. On the other hand, it is clear, from the proof of Theorem 2, that we do have the following result.

COROLLARY 2   If $Q(s)$ contains some terms with positive coefficients and some terms with negative coefficients, then the equation $Q(s) = 0$ has at least one root whose real part is positive.

For quadratic equations the necessary condition of Theorem 2 is also sufficient. For if the equation $a_0 s^2 + a_1 s + a_2 = 0$ contains no negative coefficients, then its roots

$$s = \frac{-a_1 \pm \sqrt{a_1{}^2 - 4a_0 a_2}}{2a_0}$$

are clearly either nonpositive real numbers or conjugate complex numbers with nonpositive real parts.

For cubic equations, a sufficient condition, supplementing Theorem 2, is contained in the following result.

THEOREM 3 A necessary and sufficient condition that every root of the cubic equation $a_0 s^3 + a_1 s^2 + a_2 s + a_3 = 0$ have negative real part is that all coefficients have the same sign and that

$$a_1 a_2 - a_0 a_3 > 0$$

Proof Let us assume for definiteness that the given equation has one real root r and one pair of conjugate complex roots, $p \pm iq$. The case in which the equation has three real roots can be handled in exactly the same fashion. From algebra we recall that the roots, say r_1, r_2, r_3, of any cubic equation are related to the coefficients through the equations

$$\frac{a_1}{a_0} = -(r_1 + r_2 + r_3)$$

$$\frac{a_2}{a_0} = r_1 r_2 + r_2 r_3 + r_3 r_1$$

$$\frac{a_3}{a_0} = -r_1 r_2 r_3$$

In the present case these become

(1) $$\frac{a_1}{a_0} = -(r + 2p)$$

(2) $$\frac{a_2}{a_0} = p^2 + q^2 + 2pr$$

(3) $$\frac{a_3}{a_0} = -r(p^2 + q^2)$$

From (3) and the assumption that the a's all have the same sign, it follows that $r < 0$. To prove that $p < 0$, we note that the condition $a_1 a_2 - a_0 a_3 > 0$ can be rewritten, after division by $a_0{}^2$, as

$$\frac{a_1}{a_0} \frac{a_2}{a_0} - \frac{a_3}{a_0} > 0$$

When the ratios of the a's are replaced by their equivalents from (1), (2), and (3), this becomes

$$-(r + 2p)(p^2 + q^2 + 2pr) + r(p^2 + q^2) > 0$$

or, simplifying and rearranging,

(4) $$-2p[(p^2 + q^2 + 2pr) + r^2] > 0$$

Now from (2) and the hypothesis that the a's are all of the same sign, it is evident that $p^2 + q^2 + 2pr > 0$. Hence

$$(p^2 + q^2 + 2pr) + r^2 > 0$$

and it follows from (4) that $p < 0$, as asserted. This proves the sufficiency of the conditions of Theorem 3.

The necessity that all the coefficients in the cubic equation have the same sign follows immediately from (1), (2), and (3), since the right-hand sides of these relations are all positive if $p < 0$ and $r < 0$. The necessity of the condition $a_1 a_2 - a_0 a_3 > 0$ follows by reversing the above steps and working backward to this inequality from (4), which is surely true if $p < 0$ and $r < 0$.

The extension of Theorem 3 to polynomial equations of higher degree is contained in the next theorem, which we state without proof.†

THEOREM 4   In the polynomial equation

$$Q(s) = a_0 s^n + a_1 s^{n-1} + a_2 s^{n-2} + \cdots + a_{n-1} s + a_n = 0$$

let every coefficient be positive, and construct the n quantities

$$D_1 = a_1, \; D_2 = \begin{vmatrix} a_1 & a_0 \\ a_3 & a_2 \end{vmatrix}, \; D_3 = \begin{vmatrix} a_1 & a_0 & 0 \\ a_3 & a_2 & a_1 \\ a_5 & a_4 & a_3 \end{vmatrix}, \ldots,$$

$$D_n = \begin{vmatrix} a_1 & a_0 & 0 & 0 & 0 & 0 & \cdots & \cdot \\ a_3 & a_2 & a_1 & a_0 & 0 & 0 & \cdots & \cdot \\ a_5 & a_4 & a_3 & a_2 & a_1 & a_0 & \cdots & \cdot \\ \cdots\cdots\cdots\cdots\cdots\cdots\cdots\cdots\cdots\cdots \\ a_{2n-1} & a_{2n-2} & a_{2n-3} & a_{2n-4} & a_{2n-5} & a_{2n-6} & \cdots & a_n \end{vmatrix}$$

where, in each determinant, all a's with negative subscripts or with subscripts greater than n are to be replaced by zero. Then a necessary and sufficient condition that each root of $Q(s) = 0$ have negative real part is that each D_i be positive.

The test provided by Theorem 4 is commonly known as the **Routh** or **Routh-Hurwitz stability criterion.**

EXAMPLE 1

For the equation $s^5 + s^4 + 2s^3 + s^2 + s + 2 = 0$, we have

$$D_1 = 1 \qquad D_2 = \begin{vmatrix} 1 & 1 \\ 1 & 2 \end{vmatrix} = 1 \qquad D_3 = \begin{vmatrix} 1 & 1 & 0 \\ 1 & 2 & 1 \\ 2 & 1 & 1 \end{vmatrix} = 2$$

$$D_4 = \begin{vmatrix} 1 & 1 & 0 & 0 \\ 1 & 2 & 1 & 1 \\ 2 & 1 & 1 & 2 \\ 0 & 0 & 2 & 1 \end{vmatrix} = -4 \qquad D_5 = \begin{vmatrix} 1 & 1 & 0 & 0 & 0 \\ 1 & 2 & 1 & 1 & 0 \\ 2 & 1 & 1 & 2 & 1 \\ 0 & 0 & 2 & 1 & 1 \\ 0 & 0 & 0 & 0 & 2 \end{vmatrix} = -8$$

---

† See, for instance, J. V. Uspensky, "Theory of Equations," pp. 304–309, McGraw-Hill, New York, 1948.

Since not all the D's are positive, the given equation has at least one root whose real part is nonnegative. This can be confirmed, of course, by actually finding the roots of the given equation, which are, in fact,

$$r_1 = -1 \qquad r_2, r_3 = \frac{1}{2} \pm i\frac{\sqrt{3}}{2} \qquad r_4, r_5 = -\frac{1}{2} \pm i\frac{\sqrt{7}}{2}$$

A somewhat different method of obtaining information about the location of the roots of an equation $f(z) = 0$, which has the advantage of telling exactly how many roots there are with positive real parts and moreover not being restricted to the case in which $f(z)$ is a polynomial, is based on the following theorem.

THEOREM 5   If $f(z)$ is analytic within and on a closed curve C except at a finite number of poles, and if $f(z)$ has neither poles nor zeros on C, then

$$\frac{1}{2\pi i}\int_C \frac{f'(z)}{f(z)}\, dz = N - P$$

where N is the number of zeros of $f(z)$ within C and P is the number of poles of $f(z)$ within C, each counted as many times as its multiplicity.

Proof   Suppose first that at a point $z = a_k$ within C, $f(z)$ has a zero of order n_k. Then $f(z)$ can be written in the form

$$f(z) = (z - a_k)^{n_k}\phi(z)$$

where $\phi(z)$ is nonvanishing and analytic in some neighborhood of $z = a_k$. From this,

$$f'(z) = n_k(z - a_k)^{n_k - 1}\phi(z) + (z - a_k)^{n_k}\phi'(z)$$

and thus $\dfrac{f'(z)}{f(z)} = \dfrac{n_k(z - a_k)^{n_k - 1}\phi(z) + (z - a_k)^{n_k}\phi'(z)}{(z - a_k)^{n_k}\phi(z)} = \dfrac{n_k}{z - a_k} + \dfrac{\phi'(z)}{\phi(z)}$

Since $\phi(z)$, and hence $\phi'(z)$, is analytic at $z = a_k$, and since $\phi(z)$ does not vanish at $z = a_k$, the fraction $\phi'(z)/\phi(z)$ is analytic at $z = a_k$. Hence it is clear from the last expression that $f'(z)/f(z)$ has a simple pole with residue n_k at every point a_k where $f(z)$ has a zero of order n_k. Similarly, if $f(z)$ has a pole of order p_k at the point $z = b_k$, we can write

$$f(z) = \frac{c_{-p_k}}{(z - b_k)^{p_k}} + \frac{c_{-p_k+1}}{(z - b_k)^{p_k - 1}} + \cdots + \frac{c_{-1}}{z - b_k} + c_0 + \cdots$$

Hence, putting these fractions over a common denominator, we have, in the neighborhood of $z = b_k$,

$$f(z) = \frac{1}{(z - b_k)^{p_k}}\psi(z) = (z - b_k)^{-p_k}\psi(z)$$

where $\psi(z) = c_{-p_k} + c_{-p_k+1}(z - b_k) + c_{-p_k+2}(z - b_k)^2 + \cdots$

is obviously analytic and nonvanishing at $z = b_k$. Therefore, around b_k,

$$f'(z) = -p_k(z - b_k)^{-p_k - 1}\psi(z) + (z - b_k)^{-p_k}\psi'(z)$$

and thus $\dfrac{f'(z)}{f(z)} = \dfrac{-p_k(z - b_k)^{-p_k - 1}\psi(z) + (z - b_k)^{-p_k}\psi'(z)}{(z - b_k)^{-p_k}\psi(z)} = \dfrac{-p_k}{z - b_k} + \dfrac{\psi'(z)}{\psi(z)}$

The last fraction on the right is clearly analytic; hence, $f'(z)/f(z)$ has a simple pole with residue $-p_k$ at every point where $f(z)$ has a pole of order p_k. Applying the residue theorem to $f'(z)/f(z)$ over the region bounded by C, we therefore have

$$\int_C \frac{f'(z)}{f(z)} \, dz = 2\pi i \sum \text{residues} = 2\pi i(\sum n_k - \sum p_k) = 2\pi i(N - P)$$

since $\sum n_k$ is the total multiplicity N of all the zeros of $f(z)$ within C and $\sum p_k$ is the total multiplicity P of all the poles of $f(z)$ within C. Dividing by $2\pi i$, we obtain the assertion of the theorem.

An important alternative form of the last theorem can be derived by noting that

$$\frac{1}{2\pi i} \int_C \frac{f'(z)}{f(z)} \, dz = \frac{1}{2\pi i} \int_C d[\ln f(z)]$$

Hence, performing the integration,

$$N - P = \frac{1}{2\pi i} \, [\text{variation of } \ln f(z) \equiv \ln |f(z)| + i \arg f(z)$$

$$\text{in going completely around } C]$$

Clearly, $\ln |f(z)|$ is the same at the beginning and at the end of one full circuit around any closed curve. Therefore

$$N - P = \frac{1}{2\pi i} \, [\text{variation of } i \arg f(z) \text{ around } C]$$

$$= \frac{\text{variation of } \arg f(z) \text{ around } C}{2\pi}$$

In particular, if $f(z)$ is analytic everywhere within C, so that $P = 0$, we have the following important result, commonly known as the **principle of the argument**.

COROLLARY 1 If $f(z)$ is analytic within and on a closed curve C and does not vanish on C, then the number of zeros of $f(z)$ within C is equal to $1/2\pi$ times the net variation in the argument of $f(z)$ as z traverses the curve C in the counterclockwise direction.

In geometric terms, this means that if the locus of $w = f(z)$ is plotted for values of z ranging around the given contour C, then the number of times this locus encircles the origin in the w plane is the number of zeros of $f(z)$ within C. Moreover, although the possibility that $f(z) = 0$ on C was excluded in the statement of Corollary 1, it is clear that *if* $f(z)$ has a zero on C, then the image curve passes *through* the origin in the w plane.

To use the last theorem and its corollary to determine whether or not each of the roots of a polynomial equation $Q(z) = 0$ has negative real part, we proceed as follows. In the z plane let the contour C consist of the segment of the imaginary axis between $-R$ and R together with the semicircle lying in the right half plane and having this segment as diameter (Fig. 17.11). Since a polynomial equation has only a finite number of roots, it is clear that if R is taken sufficiently large, any roots of $Q(z) = 0$ which lie in the right half plane, i.e., any roots which have positive real parts, will lie within C.

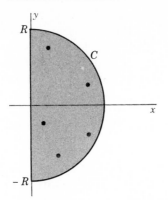

Figure 17.11
A semicircular contour enclosing all zeros of a function which lie in the right half plane.

Now let z range over the contour C, and in an auxiliary w plane let the locus of the corresponding values of $w = Q(z)$ be plotted. If this curve does not enclose the origin in the w plane, then according to the corollary of Theorem 5, $Q(z) = 0$ has no roots in the right half plane. If, further, this curve does not pass through the origin in the w plane, then $Q(z) = 0$ has no roots on the imaginary axis either; i.e., all roots of $Q(z) = 0$ have negative real parts. On the other hand, if the image curve encircles the origin in the w plane a net number of times k, then $Q(z) = 0$ has k roots in the right half plane, i.e., has k roots with positive real parts. Moreover, for every time this curve passes through the origin in the w plane there is a root of $Q(z) = 0$ lying on the imaginary axis in the z plane. Distinct imaginary roots of $Q(z) = 0$ thus give rise to a multiple point at the origin in the w plane, the tangents at the multiple point being distinct. A repeated pure imaginary root in the z plane similarly gives rise, in general, to a cusp at the origin in the w plane.

The labor of plotting the image curve in the w plane can be reduced considerably by letting $R \to \infty$. The image of the semicircular portion of C then recedes to infinity in the w plane, and without any plotting its contribution to possible encirclements of the origin can be determined in the following way. On the semicircle we have

$$z = Re^{i\theta} \qquad -\frac{\pi}{2} \le \theta \le \frac{\pi}{2}$$

For the images of these values of z, we have

$$w = Q(z) = Q(Re^{i\theta}) = a_0(Re^{i\theta})^n + a_1(Re^{i\theta})^{n-1} + a_2(Re^{i\theta})^{n-2} + \cdots + a_n$$

$$= R^n \left(a_0 e^{in\theta} + \frac{1}{R} a_1 e^{i(n-1)\theta} + \frac{1}{R^2} a_2 e^{i(n-2)\theta} + \cdots + \frac{1}{R^n} a_n \right)$$

Hence, for arbitrarily large values of R, the lengths of the terms in $Q(z)$, after the first, are vanishingly small in comparison with the length of the first term. Thus, regardless of the angles of these terms, $Q(z)$ is arbitrarily close to its leading term

$$R^n a_0 e^{in\theta}$$

Therefore, as z traverses the semicircular portion of C in the positive direction, with $\theta = \arg z$ varying from $-\pi/2$ to $\pi/2$, the argument of its image

$$w \doteq a_0 R^n e^{in\theta}$$

varies from $-n\pi/2$ to $n\pi/2$, which represents a net variation in arg w, that is, arg $Q(z)$, of $n\pi$. Hence if $w = Q(z)$ is plotted only for z varying from $i\infty$ to $-i\infty$ along the imaginary axis and the net change in the argument of w is noted, with its proper sign, of course, this change plus $n\pi$ will give the net change as the entire contour C is traversed. This change divided by 2π gives the net number of times the image curve encircles the origin in the w plane, and this number is equal to the number of roots of $Q(z) = 0$ in the right half of the z plane. The labor of plotting can be reduced still further by noting that for polynomials with real coefficients, such as we encounter as Laplace transforms, we have

$$Q(\bar{z}) = \overline{Q(z)}$$

and hence the plot of $Q(z)$ for values of z on the lower half of the imaginary axis is just the reflection in the real axis of the plot of $Q(z)$ for values of z on the upper half of the imaginary axis.

EXAMPLE 2

Discuss the stability of $y(t)$ if

$$\mathscr{L}\{y(t)\} = \frac{s^2 + 1}{s^3 + s^2 + 4s + 1}$$

As we pointed out above, the stability of $y(t)$ is determined solely by the location of the zeros of the denominator of $\mathscr{L}\{y(t)\}$. Hence, we begin by plotting

$$w = Q(s) = s^3 + s^2 + 4s + 1$$

for values of s on the imaginary s axis, i.e., for $s = i\omega$ and ω ranging from ∞ to $-\infty$. Parametric equations for the image curve are easily obtained, for

$$Q(i\omega) = -i\omega^3 - \omega^2 + 4i\omega + 1$$

and so the real and imaginary parts of $w = u + iv$ are

$$u = 1 - \omega^2 \qquad \text{and} \qquad v = 4\omega - \omega^3$$

Figure 17.12 shows a plot of this curve, together with a plot of arg w. Evidently, as s traverses the imaginary axis from $i\infty$ to $-i\infty$, arg w varies from $3\pi/2$ to $-3\pi/2$, which is a net variation of -3π. This added to the value $n\pi \equiv 3\pi$ contributed, according to our preceding discussion, by the semicircular portion of the contour C (Fig. 17.11), gives a net variation of zero as the entire contour C is traversed. Hence, $Q(s)$ has no zeros in the right half of the s plane. Moreover, since the image curve does not pass through the origin in the w plane, $Q(s)$ has no zeros on the imaginary axis. Therefore, by our earlier discussion, the inverse $y(t)$ is stable.

EXAMPLE 3

Discuss the stability of $y(t)$ if

$$\mathscr{L}\{y(t)\} = \frac{s - 2}{s^3 + s^2 + s + 4}$$

Proceeding exactly as in Example 2, we obtain from

$$Q(i\omega) = -i\omega^3 - \omega^2 + i\omega + 4$$

the parametric equations

$$u = 4 - \omega^2 \qquad \text{and} \qquad v = \omega - \omega^3$$

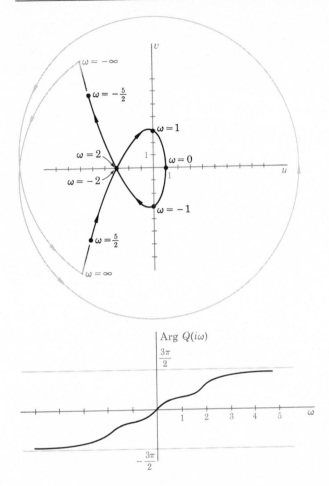

Figure 17.12
Plots of $Q(s) =$
$s^3 + s^2 + 4s + 1$ and
arg $Q(s)$ for $s = i\omega$.

and the image curve shown in Fig. 17.13. In this case, as s traverses the imaginary axis from $i\infty$ to $-i\infty$, arg w varies from $3\pi/2$, as in Example 2, to $5\pi/2$, which is a net variation of $5\pi/2 - 3\pi/2 = \pi$. Hence, adding the variation $n\pi \equiv 3\pi$ contributed by the semicircular portion of the contour C (Fig. 17.11), we obtain 4π for the net variation in arg w as the entire contour C is traversed. Dividing this by 2π, we obtain 2 as the number of zeros of $Q(s)$ in the right half plane. The inverse in this case is therefore unstable.

Theorem 5 finds its best-known application in the so-called **Nyquist stability criterion**, which is a modification of the preceding process especially well adapted to the stability analysis of closed-loop control systems. One common problem in engineering is to make the output $x_o(t)$ of a system follow quickly and accurately changes made in the input $x_i(t)$ to the system. In an **open-loop system** (Fig. 17.14a) this is often difficult to accomplish; specifically, prolonged oscillation of $x_o(t)$ about its desired value may well follow an abrupt change of the input $x_i(t)$ to some desired new value. One possible way to remedy this situation is to construct a **feedback loop**, like the one shown in Fig. 17.14b, to sample the output and feed it back to a differential device which will in turn transmit the **error signal** $x_i(t) - x_o(t)$ as a modified or corrected input to the original system. More generally, the output $x_o(t)$

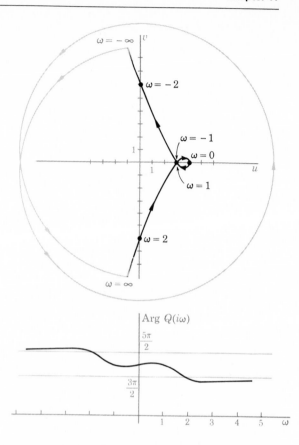

Figure 17.13
Plots of $Q(s) = s^3 + s^2 + s + 4$ and arg $Q(s)$ for $s = i\omega$.

may be (and usually is) modified by some additional device in the feedback loop to produce the **feedback signal** $x_f(t)$ before it is fed to the differential device (Fig. 17.14c).

In Fig. 17.14, let $G_1(s)$ and $G_2(s)$ be the transfer functions of the original system and the feedback loop, respectively. Then, from the definition of a transfer function as the ratio of the transformed output to the transformed input (Sec. 7.7), we can write

$$\mathcal{L}\{x_o(t)\} = G_1(s)[\mathcal{L}\{x_i(t)\} - \mathcal{L}\{x_f(t)\}]$$
$$\mathcal{L}\{x_f(t)\} = G_2(s)\mathcal{L}\{x_o(t)\}$$

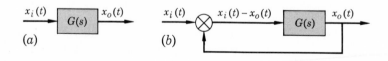

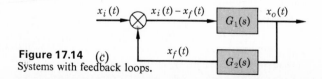

Figure 17.14   (c)
Systems with feedback loops.

If we eliminate $\mathscr{L}\{x_f(t)\}$ between these two equations, we obtain at once

$$\mathscr{L}\{x_0(t)\} = \frac{G_1(s)}{1 + G_1(s)G_2(s)} \mathscr{L}\{x_i(t)\}$$

Evidently $G_1(s)/[1 + G_1(s)G_2(s)]$ is the overall transfer function of the entire closed-loop system.

The question of the stability of a feedback system is of great importance and, as discussed above, can be answered by an examination of the Laplace transform of the output, namely,

$$\frac{G_1(s)}{1 + G_1(s)G_2(s)} \mathscr{L}\{x_i(t)\}$$

If the original system without the feedback loop is stable for the input $x_i(t)$, as we shall suppose, then the product $G_1(s)\mathscr{L}\{x_i(t)\}$ can have no poles in the right half of the s plane and the stability of the overall system depends solely on the location of the zeros of the denominator

$$1 + G_1(s)G_2(s)$$

Hence, as before, we plot the locus of the function

$$w(s) = 1 + G_1(s)G_2(s)$$

as s ranges over the contour of Fig. 17.11.

In this case, since $G_1(s)$ and $G_2(s)$ are themselves Laplace transforms, each approaches zero as R becomes infinite (Corollary 1, Theorem 5, Sec. 7.1). Hence, the image of the semicircular portion of the contour C shrinks to the single point $w = 1$ as $R \to \infty$ and therefore contributes nothing to the total variation of arg w as s traverses C. Thus, to determine stability, it is necessary only to plot $w(s) = 1 + G_1(s)G_2(s)$ for values of s on the imaginary axis and determine whether or not the resulting curve encloses the origin. Moreover, as we pointed out above, this curve can be constructed simply by plotting $1 + G_1(i\omega)G_2(i\omega)$ for positive values of ω and then reflecting the resulting arc in the real axis. In practice, instead of plotting $w = 1 + G_1(i\omega)G_2(i\omega)$ and observing whether the image curve encircles the origin, it is customary to plot $w = G_1(i\omega)G_2(i\omega)$ and observe whether it encircles the point $w = -1$. The equivalence of these two procedures is obvious.

It would take us too far afield and involve us in too many details of a purely engineering nature to discuss the applications of the Nyquist stability criterion to specific, nontrivial closed-loop systems. Such applications appear in large numbers in books on servomechanisms, to which we must turn for illustrations and further information.†

EXERCISES

1   Verify that $f(z) = z^2 + 1$ has one zero inside the circle $|z - i| = 1$ and two zeros inside the circle $|z| = 2$ by plotting $f(z)$ as z varies around the respective circles.

Using the geometric approach based on the corollary of Theorem 5, determine whether the following equations have any roots with nonnegative real parts. Check by using Theorem 4.

---

† See, for instance, G. J. Thaler and R. G. Brown, "Analysis and Design of Feedback Control Systems," 2d ed., McGraw-Hill, New York, 1960, or H. Chestnut and R. W. Mayer, "Servomechanisms and Regulating System Design," Wiley, New York, 1951.

2 $s^3 + s + 9 = 0$

3 $s^3 + 6s^2 + 10s + 6 = 0$

4 $s^4 + 2s^3 + 7s^2 + 4s + 10 = 0$

5 $s^4 + s^3 + s^2 + 10s + 10 = 0$

6 Prove Theorem 3 under the assumption that the cubic has three real roots.

7 Using the principle of the argument, prove the following result, known as **Rouché's theorem.** If $f(z)$ and $g(z)$ are analytic within and on a simple closed curve C, and if $|g(z)| < |f(z)|$ on C, then $f(z) + g(z)$ and $f(z)$ have the same number of zeros in the interior of C. *Hint:* Verify that $\arg(f + g) = \arg f + \arg[1 + (g/f)]$ and then show that the variation in $\arg[1 + (g/f)]$ as z traverses C is zero.

8 Using the result of Exercise 7, show that if $g(z)$ is analytic within and on the circle $C: |z| = 1$ and if $|g(z)| < 1$ for z on C, then there is exactly one point z_0 in the interior of C such that $g(z_0) = z_0$.

9 Using the result of Exercise 7, show that if $|a| > e$, then the equation $az^n - e^z = 0$ has exactly n roots in the interior of the circle $C: |z| = 1$.

10 Show that all the roots of the equation $z^7 - 5z^3 + 12 = 0$ are located in the annulus bounded by the circle $|z| = 1$ and the circle $|z| = 2$.

CHAPTER 18
Conformal Mapping

18.1 The Geometrical Representation of Functions of z

Although in the last section we plotted the values of a function $w = f(z)$ for certain values of z, namely, those on a particular semicircular contour, we have not yet attempted to provide a geometrical representation for $w = f(z)$ when z ranges over the entire complex plane. To do so now requires a decided departure from the conventional methods of cartesian plotting, which associate a curve with a real function of one variable $y = g(x)$ and a surface with a real function of two variables $z = h(x, y)$. In the complex domain, a functional relation $w = f(z)$, that is,

$$u + iv = f(x + iy)$$

involves *four* real variables, namely, the two independent variables x and y and the two dependent variables u and v. Hence, a space of *four* dimensions is required if we are to plot $w = f(z)$ in the cartesian fashion. To avoid the difficulties inherent in such a device, we choose to proceed as follows.

Let there be given two planes, one the z plane, in which the point $z = x + iy$ is to be plotted, and the other the w plane, in which the point $w = u + iv$ is to be plotted. A function $w = f(z)$ is now represented not by a locus of points in a space of four dimensions but by a correspondence between the points of these two cartesian planes. Whenever a point is given in the z plane, the function $w = f(z)$ determines one or more values of $u + iv$ and, hence, one or more points in the w plane. As z ranges over any configuration in the z plane, the corresponding point $u + iv$ describes some configuration in the w plane. The function $w = f(z)$ thus defines a **mapping** or a **transformation** of the z plane onto the w plane and, in turn, is represented geometrically by this mapping.

EXAMPLE 1

Discuss the way in which the z plane is mapped onto the w plane by the function $w = z^2$.

In this case we have

$$w = u + iv = z^2 = (x + iy)^2 = x^2 - y^2 + 2ixy$$

and thus

$$(1) \qquad\qquad u = x^2 - y^2 \qquad v = 2xy$$

These are the equations of the transformation from the z plane to the w plane. From them, many features of the mapping can easily be inferred.

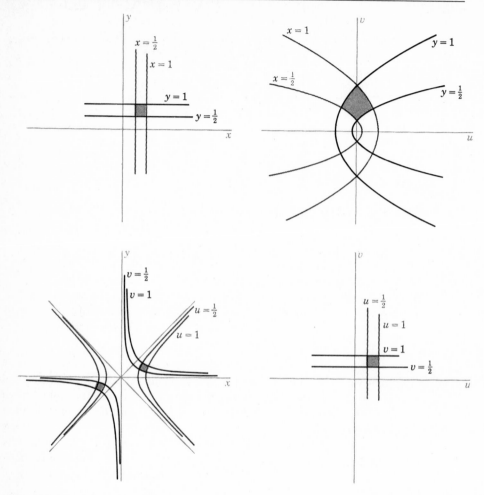

Figure 18.1
The mapping of certain lines by the function $w = z^2$.

For instance, lines parallel to the y axis, i.e., lines with equations $x = c_1$, map into curves in the w plane whose parametric equations are, from (1),

$$u = c_1{}^2 - y^2 \qquad v = 2c_1 y$$

Eliminating the parameter y, we obtain the equation

$$u = c_1{}^2 - \frac{v^2}{4c_1{}^2}$$

This defines a family of parabolas having the origin of the w plane as focus and the line $v = 0$ as axis, all opening to the left (Fig. 18.1). Similarly, lines parallel to the x axis, i.e., lines with equations $y = c_2$, map into curves in the w plane whose parametric equations are

$$u = x^2 - c_2{}^2 \qquad v = 2c_2 x$$

Eliminating x, we obtain

$$u = \frac{v^2}{4c_2{}^2} - c_2{}^2$$

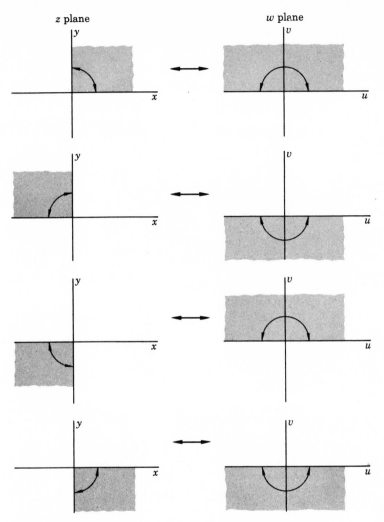

Figure 18.2
The two-valued character of the mapping defined by $z = w^{1/2}$.

which is also the equation of a family of parabolas having the origin as focus and the line $v = 0$ as axis but this time all opening to the right.

Mapping from the w plane back onto the z plane is even more immediate. From (1), it is clear that the lines $u = k_1$ correspond to the rectangular hyperbolas

$$x^2 - y^2 = k_1$$

and the lines $v = k_2$ correspond to the rectangular hyperbolas

$$xy = \tfrac{1}{2}k_2$$

The images of other curves, or regions, can be found in the same fashion with varying degrees of difficulty. For instance, to find the curve into which the line

$$y = 2x + 1$$

is transformed, we must eliminate x and y between this equation and the equations of the transformation. To do this, we first substitute for y in Eqs. (1), getting

$$u = x^2 - (2x + 1)^2 = -3x^2 - 4x - 1$$

$$v = 2x(2x + 1) = 4x^2 + 2x$$

Now if we regard these as two simultaneous equations in the quantities x and x^2, we can solve, getting

$$x = \frac{4u + 3v + 4}{-10} \qquad x^2 = \frac{u + 2v + 1}{5}$$

Hence

$$\frac{u + 2v + 1}{5} = \left(\frac{4u + 3v + 4}{-10}\right)^2$$

or, collecting terms,

$$16u^2 + 24uv + 9v^2 + 12u - 16v = 4$$

which is the equation of a parabola.

Although w is a single-valued function of z, the converse is not true. In fact, when w is given, z may be either of the two square roots of w. Because of this, the mapping from the z plane to the w plane covers the latter twice, as Fig. 18.2 suggests. This, of course, is nothing but a graphic representation of the now familiar fact that the angles of complex numbers are doubled when the numbers are squared.

EXERCISES

1 Show that the mapping $w = z^2$ transforms every straight line into a parabola.
2 What is the image of the circle $x^2 + y^2 = a^2$ under the mapping $w = z^2$?
3 Discuss the mapping between the z plane and the w plane which is defined by the function $w = \bar{z}^2$.
4 Discuss the transformation between the z plane and the w plane defined by $w = x - iy$.
5 What relation, if any, exists between the mappings defined by $w = f(z)$ and by $w = f(\bar{z})$?
6 Discuss the transformation defined by $w = 2iz + 1$.
7 Discuss the transformation defined by $w = (x^2 - y^2) + ixy$. In what significant way does it differ from the transformation defined by $w = (x^2 - y^2) + 2ixy$?
8 Discuss the transformation defined by $w = 1/z$.
9 Discuss the transformation defined by $w = z^3$. Plot the image of the line $u = 1$. What is the equation of the image of the line $x = 1$?
10 Find the equations of the transformation defined by the function $w = (z - i)/z$, and show that every circle through the origin in the z plane is transformed into a straight line.
11 Discuss the transformation defined by the function $w = e^z$.
12 Discuss the transformation defined by the function $w = \sin z$.

18.2 Conformal Mapping

In the last section we saw that every function of a complex variable maps the xy plane onto the uv plane. We now propose to investigate in more general terms the character of this transformation when the mapping function $w = u(x,y) + iv(x,y)$ is analytic.

At the outset it is important to know when the transformation equations can be solved (at least theoretically) for x and y as single-valued functions of u and v, that is, when the transformation has a single-valued inverse. The condition for this, as established in most texts on advanced calculus,† is simply that the jacobian determinant of the transformation

$$J\left(\frac{u,v}{x,y}\right) = \begin{vmatrix} \dfrac{\partial u}{\partial x} & \dfrac{\partial u}{\partial y} \\[2mm] \dfrac{\partial v}{\partial x} & \dfrac{\partial v}{\partial y} \end{vmatrix}$$

be different from zero. Since $w = f(z)$ is assumed to be analytic, u and v must satisfy the Cauchy-Riemann equations. Hence, substituting into the jacobian determinant, we have

$$J\left(\frac{u,v}{x,y}\right) = \begin{vmatrix} \dfrac{\partial u}{\partial x} & -\dfrac{\partial v}{\partial x} \\[2mm] \dfrac{\partial v}{\partial x} & \dfrac{\partial u}{\partial x} \end{vmatrix} = \left(\frac{\partial u}{\partial x}\right)^2 + \left(\frac{\partial v}{\partial x}\right)^2 = \left|\frac{\partial u}{\partial x} + i\frac{\partial v}{\partial x}\right|^2 = |f'(z)|^2$$

which establishes the following result.

THEOREM 1   If $f(z)$ is analytic, the function $w = f(z)$ will have a single-valued inverse in the neighborhood of any point where the derivative of the mapping function is different from zero.

Exceptional points where $f'(z) = 0$ are known as **critical points** of the transformation.

Now consider a value of z and its image $w = f(z)$, where $f(z)$ is analytic, and let

$$\Delta z = |\Delta z|e^{i\theta} \qquad \text{and} \qquad \Delta w = |\Delta w|e^{i\phi}$$

be corresponding increments of these quantities (Fig. 18.3). Then

$$f'(z) = \lim_{\Delta z \to 0} \frac{\Delta w}{\Delta z} = \lim_{\Delta z \to 0} \frac{|\Delta w|e^{i\phi}}{|\Delta z|e^{i\theta}} = \lim_{\Delta z \to 0} \left|\frac{\Delta w}{\Delta z}\right| e^{i(\phi - \theta)}$$

From this it is apparent that

$$\lim_{\Delta z \to 0} \left|\frac{\Delta w}{\Delta z}\right| = |f'(z)| \qquad \text{and} \qquad \lim_{\Delta z \to 0} (\phi - \theta) = \arg f'(z)$$

Figure 18.3
Δz and its image Δw under a mapping $w = f(z)$.

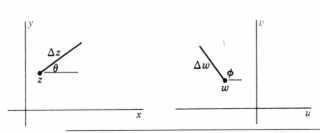

† See, for instance, R. C. Buck, "Advanced Calculus," p. 215, McGraw-Hill, New York, 1956.

or, to an arbitrary degree of approximation,

(1) $$|\Delta w| = |f'(z)| \cdot |\Delta z|$$

and $$\phi = \theta + \arg f'(z)$$

or

(2) $$\arg \Delta w = \arg \Delta z + \arg f'(z)$$

The fact that $f'(z)$ exists [which, of course, it does, since $f(z)$ is assumed to be analytic] means that both $|f'(z)|$ and $\arg f'(z)$ are independent of the manner in which $\Delta z \to 0$. In other words, they depend solely on z and not on the limiting orientation of the increment Δz. Hence, from (1) we draw the following conclusion.

THEOREM 2   In the mapping defined by an analytic function $w = f(z)$, the lengths of infinitesimal segments, regardless of their direction, are altered by a factor $|f'(z)|$ which depends only on the point from which the segments are drawn.

Since infinitesimal lengths are magnified by the factor $|f'(z)|$, it follows that infinitesimal areas are magnified by the factor $|f'(z)|^2$, that is, by $J(u,v/x,y)$.

Similarly, we conclude from (2) that, in general, the difference between the angles of an infinitesimal segment and its image is independent of the direction of the segment and depends only on the point from which the segment is drawn. In particular, two infinitesimal segments forming an angle will both be rotated in the same direction by the same amount. Hence, the measure of the angle between them will in general be left invariant by the transformation.

However, when $f'(z) = 0$, $\arg f'(z)$ is undefined and we cannot assert that angles are preserved. To investigate this case, suppose that $f'(z)$ has an n-fold zero at $z = z_0$. Then $f'(z)$ must contain the factor $(z - z_0)^n$, and hence we can write

$$f'(z) = (n + 1)a(z - z_0)^n + (n + 2)b(z - z_0)^{n+1} + \cdots$$

where $a, b, \ldots$ are complex coefficients of no concern to us and the factors $n + 1$, $n + 2, \ldots$ have been inserted for convenience in integrating $f'(z)$ to obtain $f(z)$:

$$f(z) = f(z_0) + a(z - z_0)^{n+1} + b(z - z_0)^{n+2} + \cdots$$

If in this expression we transpose $f(z_0)$, set

$$z - z_0 = \Delta z, \qquad f(z) - f(z_0) = \Delta w$$

and divide by $a(\Delta z)^{n+1}$, we obtain

$$\frac{\Delta w}{a(\Delta z)^{n+1}} = 1 + \frac{b}{a} \Delta z + \cdots$$

As $\Delta z \to 0$, the right member approaches 1. Therefore

$$\lim_{\Delta z \to 0} (\arg \Delta w) - \lim_{\Delta z \to 0} \left[\arg a(\Delta z)^{n+1}\right] = \arg 1 = 0$$

or, to an arbitrary degree of approximation,

$$\arg \Delta w = \arg a + (n + 1) \arg \Delta z$$

Now let Δz_1 and Δz_2 be two infinitesimal segments which make an angle θ with each other, and let Δw_1 and Δw_2 be their images. From the last equation we have

$$\arg \Delta w_1 = \arg a + (n + 1) \arg \Delta z_1$$

$$\arg \Delta w_2 = \arg a + (n + 1) \arg \Delta z_2$$

Hence, subtracting, we have

$$\arg \Delta w_2 - \arg \Delta w_1 = (n + 1)(\arg \Delta z_2 - \arg \Delta z_1) = (n + 1)\theta$$

Thus we have established the following theorem.

THEOREM 3 In the mapping defined by an analytic function $w = f(z)$, angles are in general preserved in magnitude and in sense. The only exception to this occurs when the vertex of the angle is an n-fold zero of $f'(z)$, in which case the measure of the angle is altered by the factor $n + 1$.

Example 1 of the last section is an excellent illustration of the behavior described by Theorem 3. The mapping function $w = f(z) = z^2$ is everywhere analytic, and, as Fig. 18.1 suggests, angle measures are in general preserved. However, the derivative $f'(z) = 2z$ has a simple zero at $z = 0$, and, as Fig. 18.2 indicates, angles with vertex at the origin are not preserved but are doubled.

A transformation which preserves the magnitudes of angles is said to be **isogonal**. A transformation which preserves the sense as well as the magnitudes of angles is said to be **conformal**. If $f(z)$ is an analytic function, it follows from Theorem 3 that in the neighborhood of any point where $f'(z) \neq 0$, the transformation defined by $w = f(z)$ is conformal. Conversely, it can be shown† that if the mapping

$$u = u(x, y) \qquad v = v(x, y)$$

is conformal, and if the first partial derivatives of u and v are continuous, then $w = u + iv = f(z)$ is an analytic function. Because of the properties guaranteed by Theorems 2 and 3, it is clear that under a conformal transformation any infinitesimal configuration and its image *conform*, in the sense of being approximately similar. This is not true, however, for large configurations, which may bear little or no resemblance to their images.

One important reason for studying conformal transformations is that solutions of Laplace's equation remain solutions of Laplace's equation when subjected to a conformal transformation. More precisely, we have the following theorem.

THEOREM 4 If $\phi(x, y)$ is a solution of the equation

$$\frac{\partial^2 \phi}{\partial x^2} + \frac{\partial^2 \phi}{\partial y^2} = 0$$

---

† See, for instance, E. G. Phillips, "Functions of a Complex Variable," pp. 35 and 36, Interscience, New York, 1945.

then when $\phi(x,y)$ is transformed into a function of u and v by a conformal transformation, it will satisfy the equation

$$\frac{\partial^2 \phi}{\partial u^2} + \frac{\partial^2 \phi}{\partial v^2} = 0$$

everywhere except possibly at the images of the points where the derivative of the mapping function is equal to zero.

Proof   Let $w = u(x,y) + iv(x,y)$ define a conformal transformation by means of which $\phi(x,y)$ is transformed into a function of u and v. Then

$$\frac{\partial \phi}{\partial x} = \frac{\partial \phi}{\partial u}\frac{\partial u}{\partial x} + \frac{\partial \phi}{\partial v}\frac{\partial v}{\partial x} \quad \text{and} \quad \frac{\partial \phi}{\partial y} = \frac{\partial \phi}{\partial u}\frac{\partial u}{\partial y} + \frac{\partial \phi}{\partial v}\frac{\partial v}{\partial y}$$

A second differentiation of each of these yields the results

$$\frac{\partial^2 \phi}{\partial x^2} = \frac{\partial \phi}{\partial u}\frac{\partial^2 u}{\partial x^2} + \left(\frac{\partial^2 \phi}{\partial u^2}\frac{\partial u}{\partial x} + \frac{\partial^2 \phi}{\partial v \partial u}\frac{\partial v}{\partial x}\right)\frac{\partial u}{\partial x} + \frac{\partial \phi}{\partial v}\frac{\partial^2 v}{\partial x^2} + \left(\frac{\partial^2 \phi}{\partial u \partial v}\frac{\partial u}{\partial x} + \frac{\partial^2 \phi}{\partial v^2}\frac{\partial v}{\partial x}\right)\frac{\partial v}{\partial x}$$

$$\frac{\partial^2 \phi}{\partial y^2} = \frac{\partial \phi}{\partial u}\frac{\partial^2 u}{\partial y^2} + \left(\frac{\partial^2 \phi}{\partial u^2}\frac{\partial u}{\partial y} + \frac{\partial^2 \phi}{\partial v \partial u}\frac{\partial v}{\partial y}\right)\frac{\partial u}{\partial y} + \frac{\partial \phi}{\partial v}\frac{\partial^2 v}{\partial y^2} + \left(\frac{\partial^2 \phi}{\partial u \partial v}\frac{\partial u}{\partial y} + \frac{\partial^2 \phi}{\partial v^2}\frac{\partial v}{\partial y}\right)\frac{\partial v}{\partial y}$$

When these are added, we obtain

$$\frac{\partial^2 \phi}{\partial x^2} + \frac{\partial^2 \phi}{\partial y^2} = \frac{\partial \phi}{\partial u}\left(\frac{\partial^2 u}{\partial x^2} + \frac{\partial^2 u}{\partial y^2}\right) + \frac{\partial^2 \phi}{\partial u^2}\left[\left(\frac{\partial u}{\partial x}\right)^2 + \left(\frac{\partial u}{\partial y}\right)^2\right]$$

$$+ 2\frac{\partial^2 \phi}{\partial u \partial v}\left(\frac{\partial u}{\partial x}\frac{\partial v}{\partial x} + \frac{\partial u}{\partial y}\frac{\partial v}{\partial y}\right) + \frac{\partial \phi}{\partial v}\left(\frac{\partial^2 v}{\partial x^2} + \frac{\partial^2 v}{\partial y^2}\right) + \frac{\partial^2 \phi}{\partial v^2}\left[\left(\frac{\partial v}{\partial x}\right)^2 + \left(\frac{\partial v}{\partial y}\right)^2\right]$$

Since $w = u + iv$ is analytic, by hypothesis, u and v themselves satisfy Laplace's equation. Hence, the first and fourth groups of terms on the right vanish identically. Moreover, u and v also satisfy the Cauchy-Riemann equations; hence the third group of terms also vanishes identically. By using the Cauchy-Riemann equations again, what remains can be written

$$\frac{\partial^2 \phi}{\partial x^2} + \frac{\partial^2 \phi}{\partial y^2} = \frac{\partial^2 \phi}{\partial u^2}\left[\left(\frac{\partial u}{\partial x}\right)^2 + \left(-\frac{\partial v}{\partial x}\right)^2\right] + \frac{\partial^2 \phi}{\partial v^2}\left[\left(\frac{\partial v}{\partial x}\right)^2 + \left(\frac{\partial u}{\partial x}\right)^2\right]$$

$$= \left[\left(\frac{\partial u}{\partial x}\right)^2 + \left(\frac{\partial v}{\partial x}\right)^2\right]\left(\frac{\partial^2 \phi}{\partial u^2} + \frac{\partial^2 \phi}{\partial v^2}\right)$$

$$= |f'(z)|^2 \left(\frac{\partial^2 \phi}{\partial u^2} + \frac{\partial^2 \phi}{\partial v^2}\right)$$

Thus, at any point where the transformation is conformal, that is, where $f'(z) \neq 0$,

$$\frac{\partial^2 \phi}{\partial x^2} + \frac{\partial^2 \phi}{\partial y^2} = 0 \quad \text{implies} \quad \frac{\partial^2 \phi}{\partial u^2} + \frac{\partial^2 \phi}{\partial v^2} = 0$$

as asserted.

     Suppose now that it is required to solve Laplace's equation, subject to certain boundary conditions, within a region R. Unless R is of a very simple shape, a direct attack upon the problem will usually be exceedingly difficult. However, it may be possible to find a conformal transformation which will convert R into some simpler region R', such as a circle or a half plane, in which Laplace's equation can be solved, subject, of course, to the transformed boundary conditions. If this is the case, the resulting solution, when carried back to R by the inverse transformation, will be the required solution of the original problem.

EXERCISES

1 (a) What is the length of the curve into which the upper half of the circle $|z| = a$ is transformed by the function $w = 1/z$?
(b) What is the length of the arc into which this function transforms the segment of $y = 1 - x$ which lies in the first quadrant?

2 Find the area of the region into which the square with vertices $z = 0, 1, 1 + i, i$ is transformed:
(a) By the function $w = z^2$ (b) By the function $w = z^3$

3 (a) What are the critical points of the transformation $w = 3z - z^3$?
(b) What is the locus of points at which the magnification of lengths is equal to 1?
(c) What is the locus of points at which infinitesimal segments are rotated through 45°?
(d) What is the locus of points at which infinitesimal segments are rotated through 90°?

4 Are there any points at which infinitesimal segments are left unchanged in length and direction:
(a) By the transformation $w = 2iz + z^2$
(b) By the transformation $w = z^2 + z^3$

5 If $u = 2x^2 + y^2$ and $v = y^2/x$, show that the curves $u =$ constant and $v =$ constant are orthogonal trajectories but that the transformation defined by $w = u + iv$ is not conformal. Give a specific illustration of the latter fact.

18.3 The Bilinear Transformation

The simplest class of conformal transformations, yet one of the most important, is the class of **bilinear** or **linear fractional** or **Möbius transformations,**† defined by the family of functions

(1)
$$w = \frac{az + b}{cz + d} \qquad ad - bc \neq 0$$

The restriction $ad - bc \neq 0$ is necessary because if $ad = bc$, then $a/c = b/d$ and the numerator and denominator of w are proportional. As a consequence, w is a constant, independent of z, and thus the entire z plane is mapped into the same point in the w plane!

It is convenient to investigate the general bilinear transformation by considering first the three special cases

     a. $w = z + \lambda$
     b. $w = \mu z$
     c. $w = 1/z$

---

† Named for the German geometer A. F. Möbius (1790–1868).

In case **a**, w is found by adding a constant vector λ to each z. Hence the transformation is just a translation in the direction defined by arg λ through a distance equal to $|\lambda|$. In particular, we note for later use that this rigid motion necessarily transforms circles into circles.

In case **b**, w is found by rotating each z through a fixed angle equal to arg μ and then multiplying its length by the factor $|\mu|$. In this case, too, circles are transformed into circles. To prove this, let us first write the equation of a general circle

$$a(x^2 + y^2) + bx + cy + d = 0 \qquad \begin{array}{l} a,b,c,d \text{ real} \\ b^2 + c^2 \geq 4ad \end{array}$$

in terms of z and $\bar{z}$ by means of the relations

$$x = \frac{z + \bar{z}}{2} \qquad y = \frac{z - \bar{z}}{2i} \qquad x^2 + y^2 = z\bar{z}$$

The result is

$$az\bar{z} + \frac{b - ic}{2} z + \frac{b + ic}{2} \bar{z} + d = 0$$

or, renaming the coefficients,

(2) $$(A + \bar{A})z\bar{z} + Bz + \bar{B}\bar{z} + (D + \bar{D}) = 0$$

where now A, B, and D are arbitrary complex numbers, subject to the condition $B\bar{B} \geq (A + \bar{A})(D + \bar{D})$, derived from the original condition $b^2 + c^2 \geq 4ad$, which ensures that the radius of the circle is real. If the substitution

$$z = \frac{w}{\mu}$$

is made in (2), we obtain the equation of the transformed curve

$$(A + \bar{A}) \frac{w}{\mu} \frac{\bar{w}}{\bar{\mu}} + B \frac{w}{\mu} + \bar{B} \frac{\bar{w}}{\bar{\mu}} + (D + \bar{D}) = 0$$

or

(3) $$(A + \bar{A})w\bar{w} + (B\bar{\mu})w + (\bar{B}\mu)\bar{w} + (D + \bar{D})\mu\bar{\mu} = 0$$

Since the coefficients of the first and last terms in (3) are real, and since the coefficients of the linear terms w and $\bar{w}$ are conjugates, this equation has the same structure as (2) and hence will also represent a circle provided its coefficients satisfy the condition necessary for the radius to be real. For the locus described by Eq. (3), this condition is

$$(B\bar{\mu})(\bar{B}\mu) \geq (A + \bar{A})(D + \bar{D})\mu\bar{\mu}$$

or, dividing through by $\mu\bar{\mu}$, which is necessarily positive,

$$B\bar{B} \geq (A + \bar{A})(D + \bar{D})$$

which is true by hypothesis. If $a = 0$, so that $A + \bar{A} = 0$, both the given circle and its image are straight lines.

In case **c** we can write

(4) $$w = \frac{1}{z} = \frac{\bar{z}}{z\bar{z}}$$

which shows that w is of length $1/|z|$ and has the direction of $\bar{z}$.

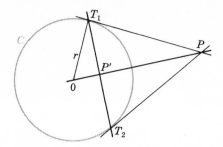

Figure 18.4
The geometrical relation between a point and its inverse.

To describe the geometrical process by which a point with these characteristics can be obtained from a given point z, we must first define the process of *inversion*. Let C be a circle with center O and radius r, and let P be any point in the plane of C. Then the **inverse** of P with respect to C is the point P' on the ray OP for which

(5)
$$OP \cdot OP' = r^2$$

From the symmetry of this relation it is clear that P is also the inverse of P'. Geometrically, a point and its inverse are related as follows. From any point outside a circle C with center O, let the two tangents to C be drawn, and let the points of contact of these tangents be joined (Fig. 18.4). The intersection of this chord with the line OP is the inverse P' of P. Conversely, let P' be any point in the interior of C. At P' erect a perpendicular to OP' and at either of the points at which this meets C let the tangent to C be drawn. The intersection of this tangent and the line OP' is the inverse P of P'. The consistency of these constructions with the definitive property (5) is evident, since in Fig. 18.4

$$\triangle OP'T_1 \sim \triangle OT_1P$$

and thus
$$\frac{OP'}{OT_1} = \frac{OT_1}{OP}$$

or
$$OP \cdot OP' = (OT_1)^2 = r^2$$

It is evident now that the construction of w from z in case **c** requires that the inverse of z in the unit circle be found and then reflected in the real axis; for the first of these steps gives a complex number whose length is $1/|z|$, and the second achieves the direction of $\bar{z}$, as required by (4).

To show that circles are also transformed into circles in case **c**, let the substitution $z = 1/w$ be made in the self-conjugate form of the equation of a circle (2). This gives

$$(A + \bar{A}) \frac{1}{w}\frac{1}{\bar{w}} + \frac{B}{w} + \frac{\bar{B}}{\bar{w}} + (D + \bar{D}) = 0$$

or
$$(D + \bar{D})w\bar{w} + \bar{B}w + B\bar{w} + (A + \bar{A}) = 0$$

which is also the equation of a circle with real radius. If $A + \bar{A} = 0$, the original circle reduces to a straight line whose image is a circle passing through the origin, since its equation contains no constant term. Conversely, any circle passing through the origin is transformed into a straight line.

The three special transformations we have just considered can be used to construct, or synthesize, the general bilinear transformation. To verify this, suppose first that

$c \neq 0$. Then the general transformation is equivalent to the following chain of special transformations:

$$w_1 = z + \frac{d}{c}$$

$$w_2 = cw_1 = cz + d$$

$$w_3 = \frac{1}{w_2} = \frac{1}{cz + d}$$

$$w_4 = \frac{bc - ad}{c} w_3 = \frac{bc - ad}{c(cz + d)}$$

$$w = w_4 + \frac{a}{c} = \frac{bc - ad}{c(cz + d)} + \frac{a}{c} = \frac{az + b}{cz + d}$$

On the other hand, if $c = 0$, it is clear from the restriction $ad - bc \neq 0$ that neither a nor d can be zero. Hence, we can write

$$w_1 = z + \frac{b}{a}$$

$$w = \frac{a}{d} w_1 = \frac{a}{d}\left(z + \frac{b}{a}\right) = \frac{az + b}{d}$$

Thus we have shown that in all cases the general bilinear transformation can be obtained as the composition of a succession of simple transformations of types **a**, **b**, and **c**. Since each of these is known to transform circles into circles, including straight lines as special cases, we have thus established the following theorem.

THEOREM 1   Under the general bilinear transformation, circles are transformed into circles.

The general bilinear transformation

(1)
$$w = \frac{az + b}{cz + d}$$

depends on three essential constants, namely, the ratios of any three of the coefficients a, b, c, d to the fourth. Hence it is evident that three conditions are necessary to determine a bilinear transformation. In particular, the requirement that three distinct values of z, say z_1, z_2, z_3, have specified distinct images w_1, w_2, w_3 leads to a unique transformation.

Although the transformation which sends three given points into three specified image points can be found by imposing these conditions on the general equation (1) and then solving the resulting three equations for the ratios of the coefficients, it is generally simpler to make use of the fact that if w_1, w_2, w_3, w_4 are, respectively, the images of z_1, z_2, z_3, z_4, then

$$\frac{(w_1 - w_2)(w_3 - w_4)}{(w_1 - w_4)(w_3 - w_2)} = \frac{(z_1 - z_2)(z_3 - z_4)}{(z_1 - z_4)(z_3 - z_2)}$$

To establish this relation, we observe that

$$w_i - w_j = \frac{az_i + b}{cz_i + d} - \frac{az_j + b}{cz_j + d} = \frac{(ad - bc)(z_i - z_j)}{(cz_i + d)(cz_j + d)}$$

Hence

$$\frac{(w_1 - w_2)(w_3 - w_4)}{(w_1 - w_4)(w_3 - w_2)} = \frac{\dfrac{(ad - bc)(z_1 - z_2)}{(cz_1 + d)(cz_2 + d)} \dfrac{(ad - bc)(z_3 - z_4)}{(cz_3 + d)(cz_4 + d)}}{\dfrac{(ad - bc)(z_1 - z_4)}{(cz_1 + d)(cz_4 + d)} \dfrac{(ad - bc)(z_3 - z_2)}{(cz_3 + d)(cz_2 + d)}}$$

$$= \frac{(z_1 - z_2)(z_3 - z_4)}{(z_1 - z_4)(z_3 - z_2)}$$

The last fraction is called the **cross ratio** or **anharmonic ratio** of the four numbers z_1, z_2, z_3, z_4. Hence the result we have just established can be formulated as the following theorem.

THEOREM 2   The cross ratio of four points is invariant under a bilinear transformation.

Suppose now that it is required to find the transformation which sends z_1, z_2, z_3 into w_1, w_2, w_3, respectively. If w is the image of a general point z under this transformation, then, according to Theorem 2, the cross ratio of w_1, w_2, w_3, and w must equal the cross ratio of z_1, z_2, z_3, and z. That is,

$$\frac{(w_1 - w_2)(w_3 - w)}{(w_1 - w)(w_3 - w_2)} = \frac{(z_1 - z_2)(z_3 - z)}{(z_1 - z)(z_3 - z_2)}$$

This equation is clearly bilinear in w and z and is satisfied by the three pairs of corresponding values (z_1, w_1), (z_2, w_2), (z_3, w_3). Moreover, everything in it is known except the variables w and z themselves. Hence it is necessary only to solve for w in terms of z to obtain the required transformation in standard form.

EXAMPLE 1

What is the bilinear transformation which sends the points $z = -1, 0, 1$ into the points $w = 0, i, 3i$, respectively?

Setting up the appropriate cross ratios, we have

$$\frac{(0 - i)(3i - w)}{(0 - w)(3i - i)} = \frac{(-1 - 0)(1 - z)}{(-1 - z)(1 - 0)}$$

or

$$\frac{3 + iw}{-2i} = \frac{-1 + z}{-1 - z}$$

Finally, solving this equation for w, we obtain without difficulty

$$w = -3i\frac{z + 1}{z - 3}$$

EXAMPLE 2

What is the most general bilinear transformation which maps the upper half of the z plane onto the interior of the unit circle in the w plane (Fig. 18.5)?

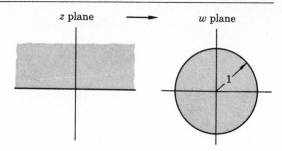

z plane ⟶ w plane

Figure 18.5
The upper half of the z plane to be mapped onto the interior of the unit circle in the w plane.

Let the required transformation be

$$w = \frac{az + b}{cz + d}$$

Since the boundaries of corresponding regions must correspond under any transformation, the unit circle in the w plane must be the image of the real axis in the z plane. Therefore, for all real values of z, we must have

$$|w| = \frac{|az + b|}{|cz + d|} = \frac{|a|}{|c|} \frac{|z + b/a|}{|z + d/c|} = 1$$

In particular, from the limiting case $|z| \to \infty$, we find

$$\frac{|a|}{|c|} = 1$$

In other words, a/c is a complex number of absolute value 1, say $e^{i\theta}$. From this, we conclude that for all real values of z

$$\left| z + \frac{b}{a} \right| = \left| z + \frac{d}{c} \right| \quad \text{or} \quad \left| z - \left(-\frac{b}{a} \right) \right| = \left| z - \left(-\frac{d}{c} \right) \right|$$

The last equation expresses the fact that the complex numbers $-b/a$ and $-d/c$ are equally far from an arbitrary point z on the real axis, which is possible if and only if the real axis in the z plane is the perpendicular bisector of the segment joining the points $-b/a$ and $-d/c$. Thus $-b/a$ and $-d/c$ must be conjugate complex numbers, say λ and $\bar{\lambda}$, and hence we can write

$$w = \frac{az + b}{cz + d}$$

$$= \frac{a}{c} \frac{z + b/a}{z + d/c}$$

$$= \frac{a}{c} \frac{z - \lambda}{z - \bar{\lambda}}$$

(6)
$$= e^{i\theta} \frac{z - \lambda}{z - \bar{\lambda}}$$

So far we have enforced only the condition that the boundaries of the two regions correspond. It is now necessary to make sure that the regions themselves correspond as required and that the upper half of the z plane has not been mapped onto the *outside* of the unit circle in the w plane. This is most easily verified by determining the image of some convenient point, say $z = \lambda$. Clearly, the point $z = \lambda$ maps into the point $w = 0$, which is certainly inside the circle $|w| = 1$. Thus if λ is restricted to lie in the *upper* half of the z plane, the solution is complete.

As a special case of some interest, let $e^{i\theta} = -1$, and let λ be a pure imaginary, say i. Then

(7)
$$w = -\frac{z - i}{z + i}$$

Now
$$\mathcal{I}(w) = \frac{w - \bar{w}}{2i} = -\frac{1}{2i}\left(\frac{z - i}{z + 1} - \frac{\bar{z} + i}{\bar{z} - i}\right)$$

or, reducing to a common denominator and simplifying,

$$\mathcal{I}(w) = \frac{z + \bar{z}}{(z + i)(\bar{z} - i)}$$

The denominator of the last fraction is the product of $z + i$ and its conjugate $\bar{z} - i$ and hence is a positive quantity. Thus the imaginary part of w will be positive if and only if $z + \bar{z}$ is positive. Since $z + \bar{z}$ is equal to twice the real part of z, this shows that the transformation (7) not only maps the upper half of the z plane onto the interior of the unit circle $|w| = 1$ but does it in such a way that the first quadrant of the z plane [where $\mathcal{R}(z) > 0$] corresponds to the upper half of the circle [where $\mathcal{I}(w) > 0$] and the second quadrant of the z plane corresponds to the lower half of the circle. In the opposite direction, the inverse transformation

(8)
$$z = -i\frac{w - 1}{w + 1}$$

maps the interior of the circle $|w| = 1$ onto the upper half of the z plane in such a way that the upper half of the circle maps onto the first quadrant of the z plane.

EXAMPLE 3

Find a transformation which will map an infinite sector of angle $\pi/4$ onto the interior of the unit circle.

Since the boundary of the sector consists of portions of *two* straight lines while its image is to be a *single* circle, it is apparent that the mapping cannot be accomplished by a bilinear transformation alone. However, a simple combination of a power function and a linear fractional transformation will define a suitable transformation. Specifically, the transformation

$$t = z^4$$

will open out the given sector in the z plane into the upper half of the auxiliary t plane (Fig. 18.6). Following this, the upper half of the t plane can be mapped onto the unit circle in the w plane by any transformation of the family (6) which we obtained in the last example, say

$$w = \frac{t - i}{t + i}$$

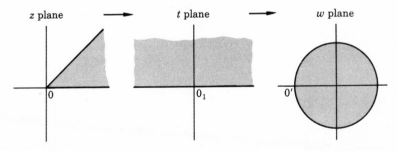

Figure 18.6
The two transformations needed to map an infinite sector onto the interior of the unit circle.

Taking the composition of these two mappings, we obtain for the required transformation

$$w = \frac{z^4 - i}{z^4 + i}$$

EXAMPLE 4

Find a transformation which will map a 60° sector of the unit circle in the z plane onto the upper half of the w plane.

At first glance it would seem that this problem could be solved simply by opening the given sector into a full circle by the transformation

$$t = z^6$$

and then mapping the circle from the t plane onto the upper half of the z plane by means of the inverse of one of the transformations of the family (6) which we obtained in Example 2, for instance the transformation (8). This method fails, however, because the circular region obtained in the t plane in this case is *not* of the type considered in Example 2. The latter consisted of a simple circular boundary plus its interior, whereas the former consists of the interior of a circle cut along a radius, since the radius $OA' \equiv OB'$ is actually the image of the two boundary radii OA and OB of the given sector (Fig. 18.7).

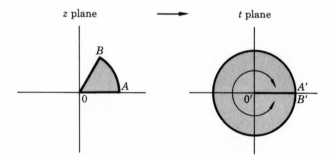

Figure 18.7
A circular sector opened out
into a circular region cut
along a radius.

To avoid this difficulty, let us first map the sector onto a semicircle by the transformation

$$t_1 = z^3$$

Then let us map the semicircle from the t_1 plane onto the first quadrant of the t_2 plane by means of the transformation (8)

$$t_2 = -i\frac{t_1 - 1}{t_1 + 1}$$

Finally (Fig. 18.8), let us open out the first quadrant of the t_2 plane into the upper half of the w plane by the transformation

$$w = t_2^2$$

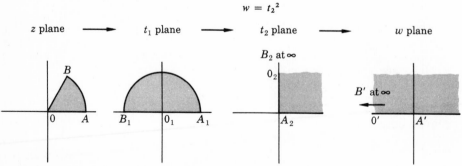

Figure 18.8
The sequence of transformations necessary to map a circular sector onto a half plane.

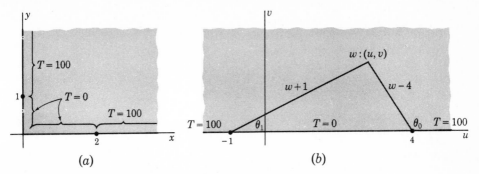

Figure 18.9
An infinite 90° sector mapped, with its boundary conditions, onto a half plane.

Taking the composition of these three transformations, we find

$$w = -\left(\frac{z^3 - 1}{z^3 + 1}\right)^2$$

as the required mapping.

EXAMPLE 5

A thin sheet of metal coincides with the first quadrant of the z plane. The upper and lower faces of the sheet are perfectly insulated so that heat flow in the sheet is strictly two-dimensional. Find the steady-state temperature at any point of the sheet if the boundary conditions are those shown in Fig. 18.9a.

Under the assumptions of the problem, the flow of heat in the sheet is two-dimensional, and we must accordingly solve Laplace's equation, i.e., the two-dimensional steady-state heat equation derived in Sec. 8.2,

$$\frac{\partial^2 T}{\partial x^2} + \frac{\partial^2 T}{\partial y^2} = 0$$

subject to the given conditions along the boundaries of the first quadrant. To do this, it is convenient to map the first quadrant of the z plane onto the upper half of the w plane by the transformation

$$w = z^2 = (x^2 - y^2) + 2ixy$$

This reduces the problem to that of finding a solution of Laplace's equation in the upper half of the w plane which assumes along the real axis the transformed boundary conditions shown in Fig. 18.9b.

We have long since discovered that both the real part and the imaginary part of any analytic function satisfy Laplace's equation. Therefore if we can find a function of w which is analytic in the upper half plane and whose real part or imaginary part takes on the given boundary values when w was real, we shall have the required solution. To obtain such a function to use in our problem, we observe first that

$$(9) \quad f(w) = iT_0 + \frac{1}{\pi}[(T_1 - T_0)\ln(w - u_0) + (T_2 - T_1)\ln(w - u_1) + \cdots$$
$$+ (T_{n+1} - T_n)\ln(w - u_n)]$$

is analytic everywhere in the upper half plane. Hence its imaginary part

$$(10) \quad \mathscr{I}[f(w)] = T_0 + \frac{1}{\pi}[(T_1 - T_0)\arg(w - u_0) + (T_2 - T_1)\arg(w - u_1) + \cdots$$
$$+ (T_{n+1} - T_n)\arg(w - u_n)]$$

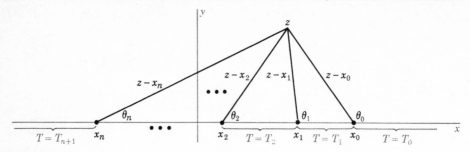

Figure 18.10
The behavior of arg $(z - x_i)$ as z varies along the real axis.

will be a solution of Laplace's equation everywhere in the upper half plane. Moreover, along the real axis, this solution takes on the boundary values shown in Fig. 18.10. To verify this, we observe from Fig. 18.10 that the complex number $w - u_i$ is represented by the vector joining the fixed point u_i to the variable point w, and thus arg $(w - u_i)$ is simply the inclination angle θ_i of this vector. Hence the function (10) can be rewritten

$$(11) \quad \mathscr{I}[f(w)] = T(w) = T_0 + \frac{1}{\pi}[(T_1 - T_0)\theta_0 + (T_2 - T_1)\theta_1 + \cdots$$
$$+ (T_{n+1} - T_n)\theta_n]$$

Again referring to Fig. 18.10, it is clear that for all values of w on the real axis to the right of u_0, each of the θ's is zero. Hence from (11) we see that T reduces to the constant value T_0 along this portion of the real axis. Furthermore, when w lies between u_1 and u_0, the angle θ_0 is equal to π while all the other θ's are still zero. Hence, along this segment the temperature (10), or (11), reduces to

$$T = T_0 + \frac{1}{\pi}[(T_1 - T_0)\pi] = T_1$$

Similarly, for values of w between u_2 and u_1, the angles θ_2 and θ_1 are each equal to π, but all other θ's are equal to zero. Hence along this segment of the real axis, we have

$$T = T_0 + \frac{1}{\pi}[(T_1 - T_0)\pi + (T_2 - T_1)\pi] = T_2$$

Continuing in this fashion, we can verify that T, as defined by either (10) or (11), not only is a solution of Laplace's equation, being the imaginary part of the analytic function (9), but also assumes along the real axis the temperature distribution shown in Fig. 18.10.

When we specialize these observations to our problem, it appears that the solution we require in the w plane is

$$T = 100 + \frac{1}{\pi}[(0 - 100)\theta_0 + (100 - 0)\theta_1]$$

$$= 100 + \frac{100}{\pi}(\theta_1 - \theta_0) = \frac{100}{\pi}[\pi + (\theta_1 - \theta_0)]$$

Multiplying by $\pi/100$ and then taking the tangent of both sides of the last equation, we have

$$\tan\frac{\pi T}{100} = \tan[\pi + (\theta_1 - \theta_0)] = \tan(\theta_1 - \theta_0)$$

$$= \frac{\tan\theta_1 - \tan\theta_0}{1 + \tan\theta_0\tan\theta_1}$$

Substituting for tan θ_0 and tan θ_1 their values, as read from Fig. 18.9*b*, we obtain from the last expression

$$\tan \frac{\pi T}{100} = \frac{v/(u+1) - v/(u-4)}{1 + v^2/(u+1)(u-4)}$$

(12)
$$= \frac{-v}{u^2 + v^2 - 3u - 4}$$

which is the solution of the transformed problem in the *w* plane. Returning to the *z* plane by means of the mapping equations

$$u = x^2 - y^2 \qquad \text{and} \qquad v = 2xy$$

we thus find, from (12), that

$$T = \frac{100}{\pi} \tan^{-1} \frac{-10xy}{(x^2 + y^2)^2 - 3x^2 + 3y^2 - 4}$$

is the solution to the original problem given in the *z* plane.

EXERCISES

1 Show that if a transformation of the form $w = (az + b)/(cz + d)$ maps z_1 into w_1 and z_2 into w_1, then either $z_1 = z_2$ or else $ad - bc = 0$.

2 What is the cross ratio of the four fourth roots of -1?

3 What is the cross ratio of the four complex sixth roots of 1?

4 Show that in general there are two points which are left invariant by a bilinear transformation, thought of as a mapping of the *z* plane onto itself. Are there any bilinear transformations which leave only one point invariant? No points invariant?

5 Find the invariant points of the transformation $z' = -(2z + 4i)/(iz + 1)$, and prove that these two points, together with an arbitrary point *z* and its image z', form a set of four points whose cross ratio is independent of *z*.

6 Find the invariant points of the transformation

$$z' = \frac{2iz + 1}{(-3 + 4i)z + 4}$$

7 Under what conditions, if any, will the transformation $z' = (az + b)/(cz + d)$ have the property that if z' is the image of *z*, then *z* is also the image of z'?

8 What is the bilinear transformation which sends the points $z = 0, -1, \infty$ into the points $w = -1, -2 - i, i$, respectively? What is the image of the circle $|z| = 1$ under this transformation?

9 What is the bilinear transformation which sends the points $z = 0, -i, 2i$ into the points $z' = 5i, \infty, -i/3$, respectively? What are the invariant points of this transformation?

10 Find the equations of the transformation of inversion in the circle $x^2 + y^2 = 1$, and show that under this transformation a circle is mapped into itself if and only if it is perpendicular to the circle defining the inversion.

11 What is the most general bilinear transformation which maps the upper half of the *z* plane onto the lower half of the *w* plane?

12 Prove that $w = z/(1 - z)$ maps the upper half of the *z* plane onto the upper half of the *w* plane. What is the image of the circle $|z| = 1$ under this transformation?

13 Find a transformation which will map an infinite sector of angle $\pi/3$ onto the interior of the unit circle.

14 Show that along the circle $|cz + d| = \sqrt{|ad - bc|}$ the transformation

$$w = \frac{(az + b)}{(cz + d)}$$

does not alter the lengths of infinitesimal segments. What happens to segments inside this circle? Outside this circle? What is the locus of points where infinitesimal segments are not rotated by the transformation?

15   Find a transformation which will map a 45° sector of the unit circle in the z plane
onto the upper half of the w plane.

16   Find a transformation which will map the upper half of the unit circle onto the
entire unit circle.

17   Show that if $|c| = |d|$, then the transformation $w = (az + b)/(cz + d)$ maps the
unit circle in the z plane onto a straight line in the w plane.

18   Verify that under the transformation

$$w = k\,\frac{z + a}{z - a} \qquad a \text{ real}$$

every circle which passes through the two points $z = -a$, a is transformed into a
straight line through the origin in the w plane. If C_1 is the circle of this family
whose y intercept is p, and if k is real, show that the image of C_1 is the straight line
through the origin whose inclination angle is $\phi = -2\cot^{-1}(p/a)$. Finally, show
that if C_2 is a circle intersecting C_1 at an angle α at $z = -a$ and at $z = a$, then the
transformation

$$w = \exp\left(2i\cot^{-1}\frac{p}{a}\right)\left(\frac{z + a}{z - a}\right)$$

maps the crescent-shaped region between C_1 and C_2 onto the interior of an angle
of measure α in standard position in the w plane.

19   Prove that four points z_1, z_2, z_3, z_4 lie on a circle if and only if their cross ratio
is real.

20   Find the steady-state temperature distribution in a sheet of metal coinciding with
the first quadrant of the z plane if $T = 100°C$ along the positive x axis and $T = 0°C$
along the positive y axis.

21   Find the steady-state temperature distribution in a sheet of metal coinciding with
the interior of a 60° angle in standard position in the z plane if $T = 0°C$ along the
horizontal side of the angle and 100°C along the other side.

22   Find the steady-state temperature distribution in a sheet of metal coinciding with the
first quadrant of the z plane if $T = 100°C$ along the positive y axis, $T = 50°C$
between $x = 0$ and $x = 3$, and $T = 0°C$ to the right of $x = 3$ on the x axis.

23   Find the steady-state temperature distribution in a sheet of metal coinciding with
the interior of the unit circle in the z plane if the upper half of the circumference
of the circle is kept at the temperature $T = 100°C$ and the lower half of the
circumference is kept at the temperature $T = 0°C$. *Hint:* Recall formula (8).

24   Find the steady-state temperature distribution in a sheet of metal coinciding with
the upper half of the unit circle in the z plane if the curved portion of the boundary
is kept at the temperature $T = 100°C$ and the bounding diameter is kept at the
temperature $T = 0°C$. *Hint:* Recall the results of Example 4.

25   Show that $w = z + 1/z$ maps the portion of the upper half of the z plane exterior
to the circle $|z| = 1$ onto the entire upper half of the w plane. Use this result to
find the steady-state temperature distribution in the upper half of the z plane
exterior to the unit circle if $T = 100°C$ along the linear portion of the boundary
and $T = 0°C$ along the circular portion of the boundary.

18.4 The Schwarz-Christoffel Transformation

In general, the conformal transformation of one given region onto another is
exceedingly difficult. The *existence* of such a transformation is assured by the
following theorem, due to Riemann.

THEOREM 1   Either of two bounded simply connected regions can be mapped
conformally onto the other.

The determination of the specific function which accomplishes a required mapping, however, is usually out of the question. In fact, in addition to the simple regions which we found could be mapped by means of the elementary functions, the only class of regions for which conformal transformations of practical interest exist are those bounded by polygons having a finite number of vertices (one or more of which may lie at infinity). These can always be mapped onto a half plane (and hence onto any region into which a half plane can be transformed) by means of a transformation which we shall now discuss.

To see how this can be done, we first recall the mapping properties of the power function

$$w = z^m$$

Since this transformation has the property (Theorem 3, Sec. 18.2) that it alters by the factor m any angle with vertex at the origin, it follows that the transformation

$$(1) \qquad w - w_1 = (z - x_1)^{\alpha_1/\pi} \qquad \frac{dw}{dz} = \frac{\alpha_1}{\pi}(z - x_1)^{(\alpha_1/\pi)-1}$$

will take a segment of the x axis containing x_1 in its interior, i.e., a straight angle with vertex at x_1, and fold it into an angle of

$$\frac{\alpha_1}{\pi}\pi = \alpha_1$$

with vertex at w_1. Clearly, if this could be done simultaneously for a number of points $x_1, x_2, \ldots, x_n$ on the x axis, the x axis would be mapped into a polygon whose angles were, respectively, $\alpha_1, \alpha_2, \ldots, \alpha_n$, and conversely, and the biggest step in the solution of our problem would be taken. This is actually possible, and the transformation which accomplishes it, suggested by the form of the derivative of the function in (1), is defined by

$$(2) \qquad \frac{dw}{dz} = K(z - x_1)^{(\alpha_1/\pi)-1}(z - x_2)^{(\alpha_2/\pi)-1}\cdots(z - x_n)^{(\alpha_n/\pi)-1}$$

To verify this, we begin with a point z on the x axis to the left of the first of the given points $x_1, x_2, \ldots, x_n$ and investigate the locus of its image as it moves to the right along the x axis (Fig. 18.11).† From (2) we obtain at once the relation

$$(3) \quad \arg dw = \arg K + \left(\frac{\alpha_1}{\pi} - 1\right)\arg(z - x_1)$$

$$+ \left(\frac{\alpha_2}{\pi} - 1\right)\arg(z - x_2) + \cdots$$

$$+ \left(\frac{\alpha_n}{\pi} - 1\right)\arg(z - x_n) + \arg dz$$

---

† In reasoning in this fashion, we are actually considering the inverse problem of mapping the upper half of the z plane onto a polygonal region in the w plane. As a consequence, in our applications we will be mapping a polygon *given in the w plane* onto the upper half of the z plane. This interchange of the usual roles of the z and w planes should cause no confusion, however.

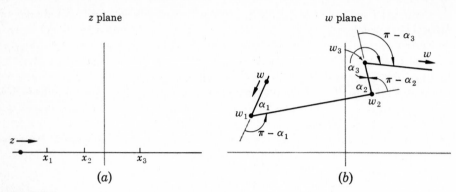

Figure 18.11
The mapping of the real axis in the z plane into a polygon with prescribed angles in the w plane.

and from this it is apparent that until z reaches x_1, every term on the right remains constant since $z - x_1, z - x_2, \ldots, z - x_n$ are all negative real numbers and hence have π for their respective arguments and since dz is positive and therefore has 0 as its argument. Thus the image point w traces a straight line, since the argument of the increment dw remains constant. However, as z passes through x_1, the difference $z - x_1$ changes abruptly from negative to positive, and thus arg $(z - x_1)$ decreases abruptly from π to 0. Hence, arg dw changes by the amount

$$\left(\frac{\alpha_1}{\pi} - 1\right)(-\pi) = \pi - \alpha_1$$

But, from Fig. 18.11b it is evident that this is the precise amount through which it is necessary to turn if w is to begin to move in the direction of the next side of the polygon. As z moves from x_1 to x_2, the same situation exists. The argument of dw remains constant, and thus w moves in a straight line until z reaches x_2. Here $z - x_2$ changes abruptly from negative to positive, its argument jumps immediately from π to 0, and, as a consequence, arg dw increases by the amount $\pi - \alpha_2$, which is just the amount of rotation required to give the direction of the next side of the polygon.

Thus as z traverses the x axis, it is clear that w moves along the boundary of a polygon whose interior angles are precisely the given angles $\alpha_1, \alpha_2, \ldots, \alpha_n$. Moreover, it is evident that the region which is mapped onto the half plane is the region which contains these angles. The required transformation will be obtained if we can ensure that the lengths of the sides of the polygon, as well as its angles, have the correct values.

Now the mapping function w, obtained by integrating (2), is

(4) $\qquad w = K \int \left[(z - x_1)^{(\alpha_1/\pi) - 1}(z - x_2)^{(\alpha_2/\pi) - 1} \cdots (z - x_n)^{(\alpha_n/\pi) - 1}\right] dz + C$

and this can be thought of as the composition of the two transformations

(5) $\qquad t = \int \left[(z - x_1)^{(\alpha_1/\pi) - 1}(z - x_2)^{(\alpha_2/\pi) - 1} \cdots (z - x_n)^{(\alpha_n/\pi) - 1}\right] dz$

(6) $\qquad w = Kt + C$

The first of these transforms the x axis into some polygon which the second then translates, rotates, and either stretches or shrinks, as the case may be. If, then, the polygon determined by (5) is similar to the given polygon, the constants in (6) can always be determined so as to make the two polygons coincide.

For two polygons to be similar, not only must corresponding angles be congruent but corresponding sides must also be proportional. For triangles this is automatically the case. For quadrilaterals one further condition is required, namely, that two pairs of corresponding sides have the same ratio. For pentagons, two such conditions are required, and, in general, for polygons of n sides, $n - 3$ conditions, over and above the congruence of corresponding angles, are necessary to ensure similarity. Hence, in mapping a polygon of n sides onto a half plane, three of the points $x_1, x_2, \ldots, x_n$, which are the images of the vertices of the polygon, can be assigned arbitrarily, following which the remaining $n - 3$ image points are determined by the conditions of similarity. In many important problems, one vertex of the polygon, often an infinite one, will correspond to $z = \infty$. This, of course, accounts for one of the three assignments of image points which can be made arbitrarily. As a consequence, there will be one less finite image point in the expression (2) for dw/dz, and only *two* of the finite image points $x_1, x_2, \ldots, x_{n-1}$ can be specified arbitrarily. In either case, the resulting transformation is known as the **Schwarz-Christoffel transformation.**[†] Obviously, since w is analytic everywhere except possibly at the points $z = x_1$, $x_2, \ldots, x_n$, the transformation is conformal over the interior of the two regions. In practice, the usefulness of the Schwarz-Christoffel transformation is often limited by the complexity of the integral which defines the mapping function w.

EXAMPLE 1

Find the transformation which maps the semi-infinite strip shown in Fig. 18.12*a* onto the half plane, as indicated.

Clearly, $\alpha_1 = \alpha_2 = \pi/2$, and therefore the required transformation is defined by

$$\frac{dw}{dz} = K(z + 1)^{(\pi/2)/\pi \, -1}(z - 1)^{(\pi/2)/\pi \, -1} = K(z + 1)^{-1/2}(z - 1)^{-1/2}$$

Hence, $\quad w = K \int \dfrac{dz}{\sqrt{z^2 - 1}} = K \cosh^{-1} z + C$

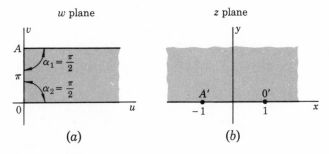

         (a)   (b)

Figure 18.12
A semi-infinite strip to be mapped onto a half plane.

---

† Named for the German mathematician H. A. Schwarz (1843–1921) and the Swiss mathematician E. B. Christoffel (1829–1900), who discovered it independently about 1865.

Since $w = 0$ is to correspond to $z = 1$ (that is, $O \to O'$, in Fig. 18.12) we have

$$0 = K \cosh^{-1} 1 + C \quad \text{or} \quad C = 0$$

Also, $w = i\pi$ (A in Fig. 18.12a) is to correspond to $z = -1$ (A' in Fig. 18.12b), and thus

$$i\pi = K \cosh^{-1}(-1) = K(i\pi) \quad \text{or} \quad K = 1$$

The required transformation is therefore $w = \cosh^{-1} z$, or

$$z = \cosh w$$

Broken down into real and imaginary parts, this becomes

$$x + iy = \cosh u \cos v + i \sinh u \sin v$$

or

$$x = \cosh u \cos v \qquad y = \sinh u \sin v$$

Eliminating u and v in turn, we have also

$$\frac{x^2}{\cosh^2 u} + \frac{y^2}{\sinh^2 u} = 1 \qquad \frac{x^2}{\cos^2 v} - \frac{y^2}{\sin^2 v} = 1$$

which, if necessary, can be solved for u and v in terms of x and y.

EXAMPLE 2

Find the transformation which maps the infinite region shown in Fig. 18.13a onto the upper half of the z plane, as indicated.

With images assigned as shown and with the angle at the finite vertex A identified as $\alpha_1 = 2\pi$ and the angle at the infinite vertex B identified as $\alpha_2 = 0$, we have

$$\frac{dw}{dz} = K(z + 1)^{(2\pi/\pi) - 1} z^{(0/\pi) - 1} = K\left(1 + \frac{1}{z}\right)$$

and

(7) $$w = K(z + \ln z) + C$$

To determine the constants K and C, we write (7) in the form

$$u + iv = (K_1 + iK_2)(x + iy + \ln |z| + i \arg z) + C_1 + iC_2$$

from which, by equating imaginary parts, we obtain

(8) $$v = K_1 y + K_2 x + K_2 \ln |z| + K_1 \arg z + C_2$$

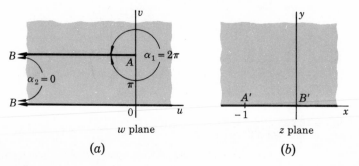

Figure 18.13
A semi-infinite channel to be mapped onto a half plane.

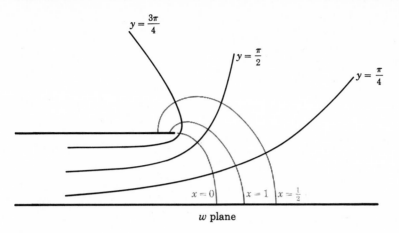

w plane

Figure 18.14
Typical streamlines for fluid flow from a long straight channel.

When w becomes infinite along AB, on which $v = \pi$, the image point z approaches zero along the negative real axis, on which $y = 0$ and arg $z = \pi$. Hence, from (8),

$$\pi = \lim_{z \to 0^-} (K_1 \cdot 0 + K_2 x + K_2 \ln |z| + K_1 \pi + C_2)$$

Obviously K_2 must be zero to keep $\ln |z|$ from making the right member infinite. Hence

(9)
$$\pi = K_1 \pi + C_2$$

Also, as w becomes infinite along OB, on which $v = 0$, the image point z approaches zero along the positive real axis, on which $y = 0$ and arg $z = 0$. Hence, using (8) again, we have

$$0 = \lim_{z \to 0^+} (K_1 \cdot 0 + C_2) = C_2$$

Therefore, $C_2 = 0$, and so from (9) we find that $K_1 = 1$. Thus (7) reduces to

$$w = z + \ln z + C_1$$

Finally, the point A, that is, $w = i\pi$, must map into the point A', that is, $z = -1$. Hence from the last equation,

$$i\pi = -1 + \ln (-1) + C_1$$

$$= -1 + i\pi + C_1$$

and so C_1 must equal 1. The required mapping function is therefore

$$w = z + \ln z + 1$$

Figure 18.14 shows the curves in the w plane which correspond to the semicircles $r = 1, 2, 3$ and the rays $\theta = \pi/4, \pi/2, 3\pi/4$. The resulting configuration can be shown to represent either the lines of equal velocity potential and the streamlines for the flow of an ideal incompressible fluid from an infinite straight channel or the lines of flux and the equipotential lines for a parallel-plate capacitor.

1   Find the transformation which will map the region shown in **Fig. 18.15a** onto the upper half of the z plane, as indicated.

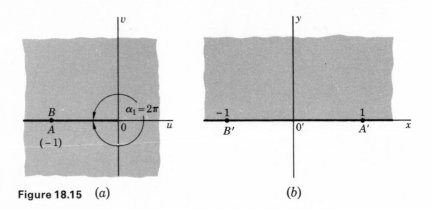

Figure 18.15 (a)   (b)

2   Using the results of Exercise 1, find the steady-state temperature distribution in the w plane if the upper side of the negative u axis is kept at the temperature $T = 100°C$ and the lower side of the negative u axis is kept at the temperature $T = 0°C$.

3   Using the results of Exercise 1, find the steady-state temperature distribution in the w plane if the portion of the negative u axis between -1 and 0 is maintained at the temperature $T = 100°C$ while the rest of the negative u axis is maintained at the temperature $T = 0°C$.

4   Using the results of Example 1, find the steady-state temperature distribution in the semi-infinite strip shown in **Fig. 18.12a** if the portions of the upper and lower boundaries to the right of the line $\mathscr{R}(w) = \cosh^{-1} 2$ are kept at the temperature $T = 0°C$ and the rest of the boundary is kept at the temperature $T = 100°C$.

5   Rework Example 1 using the mapping indicated in **Fig. 18.16**.

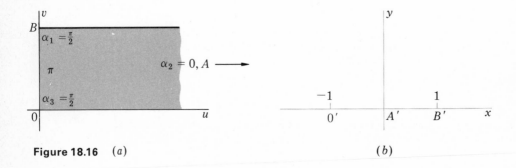

Figure 18.16 (a)   (b)

6   Rework Exercise 4 using the mapping obtained in Exercise 5.

7   Find the transformation which will map the infinite strip shown in **Fig. 18.17** onto the upper half of the z plane, as indicated.

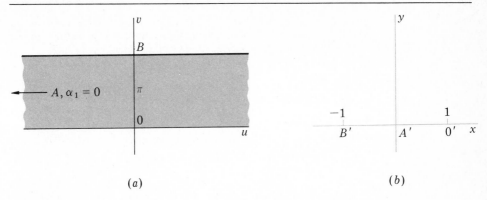

(a)

(b)

Figure 18.17

8 Using the results of Exercise 7, find the steady-state temperature distribution in the infinite strip shown in Fig. 18.17 if the lower boundary is kept at the temperature $T = 100°C$ and the upper boundary is kept at the temperature $T = 0°C$.

9 Work Exercise 8 if the left half of the lower boundary and the right half of the upper boundary are maintained at the temperature $T = 100°C$ while the rest of the boundary is kept at the temperature $T = 0°C$.

10 Find the transformation which will map the exterior of the first quadrant in the w plane onto the upper half of the z plane, as indicated in Fig. 18.18.

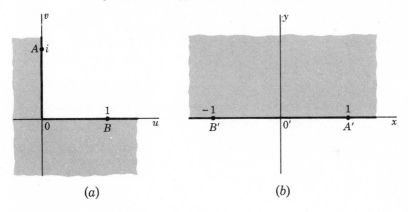

(a)

(b)

Figure 18.18

11 Using the results of Exercise 10, find the equations of the isotherms in the exterior of the first quadrant of the w plane if the positive u axis is kept at the temperature $T = 100°C$ and the positive half of the v axis is kept at the temperature $T = 0°C$.

12 Find the transformation which will map the region shown in Fig. 18.19 onto the upper half of the z plane, as indicated.

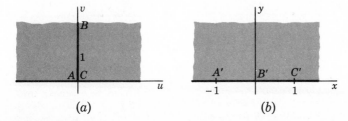

(a)

(b)

Figure 18.19

13 Using the results of Exercise 12, find the equations of the curves in the w plane which correspond to the lines $y = c$ in the z plane. (These are the streamlines of a perfect incompressible fluid flowing past an idealized vertical obstacle.)

14 Find the transformation which will map the region shown in Fig. 18.20 onto the upper half of the z plane, as indicated.

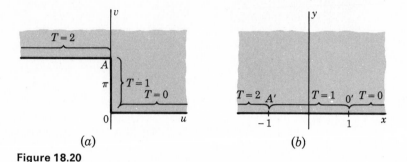

(a)   (b)

Figure 18.20

15 Find the transformation which will map the region shown in Fig. 18.21 onto the upper half of the z plane, as indicated.

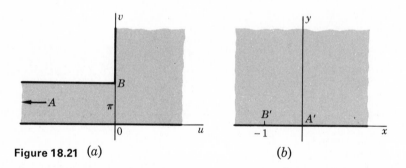

Figure 18.21 (a)   (b)

16 Find the transformation which will map the region shown in Fig. 18.22 onto the upper half of the z plane, as indicated.

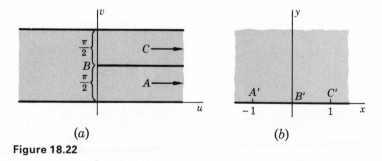

(a)   (b)

Figure 18.22

17 Find the integral which defines the transformation which maps the rectangle shown in Fig. 18.23 onto the upper half of the z plane, as indicated. Discuss, as far as possible, the determination of the constants k and C in this case. (This integral is known as an **elliptic integral of the first kind**; see Exercises 31 and 32, Sec. 2.6.)

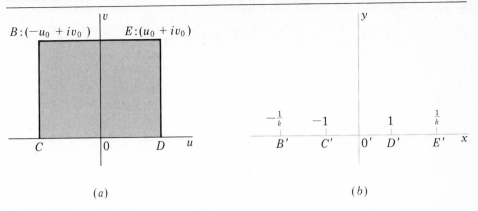

Figure 18.23

18 Find the transformation which maps the region shown in Fig. 18.24 onto the upper

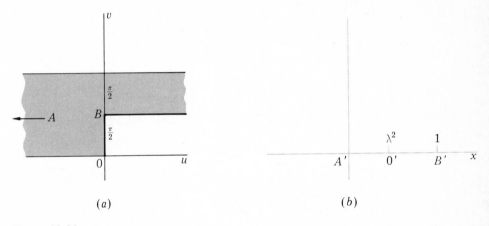

Figure 18.24

half of the z plane, as indicated. *Hint:* When the integral which defines w is set up, make the substitution

$$s^2 = \frac{z - \lambda^2}{z - 1}$$

choosing s to be the square root for which

$$\arg s = \tfrac{1}{2}[\arg(z - \lambda^2) - \arg(z - 1)]$$

Answers to Odd-numbered Exercises

Chapter 1
Sec. 1.2
p. 6

1 Second order, ordinary, linear
3 Third order, ordinary, linear
5 Second order, ordinary, linear
7 Second order, partial, linear
9 Second order, partial, nonlinear
11 Be^x, where $B = Ae^k$
13 $C \ln x$, where $C = ab$
15 $P \sin x + Q \cos x$, where $P = A \cos b + C \cos d$ and $Q = A \sin b + C \sin d$
17 $P \cosh^2 \theta + Q \sinh^2 \theta$, where $P = a + c$ and $Q = b + c$

19 $\dfrac{P}{x + 1} + \dfrac{Q}{x + 2}$, where $P = A + C$ and $Q = B - C$

31 $y'' + 4y' + 4y = 0$
33 $x^2 y'' - 2xy' + 2y = 0$
35 $y'' + a^2 y = 0$
37 $(y')^2 - 2xy' + 2y = 0$
39 The first equation is linear, and the second is nonlinear.
41 The first equation is linear, and the second is nonlinear.
43 Neither $f(x,y)$ nor $f_y(x,y)$ is defined for $x = \pm 1$.

Sec. 1.3
p. 11

1 $y = ce^{-x^2}$
3 $y = \tan (x^3 + c)$
5 $y^2 e^{2y} = cx$
7 $y^2 = 1 + c(x^2 + 1)$
9 $y - \ln (y + 1) = x^2 + c$
11 $e^y(y - 1) = c - e^{-x}$

13 $y = \dfrac{x - c(x + 1)}{1 + c(x + 1)}$
15 $y = \ln \cos (c - x) + k$

17 $y = ke^{cx}$
19 $y = 100e^{-2x}$
21 $y = 2(1 + x^2)$

23 $y = \begin{cases} (1 - x^2)^2 & x \leq -1 \\ 2(1 - x^2)^2 & -1 \leq x \leq 1 \\ 0 & 1 \leq x \end{cases}$

25 From calculus we know that
$$\int \left[\phi(u) \frac{du}{dx} \right] dx = \int \phi(u)\, du$$
Hence, writing the given equation in the form $f(x) = g(y)\, dy/dx$ and integrating both sides *with respect to x*, we have
$$\int f(x)\, dx = \int \left[g(y) \frac{dy}{dx} \right] dx = \int g(y)\, dy$$
as required.

29 $y = 1 - 2x - \ln (c - x)$
31 $y = 1 + x + \dfrac{1 + 2ce^{-3x}}{1 - ce^{-3x}}$

Sec. 1.4
p. 14

1 Homogeneous of degree -1

3 The expression as a whole is not homogeneous, although the first major term is homogeneous of degree 1 and the second is homogeneous of degree 2.

11 $(x + y)^2 = cx$

13 $y = \dfrac{x}{\ln cx}$

15 $x^2 - 2xy - 3y^2 = c$; the curves of this family are all hyperbolas.

17 $2 \tan^{-1}(y/x) = \ln c(x^2 + y^2)$; the curves of this family are all complicated transcendental curves.

19 $2y + 2\sqrt{x^2 + y^2} = x^2$

21 $y^2 = \dfrac{x^2 \ln x}{1 + \ln x}$

23 $y^3 = x^3 - x^{3/2}$

25 $(y - 3)^2 + 2(y - 3)(x + 2) - (x + 2)^2 = c$

27 If $aB = bA$, then $a = \lambda A$, $b = \lambda B$, and the equation of Exercise 24 can be written in the form

$$\frac{dy}{dx} = \frac{\lambda Ax + \lambda By + c}{Ax + By + C}$$

$$= \frac{\lambda(Ax + By + C) + (c - \lambda C)}{Ax + By + C} = f(Ax + By + C)$$

and the method of Exercise 27, Sec. 1.3 can be applied.

31 $\theta = \ln cr$ or $\tan^{-1}(y/x) = \ln c\sqrt{x^2 + y^2}$ (The solution process here should be compared with the analysis of the same problem in Exercise 17.)

33 $\sqrt{x^2 + y^2}$

Sec. 1.5
p. 19

1 $xy^2 - x + \cos y = c$

3 $x^3 - 3x^2y - y^2 = c$

5 $x^2y^4 + x \sin y = c$

7 $x^3y^4 + x^3 = c$ (multiply by x^2)

9 $y = 2 \tan^{-1}(x/y) + c$ (divide by y^2)

11 $2x/y + 3 \ln y = c$ (homogeneous, or integrating factor $1/y^2$)

13 $x^2y^2 = x^2 - y^2 + c$ (separable, or integrating factor xy)

15 $cx^2 = xy + \sqrt{x^2 + y^2}$ (homogeneous or integrating factor $1/x^2$)

Sec. 1.6
p. 22

1 $y = x^2 \ln x + cx^2$

3 $y = \sin x + c \cos x$

5 $y = \dfrac{1}{2} - \dfrac{1}{x} + \dfrac{c}{x^2}$

7 $y = \dfrac{x^2(x - 1)}{2} + c(1 - x)$

9 $y = e^{-x} + c\dfrac{e^{-x}}{x}$

11 $y = \dfrac{e^x}{2} + \dfrac{3e^{-x}}{2}$

13 $y = \dfrac{x^2 + 1}{4} + \dfrac{1}{1 + x^2}$

15 $y = x - \sqrt{\dfrac{1 + x^2}{2}}$

17 $x = \dfrac{4y^2}{5} + \dfrac{c}{y^3}$

21 $y^2 = x - \frac{1}{2} + ce^{-2x}$

23 $y^3 = \dfrac{1}{x^2 + cx}$

25 $y = \dfrac{1}{1 - x + ce^{-x}} + 1$

27 $y = 1 - x + \dfrac{1}{c - x}$

Sec. 1.7
p. 30

1 4.3 percent, 2.9 percent, 660 years

3 $P = P_0 e^{(k_b - k_d)t}$, where k_b and k_d are, respectively, the proportionality constants in the birth and death rates.

5 $\ln \dfrac{P}{P_0} = \dfrac{b_1}{b_2}(1 - e^{-b_2 t}) - \dfrac{d_1}{d_2}(1 - e^{-d_2 t})$

where $b_1 e^{-b_2 t}$ and $d_1 e^{-d_2 t}$ are, respectively, the proportionality factors in the birth and death rates.

7 $I = 26.6e^{-0.0196y}$

9 $p = \dfrac{e^{kwy} - 1}{k}$, 1.02

11 $Q = 30 - \dfrac{1,500}{t + 50}$, $t = 50$ min

13 $Q = 40 - \dfrac{4,320}{t + 108}$

15 $Q = \dfrac{20t}{t + 48}$

17 $\dfrac{7Q}{20 - Q} + \ln\left(1 - \dfrac{Q}{20}\right) = 0.133t$

19 $Q = 2(100 - t) - 150\left(\dfrac{100 - t}{100}\right)^3$; 42.3 min; $Q_{max} = 89$ lb when $t = 33\frac{1}{3}$ min

21 $2x^2 + 3y^2 = k$

23 $y^2 = -x^2 \ln kx$

25 $x^2 + y^2 = ky$

29 $H = \dfrac{k(T_1 - T_0)}{h}$, $T = T_0 - (T_0 - T_1)\dfrac{x}{h}$

31 $T = 120 - 65e^{-t/25}$

33 $T = T_a - (T_a - T_0)\left(1 + \dfrac{a - b}{P}t\right)^{-a/(a-b)}$

35 $\dot{T} = 60 + 40e^{-0.0462t}$, 21.2 min

37 $\frac{4}{3}Ry^{3/2} - \frac{2}{5}y^{5/2} = -r^2\sqrt{2g}\,t + \frac{16}{15}\sqrt{2}\,R^{5/2}$;

$\dfrac{16\sqrt{2} - 14}{15r^2\sqrt{2g}}R^{5/2}$; $\dfrac{16}{15\sqrt{g}\,r^2}R^{5/2}$

39 $y^{-1/2} = \dfrac{h\sqrt{2g}}{2R^2}t + h^{-1/2}$; the tank will never be completely empty.

41 $(2R - y)^{-3/2} = R^{3/2} + \dfrac{3\pi r^2\sqrt{2g}}{4l}t$, $t_0 = \dfrac{4l}{3\pi r^2\sqrt{2g}}[(2R)^{3/2} - R^{3/2}]$

43 $y = kx^2$

45 $\dfrac{1}{\sqrt{y}} = \dfrac{w\sqrt{2g}}{3\pi r^2}t + \dfrac{1}{\sqrt{h}}$; the tank will never be completely empty.

47 $y = h - (k_1/k_2\rho)(1 - e^{-k_2 t})$, $h < k_1/k_2\rho$, where k_1 and k_2 are, respectively, the proportionality constants in the rates of evaporation and condensation.

49 $s = \dfrac{(480 - t)^3}{-43,200} + 2,560$, $s_{max} = 2,560$ ft

51 $\omega = \begin{cases} 100\sqrt{3}\tan\left(1.40 - \dfrac{6\sqrt{3}}{5}t\right) & 0 \le t \le 0.42 \\ (20.1 - 24t)^2 & 0.42 \le t \le 0.84 \end{cases}$

$\omega = 0$ when $t = 0.84$ min

53 $y = 256t - 2,560(1 - e^{-t/8})$, $y_{max} = 55.0$ ft; the object strikes the ground at 54.8 ft/s when $t = 3.71$ s.

55 (a) 1.78 (b) 1.41 (c) 1.04 (d) 0.69
(e) 0.55 (f) 0.52 (g) 0.51

57 (a) $x = \sqrt{\dfrac{m}{k}}\,v_0\sin\sqrt{\dfrac{k}{m}}\,t$

(b) $x = \sqrt{\dfrac{m}{k}}\,v_0\sin\sqrt{\dfrac{k}{m}}\,t + x_0\cos\sqrt{\dfrac{k}{m}}\,t$

59 (a) $t = \sqrt{\dfrac{x_0 m}{2k}} \left(\sqrt{x_0 x - x^2} + \dfrac{x_0}{2} \cos^{-1} \dfrac{2x - x_0}{x_0} \right)$

   (b) $t = \sqrt{\dfrac{x_0 m}{2k}} \left(\sqrt{x^2 - x_0 x} + \dfrac{x_0}{2} \cosh^{-1} \dfrac{2x - x_0}{x_0} \right)$

61 7.0 mi/s   **63** $s = (\frac{5}{12} g \sin \alpha) t^2$

65 $Q = \dfrac{E_0 C}{1 + \omega^2 R^2 C^2} [\sin \omega t - \omega RC(\cos \omega t - e^{-t/RC})]$

67 $i = \dfrac{E}{R} (1 - e^{-Rt/L})$, $t_{1/2} = 0.693 \dfrac{L}{R}$

69 $y = \dfrac{H}{w} \cosh \dfrac{wx}{H}$ **71** $y^2 = 2cx + c^2$

73 $y = \dfrac{x^2 \omega^2}{2g} + h - \dfrac{\omega^2 r^2}{4g}$ **75** 11:23 A.M.

Chapter 2   **1** $y = c_1 \sin x + c_2 \cos x$
Sec. 2.1   **3** $y = c_1 x + c_2 x^{-4}$
p. 45   **5** $y = c_1 e^x + c_2 x e^{-x}$

Sec. 2.2   **1** Dy means the derivative of y with respect to the relevant independent
p. 51   variable. yD is an operator which means y times the derivative of a function
   as yet unspecified.
3 $(D + 1)(D + x)e^x = 3e^x + 2xe^x$, $(D + x)(D + 1)e^x = 2e^x + 2xe^x$.
   These expressions differ because in permuting the operational coefficients
   variable terms are moved across symbols of differentiation.
5 $y = c_1 e^{-x} + c_2 e^{-4x}$ **7** $y = c_1 + c_2 e^{-5x}$
9 $y = c_1 e^{2x/3} + c_2 x e^{2x/3}$ **11** $y = e^{-5x}(A \cos x + B \sin x)$
13 $y = 2 \cos 2x + 3 \sin 2x$ **15** $y = 0$
17 There is no solution meeting the given conditions.
19 If $x_1 = x_0 + n\pi$, the conditions can be met if and only if $y_1 = (-1)^n y_0$.
   In this case, the equations arising from the end conditions are dependent,
   and there is an infinite family of solutions meeting the conditions.
21 $\lambda = n$, $y = A \cos nx$
25 There are no nontrivial solutions satisfying the given conditions.
31 As $m_2 \to m_1$, y becomes an indeterminate of the form 0/0. This evaluates
   by L'Hospital's rule to $xe^{m_1 x}$, which is a second, independent solution of
   the limiting equation.

37 $y = -\dfrac{ae^x + 2be^{2x}}{x^2(ae^x + be^{2x})}$

Sec. 2.3   **1** $y = e^{-2x}(c_1 \cos x + c_2 \sin 2x) + \dfrac{e^x}{5}$
p. 59

3 $y = c_1 + c_2 e^{-x} + \dfrac{x^2}{2} + x$

5 $y = c_1 \cos x + c_2 \sin x + \dfrac{x \sin x}{2} - \sin 2x$

7 $y = c_1 + c_2 e^{-3x} + \dfrac{\sin x - \cos x}{2}$

9 $y = c_1 e^{-2x} + c_2 x e^{-2x} + x e^{-x} - 2e^{-x}$

11 $y = c_1 e^{-x} + c_2 e^{-2x} - \dfrac{3\cos 2x}{50} + \dfrac{2\sin 2x}{25} + \dfrac{1}{2}$

13 $y = c_1 e^{2x} + c_2 e^{3x} + \dfrac{7\cosh x + 5\sinh x}{24}$

15 (b) $A = 1$

17 $y = \frac{13}{2} e^{-2x} + 11 x e^{-2x} + 2x - \frac{9}{2}$

19 $y = e^{-x}(3\cos 2x + 4\sin 2x) + 2\cos x + \sin x$

21 $Y = \dfrac{(e^{\lambda t} - e^{at}) - (\lambda - a)t e^{at}}{(\lambda - a)^2}$

25 $Y = x^{1/2} + \frac{1}{4} x^{-3/2} - \dfrac{3 \cdot 5}{4^2} x^{-7/2} + \dfrac{3 \cdot 5 \cdot 7 \cdot 9}{4^3} x^{-11/2} - \cdots$. This series diverges for all values of x.

27 (a) $Y = -x e^x$ (b) $Y = -x e^x - e^x$

29 (a) $Y = \dfrac{x \sin x}{2}$ (b) $Y = \dfrac{x \sin x}{2}$

Sec. 2.4
p. 62

1 $y = c_1 e^x + c_2 x e^x + \dfrac{x^5 e^x}{20}$

3 $y = c_1 \cosh x + c_2 \sinh x + \dfrac{x \sinh x}{2}$

5 $y = c_1 e^{-x} + c_2 x e^{-x} + x e^{-x}(\ln x - 1)$

7 $y = c_1 \cos x + c_2 \sin x - 1 - x \cos x + \sin x \ln(1 + \sin x)$

9 $y = c_1 e^{-x} + c_2 x e^{-x} + \left(\dfrac{x^2}{2} \ln x - \frac{3}{4} x^2\right) e^{-x}$

11 $y = c_1 x + c_2 \dfrac{1}{x} - \left[\frac{1}{2} \ln x + \dfrac{1}{2x} - \frac{1}{2} \ln(x + 1)\right] x - \dfrac{1}{2x} \ln(x + 1)$

13 $y = c_1 x + c_2 x^2 - x - x \ln x - \dfrac{x}{2} \ln^2 x$

15 $y = c_1 e^{(-a+b)x} + c_2 e^{(-a-b)x} + \dfrac{1}{2b} \displaystyle\int_{x_0}^x f(s)[e^{(-a+b)(x-s)} - e^{(-a-b)(x-s)}]\, ds$

17 $y = e^{-ax}(A \cos bx + B \sin bx) + \dfrac{1}{b} \displaystyle\int_{x_0}^x f(s) e^{-a(x-s)} \sin b(x - s)\, ds$

19 $Y = \displaystyle\int_0^x \sqrt{s} \sin(x - s)\, ds$. Unlike the formal series obtained in Exercise 25, Sec. 2.3, this expression is meaningful for all positive values of x.

21 (b) $\dfrac{L}{a^2 + b^2}$

Sec. 2.5
p. 65

1 $y = c_1 e^{-x} + c_2 e^{-2x} + c_3 e^{-3x} + x - 3$

3 $y = c_1 e^{-2x} + e^{2x}(c_2 \cos x + c_3 \sin x) + 3 \cos x - \sin x$

5 $y = c_1 e^x + c_2 e^{-x} + c_3 \cos 3x + c_4 \sin 3x - x^2 - \dfrac{16}{9} - \dfrac{\sin 2x}{5}$

7 $y = c_1 e^x + c_2 x e^x + c_3 e^{-2x} + c_4 x e^{-2x} + \frac{1}{18} x^2 e^x$

9 $y = e^x(c_1 \cos x + c_2 \sin x) + e^{-x}(c_3 \cos x + c_4 \sin x) + \dfrac{\cos x}{5} + \dfrac{\sin 2x}{20}$

11 $y = -2 \sin x + \cos x$ **13** $y = \sin x - \sin 2x$

15 $Y = \dfrac{1}{2} \displaystyle\int_{x_0}^{x} f(s)[e^{x-s} - 2e^{2(x-s)} + e^{3(x-s)}]\, ds$

17 $Y = \dfrac{1}{2} \displaystyle\int_{x_0}^{x} f(s)[(x - s)e^{x-s} - \sinh (x - s)]\, ds$

19   Only for values of λ which satisfy the equation $\tanh \lambda = \tan \lambda$. $y_n = A_n(\sin \lambda_n \sinh \lambda_n x - \sinh \lambda_n \sin \lambda_n x)$, where λ_n is the nth one of the roots of the equation $\tanh \lambda = \tan \lambda$.

27   If y_1, y_2, and y_3 are any solutions of the homogeneous equation

$$y''' + P_1(x)y'' + P_2(x)y' + P_3(x)y = 0$$

then $y_4 = c_1 y_1 + c_2 y_2 + c_3 y_3$, where c_1, c_2, and c_3 are arbitrary constants, is also a solution.

Sec. 2.6
p. 73

1 $y = c_1 \dfrac{1}{x} + c_2 x$ **3** $y = c_1 x + c_2 x \ln x + \dfrac{x^5}{16}$

5 $y = \dfrac{c_1}{\sqrt{x}} + \dfrac{c_2}{x} + 2 + \dfrac{x}{2}$

7 $y = c_1 x + c_2 \dfrac{1}{x} + c_3 \sin (\ln x) + c_4 \cos (\ln x) + \dfrac{1}{39x^2} - \dfrac{\ln x}{3}$

11   **(a)** $\dfrac{d^2 y}{dt^2} = -\dfrac{kg}{w} y, \ \omega = \dfrac{1}{2\pi}\sqrt{\dfrac{kg}{w}}$ **(b)** $y = -2 \cos 12t$

       (c) $y = \frac{1}{4} \sin 12t$ **(d)** $y = 4 \cos 12t + \frac{2}{3} \sin 12t$

13   6.75 in

15   If the pellet starts from the midpoint of the tube with radial velocity $g/2\omega$ when the tube is horizontal, it will execute simple harmonic motion described by the relation $r = (g/2\omega^2) \sin \omega t$. Furthermore, if the pellet starts with arbitrary displacement r_0 and initial velocity $-\omega r_0 + g/2\omega$ when the tube is horizontal, it will move according to the law

$$r = r_0 e^{-\omega t} + \dfrac{g}{2\omega^2} \sin \omega t$$

which is asymptotically simple harmonic.

19 $y = x - x^{m_1}/m_1 L^{m_1 - 1}$, where m_1 is the positive root of the equation $m^2 - m - 2\sqrt{2}\, F/\pi k^4 E = 0$ and k is the proportionality constant in the formula which defines the radius of the beam cross section as a function of x.

21 $\omega_n = \dfrac{z_n^2}{L^2}\sqrt{\dfrac{EIg}{A\rho}}$

where z_n is the nth one of the roots of the equation $\cos z \cosh z = 1$. The deflection curve corresponding to ω_n is

$$y_n = (\cos z_n - \cosh z_n)\left(\cos z_n \dfrac{x}{L} - \cosh z_n \dfrac{x}{L}\right)$$

$$+ (\sin z_n + \sinh z_n)\left(\sin z_n \dfrac{x}{L} - \sinh z_n \dfrac{x}{L}\right)$$

23 $\omega = \dfrac{1}{2\pi}\sqrt{\dfrac{3kL^2 g}{wL^2 + 3Wl^2}}$ **25** $y = \dfrac{a}{2}\left(\cosh \sqrt{\dfrac{g}{L}}\, t - 1\right)$

27 $y = \dfrac{a}{2}\left(\cosh \sqrt{\dfrac{2wgR^2}{wLR^2 + Ig}}\, t - 1\right)$

29 $\omega = \dfrac{1}{2\pi} \sqrt{\dfrac{3g[4kl^2 - (2W + w)L]}{2L^2(w + 3W)}}$

provided $4k^2 > (2W + w)L$. If this is not the case, the differential equation has hyperbolic functions in its solution, and the system is unstable.

31 $\omega = \dfrac{1}{2\pi} \sqrt{\dfrac{g}{l}}$

33   True period $= 1.18 \times$ approximate period

Sec. 2.7
p. 87

3 $y = \begin{cases} \dfrac{Px}{2T} & 0 \le x \le \dfrac{L}{4} \\[2ex] \dfrac{P(L - 2x)}{4T} & \dfrac{L}{4} \le x \le \dfrac{L}{2} \\[2ex] \dfrac{3P(L - 2x)}{4T} & \dfrac{L}{2} \le x \le \dfrac{3L}{4} \\[2ex] \dfrac{3P(L - x)}{2T} & \dfrac{3L}{4} \le x \le L \end{cases}$

5 $y = \begin{cases} -\dfrac{Lx}{8T} & 0 \le x \le \dfrac{L}{2} \\[2ex] \dfrac{4x^2 - 5Lx + L^2}{8T} & \dfrac{L}{2} \le x \le L \end{cases}$

7 $G(x,s) = \begin{cases} \dfrac{e^{-s} \cos s\, e^{-x} \sin x}{e^{-2s}} & 0 \le x \le s \\[2ex] \dfrac{e^{-x} \cos x\, e^{-s} \sin s}{e^{-2s}} & s \le x \le \dfrac{\pi}{2} \end{cases}$

In this case, $G(x,s)$ is not symmetric. For the equivalent equation $e^{2x}y'' + 2e^{2x}y' + 2e^{2x}y = 0$, $G(x,s)$ is the symmetric function

$$\begin{cases} e^{-s} \cos s\, e^{-x} \sin x & 0 \le x \le s \\[1ex] e^{-x} \cos x\, e^{-s} \sin s & s \le x \le \dfrac{\pi}{2} \end{cases}$$

9   **(a)** $G(x,s) = \begin{cases} \dfrac{\cosh k(b - s) \sinh kx}{k \cosh kb} & 0 \le x \le s \\[2ex] \dfrac{\cosh k(b - x) \sinh ks}{k \cosh kb} & s \le x \le b \end{cases}$

(b) $G(x,s) = \begin{cases} \dfrac{\sinh k(b - s) \cosh kx}{k \cosh kb} & 0 \le x \le s \\[2ex] \dfrac{\sinh k(b - x) \cosh ks}{k \cosh kb} & s \le x \le b \end{cases}$

(c) $G(x,s) = \begin{cases} \dfrac{\cosh k(b - s) \cosh k(x - a)}{k \sinh k(b - a)} & a \le x \le s \\[2ex] \dfrac{\cosh k(b - x) \cosh k(s - a)}{k \sinh k(b - a)} & s \le x \le b \end{cases}$

(d) $G(x,s) = \begin{cases} \dfrac{\sinh k(b - s)[\sinh k(x - a) + k \cosh k(x - a)]}{k \sinh k(b - a) + k^2 \cosh k(b - a)} & a \le x \le s \\[2ex] \dfrac{\sinh k(b - x)[\sinh k(s - a) + k \cosh k(s - a)]}{k \sinh k(b - a) + k^2 \cosh k(b - a)} & s \le x \le b \end{cases}$

15 $\phi(x) = \dfrac{\omega^2 \rho}{E_s g} \displaystyle\int_0^s s\phi(s)\,ds + \dfrac{\omega^2 \rho x}{E_s g} \displaystyle\int_x^L \phi(s)\,ds$

$\omega_n = \dfrac{(2n+1)\pi}{2L} \sqrt{\dfrac{E_s g}{\rho}}$

17 $G(x,s) = \begin{cases} \dfrac{x^2(x-3s)}{6EI} & 0 \le x \le s \\[3mm] \dfrac{s^2(s-3x)}{6EI} & s \le x \le L \end{cases}$

19 $y = -\dfrac{x^5}{120EI} + \dfrac{L^2 x^3}{12EI} - \dfrac{L^3 x^2}{6EI}$

Chapter 3
Sec. 3.2
p. 96

1 $x = 4$
$y = -5 + \frac{1}{2}e^{-t}$

3 $x = -\frac{2}{3}c_1 e^t - \frac{1}{3}c_2 e^{-2t} - 9;\ y = c_1 e^t + c_2 e^{-2t} + \frac{15}{2} + \frac{1}{2}e^{-t}$

5 $x = 3c_2 \cos 3t - 3c_1 \sin 3t + t - \frac{34}{9}$

$y = c_1 \cos 3t + c_2 \sin 3t + \dfrac{2t}{9} - \dfrac{5}{9}$

7 $x = e^{-t}[(-c_1 + c_2)\cos t - (c_1 + c_2)\sin t] - 3\sin t + 2\cos t$
$y = e^{-t}(c_1 \cos t + c_2 \sin t) - \cos t + 3\sin t$

9 $x = (-2c_1 + c_2)e^t - 2c_2 t e^t - \frac{6}{5}c_3 e^{3t}$
$y = c_1 e^t + c_2 t e^t + c_3 e^{3t}$

11 $x = -\frac{1}{2}c_1 e^t + \frac{1}{2}c_2 e^{-t} + \frac{1}{2}c_3 e^{3t} - 2e^{2t}$
$y = -c_2 e^{-t} + c_3 e^{3t} - 2e^{2t}$
$z = c_1 e^t + c_2 e^{-t} + c_3 e^{3t} + 2e^{2t}$

13 $x = e^t;\ y = e^t;\ z = e^t$

15 $x = c_1 e^{-t} + c_2 e^t + \displaystyle\int_0^t \sinh(t-s)[-z''(s) - z'(s) + z(s)]\,ds$

$y = c_1 e^{-t} - c_2 e^t + \displaystyle\int_0^t \sinh(t-s)[-z''(s) + 2z(s)]\,ds$

$z = $ any twice differentiable function of t

17 $x = c_1 e^{-t} + c_2 e^{-2t} + c_3 \cos 2t + c_4 \sin 2t - 3$
$y = -c_1 e^{-t} - \frac{8}{3}c_2 e^{-2t} - 2(c_3 + c_4)\cos 2t + 2(c_3 - c_4)\sin 2t + 2$

19 (a) $x = 4\sin 2t - 2;\ y = -3\sin 2t - \cos 2t + 1$
(b) $x = 3e^t + 2\sin 2t + 2\cos 2t - 2$
$y = -2e^t - \sin 2t - 2\cos 2t + 1$

21 $(5D - 4)x - (4D - 5)y = 0$
$(5D^2 - 8)x - 3Dy = 0$

23 If D times the second equation in Exercise 17 is subtracted from $D - 1$ times the first equation, the result is $3y = (D^3 + D^2 + D - 2)x$, which yields y without introducing any additional arbitrary constants. In Exercise 19, if $D + 5$ times the second equation is subtracted from the first equation, the result is $24y = -4 + (D^2 - 3D - 14)x$, which yields y without introducing any additional constants. This method can be used in general.

25 $Q_1 = 100(1 - e^{-t/10});\ Q_2 = 100(1 + e^{-t/10})$

27 (a) $cp_1 \dfrac{du_1}{dt} + (k_1 + qc)u_1 - qcu_2 = h + k_1 T_1$

$-qcu_1 + cp_2 \dfrac{du_2}{dt} + (k_2 + qc)u_2 = k_2 T_2$

(b) $u_1 = -33.9e^{-t/100} + 0.5e^{-27t/100} + 83.4$
$u_2 = -20.3e^{-t/100} - 7.6e^{-27t/100} + 78.0$

(c) $u_1 = 1.4e^{-31t/100} - 29.8e^{-t/100} + 88.4$
$u_2 = -7.1e^{-31t/100} - 6.0e^{-t/100} + 73.1$

Sec. 3.3
p. 102

1 $x = ce^{-3t} - \frac{1}{2};\ y = ce^{-3t} + \frac{1}{2}$

3 $x = -b_1 e^t - 2b_2 e^{2t} + t;\ y = b_1 e^t + b_2 e^{2t} + 1$

5 $x = -2b_1 + b_2 e^{-5t} + 5e^t + e^{-t}$
$y = b_1 - 3b_2 e^{-5t} - 3e^t + e^{-t}$

7 $x = a_1 e^t + 4a_2 e^{2t} + 5a_3 e^{3t} - 5e^{-t} + 2$
$y = -a_1 e^t - 5a_2 e^{2t} - 7a_3 e^{3t} + 3e^{-t} - 1$

9 $x = 2a_1 \cos t + 2a_2 \sin t + b_1 \cos 2t + b_2 \sin 2t + 2 \sin 3t$
$y = -3a_1 \cos t - 3a_2 \sin t - 3b_1 \cos 2t - 3b_2 \sin 2t - 3 \sin 3t$

11 $x = 4ae^{-t} + 14be^{-2t} + ce^{-3t} + e^t$
$y = -9ae^{-t} - 30be^{-2t} - 2ce^{-3t}$
$z = 6ae^{-t} + 17be^{-2t} + ce^{-3t} - e^{-t}$

13 $x = 2ae^t + 3be^{2t} + te^t$
$y = (-3a + \frac{1}{4})e^t - 5be^{2t} - \frac{3}{2}te^t$

15 $\ddot{y}_1 = -\lambda^2(2y_1 - y_2)$

$\ddot{y}_2 = -\lambda^2(-y_1 + 2y_2 - y_3)$ $\lambda^2 = \dfrac{4Tg}{wl}$

$\ddot{y}_3 = -\lambda^2(-y_2 + 2y_3)$

$\omega_1 = \sqrt{2 - \sqrt{2}}\,\lambda,\quad 1:\sqrt{2}:1$

$\omega_2 = \sqrt{2}\,\lambda,\quad 1:0:-1$

$\omega_3 = \sqrt{2 + \sqrt{2}}\,\lambda,\quad 1:-\sqrt{2}:1$

Chapter 4
Sec. 4.1
p. 113

5 (a) $(x)^{(3)} + 3(x)^{(2)} + 1$ (b) $(x)^{(4)} + 4(x)^{(3)} + (x)^{(2)} - 2(x)$
(c) $(x)^{(5)} + 8(x)^{(4)} + 17(x)^{(3)} + 13(x)^{(2)} + 2(x) + 6$

11 (a) $\dfrac{n^4 + 6n^3 + 11n^2 + 6n}{4}$ (b) $\dfrac{n(n + 1)(n + 2)(3n + 13)}{12}$

15 (a) $(x)^{-(1)} - 2(x)^{-(2)}$ (b) $(x)^{-(2)} - 2(x)^{-(3)}$
(c) $(x)^{-(1)} - 4(x)^{-(2)} + 4(x)^{-(3)}$

21 (a) $(n - 1)2^{n+1} + 2$ (b) $\dfrac{3^{n+1}}{2}(n^2 - n + 1) - \frac{3}{2}$

(c) $\dfrac{(n + 1)\sin (n + \frac{1}{2})}{2 \sin \frac{1}{2}} - \dfrac{1}{2} + \dfrac{\cos (n + 1) - \cos 1}{4 \sin^2 \frac{1}{2}}$

(d) $-\dfrac{(n + 1)\cos (an + a/2)}{2 \sin (a/2)} + \dfrac{\sin a(n + 1)}{4 \sin^2 (a/2)}$

Sec. 4.2
p. 123

3 (a) 7.08520   (b) 7.47663

5 $x^3 - 3x^2 + 17$

11 $f_0 - \dfrac{(f_{01} - f_{12} + f_{02})^2}{4f_{012}}$

15 (a) 0.000000035   (b) 0.000000017

Sec. 4.3   **1** $f'(200) = 0.00500000$, exact value $= 0.00500000$
p. 131 $f''(200) = -0.00002499$, exact value $= -0.00002500$
               $f'''(200) = 0.00000024$, exact value $= 0.00000025$
               $f'(205) = 0.00487806$, exact value $= 0.00487805$
               $f''(205) = -0.00002377$, exact value $= -0.00002380$
               $f'''(205) = 0.00000025$, exact value $= 0.00000023$
3 $f'(0) = 0.2005$; $f'(10) = 1.170$
5 When the first difference correction terms are taken into account, the value
of the integral is 0.31028. When correction terms through the third differences are taken into account, the value of the integral is 0.31027.

7 $\dfrac{n(n + 1)(2n + 1)(3n^2 + 3n - 1)}{30}$

9

x	$\int_x^1 \dfrac{\sin x}{x} dx$	x	$\int_x^1 \dfrac{\sin x}{x} dx$
0.0	0.946	0.6	0.358
0.1	0.846	0.7	0.265
0.2	0.746	0.8	0.174
0.3	0.647	0.9	0.086
0.4	0.549	1.0	0.000
0.5	0.453		

11 $I = \dfrac{hk}{9} \displaystyle\sum_{i,j} w_{ij} f_{ij}$

where

$$w_{ij} = \begin{cases} 1 & i = j = 0 \\ 2 & i = 0, j \text{ even}; i \text{ even}, j = 0 \\ 4 & i = 0, j \text{ odd}; i \text{ odd}, j = 0; i, j \text{ both even and } > 0 \\ 8 & i \text{ odd}, j \text{ even and } > 0; i \text{ even and } > 0, j \text{ odd} \\ 16 & i, j \text{ both odd} \end{cases}$$

13 With $c = 5$, the value of the sum is 1.64493 which is correct to five decimal
places. With $c = 10$ the sum is 1.63382, which is significantly less accurate.
15 7.4855

17 $f'(0) = \dfrac{1}{h}\left(\delta f_0 - \dfrac{\delta^3 f_0}{24} + \dfrac{3}{640}\delta^5 f_0 - \cdots\right)$

19 $c_0 = -\dfrac{(x_0 - x_2)(2x_0 - 3x_1 + x_2)}{6(x_0 - x_1)}$

$c_1 = \dfrac{-(x_2 - x_0)^3}{6(x_1 - x_0)(x_1 - x_2)}$

$c_2 = \dfrac{(x_2 - x_0)(x_0 - 3x_1 + 2x_2)}{6(x_2 - x_1)}$

If the x's are equally spaced, so that $x_1 = x_0 + h$ and $x_2 = x_0 + 2h$, then
$c_0 = h/3$, $c_1 = 4h/3$, and $c_2 = h/3$, which are precisely the weights in
Simpson's rule.

Sec. 4.4   **1** Replace the given third-order equation by the system of first-order equations
p. 140 $v' = f(x,y,u,v)$, $u' = v$, $y' = u$ with the initial conditions $y = y_0$, $u = u_0 = y_0'$, $v = v_0 = y_0''$.
3 $y_1 = 1.00483$, $y_2 = 1.01872$, $y_3 = 1.04080$, $y_4 = 1.07030$, $y_5 = 1.10651$

5	x	e^{-x^2}	Exact value
	0.0	1.00000	1.00000
	0.1	0.99005	0.99005
	0.2	0.96079	0.96079
	0.3	0.91393	0.91393
	0.4	0.85214	0.85214
	0.5	0.77880	0.77880

7	x	y	z
	0.0	0.00000	1.00000
	0.1	0.10500	0.99983
	0.2	0.21994	0.99866
	0.3	0.34467	0.99550

9	x	y
	1.0	1.00000
	1.1	1.00483
	1.2	1.01872
	1.3	1.04080
	1.4	1.07030
	1.5	1.10651
	1.6	1.14879

13	x	y	Exact value, $y = e^x - x$
	0.0	1.00000	1.00000
	0.1	1.00517	1.00517
	0.2	1.02140	1.02140
	0.3	1.04986	1.04986
	0.4	1.09183	1.09183
	0.5	1.14873	1.14872
	0.6	1.22213	1.22212

15 $\Delta y_1 \equiv k_1 = f(x_0, y_0)h$

$\Delta y_2 \equiv k_2 = f(x_0 + h/2, y_0 + \frac{1}{2}k_1)h$

$\Delta y_3 \equiv k_3 = f(x_0 + h, y_0 + 2k_2 - k_1)h$

$$\Delta y = \tfrac{1}{6}k_1 + \tfrac{4}{6}k_2 + \tfrac{1}{6}k_3$$

17 $y = y_0 + hy_0' \dfrac{x}{h} + (-3y_0 + 3y_1 - 2hy_0' - hy_1')\left(\dfrac{x}{h}\right)^2$

$$+ (2y_0 - 2y_1 + hy_0' + hy_1')\left(\dfrac{x}{h}\right)^3$$

One way to obtain a closed formula is to read y_2 from the polynomial of minimum degree which takes on the values y_0, y_0', y_1', and y_2'. The resulting formula is

$$y_2 = y_0 + \frac{h}{3}(y_0' + 4y_1' + y_2')$$

Sec. 4.5
p. 151

1 (a) $y = c_1(-3)^x + c_2(-4)^x$ (b) $y = c_1(-3)^x + c_2 x(-3)^x$

(c) $y = 2^{x/2}\left(A\cos\dfrac{3\pi x}{4} + B\sin\dfrac{3\pi x}{4}\right)$

(d) $y = c_1 2^x + c_2 3^x$

3 (a) $y = c_1(-2)^x + c_2 3^x - \dfrac{x}{6} - \dfrac{1}{36} + \dfrac{x3^x}{15}$

(b) $y = c_1 \cos \dfrac{\pi x}{2} + c_2 \sin \dfrac{\pi x}{2} + \dfrac{\sin x + \sin (x - 2)}{2(1 + \cos 2)}$

7 $V_x = V_0 \dfrac{\sinh (n - x)\mu}{\sinh n\mu}$; $\cosh \mu = 1 + \dfrac{1}{2k}$

9

$$D_n = \begin{cases} \dfrac{\sinh (n + 1)\mu}{\sinh \mu}; \cosh \mu = \dfrac{a}{2} & a > 2 \\[2mm] 1 + n & a = 2 \\[2mm] \dfrac{\sin (n + 1)\mu}{\sin \mu}, \cos \mu = \dfrac{a}{2} & -2 < a < 2 \\[2mm] (-1)^n(1 + n) & a = -2 \\[2mm] (-1)^n \dfrac{\sinh (n + 1)\mu}{\sinh \mu}, \cosh \mu = -\dfrac{a}{2} & a < -2 \end{cases}$$

15 (a) The only solution is $y = 0$.
(b) $y = \phi(x - 1)$
(c) Because $a_2 = 0$, the equation is actually of the first order, and its complete solution contains only one arbitrary constant:

$$y = A \left(-\dfrac{a_1}{a_0} \right)^x$$

(d) $y = A(-a_1/a_0)^x + \Phi(x - 1)$, where $\Phi(x)$ is a particular solution of the nonhomogeneous equation $(a_0 E + a_1)y = \phi(x)$.

19 The formula is numerically stable for the equation $y' = Ay$ for all values of A.

21 The formula is numerically stable for the equation $y' = Ay$ only if $A > 0$.

Sec. 4.6
p. 167

3 (a) $y = \dfrac{2 + 133x}{102} = 0.0196 + 1.304x$

(b) $y = \dfrac{68 + 187x}{133} = 0.511 + 1.406x$

5 (a) $x = 1.683; y = -1.847$ (b) $x = 1.739; y = -1.811$
The factor 4 acts as a weight attaching relatively more significance to the third equation.

9 $y = 0.8190 + 2.1786x; E_1 = 0.1305$
$y = 1.0124 - 0.7280x + 1.4500x^2; E_2 = 0.0047$

11 (a) $A = 1.000; a = 0.499$ (b) $A = 1.005; a = 0.498$

15 After a has been found, A can be determined by applying the method of least squares to the equations

$$y_1 = Ae^{ax_1}, y_2 = Ae^{ax_2}, \ldots, y_n = Ae^{ax_n}$$

in which A is the only unknown. This method is generally to be preferred to linearizing by taking logarithms because it does not introduce any unwarranted weighting of the data. It is clearly preferable to both linearizing by taking logarithms and using Taylor's series.

17 $A = 1.008$, $a = 0.496$, $E = 0.002081$. For part (a) of Exercise 11, $E = 0.001272$; for part (b) (beginning with $A = 1$ and $a = \frac{1}{2}$), $E = 0.001257$.

19 (a) $y = 0.980 - 0.417x^2$ (b) $y = 0.989x - 0.145x^3$

21		Derivative of ln x		
	x	**(a)**	**(b)**	Correct value
	2.10	0.49340	0.5852	0.47619
	2.20	0.44820	0.3632	0.45455
	2.30	0.43638	0.3520	0.43478

23 $A = 0.274$

25 $p = \bar{x} \cos \theta + \bar{y} \sin \theta$

Chapter 5

Sec. 5.2
p. 177

1 $k = \dfrac{k_1 k_2}{k_1 + k_2}$

3 $\frac{1}{3} M_s$

5 In equilibrium, the upper spring is stretched an amount equal to

$$\frac{w + k_2(l - l_1 - l_2)}{k_1 + k_2}$$

7 $CL \dfrac{d^2 i_l}{dt^2} + \dfrac{L}{R} \dfrac{d i_l}{dt} + i_l = I_0 \cos \omega t$

$CL \dfrac{d^2 i_r}{dt^2} + \dfrac{L}{R} \dfrac{d i_r}{dt} + i_r = -\dfrac{\omega L}{R} I_0 \sin \omega t$

$CL \dfrac{d^2 i_c}{dt^2} + \dfrac{L}{R} \dfrac{d i_c}{dt} + i_c = -\omega^2 LCI_0 \cos \omega t$

9 $\dfrac{d^2 X}{dT^2} + \dfrac{c}{I v_2} \dfrac{dX}{dT} + \dfrac{k}{I v_2{}^2} X = \dfrac{T_0}{I a v_2{}^2} \cos \dfrac{\omega}{v_2} T$

11 $\dfrac{d^2 X}{dT^2} + \dfrac{1}{CR v_4} \dfrac{dX}{dT} + \dfrac{1}{LC v_4{}^2} X = \dfrac{I_0}{C \varepsilon v_4{}^2} \cos \dfrac{\omega}{v_4} T$

where v_4 is an arbitrary frequency, ε is an arbitrary value of $U = \int^t e\, dt$, $X = U/\varepsilon$, and $T = v_4 t$.

Sec. 5.3
p. 191

5 The first integer equal to or greater than $\dfrac{\ln 2}{2\pi} \dfrac{\sqrt{1 - (c/c_c)^2}}{c/c_c}$

7 $Y_{ss} = 0.91 \sin (15t - 48°34')$

9 $y = -e^{-6t}(2 \cos 8t + \frac{3}{2} \sin 8t) + 2$

11 $y = \dfrac{F_0}{241} [e^{-4.2t}(-3 \cos 14.4t + \frac{29}{24} \sin 14.4t) + 3 e^{-10t}]$

17 The magnification ratio is

$$M = \frac{(\omega/\omega_n)^2}{\sqrt{[1 - (\omega/\omega_n)^2]^2 + [2(c/c_c)(\omega/\omega_n)]^2}}$$

and is to be applied to the static deflection produced in the system by a force equal to the amplitude of the disturbing force at the frequency ω_n. The phase angle is the same [Eq. (12)] whether the amplitude of the excitation is constant or proportional to ω^2.

19 Period $= \tau = 2\pi \sqrt{a/\mu g}$. Hence $\mu = 4\pi^2 a/g\tau^2$.

23 $2 \dfrac{w}{g} v_0{}^2 + c v_0 y_0 > 0$

29 $Y_{ss} = \dfrac{F_0}{50}[2.28 \cos(15t - 1°) + 2.77 \cos(16t - 1°30')]$

Because of the near equality of the frequencies and amplitudes of the impressed forces, the phenomenon of beats is to be expected.

Sec. 5.4
p. 200

1   Across the resistance: $E = iR = 40(e^{-200t} - e^{-800t})$

Across the inductance: $E = L\dfrac{di}{dt} = -8e^{-200t} + 32e^{-800t}$

Across the capacitor: $E = \dfrac{1}{C}\displaystyle\int_0^t i\,dt = -32e^{-200t} + 8e^{-800t} + 24$

3 $i = \frac{1}{15}e^{-20,000t} - \frac{1}{15}e^{-5,000t}$

5 $i = -\frac{1}{8}e^{-6,000t}\sin 8,000t$

7 $i = \frac{625}{4}te^{-2,500t}$

9 $i = -50\pi A e^{-500t} + 500(50\pi A + 500B)te^{-500t}$
$$+ 50\pi A \cos 50\pi t - 50\pi B \sin 50\pi t$$

where
$$A = \frac{110[\frac{1}{4} \times 10^6 - (50\pi)^2]}{[\frac{1}{4} \times 10^6 - (50\pi)^2]^2 + (50,000\pi)^2}$$

$$B = -\frac{110(50,000\pi)}{[\frac{1}{4} \times 10^6 - (50\pi)^2]^2 + (50,000\pi)^2}$$

11 $t = 0.00039$

13 $E_{ss} = \frac{150}{109}\sin(150t + \delta)$, where $\delta = \tan^{-1}\frac{91}{60}$

19 $|Z|$ is a minimum for the undamped natural frequency $\Omega_n = 1/\sqrt{LC}$. For the magnification ratio the maximum always occurs at a frequency below the undamped natural frequency. This involves no contradiction, since the magnification ratio M relates F and y, whose electrical analogs are E and Q, whereas the impedance relates E and $i = dQ/dt$.

21 $|Z| = R\sqrt{1 + Q^2\left(\dfrac{\Omega}{\Omega_n} - \dfrac{\Omega_n}{\Omega}\right)^2}$; $\delta = \text{Tan}^{-1}Q\left(\dfrac{\Omega}{\Omega_n} - \dfrac{\Omega_n}{\Omega}\right)$

23 $L = 0.12$, $R = 66.2$, $C = 1.56 \times 10^{-6}$, $E_0 = 3.69 \times 10^5 \, \gamma/\sigma$, $v_3/v_1 = 200$, and $\gamma/\sigma = 0.3 \times 10^{-6}$.

Sec. 5.5
p. 209

1 $y_1 = \frac{5}{2}\sin t - 2\sin 2t + \frac{1}{2}\sin 3t$
$y_2 = 5\sin t + 2\sin 2t - 3\sin 3t$

3 $Y_1 = \dfrac{F_0}{2(\omega^2 - 1)(\omega^2 - 4)}\sin \omega t$

$Y_2 = \dfrac{-F_0(\omega^2 - 3)}{2(\omega^2 - 1)(\omega^2 - 4)}\sin \omega t$

The amplitudes of the steady-state motion vary with the impressed frequency ω and are unbounded in the neighborhood of either natural frequency $\omega_1 = 1$ or $\omega_2 = 2$.

5 $\omega_1^2 = 1$, $a_1{:}a_2 = 1{:}1$; $\omega_2^2 = 5$, $a_1{:}a_2 = 3{:}{-1}$

7 $x_1 = \frac{1}{3}\cos t + \frac{2}{3}\cos 2t$; $x_2 = \frac{1}{3}\cos t - \frac{1}{3}\cos 2t$

9 $k_2 = \omega^2 M_2$, $a_2 = -F_0/k_2$. Friction introduced to reduce the amplitude a_2 makes it impossible to keep M_1 strictly at rest, although its motion will still be within acceptable limits.

11 $x_1 = \cos 9t + \cos 11t$; $x_2 = \cos 9t - \cos 11t$. Since these can be rewritten in the form $x_1 = 2\cos 10t \cos t$ and $x_2 = 2\sin 10t \sin t$, it is clear that both x_1 and x_2 appear to vary with frequency $(\omega_1 + \omega_2)/2 = 10$ with the slowly varying amplitudes $2\cos t$ and $2\sin t$, respectively. Thus the system exhibits the phenomenon of beats.

15 $\omega_1 = 6\sqrt{6}$, $\omega_2 = 12\sqrt{5}$

17 $i_1 = -\dfrac{Q_0}{60\sqrt{LC}}\left(16\sin\dfrac{t}{3\sqrt{LC}} + \sin\dfrac{t}{12\sqrt{LC}}\right)$

$i_2 = -\dfrac{Q_0}{20\sqrt{LC}}\left(4\sin\dfrac{t}{3\sqrt{LC}} - \sin\dfrac{t}{12\sqrt{LC}}\right)$

19 $i_1 = E_0\sqrt{\dfrac{C}{L}}\left(\dfrac{243}{80}\sin\dfrac{3t}{\sqrt{LC}} - \dfrac{1}{240}\sin\dfrac{t}{3\sqrt{LC}} - \sin\dfrac{t}{\sqrt{LC}}\right)$

$i_2 = E_0\sqrt{\dfrac{C}{L}}\left(\dfrac{27}{10}\sin\dfrac{3t}{\sqrt{LC}} + \dfrac{1}{30}\sin\dfrac{t}{3\sqrt{LC}} - \sin\dfrac{t}{\sqrt{LC}}\right)$

21 $\omega_N = \dfrac{2}{\sqrt{LC}}\sin\dfrac{N\pi}{2n}$ $N = 1, 2, \ldots, n-1$

For the kth amplitude when the system is vibrating at the Nth natural frequency, we have

$$(a_k)_N = \sin\dfrac{N\pi k}{n} - \sin\dfrac{N\pi(k-1)}{n}$$

23 $\omega_N = 2\sqrt{\dfrac{k}{m}}\sin\left(\dfrac{N}{n+1}\dfrac{\pi}{2}\right)$ $N = 1, 2, \ldots, n$

$(a_j)_N = \sin\dfrac{N\pi j}{n+1}$

25 $\omega_N = 2\sqrt{\dfrac{k}{m}}\sin\dfrac{(2N+1)\pi}{(2n+1)2}$ $N = 0, 1, \ldots, n-1$

$(a_{n-j})_N = \sin\left(\dfrac{2N+1}{2n+1}\pi j\right) - \sin\left[\dfrac{2N+1}{2n+1}\pi(j-1)\right]$

27 (a) $Q_k = \dfrac{E_0}{\omega}\sqrt{\dfrac{C}{L}}\dfrac{\sin(n+1-k)\mu}{\cos[(2n+1)/2]\mu}\cos\omega t; \quad \mu = \text{Cos}^{-1}\left(1 - \dfrac{LC\omega^2}{2}\right)$

(b) $Q_k = (-1)^k\dfrac{E_0}{\omega}\sqrt{\dfrac{C}{L}}\dfrac{\sinh(n+1-k)\mu}{\sinh[(2n+1)/2]\mu}\cos\omega t$

$\mu = \text{Cosh}^{-1}\left(\dfrac{LC\omega^2}{2} - 1\right)$

29 Let s_1 and s_2 be arbitrary lengths, let v_1 be an arbitrary frequency, and let $Y_1 = y_1/s_1$, $Y_2 = y_2/s_2$, $T = v_1 t$. The mechanical system is governed by the differential equations

$$\dfrac{d^2 Y_1}{dT^2} + \dfrac{c}{m_1 v_1}\dfrac{dY_1}{dT} + \dfrac{k_1 + k_2}{m_1 v_1^2}Y_1 - \dfrac{cs_2}{m_1 s_1 v_1}\dfrac{dY_2}{dT} - \dfrac{k_2 s_2}{m_1 s_1 v_1^2}Y_2 = 0$$

$$-\dfrac{cs_1}{v_1 s_2 m_2}\dfrac{dY_1}{dT} - \dfrac{k_2 s_1}{m_2 s_2 v_1^2}Y_1 + \dfrac{d^2 Y_2}{dT^2} + \dfrac{c}{m_2 v_1}\dfrac{dY_2}{dT} + \dfrac{k_2}{m_2 v_1^2}Y_2 = 0$$

Chapter 6
Sec. 6.2
p. 221

1 $f(t) = \dfrac{1}{2} + \dfrac{2}{\pi}\left(\dfrac{\sin\pi t}{1} + \dfrac{\sin 3\pi t}{3} + \dfrac{\sin 5\pi t}{5} + \cdots\right)$

3 $a_n \equiv 0, \; b_n = \begin{cases} -\dfrac{2}{n\pi} & n = 1, 3, 5, 7, \ldots \\[2mm] -\dfrac{4}{n\pi} & n = 2, 6, 10, 14, \ldots \\[2mm] 0 & n = 4, 8, 12, 16, \ldots \end{cases}$

5 $f(t) = \dfrac{1}{2} - \dfrac{1}{\pi}\left(\dfrac{\sin 2\pi t}{1} + \dfrac{\sin 4\pi t}{2} + \dfrac{\sin 6\pi t}{3} + \cdots\right)$

7 $a_0 = 2,\ a_n \equiv 0 \qquad n \neq 0$

$$b_n = \begin{cases} \dfrac{3}{n\pi} & n = 1, 2, 4, 5, 7, 8, \ldots \\ 0 & n = 3, 6, 9, 12, \ldots \end{cases}$$

9 $a_n = \dfrac{1 - e^{-2}}{1 + n^2\pi^2} \qquad b_n = \dfrac{n\pi(1 - e^{-2})}{1 + n^2\pi^2}$

11 $a_0 = \dfrac{\pi}{2},\ a_n = \begin{cases} -\dfrac{2}{n^2\pi^2} & n = 1, 3, 5, \ldots \\ 0 & n = 2, 4, 6, \ldots \end{cases}$

$b_n = \dfrac{(-1)^{n+1}}{n}$

13 $a_n \equiv 0,\ b_n = (-1)^{n+1}\dfrac{12}{n^3\pi^3}$

Sec. 6.3
p. 227

3 $f(t) = \dfrac{2p}{\pi}\left(\sin\dfrac{\pi t}{p} - \dfrac{1}{2}\sin\dfrac{2\pi t}{p} + \dfrac{1}{3}\sin\dfrac{3\pi t}{p} - \dfrac{1}{4}\sin\dfrac{4\pi t}{p} + \cdots\right)$

$f(t) = -\dfrac{4p}{\pi^2}\left(\cos\dfrac{\pi t}{p} + \dfrac{1}{9}\cos\dfrac{3\pi t}{p} + \dfrac{1}{25}\cos\dfrac{5\pi t}{p} + \dfrac{1}{49}\cos\dfrac{7\pi t}{p} + \cdots\right)$

5 $f(t) = \dfrac{4}{\pi}\left(\dfrac{1}{1^2 - 4}\sin\dfrac{t}{2} + \dfrac{3}{3^2 - 4}\sin\dfrac{3\pi}{2} + \dfrac{5}{5^2 - 4}\sin\dfrac{5t}{2} + \cdots\right)$

$f(t) = \cos t$

7 Half-range sine expansion: $b_n = \dfrac{2n\pi}{a^2 + n^2\pi^2}[1 - (-1)^n e^{-a}]$

Half-range cosine expansion: $a_n = \dfrac{2a}{a^2 + n^2\pi^2}[1 - (-1)^n e^{-a}]$

9 Half-range sine expansion: $b_n = \dfrac{2n}{\pi(n^2 - a^2)}[1 - (-1)^n \cos a\pi]$

Half-range cosine expansion: $a_n = \dfrac{(-1)^{n+1}2a \sin a\pi}{\pi(n^2 - a^2)}$

13 $t^3 = \dfrac{p^3}{4} + 6p^3\left[\left(\dfrac{4}{\pi^4} - \dfrac{1}{\pi^2}\right)\cos\dfrac{\pi t}{p} + \dfrac{1}{2^2\pi^2}\cos\dfrac{2\pi t}{p}\right.$

$+ \left(\dfrac{4}{3^4\pi^4} - \dfrac{1}{3^2\pi^2}\right)\cos\dfrac{3\pi t}{p} + \dfrac{1}{4^2\pi^2}\cos\dfrac{4\pi t}{p}$

$\left. + \left(\dfrac{4}{5^4\pi^4} - \dfrac{1}{5^2\pi^2}\right)\cos\dfrac{5\pi t}{p} + \dfrac{1}{6^2\pi^2}\cos\dfrac{6\pi t}{p} + \cdots\right]$

17 Any extension of $f(t) = t - t^2$ whose derivative is continuous over the interval $(-1, 0)$, which takes on the value 0 at $t = -1$ and at $t = 0$, and whose derivative takes on the values -1 and 1, respectively, at these points, will give rise to a series whose coefficients decrease at least as fast as $1/n^3$. Moreover, if the second derivative of such a function is different from 0 at either $t = -1$ or $t = 0$, the coefficients can decrease no faster than $1/n^3$. Examples of such extensions are $e^{t(t+1)} - 1$, $t(t + 1) + \lambda t^2(t + 1)^2$, λ arbitrary, and $(\sin \pi t)/\pi$.

19   One such function is $f(t) = (1 - t^2)^2$, $0 \le t \le 1$, whose half-range cosine expansion has the coefficients

$$a_0 = \tfrac{16}{15} \qquad a_n = \frac{(-1)^{n+1}48}{n^4\pi^4} \qquad n \ne 0$$

23   The coefficients are

$$a_0 = \frac{a+1}{2} \qquad a_n = \frac{-4}{(1-a)n^2\pi^2}\left(\cos\frac{n\pi}{2} - \cos\frac{n\pi a}{2}\right) \qquad n \ne 0$$

$$b_n \equiv 0$$

Clearly, for fixed a (< 1) and n sufficiently large, a_n decreases as $1/n^2$, as expected. However, by converting the difference of the cosine terms in the expression for a_n into a product, we can write

$$a_n = \frac{2}{n\pi}\sin\frac{(a+1)n\pi}{4}\left\{\frac{\sin\left[(1-a)n\pi/4\right]}{(1-a)n\pi/4}\right\}$$

For arbitrarily large but fixed n, the fraction in braces is arbitrarily close to 1 for a sufficiently close to 1. Hence, by taking a close enough to 1, the coefficients can be made to decrease only as fast as $1/n$ through as large a range of n as desired. Descriptively, since the function is continuous (as long as $0 < a < 1$) but has a discontinuous derivative, its coefficients must eventually decrease as $1/n^2$. However, by making one segment of the graph arbitrarily close to the vertical, i.e., by approximating a finite jump in the function, the onset of the inevitable asymptotic behavior of the coefficients can be postponed to arbitrarily large values of n.

25 a_n will decrease faster than $1/n^2$ provided that for all values of n,

$$\sum_i [f'(t_i^+) - f'(t_i^-)]\cos\frac{n\pi t_i}{p} = 0$$

where $\{t_i\}$ is the set of points of discontinuity of f'. b_n will decrease faster than $1/n^2$ provided that for all values of n

$$\sum_i [f'(t_i^+) - f'(t_i^-)]\sin\frac{n\pi t_i}{p} = 0$$

where $\{t_i\}$ is the set of points of discontinuity of f'.

Sec. 6.4
p. 232

1 $A_0 = \tfrac{1}{3},\ A_n = \dfrac{2}{n^2\pi^2}\sqrt{n^2\pi^2 + 4}$

3 $A_n = \dfrac{2\sinh 1}{\sqrt{1 + n^2\pi^2}}$ **5** $c_0 = \tfrac{1}{2};\ c_n = \dfrac{1 - e^{-ni\pi}}{2ni\pi}$

7 $c_0 = \tfrac{1}{2};\ c_n = \dfrac{(1 + 2ni\pi)e^{-2ni\pi} - 1}{4n^2\pi^2} = \dfrac{i}{2n\pi}$

9 $c_n = \dfrac{2(-1)^n}{\pi(1 - 4n^2)}$ **11** $c_n = \dfrac{(-1)^n\sinh 1}{1 + n^2\pi^2}$

Sec. 6.5
p. 238

1   Yes

5 $i_{ss} = \displaystyle\sum_{n=-\infty}^{\infty} - \frac{2iE_0e^{200ni\pi t}}{250n\pi + i(4n^2\pi^2 - 2,500)}$

7 $y_{ss} = F_0[0.011\sin(2\pi t - 1°34') - 0.081\sin(4\pi t - 124°37')$
$\qquad\qquad\qquad\qquad + 0.002\sin(6\pi t - 177°21') - \cdots]$

9 $y_{ss} = F_0[\tfrac{1}{120} + 0.011\sin(\pi t - 1°17') + 0.009\sin(3\pi t - 9°19')$
$\qquad\qquad\qquad\qquad + 0.003\sin(5\pi t - 171°39') + \cdots]$

11 $i_{ss} = E_0 \displaystyle\sum_{n=-\infty}^{\infty} \frac{ie^{200ni\pi t}}{200[600n\pi + i(200n^2\pi^2 - 1,250)]}$

13 The complete solution will originally appear in the form $y = c_1 y_1 + c_2 y_2 + Y$, where Y is the Fourier series obtained as the answer to Example 2. Imposing initial conditions of displacement and velocity will thus lead to a pair of simultaneous linear equations in c_1 and c_2 in which the constant terms will involve the infinite series which result from the evaluation of Y and Y' when $t = 0$. Although there is no theoretical problem in determining c_1 and c_2 from these equations, the arithmetical complications are obvious.

15 (a) $y = \left[\dfrac{\pi}{2} - \dfrac{4}{\pi} \displaystyle\sum_{n=1}^{\infty} \dfrac{1}{(2n-1)^4 + (2n-1)^2} \right] \cosh\left(t - \dfrac{\pi}{2} \right)$

$$+ \dfrac{4}{\pi} \sum_{n=1}^{\infty} \dfrac{\cos(2n-1)t}{(2n-1)^4 + (2n-1)^2}$$

(b) $y = \left[-\dfrac{\pi}{8} - \dfrac{4}{\pi} \displaystyle\sum_{n=1}^{\infty} \dfrac{1}{(2n-1)^4 - 4(2n-1)^2} \right] \cos 2t$

$$+ \dfrac{\pi}{8} + \dfrac{4}{\pi} \sum_{n=1}^{\infty} \dfrac{\cos(2n-1)t}{(2n-1)^4 - 4(2n-1)^2}$$

(c) $y = \left[-\dfrac{\pi}{18} + \dfrac{1}{2\pi} - \dfrac{4}{\pi} \displaystyle\sum_{n=1}^{\infty} \dfrac{1}{(2n-1)^4 - 9(2n-1)^2} \right] \cos 3t$

$$+ \dfrac{\pi}{18} - \dfrac{4}{\pi} \left[\dfrac{\cos t}{8} + \dfrac{t \sin t}{54} - \sum_{n=1}^{\infty} \dfrac{\cos(2n-1)t}{(2n-1)^4 - 9(2n-1)^2} \right]$$

Sec. 6.6
p. 249

1 (a) $a_n \equiv 0,\ b_n = \dfrac{2}{\pi}\dfrac{1 - \cos \omega_n}{\omega_n}\ \Delta\omega;\ \omega_n = \dfrac{n\pi}{p},\ \Delta\omega = \dfrac{\pi}{p}$

(b) $a_n = \dfrac{2}{\pi}\dfrac{1 - \cos \omega_n}{\omega_n^2}\ \Delta\omega,\ b_n \equiv 0;\ \omega_n = \dfrac{n\pi}{p},\ \Delta\omega = \dfrac{\pi}{p}$

(c) $a_n = \dfrac{4}{\pi}\dfrac{\sin \omega_n - \omega_n \cos \omega_n}{\omega_n^3}\ \Delta\omega,\ b_n \equiv 0;\ \omega_n = \dfrac{n\pi}{p},\ \Delta\omega = \dfrac{\pi}{p}$

(d) $a_n \equiv 0,\ b_n = \dfrac{2\sin \omega_n}{\pi^2 - \omega_n^2}\ \Delta\omega;\ \omega_n = \dfrac{n\pi}{p},\ \Delta\omega = \dfrac{\pi}{p}$

3 (a) $a_n = \dfrac{2}{\pi}\dfrac{1 - (-1)^n e^{-p}}{1 + \omega_n^2}\ \Delta\omega,\ b_n \equiv 0;\ \omega_n = \dfrac{n\pi}{p},\ \Delta\omega = \dfrac{\pi}{p}$

(b) $a_n \equiv 0,\ b_n = \dfrac{2\omega_n[1 - (-1)^n e^{-p}]}{\pi(1 + \omega_n^2)}\ \Delta\omega;\ \omega_n = \dfrac{n\pi}{p},\ \Delta\omega = \dfrac{\pi}{p}$

In Exercise 2, the coefficient formulas involved only ω_n and $\Delta\omega$, and hence the amplitude envelope did not change its shape as p became infinite. In the present example, the coefficient formulas involve p as well as ω_n and $\Delta\omega$, and hence the amplitude envelope changes its shape as p becomes infinite.

5 (a) $f(t) = \dfrac{2a}{\pi} \displaystyle\int_0^{\infty} \dfrac{\cos \omega t}{a^2 + \omega^2}\ d\omega$

(b) $f(t) = \dfrac{1}{\pi} \displaystyle\int_0^{\infty} \dfrac{a \cos \omega t + \omega \sin \omega t}{a^2 + \omega^2}\ d\omega$

(c) $f(t) = \dfrac{2}{\pi} \displaystyle\int_0^{\infty} \dfrac{\sin \omega \pi \sin \omega t}{1 - \omega^2}\ d\omega$

(d) $f(t) = \dfrac{2}{\pi} \displaystyle\int_0^{\infty} \dfrac{\cos(\omega \pi / 2) \cos \omega t}{1 - \omega^2}\ d\omega$

(e) $f(t) = \dfrac{1}{\pi} \displaystyle\int_0^{\infty} \dfrac{\sin \omega(1 - t) + \sin \omega t}{\omega}\ d\omega$

(f) $f(t) = \dfrac{4}{\pi} \displaystyle\int_0^{\infty} \dfrac{\sin \omega - \omega \cos \omega}{\omega^3}\ \cos \omega t\ d\omega$

7 $f(t) = \dfrac{2}{\pi} \displaystyle\int_0^\infty \dfrac{1 - \cos \omega}{\omega} \sin \omega t \, d\omega$

$$\doteq -\dfrac{1}{\pi}[\text{Si}\,(t - 1)\omega_0 - 2\,\text{Si}\,(t\omega_0) + \text{Si}\,(t + 1)\omega_0]$$

9 $f(t) \doteq \dfrac{1}{\pi}\left[\dfrac{\cos (t + 1)\omega_0}{\omega_0} - \dfrac{\cos (t - 1)\omega_0}{\omega_0}\right.$

$$\left. + t\,\text{Si}\,(t + 1)\omega_0 - t\,\text{Si}\,(t - 1)\omega_0\right]$$

13 $Y = \dfrac{2}{\pi} \displaystyle\int_0^\infty \dfrac{(b - \omega^2)\cos \omega t + a\omega \sin \omega t}{(b - \omega^2)^2 + (a\omega)^2} \dfrac{\sin \omega}{\omega} \, d\omega$

15 $Y = \dfrac{2}{\pi} \displaystyle\int_0^\infty \dfrac{-a\omega \cos \omega t + (b - \omega^2)\sin \omega t}{(b - \omega^2)^2 + (a\omega)^2} \dfrac{\sin \omega - \omega \cos \omega}{\omega^2} \, d\omega$

Sec. 6.7
p. 256

1 The transform of $e^{-at}t$ is $g(\omega) = 1/(a + i\omega)^2$.

3 No; yes; $g(\omega) = \dfrac{a + i\omega}{(a + i\omega)^2 + k^2}$

Chapter 7
Sec. 7.1
p. 262

1 (a) Yes   (b) No   (c) No
(d) Yes   (e) Yes   (f) Yes
3 (a) 0   (b) 0   (c) 0   (d) k (e) k (f) 0

Sec. 7.2
p. 267

1 $\mathscr{L}\{f^{(n)}\} = s^n \mathscr{L}\{f\} - \displaystyle\sum_{j=0}^{n-1} s^{n-1-j}f^{(j)}(0^+)$

9

$$\mathscr{L}\{y\} = \dfrac{\begin{vmatrix} \mathscr{L}\{f_1\} + a_1y_0 + c_1z_0 & c_1s + d_1 \\ \mathscr{L}\{f_2\} + a_2y_0 + c_2z_0 & c_2s + d_2 \end{vmatrix}}{\begin{vmatrix} a_1s + b_1 & c_1s + d_1 \\ a_2s + b_2 & c_2s + d_2 \end{vmatrix}}$$

$$\mathscr{L}\{z\} = \dfrac{\begin{vmatrix} a_1s + b_1 & \mathscr{L}\{f_1\} + a_1y_0 + c_1z_0 \\ a_2s + b_2 & \mathscr{L}\{f_2\} + a_2y_0 + c_2z_0 \end{vmatrix}}{\begin{vmatrix} a_1s + b_1 & c_1s + d_1 \\ a_2s + b_2 & c_2s + d_2 \end{vmatrix}}$$

11 $C(f'') = -n^2C(f) + (-1)^n f'(\pi) - f'(0)$

13 (a) $b_n = -\dfrac{4}{\pi n^3} + (-1)^n\left(\dfrac{4}{\pi n^3} - \dfrac{2\pi}{n}\right)$

(b) $a_n = \dfrac{12}{\pi n^4} - (-1)^n\left(\dfrac{12}{\pi n^4} - \dfrac{6\pi}{n^2}\right)$

15 If $f(t), f'(t), \ldots, f^{(k)}(t)$ are all piecewise regular and of exponential order, and if $f^{(k)}(t)$ is the first of these quantities which is not everywhere continuous, which requires, of course, that

$$f(0^+) = f'(0^+) = \cdots = f^{(k-1)}(0^+) = 0$$

then

$$\mathscr{L}\{f\} \sim \dfrac{A}{s^{k+1}} \qquad \text{as } s \to \infty$$

It is interesting to compare this result with Theorem 3, Sec. 6.3.

Sec. 7.3
p. 273

3 $\dfrac{s}{s^2 - k^2}$

5 $\dfrac{1}{2}\left[\dfrac{1}{s} + \dfrac{s}{s^2 + 4b^2}\right]$

7 (a) e^{-3t} (b) $\dfrac{1}{3!} t^3$ (c) $\tfrac{1}{3} \sin 3t$

(d) $2 \cos 3t + \sin 3t$ (e) $-\tfrac{1}{2}e^{-t} + \tfrac{3}{2}e^{3t}$

9 $z = 4e^{-4t} + 2e^t$

11 $y = -\tfrac{2}{3} \cos t - \sin t + \tfrac{8}{3} \cos 2t + 2 \sin 2t$
$z = \tfrac{1}{3} \cos t + \tfrac{1}{3} \sin t - \tfrac{4}{3} \cos 2t - \tfrac{2}{3} \sin 2t$

13 (a) $\dfrac{1}{(\ln c)^{c+1}} \Gamma(c + 1)$ (b) $\Gamma(\tfrac{1}{2}) = \sqrt{\pi}$

(c) $\dfrac{1}{(m + 1)^{n+1}} \Gamma(n + 1)$

15 $a\sqrt{\dfrac{m\pi}{k}}$ **17** $\dfrac{\sqrt{\pi}}{2} \dfrac{\Gamma[(k + 1)/2]}{\Gamma(k/2 + 1)}$ $k > -1$

19 (a) $\dfrac{\sqrt{\pi}}{2} \dfrac{\Gamma(\tfrac{3}{4})}{\Gamma(\tfrac{5}{4})}$ (b) $\tfrac{1}{2}\Gamma(\tfrac{3}{4})\Gamma(\tfrac{1}{4})$

(c) $\dfrac{\sqrt{\pi}}{4} \dfrac{\Gamma(\tfrac{1}{4})}{\Gamma(\tfrac{3}{4})}$ (d) $\dfrac{1}{k} \dfrac{\Gamma(1 + 1/k)\Gamma(1/k)}{\Gamma(1 + 2/k)}$

(e) $\dfrac{a^{(n+2)/2}}{n} \dfrac{\Gamma(1/n)\Gamma(\tfrac{3}{2})}{\Gamma(1/n + \tfrac{3}{2})}$

Sec. 7.4
p. 286

1 $\dfrac{1}{s} e^{-as}$ **3** $\dfrac{1}{s}$

5 $\left(\dfrac{2}{s^3} + \dfrac{4}{s^2} + \dfrac{4}{s}\right) e^{-2s}$ **7** $\left(\dfrac{s \cos 1 - \sin 1}{s^2 + 1}\right) e^{-s}$

9 $\dfrac{1 + e^{-\pi s}}{s^2 + 1}$ **11** $\dfrac{b}{s}(e^{-as} - 2e^{-2as} + e^{-3as})$

13 $\dfrac{4(s + 3)}{(s^2 + 6s + 13)^2}$ **15** $\dfrac{4}{[(s + 3)^2 + 4]^2}$

17 $\ln \dfrac{\sqrt{s^2 + 9}}{s}$ **19** $\cot^{-1} \dfrac{s + 3}{2}$

21 $\dfrac{1}{s + 3} \cot^{-1} \dfrac{s + 3}{2}$ **23** $\dfrac{1}{3!} t^3 e^{-2t}$

25 $\dfrac{1}{9}\left(\cos \dfrac{2t}{3} + \sin \dfrac{2t}{3}\right) e^{-t/3}$

27 $t - 1 + e^{-t}$ **29** $\tfrac{1}{3} \sin t - \tfrac{1}{6} \sin 2t$

31 $\tfrac{1}{3} \sinh 3(t - 3) u(t - 3)$

33 $(e^{2(t-1)} - e^{(t-1)})u(t - 1) + (e^{2(t-2)} - e^{(t-2)})u(t - 2)$

35 $\dfrac{2(1 - \cosh t)}{t}$

37 $\dfrac{2(\sinh t - t \cosh t)}{t^2}$

The number 2 in the given transform, though it appears to play no role in the determination of the inverse, is necessary in order that $\lim\limits_{s \to \infty} \phi(s) = 0$, as required by Corollary 1, Theorem 5, Sec. 7.1.

39 $\dfrac{te^{-2t}\sin t}{2}$

41 $\displaystyle\int_0^t \dfrac{\sin t}{t}\,dt$ or Si (t)

43 $\dfrac{(-t\cos t + \sin t)e^{-t}}{2}$

45 $y = \dfrac{(7 + 2t)e^{-t}}{4} - \dfrac{3e^{-3t}}{4}$

47 $y = e^{-t} - e^{-2t} + (-e^{-(t-1)} + \frac{1}{2}e^{-2(t-1)} + \frac{1}{2})u(t-1)$

49 $y = \dfrac{t\sin t}{2}$

51 $f(t)$, $f'(t)$, and $f''(t)$ are each piecewise regular and of exponential order; $f(t)$, $f'(t)$, $f''(t)$, and $f'''(t)$ are each piecewise regular and of exponential order.

$$f^{(n)}(0^+) = \lim_{s\to\infty} s\left[s^n\mathcal{L}\{f\} - \sum_{j=0}^{n-1} s^{n-1-j}f^{(m)}(0^+) \right]$$

53 ln 2

55 $\frac{1}{2}[\ln(p^2 + q^2) - \ln(a^2 + b^2)]$

57 $y = y_0e^{-t} + \dfrac{c}{6}t^3e^{-t}$

Sec. 7.5
p. 294

1 $\frac{5}{2}e^{-t} - 9e^{-2t} + \frac{15}{2}e^{-3t}$

3 $\frac{1}{25}(3\cos t + 4\sin t - 3e^{-2t} - 10te^{-2t})$

5 $\dfrac{t}{4}(e^t - e^{-t})$

7 $\frac{3}{20}e^t - \frac{1}{4}e^{-t} + \dfrac{(\cos t - 2\sin t)e^{-2t}}{10}$

9 $\left(\dfrac{4}{25} + \dfrac{2t}{5} + \dfrac{2t^2}{5} + \dfrac{t^3}{6}\right)\dfrac{e^{-2t}}{5^4} + \left(-\dfrac{4}{25} + \dfrac{2t}{5} - \dfrac{2t^2}{5} + \dfrac{t^3}{6}\right)\dfrac{e^{3t}}{5^4}$

11 $y = \frac{1}{3}e^{2t} - \frac{1}{2}e^t + \frac{1}{6}e^{-t} + \frac{1}{2}u(t-2) + \frac{1}{6}e^{2(t-2)}u(t-2)$
$\qquad\qquad\qquad\qquad\qquad - \frac{1}{2}e^{t-2}u(t-2) - \frac{1}{6}e^{-(t-2)}u(t-2)$

13 $y = \dfrac{t^2 + 2t + 1}{4}e^{-t} - \dfrac{\cos t + \sin t}{4}$

15 $x = \dfrac{-28 + 20t - 3t^2}{6} + \dfrac{14 + 4t}{3}e^{-t}$

$\quad\ \ y = \dfrac{-6 + 8t - 3t^2}{6} + \dfrac{6 + 2t}{3}e^{-t}$

19 Yes

Sec. 7.6
p. 307

3 $\dfrac{\coth(\pi s/3)}{1 + s^2}$

5 $\dfrac{1 - (1 + as)e^{-as}}{s^2(1 - e^{-2as})}$

7 $2[\phi_5(\tau,2,2) - \phi_5(t,2,2)] - [\phi_5(\tau,1,2) - \phi_5(t,1,2)]$

9 $(-1)^n\phi_6(\tau,1,1) + \phi_6(t,1,1) - [(-1)^n\phi_6(\tau,2,1) + \phi_6(t,2,1)]$
$\qquad\qquad\qquad\qquad\qquad - [(-1)^n\phi_{12}(\tau,2,1) + \phi_{12}(t,2,1)]$

11 $y = -e^{-t} + 2e^{-3t} + (-1)^n\phi_6(\tau,1,2) + \phi_6(t,1,2)$
$\qquad\qquad\qquad\qquad\qquad - (-1)^n\phi_6(\tau,3,2) - \phi_6(t,3,2)$

13 $y = \phi_2(t,\pi) - \phi_1(t,\pi)\cos t$

15 Because the corresponding ϕ's are undefined for these sets of values, which, in turn, happens because in the finite geometric series which are summed in the derivation of the inverses the common ratio is either 1 or -1.

19 (b) $\dfrac{2!\,(e^s - 1)}{s(e - c)^3}$

21 (a) $\dfrac{e^a \sin b}{e^{2a} - 2 \cos b\, e^{a+s} + e^{2s}} \quad \dfrac{e^s - 1}{s}$

 (b) $\dfrac{e^s - e^a \cos b}{e^{2a} - 2 \cos b\, e^{a+s} + e^{2s}} \quad \dfrac{e^s - 1}{s}$

Sec. 7.7
p. 319

1 $\dfrac{\sin 2t}{16} - \dfrac{t \cos 2t}{8}$ **3** $\delta(t) - 2e^{-2t}$

5 $\frac{1}{6}(3t \cos 3t + \sin 3t)e^{-2t}$

7 $Y = \displaystyle\int_0^t (t - \lambda)e^{-a(t-\lambda)}f(\lambda)\,d\lambda = \int_0^t \lambda e^{-a\lambda}f(t - \lambda)\,d\lambda$

9 (b) $\dfrac{2}{\sqrt{\pi}} \displaystyle\int_0^{\sqrt{t}} e^{-\tau^2}\,d\tau$

13 (a) $\left(\dfrac{1 - e^{-as}}{as}\right)^2 ; 1$

 (b) $\dfrac{3}{4}\,\dfrac{(2as - 2) + (2as + 2)e^{-2as}}{(as)^3} ; 1$

15 $A(t) = \frac{1}{2} - e^{-t} + \frac{1}{2}e^{-2t}; \; h(t) = e^{-t} - e^{-2t}$

17 $A(t) = \dfrac{3 \times 10^{-3}}{11}\,(2e^{-2,000t} - e^{-1,000t/6})$

 $h(t) = \dfrac{1}{2 \times 10^3}\,[I(t) - \frac{12}{11}e^{-2,000t} + \frac{1}{22}e^{-1,000t/6}]$

Note: The term $I(t)$ arises because there is no inductance in the circuit and hence the current builds up *instantaneously* to a nonzero value.

 $i_2(t) = \displaystyle\int_0^t h(\lambda)E(t - \lambda)\,d\lambda$

23 (a) $\begin{cases} 0 & t < a \\ f(t - a) & 0 \le a \le t \end{cases}$

 (b) $\begin{cases} 0 & t < a \\ \displaystyle\int_a^t f(t - \lambda)\,d\lambda & 0 \le a \le t \end{cases}$

 (c) $\dfrac{m!\,n!}{(m + n + 1)!}\,t^{m+n+1}$

25 It depends on what kinds of functions are acceptable as solutions. Since

$$f * x = g \rightarrow \mathscr{L}\{x\} = \dfrac{\mathscr{L}\{g\}}{\mathscr{L}\{f\}}$$

it is possible that $\mathscr{L}\{x\}$ will not be the transform of any "respectable" function. This is the case, for example, if $f = g$.

Chapter 8

Sec. 8.2
p. 331

7 $$\frac{\partial^2[EI(x)\,\partial^2 y/\partial x^2]}{\partial x^2} = -\frac{p}{g}A(x)\frac{\partial^2 y}{\partial t^2} - f(x,y,\dot{y},t) + \frac{p}{g}\frac{\partial}{\partial x}\left[I(x)\frac{\partial^3 y}{\partial^2 t\,\partial x}\right]$$

9 (a) Assume a particular integral of the form $\Phi(x,y) = Ae^{mx}e^{ny}$
(b) Assume a particular integral of the form $\Phi(x,y) = A\sin(mx + ny)$
(c) Assume a particular integral of the form $\Phi(x,y) = A\cos(mx + ny)$
(d) Assume a particular integral of any one of the forms
$$\Phi(x,y) = x^2(Ax^2 + Bxy + Cy^2)$$
$$\Phi(x,y) = xy(Ax^2 + Bxy + Cy^2)$$
$$\Phi(x,y) = y^2(Ax^2 + Bxy + Cy^2)$$
(e) Assume $\Phi(x,y)$ equal to a homogeneous polynomial of degree k in x and y multiplied by either x^2, xy, or y^2.

11 (b) It is not.

Sec. 8.3
p. 339

1 $y(x,t) = \frac{1}{2}(1 - |x - at|)[u(x - at + 1) - u(x - at - 1)]$
$$+ \frac{1}{2}(1 - |x + at|)[u(x + at + 1) - u(x + at - 1)]$$

$$v(0,t) = \begin{cases} -a & 0 \le at < 1 \\ 0 & 1 < at \end{cases}$$

3 $y(x,t) = \frac{1}{2}\cos(x - at)\left[u\left(x - at + \frac{\pi}{2}\right) - u\left(x - at - \frac{\pi}{2}\right)\right]$
$$+ \frac{1}{2}\cos(x + at)\left[u\left(x + at + \frac{\pi}{2}\right) - u\left(x + at - \frac{\pi}{2}\right)\right]$$

$$v(0,t) = \begin{cases} -a\sin at & 0 \le at < \pi/2 \\ 0 & \pi/2 < at \end{cases}$$

5 $y(x,t) = \dfrac{1}{2a}[(x + at + 1)u(x + at + 1) - (x - at + 1)u(x - at + 1)$
$$- (x + at - 1)u(x + at - 1) + (x - at - 1)u(x - at - 1)]$$

7 $y(x,t) = \frac{1}{2}[(x - at)e^{-(x-at)}u(x - at) + (x - at)e^{x-at}u(-x + at)$
$$+ (x + at)e^{-(x+at)}u(x + at) + (x + at)e^{x+at}u(-x - at)]$$

11 $\dot{y}(x,0) = \dfrac{2ax}{(1 + x^2)^2}$

17 (a) Parabolic at points on the parabola $x^2 - y - 1 = 0$; elliptic at points in the interior of this parabola; hyperbolic at points outside this parabola.
(b) Parabolic at points on the hyperbola $x^2 - xy - y = 0$; elliptic at points in the interior of this hyperbola; hyperbolic at points outside this hyperbola.
(c) Parabolic at points on the circle $(x - 1)^2 + y^2 = 1$; elliptic at points in the interior of this circle; hyperbolic at points outside this circle.

19 (a) $u = f(y - x) + g(y - 2x)$
(b) $u = xf(y - 2x) + g(y - 2x)$
(c) $u = f(y - \{2 + i\}x) + g(y - \{2 - i\}x)$
(d) $u = f(x) + xg(y/x)$
(e) $u = f(xy) + g(x)$
(f) $u = f(y - x^2) + g(y + x^2)$

Sec. 8.4
p. 349

1 The restrictions are $\int_0^l f(x)\,dx = 0$ and $\int_0^l g(x)\,dx = 0$. Physically speaking, the first restriction implies that the integrated initial angular displacement is zero, which will always be the case if the origin of θ is suitably chosen. Since the shaft is of uniform cross section, the second condition implies that

$$\int_0^l I\dot{\theta}(x,0)\,dx = 0$$

which is precisely the statement that the total angular momentum of the shaft is initially (and hence permanently) zero. In other words, the restriction on $g(x)$ implies that the vibration being studied is not superposed on a uniform rotation.

3 (a) Yes   (b) Yes   (c) Yes   (d) Yes   (e) No   (f) Yes

5 $\theta(x,t) = \dfrac{8l^3}{\pi^3} \displaystyle\sum_{m=1}^{\infty} \dfrac{1}{(2m-1)^3} \sin \dfrac{(2m-1)\pi x}{l} \cos \dfrac{(2m-1)\pi a t}{l}$

7 Doubling the tension multiplies the frequency by $\sqrt{2}$. Because it is easier to change the length quickly and accurately than it is to change the tension.

13 $y(x,t) = \sin \dfrac{\pi x}{l} \cos \dfrac{\pi a t}{l} + \dfrac{4l}{\pi^2} \displaystyle\sum_{n=1}^{\infty} \dfrac{1}{n^2} \sin \dfrac{n\pi}{2} \sin \dfrac{n\pi}{4} \sin \dfrac{n\pi a t}{l}$

15 $u\left(\dfrac{l}{2}, t\right) = \dfrac{400}{\pi}\left(\dfrac{z}{1} - \dfrac{z^9}{3} + \dfrac{z^{25}}{5} - \dfrac{z^{49}}{7} + \cdots\right)$ where $z = \exp(\pi^2 t/a^2 l^2)$

This seems to assert that thermal disturbances propagate with infinite velocity.

17 (a) $u(x,t) = \displaystyle\sum_{n=0}^{\infty} A_n \cos \dfrac{(2n+1)\pi x}{2l} \exp\left[-(2n+1)^2\pi^2 t/4a^2 l^2\right]$

where $A_n = (-1)^n \dfrac{400}{(2n+1)\pi}$

(b) $u(x,t) = 100 - \displaystyle\sum_{n=0}^{\infty} A_n \cos \dfrac{(2n+1)\pi x}{2l} \exp\left[-(2n+1)^2\pi^2 t/4a^2 l^2\right]$

where $A_n = (-1)^n \dfrac{400}{(2n+1)\pi}$

19 (a) The amount of heat in the rod when $t = \alpha a^2 l^2$ is

$$\dfrac{800 c\rho A l}{g\pi^2} \sum_{n=1}^{\infty} \dfrac{1}{n^2} \exp(-n^2\pi^2\alpha) \qquad n \text{ odd}$$

where A is the cross-sectional area of the rod.

(b) Heat loss through left end between $t = 0$ and $t = \alpha a^2 l^2$ is

$$100\alpha a^2 lk A + \dfrac{200 k A a^2 l}{\pi^2} \sum_{n=1}^{\infty} \dfrac{(-1)^n[1 - \exp(-n^2\pi^2\alpha)]}{n^2}$$

Heat gain through right end between $t = 0$ and $t = \alpha a^2 l^2$ is

$$100\alpha a^2 lk A + \dfrac{200 k A a^2 l}{\pi^2} \sum_{n=1}^{\infty} \dfrac{1 - \exp(-n^2\pi^2\alpha)}{n^2}$$

Heat stored in rod at $t = \alpha a^2 l^2$ is

$$\dfrac{50\rho c A l}{g} + \dfrac{400\rho c A l}{g\pi^2} \sum_{n=1}^{\infty} \dfrac{(-1)^n \exp(-n^2\pi^2\alpha)}{n^2}$$

where, in each expression, n is odd.

Sec. 8.5
p. 367

5 $\cot z = \alpha z$, where $z = \lambda l$ and $\alpha = 1/hl$

7 $\tan z = 2\alpha z/(z^2 - \alpha^2)$, where $z = \lambda l$ and $\alpha = hl$; or, equivalently, $\tan (z/2) = -\beta z$, where $\beta = 1/\alpha$

9 $u(x,t) = 100 - \dfrac{100x}{1 + hl} + \displaystyle\sum_{n=1}^{\infty} B_n \sin \lambda_n x \exp(-\lambda_n t/a^2)$

where $B_n = \dfrac{2}{l(1 + \alpha \cos^2 z_n)} \displaystyle\int_0^l \phi(x) \sin \lambda_n x \, dx$

and the values of λ_n are determined by the same equation as in Example 1.

11 $\cot z = \alpha z$, where $z = \lambda l/a$ and $\alpha = E_s J/kl$

17 (a) $y_n = \sin(n\pi \ln x)$ (b) $y_n = \cos(n\pi \ln x)$
(c) $y_n = \cos[(2n+1)\pi \ln x]$
(d) $y_n = \sin(\lambda_n \ln x)$, where λ_n is the nth one of the roots of the equation $\lambda \cos \lambda = e \sin \lambda$.

19 (a) $\sin z = 0$; $X_n = \sin\left(z_n \dfrac{x}{l}\right)$

(b) $\cos z \cosh z = 1$

$$X_n = (\sin z_n - \sinh z_n)\left[\cos\left(z_n \frac{x}{l}\right) - \cosh\left(z_n \frac{x}{l}\right)\right]$$
$$- (\cos z_n - \cosh z_n)\left[\sin\left(z_n \frac{x}{l}\right) - \sinh\left(z_n \frac{x}{l}\right)\right]$$

(c) $\cos z \cosh z = 1$

$$X_n = (\sin z_n - \sinh z_n)\left[\cos\left(z_n \frac{x}{l}\right) + \cosh\left(z_n \frac{x}{l}\right)\right]$$
$$- (\cos z_n - \cosh z_n)\left[\sin\left(z_n \frac{x}{l}\right) + \sinh\left(z_n \frac{x}{l}\right)\right]$$

(d) $\tan z = \tanh z$

$$X_n = (\sin z_n - \sinh z_n)\left[\cos\left(z_n \frac{x}{l}\right) - \cosh\left(z_n \frac{x}{l}\right)\right]$$
$$- (\cos z_n - \cosh z_n)\left[\sin\left(z_n \frac{x}{l}\right) - \sinh\left(z_n \frac{x}{l}\right)\right]$$

(e) $\tan z = \tanh z$

$$X_n = (\sin z_n + \sinh z_n)\left[\cos\left(z_n \frac{x}{l}\right) + \cosh\left(z_n \frac{x}{l}\right)\right]$$
$$- (\cos z_n + \cosh z_n)\left[\sin\left(z_n \frac{x}{l}\right) + \sinh\left(z_n \frac{x}{l}\right)\right]$$

21 $\tan^2 z = (r_1/r_2)^4$, where $z = \lambda l/a$

23 $\cot z = \alpha z$, where $z = \lambda l/a$ and $\alpha = I_p/I_s$, I_s being the polar moment of inertia of the shaft

25 $\sin z[2 \cos z \cosh z + rz(\cos z \sinh z - \sin z \cosh z)] = 0$ where $r = M/M_b$, M being the concentrated mass and M_b being the mass of the beam.

Sec. 8.6
p. 377

1 $y(x,t) = \displaystyle\sum_{n=1}^{\infty} C_n \sin\frac{n\pi x}{l}\left[-\frac{\omega l}{n\pi a(\omega_n^2 - \omega^2)}\sin\frac{n\pi a t}{l} + \frac{1}{\omega_n^2 - \omega^2}\sin \omega t\right]$

where $C_n = \dfrac{4}{n\pi}\sin\dfrac{n\pi}{2}\sin\dfrac{n\pi}{4}$

3 $y(x,t) = \displaystyle\sum_{n=1}^{\infty} C_n \sin\frac{n\pi x}{l}\left[-\frac{\omega l}{n\pi a(\omega_n^2 - \omega^2)}\sin\frac{n\pi a t}{l} + \frac{1}{\omega_n^2 - \omega^2}\sin \omega t\right]$

where $C_n = \dfrac{4}{n^2\pi^2}\sin\dfrac{n\pi}{2}$

5 If $\omega = n\pi a/l$, the boundary conditions on Φ cannot be imposed.

7 $y(x,t) = \dfrac{4}{p}\displaystyle\sum_{n=1}^{\infty}\frac{1}{\omega_1^2 - \Omega_n^2}\sin\frac{\pi x}{l}\left(\frac{\sin \Omega_n t}{\Omega_n} - \frac{\sin \omega_1 t}{\omega_1}\right)$, $\Omega_n = \dfrac{(2n-1)\pi}{p}$

9 Yes. If the frequency of the impressed force is, say, $\omega_m = m\pi a/l$, then, for the term $C_m \sin(m\pi x/l)$ in $\phi(x)\sin(m\pi a t/l)$, assume a term of the form

$$D_m \sin\frac{m\pi x}{l}\left(t\cos\frac{m\pi a t}{l}\right)$$

in the series of particular integrals used in the second method.

11 (a) $y(x,t) = \sum_{n=1}^{\infty} C_n \sin \frac{(2n-1)n\pi}{l} \left[-\frac{2}{2n-1} \sin \frac{(2n-1)\pi at}{l} \right.$

$$\left. + \sin \frac{2\pi at}{l} \right]$$

where $\qquad C_n = \dfrac{(-1)^{n-1}l^3}{a^2\pi^4[(2n-1)^4 - 4(2n-1)^2]}$

(b) $y(x,t) = \dfrac{l^3}{2a^2\pi^4} \sin \dfrac{\pi x}{l} \left(3 \sin \dfrac{\pi at}{l} - \sin \dfrac{3\pi at}{l} \right)$

$$+ \frac{2l^2}{27\pi a^3} \sin \frac{3\pi x}{l} \left(t \cos \frac{3\pi at}{l} - \frac{l}{3a\pi} \sin \frac{3\pi at}{l} \right)$$

$$- \sum_{n=3}^{\infty} C_n \sin \frac{(2n-1)\pi x}{l} \left[\frac{3}{2n-1} \sin \frac{(2n-1)\pi at}{l} - \sin \frac{3\pi at}{l} \right]$$

where $\qquad C_n = \dfrac{(-1)^{n-1}4l^3}{a^2\pi^4[(2n-1)^4 - 9(2n-1)^2]}$

19 $u(x,t) = \sin \left(\omega t - a \sqrt{\dfrac{\omega}{2}} x \right) \exp \left(-\dfrac{a\sqrt{\omega} x}{\sqrt{2}} \right)$

21 $u(x,t) = \sum_{n=1}^{\infty} C_n e^{-n^2\pi^2 t/a^2 l} \sin \dfrac{n\pi x}{l} + \Phi(x)$

where $\qquad \Phi(x) = \dfrac{x}{kl} \int_0^l \int_0^s \phi(r) \, dr \, ds - \dfrac{1}{k} \int_0^x \int_0^s \phi(r) \, dr \, ds$

and $\qquad C_n = \dfrac{2}{l} \int_0^l [g(x) - \Phi(x)] \sin \dfrac{n\pi x}{l} \, dx$

25 $u(x,y,t) = \sum_{m=1}^{\infty} \sum_{n=1}^{\infty} E_{mn} \sin \dfrac{(2m-1)\pi x}{2} \sin \dfrac{(2n-1)\pi y}{2}$

$$\times \, e^{-[(2m-1)^2+(2n-1)^2]\pi^2 t/4a^2}$$

where $\quad E_{mn} = 4 \int_0^1 \int_0^1 f(x,y) \sin \dfrac{(2m-1)\pi x}{2} \sin \dfrac{(2n-1)\pi y}{2} \, dx \, dy$

27 $u(x,y) = \sum_{n=1}^{\infty} A_n \sinh n\pi(1-y) \sin n\pi x$

where $\qquad A_n = \dfrac{2}{\sinh n\pi} \int_0^1 f(x) \sin n\pi x \, dx$

29 The problem can be solved by superimposing the solutions to the problems
defined by the following sets of boundary conditions:

$u(x,0) = f_1(x) \qquad u(x,1) = u(0,y) = u(1,y) = 0$

$u(x,1) = f_2(x) \qquad u(x,0) = u(0,y) = u(1,y) = 0$

$u(0,y) = g_1(y) \qquad u(1,y) = u(x,0) = u(x,1) = 0$

$u(1,y) = g_2(y) \qquad u(0,y) = u(x,0) = u(x,1) = 0$

31 (a) $u(x,y,z) = \sum_{n=1}^{\infty} \sum_{m=1}^{\infty} E_{mn} \cosh [\sqrt{m^2+n^2}\,\pi(1-z)] \sin m\pi x \sin n\pi y$

where

$$E_{mn} = \frac{4}{\cosh [\sqrt{m^2+n^2}\,\pi]} \int_0^1 \int_0^1 f(x,y) \sin m\pi x \sin n\pi y \, dx \, dy$$

(b) $u(x,y,z) = \sum_{n=0}^{\infty} \sum_{m=0}^{\infty} E_{mn} \sinh [\sqrt{m^2+n^2}\,\pi(1-z)] \cos m\pi x \cos n\pi y$

where

$$E_{mn} = \frac{k_{mn}}{\sinh [\sqrt{m^2+n^2}\,\pi]} \int_0^1 \int_0^1 f(x,y) \cos m\pi x \cos n\pi y \, dx \, dy$$

and $k_{mn} = 4$ if $m, n \neq 0$; $k_{mn} = 2$ if either m or n, but not both, is equal to zero; and $k_{mn} = 1$ if $m = n = 0$.

(c) $u(x,y,z) = \sum\limits_{n=1}^{\infty} \sum\limits_{m=0}^{\infty} E_{mn} \sinh [\sqrt{m^2 + n^2}\, \pi(1 - z)] \cos m\pi x \sin n\pi y$

where

$$E_{mn} = \frac{k_{mn}}{\sinh [\sqrt{m^2 + n^2}\, \pi]} \int_0^1 \int_0^1 f(x,y) \cos m\pi x \sin n\pi y \, dx \, dy$$

and
$$k_{mn} = \begin{cases} 4 & m \neq 0 \\ 2 & m = 0 \end{cases}$$

33 (a) $u(x,y) = 100$ (as should have been obvious)
    (b) $u(x,y) = 100x$ (as should have been obvious)

(c) $u(x,y) = \dfrac{400}{\pi} \sum\limits_{n=1}^{\infty} \dfrac{1}{2n - 1} \sin \dfrac{(2n - 1)\pi x}{2}\, e^{-(2n-1)\pi y/2}$

(d) $u(x,y) = 100x + \dfrac{200}{\pi} \sum\limits_{n=1}^{\infty} \dfrac{(-1)^n}{n} \sin n\pi x\, e^{-n\pi y}$

35 $u(x,y,t) = \sum\limits_{m=1}^{\infty} \sum\limits_{n=1}^{\infty} E_{mn} \sin m\pi x \sin n\pi y\, e^{-(m^2+n^2)t/a^2}$

$$+ \sum\limits_{n=1}^{\infty} B_n \sin n\pi x \sinh n\pi y$$

where
$$B_n = \frac{2}{\sinh n\pi} \int_0^1 f(x) \sin n\pi x \, dx$$

and

$$E_{mn} = -4 \int_0^1 \int_0^1 \left(\sum\limits_{n=1}^{\infty} B_n \sin n\pi x \sinh n\pi y \right) \sin m\pi x \sin n\pi y \, dx \, dy$$

37 $y(x,t) = X(x) \sin \omega t + \dfrac{a^2 x(l - x)}{\omega^2 EI} \sin \omega t$

where

$$X(x) = \frac{a^3}{\omega^3 EI} \left\{ - \frac{\cos [\sqrt{\omega/a}(x - l/2)]}{\cos (\sqrt{\omega/a}\, l/2)} + \frac{\cosh [\sqrt{\omega/a}(x - l/2)]}{\cosh (\sqrt{\omega/a}\, l/2)} \right\}$$

39 $y(x,t) = X(x) \sin \omega t$

where

$$X(x) = \left[(\sin z + \sinh z) \left(- \cos \frac{zx}{l} + \cosh \frac{zx}{l} \right) \right.$$
$$+ (\cos z + \cosh z) \left(\sin \frac{zx}{l} - \sinh \frac{zx}{l} \right) \left. \right]$$

$$\times \frac{A}{\sin z \cosh z - \cos z \sinh z}$$

where
$$z = \sqrt{\frac{\omega}{a}}\, l$$

41 $(\cosh z \cos z + 1) + \dfrac{kl^3}{EIz^3} (\sin z \cosh z - \cos z \sinh z) = 0$

where
$$z = \sqrt{\frac{\lambda}{a}}\, l$$

47 $X(x) = \dfrac{\lambda^2 \rho}{E_s g} \left[\int_0^x \dfrac{1}{J(s)} \int_s^l J(r)X(r)\, dr\, ds + \dfrac{IX(l)g}{\rho} \int_0^x \dfrac{ds}{J(s)} \right]$

49 $X(x) = \dfrac{\lambda^2 \rho}{Eg} \int_0^x \int_0^v \dfrac{1}{I(u)} \int_u^l \int_s^l A(r)X(r)\, dr\, ds\, du\, dv$

Sec. 8.7
p. 386

3 (a) $e(x,t) = \dfrac{ax \exp\left(-a^2 x^2/4t\right)}{2\sqrt{\pi}\, t^{3/2}}$

(b) $e(x,t) = \displaystyle\int_0^t \left(1 - \operatorname{erf} \dfrac{ax}{2\sqrt{\lambda}}\right) E'(t - \lambda)\, d\lambda + \left(1 - \operatorname{erf} \dfrac{ax}{2\sqrt{t}}\right) E(0)$

5 $y(x,t) = \dfrac{g}{w(n^2\pi^2 a^2/l^2 - \omega^2)}\left(\sin \omega t \sin \dfrac{n\pi x}{l} - \dfrac{\omega l}{n\pi a}\sin \dfrac{n\pi a t}{l}\sin \dfrac{n\pi x}{l}\right)$

The second term, whose frequency is different from that of the impressed force, is present only because friction has been neglected. Actually, it will die away rapidly in any realistic physical system. The response of the string to a disturbed force $f(x,t) = g(x)\sin \omega t$ or $F(x,t) = (\sin n\pi x/l)h(t)$, where $h(t)$ is periodic, can be found by first expanding $g(x)$ or $h(t)$, as the case may be in a Fourier series and then applying the result of the first part of this exercise to each term in the series.

7 $y(x,t) = \begin{cases} \dfrac{g}{2a^2}(x^2 - 2axt) & x < at \\[2mm] -\tfrac{1}{2}gt^2 & x \geq at \end{cases}$

9 $$\mathcal{L}\{\theta\} = \dfrac{aT_0 \sinh (sx/a)}{E_s Js \cosh (sl/a)}$$

When $x = l$, this becomes

$$\dfrac{aT_0}{E_s J}\dfrac{1}{s}\tanh \dfrac{sl}{a}$$

hence, in this case θ is the Morse dot function of period $4l/a$.

Chapter 9
Sec. 9.1
p. 393

1 (a) There are no singular points.
(b) There are no singular points.
(c) $x = 0$, regular   (d) $x = 0$, irregular
(e) $x = 1$, regular; $x = -1$, regular.
(f) $x = 0$, irregular; $x = 1$, regular.

3 $y_1 = 1 - \dfrac{x^2}{2!} + \dfrac{x^4}{4!} - \dfrac{x^6}{6!} + \cdots = \cos x$

$y_2 = x - \dfrac{x^3}{3!} + \dfrac{x}{5!} - \dfrac{x^7}{7!} + \cdots = \sin x$

5 $y_1 = 1 - \dfrac{x^3}{2\cdot 3} + \dfrac{x^6}{2\cdot 3\cdot 5\cdot 6} - \dfrac{x^7}{2\cdot 3\cdot 5\cdot 6\cdot 8\cdot 9} + \cdots$

$y_2 = x - \dfrac{x^4}{3\cdot 4} + \dfrac{x^7}{3\cdot 4\cdot 6\cdot 7} - \dfrac{x^{10}}{3\cdot 4\cdot 6\cdot 7\cdot 9\cdot 10} + \cdots$

7 $y_1 = \displaystyle\sum_{n=0}^{\infty} \dfrac{(-1)^n x^{2n-1}}{2^{3n}n!\,\Gamma(n + \frac{1}{4})}$

$y_2 = \displaystyle\sum_{n=0}^{\infty} \dfrac{(-1)^n x^{(4n+1)/2}}{2^{3n}n!\,\Gamma(n + \frac{7}{4})}$

9 $y_1 = \displaystyle\sum_{n=0}^{\infty} \dfrac{(-1)^n x^{n+2}}{n!\,(n + 3)!}$. A second solution cannot be found by the method of Example 2.

11 $x = a$ is an ordinary point if $P(x)$, $Q(x)$, and $R(x)$ are analytic at $x = a$. $x = a$ is a regular singular point if at least one of the functions $P(x)$, $Q(x)$, $R(x)$ is not analytic at $x = a$ but each of the functions $(x - a)P(x)$, $(x - a)^2 Q(x)$, $(x - a)^3 R(x)$ is analytic at $x = a$. $x = a$ is an irregular singular point if at least one of the functions $(x - a)P(x)$, $(x - a)^2 Q(x)$, $(x - a)^3 R(x)$ is not analytic at $x = a$.

13 $y_1 = \sum\limits_{n=0}^{\infty} \dfrac{(-1)^n x^{n+2}}{(n+2)(n+1)^2(n!)^3}$. There are no other power-series solutions

around the origin.

15 The point at infinity is a regular singular point.

17 The conditions $b_2 = a_1 b_1/2$ and $c_2 = b_1{}^2/4$ are sufficient to ensure that the given equation can be solved in terms of elementary functions.

19 If the roots of the indicial equation are r and $r - 1$, then two independent power-series solutions can be found if and only if $b_1(r-1) + c_1 = 0$. If the roots of the indicial equation are r and $r - 2$, then two independent power-series solutions can be found if and only if

$$\begin{vmatrix} -1 & b_1(r-2) + c_1 \\ b_1(r-1) + c_1 & b_2(r-2) + c_2 \end{vmatrix} = 0$$

Similar determinantal conditions of higher and higher order are involved if the roots of the indicial equation differ by an integer more than 2.

Sec. 9.2
p. 400

3 Yes, since J_n and J_{-n} are dependent and therefore have a vanishing wronskian.

11 $\dfrac{2}{\pi x\, Y_\nu(x) J_\nu(x)}$

Sec. 9.3
p. 407

5 $\dfrac{I_\nu(x)}{K_\nu(x)}$

Sec. 9.4
p. 410

3 $y(x) = \sqrt{x}\,[c_1 J_{1/3}(\tfrac{2}{3}\sqrt{x}) + c_2 J_{-1/3}(\tfrac{2}{3}\sqrt{x})]$

5 No

7 $y(x) = \sqrt{x}\,[c_1 J_0(2\sqrt{x}) + c_2 Y_0(2\sqrt{x})]$

9 $y(x) = \sqrt{x}\left[c_1 J_{1/(2+m)}\left(\dfrac{2}{2+m}\, x^{(2+m)/2} \right) \right.$

$$\left. + c_2 Y_{1/(2+m)}\left(\dfrac{2}{2+m}\, x^{(2+m)/2} \right) \right]$$

11 $y(x) = x[c_1 J_{1/3}(\tfrac{2}{3}x^3) + c_2 J_{-1/3}(\tfrac{2}{3}x^3)]$

13 $y(x) = \sqrt{x}\, e^{-x}[c_1 J_{3/4}(\tfrac{1}{2}x^2) + c_2 J_{-3/4}(\tfrac{1}{2}x^2)]$

15 $y(x) = c_1 I_0\left(2\dfrac{\sqrt{a+bx}}{|b|} \right) + c_2 K_0\left(2\dfrac{\sqrt{a+bx}}{|b|} \right)$

17 $y(x) = c_1 J_0\left(2\sqrt{\dfrac{a}{m^2}}\, e^{mx/2} \right) + c_2 Y_0\left(2\sqrt{\dfrac{a}{m^2}}\, e^{mx/2} \right)$. If $a < 0$, J_0, and Y_0

are to be replaced by I_0 and K_0, respectively.

21 $y(x) = c_1 J_0(2\sqrt{3x}) + c_2 Y_0(2\sqrt{3x}) + c_3 I_0(2\sqrt{3x}) + c_4 K_0(2\sqrt{3x})$

Sec. 9.5
p. 416

1 $J_5(x) = \left(\dfrac{384}{x^4} - \dfrac{72}{x^2} + 1 \right)' J_1(x) - \left(\dfrac{192}{x^3} - \dfrac{12}{x} \right) J_0(x)$

3 $-x J_3(2x) + 2x^2 J_2(2x)$

15 $\dfrac{2(1 - \cos x)}{x^2} - \dfrac{\sin x}{x}$ **21** $-\dfrac{2}{\pi x}\sin \nu\pi$

27 $x J_1(x)\cos x - J_0(x)(x\sin x + \cos x)$

31 (a) $\tfrac{1}{3}\{x^2[J_0(x)\cos x + J_1(x)\sin x] + x J_1(x)\cos x\}$

(b) $\tfrac{1}{3}\{x^2[J_0(x)\cos x + J_1(x)\sin x] - 2x J_1(x)\cos x\}$

37 (a) $a/(a^2 + \lambda^2)^{3/2}$ (b) $\lambda/(a^2 + \lambda^2)^{3/2}$

39 (a) $2\sqrt{x}\, J_1(\sqrt{x}) + c$ (b) $-4J_0(\sqrt{x}) - 2\sqrt{x}\, J_1(\sqrt{x}) + c$

41 $\dfrac{1}{\lambda}[x \ln x\, J_1(\lambda x)] + \dfrac{1}{\lambda^2} J_0(\lambda x) + c$

Sec. 9.6
p. 424

1 $\displaystyle\sum_{n=1}^{\infty} C_n J_1(\lambda_n x)$ where $C_n = -\dfrac{32(\lambda_n^2 - 1)}{\lambda_n^2(4\lambda_n^2 - 1)J_1(2\lambda_n)}$

3 $\displaystyle\sum_{n=1}^{\infty} \dfrac{2}{3\lambda_n J_1(3\lambda_n)} J_0(\lambda_n x)$

5 $2 + \displaystyle\sum_{n=1}^{\infty} \dfrac{4}{\lambda_n^2 J_0(2\lambda_n)} J_0(\lambda_n x)$

7 $\displaystyle\sum_{n=1}^{\infty} \dfrac{10}{\lambda_n J_3(5\lambda_n)} J_2(\lambda_n x)$

9 $J_\nu(\lambda)J_{-\nu}(2\lambda) - J_\nu(2\lambda)J_{-\nu}(\lambda) = 0$. If ν is an integer, $J_{-\nu}$ must be replaced by Y_ν.

11 $J_\nu(\lambda)J'_{-\nu}(2\lambda) - J'_\nu(2\lambda)J'_{-\nu}(\lambda) = 0$. If ν is an integer, $J_{-\nu}$ must be replaced by Y_ν.

13 $J_\nu(\lambda)J'_{-\nu}(2\lambda) - J'_\nu(2\lambda)J_{-\nu}(\lambda) = 0$. If ν is an integer, $J_{-\nu}$ must be replaced by Y_ν.

15 $\displaystyle\sum_{n=1}^{\infty} \dfrac{2}{\lambda_n[3J_1(3\lambda_n) + J_1(\lambda_n)]} J_0(\lambda_n x)$

17 $\displaystyle\sum_{n=1}^{\infty} C_n x J_1(\lambda_n x^2)$ where $C_n = \dfrac{2}{\lambda_n J_0^2(\lambda_n)}\left[-J_0(\lambda_n) + \int_0^1 J_0(\lambda_n t)\, dt\right]$

Sec. 9.7
p. 436

1 $\dfrac{s}{(s^2 + \lambda^2)^{3/2}}$ **3** $\dfrac{\lambda^2}{\sqrt{s^2 + \lambda^2}\,(s + \sqrt{s^2 + \lambda^2})^2}$

5 $\dfrac{1}{\lambda}$ **7** $\dfrac{1}{\sqrt{a^2 + \lambda^2}}$

9 $e^{-2t}J_0(3t)$ **11** $e^{-at}\displaystyle\int_0^t e^{ax}J_0(bx)\, dx$

15 $\dfrac{\lambda}{\sqrt{s^2 - \lambda^2}\,(s + \sqrt{s^2 - \lambda^2})}$

19 $\displaystyle\int_0^t J_0(2\lambda)J_0(t - \lambda)\, d\lambda$

29 (a) $-J_1(2\sqrt{t})/\sqrt{t}$ (b) $I_1(2\sqrt{t})/\sqrt{t}$

35 $\dfrac{\sqrt{\lambda} I_1(2\alpha\sqrt{\lambda a})}{I_1(2\alpha\sqrt{a})}$

37 (a) $u(x) = u_0 + (u_w - u_0)\dfrac{\cosh \alpha x}{\cosh \alpha a}$ where $\alpha^2 = \dfrac{2h}{kw}$

     (b) $u(x) = u_0 + (u_w - u_0)\dfrac{\alpha \cosh \alpha x + \sinh \alpha x}{\alpha \cosh \alpha a + \sinh \alpha a}$

39 $u(x) = u_0 + \dfrac{(u_c - u_0)[K_1(\alpha R)I_0(\alpha x) + I_1(\alpha R)K_0(\alpha x)]}{I_0(\alpha r)K_1(\alpha R) - I_1(\alpha R)K_0(\alpha r)}$

41 $\omega_1 = 2{,}430$ Hz; $\omega_2 = 5{,}590$ Hz

43 $\theta(x,t) = \sum\limits_{n=1}^{\infty} A_n J_0\left(\dfrac{\lambda_n x}{a}\right)\cos \lambda_n t$ where $a^2 = \dfrac{E_s g}{\rho}$

and $A_n = \dfrac{a^2}{4{,}000\lambda_n{}^2 J_1{}^2(20\lambda_n/a)} \displaystyle\int_0^{20} J_0\left(\dfrac{\lambda_n x}{a}\right) dx$

45 $x(t) = \dfrac{x_0\{Y_1(\lambda)J_0[\lambda(1+\alpha t)^{1/2}] - J_1(\lambda)Y_0[\lambda(1+\alpha t)^{1/2}]\}}{J_0(\lambda)Y_1(\lambda) - J_1(\lambda)Y_0(\lambda)}$

where $\lambda = 2k/\alpha\sqrt{m_0}$ and k^2 is the proportionality constant in the force law.

47 $x(t) = \dfrac{x_0\{K_1(\lambda)I_0[\lambda(1+\alpha t)^{1/2}] + I_1(\lambda)K_0[\lambda(1+\alpha t)^{1/2}]\}}{I_0(\lambda)K_1(\lambda) + I_1(\lambda)K_0(\lambda)}$

where $\lambda = 2k/\alpha\sqrt{m_0}$ and k^2 is the proportionality constant in the force law.

49 $J_{1/3}(\lambda)J_{-1/3}(\lambda s^{3/2}) - J_{1/3}(\lambda s^{3/2})J_{-1/3}(\lambda) = 0$

where $s = 1 + \alpha l$ and $\lambda = \dfrac{2\omega}{3\alpha}\sqrt{\dfrac{w_0}{Tg}}$

51 $J_1(2\lambda\sqrt{l})I_2(2\lambda\sqrt{l}) + I_1(2\lambda\sqrt{l})J_2(2\lambda\sqrt{l}) = 0$
where $\lambda^4 = 12\rho\omega^2/Egk^2$ and k is the proportionality constant in the expression for the depth of the beam.

53 $J_{5/2}(\lambda R)J_{-5/2}(\lambda r) - J_{5/2}(\lambda r)J_{-5/2}(\lambda R) = 0$ where $\lambda = \dfrac{\omega l}{a(R-r)}$

55 $y(x) = \tan\theta\left[\dfrac{\sqrt{x}\,I_1(2a\sqrt{x})}{aI_0(2a\sqrt{l})} - x\right]$ where $a^2 = \dfrac{12F\cos\theta}{Ebk^3}$

57 $J_0\left(\dfrac{\omega b}{a}\right) = 0$

59 $z(r,t) = \dfrac{gF_0}{w\omega^2}\dfrac{J_0(\omega r/a) - 1}{J_0(\omega b/a)}$

61 $e_n(t) = \dfrac{1}{(LC)^n}\dfrac{\Gamma[(2n+1)/2]}{\Gamma(2n)}\left(\dfrac{2t}{\beta}\right)^{(2n-1)/2} e^{-\alpha t}J_{(2n-1)/2}(\beta t)$

where $\alpha = \dfrac{R}{2L}$ and $\beta^2 = \dfrac{1}{LC} - \dfrac{R^2}{4L^2}$

63 $u(r,\theta,z) = \sum\limits_{n=1}^{\infty}\sum\limits_{m=1}^{\infty} A_{nm}J_n\left(\dfrac{m\pi r}{h}\right)\sin\dfrac{m\pi z}{h}\sin n\theta$

where $A_{nm} = \dfrac{2}{hJ_n(m\pi b/h)}\displaystyle\int_0^h F_n(z)\sin\dfrac{m\pi z}{h}\,dz$

and $F_n(z) = \dfrac{2}{\pi}\displaystyle\int_0^{\pi} g(\theta,z)\sin n\theta\,d\theta$

65 $u(r,z) = \sum\limits_{n=1}^{\infty} B_n I_0\left(\dfrac{n\pi r}{h}\right)\sin\dfrac{n\pi z}{h}$

where $B_n = \dfrac{2}{hI_0(n\pi b/h)}\displaystyle\int_0^h f(z)\sin\dfrac{n\pi z}{h}\,dz$

67 $u(r,z) = \dfrac{100(h-z)}{h}$

69 $u(r,z) = \sum\limits_{n=1}^{\infty} A_n J_0(\lambda_n r)\sinh\lambda_n(h-z)$

where $A_n = \dfrac{200\sigma}{\sinh\lambda_n h\,J_0(\lambda_n b)b(\lambda_n{}^2 + \sigma^2)}$

and σ is the proportionality constant in the cooling law for the surface.

71 $u(r,\theta,t) = \sum\limits_{n=1}^{\infty} \sum\limits_{m=1}^{\infty} A_{nm} J_{2n}(\lambda_{nm} r) \sin 2n\theta \, \exp\left(-\lambda^2_{nm} t/a^2\right)$

where λ_{nm} is the mth one of the roots of the equation $J_{2n}(\lambda b) = 0$ and

$$A_{nm} = \frac{800 \int_0^b r J_{2n}(\lambda_{nm} r) \, dr}{n\pi b^2 J^2_{2n+1}(\lambda_{nm} b)}$$

73 $u(r,\theta,t) = 100 \left(\dfrac{r}{b}\right)^n + \sum\limits_{n=1}^{\infty} \sum\limits_{m=1}^{\infty} A_{nm} J_n(\lambda_{nm} r) \sin n\theta \, \exp\left(-\lambda_{nm}^2 t/a^2\right)$

where n ranges over the odd positive integers, m ranges over all the positive integers, λ_{nm} is the mth one of the roots of the equation $J_n(\lambda b) = 0$, and

$$A_{nm} = \frac{800 \int_0^b [1 - (r/b)^n] r J_n(\lambda_{nm} r) \, dr}{\pi b^2 J^2_{n+1}(\lambda_{nm} b)}$$

75 $u(r,t) = 100 \dfrac{\ln r - \ln r_2}{\ln r_1 - \ln r_2}$

$$+ \sum\limits_{n=1}^{\infty} A_n [Y_0(\lambda_n r_2) J_0(\lambda_n r) - J_0(\lambda_n r_2) Y_0(\lambda_n r)] e^{-\lambda_n^2 t/a^2}$$

where λ_n is the nth one of the roots of the equation

$Y_0(\lambda r_1) J_0(\lambda r_2) - J_0(\lambda r_1) Y_0(\lambda r_2) = 0$

and

$$A_n = \frac{-100 \int_{r_1}^{r_2} \dfrac{\ln r - \ln r_2}{\ln r_1 - \ln r_2} [Y_0(\lambda_n r_2) J_0(\lambda_n r) - J_0(\lambda_n r_2) Y_0(\lambda_n r)] r \, dr}{(r_2{}^2/2)[Y_0(\lambda_n r_2) J_1(\lambda_n r_2) - J_0(\lambda_n r_2) Y_1(\lambda_n r_2)]^2 \\ -(r_1{}^2/2)[Y_0(\lambda_n r_2) J_1(\lambda_n r_1) - J_0(\lambda_n r_2) Y_1(\lambda_n r_1)]^2}$$

Sec. 9.8
p. 451

3 $x^2 = \dfrac{P_0(x) + 2P_2(x)}{3}$ $\qquad x^3 = \dfrac{3P_1(x) + 2P_3(x)}{5}$

$x^4 = \dfrac{7P_0(x) + 20P_2(x) + 8P_4(x)}{35}$

17 $u(r,\theta) = 50 + \sum\limits_{n=1}^{\infty} \dfrac{50}{b^{2n-1}} [P_{2n-2}(0) - P_{2n}(0)] r^{2n-1} P_{2n-1}(\cos\theta)$

19 $u(r,\theta) = \sum\limits_{n=1}^{\infty} \left(A_n r^n + \dfrac{B_n}{r^{n+1}}\right) P_n(\cos\theta)$, where A_n and B_n are determined by

the equations

$$A_n b_1{}^n + \frac{B_n}{b_1^{n+1}} = \frac{2n+1}{2} \int_0^{\pi} f_1(\theta) \sin\theta \, P_n(\cos\theta) \, d\theta$$

$$A_n b_2{}^n + \frac{B_n}{b_2^{n+1}} = \frac{2n+1}{2} \int_0^{\pi} f_2(\theta) \sin\theta \, P_n(\cos\theta) \, d\theta$$

21 $u(r,\theta) = \sum\limits_{n=1}^{\infty} A_{2n-1} r^{2n-1} P_{2n-1}(\cos\theta)$

where $\qquad A_{2n-1} = \dfrac{4n-1}{2b^{2n-1}} \int_0^{\pi/2} f(\theta) \sin\theta \, P_{2n-1}(\cos\theta) \, d\theta$

25 $L_n(x) = a_0 \left[1 - \dfrac{n}{(1!)^2} x + \dfrac{n(n-1)}{(2!)^2} x^2 - \dfrac{n(n-1)(n-2)}{(3!)^2} x^3 + \cdots \right]$

where, customarily, a_0 is taken equal to 1. Since Laguerre's equation can be written in the form $(xe^{-x}y')' + ne^{-x}y = 0$, it follows from Theorem 4, Sec. 8.5, that solutions corresponding to different values of n are orthogonal with respect to e^{-x} over the interval $(0,\infty)$ without the necessity of boundary conditions, since $r(x) \equiv xe^{-x}$ vanishes at both $x = 0$ and $x = \infty$.

Chapter 10

1 (a) 80   (b) 0   (c) 4

5 The values of the determinants in (a), (b), and (c) are, respectively, 3, $-13, -13$.

7 If n is odd, $|A| = 0$.

13 $D_n = 2^{n+1} - 1$

15 Yes. In this case the equation reduces to the equation of the line on which the three points lie.

3 (a) $\begin{bmatrix} 4 & 6 \\ -12 & 10 \end{bmatrix}$ (b) $\begin{bmatrix} 0 & 0 & 0 \\ 0 & 0 & 0 \\ 0 & 0 & 0 \end{bmatrix}$ (c) $\begin{bmatrix} 12 & 0 & -8 \\ 8 & 12 & 0 \\ 0 & 8 & 12 \end{bmatrix}$

5 $A^3 = \begin{bmatrix} -9 & 22 \\ -11 & 13 \end{bmatrix}$ $A^4 = \begin{bmatrix} -31 & 48 \\ -24 & 17 \end{bmatrix}$ $A^5 = \begin{bmatrix} -79 & 82 \\ -41 & 3 \end{bmatrix}$

7 (a) $X = \begin{bmatrix} -3 & -3 \\ \frac{5}{2} & 2 \end{bmatrix}$ (b) $X = \begin{bmatrix} \frac{4}{3} & 2 \\ -\frac{5}{3} & -2 \end{bmatrix}$

(c) $X = \begin{bmatrix} 0 & 0 \\ \frac{1}{2} & 1 \end{bmatrix}$ (d) $X = \begin{bmatrix} 1 & 2 \\ -1 & -2 \end{bmatrix}$

13 Yes   **17** No

19 No. The submatrices must also be transposed.

21 $(ABC)' = A'BC + AB'C + ABC'$. No, unless A and A' commute.

23 $\dfrac{dA}{dT} = D \begin{bmatrix} \cos \lambda_1 t & 0 & \cdots & 0 \\ 0 & \cos \lambda_2 t & \cdots & 0 \\ \cdots & \cdots & \cdots & \cdots \\ 0 & 0 & \cdots & \cos \lambda_n t \end{bmatrix}$ $\dfrac{d^2 A}{dt^2} = -D^2 A$

27 $P = \dfrac{1}{8} \begin{bmatrix} 4 & 3 & 1 \\ 3 & 4 & 1 \\ 3 & 3 & 2 \end{bmatrix}$ $P^2 = \dfrac{1}{8^2} \begin{bmatrix} 28 & 27 & 9 \\ 27 & 28 & 9 \\ 27 & 27 & 10 \end{bmatrix}$

$P^3 = \dfrac{1}{8^3} \begin{bmatrix} 220 & 219 & 73 \\ 219 & 220 & 73 \\ 219 & 219 & 74 \end{bmatrix}$ $P^4 = \dfrac{1}{8^4} \begin{bmatrix} 1,756 & 1,755 & 585 \\ 1,755 & 1,756 & 585 \\ 1,755 & 1,755 & 586 \end{bmatrix}$

1 (a) adj $A = \begin{bmatrix} 4 & -2 \\ -3 & 1 \end{bmatrix}$ $A^{-1} = -\dfrac{1}{2} \begin{bmatrix} 4 & -2 \\ -3 & 1 \end{bmatrix}$

(b) adj $A = \begin{bmatrix} 3 & 11 & 1 \\ -11 & -5 & 14 \\ 12 & -9 & 4 \end{bmatrix}$ $A^{-1} = \dfrac{1}{53} \begin{bmatrix} 3 & 11 & 1 \\ -11 & -5 & 14 \\ 12 & -9 & 4 \end{bmatrix}$

(c) adj $A = \begin{bmatrix} -4 & 1 & 1 \\ 8 & -2 & -2 \\ -4 & 1 & 1 \end{bmatrix}$ Since $|A| = 0$, A^{-1} does not exist.

(d) adj $A = \begin{bmatrix} -14 & 21 & 7 \\ 6 & -9 & -3 \\ 10 & -15 & -5 \end{bmatrix}$ Since $|A| = 0$, A^{-1} does not exist.

3 (a) $|A| = 1$ $A^{-1} = \begin{bmatrix} 3 & -1 \\ -2 & 1 \end{bmatrix}$

(b) $|A| = -2$ $A^{-1} = \begin{bmatrix} -\frac{5}{2} & \frac{3}{2} \\ 2 & -1 \end{bmatrix}$

(c) $|A| = -1$ $A^{-1} = \begin{bmatrix} -4 & 3 & -1 \\ 5 & -3 & 1 \\ -2 & 1 & 0 \end{bmatrix}$

5 $x_1 = \frac{5}{4}; x_2 = \frac{5}{4}; x_3 = \frac{1}{2}$

7   The inverse of a nonsingular triangular matrix is a triangular matrix whose diagonal elements are the reciprocals of the corresponding diagonal elements in the original matrix.

9   Yes   **13**   No

17 $K = \begin{bmatrix} -(k_1 + k_{12}) & k_{12} & 0 \\ k_{12} & -(k_{12} + k_{23} + k_2) & k_{23} \\ 0 & k_{23} & -(k_{23} + k_3) \end{bmatrix}$

19 $PE = \frac{1}{2}k_1x_1{}^2 + \frac{1}{2}k_{12}(x_1 - x_2)^2 + \frac{1}{2}k_{13}(x_1 - x_3)^2 + \frac{1}{2}k_{23}(x_2 - x_3)^2 + \frac{1}{2}k_3x_3{}^2$

Sec. 10.4
p. 497

1   No, for any minor of order $r > p$ can itself be expanded in terms of minors of order p, and if all of the latter vanish, so must all of the former.

9 $PN_1N_2N_3N_4N_5N_6 = I$, where

$$N_1 = \begin{bmatrix} 1 & -2 & 0 \\ 0 & 1 & 0 \\ 0 & 0 & 1 \end{bmatrix} \quad N_2 = \begin{bmatrix} 1 & 0 & 0 \\ 0 & 1 & -1 \\ 0 & 0 & 1 \end{bmatrix} \quad N_3 = \begin{bmatrix} 1 & 0 & 0 \\ 2 & 1 & 0 \\ 0 & 0 & 1 \end{bmatrix}$$

$$N_4 = \begin{bmatrix} 1 & 0 & 0 \\ 0 & 1 & 0 \\ -1 & 0 & 1 \end{bmatrix} \quad N_5 = \begin{bmatrix} 1 & 0 & 0 \\ 0 & 1 & 0 \\ 0 & -1 & 0 \end{bmatrix} \quad N_6 = \begin{bmatrix} 1 & 0 & 0 \\ 0 & -1 & 0 \\ 0 & 0 & 1 \end{bmatrix}$$

11   **(a)** Rank $= \begin{cases} 3 & \lambda \neq \frac{1}{2}, 1, 2 \\ 2 & \lambda = \frac{1}{2}, 1, 2 \end{cases}$ **(b)** Rank $= \begin{cases} 3 & \lambda \neq 1, 6 \\ 2 & \lambda = 1, 6 \end{cases}$

       (c) Rank $= \begin{cases} 3 & \lambda \neq 1, \frac{19}{2} \\ 2 & \lambda = \frac{19}{2} \\ 1 & \lambda = 1 \end{cases}$ **(d)** Rank $= \begin{cases} 3 & \lambda \neq 1 \\ 1 & \lambda = 1 \end{cases}$

13 $B = PAQ$ where $P = \begin{bmatrix} -2 & 1 \\ 1 & 0 \end{bmatrix}$ and $Q = \begin{bmatrix} 1 & 0 & -1 \\ 0 & 1 & 1 \\ 0 & 0 & 1 \end{bmatrix}$

15 $B = PAQ$ where $P = \begin{bmatrix} 1 & 0 & 0 \\ 0 & 1 & 0 \\ -2 & 0 & 1 \end{bmatrix}$ and $Q = \begin{bmatrix} 1 & 0 & 0 \\ 0 & 1 & 0 \\ 0 & -1 & 2 \end{bmatrix}$

Sec. 10.5
p. 513

3 $3X_1 + 4X_2 - 3X_3 + X_4 = 0$

5   **(a)** Linearly independent   **(b)** Linearly independent

9   No

11   **(a)** $X = c_1 \begin{bmatrix} 0 \\ 3 \\ -2 \\ 0 \\ 1 \end{bmatrix} + c_2 \begin{bmatrix} 5 \\ -6 \\ 2 \\ 1 \\ 0 \end{bmatrix} + \begin{bmatrix} 1 \\ 1 \\ 0 \\ 0 \\ 0 \end{bmatrix}$

     (b) $X = c \begin{bmatrix} 5 \\ -6 \\ 2 \\ 1 \\ 0 \end{bmatrix} + \begin{bmatrix} 1 \\ 1 \\ 0 \\ 0 \\ 0 \end{bmatrix}$

13   **(a)** $x = -\frac{12}{11}; y = \frac{35}{11}; z = -\frac{4}{11}$
     (b) $x = 1; y = 2; z = -1; w = 3$

15 (a) $\lambda = -1$, $X = \begin{bmatrix} 1 \\ -1 \end{bmatrix}$; $\lambda = 4$, $X = \begin{bmatrix} 2 \\ 3 \end{bmatrix}$

(b) $\lambda = \frac{4}{3}$, $X = \begin{bmatrix} 1 \\ 2 \end{bmatrix}$; $\lambda = 4$, $X = \begin{bmatrix} 3 \\ -2 \end{bmatrix}$

(c) $\lambda = -1$, $X = \begin{bmatrix} 1 \\ 0 \\ 1 \end{bmatrix}$; $\lambda = 1$, $X = \begin{bmatrix} 1 \\ 0 \\ -1 \end{bmatrix}$; $\lambda = 3$, $X = \begin{bmatrix} 11 \\ 16 \\ -1 \end{bmatrix}$

(d) $\lambda = 1$, $X = c_1 \begin{bmatrix} 1 \\ 0 \\ 2 \end{bmatrix} + c_2 \begin{bmatrix} 0 \\ 1 \\ 2 \end{bmatrix}$; $\lambda = \frac{19}{2}$, $X = \begin{bmatrix} 4 \\ 4 \\ -1 \end{bmatrix}$

17 No

23 (a) $U_1 = \frac{1}{5} \begin{bmatrix} 3 \\ 4 \end{bmatrix}$ $U_2 = \frac{1}{5} \begin{bmatrix} -4 \\ 3 \end{bmatrix}$

(b) $U_1 = \frac{1}{\sqrt{2}} \begin{bmatrix} 1 \\ 1 \\ 0 \end{bmatrix}$ $U_2 = \frac{1}{\sqrt{6}} \begin{bmatrix} 1 \\ -1 \\ 2 \end{bmatrix}$ $U_3 = \frac{1}{\sqrt{3}} \begin{bmatrix} -1 \\ 1 \\ 1 \end{bmatrix}$

31 (a) The set of points for which $|x| > 2$ and $|y| > 2$.
(b) The locus of points for which the determinant is equal to zero is the hyperbola $y = (x + 1)/(3x - 1)$.

Sec. 10.6
p. 522

1 $\begin{bmatrix} x \\ y \end{bmatrix} = c_1 \begin{bmatrix} 2 \\ -1 \end{bmatrix} e^t + c_2 \begin{bmatrix} 1 \\ 1 \end{bmatrix} e^{-2t} + \frac{1}{2} \begin{bmatrix} 2 \\ 1 \end{bmatrix} e^{-t}$

3 $\begin{bmatrix} x \\ y \end{bmatrix} = c_1 \begin{bmatrix} 4 \\ -3 \end{bmatrix} e^t + c_2 \begin{bmatrix} 5 \\ -3 \end{bmatrix} e^{-2t} + \begin{bmatrix} 1 \\ 0 \end{bmatrix} e^t$

5 $\begin{bmatrix} x \\ y \end{bmatrix} = c_1 e^{-2t} \left(\begin{bmatrix} 0 \\ 1 \end{bmatrix} \cos t - \begin{bmatrix} 1 \\ -1 \end{bmatrix} \sin t \right)$

$+ c_2 e^{-2t} \left(\begin{bmatrix} 1 \\ -1 \end{bmatrix} \cos t + \begin{bmatrix} 0 \\ 1 \end{bmatrix} \sin t \right) + \begin{bmatrix} 1 \\ -1 \end{bmatrix} e^t$

7 $\begin{bmatrix} x \\ y \end{bmatrix} = c_1 e^{-t} \left(\begin{bmatrix} 1 \\ 1 \end{bmatrix} \cos 2t - \begin{bmatrix} 2 \\ -4 \end{bmatrix} \sin 2t \right)$

$+ c_2 e^{-t} \left(\begin{bmatrix} 2 \\ -4 \end{bmatrix} \cos 2t + \begin{bmatrix} 1 \\ 1 \end{bmatrix} \sin 2t \right) + \frac{1}{6} \begin{bmatrix} -1 \\ 5 \end{bmatrix} e^{-t}$

13 $D^r(t^3 e^{mt}) = m^r t^3 e^{mt} + 3rm^{r-1} t^2 e^{mt} + 3r(r-1)te^{mt} + r(r-1)(r-2)e^{mt}$
$p(D)(t^3 e^{mt}) = p(m)t^3 e^{mt} + 3p'(m)t^2 e^{mt} + 3p''(m)te^{mt} + p'''(m)e^{mt}$
$P(D)(t^3 e^{mt}) = P(m)t^3 e^{mt} + 3P'(m)t^2 e^{mt} + 3P''(m)te^{mt} + P'''(m)e^{mt}$

Chapter 11
Sec. 11.1
p. 531

1 (a) Positive-definite   (b) Negative-definite   (c) Indefinite
(d) Positive-semidefinite   (e) Indefinite

3 $y_1 = -\frac{1}{2}x_1 + 2x_2 + 3x_3 + 2x_4$
$y_2 = \frac{3}{2}x_1 \qquad\quad + x_3 + 2x_4$
$y_3 = x_1$
$y_4 = \qquad\qquad\qquad x_3 \qquad$ and $Q = y_1^2 + y_2^2 - \frac{3}{2}y_3^2 + y_4^2$
or, equally well,
$z_1 = \frac{1}{2}x_1 + x_2 + 2x_3 + 2x_4$
$z_2 = -\frac{1}{2}x_1 + 4x_2 + 4x_3$
$z_3 = x_2 + x_3$
$z_4 = \qquad\qquad\qquad x_3 \qquad$ and $Q = 2z_1^2 + \frac{1}{2}z_2^2 - 6z_3^2 + z_4^2$
In each case, after transformation, Q contains three terms with positive coefficients and one term whose coefficient is negative. (This is an example

of **Sylvester's law of inertia**: if a quadratic form is reduced to a sum of squares by each of two nonsingular linear transformations, the number of terms with positive coefficients will be the same in each case.)

5 (a) Positive-semidefinite   (b) Indefinite
    (c) Positive-semidefinite   (d) Indefinite
    (e) Indefinite   (f) Indefinite

13 $PE = \frac{1}{2}k_1x_1{}^2 + \frac{1}{2}k_{12}(x_2 - x_1)^2 + \frac{1}{2}k_{13}(x_3 - x_1)^2$
$$+ \frac{1}{2}k_{23}(x_3 - x_2)^2 + \frac{1}{2}k_3x_3{}^2$$

The matrix of this quadratic form is just $-\frac{1}{2}$ the matrix K.

15 $U_1 = \begin{bmatrix} 1 \\ 0 \\ 0 \end{bmatrix}$ $U_2 = \frac{1}{\sqrt{2}}\begin{bmatrix} 0 \\ 1 \\ 0 \end{bmatrix}$ $U_3 = \frac{1}{\sqrt{3}}\begin{bmatrix} 0 \\ 0 \\ 1 \end{bmatrix}$

Sec. 11.2
p. 546   **1**   (a) $\lambda_1 = 1$, $X_1 = \begin{bmatrix} 4 \\ 1 \\ -3 \end{bmatrix}$; $\lambda_2 = 2$, $X_2 = \begin{bmatrix} 3 \\ 1 \\ -2 \end{bmatrix}$

The characteristic vectors are not orthogonal.

(b) $\lambda_1 = 1$, $X_1 = \begin{bmatrix} 1 \\ 3 \\ 0 \end{bmatrix}$; $\lambda_2 = 2$, $X_2 = \begin{bmatrix} 2 \\ 1 \\ 2 \end{bmatrix}$

The characteristic vectors are not orthogonal.

(c) $\lambda_1 = -2$, $X_1 = \begin{bmatrix} 1 \\ -1 \\ 0 \end{bmatrix}$; $\lambda_2 = 9$, $X_2 = \begin{bmatrix} 2 \\ 2 \\ -1 \end{bmatrix}$

$$\lambda_3 = -18,\ X_3 = \begin{bmatrix} 1 \\ 1 \\ 4 \end{bmatrix}$$

The characteristic vectors are orthogonal.

(d) $\lambda_1 = 0$, $X_1 = \begin{bmatrix} 1 \\ 1 \\ 1 \end{bmatrix}$; $\lambda_2 = 1$, $X_2 = \begin{bmatrix} 1 \\ -1 \\ 2 \end{bmatrix}$; $\lambda_3 = 2$, $X_3 = \begin{bmatrix} 2 \\ 1 \\ 2 \end{bmatrix}$

The characteristic vectors are not orthogonal.

(e) $\lambda_1 = -1$, $X_1 = \begin{bmatrix} 6 \\ 2 \\ -7 \end{bmatrix}$; $\lambda_2 = 1$, $X_2 = \begin{bmatrix} 0 \\ 1 \\ -1 \end{bmatrix}$;

$$\lambda_3 = 4,\ X_3 = \begin{bmatrix} 3 \\ 1 \\ -1 \end{bmatrix}$$

The characteristic vectors are not orthogonal.

(f) $\lambda_1 = 1$, $(X_1)_1 = \begin{bmatrix} 1 \\ 1 \\ 0 \end{bmatrix}$, $(X_1)_2 = \begin{bmatrix} 1 \\ 0 \\ 1 \end{bmatrix}$; $\lambda_2 = 6$, $X_2 = \begin{bmatrix} 1 \\ 1 \\ 1 \end{bmatrix}$

X_2 is orthogonal to $(X_1)_1$ and $(X_1)_2$ since they correspond to different characteristic values and the given matrix is symmetric. $(X_1)_1$ and $(X_1)_2$ are not orthogonal since they arise from the same characteristic value. They can be replaced by equivalent, orthogonal characteristic vectors, however, by means of the Schmidt process.

3 (a) $\lambda_1 = 1$, $(X_1)_1 = \dfrac{1}{2\sqrt{2}}\begin{bmatrix}1\\0\\1\end{bmatrix}$, $(X_1)_2 = \dfrac{1}{6\sqrt{2}}\begin{bmatrix}-1\\4\\3\end{bmatrix}$;

$$\lambda_2 = 2,\ X_2 = \frac{1}{6}\begin{bmatrix}1\\2\\-3\end{bmatrix}$$

(b) $\lambda_1 = 1$, $(X_1)_1 = \dfrac{1}{3}\begin{bmatrix}1\\1\\0\end{bmatrix}$, $(X_1)_2 = \dfrac{1}{6}\begin{bmatrix}1\\-2\\3\end{bmatrix}$; $\lambda_2 = 2,\ X_2 = \dfrac{1}{6}\begin{bmatrix}-1\\2\\3\end{bmatrix}$

5 Unless B is positive-definite, the n (orthogonal) characteristic vectors cannot be normalized with respect to B.

9 If $A = \begin{bmatrix}1 & 2\\2 & -1\end{bmatrix}$ and $B = \begin{bmatrix}1 & 0\\0 & -1\end{bmatrix}$, the roots of the equation $A - \lambda B = 0$ are $1 \pm 2i$.

13 No. For example, $A = \begin{bmatrix}1 & 1\\-1 & -1\end{bmatrix}$ is a nonnull *unsymmetrical* matrix whose characteristic values are all zero.

19 $\lambda_1 = \frac{7}{4},\quad X_1 = \begin{bmatrix}1\\-4\\2\end{bmatrix}$

21 (a) $\lambda_2 = 1.56$, $X_2 = \begin{bmatrix}4.96\\-2.60\\1.00\end{bmatrix}$; $\lambda_3 = 0.033$, $X_3 = \begin{bmatrix}1.27\\1.72\\1.00\end{bmatrix}$

(b) $\lambda_2 = 1$, $X_2 = \begin{bmatrix}-1\\1\\1\end{bmatrix}$; $\lambda_3 = \frac{1}{4}$, $X_3 = \begin{bmatrix}2\\4\\1\end{bmatrix}$

23 $[1\quad r_1\quad r_1^2\quad r_1^3\quad \cdots\quad r_1^{n-1}]$

27 If and only if $\lambda_1 = \lambda_2$.

29 When $\lambda = \lambda_i$, a solution of the nonhomogeneous equation will exist when and only when $X_i^T V = 0$.

Sec. 11.3
p. 560

1 (a) $P = \begin{bmatrix}1 & 0\\-3 & 1\end{bmatrix}$ $Q = \begin{bmatrix}1 & -2\\0 & 1\end{bmatrix}$

(b) $P = \begin{bmatrix}1 & 0\\3 & 1\end{bmatrix}$ $Q = \begin{bmatrix}0 & 1\\1 & 1\end{bmatrix}$

(c) $P = \begin{bmatrix}3 & 0 & 0\\-6 & 3 & 0\\2 & -1 & 3\end{bmatrix}$ $Q = \begin{bmatrix}1 & 1 & -1\\0 & 1 & 0\\0 & 0 & 1\end{bmatrix}$

(d) $P = \begin{bmatrix}1 & 0 & 0\\-1 & 1 & 0\\-2 & 3 & 1\end{bmatrix}$ $Q = \begin{bmatrix}1 & 0 & -3\\0 & 1 & -2\\0 & 0 & 1\end{bmatrix}$

3 (a) $\begin{bmatrix}x_1\\x_2\end{bmatrix} = \begin{bmatrix}\dfrac{1}{\sqrt{3}} & \dfrac{2}{\sqrt{6}}\\[2mm] \dfrac{1}{\sqrt{3}} & -\dfrac{1}{\sqrt{6}}\end{bmatrix}\begin{bmatrix}y_1\\y_2\end{bmatrix}$

(b) $\begin{bmatrix}x_1\\x_2\end{bmatrix} = \begin{bmatrix}\dfrac{1}{\sqrt{6}} & \dfrac{1}{\sqrt{3}}\\[2mm] -\dfrac{2}{\sqrt{6}} & \dfrac{1}{\sqrt{3}}\end{bmatrix}\begin{bmatrix}y_1\\y_2\end{bmatrix}$

(c) $\begin{bmatrix} x_1 \\ x_2 \end{bmatrix} = \begin{bmatrix} \dfrac{1}{2} & \dfrac{3}{\sqrt{12}} \\ -\dfrac{1}{2} & \dfrac{1}{\sqrt{12}} \end{bmatrix} \begin{bmatrix} y_1 \\ y_2 \end{bmatrix}$

(d) $\begin{bmatrix} x_1 \\ x_2 \end{bmatrix} = \begin{bmatrix} 1 & 1 \\ -1 & 0 \end{bmatrix} \begin{bmatrix} y_1 \\ y_2 \end{bmatrix}$

(e) $\begin{bmatrix} x_1 \\ x_2 \\ x_3 \end{bmatrix} = \begin{bmatrix} \dfrac{1}{2\sqrt{6}} & \dfrac{1}{4} & \dfrac{1}{4\sqrt{3}} \\ \dfrac{1}{\sqrt{6}} & 0 & -\dfrac{1}{\sqrt{3}} \\ \dfrac{1}{\sqrt{6}} & -\dfrac{1}{2} & \dfrac{1}{2\sqrt{3}} \end{bmatrix} \begin{bmatrix} y_1 \\ y_2 \\ y_3 \end{bmatrix}$

(f) $\begin{bmatrix} x_1 \\ x_2 \\ x_3 \end{bmatrix} = \begin{bmatrix} \frac{1}{3} & -\frac{1}{3} & \frac{1}{6} \\ \frac{2}{3} & \frac{1}{3} & -\frac{2}{3} \\ \frac{1}{6} & \frac{1}{3} & \frac{1}{3} \end{bmatrix} \begin{bmatrix} y_1 \\ y_2 \\ y_3 \end{bmatrix}$

(g) $\begin{bmatrix} x_1 \\ x_2 \\ x_3 \end{bmatrix} = \dfrac{1}{2} \begin{bmatrix} 0 & 1 & 1 \\ 1 & 0 & -1 \\ 1 & -1 & 0 \end{bmatrix} \begin{bmatrix} y_1 \\ y_2 \\ y_3 \end{bmatrix}$

(h) $\begin{bmatrix} x_1 \\ x_2 \\ x_3 \end{bmatrix} = \begin{bmatrix} \dfrac{1}{2\sqrt{5}} & \dfrac{4}{9\sqrt{5}} & \dfrac{1}{3} \\ -\dfrac{1}{2\sqrt{5}} & \dfrac{1}{9\sqrt{5}} & \dfrac{1}{3} \\ 0 & -\dfrac{5}{9\sqrt{5}} & \dfrac{1}{3} \end{bmatrix} \begin{bmatrix} y_1 \\ y_2 \\ y_3 \end{bmatrix}$

5 (a) $S = \begin{bmatrix} 1 & 2 \\ 2 & 5 \end{bmatrix} \qquad S^{-1} = \begin{bmatrix} 5 & -2 \\ -2 & 1 \end{bmatrix}$

(b) $S = \begin{bmatrix} 1 & 1 \\ 1 & -1 \end{bmatrix} \qquad S^{-1} = \dfrac{1}{2}\begin{bmatrix} 1 & 1 \\ 1 & -1 \end{bmatrix}$

(c) $S = \begin{bmatrix} 1 & 1 \\ -2 & 1 \end{bmatrix} \qquad S^{-1} = \dfrac{1}{3}\begin{bmatrix} 1 & -1 \\ 2 & 1 \end{bmatrix}$

(d) $S = \begin{bmatrix} 1 & 1 & 1 \\ 1 & 1 & -1 \\ 1 & -1 & 1 \end{bmatrix} \qquad S^{-1} = \dfrac{1}{2}\begin{bmatrix} 0 & 1 & 1 \\ 1 & 0 & -1 \\ 1 & -1 & 0 \end{bmatrix}$

(e) $S = \begin{bmatrix} 1 & 0 & 3 \\ 0 & 1 & -1 \\ 0 & 1 & 1 \end{bmatrix} \qquad S^{-1} = \dfrac{1}{2}\begin{bmatrix} 2 & 3 & -3 \\ 0 & 1 & 1 \\ 0 & -1 & 1 \end{bmatrix}$

(f) $S = \begin{bmatrix} 1 & 0 & 1 \\ 1 & 1 & -1 \\ 0 & -1 & 1 \end{bmatrix} \qquad S^{-1} = \begin{bmatrix} 0 & 1 & 1 \\ 1 & -1 & -2 \\ 1 & -1 & -1 \end{bmatrix}$

7 $\quad X = \dfrac{5}{3}\begin{bmatrix} 1 \\ 1 \end{bmatrix}\cos t - \dfrac{1}{3}\begin{bmatrix} 2 \\ -1 \end{bmatrix}\cos 2t + \dfrac{1}{2}\begin{bmatrix} 2 \\ -1 \end{bmatrix}\sin 2t$

9 $\quad \Theta = \dfrac{16}{45}\begin{bmatrix} 3 \\ 3 \\ 2 \end{bmatrix}\cos\dfrac{t}{2} - \dfrac{5}{45}\begin{bmatrix} 3 \\ 0 \\ -10 \end{bmatrix}\cos t + \dfrac{4}{45}\begin{bmatrix} 3 \\ -12 \\ 2 \end{bmatrix}\cos 2t$

$\qquad\qquad + \dfrac{208}{180}\begin{bmatrix} 3 \\ 3 \\ 2 \end{bmatrix}\sin\dfrac{t}{2} + \dfrac{5}{180}\begin{bmatrix} 3 \\ 0 \\ -10 \end{bmatrix}\sin t + \dfrac{11}{360}\begin{bmatrix} 3 \\ -12 \\ 2 \end{bmatrix}\sin 2t$

11 $\omega_1{}^2 = \frac{7}{4}$, $A_1 = \begin{bmatrix} 1 \\ 4 \\ 4 \end{bmatrix}$; $\omega_2{}^2 = 4$, $(A_2)_1 = \begin{bmatrix} 1 \\ -1 \\ 0 \end{bmatrix}$, $(A_2)_2 = \begin{bmatrix} 1 \\ 0 \\ -1 \end{bmatrix}$

13 $X = \begin{bmatrix} 1 \\ 1 \\ 1 \end{bmatrix} (2 \cos t + \frac{1}{3} \sin t) - \begin{bmatrix} 1 \\ 0 \\ -1 \end{bmatrix} \cos 2t + \frac{1}{6} \begin{bmatrix} 2 \\ -1 \\ -1 \end{bmatrix} \sin 2t$

15 $X = \begin{bmatrix} 1 \\ -2 \\ 1 \end{bmatrix} \cos t + \begin{bmatrix} 8 \\ -3 \\ -5 \end{bmatrix} \sin t$

23 $X = \begin{bmatrix} 1 \\ -4 \\ -2 \end{bmatrix} \cos 2t$

Sec. 11.4
p. 573

5 Yes

9 For all values of a and for all nonzero values of b, the given equations are satisfied, respectively, by the following matrices:

(a) $X = \begin{bmatrix} a & b \\ \dfrac{-a^2 + 2a + 3}{b} & 2 - a \end{bmatrix}$

(b) $X = \begin{bmatrix} a & b \\ \dfrac{-4a^2 + 4a - 3}{b} & 4 - a \end{bmatrix}$

(c) $X = \begin{bmatrix} a & b \\ \dfrac{-a^2 + 4a + 5}{b} & 4 - a \end{bmatrix}$

(d) $X = \dfrac{1}{b} \begin{bmatrix} b & ab & 0 \\ 0 & 2b & 0 \\ -2 & -2a & 3b \end{bmatrix}$

15 (a) $X_1 = \begin{bmatrix} 0 & 2 \\ -1 & 3 \end{bmatrix}$, $X_2 = \begin{bmatrix} -1 & 4 \\ -2 & 5 \end{bmatrix}$, $X_3 = \begin{bmatrix} 6 & -4 \\ 2 & 0 \end{bmatrix}$, $X_4 = \begin{bmatrix} 5 & -2 \\ 1 & 2 \end{bmatrix}$

(b) $X_1 = \begin{bmatrix} -4 & 3 \\ -2 & 1 \end{bmatrix}$, $X_2 = \begin{bmatrix} 4 & -9 \\ 6 & -11 \end{bmatrix}$, $X_3 = \begin{bmatrix} -10 & 9 \\ -6 & 5 \end{bmatrix}$

$X_4 = \begin{bmatrix} -2 & -3 \\ 2 & -7 \end{bmatrix}$

(c) $X_1 = \begin{bmatrix} 0 & 2 \\ -1 & 3 \end{bmatrix}$ (d) $X_1 = \begin{bmatrix} 2 & -1 & -1 \\ -3 & 4 & 5 \\ 3 & -3 & -4 \end{bmatrix}$

17 (a) $\begin{bmatrix} 60 & 19 \\ -57 & -16 \end{bmatrix}$ (b) $\begin{bmatrix} -3 & -6 \\ 9 & 12 \end{bmatrix}$ (c) $\begin{bmatrix} 0 & -6 \\ 3 & 9 \end{bmatrix}$

(d) $\begin{bmatrix} -12 & -27 & -9 \\ 3 & 12 & 3 \\ 9 & 9 & 6 \end{bmatrix}$ (e) $\begin{bmatrix} 6 & 35 & 35 \\ 3 & 76 & 73 \\ -3 & -35 & -32 \end{bmatrix}$

Sec. 11.5
p. 582

3 (a) $A^2 - 3A = 0$ (b) $A^2 - I = 0$
(c) $A^2 - 3A + 2I = 0$ (d) $A^2 - 4A + 3I = 0$
5 (a) $\frac{1}{35}(-A^2 + 8A)$ (b) $\frac{1}{30}(-A^2 + 7A)$
(c) $\frac{1}{10}(-A^2 + 3A)$ (d) $\frac{1}{90}(A^3 - 6A^2 + 23A)$
(e) $\frac{1}{210}(A^3 - 10A^2 + 51A)$

Sec. 11.6
p. 589

3 $e^{-A} = \begin{bmatrix} 2e^{-1} - e^{-2} & 2e^{-1} - 2e^{-2} \\ -e^{-1} + e^{-2} & -e^{-1} + 2e^{-2} \end{bmatrix}$

7 (a) $\sin A = A \sin 1$, $\cos A = (\cos 1 - 1)A + I$, $\sin 2A = A \sin 2$
(b) $\sin A = A \sin 1$, $\cos A = I \cos 1$, $\sin 2A = A \sin 2$

(c) $\sin A = \dfrac{A \sin 3}{3}$ $\cos A = I - \dfrac{A}{3}(1 - \cos 3)$ $\sin 2A = \dfrac{A \sin 6}{3}$

(d) $\sin A = A \sin 1$, $\cos A = I \cos 1$, $\sin 2A = A \sin 2$
9 $\sin A = A \sin 1$, $\cos A = A(\cos 1 - 1) + I$
$\sin B = B \sin 1$, $\cos B = I \cos 1$

$\sin (A + B) = \dfrac{A + B}{3}(\sin 1 + \sin 2) - \dfrac{I}{3}(2 \sin 1 - \sin 2)$

$\cos (A + B) = \dfrac{A + B}{3}(-\cos 1 + \cos 2) + \dfrac{I}{3}(2 \cos 1 + \cos 2)$

$\sin (A - B) = (A - B) \sin 1$
$\cos (A - B) = (A - B)(\cos 1 - 1) + I$

11 $\cos A = A^2 \dfrac{\cos 3 - 1}{9} + I$

13 $e^{At} = \begin{bmatrix} -5e^{-2t} + 6e^{-3t} & 3e^{-2t} - 3e^{-3t} \\ -10e^{-2t} + 10e^{-3t} & 6e^{-2t} - 5e^{-3t} \end{bmatrix}$ $\begin{aligned} x_1 &= -8e^{-2t} + 9e^{-3t} \\ x_2 &= -16e^{-2t} + 15e^{-3t} \end{aligned}$

15 $x_1 = \cosh t + 2 \sinh t$ $x_2 = \cosh 2t - \dfrac{\sinh 2t}{2}$

19 $x_1 = -\frac{3}{8} \sinh t + \cosh t + \frac{1}{8} \sinh 3t$
$x_2 = \frac{3}{8} \sinh t - \cosh t + \frac{5}{24} \sinh 3t$
$x_1 = -\frac{3}{8} \sin t + \cos t + \frac{1}{8} \sin 3t$
$x_2 = \frac{3}{8} \sin t - \cos t + \frac{5}{24} \sin 3t$

Chapter 12

Sec. 12.2
p. 595

1 Minimum at $(0, -3)$ **3** Maximum at $(3,2,3)$
5 Minimum at $(1,0)$; critical point which is neither a maximum nor minimum at $(-1,0)$
7 Minimum at $(0,0)$; critical point which is neither a maximum nor minimum at $(1,-1)$
9 Minimum at $(0,0,0)$; critical point which is neither a maximum nor minimum at $(-2,0,0)$

Sec. 12.3
p. 599

3 $(2,-2,1)$, $(-2,2,1)$ **5** $r = h = \sqrt{\dfrac{S}{3\pi}}$

7 $h = \left(1 - \dfrac{2\sqrt{5}}{15}\right) r$ $s = \dfrac{2\sqrt{5}}{5} r$, where s is the altitude of the cone

11 $A_0 = \frac{17}{35}$, $A_1 = A_{-1} = \frac{12}{35}$, $A_2 = A_{-2} = -\frac{3}{35}$

Sec. 12.4
p. 607

1 (a) The exact values are $\lambda_1 = \frac{1}{2}$, $\lambda_2 = 1$, $\lambda_3 = 2$.
(b) The exact values are $\lambda_1 = \frac{1}{6}$, $\lambda_2 = \frac{2}{3}$, $\lambda_3 = 1$.
(c) The exact values are the roots of the equation
$\lambda^3 - 10\lambda^2 + 17\lambda - 3 = 0$, approximately $\lambda_1 = 0.200$, $\lambda_2 = 1.90$, $\lambda_3 = 7.90$.
3 In each case the frequencies are the roots of the equation $\omega^4 - 3\omega^2 + 1 = 0$, or, approximately, $\omega_1{}^2 = 0.382$, $\omega_2{}^2 = 2.618$.

Sec. 12.5
p. 612

3 $2y' - y = 0$; $y'' - y = 0$; $y = \dfrac{\sinh (1 - x) + 2 \sinh x}{\sinh 1}$;

$y = \dfrac{\sinh (1 - x) - 2 \cosh (1 - x)}{\sinh 1 - 2 \cosh 1}$; $y = \dfrac{4 \cosh x + 2 \sinh x}{2 \cosh 1 + \sinh 1}$; $y = 0$

5 $y = c \cosh^{-1}\left(\dfrac{1 + x}{c}\right) + d$ **7** $(x + 1)^2 + (y - \tfrac{17}{6})^2 = \tfrac{325}{36}$

Sec. 12.6
p. 616

1 $\Delta F = (4nx^n + x^{n+1})\varepsilon + n^2 x^{2n+1}\varepsilon^2$;
$\delta F = (4nx^n + x^{n+1})\varepsilon$

Sec. 12.7
p. 620

1 $(x - c)^2 + (y - d)^2 = \lambda^2$, where $\lambda^2 = d^2 + a^2$, $c = (x_1 + x_2)/2$,
$\sqrt{d^2 + a^2} \cot^{-1} (-d/a) = L/2$, and $a = (x_2 - x_1)/2$.

3 $y = c \cosh \dfrac{x}{c} - \lambda$ where $c = \dfrac{a}{\sinh^{-1} (L/2)}$ and $\lambda = \dfrac{c}{2} \sqrt{4 + L^2} - b$

5 The path is the circular arc of the prescribed length which joins the given points and is concave upward.

7 The particle moves along a meridian circle according to the laws $x = a \sin (\pi t/T)$, $y = b \sin (\pi t/T)$, $z = \cos (\pi t/T)$, where $a^2 + b^2 = 1$. The minimum value of the integrated kinetic energy is $m\pi^2/2T$.

Sec. 12.8
p. 625

3 $\delta \displaystyle\int_a^b [P(y')^2 - Ry^2 + \lambda Qy^2]\, dx = 0$

where $P = \exp\left(\displaystyle\int \dfrac{q}{p}\, dx\right)$ $R = \dfrac{r}{p} P$ $Q = \dfrac{P}{p}$

The boundary conditions are $Py'\, \delta y|_a^b = 0$.

5 $\delta \displaystyle\int_a^b [p(y'')^2 - q(y')^2 + ry^2 + \lambda y^2]\, dx = 0$, with boundary conditions $[(py'')' + qy']\, \delta y|_a^b = 0$ and $py''\, \delta y'|_a^b = 0$.

7 $y = \dfrac{25x(x - 2)}{14}$

9 $\lambda = 0.503$; exact value is $\lambda = \tfrac{1}{2}$.

Sec. 12.9
p. 629

3 $y = \dfrac{a}{1 - \omega^2 m/k}\left[\sin \omega t - \omega \sqrt{\dfrac{m}{k}} \sin \left(\sqrt{\dfrac{k}{m}}\, t\right)\right]$

7 $PE = mgl(1 - \cos \theta) + kx^2$ $KE = \dfrac{m}{2} (\dot{x}^2 + 2\dot{x}\dot{\theta}l \cos \theta + l^2\dot{\theta}^2)$

$\ddot{x} \cos \theta + l\ddot{\theta} + g \sin \theta = 0$ $m\ddot{x} + m\ddot{\theta}l \cos \theta - m(\dot{\theta})^2 l \sin \theta + 2kx = 0$

There is a single natural frequency, $\omega^2 = 2kgl/(mg + 2kl)$.

9 $mr - mr(\dot{\theta})^2 + f'(r) = 0$, $\dfrac{d}{dt} (mr^2\dot{\theta}) = 0$

Chapter 13
Sec. 13.1
p. 642

1 (a) 3, 9, 13; -18; $10i + 18j + 16k$; $-\tfrac{8}{9}$; $\cos^{-1} (-\tfrac{2}{3})$; -90;
     $245i + 160j - 170k$; $45i + 216j - 108k$; -180; 0

(b) 7, 15, 11; 80; $72i + 24j - 12k$; $-\tfrac{64}{15}$; $\cos^{-1} \tfrac{16}{21}$; -636;
     $-210i + 710j + 425k$; $78i - 202j + 329k$; $-1,272$; 0

(c) 15, 15, 9; -40; $-75i + 60j + 30k$; $\tfrac{62}{15}$; $\cos^{-1} (-\tfrac{8}{45})$; 915;
     $610i + 100j - 1,420k$; $-170i - 800j - 1,570k$; 1,830; 0

7 Not necessarily; not necessarily; not necessarily; yes

9 $(17\mathbf{i} - 13\mathbf{j} + 8\mathbf{k})/\sqrt{522}$

11 No. In fact, if $\mathbf{A} = \mathbf{0}$ and $\mathbf{B} = \mathbf{C} \neq \mathbf{0}$, then $\mathbf{A} \times \mathbf{B} = \mathbf{B} \times \mathbf{C} = \mathbf{C} \times \mathbf{A} = \mathbf{0}$
   whereas $\mathbf{A} + \mathbf{B} + \mathbf{C} = 2\mathbf{C} \neq \mathbf{0}$.

23 $\mathbf{i} + 2\mathbf{j} + 3\mathbf{k} = (-17\mathbf{A} + 14\mathbf{B} + 3\mathbf{C})/33 = (-11\mathbf{U} + 319\mathbf{V} + 143\mathbf{W})/11$

25 No, because $\mathbf{F} \times \mathbf{R}$ is opposite to the direction in which $\mathbf{F}$ would cause a
   right-hand screw to advance.

Sec. 13.2
p. 648

3 $\mathbf{V} \cdot \dfrac{d\mathbf{V}}{dt} \times \dfrac{d^3\mathbf{V}}{dt^3}$ (b) $\dfrac{d\mathbf{V}}{dt} \times \left(\dfrac{d\mathbf{V}}{dt} \times \dfrac{d^2\mathbf{V}}{dt^2} \right) + \mathbf{V} \times \left(\dfrac{d\mathbf{V}}{dt} \times \dfrac{d^3\mathbf{V}}{dt^3} \right)$

7 Yes; the tangents at $t = 0$ and $t = 1$ are parallel. Yes; for all values of t
   the tangents at t and $-t$ are parallel.

9 $\mathbf{R} = \left(\dfrac{t^2}{2} + \dfrac{t^3}{6} \right) \mathbf{i} + \left(\dfrac{t^5}{20} + 1 \right) \mathbf{j} + \left(\dfrac{t^3}{3} - \dfrac{t^4}{12} + 2 \right) \mathbf{k}$

17 $\begin{vmatrix} x - t^4 & y - t^2 & z - t^3 \\ 4t^3 & 2t & 3t^2 \\ 12t^2 & 2 & 6t \end{vmatrix} = 0$

Sec. 13.3
p. 658

1 (a) $-xy, -x(2x + z)\mathbf{i} + yz\mathbf{j} + 2(2xz - y)\mathbf{k}$
   (b) $yz + 3x^2 + (2xz - y^2), -2yz\mathbf{i} + (xy - z^2)\mathbf{j} + x(6y - z)\mathbf{k}$

3 $\frac{14}{3}$

5 $-1/\sqrt{22}$

11 If $\mathbf{v} = y\mathbf{i} + z\mathbf{j} + x\mathbf{k}$, then $\nabla \times \mathbf{v} \cdot \mathbf{v} = -y - z - x \neq 0$.

13 $\nabla \phi = \dfrac{\partial \phi}{\partial u} \nabla u + \dfrac{\partial \phi}{\partial v} \nabla v + \dfrac{\partial \phi}{\partial w} \nabla w$

15 A   17 $n = -1, 0$ 19 $n = -3$

Sec. 13.4
p. 670

1 Along the parabola, $2 \ln \frac{3}{2}$; along AB, $\frac{3}{4} \ln 3$; along APB, $\ln \frac{5}{2}$;
   along AQB, $\ln 2$

3 (a) 0   (b) $-\frac{8}{15}$ 5 $\dfrac{\pi}{4}$ 7 $n = \frac{1}{2}$ 9 3   11 $\dfrac{\pi a^3}{24}$

13 The common value of the integrals is $\frac{2}{3}$.

15 $\dfrac{k}{n + 1} [(x_1^2 + y_1^2)^{(n+1)/2} - (x_0^2 + y_0^2)^{(n+1)/2}]$

17 Yes

19 $I_p = \displaystyle\int_C -x^2 y \, dx + xy^2 \, dy$

Sec. 13.5
p. 684

1 (a) $\frac{4}{3}$ (b) $\frac{11}{6}$ (c) $\frac{11}{6}$ (d) $\frac{11}{6}$

3 (a) 1   (b) 6   (c) $\dfrac{\pi}{2}$ (d) $\dfrac{\pi}{3}$

5 $\displaystyle\iiint_V \left[u \left(\dfrac{\partial^2 v}{\partial x^2} + \dfrac{\partial^2 v}{\partial y^2} + \dfrac{\partial^2 v}{\partial z^2} \right) - v \left(\dfrac{\partial^2 u}{\partial x^2} + \dfrac{\partial^2 u}{\partial y^2} + \dfrac{\partial^2 u}{\partial z^2} \right) \right] dV$
   $= \displaystyle\iint_S \left(u \dfrac{dv}{dn} - v \dfrac{du}{dn} \right) dS$

7
$$\int_C F_1 \, dx + F_2 \, dy + F_3 \, dz = \iint_S \left[\left(\frac{\partial F_3}{\partial y} - \frac{\partial F_2}{\partial z} \right) \cos \alpha \right.$$
$$\left. + \left(\frac{\partial F_1}{\partial z} - \frac{\partial F_3}{\partial x} \right) \cos \beta + \left(\frac{\partial F_2}{\partial x} - \frac{\partial F_1}{\partial y} \right) \cos \gamma \right] dS$$

9 The length of the curve. No, because the vector function **T** is defined *only* on the curve *C*.

19 The common value of the integrals is $5\pi a^4/4$.

21 The common value of the integrals is $\frac{3}{2}$.

23 -2π

29 If 0 is a point at which the surface has a tangent plane, the integral in Gauss' theorem is equal to 2π. If 0 is a singular point on S, the integral may have any value between 0 and 4π.

Sec. 13.6
p. 694

1 $-\dfrac{kr^3}{3}, \ -k \ln r$ **3** $\dfrac{2M}{a^2} (\sqrt{a^2 + z^2} - z)$

9 $\mathbf{E} = \mathbf{i} \displaystyle\sum_{n=1}^{\infty} \left(\frac{4}{n\pi} \cos \frac{n\pi a t}{l} + \frac{4}{n^2 \pi^2 a} \sin \frac{n\pi a t}{l} \right) \sin \frac{n\pi y}{l}$

$\qquad + \mathbf{k} \displaystyle\sum_{n=1}^{\infty} \left(\frac{4}{n\pi} \cos \frac{n\pi a t}{l} - \frac{4l}{n^2 \pi^2 a} \sin \frac{n\pi a t}{l} \right) \sin \frac{n\pi y}{l} \qquad a^2 = \frac{1}{\mu \varepsilon}$

the summations extending only over the odd positive integers

Chapter 14
Sec. 14.2
p. 705

3 (a) $\bar{\mathbf{e}}_1 = \frac{1}{3}(\mathbf{i} - 2\mathbf{j} + \mathbf{k}), \ \bar{\mathbf{e}}_2 = \frac{1}{3}(-\mathbf{i} - \mathbf{j} + 2\mathbf{k}), \ \bar{\mathbf{e}}_3 = \frac{1}{3}(2\mathbf{i} + 5\mathbf{j} - 4\mathbf{k})$
$\bar{\mathbf{e}}^1 = 2\mathbf{i} + \mathbf{k}, \ \bar{\mathbf{e}}^2 = \mathbf{i} + 2\mathbf{j} + 3\mathbf{k}, \ \bar{\mathbf{e}}^3 = \mathbf{i} + \mathbf{j} + \mathbf{k}$
(b) $\bar{\mathbf{e}}_1 = 2\mathbf{i} - \mathbf{j} - \mathbf{k}, \ \bar{\mathbf{e}}_2 = 3\mathbf{i} - 2\mathbf{j} - \mathbf{k}, \ \bar{\mathbf{e}}_3 = -7\mathbf{i} + 5\mathbf{j} + 3\mathbf{k}$
$\bar{\mathbf{e}}^1 = \mathbf{i} + 2\mathbf{j} - \mathbf{k}, \ \bar{\mathbf{e}}^2 = 2\mathbf{i} + \mathbf{j} + 3\mathbf{k}. \ \bar{\mathbf{e}}^3 = \mathbf{i} + \mathbf{j} + \mathbf{k}$

5 (b) $\qquad \dfrac{\bar{U}^T \bar{G}^{-1} \bar{V}}{\sqrt{\bar{U}^T \bar{G}^{-1} \bar{U}} \sqrt{\bar{V}^T \bar{G}^{-1} \bar{V}}}$

Sec. 14.3
p. 718

1 (a) $f(x_1) \, \Delta_1 x + f(x_2) \, \Delta x_2 + f(x_3) \, \Delta x_3$

(b) $\quad a_{11}x_1 x_1 + a_{12}x_1 x_2 + a_{13}x_1 x_3$
$\quad + a_{21}x_2 x_1 + a_{22}x_2 x_2 + a_{23}x_2 x_3$
$\quad + a_{31}x_3 x_1 + a_{32}x_3 x_2 + a_{33}x_3 x_3$

(c) $\quad a_{11}x_1 y_1 + a_{12}x_1 y_2 + a_{13}x_1 y_3$
$\quad + a_{21}x_2 y_1 + a_{22}x_2 y_2 + a_{23}x_2 y_3$
$\quad + a_{31}x_3 y_1 + a_{32}x_3 y_2 + a_{33}x_3 y_3$

(d) $a_{11}x_1 x_1 + a_{22}x_2 x_2 + a_{33}x_3 x_3$

(e) $(a_1 x^1 + a_2 x^2 + a_3 x^3)^2 = (a_1 x^1)^2 + (a_2 x^2)^2 + (a_3 x^3)^2$
$\qquad\qquad\qquad\qquad\qquad + 2a_1 a_2 x^1 x^2 + 2a_1 a_3 x^1 x^3 + 2a_2 a_3 x^2 x^3$

(f) $\dfrac{\partial x^t \, \partial y^1}{\partial y^1 \, \partial x^1} z^1 + \dfrac{\partial x^t \, \partial y^2}{\partial y^2 \, \partial x^1} z^1 + \dfrac{\partial x^t \, \partial y^3}{\partial y^3 \, \partial x^1} z^1 +$

$\dfrac{\partial x^t \, \partial y^1}{\partial y^1 \, \partial x^2} z^2 + \dfrac{\partial x^t \, \partial y^2}{\partial y^2 \, \partial x^2} z^2 + \dfrac{\partial x^t \, \partial y^3}{\partial y^3 \, \partial x^2} z^2 +$

$\dfrac{\partial x^t \, \partial y^1}{\partial y^1 \, \partial x^3} z^3 + \dfrac{\partial x^t \, \partial y^2}{\partial y^2 \, \partial x^3} z^3 + \dfrac{\partial x^t \, \partial y^3}{\partial y^3 \, \partial x^3} z^3$

(g) $\dfrac{\partial x^1 \, \partial y^1}{\partial y^1 \, \partial x^k} z_1 + \dfrac{\partial x^1 \, \partial y^2}{\partial y^2 \, \partial x^k} z_1 + \dfrac{\partial x^1 \, \partial y^3}{\partial y^3 \, \partial x^k} z_1 +$

$\dfrac{\partial x^2 \, \partial y^1}{\partial y^1 \, \partial x^k} z_2 + \dfrac{\partial x^2 \, \partial y^2}{\partial y^2 \, \partial x^k} z_2 + \dfrac{\partial x^2 \, \partial y^3}{\partial y^3 \, \partial x^k} z_2 +$

$\dfrac{\partial x^3 \, \partial y^1}{\partial y^1 \, \partial x^k} z_3 + \dfrac{\partial x^3 \, \partial y^2}{\partial y^2 \, \partial x^k} z_3 + \dfrac{\partial x^3 \, \partial y^3}{\partial y^3 \, \partial x^k} z_3$

9 The tangents to the parametric curves are coplanar at any point on any one of the coordinate planes. In the first octant

$$x = \sqrt{\frac{x^1 + x^2}{2}} + 2\sqrt{x^3 - x^1} \qquad y = \sqrt{\frac{x^1 + x^2}{2}} \qquad z = \sqrt{x^3 - x^1}$$

In the octant in which $x < 0$, $y < 0$, $z < 0$,

$$x = -\sqrt{\frac{x^1 + x^2}{2}} - 2\sqrt{x^3 - x^1} \qquad y = -\sqrt{\frac{x^1 + x^2}{2}} \qquad z = -\sqrt{x^3 - x^1}$$

11 (a) $(ds)^2 = (dr)^2 + (r\, d\theta)^2 + (dz)^2$

(b) $G = \begin{bmatrix} 1 & 0 & 0 \\ 0 & r^2 & 0 \\ 0 & 0 & 1 \end{bmatrix}$ $\qquad G^{-1} = \begin{bmatrix} 1 & 0 & 0 \\ 0 & \dfrac{1}{r^2} & 0 \\ 0 & 0 & 1 \end{bmatrix}$

13 $\sqrt{41}$; 3; $\cos \theta = -\dfrac{8}{3\sqrt{41}}$

15 At $(2,0,1)$, $\mathbf{V} = \mathbf{e}^1 + 2\sqrt{3}\,\mathbf{e}^2$; at $(2,\pi/3,1)$, $\mathbf{V} = \mathbf{e}^1 + 2\sqrt{3}\,\mathbf{e}^2$

Sec. 14.4
p. 724

5 (a) No
(b) No

Sec. 14.5
p. 728

3 (a) For a contravariant vector $\mathbf{V} = \xi^i \mathbf{e}_i$, the divergence is

$$\frac{1}{r^2 \sin \theta} \left[\frac{\partial(r^2 \sin \theta\, \xi^1)}{\partial r} + \frac{\partial(r^2 \sin \theta\, \xi^2)}{\partial \theta} + \frac{\partial(r^2 \sin \theta\, \xi^3)}{\partial \phi} \right]$$

If we let V^1, V^2, V^3 be the components of $\mathbf{V}$ along *unit* vectors in the directions of $\mathbf{e}_1$, $\mathbf{e}_2$, and $\mathbf{e}_3$, respectively, so that $V^1 = \xi^1$, $V^2 = r\xi^2$, $V^3 = r \sin \theta \xi^3$, the divergence appears in the more usual form

$$\frac{1}{r^2} \frac{\partial(r^2 V^1)}{\partial r} + \frac{1}{r \sin \theta} \frac{\partial(\sin \theta\, V^2)}{\partial \theta} + \frac{1}{r \sin \theta} \frac{\partial V^3}{\partial \phi}$$

(b) For a covariant vector $\mathbf{V} = \xi_i \mathbf{e}^i$, the divergence is

$$\frac{1}{r^2 \sin \theta} \left[\frac{\partial(r^2 \sin \theta\, \xi_1)}{\partial r} + \frac{\partial(\sin \theta\, \xi_2)}{\partial \theta} + \frac{\partial[(1/\sin \theta)\, \xi_3]}{\partial \phi} \right]$$

If we let V_1, V_2, V_3 be the components of $\mathbf{V}$ along *unit* vectors in the directions of $\mathbf{e}^1$, $\mathbf{e}^2$, $\mathbf{e}^3$, respectively, so that $V_1 = \xi_1$, $V_2 = \xi_2/r$, $V_3 = \xi_3/(r \sin \theta)$, the divergence appears in the more usual form

$$\frac{1}{r^2} \frac{\partial(r^2 V_1)}{\partial r} + \frac{1}{r \sin \theta} \frac{\partial(\sin \theta\, V_2)}{\partial \theta} + \frac{1}{r \sin \theta} \frac{\partial V_3}{\partial \phi}$$

Chapter 15
Sec. 15.2
p. 735

9 $x^3 - 3xy^2 - 2x$, $3x^2y - y^3 - 2y$

11 $3 + 2i$ $\qquad\qquad$ **13** $120 - 119i$

15 $\dfrac{-3 + 4i}{25}$ $\qquad\qquad$ **17** $(1,2)$, $(4,\tfrac{1}{8})$

19 The radius of the circle is $\dfrac{\sqrt{B\bar{B} - (A + \bar{A})(C + \bar{C})}}{A + \bar{A}}$. The center of the circle is $-\dfrac{B + \bar{B}}{2(A + \bar{A})}$, $\dfrac{B - \bar{B}}{2i(A + \bar{A})}$.

Sec. 15.3
p. 740

1 Rotation through $-90°$; rotation through $45°$

3 $\dfrac{1+i}{\sqrt{2}}, \dfrac{-1+i}{\sqrt{2}}, \dfrac{-1-i}{\sqrt{2}}, \dfrac{1-i}{\sqrt{2}}$

5 $\dfrac{\sqrt{3}+i}{2}, \dfrac{-\sqrt{3}+i}{2}, -i$

7 $2^{1/6}(\cos 15° + i \sin 15°) = 1.084 + 0.291i$
$2^{1/6}(\cos 135° + i \sin 135°) = -0.794 + 0.794i$
$2^{1/6}(\cos 255° + i \sin 255°) = -0.291 - 1.084i$

9 $2^{2/5}(\cos \pi + i \sin \pi)$

$2^{2/5}\left(\cos \dfrac{3\pi}{5} + i \sin \dfrac{3\pi}{5}\right)$

$2^{2/5}\left(\cos \dfrac{\pi}{5} + i \sin \dfrac{\pi}{5}\right)$

$2^{2/5}\left(\cos \dfrac{9\pi}{5} + i \sin \dfrac{9\pi}{5}\right)$

$2^{2/5}\left(\cos \dfrac{7\pi}{5} + i \sin \dfrac{7\pi}{5}\right)$

13 At the point $\dfrac{m_1 z_1 + m_2 z_2 + m_3 z_3}{m_1 + m_2 + m_3}$

Sec. 15.4
p. 744

1 No

3 If and only if z_1 and z_2 have the same argument (or arguments differing by an integral multiple of 2π)

7 The 45° sector of the first quadrant bounded by the y axis and the line $y = x$. The points on the y axis are not included; the points on the line $y = x$ are included.

15 $z_k, \bar{z}_k$

Sec. 15.5
p. 749

1 $-2 - 3i$ **3** $-iz^2 + 2z - 1$

5 (a) The upper half plane, excluding the real axis; unbounded, open, simply connected
(b) The annulus bounded by the circles with center at the origin and radii 2 and 3, including the boundary circles; bounded, closed, multiply connected
(c) The exterior of the circle with center at $z = 1$ and radius 4, excluding the boundary circle; unbounded, open, multiply connected
(d) The infinite vertical strip between the y axis and the line $x = 1$ including the boundaries; unbounded, closed, simply connected
(e) The interior of the circle with center at $z = 1$ and radius $\frac{3}{4}$, including the boundary circle; bounded, closed, simply connected
(f) The portion of the first quadrant below the line $y = x$, including the positive real axis but excluding the positive half of the line $y = x$; unbounded, neither open nor closed, simply connected

7 (a) $z = \pm i$; $f(z)$ is undefined.
(b) $z = 1$; $f(z)$ is undefined. [If $f(1)$ is defined to be 2, then $f(z)$ is continuous everywhere.]
(c) $f(z)$ is continuous everywhere.
(d) $f(z)$ is continuous everywhere.
(e) $z = i$; $\lim_{z \to i} f(z) = 0$ is not equal to $f(i) = i$. [If $f(i)$ is defined to be 0, then $f(z)$ is continuous everywhere.]

Sec. 15.6 **3** Only at the origin; only at the origin; $f(z)$ is nowhere analytic.
p. 756 **5** Only if u and v are constants.
7 No, since $u = x^2 + y$ does not satisfy Laplace's equation.
11 The values all lie on the circle $x^2 + y^2 = 1$.

Sec. 15.7 **7** (a) $\sin 2 \cosh 1 - i \cos 2 \sinh 1 = 1.403 - 0.489i$
p. 764 (b) $\cosh 1 \cos 1 + i \sinh 1 \sin 1 = 0.834 + 0.989i$
(c) $\sinh 2 \cos 3 + i \cosh 2 \sin 3 = -3.591 + 0.531i$
(d) $i \cosh^{-1} 2 = 1.317i$
(e) $i \tanh 1 = 0.752i$

(f) $\dfrac{1}{2}\left(\ln 3 + \dfrac{i\pi}{2}\right) = 0.549 + 0.785i$

(g) $\ln 5 + i \tan^{-1}\left(\dfrac{4}{-3}\right) = 1.609 + 2.214i$

(h) $\cos(\ln 2) + i \sin(\ln 2) = 0.769 + 0.639i$

(i) $\exp\left[\left(2\ln\sqrt{2} - \dfrac{3\pi}{4}\right) - i\left(3\ln\sqrt{2} + \dfrac{\pi}{2}\right)\right] = e^{-(1.663 + 2.611i)}$

$= -0.163 - 0.096i$

13 Yes
17 $z = \cosh^{-1} 2 + i(2m + 1)\pi$
19 The complex numbers are not ordered (see Exercise 20, Sec. 15.2), and hence the inequalities implicit in the property of real continuous functions referred to in the exercise are meaningless for complex quantities.

21 (c) $\dfrac{1}{1 + z^2}$

23 $|\sinh z|^2 = \sinh^2 x + \sin^2 y$; $|\cosh z|^2 = \sinh^2 x + \cos^2 y$
29 The plot consists of the family of all circles having their center at the origin and the family of all straight lines which pass through the origin.

Sec. 15.8 **1** (a) $6 + \frac{2}{3}{}^6 i$ (b) $6 + \frac{2}{3}{}^6 i$ (c) $6 + \frac{2}{3}{}^6 i$
p. 776 **3** Along $y = x$, $-\frac{1}{6} + \frac{5}{6}i$; along $y = x^2$, $-\frac{1}{6} + \frac{5}{6}i$
5 $3e^6/8, 2e/3$. (Each of these is a very crude estimate, because when the methods of Chap. 17 are used, the value of the integral around $|z| = 3$ is sin 2, and by Cauchy's theorem the value of the integral around $|z| = \frac{1}{2}$ is 0.)

7 (a) 0   (b) $\dfrac{3 + 2i}{2}\pi$ (c) $\dfrac{-3 + 2i}{2}\pi$

9 (a) $-\dfrac{3\pi i}{2}$ (b) $\dfrac{3\pi i}{2}$ (c) 0

17 $I_k(2x) = \dfrac{1}{2\pi}\displaystyle\int_0^{2\pi} \exp(2x \cos\theta - ik\theta)\, d\theta$

$J_k(2x) = \dfrac{1}{2\pi}\displaystyle\int_0^{2\pi} \exp[i(2x \sin\theta - k\theta)]\, d\theta$

Chapter 16
Sec. 16.1 **1** (a) The series converges to the sum $1/(1 + i - z)$ in the interior of the
p. 787 circle of unit radius whose center is $z = i$.
(b) The series converges to the sum

$$S = \begin{cases} z & |z| < 1 \\ 0 & z = 1 \end{cases}$$

(c) The series converges to the sum $(z + 1)/(z - 3)$ in the interior of the circle whose radius is $\frac{4}{3}$ and whose center is the point $(\frac{5}{3},0)$.

(d) The series converges to the sum $1/(2z - 1 + 2i)$ in the interior of the circle $|z + i| = \frac{1}{2}$.

5 (b) Yes

7 (a) The series converges to the sum 1 for all values of z except $z = 1/n$, $n = 1, 2, 3, \ldots$.

(b) The series converges to the sum $1/z$ for all values of z except $z = 0$, $-1, -2, \ldots$. The convergence is uniform over the entire region of convergence.

9 The series

$$- \frac{x}{1 + x^2} + \left[\frac{x}{1 + x^2} - \frac{2x}{1 + (2x)^2} \right] + \left[\frac{2x}{1 + (2x)^2} - \frac{3x}{1 + (3x)^2} \right] + \cdots$$

converges to the continuous function 0 for all values of x, although the convergence is not uniform in any interval containing the origin.

Sec. 16.2
p. 794

1 (a) $f(z) = -1 + 2z - 2z^2 + 2z^3 - \cdots$ $|z| < 1$

(b) $f(z) = \dfrac{z - 1}{2} - \dfrac{(z - 1)^2}{4} + \dfrac{(z - 1)^3}{8} - \dfrac{(z - 1)^4}{16} + \cdots$

$$|z - 1| < 2$$

3 $\cosh z = -1 - \dfrac{(z - i\pi)^2}{2!} - \dfrac{(z - i\pi)^4}{4!} - \cdots$ $|z| < \infty$

5 (a) $\sqrt{2}$ (b) $2\pi - 4$ (c) $\sqrt{10}$ (d) $\sqrt{17}$

11 $f(z) = 2\pi i$

Sec. 16.3
p. 800

1 (a) $f(z) = \frac{1}{2} + \frac{3}{4}z + \frac{7}{8}z^2 + \frac{15}{16}z^3 + \cdots$

(b) $f(z) = \cdots - \dfrac{1}{z^3} - \dfrac{1}{z^2} - \dfrac{1}{z} - \dfrac{1}{2} - \dfrac{z}{4} - \dfrac{z^2}{8} - \dfrac{z^3}{16} - \cdots$

(c) $f(z) = \cdots + \dfrac{15}{z^5} + \dfrac{7}{z^4} + \dfrac{3}{z^3} + \dfrac{1}{z^2}$

(d) $f(z) = -\dfrac{1}{z - 1} - 1 - (z - 1) - (z - 1)^2 - \cdots$

(e) $f(z) = \cdots + \dfrac{1}{(z - 1)^4} + \dfrac{1}{(z - 1)^3} + \dfrac{1}{(z - 1)^2}$

(f) $f(z) = \dfrac{1}{z - 2} - 1 + (z - 2) - (z - 2)^2 + (z - 2)^3 - \cdots$

(g) $f(z) = \cdots + \dfrac{1}{(z - 2)^4} - \dfrac{1}{(z - 2)^3} + \dfrac{1}{(z - 2)^2}$

3 $f(z) = -\dfrac{1}{z - 1} - 2i + 3(z - i) - 4i(z - i)^2 + 5(z - i)^3 - \cdots$

$$0 < |z - i| < 1$$

$f(z) = \cdots + \dfrac{4i}{(z - i)^6} - \dfrac{3}{(z - i)^5} - \dfrac{2i}{(z - i)^4} + \dfrac{1}{(z - i)^3}$ $|z - i| > 1$

5 (a) 0   (b) $2\pi i$ (c) 0   (d) 0   (e) $2\pi i$ (f) $-\dfrac{i\pi}{6}$

7 The first series converges only for $|z| < 1$, the second converges only for $|z| > 1$. Hence there is no value of z for which the two series are simultaneously valid.

9 **(c)** $a_n = \dfrac{1}{2\pi} \displaystyle\int_0^{2\pi} \cosh\,(2\cos\theta)\cos n\theta\,d\theta$

(d) $a_n = \dfrac{1}{2\pi} \displaystyle\int_0^{2\pi} \cos\,(2\cos\theta)\cos n\theta\,d\theta$

Chapter 17

Sec. 17.1
p. 808

1 **(a)** $\frac{1}{2}$ **(b)** $\frac{1}{2}$

3 At $z = -1 + 2i$ the residue is $(2 + i)/4$. At $z = -1 - 2i$ the residue is $(2 - i)/4$.

5 -1 **7** $\frac{3}{10}$

9 **(a)** $2\pi i$ **(b)** 0 **(c)** 0

(d) 0 **(e)** 0 **(f)** $\dfrac{(5 - 3\sqrt{5})i\pi}{20}$

15 **(a)** $\dfrac{\pi \coth \pi a}{a}$ **(b)** $\dfrac{\pi \sinh 2\pi b}{2b(\cosh^2 \pi b - \cos^2 \pi a)}$

(c) $\dfrac{\pi^3 \cos \pi a}{\sin^3 \pi a}$ **(d)** $\dfrac{\pi(\sinh 2\pi a + \sin 2\pi a)}{4a^3(\cosh 2\pi a - \cos 2\pi a)}$

Sec. 17.2
p. 816

1 $\dfrac{2\pi}{1 - p^2}$ **3** π

5 $\dfrac{2\pi}{b^2}(a - \sqrt{a^2 - b^2})$ **7** $\dfrac{\pi}{\sqrt{2}\,a^3}$

9 $\dfrac{\pi}{3}$ **11** $\dfrac{\pi}{4a^3}$

13 $\dfrac{\pi e^{-mb} \cos ma}{b}$ **15** $\dfrac{\pi(1 + am)e^{-ma}}{4a^3}$

17 $\dfrac{\pi}{a^2 - b^2}\left(\dfrac{e^{-bm}}{b} - \dfrac{e^{-am}}{a}\right)$ **19** $\pi \exp\left(-\dfrac{m}{\sqrt{2}}\right)\sin\dfrac{m}{\sqrt{2}}$

21 $\dfrac{\pi}{\sin a\pi}\dfrac{r^{a-1}\sin(a-1)\theta}{c}$ where $r = \sqrt{a^2 + b^2}$
$\theta = \tan^{-1}(-c/-b)$

23 $\dfrac{\pi}{3\sin a\pi}\left(1 + 2\cos\dfrac{2\pi a}{3}\right)$

31 $a_n = \dfrac{2(-1)^n}{\sqrt{a^2 - b^2}}\left[\tan\left(\dfrac{1}{2}\csc^{-1}\dfrac{a}{b}\right)\right]^n$. As $n \to \infty$, a_n approaches zero more rapidly than the reciprocal of any fixed power of n. This, of course, is implied by Theorem 3, Sec. 6.3, since all derivatives of the given function are everywhere continuous.

33 $\dfrac{\pi^3}{8a} + \pi\dfrac{\ln^2 a}{2a}$; $\pi\dfrac{\ln a}{2a}$

Sec. 17.3
p. 824

1 $\dfrac{e^{-t} - e^{-3t}}{2}$ **3** $\dfrac{\sin 2t}{2}$

5 $1 - \cos t$ **7** $\dfrac{t \sin 2t}{4}$

9 $(2 - t)e^{-2t} - 2e^{-3t}$

13 $\dfrac{t}{a} + \dfrac{2}{\pi} \sum\limits_{n=1}^{\infty} \dfrac{(-1)^n}{n} \sin \dfrac{n\pi t}{a}$

15 $x + \dfrac{2}{\pi} \sum\limits_{n=1}^{\infty} \dfrac{(-1)^n}{n} \exp\left(-n^2\pi^2 t\right) \sin n\pi x$

Sec. 17.4
p. 835

3 $D_1 = 6$, $D_2 = 4$, $D_3 = 324$. Hence each root has negative real part.
5 $D_1 = 1$, $D_2 = -9$, $D_3 = -100$, $D_4 = -1{,}000$. Hence there is at least one root with nonnegative real part.

Chapter 18
Sec. 18.1
p. 840

5 The mapping is equivalent to the mapping defined by the function $w = z^2$, discussed in Example 1, followed by reflection in the real axis.
3 The transformation $w = f(\bar{z})$ is equivalent to the transformation $w = f(z)$ followed by a reflection in the real axis.
7 Angle measures are not preserved.
9 The equations of the transformation are

$$u = x^3 - 3xy^2 \qquad v = 3x^2y - y^3$$

The image of the line $x = 1$ is the cubic curve

$$27v^2 = (1 - u)(u + 8)^2$$

11 The equations of the transformation are

$$u = e^x \cos y \qquad v = e^x \sin y$$

The vertical lines $x = c$ map into the concentric circles $u^2 + v^2 = e^{2c}$. The horizontal lines $y = k$ map into the radial lines $v = (\tan k)u$. The infinite strip $0 < y < \pi$, $-\infty < x < \infty$, maps onto the upper half of the w plane.

Sec. 18.2
p. 845

1 (a) $\dfrac{\pi}{a}$ **(b)** $\dfrac{\pi}{\sqrt{2}}$

3 (a) $z = \pm 1$ **(b)** $z^2 - 1 = \frac{1}{3}$ or $(x^2 - y^2 - 1)^2 + 4x^2y^2 = \frac{1}{9}$
(c) $x^2 - 2xy - y^2 = 1$ **(d)** $x^2 - y^2 = 1$
5 The images of the perpendicular lines $x = 1$ and $y = 0$ intersect at an angle of $45°$.

Sec. 18.3
p. 855

3 $-\frac{1}{3}$.
5 The invariant points are $-i$, $4i$. The constant value of the cross ratio is $-\frac{3}{2}$.
7 If and only if $a = d$.

9 $w = \dfrac{-3z + i}{-iz + 1}$. The invariant points are i, $-5i$.

11 $w = \dfrac{az + b}{cz + d}$, where a, b, c, d are all real and $ad - bc < 0$.

13 $w = \dfrac{z^3 - i}{z^3 + i}$ **15** $w = -\left(\dfrac{z^4 - 1}{z^4 + 1}\right)^2$

21 $T = \dfrac{100}{\pi}\tan^{-1}\dfrac{3x^2y - y^3}{x^3 - 3xy^2}$

23 $T = \dfrac{100}{\pi}\tan^{-1}\dfrac{x^2 + y^2 - 1}{2y}$

25 $T = \dfrac{100}{\pi}\tan^{-1}\dfrac{4y(1 - x^2 - y^2)}{(x^2 + y^2)^2 - 2x^2 - 6y^2 + 1}$

Sec. 18.4   **1** $w = -z^2$
p. 862

3 $\cos \dfrac{\theta}{2} = \dfrac{r-1}{2\sqrt{r}} \tan \dfrac{\pi T}{100}$, where (r,θ) are polar coordinates in the w plane.

7 $z = e^w$ **9** $\tan \dfrac{\pi T}{100} = \dfrac{\tan v}{\tanh u}$

11 $T = \dfrac{100}{\pi} (\pi + \tfrac{2}{3}\theta)$, where θ is the polar coordinate angle in the w plane.

13 $v^2 = c^2 + \dfrac{c^2}{u^2 + c^2}$

15 $w = 2\sqrt{z+1} + \ln \dfrac{\sqrt{z+1}-1}{\sqrt{z+1}+1}$

17 $w = A \displaystyle\int_0^z \dfrac{dz}{\sqrt{(1-z^2)(1-k^2 z^2)}}$

If E is the point $w = u_0$, then

$$u_0 = A \int_0^1 \frac{dz}{\sqrt{(1-z^2)(1-k^2 z^2)}} = AK(k),$$

where $K(k)$ denotes the complete elliptic integral of the first kind of modulus k. Therefore $A = u_0/K(k)$.

INDEX

INDEX

Page numbers followed by an *e* indicate exercise.